Fundamental Equations of Dynamics

KINEMATICS

Particle Rectilinear Motion

Variable a	*Constant $a = a_c$*
$a = \dfrac{dv}{dt}$	$v = v_0 + a_c t$
$v = \dfrac{ds}{dt}$	$s = s_0 + v_0 t + \frac{1}{2} a_c t^2$
$v\,dv = a\,ds$	$v^2 = v_0^2 + 2a_c(s - s_0)$

Particle Curvilinear Motion

x, y, z Coordinates		*r, θ, z Coordinates*	
$v_x = \dot{x}$	$a_x = \ddot{x}$	$v_r = \dot{r}$	$a_r = \ddot{r} - r\dot{\theta}^2$
$v_y = \dot{y}$	$a_y = \ddot{y}$	$v_\theta = r\dot{\theta}$	$a_\theta = r\ddot{\theta} + 2\dot{r}\dot{\theta}$
$v_z = \dot{z}$	$a_z = \ddot{z}$	$v_z = \dot{z}$	$a_z = \ddot{z}$

n, t, Coordinates

$v = \dot{s}$ | $a_t = \dot{v} = v\dfrac{dv}{ds}$, $a_n = \dfrac{v^2}{\rho}$, $\rho = \left|\dfrac{[1 + (dy/dx)^2]^{3/2}}{d^2y/dx^2}\right|$

Relative Motion

$\mathbf{v}_B = \mathbf{v}_A + \mathbf{v}_{B/A}$ $\quad \mathbf{a}_B = \mathbf{a}_A + \mathbf{a}_{B/A}$

Rigid Body Motion About a Fixed Axis

Variable α	*Constant $\alpha = \alpha_c$*
$\alpha = \dfrac{d\omega}{dt}$	$\omega = \omega_0 + \alpha_c t$
$\omega = \dfrac{d\theta}{dt}$	$\theta = \theta_0 + \omega_0 t + \frac{1}{2}\alpha_c t^2$
$\omega\,d\omega = \alpha\,d\theta$	$\omega^2 = \omega_0^2 + 2\alpha_c(\theta - \theta_0)$

For Point P

$s = \theta r \quad v = \omega r \quad a_t = \alpha r \quad a_n = \omega^2 r$

Relative General Plane Motion-Translating Axes

$\mathbf{v}_B = \mathbf{v}_A + \mathbf{v}_{B/A(\text{pin})}$ $\quad \mathbf{a}_B = \mathbf{a}_A + \mathbf{a}_{B/A(\text{pin})}$

Relative General Plane Motion-Trans. and Rot. Axis

$\mathbf{v}_B = \mathbf{v}_A + (\mathbf{v}_{B/A})_{\text{rel}}$

$\mathbf{a}_B = \mathbf{a}_A + (\mathbf{a}_{B/A})_{\text{rel}} + \mathbf{a}_{\text{cor}}$

KINF

Moment of Inertia $\quad I$

Parallel-Axis Theorem $\quad I =$

Radius of Gyration $\quad k =$

Equations of Motion

Particle	$\Sigma\mathbf{F} = m\mathbf{a}$
Rigid Body (Plane Motion)	$\Sigma F_x = m(a_G)_x$ $\Sigma F_y = m(a_G)_y$ $\Sigma M_G = I_G\alpha \;/\; \Sigma M_P = \Sigma(\mathcal{M}_k)_P$

Principle of Work and Energy

$T_1 + U_{1-2} = T_2$

Kinetic Energy

Particle	$T = \frac{1}{2}mv^2$
Rigid Body (Plane Motion)	$T = \frac{1}{2}mv_G^2 + \frac{1}{2}I_G\omega^2$

Work

Variable force	$U_F = \int F \cdot ds$
Constant force	$U_{F_c} = F_c \cos\theta(s_2 - s_1)$
Weight	$U_W = W(y_2 - y_1)$
Force on spring	$U_s = \frac{1}{2}kx_2^2 - \frac{1}{2}kx_1^2$
Couple moment	$U_M = M(\theta_2 - \theta_1)$

Power and Efficiency

$P = \dfrac{dU}{dt} = (F\cos\theta)v, \; \varepsilon = \dfrac{P_{\text{out}}}{P_{\text{in}}}$

Conservation of Energy Theorem

$T_1 + V_1 = T_2 + V_2$

Potential Energy

$V = V_g + V_e$, where $V_g = \pm Wy$, $V_e = +\frac{1}{2}kx^2$

Principle of Linear Impulse and Momentum

Particle	$m\mathbf{v}_1 + \Sigma\int \mathbf{F}\,dt = m\mathbf{v}_2$
Rigid Body	$m(\mathbf{v}_G)_1 + \Sigma\int \mathbf{F}\,dt = m(\mathbf{v}_G)_2$

Conservation of Linear Momentum

$\Sigma(\text{syst. } m\mathbf{v})_1 = \Sigma(\text{syst. } m\mathbf{v})_2$

Coefficient of Restitution $\quad e = \dfrac{(v_B)_2 - (v_A)_2}{(v_A)_1 - (v_B)_1}$

Principle of Angular Impulse and Momentum

Particle	$(\mathbf{H}_O)_1 + \Sigma\int \mathbf{M}_O\,dt = (\mathbf{H}_O)_2$, where $H_O = (d)(mv)$
igid Body 'lane Motion)	$(\mathbf{H}_G)_1 + \Sigma\int \mathbf{M}_G\,dt = (\mathbf{H}_G)_2$, where $H_G = I_G\omega$ $(\mathbf{H}_O)_1 + \Sigma\int \mathbf{M}_O\,dt = (\mathbf{H}_O)_2$, where $H_O = I_G\omega + (d)(mv_G)$

Conservation of Angular Momentum

$\Sigma(\text{syst. }\mathbf{H})_1 = \Sigma(\text{syst. }\mathbf{H})_2$

Mechanics for Engineers
STATICS AND DYNAMICS

DEON STOCKERT

MECHANICS FOR ENGINEERS

STATICS AND DYNAMICS

Russell C. Hibbeler

Macmillan Publishing Company
New York
Collier Macmillan Publishers
London

Printed in the United States of America

Macmillan Publishing Company
866 Third Avenue, New York, New York 10022

Collier Macmillan Canada, Inc.

Library of Congress Cataloging in Publication Data

Hibbeler, R. C.
Mechanics for engineers.

Rev. ed. of: Engineering mechanics—statics and dynamics. 3rd ed. 1983.
Includes index.
1. Mechanics, Applied. I. Hibbeler, R. C. Engineering mechanics—statics and dynamics. II. Title.
TA350.H483 1985 620.1 84-23057
ISBN 0-02-354410-4

Printing: 2 3 4 5 6 7 8 Year: 6 7 8 9 0 1 2

ISBN 0-02-354410-4

Preface

The purpose of this book is to provide the student with a clear and thorough presentation of the theory and application of the principles of engineering mechanics. Emphasis is placed on developing the student's ability to analyze problems—a most important skill for any engineer. In order to achieve this objective the contents of each chapter are organized into well-defined sections. Selected groups of sections contain an explanation of specific topics, a "procedure for analysis" which provides a systematic approach for applying the theory, illustrative example problems, and a set of homework problems.

Numerous problems in the book depict realistic situations encountered in engineering practice. It is hoped that this realism will both stimulate the student's interest in engineering mechanics and provide a means for developing the skill to reduce any such problem from its physical description to a model or symbolic representation to which the principles of mechanics may be applied. Both SI and FPS units are used equally throughout the book. Furthermore, in any set, the problems are arranged in order of increasing difficulty* and the answers to all but every fourth problem, which is indicated by an asterisk, are listed in the back of the book.

Most of the text material has been organized so that topics within each section are placed into subgroups defined by boldface titles. The purpose of this is to present a structured method for introducing each new definition or concept, and to provide a convenient means for later reference or review. As stated above, a "procedure for analysis" is used in many sections of the book. These guides provide the student with a logical and orderly method to follow when applying the theory. Most often the first step in the procedure will require drawing a diagram. In doing so, the student forms the habit of tabulating

*Review problems, at the end of each chapter, are presented in random order.

the necessary data while focusing on the physical aspects of the problem and its associated geometry. If this step is correctly performed, applying the relevant equations of mechanics becomes somewhat methodical, since the data can be taken directly from the diagram. The example problems are solved using this outlined method in order to clarify its numerical application.

Since mathematics provides a systematic means of applying the principles of mechanics, the student is expected to have prior knowledge of algebra, geometry, trigonometry, and, for complete coverage, some calculus. Occasionally, the example problems are solved using several different methods of analysis so that the student develops the ability to use mathematics as a tool whereby the solution of any problem may be carried out in the most direct and effective manner.

Contents: *Statics*. The subject of statics is presented in 11 chapters, in which the principles introduced are first applied to simple situations. Specifically, each principle is applied first to a particle, then to a rigid body subjected to a coplanar system of forces, and finally to the most general case of three-dimensional force systems acting on a rigid body.

In particular, an introduction to mechanics and a discussion of units is outlined in Chapter 1. The notion of a vector and the properties of a concurrent force system are introduced in Chapter 2. This theory is then applied to the equilibrium of particles in Chapter 3. Chapter 4 contains a general discussion of both concentrated and distributed force systems and the methods used to simplify them. The principles of rigid-body equilibrium are developed in Chapter 5. These principles are applied to specific problems involving the equilibrium of trusses, frames, and machines in Chapter 6, and to the analysis of internal forces in beams in Chapter 7. Applications to problems involving frictional forces are discussed in Chapter 8; and topics related to the centroid and the center of gravity are given in Chapter 9. If time permits, sections concerning more advanced topics, indicated by stars, may be covered. Some topics in Chapter 10 and Appendix B (Moments of Inertia) and Chapter 11 (Virtual Work) may be omitted from the basic course. Note, however, that this more advanced material provides a suitable reference for basic principles when it is discussed in more advanced courses.

At the discretion of the instructor, some of the material may be presented in a different sequence with no loss in continuity. For example, it is possible to introduce the concept of a force by first covering Chapter 2. Then, after covering Chapter 4 (force and moment systems), the equilibrium methods in Chapters 3 and 5 can be discussed. Furthermore, Chapter 9 may be covered after Sec. 4-8 (distributed force systems), since understanding of this material does not depend upon the methods of equilibrium.

Contents: *Dynamics*. The subject of dynamics is presented in the last 10 chapters. In particular, the kinematics of a particle is discussed in Chapter 12, followed by a discussion of particle kinetics in Chapter 13 (equation of motion), Chapter 14 (work and energy), and Chapter 15 (impulse and momentum). A similar sequence of presentation is given for the planar motion of a rigid body: Chapter 16 (planar kinematics), Chapter 17 (equation of motion), Chapter 18 (work and energy), and Chapter 19 (impulse and momentum). If desired, it is possible to cover Chapters 12 through 19 in the following order with no loss in continuity: Chapters 12 and 16 (kinematics), Chapters 13 and 17 (equation of motion), Chapters 14 and 18 (work and energy), and Chapters 15 and 19 (impulse and momentum).

Time permitting, some of the material included in Chapter 20 (special applications) may be covered in the course at points where the instructor finds it beneficial. Chapter 21 (vibrations) may be included if the student has the necessary mathematical background. Sections of the book which are considered to be beyond the scope of the basic dynamics course are indicated by a star and may be omitted. Note, however, that this more advanced material provides a suitable reference for basic principles when it is covered in more advanced courses.

Acknowledgments. I have endeavored to write this book so that it will appeal to both the student and the instructor. Many people helped in its development. I wish to acknowledge the valuable suggestions and comments made by Richard G. Camp, Jr., California State Polytechnic University, Pomona; Thomas E. Kirk, Purdue University; Karl D. Lilje, California Polytechnic State University, San Luis Obispo; and Grover J. Trammell, Louisiana Tech University. Many thanks are also extended to all of the author's students and to those in the teaching profession who have taken the time to send me their suggestions and comments. Although the list is too long to mention, it is hoped that those who have given help will accept this anonymous recognition. Lastly, I should like to acknowledge the assistance of my wife, Conny, who has once again been quite helpful in preparing the manuscript for publication.

Russell Charles Hibbeler

Contents

STATICS

Mechanics for Engineers
STATICS AND DYNAMICS

1 General Principles

1.1 Mechanics

Mechanics can be defined as that branch of the physical sciences concerned with the state of rest or motion of bodies that are subjected to the action of forces. A thorough understanding of this subject is required for the study of structural engineering, machine design, fluid flow, electrical instrumentation, and even the molecular and atomic behavior of elements.

In general, mechanics is subdivided into three branches: *rigid-body mechanics, deformable-body mechanics,* and *fluid mechanics*. In this book only rigid-body mechanics will be studied, since this subject forms a suitable basis for the design and analysis of many engineering problems, and it provides the necessary background for the study of the mechanics of deformable bodies and the mechanics of fluids.

Rigid-body mechanics is divided into two areas: statics and dynamics. *Statics* deals with the equilibrium of bodies, that is, those which are either at rest or moving with a constant velocity; whereas *dynamics* is concerned with the accelerated motion of bodies. Although statics can be considered as a special case of dynamics, in which the acceleration is zero, statics deserves separate treatment in engineering education, since most structures are designed with the intention that they remain in equilibrium.

Historical Development. The subject of statics developed very early in history, because the principles involved could be formulated simply from measurements of geometry and force. For example, the writings of Archimedes (287–212 B.C.) provide an explanation of the equilibrium of the lever. Studies of the pulley, inclined plane, and wrench are also recorded in ancient writings—at times when the requirements of engineering were limited primarily to building construction.

Since the principles of dynamics depend upon an accurate measurement of time, this subject developed much later. Galileo Galilei (1564–1642) was one of the first major contributors to this field. His work consisted of experiments using pendulums and falling bodies. The most significant contributions in dynamics, however, were made by Isaac Newton (1642–1727), who is noted for his formulation of the three fundamental laws of motion and the law of universal gravitational attraction. Shortly after these laws were postulated, important techniques for their application were developed by Euler, D'Alembert, Lagrange, and others.

Newton's Three Laws of Motion. The entire subject of rigid-body mechanics is formulated on the basis of Newton's three laws of motion. These laws, which apply to the motion of a particle, may be briefly stated as follows:

First Law. A particle originally at rest, or moving in a straight line with constant velocity, will remain in this state provided the particle is not subjected to an unbalanced force.

Second Law. A particle acted upon by an unbalanced force **F** experiences an acceleration **a** that has the same direction as the force and a magnitude that is directly proportional to the force.* If **F** is applied to a particle of mass m, this law may be expressed mathematically as

$$\mathbf{F} = m\mathbf{a} \tag{1–1}$$

Third Law. For every force acting on a particle, the particle exerts an equal, opposite, and collinear reactive force.

Newton's Law of Gravitational Attraction. Shortly after formulating his three laws of motion, Newton postulated a law governing the gravitational attraction between any two particles. This law can be expressed mathematically as

$$F = G\frac{m_1 m_2}{r^2} \tag{1–2}$$

where F = force of gravitation between the two particles
G = universal constant of gravitation; according to experimental evidence, $G = 6.673(10^{-11})\ \text{m}^3/(\text{kg}\cdot\text{s}^2)$
m_1, m_2 = mass of each of the two particles
r = distance between the centers of mass of the two particles

Any two particles or bodies have a mutual attractive (gravitational) force acting between them. In the case of a particle located at or near the surface of

*Stated another way, the unbalanced force acting on the particle is proportional to the time rate of change of the particle's linear momentum.

the earth, however, the only attractive force having any sizable magnitude is that of the earth's gravitation. Consequently, this force, termed the *weight,* will be the only gravitational force considered.

Fundamental Idealizations 1.2

In mechanics models or idealizations are used in order to simplify application of the theory. A few of the more important idealizations will now be defined. Others that are of importance will be discussed at points where they are needed.

Particle. A *particle* has a mass but essentially has no size or shape. When a body is idealized as a particle, the principles of mechanics reduce to a rather simplified form since the geometry of the body will not be involved in the analysis of the problem.

Rigid Body. A *rigid body* can be considered as a combination of a large number of particles in which all the particles remain at a fixed distance from one another both before and after a load is applied. This idealization eliminates the need for any experimental testing, since no deformation occurs and consequently a body's material properties are not considered in the analysis of the forces acting on the body. In most cases the actual deformations occurring in structures, machines, mechanisms, and the like are relatively small, and the rigid-body assumption is suitable for analysis.

Concentrated Force. A concentrated force represents the effect of a loading which is assumed to act at a *point* on a body. Most often we can represent the effect of a loading by a concentrated force, provided the area over which the load is applied is *small* compared to the over-all size of the body.

Units of Measurement 1.3

Fundamental Quantities. Four fundamental quantities commonly used in mechanics are length, time, mass, and force. In general, the magnitude of each of these quantities is defined by an arbitrarily chosen *unit* or "standard."

Length. The concept of *length* is needed to locate the position of a point in space and thereby to describe the size of a physical system. The standard unit of length measurement is the *meter* (m), which is defined as the distance light travels through a vacuum in 1/299,792,458 of a second. All other units of length are defined in terms of this standard. For example, 1 foot (ft) is equal to 0.3048 m.

Time. The concept of *time* is conceived by a succession of events. The standard unit used for its measurement is the second (s), which is based on the duration of 9 192 631 770 cycles of vibration of an isotope cesium 133.

Mass. The *mass* of a body is regarded as a quantitative property of matter used to measure the resistance of matter to a change in velocity. The standard unit of mass is the *kilogram* (kg), defined by a bar of platinum–iridium alloy kept at the International Bureau of Weights and Measures in Sèvres, France.

Force. In general, *force* is considered as a "push" or "pull" exerted by one body on another. This interaction can occur when there is either direct contact between bodies, such as a person pushing on a wall, or it can occur through a distance by which the bodies are physically separated. Examples of the latter type include gravitational, electrical, and magnetic forces. In any case, a force is completely characterized by its magnitude, direction, and point of application. Most often, engineers define the standard unit of force using either the newton (N) or the pound (lb).

Systems of Units. The four fundamental quantities—length, time, mass, and force—are not all independent from one another; in fact, they are *related* by Newton's second law of motion, $\mathbf{F} = m\mathbf{a}$. Hence, the units used to define force, mass, length, and time cannot *all* be selected arbitrarily. The equality $\mathbf{F} = m\mathbf{a}$ is maintained only if three of the four units, called *base units,* are *arbitrarily defined* and the fourth unit is *derived* from the equation.

Table 1–1 Systems of Units

Name	*Length*	*Time*	*Mass*	*Force*
International System of Units (SI)	meter (m)	second (s)	kilogram (kg)	newton* (N) $\left(\frac{\text{kg}\cdot\text{m}}{\text{s}^2}\right)$
U.S. Customary (FPS)	foot (ft)	second (s)	slug* $\left(\frac{\text{lb}\cdot\text{s}^2}{\text{ft}}\right)$	pound (lb)

*Derived unit.

SI Units. The International System of units, abbreviated SI after the French "Système International d'Unités," is a modern version of the metric system which has received worldwide recognition. As shown in Table 1–1, the SI system specifies length in meters (m), time in seconds (s), and mass in kilograms (kg). The unit of force, called a newton (N), is *derived* from $\mathbf{F} = m\mathbf{a}$. Thus, 1 newton is equal to a force required to give 1 kilogram of mass an acceleration of 1 m/s^2 ($\text{N} = \text{kg}\cdot\text{m/s}^2$).

U.S. Customary. In the U.S. Customary system of units (FPS), Table 1–1, length is in feet (ft), time is in seconds (s), and force is in pounds (lb). The unit of mass, called a *slug,* is *derived* from $\mathbf{F} = m\mathbf{a}$. Hence, 1 slug is equal to the amount of matter accelerated at 1 ft/s^2 when acted upon by a force of 1 lb ($\text{slug} = \text{lb}\cdot\text{s}^2/\text{ft}$).

The International System of Units 1.4

The SI system of units is used in this book since it is intended to become the worldwide standard for measurement. Consequently, the rules for its use and some of its terminology relevant to mechanics will now be presented.

Prefixes. When a numerical quantity is either very large or very small the units used to define its size may be modified by using a prefix. Some of the prefixes used in the SI system are shown in Table 1–2. Each represents a multiple or submultiple of a unit which, if applied successively, moves the decimal point of a numerical quantity to every third place.* For example, 4 000 000 N = 4 000 kN (kilo-newton) = 4 MN (mega-newton), or 0.005 m = 5 mm (milli-meter). Notice that the SI system does not include the multiple deca (10) or the submultiple centi (0.01), which form part of the old metric system. Except for some volume and area measurements, the use of these prefixes is to be avoided in science and engineering.

Table 1–2 Prefixes

	Exponential Form	*Prefix*	*SI Symbol*
Multiple			
1 000 000 000	10^9	giga	G
1 000 000	10^6	mega	M
1 000	10^3	kilo	k
Submultiple			
0.001	10^{-3}	milli	m
0.000 001	10^{-6}	micro	μ
0.000 000 001	10^{-9}	nano	n

Attaching a prefix to a unit in effect creates a new unit. Thus, if a multiple or submultiple is raised to a power, the power applies to this new unit, not just to the original unit *without* the multiple or submultiple. For example, $(2\text{ km})^2 = (2)^2(\text{km})^2$ and is represented symbolically as 4 km^2. Also, 1 mm^2 represents $1\ (\text{mm})^2$, not $1\ \text{m}(\text{m}^2)$.

Rules for Use. The following rules are given for the proper use of the various SI symbols:

1. A symbol is *never* written with a plural "s," since it may be confused with the unit for second (s).
2. Symbols are always written in lower-case letters, with two exceptions: symbols for the two largest prefixes shown in Table 1–2, giga and mega,

*The kilogram is the only base unit that is defined with a prefix.

are capitalized as G and M, respectively; and symbols named after an individual are capitalized, e.g., N.

3. Quantities defined by several units which are multiples of one another are separated by a *dot* to avoid confusion with prefix notation, as illustrated by $N = kg \cdot m/s^2 = kg \cdot m \cdot s^{-2}$. Also, m · s (meter-second); whereas ms (milli-second).
4. Physical constants or numbers having several digits on either side of the decimal point should be reported with a *space* between every three digits rather than with a comma, e.g., 73 569.213 427. In the case of four digits on either side of the decimal, the spacing is optional, e.g., 8537 or 8 537. Furthermore, always try to use decimals and avoid fractions; that is, write 15.25, *not* $15\frac{1}{4}$.
5. When performing calculations, represent the numbers in terms of their *base or derived units* by converting all prefixes to powers of 10. The final result should then be expressed using a *single prefix*. Also, after calculation, it is best to keep numerical values between 0.1 and 1000; otherwise, a suitable prefix should be chosen. For example,

$$\begin{aligned}(50\ \text{kN})(60\ \text{nm}) &= [50(10^3)\ \text{N}][60(10^{-9})\ \text{m}] \\ &= 3000(10^{-6})\ \text{N} \cdot \text{m} = 3\ \text{mN} \cdot \text{m}\end{aligned}$$

6. Compound prefixes should not be used, e.g., kμs (kilo-micro-second) should be expressed as ms (milli-second) since $1\ \text{k}\mu\text{s} = 1(10^3)(10^{-6})\ \text{s} = (10^{-3})\ \text{s} = 1\ \text{ms}$.
7. With the exception of the base unit the kilogram, in general avoid the use of a prefix in the denominator of a composite unit. For example, do not write N/mm, but rather kN/m.
8. Although not expressed in multiples of 10, the minute, hour, etc., are retained for practical purposes as multiples of the second. Furthermore, plane angular measurement is made using radians (rad). In this book, degrees will sometimes be used, where $180° = \pi$ rad. Fractions of a degree, however, should be expressed in decimal form rather than in minutes, as in 10.4°, not 10°24′.

When learning to use SI units, it is generally agreed that one should *not* think in terms of conversion factors between systems. Instead, it is better to think *only* in terms of SI units. A "feeling" for these units can only be gained through experience. As a memory aid, it might be helpful to recall that a standard flashlight battery or a small apple weighs about 1 newton. Your body is a suitable reference for small distances. For example, the millimeter scale in Fig. 1–1 can be used to measure, say, the width of three or four fingers

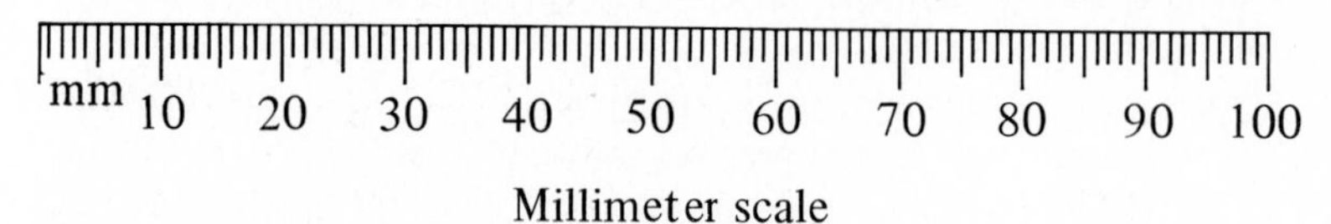

Millimeter scale

Fig. 1–1

pressed together, about 50 mm, or the width of the small fingernail, about 10 mm. For most people, a stretched walking pace is about 1 meter long.

Mass and Weight 1.5

The *mass* of a body is an *absolute* quantity since its measurement can be made at any location. The *weight* of a body, however, is *not absolute* since it is measured in a gravitational field, and hence its magnitude depends upon where the measurement is made.

The mass and weight of a body are measured differently in the SI and FPS systems, and the method of defining the units should be thoroughly understood.

SI System of Units. In the SI system the mass of the body is specified in kg and the weight must be calculated using the equation $\mathbf{F} = m\mathbf{a}$. Hence, if a body has a mass of m (kg) and is located at a point where the acceleration due to gravity is g (m/s^2), then the weight is expressed in *newtons* as $W = mg$ (N). In particular, if the body is located on the earth at sea level and at a latitude of 45° (considered the ''standard location''), the acceleration due to gravity is $g = 9.806\ 65$ m/s^2. For calculations, the value $g = 9.81$ m/s^2 will be used, so that

$$W = mg \text{ (N)} \quad (g = 9.81 \text{ m/s}^2) \tag{1–3}$$

Therefore, a body of mass 1 kg has a weight of 9.81 N; a 2-kg body weighs 19.62 N; and so on.

FPS System of Units. In the FPS system the weight of the body is specified in lb and the mass must be calculated from $\mathbf{F} = m\mathbf{a}$. Hence, if a body has a weight of W (lb) and is located at a point where the acceleration due to gravity is g (ft/s^2), then the mass is expressed in *slugs* as $m = W/g$ (slug). Since the acceleration of gravity at the standard location is approximately 32.2 ft/s^2 (= 9.81 m/s^2), the mass of the body measured in slugs is

$$m = \frac{W}{g} \text{(slug)} \quad (g = 32.2 \text{ ft/s}^2) \tag{1–4}$$

Therefore, a body weighing 32.2 lb has a mass of 1 slug; a 64.4-lb body has a mass of 2 slugs; and so on.

Dimensional Quantities 1.6

Dimensional Homogeneity. The terms of any equation used to describe a physical process must be *dimensionally homogeneous,* that is, each term must be expressed in the same units. Provided this is the case, all the terms of an equation can then be combined if numerical values are substituted for the variables. Consider, for example, the equation $s = vt + \frac{1}{2}at^2$, where, in SI units, s is the position in meters (m), t is time in seconds (s), v is velocity in m/s, and a is acceleration in m/s^2. Regardless of how this equation is evaluated, it maintains its dimensional homogeneity. In the form stated each of the three terms is expressed in meters [m, (m/$\not{s}$)$\not{s}$, (m/$\not{s}^{\not{2}}$)$\not{s}^{\not{2}}$], or solving for a, $a = 2s/t^2 - 2v/t$, the terms are each expressed in units of m/s^2 [m/s^2, m/s^2, (m/s)1/s].

Since mechanics problems involve the solution of dimensionally homogeneous equations, the fact that all terms of an equation are represented by a consistent set of units can be used as a partial check for algebraic manipulations of an equation.

Accuracy. In practice, forces are estimated to be about 90 per cent accurate, and often the mass of a body and its measurements are approximated. Hence, the accuracy obtained from the solution of a problem generally can never be better than the accuracy of the problem data. This, of course, is to be expected, but often computers or hand calculators involve more figures in the answer than the numbers used for the data. Since accuracy is often lost when subtracting numbers that are approximately equal, numerical work for problem solving should be carried out as accurately as possible, and the final answers should be rounded off to a value that reflects the accuracy of the original data.

Conversion of Units. In some cases it may be necessary to convert from one system of units to another. In this regard, Table 1–3 provides a set of direct conversion factors between FPS and SI units for the fundamental quantities. Also, in the FPS system, recall that 1 ft = 12 in. (inches), 5280 ft = 1 mi (mile); and 1000 lb = 1 kip (kilo-pound), 2000 lb = 1 ton.

Table 1–3 Conversion Factors

Quantity	*Unit of Measurement (FPS)*	*Equals*	*Unit of Measurement (SI)*
Force	lb		4.4482 N
Mass	slug		14.5938 kg
Length	ft		0.3048 m

When derived units are present, a general procedure using a simple cancellation technique should be applied. The following examples illustrate this method.

Example 1–1

Convert the speed of 2 km/h to m/s.

Solution

Since 1 km = 1000 m and 1 h = 3600 s, the factors of conversion are arranged as follows:

$$2\text{ km/h} = \frac{2\ \not{\text{km}}}{\not{\text{h}}}\left(\frac{1000\text{ m}}{\not{\text{km}}}\right)\left(\frac{1\ \not{\text{h}}}{3600\text{ s}}\right)$$

$$= \frac{2000\text{ m}}{3600\text{ s}} = 0.556\text{ m/s} \qquad \textit{Ans.}$$

Example 1–2

Convert the "impulse" 300 lb · s to appropriate SI units.

Solution

Using Table 1–3, 1 lb = 4.4482 N, then

$$300\text{ lb}\cdot\text{s} = 300\ \not{\text{lb}}\cdot\text{s}\left(\frac{4.4482\text{ N}}{\not{\text{lb}}}\right)$$

$$= 1334.4\text{ N}\cdot\text{s} = 1.33\text{ kN}\cdot\text{s} \qquad \textit{Ans.}$$

Example 1–3

Evaluate each of the following and express with SI units having an appropriate prefix: (a) (50 mN)(6 GN), (b) (400 mm)(0.6 MN).

Solution

First convert each number to base units, perform the indicated operations, then choose an appropriate prefix.

Part (a):

$$(50\text{ mN})(6\text{ GN}) = [50(10^{-3})\text{ N}][6(10^{9})\text{ N}]$$

$$= 300(10^{6})\text{ N}^2$$

$$= 300(10^{6})\ \not{\text{N}}^{\not{2}}\ (10^{-3}\text{ mN}/\not{\text{N}})(10^{-3}\text{ mN}/\not{\text{N}})$$

$$= 300\text{ mN}^2 \qquad \textit{Ans.}$$

Part (b):

$$(400\text{ mm})(0.6\text{ MN}) = [400(10^{-3})\text{ m}][0.6(10^{6})\text{ N}]$$

$$= 240(10^{3})\text{ m}\cdot\text{N}$$

$$= 240\text{ km}\cdot\text{N} \qquad \textit{Ans.}$$

1.7 General Procedure for Analysis

The most effective way of learning the principles of engineering mechanics is to *solve problems*. To be successful at this, it is important to present the work in a *logical* and *orderly manner*, as suggested by the following sequence of steps:

1. Read the problem carefully and try to correlate the actual physical situation with the theory studied.
2. Draw any necessary diagrams and tabulate the problem data.
3. List all the relevant principles, generally in mathematical form.
4. Solve the necessary equations algebraically as far as practical, then making sure they are dimensionally homogeneous, use a consistent set of units and complete the solution numerically.
5. Study the answer with technical judgment and common sense to determine whether or not it seems reasonable.
6. Once the solution has been completed, review the problem. Try to think of other ways of obtaining the same solution.

In applying this general procedure, do the work as neatly as possible. Being neat generally stimulates clear and orderly thinking, and vice versa.

Problems

1–1. What is the weight in SI units of an object that has a mass of: (a) 10 kg; (b) 0.5 g; (c) 4.50 Mg?

1–2. Water has a density of 1.94 slug/ft^3. What is its density expressed in SI units?

1–3. Using Table 1–3, determine your mass in kilograms, weight in newtons, and height in meters.

***1–4.** Represent each of the following with SI units having an appropriate prefix: (a) 8653 ms; (b) 8368 N; (c) 0.893 kg.

1–5. Evaluate each of the following and express with SI units having an appropriate prefix: (a) $(200 \text{ kN})^2$, (b) $(0.005 \text{ mm})^2$, (c) $(400 \text{ m})^3$.

1–6. Represent each of the following combinations of units in the correct SI form: (a) μMN; (b) N/μm; (c) MN/ks^2; (d) kN/ms.

1–7. Represent each of the following combinations of units in the correct SI form: (a) Mg/ms; (b) N/mm; (c) mN/(kg $\cdot$ μs).

***1–8.** Convert: (a) 200 lb $\cdot$ ft to N $\cdot$ m; (b) 350 lb/ft^3 to kN/m^3; (c) 8 ft/h to mm/s.

1–9. The *pascal* (Pa) is actually a very small unit of pressure. To show this, convert 1 Pa = 1 N/m^2 to lb/ft^2. Atmospheric pressure at sea level is 14.7 lb/in.^{2}. How many pascals is this?

1–10. Using the base units of the SI system, show that Eq. 1–2 is a dimensionally homogeneous equation which gives F in newtons. Compute the gravitational force acting between two spheres that are touching each other. The mass of each sphere is 200 kg and the radius is 300 mm.

1–11. If a body weighs 600 lb on earth, specify: (a) its mass in slugs, (b) its mass in kilograms, and (c) its weight in newtons. If the 600-lb body is placed on the moon, where the acceleration due to gravity is $g_m = 5.30$ ft/s^2, determine: (d) its weight in pounds and (e) its mass in kilograms.

Force Vectors

2

Force Resultants in Two Dimensions

In this chapter we will introduce the concept of a concentrated force and give the procedures for adding forces and resolving them into components. In order to better understand the methods of analysis, application will first be made to force systems in two dimensions, followed by that in three dimensions. Since force is a vector quantity, we must use the rules of vector algebra whenever forces are added. In this regard, we will begin our study by defining scalar and vector quantities and then develop some of the basic rules of vector addition.

Scalars and Vectors 2.1

Most of the physical quantities in mechanics can be expressed mathematically by means of scalars and vectors.

Scalar. A quantity possessing only a magnitude is called a *scalar*. Mass, volume, and length are scalar quantities often used in statics. In this book, scalars are indicated by letters in italic type, such as the scalar A. The mathematical operations involving scalars follow the same rules as those of elementary algebra.

Vector. A *vector* is a quantity that has both a magnitude and direction and "adds" according to the parallelogram law. This law, which will be described later, utilizes a form of construction that accounts for the combined magnitude and direction of the vector. Vector quantities commonly used in statics are position, force, and moment vectors.

For handwritten work, a vector is generally represented by a letter with an arrow written over it, such as $\vec{A}$. The magnitude is designated by $|\vec{A}|$, or simply A. In this book vectors are symbolized in boldface type; for example, **A** is used to designate the vector "A", and its magnitude is symbolized in italic type, A.

A vector is represented graphically by an arrow, which is used to define its magnitude, line of action, and direction. The *magnitude* of the vector is indicated by the *length* of the arrow, and its *direction* is indicated by an *arrowhead*. For example, the vector **A** shown in Fig. 2–1 has a magnitude of 4 units and is directed upward, along its line of action, which is 20° above the horizontal. The point O is called the *tail* of the vector; the point P is the *tip*.

Fig. 2–1

Types of Vectors. In statics various types of vectors occur and it is important to know how they are described.

1. A *fixed vector* acts at a fixed point in space, Fig. 2–2*a*.
2. A *sliding vector* may be applied at any point along its line of action, Fig. 2–2*b*.
3. A *free vector* can act anywhere in space; it is only necessary that it preserve its magnitude and direction, Fig. 2–2*c*.
4. *Equal vectors* have the same magnitude and direction, Fig. 2–2*d*.

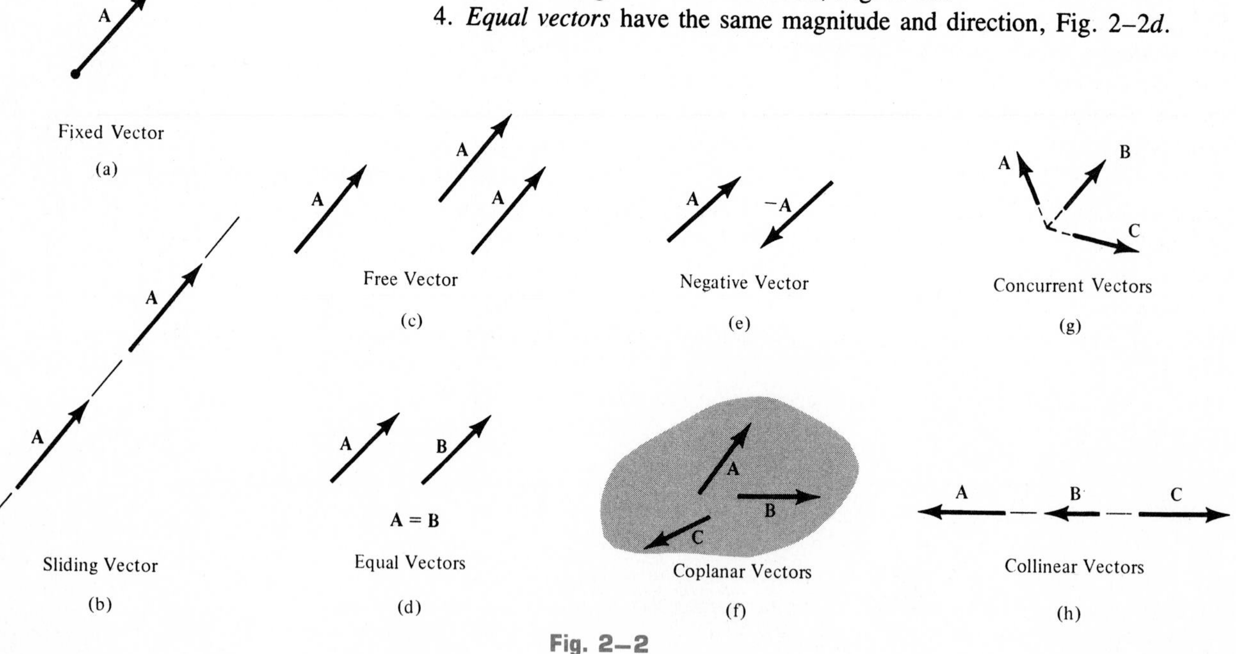

Fig. 2–2

pg 12 & 13

5. A *negative vector* has a direction opposite to its positive counterpart, but has the same magnitude, Fig. 2–2*e*.
6. *Coplanar vectors* lie in the same plane, Fig. 2–2*f*.
7. *Concurrent vectors* have lines of action that pass through the same point, Fig. 2–2*g*.
8. *Collinear vectors* have the same line of action, Fig. 2–2*h*.

Vector Addition 2.2

In this section we will discuss the addition of two *component vectors* **A** and **B** which yield a *resultant vector* **R.** In general, two vectors are added together using the *parallelogram law*. This method of addition will now be described for two types of problems which often occur in practice.

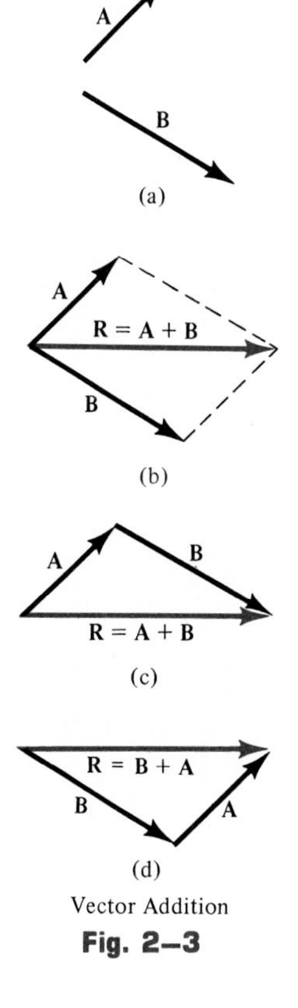

Vector Addition

Fig. 2–3

Known Components, Unknown Resultant. When the two components shown in Fig. 2–3*a* are added to form the resultant **R,** they are joined at their tails, forming adjacent sides of a parallelogram constructed as shown in Fig. 2–3*b*. The resultant **R** is then determined by extending an arrow from the tails of **A** and **B** along the diagonal of the parallelogram to the opposite corner.

We can also add **A** to **B** using a *triangle construction,* which is a special case of the parallelogram law, whereby vector **B** is added to vector **A** in a "tip-to-tail" fashion, i.e., by placing the tail of **B** at the tip of **A,** Fig. 2–3*c*. The resultant **R** extends from the tail of **A** to the tip of **B.** In a similar manner, **R** can also be obtained by adding **A** to **B,** Fig. 2–3*d*. Hence, it is seen that vector addition is commutative; in other words, the component vectors can be added in *any* order, i.e., **R** = **A** + **B** = **B** + **A.** Once the triangle has been formed, the magnitude and direction of the resultant can then be determined using trigonometry.

Known Resultant, Unknown Components. In some problems the magnitude and direction of the resultant force will be known and it will have to be resolved into two components having known lines of action. Again, the parallelogram law must be used for the construction. For example, if **R** in Fig. 2–4 is to be resolved along the dashed axes *a* and *b*, one would start at the *tip*

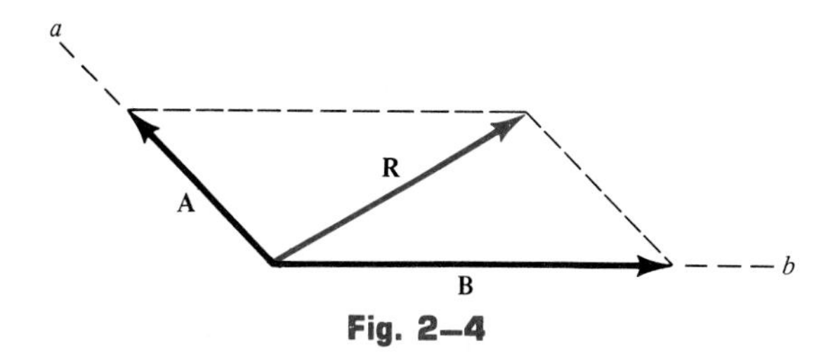

Fig. 2–4

of **R** and extend a dashed line drawn *parallel* to the *a* axis until it intersects the *b* axis. Likewise, a dashed line *parallel* to the *b* axis is drawn from the tip of **R** to the point of intersection with the *a* axis. The two components **A** and **B** are then scaled from the tail of **R** to the points of intersection, as shown in the figure. From this, one can then construct the triangle showing a tip-to-tail addition, and determine the unknown magnitudes of **A** and **B** by trigonometry.

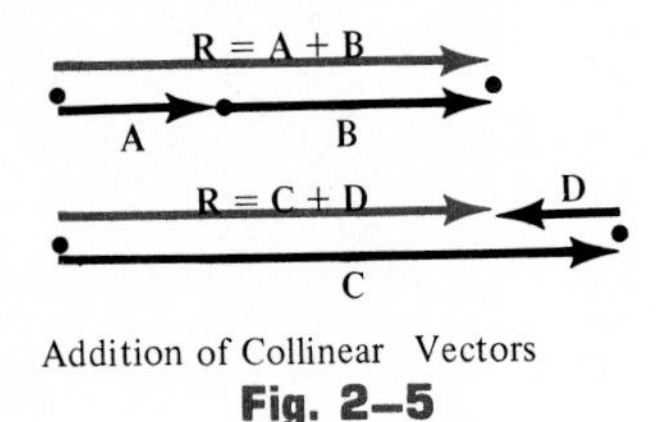

Addition of Collinear Vectors
Fig. 2–5

Collinear Vectors. If the two component vectors are collinear, i.e., both lie along the same line of action, the parallelogram law reduces to an algebraic addition of the components such as shown in Fig. 2–5.

2.3 Multiplication and Division of a Vector by a Scalar

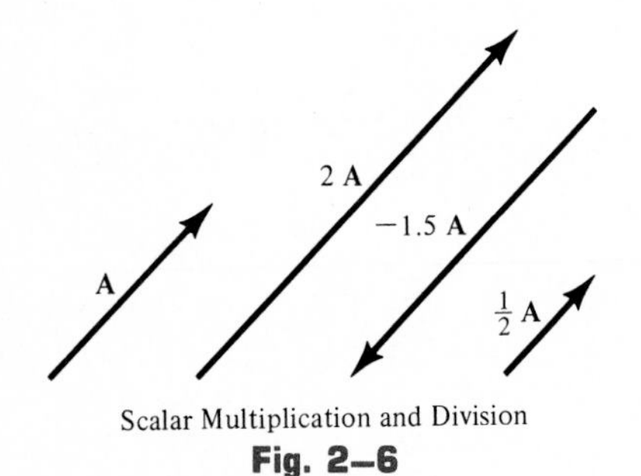

Scalar Multiplication and Division
Fig. 2–6

The product of vector **A** and scalar m, yielding $m\mathbf{A}$, is defined as a vector having a magnitude mA. The direction of $m\mathbf{A}$ is the same as **A** provided m is positive; it is opposite to **A** if m is negative. Consequently, the negative of a vector is formalized by multiplying the vector by the scalar (-1). Since the vector magnitude is always positive, a minus sign in front of a vector simply means that its direction is reversed.

Division of a vector by a scalar can be defined using the laws of multiplication, since $\mathbf{A}/m = (1/m)\mathbf{A}$, $m \neq 0$. Graphic examples of these operations are shown in Fig. 2–6.

2.4 Vector Addition of Forces Using the Parallelogram Law

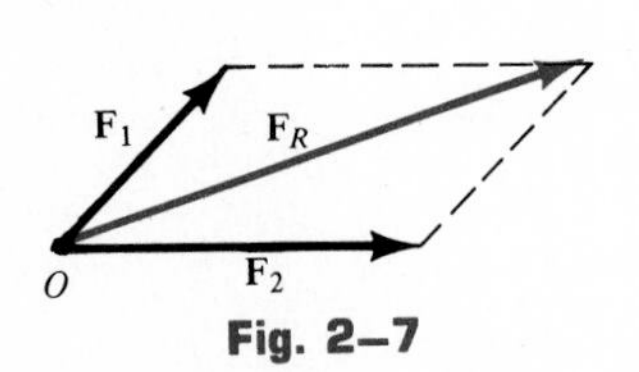

Fig. 2–7

Experimental evidence has shown that a force is a vector quantity since it has a specified magnitude and direction, and it adds according to the parallelogram law. Two common problems in statics involve either the determination of the magnitude and direction of the resultant force $\mathbf{F}_R$, given the component forces $\mathbf{F}_1$ and $\mathbf{F}_2$ acting at a point O, or the determination of the magnitudes and/or directions of the components $\mathbf{F}_1$ and $\mathbf{F}_2$, given the resultant force $\mathbf{F}_R$. Both of these problems require application of the parallelogram law, Fig. 2–7.

If more than two forces act at point O, successive applications of the parallelogram law can be carried out in order to obtain the resultant force. For example, if three coplanar forces $\mathbf{F}_1$, $\mathbf{F}_2$, $\mathbf{F}_3$ act at O, Fig. 2–8, the resultant of any two of the forces is found, say $\mathbf{F}_1 + \mathbf{F}_2$, and then this resultant is added to the third force, yielding the resultant of all three forces, i.e., $\mathbf{F}_R = (\mathbf{F}_1 + \mathbf{F}_2) + \mathbf{F}_3$. Using the parallelogram law to add more than two forces most often requires extensive geometric and trigonometric calculation. Problems of this type, however, are easily solved by using the "rectangular-component method," which is explained in the next section.

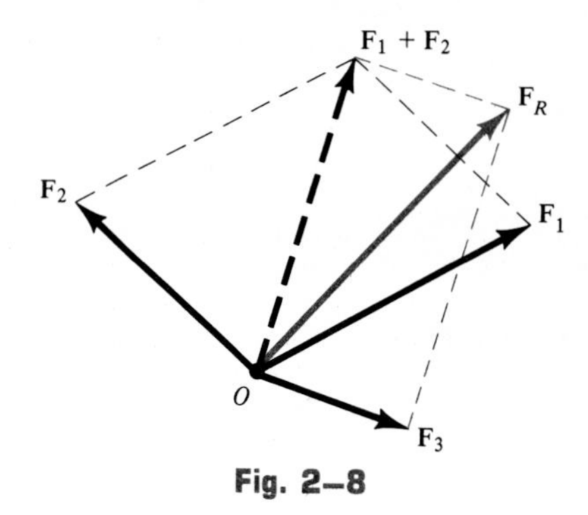

Fig. 2–8

PROCEDURE FOR ANALYSIS

Problems that involve the addition of two forces and contain at most *two unknowns* can be solved by using the following procedure:

Parallelogram Law. Make a sketch showing the vector addition using the parallelogram law. If possible, determine the interior angles of the parallelogram from the geometry of the problem. Recall that the sum total of the angles is to be 360°. Unknown angles, along with known and unknown force magnitudes, should clearly be labeled on this sketch. Redraw a portion of the constructed parallelogram to illustrate the triangular tip-to-tail addition of the components.

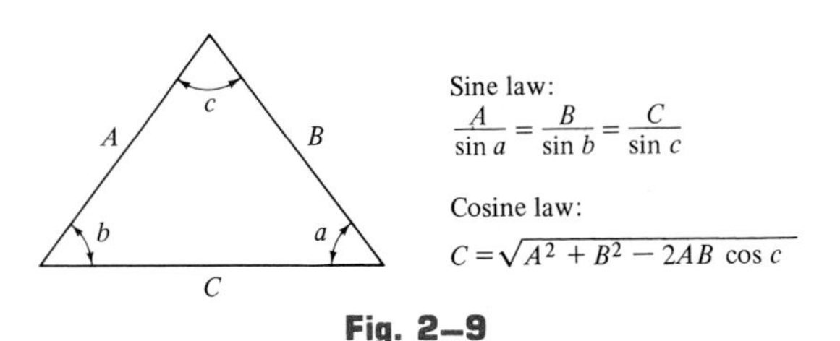

Fig. 2–9

Trigonometry. By using trigonometry, the two unknowns can be determined from the data listed on the triangle. If the triangle does *not* contain a 90° angle, the law of sines and/or the law of cosines may be used for the solution. These formulas are given in Fig. 2–9 for the triangle shown.

The following examples numerically illustrate this method for solution.

Example 2–1

The screw eye in Fig. 2–10*a* is subjected to two forces, $\mathbf{F}_1$ and $\mathbf{F}_2$. Determine the magnitude and direction of the resultant force.

Solution

Parallelogram Law. The parallelogram law of addition is shown in Fig. 2–10*b*. The two unknowns are the magnitude of $\mathbf{F}_R$ and the angle θ (theta). From Fig. 2–10*b*, the force triangle, Fig. 2–10*c*, is constructed.

Trigonometry. F_R is determined by using the law of cosines:

$$\begin{aligned} F_R &= \sqrt{(100)^2 + (150)^2 - 2(100)(150)\cos 115^\circ} \\ &= \sqrt{10\,000 + 22\,500 - 30\,000(-0.423)} \\ &= 212.6 \text{ N} \end{aligned}$$ ***Ans.***

The angle θ is determined by applying the law of sines, using the computed value of F_R.

$$\frac{150}{\sin\theta} = \frac{212.6}{\sin 115^\circ}$$

$$\sin\theta = \frac{150}{212.6}(0.906) = 0.639$$

$$\theta = 39.7^\circ$$

Thus, the direction ϕ (phi) of $\mathbf{F}_R$, measured from the horizontal, is

$$\phi = 39.7^\circ + 15.0^\circ = 54.7^\circ \ \angle^{\phi}$$ ***Ans.***

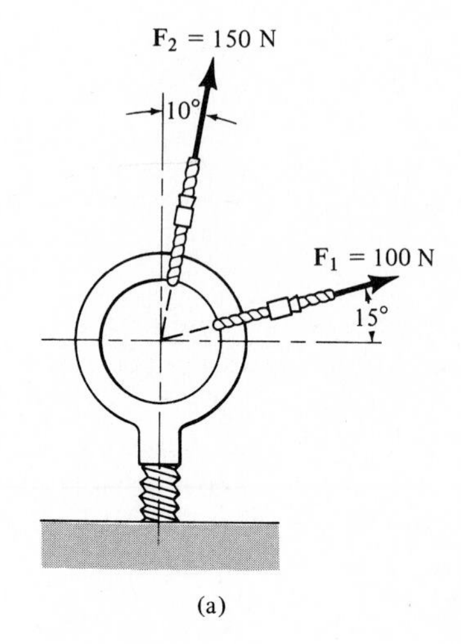

(a)

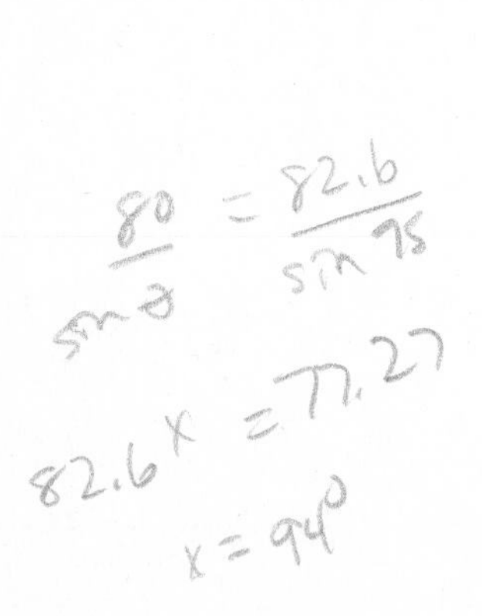

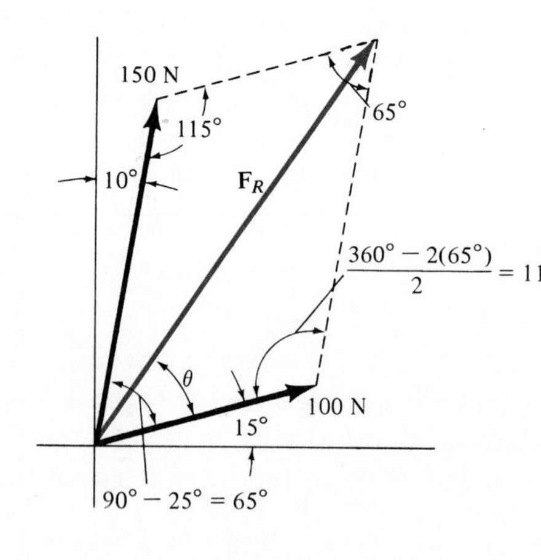

(b)

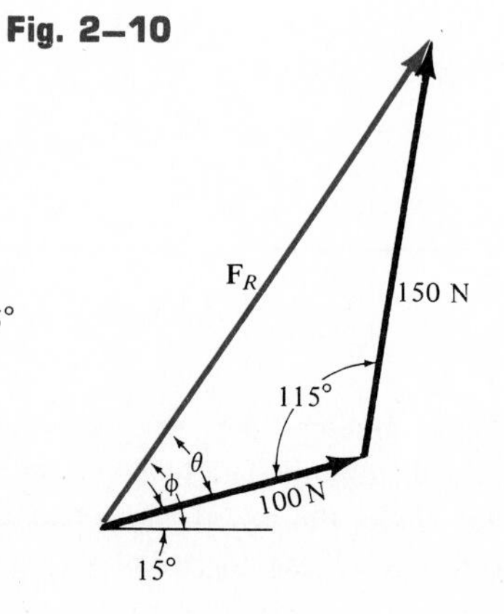

(c)

Fig. 2–10

Example 2–2

Resolve the 200-lb force shown acting on the pin, Fig. 2–11a, into components in the (a) x and y directions and (b) x' and y directions.

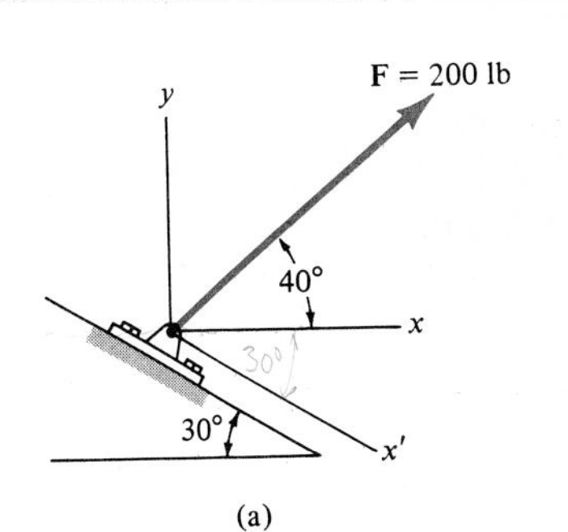

(a)

Solution

In each case the parallelogram law is used to resolve **F** into its two components, and then the triangle law is applied to determine the numerical results by trigonometry.

Part (a). The vector addition $\mathbf{F} = \mathbf{F}_x + \mathbf{F}_y$ is shown in Fig. 2–11b. In particular, note that the length of the components is scaled along the x and y axes by first constructing dashed lines parallel to the axes in accordance with the parallelogram law. From the vector triangle, Fig. 2–11c,

$$F_x = 200 \cos 40° = 153.2 \text{ lb} \qquad \textit{Ans.}$$

$$F_y = 200 \sin 40° = 128.6 \text{ lb} \qquad \textit{Ans.}$$

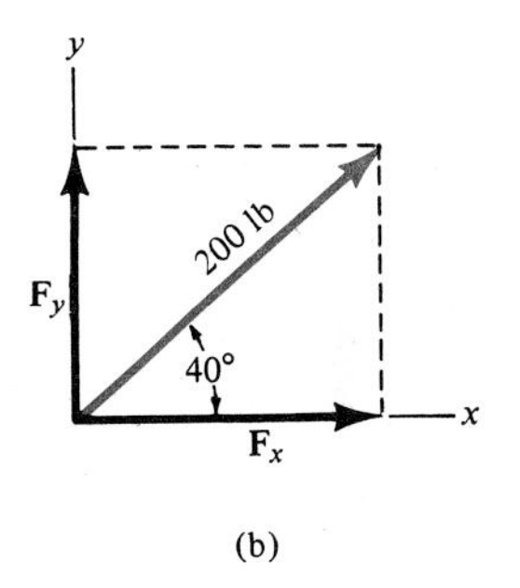

(b)

Part (b). The vector addition $\mathbf{F} = \mathbf{F}_{x'} + \mathbf{F}_y$ is shown in Fig. 2–11d. Note carefully how the parallelogram is constructed. Applying the law of sines and using the data listed on the vector triangle, Fig. 2–11e, yields

$$\frac{F_{x'}}{\sin 50°} = \frac{200}{\sin 60°}$$

$$F_{x'} = 200\left(\frac{\sin 50°}{\sin 60°}\right) = 176.9 \text{ lb} \qquad \textit{Ans.}$$

$$\frac{F_y}{\sin 70°} = \frac{200}{\sin 60°}$$

$$F_y = 200\left(\frac{\sin 70°}{\sin 60°}\right) = 217.0 \text{ lb} \qquad \textit{Ans.}$$

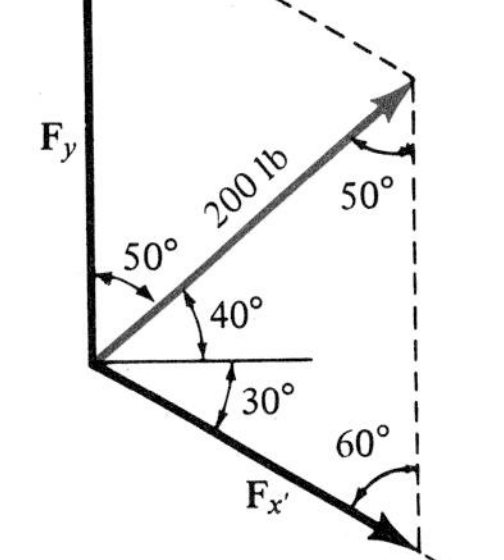

(d)

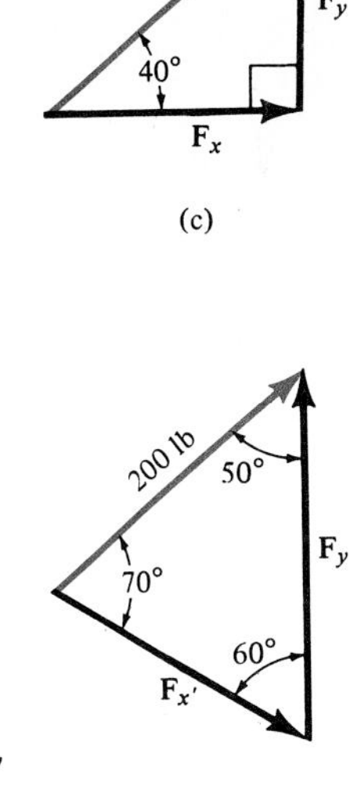

(e)

Fig. 2–11

Example 2–3

The force **F** acting on the frame shown in Fig. 2–12*a* has a magnitude of 500 N and is to be resolved into two components acting along struts *AB* and *AC*. Determine the angle θ ($0° \leq \theta \leq 90°$) so that the component $\mathbf{F}_{AC}$ is directed from *A* toward *C* and has a magnitude of 400 N.

Solution

By using the parallelogram law, the vector addition of the two components yielding the resultant is shown in Fig. 2–12*b*. Note carefully how the resultant force is resolved into the two components $\mathbf{F}_{AB}$ and $\mathbf{F}_{AC}$, which have specified lines of action. The corresponding vector triangle is shown in Fig. 2–12*c*. The angle ϕ can be determined by using the law of sines:

$$\frac{400}{\sin \phi} = \frac{500}{\sin 60°}$$

$$\sin \phi = \left(\frac{400}{500}\right) \sin 60° = 0.693$$

$$\phi = 43.9°$$

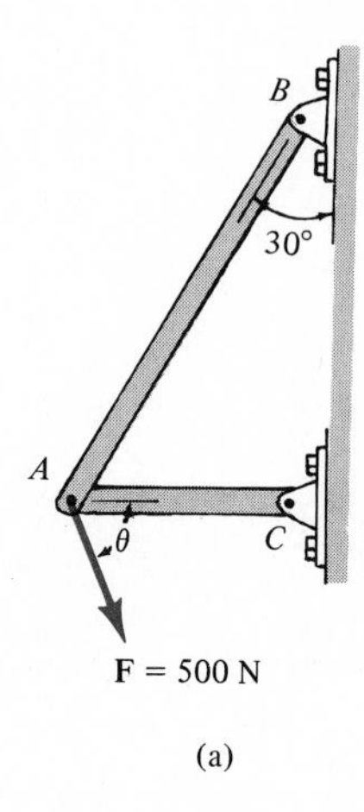

(a)

Hence,

$$\theta = 180° - 60° - 43.9° = 76.1° \quad \text{Ans.}$$

Show that $\mathbf{F}_{AB}$ has a magnitude of 560.4 N. Note that this component is *larger* in magnitude than the applied force **F**.

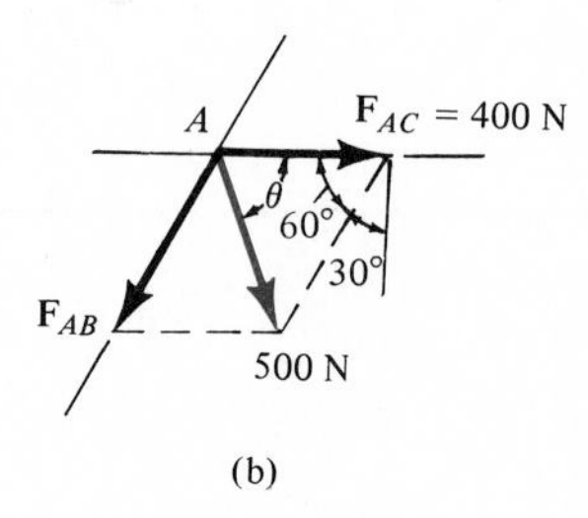

(b)

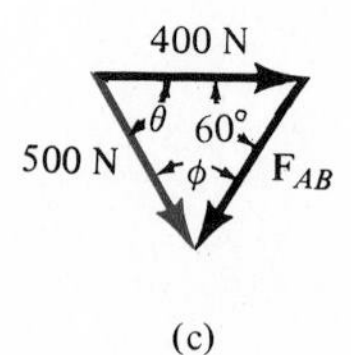

(c)

Fig. 2–12

Example 2–4

The ring shown in Fig. 2–13*a* is subjected to two forces, $\mathbf{F}_1$ and $\mathbf{F}_2$. If it is required that the resultant force have a magnitude of 1 kN and be directed vertically downward, determine: (a) the magnitudes of $\mathbf{F}_1$ and $\mathbf{F}_2$ provided $\theta = 30°$; (b) the magnitudes of $\mathbf{F}_1$ and $\mathbf{F}_2$ if F_2 is to be a minimum.

Solution

Part (a). A sketch of the vector addition, according to the parallelogram law, is shown in Fig. 2–13*b*. From the vector triangle constructed in Fig. 2–13*c* the unknown magnitudes F_1 and F_2 can be determined by using the law of sines.

$$\frac{F_1}{\sin 30°} = \frac{1000}{\sin 130°}$$

$$F_1 = 652.7 \text{ N} \qquad \textit{Ans.}$$

$$\frac{F_2}{\sin 20°} = \frac{1000}{\sin 130°}$$

$$F_2 = 446.5 \text{ N} \qquad \textit{Ans.}$$

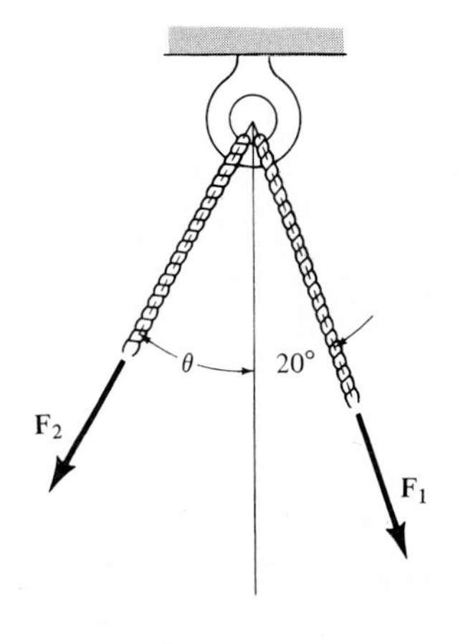

(a)

Part (b). By the vector triangle, $\mathbf{F}_2$ may be added to $\mathbf{F}_1$ in various ways to yield $\mathbf{F}_R$, Fig. 2–13*d*. In particular, the *minimum* length or magnitude of $\mathbf{F}_2$ will occur when its line of action is *perpendicular* to $\mathbf{F}_1$. Any other direction, such as *OA* or *OB*, yields a larger value for F_2. Hence, when $\theta = 90° - 20° = 70°$, F_2 is minimum. From the triangle shown in Fig. 2–13*e*, it is seen that

$$F_1 = 1000 \sin 70° = 939.7 \text{ N} \qquad \textit{Ans.}$$

$$F_2 = 1000 \sin 20° = 342.0 \text{ N} \qquad \textit{Ans.}$$

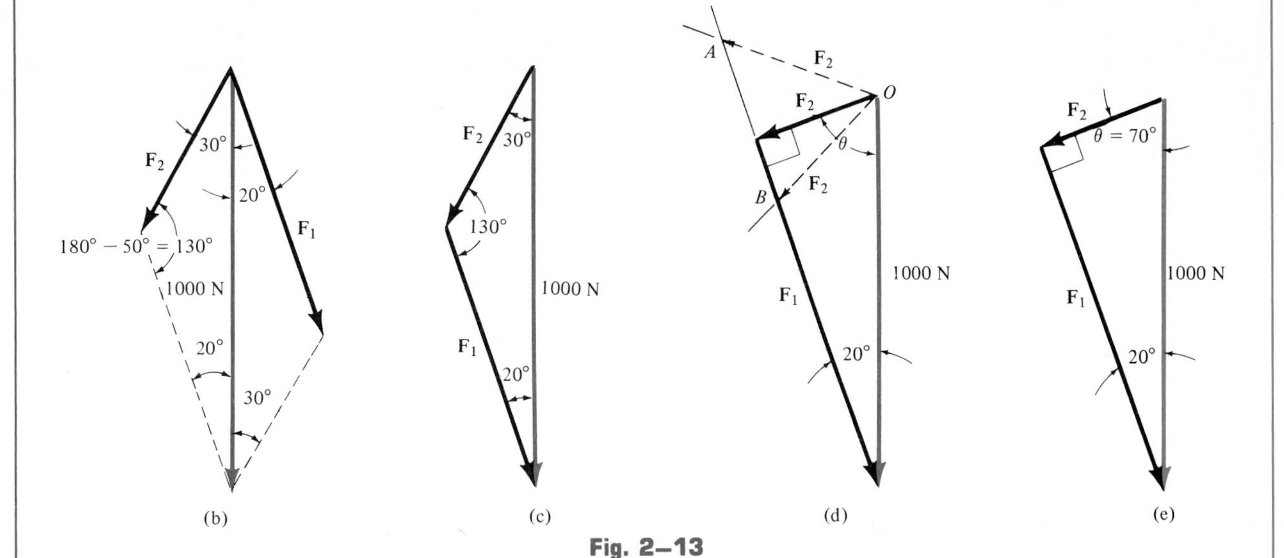

Fig. 2–13

Problems

2–1. Determine the magnitude of the resultant force and its direction measured from the positive x axis.

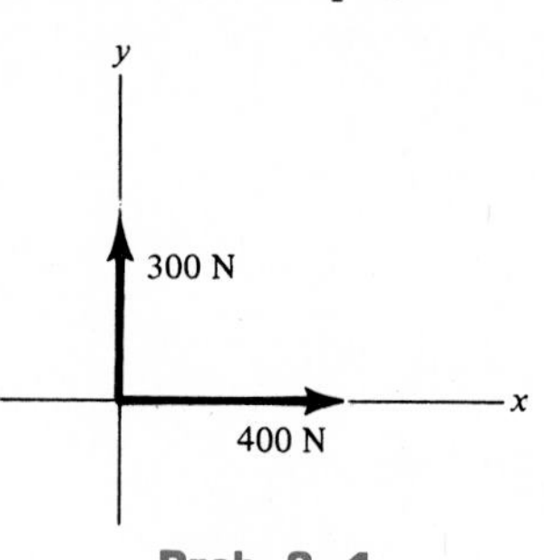
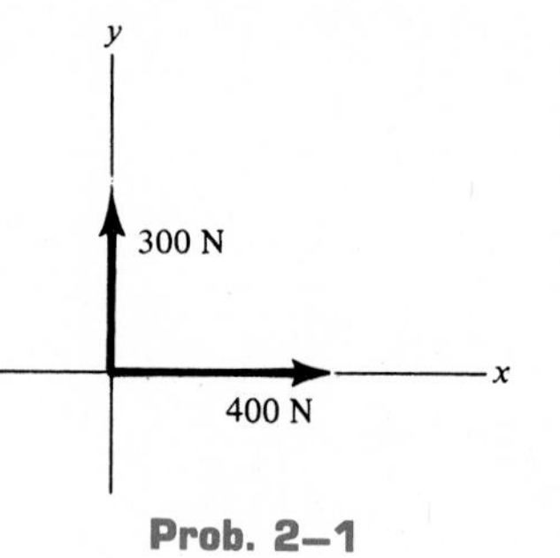

Prob. 2–1

2–2. Determine the magnitude of the resultant force and its direction measured from the positive x axis.

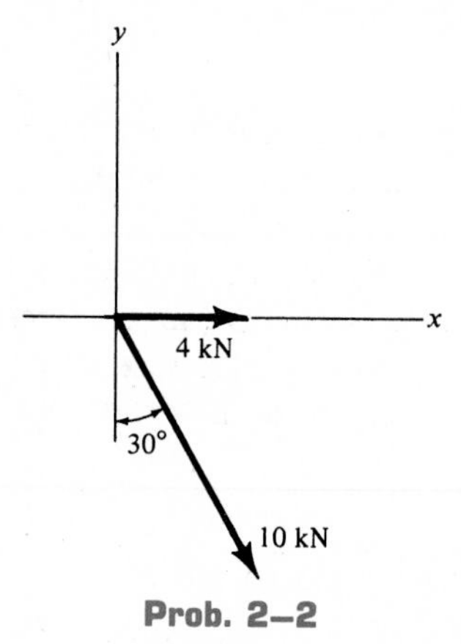

Prob. 2–2

2–3. Determine the magnitude of the resultant force and its direction measured from the positive x axis.

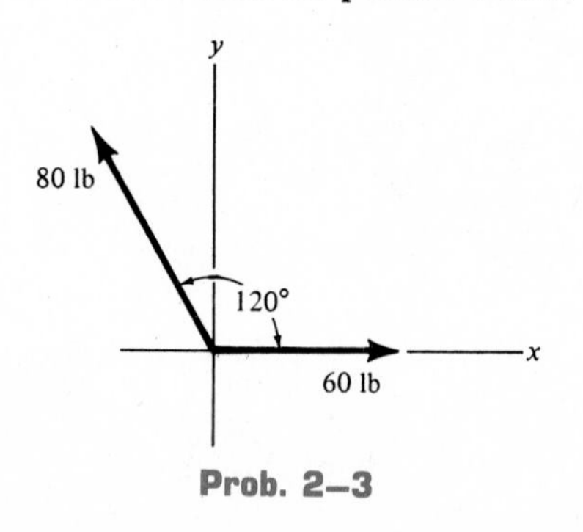

Prob. 2–3

***2–4.** Determine the magnitudes of the x and y components of the 150-N force.

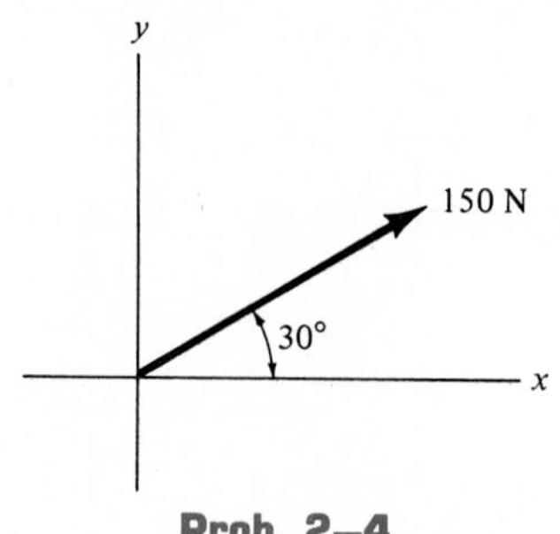

Prob. 2–4

2–5. Determine the magnitudes of the x and y components of the 2-kN force.

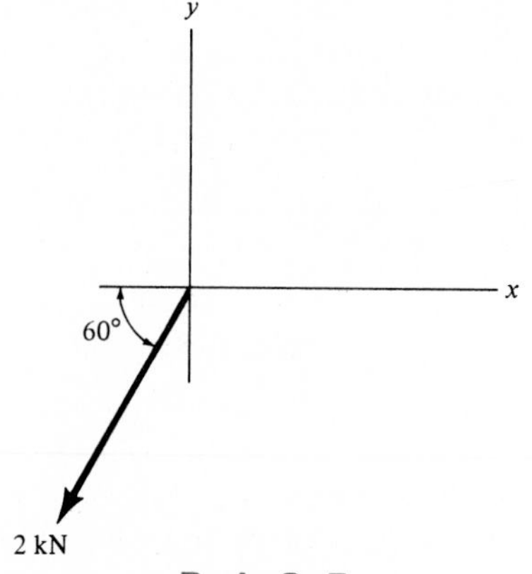

Prob. 2–5

2–6. Determine the magnitudes of the x and y components of the 700-lb force.

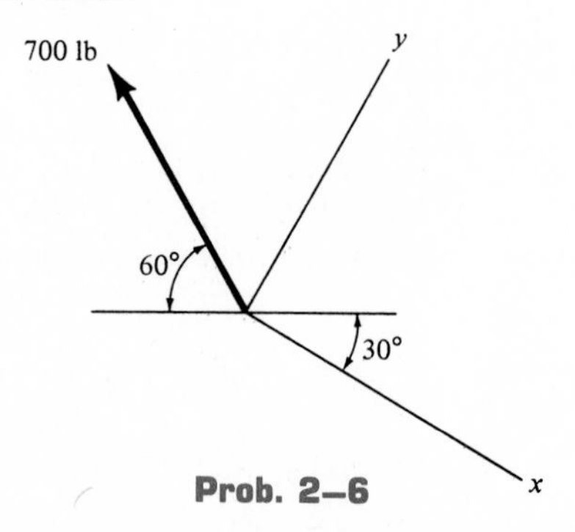

Prob. 2–6

2–7. Resolve the 200-N force into components acting along the u and v axes and determine the magnitudes of the components.

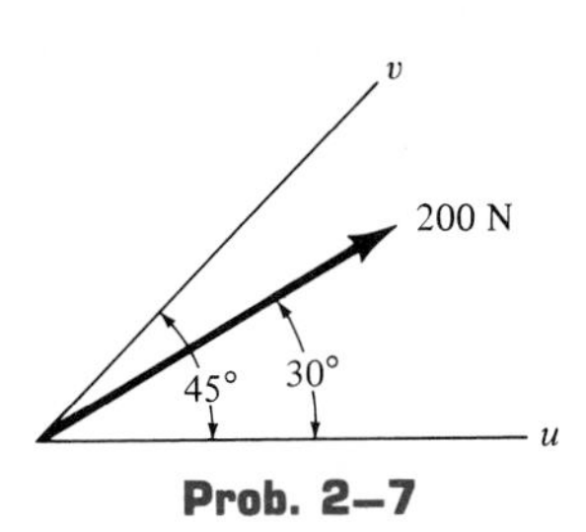

Prob. 2–7

***2–8.** Resolve the 40-kN force into components acting along the u and v axes and determine the magnitudes of the components.

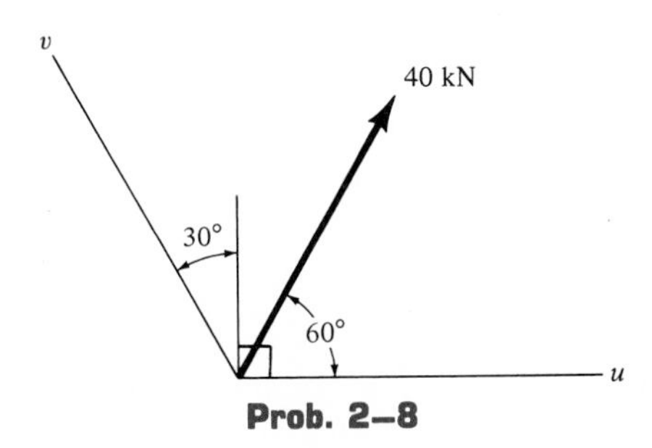

Prob. 2–8

2–9. Resolve the 25-lb force into components acting along the u and v axes and determine the magnitudes of the components. *Hint:* One of the components will act in the $-v$ direction.

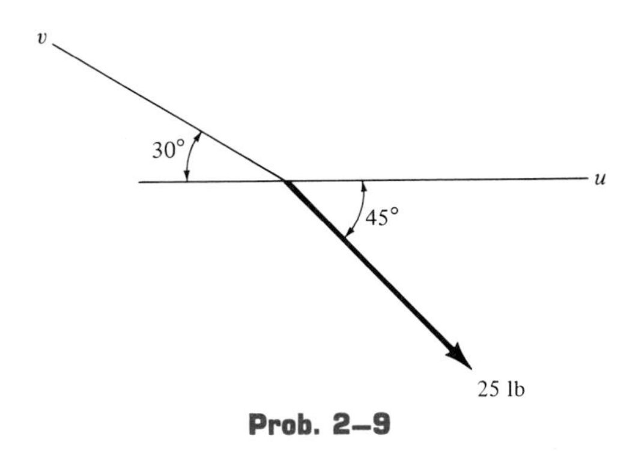

Prob. 2–9

2–10. The girl exerts a force of 75 N along the handle AB of the lawn mower. Resolve this force into x and y components which act parallel and perpendicular to the ground.

Prob. 2–10

2–11. The wind is deflected by the sail of a boat such that it exerts a resultant force of $F = 80$ lb perpendicular to the sail. Resolve this force into two components, one parallel and one perpendicular to the keel aa of the boat. *Note:* The ability to sail into the wind is known as tacking, made possible by the force parallel to the boat's keel. The perpendicular component tends to tip the boat or push it over.

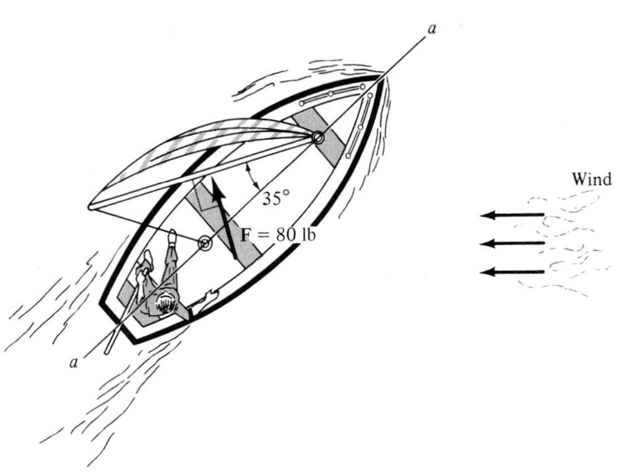

Prob. 2–11

*2–12. The leg is held in position by the quadriceps AB, which is attached to the pelvis at A. If the force exerted on this muscle by the pelvis is 85 N, in the direction shown, determine the stabilizing force component acting along the positive y axis and the supporting force component acting along the negative x axis.

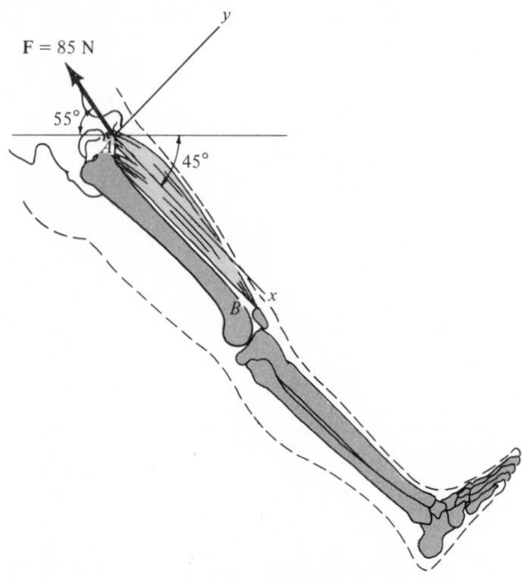

Prob. 2–12

2–13. The needle of the tone arm of a record player is subjected to a frictional force of $F = 7$ mN acting tangent to the spiral groove of the record. Resolve this force into components along the u and v axes and compute the magnitudes of these components. *Note:* The u component of force is actually supported by the tone-arm bearing B, while the v component represents the "skating" force or tendency of the tone arm to move toward the center of the record, faster than the spiral of the groove would normally move it.

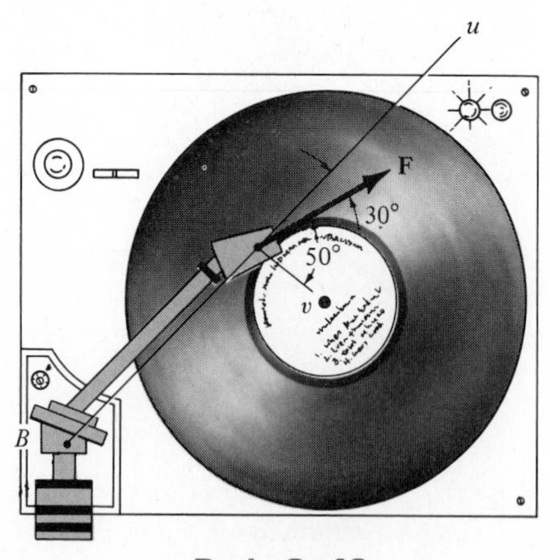

Prob. 2–13

2–14. Using chains, two tractors exert forces of $\mathbf{F}_A$ and $\mathbf{F}_B$ on the trunk of a tree as shown. Determine the magnitude of the resultant force. If the same magnitudes of forces $\mathbf{F}_A$ and $\mathbf{F}_B$ are maintained, in what direction should the tractors pull on the trunk so as to cause the *maximum* resultant force? What is the magnitude of this maximum force?

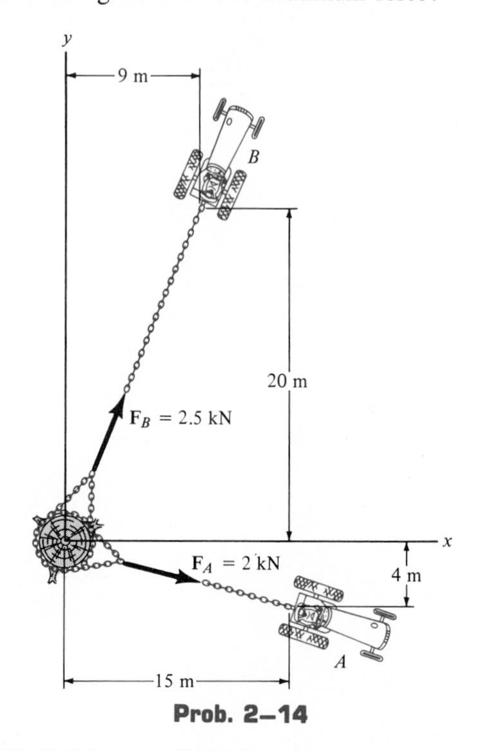

Prob. 2–14

2–15. Determine the magnitude and direction θ of $\mathbf{F}$ so that this force has components of 40 lb acting from A toward B and 60 lb acting from A toward C on the frame.

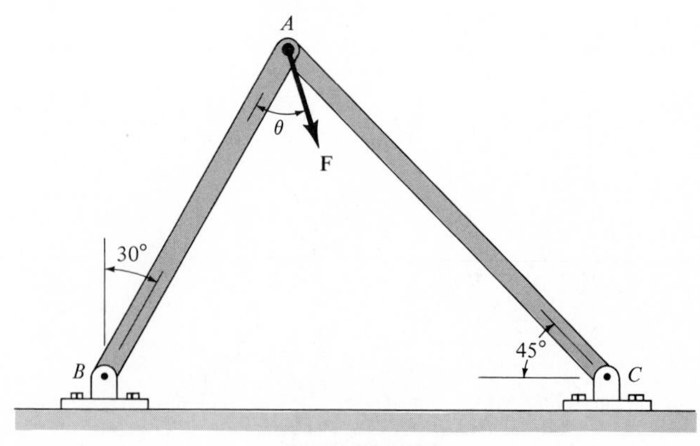

Prob. 2–15

***2–16.** Resolve the 50-lb force into components along the (a) x and y axes and (b) x and y' axes.

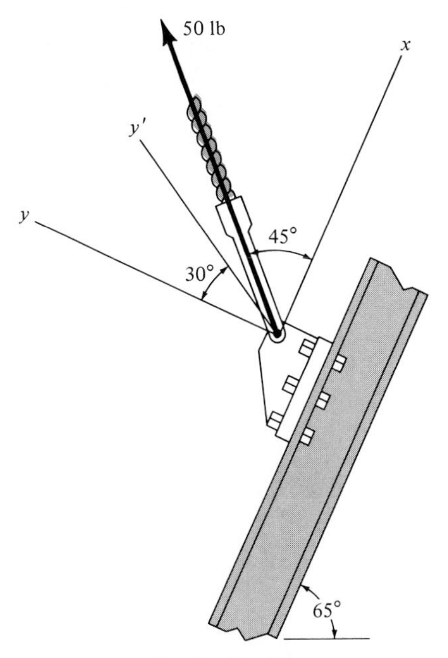

Prob. 2–16

2–17. Determine the design angle θ ($\theta < 90°$) between the two struts so that the 500-lb horizontal force has a component of $F_{AC} = 600$ lb directed from A toward C. What is the magnitude of force $\mathbf{F}_{AB}$ acting along member BA?

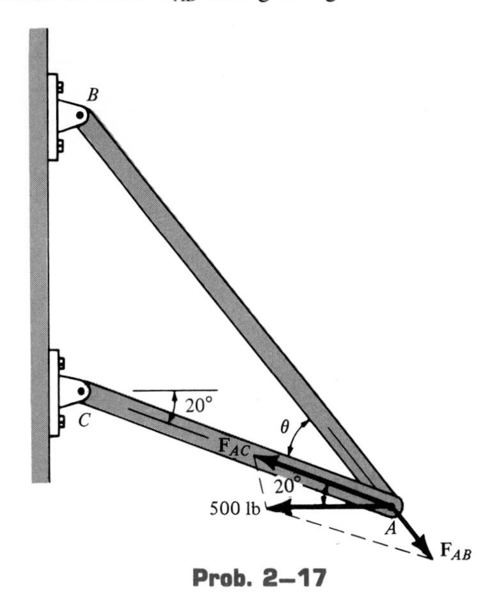

Prob. 2–17

2–18. Determine the magnitudes of the two components of the 600-N force, one directed along the cable AC and the other along the axis of strut AB.

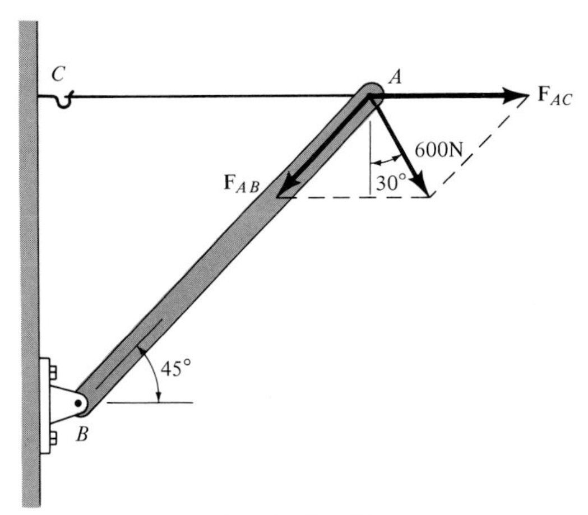

Prob. 2–18

2–19. If $\theta = 20°$ and $\phi = 35°$, determine the magnitudes of $\mathbf{F}_1$ and $\mathbf{F}_2$ so that the resultant force has a magnitude of 20 lb and is directed along the positive x axis.

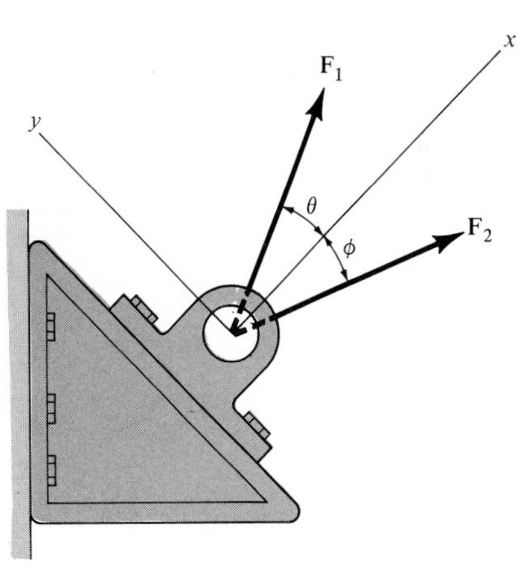

Probs. 2–19 / 2–20

***2–20.** If $F_1 = F_2 = 30$ lb, determine the angles θ and ϕ so that the resultant force is directed along the positive x axis and has a magnitude of $F_R = 20$ lb.

2–21. The *gusset plate G* of a bridge joint is subjected to the two member forces at A and B. If the force at B is horizontal and the force at A is directed at $\theta = 30°$, determine the magnitude and direction of the resultant force.

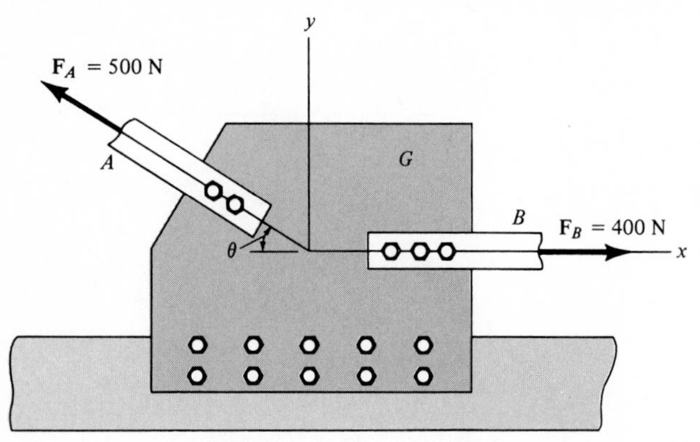

Probs. 2–21 / 2–22

2–22. Determine the design angle θ for connecting member A to the plate if the resultant force is to be directed vertically upward. Also, what is the magnitude of the resultant?

2–23. The log is being towed by two tractors A and B. Determine the magnitudes of the two towing forces $\mathbf{F}_A$ and $\mathbf{F}_B$ if it is required that the resultant force have a magnitude of $F_R = 8$ kN and be directed along the x axis. Set $\theta = 20°$.

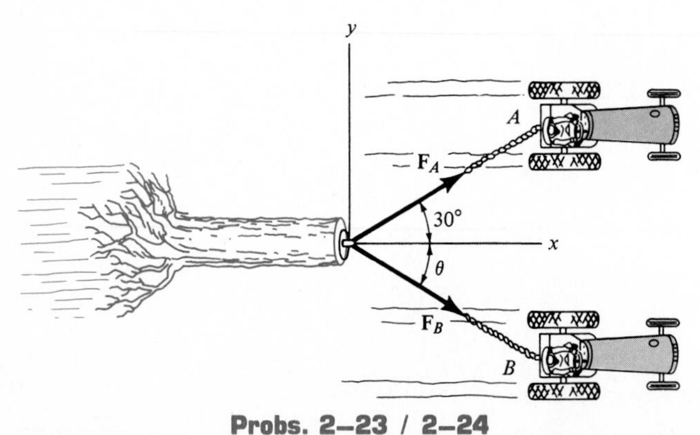

Probs. 2–23 / 2–24

***2–24.** If the resultant $\mathbf{F}_R$ of the two forces acting on the log is to be directed along the x axis and have a magnitude of 8 kN, determine the angle θ of the cable attached to B such that the force $\mathbf{F}_B$ in this cable is a *minimum*. What is the magnitude of force in each cable for this situation?

2–25. The power line BC can be hoisted over the pole provided the resultant force acting on it is 200 lb. In order to develop this loading, three cables are attached to the line at A. If two of the cables are subjected to known forces, as shown, determine the direction θ ($0 \leq \theta \leq 90°$) of the third cable so that the magnitude of force $\mathbf{F}_3$ is a minimum. What is the magnitude of $\mathbf{F}_3$? All forces lie in the same plane. *Hint:* First determine the resultant of $\mathbf{F}_1$ and $\mathbf{F}_2$. $\mathbf{F}_3$ acts in the *same direction* θ as this resultant.

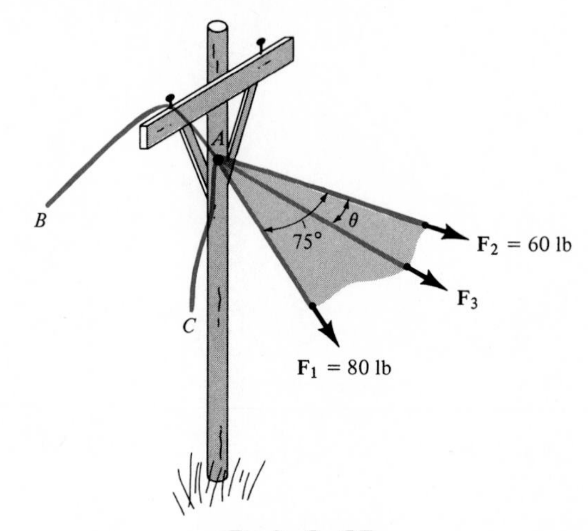

Prob. 2–25

2–26. Determine the magnitude and direction of the resultant $\mathbf{F}_R = \mathbf{F}_1 + \mathbf{F}_2 + \mathbf{F}_3$ of the three forces by first finding the resultant $\mathbf{F}' = \mathbf{F}_1 + \mathbf{F}_2$ and then forming $\mathbf{F}_R = \mathbf{F}' + \mathbf{F}_3$.

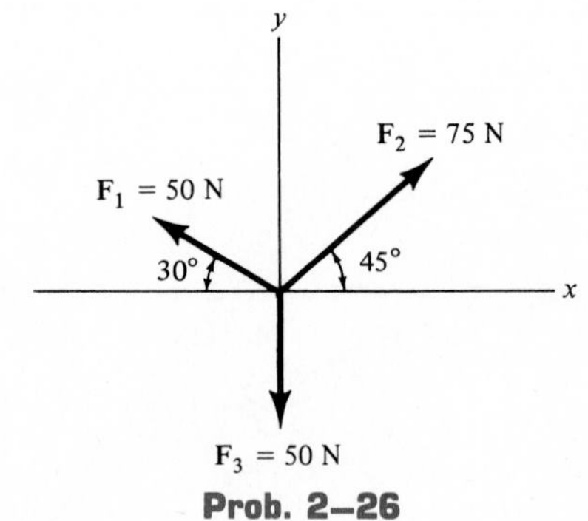

Prob. 2–26

Addition of Rectangular Force Components 2.5

If a force **F** lies in the x-y plane, Fig. 2–14*a*, the resolution of the force along the x and y axes yields its two *rectangular components* $\mathbf{F}_x$ and $\mathbf{F}_y$. By the parallelogram law,

$$\mathbf{F} = \mathbf{F}_x + \mathbf{F}_y$$

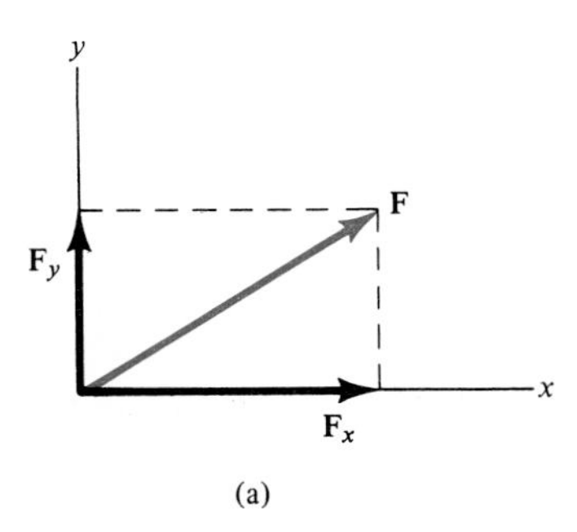

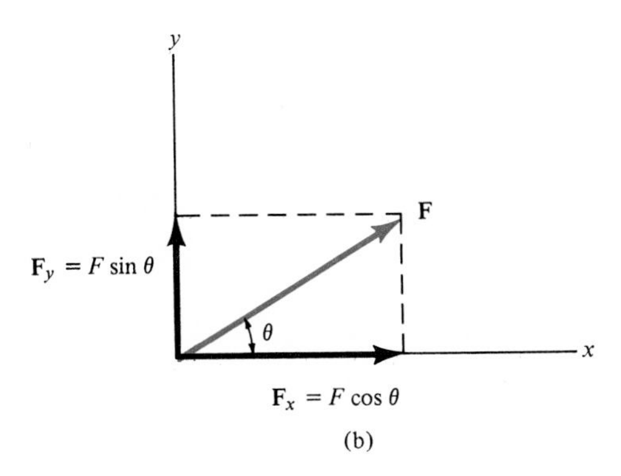

Fig. 2–14

The *magnitudes* of $\mathbf{F}_x$ and $\mathbf{F}_y$ are always positive and can be determined by trigonometry provided both the magnitude and direction of **F** are specified. For example, if the line of action of **F** is inclined θ from the x axis, Fig. 2–14*b*, then $F_x = F \cos \theta$ and $F_y = F \sin \theta$. For purposes of calculation the *sense of direction* of $\mathbf{F}_x$ and $\mathbf{F}_y$ will be designated by a plus or minus sign, depending upon whether these vectors are pointing along the positive or negative x or y axis, respectively. In the example above, Fig. 2–14*b*, both components are positive. Why?

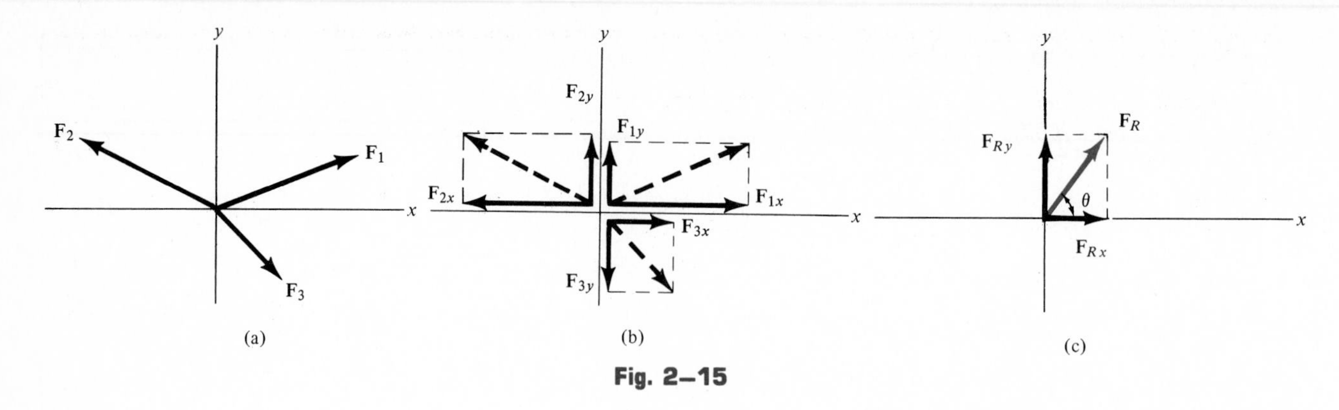

Fig. 2–15

This method of resolving a force into its x and y components can now be used to determine the resultant of several coplanar forces. To do this, each force is first resolved into its x and y components, the magnitude and direction of each component is then determined, and finally the respective x and y components are added *algebraically* since they are collinear. For example, consider the three forces in Fig. 2–15*a*. Here $\mathbf{F}_{2x}$ points in the $-x$ direction, Fig. 2–15*b*, and $\mathbf{F}_{3y}$ is in the $-y$ direction. The magnitudes of the x and y components of the resultant force are therefore

$$F_{Rx} = F_{1x} - F_{2x} + F_{3x}$$
$$F_{Ry} = F_{1y} + F_{2y} - F_{3y}$$

Thus, *the x and y components of the resultant force equal the algebraic sum of the x and y components of the forces being added.*

In the general case, the resultant of any number of forces can be represented symbolically as

$$\mathbf{F}_R = \Sigma\mathbf{F} = \mathbf{F}_{Rx} + \mathbf{F}_{Ry}$$

where the components have magnitudes of

$$F_{Rx} = \Sigma F_x \qquad F_{Ry} = \Sigma F_y \tag{2–1}$$

Once the x and y components of $\mathbf{F}_R$ have been determined, the *magnitude* of $\mathbf{F}_R$ can be calculated from the Pythagorean theorem, Fig. 2–15*c*, that is,

$$F_R = \sqrt{F_{Rx}^2 + F_{Ry}^2}$$

If the angle θ is used to specify the direction of $\mathbf{F}_R$, then by trigonometry,

$$\theta = \tan^{-1}\left(\frac{F_{Ry}}{F_{Rx}}\right)$$

The above concepts are illustrated numerically in the following examples.

Example 2–5

Determine the x and y components of each of the concurrent forces $\mathbf{F}_1$, $\mathbf{F}_2$, and $\mathbf{F}_3$ shown in Fig. 2–16*a*.

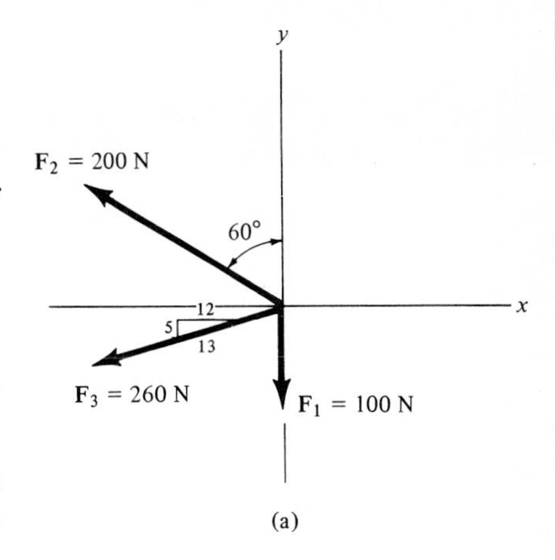

(a)

Solution

Force $\mathbf{F}_1$ has only a y component. Its *magnitude* is 100 N and its *sense of direction* is specified by a minus sign since it acts along the negative y axis. Thus,

$$F_{1x} = 0 \qquad \textit{Ans.}$$
$$F_{1y} = -100 \qquad \textit{Ans.}$$

By the parallelogram law, $\mathbf{F}_2$ is resolved into x and y components, Fig. 2–16*b*. Notice that $\mathbf{F}_{2x}$ acts in the $-x$ direction and $\mathbf{F}_{2y}$ acts in the $+y$ direction. The magnitude of each component is determined by trigonometry. Thus, the components are

$$F_{2x} = -200 \sin 60° = -173.2 \text{ N} \qquad \textit{Ans.}$$
$$F_{2y} = 200 \cos 60° = 100.0 \text{ N} \qquad \textit{Ans.}$$

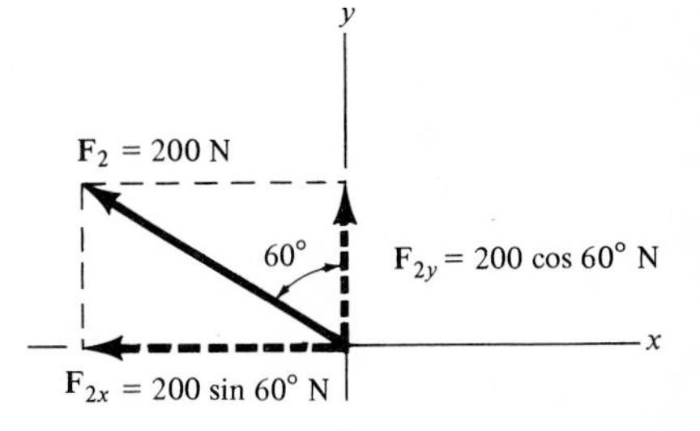

(b)

Both components of $\mathbf{F}_3$ act in a negative direction, Fig. 2–16*c*. Their magnitudes can be obtained by solving for the angle θ using the "slope triangle," i.e., $\theta = \tan^{-1} \frac{5}{12}$, and then proceeding in the same manner as for $\mathbf{F}_2$. An easier method, however, consists of using proportional parts of similar triangles, i.e.,

$$\frac{F_{3x}}{260} = \frac{12}{13} \qquad F_{3x} = 260\left(\frac{12}{13}\right) = 240 \text{ N}$$

Similarly,

$$F_{3y} = 260\left(\frac{5}{13}\right) = 100 \text{ N}$$

Notice that the magnitude of the *horizontal component*, F_{3x}, was obtained by multiplying the force magnitude by the ratio of the *horizontal leg* of the slope triangle divided by the hypotenuse; whereas the magnitude of the *vertical component*, F_{3y}, was obtained by multiplying the force magnitude by the ratio of the *vertical leg* divided by the hypotenuse. Hence, the components are

$$F_{3x} = -240 \text{ N} \qquad \textit{Ans.}$$
$$F_{3y} = -100 \text{ N} \qquad \textit{Ans.}$$

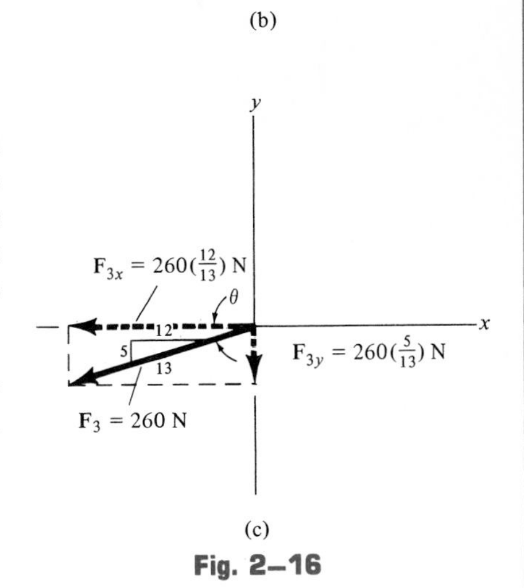

(c)

Fig. 2–16

Example 2–6

The pin in Fig. 2–17*a* is subjected to two forces $\mathbf{F}_1$ and $\mathbf{F}_2$. Determine the magnitude and direction of the resultant force.

Solution

This problem can be solved by using the parallelogram law; however, here we will resolve each force into its *x* and *y* components, Fig. 2–17*b*, and apply Eqs. 2–1. Indicating the "positive" sense of direction of the force components, i.e., the $+x$ and $+y$ directions, alongside the equations, we have

$$\overset{+}{\rightarrow} F_{Rx} = \Sigma F_x; \qquad F_{Rx} = 600 \cos 30^\circ - 400 \sin 45^\circ = 236.8 \text{ N}$$

$$+\uparrow F_{Ry} = \Sigma F_y; \qquad F_{Ry} = 600 \sin 30^\circ + 400 \cos 45^\circ = 582.8 \text{ N}$$

The resultant force, shown in Fig. 2–17*c*, has a *magnitude* of

$$F_R = \sqrt{(236.8)^2 + (582.8)^2} = 629.1 \text{ N} \qquad \textit{Ans.}$$

The *direction*, defined by the angle θ in Fig. 2–17*c*, is

$$\theta = \tan^{-1}\left(\frac{582.8}{236.8}\right) = 67.9^\circ \qquad \textit{Ans.}$$

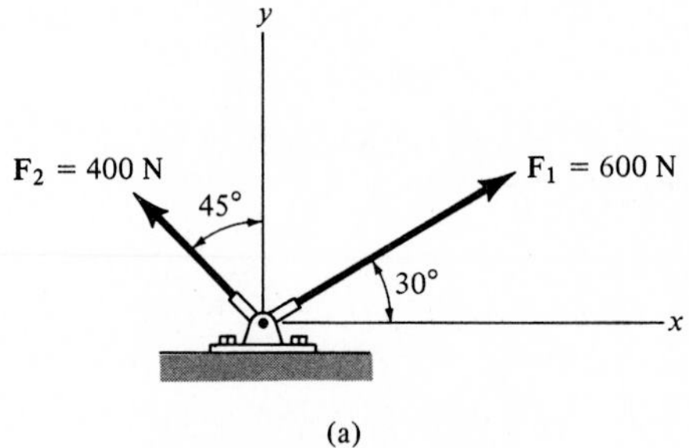

Fig. 2–17

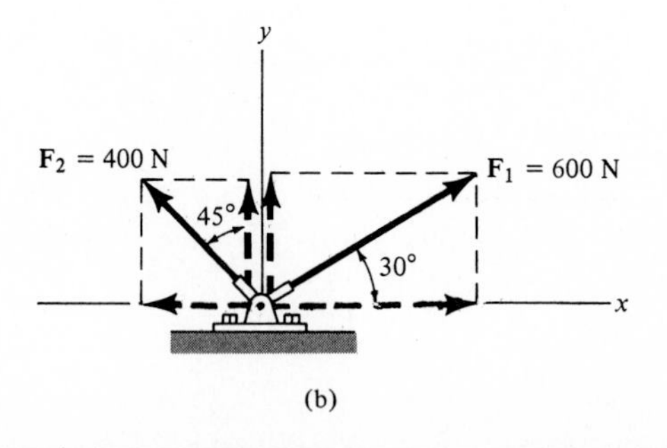

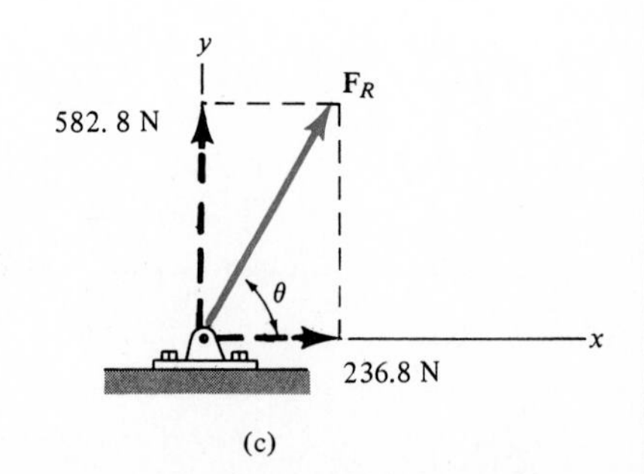

Example 2–7

The end of the boom O in Fig. 2–18a is subjected to three concurrent and coplanar forces. Determine the magnitude and direction of the resultant force.

Solution

Each force is resolved into its x and y components as shown in Fig. 2–18b. Applying Eqs. 2–1, we have

$$\overset{+}{\rightarrow} F_{Rx} = \Sigma F_x; \qquad F_{Rx} = -400 + 250 \sin 45^\circ - 200(\tfrac{4}{5})$$
$$= -383.2 \text{ N}$$

The negative sign indicates that F_{Rx} acts to the left, i.e., in the $-x$ direction.

$$+\uparrow F_{Ry} = \Sigma F_y; \qquad F_{Ry} = 250 \cos 45^\circ + 200(\tfrac{3}{5})$$
$$= 296.8 \text{ N}$$

The *magnitude* of the resultant force is

$$F_R = \sqrt{(-383.2)^2 + (296.8)^2}$$
$$= 484.7 \text{ N} \qquad \textit{Ans.}$$

The direction, defined by the angle θ in Fig. 2–18c, is

$$\theta = \tan^{-1}\left(\frac{296.8}{383.2}\right) = 37.8^\circ \qquad \textit{Ans.}$$

Note that $\mathbf{F}_R$ creates the *same effect* on the boom as the three forces in Fig. 2–18a.

(a)

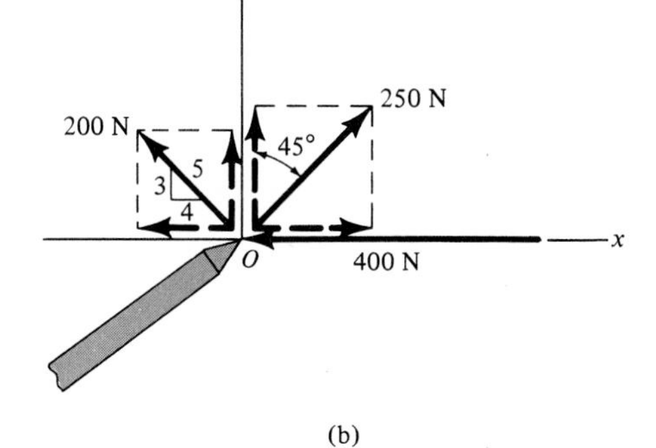

(b)

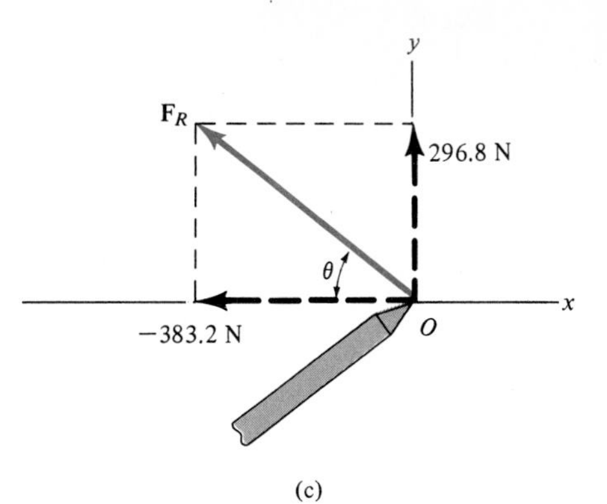

(c)

Fig. 2–18

Problems

2–27. Determine the x and y components of the 95-N force.

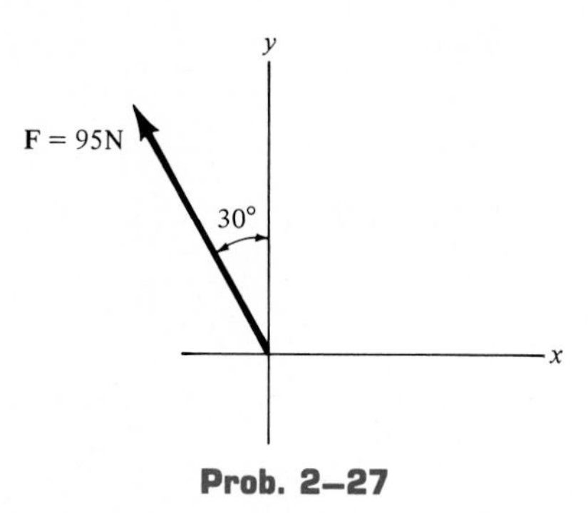

Prob. 2–27

***2–28.** Determine the x and y components of the 25-kN force.

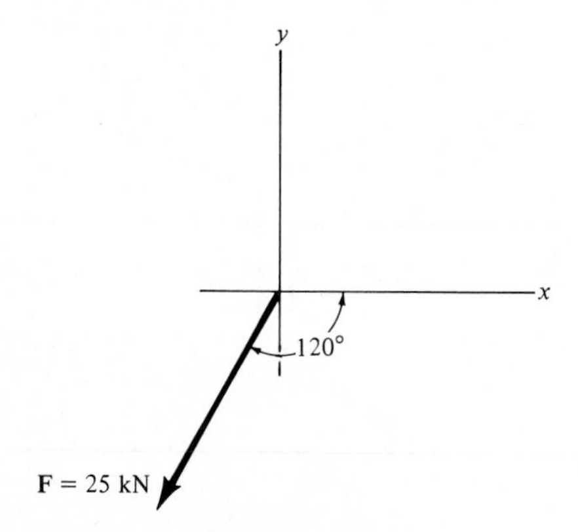

Prob. 2–28

2–29. Determine the x and y components of the 250-lb force.

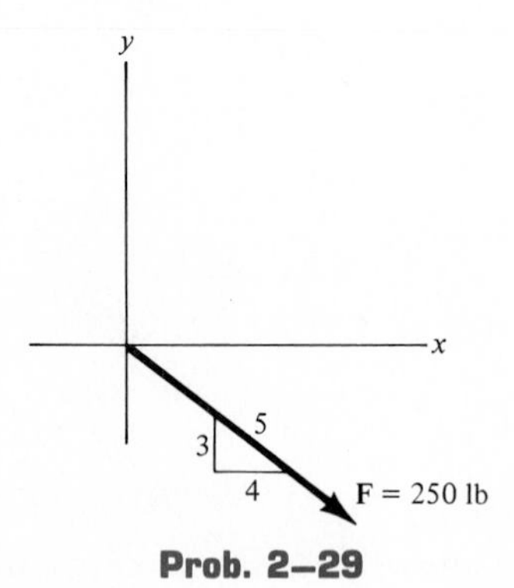

Prob. 2–29

2–30. Determine the magnitude of the resultant force acting on the column.

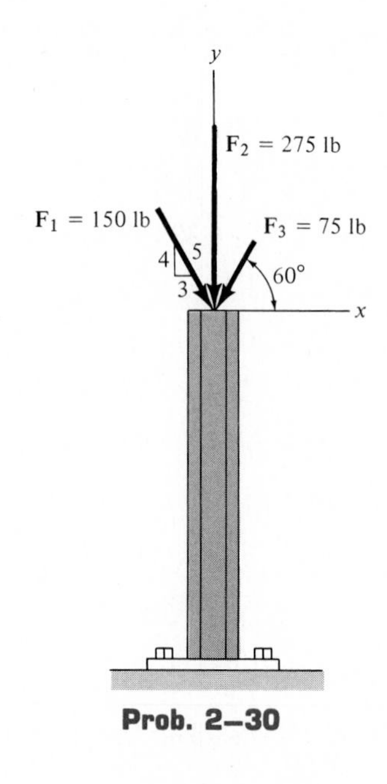

Prob. 2–30

2–31. Solve Prob. 2–19 by adding the rectangular components of the forces and requiring $F_R = \Sigma F_x$ and $0 = \Sigma F_y$.

***2–32.** Solve Prob. 2–21 by adding the rectangular components of the two forces to determine the resultant.

2–33. Solve Prob. 2–23 by adding the rectangular components of the two forces and requiring $F_R = \Sigma F_x$ and $0 = \Sigma F_y$.

2–34. Solve Prob. 2–26 by adding the rectangular components of the three forces to determine the resultant.

2–35. Determine the x and y components of each force.

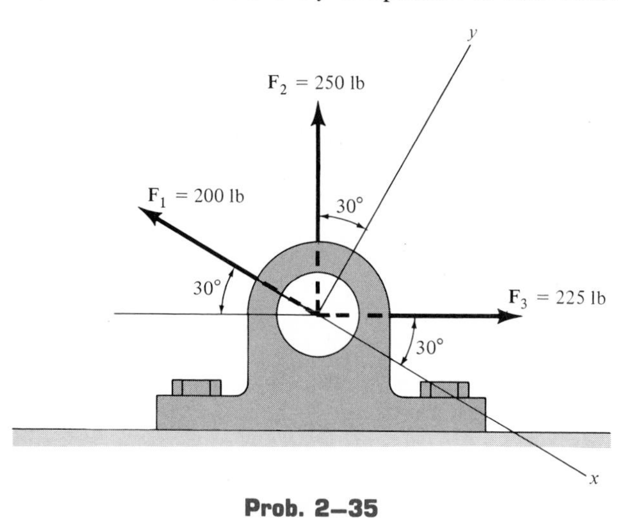

Prob. 2–35

***2–36.** Three concurrent forces act on the ring as shown. If each has a magnitude of 80 N, determine the x and y components of each force and find the resultant force.

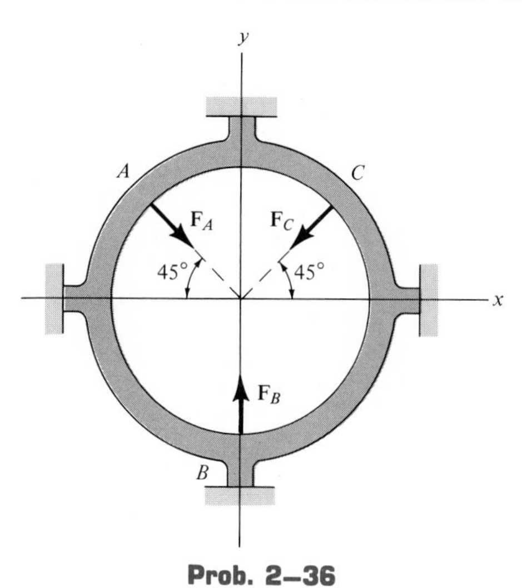

Prob. 2–36

2–37. Determine the x and y components of each force and then compute the magnitude and direction θ of $\mathbf{F}_1$ so that the resultant force is directed along the positive x' axis and has a magnitude of $F_R = 600$ N.

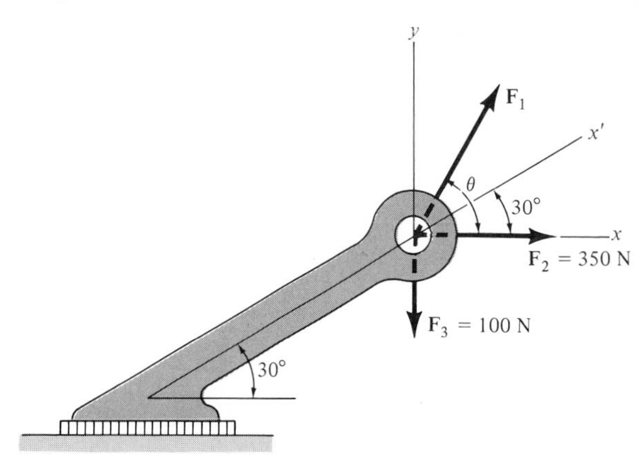

Prob. 2–37

Force Resultants in Three Dimensions

2.6 The Rectangular Components of a Force in Three Dimensions

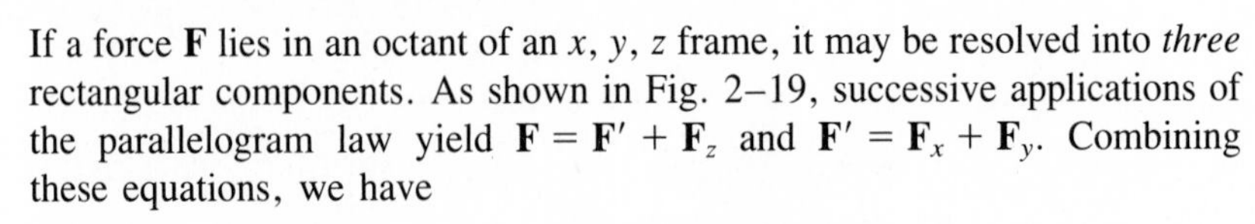

If a force **F** lies in an octant of an x, y, z frame, it may be resolved into *three* rectangular components. As shown in Fig. 2–19, successive applications of the parallelogram law yield $\mathbf{F} = \mathbf{F}' + \mathbf{F}_z$ and $\mathbf{F}' = \mathbf{F}_x + \mathbf{F}_y$. Combining these equations, we have

$$\mathbf{F} = \mathbf{F}_x + \mathbf{F}_y + \mathbf{F}_z$$

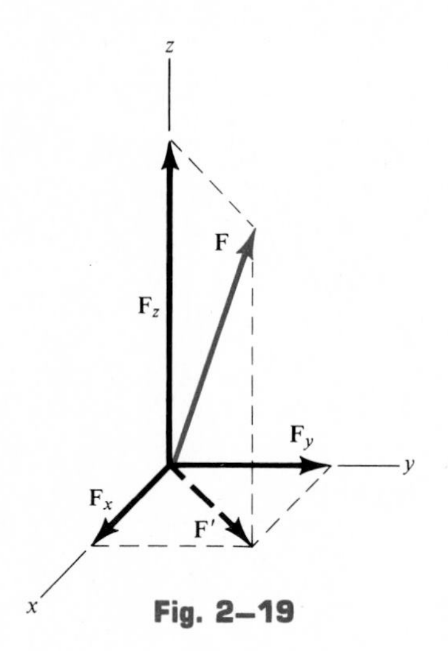

Fig. 2–19

We will now show that if one knows the magnitudes and directions of the three components $\mathbf{F}_x$, $\mathbf{F}_y$, $\mathbf{F}_z$ one can then specify the magnitude and direction of the force **F**.

Magnitude of Force. As shown in Fig. 2–20, from the shaded right triangle, the magnitude of $F' = \sqrt{F_x^2 + F_y^2}$. Similarly, from the colored right triangle, $F = \sqrt{F'^2 + F_z^2}$. Combining these equations yields

$$F = \sqrt{F_x^2 + F_y^2 + F_z^2} \qquad (2\text{–}2)$$

Hence, the magnitude of **F** *is equal to the square root of the sum of the squares of the magnitudes of its components.*

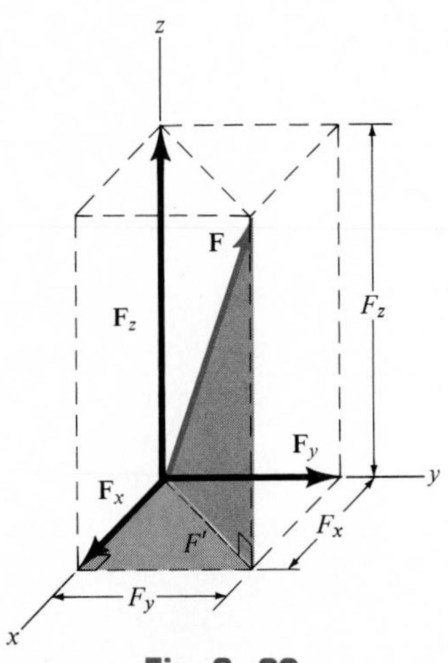

Fig. 2–20

Direction of Force. In three dimensions the direction of a force **F** is defined by three coordinate direction angles α (alpha), β (beta), and γ (gamma), which are measured from the *positive* x, y, z axes to the tail of **F**, Fig. 2–21.

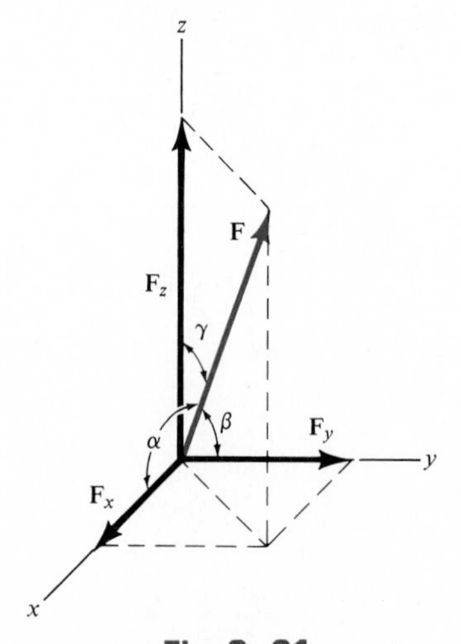

Fig. 2–21

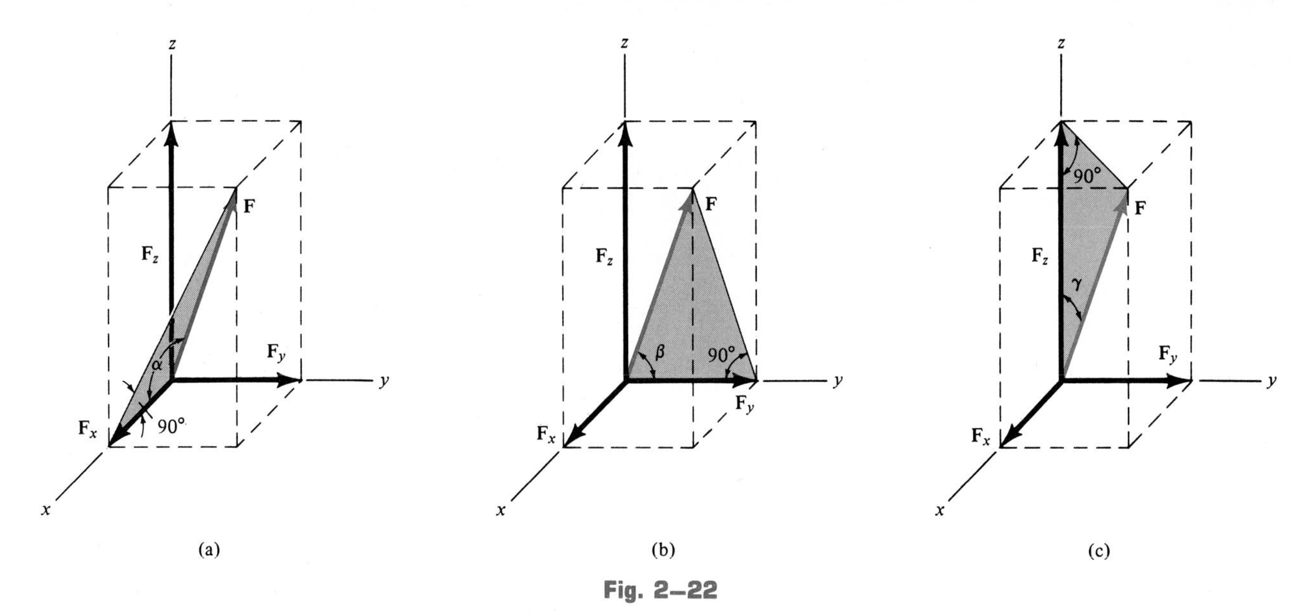

Fig. 2–22

Note that regardless of where **F** is directed, none of these angles will be greater than 180°. To determine α, β, and γ, consider the projection of **F** onto the x, y, z axes, Fig. 2–22. Referring to the colored right triangles shown in each figure, we have

$$\cos \alpha = \frac{F_x}{F} \qquad \cos \beta = \frac{F_y}{F} \qquad \cos \gamma = \frac{F_z}{F} \tag{2–3}$$

These numbers are known as the *direction cosines* of **F.** Once they have been obtained, the direction angles α, β, γ can then be determined from the inverse cosines.

If the magnitude of **F** and its direction cosines α, β, γ are known, Eqs. 2–3 can be used to determine the components of F, namely,

$$F_x = F \cos \alpha \qquad F_y = F \cos \beta \qquad F_z = F \cos \gamma \tag{2–4}$$

If these equations are substituted into Eq. 2–2, we obtain an important relation between the direction cosines, i.e.,

$$\cos^2 \alpha + \cos^2 \beta + \cos^2 \gamma = 1 \tag{2–5}$$

This equation is useful for determining one of the coordinate direction angles if the other two are known.

2.7 Addition of Several Forces

As shown in Sec. 2–5, the vector addition of several forces is greatly simplified if the forces are first expressed in terms of their rectangular components. When this is done in three dimensions, the resultant force will have x, y, z components which represent the *algebraic sum* of the x, y, z components of all forces added. We can write this in symbolic form as

$$\begin{aligned} F_{Rx} &= \Sigma F_x \\ F_{Ry} &= \Sigma F_y \\ F_{Rz} &= \Sigma F_z \end{aligned} \qquad (2\text{–}6)$$

Once the three components of $\mathbf{F}_R$ have been determined, its magnitude and direction angles can be calculated using Eqs. 2–2 and 2–3.

The following examples numerically illustrate the above methods for the solution of problems involving force as a three-dimensional vector quantity.

Example 2–8

The components of $\mathbf{F}_1$ and $\mathbf{F}_2$ are shown in Fig. 2–23*a*. Determine the magnitudes and coordinate direction angles of $\mathbf{F}_1$ and $\mathbf{F}_2$, and also find the components of the resultant force.

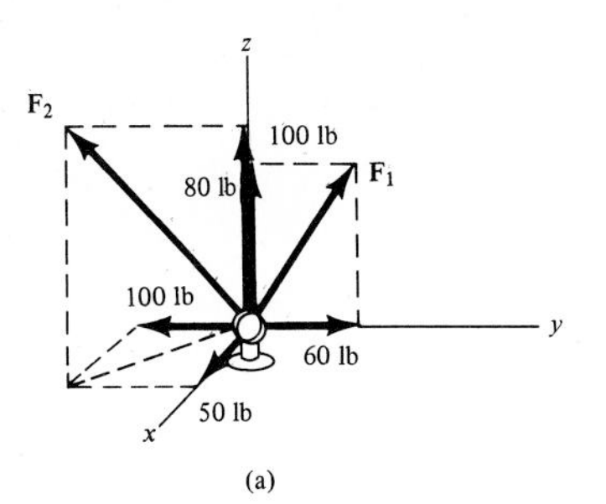

Solution

In each case the magnitude of the force is determined from Eq. 2–2 and the coordinate direction angles are determined from Eqs. 2–3. For $\mathbf{F}_1$, Fig. 2–23*a*, since $F_{1x} = 60$ lb, $F_{1y} = 80$ lb, we have

$$F_1 = \sqrt{(60)^2 + (80)^2} = 100 \text{ lb} \qquad \textit{Ans.}$$

$$\alpha_1 = \cos^{-1}\left(\frac{0}{100}\right) = 90^\circ \qquad \textit{Ans.}$$

$$\beta_1 = \cos^{-1}\left(\frac{60}{100}\right) = 53.1^\circ \qquad \textit{Ans.}$$

$$\gamma_1 = \cos^{-1}\left(\frac{80}{100}\right) = 36.9^\circ \qquad \textit{Ans.}$$

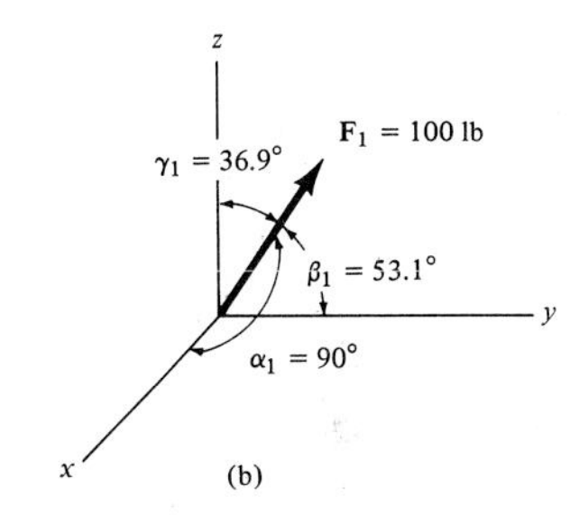

The results are shown in Fig. 2–23*b*.

In a similar manner $\mathbf{F}_2$ in Fig. 2–23*a* has components of $F_x = 50$ lb, $F_y = -100$ lb, $F_z = 100$ lb, so that

$$F_2 = \sqrt{(50)^2 + (-100)^2 + (100)^2} = 150 \text{ lb} \qquad \textit{Ans.}$$

$$\alpha_2 = \cos^{-1}\left(\frac{50}{150}\right) = 70.5^\circ \qquad \textit{Ans.}$$

$$\beta_2 = \cos^{-1}\left(\frac{-100}{150}\right) = 131.8^\circ \qquad \textit{Ans.}$$

$$\gamma_2 = \cos^{-1}\left(\frac{100}{150}\right) = 48.2^\circ \qquad \textit{Ans.}$$

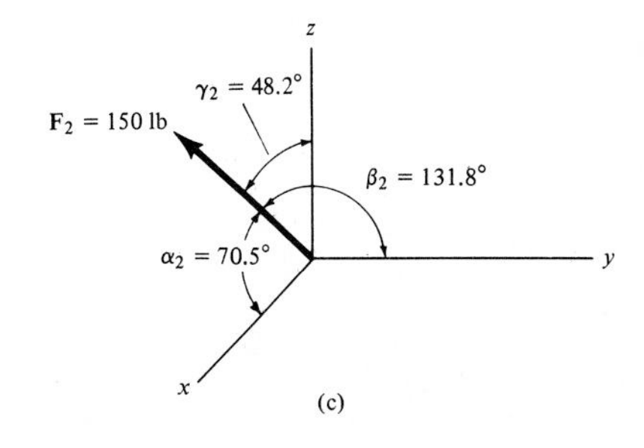

The results are shown in Fig. 2–23*c*.

Using Eqs. 2–6, the components of the resultant force are

$F_{Rx} = \Sigma F_x;$ $\qquad F_{Rx} = 50$ lb $\qquad$ *Ans.*

$F_{Ry} = \Sigma F_y;$ $\qquad F_{Ry} = 60 - 100 = -40$ lb $\qquad$ *Ans.*

$F_{Rz} = \Sigma F_z;$ $\qquad F_{Rz} = 80 + 100 = 180$ lb $\qquad$ *Ans.*

The resultant force is shown in Fig. 2–23*d*. Show that it has a magnitude of 191.0 lb and coordinate direction angles $\alpha = 74.8^\circ$, $\beta = 102.1^\circ$, $\gamma = 19.6^\circ$. Indicate these angles on the figure.

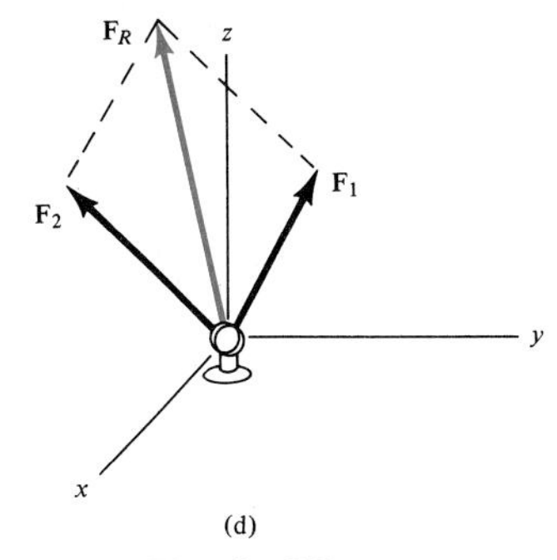

Fig. 2–23

Example 2–9

Determine the rectangular components of **F** shown in Fig. 2–24.

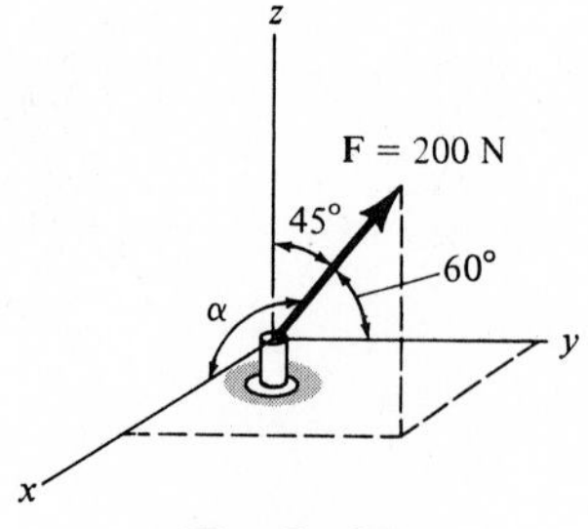

Fig. 2–24

Solution

Since only two coordinate direction angles are specified, Fig. 2–24, the third angle α is determined from Eq. 2–5, i.e.,

$$\cos^2 \alpha + \cos^2 \beta + \cos^2 \gamma = 1$$
$$\cos^2 \alpha + \cos^2 60° + \cos^2 45° = 1$$
$$\cos \alpha = \sqrt{1 - (0.707)^2 - (0.5)^2} = \pm 0.50$$

Hence,

$$\alpha = \cos^{-1}(0.5) = 60° \qquad \text{or} \qquad \alpha = \cos^{-1}(-0.5) = 120°$$

By inspection of Fig. 2–24, however, it is necessary that $\alpha = 60°$.

Knowing both the magnitude and coordinate direction angles of **F**, we can determine its rectangular components using Eqs. 2–4.

$F_x = F \cos \alpha;$ $\quad F_x = 200 \cos 60° = 100.0 \text{ N}$ *Ans.*

$F_y = F \cos \beta;$ $\quad F_y = 200 \cos 60° = 100.0 \text{ N}$ *Ans.*

$F_z = F \cos \gamma;$ $\quad F_z = 200 \cos 45° = 141.4 \text{ N}$ *Ans.*

By applying Eq. 2–2, note that indeed the magnitude of $F = 200$ N.

$$F = \sqrt{(F_x)^2 + (F_y)^2 + (F_z)^2}$$
$$= \sqrt{(100.0)^2 + (100.0)^2 + (141.4)^2} = 200 \text{ N}$$

Example 2–10

Determine the rectangular components of **F** shown in Fig. 2–25*a*.

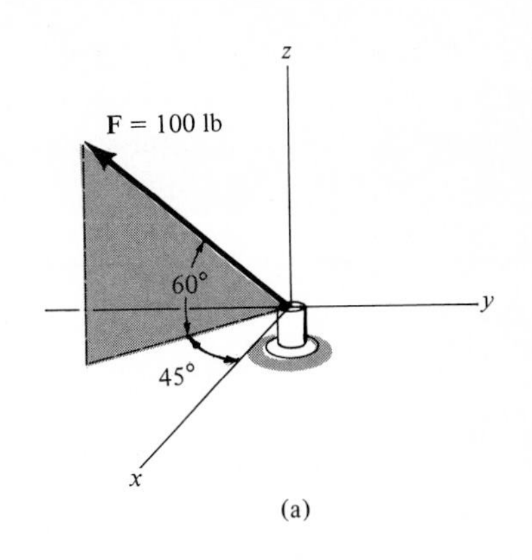

(a)

Solution

In this case, the angles of 60° and 45° defining the direction of **F** are *not* coordinate direction angles. Why? By two successive applications of the parallelogram law, however, **F** can be resolved into its x, y, z components as shown in Fig. 2–25*b*. By trigonometry, the magnitudes of the components are

$$F_z = 100 \sin 60° = 86.6 \text{ lb}$$
$$F' = 100 \cos 60° = 50 \text{ lb}$$
$$F_x = 50 \cos 45° = 35.4 \text{ lb}$$
$$F_y = 50 \sin 45° = 35.4 \text{ lb}$$

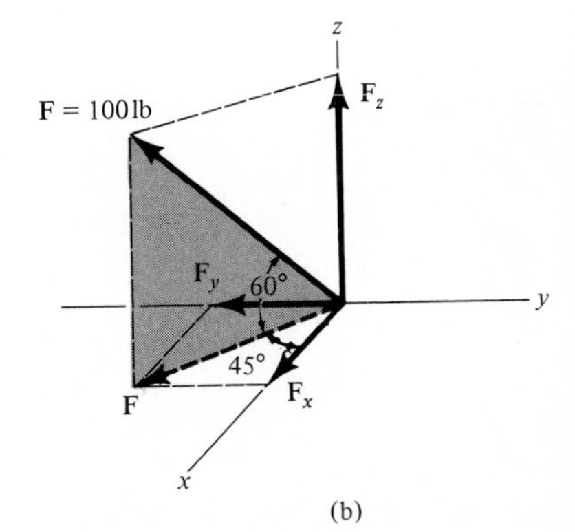

(b)

Realizing that $\mathbf{F}_y$ acts in the $-y$ direction, we have

$$F_x = 35.4 \text{ lb} \qquad \textit{Ans.}$$
$$F_y = -35.4 \text{ lb} \qquad \textit{Ans.}$$
$$F_z = 86.6 \text{ lb} \qquad \textit{Ans.}$$

To show that the magnitude of this vector is indeed 100 lb, apply Eq. 2–2,

$$F = \sqrt{F_x^2 + F_y^2 + F_z^2}$$
$$= \sqrt{(35.4)^2 + (-35.4)^2 + (86.6)^2} = 100 \text{ lb}$$

If needed, the coordinate direction angles of **F** can be determined using Eqs. 2–3, i.e.,

$$\cos \alpha = \frac{35.4}{100} = 0.354$$

$$\cos \beta = \frac{-35.4}{100} = -0.354$$

$$\cos \gamma = \frac{86.6}{100} = 0.866$$

or

$$\alpha = \cos^{-1} (0.354) = 69.3°$$
$$\beta = \cos^{-1} (-0.354) = 110.7°$$
$$\gamma = \cos^{-1} (0.866) = 30.0°$$

These results are shown in Fig. 2–25*c*.

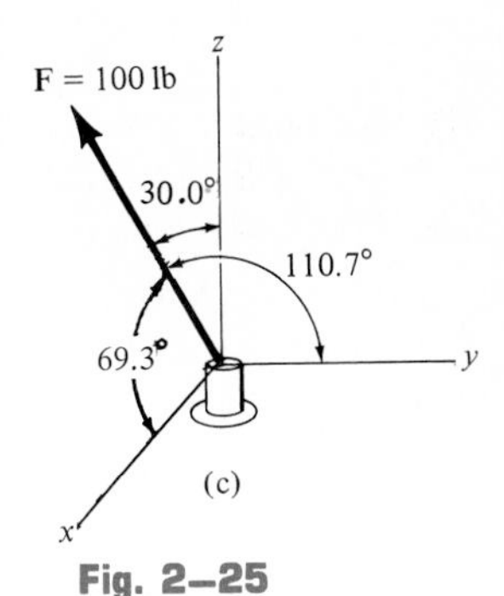

(c)

Fig. 2–25

Example 2–11

Two forces act on the pipe shown in Fig. 2–26*a*. Specify the coordinate direction angles of $\mathbf{F}_2$ so that the resultant force $\mathbf{F}_R$ acts in the positive y direction and has a magnitude of 800 N.

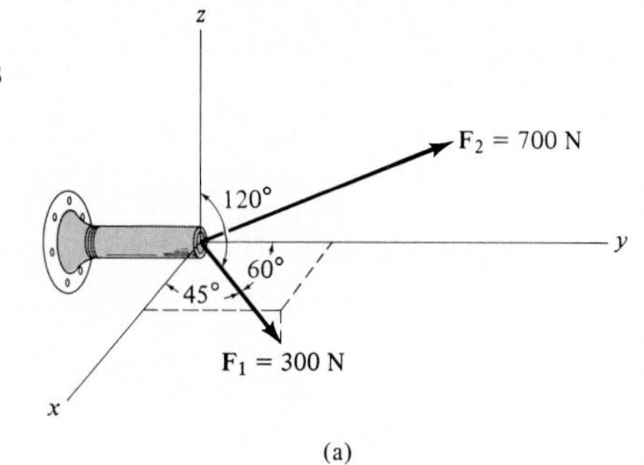

(a)

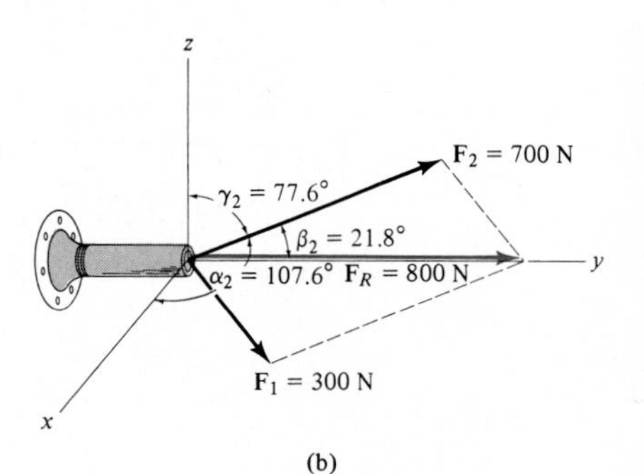

(b)

Fig. 2–26

Solution

We require that $\mathbf{F}_R = \mathbf{F}_1 + \mathbf{F}_2$. The components of $\mathbf{F}_1$ are computed as follows:

$$F_{1x} = 300 \cos 45° = 212.1 \text{ N}$$
$$F_{1y} = 300 \cos 60° = 150 \text{ N}$$
$$F_{1z} = 300 \cos 120° = -150 \text{ N}$$

Since the coordinate direction angles for $\mathbf{F}_2$ are unknown, we have

$$F_{2x} = 700 \cos \alpha_2$$
$$F_{2y} = 700 \cos \beta_2$$
$$F_{2z} = 700 \cos \gamma_2$$

Finally, $\mathbf{F}_R$ is to have a magnitude of 800 N and act in the positive y direction. Therefore,

$$F_{Rx} = 0$$
$$F_{Ry} = 800 \text{ N}$$
$$F_{Rz} = 0$$

Since $\mathbf{F}_R = \mathbf{F}_1 + \mathbf{F}_2$, the corresponding x, y, z components of these forces must satisfy Eqs. 2–6. Thus,

$$F_{Rx} = \Sigma F_x; \qquad 0 = 212.1 + 700 \cos \alpha_2$$
$$F_{Ry} = \Sigma F_y; \qquad 800 = 150 + 700 \cos \beta_2$$
$$F_{Rz} = \Sigma F_z; \qquad 0 = -150 + 700 \cos \gamma_2$$

Solving, we obtain

$$\alpha_2 = \cos^{-1}\left(\frac{-212.1}{700}\right) = 107.6° \qquad \textit{Ans.}$$

$$\beta_2 = \cos^{-1}\left(\frac{650}{700}\right) = 21.8° \qquad \textit{Ans.}$$

$$\gamma_2 = \cos^{-1}\left(\frac{150}{700}\right) = 77.6° \qquad \textit{Ans.}$$

The results are shown in Fig. 2–26*b*.

Problems

2–38. Determine the x, y components of each force.

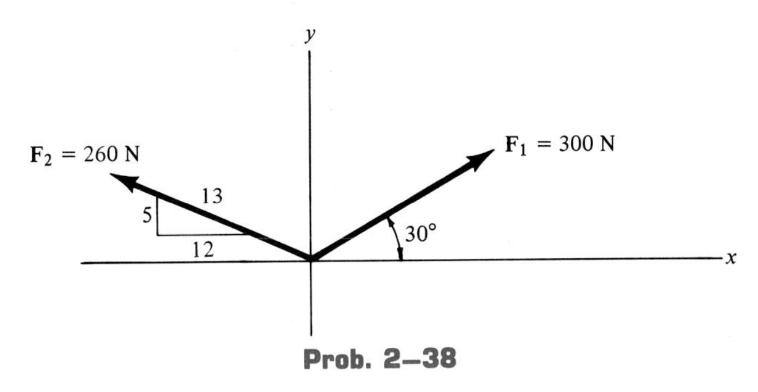

Prob. 2–38

2–39. Determine the x, y, z components of each force.

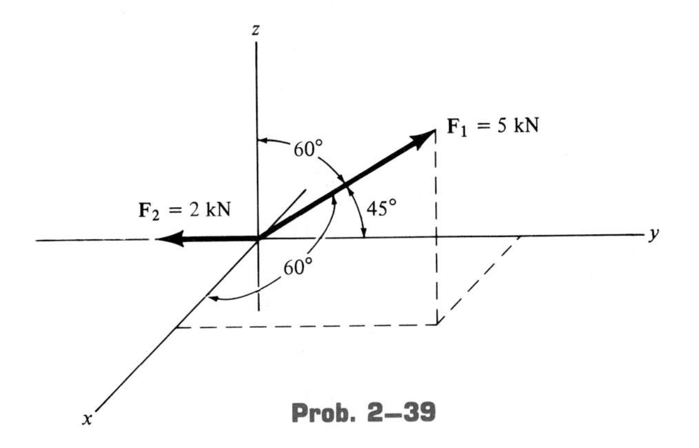

Prob. 2–39

***2–40.** Determine the x, y, z components of each force.

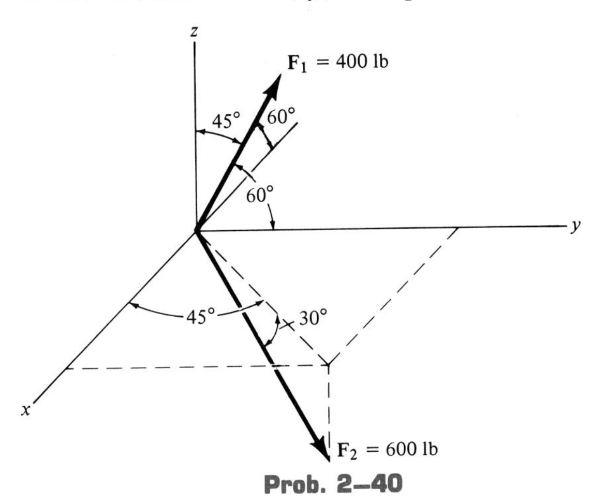

Prob. 2–40

2–41. The ball joint is subjected to the three forces shown. Determine the x, y, z components of each force and compute the magnitude and coordinate direction angles of the resultant force.

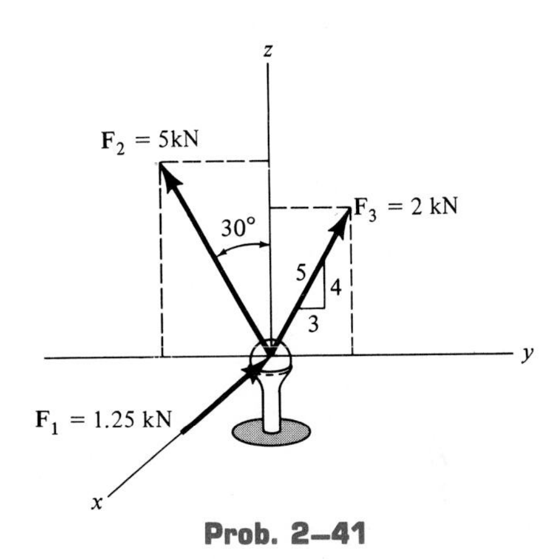

Prob. 2–41

2–42. The beam is subjected to the two forces shown. Determine the x, y, z components of each force and compute the magnitude and coordinate direction angles of the resultant force.

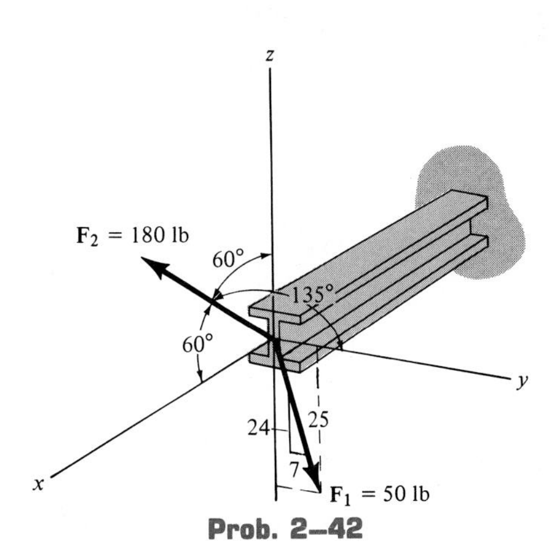

Prob. 2–42

2–43. The pith ball is subjected to an electrostatic repulsive force **F** which has measured components of $F_x = 40$ mN and $F_z = 20$ mN. If the angle $\beta = 135°$, determine the magnitudes of **F** and $\mathbf{F}_y$.

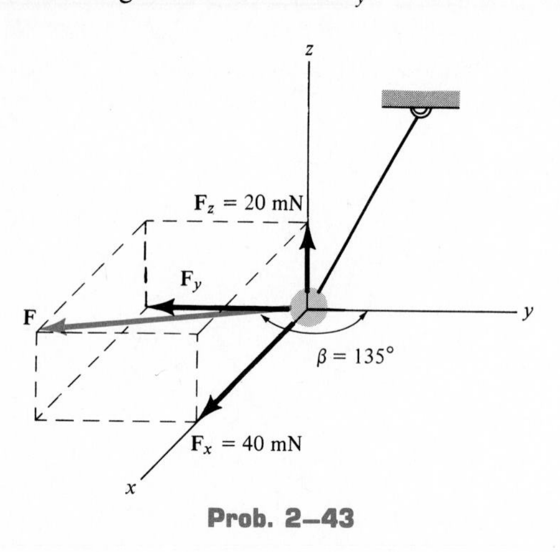

Prob. 2–43

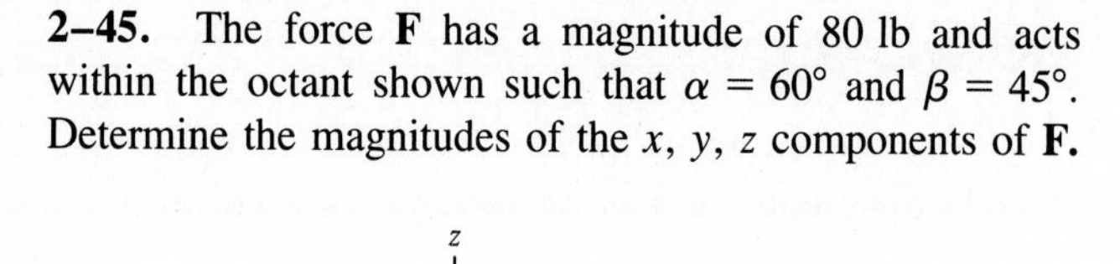

2–45. The force **F** has a magnitude of 80 lb and acts within the octant shown such that $\alpha = 60°$ and $\beta = 45°$. Determine the magnitudes of the x, y, z components of **F**.

Prob. 2–45

***2–44.** Determine the magnitude and direction angles of $\mathbf{F}_2$ so that the resultant of the two forces acts upward along the z axis of the pole and has a magnitude of 275 N.

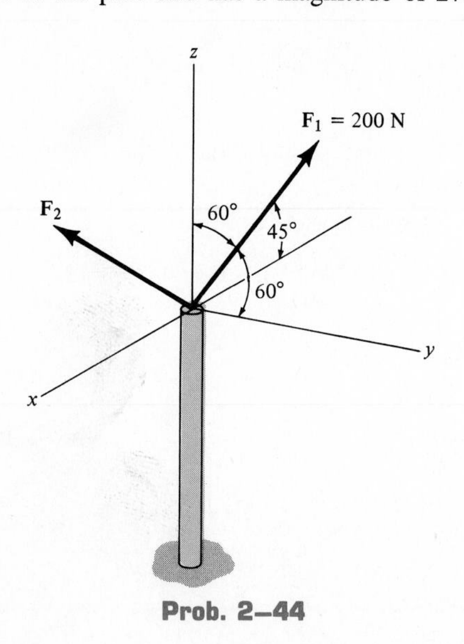

Prob. 2–44

2–46. Specify the magnitude of $\mathbf{F}_1$ and its coordinate direction angles α_1, β_1, and γ_1 so that the resultant of the three forces acting on the post has a magnitude of 350 lb and is directed along the negative z axis. Note that $\mathbf{F}_3$ lies in the x-y plane.

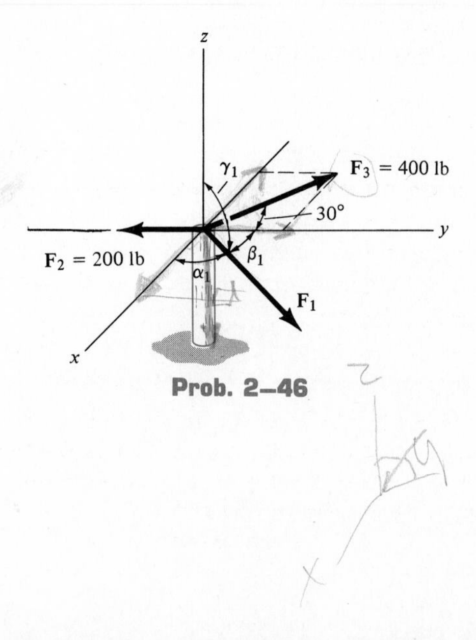

Prob. 2–46

2–47. Force **F** acts on peg A such that one of its components, lying in the x-y plane, has a magnitude of 50 lb. Determine the magnitude of **F** and its x, y, z components.

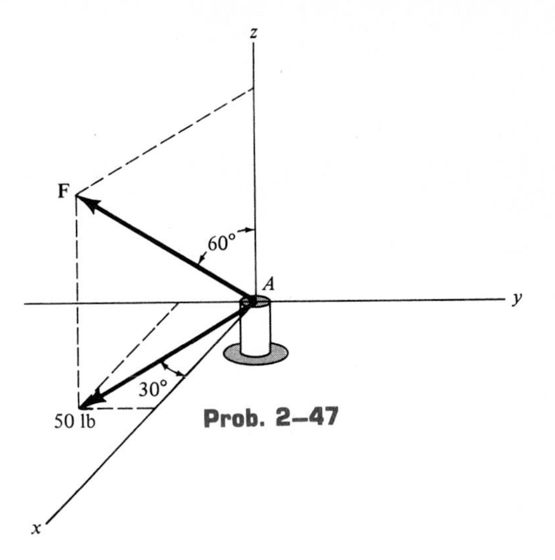

Prob. 2–47

Position Vectors 2.8

In this section we will introduce the concept of a position vector. It will be shown in the next section that this vector is of importance in formulating the x, y, z components of a force directed between any two points in space.

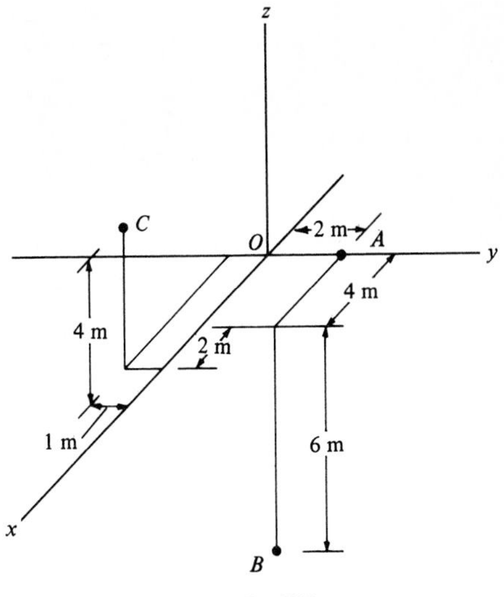

Fig. 2–27

x, y, z Coordinates. In order to formulate a position vector, we must first be able to define the x, y, z coordinates of a point in space. In this text we will follow the convention used in the majority of mathematics and technical books, and that is to require the z axis to be directed upward so that it measures the height of an object or the altitude of a point. The x, y axes then refer to the horizontal plane, Fig. 2–27. Points in space are then located relative to the origin of coordinates, O, by successive measurements along the x, y, z axes. For example, in Fig. 2–27 the coordinates of point A are obtained by starting at O and measuring $x_A = 0$ along the x axis, $y_A = +2$ m along the y axis, and $z_A = 0$ along the z axis. Thus, $A(0, 2, 0)$. In a similar manner, measurements along the x, y, z axes from O to B yield the coordinates of B, i.e., $B(4, 2, -6)$. Also notice that $C(6, -1, 4)$.

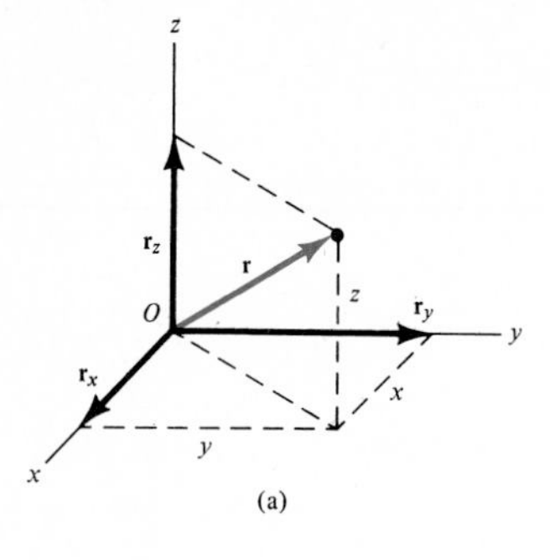

(a)

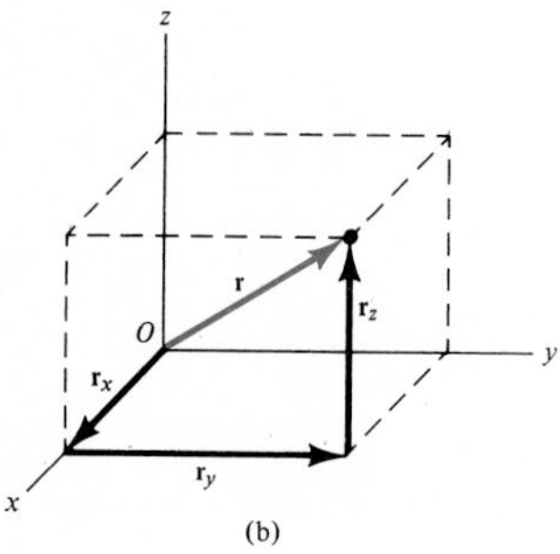

(b)

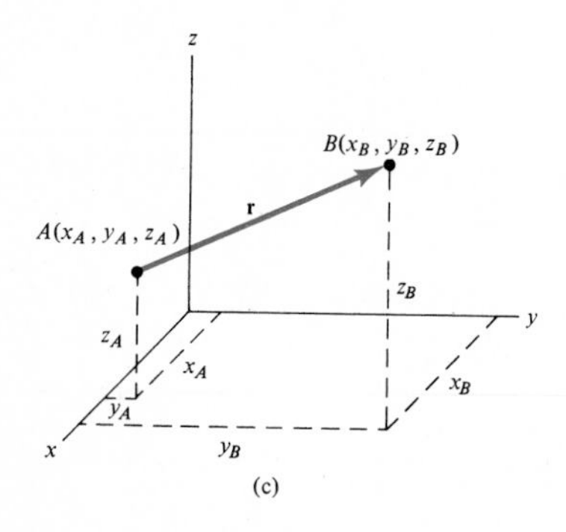

(c)

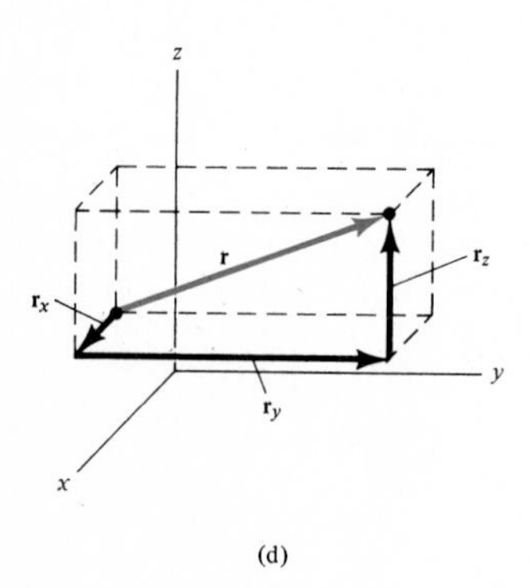

(d)

Fig. 2–28

Position Vector. The *position vector* **r** is defined as a fixed vector which locates a point in space relative to another point. For example, if **r** extends from the origin of coordinates O to point $P(x, y, z)$, Fig. 2–28*a*, then **r** can be expressed in vector form as

$$\mathbf{r} = \mathbf{r}_x + \mathbf{r}_y + \mathbf{r}_z$$

where the magnitudes of the components are

$$r_x = x \qquad r_y = y \qquad r_z = z$$

In particular, note how the tip-to-tail vector addition of the three components, going from O to P, yields vector **r**, as shown in Fig. 2–28*b*.

In the more general case, the position vector is directed from point $A(x_A, y_A, z_A)$ to point $B(x_B, y_B, z_B)$ in space, Fig. 2–28*c*. The x, y, z components of **r**, shown in Fig. 2–28*d*, represent the measured lengths of the x, y, z distances between points A and B. These components can therefore be determined as the difference in the coordinates of points A and B, Fig. 2–28*c*, i.e.,

$$\begin{aligned} r_x &= (x_B - x_A) \\ r_y &= (y_B - y_A) \\ r_z &= (z_B - z_A) \end{aligned} \qquad (2\text{–}7)$$

In other words, *the x, y, z components of the position vector* **r** *are determined by taking the coordinates of the tip of the vector, $B(x_B, y_B, z_B)$, and subtracting from them the corresponding coordinates of the tail, $A(x_A, y_A, z_A)$.* Again note how the tip-to-tail addition of these three components, going from A to B, yields **r**, Fig. 2–28*d*.

Example 2–12

Determine the magnitude and coordinate direction angles of the position vector extending from A to B in Fig. 2–29a.

Solution

The coordinates of A and B are $A(1, 0, -3)$, $B(-2, 2, 3)$. Hence, using Eqs. 2–7, the components of $\mathbf{r}$ are

$$r_x = -2 - 1 = -3 \text{ m}$$
$$r_y = 2 - 0 = 2 \text{ m}$$
$$r_z = 3 - (-3) = 6 \text{ m}$$

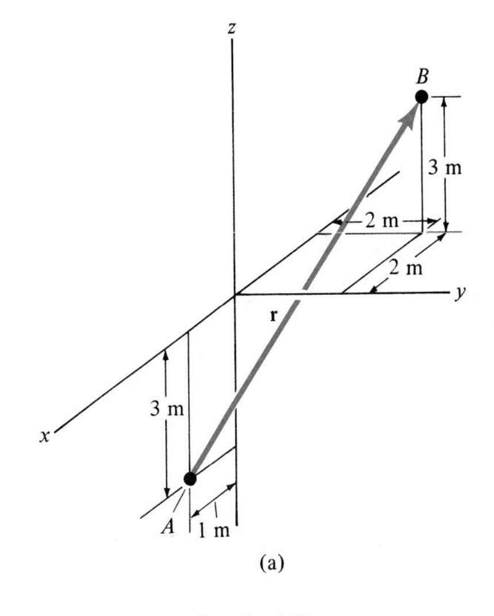

(a)

Fig. 2–29

As shown in Fig. 2–29b, the three components of $\mathbf{r}$ can be obtained in a more *direct manner* by realizing that in going from A to B, Fig. 2–29a, one must move along the x axis -3 m, then parallel to the y axis $+2$ m, and finally parallel to the z axis $+6$ m.

The *magnitude of* $\mathbf{r}$ is thus

$$r = \sqrt{(-3)^2 + (2)^2 + (6)^2} = 7 \text{ m} \qquad \textit{Ans.}$$

The *coordinate direction angles* of $\mathbf{r}$ are

$$\alpha = \cos^{-1}\left(\frac{-3}{7}\right) = 115.4° \qquad \textit{Ans.}$$

$$\beta = \cos^{-1}\left(\frac{2}{7}\right) = 73.4° \qquad \textit{Ans.}$$

$$\gamma = \cos^{-1}\left(\frac{6}{7}\right) = 31.0° \qquad \textit{Ans.}$$

These angles are measured from a localized coordinate system placed at the tail of $\mathbf{r}$ as shown in Fig. 2–29c.

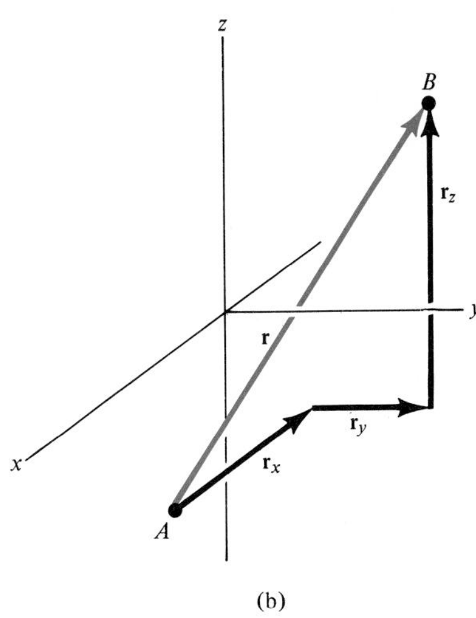

(b)

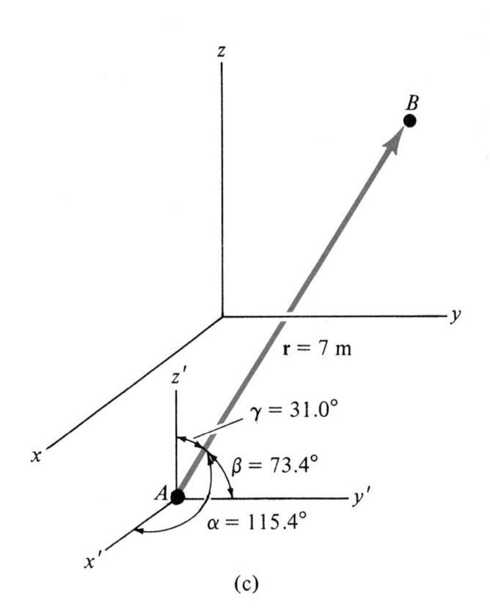

(c)

2.9 Force Vector Directed Along a Line

Quite often in three-dimensional statics problems the direction of a force is specified by two points through which its line of action passes. Such a situation is shown in Fig. 2–30, where the force **F** is directed along the cord AB. We can determine the components of **F** in this case by realizing that **F** acts in the *same direction* as the position vector **r** directed from point A to point B on the cord. This common direction can be specified by the direction cosines of **r**. In particular, note that *the force does not extend from A to B; only the position vector does*. This is because **r** has units of length, whereas **F** has units of force.

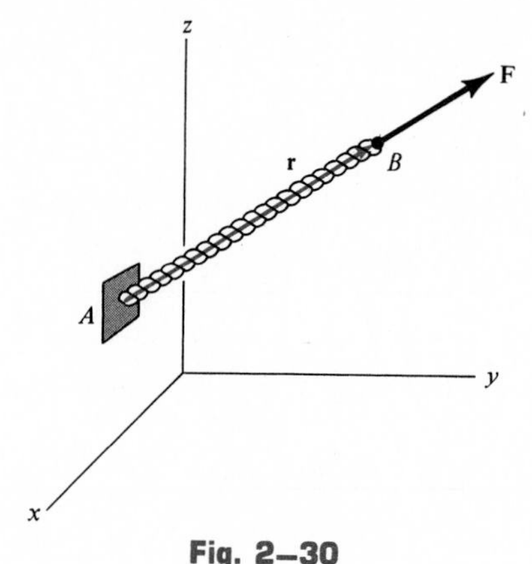

Fig. 2–30

PROCEDURE FOR ANALYSIS

When **F** is directed along a line which extends from point A to point B, then the components of **F** can be determined as follows:

Position Vector. Determine the components of the position vector **r** directed from A to B, and compute the magnitude of **r**.

Direction Cosines. Determine the direction cosines of **r**, i.e., $\cos \alpha = r_x/r$, $\cos \beta = r_y/r$, $\cos \gamma = r_z/r$. These angles define the *direction* of *both* **r** and **F**.

Force Vector. Determine the components of **F** by applying Eqs. 2–4, i.e., $F_x = F \cos \alpha$, $F_y = F \cos \beta$, $F_z = F \cos \gamma$ or $F_x = F(r_x/r)$, $F_y = F(r_y/r)$, $F_z = F(r_z/r)$.

This procedure is illustrated numerically in the following example problems.

Example 2–13

The man shown in Fig. 2–31*a* pulls on the cord with a force of 70 lb. Determine the components of this force, acting on the support at *A*.

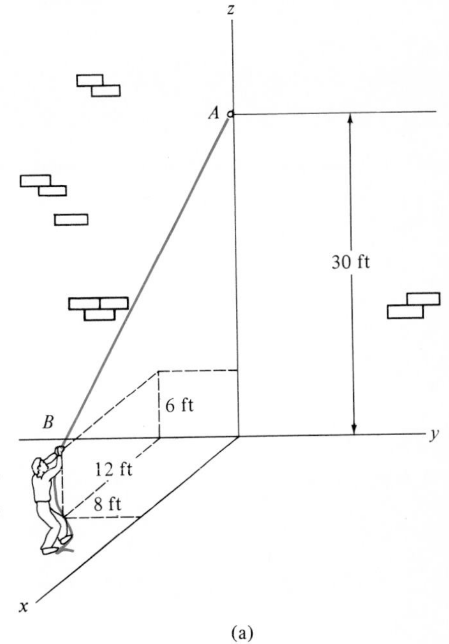

(a)

Solution

Force **F** is shown in Fig. 2–31*b*. The direction of **F** is the *same* as that of the position vector **r**, which extends from *A* to *B*. To formulate the components of **F** we use the following procedure.

Position Vector. The coordinates of the end points of the cord are $A(0, 0, 30)$ and $B(12, -8, 6)$. Thus, the components of **r** are obtained by subtracting the corresponding *x*, *y*, *z* coordinates of *A* from those of *B*. We have

$$r_x = 12 - 0 = 12 \text{ ft}$$
$$r_y = -8 - 0 = -8 \text{ ft}$$
$$r_z = 6 - 30 = -24 \text{ ft}$$

Show on Fig. 2–31*a* how one can calculate these components *directly* by going from A $r_z = -24$ ft along the *z* axis, then $r_y = -8$ ft parallel to the *y* axis, and finally $r_x = +12$ ft parallel to the *x* axis to get to *B*.

The magnitude of **r**, which represents the *length* of cord *AB*, is

$$r = \sqrt{(12)^2 + (-8)^2 + (-24)^2} = 28 \text{ ft}$$

Direction Cosines. Here we have

$$\cos\alpha = \frac{12}{28} \qquad \cos\beta = \frac{-8}{28} \qquad \cos\gamma = \frac{-24}{28}$$

Force Vector. Since **F** has a magnitude of 70 lb and the same direction as **r**, then the components of **F** are

$$F_x = F\cos\alpha = 70\left(\frac{12}{28}\right) = 30 \text{ lb} \qquad \textit{Ans.}$$

$$F_y = F\cos\beta = 70\left(\frac{-8}{28}\right) = -20 \text{ lb} \qquad \textit{Ans.}$$

$$F_z = F\cos\gamma = 70\left(\frac{-24}{28}\right) = -60 \text{ lb} \qquad \textit{Ans.}$$

As shown in Fig. 2–31*b*, the coordinate direction angles are measured between **r** (or **F**) and the *positive axes* of a localized coordinate system with origin placed at *A*. From the direction cosines, we have

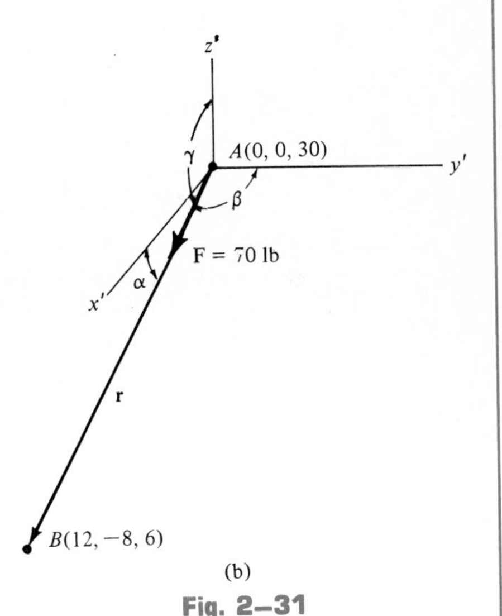

(b)

Fig. 2–31

$$\alpha = \cos^{-1}\left(\frac{12}{28}\right) = 64.6^\circ \qquad \textit{Ans.}$$

$$\beta = \cos^{-1}\left(\frac{-8}{28}\right) = 106.6^\circ \qquad \textit{Ans.}$$

$$\gamma = \cos^{-1}\left(\frac{-24}{28}\right) = 149.0^\circ \qquad \textit{Ans.}$$

Example 2–14

The circular plate shown in Fig. 2–32*a* is partially supported by the cable *AB*. If the magnitude of force in the cable is $F_B = 500$ N, determine the *x*, *y*, *z* components of this force acting on the hook at *A*.

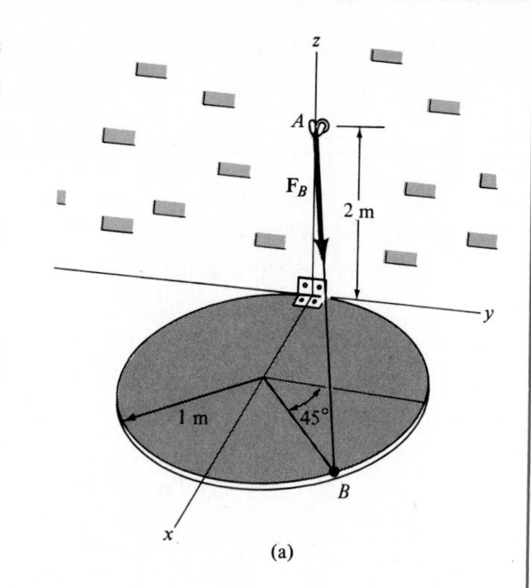

(a)

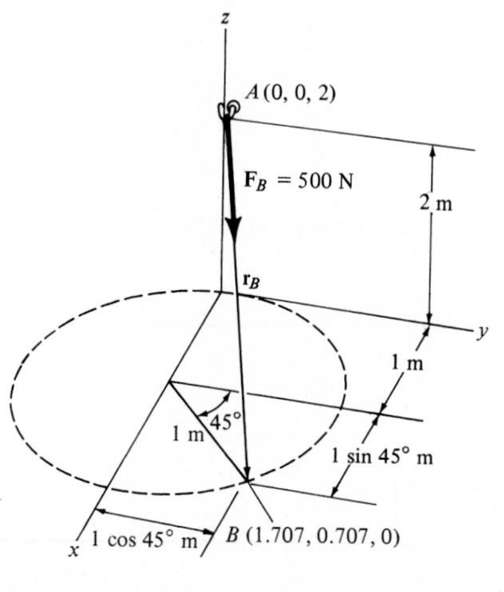

(b)

Fig. 2–32

Solution

As shown in Fig. 2–32*b*, **F** acts in the same direction as the position vector $\mathbf{r}_B$, which extends from *A* to *B*.

Position Vector. The coordinates of the end points of the cable are *A*(0, 0, 2) and *B*(1.707, 0.707, 0), as indicated in the figure. Thus, the components of $\mathbf{r}_B$ are

$$r_x = 1.707 - 0 = 1.707 \text{ m}$$
$$r_y = 0.707 - 0 = 0.707 \text{ m}$$
$$r_z = 0 - 2 = -2 \text{ m}$$

Show how one can calculate these components directly, by going from *A* $r_z = -2$ m along the *z* axis, then $r_x = +1.707$ m along the *x* axis, and finally $r_y = +0.707$ m parallel to the *y* axis to get to *B*.

The magnitude of **r** is

$$r = \sqrt{(1.707)^2 + (0.707)^2 + (-2)^2} = 2.72 \text{ m}$$

Direction Cosines. Thus,

$$\cos\alpha = \frac{1.707}{2.72}$$
$$\cos\beta = \frac{0.707}{2.72}$$
$$\cos\gamma = \frac{-2}{2.72}$$

Force Vector. Since $F = 500$ N and **F** has the same direction cosines as **r**, we have

$$F_x = F\cos\alpha = 500\left(\frac{1.707}{2.72}\right) = 313.5 \text{ N} \qquad \textit{Ans.}$$

$$F_y = F\cos\beta = 500\left(\frac{0.707}{2.72}\right) = 129.8 \text{ N} \qquad \textit{Ans.}$$

$$F_z = F\cos\gamma = 500\left(\frac{-2}{2.72}\right) = -367.3 \text{ N} \qquad \textit{Ans.}$$

Using these components, notice that indeed the magnitude of $F = 500$ N, i.e.,

$$F = \sqrt{(313.5)^2 + (129.8)^2 + (-367.3)^2} = 500.0 \text{ N}$$

Example 2–15

The chains exert forces of $F_B = 100$ N and $F_C = 120$ N on the ring at A as shown in Fig. 2–33a. Determine the magnitude of the resultant force acting at A.

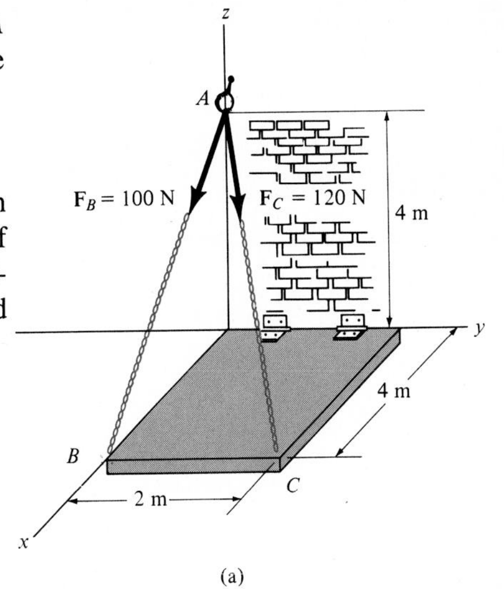

(a)

Solution

The resultant force $\mathbf{F}_R$ is shown graphically in Fig. 2–33b. We can determine its x, y, z components by first finding the x, y, z components of $\mathbf{F}_B$ and $\mathbf{F}_C$ and then using Eqs. 2–6. Proceeding in this manner, the coordinates of the end points of the position vectors $\mathbf{r}_B$ and $\mathbf{r}_C$ are determined and shown in the figure. Thus, for $\mathbf{F}_B$ we have

$$r_{B_x} = 4 - 0 = 4 \text{ m} \qquad r_{B_y} = 0 \qquad r_{B_z} = 0 - 4 = -4 \text{ m}$$

$$r_B = \sqrt{(4)^2 + (0)^2 + (-4)^2} = 4\sqrt{2} \text{ m}$$

Using this data to formulate the direction cosines for $\mathbf{F}_B$ yields

$$F_{B_x} = F_B\left(\frac{r_{B_x}}{r_B}\right) = 100\left(\frac{4}{4\sqrt{2}}\right) = 70.7 \text{ N}$$

$$F_{B_y} = F_B\left(\frac{r_{B_y}}{r_B}\right) = 100\left(\frac{0}{4\sqrt{2}}\right) = 0$$

$$F_{B_z} = F_B\left(\frac{r_{B_z}}{r_B}\right) = 100\left(\frac{-4}{4\sqrt{2}}\right) = -70.7 \text{ N}$$

For $\mathbf{F}_C$ we have

$$r_{C_x} = 4 - 0 = 4 \text{ m} \qquad r_{C_y} = 2 - 0 = 2 \text{ m} \qquad r_{C_z} = 0 - 4 = -4 \text{ m}$$

$$r_C = \sqrt{(4)^2 + (2)^2 + (4)^2} = 6 \text{ m}$$

so that

$$F_{C_x} = F_C\left(\frac{r_{C_x}}{r_C}\right) = 120\left(\frac{4}{6}\right) = 80 \text{ N}$$

$$F_{C_y} = F_C\left(\frac{r_{C_y}}{r_C}\right) = 120\left(\frac{2}{6}\right) = 40 \text{ N}$$

$$F_{C_z} = F_C\left(\frac{r_{C_z}}{r_C}\right) = 120\left(\frac{-4}{6}\right) = -80 \text{ N}$$

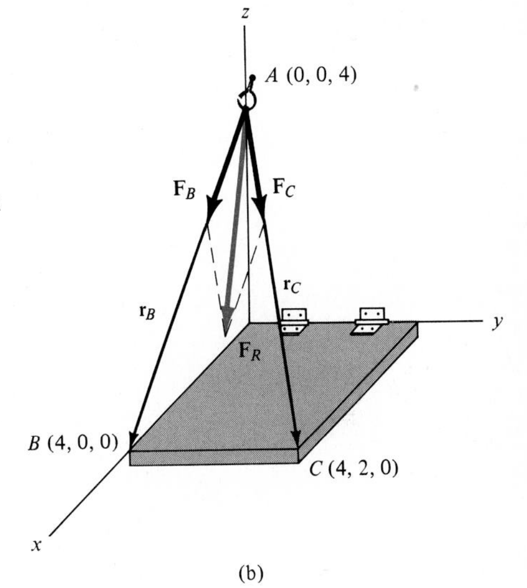

(b)

Fig. 2–33

Applying Eqs. 2–6, the components of $\mathbf{F}_R$ are

$$F_{R_x} = \Sigma F_x; \qquad F_{R_x} = 70.7 + 80 = 150.7 \text{ N}$$

$$F_{R_y} = \Sigma F_y; \qquad F_{R_y} = 0 + 40 = 40 \text{ N}$$

$$F_{R_z} = \Sigma F_z; \qquad F_{R_z} = -70.7 - 80 = -150.7 \text{ N}$$

The magnitude of the resultant force is therefore

$$F_R = \sqrt{(150.7)^2 + (40)^2 + (-150.7)^2}$$

$$= 216.8 \text{ N} \qquad \textit{Ans.}$$

Problems

***2–48.** Determine the magnitude and coordinate direction angles of **r**.

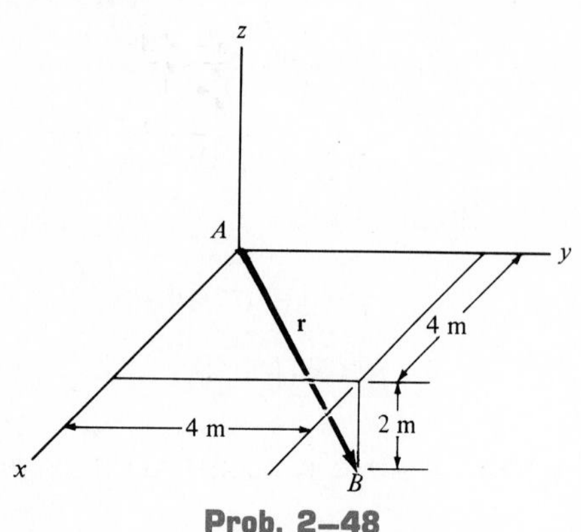

Prob. 2–48

2–49. Determine the magnitude and coordinate direction angles of **r**.

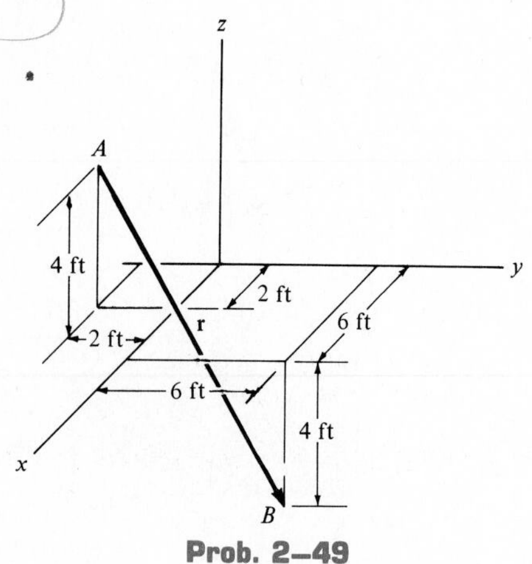

Prob. 2–49

2–50. Determine the magnitude and coordinate direction angles of **r**.

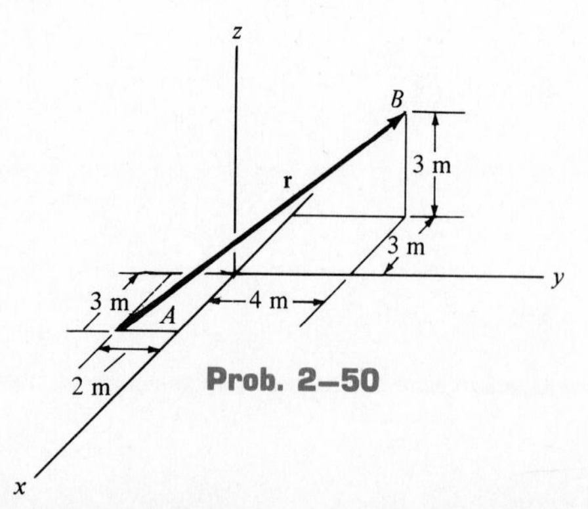

Prob. 2–50

2–51. Determine the components of a position vector extending from *A* to *B* on the crankshaft; then compute the length of the crankshaft *AB*.

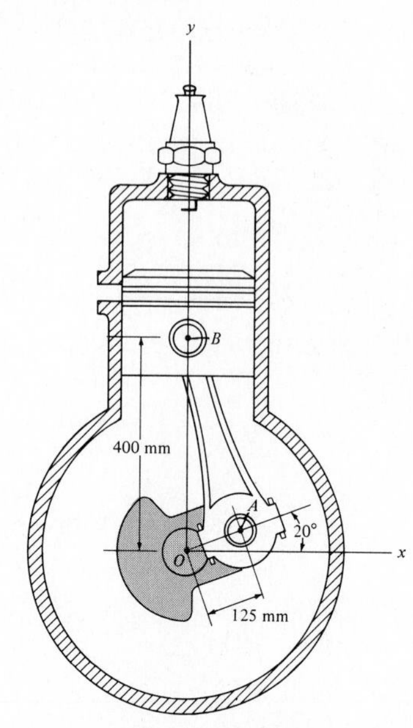

Prob. 2–51

***2–52.** The position vector directed from *B* to *A* along the pipe *BA* has components of $(r_{BA})_x = 0$, $(r_{BA})_y = 2$ ft, $(r_{BA})_z = -1$ ft. Determine the distance from *O* to *A*. *Hint:* First determine the components of the position vector $\mathbf{r}_{OA}$ and then compute its magnitude.

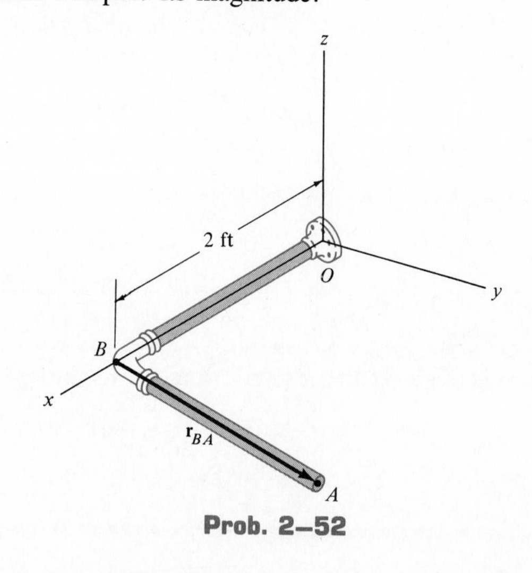

Prob. 2–52

2–53. The cord is attached between two walls. If it is 8 m long, determine the distance x to the point of attachment at B.

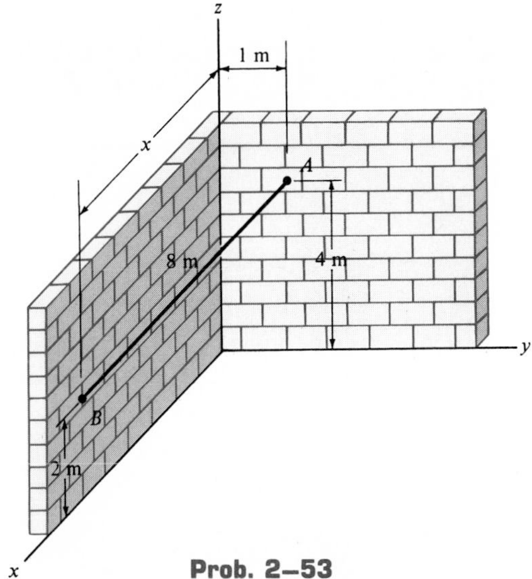

Prob. 2–53

2–54. At a given instant, the positions of a plane at A and a train at B are measured relative to a radar antenna at O. Determine the distance d between A and B at this instant. To solve the problem, formulate a position vector, directed from A to B, and then determine its magnitude.

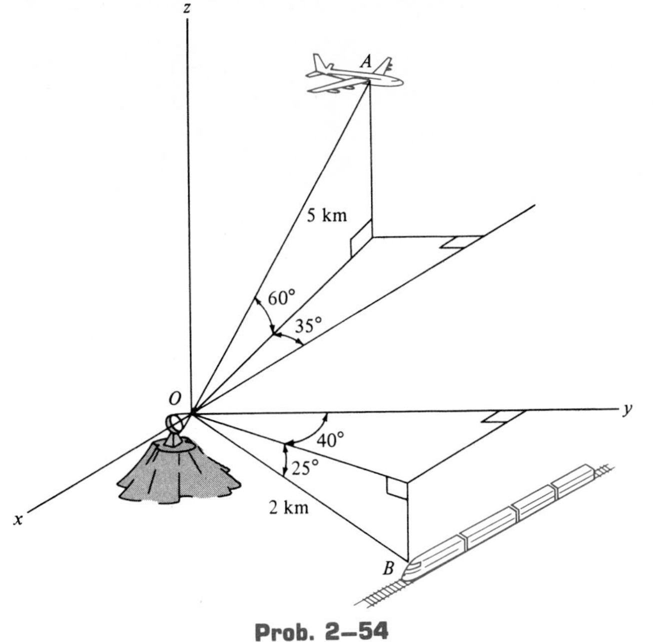

Prob. 2–54

2–55. Determine the x, y components of the force **F**.

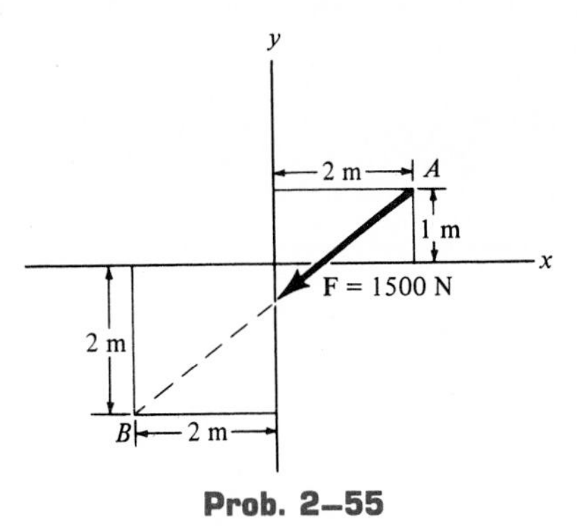

Prob. 2–55

***2–56.** Determine the x, y, z components of the force **F**.

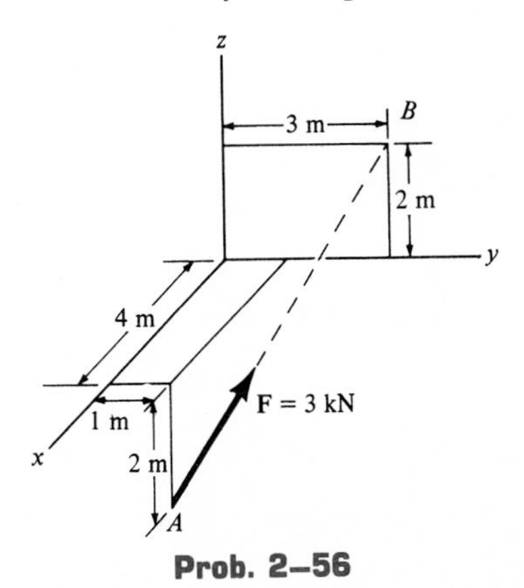

Prob. 2–56

2–57. Determine the x, y, z components of the force **F**.

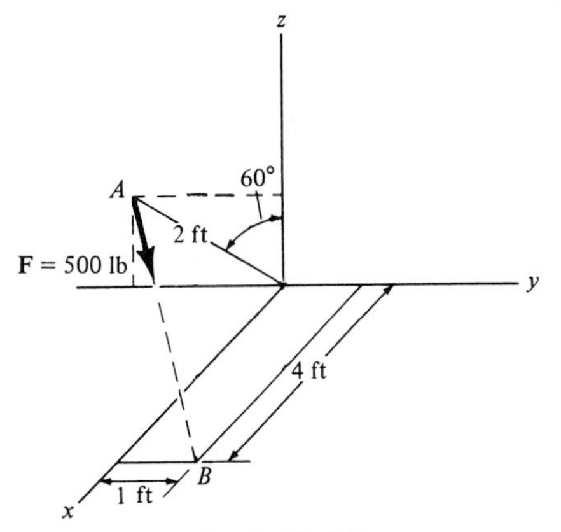

Prob. 2–57

2–58. The hinged plate is supported by the cord AB. If the force in the cord is $F = 340$ lb, determine the x, y, z components of **F**. What is the length of the cord?

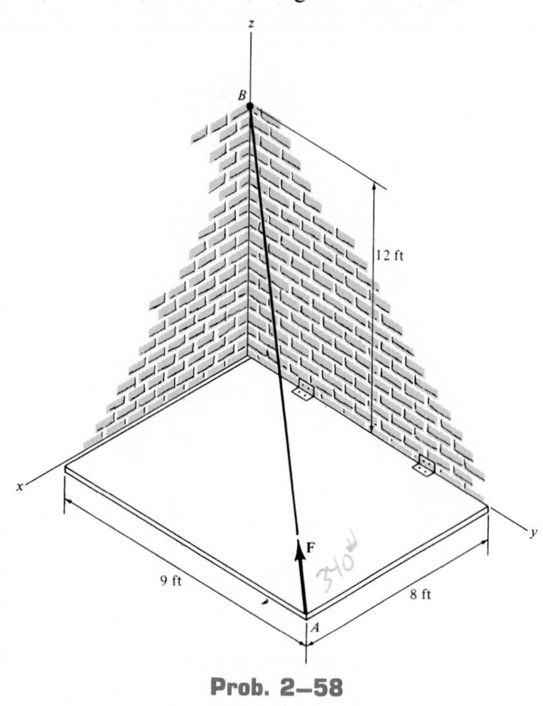

Prob. 2–58

*2–60. The two mooring cables exert forces on the stern of a ship as shown. Determine the x, y, z components of each force and determine the magnitude and coordinate direction angles of the resultant.

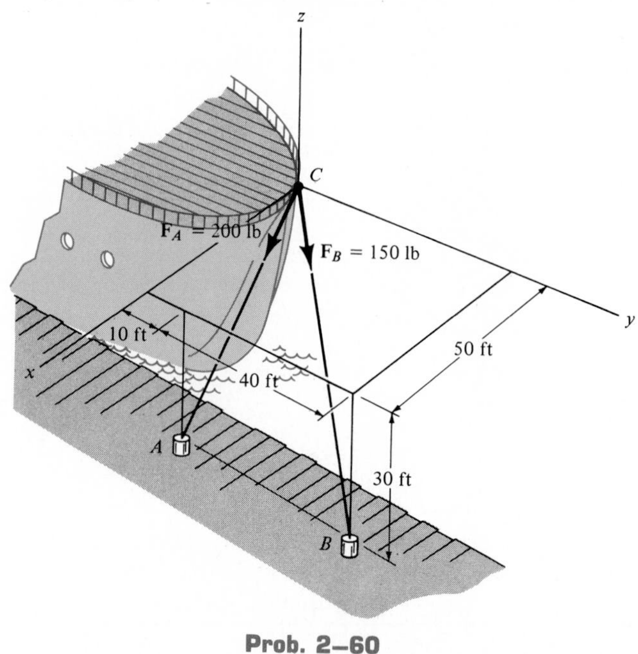

Prob. 2–60

2–59. The antenna tower is supported by three cables. If the forces in these cables are as follows: $F_B = 520$ N, $F_C = 680$ N, and $F_D = 560$ N, determine the magnitude and coordinate direction angles of the resultant force acting at A.

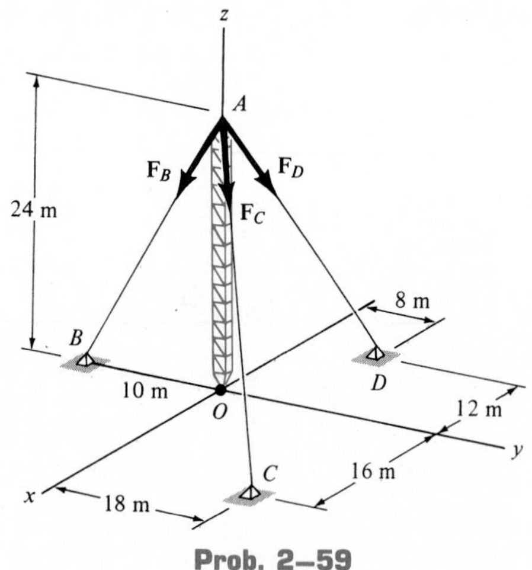

Prob. 2–59

2–61. The rope AB exerts a tie-down tension force of $F_B = 60$ N on the wing of the plane. Determine the x, y, z components of this force.

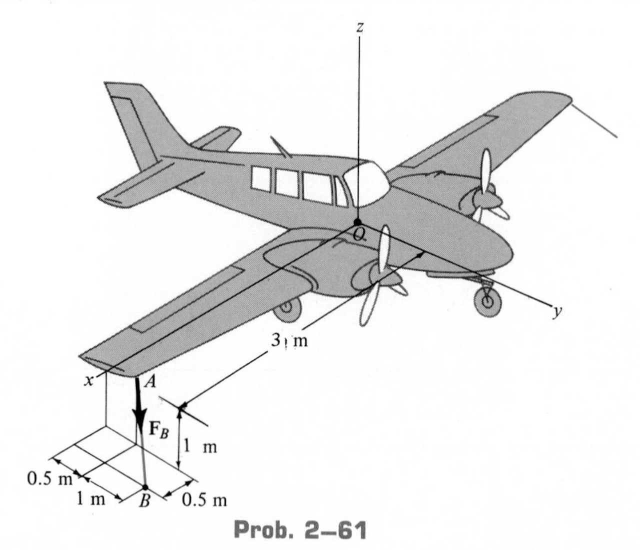

Prob. 2–61

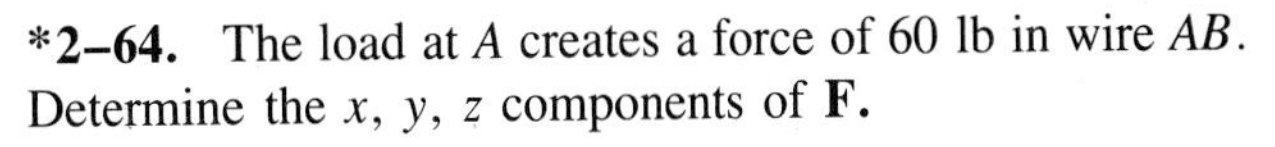

2–62. Each of the four forces acting at E has a magnitude of 28 kN. What is the magnitude of the resultant force?

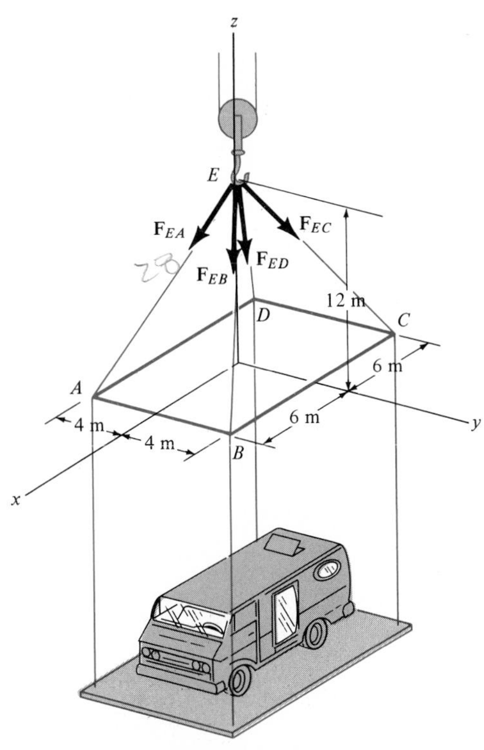

Prob. 2–62

2–63. The cable AO exerts a force on the top of the pole having components of $F_x = -120$ lb, $F_y = -90$ lb, $F_z = -80$ lb. If the cable has a length of 34 ft, determine the height z of the pole and the location (x, y) of its base.

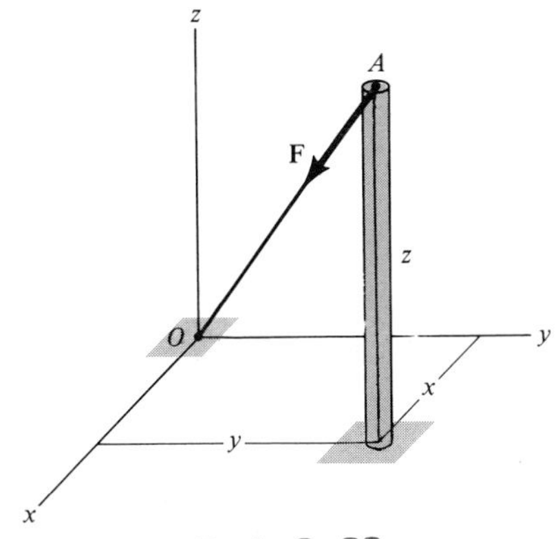

Prob. 2–63

***2–64.** The load at A creates a force of 60 lb in wire AB. Determine the x, y, z components of $\mathbf{F}$.

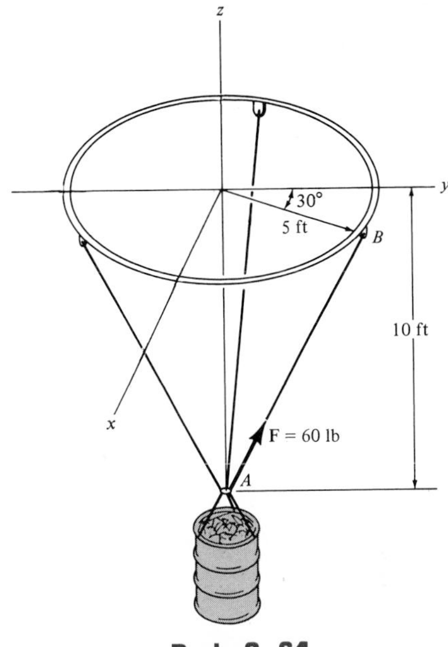

Prob. 2–64

2–65. The window is held open by cable AB. Determine the length of the cable and the x, y, z components of the 30-N force which acts at A and is directed along the cable.

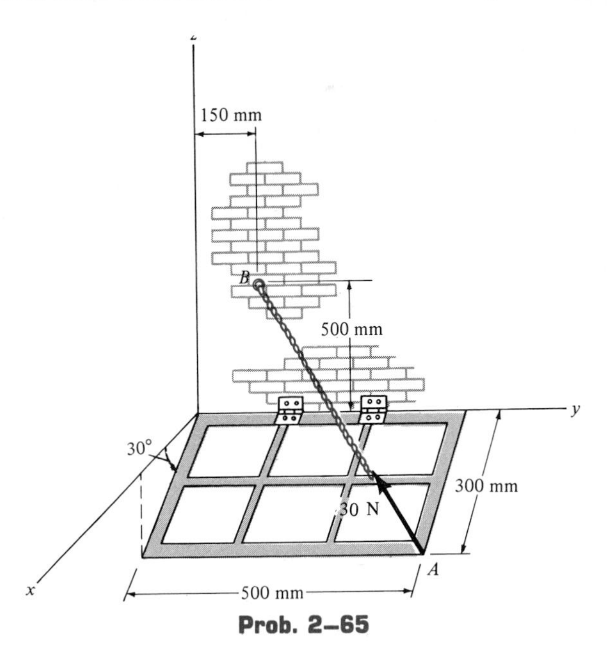

Prob. 2–65

2–66. The force **F** acting on the man, caused by his pulling on the anchor cord, has components of $F_x = 40$ N, $F_y = 20$ N, $F_z = -50$ N. If the length of the cord is 25 m, determine the coordinates $A(x, y, -z)$ of the anchor. Assume that the cord OA is along a straight line.

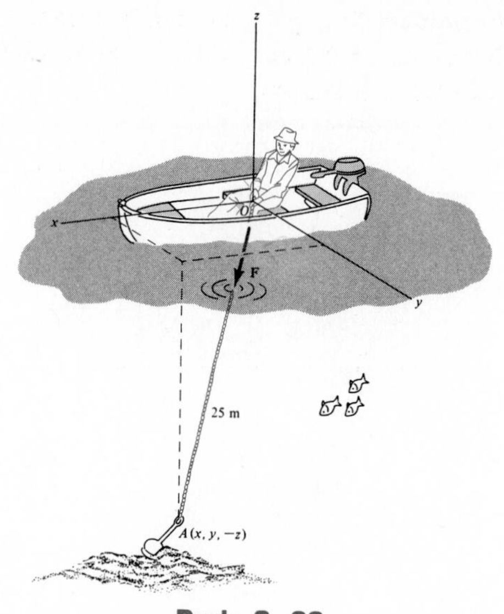

Prob. 2–66

Review Problems

2–67. Use the parallelogram law to determine the magnitude of the resultant force acting on the pin. Specify the resultant's direction, measured from the x axis.

***2–68.** Determine the x, y components of each force acting on the pin, and then determine the magnitude and direction of the resultant force.

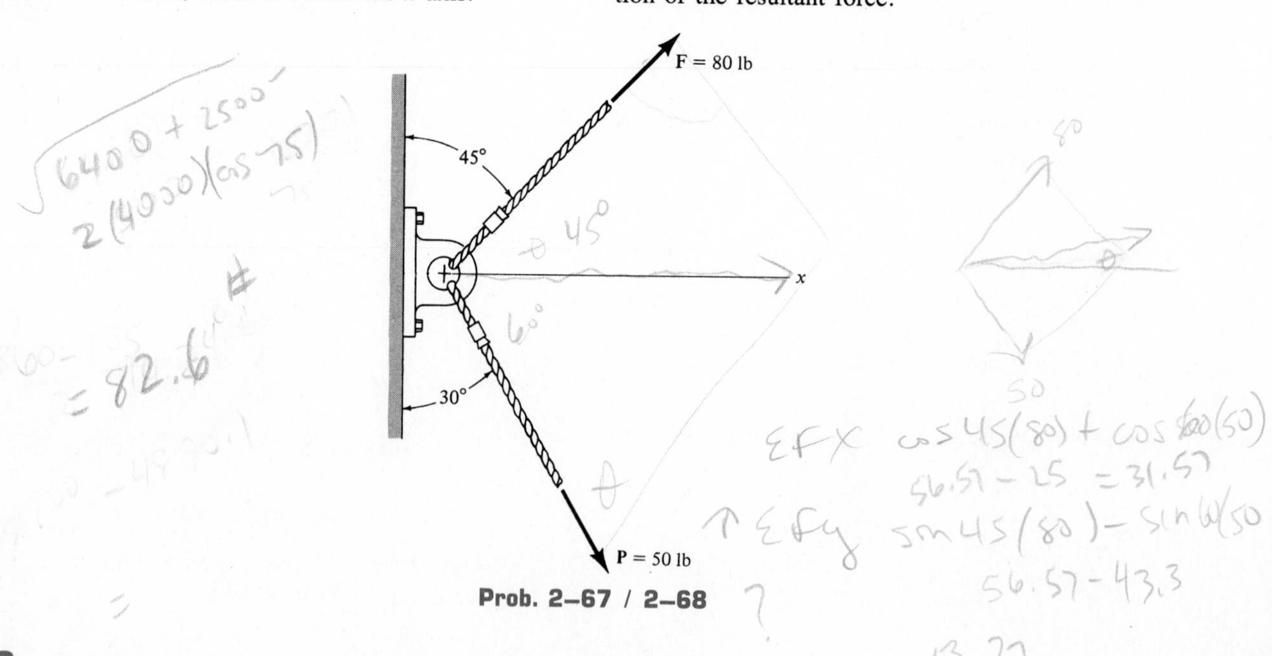

Prob. 2–67 / 2–68

2–69. Determine the x, y components of each force acting on the screw eye, and then determine the magnitude and direction of the resultant force.

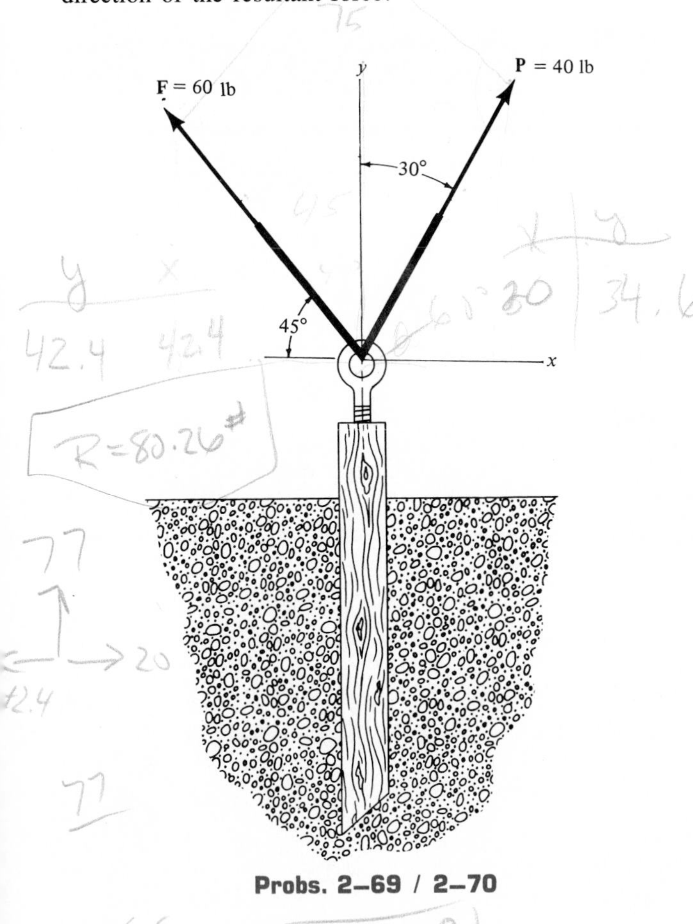

Probs. 2–69 / 2–70

2–70. Use the parallelogram law to determine the magnitude of the resultant force acting on the screw eye. Determine its direction, measured from the x axis.

2–71. Determine the x, y, z components of $\mathbf{F}_{AB}$ and $\mathbf{F}_{AC}$ and then determine the magnitude of the resultant force.

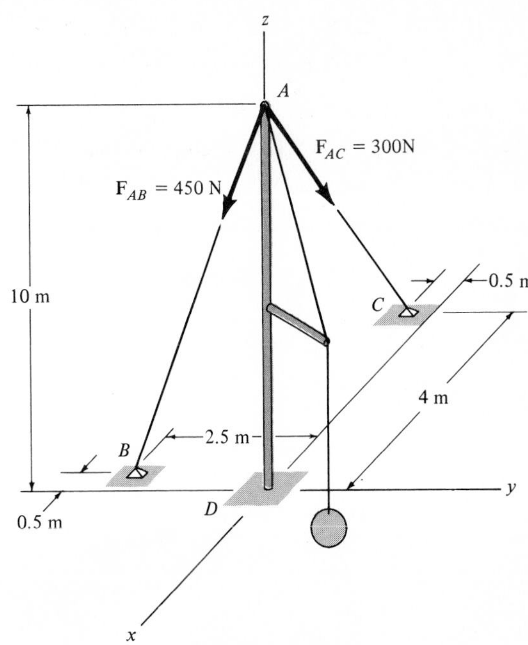

Prob. 2–71

***2–72.** The bracket supports two forces. Determine the angle θ so that the resultant force is directed to the left along the horizontal. What is the magnitude of the resultant force?

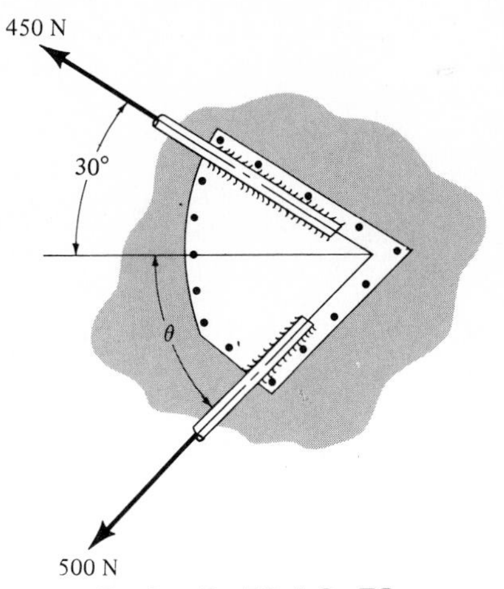

Probs. 2–72 / 2–73

2–73. If $\theta = 45°$, determine the magnitude of the resultant force acting on the bracket and its direction measured from the horizontal.

2–74. The *gusset plate* is subjected to four forces which are concurrent at point O. Determine the magnitude and direction of the resultant force.

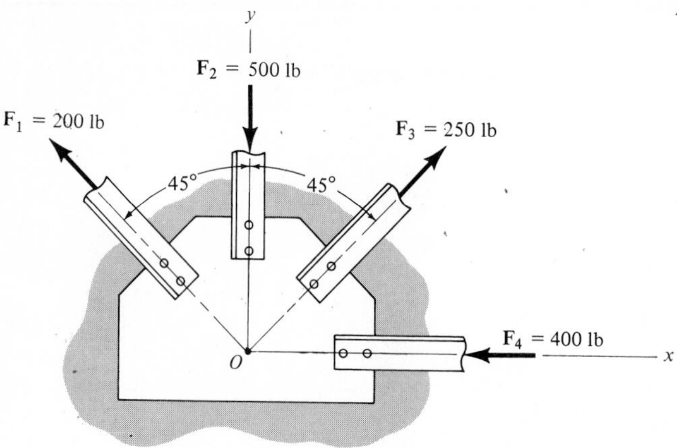

Prob. 2–74

2–75. The shaft S exerts three force components on the die D. Find the magnitude and coordinate direction angles of the resultant force. Force $\mathbf{F}_2$ acts in the octant shown.

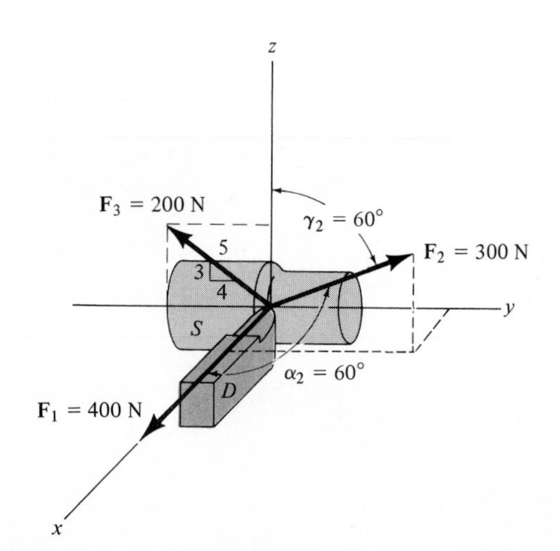

Prob. 2–75

***2–76.** The vertical force of 60 lb acts downward at A on the two-member frame. Determine the magnitudes of the two components of $\mathbf{F}$ directed along the axes of members AB and AC. Set $\theta = 45°$.

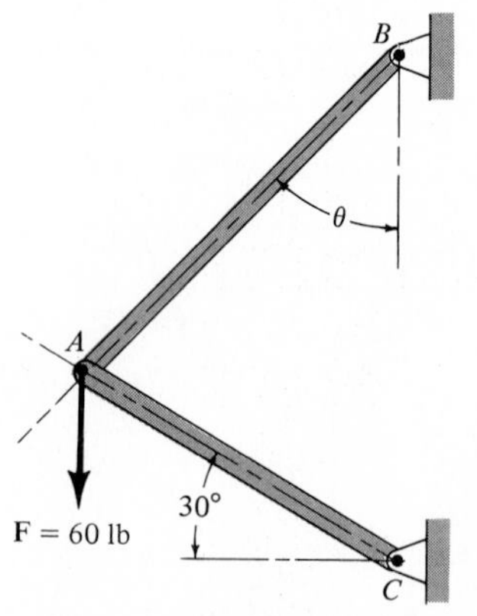

Probs. 2–76 / 2–77

2–77. Determine the angle θ $(0° \leq \theta \leq 90°)$ of member AB so that the force acting along the axis of AB is 80 lb. What is the magnitude of force acting along the axis of member AC?

2–78. The 6-ft-long guy line OA used to hold the umbrella tent in place exerts a force having components of $F_x = 4$ lb, $F_y = -3$ lb, $F_z = -12$ lb on the tent. Determine the height h of point A from the ground.

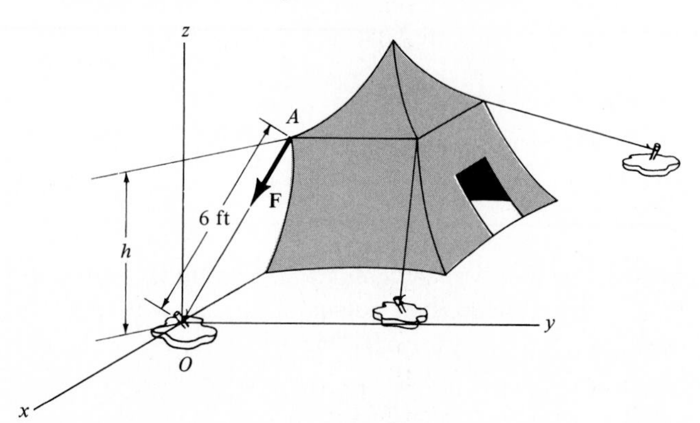

Prob. 2–78

Equilibrium of a Particle

In this chapter the methods of resolving a force into components will be used to solve problems involving the equilibrium of a particle. Recall that the dimensions or size of a particle are assumed to be neglected, and therefore a particle can be subjected *only* to a system of *concurrent forces*. In order to simplify the discussion, the equilibrium of a particle subjected to a coplanar force system will be considered first. Then, in the last part of the chapter, we will solve equilibrium problems involving three-dimensional force systems.

3.1 Condition for the Equilibrium of a Particle

The condition for particle equilibrium is based upon a balance of force. Formally stated as Newton's first law of motion: If the *resultant force* acting on a particle is *zero,* then the particle is in equilibrium. Hence, equilibrium requires that a particle either be at rest, if originally at rest, or move with constant velocity, if originally in motion. Most often, however, the term "equilibrium" or more specifically "static equilibrium" is used to describe an object at rest.

The above condition for particle equilibrium may be stated mathematically as

$$\Sigma \mathbf{F} = \mathbf{0} \tag{3–1}$$

where $\Sigma \mathbf{F}$ is the *vector sum* of *all the forces* acting on the particle. When satisfied, this equation provides both the necessary and sufficient condition for equilibrium.

*3.2 The Free-Body Diagram

A *free-body diagram* of a particle is a drawing which shows *all* the forces that act on the particle. If this diagram is correctly drawn, it will be easy to account for all the terms in Eq. 3–1.

PROCEDURE FOR DRAWING A FREE-BODY DIAGRAM

The following three steps are necessary to construct a free-body diagram.

Step 1. Imagine the particle to be *isolated* from its surroundings by drawing (sketching) an outlined shape of the particle.

Step 2. Indicate on this sketch *all* the forces that act *on the particle*. These forces will either be *active forces,* which tend to set the particle in motion, e.g., weight, or magnetic and electrostatic interaction; or *reactive forces,* such as those caused by the constraints or supports that tend to prevent motion.

Step 3. The forces that are *known* should be labeled with their proper magnitudes and directions. Letters are used to represent the magnitudes and direction angles of forces that are unknown. In particular, if a force has a known line of action but unknown magnitude, the ''arrowhead,'' which defines the directional sense of the force, can be *assumed*. The correctness of the directional sense will become apparent after solving the equilibrium equations for the unknown magnitude. By definition, the *magnitude* of a force is *always positive* so that, if the solution yields a ''negative'' magnitude, the *minus sign* indicates that the arrowhead or directional sense of the force is opposite to that which was originally assumed.

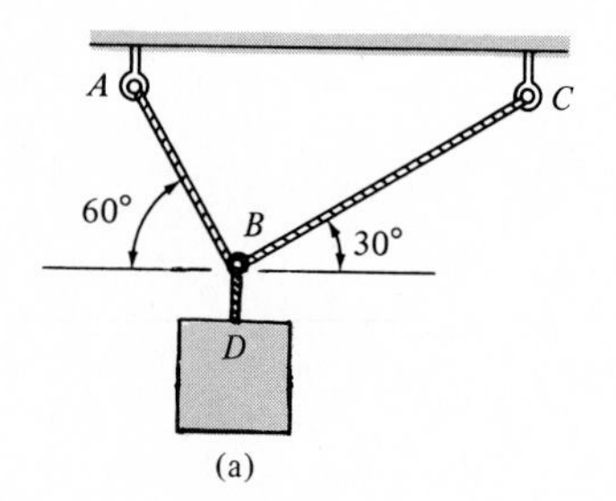

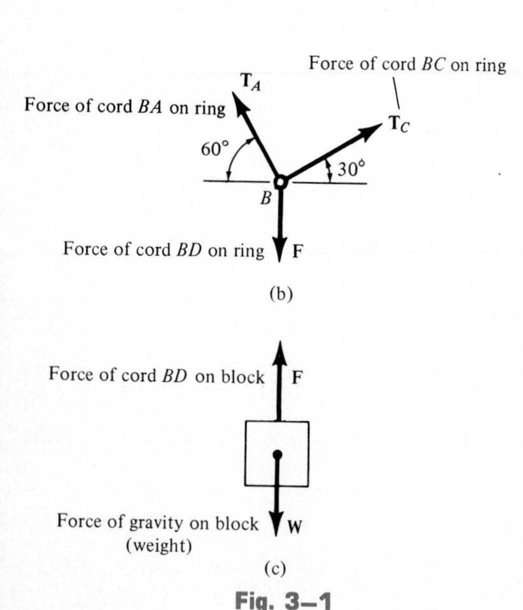

Fig. 3–1

When drawing a free-body diagram it is most important to *show only the forces acting on the object being considered*. For example, following the above procedure, the free-body diagram of the block in Fig. 3–1*a* is shown in Fig. 3–1*c*. There are only two forces acting *on the block,* its weight **W** and the force **F**, which represents the force of cord *BD* on the block. Note that **F** acts where the cord is attached. Since the cords *BA* and *BC* are not attached to the block, they do *not* exert forces *on the block* and are therefore not represented on the block's free-body diagram. Instead, these forces are shown on the free-body diagram of the ring at *B*, Fig. 3–1*b*.

Types of Connections. The following types of supports and connections are often encountered in particle equilibrium problems.

Cables and Pulleys. Throughout this book all cables are assumed to have negligible weight and they cannot be stretched. A cable can support only a tension or "pulling" force, and this force always acts in the direction of the cable. In Chapter 5 it will be shown that the tension force developed in a *continuous cable* which passes over a frictionless pulley must have a *constant* magnitude to keep the cable in equilibrium. Hence, for any angle θ, shown in Fig. 3–2, the cable is subjected to a constant tension **T** throughout its length.

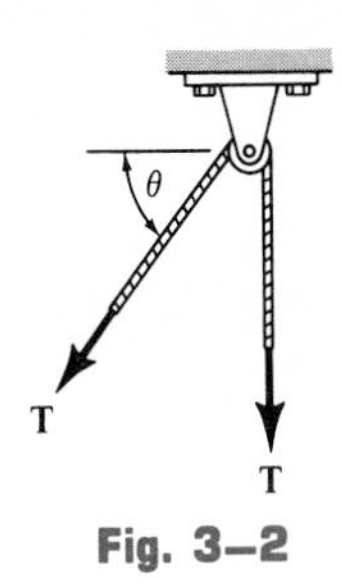

Fig. 3–2

Springs. If a *linear elastic spring* is used for a connection, the length of the spring will change in direct proportion to the force acting on it. A characteristic that defines the "elasticity" of a spring is the *stiffness k*. Specifically, the magnitude of force developed by a linear elastic spring which has a stiffness k, and is deformed (compressed or elongated) a distance x measured from its unloaded position, is

$$F = kx \tag{3–2}$$

For example, the spring shown in Fig. 3–3 has a stiffness of $k = 500$ N/m so that to stretch or compress it a distance of $x = \pm 0.2$ m, a force of $F = kx = 500$ N/m$(0.2$ m$) = 100$ N is needed.

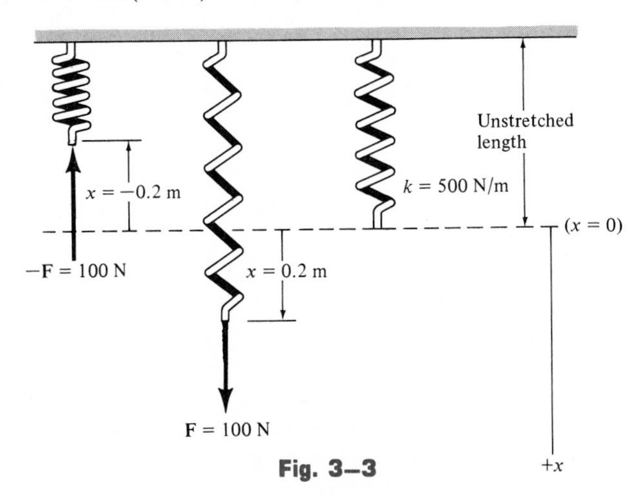

Fig. 3–3

Concurrent Forces and Weight of a Rigid Body. A special case of equilibrium occurs when a rigid body is subjected to a system of *concurrent forces*. The free-body diagram of the suspended sphere, Fig. 3–4, illustrates this situation. In particular, the weight **W** of the sphere, which represents the effect of gravity, acts downward toward the center of the earth. The weight of each particle of the sphere contributes to the total weight and provided the material is the same throughout, i.e., homogeneous, the total weight may be represented as a concentrated force acting through the sphere's geometric center O.* As shown, the lines of action of the three forces intersect at the common point O. Hence, for equilibrium of the sphere, the force system may be considered *concurrent* and coplanar at point O and must therefore satisfy the same equilibrium condition as for a particle, i.e., $\Sigma \mathbf{F} = \mathbf{0}$.

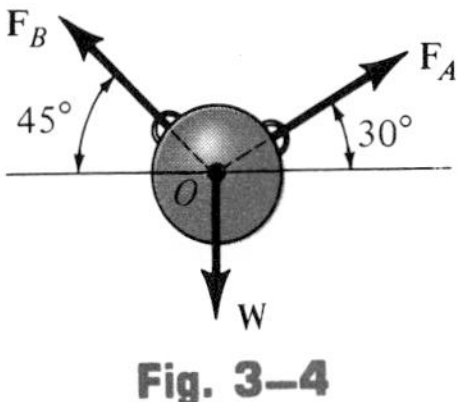

Fig. 3–4

*The methods used to determine the geometric center of an object will be discussed in Chapter 9.

Example 3–1

The crate in Fig. 3–5*a* has a weight of 20 lb. If each of the two members *AB* and *AC* supports a force directed along its axis, draw a free-body diagram of (a) the crate, and (b) the pin at *A*. Assume the cord is attached to the pin.

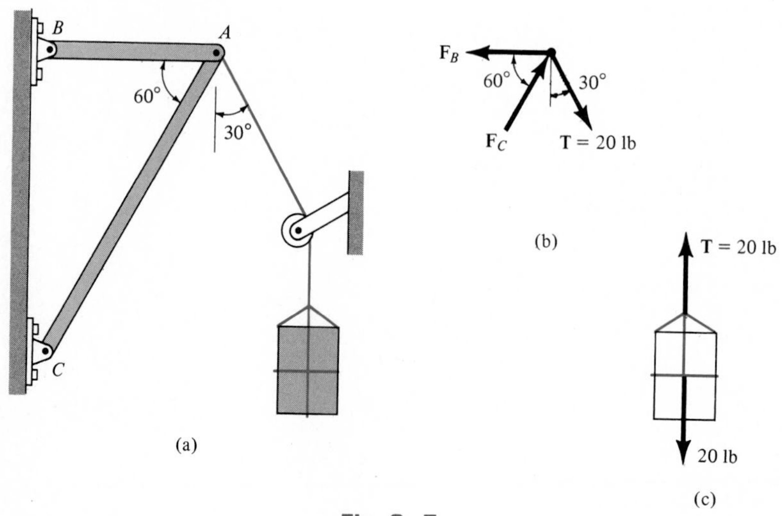

Fig. 3–5

Solution

Part (a). By inspection, two forces act on the crate, i.e., its weight of 20 lb and the cord tension **T**. The free-body diagram is shown in Fig. 3–5*b*. Notice that **T** is applied at the point where the cord is attached to the crate. Since the forces are collinear, they must be equal but opposite for equilibrium, i.e., $T = 20$ lb.

Part (b). By inspection of Fig. 3–5*a*, there are *three* forces which act on the pin at *A*, namely the forces of members *BA* and *CA* and the force of the cord. The free-body diagram is shown in Fig. 3–5*c*. As stated in the problem, the member forces act along the axis of each member. The "arrowhead" directions of $\mathbf{F}_B$ and $\mathbf{F}_C$ are assumed. Since the cord is continuous and passes over a pulley, it maintains the same constant tension $T = 20$ lb. This force "pulls" on the pin in the direction of the cord in the same manner as it "pulls" on the crate, Fig. 3–5*b*.

Coplanar Force Systems 3.3

Many particle equilibrium problems involve a coplanar force system. If the forces lie in the x-y plane, they can each be resolved into their respective x and y components and Eq. 3–1, $\Sigma \mathbf{F} = \mathbf{0}$, is satisfied provided

$$\Sigma F_x = 0 \qquad\qquad (3\text{–}3)$$
$$\Sigma F_y = 0$$

These equilibrium equations require that the algebraic sum of the x and y components of all the forces acting on the particle be equal to zero. As a result, Eqs. 3–3 can be solved for at most two unknowns, generally represented as angles and magnitudes of forces shown on the particle's free-body diagram.

PROCEDURE FOR ANALYSIS

The following procedure provides a method for solving coplanar force equilibrium problems:

Free-Body Diagram. Draw a free-body diagram of the particle. As outlined in Sec. 3–2, this requires that all the known and unknown force magnitudes and angles be labeled on the diagram. The directional sense of a force having an unknown magnitude and known line of action can be assumed.

Equations of Equilibrium. Apply the two equations of equilibrium, $\Sigma F_x = 0$ and $\Sigma F_y = 0$, to the force system shown on the free-body diagram. If more than two unknowns exist and the problem involves a spring, apply $F = kx$ (Eq. 3–2) to relate the spring force to the deformation x of the spring.

If the solution of the equations yields a *negative* force magnitude, it indicates that the direction of the force shown on the free-body diagram is *opposite* to that which was *assumed*.

The following example problems numerically illustrate this solution procedure.

Example 3–2

Determine the tension in cords AB and AD for equilibrium of the 10-kg block shown in Fig. 3–6a.

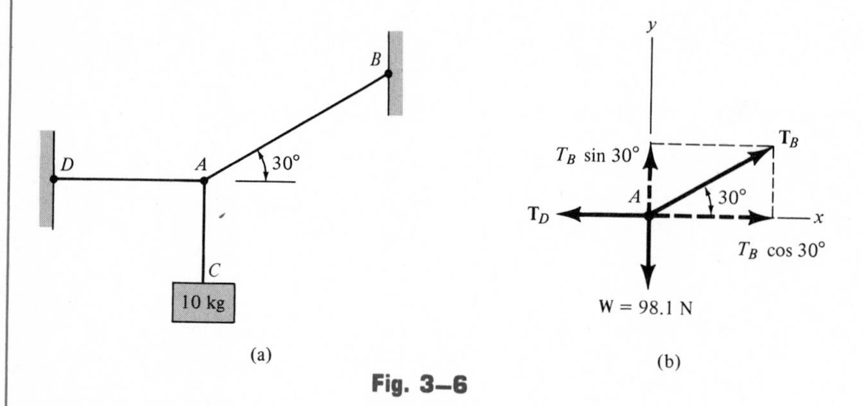

Fig. 3–6

Solution

The cord tensions can be obtained by investigating the equilibrium of point A.

Free-Body Diagram. As shown in Fig. 3–6b, there are three concurrent forces *acting on point A*. The cord tension forces $\mathbf{T}_B$ and $\mathbf{T}_D$ have unknown magnitudes but known directions. Cord AC exerts a downward force on A equal to the weight of the block, $W = (10 \text{ kg})(9.81 \text{ m/s}^2) = 98.1 \text{ N}$.

Equations of Equilibrium. Since the equilibrium equations require a summation of the x and y components of each force, $\mathbf{T}_B$ must be resolved into x and y components. These components, shown dashed on the free-body diagram, have magnitudes of $T_B \cos 30°$ and $T_B \sin 30°$, respectively. Equations 3–3 will be applied by assuming that "positive" force components act along the positive x and y axes. Indicating these directions alongside the equations, we have

$$\overset{+}{\rightarrow}\Sigma F_x = 0; \qquad T_B \cos 30° - T_D = 0 \qquad (1)$$

$$+\uparrow \Sigma F_y = 0; \qquad T_B \sin 30° - 98.1 = 0 \qquad (2)$$

Solving Eq. (2) for T_B and substituting into Eq. (1) to obtain T_D yields

$$T_B = 196.2 \text{ N} \qquad \textit{Ans.}$$

$$T_D = 169.9 \text{ N} \qquad \textit{Ans.}$$

Example 3–3

If the cylinder at A in Fig. 3–7a has a weight of 20 lb, determine the weight of B and the force in each cord needed to hold the system in the equilibrium position shown.

Solution

Since the weight of A is known, the unknown tension in cables EG and EC can be determined by investigating the equilibrium of point E.

Free-Body Diagram. There are three forces acting on point E, as shown in Fig. 3–7b.

Equations of Equilibrium. Resolving each force into its x and y components using trigonometry, and applying the equations of equilibrium, we have

$$\overset{+}{\rightarrow}\Sigma F_x = 0; \qquad T_{EG} \sin 30° - T_{EC} \cos 45° = 0 \qquad (1)$$

$$+\uparrow \Sigma F_y = 0; \qquad T_{EG} \cos 30° - T_{EC} \sin 45° - 20 = 0 \qquad (2)$$

Solving Eq. (1) for T_{EG} in terms of T_{EC} and substituting the result into Eq. (2) allows a solution for T_{EC}. One then obtains T_{EG} from Eq. (1). The results are

$$T_{EC} = 38.6 \text{ lb} \qquad \textit{Ans.}$$

$$T_{EG} = 54.6 \text{ lb} \qquad \textit{Ans.}$$

Using the calculated result for T_{EC}, the equilibrium of point C can now be investigated to determine the tension in CD and the weight of B.

Free-Body Diagram. As shown in Fig. 3–7c, $T_{EC} = 38.6$ lb "pulls" on C. The reason for this becomes clear when one draws the free-body diagram of cord CE and applies the principle of action, equal but opposite force reaction (Newton's third law), Fig. 3–7d.

Equations of Equilibrium. Noting the components of T_{CD} are proportional to the slope triangle, we have

$$\overset{+}{\rightarrow}\Sigma F_x = 0; \qquad 38.6 \cos 45° - (\tfrac{4}{5})T_{CD} = 0 \qquad (3)$$

$$+\uparrow \Sigma F_y = 0; \qquad (\tfrac{3}{5})T_{CD} + 38.6 \sin 45° - W_B = 0 \qquad (4)$$

Solving Eq. (3) and substituting the result into Eq. (4) yields

$$T_{CD} = 34.1 \text{ lb} \qquad \textit{Ans.}$$

$$W_B = 47.8 \text{ lb} \qquad \textit{Ans.}$$

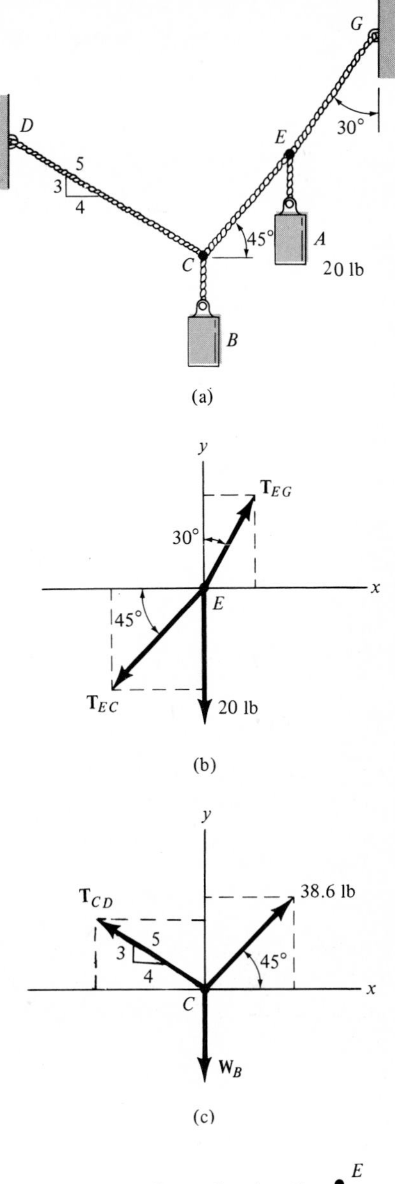

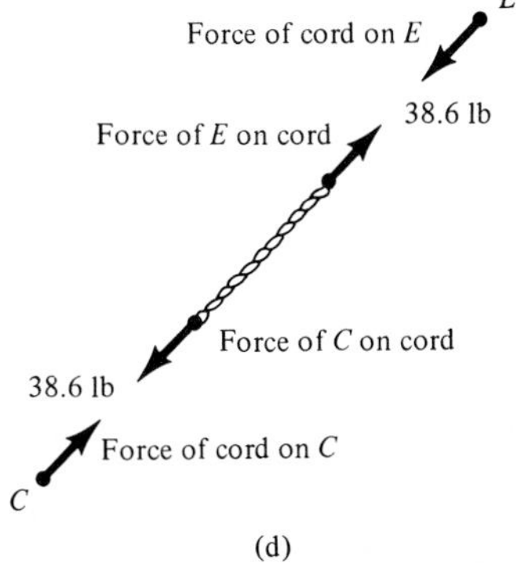

Fig. 3–7

Example 3–4

Determine the required length of cord AC in Fig. 3–8a so that the 8-kg lamp is suspended in the position shown. The *unstretched* length of the spring AB is $l'_{AB} = 0.4$ m, and the spring has a stiffness of $k_{AB} = 300$ N/m.

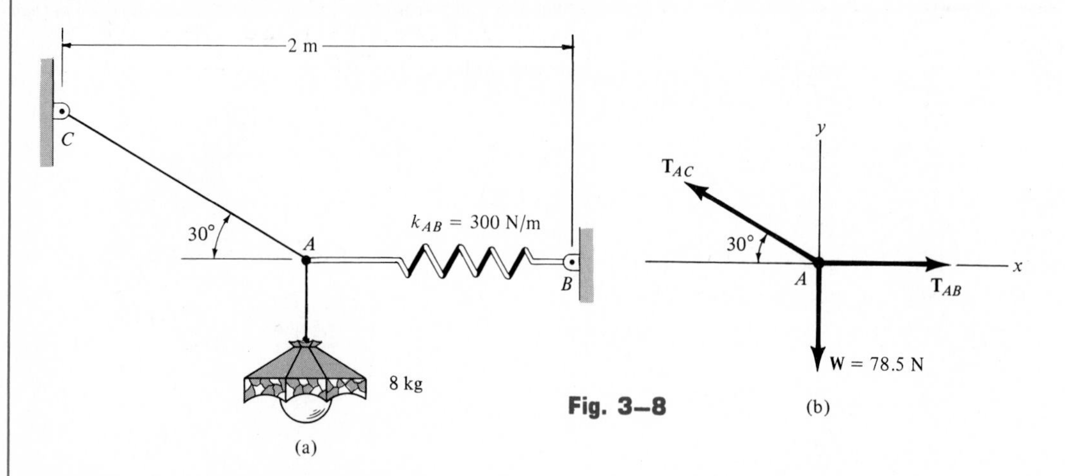

Fig. 3–8

Solution

If the force in spring AB is known, the stretch in the spring can be found ($F = kx$). Using the problem geometry, it is then possible to calculate the required length of AC.

Free-Body Diagram. The lamp has a weight of $W = 8(9.81) = 78.5$ N. The free-body diagram of point A is shown in Fig. 3–8b.

Equations of Equilibrium.

$$\stackrel{+}{\rightarrow}\Sigma F_x = 0; \qquad T_{AB} - T_{AC}\cos 30^\circ = 0$$

$$+\uparrow \Sigma F_y = 0; \qquad T_{AC}\sin 30^\circ - 78.5 = 0$$

Solving, we obtain

$$T_{AC} = 157.0 \text{ N}$$
$$T_{AB} = 136.0 \text{ N}$$

The stretch in spring AB is therefore

$$T_{AB} = k_{AB}x_{AB}; \qquad 136.0 \text{ N} = 300 \text{ N/m}(x_{AB})$$
$$x_{AB} = 0.453 \text{ m}$$

so that the stretched length of spring AB is

$$l_{AB} = l'_{AB} + x_{AB}$$
$$l_{AB} = 0.4 \text{ m} + 0.453 \text{ m} = 0.853 \text{ m}$$

The horizontal distance from C to B, Fig. 3–8a, requires

$$2 \text{ m} = l_{AC}\cos 30^\circ + 0.853 \text{ m}$$
$$l_{AC} = 1.32 \text{ m} \qquad \textit{Ans.}$$

Problems

3–1. Determine the magnitudes of $\mathbf{F}_1$ and $\mathbf{F}_2$ so that particle P is in equilibrium.

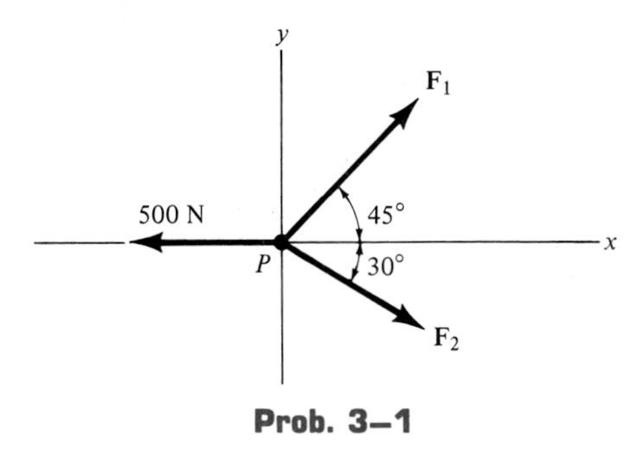

Prob. 3–1

3–2. Determine the magnitude and direction θ of $\mathbf{F}_1$ so that particle P is in equilibrium.

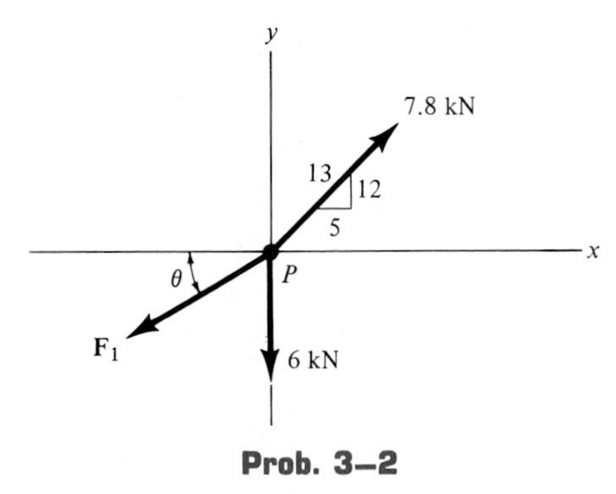

Prob. 3–2

3–3. Determine the magnitude and direction θ of $\mathbf{F}_1$ so that particle P is in equilibrium.

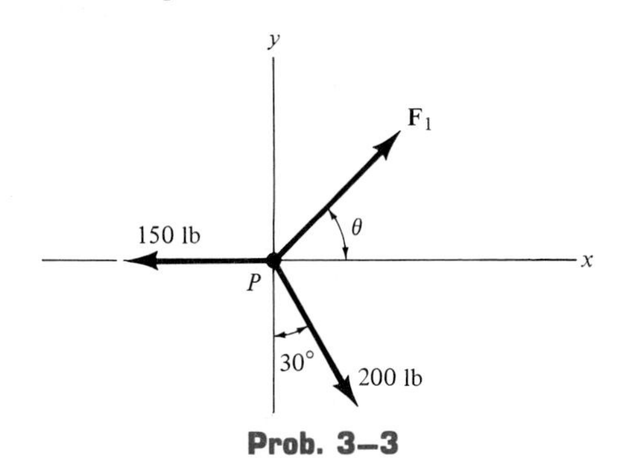

Prob. 3–3

***3–4.** The members of a truss are connected to the gusset plate. If the forces are concurrent at point O, determine the magnitudes of $\mathbf{F}$ and $\mathbf{T}$ for equilibrium.

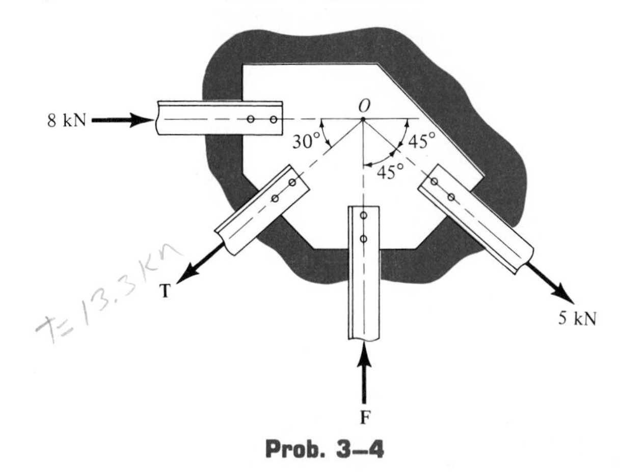

Prob. 3–4

3–5. The patella P located in the human knee joint is subjected to tendon forces $\mathbf{T}_1$ and $\mathbf{T}_2$ and a force $\mathbf{F}$ exerted on the patella by the femoral articular A. If the directions of these forces are estimated from an X-ray as shown, determine the magnitudes of $\mathbf{T}_1$ and $\mathbf{F}$ when the tendon force $T_2 = 6$ lb. The forces are concurrent at point O.

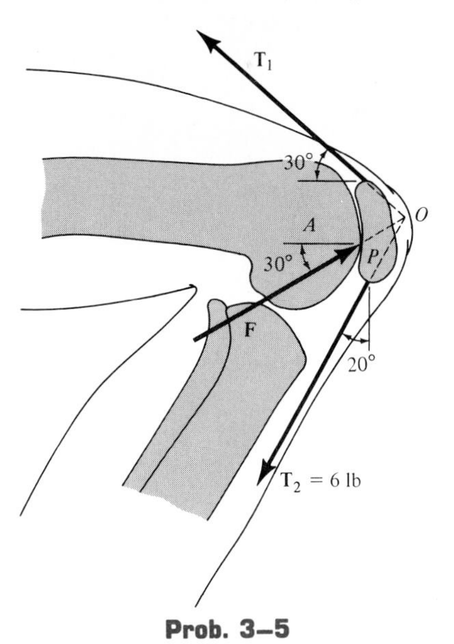

Prob. 3–5

3–6. The 75-lb vertical force at A is supported by members AB and AC of the frame. If each member supports a force which is directed along its axis, determine these forces for equilibrium of pin A.

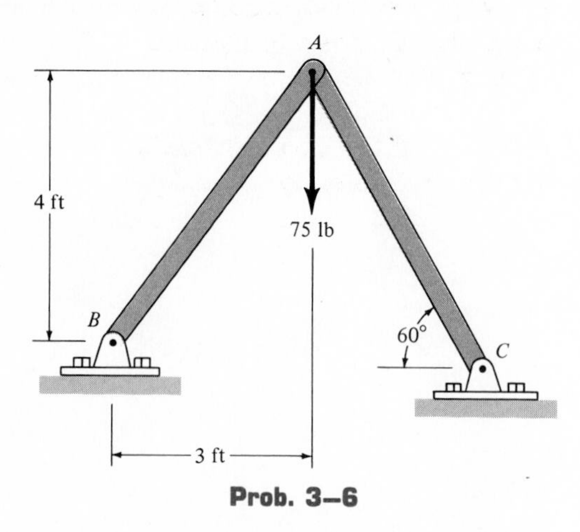

Prob. 3–6

3–7. The engine E of an automobile has a mass of 120 kg. Determine the force which the man at B must exert on the horizontal rope to position the engine over the truck bed as shown. What force does the man at C exert on the cord? Neglect the size of the pulley at A.

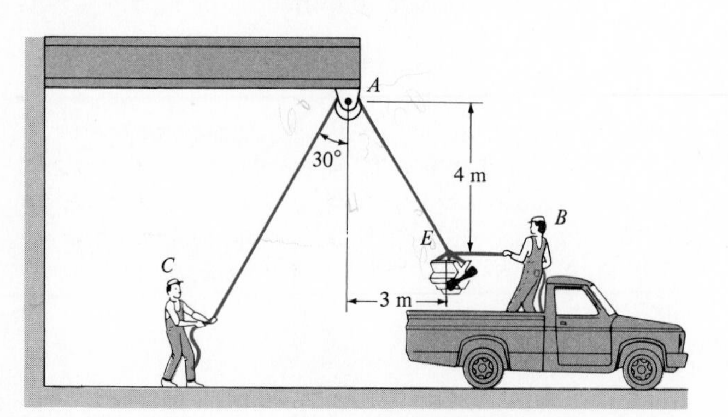

Prob. 3–7

***3–8.** Determine the magnitude and direction θ of the resultant force $\mathbf{F}_{AB}$ exerted along link AB by the tractive apparatus shown. The suspended mass is 10 kg. Neglect the size of the pulley at A and treat it as a particle.

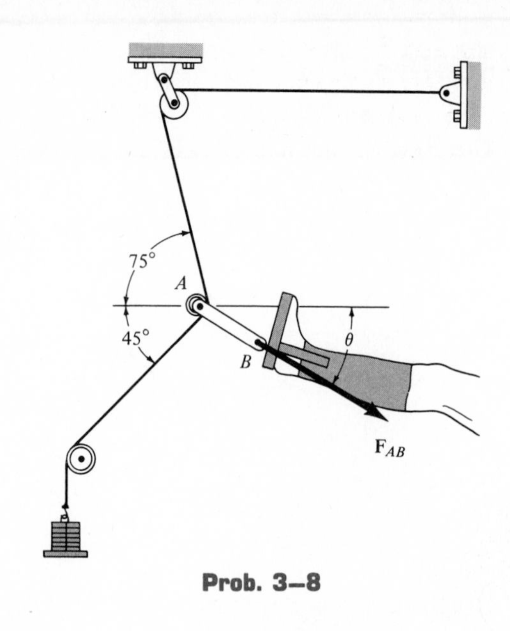

Prob. 3–8

3–9. The motor at B winds up the cord attached to the 65-lb crate with a constant speed. Determine the force in rope CD supporting the pulley and the angle θ for equilibrium. Neglect the size of the pulley at C and treat it as a particle.

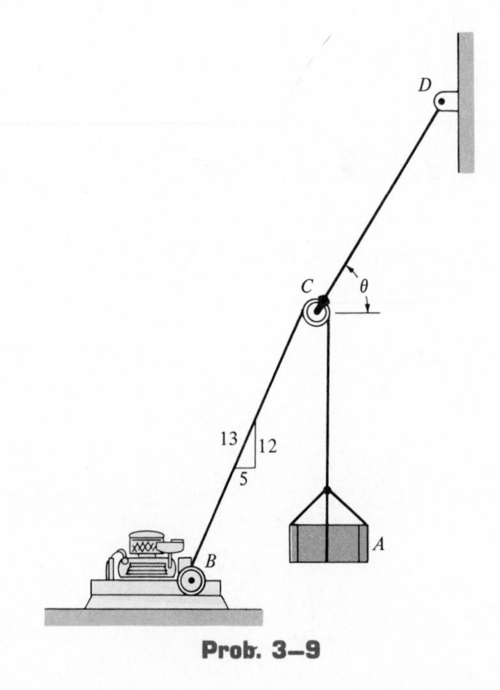

Prob. 3–9

3–10. The gusset plate P is subjected to the forces of three members as shown. Determine the force in member C and its proper orientation θ for equilibrium. The forces are concurrent at point O.

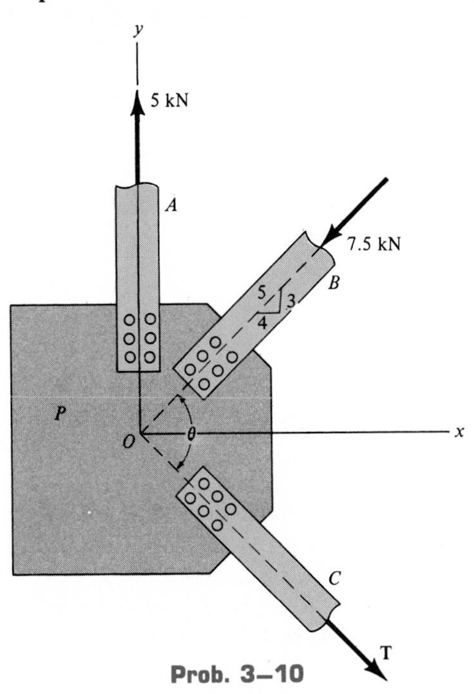

Prob. 3–10

3–11. A child's toy doll has a mass of 0.5 kg and is suspended from a cord ABC. If the distance the doll hangs downward from the ceiling is $d = 150$ mm and the ring at B is free to slide over the cord, i.e., it acts like a pulley, show that cord length AB must be equal to CB and determine the tension in the cord.

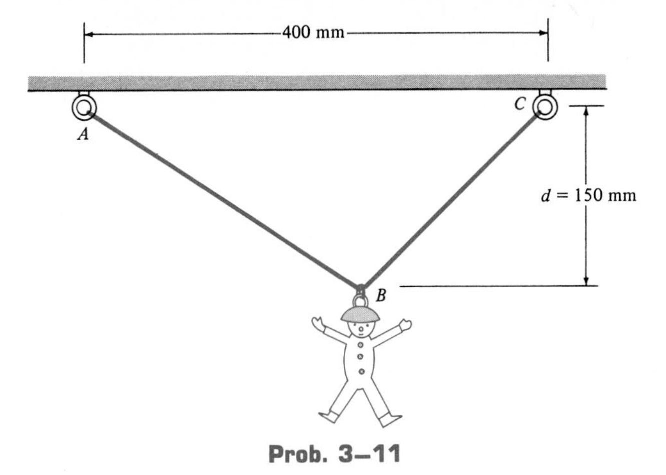

Prob. 3–11

***3–12.** Romeo tries to reach Juliet by climbing with constant velocity up a rope which is knotted at point A. Any of the three segments of the rope can sustain a maximum force of 1.5 kN before it breaks. Perform an equilibrium analysis of point A to determine if Romeo, who has a mass of 70 kg, can climb the rope. If this is possible, can he along with his Juliet, who has a mass of 60 kg, climb down with constant velocity?

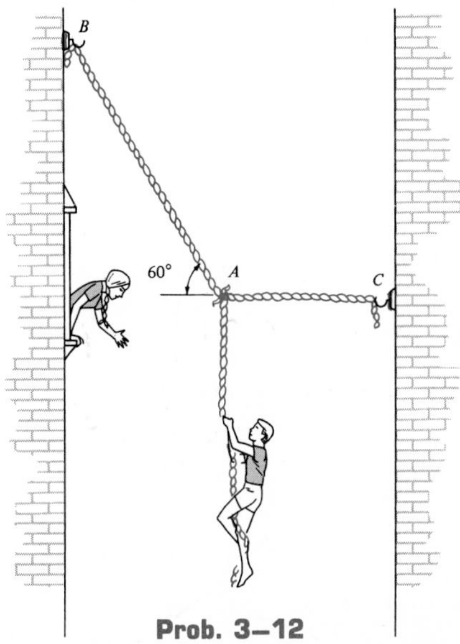

Prob. 3–12

3–13. The ring of negligible size is subjected to a vertical force of 200 lb. Determine the required length l of cord AC such that the tension acting in AC is 160 lb. Also, what is the force acting in cord AB? *Hint:* Use the equilibrium condition to determine the required angle θ for attachment, then determine l using trigonometry applied to ΔABC.

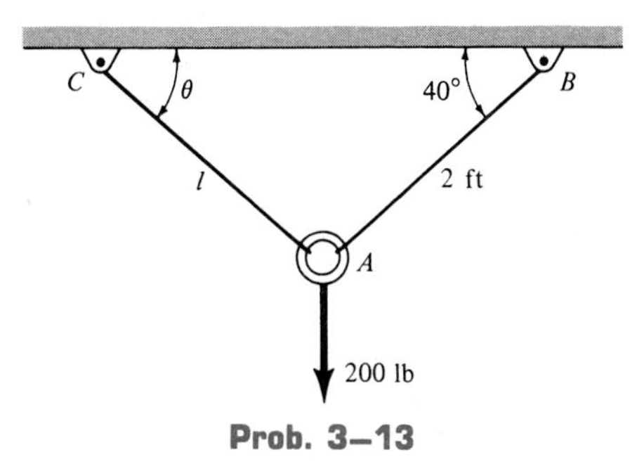

Prob. 3–13

3–14. A "scale" is constructed with a 4-ft-long cord and the 10-lb block D. The cord is fixed to a pin at A and passes over two *small* pulleys at B and C. Determine the weight of the suspended block E if the system is in equilibrium when $s = 1.5$ ft.

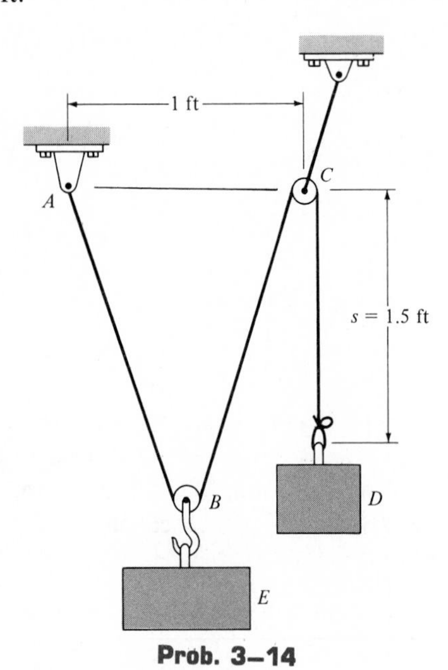

Prob. 3–14

3–15. Determine the stiffness k_T of the single spring such that the force **F** will stretch it by the same amount x as the force **F** stretches the two springs. Express the result in terms of the stiffness k_1 and k_2 of the two springs.

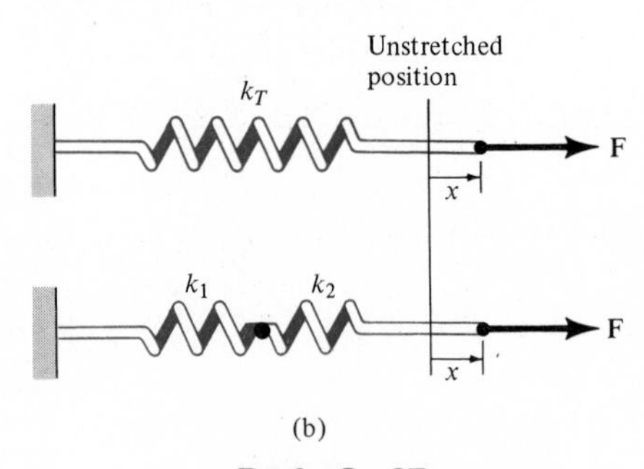

Prob. 3–15

***3–16.** The 30-kg block is supported by two springs having the stiffness shown. Determine the unstretched length of each spring after the block is removed.

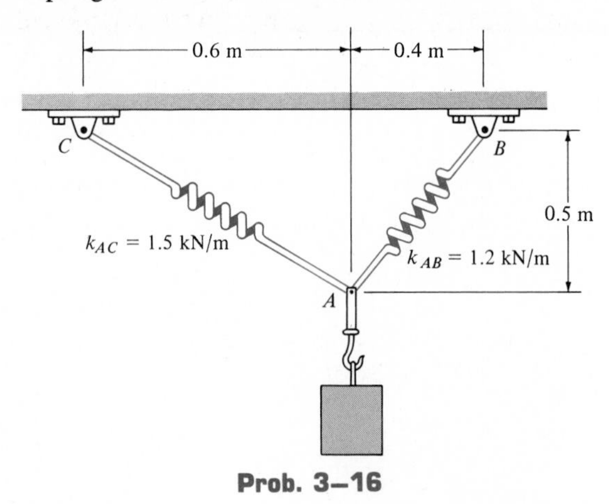

Prob. 3–16

3–17. The spider has a mass of 0.7 g and is suspended from a portion of its web as shown. Determine the force which each of the three web "strings" exerts on the twigs at C, D, and E. String CB is horizontal. *Hint:* First analyze the equilibrium at point A; then, using the result for the force in AB, analyze the equilibrium at B.

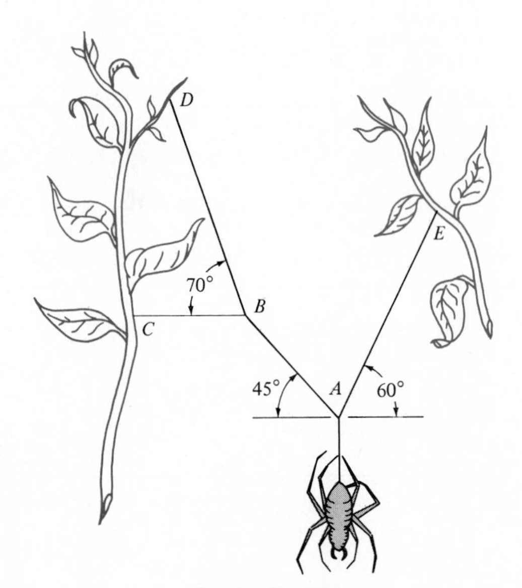

Prob. 3–17

3–18. The sack has a weight of 15 lb and is supported by the six cords tied together as shown. Determine the tension in each cord and the angle θ for equilibrium of DC. Cord BC is horizontal. *Hint:* Analyze the equilibrium of the knots in the following sequence: A, B, C, using the results previously calculated.

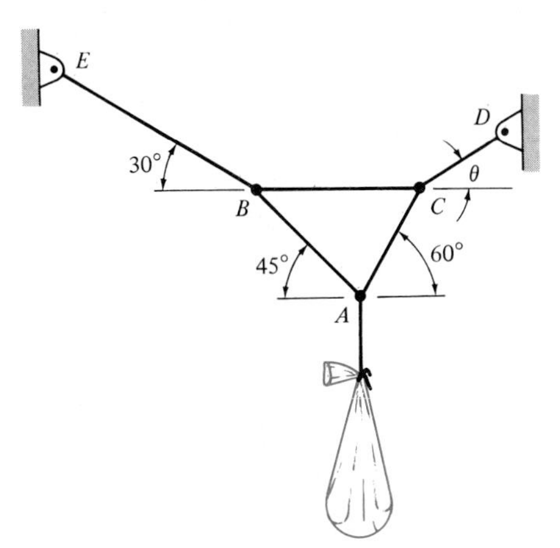

Prob. 3–18

3–19. Determine the force in each cable and the force **F** needed to hold the 4-kg lamp in the position shown. *Hint:* First analyze the equilibrium at B; then, using the result for the force in BC, analyze the equilibrium at C.

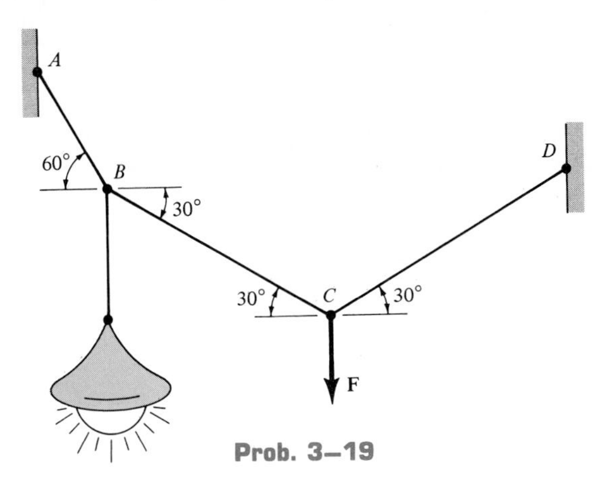

Prob. 3–19

***3–20.** The bulldozer is used to pull down the chimney using the cable and *small* pulley arrangement shown. If the tension force in AB is 500 lb, determine the tension in cable CAD and the angle θ which the cable makes at the pulley.

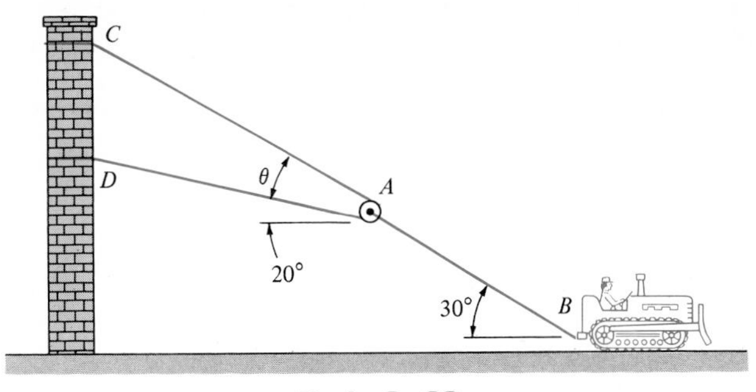

Prob. 3–20

3–21. The ends of the 7-m-long cable AB are attached to the fixed walls. If a bucket and its contents have a mass of 10 kg and are suspended from the cable by means of a *small* pulley as shown, determine the location x of the pulley for equilibrium.

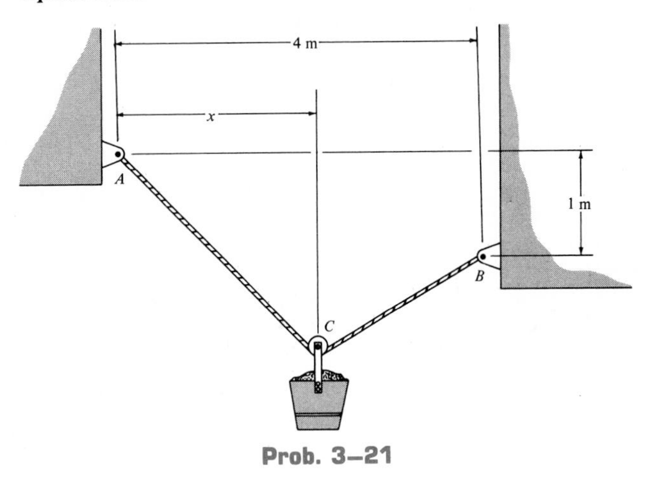

Prob. 3–21

3.4 Three-Dimensional Force Systems

It was shown in Sec. 3–1 that particle equilibrium requires $\Sigma \mathbf{F} = \mathbf{0}$. If all the forces acting on the particle are resolved into their x, y, z components, then for equilibrium we require the following three scalar component equations be satisfied:

$$\Sigma F_x = 0 \qquad \Sigma F_y = 0 \qquad \Sigma F_z = 0 \tag{3–4}$$

These equations represent the *algebraic sums* of the x, y, z force components acting on the particle. Using them we can solve for at most three unknowns, generally represented as angles or magnitudes of forces shown on the free-body diagram of the particle. The equilibrium analysis is similar to that used for coplanar force systems.

PROCEDURE FOR ANALYSIS

The following procedure provides a method for solving three-dimensional force equilibrium problems.

Free-Body Diagram. Draw a free-body diagram of the particle and label all the known and unknown forces on this diagram.

Equations of Equilibrium. Establish the x, y, z coordinate axes with origin located at the particle and then determine the x, y, z components of each force shown on the free-body diagram. After this is done, the three equations of equilibrium can be applied. If more than three unknowns exist and the problem involves a spring, consider using $F = kx$ to relate the spring force to the deformation x of the spring.

The following example problems numerically illustrate this solution procedure.

Example 3–5

The 90-lb cylinder shown in Fig. 3–9a is supported by two cables and a spring having a stiffness of $k = 500$ lb/ft. Determine the force in the cables and the stretch of the spring for equilibrium. Cord AD lies in the x-y plane and cord AC lies in the x-z plane.

Solution

The stretch of the spring can be determined once the force in AB is determined.

Free-Body Diagram. Point A is chosen for the analysis since the cable forces are concurrent at this point, Fig. 3–8b.

Equations of Equilibrium. By inspection, each force can easily be resolved into its x, y, z components. Considering components directed along the positive axes as "positive," we have

$$\Sigma F_x = 0; \qquad F_D \sin 30^\circ - \tfrac{4}{5}F_C = 0 \qquad (1)$$

$$\Sigma F_y = 0; \qquad -F_D \cos 30^\circ + F_B = 0 \qquad (2)$$

$$\Sigma F_z = 0; \qquad \tfrac{3}{5}F_C - 90 \text{ lb} = 0 \qquad (3)$$

Solving Eq. (3) for F_C, then Eq. (1) for F_D, and finally Eq. (2) for F_B, we get

$$F_C = 150.0 \text{ lb} \qquad \textit{Ans.}$$

$$F_D = 240.0 \text{ lb} \qquad \textit{Ans.}$$

$$F_B = 207.8 \text{ lb} \qquad \textit{Ans.}$$

The stretch of the spring is therefore

$$F_B = kx_{AB}$$

$$207.8 \text{ lb} = 500 \text{ lb/ft}\,(x_{AB})$$

$$x_{AB} = 0.416 \text{ ft} \qquad \textit{Ans.}$$

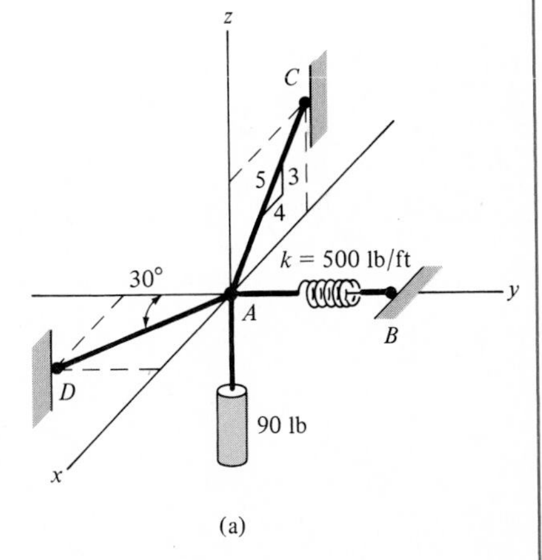

(a)

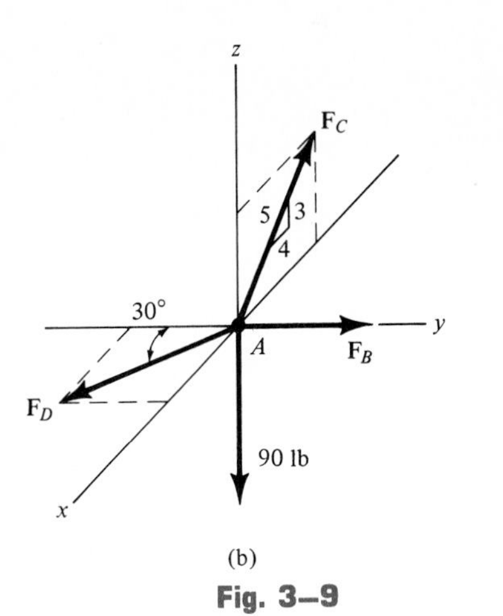

(b)

Fig. 3–9

Example 3–6

Determine the magnitude and coordinate direction angles of **F** in Fig. 3–10*a* that are required for equilibrium of point *O*.

Solution

Free-Body Diagram. The four forces acting on point *O* are shown in Fig. 3–10*b*.

Equations of Equilibrium. Before applying the equations of equilibrium we must first determine the components of each force.

For $\mathbf{F}_1$ $\quad F_{1x} = 0 \quad F_{1y} = 400 \text{ N} \quad F_{1z} = 0$

For $\mathbf{F}_2$ $\quad F_{2x} = 0 \quad F_{2y} = 0 \quad F_{2z} = -800 \text{ N}$

The components of $\mathbf{F}_3$ are determined by first finding a position vector $\mathbf{r}_B$, extending from *O* to *B*. Note that the coordinates of point *B* are $B(-2, -3, 6)$ so that $r_{Bx} = -2$ m, $r_{By} = -3$ m, $r_{Bz} = 6$ m. Also $r_B = \sqrt{(-2)^2 + (-3)^2 + (6)^2} = 7$ m. Using the direction cosines of $\mathbf{r}_B$, the components of $\mathbf{F}_3$ are

$$F_{3x} = 700\left(\frac{-2}{7}\right) = -200 \text{ N} \qquad F_{3y} = 700\left(\frac{-3}{7}\right) = -300 \text{ N}$$

$$F_{3z} = 700\left(\frac{6}{7}\right) = 600 \text{ N}$$

Lastly, **F** has unknown components.

$$F_x = ? \qquad F_y = ? \qquad F_z = ?$$

Equilibrium requires

$\Sigma \mathbf{F} = \mathbf{0}; \qquad \mathbf{F}_1 + \mathbf{F}_2 + \mathbf{F}_3 + \mathbf{F} = \mathbf{0}$

or, using the above data, we have

$\Sigma F_x = 0; \qquad -200 + F_x = 0 \qquad F_x = 200 \text{ N}$

$\Sigma F_y = 0; \qquad 400 - 300 + F_y = 0 \qquad F_y = -100 \text{ N}$

$\Sigma F_z = 0; \qquad -800 + 600 + F_z = 0 \qquad F_z = 200 \text{ N}$

Hence, the magnitude of **F** is

$$F = \sqrt{(200)^2 + (-100)^2 + (200)^2} = 300 \text{ N} \qquad \textit{Ans.}$$

The coordinate direction angles are

$$\alpha = \cos^{-1}\left(\frac{200}{300}\right) = 48.2° \qquad \textit{Ans.}$$

$$\beta = \cos^{-1}\left(\frac{-100}{300}\right) = 109.5° \qquad \textit{Ans.}$$

$$\gamma = \cos^{-1}\left(\frac{200}{300}\right) = 48.2° \qquad \textit{Ans.}$$

The magnitude and correct direction of **F** are shown in Fig. 3–10*c*.

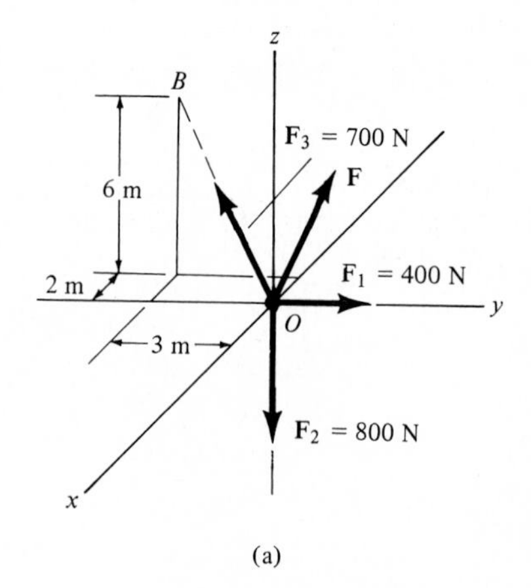

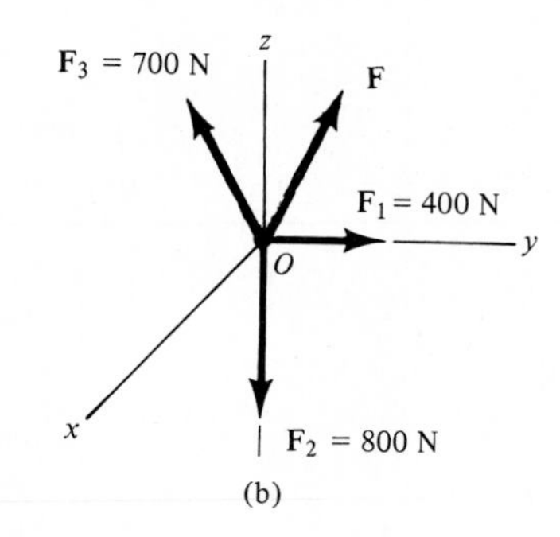

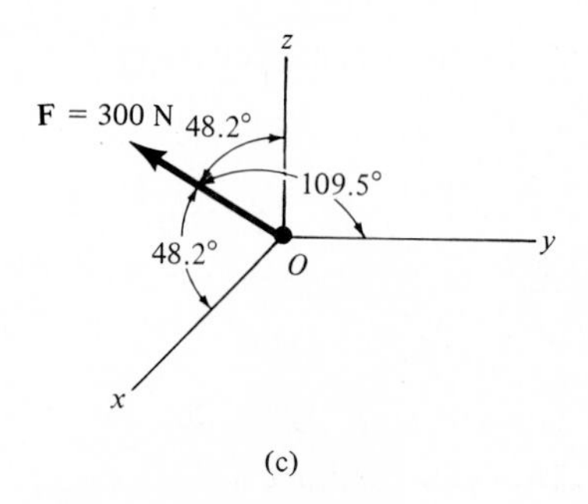

Fig. 3–10

Example 3–7

The 40-lb crate is supported by three struts AB, AC, and AD as shown in Fig. 3–11a. Determine the force developed in each strut if it passes along the axis of the strut. These forces may be either tensile or compressive.

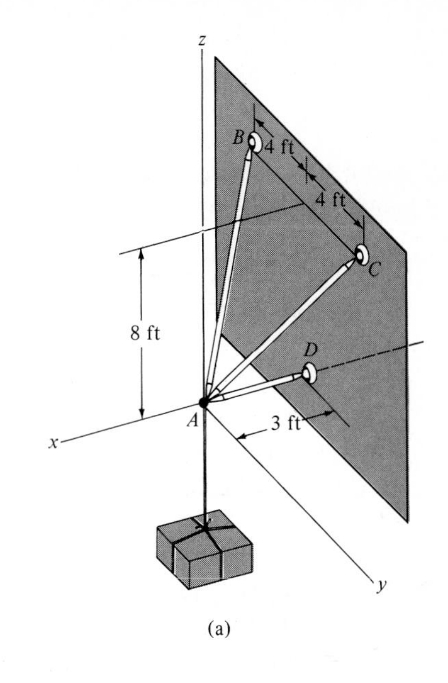

(a)

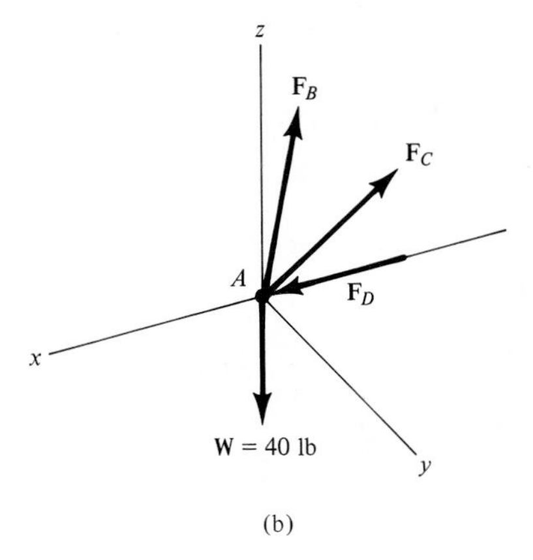

(b)

Fig. 3–11

Solution

Free-Body Diagram. As shown in Fig. 3–11b, point A is considered in order to ''expose'' the three unknown forces in the struts. Here $\mathbf{F}_B$ and $\mathbf{F}_C$ are assumed to be *tensile* forces since they *pull* on A, and $\mathbf{F}_D$ is assumed to be *compressive* since it *pushes* on A.

Equations of Equilibrium. The components of each force are first determined. For $\mathbf{F}_B$, note that point B has coordinates of $B(-3, -4, 8)$. Thus, $r_{Bx} = -3$ ft, $r_{By} = -4$ ft, $r_{Bz} = 8$ ft. Also, $r_B = 9.43$. Since F_B is unknown,

$$F_{Bx} = F_B\left(\frac{-3}{9.43}\right) = -0.318F_B$$

$$F_{By} = F_B\left(\frac{-4}{9.43}\right) = -0.424F_B$$

$$F_{Bz} = F_B\left(\frac{8}{9.43}\right) = 0.848F_B$$

For $\mathbf{F}_C$, point C has coordinates of $C(-3, 4, 8)$ so that $r_{Cx} = -3$ ft, $r_{Cy} = 4$ ft, $r_{Cz} = 8$ ft, and $r_C = 9.43$ ft. Since F_C is unknown,

$$F_{Cx} = F_C\left(\frac{-3}{9.43}\right) = -0.318F_C$$

$$F_{Cy} = F_C\left(\frac{4}{9.43}\right) = 0.424F_C$$

$$F_{Cz} = F_C\left(\frac{8}{9.43}\right) = 0.848F_C$$

Also,

$$F_{Dx} = F_D \qquad F_{Dy} = 0 \qquad F_{Dz} = 0$$

$$W_x = 0 \qquad W_y = 0 \qquad W_z = -40 \text{ lb}$$

For equilibrium we require that

$$\Sigma \mathbf{F} = \mathbf{0}; \qquad \mathbf{F}_B + \mathbf{F}_C + \mathbf{F}_D + \mathbf{W} = \mathbf{0}$$

or

$$\Sigma F_x = 0; \qquad -0.318F_B - 0.318F_C + F_D = 0 \qquad (1)$$

$$\Sigma F_y = 0; \qquad -0.424F_B + 0.424F_C = 0 \qquad (2)$$

$$\Sigma F_z = 0; \qquad 0.848F_B + 0.848F_C - 40 = 0 \qquad (3)$$

Equation (2) states $F_B = F_C$. Thus, solving Eq. (3) for F_B and F_C and substituting the result into Eq. (1) to obtain F_D, we have

$$F_B = F_C = 23.6 \text{ lb} \qquad \textit{Ans.}$$

$$F_D = 15.0 \text{ lb} \qquad \textit{Ans.}$$

Example 3–8

The 100-kg cylinder shown in Fig. 3–12*a* is supported by three cords, one of which is connected to a spring. Determine the tension in each cord for equilibrium and the stretch in the spring.

Solution

Free-Body Diagram. The force in each cord can be determined by investigating the free-body diagram of point A, Fig. 3–12*b*. The weight of the cylinder is $W = 100(9.81) = 981$ N.

Equations of Equilibrium. The components of each force are first determined from the geometric data shown in Fig. 3–12*a*.
For $\mathbf{F}_B$, by inspection,

$$F_{Bx} = F_B \qquad F_{By} = 0 \qquad F_{Bz} = 0$$

For $\mathbf{F}_C$ use Eqs. 2–4:

$$F_{Cx} = F_C \cos 120° = -0.5F_C \qquad F_{Cy} = F_C \cos 135° = -0.707F_C \qquad F_{Cz} = F_C \cos 60° = 0.5F_C$$

For $\mathbf{F}_D$, note that point D has coordinates of $D(-1, 2, 2)$, so that $r_{Dx} = -1$ m, $r_{Dy} = 2$ m, $r_{Dz} = 2$ m. Also, $r_D = 3$ m. Thus

$$F_{Dx} = F_D\left(\frac{-1}{3}\right) = -0.333F_D \qquad F_{Dy} = F_D\left(\frac{2}{3}\right) = 0.667F_D \qquad F_{Dz} = F_D\left(\frac{2}{3}\right) = 0.667F_D$$

For $\mathbf{W}$, by inspection,

$$W_x = 0 \qquad W_y = 0 \qquad W_z = -981 \text{ N}$$

For equilibrium,

$\Sigma \mathbf{F} = \mathbf{0};$ $\qquad \mathbf{F}_B + \mathbf{F}_C + \mathbf{F}_D + \mathbf{W} = \mathbf{0}$

or

$\Sigma F_x = 0;$ $\qquad F_B - 0.5F_C - 0.333F_D = 0 \qquad (1)$

$\Sigma F_y = 0;$ $\qquad -0.707F_C + 0.667F_D = 0 \qquad (2)$

$\Sigma F_z = 0;$ $\qquad 0.5F_C + 0.667F_D - 981 = 0 \qquad (3)$

Solving Eq. (2) for F_D in terms of F_C and substituting into Eq. (3) yields F_C. F_D is determined from Eq. (2). Substituting the results into Eq. (1) yields F_B. Hence,

$$F_C = 812.7 \text{ N} \qquad \textit{Ans.}$$
$$F_D = 862.0 \text{ N} \qquad \textit{Ans.}$$
$$F_B = 693.7 \text{ N} \qquad \textit{Ans.}$$

The stretch in the spring is therefore

$F = kx;$ $\qquad 693.7 = 1500x$

$$x = 0.462 \text{ m} \qquad \textit{Ans.}$$

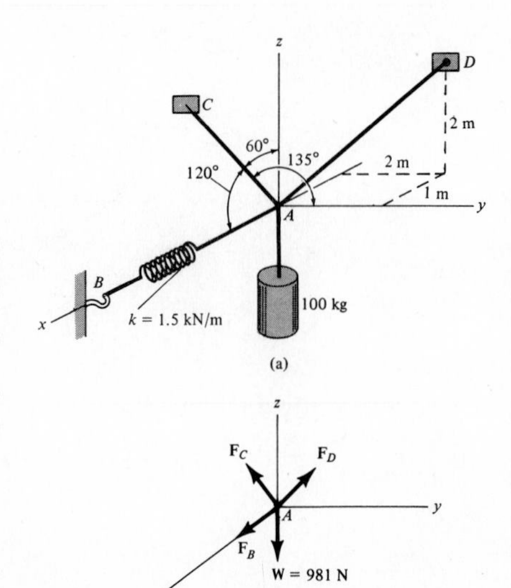

Fig. 3–12

Problems

3–22. Determine the magnitudes of $\mathbf{F}_1$, $\mathbf{F}_2$, and $\mathbf{F}_3$ for equilibrium of particle P.

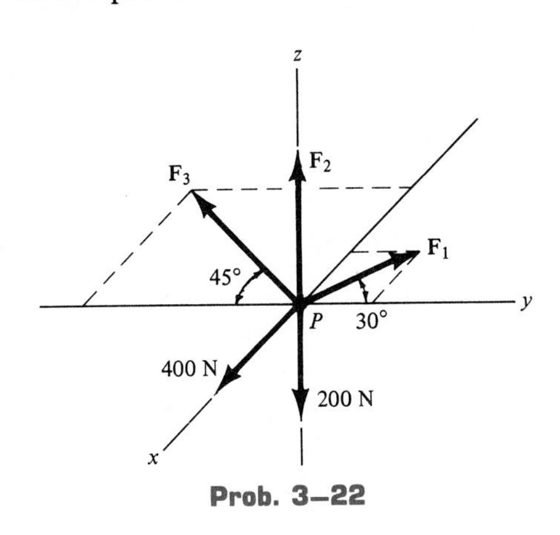

Prob. 3–22

3–23. Determine the magnitudes of $\mathbf{F}_1$, $\mathbf{F}_2$, and $\mathbf{F}_3$ for equilibrium of particle P.

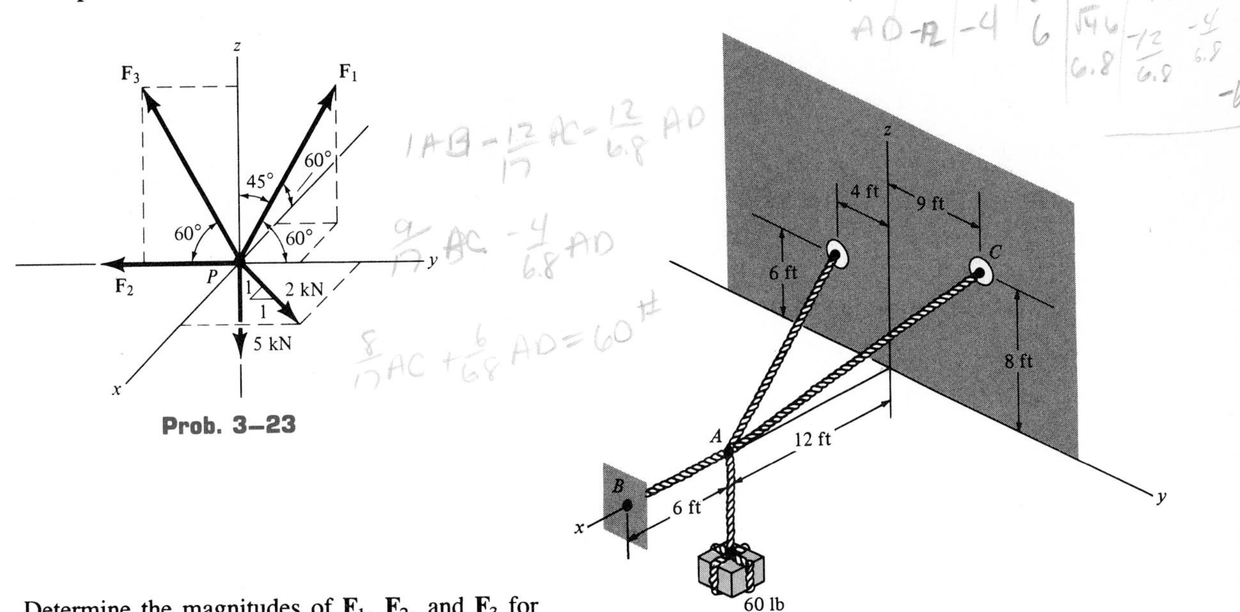

Prob. 3–23

***3–24.** Determine the magnitudes of $\mathbf{F}_1$, $\mathbf{F}_2$, and $\mathbf{F}_3$ for equilibrium of particle P.

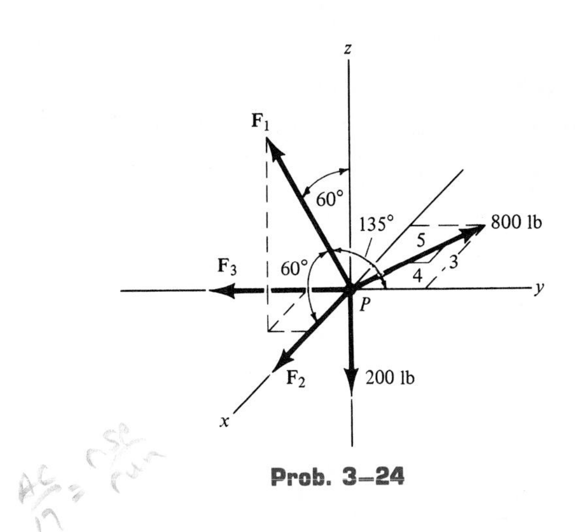

Prob. 3–24

3–25. Determine the tension in cables AB, AC, and AD, required to hold the 60-lb crate in equilibrium.

Prob. 3–25

3–26. Determine the force acting along the axis of each strut necessary to hold the 20-lb flowerpot in equilibrium.

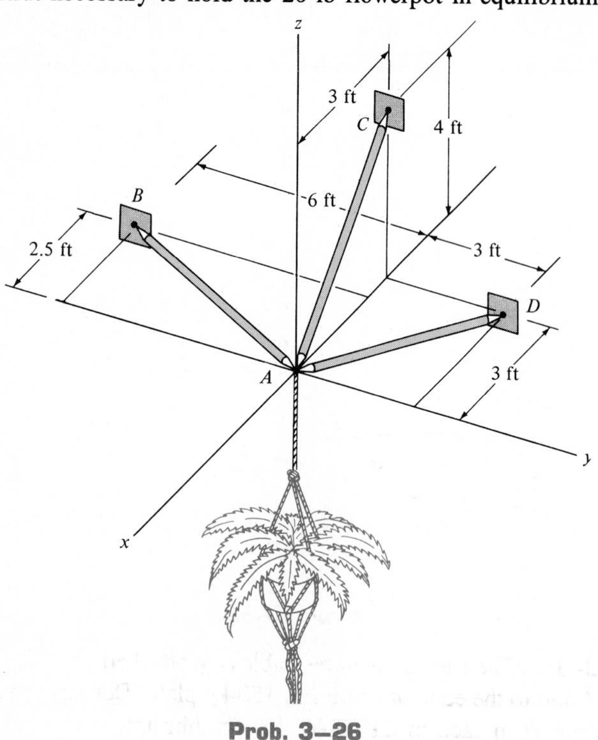

Prob. 3–26

3–27. The boom supports a bucket and its contents, which have a total mass of 300 kg. Determine the forces developed in struts AD and AE and the tension in cable AB for equilibrium. The force in each strut acts along its axis.

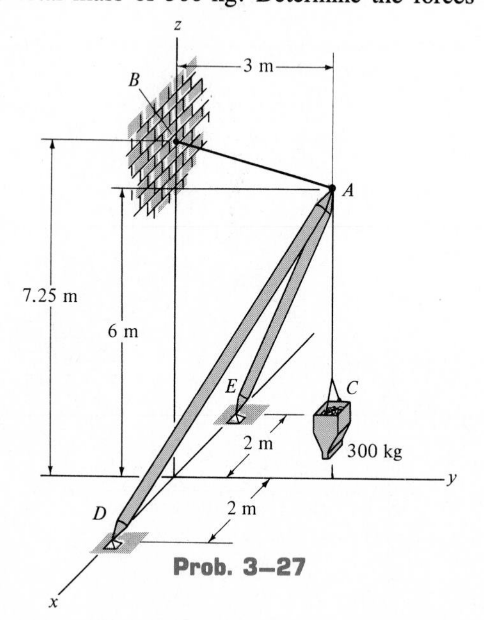

Prob. 3–27

***3–28.** Determine the stretch in each of the two springs required to hold the 20-kg crate in the equilibrium position shown. Each spring has an unstretched length of 2 m and a stiffness of $k = 300$ N/m.

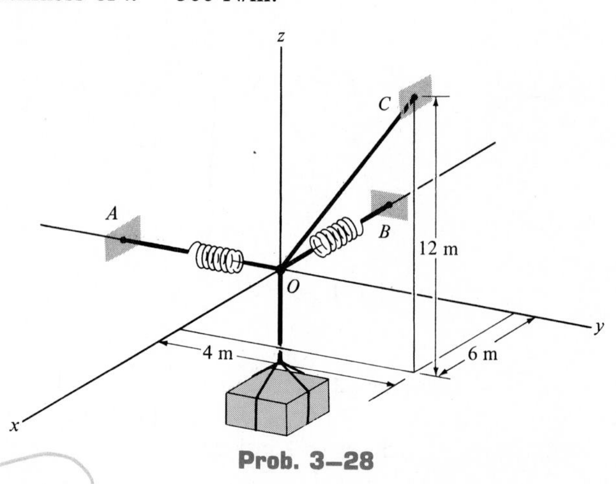

Prob. 3–28

3–29. The shear leg derrick is used to haul the 200-kg net of fish onto the dock. Determine the compressive force in each of the legs AB and CB and the tension in cable DB. Assume the force in each leg acts along its axis.

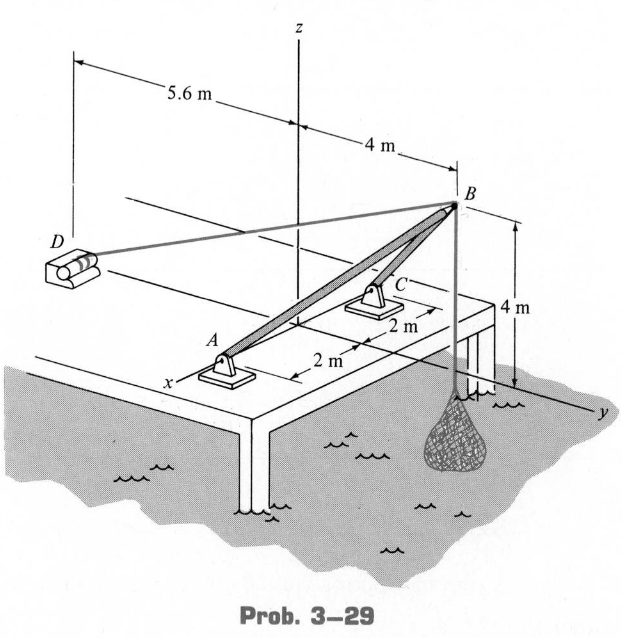

Prob. 3–29

3–30. The block having a weight of 150 lb is suspended from point A by the three struts AB, AC, and AD. Determine the force acting along the axis of each strut for equilibrium.

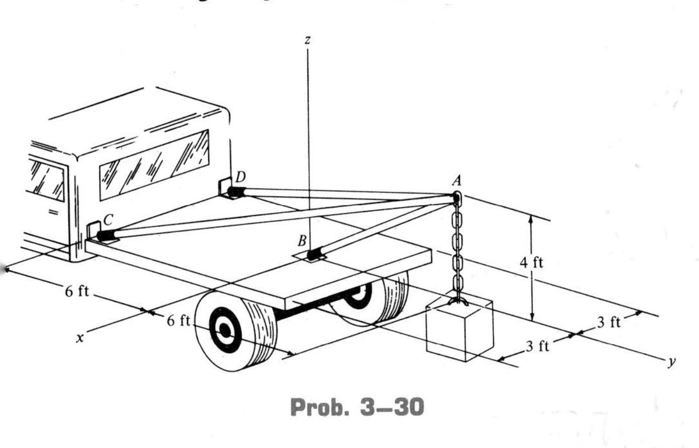

Prob. 3–30

3–31. If the bucket and its contents have a total weight of 35 lb, determine the force in the supporting cables DA and DC and the force acting along strut DB.

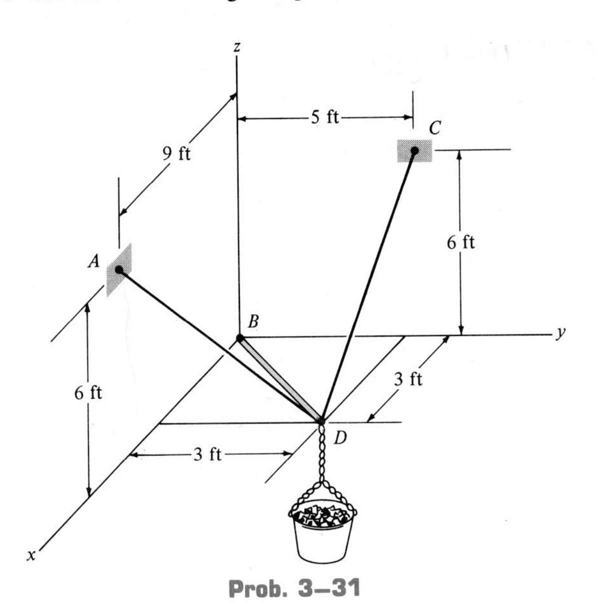

Prob. 3–31

***3–32.** Determine the tensile forces acting along the axes of struts DE and DF required to support the lamp, which has a weight of 20 lb. *Hint:* First determine the forces at C, then determine the forces at D.

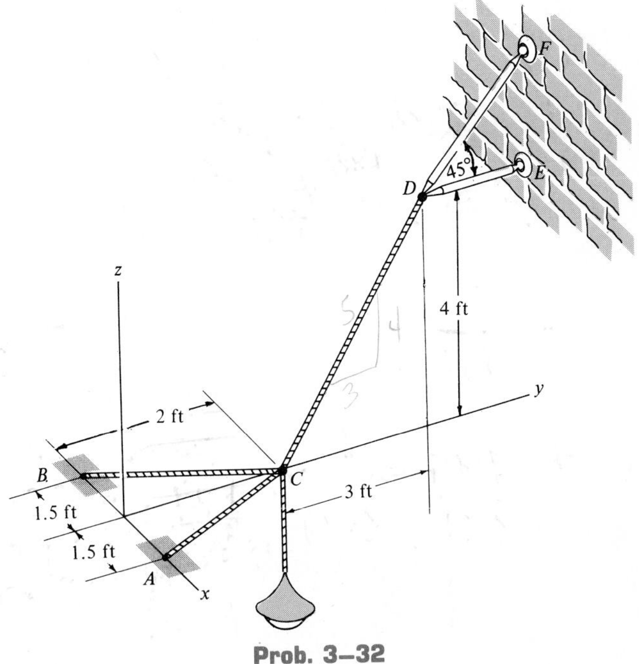

Prob. 3–32

3–33. The ends of the three cables are attached to a ring at A and to the edge of a uniform 150-kg plate. Determine the tension in each of the cables for equilibrium.

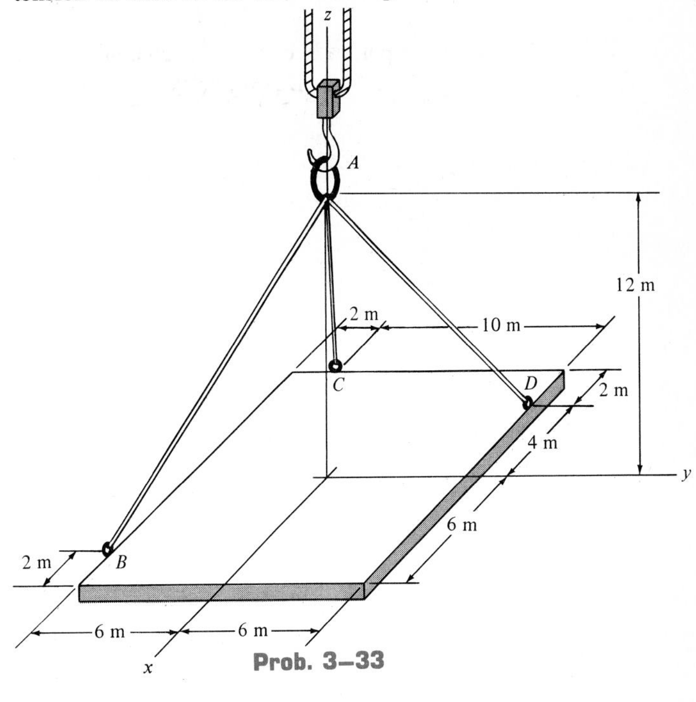

Prob. 3–33

3–34. The lamp has a mass of 15 kg and is supported by a pole AO and cables AB and AC. If the force in the pole acts along its axis, determine the forces in AO, AB, and AC for equilibrium.

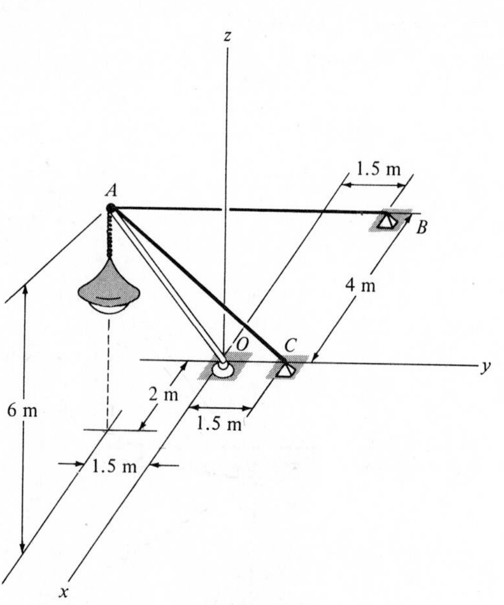

Prob. 3–34

3–35. Determine the tension developed in the three cables required to support the traffic light, which has a mass of 10 kg.

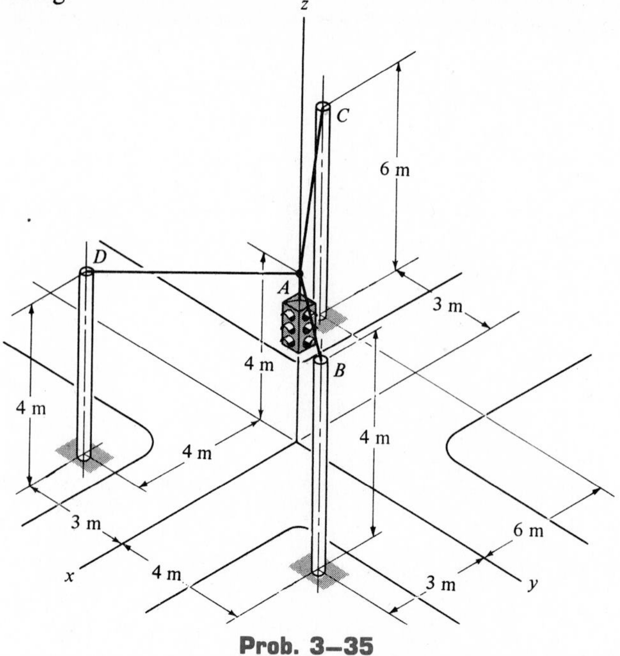

Prob. 3–35

Review Problems

***3–36.** The joint of a light metal truss is formed by bolting four angles to the *gusset plate*. Knowing the force in members A and C, determine the forces $\mathbf{F}_B$ and $\mathbf{F}_D$ acting in members B and D required for equilibrium. The force system is concurrent at point O.

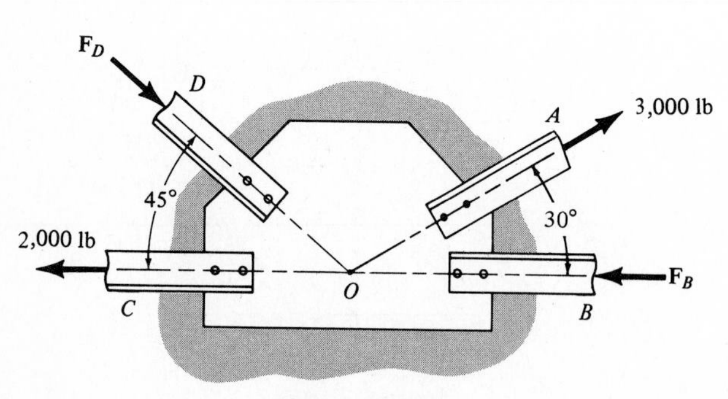

Prob. 3–36

3–37. Blocks D and F weigh 5 lb each and block E weighs 8 lb. If the cord connecting the blocks passes over small frictionless pulleys, determine the sag s for equilibrium.

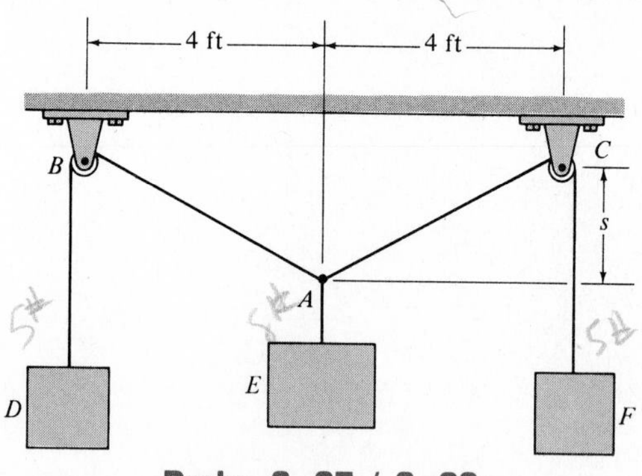

Probs. 3–37 / 3–38

3–38. If blocks D and F weigh 5 lb each, determine the weight of block E if $s = 3$ ft.

3–39. The joint O of a space frame is subjected to four forces. Strut OA lies in the x-y plane and strut OB lies in the y-z plane. Determine the force acting in each of the three struts required for equilibrium of the joint. Set $\theta = 45°$.

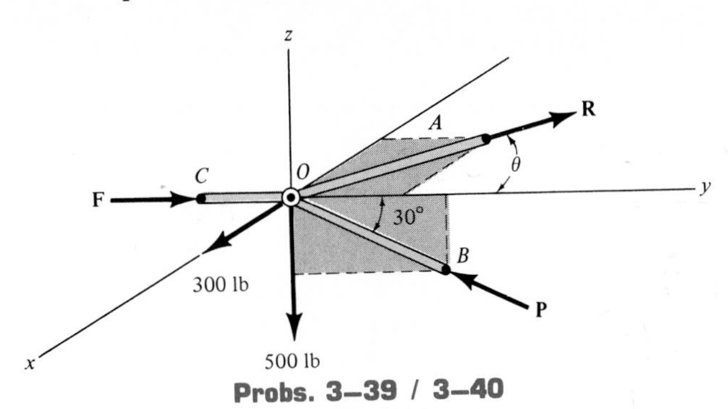

Probs. 3–39 / 3–40

***3–40.** The joint O of the space frame is subjected to four forces. Determine the angle θ so that the force in member OA has a magnitude of $R = 400$ lb. What are the forces in members OC and OB?

3–41. The uniform 200-lb crate is suspended by using a 6-ft-long cord that is attached to the sides of the crate and passes over the small pulley located at O. If the cord can be attached at either points A and B, or C and D, determine which attachment produces the least amount of tension in the cord and specify the cord tension in this case.

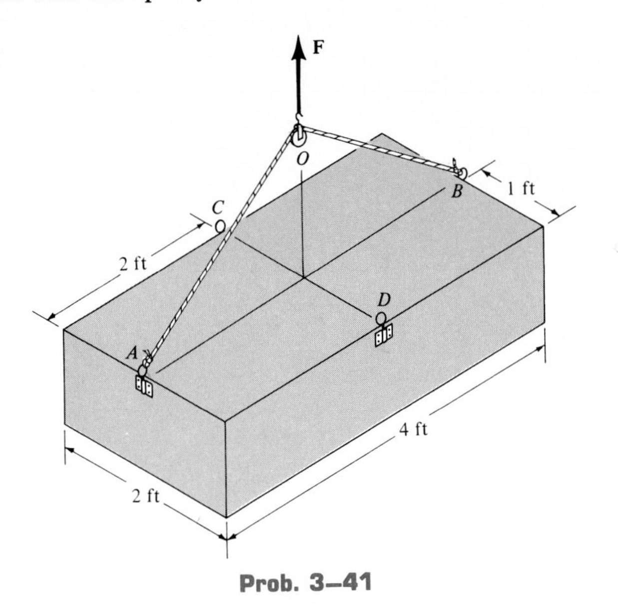

Prob. 3–41

3–42. Determine the tension developed in cables OD and OB and the strut OC, required to support the 500-lb crate. The spring OA has an unstretched length of 0.2 ft and a stiffness of $k_{OA} = 350$ lb/ft. The force in the strut acts along the axis of the strut.

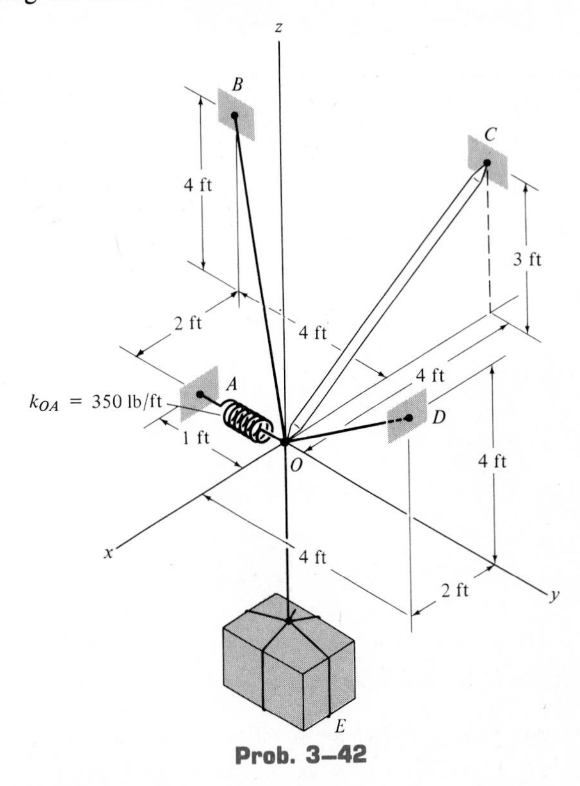

Prob. 3–42

3–43. Determine the magnitudes of the forces **P**, **R**, and **F** required for equilibrium of point O.

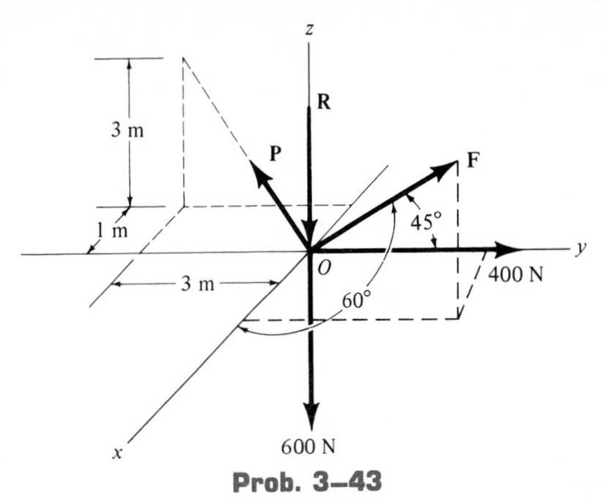

Prob. 3–43

***3–44.** The links AB, BC, and CD support the 40-lb sack at E and the 20-lb crate at B. Determine the tension in AB and the angle θ for equilibrium. The force in each link acts along the axis of the link. *Hint:* Analyze the forces at C to obtain $\mathbf{F}_{BC}$, then analyze the forces at B.

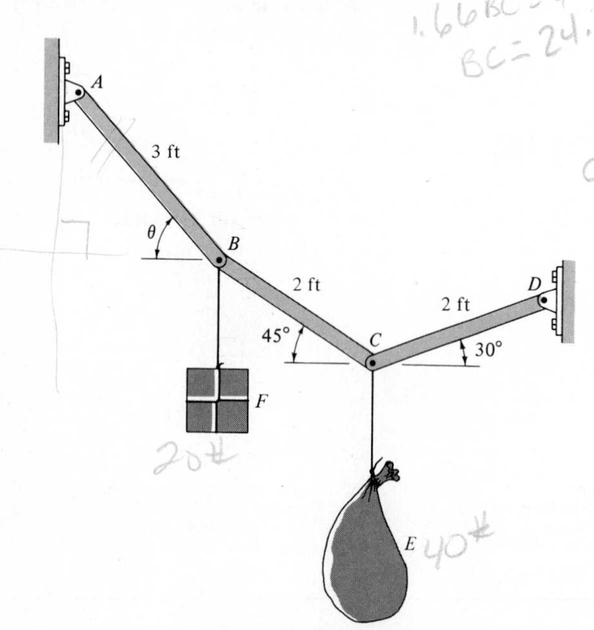

Probs. 3–44 / 3–45

3–45. If the sack at E weighs 40 lb, determine the weight of the crate at F if it is observed that the angle $\theta = 60°$. *Hint:* First analyze the forces at C, then analyze the forces at B.

3–46. A small peg P rests on a spring that is contained inside the smooth pipe. When the spring is compressed so that $s = 0.15$ m, the spring exerts an upward force of 60 N on the peg. Determine the point of attachment $A(x, y, 0)$ of cord PA so that the tension in cords PB and PC equals 30 N and 50 N, respectively.

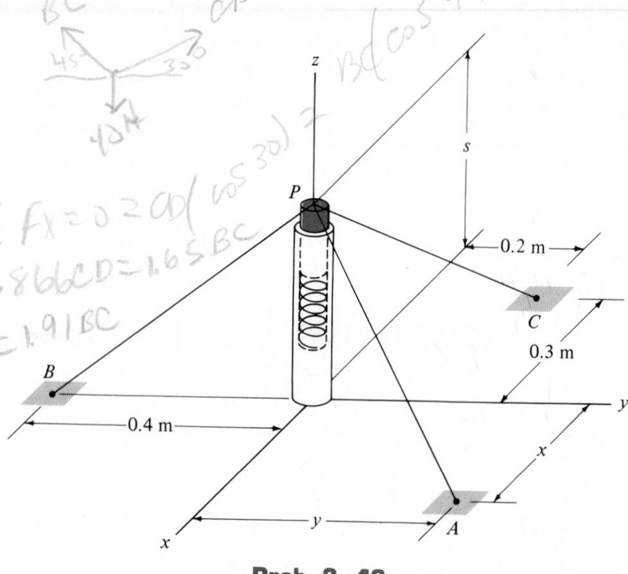

Prob. 3–46

3–47. A continuous cable of total length 4 m is wrapped around the *small* frictionless pulleys at A, B, C, and D. If the stiffness of each spring is $k = 500$ N/m and each spring is stretched 300 mm, determine the mass m of each block. Neglect the weight of the pulleys and cords. The springs are unstretched when $d = 2$ m.

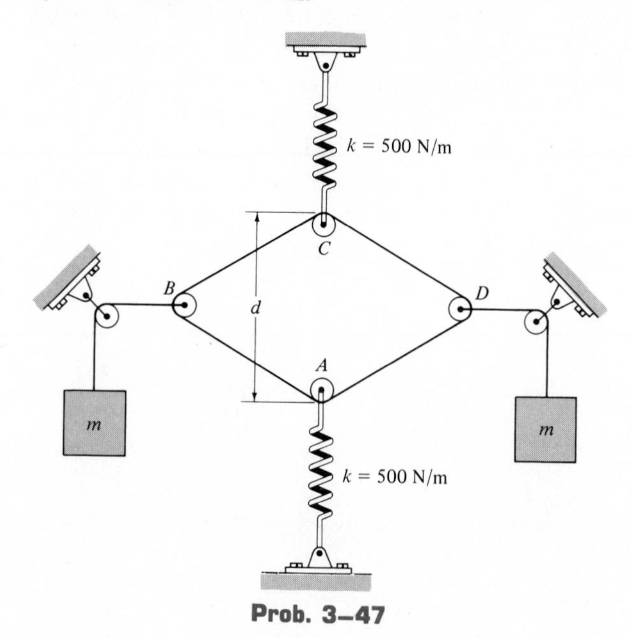

Prob. 3–47

4 Equivalent Force Systems

In Chapter 3 it was shown that a particle is in equilibrium if the resultant of the force system acting on it is equal to zero. Chapter 5 shows that such a restriction is necessary but not sufficient for the equilibrium of a rigid body. A further restriction must be made with regard to the nonconcurrency of the applied force system, giving rise to the concept of moment. In this chapter we will develop a formal definition of a moment and discuss ways of finding the moments caused by concentrated forces and distributed loadings about points and axes. We will also develop methods of reducing nonconcurrent concentrated force systems and distributed loadings to equivalent yet simpler systems. In order to clarify the concepts involved, we will first apply the principles to coplanar force systems, followed by the more difficult applications involving three-dimensional force systems.

Coplanar Force Systems

4.1 Moment of a Force

A force can provide the effect of turning about an axis if it does not act parallel to the axis or have a line of action which passes through the axis. For example, consider the force **F** in Fig. 4–1, which acts on the handle of the wrench and is located a distance *d* from the vertical axis of the pipe. If the

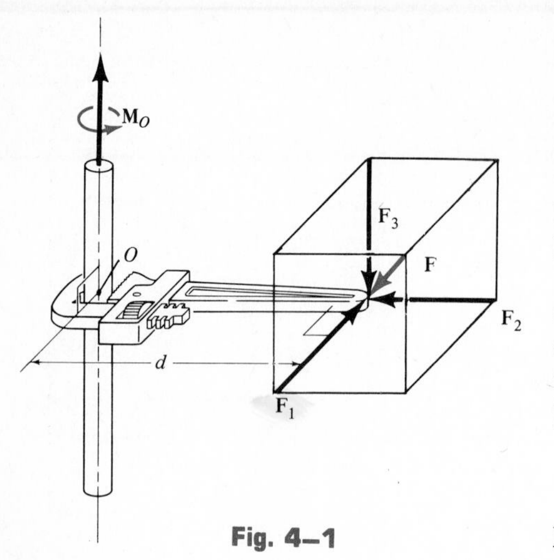

Fig. 4–1

force is resolved into three perpendicular components, it is seen that *only* $\mathbf{F}_1$ tends to rotate the wrench and pipe about the axis. Obviously, $\mathbf{F}_2$ and $\mathbf{F}_3$ do not cause the wrench to turn about this axis since $\mathbf{F}_2$ intersects the axis and $\mathbf{F}_3$ is parallel to it. This rotational effect of $\mathbf{F}_1$ is called the *moment of a force* or simply the *moment* $\mathbf{M}_O$. In particular, note that the pipe axis is perpendicular to a plane containing both $\mathbf{F}_1$ and d, and that the axis intersects this plane at point O.

In the more general case, we consider the force $\mathbf{F}$ lying in the shaded plane shown in Fig. 4–2*a*. The moment $\mathbf{M}_O$ about an axis perpendicular to the plane and intersecting the plane at point O is a *vector quantity*, since it has a specified magnitude and direction and adds according to the parallelogram law.

Magnitude. The magnitude of $\mathbf{M}_O$ is

$$M_O = Fd \tag{4–1}$$

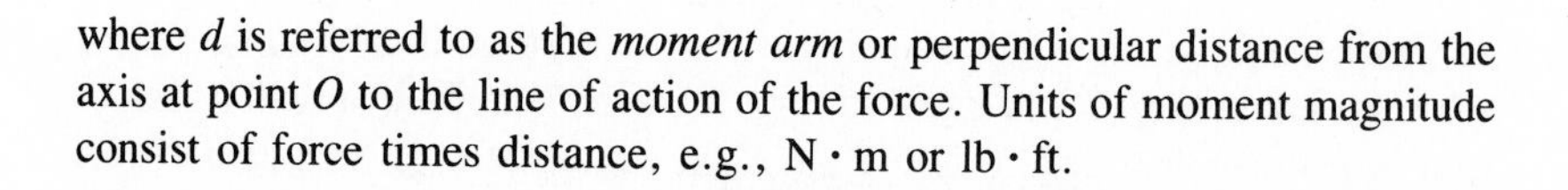

where d is referred to as the *moment arm* or perpendicular distance from the axis at point O to the line of action of the force. Units of moment magnitude consist of force times distance, e.g., N · m or lb · ft.

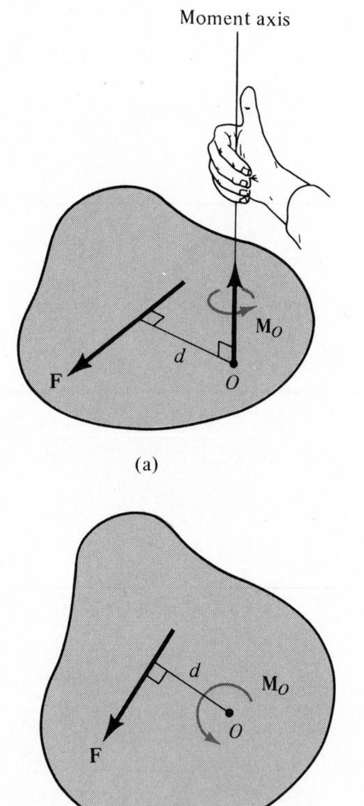

Fig. 4–2

Direction. The *direction* of $\mathbf{M}_O$ will be specified by using the "right-hand rule." The fingers of the right hand are curled such that they follow the sense of rotation as caused by the force acting about the axis; the *thumb* then *points* in the direction of the moment vector along the *moment axis,* which is *perpendicular* to the shaded plane containing $\mathbf{F}$ and d, Fig. 4–2*a*. By this definition, the moment $\mathbf{M}_O$ can be considered as a *sliding vector* and therefore acts at any point along the moment axis.

In three dimensions, $\mathbf{M}_O$ is illustrated by a regular vector with a curl on it to *distinguish* it from a force vector, Fig. 4–2*a*. Many problems in mechanics, however, involve coplanar force systems that may be conveniently viewed in two dimensions. For example, a two-dimensional view of Fig. 4–2*a* is given in Fig. 4–2*b*. Here $\mathbf{M}_O$ is simply represented by the (counterclockwise) curl, which indicates the action of $\mathbf{F}$. This curl is used to show the *sense of rotation* caused by $\mathbf{F}$. Using the right-hand rule, however, note that the *direction* of the moment vector in Fig. 4–2*b* is indicated by the thumb and thus points *out* of the page, since the fingers follow the curl. In two dimensions we will often refer to finding the moment of a force "about a point" (O). Keep in mind that the moment actually acts about an axis which is perpendicular to the plane of $\mathbf{F}$ and d and intersects the plane at the point (O), Fig. 4–2.

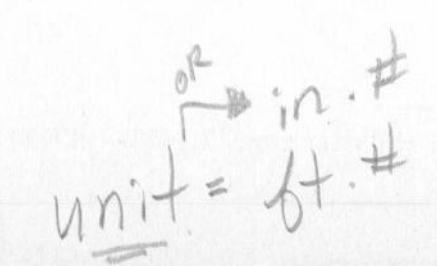

Example 4–1

Determine the moment of force **F** about points A and B of the beam shown in Fig. 4–3a.

Solution

For the moment about A, the moment arm $d = 2$ m, since it is the *perpendicular distance* from A to the line of action of **F**. Hence, the magnitude of the moment of **F** about A is

$$M_A = Fd = 200 \text{ N}(2 \text{ m}) = 400 \text{ N} \cdot \text{m} \qquad \textit{Ans.}$$

By the right-hand rule, the moment is directed *into* the page, Fig. 4–3b, since the force *tends* to rotate the beam in a clockwise direction about an axis passing through A. (In reality, this rotation is *prevented* by the roller constraint at B.) In a similar manner, the magnitude of moment at B is

$$M_B = Fd = (200 \text{ N})(0.5 \text{ m}) = 100 \text{ N} \cdot \text{m} \qquad \textit{Ans.}$$

This moment is counterclockwise and hence directed out of the page. A three-dimensional view of the moment vectors is shown in Fig. 4–3c.

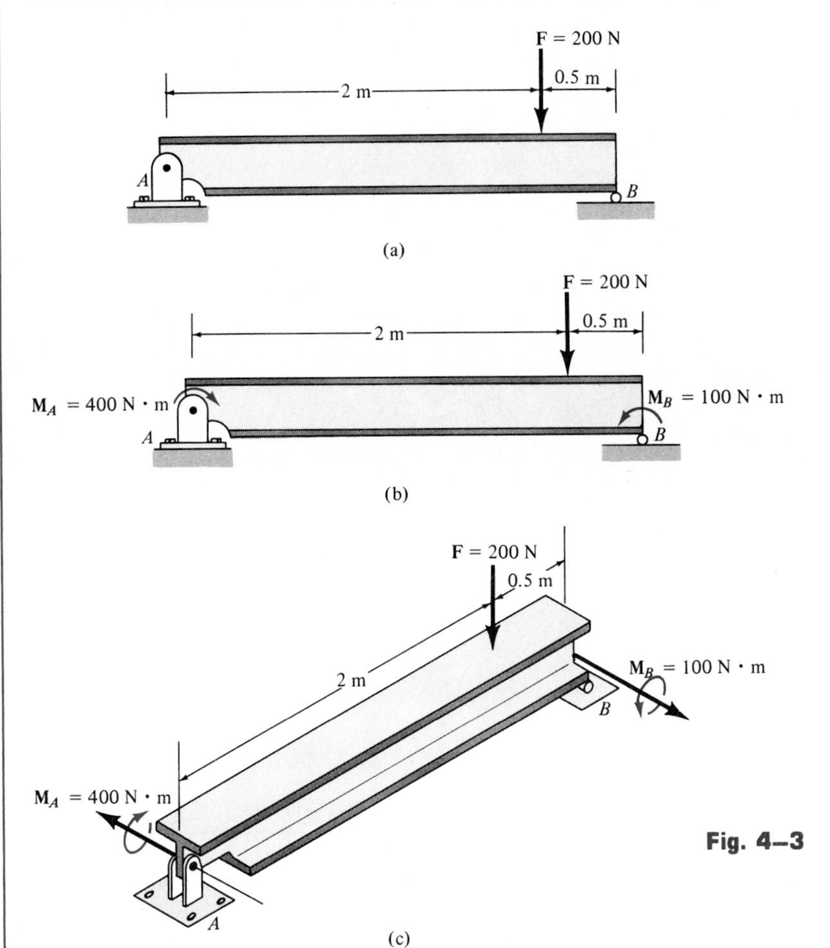

Fig. 4–3

Example 4–2

Determine the moment of the 800-N force acting on the frame in Fig. 4–4 about points A, B, C, and D.

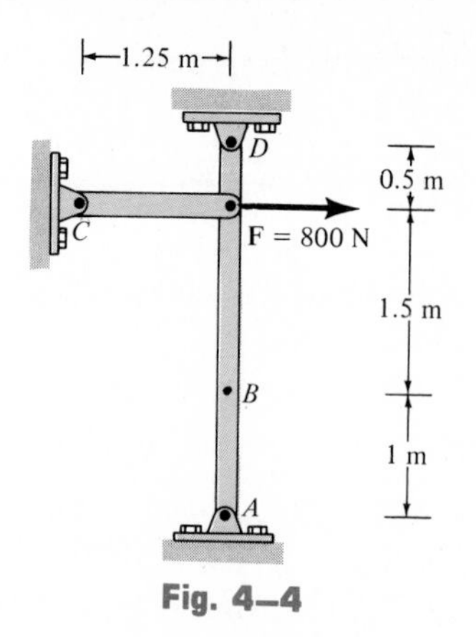

Fig. 4–4

Solution

In general, $M = Fd$, where d is the moment arm or *perpendicular distance* from the *point* on the moment axis to the *line of action* of the force. Hence,

$M_A = 800 \text{ N}(2.5 \text{ m}) = 2000 \text{ N} \cdot \text{m} \circlearrowright$ *Ans.*

$M_B = 800 \text{ N}(1.5 \text{ m}) = 1200 \text{ N} \cdot \text{m} \circlearrowright$ *Ans.*

$M_C = 800 \text{ N}(0) = 0$ (line of action of **F** passes through C) *Ans.*

$M_D = 800 \text{ N}(0.5 \text{ m}) = 400 \text{ N} \cdot \text{m} \circlearrowleft$ *Ans.*

Example 4–3

Determine the point of application P and the direction of a 20-lb force that lies in the plane of the square plate shown in Fig. 4–5a, so that this force creates the greatest counterclockwise moment about point O. What is this moment?

Solution

Since the maximum moment created by the force is required, the force must act on the plate at a distance *farthest* from point O. As shown in Fig. 4–5b, the point of application must therefore be at the diagonal corner. In order to produce *counterclockwise* rotation of the plate about O, **F** must act at an angle $45° < \phi < 225°$. The greatest moment is produced when the line of action of **F** is *perpendicular* to d, i.e., $\phi = 135°$, Fig. 4–5c. The maximum moment is therefore

$$M_O = Fd = (20 \text{ lb})(2\sqrt{2} \text{ ft}) = 56.6 \text{ lb} \cdot \text{ft} \circlearrowleft \qquad \textit{Ans.}$$

By the right-hand rule, $\mathbf{M}_O$ is directed out of the page.

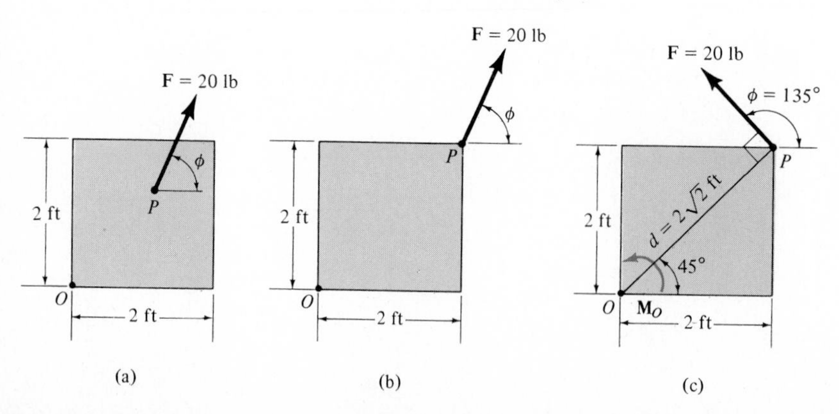

Fig. 4–5

Principle of Moments 4.2

A concept often used in mechanics is the *principle of moments,* which is sometimes referred to as Varignon's theorem since it was originally developed by the French mathematician Varignon (1654–1722). This theorem states that *the moment of a force about a point is equal to the sum of the moments of the force's components about the point.*

Proof. Consider the force **F** and its two rectangular components $\mathbf{F}_x$ and $\mathbf{F}_y$ shown in Fig. 4–6*a*. If moments are summed about point O, assuming positive moments to be counterclockwise we require

$$\circlearrowleft + F(d) = F_y(x) - F_x(y) \qquad (4\text{–}2)$$

However, $F_x = F \cos \theta$ and $F_y = F \sin \theta$. Thus,

$$\circlearrowleft + F(d) = F \sin \theta(x) - F \cos \theta(y)$$

Factoring out F,

$$d = x \sin \theta - y \cos \theta$$

which is evident from the geometrical construction in Fig. 4–6.

The theorem is also valid for nonrectangular components $\mathbf{F}_1$ and $\mathbf{F}_2$ as shown in Fig. 4–6*b*. If each force is resolved into its rectangular components and Eq. 4–2 is applied, we require that

$$\circlearrowleft + F(d) = F_1(d_1) - F_2(d_2)$$
$$F_y(x) - F_x(y) = (F_{1y}x - F_{1x}y) + (F_{2y}x - F_{2x}y)$$
$$F_y(x) - F_x(y) = (F_{1y} + F_{2y})x - (F_{1x} + F_{2x})y$$

This equation is valid since $F_x = F_{1x} + F_{2x}$ and $F_y = F_{1y} + F_{2y}$ (Eqs. 2–1). Again the theorem is proved.

Throughout this book it will be shown that application of the principle of moments considerably simplifies the computation of the moment of a force about a point or axis, since it allows one to consider the moments of the force's components rather than the force itself.

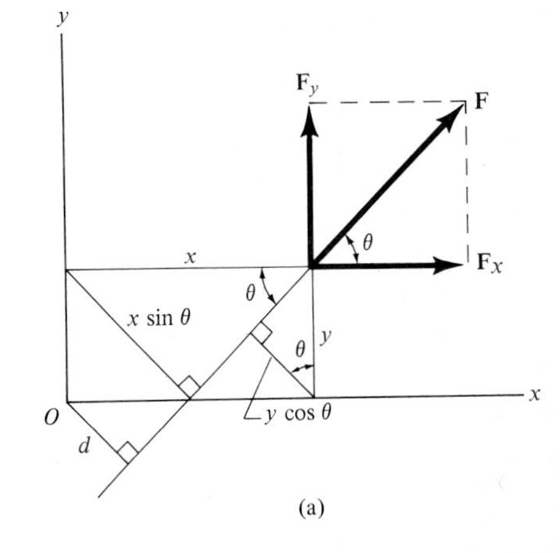

(a)

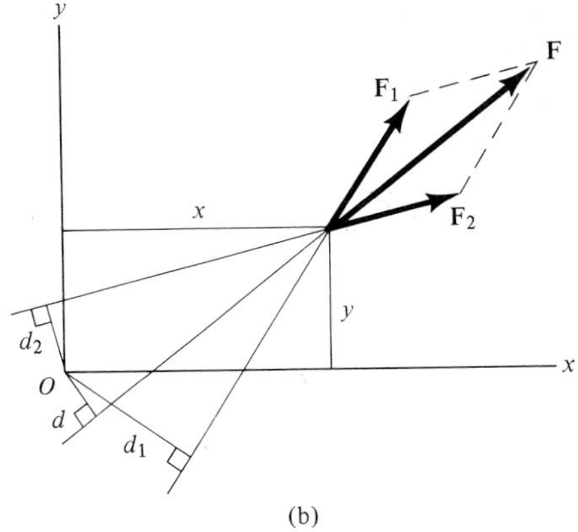

(b)

Fig. 4–6

Example 4–4

A 200-N force acts on the bracket shown in Fig. 4–7*a*. Determine the moment of the force about point *A*.

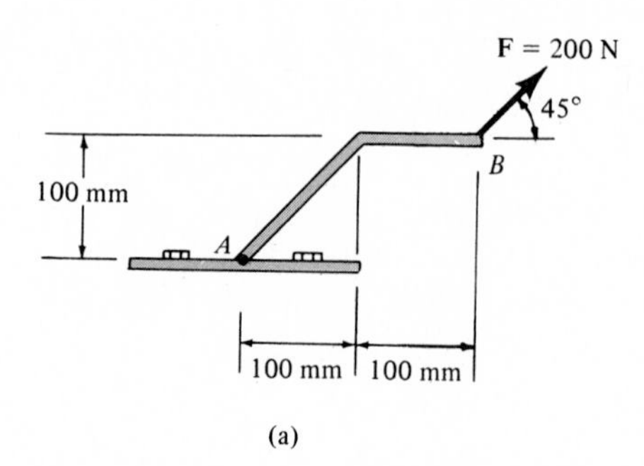

(a)

Fig. 4–7

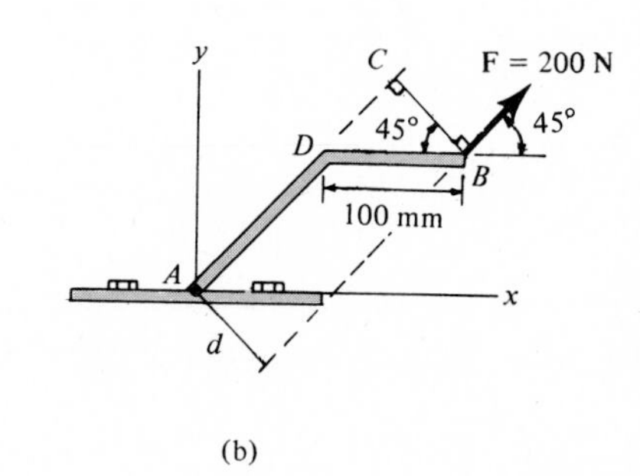

(b)

Solution I

The moment arm *d* can be computed by trigonometry, using the construction shown in Fig. 4–7*b*. From triangle *BCD*,

$$CB = d = 100 \cos 45° = 70.71 \text{ mm} = 0.070\ 71 \text{ m}$$

Thus,

$$M_A = Fd = 200 \text{ N}(0.070\ 71 \text{ m}) = 14.14 \text{ N} \cdot \text{m} \qquad \textit{Ans.}$$

According to the right-hand rule, $\mathbf{M}_A$ is directed out of the page since the force tends to rotate *counterclockwise* about point *A*.

Solution II

The 200-N force may be resolved into *x* and *y* components, as shown in Fig. 4–7*c*. In accordance with the principle of moments, the moment computed about point *A* is equivalent to the sum of the moments produced by the two force components. Assuming counterclockwise rotation as positive, i.e., out of the page, we can add the moments of both forces algebraically, since the moments act along the same axis (perpendicular to the *x*-*y* plane, passing through *A*).

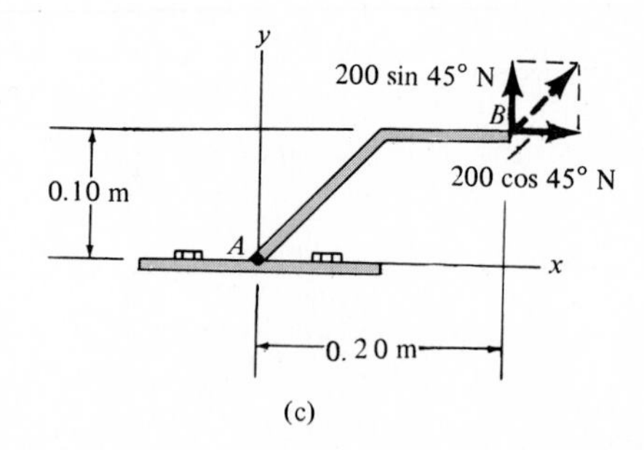

(c)

$$\circlearrowleft +M_A = (200 \sin 45°)(0.20) - (200 \cos 45°)(0.10)$$
$$= 14.14 \text{ N} \cdot \text{m} \circlearrowleft \qquad \textit{Ans.}$$

By comparison, it is seen that solution II provides a more *convenient method* for analysis than solution I since the moment arm for each component force is easier to establish.

Example 4–5

Two coplanar forces act on the beam shown in Fig. 4–8a. Determine the moment of each force about point O.

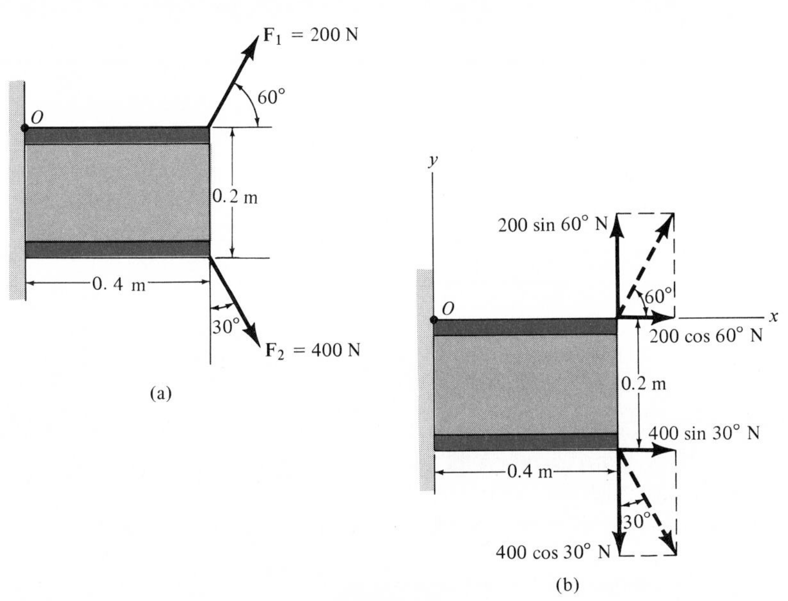

Fig. 4–8

Solution

Each force is resolved into its x and y components as shown in Fig. 4–8b, and the moments of the components are computed about O. Assuming that positive moments act counterclockwise, then for $\mathbf{F}_1$,

$$\circlearrowleft + (M_O)_1 = 200 \cos 60°(0) + 200 \sin 60°(0.4)$$
$$= 69.3 \text{ N} \cdot \text{m} \circlearrowleft \qquad \textit{Ans.}$$

For $\mathbf{F}_2$,

$$\circlearrowleft + (M_O)_2 = 400 \sin 30°(0.2) - 400 \cos 30°(0.4)$$
$$= -98.6 \text{ N} \cdot \text{m} \qquad \textit{Ans.}$$

The negative sign indicates $(\mathbf{M}_O)_2$ acts clockwise. Hence,

$$(M_O)_2 = 98.6 \text{ N} \cdot \text{m} \circlearrowright \qquad \textit{Ans.}$$

As stated in the previous example, the above method provides the most direct solution to this problem, since the moment arms for the *force components* are easily obtained.

4.3 Moment of a Couple

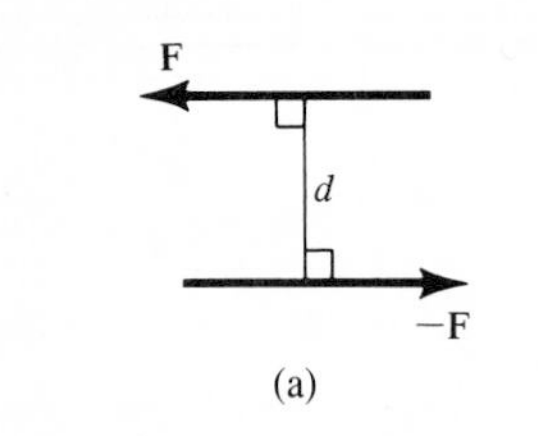

(a)

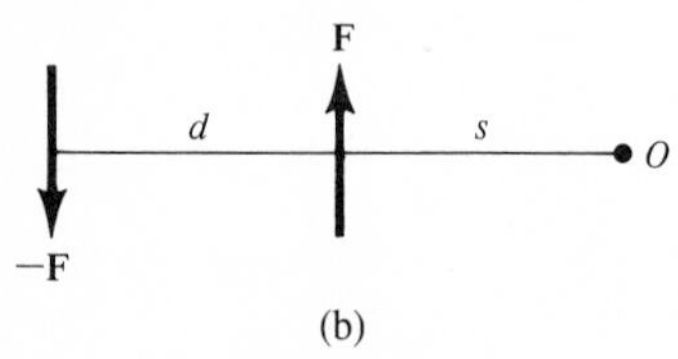

(b)

Fig. 4–9

A *couple* is defined as two parallel forces that have the same magnitude, opposite directions, and are separated by a perpendicular distance d, Fig. 4–9*a*. Since the resultant force of the two forces composing the couple is zero, the only effect of a couple is to produce a rotation or tendency of rotation in a specified direction. This concept has important applications throughout this chapter regarding the simplification of force systems.

The moment produced by a couple, called a *couple moment*, is equivalent to the sum of the moments of both couple forces, computed about *any arbitrary point*. For example, consider the couple in Fig. 4–9*b*. If the moments of the two forces are computed about point O, then, assuming positive moments are directed out of the page, i.e., counterclockwise by the right-hand rule, we have

$$\circlearrowleft + M_C = -Fs + F(d + s)$$
$$= Fd \circlearrowleft$$

This result indicates that a couple moment is a *free vector* since $\mathbf{M}_C$ depends only on the distance *between* the forces and *not* on the location s of point O. Consequently, a couple is unlike the moment of a force, which requires a definite point (or axis) about which it is computed.

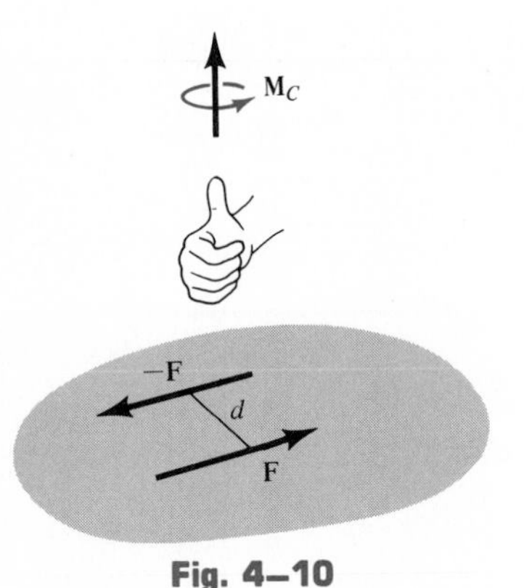

Fig. 4–10

Magnitude. The moment of a couple $\mathbf{M}_C$, Fig. 4–10, is defined as having a *magnitude* of

$$M_C = Fd \tag{4–3}$$

where F is the magnitude of one of the forces and d is the perpendicular distance or moment arm between the forces.

Direction. The *direction* of the couple moment is determined by the right-hand rule, where the thumb indicates the direction when the fingers are curled with the sense of rotation caused by the forces. In all cases, $\mathbf{M}_C$ acts perpendicular to the plane containing the two forces, Fig. 4–10.

Equivalent Couples. Two couples are said to be equivalent if they produce the same moment. Since the moment produced by a couple is always perpendicular to the plane containing the couple forces, it is therefore necessary that the forces of equal couples lie either in the same plane or in planes that are *parallel* to one another. In this way, the line of action of each couple moment will be the same, that is, perpendicular to the parallel planes. For example, the two couples shown in Fig. 4–11 are equivalent. One couple is produced by a pair of 100-N forces separated by a distance of $d = 0.5$ m, and the other is produced by a pair of 200-N forces separated by a distance of 0.25 m. Since

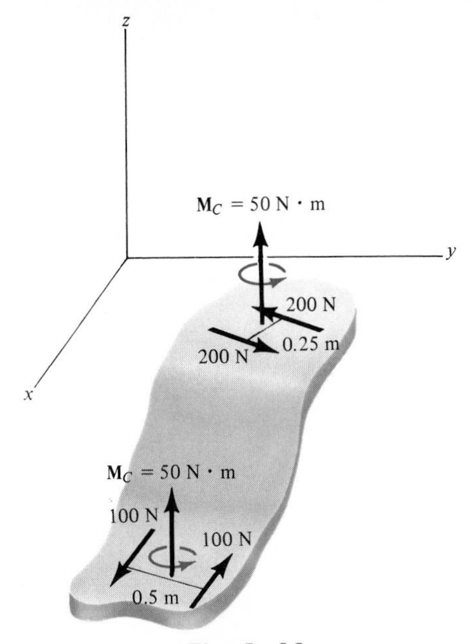

Fig. 4–11

the planes in which the forces act are parallel to the *x*-*y* plane, the moment produced by each of the couples has a magnitude of $M_C = 50$ N · m and is in the direction of the positive *z* axis.

Resultant Couple. Since couple moments are free vectors, they may be applied at any point *P* on a body and added vectorially. For example, the two couples acting on different planes of the rigid body in Fig. 4–12*a* may be replaced by their corresponding moments $\mathbf{M}_{C_1}$ and $\mathbf{M}_{C_2}$, shown in Fig. 4–12*b*. These free vectors may be moved to the *arbitrary point P* and added to obtain their resultant vector sum $\mathbf{M}_{C_R} = \mathbf{M}_{C_1} + \mathbf{M}_{C_2}$, shown in Fig. 4–12*c*. Hence, the effect of the two couples in Fig. 4–12*a* imparts a "total twist" or moment $\mathbf{M}_{C_R}$ to the body about an axis that is parallel to the direction of $\mathbf{M}_{C_R}$.

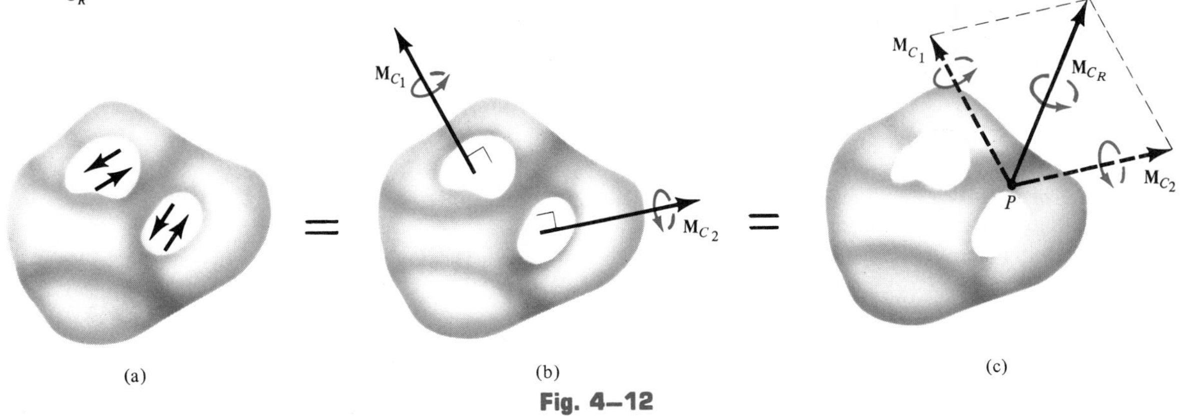

Fig. 4–12

Example 4–6

Determine the moment of the couple acting on the beam shown in Fig. 4–13*a*.

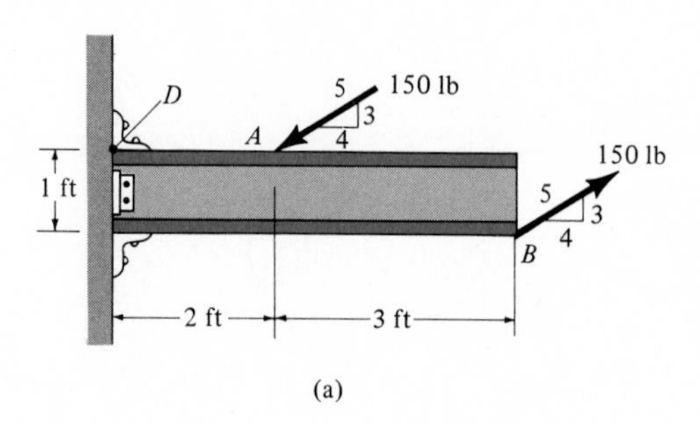

(a)

Fig. 4–13

Solution

Here it is somewhat difficult to determine the perpendicular distance between the forces and compute the couple moment as $M = Fd$. Instead, we can resolve each force into its horizontal and vertical components, $F_x = \frac{4}{5}(150) = 120$ lb and $F_y = \frac{3}{5}(150) = 90$ lb, Fig. 4–13*b*, and then use the principle of moments. The couple moment can be computed about any point. For example, if point *D* is chosen, we have

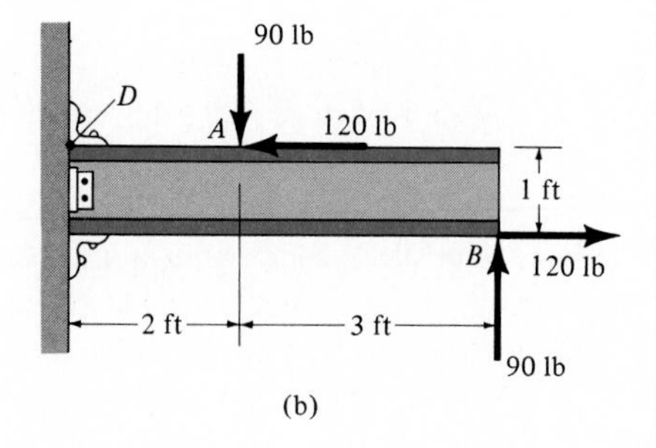

(b)

$$\circlearrowleft\!+M_C = 120\text{ lb}(0\text{ ft}) - 90\text{ lb}(2\text{ ft}) + 90\text{ lb}(5\text{ ft}) + 120\text{ lb}(1\text{ ft})$$
$$= 390\text{ lb}\cdot\text{ft}\circlearrowleft \qquad \textit{Ans.}$$

It is easier to compute the moments about point *A* or *B* in order to *eliminate* the moment of the force acting at the moment point. For point *A*, Fig. 4–13*b*, we have

$$\circlearrowleft\!+M_C = 90\text{ lb}(3\text{ ft}) + 120\text{ lb}(1\text{ ft})$$
$$= 390\text{ lb}\cdot\text{ft}\circlearrowleft \qquad \textit{Ans.}$$

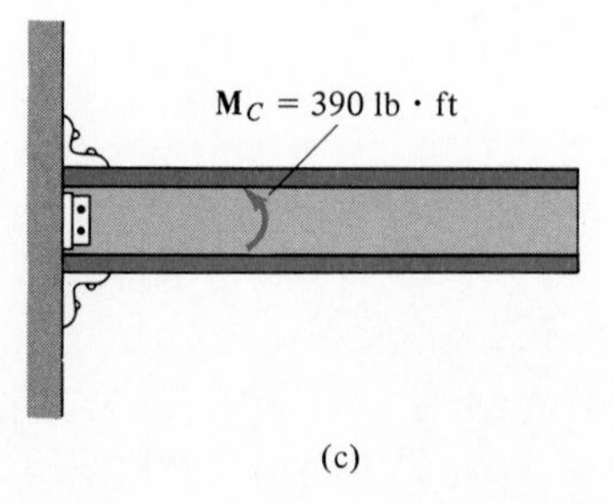

(c)

Show that one obtains this same result if moments are summed about point *B*. Notice also that the couple in Fig. 4–13*a* has been replaced by *two* couples in Fig. 4–13*b*. Using $M_C = Fd$, one couple has a moment of $M_{C_1} = 90\text{ lb}(3\text{ ft}) = 270\text{ lb}\cdot\text{ft}$ and the other has a moment of $M_{C_2} = 120\text{ lb}(1\text{ ft}) = 120\text{ lb}\cdot\text{ft}$. By the right-hand rule both couple moments are counterclockwise and are therefore directed out of the page. Since these couples are free vectors they can be moved to a point and added, which yields $M_C = 270\text{ lb}\cdot\text{ft} + 120\text{ lb}\cdot\text{ft} = 390\text{ lb}\cdot\text{ft}\circlearrowleft$, the same result computed above. $\mathbf{M}_C$ is a free vector and can therefore act at any point on the beam, Fig. 4–13*c*.

Example 4–7

A couple acts at the end of the beam shown in Fig. 4–14*a*. Replace it by an equivalent one having a pair of forces that act through (a) points A and B; (b) points D and E.

Solution

The couple has a magnitude of $M_C = Fd = 400(0.2) = 80\ \text{N}\cdot\text{m}$ and a direction that is into the page since the forces tend to rotate clockwise. $\mathbf{M}_C$ is a free vector so that it can be placed at any point on the beam, Fig. 4–14*b*.

Part (a). To preserve the direction of $\mathbf{M}_C$, *horizontal* forces acting through points A and B must be directed as shown in Fig. 4–14*c*. The magnitude of each force is

$$M_C = Fd$$
$$80 = F(0.25)$$
$$F = 320\ \text{N} \qquad \textit{Ans.}$$

Part (b). To generate the required clockwise rotation, forces acting through points D and E must be *vertical* and directed as shown in Fig. 4–14*d*. The magnitude of each force is

$$M_C = Pd$$
$$80 = P(0.1)$$
$$P = 800\ \text{N} \qquad \textit{Ans.}$$

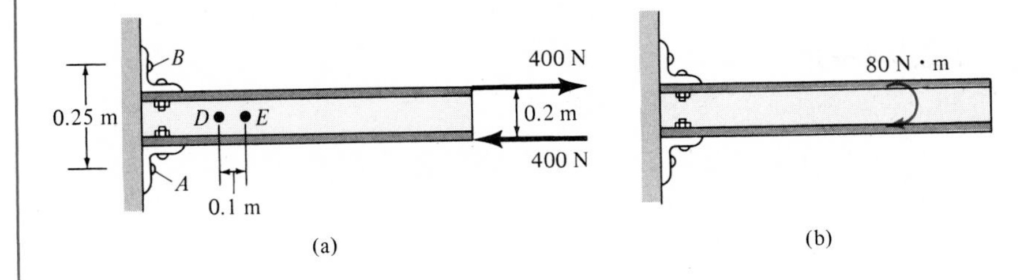

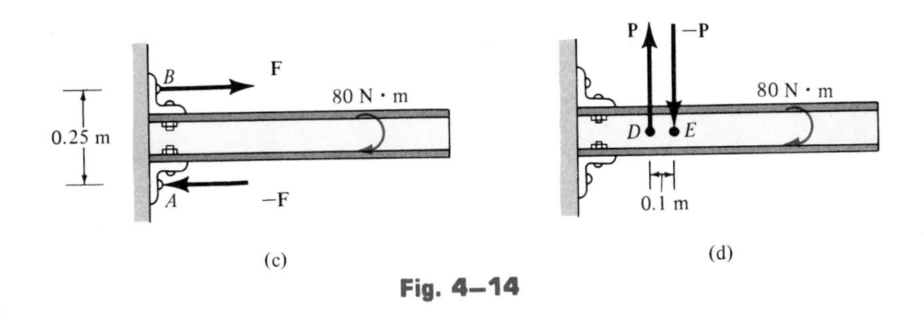

Fig. 4–14

Problems

4–1. In each case, determine the magnitude and direction of the moment of the force at A about point P.

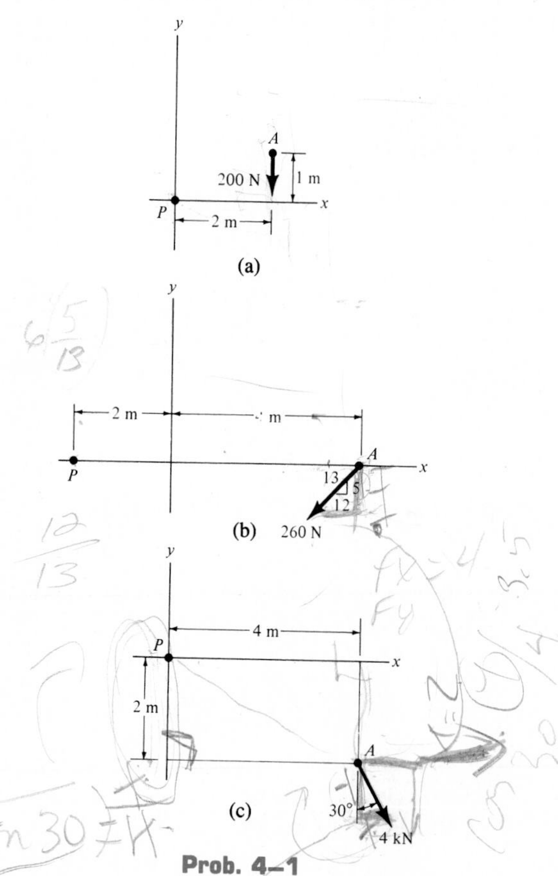

Prob. 4–1

4–2. In each case, determine the magnitude and direction of the moment of the force at A about point P.

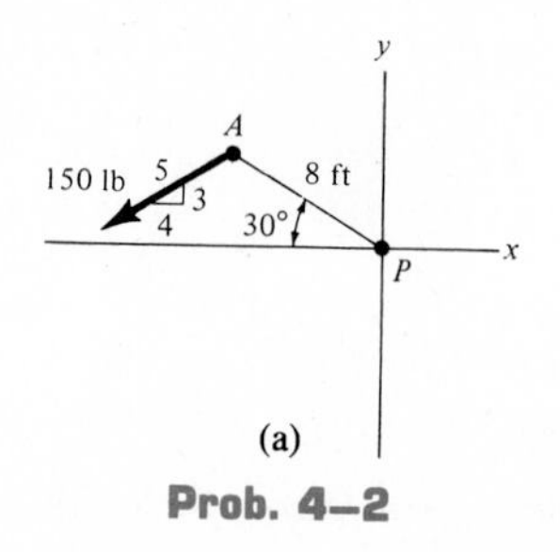

Prob. 4–2

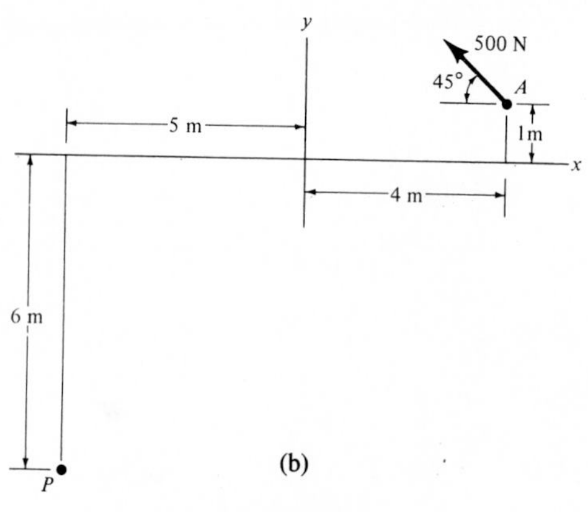

Prob. 4–2

4–3. The wrench is used to loosen the bolt. Determine the moment of each force about the bolt's axis passing through point O.

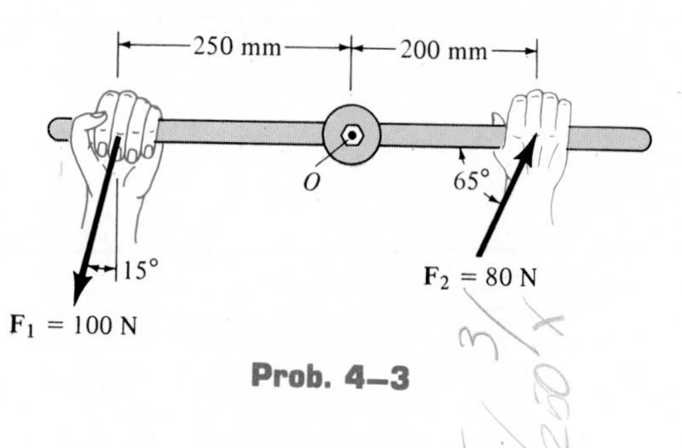

Prob. 4–3

***4–4.** Determine the moment of each of the three forces about point A on the beam.

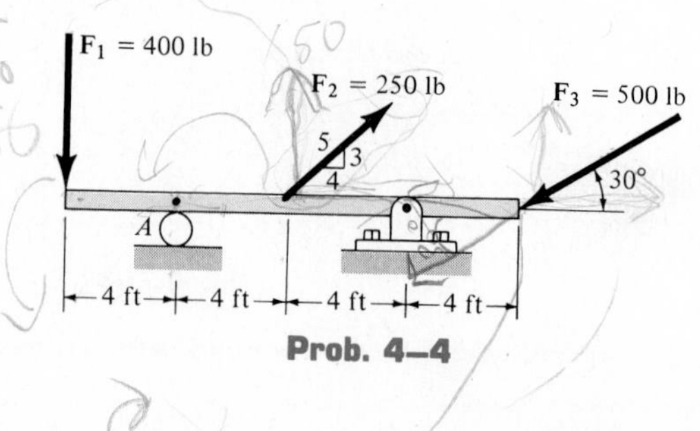

Prob. 4–4

4–5. The 500-N force acts on the end of the pipe at B. Determine: (a) the moment of this force about point A and (b) the magnitude and direction of a horizontal force, applied at C, which produces the same moment.

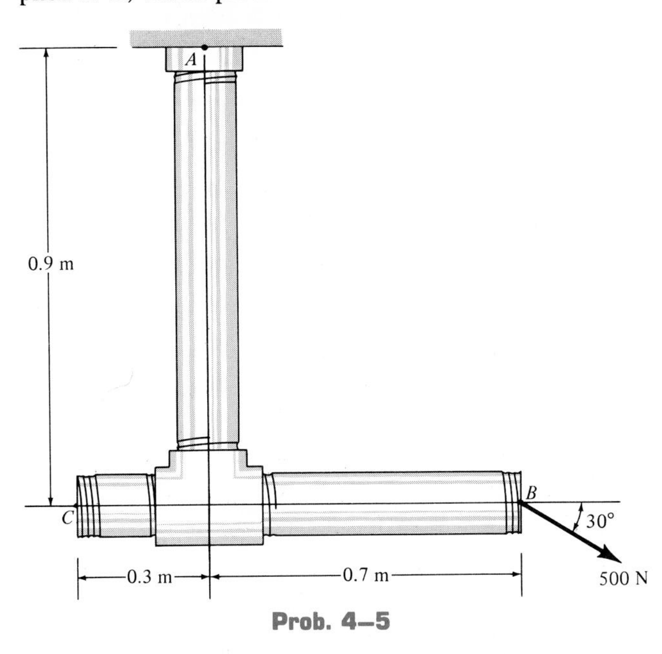

Prob. 4–5

4–6. A force of 90 N acts on the handle of the paper cutter at A. Determine the moment created by this force about the hinge at O if $\theta = 60°$. At what angle θ should the force be applied so that the clockwise moment it creates about point O is a maximum?

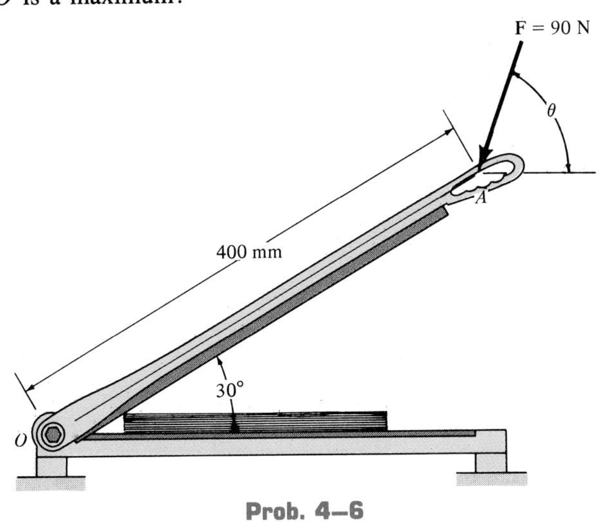

Prob. 4–6

4–7. Determine the moment of each force about the bolt axis passing through point A.

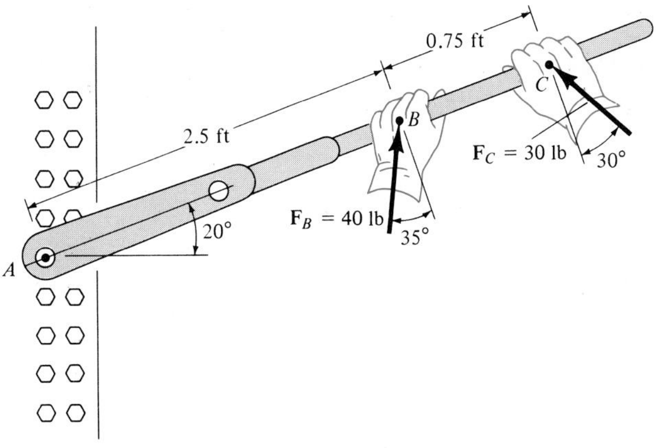

Prob. 4–7

***4–8.** As part of an acrobatic stunt, a man supports a girl who has a weight of 120 lb and is seated in a chair on top of a pole as shown. If her center of gravity is at G, and if the maximum counterclockwise moment the man can exert on the pole at A is 250 lb · ft, determine the maximum angle of tilt, θ, which will not allow the girl to fall.

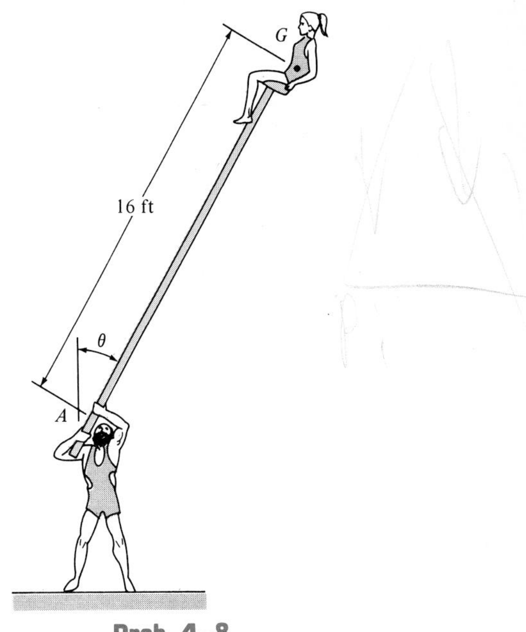

Prob. 4–8

4–9. The power pole supports the three lines, each line exerting a vertical force on the pole due to its weight as shown. Determine the resultant moment at the base D due to all of these forces. If it is possible for wind or ice to snap the lines, determine which line(s) when removed create(s) a condition for the greatest moment about the base. What is this resultant moment?

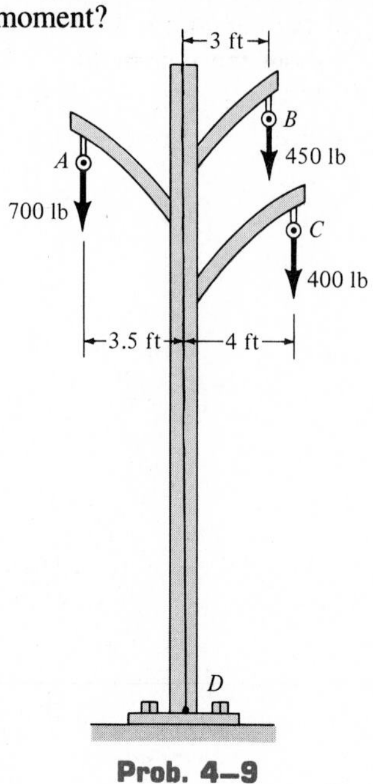

Prob. 4–9

4–10. The crane can be adjusted for any angle $0° \leq \theta \leq 90°$ and any extension $0 \leq x \leq 5$ m. For a suspended mass of 150 kg, determine the moment developed at A as a function of x and θ. What values of both x and θ develop the maximum possible moment at A? Compute this moment. Neglect the size of the pulley at B.

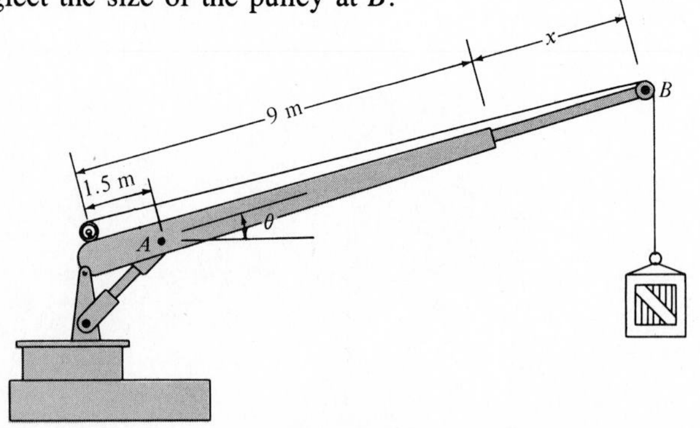

Prob. 4–10

4–11. The crane is used to lift the block that has a mass of 200 kg. Determine the clockwise moment of the block's weight about point A at the base of the boom. What magnitude of cable tension $\mathbf{T}$ is needed to create a counterclockwise moment of the same magnitude about point A?

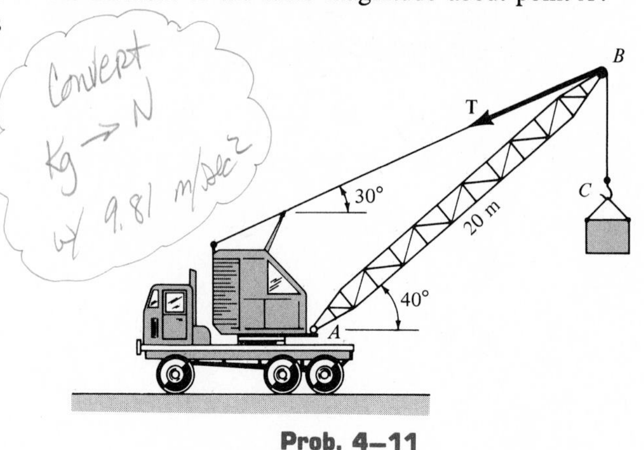

Prob. 4–11

***4–12.** The hammer is subjected to a horizontal force of $P = 25$ lb at the grip, whereas it takes a force of $F = 130$ lb at the claw to pull the nail out. Find the moment of each force about point A and determine if $\mathbf{P}$ is sufficient to pull out the nail. The hammer head contacts the board at point A.

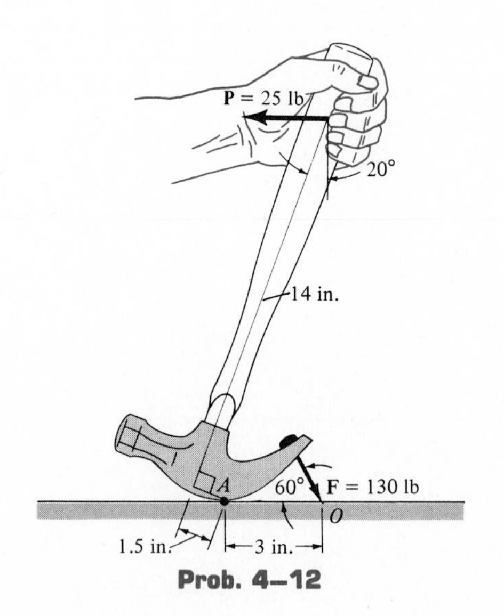

Prob. 4–12

4–13. Determine the direction θ ($0° \leq \theta \leq 180°$) of the 30-lb force **F** so that **F** produces (a) the maximum moment about point A and (b) the minimum moment about point A. Compute the moment in each case.

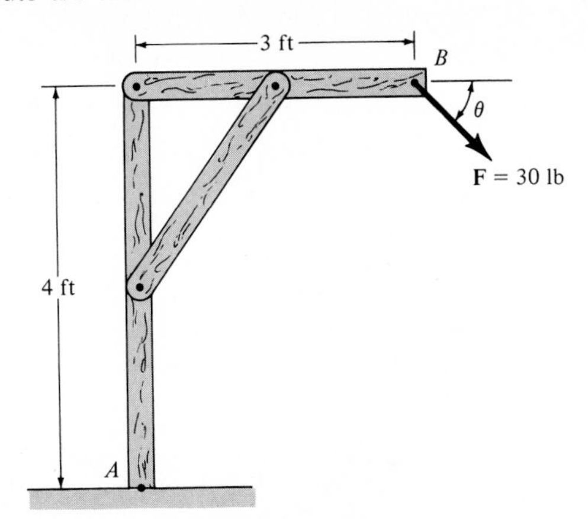

Prob. 4–13

4–14. In each case, determine the magnitude and direction of the couple moment.

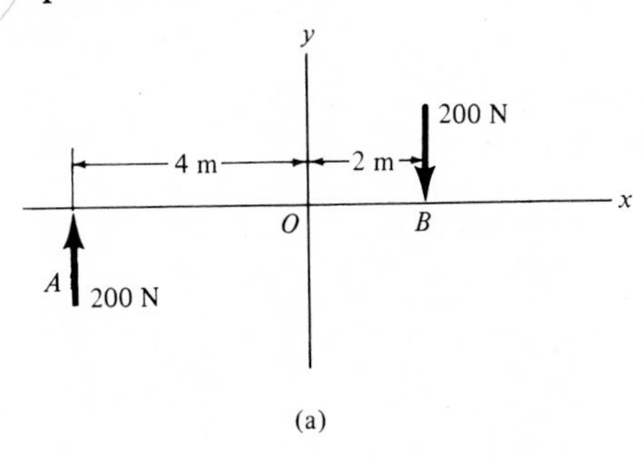

(a)

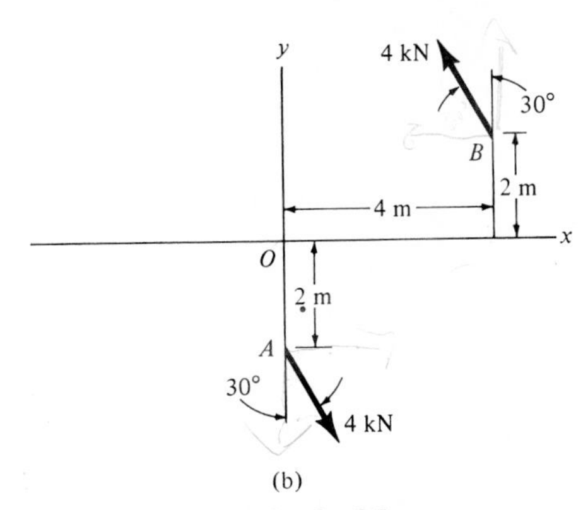

(b)

Prob. 4–14

4–15. Segments of drill pipe D for an oil well are tightened a prescribed amount by using a set of tongs T, which grip the pipe, and a hydraulic cylinder (not shown) to regulate the force **F** applied to the tongs. This force acts along the cable which passes around the small pulley P. If the cable is originally perpendicular to the tongs as shown, determine the magnitude of force **F** which must be applied so that the moment about the pipe is $M = 30{,}000$ lb · ft. In order to maintain this same moment what magnitude of **F** is required when the tongs rotate 30° to the dashed position?

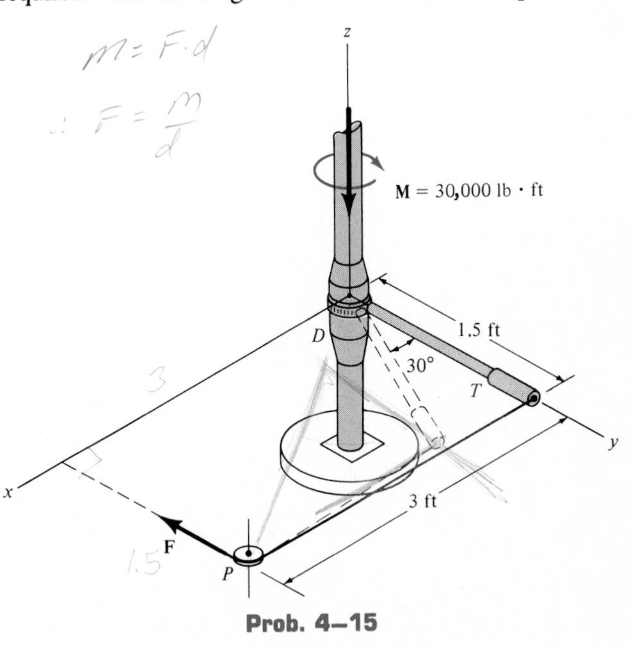

Prob. 4–15

***4–16.** Determine the magnitude and direction of the couple moment.

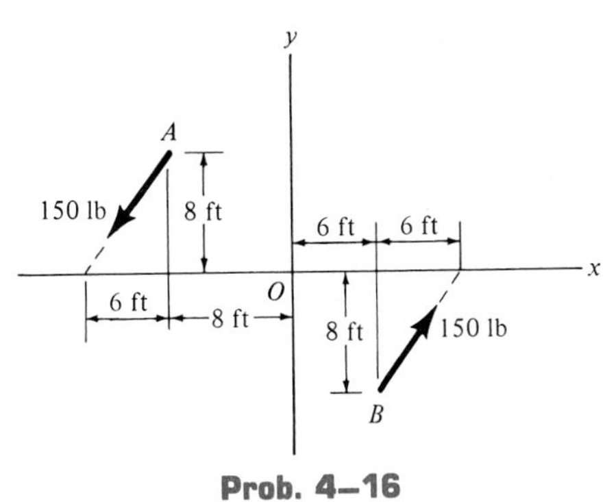

Prob. 4–16

4–17. A twist of 4 N · m is applied to the handle of the screwdriver. Resolve this couple moment into a pair of couple forces **F** exerted on the handle. Also resolve this couple moment into a pair of forces **P** exerted on the blade.

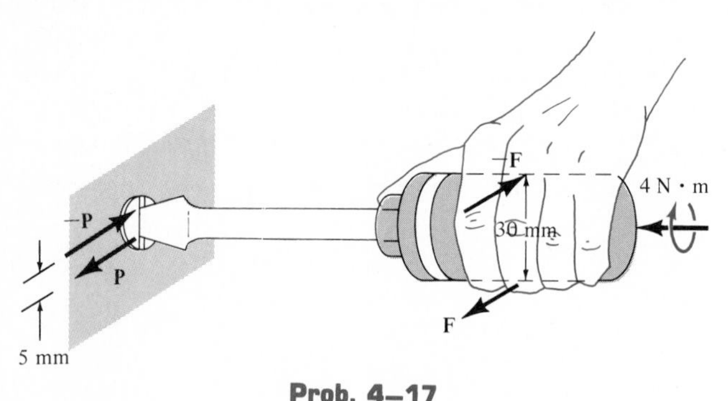

Prob. 4–17

4–18. The crossbar wrench is used to remove a lug nut from the automobile wheel. The mechanic applies a couple to the wrench such that his hands are a constant distance apart. Is it necessary that $a = b$ in order to produce the most effective turning of the nut? Explain. Also, what is the effect of changing the shaft dimension c in this regard? The forces act in the vertical plane.

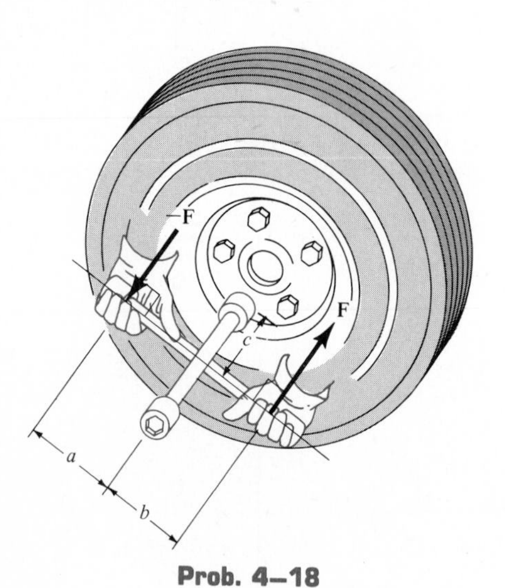

Prob. 4–18

4–19. The main beam along the wing of an airplane is swept back at an angle of 25°. From load calculations it is determined that the beam is subjected to couple moments $M_x = 25{,}000$ lb · ft and $M_y = 17{,}000$ lb · ft. Determine the equivalent couple moments created about the x' and y' axes.

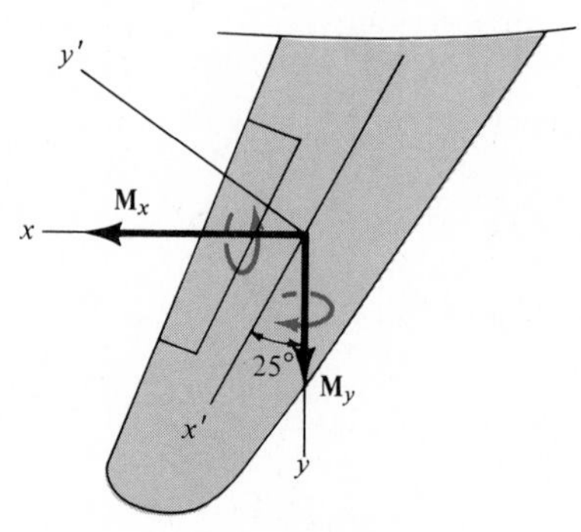

Prob. 4–19

***4–20.** When the engine of the plane is running, the vertical reaction that the ground exerts on the wheel at A is measured as 650 lb. When the engine is turned off, however, the vertical reactions at A and B are 575 lb each. The difference in readings at A is caused by a couple acting on the propeller when the engine is running. This couple tends to overturn the plane counterclockwise, which is opposite to the propeller's clockwise rotation. Determine the magnitude of this couple and the magnitude of the vertical force exerted at B when the engine is running.

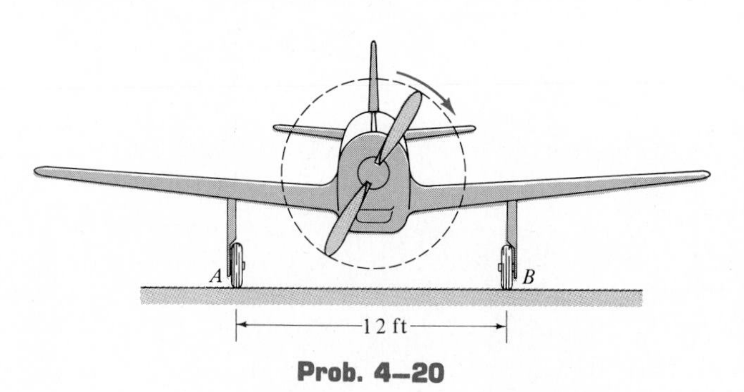

Prob. 4–20

The Principle of Transmissibility 4.4

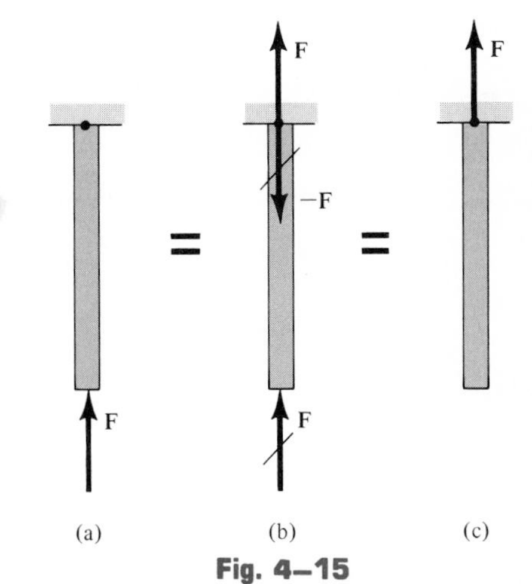

Fig. 4–15

The *principle of transmissibility* is an important concept often used in mechanics for studying the action of a force on a rigid body. *This principle states that the external effects on a rigid body remain unchanged when a force, acting at a given point on the body, is applied to another point lying on the line of action of the force.* In other words, the force can be considered as a *sliding vector.*

To illustrate this concept, consider the bar shown in Fig. 4–15*a*, which is subjected to a force **F** acting at its bottom end. If equal but opposite forces **F** and −**F** are applied at the top, Fig. 4–15*b*, the state of rest or motion will not be altered. As a consequence, two of the forces, indicated with a slash across them, cancel and the force acting at the top end of the bar remains, Fig. 4–15*c*. The force, therefore, has been "transmitted" along its line of action from the bottom to the top of the bar.

A further illustration is shown in Fig. 4–16, where a beam is subjected to a force at point P_1. Rather than applying **F** at the beam's top surface, it can be applied from the bottom of the beam at P_2. Since P_1 and P_2 lie on the line of action of **F,** the principle of transmissibility states that **F** may be applied at *any* of these two points and the *external reactive forces* developed at the supports *A* and *B* will remain the same. The *internal forces* developed in the beam, however, will depend upon the location of **F.** For example, if **F** acts at P_1, the internal forces in the beam have a high intensity around P_1; whereas if **F** acts at P_2, the effect of **F** on generating internal forces at P_1 will be less.

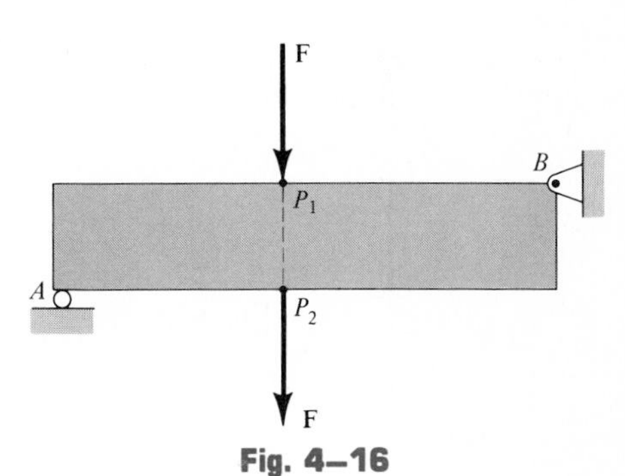

Fig. 4–16

Resolution of a Force into a Force and a Couple 4.5

Many problems in statics deal with a rigid body subjected to a system of forces. In order to understand fully what external effects these forces have on the body, it is best to reduce the system to the simplest resultant possible. In the next section we will show how this is accomplished for a coplanar force system. Before doing this, however, it will be necessary to determine the effect of moving a force from one point to another on a rigid body.

For example, consider moving the force **F** acting at point *A* on the constrained rigid body in Fig. 4–17*a* to the arbitrary point *O*, which does *not* lie along the line of action of **F.** Using the principle of transmissibility, **F** may first be applied at point *A*′, which lies on the line of action of the force at a

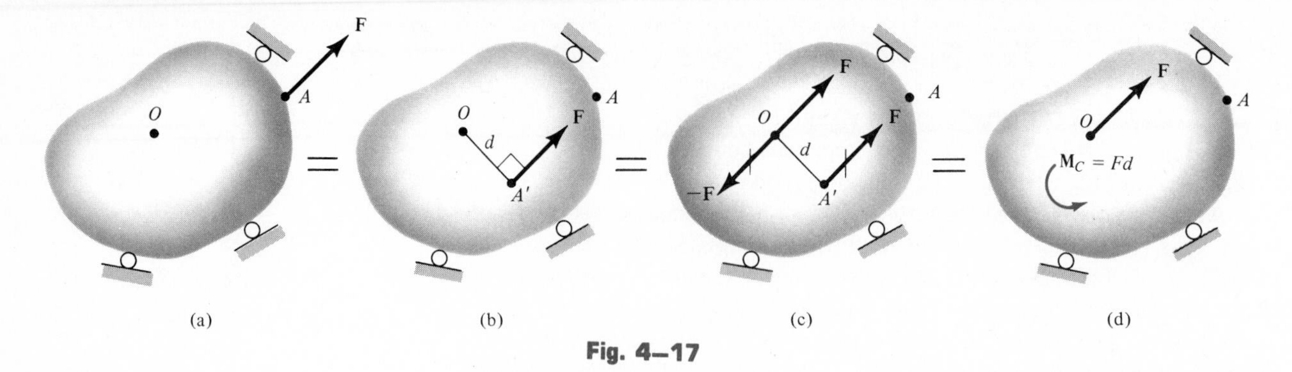

Fig. 4–17

perpendicular distance d from O, Fig. 4–17b. Applying equal but opposite forces **F** and −**F** at O, as shown in Fig. 4–17c, in no way alters the external effects on the body; however, the two forces indicated by a slash across them form a couple which has a magnitude $M_C = Fd$ and tends to rotate the body in a counterclockwise direction. Since the couple is a free vector, it may be applied at *any point* on the body, as shown in Fig. 4–17d. In addition to this couple, **F** now acts at point O. By using this construction procedure, an equivalent system has been maintained between each of the diagrams, as indicated by the "equal signs." In other words, when **F** acts at A, Fig. 4–17a, it will produce the *same reactions* at the three roller supports as when **F** is applied at O and a couple moment $\mathbf{M}_C$ is also applied to the body, Fig. 4–17d. From the construction, *note that the magnitude and direction of the couple moment can also be determined by taking the moment of* **F** *about O when the force is located at its original point A (or point A'). The line of action of* $\mathbf{M}_C$ *is always perpendicular to the plane containing* **F** *and d.*

The foregoing concept regarding the movement of a force to any point on a body may be summarized by the following two statements:

1. If the force is to be moved to a *point O located on its line of action,* by the principle of transmissibility, simply move the force to the point.
2. If the force is to be moved to a *point O that is not located on its line of action,* an equivalent system is maintained when the force is moved to point O and a couple moment is placed on the body. The magnitude and direction of the couple moment are determined by finding the moment of the force about point O.

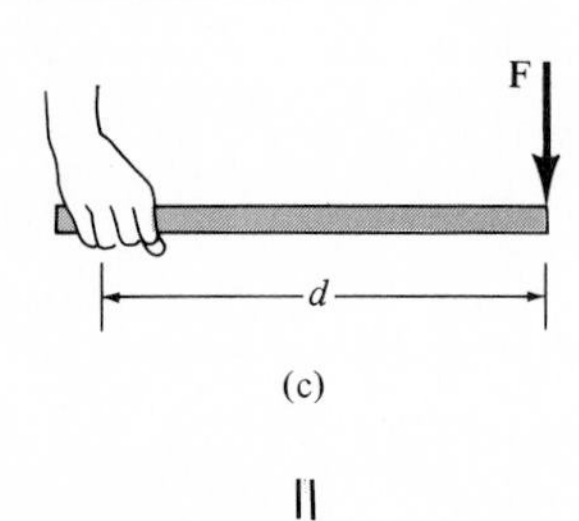

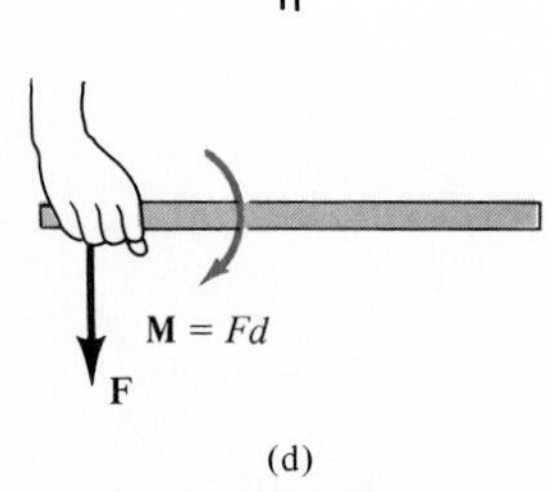

Fig. 4–18

As a physical illustration of these two statements, consider holding the end of a stick of negligible weight. If a vertical force **F** is applied at the other end, and the stick is held in the vertical position, Fig. 4–18a, then, by the principle of transmissibility, the same force is felt at the grip, Fig. 4–18b. When the stick is held in the horizontal position, Fig. 4–18c, the force has the effect of producing *both* a downward force at the grip and a clockwise twist, Fig. 4–18d. The twist can be thought of as being caused by a *couple moment* that is produced when **F** is moved to the grip. This couple moment has the *same* magnitude and direction as the moment of **F** about the grip, Fig. 4–18c.

Example 4–8

Replace the force **F** acting on the end of the beam shown in Fig. 4–19*a* by an equivalent force and couple system acting at point *O*.

Solution

Since point *O* is not located on the line of action of **F**, in order to maintain an equivalent loading the force is moved to *O* and a couple moment is applied to the beam. Since the couple moment is equivalent to the moment of **F** about *O*, it is easy to determine $\mathbf{M}_O$ by using the principle of moments. In this regard, **F** is resolved into its *x* and *y* components, and each component is moved to *O*. From the data shown in Fig. 4–19*b*, assuming positive moments to be counterclockwise, i.e., out of the page, we have

$$\circlearrowleft + M_O = 150(0.2) - 260(0.5)$$
$$\mathbf{M}_O = -100 \text{ N} \cdot \text{m} \qquad \textit{Ans.}$$

The results are shown in Fig. 4–19*c*.

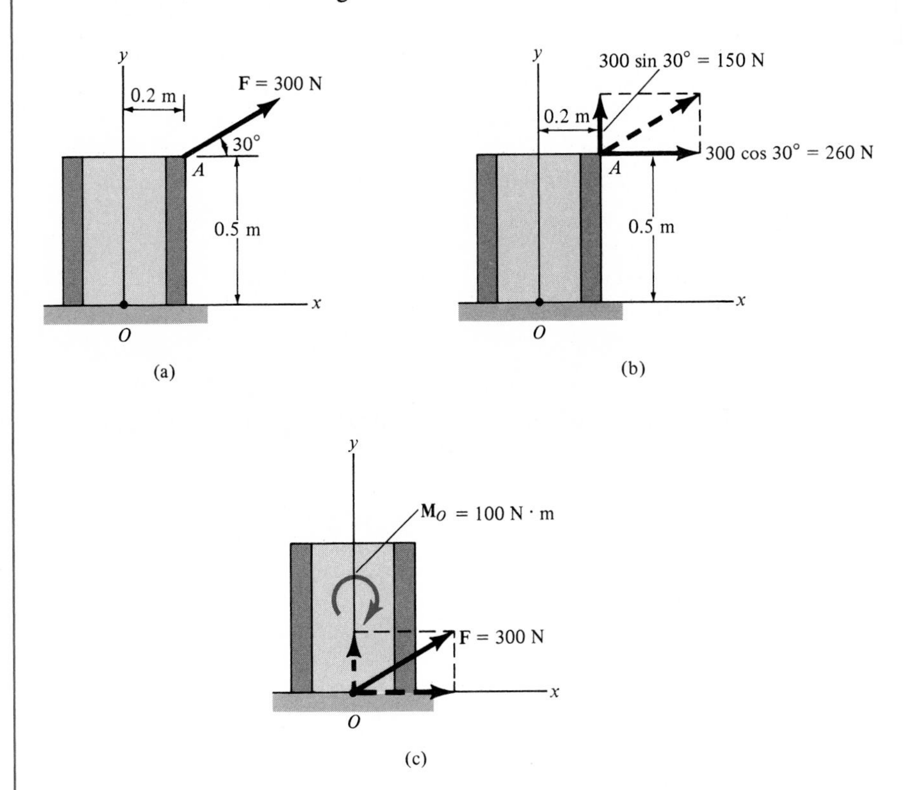

Fig. 4–19

Example 4–9

Replace the force **F** acting at the end of the beam shown in Fig. 4–20*a* by an equivalent force and couple system acting at point *A*.

Solution

The force is resolved into its components $\mathbf{F}_x$ and $\mathbf{F}_y$, Fig. 4–20*b*, and each component is moved separately to point *A*. Since the line of action of $\mathbf{F}_x$ passes through *A*, this force is simply moved to the point—principle of transmissibility, Fig. 4–20*c*. $\mathbf{F}_y$ is moved to *A* and a couple $\mathbf{M}_A$ is added to the beam. The moment of $\mathbf{F}_y$ about point *A* has a magnitude of

$$M_A = (500 \sin 60°)(3) = 1299.0 \text{ N} \cdot \text{m} \qquad \textit{Ans.}$$

Since $\mathbf{F}_y$ tends to rotate the beam clockwise about *A*, by the right-hand rule $\mathbf{M}_A$ acts into the page. Adding $\mathbf{F}_x$ and $\mathbf{F}_y$ yields **F**. The results are shown in Fig. 4–20*d*.

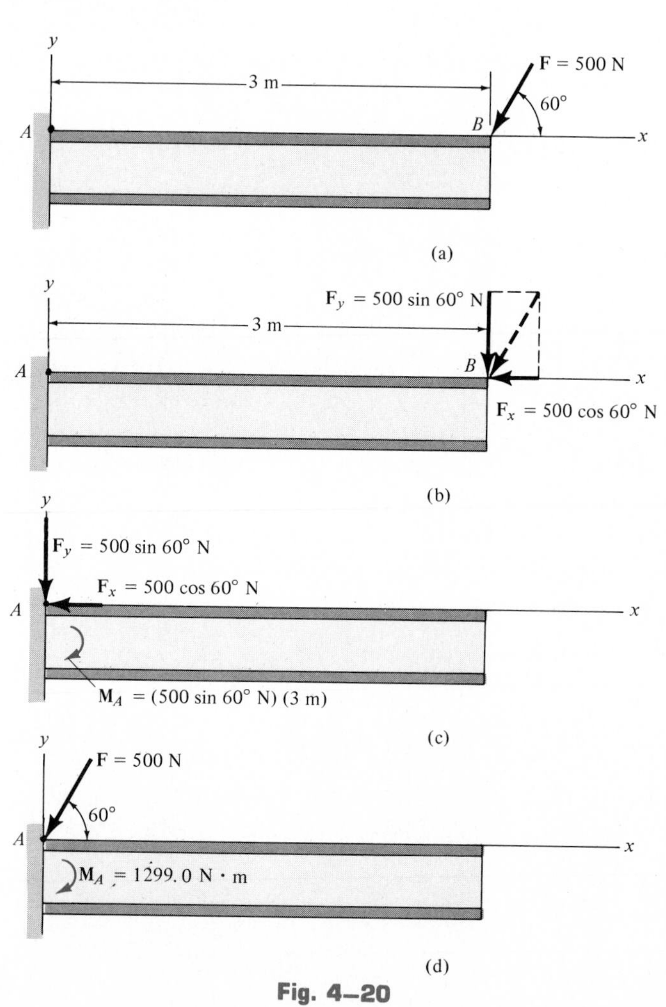

Fig. 4–20

Simplification of a Coplanar Force System 4.6

By moving each force of a system of coplanar forces to an arbitrary point O, it is possible to simplify the system to a *single* resultant force acting at O and a *single* resultant couple. To demonstrate this, consider the rigid body subjected to the coplanar force and couple system in Fig. 4–21*a*.

1. Couple moments $\mathbf{M}_{C_1}$ and $\mathbf{M}_{C_2}$ can simply be moved to point O since they are *free vectors,* Fig. 4–21*b*.
2. Point O is located on the line of action of $\mathbf{F}_1$. This *sliding vector* is moved to O by the principle of transmissibility, Fig. 4–21*b*.
3. Since point O is not on the line of action of $\mathbf{F}_2$ and $\mathbf{F}_3$, each of these forces must be placed at O in accordance with the procedure outlined in Sec. 4–5. For example, when $\mathbf{F}_2$ is applied at O, a corresponding couple $M_2 = F_2 d_2$ must also be applied to the body, Fig. 4–21*b*.

Using vector addition, this equivalent concurrent force and couple system at O can now be summed to a single resultant force and resultant couple, where

$$\mathbf{F}_R = \Sigma \mathbf{F} = \mathbf{F}_1 + \mathbf{F}_2 + \mathbf{F}_3$$
$$\mathbf{M}_{R_O} = \Sigma \mathbf{M}_O = \mathbf{M}_{C_1} + \mathbf{M}_{C_2} + \mathbf{M}_3 + \mathbf{M}_4$$

The results are shown in Fig. 4–21*c*. Note that both the magnitude and direction of $\mathbf{F}_R$ are independent of the location of point O. However, $\mathbf{M}_{R_O}$ depends upon this location, since the moments $\mathbf{M}_2$ and $\mathbf{M}_3$ are computed using the moment arms d_2 and d_3. Also, $\mathbf{M}_{R_O}$ will always be *perpendicular* to $\mathbf{F}_R$ since by the right-hand rule the moments of a coplanar force system are perpendicular to the plane containing the forces and moment arms.

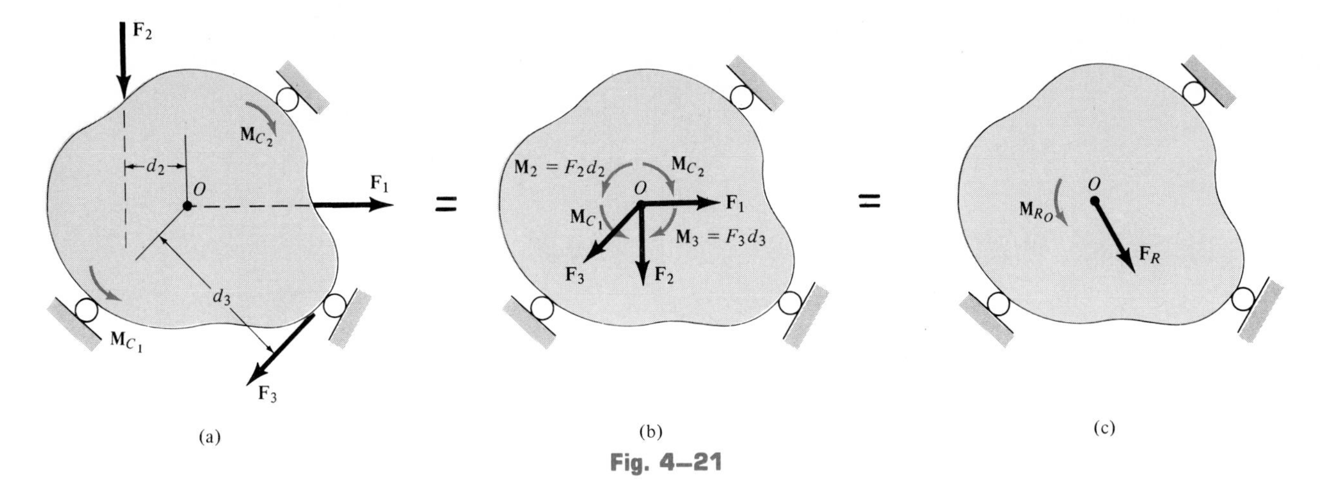

Fig. 4–21

PROCEDURE FOR ANALYSIS

The method just described for simplifying a coplanar force and couple system to an equivalent single resultant force acting at point O and a resultant couple may be generalized as follows:

Force Summation. The *resultant force* $\mathbf{F}_R$ is equivalent to the vector sum of its two components $\mathbf{F}_{R_x}$ and $\mathbf{F}_{R_y}$. Each component is found from the scalar (algebraic) sum of the components of all the forces in the system, i.e.,

$$F_{R_x} = \Sigma F_x$$
$$F_{R_y} = \Sigma F_y$$

Moment Summation. The *resultant couple moment* $\mathbf{M}_{R_O}$ is perpendicular to the plane containing the forces and is equivalent to the scalar (algebraic) sum of all the couples in the system *plus* the moments about point O of all the forces in the system. This can be expressed mathematically as

$$M_{R_O} = \Sigma M_O$$

When computing the moments of the forces about O, it is generally advantageous to use the *principle of moments,* i.e., compute the moments of the *components* of each force rather than compute the moment of the force.

The following example numerically illustrates this procedure.

Example 4–10

Replace the forces acting on the pipe shown in Fig. 4–22*a* by an equivalent single force and couple system acting at point *A*.

Solution

The principle of moments will be applied to the 400-N force, whereby the moments of its two rectangular components will be considered.

Force Summation. The resultant force has x and y components of

$$\overset{+}{\rightarrow} F_{R_x} = \Sigma F_x; \quad F_{R_x} = -100 - 400 \cos 45° = -382.8 \text{ N} = 382.8 \text{ N} \leftarrow$$

$$+\uparrow F_{R_y} = \Sigma F_y; \quad F_{R_y} = -600 - 400 \sin 45° = -882.8 \text{ N} = 882.8 \text{ N} \downarrow$$

As shown in Fig. 4–22*b*, $\mathbf{F}_R$ has a magnitude of

$$F_R = \sqrt{(F_{R_x})^2 + (F_{R_y})^2} = \sqrt{(382.8)^2 + (882.8)^2} = 962.2 \text{ N} \qquad \textit{Ans.}$$

and a direction defined by

$$\theta = \tan^{-1}\left(\frac{F_{R_y}}{F_{R_x}}\right) = \tan^{-1}\left(\frac{882.8}{382.8}\right) = 66.6° \qquad \textit{Ans.}$$

Moment Summation. The resultant couple moment $\mathbf{M}_{R_A}$ is determined by summing moments about point *A*. Noting that the moment of each force in the system is perpendicular to the plane of the forces and assuming that positive moments act counterclockwise, we have

$$\circlearrowleft + M_{R_A} = \Sigma M_A;$$

$$M_{R_A} = 100(0) - 600(0.4) - (400 \sin 45°)(0.8) - (400 \cos 45°)(0.3)$$

$$= -551.1 \text{ N} \cdot \text{m}$$

The negative sign indicates that $\mathbf{M}_{R_A}$ acts clockwise as shown in Fig. 4–22*b*.

In conclusion, when $\mathbf{M}_{R_A}$ and $\mathbf{F}_R$ act on the pipe, Fig. 4–22*b*, they will produce the *same* reaction at the support *A* as that produced by the force system in Fig. 4–22*a*.

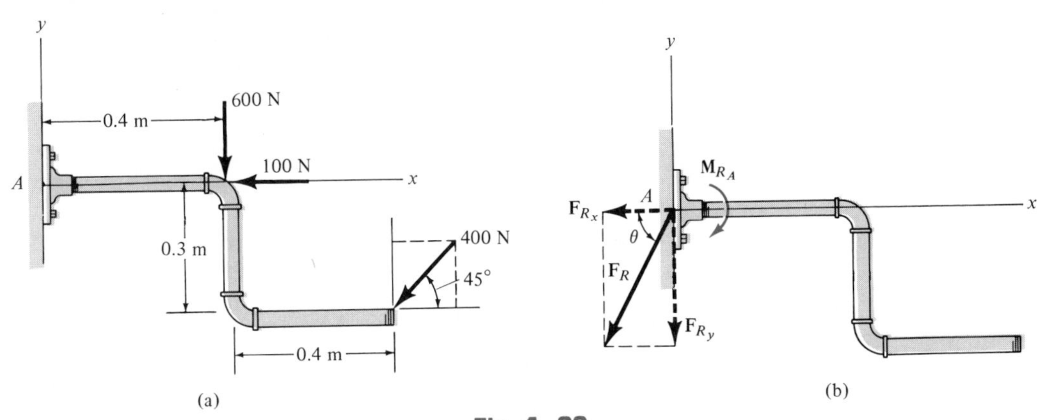

Fig. 4–22

4.7 Further Simplification of a Coplanar Force System

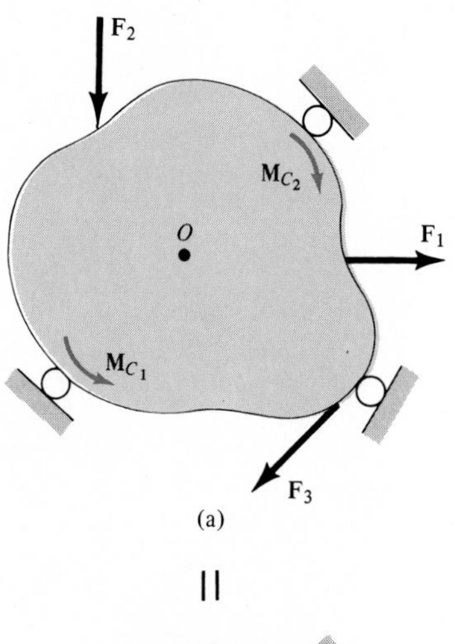

(a)

=

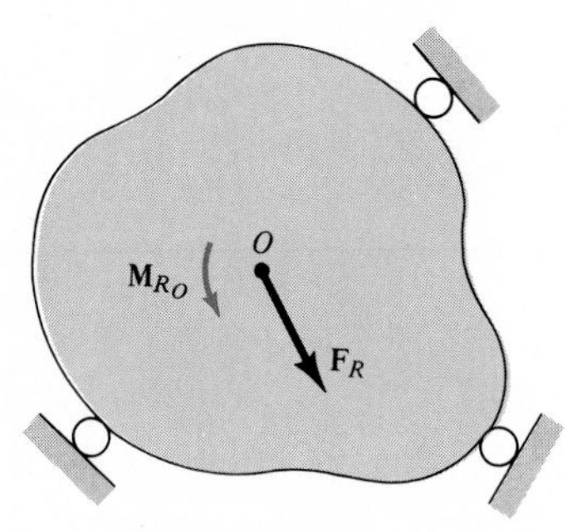

(b)

=

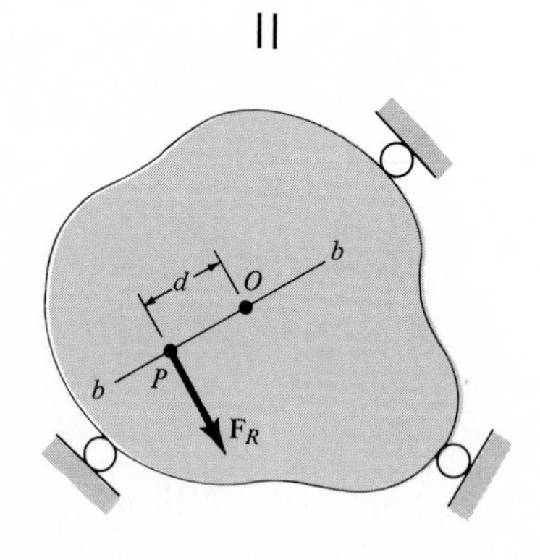

(c)

Fig. 4–23

Since the couple moments developed by moving a system of coplanar forces to some arbitrary point are *always perpendicular* to the plane containing the forces, we can further simplify this system by eliminating the resultant couple $\mathbf{M}_{R_O}$. To see how this is done, consider again the rigid body subjected to the coplanar force and couple system shown in Fig. 4–23*a*. The system is reduced to an equivalent single resultant force $\mathbf{F}_R$ acting at point O and a (perpendicular) resultant couple moment $\mathbf{M}_{R_O}$, Fig. 4–23*b*. The couple moment can be eliminated by moving $\mathbf{F}_R$ to point P located on line *bb*, which is perpendicular to both $\mathbf{F}_R$ and $\mathbf{M}_{R_O}$, Fig. 4–23*c*. To maintain an equivalent loading, P is chosen so that the moment arm d satisfies the scalar equation $M_{R_O} = F_R d$ or $d = M_{R_O}/F_R$. Furthermore, P must be located to the *left* of O in order to preserve the correct direction of $\mathbf{M}_{R_O}$ if $\mathbf{F}_R$ were to be moved back to O, Fig. 4–23*b*.

PROCEDURE FOR ANALYSIS

The techniques used to simplify a coplanar force system to an equivalent single resultant force follow the general procedure outlined in the preceding section.

Force Summation. The *resultant force* $\mathbf{F}_R$ is equivalent to the vector sum of its two components $\mathbf{F}_{R_x}$ and $\mathbf{F}_{R_y}$. Each component is found from the scalar (algebraic) sum of the components of all the forces in the system, i.e.,

$$F_{R_x} = \Sigma F_x$$
$$F_{R_y} = \Sigma F_y$$

Moment Summation. The location of $\mathbf{F}_R$, specified by the moment arm d measured from an arbitrary point O, is determined from the moment equation

$$M_{R_O} = F_R d = \Sigma M_O$$

which states that the moment of the resultant force about point O, $F_R d$, is equivalent to the sum of the moments about point O of all the forces in the system, ΣM_O. Again, when computing the moments of coplanar forces it is generally best to use the principle of moments.

The following examples numerically illustrate this procedure.

Example 4–11

Replace the system of forces acting on the beam shown in Fig. 4–24*a* by an equivalent single resultant force. Specify the distance the force acts from point *A*.

Solution

Force Summation. If "positive" forces are assumed to act upward, then from Fig. 4–24*a* the force resultant $\mathbf{F}_R$ is

$$+\uparrow F_R = \Sigma F; \quad F_R = -100 \text{ N} + 400 \text{ N} - 200 \text{ N} = 100 \text{ N}\uparrow \qquad \textit{Ans.}$$

Moment Summation. Moments will be summed about point *A*. Considering counterclockwise rotations as positive, i.e., positive moment vectors are directed out of the page, then from Fig. 4–24*a* and *b* we require the moment of $\mathbf{F}_R$ about *A* to equal the moments of the force system about *A*, i.e.,

$$\circlearrowleft +M_{R_A} = \Sigma M_A;$$

$$100 \text{ N}(d) = -(100 \text{ N})(3 \text{ m}) + (400 \text{ N})(5 \text{ m}) - (200 \text{ N})(8 \text{ m})$$

$$(100)d = 100$$

$$d = 1 \text{ m} \qquad \textit{Ans.}$$

Since *d* is *positive*, $\mathbf{F}_R$ acts indeed to the right of *A* as shown in Fig. 4–24*b*.

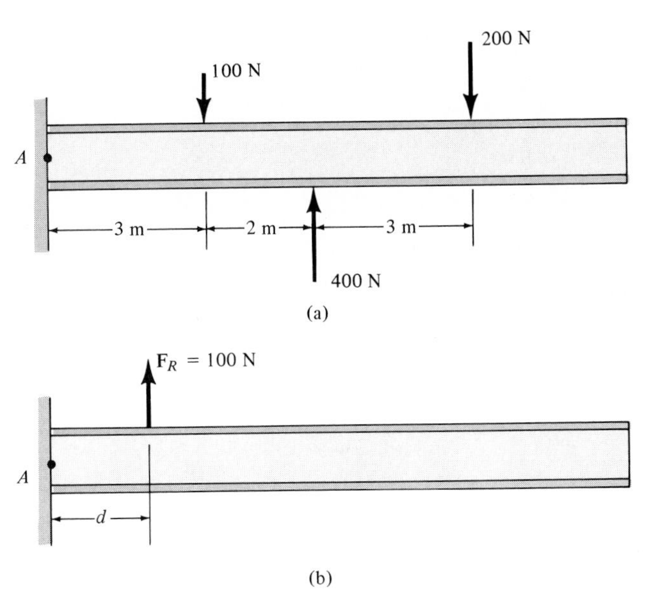

Fig. 4–24

Example 4–12

The beam AE in Fig. 4–25a is subjected to a system of coplanar forces. Determine the magnitude, direction, and location on the beam of a single resultant force which is equivalent to the given system of forces.

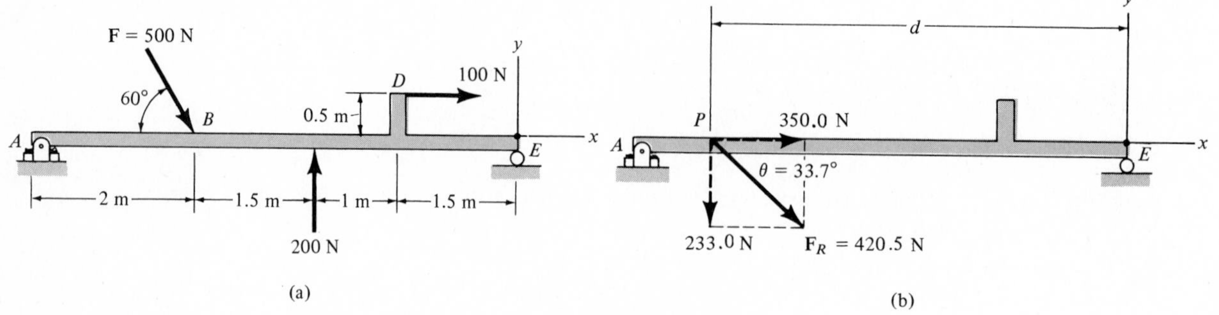

Fig. 4–25

Solution

Force Summation. The origin of coordinates is arbitrarily located at point E as shown in Fig. 4–25a. Resolving the 500-N force into x and y components, and summing the force components, yields

$$\overset{+}{\rightarrow} F_{R_x} = \Sigma F_x; \qquad F_{R_x} = 500 \cos 60° + 100 = 350.0 \text{ N} \rightarrow$$

$$+\uparrow F_{R_y} = \Sigma F_y; \qquad F_{R_y} = -500 \sin 60° + 200 = -233.0 \text{ N} = 233.0 \text{ N} \downarrow$$

The magnitude and direction of the resultant force shown in Fig. 4–25b are, therefore,

$$F_R = \sqrt{(350.0)^2 + (233.0)^2} = 420.5 \text{ N} \qquad \textit{Ans.}$$

$$\theta = \tan^{-1}\left(\frac{233.0}{350.0}\right) = 33.7° \qquad \textit{Ans.}$$

Moment Summation. Moments will be summed about point E. Since P lies on the x axis, $\mathbf{F}_{R_x}$ (350 N) does not create a moment about E (principle of transmissibility), only $\mathbf{F}_{R_y}$ (233.0 N) does. Hence, from Fig. 4–25a and b, we require that the moment of $\mathbf{F}_R$ about point E equal the moments of the force system about E, i.e.,

$$\circlearrowleft + M_{R_E} = \Sigma M_E;$$

$$233.0(d) = (500 \sin 60°)(4) + (500 \cos 60°)(0) - (100)(0.5) - (200)(2.5)$$

$$d = \frac{1182.1}{233.0} = 5.07 \text{ m} \qquad \textit{Ans.}$$

Example 4–13

The frame shown in Fig. 4–26*a* is subjected to three coplanar forces. Replace this loading by a single resultant force and specify where the resultant's line of action intersects members *AB* and *BC*.

Solution

Force Summation. Resolving the 75-lb force into *x* and *y* components and summing the force components yields

$$\overset{+}{\leftarrow} F_{R_x} = \Sigma F_x; \qquad F_{R_x} = 75(\tfrac{3}{5}) + 15 = 60 \text{ lb} \leftarrow$$

$$+\downarrow F_{R_y} = \Sigma F_y; \qquad F_{R_y} = 75(\tfrac{4}{5}) + 90 = 150 \text{ lb} \downarrow$$

As shown in Fig. 4–26*b*,

$$F_R = \sqrt{(60)^2 + (150)^2} = 161.6 \text{ lb} \qquad \textit{Ans.}$$

$$\theta = \tan^{-1}\left(\frac{150}{60}\right) = 68.2° \qquad \textit{Ans.}$$

Moment Summation. Moments will be summed about point *A*. If the line of action of $\mathbf{F}_R$ intersects *AB*, Fig. 4–26*b*, we require the moment of $\mathbf{F}_R$ in Fig. 4–26*b* about *A* to equal the moments of the force system in Fig. 4–26*a* about *A*, i.e.,

$$\circlearrowleft + M_{R_A} = \Sigma M_A;$$

$$0(161.6 \sin 68.2°) + y(161.6 \cos 68.2°) = 4(15) + 0(90) + 7(75)(\tfrac{3}{5}) - 2(75)(\tfrac{4}{5})$$

$$y = 4.25 \text{ ft} \qquad \textit{Ans.}$$

By the principle of transmissibility, $\mathbf{F}_R$ can also intersect *BC*, Fig. 4–26*b*, in which case we have

$$\circlearrowleft + M_{R_A} = \Sigma M_A;$$

$$7(161.6 \cos 68.2°) - x(161.6 \sin 68.2°) = 4(15) + 0(90) + 7(75)(\tfrac{3}{5}) - 2(75)(\tfrac{4}{5})$$

$$x = 1.10 \text{ ft} \qquad \textit{Ans.}$$

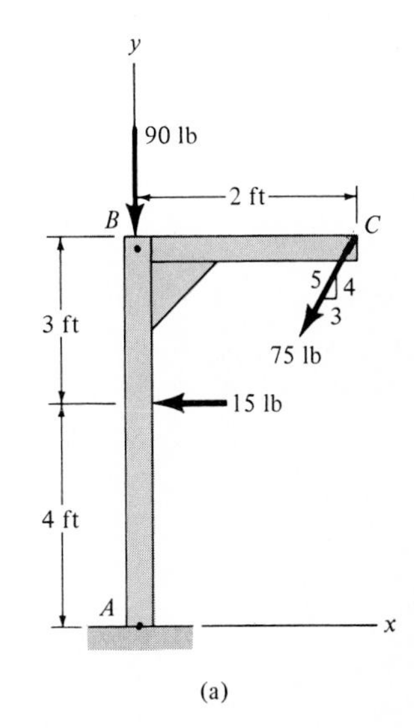

(a)

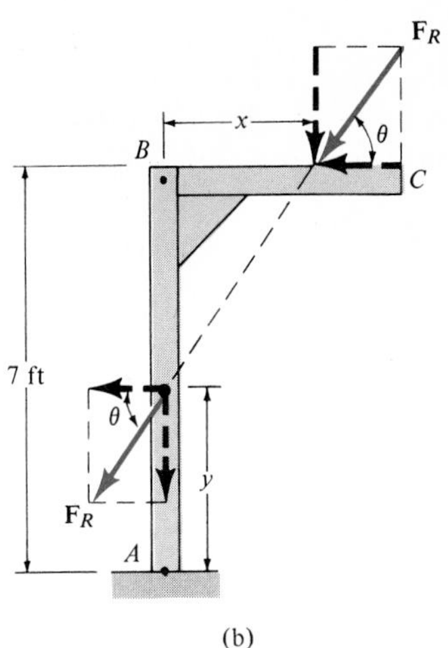

(b)

Fig. 4–26

Problems

4–21. Replace the force at *A* by an equivalent force and couple moment at *P*.

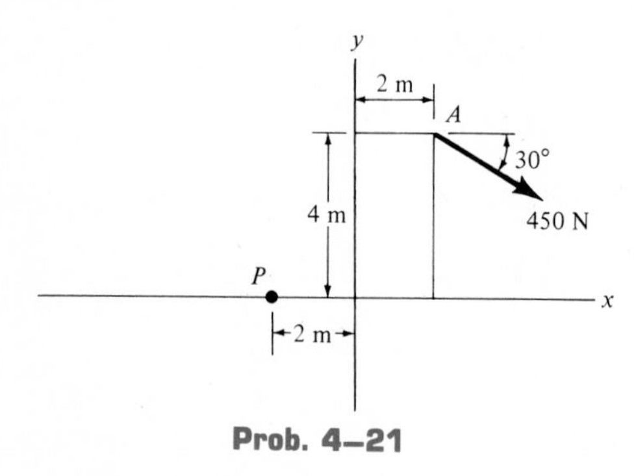

Prob. 4–21

4–22. Replace the force at *A* by an equivalent force and couple moment at *P*.

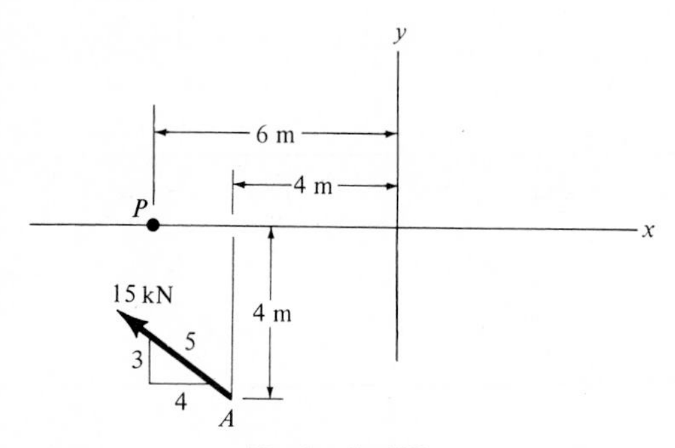

Prob. 4–22

4–23. Replace the force at *A* by an equivalent force and couple moment at *P*.

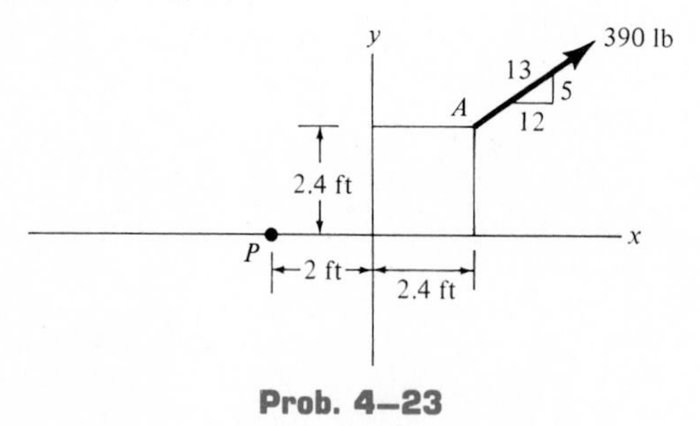

Prob. 4–23

***4–24.** The structural connection is subjected to the 8000-lb force. Replace this force by an equivalent force and couple acting at the center of the bolt group, *O*.

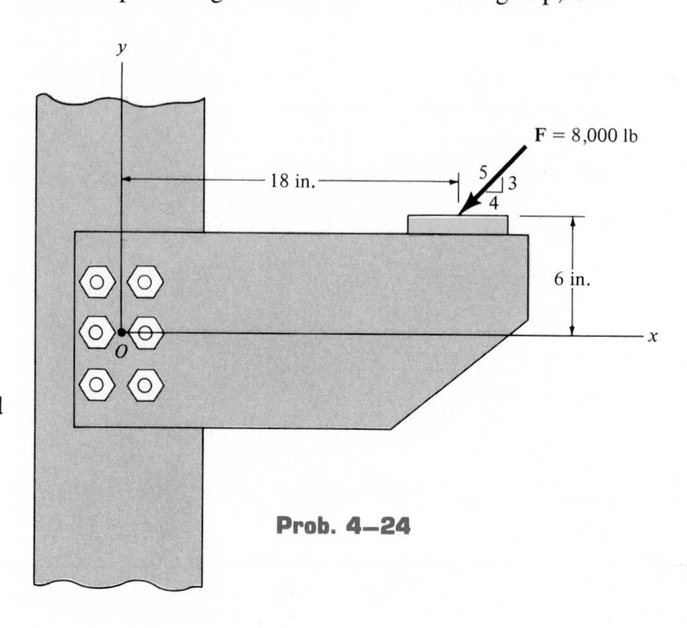

Prob. 4–24

4–25. The force of 50 N is exerted on the handle of the monkey wrench. Replace this force by an equivalent force and couple acting through the axis of the bolt at *O*.

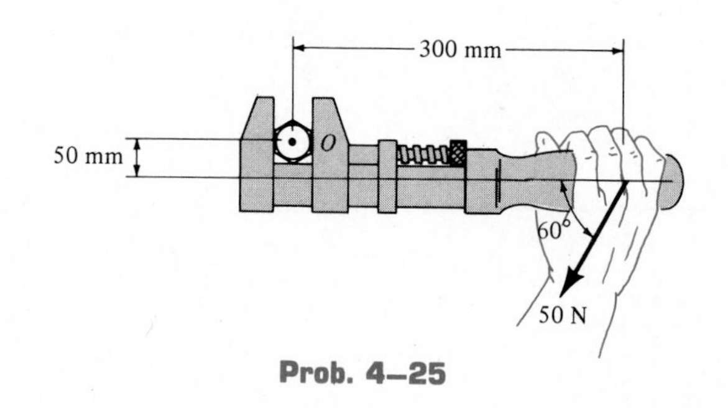

Prob. 4–25

4–26. Replace the force and couple system by an equivalent single force and couple acting at point P.

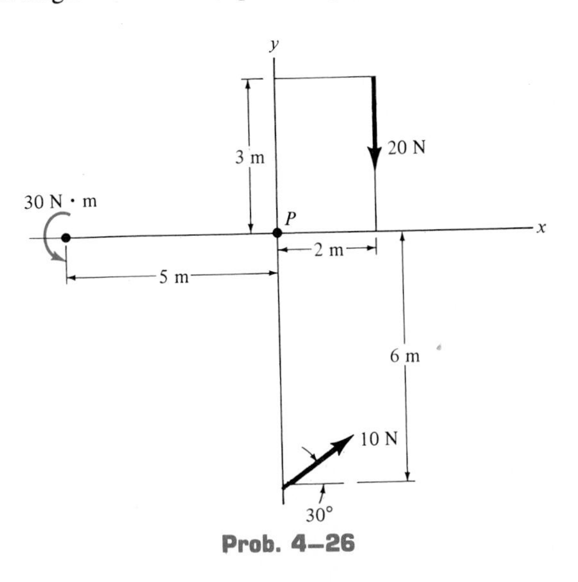

Prob. 4–26

4–27. Replace the force and couple system by an equivalent single force and couple acting at point P.

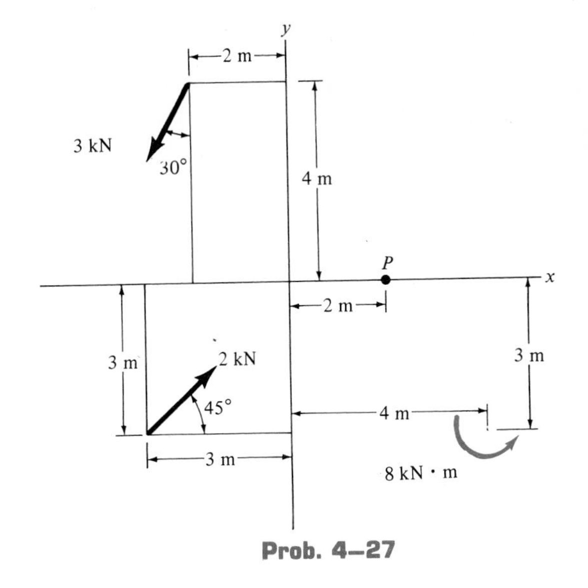

Prob. 4–27

***4–28.** Replace the force and couple system by an equivalent single force and couple acting at point P.

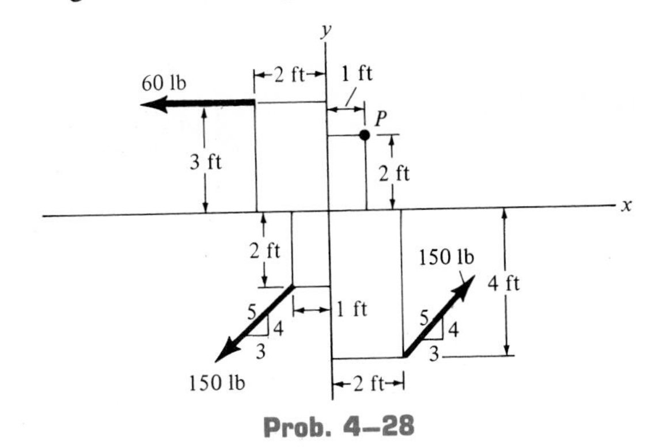

Prob. 4–28

4–29. Replace the loading system acting on the beam by an equivalent force and couple system at point O.

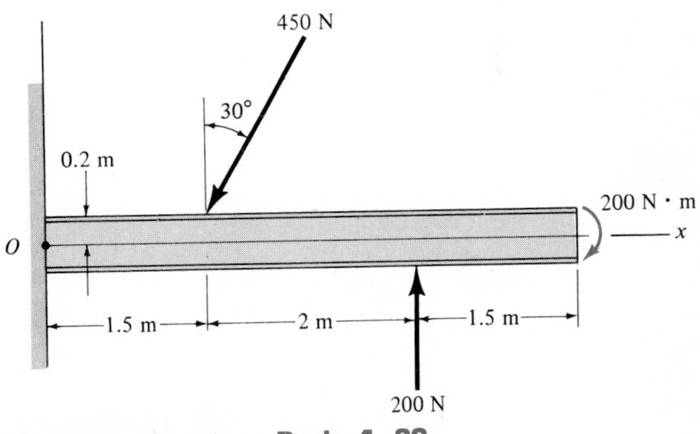

Prob. 4–29

4–30. Determine the magnitude and direction θ of force **F** and the couple moment **M** such that the loading system is equivalent to a resultant force of 600 N, acting vertically downward at O, and a clockwise couple moment of 4000 N · m.

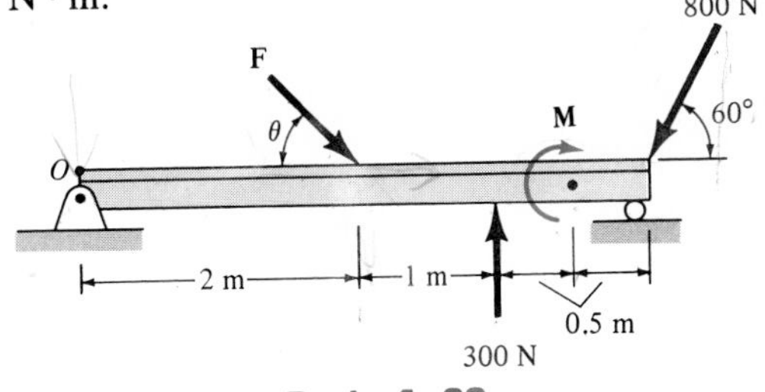

Prob. 4–30

4–31. Replace the loading system acting on the post by an equivalent force and couple system at point O.

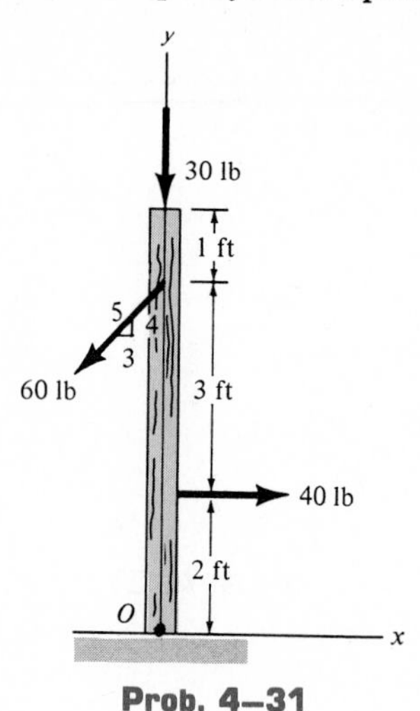

Prob. 4–31

***4–32.** Replace the force system by a single force resultant and specify its point of application measured from point O.

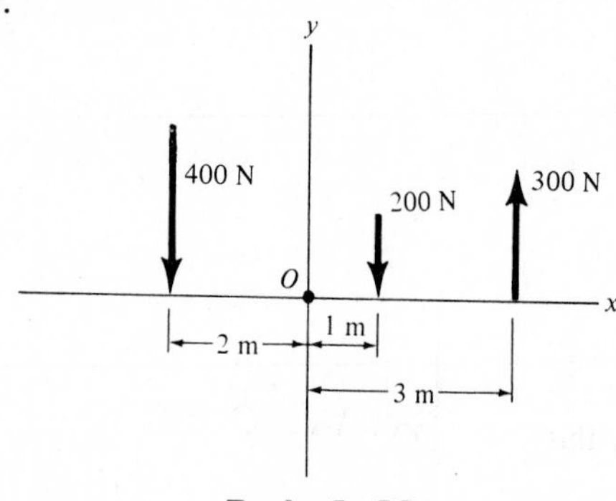

Prob. 4–32

4–33. The tires of a truck exert the forces shown on the deck of the bridge. Replace this system of forces by a single resultant force and specify its location measured from point A.

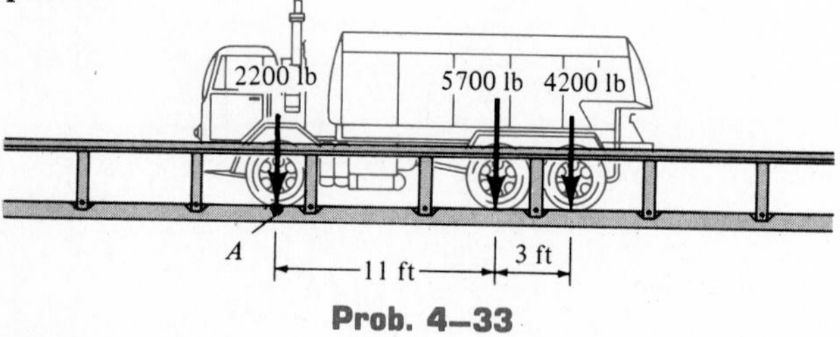

Prob. 4–33

4–34. Replace the force system by a single force resultant and specify its point of application measured from point O.

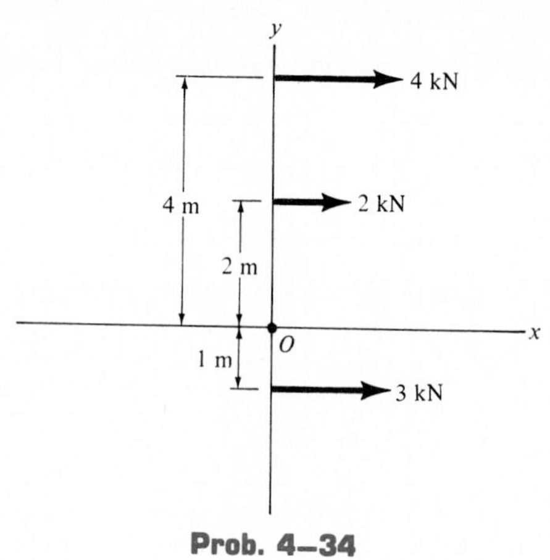

Prob. 4–34

4–35. The three forces acting on the water tank represent the effect of the wind. Replace this system by a *single resultant force* and specify its vertical location from point O.

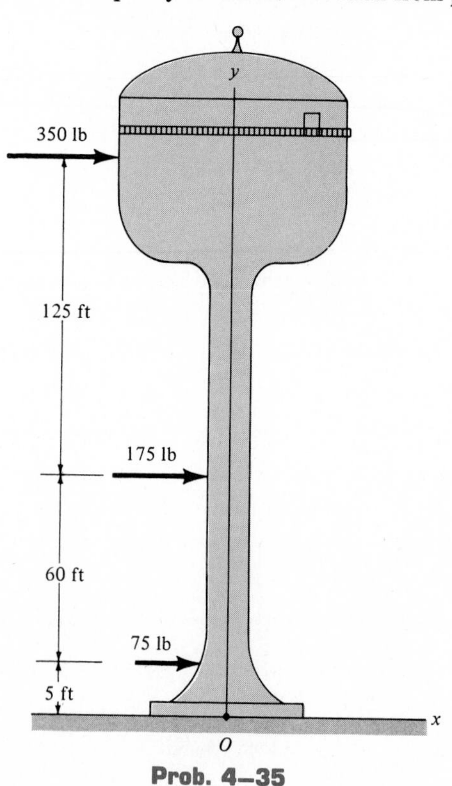

Prob. 4–35

***4–36.** The system of parallel forces acts on the top surface of the *Warren truss*. Determine the resultant force of the system and specify its location measured from point *A*.

4–37. The system of four forces acts on the roof truss. Determine the resultant force and specify its location along *AB*, measured from point *A*.

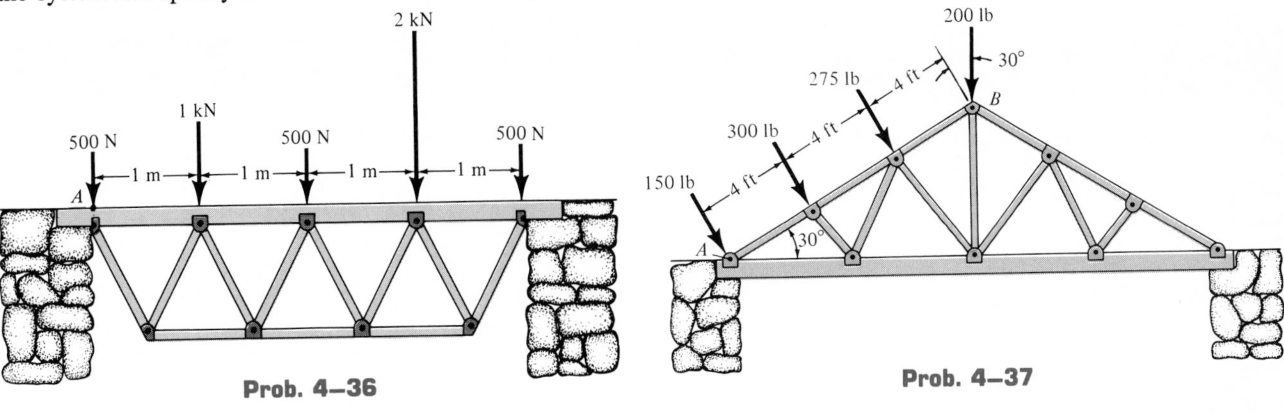

Prob. 4–36

Prob. 4–37

4.8 Reduction of a Simple Distributed Loading

In some situations the surface of a body may be subjected to distributed loadings such as those caused by wind, fluids, or simply the weight of material supported over the body's surface. If this loading is *symmetric* along one axis of a flat surface, it may be represented as a force per unit length distributed along the other axis.* For example, the distribution of load along the length of a beam is shown in Fig. 4–27*a*. The loading function $w = w(x)$ defines the intensity of the load at each point x measured in units of lb/ft or N/m. The direction of the load intensity is indicated by arrows shown on the diagrams. The entire loading on the beam is therefore a system of coplanar parallel forces, infinite in number and each acting at a specified point x on the beam. Using the methods of the previous section, we will now simplify this system to a single resultant force $\mathbf{F}_R$ and specify its location $\bar{x}$.

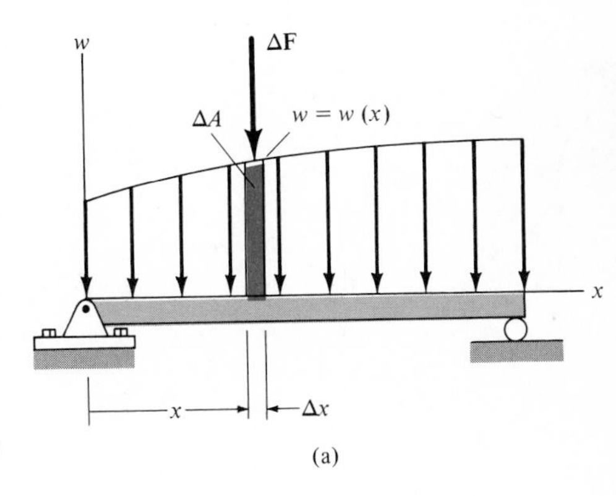

(a)

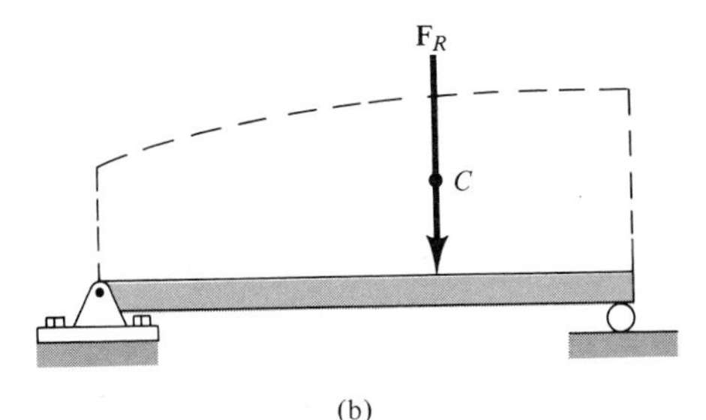

(b)

Fig. 4–27

Magnitude of Resultant Force. The magnitude of $\mathbf{F}_R$ is equivalent to the sum of all the forces in the system. Since w measures the force per unit length, then the force $\Delta\mathbf{F}$ acting on a thin segment Δx of the beam, Fig. 4–27*b*, is $\Delta F = w(x)[\Delta x] = \Delta A$. In other words, the magnitude of $\Delta\mathbf{F}$ is represented by the colored area segment ΔA under the loading curve. For the entire length of the beam, we require

$$+\downarrow F_R = \Sigma F; \qquad F_R = \Sigma w(x)\,\Delta x = \Sigma \Delta A = A \tag{4–4}$$

Hence, the magnitude of the resultant force is equal to the total area under the loading diagram $w = w(x)$.

*The more general case of a nonuniform surface loading acting on a body is considered in Sec. 9–5.

Location of Resultant Force. The location $\bar{x}$ of the line of action of $\mathbf{F}_R$ is determined by applying the principle of moments, which requires that the moment of $\mathbf{F}_R$ about point O, Fig. 4–27*b*, be equal to the sum of the moments of all the forces $\Delta\mathbf{F}$ about O, Fig. 4–27*a*. Since $\Delta\mathbf{F}$ produces a moment of $x\ \Delta F = x\ \Delta A$ about O, then for the entire beam,

$$\circlearrowleft + M_{R_O} = \Sigma M_O; \qquad \bar{x}F_R = \Sigma x(\Delta F)$$

or

$$\bar{x} = \frac{\Sigma x(\Delta F)}{F_R} = \frac{\Sigma x(\Delta A)}{A} \tag{4–5}$$

This equation locates the x coordinate for the geometric center of the area under the distributed-loading diagram. To better understand why this is so, consider the equation written in the form $\bar{x}(F_R) = \Sigma x(\Delta F)$ or $xA = \Sigma x\ \Delta A$.

Example 4–14

In each case, determine the magnitude and location of the resultant of the distributed load acting on the beams in Fig. 4–28.

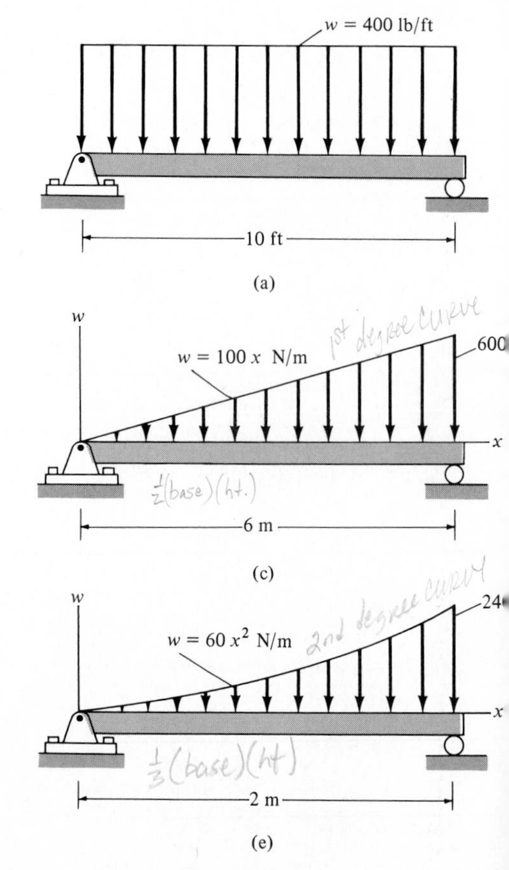

Solution

Uniform Loading. As indicated $w = 400$ lb/ft, which is constant over the entire beam, Fig. 4–28*a*. This loading forms a rectangle, the area of which is equal to the resultant force, Fig. 4–28*b*, i.e.,

$$F_R = (400\ \text{lb/ft})(10\ \text{ft}) = 4000\ \text{lb} \qquad \textit{Ans.}$$

The location of $\mathbf{F}_R$ passes through the center or centroid C of the area, so that

$$\bar{x} = 5\ \text{ft} \qquad \textit{Ans.}$$

Triangular Loading. Here the loading varies uniformly in intensity from 0 to 600 N/m, Fig. 4–28*c*. These values can be verified by substitution of $x = 0$ and $x = 6$ m into the loading function $w = 100x$ N/m. The area of this triangular loading is equal to $\mathbf{F}_R$, Fig. 4–28*d*. From the table on the inside back cover, $A = \frac{1}{2}bh$, so that

$$F_R = \tfrac{1}{2}(6\ \text{m})(600\ \text{N/m}) = 1800\ \text{N} \qquad \textit{Ans.}$$

The line of action of $\mathbf{F}_R$ passes through the centroid C of the triangle. Using the table on the inside back cover, this point lies at a distance of one third the length of the beam, measured from the right side. Hence,

$$\bar{x} = 6\ \text{m} - \tfrac{1}{3}(6\ \text{m}) = 4\ \text{m} \qquad \textit{Ans.}$$

By comparison, it can be seen that the "moment" of the entire area, $\bar{x}A$, about O must be equal to the sum of the moments of all the area strips, ΔA, about O, $\Sigma x\,\Delta A$. In this regard the total area (or $\mathbf{F}_R$) can be thought of as concentrated at the geometric center, a distance $\bar{x}$ from O. We define the geometric center of an area as the *centroid* C of the area. Therefore, *the resultant force has a line of action which passes through the centroid* C *(geometric center) of the area defined by the distributed loading diagram* $w(x)$, *Fig. 4–27b.*

Detailed treatment of the techniques for computing the centroid of an area is given in Chapter 9. In most cases, the distributed-loading diagram is of a simple geometric form so that the centroids for such common shapes can be obtained directly from tables such as the one given on the inside back cover of this book. The following examples numerically illustrate the concepts developed above using this table.

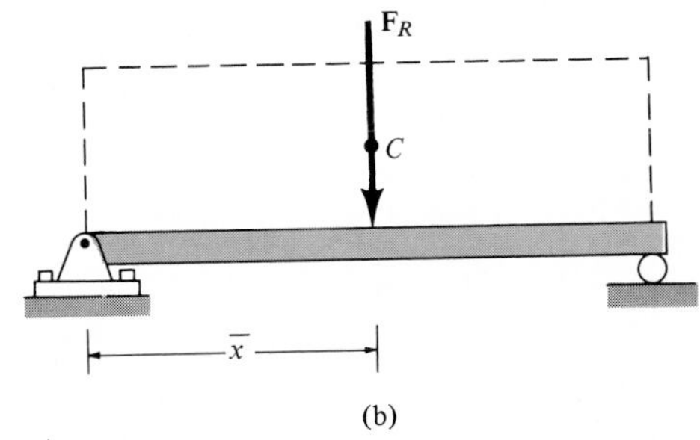

Parabolic Loading. In this case the loading function is described by $w = 60x^2$, so that $w = 0$ at $x = 0$ and $w = 240$ N/m at $x = 2$ m as shown in Fig. 4–28*e*. Using the table on the inside back cover, $A = ab/3$, so that we have

$$F_R = \tfrac{1}{3}(240\text{ N/m})(2\text{ m}) = 160\text{ N} \qquad \textit{Ans.}$$

Since $\bar{x} = \frac{3}{4}b$ for the area, Fig. 4–28*f*, then

$$\bar{x} = \tfrac{3}{4}(2\text{ m}) = 1.5\text{ m} \qquad \textit{Ans.}$$

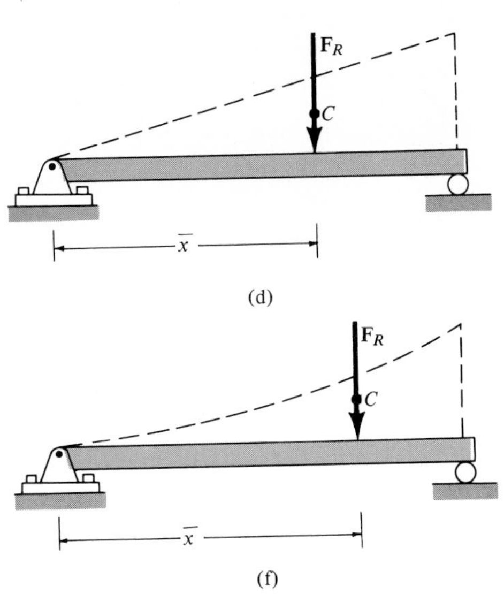

Fig. 4–28

Example 4–15

Determine the magnitude and location of the resultant of the distributed load acting on the beam shown in Fig. 4–29*a*.

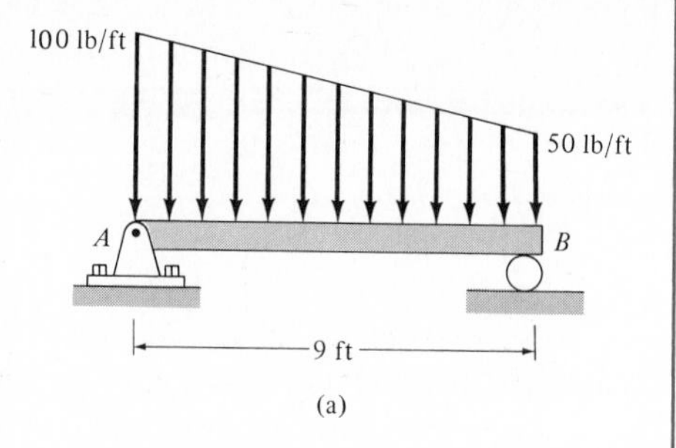

(a)

Solution

The area of the loading diagram is a *trapezoid*, and therefore the solution can be obtained directly from the area and centroid formulas for a trapezoid listed on the inside back cover. Since these formulas are not easily remembered, instead we will solve this problem by using "composite" areas. In this regard, we can divide the trapezoidal loading into a rectangular and triangular loading as shown in Fig. 4–29*b*. The magnitude of the force represented by each of these loadings is equal to its associated *area*,

$$F_1 = \tfrac{1}{2}(9 \text{ ft})(50 \text{ lb/ft}) = 225 \text{ lb}$$
$$F_2 = (9 \text{ ft})(50 \text{ lb/ft}) = 450 \text{ lb}$$

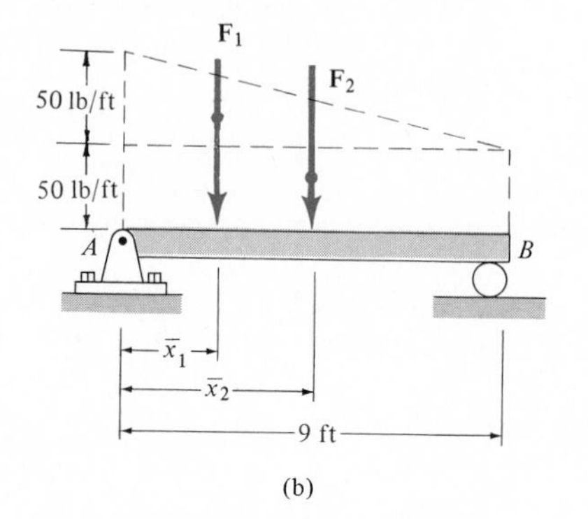

(b)

The lines of action of these parallel forces act through the *centroid* of their associated areas and therefore intersect the beam at

$$\bar{x}_1 = \tfrac{1}{3}(9 \text{ ft}) = 3 \text{ ft}$$
$$\bar{x}_2 = \tfrac{1}{2}(9 \text{ ft}) = 4.5 \text{ ft}$$

The two parallel forces $\mathbf{F}_1$ and $\mathbf{F}_2$ can be reduced to a single resultant $\mathbf{F}_R$. The magnitude of $\mathbf{F}_R$ is

$+\downarrow F_R = \Sigma F;$ $\qquad F_R = 225 + 450 = 675 \text{ lb}$ ***Ans.***

Applying the principle of moments with reference to point A, Fig. 4–29*b* and *c*, we can define the location of $\mathbf{F}_R$,

$\circlearrowright + M_{R_A} = \Sigma M_A;$ $\qquad \bar{x}(675) = 3(225) + 4.5(450)$

$$\bar{x} = 4.0 \text{ ft}$$ ***Ans.***

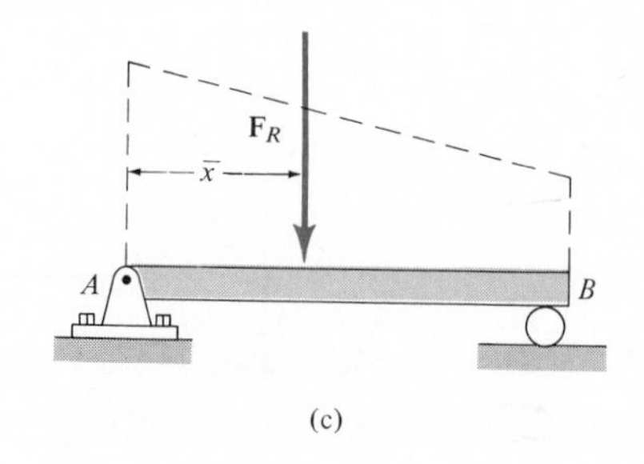

(c)

Note: The trapezoidal area can also be divided into two triangular areas as shown in Fig. 4–29*d*. In this case

$$F_1 = \tfrac{1}{2}(9 \text{ ft})(100 \text{ lb/ft}) = 450 \text{ lb}$$
$$F_2 = \tfrac{1}{2}(9 \text{ ft})(50 \text{ lb/ft}) = 225 \text{ lb}$$

and

$$\bar{x}_1 = \tfrac{1}{3}(9 \text{ ft}) = 3 \text{ ft}$$
$$\bar{x}_2 = \tfrac{1}{3}(9 \text{ ft}) = 3 \text{ ft}$$

Using these results, show that $F_R = 675$ lb and $\bar{x} = 4.0$ ft as obtained above.

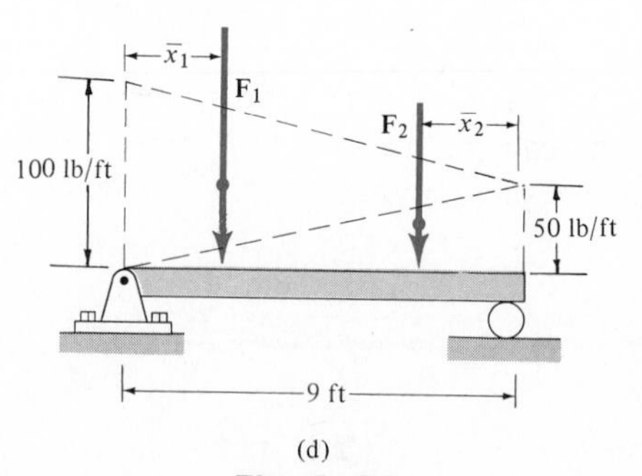

(d)

Fig. 4–29

Problems

4–38. Replace the loading by an equivalent force and couple moment acting at point O.

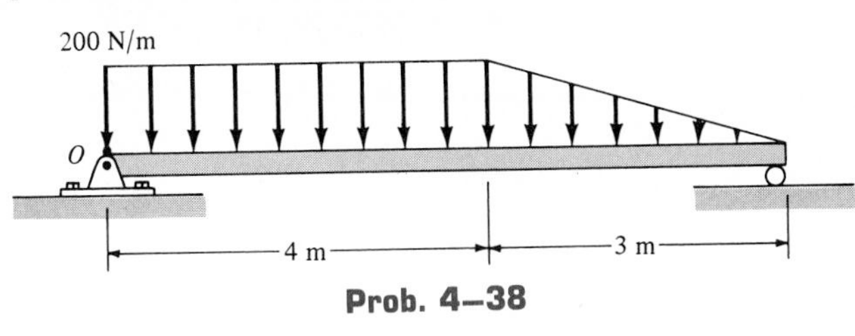

Prob. 4–38

4–39. Replace the loading by an equivalent force and couple moment acting at point O.

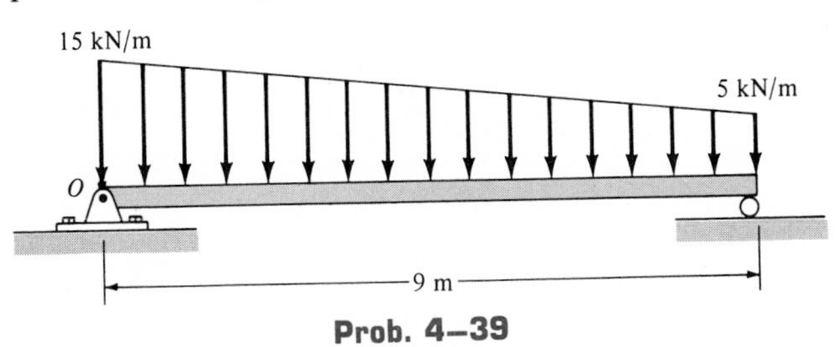

Prob. 4–39

***4–40.** Replace the loading by an equivalent force and couple moment acting at point O.

ANSWER: Ma = 1350 ft# ↺ F = 75# ↓

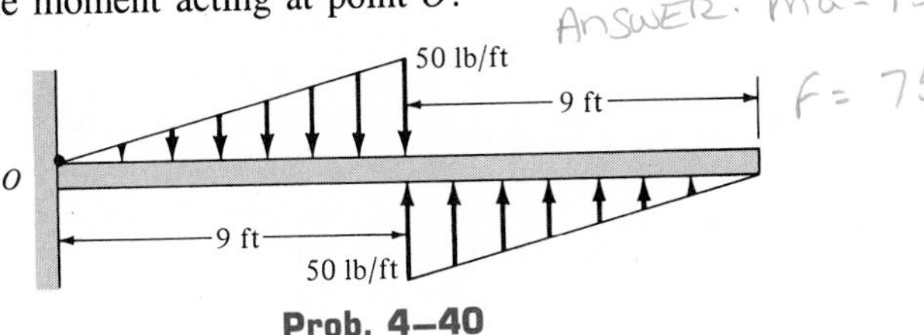

Prob. 4–40

4–41. The masonry support creates the loading distribution acting on the end of the beam. Simplify this load to a single resultant force and specify its location measured from point O.

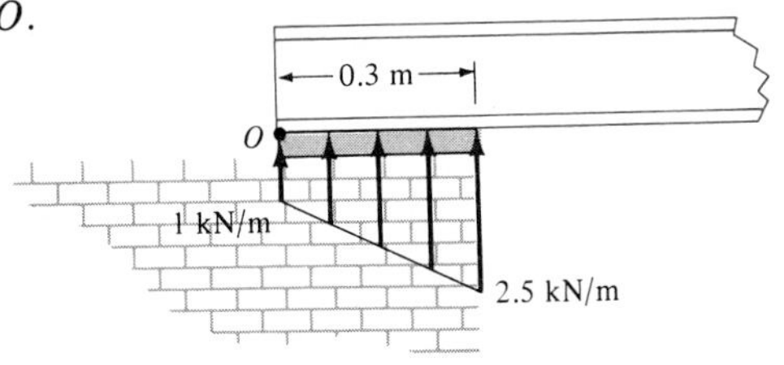

Prob. 4–41

150.45 ft to Right of A

4–42. The distribution of loading along the deck of a bridge is approximated as shown. Replace this loading by a single resultant force and specify its location on the bridge, measured from point A.

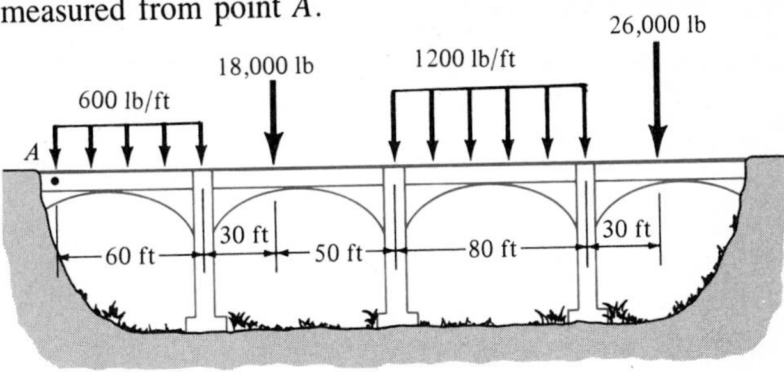

Prob. 4–42

4–43. The distributed loadings of soil pressure on the sides and bottom of a spread footing are shown. Simplify this system to a single resultant force and couple moment acting at A.

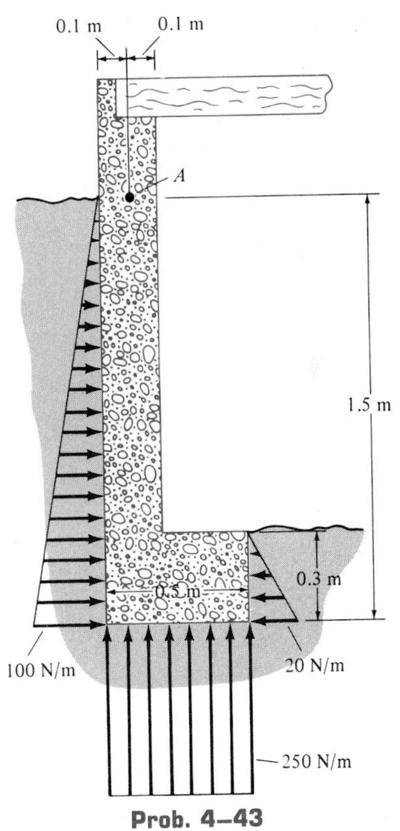

Prob. 4–43

***4–44.** Determine the length b of the uniform load and its position a on the beam such that the resultant force and resultant couple acting on the beam are zero.

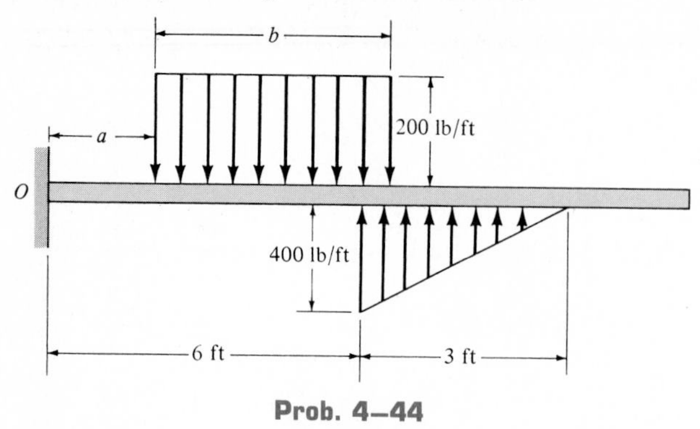

Prob. 4–44

4–45. The gabled frame is subjected to the uniform wind loadings shown. Note that member AB is subjected to a "suction" since it is on the leeward side of the roof. Reduce this loading to a single resultant force and couple moment acting at D.

4–46. The bricks on top of the beam and the supports at the bottom create the distributed loading shown in the second figure. Determine the required intensity w and dimension d of the right support so that the resultant force and couple moment about point A of the system at both zero.

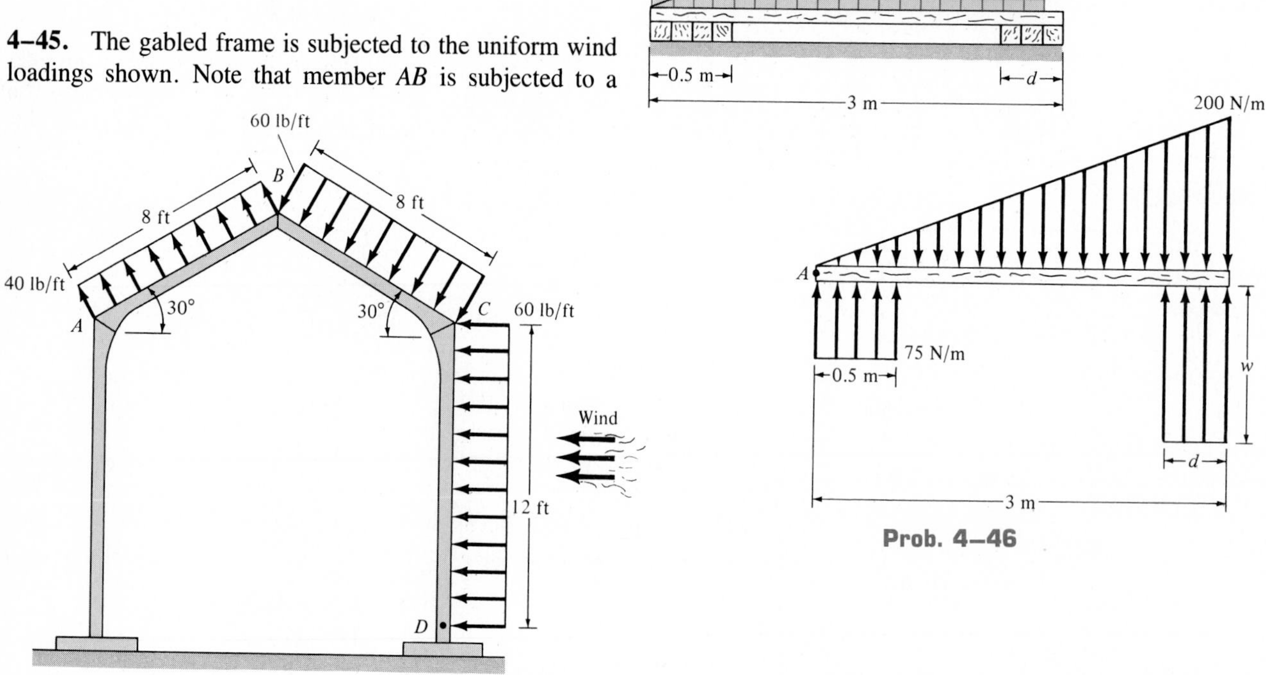

Prob. 4–45

Prob. 4–46

Three-Dimensional Force Systems

4.9 Moment of a Force in Three Dimensions

Recall that when the moment of a force is computed about an axis the moment and the axis are always perpendicular to the plane containing the force **F** and moment arm d. See Fig. 4–2. The magnitude of the moment is determined using the formula $M_O = Fd$. In three dimensions, however, it is often difficult to determine the moment arm d, since it must be the perpendicular distance from a point (O) on the axis to the line of action of the force, Fig. 4–30. An easier method to determine $\mathbf{M}_O$ consists of first finding the components of the moment about the x, y, z axes and then computing the *magnitude* of the moment from

$$M_O = \sqrt{M_x^2 + M_y^2 + M_z^2} \tag{4–6}$$

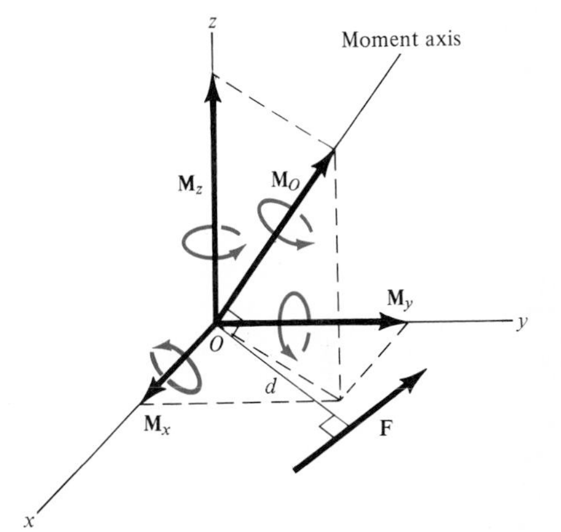

Fig. 4–30

and its *coordinate direction angles* from

$$\alpha = \cos^{-1}\left(\frac{M_x}{M_O}\right)$$
$$\beta = \cos^{-1}\left(\frac{M_y}{M_O}\right) \tag{4–7}$$
$$\gamma = \cos^{-1}\left(\frac{M_z}{M_O}\right)$$

This procedure is a direct application of the *principle of moments, which states that the moment of a force about a point lying on the moment axis is equal to the sum of the moments of the force's components about the point.* (See Sec. 4–2).

VARignon's thm.

If the origin of coordinates, O, is chosen as a point on the moment axis, Fig. 4–30, then the moment's components $\mathbf{M}_x$, $\mathbf{M}_y$, $\mathbf{M}_z$ can always be computed provided both the x, y, z components of the force and the x, y, z coordinates of a point on the line of action of the force are known. To illustrate how this can be done, let us first consider finding the moment components of $\mathbf{F}_x$, $\mathbf{F}_y$, and $\mathbf{F}_z$ in Fig. 4–31. In Fig. 4–31*a* the force $\mathbf{F}_x$ creates no moment (or tendency for rotation) about the x axis since the force is *parallel* to the x axis. The perpendicular distance (moment arm) from the line of action of $\mathbf{F}_x$ to the y axis is z; hence, the moment of $\mathbf{F}_x$ about the y axis is $M_y = F_x z$. This moment is positive since, by the right-hand rule, the tendency for rotation as indicated by the curl of the fingers directs the thumb in the positive y direction. In a similar manner, the moment of $\mathbf{F}_x$ about the z axis is $M_z = -F_x y$. Here the moment is negative as illustrated by the right-hand rule, shown in Fig. 4–31*a*.

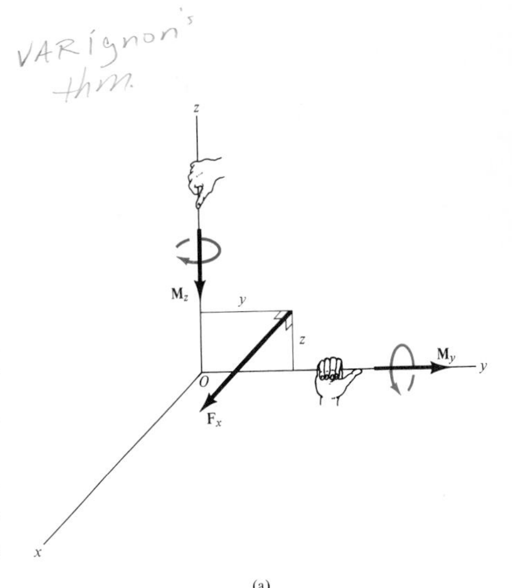

(a)

Fig. 4–31a

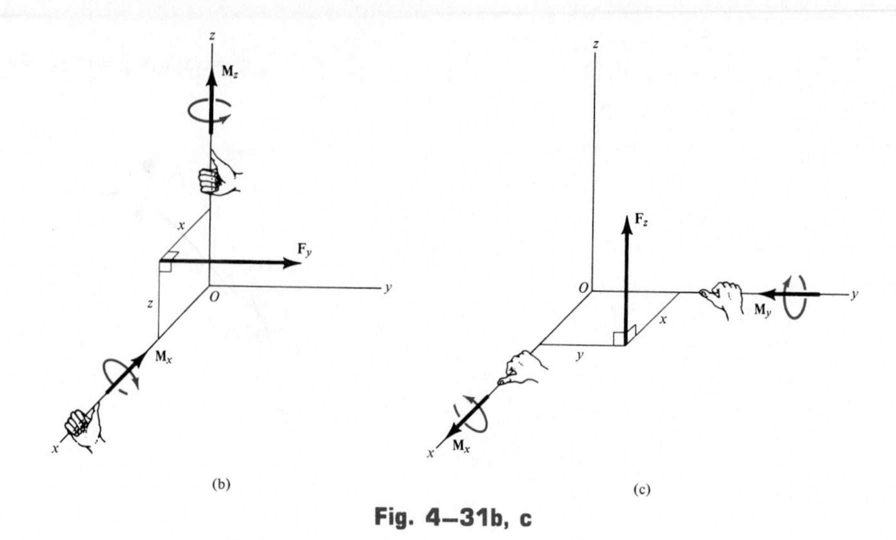

Fig. 4–31b, c

In Fig. 4–31*b* force $\mathbf{F}_y$ creates no moment about the *y* axis since the force is parallel to the *y* axis. Using the coordinates *x* and *z*, which are measured to a point lying on the line of action of $\mathbf{F}_y$, the moments about the *x* and *z* axes are $M_x = -F_y z$ and $M_z = F_y x$, respectively. As before, the signs of these moments are determined by the right-hand rule.

In Fig. 4–31*c*, $\mathbf{F}_z$ creates no moment about the *z* axis. Why? Using the moment arms *x* and *y*, the moments about the *x* and *y* axes are $M_x = F_z y$ and $M_y = -F_z x$, respectively. These results should *not* be memorized; rather, it should be clearly understood how each is obtained by using the right-hand rule and the basic definition of the moment of a force.

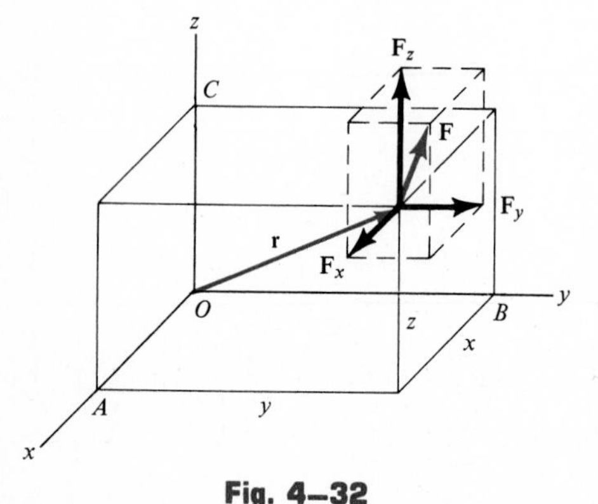

Fig. 4–32

The above results can now be combined to illustrate how to compute the moment components for the general case of a force $\mathbf{F}$ having components $\mathbf{F}_x$, $\mathbf{F}_y$, $\mathbf{F}_z$ and acting at a point *x*, *y*, *z* in space, Fig. 4–32. The moment of $\mathbf{F}$ about the *x* axis is computed from the moments of forces $\mathbf{F}_z$ and $\mathbf{F}_y$. Treating these forces as sliding vectors, the moment of $\mathbf{F}_z$ about the *x* axis is $+F_z y$. (By the right-hand rule this component acts along the positive *x* axis.) Likewise, $\mathbf{F}_y$ contributes a moment component of $-F_y z$. Force component $\mathbf{F}_x$ does *not* create a moment about the *x* axis since this force is *parallel* to the *x* axis. Hence, the *x* component of the moment of $\mathbf{F}$ is

$$M_x = F_z y - F_y z$$

As an exercise, show that

$$M_y = F_x z - F_z x$$
$$M_z = F_y x - F_x y$$

These results should not be memorized; instead, one should clearly understand how each component was obtained. The following examples numerically illustrate application of the same technique.

Example 4–16

The window is subjected to a 200-N force at its corner A as shown in Fig. 4–33a. Compute the moment of this force about the window's hinged axis y'. The force lies in the x-y plane.

Solution

If the force is resolved into its rectangular components, Fig. 4–33b, it is noted that only $\mathbf{F}_x$ produces a moment about the y' axis. (The line of action of $\mathbf{F}_y$ is *parallel* to this axis, and hence the moment about y' is zero.) The moment arm is the perpendicular distance from the line of action of $\mathbf{F}_x$ to the axis. This distance is $d_{CA} = 0.3$ m. Hence, the magnitude of moment is $M_{y'} = F_x d_{CA} = (200 \sin 60° \text{ N})(0.3 \text{ m}) = 52.0 \text{ N} \cdot \text{m}$. By the right-hand rule, Fig. 4–33b, $\mathbf{F}_x$ tends to rotate the window out and around so that the thumb points in the negative y' direction. Thus,

$$M_{y'} = -52 \text{ N} \cdot \text{m} \qquad \textit{Ans.}$$

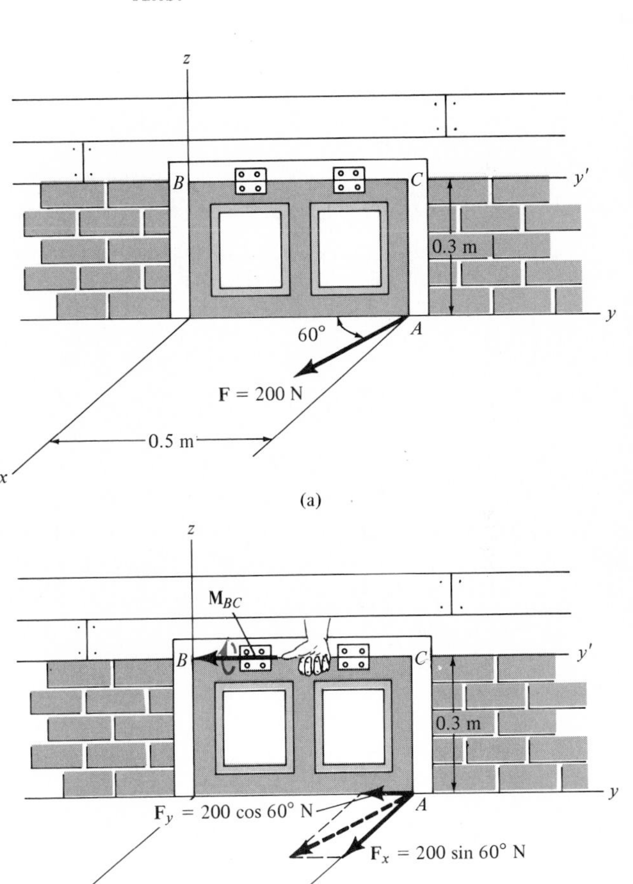

Fig. 4–33

Example 4–17

Determine the magnitude of the moment about point O of the three components of the force **F** acting at the end A of the pipe in Fig. 4–34*a*. Specify the coordinate direction angles of the moment axis of the resultant moment.

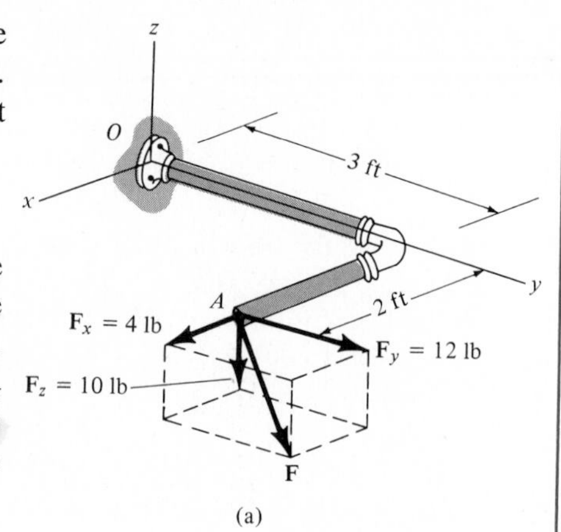

(a)

Solution

The coordinate or perpendicular distances from the line of action of the force components to the x, y, z axes are indicated in the figure. Using the right-hand rule, we will consider each moment component separately.

For $\mathbf{M}_x$: Note that $\mathbf{F}_x$ does not create a moment about the x axis since $\mathbf{F}_x$ is parallel to the x axis. Also, $\mathbf{F}_y$ does not create a moment about the x axis since the line of action of $\mathbf{F}_y$ passes through a point lying on the axis. The moment of $\mathbf{F}_z$ about the x axis is $-(10 \text{ lb})(3 \text{ ft})$. Thus,

$$M_x = 0 + 0 - (10 \text{ lb})(3 \text{ ft}) = -30 \text{ lb} \cdot \text{ft}$$

For $\mathbf{M}_y$: $\mathbf{F}_x$ and $\mathbf{F}_y$ do not create moments about the y axis. Why? The moment of $\mathbf{F}_z$ about the y axis is $+(10 \text{ lb})(2 \text{ ft})$. Thus

$$M_y = 0 + 0 + (10 \text{ lb})(2 \text{ ft}) = 20 \text{ lb} \cdot \text{ft}$$

For $\mathbf{M}_z$: Here $\mathbf{F}_x$ creates a moment about the z axis of $-(4 \text{ lb})(3 \text{ ft})$ and $\mathbf{F}_y$ creates a moment of $+(12 \text{ lb})(2 \text{ ft})$. $\mathbf{F}_z$ does not contribute moment about the z axis. Thus,

$$M_z = -(4 \text{ lb})(3 \text{ ft}) + (12 \text{ lb})(2 \text{ ft}) + 0 = 12 \text{ lb} \cdot \text{ft}$$

Using Eq. 4–6, the magnitude of the resultant moment at O is thus

$$M_O = \sqrt{(-30)^2 + (20)^2 + (12)^2} = 38.0 \text{ lb} \cdot \text{ft} \qquad \textit{Ans.}$$

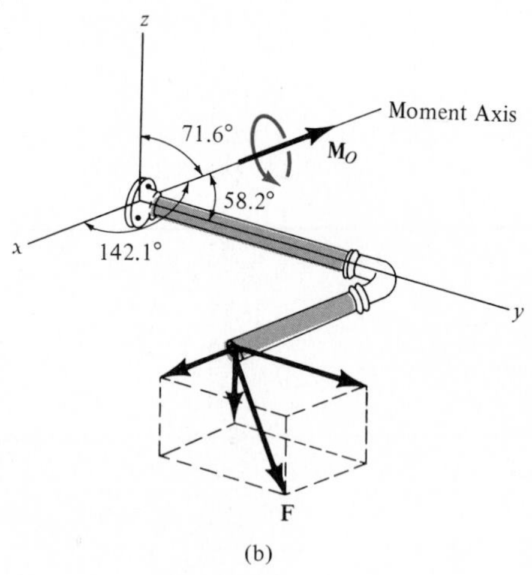

(b)

Fig. 4–34

The coordinate direction angles of the moment axis are determined from Eqs. 4–7, i.e.,

$$\alpha = \cos^{-1}\left(\frac{-30}{38.0}\right) = 142.1° \qquad \textit{Ans.}$$

$$\beta = \cos^{-1}\left(\frac{20}{38.0}\right) = 58.2° \qquad \textit{Ans.}$$

$$\gamma = \cos^{-1}\left(\frac{12}{38.0}\right) = 71.6° \qquad \textit{Ans.}$$

The results are shown in Fig. 4–34*b*. Realize that when the moment M_O of a force is to be determined about a point, this actually implies finding the moment of the force about a *point lying on the moment axis*.

Example 4–18

The pole in Fig. 4–35*a* is subjected to a 60-N force that is directed from C to B. Determine the magnitude of the moment of this force about point A.

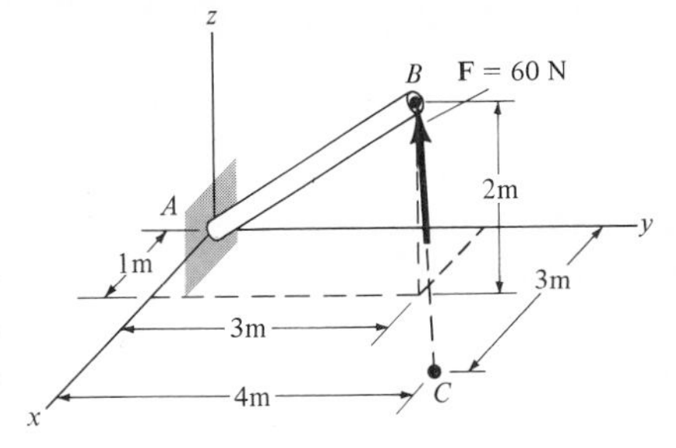

Solution

We will solve this problem by finding the x, y, z components of the moment at A. To do this the x, y, z components of the force must be determined first. Since a position vector, directed from $C(3, 4, 0)$ to $B(1, 3, 2)$, has components of $r_{CB_x} = -2$ m, $r_{CB_y} = -1$ m, $r_{CB_z} = 2$ m, and $r_{CB} = 3$ m, then

$$F_x = 60\left(\frac{-2}{3}\right) = -40 \text{ N},\ F_y = 60\left(\frac{-1}{3}\right) = -20 \text{ N},\ F_z = 60\left(\frac{2}{3}\right) = 40 \text{ N}$$

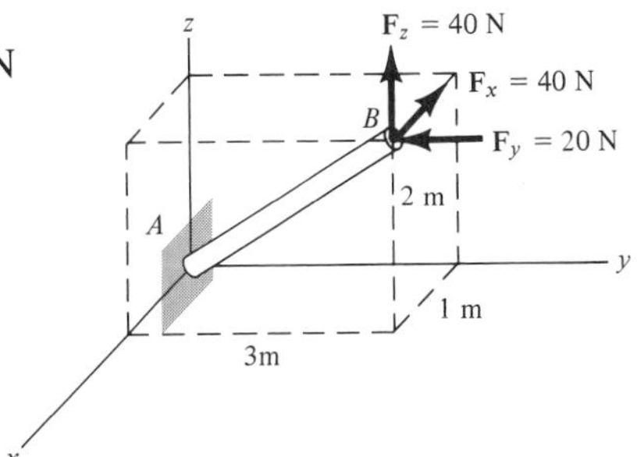

(b)

These components acting in their appropriate directions are shown on the pole in Fig. 4–35*b*. The moment components of these forces about the x, y, and z axes can now be computed. The moment of $\mathbf{F}_x$ about the x axis is zero since $\mathbf{F}_x$ is parallel to the x axis. $\mathbf{F}_y$ contributes a moment of $M_x = +(20 \text{ N})(2 \text{ m})$ and the moment of $\mathbf{F}_z$ is $M_x = +(40 \text{ N})(3 \text{ m})$. The total moment is thus

$$M_x = +(20 \text{ N})(2 \text{ m}) + (40 \text{ N})(3 \text{ m}) = 160 \text{ N} \cdot \text{m}$$

In a similar manner,

$$M_y = -(40 \text{ N})(2 \text{ m}) - (40 \text{ N})(1 \text{ m}) = -120 \text{ N} \cdot \text{m}$$
$$M_z = (40 \text{ N})(3 \text{ m}) - (20 \text{ N})(1 \text{ m}) = 100 \text{ N} \cdot \text{m}$$

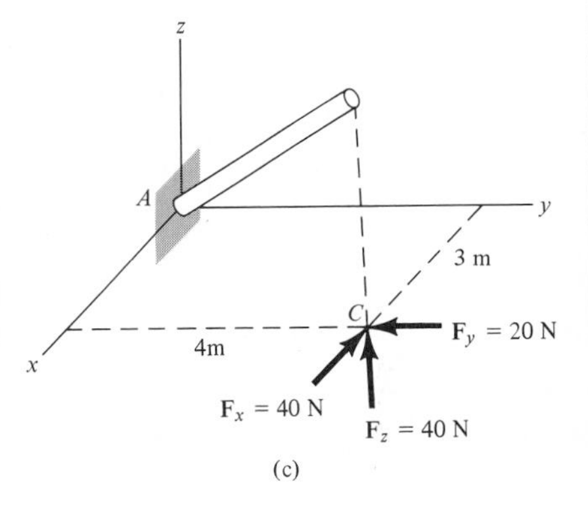

(c)

We can also solve for these components by using the principle of transmissibility and place the force (or its components) at point C, Fig. 4–35*c*. As usual $\mathbf{F}_x$ creates no moment about the x axis. Furthermore, the moment of $\mathbf{F}_y$ about the x axis is also zero since its line of action passes through a point on the axis. The moment of $\mathbf{F}_z$ about the x axis is

$$M_x = +(40 \text{ N})(4 \text{ m}) = +160 \text{ N} \cdot \text{m}$$

In a similar manner,

$$M_y = -(40 \text{ N})(3 \text{ m}) = -120 \text{ N} \cdot \text{m}$$
$$M_z = +(40 \text{ N})(4 \text{ m}) - (20 \text{ N})(3 \text{ m}) = +100 \text{ N} \cdot \text{m}$$

Thus,

$$M_A = \sqrt{(160)^2 + (-120)^2 + (100)^2} = 223.6 \text{ N} \cdot \text{m} \qquad \textbf{\textit{Ans.}}$$

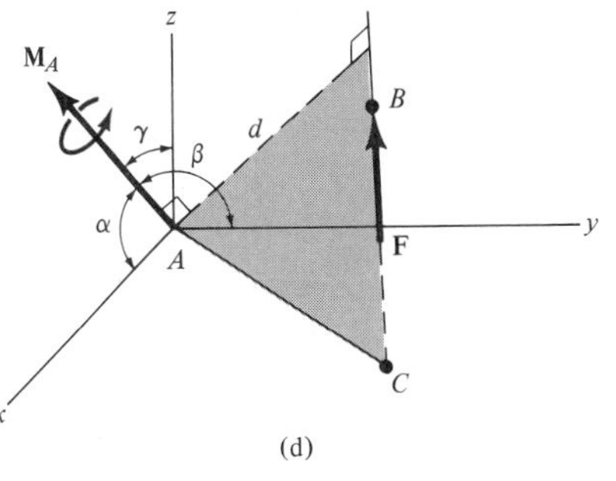

(d)

Fig. 4–35

Note that $\mathbf{M}_A$ acts perpendicular to the shaded plane containing $\mathbf{F}$ and the moment arm d, Fig. 4–35*d*. (How would you find the coordinate direction angles $\alpha = 44.3°$, $\beta = 122.5°$, $\gamma = 63.4°$?) Had this problem been worked using $M_A = Fd$, notice the difficulty that would arise in obtaining the moment arm d.

Problems

4–47. Determine the x, y, z components of the moment of the force at A about point P. The force has components of $F_x = 40$ N, $F_y = 60$ N, $F_z = -20$ N.

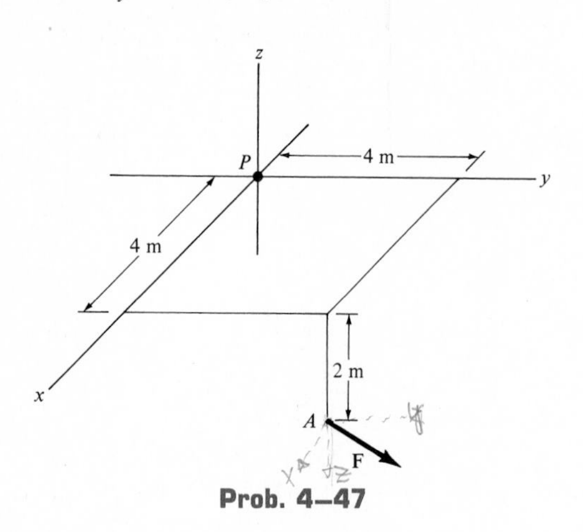

Prob. 4–47

***4–48.** Determine the x, y, z components of the moment of the force at A about point P. The force has components of $F_x = -6$ kN, $F_y = 4$ kN, $F_z = 3$ kN.

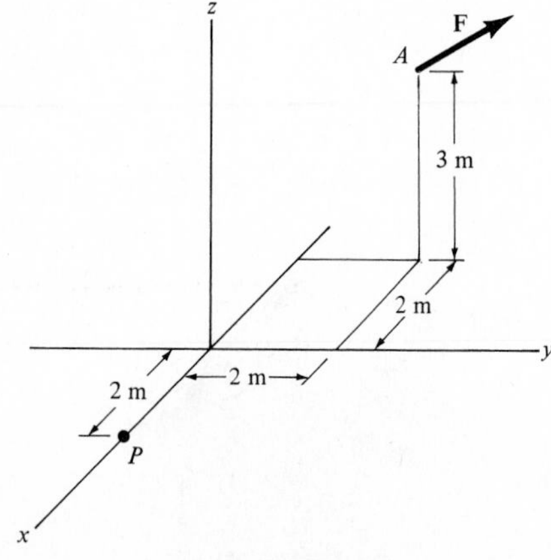

Prob. 4–48

4–49. Determine the x, y, z components of the moment of the 300-lb force at A about point P.

Prob. 4–49

4–50. The x-ray machine is used for medical diagnosis. If the equipment and housing at C have a mass of 150 kg and a mass center at G, determine the x, y, z components of the moment of its weight about point O when it is in the position shown.

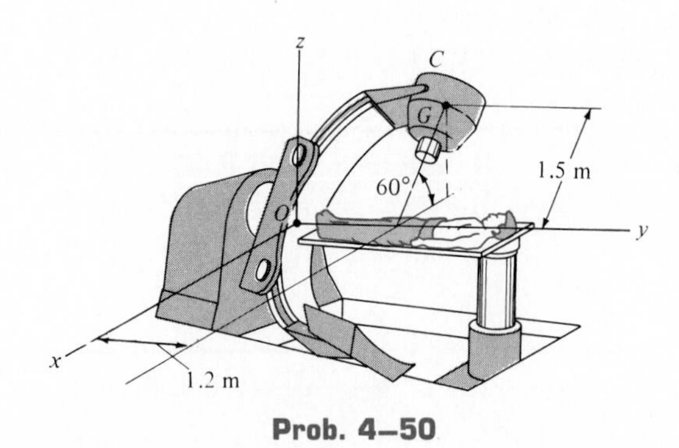

Prob. 4–50

4–51. The man at B exerts a force of 140 N on the rope attached to the end of the beam as shown. Determine the magnitude of the moment of this force at the wall A.

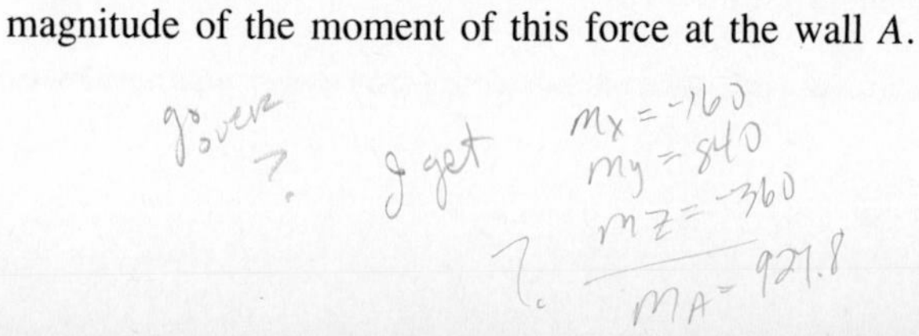

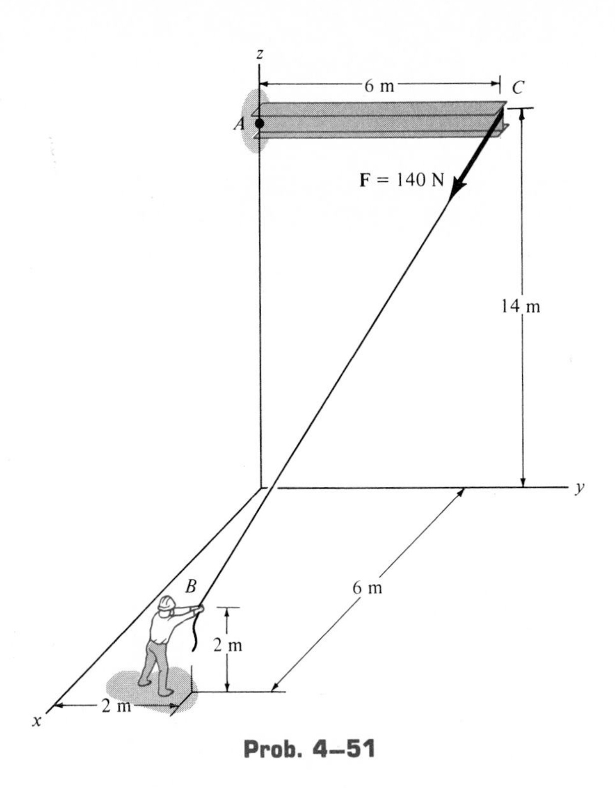

Prob. 4–51

4–53. The pole supports a 22-lb traffic light. Determine the magnitude of the moment of the weight of the traffic light about the base of the pole at A.

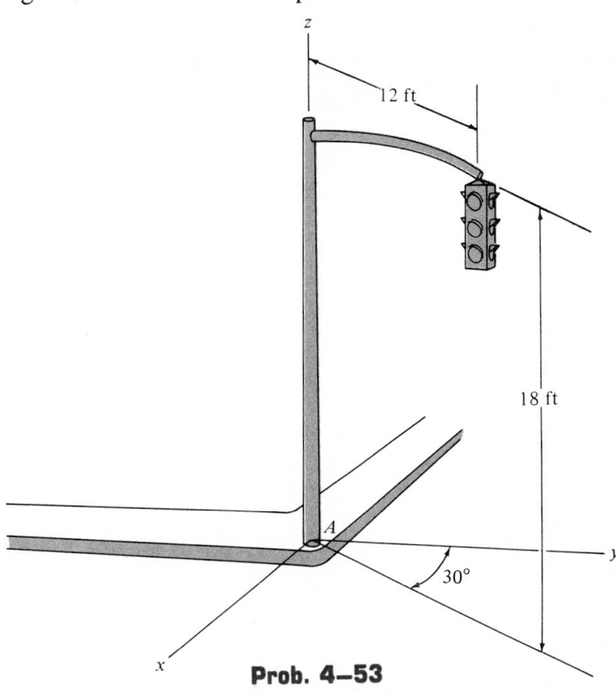

Prob. 4–53

***4–52.** The curved beam has a radius of 4 ft and is supported by the cable AB, which exerts a force of 80 lb on the beam. Determine the magnitude of the moment of this force about point C.

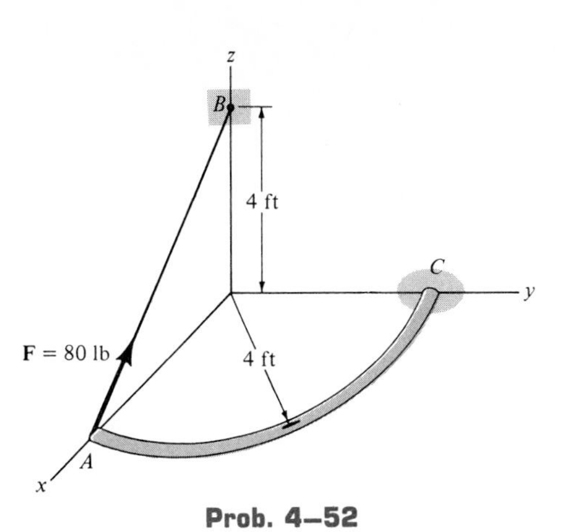

Prob. 4–52

4–54. Each of the four traffic lights has a weight of 25 lb. Determine the magnitude of the resultant moment created by the weight of all of them at the base of the post A.

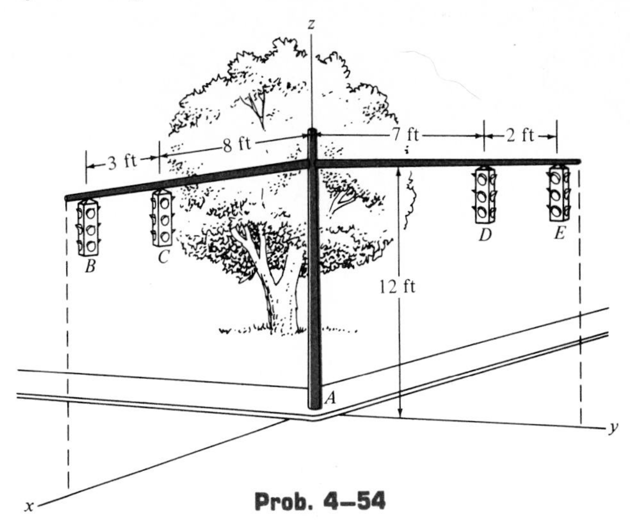

Prob. 4–54

4–55. The 5-m-long boom AB lies in the y-z plane. If a tension force of $F = 600$ N acts in cable BC, determine the moment of this force about point A.

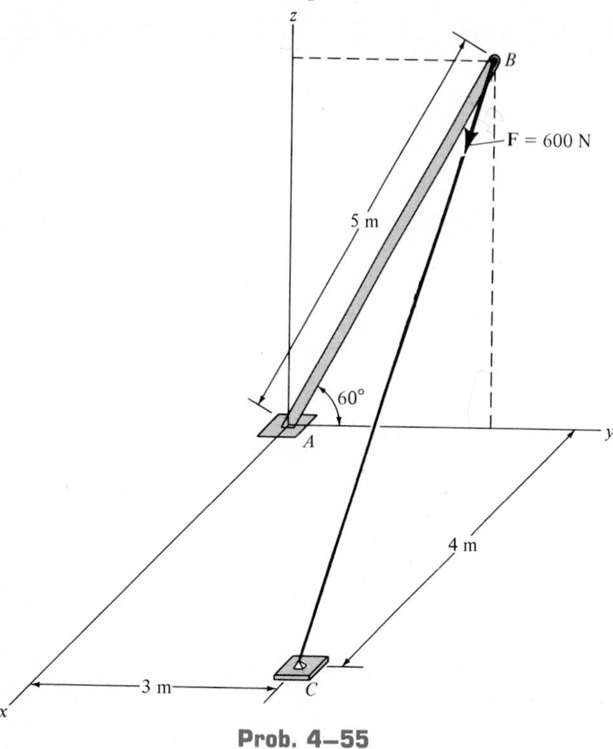

Prob. 4–55

*4–56. The hatch door is held open by the two force components $\mathbf{F}_y$ and $\mathbf{F}_z$ shown. Determine the magnitudes of the moments of these forces about point O.

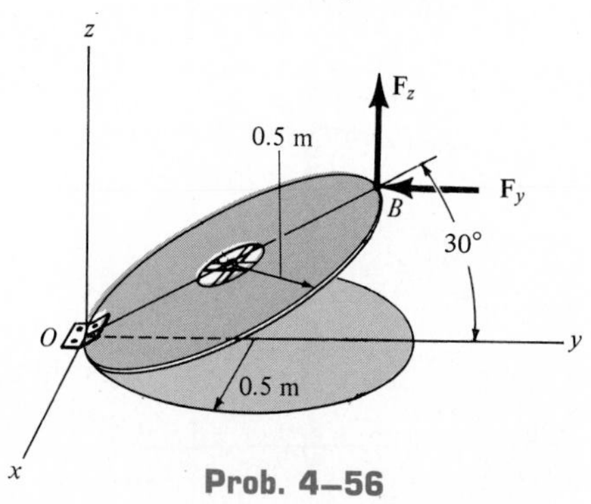

Prob. 4–56

4–57. The hood of the automobile is supported by the strut AB, which exerts a force of 12 lb on the hood. Determine the moment of this force about the hinged axis y.

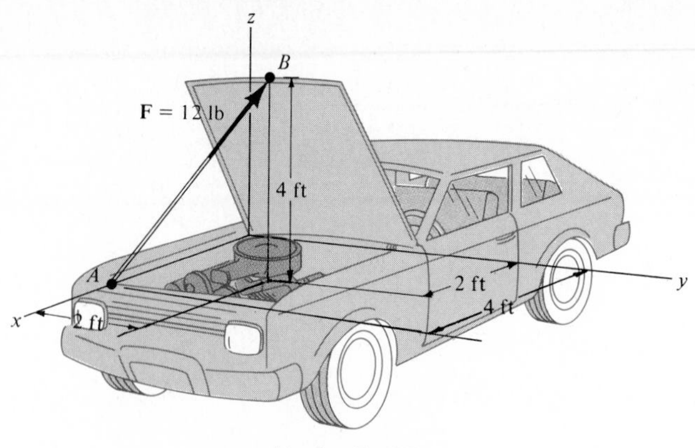

Prob. 4–57

4–58. The A frame is being hoisted into an upright position by the *vertical* towing force of $F = 80$ lb. Determine the moment of this force about the x' axis passing through points A and B when the frame is in the position shown.

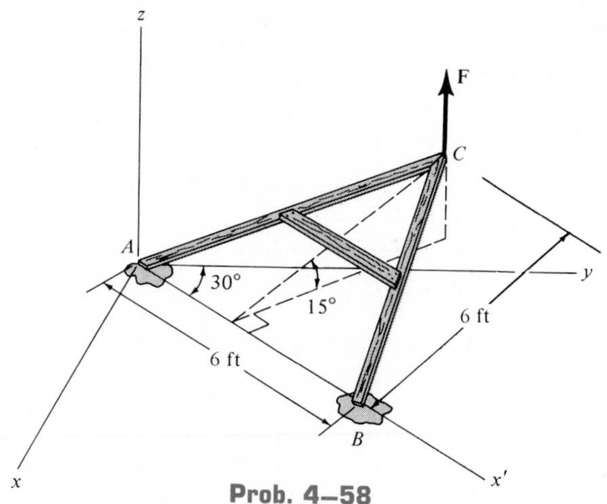

Prob. 4–58

4–59. The hatch door is held open by the two force components $\mathbf{F}_y$ and $\mathbf{F}_z$ shown. Determine the moments of these forces about the x axis of the hinge located at O.

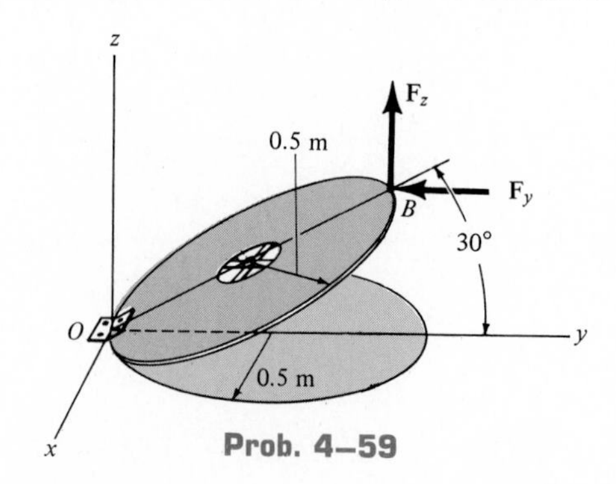

Prob. 4–59

***4–60.** A force of 50 N is applied to the handle of the door as shown. Determine the moment of this force about the hinged axis z. Neglect the size of the doorknob.

Prob. 4–60

4–61. A vertical force of $F = 60$ N is applied to the handle of the pipe wrench. Determine the moment that this force exerts along the axis AB (x axis) of the pipe assembly. Both the wrench and pipe assembly ABC lie in the x-y plane.

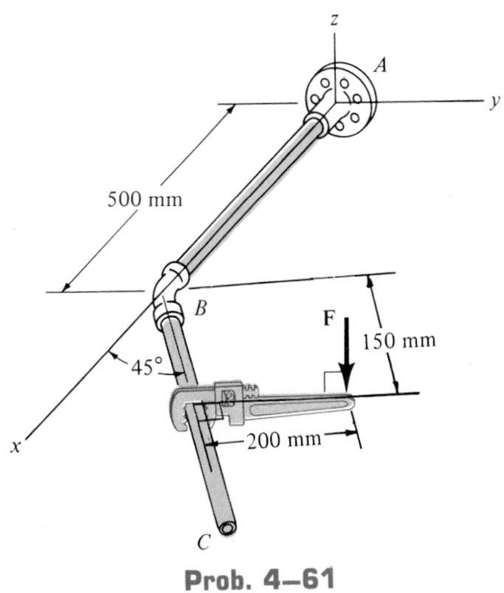

Prob. 4–61

4–62. Determine the magnitude of the vertical force **F** acting on the handle of the wrench so that this force produces a component of moment along the AB axis (x axis) of the pipe assembly of $\mathbf{M}_{Ax} = -5\ \text{N} \cdot \text{m}$. Both the pipe assembly ABC and the wrench lie in the x-y plane.

Moment of a Couple in Three Dimensions 4.10

Recall that a couple consists of two noncollinear parallel forces that have the same magnitude but opposite directions. The moment produced by a couple is equivalent to the sum of the moments of its two forces, computed about *any* arbitrary point O in space. In this regard, it is usually best to choose point O lying on the line of action of one of the forces, since then the moment of this force would be zero about point O and only the moment of the other force would have to be computed, Fig. 4–36.

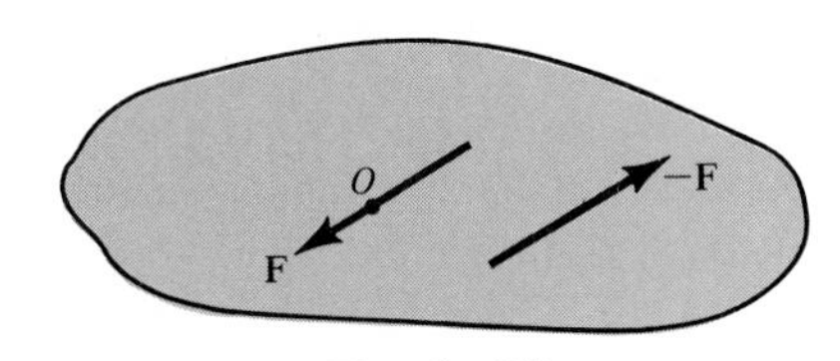

Fig. 4–36

The following examples numerically illustrate this concept.

Example 4–19

Determine the x, y, z components of the couple moment acting on the shaft shown in Fig. 4–37a. Segment AB is directed 30° below the x-y plane.

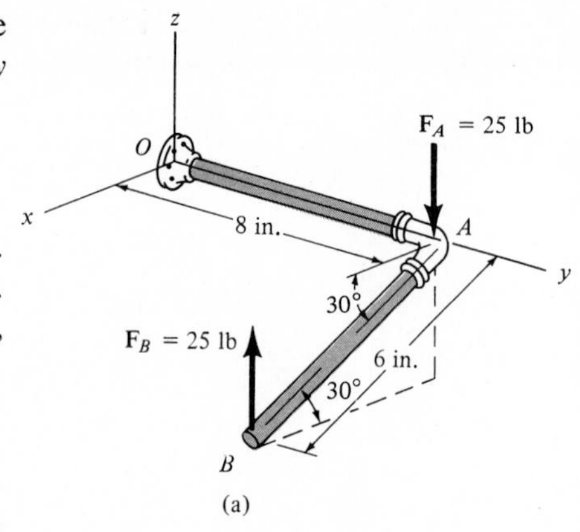

(a)

Solution I

The moments of the two couple forces can be computed about *any point*. Here we will use point O. With the x, y, z axes established at O, Fig. 4–37a, force $\mathbf{F}_A$ creates only a moment component about the x axis, namely

$$M_x = -25 \text{ lb}(8 \text{ in.}) = -200 \text{ lb} \cdot \text{in.}$$

The moment components $M_y = M_z = 0$. Why?

Force $\mathbf{F}_B$ creates no moment component about the z axis because it is parallel to the axis. The x and y moment components are computed as follows:

$$M_x = 25 \text{ lb}(8 \text{ in.}) = 200 \text{ lb} \cdot \text{in.}$$

$$M_y = -25 \text{ lb}(6 \cos 30° \text{ in.}) = -129.9 \text{ lb} \cdot \text{in.}$$

Adding the x components of both forces algebraically yields $M_x = 0$. Hence,

$$M_x = 0 \qquad M_y = -129.9 \text{ lb} \cdot \text{in.} \qquad M_z = 0 \qquad \textit{Ans.}$$

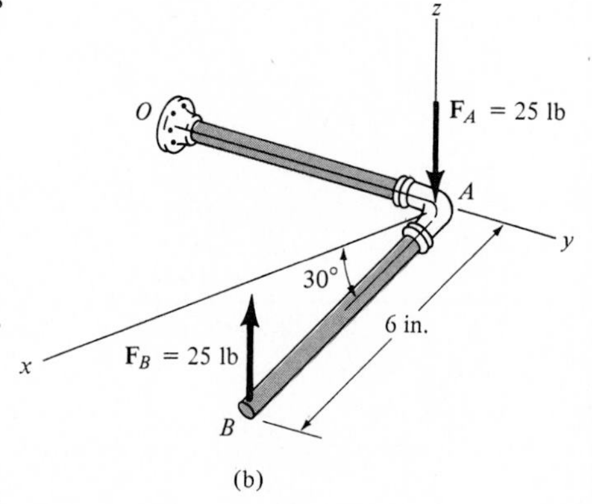

(b)

Solution II

It is easier to compute the moment components about point A, Fig. 4–37b. Why? With the origin of the coordinates located there, the moment components of $\mathbf{F}_A$ about the axes are all zero. $\mathbf{F}_B$ contributes only a component of moment about the y axis, namely, $M_y = -25 \text{ lb}(6 \cos 30° \text{ in.}) = -129.9 \text{ lb} \cdot \text{in.}$ (The moment of $\mathbf{F}_B$ about the x axis is zero since the line of action of $\mathbf{F}_B$ passes through a point lying on the axis. Why is the moment of $\mathbf{F}_B$ about the z axis equal to zero?) Hence,

$$M_x = 0 \qquad M_y = -129.9 \text{ lb} \cdot \text{in.} \qquad M_z = 0 \qquad \textit{Ans.}$$

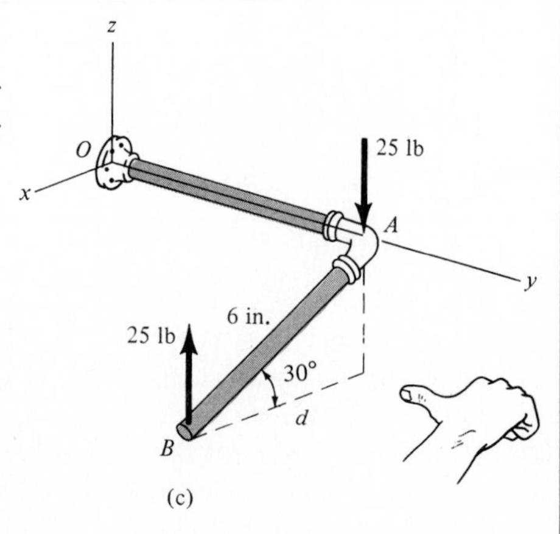

(c)

Fig. 4–37

Solution III

Although this problem is shown in three dimensions, the geometry is simple enough to use the scalar equation $M_C = Fd$ (Eq. 4–3). The perpendicular distance between the lines of action of the forces is $d = 6 \cos 30° \text{ in.} = 5.20 \text{ in.}$, Fig. 4–37$c$. Hence, taking moments of the forces about either A or B and using the right-hand rule yields

$$M_C = M_y = Fd = -25 \text{ lb}(5.20 \text{ in.}) = -129.9 \text{ lb} \cdot \text{in.}$$

Example 4–20

Replace the two couples acting on the triangular block in Fig. 4–38*a* by a single resultant couple. Specify the *x*, *y*, *z* components of this resultant.

Solution

The couple M_{C_1} caused by the forces at *A* and *B* can easily be determined from the equation $M_{C_1} = Fd$ in conjunction with the right-hand rule. We have

$$(M_{C_1})_x = 200 \text{ N}(0.3 \text{ m}) = 60 \text{ N}\cdot\text{m} \qquad (M_{C_1})_y = 0 \qquad (M_{C_1})_z = 0$$

To obtain the moment components of the couple $\mathbf{M}_{C_2}$ caused by the forces at *C* and *D*, we will take moments about point *D*. Only the moment of $\mathbf{F}_C$ has to be considered. Why? $\mathbf{F}_C$ has components $F_{C_x} = 0$, $F_{C_y} = 100 \text{ N}(\frac{4}{5}) = 80$ N, $F_{C_z} = -100 \text{ N}(\frac{3}{5}) = -60$ N. From Fig. 4–38*a*, note that $\mathbf{F}_{C_y}$ contributes only a moment component about the *z* axis, namely

$$(M_{C_2})_z = 80 \text{ N}(0.2 \text{ m}) = +16 \text{ N}\cdot\text{m}$$

And $\mathbf{F}_{C_z}$ creates only a moment component about the *y* axis, i.e.,

$$(M_{C_2})_y = 60 \text{ N}(0.2 \text{ m}) = +12 \text{ N}\cdot\text{m}$$

These components, when added by the parallelogram law, yield the resultant $\mathbf{M}_{C_2}$ shown in Fig. 4–38*b*. As expected, this vector is perpendicular to the plane *ACDE* which contains the forces $\mathbf{F}_C$ and $\mathbf{F}_D$.

Since $\mathbf{M}_{C_1}$ and $\mathbf{M}_{C_2}$ are free vectors, they may be moved to some arbitrary point *P* on the block and added vectorially, Fig. 4–38*c*. The resultant couple moment then has components of

$$(\mathbf{M}_{C_R})_x = 60 \text{ N}\cdot\text{m} \qquad (\mathbf{M}_{C_R})_y = 12 \text{ N}\cdot\text{m} \qquad (\mathbf{M}_{C_R})_z = 16 \text{ N}\cdot\text{m} \qquad \textbf{Ans.}$$

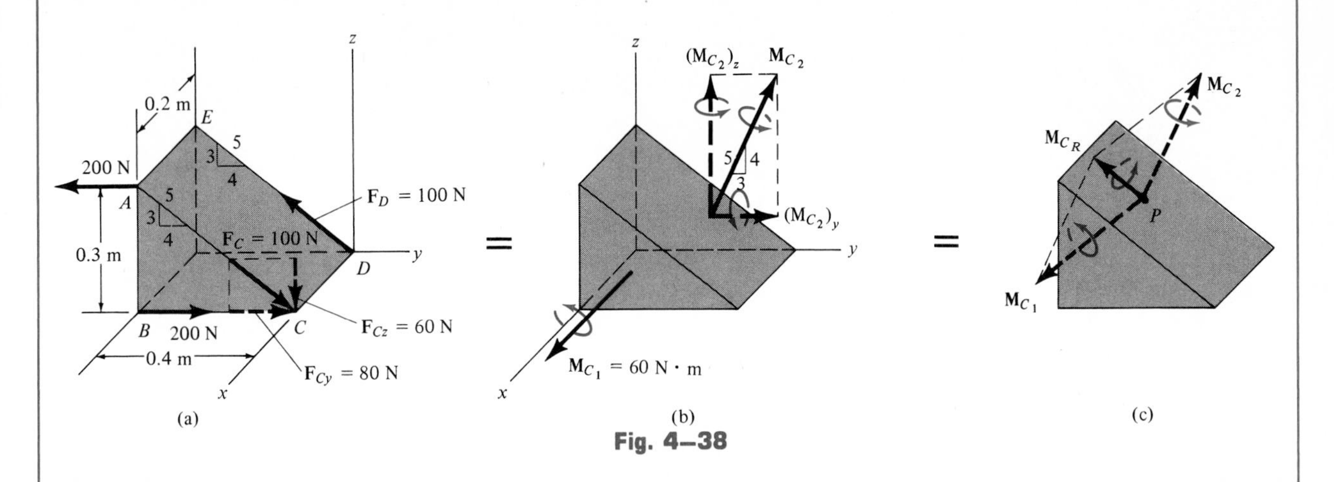

Fig. 4–38

4.11 Simplification of a Three-Dimensional Force System

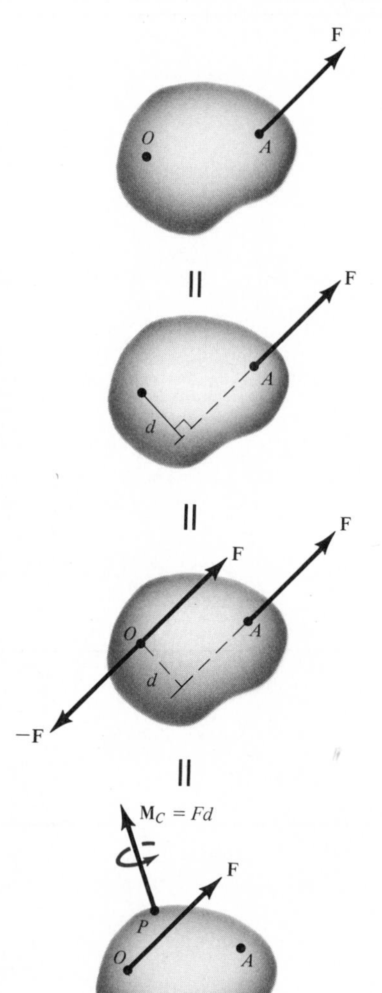

Fig. 4–39

The procedures discussed in Secs. 4–5 to 4–7 may be generalized and applied to three-dimensional force systems.

Resolution of a Force into a Force and a Couple. When a force is moved from one point to another on a rigid body, it is necessary to add a couple moment to the body in order to maintain an equivalency of loading. This should be evident from the constructions shown in Fig. 4–39, which are identical to those used for a planar analysis, Fig. 4–17. In this case $\mathbf{F}$ is moved from point A to point O and a couple moment having a magnitude of $M_C = Fd$ is added to the body. According to the right-hand rule, the couple moment is perpendicular to the plane containing $\mathbf{F}$ and d. Furthermore, since $\mathbf{M}_C$ is a free vector, it can act at *any point P* on the body.

Simplification of a Force and Couple System. Consider the force and couple-moment system acting on the rigid body shown in Fig. 4–40*a*. We can simplify this system to a single resultant force $\mathbf{F}_R$ acting at point O and a resultant couple moment $\mathbf{M}_{R_O}$. To do this, the couple moment $\mathbf{M}_C$ is simply moved to point O since it is a free vector. Forces $\mathbf{F}_1$ and $\mathbf{F}_2$ are sliding vectors and since O does not lie on the line of action of these forces, each must be moved to O in accordance with the procedure outlined above. For example, when $\mathbf{F}_1$ is applied at O, a corresponding couple moment having a magnitude of $M_1 = F_1 d_1$ must also be applied to the body, Fig. 4–40*b*. By vector addition, the equivalent system of forces and couple moments is reduced to a single resultant force $\mathbf{F}_R = \mathbf{F}_1 + \mathbf{F}_2$ and resultant couple moment $\mathbf{M}_{R_O} = \mathbf{M}_C + \mathbf{M}_1 + \mathbf{M}_2$, as shown in Fig. 4–40*c*.

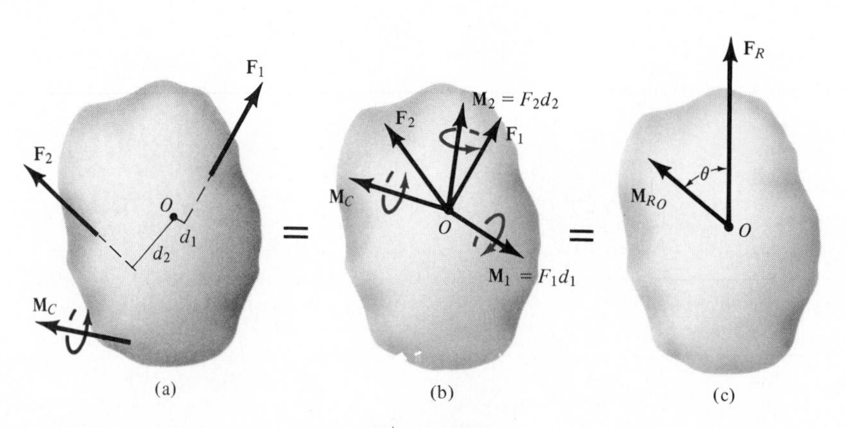

Fig. 4–40

PROCEDURE FOR ANALYSIS

The preceding method for simplifying any force and couple system to a single resultant force acting at point O and a resultant couple moment may be generalized to problems in three dimensions as follows.

Force Summation. The resultant force has x, y, z components that are equivalent to the respective algebraic sums of the x, y, z components of all the forces in the system, i.e.,

$$
\begin{aligned}
F_{R_x} &= \Sigma F_x \\
F_{R_y} &= \Sigma F_y \\
F_{R_z} &= \Sigma F_z
\end{aligned}
\tag{4–8}
$$

Moment Summation. The resultant couple moment has x, y, z components that are equivalent to the respective algebraic sums of all the x, y, z components of all the couples in the system plus the respective x, y, z components of the moments about point O of all the forces in the system, i.e.,

$$
\begin{aligned}
M_{R_x} &= \Sigma M_x \\
M_{R_y} &= \Sigma M_y \\
M_{R_z} &= \Sigma M_z
\end{aligned}
\tag{4–9}
$$

Example 4–21 illustrates numerical application of the above equations.

Simplification to a Wrench. In the general case, the force and couple system acting on a body, Fig. 4–40*a*, will simplify to a single resultant force $\mathbf{F}_R$ and couple $\mathbf{M}_{R_O}$ at O which are *not* perpendicular. Instead, $\mathbf{F}_R$ will act at an angle θ from $\mathbf{M}_{R_O}$, Fig. 4–40*c*. As shown in Fig. 4–41*a*, however, $\mathbf{M}_{R_O}$ may be resolved into two components: one perpendicular, $\mathbf{M}_\perp$, and the other parallel, $\mathbf{M}_\parallel$, to the line of action of $\mathbf{F}_R$. The perpendicular component $\mathbf{M}_\perp$ may be *eliminated* by moving $\mathbf{F}_R$ to point P, as shown in Fig. 4–41*b*. This point lies on axis *bb*, which is perpendicular to both $\mathbf{M}_{R_O}$ and $\mathbf{F}_R$. In order to maintain an equivalency of loading, the distance from O to P is $d = M_\perp / F_R$. Furthermore, when $\mathbf{F}_R$ is applied at P, the moment of $\mathbf{F}_R$ tending to cause rotation of the body *about O* is in the *same direction* as $\mathbf{M}_\perp$, Fig. 4–41*b*. Finally, since $\mathbf{M}_\parallel$ is a free vector, it may be moved to P so that it coincides with $\mathbf{F}_R$, Fig. 4–41*c*. This combination of a *collinear* force and couple is called a *wrench*. The *axis of the wrench* has the same line of action as the force. Hence, the wrench tends to cause both a translation along and a rotation about this axis. Comparing Fig. 4–40*a* to Fig. 4–41*c*, it is seen that a general force and couple system acting on a body can be simplified to a wrench. The axis of the wrench and a point through which this axis passes are unique and can always be determined.

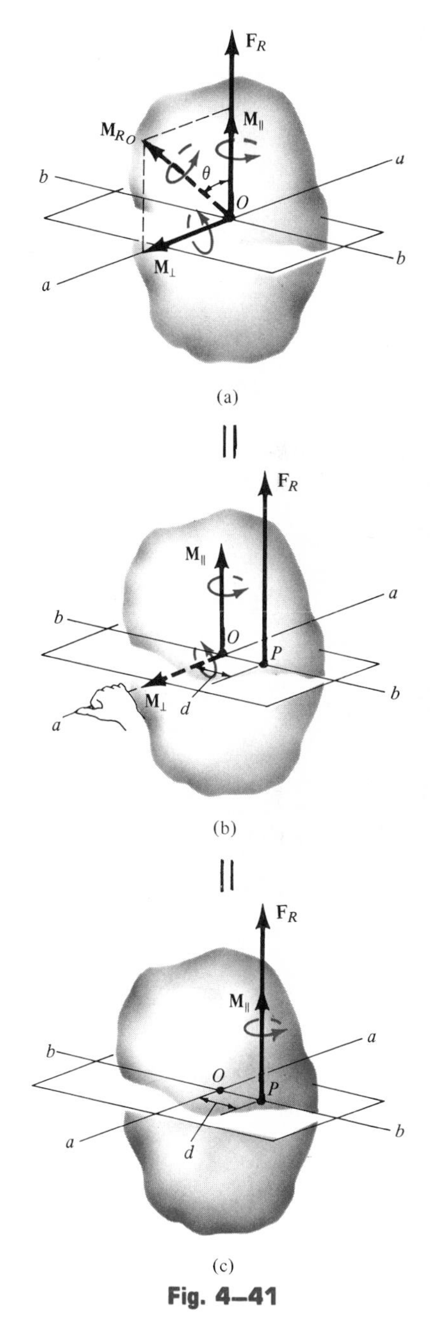

Fig. 4–41

Parallel Force System. A system of parallel forces, not necessarily all lying in the same plane, can be reduced to a *single resultant force* which acts through a unique point.* This is possible because, when all the forces in the system are moved to an arbitrary point O, they produce couple moments which are *all perpendicular* to the common line of action of the forces. Since $\mathbf{M}_{R_O}$ will be perpendicular to $\mathbf{F}_R$, then, like $\mathbf{M}_\perp$ in Fig. 4–41a ($\theta = 90°$), $\mathbf{M}_{R_O}$ can be eliminated. This requires that the moment of the resultant force about

Example 4–21

The beam is subjected to a couple moment $\mathbf{M}_C$ and forces $\mathbf{F}_1$ and $\mathbf{F}_2$ as shown in Fig. 4–42a. Replace this system by an equivalent single resultant force and couple moment acting at point O. Specify the x, y, z components of the resultants.

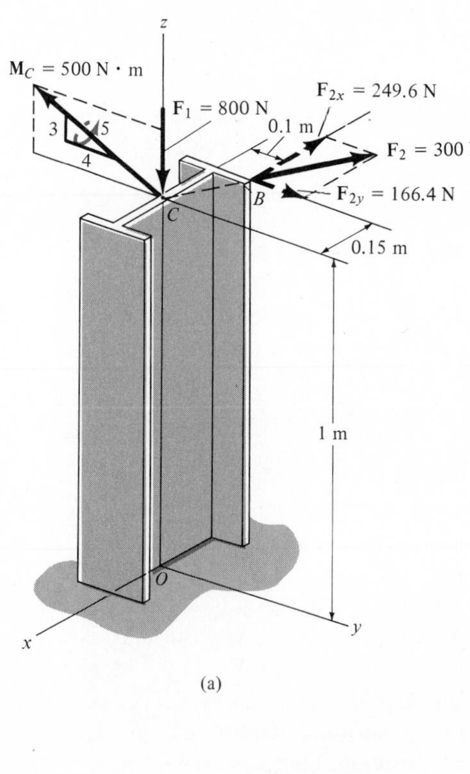

(a)

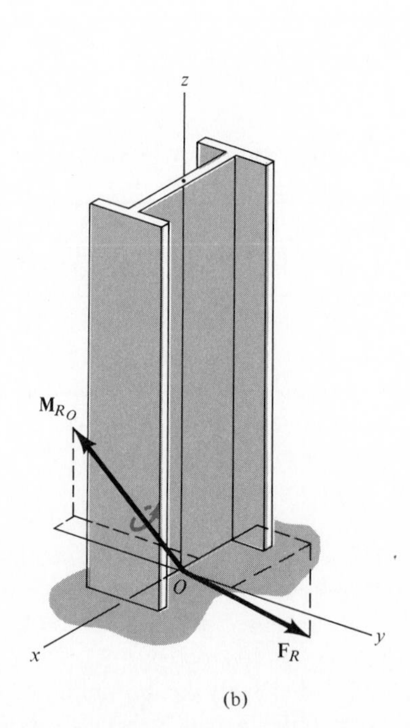

(b)

Fig. 4–42

point O be equivalent to the sum of the moments about O of all the forces in the system. Numerical application of this simplification, using the procedure outlined above, is given in Examples 4–22 and 4–23.

*It is necessary that $\mathbf{F}_R = \Sigma\mathbf{F} \neq \mathbf{0}$. However, if $\mathbf{F}_R = \mathbf{0}$, the simplest resultant one can guarantee is a couple moment.

Solution

The x, y, z components of the forces and couple can be determined directly from the figure. They are

$$F_{1_x} = 0 \qquad F_{1_y} = 0 \qquad F_{1_z} = -800 \text{ N}$$

$$F_{2_x} = -300 \text{ N}\left(\frac{0.15}{\sqrt{(0.15)^2 + (0.1)^2}}\right) = -249.6 \text{ N}, \qquad F_{2_y} = 300 \text{ N}\left(\frac{0.1}{\sqrt{(0.15)^2 + (0.1)^2}}\right) = 166.4 \text{ N}, \qquad F_{2_z} = 0$$

$$M_{C_x} = 0 \qquad M_{C_y} = -500 \text{ N}(\tfrac{4}{5}) = -400 \text{ N}\cdot\text{m} \qquad M_{C_z} = -500 \text{ N}(\tfrac{3}{5}) = 300 \text{ N}\cdot\text{m}$$

Force Summation

$F_{R_x} = \Sigma F_x;$ $\qquad F_{R_x} = 0 - 249.6 \text{ N} = -249.6 \text{ N}$ *Ans.*

$F_{R_y} = \Sigma F_y;$ $\qquad F_{R_y} = 0 + 166.4 \text{ N} = 166.4 \text{ N}$ *Ans.*

$F_{R_z} = \Sigma F_z;$ $\qquad F_{R_z} = -800 \text{ N} + 0 = -800 \text{ N}$ *Ans.*

Moment Summation. From Fig. 4–42*a* notice that the moment of $\mathbf{F}_1$ about O is zero since O lies on the line of action of $\mathbf{F}_1$. Also, the couple $\mathbf{M}_C$ can be moved directly to point O. Why? The moment of $\mathbf{F}_2$ about O is determined from the moments of its components. The computations are as follows:

$M_{R_x} = \Sigma M_x;$

$$M_{R_x} = 0 - 166.4 \text{ N}(1 \text{ m}) = -166.4 \text{ N}\cdot\text{m} \qquad \textit{Ans.}$$

$M_{R_y} = \Sigma M_y;$

$$M_{R_y} = -(\tfrac{4}{5})(500 \text{ N}\cdot\text{m}) - (249.6 \text{ N})(1 \text{ m}) = -649.6 \text{ N}\cdot\text{m} \qquad \textit{Ans.}$$

$M_{R_z} = \Sigma M_z;$

$$M_{R_z} = (\tfrac{3}{5})(500 \text{ N}\cdot\text{m}) = 300 \text{ N}\cdot\text{m} \qquad \textit{Ans.}$$

The resultant couple and force acting at O can be found by vector addition of their components. The results are shown graphically in Fig. 4–42*b*.

Example 4–22

The slab in Fig. 4–43a is subjected to a series of four parallel forces. Determine the magnitude and direction of a single resultant force equivalent to the given force system, and locate its point of application on the slab.

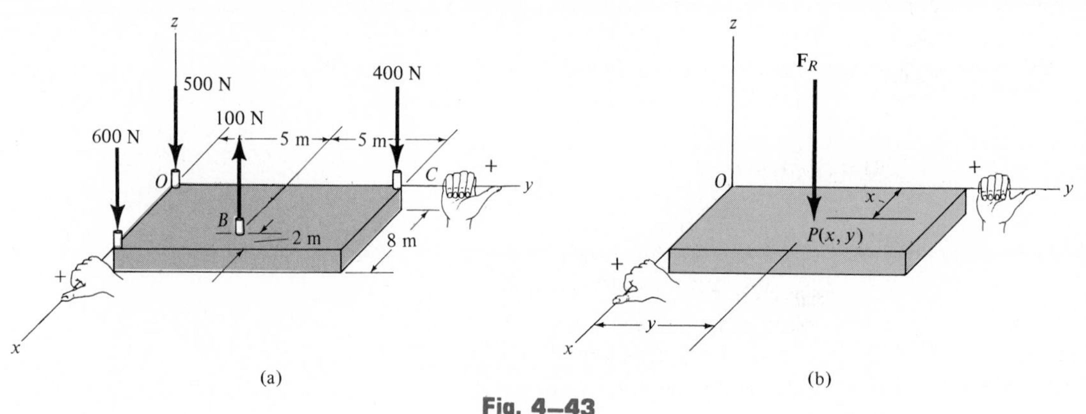

Fig. 4–43

Solution

Force Summation. From Fig. 4–43a, the resultant force is

$$+\downarrow F_{R_z} = \Sigma F_z;\; F_R = 600\text{ N} - 100\text{ N} + 400\text{ N} + 500\text{ N}$$
$$= 1400\text{ N}(\downarrow) \qquad \textit{Ans.}$$

Moment Summation. We require the moment about the x axis of the resultant force, Fig. 4–43b, to be equal to the sum of the moments about the x axis of all the forces in the system, Fig. 4–43a. The moment arms are determined from the y coordinates since these coordinates represent the *perpendicular distances* from the x axis to the lines of action of the forces. Using the right-hand rule, where positive moments act in the $+x$ direction, we have

$$M_{R_x} = \Sigma M_x;$$
$$-1400\text{N}\, y = 600\text{ N}(0\text{ m}) + 100\text{ N}(5\text{ m}) - 400\text{ N}(10\text{ m}) + 500\text{ N}(0\text{ m})$$
$$-1400y = -3500 \qquad y = 2.50\text{ m} \qquad \textit{Ans.}$$

In a similar manner, assuming that positive moments act in the $+y$ direction, a moment equation can be written about the y axis using moment arms defined by the x coordinates of each force.

$$M_{R_y} = \Sigma M_y;$$
$$1400\text{N}\, x = 600\text{ N}(8\text{ m}) - 100\text{ N}(6\text{ m}) + 400\text{ N}(0\text{ m}) + 500\text{ N}(0\text{ m})$$
$$1400x = 4200 \qquad x = 3.00\text{ m} \qquad \textit{Ans.}$$

Hence, a force of $F_R = 1400$ N placed at point $P(3.00\text{ m}, 2.50\text{ m})$ on the slab, Fig. 4–43b, is equivalent to the parallel force system acting on the slab in Fig. 4–43a.

Example 4–23

Three parallel forces act on the rim of the circular plate in Fig. 4–44*a*. Determine the magnitude and direction of a single resultant force equivalent to the given force system and locate its point of application, *P*, on the plate.

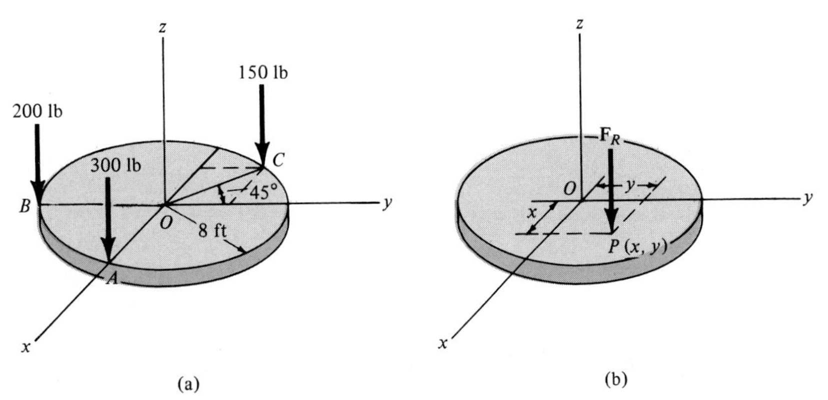

Fig. 4–44

Solution

Force Summation. From Fig. 4–44*a*, the force resultant $\mathbf{F}_R$ is

$$+\downarrow F_R = \Sigma F_z; \qquad F_R = -300 \text{ lb} - 200 \text{ lb} - 150 \text{ lb}$$
$$= -650 \text{ lb}(\downarrow) \qquad \textit{Ans.}$$

Moment Summation. Choosing point O as a reference for computing moments and assuming that $\mathbf{F}_R$ acts at a point $P(x, y)$, Fig. 4–44*b*, we require the moment about the x axis of $\mathbf{F}_R$, Fig. 4–44*b*, to be equal to the sum of the moments about the x axis of all the forces in the system, i.e.,

$$M_{R_x} = \Sigma M_x;$$
$$-650 \text{ lb } y = -150 \text{ lb}(8 \cos 45° \text{ ft}) - 300 \text{ lb}(0 \text{ ft}) + 200 \text{ lb}(8 \text{ ft})$$
$$y = -1.16 \text{ ft} \qquad \textit{Ans.}$$

The negative sign indicates that it was wrong to have assumed a $+y$ position for $\mathbf{F}_R$ as shown in Fig. 4–44*b*.

In a similar manner, a balance of moments about the y axis yields

$$M_{R_y} = \Sigma M_y;$$
$$650 \text{ lb } x = -150 \text{ lb}(8 \sin 45° \text{ ft}) + 300 \text{ lb}(8 \text{ ft}) + 200 \text{ lb}(0 \text{ ft})$$
$$x = 2.39 \text{ ft} \qquad \textit{Ans.}$$

Problems

4–63. Determine the x, y, z components of the couple moment. The forces shown in color have the components shown.

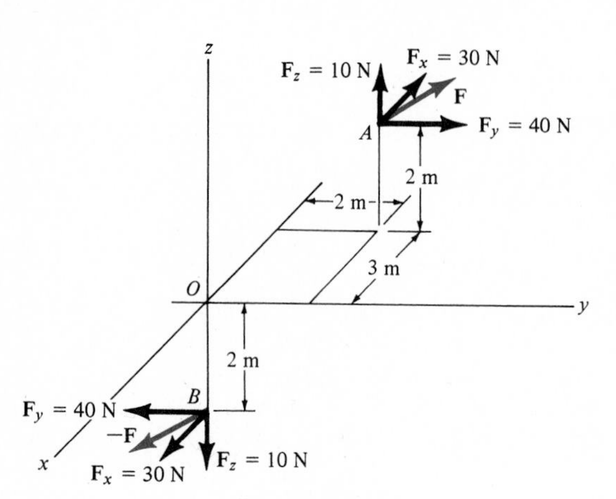

Prob. 4–63

***4–64.** Determine the x, y, z components of the couple moment. The forces shown in color have the components shown.

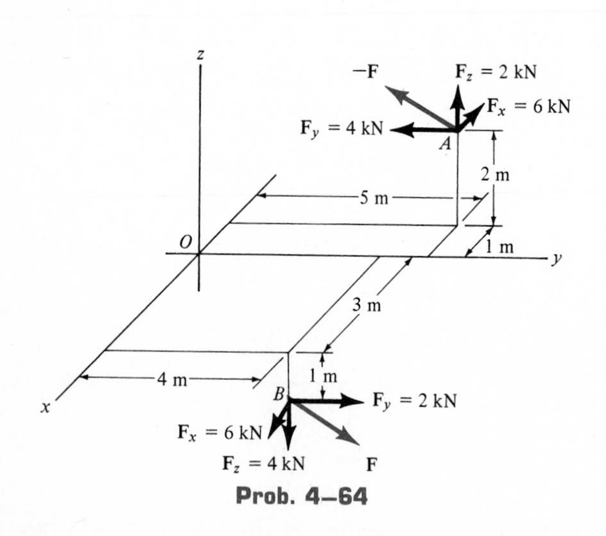

Prob. 4–64

4–65. Determine the x, y, z components of the couple moment.

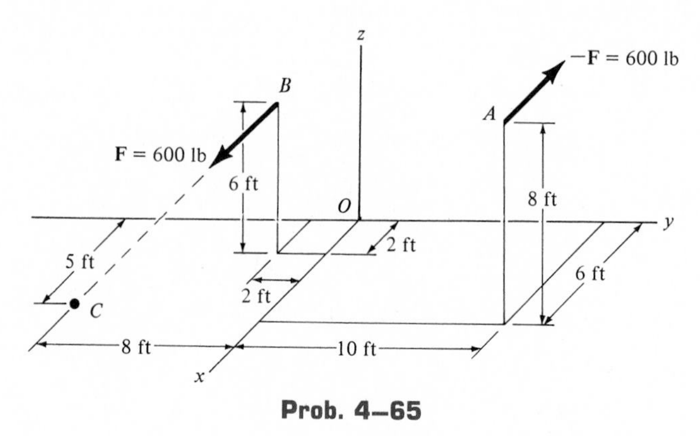

Prob. 4–65

4–66. A wire cable passes through the bearing support. If the cable ends are subjected to the couple moments shown, determine the resultant couple moment on the bearing. Specify its magnitude and coordinate direction angles.

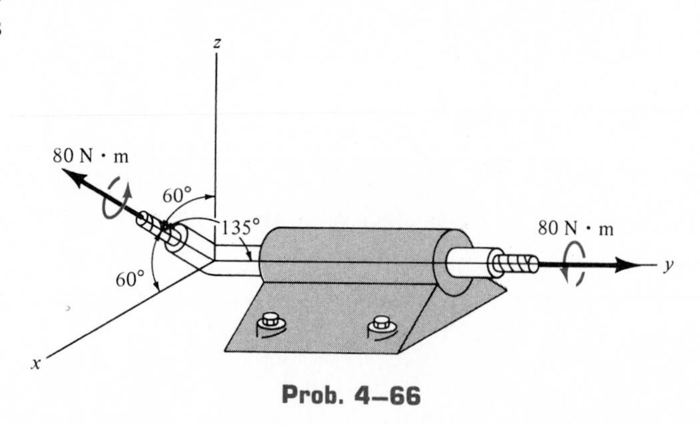

Prob. 4–66

4–67. The three couples act on the fire hydrant as shown. Determine the magnitude of the resultant couple moment and its coordinate direction angles α, β, γ.

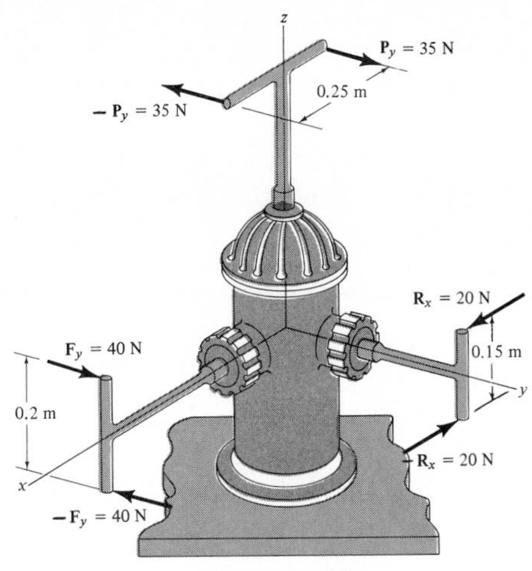

Prob. 4–67

***4–68.** Determine the x, y, z components of the couple moment acting on the pipe.

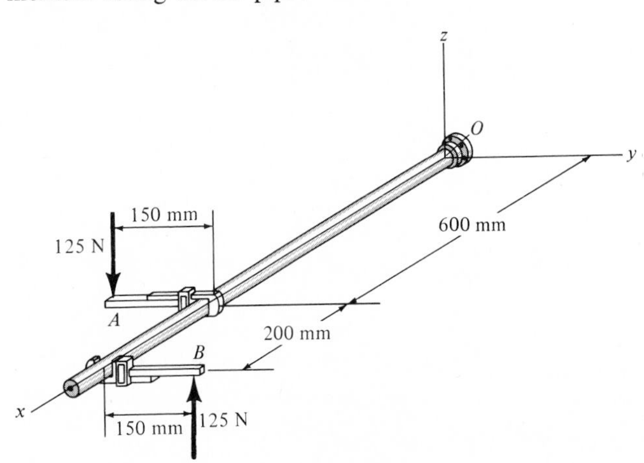

Prob. 4–68

4–69. Replace the force at A by an equivalent force and couple moment at P. Express the results in terms of their x, y, z components. The force has components of $F_x = 50$ N, $F_y = -20$ N, $F_z = -30$ N.

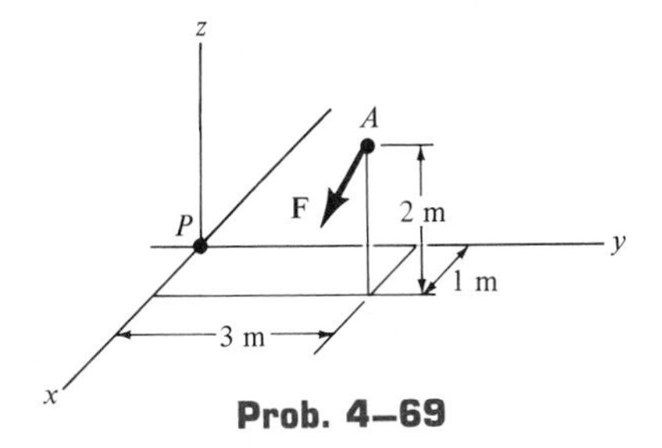

Prob. 4–69

4–70. Determine the resultant couple of the two couples that act on the assembly. Member OB lies in the x-z plane. Specify the x, y, z components of the resultant.

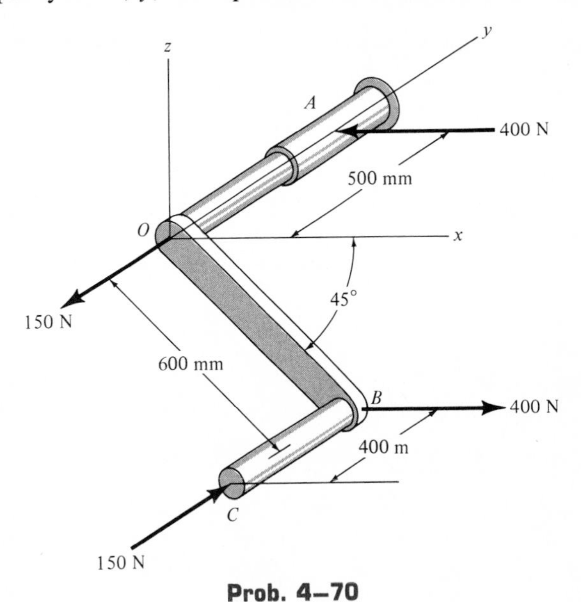

Prob. 4–70

4–71. Replace the force at A by an equivalent force and couple moment at P. Express the results in terms of their x, y, z components.

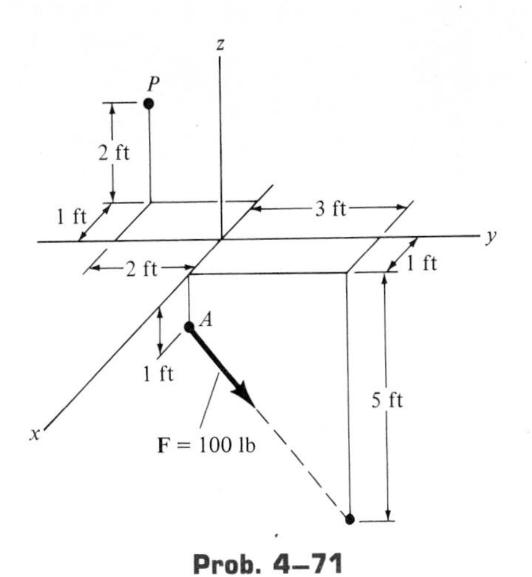

Prob. 4–71

***4–72.** Replace the force at A by an equivalent force and couple moment at P. Express the results in terms of their x, y, z components. The force has components of $F_x = -3$ kN, $F_y = 2$ kN, $F_z = 4$ kN.

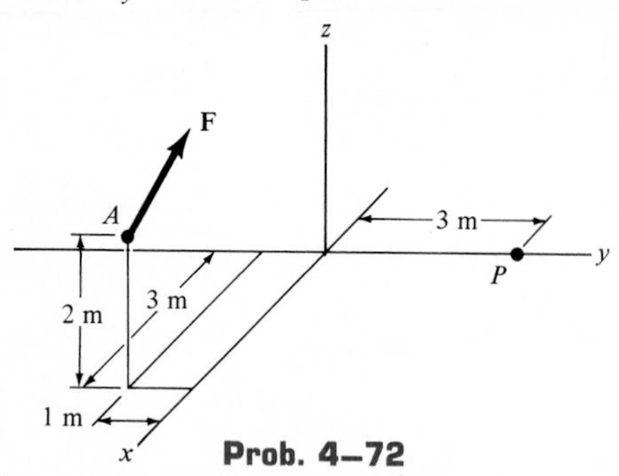

Prob. 4–72

4–73. The resultant force of a wind loading acts perpendicular to the face of the sign as shown. Replace this force by an equivalent force and couple moment acting at point O. Express the results in terms of their x, y, z components.

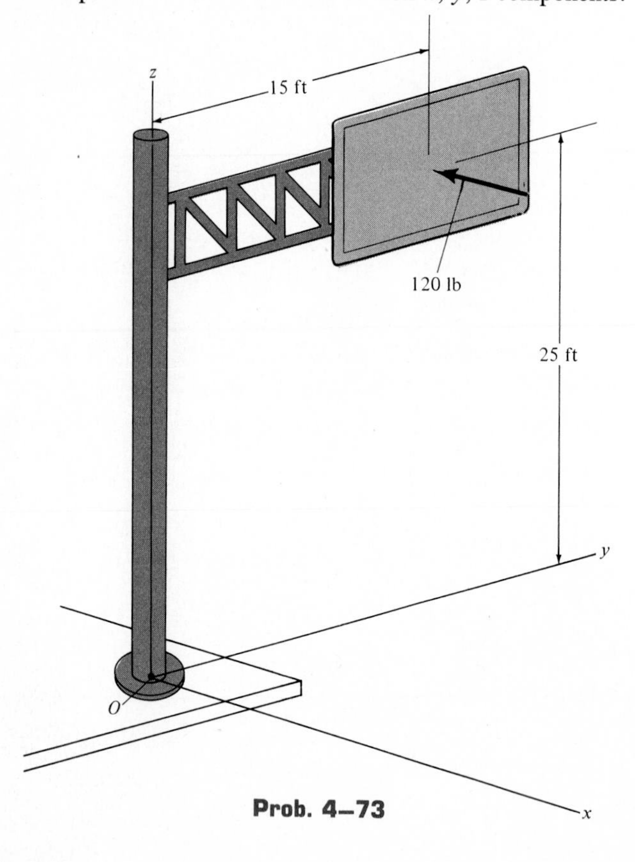

Prob. 4–73

4–74. The building column is subjected to a force of 1,500 lb acting at the end of the supporting bracket. Replace this force by an equivalent force and couple moment acting at (a) point O and (b) point A. Express the results in terms of their x, y, z components.

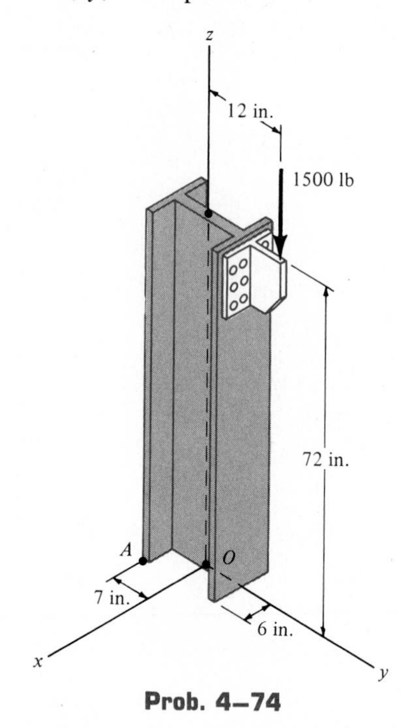

Prob. 4–74

4–75. A fish exerts a momentary pull on the line of a fishing pole of 70 N. Replace this force by an equivalent

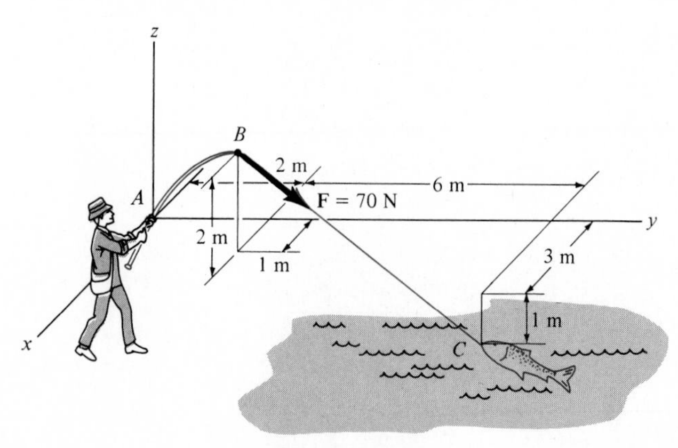

Prob. 4–75

force and couple moment at the fisherman's grip A. Solve the problem two ways, a) by placing the force **F** at the fish C, and b) by keeping it at B. Express the results in terms of their x, y, z components.

results in terms of their x, y, z components. **F** has components of $F_x = 20$ N, $F_y = 30$ N, $F_z = -20$ N, and $\mathbf{M}_C$ has components of $M_x = -10$ N · m, $M_y = 40$ N · m, $M_z = 10$ N · m.

***4–76.** Replace the force **F**, having a magnitude of $F = 40$ lb and acting at B, by an equivalent force and couple moment at A. Express the results in term of their x, y, z components.

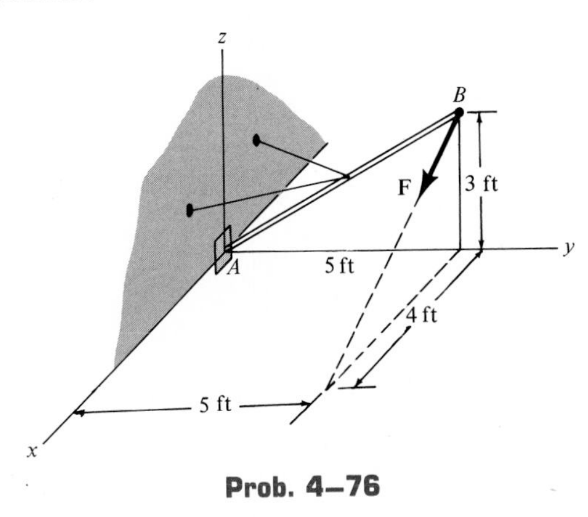

Prob. 4–76

4–78. Replace the force and couple system by an equivalent force and couple moment at point P. Express the results in terms of their x, y, z components.

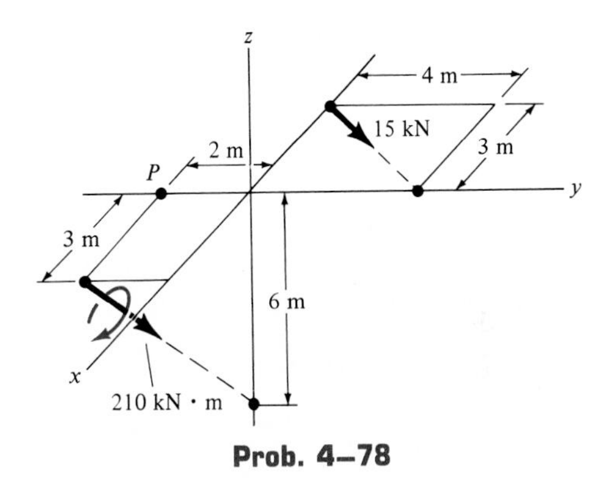

Prob. 4–78

4–77. Replace the force and couple system by an equivalent force and couple moment at point P. Express the results in terms of their x, y, z components.

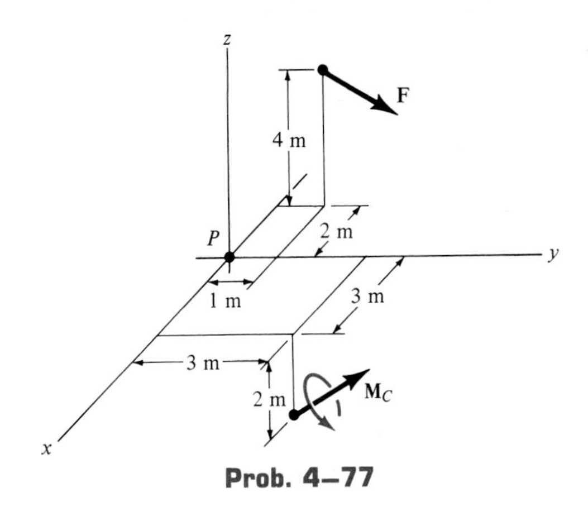

Prob. 4–77

4–79. Replace the force and couple system by an equivalent force and couple moment at point P. Express the results in terms of their x, y, z components.

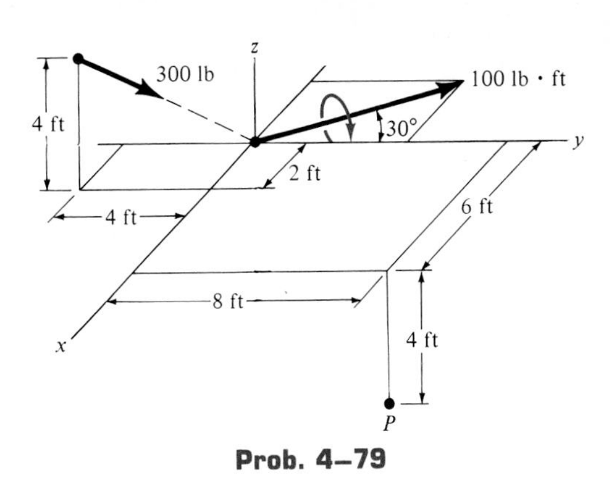

Prob. 4–79

***4–80.** Replace the force acting on the tree branch by an equivalent force and couple moment acting at point O. Express the results in terms of their x, y, z components.

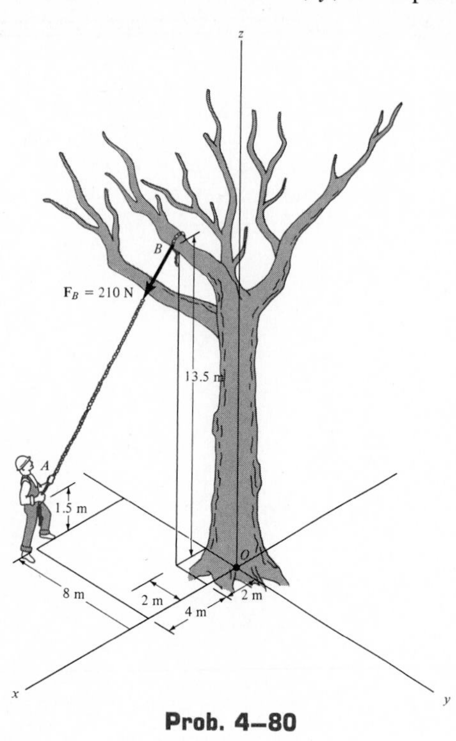

Prob. 4–80

4–81. Replace the wrench $F_1 = 60$ lb, $M_1 = 50$ lb · ft and the force $F_2 = 20$ lb, acting on the pipe assembly, by an equivalent force and couple moment acting at point O. Express the results in terms of their x, y, z components.

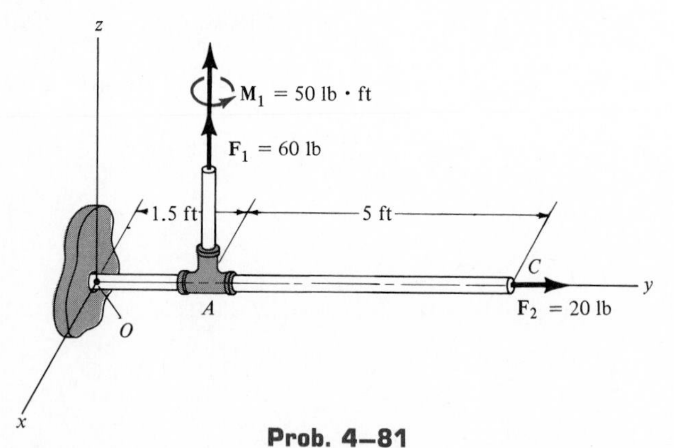

Prob. 4–81

4–82. The boys A, B, and C stand near the edges of a raft as shown. Determine the location (x, y) of boy D so that all four boys create a single resultant force acting through the raft's center O. Provided the raft itself is symmetric, this would keep the raft afloat in a horizontal plane. The mass of each boy is indicated on the diagram.

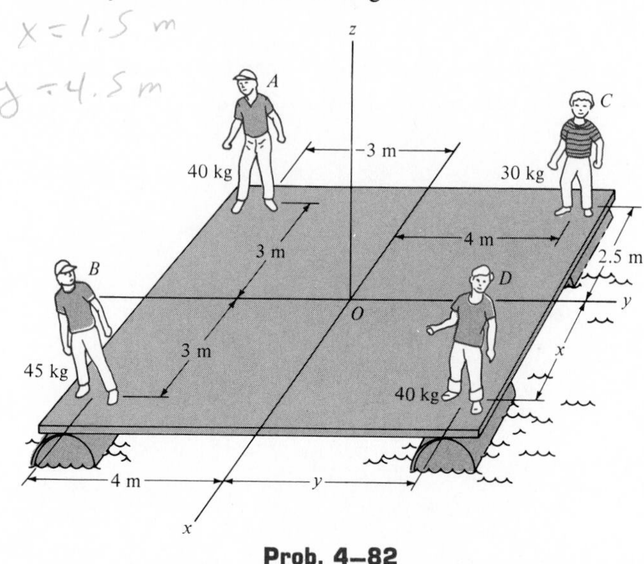

Prob. 4–82

4–83. A force and couple act on the pipe assembly. Replace this system by an equivalent single resultant force. Specify the location of the resultant force along the y axis, measured from A. The pipe lies in the x-y plane.

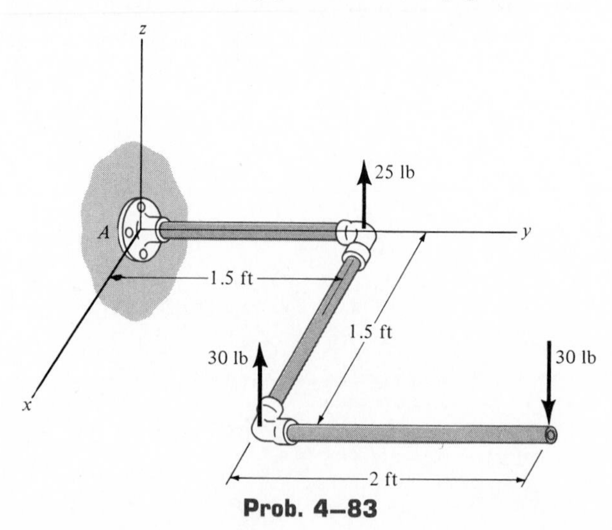

Prob. 4–83

*4–84. The hollow concrete column supports the four parallel forces shown. Determine the magnitudes of forces $\mathbf{F}_C$ and $\mathbf{F}_D$ acting at C and D so that the resultant force of the system acts through the midpoint O of the column. *Hint:* This requires $\Sigma M_x = 0$ and $\Sigma M_y = 0$. Why?

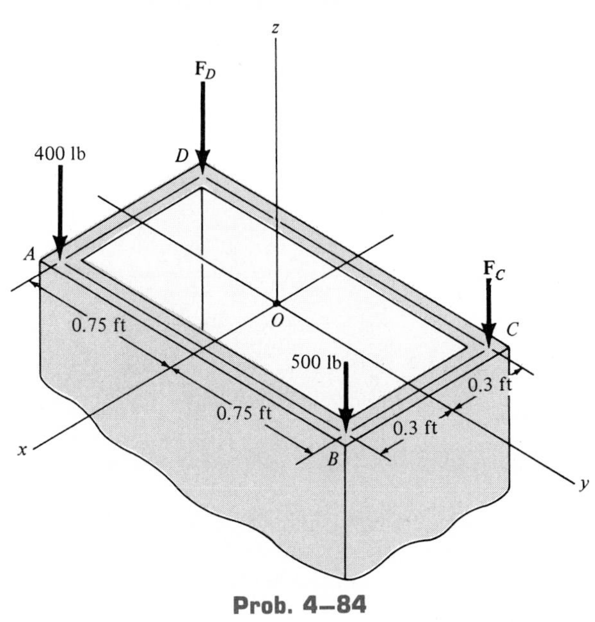

Prob. 4–84

4–85. Three parallel forces act on the circular slab. Determine the resultant force, and specify its location (x, y) on the slab.

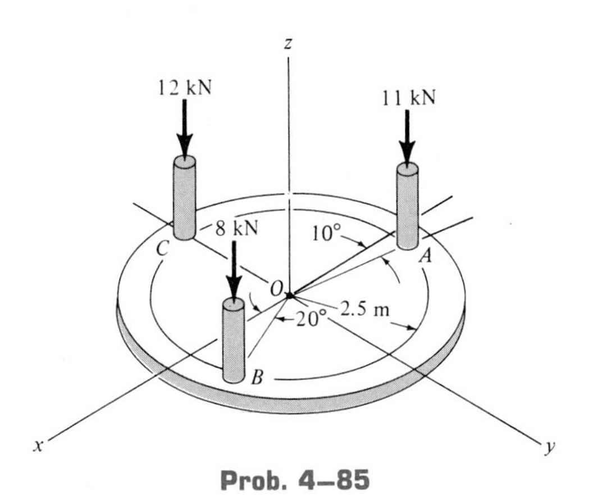

Prob. 4–85

4–86. The pipe assembly is subjected to the action of a wrench at B and a couple at A. Simplify this system to a single resultant wrench and specify the location of the wrench along the axis of pipe CD, measured from point C.

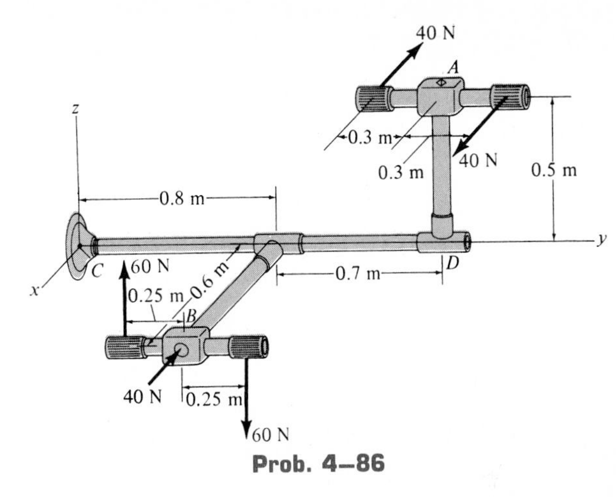

Prob. 4–86

4–87. Three parallel forces act on the rim of the tube. If it is required that the resultant force $\mathbf{F}_R$ of the system have a line of action that coincides with the central z axis, determine the magnitude of $\mathbf{F}_C$ and its location θ on the rim. What is the magnitude of the resultant force $\mathbf{F}_R$? *Hint:* This requires $\Sigma M_x = 0$ and $\Sigma M_y = 0$. Why?

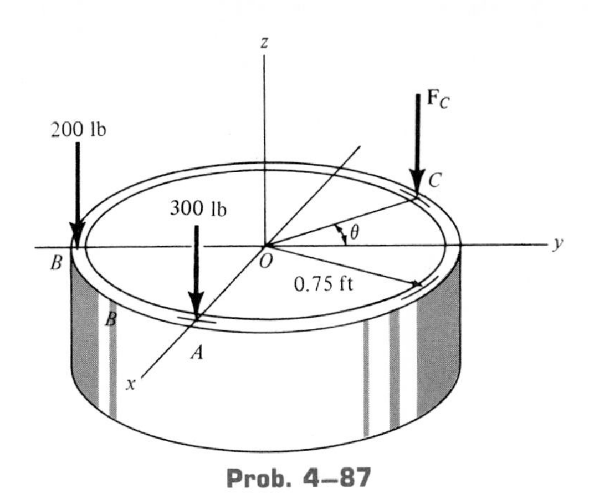

Prob. 4–87

Review Problems

***4–88.** Replace the loading acting on the beam by an equivalent force and couple system at point A.

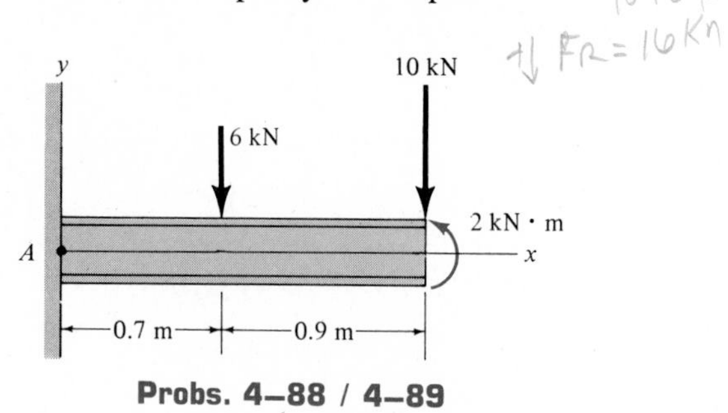

Probs. 4–88 / 4–89

4–89. Replace the loading acting on the beam by a single resultant force. Specify the location of this force measured from A.

4–90. Replace the force system acting on the bracket by an equivalent force and couple system at point O.

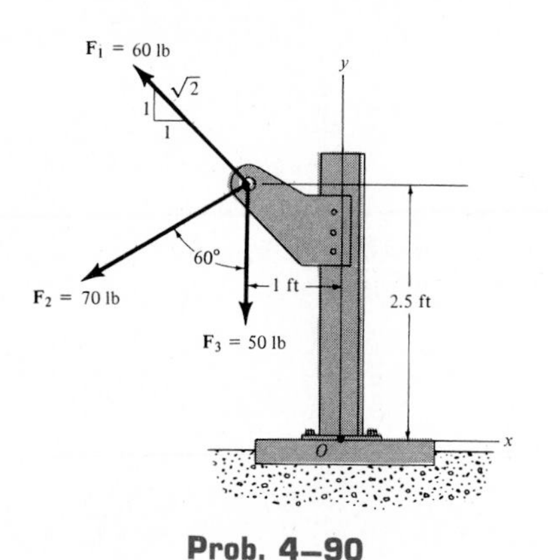

Prob. 4–90

4–91. A clockwise couple having a magnitude of $M = 40$ lb · ft is resisted by the shaft of an electric motor. Determine the magnitudes of the reactive forces **R** and $-$**R**, which act at the supports A and B, so that the resultant of the two couples is zero.

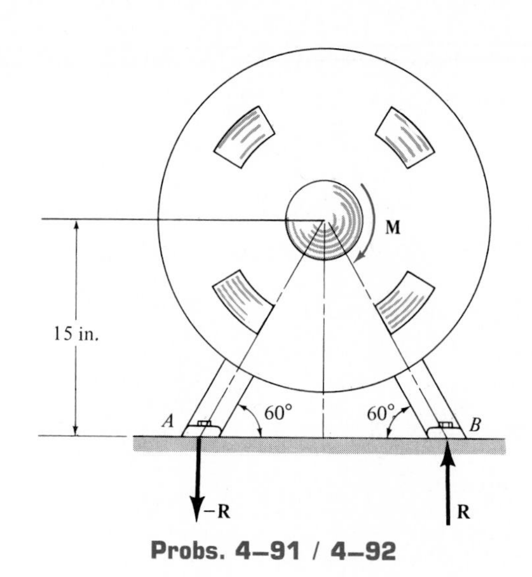

Probs. 4–91 / 4–92

***4–92.** Determine the magnitude of the couple **M** acting on the electric motor if the reactive forces **R** and $-$**R** each have a magnitude of 20 lb and the resultant of the two couples is zero.

4–93. Determine the moment of force **F** about point O. Set $F = 12$ lb.

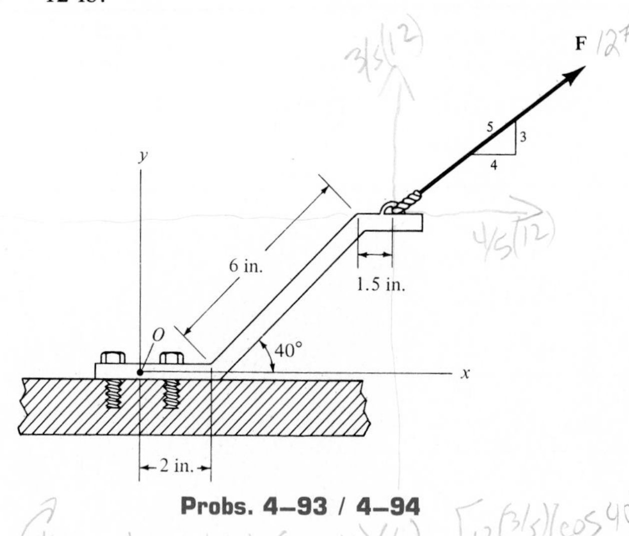

Probs. 4–93 / 4–94

4–94. Replace the force **F** by a resultant force and couple moment at point O. Set $F = 20$ lb.

4–95. Determine the resultant moment of both the 100-lb force and the triangular distributed load about point O.

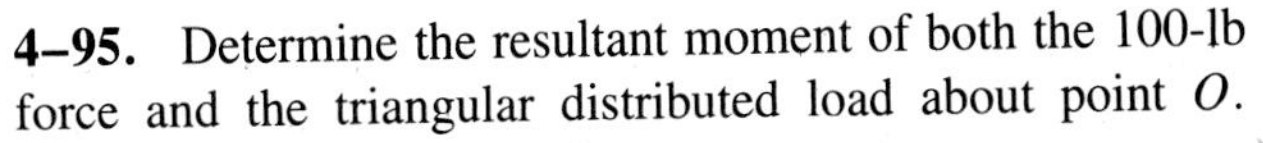

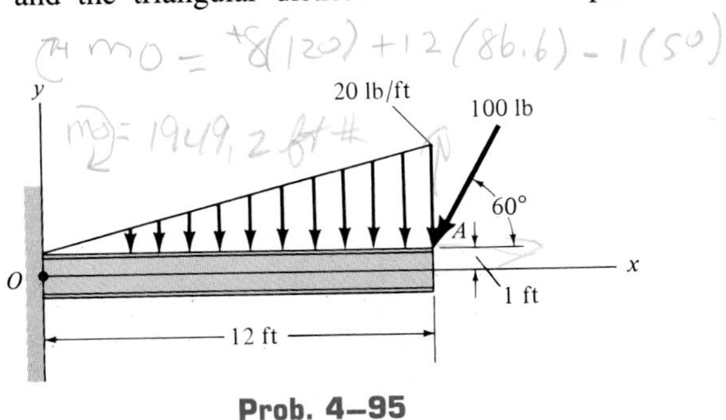

Prob. 4–95

***4–96.** The pressure of water flowing from the fire hose exerts a force of 18 lb on the nozzle in the direction shown. Determine the moment of this force about point A.

18 lb
40°
4 ft
A
2.5 ft

Prob. 4–96

4–97. Replace the two forces acting on the block by an equivalent force and couple-moment system acting at point A. Express the results in terms of their x, y, z components.

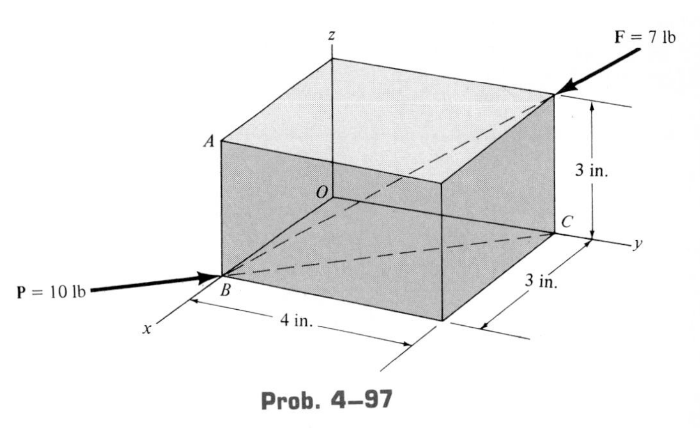

Prob. 4–97

4–98. The force **F** acts at the end B of the beam. If $F_x = 600$ N, $F_y = 300$ N, $F_z = -600$ N, determine the x, y, z components of the moment of this force about point O.

z
O
x
B
1.2 m
F
0.4 m
0.2 m
y

Prob. 4–98

4–99. The horizontal 30-N force acts on the handle of the wrench. Determine the moment of this force about point O. Specify the magnitude of the moment and the coordinate direction angles α, β, γ of the moment axis.

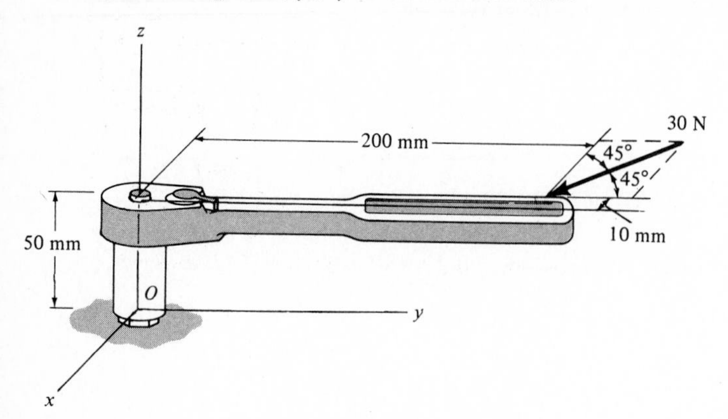

Probs. 4–99 / 4–100

***4–100.** The horizontal 30-N force acts on the handle of the wrench. What is the moment of this force about the z axis?

4–101. Determine the coordinate direction angles α, β, γ of **F**, which is applied to the end A of the pipe assembly, so that the moment of **F** about O is zero. Require $0° \leq \alpha \leq 180°$.

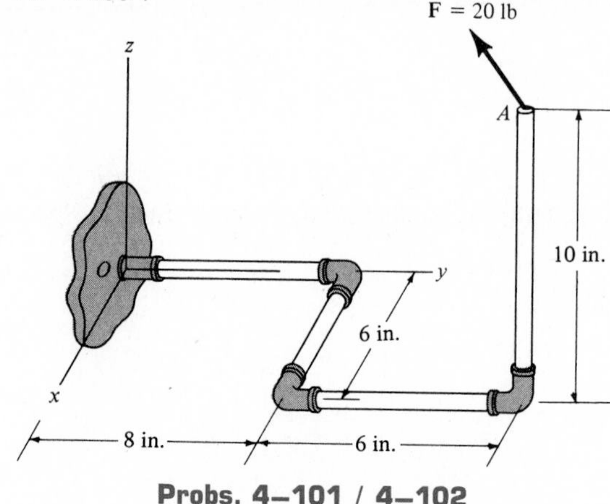

Probs. 4–101 / 4–102

4–102. Determine the x, y, z components of the moment of the force **F** about point O. The force has coordinate direction angles of $\alpha = 60°$, $\beta = 120°$, $\gamma = 45°$.

Equilibrium of a Rigid Body

In this chapter the fundamental concepts of rigid-body equilibrium will be discussed. It will be shown that equilibrium requires both a *balance of forces,* to prevent the body from translating or moving along a straight or curved path, and a *balance of moments,* to prevent the body from rotating.

Many types of engineering problems involve symmetric loadings and can be solved by projecting all the forces acting on a body onto a single plane. Hence, in the first part of this chapter, the equilibrium of a body subjected to a *coplanar* or *two-dimensional force system* will be considered. The more general discussion of rigid bodies subjected to *three-dimensional force systems* is given in the second part of the chapter.

Equilibrium in Two Dimensions

5.1 Support Reactions and Free-Body Diagrams

The free-body diagram concept, introduced in Chapter 3 in connection with problems involving the equilibrium of a particle, plays an important role in the solution of rigid-body equilibrium problems. Before discussing how to draw the free-body diagram, however, it is first necessary to consider the various types of *reactions* that can occur at supports and points of contact between bodies.

Support Reactions. For two-dimensional problems, i.e., bodies subjected to coplanar force systems, the types of supports most commonly encountered

Table 5–1 Supports for Rigid Bodies Subjected to Two-Dimensional Force Systems

Types of Connection	*Reaction*	*Number of Unknowns*
(1) light cable		One unknown. The reaction is a force which acts away from the member in the direction of the cable.
(2) weightless link	or	One unknown. The reaction is a force which acts along the axis of the link.
(3) roller		One unknown. The reaction is a force which acts perpendicular to the surface at the point of contact.
(4) roller or pin in confined smooth slot	or	One unknown. The reaction is a force which acts perpendicular to the slot.
(5) rocker		One unknown. The reaction is a force which acts perpendicular to the surface at the point of contact.
(6) smooth contacting surface		One unknown. The reaction is a force which acts perpendicular to the surface at the point of contact.
(7) collar on smooth rod	or	One unknown. The reaction is a force which acts perpendicular to the rod.

Table 5–1 (Contd.)

Types of Connection	Reaction	Number of Unknowns
(8) smooth pin or hinge	F_y, F_x or F, ϕ	Two unknowns. The reactions are two components of force, or the magnitude and its direction ϕ of the resultant force. Note that ϕ and θ are not necessarily equal (usually not, unless the rod shown is a link as in (2)).
(9) fixed support	M, F_x, F_y or M, F, ϕ	Three unknowns. The reactions are the couple and the two force components, or the couple and the magnitude and direction ϕ of the resultant force.

are given in Table 5–1. It is important to carefully study each of the symbols used to represent these supports and the types of reactions that occur. Notice that a support develops a *force* on a member to which it is attached if it *prevents translation* of the member, and it develops a *couple* if it *prevents rotation* of the member. For example, a member in contact with the smooth surface (6) is prevented from translating *only* in the downward direction, perpendicular or normal to the surface. Hence, the surface exerts only a *normal force* **F** on the member at the point of contact. The magnitude of this force represents *one unknown*. Since the member is free to rotate on the surface, a couple cannot be developed by the surface on the member at the point of contact. The pin or hinge support (8) prevents translation of the connecting member at its point of connection. In this case, unlike that of the smooth surface, translation is prevented in any direction. Hence, a force **F** must be developed at the support such that it has *two unknowns*, its magnitude F and direction ϕ, or, equivalently, the magnitudes of its two components $\mathbf{F}_x$ and $\mathbf{F}_y$. Since the connecting member is allowed to rotate freely in the plane about the pin, a pin support does not resist a couple acting perpendicular to this plane. The fixed support (9), however, prevents *both* planar translation and rotation of the connecting member at the point of connection. Therefore, this type of support exerts both a force and a couple on the member. Note that the couple acts *perpendicular* to the plane of the page since rotation is prevented in the plane. Hence, there are *three unknowns* at a fixed support.

In reality, all supports actually exert *distributed surface loads* on their contacting members. The concentrated forces and couples shown in Table 5–1 represent the *resultants* of this load distribution. This representation is, of course, an idealization. However, it can be used provided the surface area over which the distributed load acts is considerably *smaller* than the *total* surface area of the connecting member. (See Prob. 5–37.)

Free-Body Diagrams. The ability to draw a correct free-body diagram is extremely important for solving equilibrium problems. *If the free-body diagram is correctly drawn, the effects of all the forces and couples acting on the rigid body can be accounted for when the equations of equilibrium are applied.* For this reason, a thorough understanding of how to draw a free-body diagram is of primary importance for solving problems in mechanics. It is the preliminary step before applying the equations of equilibrium.

PROCEDURE FOR DRAWING A FREE-BODY DIAGRAM

To construct a free-body diagram for a rigid body or group of bodies considered as a single system, the following steps should be performed:

Step 1. Imagine the body to be *isolated* from its surroundings, and draw (sketch) its outlined shape.

Step 2. Identify all the forces and couples that act on the body. Forces and couples generally encountered are those due to (1) applied *external* loadings, (2) reactions occurring at the supports or at points of contact with other bodies, and (3) the weight of the isolated body. Forces that are *internal* to the body, such as the forces of contact between any two particles of the body, are *not shown* on the free-body diagram of the entire body. These forces always occur in equal and opposite collinear pairs, and therefore their *net effect* on the body will be zero.

Step 3. Indicate the dimensions of the body necessary for computing the moments of forces. The forces and couples that are known should be labeled with their proper magnitudes and directions. Letters are used to represent the magnitudes and direction angles of forces and couples that are *unknown*. In particular, if a force or couple has a known line of action but unknown magnitude, the arrowhead which defines the directional sense of the vector can be assumed. The correctness of the assumed direction will become apparent after solving the equilibrium equations for the unknown magnitude. By definition, the *magnitude* of a vector is *always positive,* so that if the solution yields a ''negative'' magnitude, the *minus sign* indicates that the vector's ''arrowhead'' direction is *opposite* to that which was originally assumed.

Before proceeding, carefully review this section; then attempt to draw the free-body diagrams for the following example problems using the above procedure before ''looking'' at the solutions. Further practice in drawing free-body diagrams should be gained by consulting Figs. 5–6 to 5–12 and then solving *all* the problems given at the end of this section.

Example 5–1

Draw the free-body diagram of the uniform beam shown in Fig. 5–1*a*. The beam has a mass of 100 kg.

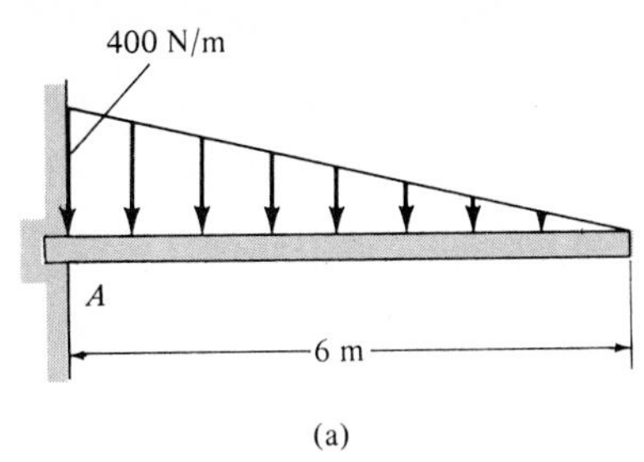

(a)

Fig. 5–1a

Solution

The free-body diagram of the beam is shown in Fig. 5–1*b*. Since the support at A is a fixed wall, there are three reactions acting *on the beam* at A, denoted as $\mathbf{A}_x$, $\mathbf{A}_y$, and $\mathbf{M}_A$. The magnitudes of these vectors are *unknown,* and their directions have all been *assumed.* (How does one obtain the *correct* directional sense of these vectors?) For convenience in computing the reactions at A, the distributed loading is reduced to a concentrated force. Using the methods of Sec. 4–8, the resultant of the distributed loading is equal to the area under the triangular loading diagram, i.e., $\frac{1}{2}(6\text{ m})(400\text{ N/m}) = 1200\text{ N}$. This force acts through the centroid of the triangle, i.e., $\frac{1}{3}(6\text{ m}) = 2\text{ m}$ from A. The weight of the beam, $W = 100(9.81) = 981\text{ N}$, acts 3 m from A since the beam is uniform.

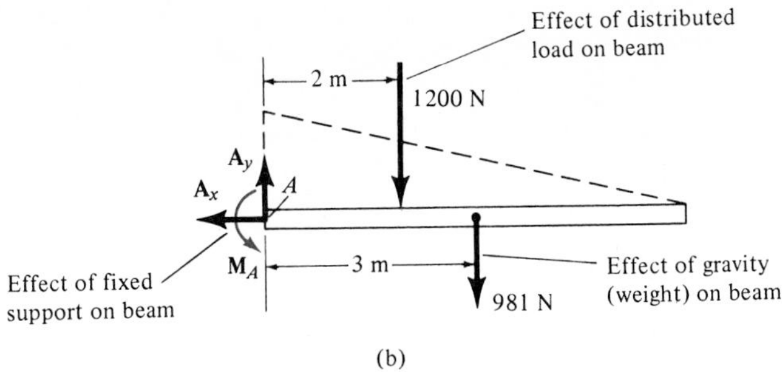

(b)

Fig. 5–1b

Example 5–2

Draw the free-body diagram for the bell crank *ABC* shown in Fig. 5–2*a*.

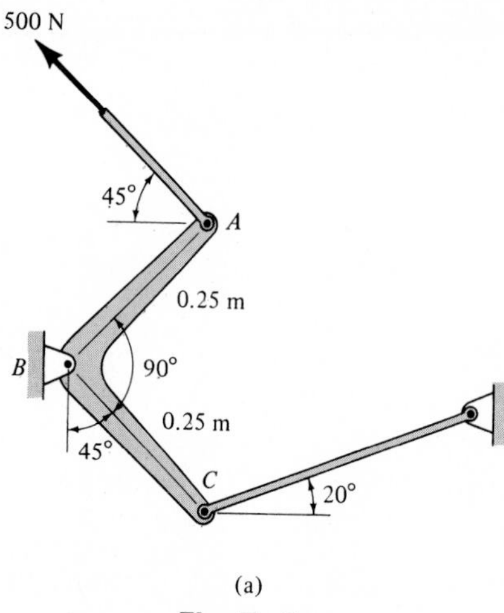

Fig. 5–2a

Solution

The free-body diagram is shown in Fig. 5–2*b*. The pin support at *B* exerts force components $\mathbf{B}_x$ and $\mathbf{B}_y$ *on the crank,* each having a known line of action but unknown magnitude. The link at *C* exerts a force $\mathbf{F}_C$ acting in the direction of the link and having an unknown magnitude. The dimensions of the crank are also labeled on the free-body diagram, since this information will be useful in computing the moments of the forces. As usual, the directions of the three unknown forces have been assumed. The correct directions will become apparent after solving the equilibrium equations.

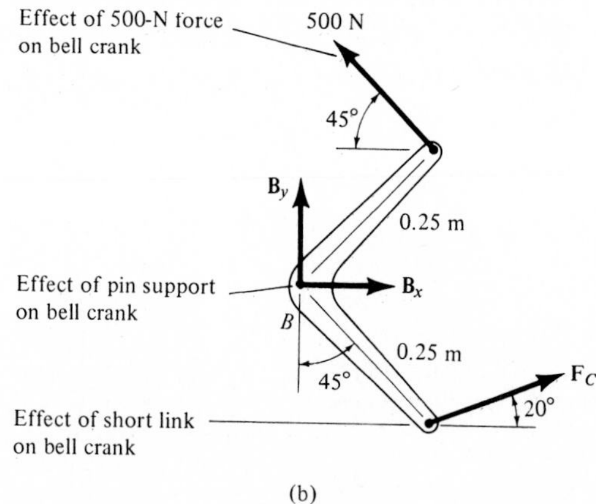

Fig. 5–2b

Example 5–3

Two smooth balls A and B, each having a mass of 2 kg, rest between the inclined planes shown in Fig. 5–3*a*. Draw the free-body diagram for ball A.

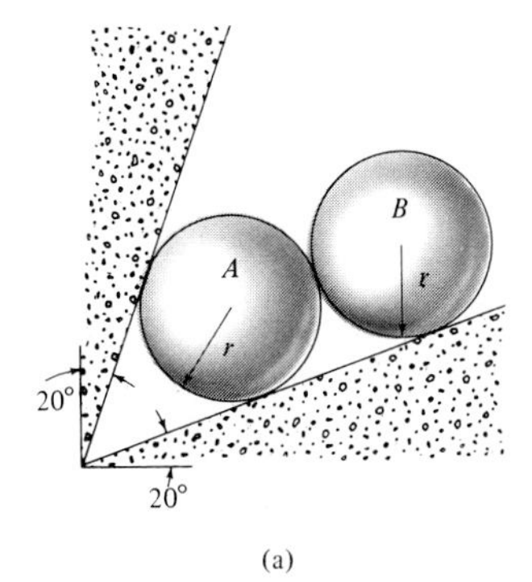

(a)

Fig. 5–3a

Solution

The free-body diagram is shown in Fig. 5–3*b*. Note that the weight of the ball is calculated as $W = 2(9.81) = 19.62$ N. Since all contacting surfaces are *smooth,* the reactive forces **T, F, R** act in a direction *normal* to the tangent at their surfaces of contact.

Although not required here, the free-body diagram of ball B is shown in Fig. 5–3*c*. Can you identify each of the three forces acting *on the ball?* In particular, note that **R** representing the force of ball A on ball B, Fig. 5–3*c*, is equal and opposite to **R** representing the force of ball B on ball A, Fig. 5–3*b*. This is a consequence of Newton's third law of motion.

The free-body diagram of both balls combined is shown in Fig. 5–3*d*. Here the contact force **R**, which acts between A and B, is considered as an *internal* force and hence is not shown on the free-body diagram. That is, it represents an equal but opposite pair of collinear forces which cancel each other.

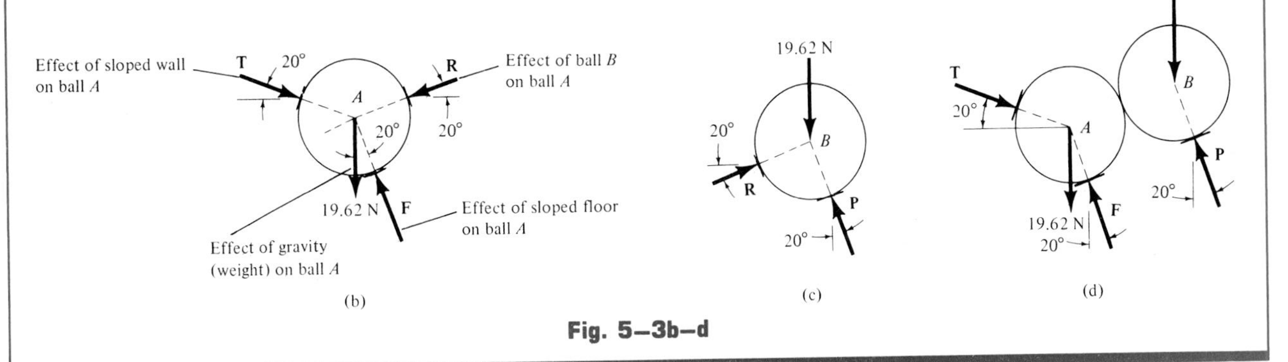

Fig. 5–3b–d

Example 5–4

The highway sign shown in Fig. 5–4*a* has a uniform mass of 100 kg with a center of gravity at *G*. In order to provide roadway clearance, it is pin-supported at *C* and *D* and held over a traffic lane by means of cable *AB*. Draw a free-body diagram of the sign and the supporting frame. Neglect the weight of the frame.

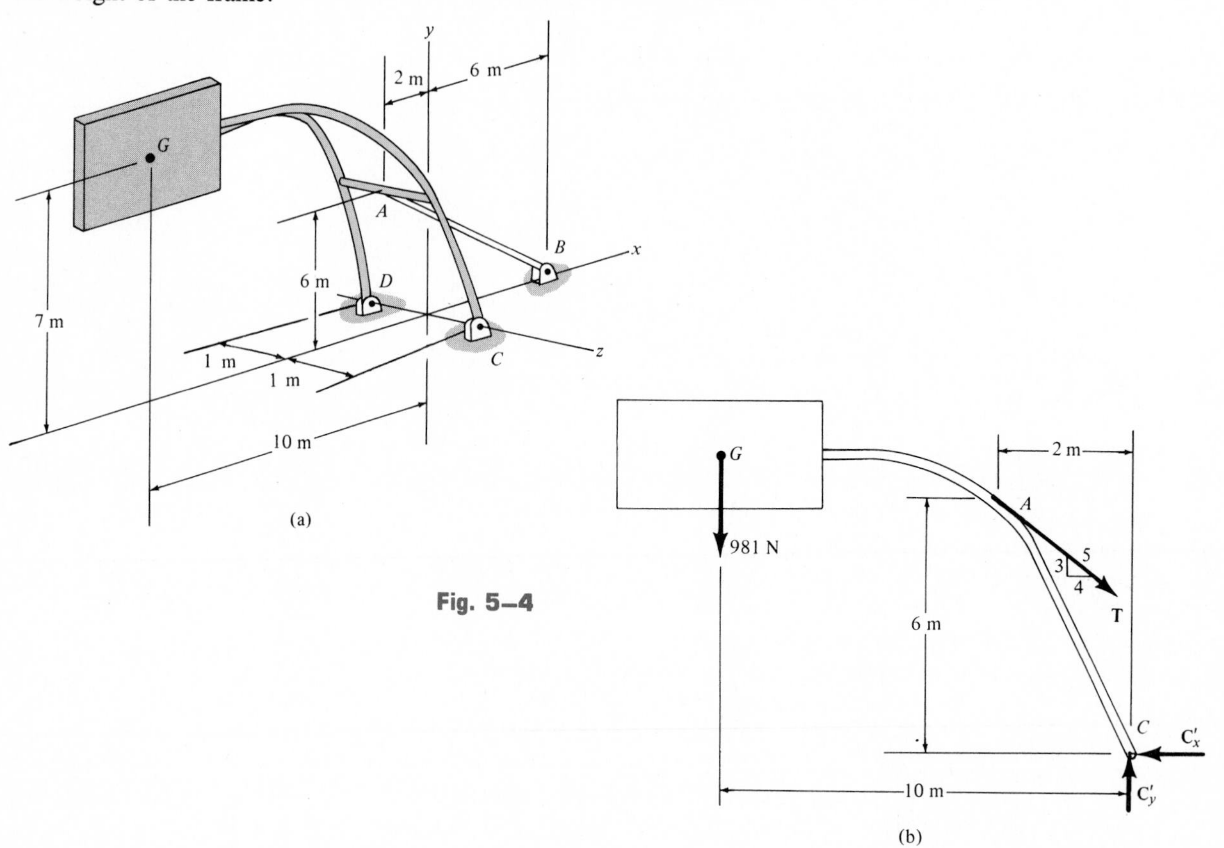

Fig. 5–4

Solution

By observation, the frame and sign and the loading are all symmetrical about the vertical *x-y* plane, hence the problem may be analyzed as a system of *coplanar forces*. The free-body diagram is shown in Fig. 5–4*b*. Can you identify each of the forces acting *on the sign and frame?* Force **T** created by the cable has a known line of action indicated by the 3–4–5 triangle. The force components $\mathbf{C}'_x$ and $\mathbf{C}'_y$ represent the horizontal and vertical reactions of *both* pins *C* and *D*. Consequently, after solving for these reactions, *half* of their magnitude is applied at *C* and half at *D*.

Problems

Draw the free-body diagram in each of the following problems and determine the total number of unknown force and couple magnitudes and/or directions. Neglect the weight of the members unless otherwise stated.

5–1. The uniform 4-m-long beam pinned at A and supported by a cord at B; the mass of the beam is 100 kg.

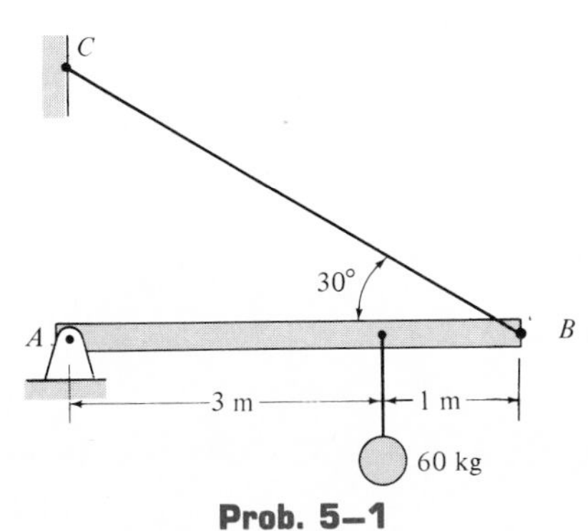

Prob. 5–1

5–2. The holding block B, which is subjected to the 40-lb force of the clamp screw at A and the forces of the smooth pipe at C and D.

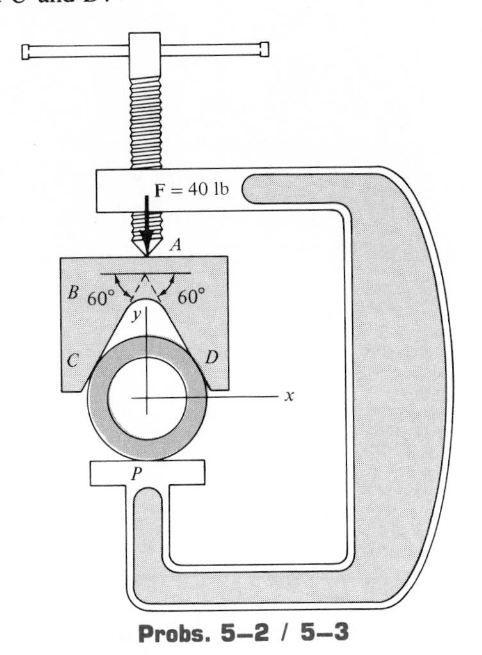

Probs. 5–2 / 5–3

5–3. The smooth pipe in Prob. 5–2. Neglect its weight.

***5–4.** The jib crane, which is fixed to the floor at A and carries a load of 3 kN at the hoist.

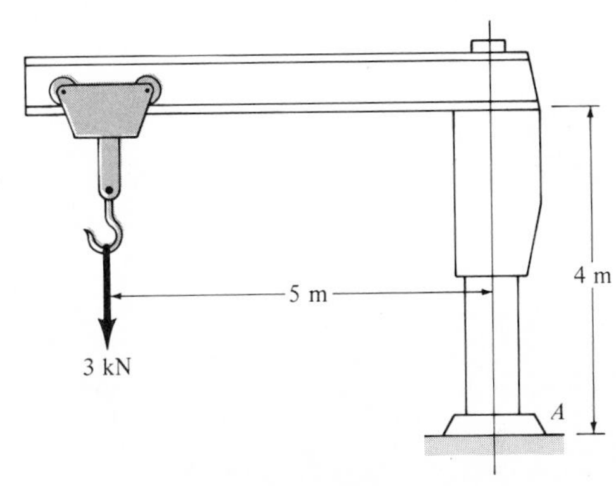

Prob. 5–4

5–5. The truss supported by cable AB and pin C.

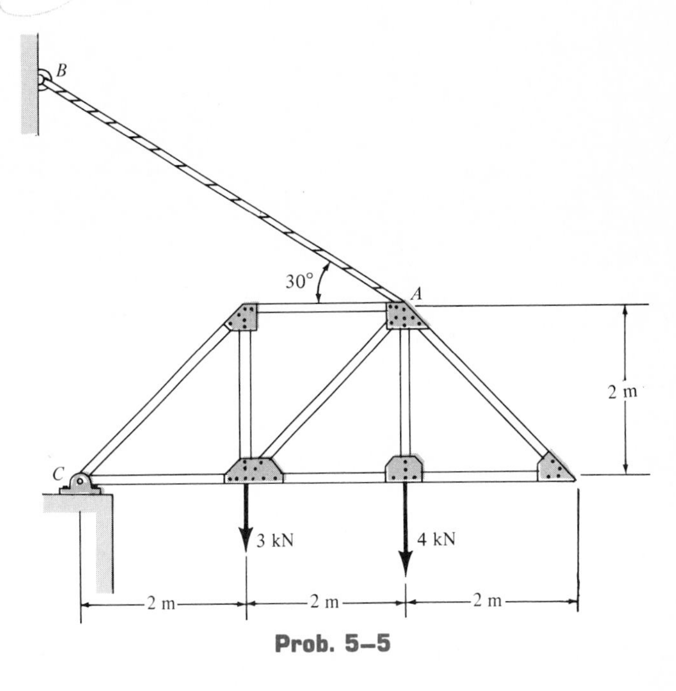

Prob. 5–5

5–6. The arm AB of a simple compression machine, which is pinned at B, supports a 20-kg mass at C, and subjects the specimen S only to a compressive force.

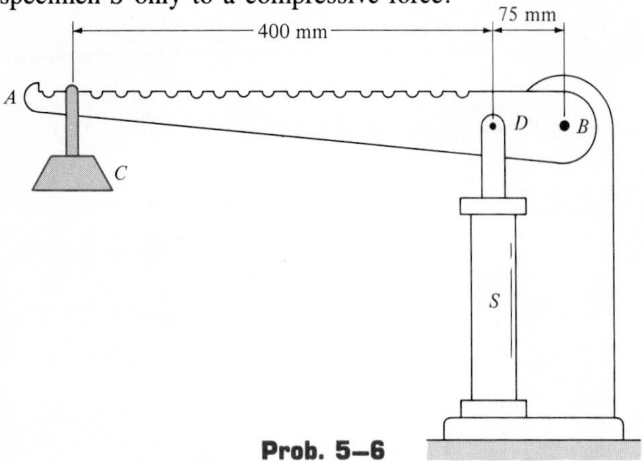

Prob. 5–6

5–7. The smooth 20-g rod which rests inside the glass.

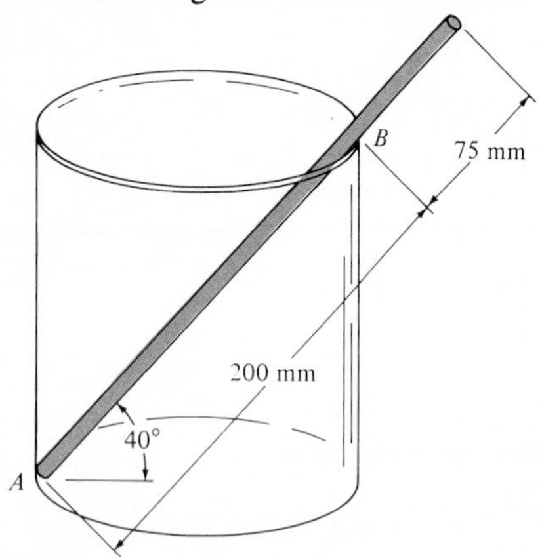

Prob. 5–7

***5–8.** The pulley, pin-connected at its center and in contact with a cable sustaining a tension **T**.

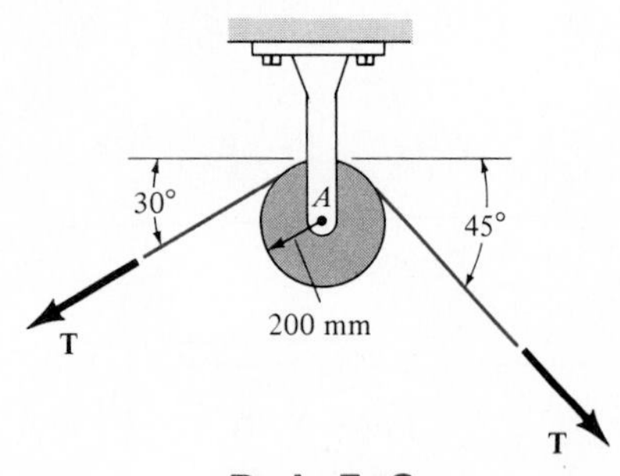

Prob. 5–8

5–9. The beam AB subjected to the triangular distributed load and supported by a pin at A and cable wrapped over a smooth pulley.

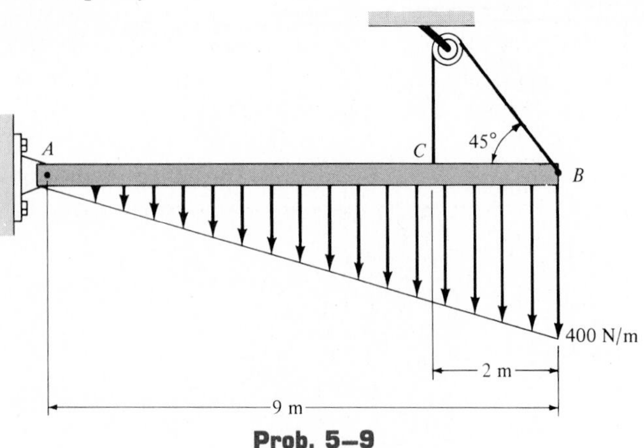

Prob. 5–9

5–10. The two-hinged gate, which has a weight of 30 lb and center of gravity at G. Due to the contact, the hinge at A is designed to carry the entire vertical load, whereas the hinge at B acts as a collar.

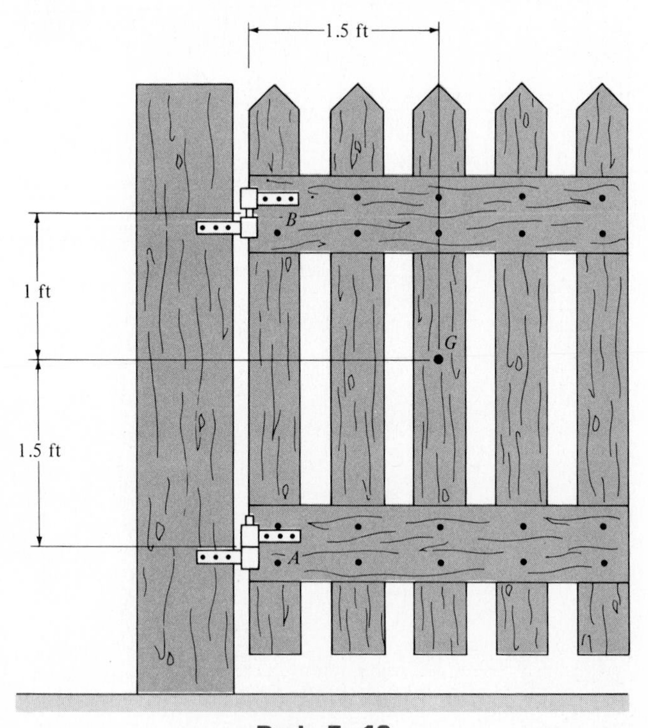

Prob. 5–10

5–11. The jib crane AB, which is pin-connected at A and supported by member (link) BC.

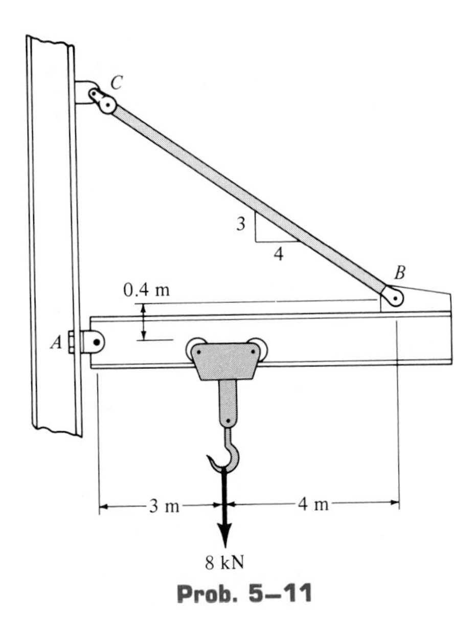

Prob. 5–11

5–13. The spanner wrench subjected to the 200-N force; the support at A acts as a pin and the surface of contact at B is smooth.

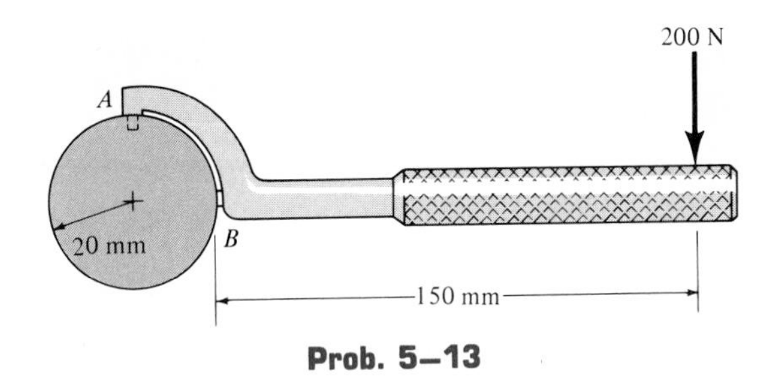

Prob. 5–13

5–14. The beam supported at A by a fixed support and at B by a roller.

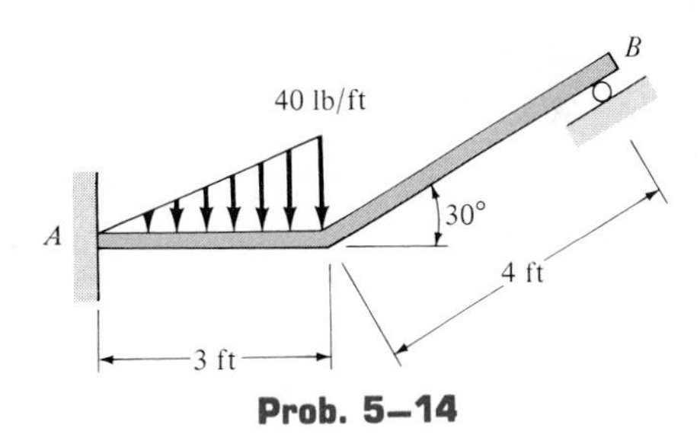

Prob. 5–14

***5–12.** The uniform beam that has a mass of 100 kg and is supported at the smooth surfaces A, B, and C.

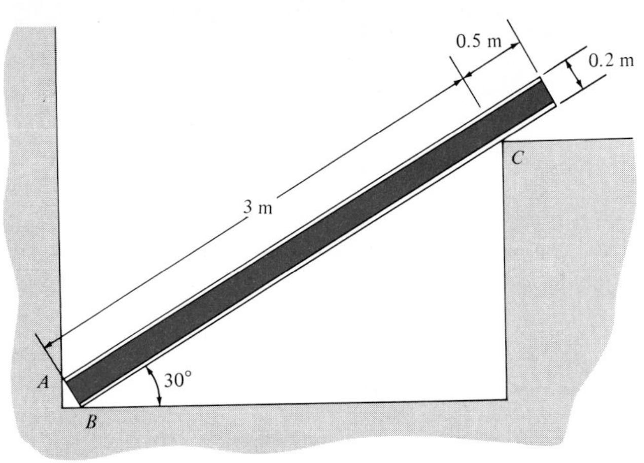

Prob. 5–12

5–15. The uniform rod ABC supported by a pin at A and a short link BD.

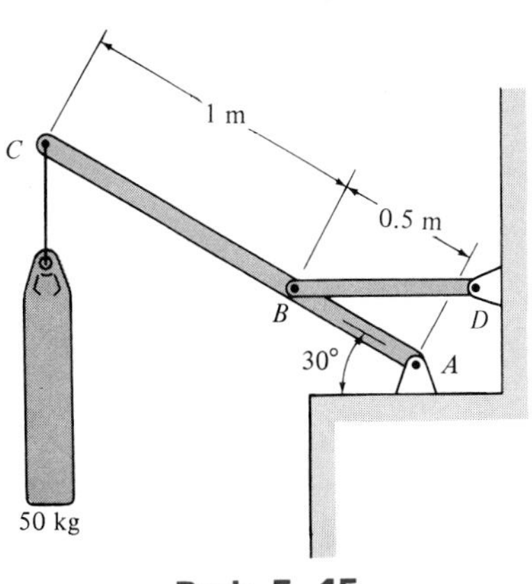

Prob. 5–15

5.2 Equations of Equilibrium

It was shown in Sec. 4–6 that a system of *coplanar forces* acting on a body can always be simplified to a resultant force $\mathbf{F}_R$ acting through an arbitrary point O and a resulant couple moment $\mathbf{M}_{R_O}$ which acts perpendicular to $\mathbf{F}_R$. In particular, if the applied forces lie in the x-y plane, then the magnitudes of the components of the resultant force can be found from $F_{R_x} = \Sigma F_x$, $F_{R_y} = \Sigma F_y$. Also, the resultant couple moment has a magnitude which is determined from $M_{R_O} = \Sigma M_O$.

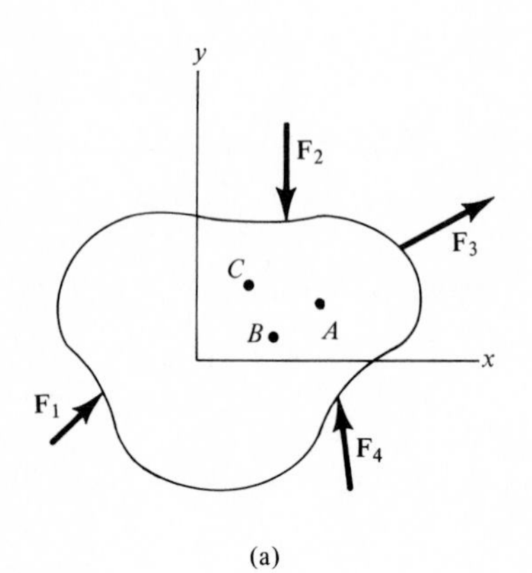

(a)

If $\mathbf{F}_R$ and $\mathbf{M}_{R_O}$ both equal zero, the body is said to be in *statical equilibrium* since physically this situation maintains a balance of both force and moment. Three scalar equations that express this condition are

$$\Sigma F_x = 0 \quad \Sigma F_y = 0 \quad \Sigma M_O = 0 \qquad (5\text{–}1)$$

Here ΣF_x and ΣF_y represent, respectively, the algebraic sums of the x and y components of all the forces acting on the body, and ΣM_O represents the algebraic sum of the moments of all these force components about an axis perpendicular to the x-y plane and passing through point O.

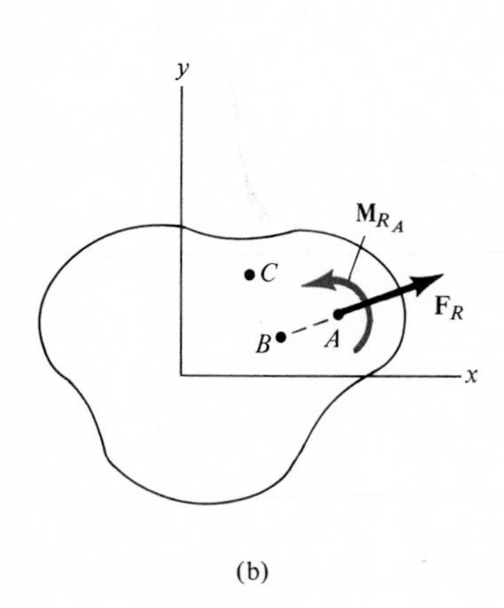

(b)

Equations 5–1 provide both the *necessary and sufficient conditions* for determining the minimum number of support reactions which are required to hold a body in equilibrium. In this book only those types of bodies that meet the conditions of proper support will be considered. Situations for which bodies are improperly supported, or have more supports than needed for equilibrium, are discussed further in Sec. 5–6.

Alternative Sets of Scalar Equilibrium Equations. Although Eqs. 5–1 are *most often* used for solving equilibrium problems involving coplanar force systems, two *alternative* sets of three independent equilibrium equations may also be used. One such set is

$$\Sigma F_y = 0 \quad \Sigma M_A = 0 \quad \Sigma M_B = 0 \qquad (5\text{–}2)$$

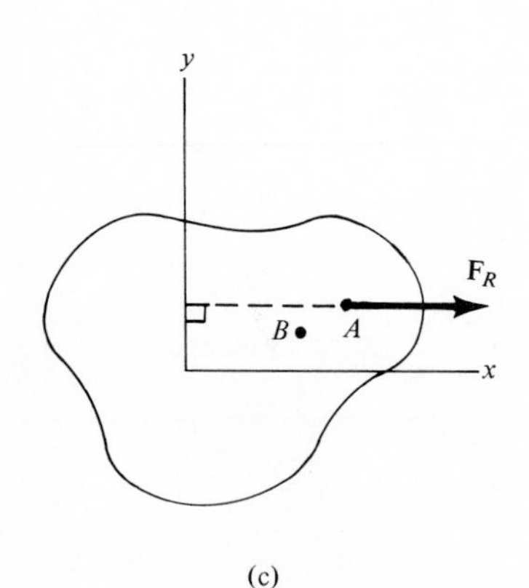

(c)

Fig. 5–5

When using these equations it is required that the moment points A and B do *not* lie on a line that is *perpendicular* to the y axis. To prove that Eqs. 5–2 provide *necessary conditions* for equilibrium, consider the free-body diagram of an arbitrarily shaped body shown in Fig. 5–5a. Using the methods of Sec. 4–6, the system of forces may be replaced by a single resultant force $\mathbf{F}_R = \Sigma\mathbf{F}$, acting at point A, and a resultant couple $\mathbf{M}_{R_A} = \Sigma\mathbf{M}_A$, Fig. 5–5$b$. For equilibrium, $\mathbf{F}_R = \mathbf{0}$ and $\mathbf{M}_{R_A} = \mathbf{0}$. Hence, if it is required that $\Sigma M_A = 0$ (Eq. 5–2), it is necessary that $\mathbf{M}_{R_A} = \mathbf{0}$. Furthermore, in order that $\mathbf{F}_R$ satisfy

$\Sigma F_y = 0$, it must have no component along the y axis, and therefore its line of action must be perpendicular to the y axis, Fig. 5–5c. Finally, if it is required that $\Sigma M_B = 0$, where B does not lie on the line of action of $\mathbf{F}_R$, then $\mathbf{F}_R = \mathbf{0}$, and indeed the body shown in Fig. 5–5a must be in equilibrium.

A second alternative set of equilibrium equations is

$$\Sigma M_A = 0$$
$$\Sigma M_B = 0 \qquad (5\text{–}3)$$
$$\Sigma M_C = 0$$

Here it is necessary that points A, B, and C do not lie on the same line. To prove that these equations when satisfied ensure equilibrium, consider again the resultant force and moment in Fig. 5–5b. If it is required that $\Sigma M_A = 0$, then the couple $\mathbf{M}_{R_A} = \mathbf{0}$. $\Sigma M_B = 0$ is satisfied if the line of action of $\mathbf{F}_R$ passes through point B as shown, and finally, if $\Sigma M_C = 0$, where C does not lie on line AB, it is necessary that $\mathbf{F}_R = \mathbf{0}$, and the body in Fig. 5–5a must then be in equilibrium.

PROCEDURE FOR ANALYSIS

The following procedure provides a method for solving coplanar force equilibrium problems:

Free-Body Diagram. Draw a free-body diagram of the body as discussed in Sec. 5–1. Briefly this requires showing all the external forces and couple moments acting *on the body*. The magnitudes of these vectors must be labeled and their directions specified. The "arrowhead" direction of a force or couple having an *unknown* magnitude but known line of action can be *assumed*. Dimensions of the body, necessary for computing the moments of forces, are also included on the free-body diagram.

Equations of Equilibrium. Apply the equations of equilibrium: $\Sigma F_x = 0$, $\Sigma F_y = 0$, $\Sigma M_O = 0$ (or the alternative sets of Eqs. 5–2 or 5–3). To *avoid* having to solve simultaneous equations, apply the moment equation $\Sigma M_O = 0$ about a point (O) lying at the intersection of the *lines of action of two of the three unknown forces*. In this way, the moments of these unknowns are *zero* about O, and one can obtain a *direct solution* for the third unknown. When applying the force equations $\Sigma F_x = 0$ and $\Sigma F_y = 0$, orient the x and y axes along lines that will provide the simplest reduction of the forces into their x and y components. If the solution of the equilibrium equations yields a *negative* magnitude for an unknown force or couple, it indicates that its "arrowhead" direction is *opposite* to that which was assumed on the free-body diagram.

The following example problems numerically illustrate this procedure.

Example 5–5

Determine the horizontal and vertical components of reaction for the beam loaded as shown in Fig. 5–6*a*. Neglect the weight of the beam in the calculations.

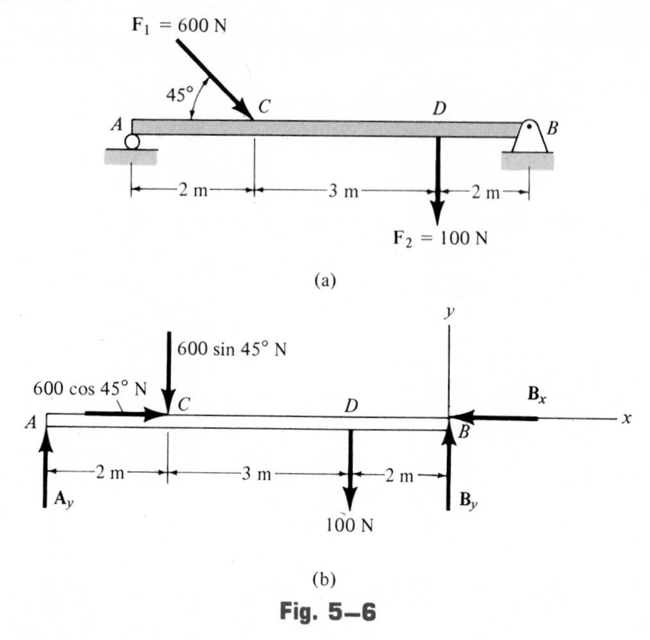

Fig. 5–6

Solution

Free-Body Diagram. Can you identify each of the forces shown on the free-body diagram of the beam, Fig. 5–6*b*? For simplicity in applying the equilibrium equations, $\mathbf{F}_1$ is represented by its x and y components as shown.

Equations of Equilibrium. Applying the equilibrium equation $\Sigma F_x = 0$ to the force system on the free-body diagram in Fig. 5–6*b* yields

$\overset{+}{\rightarrow}\Sigma F_x = 0; \qquad 600 \cos 45° \text{ N} - B_x = 0$

$$B_x = 424.3 \text{ N} \qquad \textit{Ans.}$$

A direct solution for $\mathbf{A}_y$ can be obtained by applying the moment equation $\Sigma M_B = 0$ about point B. For the calculation, it should be apparent that forces 600 cos 45° N, $\mathbf{B}_x$, and $\mathbf{B}_y$ create zero moment about B. Assuming counterclockwise rotation about B to be positive, Fig. 5–6*b*, we have

$\circlearrowleft +\Sigma M_B = 0; \quad 100 \text{ N}(2 \text{ m}) + (600 \sin 45° \text{ N})(5 \text{ m}) - A_y(7 \text{ m}) = 0$

$$A_y = 331.6 \text{ N} \qquad \textit{Ans.}$$

Summing forces in the y direction, using the result $A_y = 331.6$ N, gives

$+\uparrow \Sigma F_y = 0; \quad 331.6 \text{ N} - 600 \sin 45° \text{ N} - 100 \text{ N} + B_y = 0$

$$B_y = 192.6 \text{ N} \qquad \textit{Ans.}$$

Example 5–6

The cord shown in Fig. 5–7*a* supports a force of 100 lb and wraps over the frictionless pulley. Determine the tension in the cord and the horizontal and vertical components of reaction at pin A.

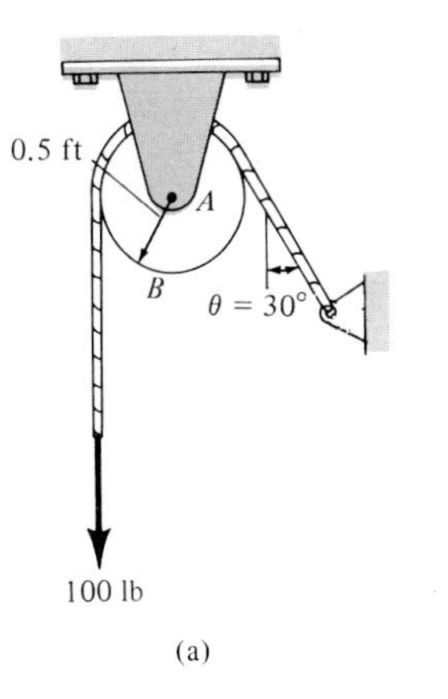

Solution

Free-Body Diagrams. The free-body diagrams of the cord and pulley are shown in Fig. 5–7*b*. Note that the principle of action, equal but opposite reaction, must be carefully observed when drawing each of these diagrams: the cord exerts an unknown pressure distribution p along part of the pulley surface, whereas the pulley exerts an equal but opposite effect on the cord. For the solution, however, it is simpler to *combine* the free-body diagrams of the pulley and a portion of the cord, so that the pressure distribution becomes *internal* to the system and is therefore eliminated from the analysis, Fig. 5–7*c*.

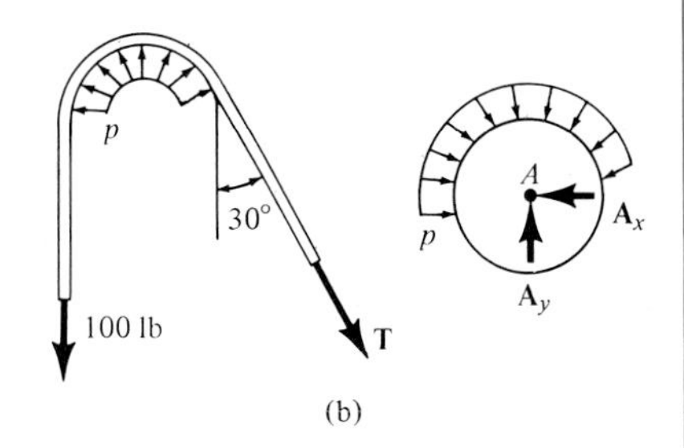

Equations of Equilibrium. Summing moments about point A to eliminate $\mathbf{A}_x$ and $\mathbf{A}_y$, Fig. 5–7*c*, we have

$$\circlearrowleft + \Sigma M_A = 0; \qquad 100\text{ lb}(0.5\text{ ft}) - T(0.5\text{ ft}) = 0$$

$$T = 100\text{ lb} \qquad \textit{Ans.}$$

It is seen that the tension remains *constant* as the cord passes over the pulley. (This of course is true for *any angle* θ at which the cord is directed and for *any radius* r of the pulley.) Using the result for T, a force summation is applied to determine the reaction at pin A.

$$\overset{+}{\rightarrow}\Sigma F_x = 0; \qquad -A_x + 100\sin 30^\circ\text{ lb} = 0$$

$$A_x = 50.0\text{ lb} \qquad \textit{Ans.}$$

$$+\uparrow \Sigma F_y = 0; \qquad A_y - 100\text{ lb} - 100\cos 30^\circ\text{ lb} = 0$$

$$A_y = 186.6\text{ lb} \qquad \textit{Ans.}$$

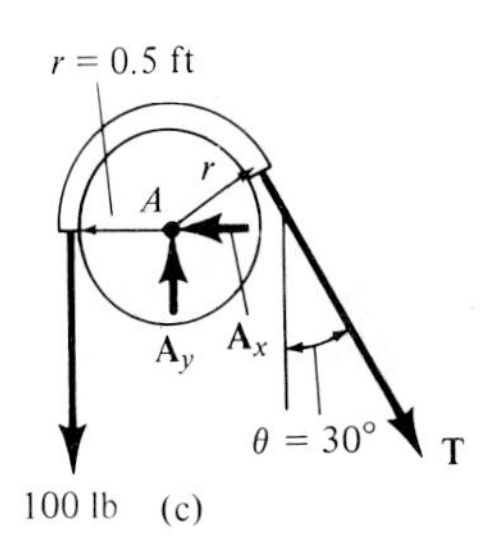

Fig. 5–7

Example 5–7

Determine the reactions at the fixed support A for the loaded frame shown in Fig. 5–8a.

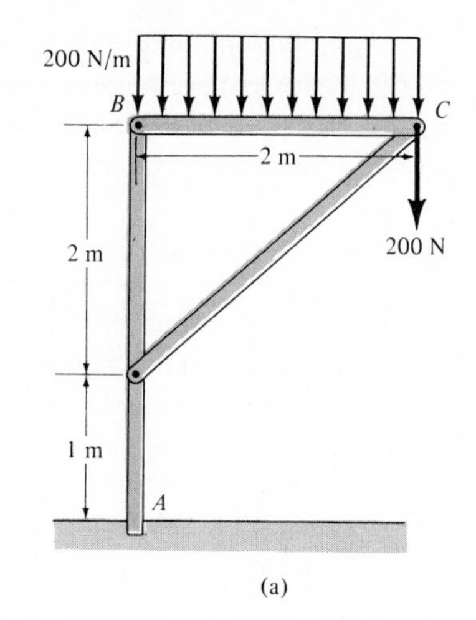

(a)

Solution

Free-Body Diagram. As shown in Fig. 5–8b, there are three unknowns at the fixed support, represented by the magnitudes of $\mathbf{A}_x$, $\mathbf{A}_y$, and $\mathbf{M}_A$. The distributed load is simplified to a resultant force equal to the area under the loading curve, i.e., (2 m)(200 N/m) = 400 N (Sec. 4–8). This force acts through the centroid or geometric center of the loading area, 1 m from B or C.

Equations of Equilibrium.

$\xrightarrow{+}\Sigma F_x = 0;\qquad A_x = 0\qquad$ ***Ans.***

$+\uparrow \Sigma F_y = 0;\qquad A_y - 400\text{ N} - 200\text{ N} = 0$

$A_y = 600\text{ N}\qquad$ ***Ans.***

$\circlearrowleft +\Sigma M_A = 0;\qquad M_A - 400\text{ N}(1\text{ m}) - 200\text{ N}(2\text{ m}) = 0$

$M_A = 800\text{ N}\cdot\text{m}\qquad$ ***Ans.***

Point A was chosen for summing moments since the lines of action of the *unknown* forces $\mathbf{A}_x$ and $\mathbf{A}_y$ pass through this point, and therefore these forces were not included in the moment summation. Note, however, that $\mathbf{M}_A$ must be *included* in the moment summation. This couple is a free vector and represents the effect of the fixed support on the frame.

Although only *three* independent equilibrium equations can be written for a rigid body, it is a good practice to *check* all calculations using a fourth equilibrium equation. The latter equation may be obtained from one of the other sets of equilibrium equations, 5–2 or 5–3. For example, the above computations may be verified by summing moments about point C:

$\circlearrowleft +\Sigma M_C = 0;\qquad 400\text{ N}(1\text{ m}) - 600\text{ N}(2\text{ m}) + 800\text{ N}\cdot\text{m} \equiv 0$

$400\text{ N}\cdot\text{m} - 1200\text{ N}\cdot\text{m} + 800\text{ N}\cdot\text{m} \equiv 0$

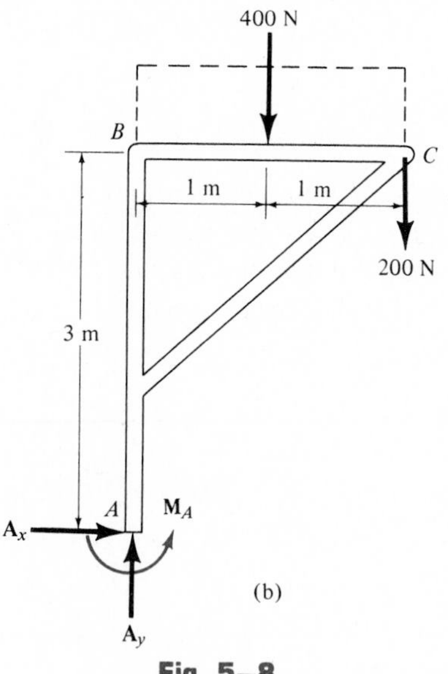

(b)

Fig. 5–8

Example 5–8

The uniform beam shown in Fig. 5–9*a* is subjected to a force and couple. If the beam is supported at *A* by a smooth wall and at *B* and *C* either at the top or bottom by smooth contacts, determine the reactions at these supports. Neglect the weight of the beam.

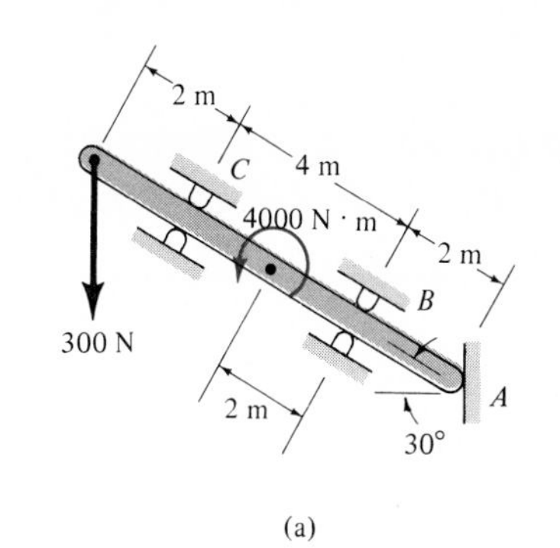

(a)

Solution

Free-Body Diagram. As shown in Fig. 5–9*b*, all the support reactions act normal to the surface of contact since the contacting surfaces are smooth. The reactions at *B* and *C* are shown acting in the positive *y'* direction. This assumes that only the contacts located on the bottom of the beam are used for support.

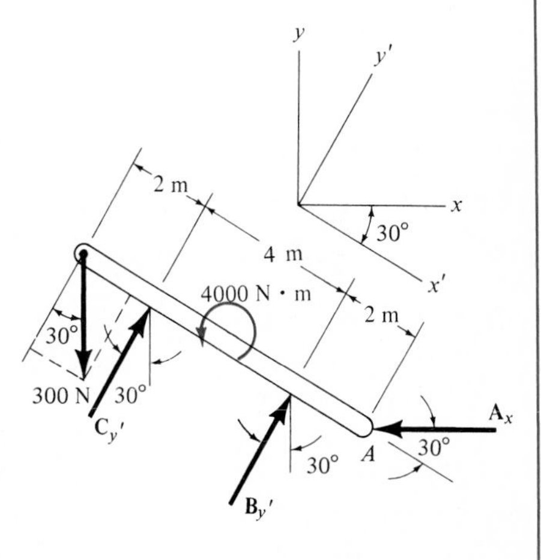

(b)

Fig. 5–9

Equations of Equilibrium. Using the *x*, *y* coordinate system in Fig. 5–9*b*, we have

$$\overset{+}{\rightarrow}\Sigma F_x = 0; \qquad C_{y'} \sin 30° + B_{y'} \sin 30° - A_x = 0 \qquad (1)$$

$$+\uparrow \Sigma F_y = 0; \quad -300 \text{ N} + C_{y'} \cos 30° + B_{y'} \cos 30° = 0 \qquad (2)$$

$$\circlearrowleft + \Sigma M_A = 0; \quad -B_{y'}(2 \text{ m}) + 4000 \text{ N} \cdot \text{m} - C_{y'}(6 \text{ m}) + (300 \cos 30° \text{ N})(8 \text{ m}) = 0 \qquad (3)$$

When writing the moment equation, it should be noticed that the line of action of the force component 300 sin 30° N passes through point *A*, and therefore this force is not included in the moment equation.

Solving Eqs. (2) and (3) simultaneously, we obtain

$$B_{y'} = -1000.0 \text{ N} \qquad \textit{Ans.}$$

$$C_{y'} = 1346.4 \text{ N} \qquad \textit{Ans.}$$

Since $B_{y'}$ is a negative quantity, the direction of $\mathbf{B}_{y'}$ is opposite to that shown on the free-body diagram in Fig. 5–9*b*. Therefore, the top contact at *B* serves as the support rather than the bottom one. Retaining the negative sign for $B_{y'}$ and substituting the results into Eq. (1), we obtain

$$1346.4 \sin 30° \text{ N} - 1000.0 \sin 30° \text{ N} - A_x = 0$$

$$A_x = 173.2 \text{ N} \qquad \textit{Ans.}$$

Example 5–9

The beam shown in Fig. 5–10a is pin-connected at A and rests against a roller at B. Compute the horizontal and vertical components of reaction at the pin A.

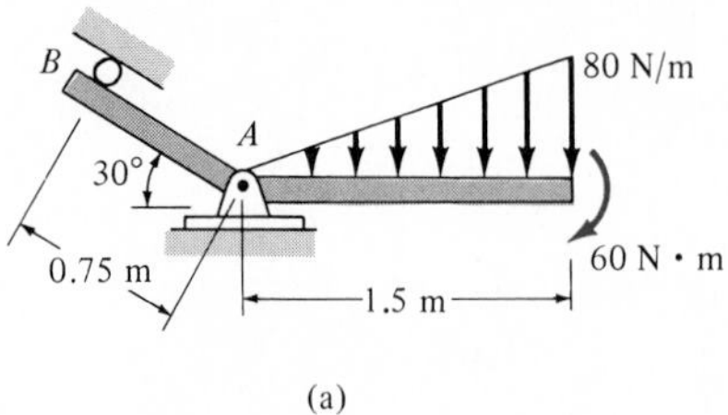

(a)

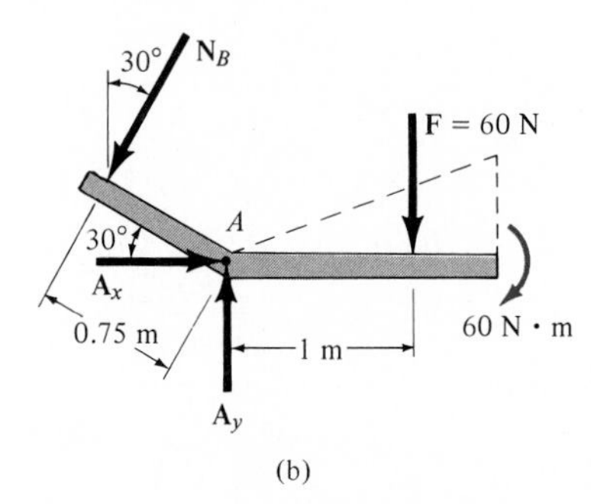

(b)

Fig. 5–10

Solution

Free-Body Diagram. As shown in Fig. 5–10b, the triangular distributed load is reduced to a single concentrated force,

$$F = \tfrac{1}{2}(1.5 \text{ m})(80 \text{ N/m}) = 60 \text{ N}$$

which acts $\tfrac{2}{3}(1.5) = 1$ m from A. The reaction $\mathbf{N}_B$ is perpendicular to the beam at B, since the support is a roller.

Equations of Equilibrium. Summing moments about A, we obtain a direct solution for N_B,

$$\circlearrowright + \Sigma M_A = 0; \quad -60 \text{ N} \cdot \text{m} - 60 \text{ N}(1 \text{ m}) + N_B(0.75 \text{ m}) = 0$$

$$N_B = 160 \text{ N}$$

Using this result,

$$\overset{+}{\rightarrow}\Sigma F_x = 0; \quad A_x - 160 \sin 30° \text{ N} = 0$$

$$A_x = 80.0 \text{ N} \qquad \textit{Ans.}$$

$$+\uparrow \Sigma F_y = 0; \quad A_y - 160 \cos 30° \text{ N} - 60 \text{ N} = 0$$

$$A_y = 198.6 \text{ N} \qquad \textit{Ans.}$$

Example 5–10

The man has a weight of 150 lb and stands at the end of the beam shown in Fig. 5–11*a*. Determine the magnitude and direction of the reaction at the pin *A* and the tension in the cable.

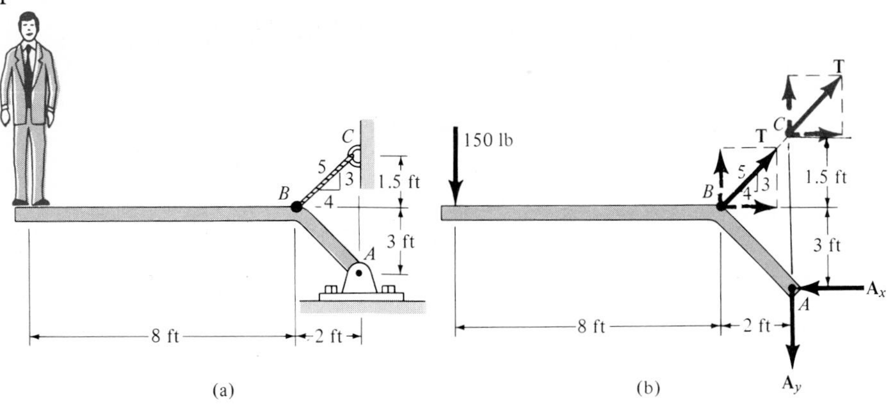

Fig. 5–11

Solution

Free-Body Diagram. The forces acting on the beam are shown in Fig. 5–11*b*.

Equations of Equilibrium. Summing moments about point *A* to obtain a direct solution for the cable tension yields

$$\circlearrowleft + \Sigma M_A = 0; \quad -(\tfrac{3}{5}T)(2 \text{ ft}) - (\tfrac{4}{5}T)(3 \text{ ft}) + 150 \text{ lb}(10 \text{ ft}) = 0$$

$$3.6T - 150 \text{ lb}(10 \text{ ft}) = 0 \qquad (1)$$

$$T = 416.7 \text{ lb} \qquad \textbf{\textit{Ans.}}$$

Note that by the principle of transmissibility it is also possible to locate **T** at *C*, even though this point is not on the beam, Fig. 5–11*b*. In this case, the vertical component of **T** creates *zero moment* about *A* and the moment arm of the horizontal component ($\frac{4}{5}T$) becomes 4.5 ft. Hence, $\Sigma M_A = 0$ yields Eq. (1) directly since $(\frac{4}{5}T)(4.5) = 3.6T$.

Summing forces to obtain A_x and A_y, using the result for *T*, we have

$$\overset{+}{\leftarrow} \Sigma F_x = 0; \qquad A_x - (\tfrac{4}{5})(416.7 \text{ lb}) = 0$$

$$A_x = 333.3 \text{ lb}$$

$$+\uparrow \Sigma F_y = 0; \qquad (\tfrac{3}{5})416.7 \text{ lb} - 150 \text{ lb} - A_y = 0$$

$$A_y = 100 \text{ lb}$$

Thus,

$$F_A = \sqrt{(333.3 \text{ lb})^2 + (100 \text{ lb})^2}$$

$$= 348.0 \text{ lb} \qquad \textbf{\textit{Ans.}}$$

$$\theta = \tan^{-1}\frac{100 \text{ lb}}{333.3 \text{ lb}} = 16.7° \qquad \textbf{\textit{Ans.}}$$

Example 5–11

The 100-kg uniform beam AB shown in Fig. 5–12a is supported at A by a pin and at B and C by a continuous cable which wraps around a frictionless pulley located at D. If a maximum tension force of 800 N can be developed in the cable before it breaks, determine the greatest length b of a uniform 2.5-kN/m distributed load that can be placed on the beam. The load is applied from the left support. What are the horizontal and vertical components of reaction at A just before the cable breaks?

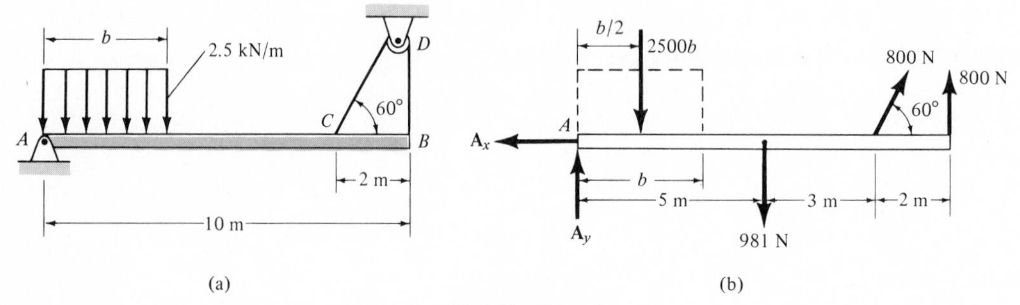

Fig. 5–12

Solution

Free-Body Diagram. Since the cable is continuous and passes over a frictionless pulley, the entire cable is subjected to its maximum tension of 800 N when the maximum loading is on the beam. Hence, the cable exerts an 800-N force at points C and B *on the beam* in the direction of the cable, Fig. 5–12b.

The distributed load is reduced to a concentrated force in accordance with the methods of Sec. 4–8. The magnitude of this force is equivalent to the area under the diagram, i.e., (2500b) N. The force acts through the centroid of the area, a distance of $b/2$ from point A.

Equations of Equilibrium. By summing moments of the force system about point A it is possible to obtain a direct solution for the dimension b. Why?

$$\circlearrowleft +\Sigma M_A = 0; \qquad -(2500b\ \text{N})\left(\frac{b}{2}\,\text{m}\right) - 981\ \text{N}(5\ \text{m})$$

$$+ (800 \sin 60^\circ\ \text{N})(8\ \text{m}) + 800\ \text{N}(10\ \text{m}) = 0$$

$$b = 2.63\ \text{m} \qquad \textit{Ans.}$$

Using this result and summing forces in the x and y directions, we have

$$\xrightarrow{+} \Sigma F_x = 0; \qquad -A_x + 800 \cos 60^\circ\ \text{N} = 0$$

$$A_x = 400\ \text{N} \qquad \textit{Ans.}$$

$$+\uparrow \Sigma F_y = 0;$$

$$A_y - 2500\ \text{N/m}(2.63\ \text{m}) - 981\ \text{N} + 800 \sin 60^\circ\ \text{N} + 800\ \text{N} = 0$$

$$A_y = 6059.9\ \text{N} \qquad \textit{Ans.}$$

Two- and Three-Force Members 5.3

The solution to some equilibrium problems can be simplified if one is able to recognize members that are subjected to either two or three forces.

Two-Force Members. When forces are applied at only two points on a member, the member is called a *two-force member*. An example of this situation is shown in Fig. 5–13*a*. When the forces at A and B are summed to obtain their respective *resultants* $\mathbf{F}_A$ and $\mathbf{F}_B$, Fig. 5–13*b*, then *force equilibrium* is satisfied provided $\mathbf{F}_A$ is of *equal magnitude* and *opposite direction* to $\mathbf{F}_B$. Furthermore, *moment equilibrium* is satisfied if $\mathbf{F}_A$ is *collinear* with $\mathbf{F}_B$. Hence, the line of action of both forces is *known* since it passes through A and B, and only the force magnitude must be determined or stated. Other examples of two-force members are given in Fig. 5–14.

Three-Force Members. *If a member is subjected to three coplanar forces, then it is necessary that the forces be either concurrent or parallel if the member is to be in equilibrium.* To show this, consider the body in Fig. 5–15 and suppose that any two of the three forces acting on the body have lines of action that intersect at point O. To satisfy moment equilibrium about O, i.e., $\Sigma M_O = 0$, the third force must also pass through O, which then makes the force system *concurrent*. If two of the three forces are parallel, the point of concurrency, O, is considered to be at "infinity" and the third force must be parallel to the other two forces to intersect at this "point." Since the three forces are concurrent at a point, only the force equilibrium equations ($\Sigma F_x = 0$, $\Sigma F_y = 0$) must be satisfied.

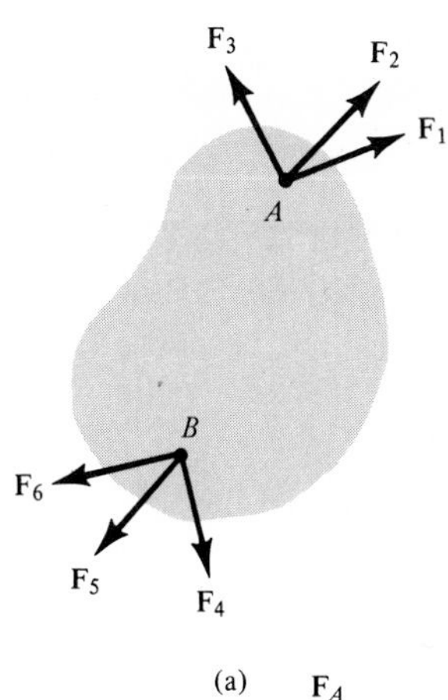

(a)

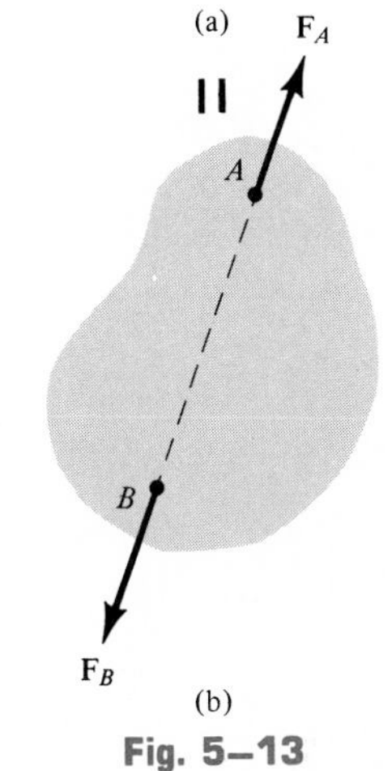

(b)

Fig. 5–13

Fig. 5–14

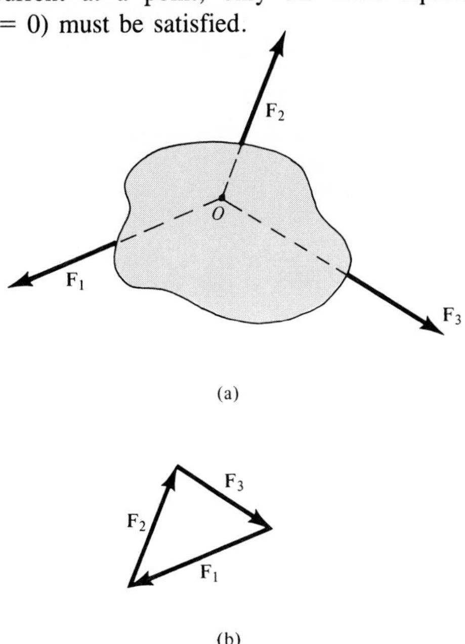

(b)

Fig. 5–15

Example 5–12

The lever ABC is pin-supported at A and connected to a short link BD as shown in Fig. 5–16a. If the weight of the members is negligible, determine the force developed on the lever at A.

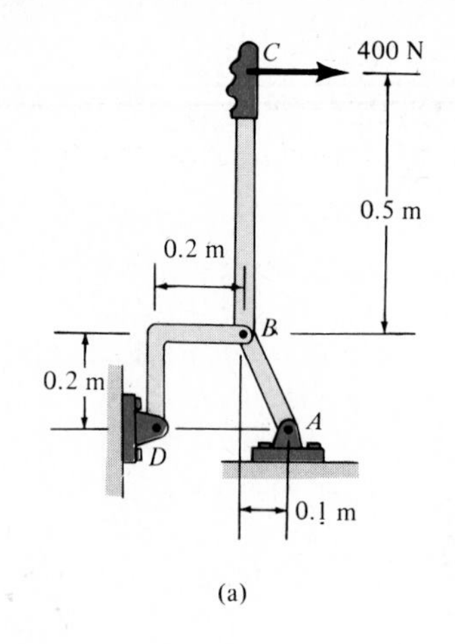

(a)

Solution

Free-Body Diagrams. As shown by the free-body diagram, Fig. 5–16b, the short link BD is a *two-force member,* so the *resultant forces* at pins D and B must be equal, opposite, and collinear. Although the magnitude of the force is unknown, the line of action is known, since it passes through B and D.

Lever ABC is a *three-force member,* and therefore the three nonparallel forces acting on it must be concurrent at O, Fig. 5–16c. In particular, note that the force $\mathbf{F}$ on the lever is equal but opposite to $\mathbf{F}$ acting at B on the link. Why? The distance CO must be 0.5 m, since the lines of action of $\mathbf{F}$ and the 400-N force are known.

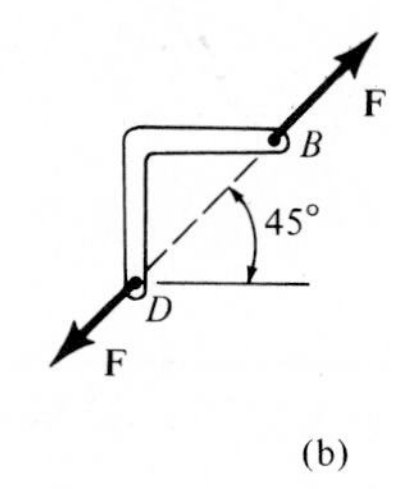

(b)

Equations of Equilibrium. Requiring the force system to be concurrent at O necessitates $\Sigma M_O = 0$. Hence, the angle θ which defines the line of action of $\mathbf{F}_A$ can be determined from trigonometry,

$$\theta = \tan^{-1}\left(\frac{0.7}{0.4}\right) = 60.3^\circ \qquad \textit{Ans.}$$

Applying the force equilibrium equations yields

$$\overset{+}{\rightarrow}\Sigma F_x = 0; \qquad F_A \cos 60.3^\circ - F \cos 45^\circ + 400 \text{ N} = 0$$

$$+\uparrow \Sigma F_y = 0; \qquad F_A \sin 60.3^\circ - F \sin 45^\circ = 0$$

Solving, we get

$$F_A = 1075.0 \text{ N} \qquad \textit{Ans.}$$

$$F = 1320.0 \text{ N}$$

We can also solve this problem by representing the force at A by its two components $\mathbf{A}_x$ and $\mathbf{A}_y$ and applying $\Sigma M_A = 0$, $\Sigma F_x = 0$, $\Sigma F_y = 0$. Once A_x and A_y are determined, how would you find F_A and θ?

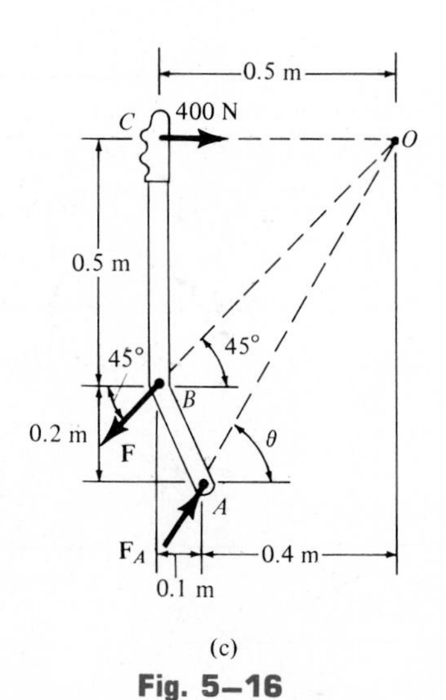

(c)

Fig. 5–16

Problems

***5–16.** Determine the reactions at the support.

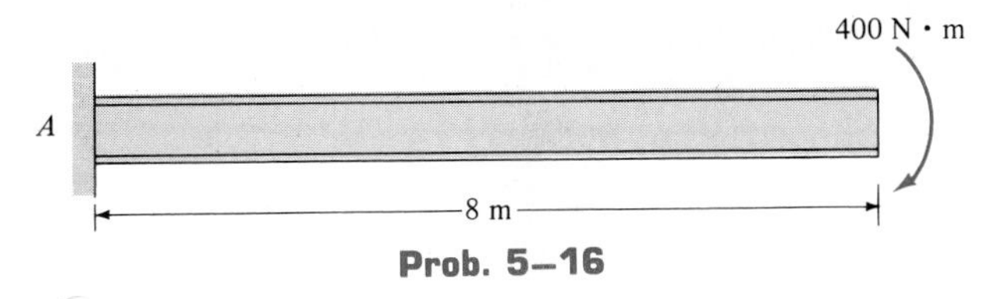

Prob. 5–16

5–17. Determine the reactions at the supports.

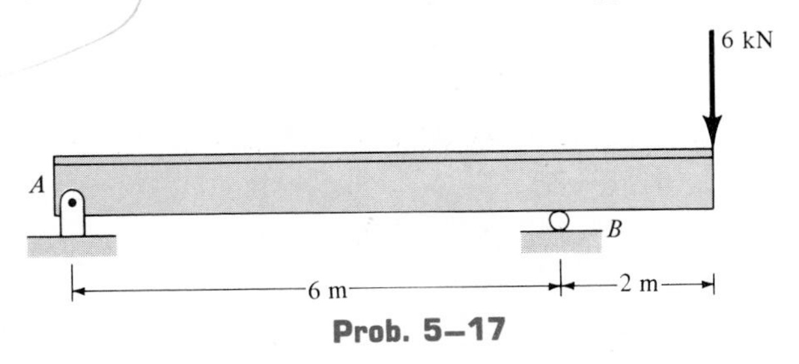

Prob. 5–17

5–18. Determine the reactions at the supports.

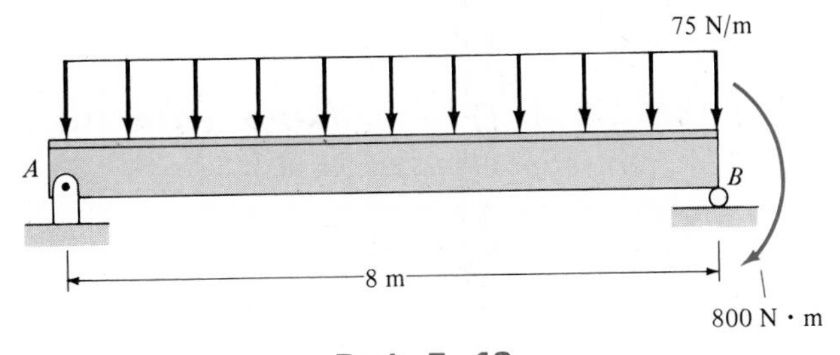

Prob. 5–18

5–19. Determine the reactions at the supports.

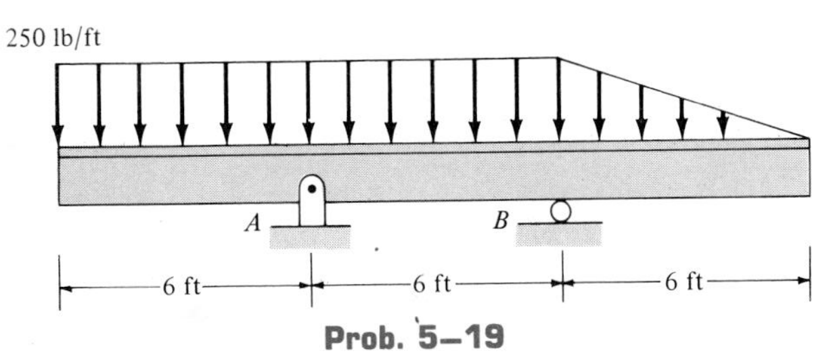

Prob. 5–19

***5–20.** One method of determining the area of a plate or surface area of an unknown shape B is by suspending it from the end of a stick along with a plate A of known area and made with the same material and thickness. Determine the area of B if A has an area of 0.23 m^2 and the dimensions measured from the center C of the stick to each supported area are as shown.

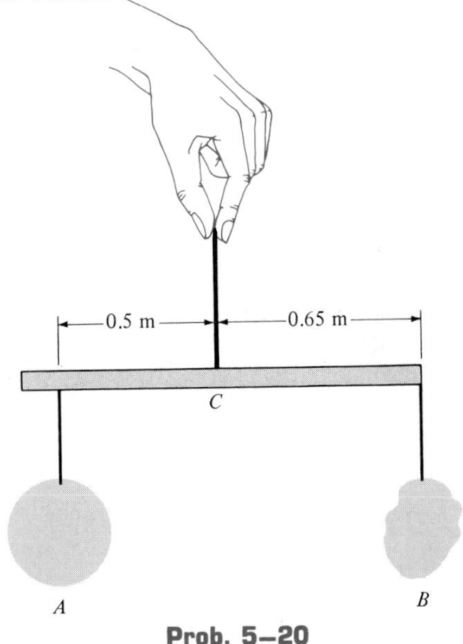

Prob. 5–20

5–21. Compute the horizontal and vertical components of force at pin B. The belt is subjected to a tension of $T = 100$ N and passes over each of the three pulleys.

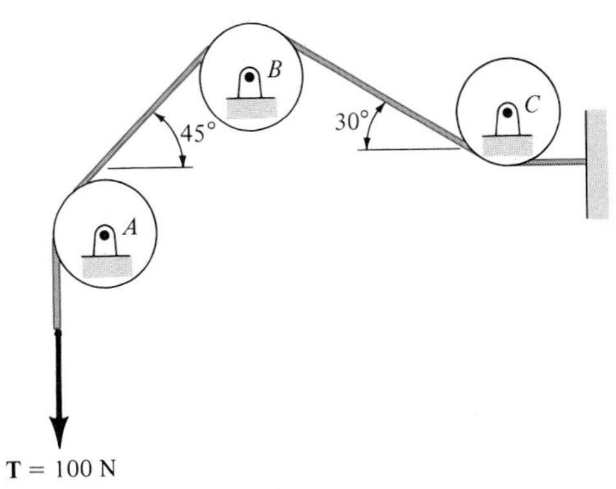

Prob. 5–21

5–22. Determine the horizontal and vertical components of reaction at the pin A and the reaction at the roller support B required for equilibrium of the truss.

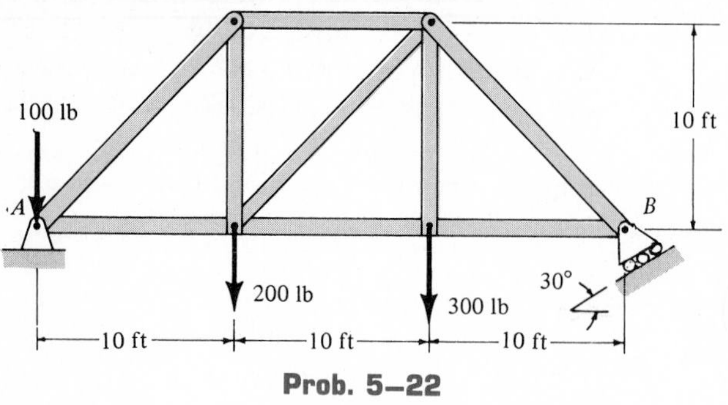

Prob. 5–22

5–23. When holding the 5-lb stone in equilibrium, the humerus H, assumed to be smooth, exerts normal forces $\mathbf{F}_C$ and $\mathbf{F}_A$ on the radius C and ulna A as shown. Determine these forces and the force $\mathbf{F}_B$ that the biceps B exerts on the radius for equilibrium. The stone has a center of mass at G.

Prob. 5–23

***5–24.** The forces acting on the plane while it is flying at constant velocity are shown. If the engine thrust is $F_T = 110$ kip and the plane's weight is $W = 170$ kip, determine the atmospheric drag $\mathbf{F}_D$ and the wing lift $\mathbf{F}_L$. Also, determine the distance s to the line of action of the drag force.

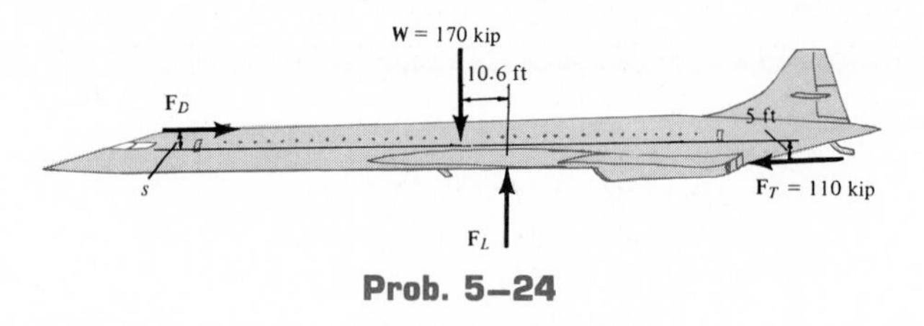

Prob. 5–24

5–25. The oil rig is supported on the trailer by the pin or axle at A and the frame at B. If the rig has a weight of 115,000 lb and center of gravity at G, determine the force $\mathbf{F}$ that must be developed along the hydraulic cylinder CD in order to *start* lifting the rig (slowly) off B toward the vertical. Also compute the horizontal and vertical components of reaction at the pin A.

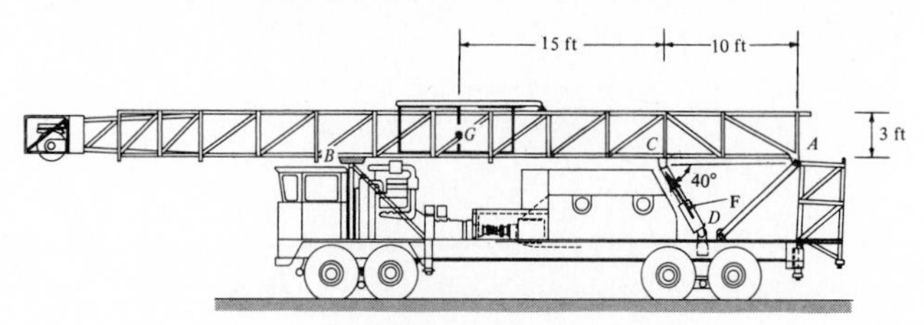

Prob. 5–25

Hint: Take Moment @ "C" due to unknowns @ C

5–26. As an airplane's brakes are applied, the nose wheel, shown dashed, exerts two forces and a moment on the end D of the member CAD. Determine the x and y components of reaction at the pin C and the force $\mathbf{F}$ in strut AB.

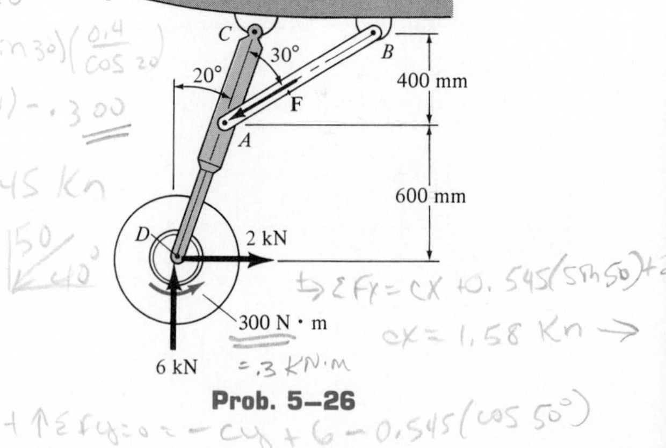

Prob. 5–26

5–27. While *slowly* walking, a man having a total weight of 150 lb places all his weight on *one foot*. Assuming that the normal force $\mathbf{N}_C$ of the ground acts on his foot at C, determine the resultant vertical compressive force $\mathbf{F}_B$ which the tibia T exerts on the astragalus B, and the vertical tension $\mathbf{F}_A$ in the achilles tendon A at the instant shown.

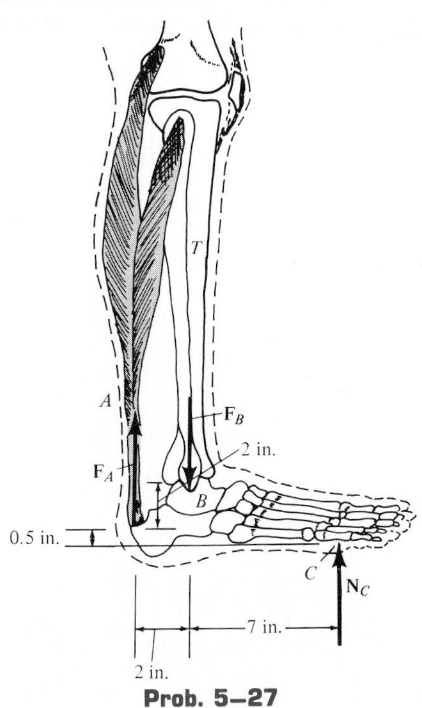

Prob. 5–27

***5–28.** If the wheelbarrow and its contents have a mass of 60 kg and center of mass at G, determine the magnitude of the resultant force which the man must exert on each of the two handles in order to hold the wheelbarrow in equilibrium. The tire at A acts as a smooth roller. *Suggestion:* Determine the horizontal and vertical components of force at B and then add them to determine the resultant force.

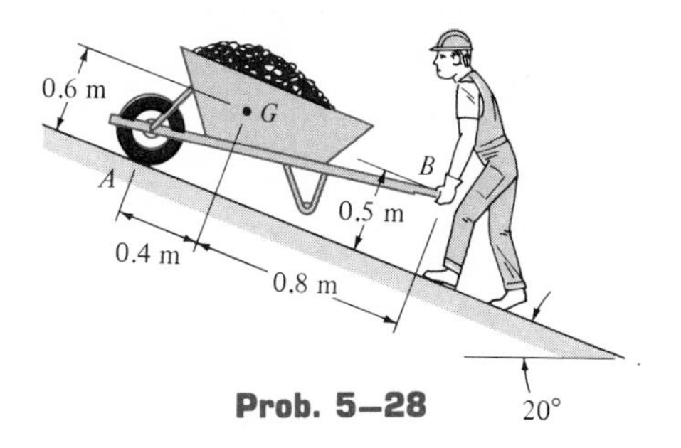

Prob. 5–28

5–29. A sack D having a weight of 8 lb is supported on the scale. Determine the largest number of 0.5-lb weights which should be suspended from the hook at A and the distance x the 0.25-lb bulb B must be placed to keep the scale beam horizontal. What is the reaction at the hook C? The beam remains horizontal when the weights at D, A, and B are removed.

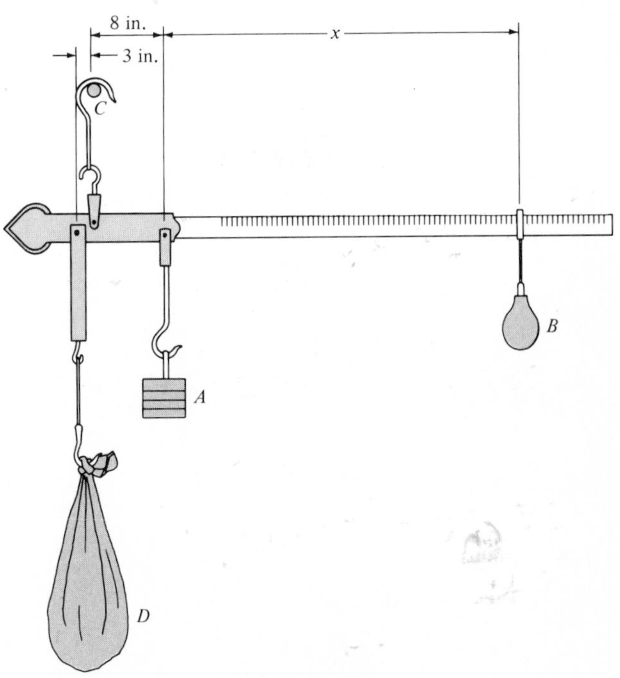

Prob. 5–29

5–30. The lawn roller has a weight of 200 lb and is to be lifted over the 0.2-ft-high step. Determine the magnitude of force $\mathbf{P}$ required to begin to (a) push it and (b) pull it over the step if in each case the force is directed at $\theta = 30°$ along the linkage AB as shown. *Hint:* The normal reaction at the ground becomes zero.

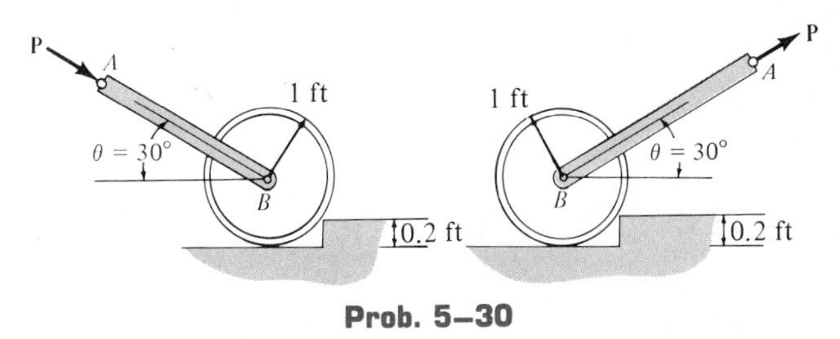

Prob. 5–30

5–31. The support is connected to three springs, which are in the unstretched position when $s = 30$ mm and $\mathbf{P} = \mathbf{0}$. The two inclined springs are attached to pivots B, C and D, E at their ends, which allows free rotation of the springs. Determine the load **P** that must be applied to the support A in order that $s = 0$. Also, compute the force in each spring when $s = 0$.

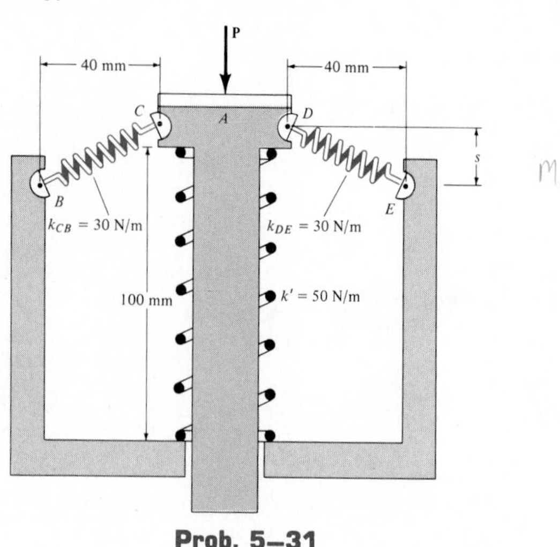

Prob. 5–31

***5–32.** Determine the force **P** in Prob. 5–31, required to hold the support at $s = 10$ mm.

5–33. The cantilever footing is used to support a wall near its edge A such that it causes a uniform soil pressure under the footing. Determine the uniform distribution of force, w_A and w_B, measured in lb/ft at pads A and B, necessary to balance the wall forces of 8,000 lb and 20,000 lb as shown.

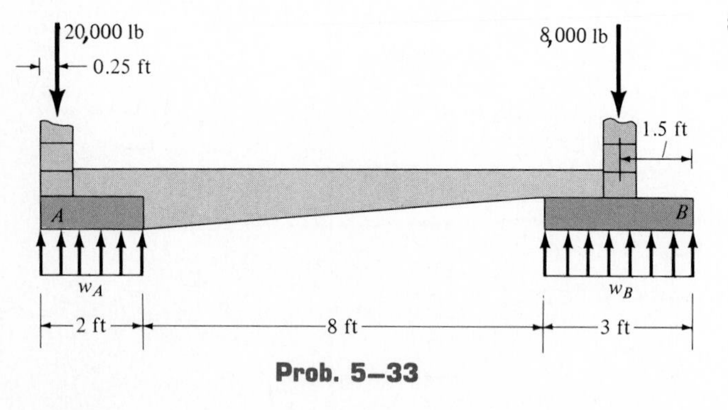

Prob. 5–33

5–34. The crane provides a long-reach capacity by using the telescopic boom segment DE. The entire boom is supported by a pin at A and by the hydraulic cylinder BC, which can be considered as a two-force member. The rated load capacity of the crane is measured by a maximum force **F** developed in the hydraulic cylinder. Determine this maximum force if it is developed when the boom is in the position shown, such that its length is $l = 40$ m, $\theta = 60°$, and it supports a mass of $m = 6$ Mg. Neglect the weight of the boom and the size of the pulley at E.

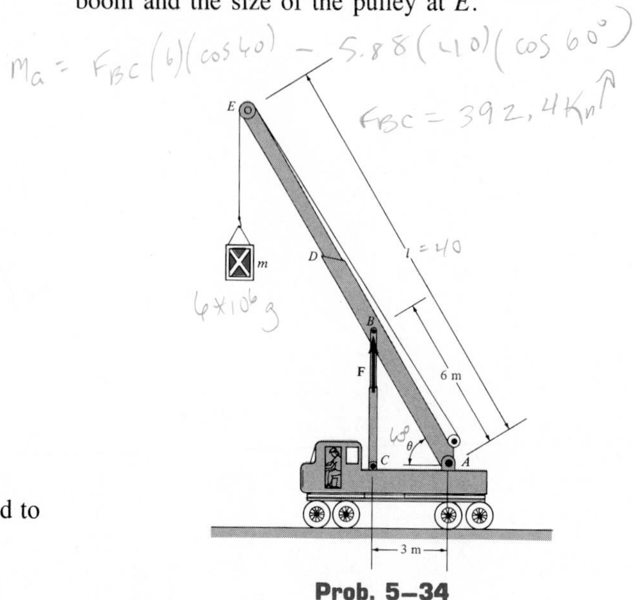

Prob. 5–34

5–35. The uniform ladder, having a length of 18 ft, rests along the wall of a building at A and on the roof at B. If the ladder has a weight of 25 lb and the surfaces at A and B are smooth, determine the angle θ for equilibrium.

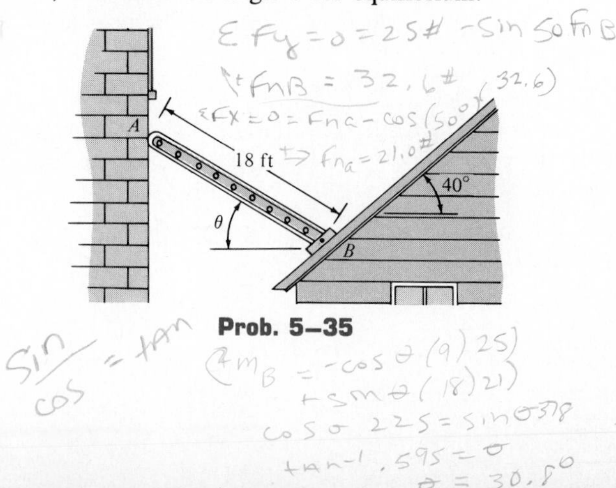

Prob. 5–35

***5–36.** The sports car has a mass of 1.5 Mg and mass center at G. If the front two springs each have a stiffness of $k_A = 58$ kN/m and the rear two springs each have a stiffness of $k_B = 65$ kN/m, determine their compression when the car is parked on the 30° incline. Also, what frictional force $\mathbf{F}_B$ must be applied to each of the rear wheels to hold the car in equilibrium? *Hint:* First determine the normal force at A and B, then determine the compression in the springs.

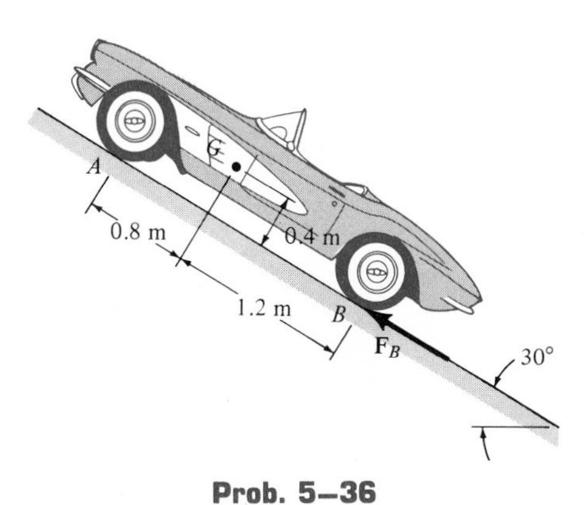

Prob. 5–36

5–37. A cantilever beam, having an extended length of 3 m, is subjected to a vertical force $F = 500$ N acting at its end. Assuming that the wall resists this load with linearly varying distributed loads over the 0.15-m length of the beam, determine the intensities w_1 and w_2 for equilibrium. *Note:* For $b \ll a$, this distribution of load is replaced by a resultant force $\mathbf{F}_y$ and couple moment $\mathbf{M}$ as shown in Table 5–1 (9).

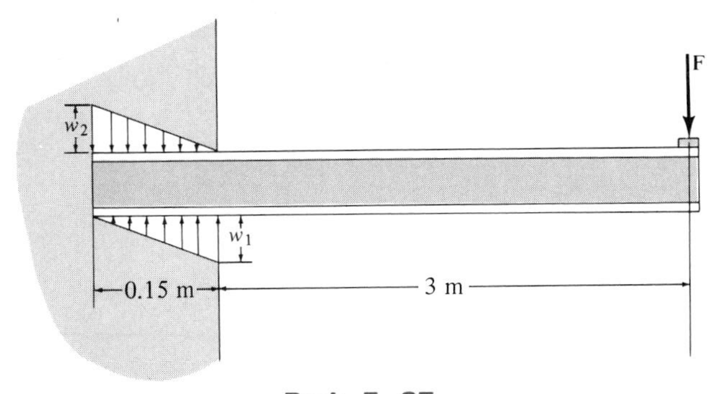

Prob. 5–37

5–38. The *flying boom B* is used with a crane to position construction materials in coves and under overhangs. The horizontal "balance" of the boom is controlled by a 250-kg block D, which has a center of gravity at G and moves by internal sensing devices along the bottom flange F of the beam. Determine the position x of the block when the boom is used to lift the stone S, which has a mass of 60 kg. The boom is uniform and has a mass of 80 kg.

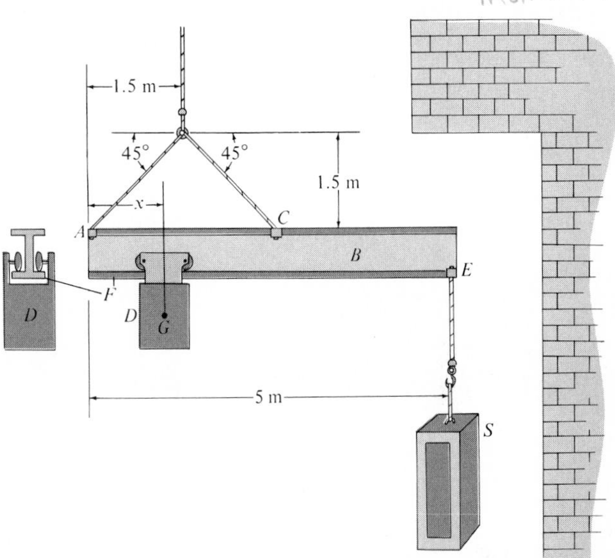

Prob. 5–38

5–39. The wooden plank resting between the buildings deflects slightly when it supports the 50-kg boy. This deflection causes a triangular distribution of load at its ends, having maximum intensities of w_A and w_B. Determine w_A and w_B, each measured in N/m, when the boy is standing 3 m from one end as shown. Neglect the mass of the plank.

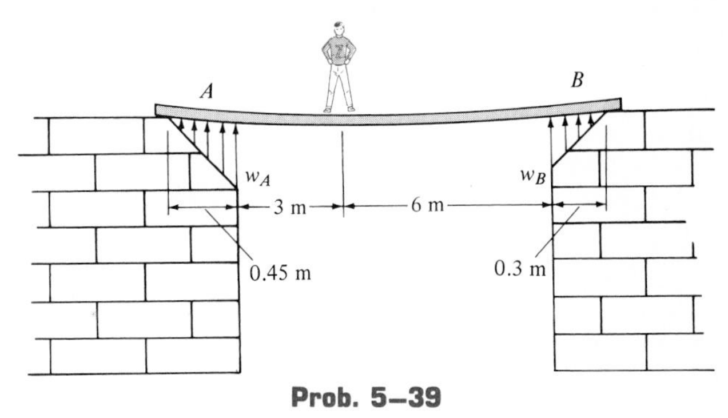

Prob. 5–39

***5–40.** The uniform concrete slab is suspended in the position shown. If it has a weight of 11 kips, determine the resultant normal force $\mathbf{N}_D$ and frictional force $\mathbf{F}_D$ at the ground required for equilibrium. Also, what is the tension in the two supporting crane cables when the slab is in the position shown? Neglect the slab thickness and the size of the rings at A and B. *Suggestion:* The geometry and loading are symmetrical, so draw the free-body diagram in two dimensions.

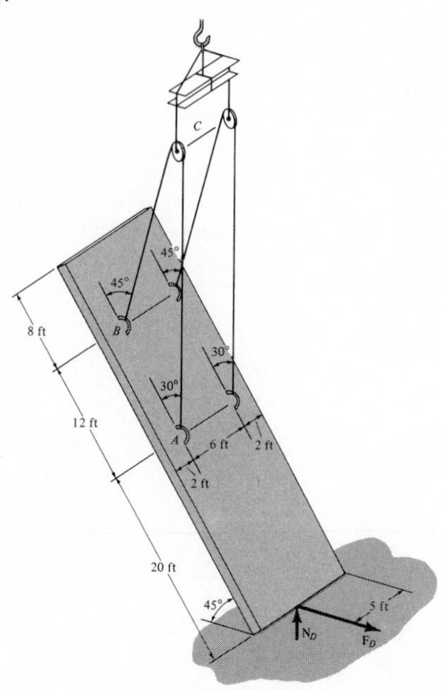

Prob. 5–40

5–41. The wench cable on a tow truck is subjected to a force of $T = 6$ kN when the cable is directed at $\theta = 60°$. Determine the magnitudes of the brake frictional force $\mathbf{F}$ on the rear wheels B and the normal forces at the front wheels A and rear wheels B for equilibrium. The truck has a total mass of 4 Mg and mass center at G.

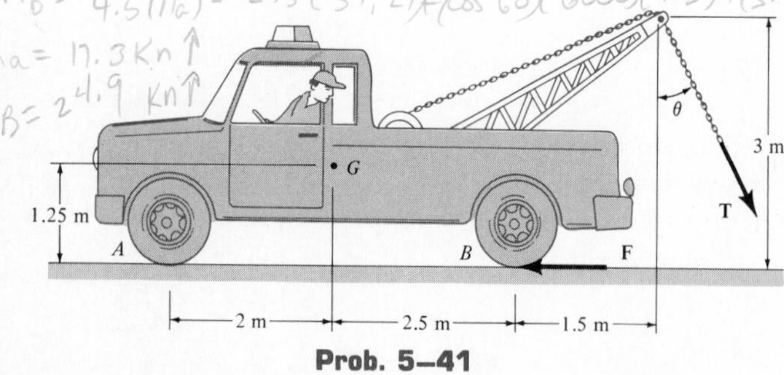

Prob. 5–41

5–42. The 300-kg beam is hoisted by two cables at A and B. If the cables are directed as shown when the beam is supported horizontally, determine the location $\bar{x}$ of the center of gravity G of the beam and compute the tension, $\mathbf{T}_1$ and $\mathbf{T}_2$, in each cable.

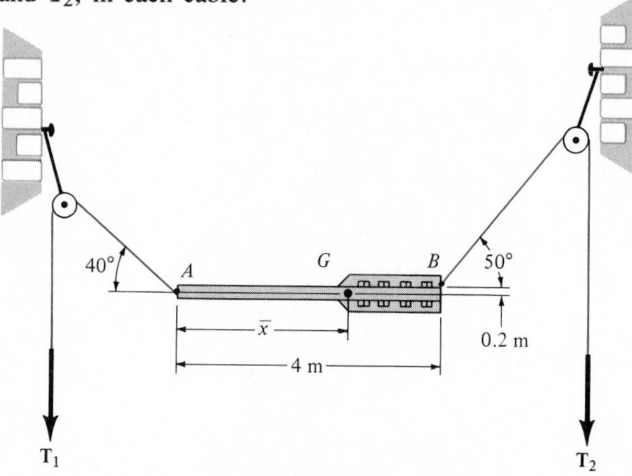

Prob. 5–42

5–43. A man stands out at the end of the diving board, which is supported by two springs A and B, each having a stiffness of $k = 15$ kN/m. In the position shown the board is horizontal. If the man has a mass of 40 kg, determine the angle of tilt which the board makes with the horizontal after he jumps off. Neglect the weight of the board.

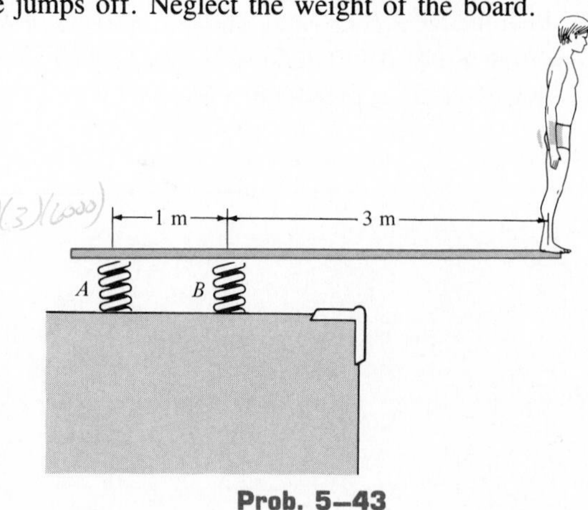

Prob. 5–43

***5–44.** Solve Prob. 5–41 if $T = 6$ kN and $\theta = 30°$.

Equilibrium in Three Dimensions

Support Reactions 5.4

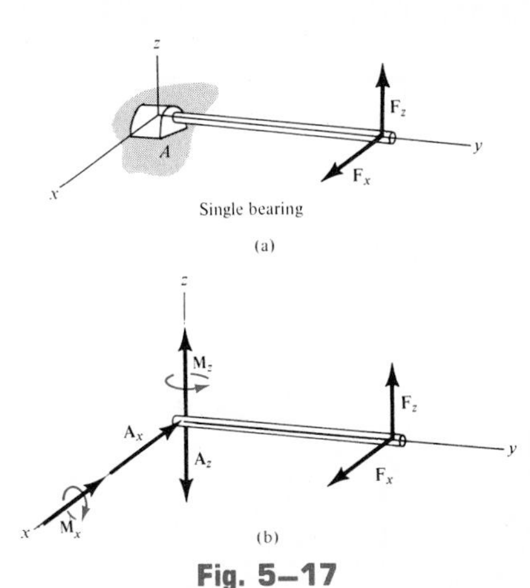

Fig. 5–17

The first step in solving three-dimensional equilibrium problems, as in the case of two dimensions, is to draw a free-body diagram of the body (or group of bodies considered as a system). When drawing this diagram, it is necessary to include *all the reactions* that can occur at each point of support.

The reactive forces and couples acting at various types of supports and connections, when the members are viewed in three dimensions, are listed in Table 5–2. It is important to recognize the symbols used to represent each of the supports and to clearly understand how the forces and couples are developed by each support. As in the two-dimensional case, a *force* is developed by a support that restricts the *translation* of the attached member, whereas a *couple* is developed when *rotation* of the attached member is prevented. For example, in Table 5–2, the ball-and-socket joint (4) prevents any translation of the connecting member; therefore, a force must act on the member at the point of connection. This force has three components having unknown magnitudes, F_x, F_y, F_z. Provided these components are known, one can obtain the magnitude of force, $F = \sqrt{F_x^2 + F_y^2 + F_z^2}$, and the force's orientation defined by the coordinate direction angles α, β, γ (Eqs. 2–3).* Since the connecting member is allowed to rotate freely about *any* axis, no moment is resisted by a ball-and-socket joint.

The *single smooth bearing* (5), Fig. 5–17*a*, is used to support the shaft subjected to $\mathbf{F}_x$ and $\mathbf{F}_z$. From the free-body diagram, Fig. 5–17*b*, it is seen that two force and two couple reactions are developed on the shaft by the support [Table 5–2 (5)]. The force reactions $\mathbf{A}_x$ and $\mathbf{A}_z$ prevent translation of the shaft in the x and z directions, and the couples $\mathbf{M}_x$ and $\mathbf{M}_z$ prevent rotation about these axes. Furthermore, observe that the support cannot prevent translation or rotation of the shaft along the y axis. Consider now the same shaft supported by *two* smooth bearings A and B which are *properly aligned* on the shaft, Fig. 5–18*a*. In this case *only force reactions* are exerted on the shaft by the bearings, Fig. 5–18*b*. Moment reactions *cannot* occur, since the force reactions alone provide the necessary constraint. In other words, it is *not possible* for the applied forces $\mathbf{F}_x$ and $\mathbf{F}_y$ to cause the shaft to rotate about one of the bearings; this is *prevented* by constraining forces developed at the *other* bearing. For example, rotation of the shaft about the x axis caused by $\mathbf{F}_z$, Fig. 5–18*b*, is prevented by the reaction $\mathbf{A}_z$, whereas a rotation about the z axis caused by $\mathbf{F}_x$ is prevented by $\mathbf{A}_x$. The same sort of situation applies for a body connected to two hinges. A *single hinge* [Table 5–2 (7)] acting on a plate,

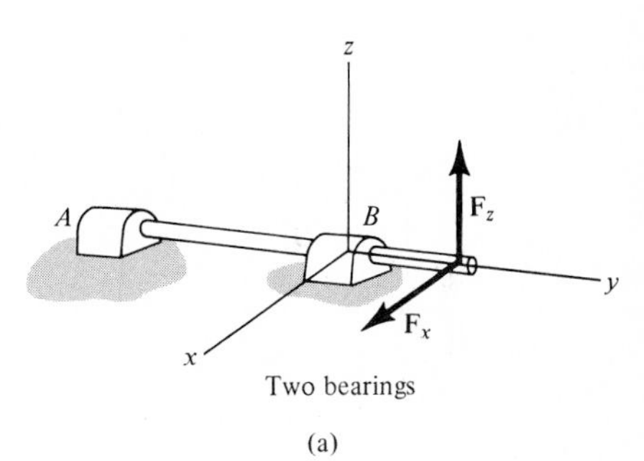

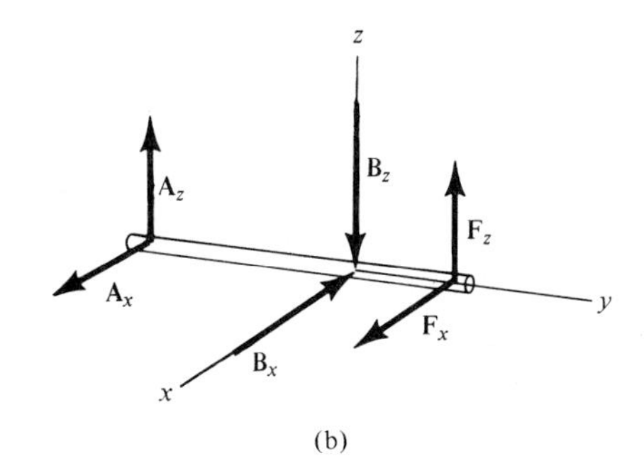

Fig. 5–18

*The three unknowns may also be represented as an unknown force magnitude F and two unknown coordinate direction angles. The third direction angle is obtained using the identity $\cos^2\alpha + \cos^2\beta + \cos^2\gamma = 1$ (Eq. 2–5).

Table 5–2 Supports for Rigid Bodies Subjected to Three-Dimensional Force Systems

Types of Connection	*Reaction*	*Number of Unknowns*
(1) 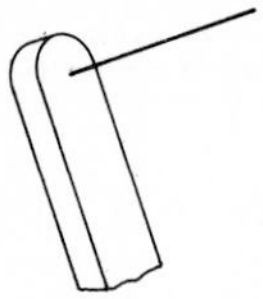 light cable	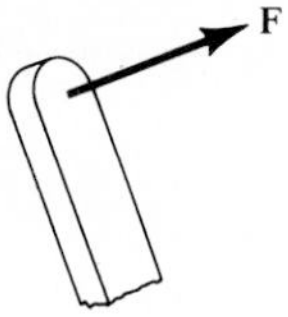 **F**	One unknown. The reaction is a force which acts away from the member in the direction of the cable.
(2) 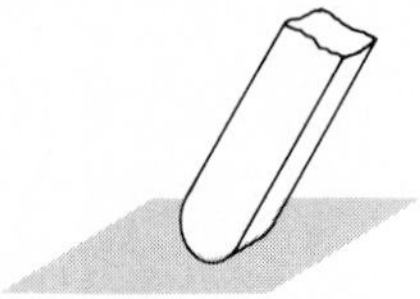 smooth surface support	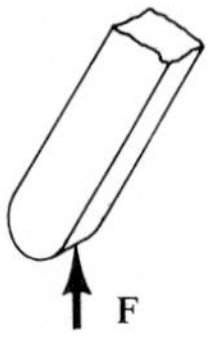 **F**	One unknown. The reaction is a force which acts perpendicular to the surface at the point of contact.
(3) 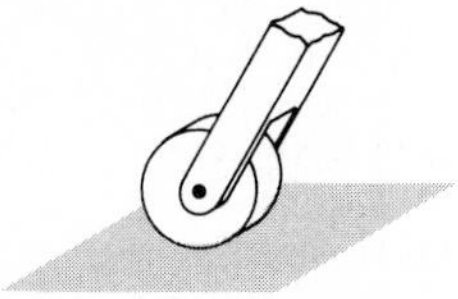 roller on a smooth surface	 **F**	One unknown. The reaction is a force which acts perpendicular to the surface at the point of contact.
(4)  ball and socket	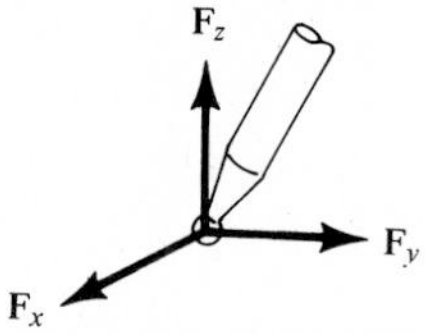 $\mathbf{F}_z$, $\mathbf{F}_y$, $\mathbf{F}_x$	Three unknowns. The reactions are three rectangular force components.
(5) 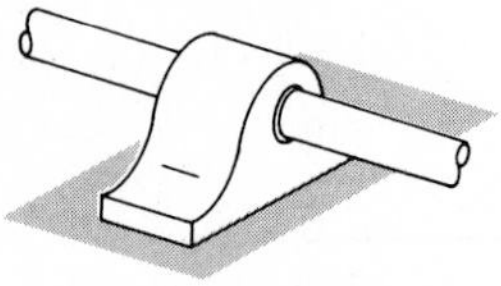 single smooth bearing	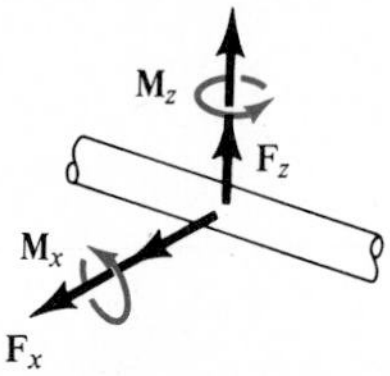 $\mathbf{M}_z$, $\mathbf{F}_z$, $\mathbf{M}_x$, $\mathbf{F}_x$	Four unknowns. The reactions are two force and two couple components which act perpendicular to the shaft.
(6)  single smooth pin	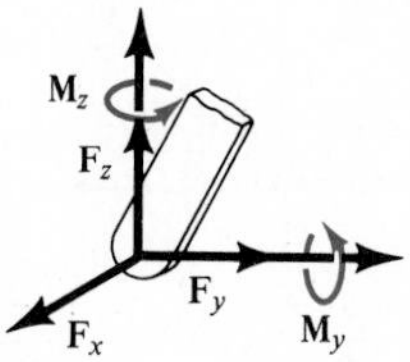 $\mathbf{M}_z$, $\mathbf{F}_z$, $\mathbf{F}_y$, $\mathbf{F}_x$, $\mathbf{M}_y$	Five unknowns. The reactions are three force and two couple components.

Table 5–2 (Contd.)

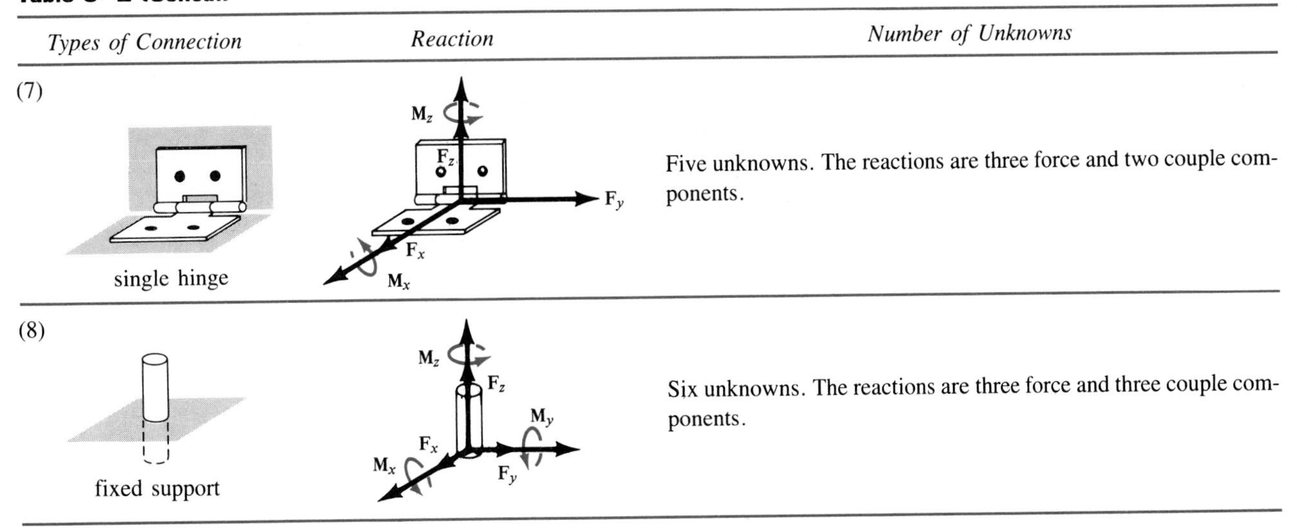

Types of Connection	Reaction	Number of Unknowns
(7) single hinge	$\mathbf{M}_z$, $\mathbf{F}_z$, $\mathbf{F}_y$, $\mathbf{F}_x$, $\mathbf{M}_x$	Five unknowns. The reactions are three force and two couple components.
(8) fixed support	$\mathbf{M}_z$, $\mathbf{F}_z$, $\mathbf{M}_y$, $\mathbf{F}_x$, $\mathbf{M}_x$, $\mathbf{F}_y$	Six unknowns. The reactions are three force and three couple components.

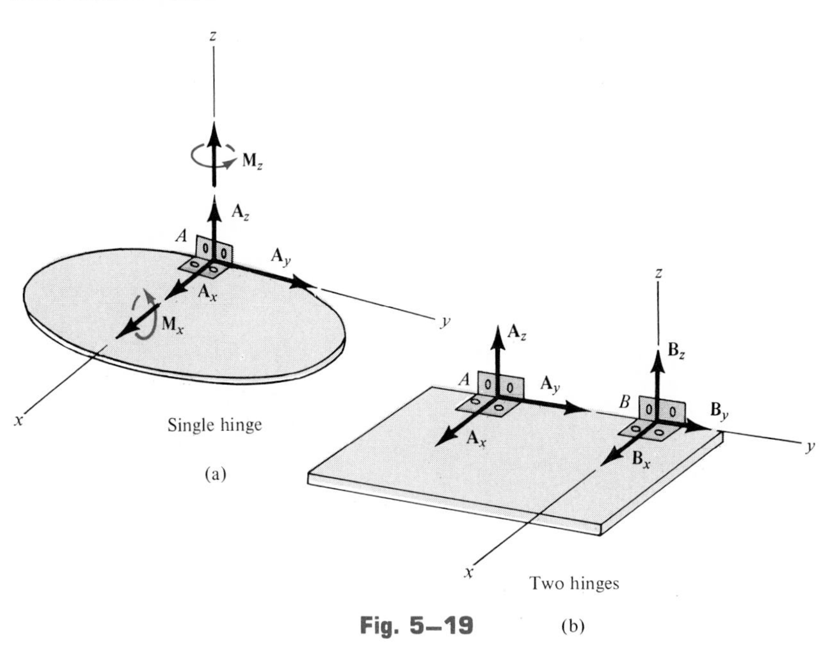

Fig. 5–19

Fig. 5–19*a*, may be subjected to "twisting" around the x and z axes due to an applied loading (not shown). To resist this, couples $\mathbf{M}_x$ and $\mathbf{M}_z$ are developed at the connection, Fig. 5–19*a*. If two *properly aligned* hinges A and B act on a plate, Fig. 5–19*b*, the hinge B, for example, cannot be twisted about the x or z axis, since there are constraining forces developed by hinge A. Therefore, no couples are developed at the hinges.

5.5 Equations of Equilibrium

A necessary condition for the equilibrium of a rigid body subjected to a three-dimensional force system is that both the *resultant* force and *resultant* couple acting on the body be equal to *zero*. From Eqs. 4–8 and 4–9 this requires

$$\Sigma F_x = 0$$
$$\Sigma F_y = 0 \qquad (5\text{–}4)$$
$$\Sigma F_z = 0$$

and

$$\Sigma M_x = 0$$
$$\Sigma M_y = 0 \qquad (5\text{–}5)$$
$$\Sigma M_z = 0$$

These six *equilibrium equations* may be used to solve for at most six unknowns shown on the free-body diagram. Equations 5–4 express the fact that the sum of the external force components acting in the x, y, and z directions must be zero, and Eqs. 5–5 require the sum of the moment components directed along the x, y, and z *axes* to be zero.

PROCEDURE FOR ANALYSIS

The following procedure provides a method for solving three-dimensional equilibrium problems.

Free-Body Diagram. Construct the free-body diagram for the body. When drawing this diagram, it is important to include *all* the forces and couples that act *on* the body. These interactions are commonly caused by the externally applied loadings, contact forces exerted by adjacent bodies, support reactions, and the weight of the body if it is significant compared to the magnitudes of the other applied forces. Dimensions of the body, necessary for computing the moments of forces, are also included on the free-body diagram. If the unknown components of forces and/or moments are *assumed* to act in the "positive" coordinate directions, then negative values obtained from the solution of equilibrium equations would indicate the components act in the "negative" coordinate directions.

Equations of Equilibrium. Establish the x, y, z coordinate axes and apply the equations of equilibrium: $\Sigma F_x = 0$, $\Sigma F_y = 0$, $\Sigma F_z = 0$, $\Sigma M_x = 0$,

$\Sigma M_y = 0$, $\Sigma M_z = 0$. Here it is *not necessary* that the set of axes chosen for force summation *coincide* with the set of axes chosen for moment summation. Instead, it is recommended that one *choose the direction of a moment axis such that it intersects the lines of action of as many unknown forces as possible*. The moments of forces passing through points on this axis or forces which are parallel to the axis will then be zero. Furthermore, the axes chosen for either the force or moment summation do not have to be perpendicular to one another. These axes must, however, *not be parallel*, or linearly dependent equations will result. By the proper choice of axes, it may be possible to solve directly for an unknown quantity, or at least reduce the need for solving a large number of simultaneous equations for the unknowns.

5.6 Sufficient Conditions for Equilibrium

To ensure the equilibrium of a rigid body, it is not only necessary to satisfy the equations of equilibrium, but the body must also be properly held or constrained by its supports. Three situations may occur where the conditions for proper constraint have not been met.

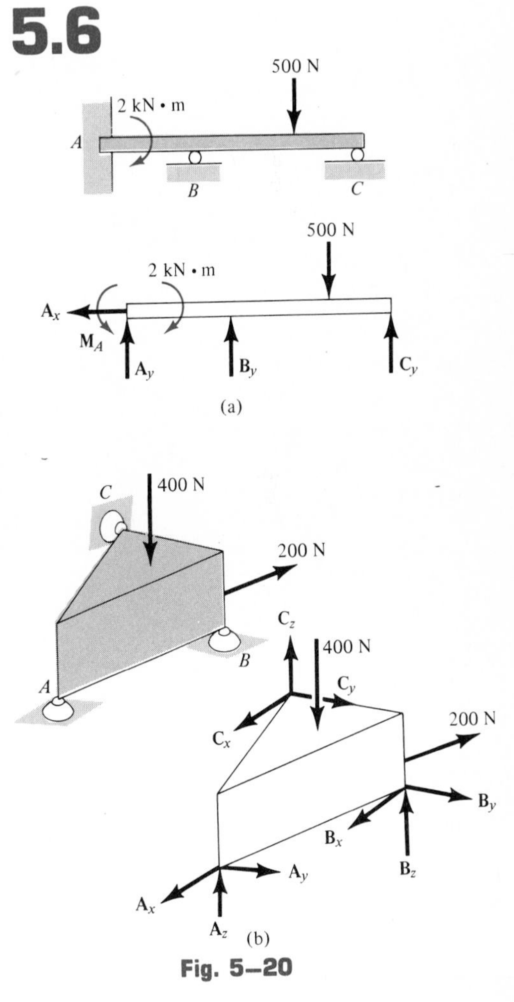

Fig. 5–20

Redundant Constraints. When all the reactive forces on a body can be determined from the equations of equilibrium, the problem is called *statically determinate*. Problems having more unknown forces than equilibrium equations are called *statically indeterminate*. Statical indeterminacy can arise if a body has redundant constraints, i.e., more than are necessary to maintain equilibrium. For example, the two-dimensional problem, Fig. 5–20*a*, and the three-dimensional problem, Fig. 5–20*b*, shown together with their free-body diagrams, are both statically indeterminate because of the redundancy in the number of support reactions. In the two-dimensional case, there are five unknowns, that is, M_A, A_x, A_y, B_y, and C_y, for which only three equilibrium equations can be written ($\Sigma F_x = 0$, $\Sigma F_y = 0$, and $\Sigma M_O = 0$, Eqs. 5–1). Two more equations are needed for a complete solution. Consequently, this problem is termed "indeterminate to the second degree" (five unknowns minus three equations). The three-dimensional problem has nine unknowns, for which only six equilibrium equations can be written (Eqs. 5–4 and 5–5). It is "indeterminate to the third degree." The additional equations needed to solve indeterminate problems of the type shown in Fig. 5–20 may be obtained from the conditions of *deformation* which occur between the loads and the internal movements of the body. These relations involve the physical properties of the body which are studied in subjects dealing with the mechanics of deformation, such as "strength of materials."

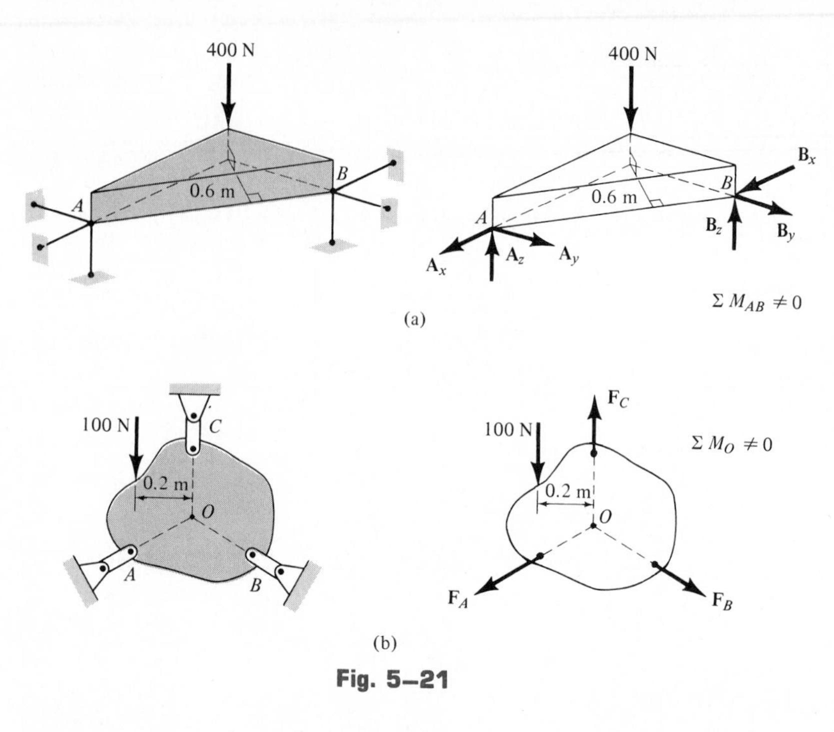

Fig. 5–21

Improper Constraints. In some cases, there may be as many unknown forces as there are equations of equilibrium; however, *instability* of the body can develop because of *improper constraining* action by the supports. In the case of three-dimensional problems, the body is improperly constrained if the support reactions *all intersect a common axis*. For two-dimensional problems, this axis is *perpendicular* to the plane of the forces and therefore appears as a point. Hence, when all the reactive forces are *concurrent* at this point, the body is improperly constrained. Examples of both cases are given in Fig. 5–21. From the free-body diagrams it is seen that the summation of moments about axis *AB*, Fig. 5–21*a*, or point *O*, Fig. 5–21*b*, will *not* be equal to zero; thus rotation about axis *AB* or point *O* will take place.* Furthermore, in both cases it becomes *impossible* to solve *completely* for all the unknowns, since one can write a moment equation that does not involve any of the unknown support reactions. (This limits by one the number of available equilibrium equations.)

Another way in which improper constraining leads to instability occurs when the *reactive forces* are all *parallel*. Three- and two-dimensional examples of this are shown in Fig. 5–22. In both cases, the summation of forces along the horizontal *aa* axis will not equal zero.

*For the three-dimensional problem, $\Sigma M_{AB} = (400 \text{ N})(0.6 \text{ m}) \neq 0$, and for the two-dimensional problem, $\Sigma M_O = (100 \text{ N})(0.2 \text{ m}) \neq 0$.

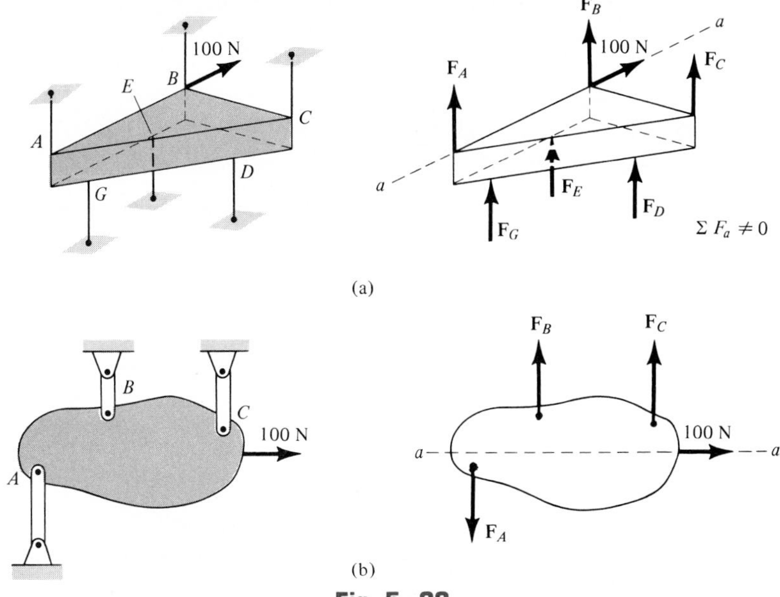

Fig. 5–22

Partial Constraints. In some cases, a body may have *fewer* reactive forces than equations of equilibrium that must be satisfied. The body then becomes only *partially constrained*. For example, consider the body shown in Fig. 5–23*a* with its corresponding free-body diagram in Fig. 5–23*b*. If O is a point not on the line of action of AB, the equations $\Sigma F_x = 0$ and $\Sigma M_O = 0$ will be satisfied by proper choice of the reactions $\mathbf{F}_A$ and $\mathbf{F}_B$. The equation $\Sigma F_y = 0$, however, will not be satisfied for the loading conditions and therefore equilibrium will not be maintained.

Proper constraining requires that (1) the lines of action of the reactive forces do not intersect points on a common axis, and (2) the reactive forces must not all be parallel to one another. When the number of reactive forces needed to properly constrain the body in question is a *minimum*, the equations of equilibrium can be used to determine *all* the reactive forces.

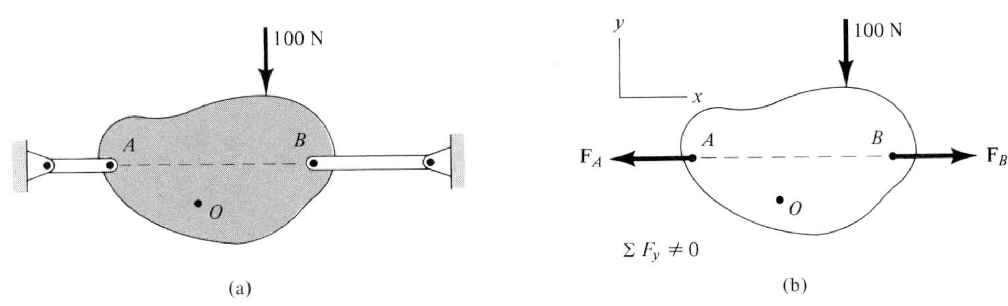

Fig. 5–23

Example 5–13

The homogeneous plate shown in Fig. 5–24*a* has a mass of 100 kg and is subjected to a distributed load along its edge and a couple moment. If it is supported in the horizontal plane by means of a roller at *A*, a ball-and-socket joint at *B*, and a cord at *C*, determine the components of reaction at the supports.

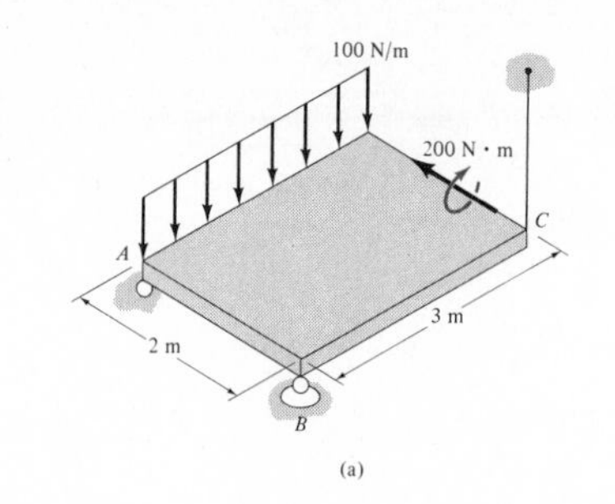

(a)

Solution

Free-Body Diagram. Can you identify each of the forces shown on the free-body diagram of the plate, Fig. 5–24*b*? Refer to Table 5–2 (1, 3, 4). Notice that the unknown components at *A* and *B* are assumed to act in the positive coordinate directions.

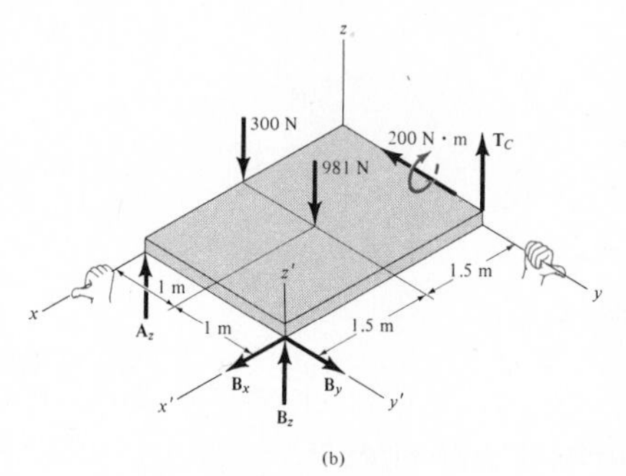

(b)

Fig. 5–24

Equations of Equilibrium. A force summation yields

$\Sigma F_x = 0;\qquad B_x = 0$ *Ans.*

$\Sigma F_y = 0;\qquad B_y = 0$ *Ans.*

$\Sigma F_z = 0;\qquad A_z + B_z + T_C - 300\text{ N} - 981\text{ N} = 0 \qquad (1)$

Recall that the moment of a force about an axis is equal to the product of the force magnitude and the perpendicular distance (moment arm) from the line of action of the force to the axis. The direction of the moment is determined by the right-hand rule. Hence, summing moments of the forces on the free-body diagram, with positive moments acting along the positive *x* or *y* axis, we have

$\Sigma M_x = 0;\quad T_C(2\text{ m}) - 981\text{ N}(1\text{ m}) + B_z(2\text{ m}) = 0 \qquad (2)$

$\Sigma M_y = 0;\quad 300\text{ N}(1.5\text{ m}) + 981\text{ N}(1.5\text{ m}) - B_z(3\text{ m}) - A_z(3\text{ m}) - 200\text{ N}\cdot\text{m} = 0 \qquad (3)$

The components of force at *B* can be eliminated if the x', y', z' axes are used. We obtain

$\Sigma M_{x'} = 0;\quad 981\text{ N}(1\text{ m}) + 300\text{ N}(2\text{ m}) - A_z(2\text{ m}) = 0 \qquad (4)$

$\Sigma M_{y'} = 0;\quad -300\text{ N}(1.5\text{ m}) - 981\text{ N}(1.5\text{ m}) - 200\text{ N}\cdot\text{m} + T_C(3\text{ m}) = 0 \qquad (5)$

Solving Eqs. (1) to (3) or the more convenient Eqs. (1), (4), and (5) yields

$A_z = 790.5\text{ N}\qquad B_z = -216.7\text{ N}\qquad T_C = 707.2\text{ N}$ *Ans.*

The negative sign indicates that $\mathbf{B}_z$ acts downward.

Example 5–14

The windlass shown in Fig. 5–25*a* is supported by two smooth bearings *A* and *B* which are properly aligned on the shaft. Determine the magnitude of the vertical force **P** that must be applied to the handle to maintain equilibrium of the 100-kg crate. Also calculate the reactions at the bearings.

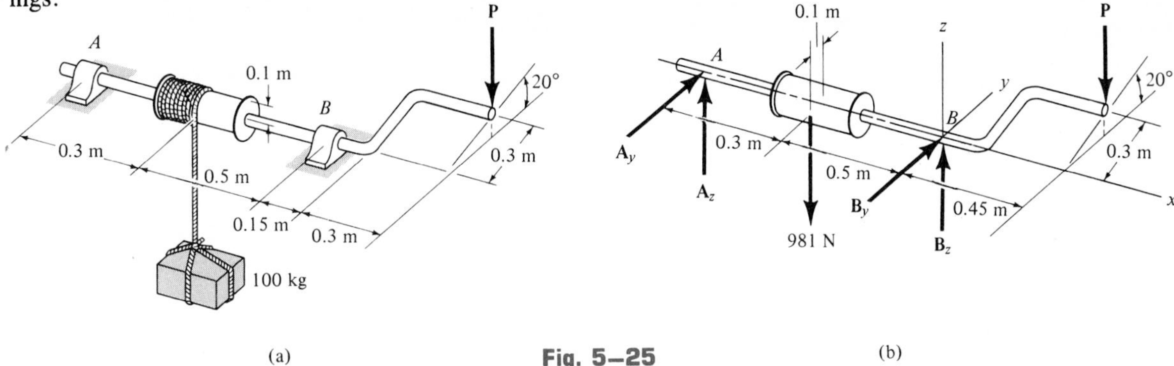

Fig. 5–25

Solution

Free-Body Diagram. Since the bearings at *A* and *B* are aligned correctly, *only* force reactions occur at these supports, Fig. 5–25*b*? Why are there no moment reactions?

Equations of Equilibrium. Summing moments about the *x* axis yields a direct solution for **P.** Why? For a moment summation, it is necessary to compute the moment of each force as the product of the force magnitude and the *perpendicular distance* from the *x* axis to the line of action of the force. Using the right-hand rule and assuming positive moments act in the $+x$ direction, we have

$\Sigma M_x = 0;\qquad 981\text{ N}(0.1\text{ m}) - P(0.3\cos 20°\text{ m}) = 0$

$$P = 348.0\text{ N}\qquad \textit{Ans.}$$

Using this result and summing moments about the *y* and *z* axes yields

$\Sigma M_y = 0;$

$$-981\text{ N}(0.5\text{ m}) + A_z(0.8\text{ m}) + (348.0\text{ N})(0.45\text{ m}) = 0$$

$$A_z = 417.4\text{ N}\qquad \textit{Ans.}$$

$\Sigma M_z = 0;\qquad -A_y(0.8\text{ m}) = 0\qquad A_y = 0\qquad \textit{Ans.}$

The reactions at *B* are obtained by a force summation, using the results computed above:

$\Sigma F_y = 0;\qquad 0 + B_y = 0\qquad B_y = 0\qquad \textit{Ans.}$

$\Sigma F_z = 0;\quad 417.4 - 981 + B_z - 348.0 = 0\qquad B_z = 911.6\text{ N}\qquad \textit{Ans.}$

As shown on the free-body diagram, the *supports* do not provide resistance against translation in the *x* direction. Hence, the windlass is only partially constrained.

Example 5–15

Determine the tension in cables BC and BD and the reactions at the ball-and-socket joint A for the mast shown in Fig. 5–26a.

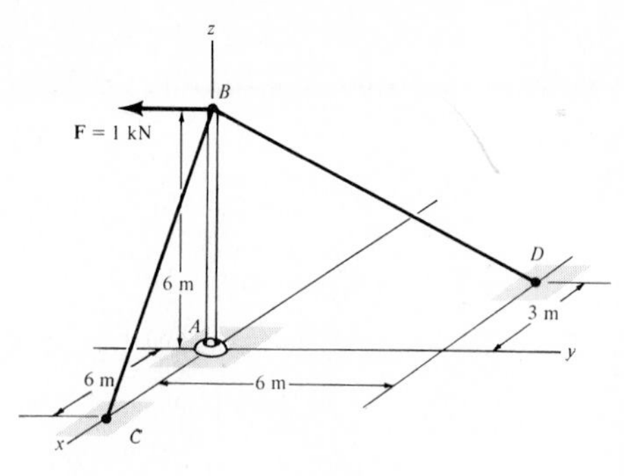

(a)

Solution

Free-Body Diagram. There are five unknown force magnitudes shown on the free-body diagram, Fig. 5–26b.

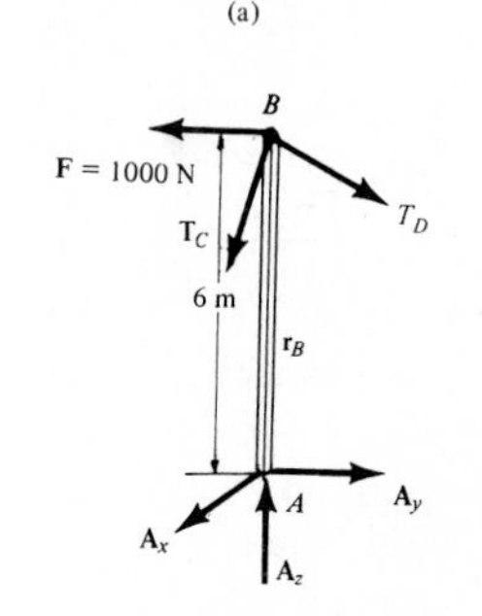

(b)

Equations of Equilibrium. The x, y, z components of the forces are first determined. From the geometry of Fig. 5–26a, we have

$$F_x = 0 \qquad F_y = -1000\text{ N} \qquad F_z = 0$$
$$A_x = ? \qquad A_y = ? \qquad A_z = ?$$
$$T_{C_x} = 0.707T_C \qquad T_{C_y} = 0 \qquad T_{C_z} = -0.707T_C$$

Since the components of a position vector directed from B to D are $r_x = -3$ m, $r_y = 6$ m, $r_z = -6$ m, and $r_{BD} = 9$ m, then

$$T_{D_x} = T_D\left(\frac{-3}{9}\right) = -0.333T_D \qquad T_{D_y} = T_D\left(\frac{6}{9}\right) = 0.667T_D$$

$$T_{D_z} = T_D\left(\frac{-6}{9}\right) = -0.667T_D$$

These components are shown on the free-body diagram in Fig. 5–26c. Note that the negative signs are *not* included here since the component arrowheads indicate the directional sense.

Applying the equations of force equilibrium gives

$$\Sigma F_x = 0; \qquad A_x + 0.707T_C - 0.333T_D = 0$$
$$\Sigma F_y = 0; \qquad A_y + 0.667T_D - 1000 = 0$$
$$\Sigma F_z = 0; \qquad A_z - 0.707T_C - 0.667T_D = 0$$

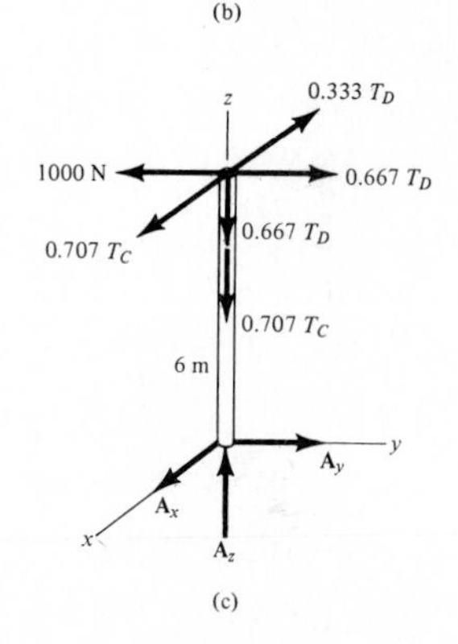

(c)

Fig. 5–26

Moments will be summed about the x, y, z axes located at A, Fig. 5–26c. We have

$$\Sigma M_x = 0; \qquad -0.667T_D(6) + 1000(6) = 0$$
$$\Sigma M_y = 0; \qquad 0.707T_C(6) - 0.333T_D(6) = 0$$
$$\Sigma M_z = 0; \qquad 0 = 0$$

Solving the above equations yields

$$T_C = 707.2\text{ N} \qquad T_D = 1500.0\text{ N} \qquad \textit{Ans.}$$
$$A_x = 0.0\text{ N} \qquad A_y = 0.0\text{ N} \qquad A_z = 1500.0\text{ N} \qquad \textit{Ans.}$$

Since the mast is a two-force member, note that the values of $A_x = A_y = 0.0$ could have been determined by inspection.

Problems

5–45. Determine the unknown force components acting at the supports A, B, and C.

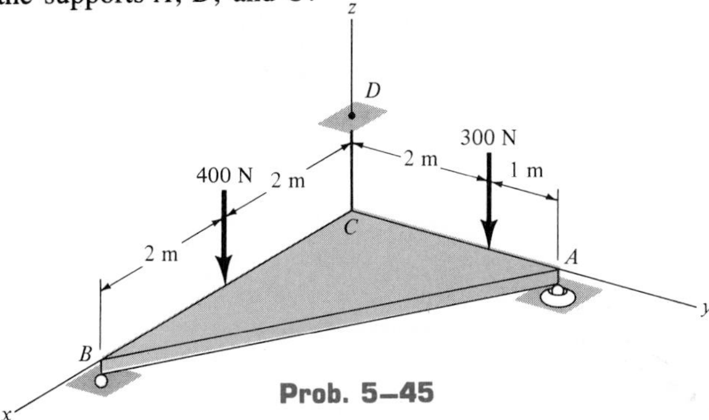

Prob. 5–45

5–46. Determine the unknown force components acting at the supports A and B. The bearing at B is properly aligned and therefore only exerts force components on the shaft.

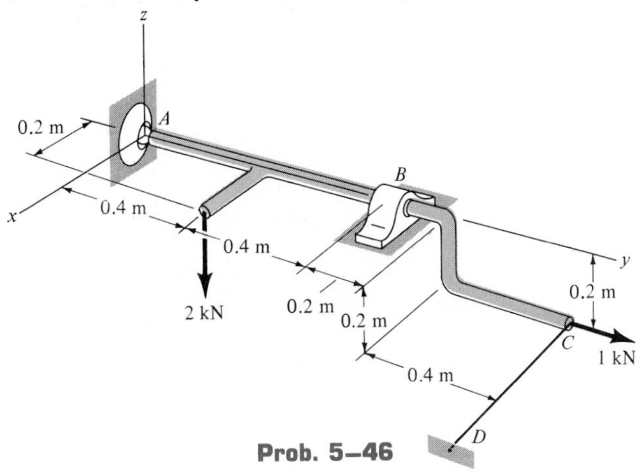

Prob. 5–46

5–47. Determine the unknown force components acting at the supports A, B, and C.

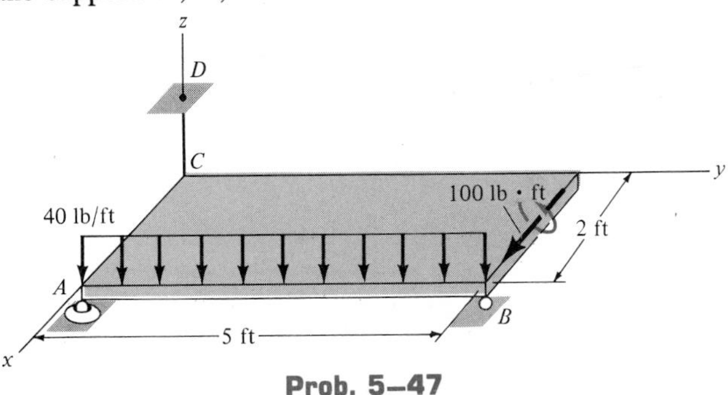

Prob. 5–47

***5–48.** The uniform concrete slab has a weight of 5500 lb. Determine the tension in each of the three parallel supporting cables when the slab is held in the horizontal plane as shown. The weight acts through the center of the slab.

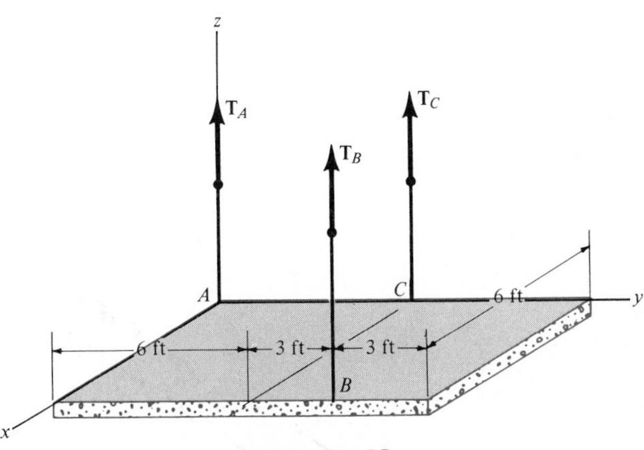

Prob. 5–48

Show on Board

5–49. The air-conditioning unit is hoisted to the roof of a building using the three cables. If the unit has a weight of 800 lb and a center of gravity at G, determine the tension in each of the cables for equilibrium.

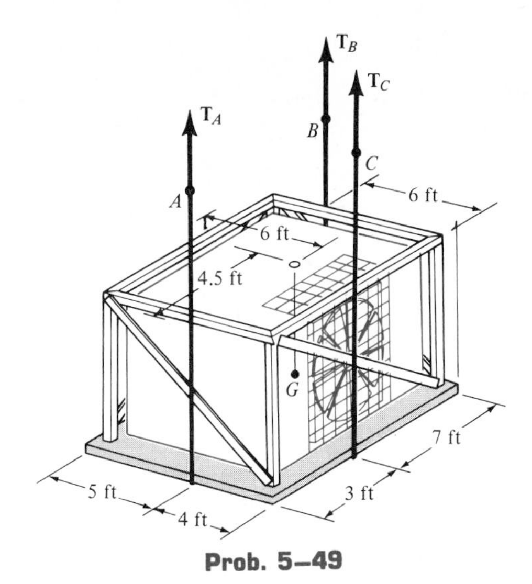

Prob. 5–49

5–50. A vertical force of 80 lb acts on the crankshaft. Determine the horizontal equilibrium force **P** that must be applied to the handle and the x, y, z components of reaction at the smooth bearings A and B. The bearings are properly aligned and therefore only exert force components on the shaft.

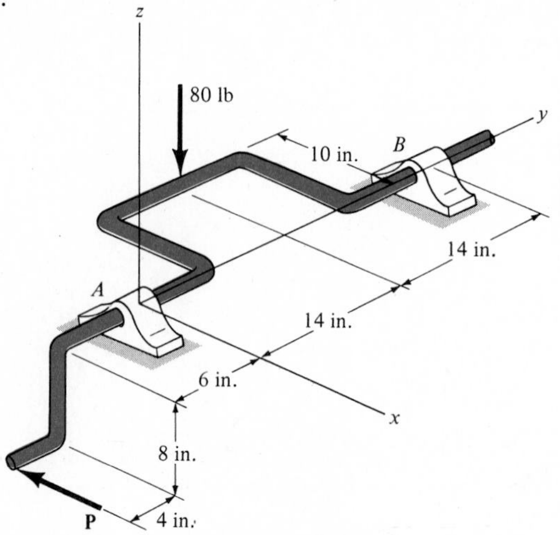

Prob. 5–50

5–51. The space truss is supported by a ball-and-socket joint at A and short links, two at C and one at D. Determine the x, y, z components of reaction at A and the force acting along each link. Note that the links are two-force members.

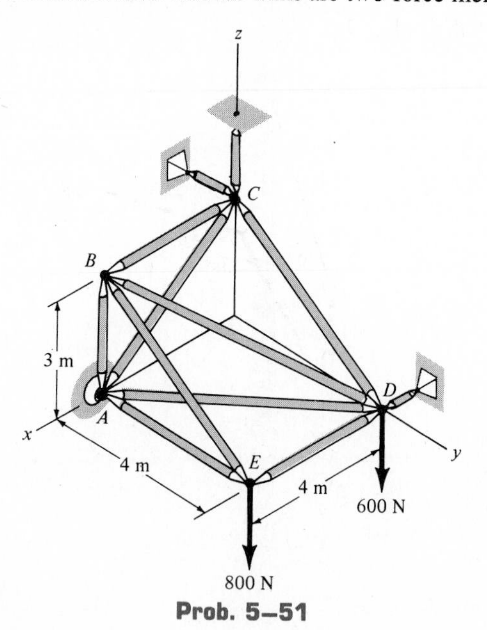

Prob. 5–51

***5–52.** Determine the x, y, z components of reaction at the fixed wall A. The triangular distributed load lies in the y-z plane. The 200-N force is parallel to the y axis.

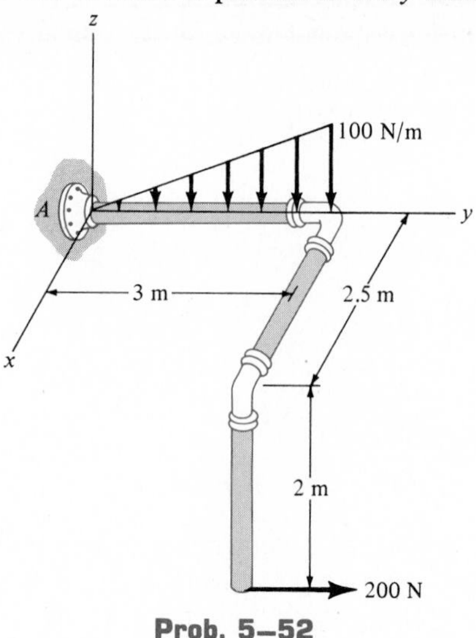

Prob. 5–52

5–53. The cable of the tower crane is subjected to a force of 560 N. Determine the x, y, z components of reaction at the fixed base A.

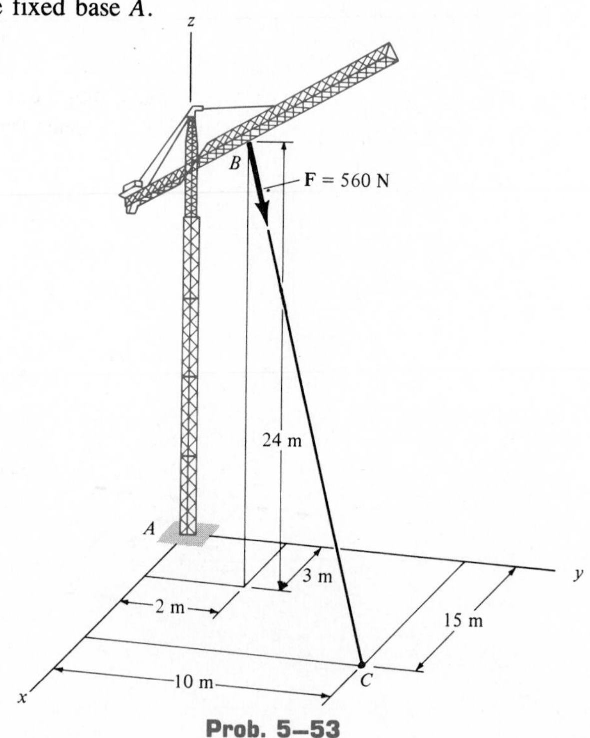

Prob. 5–53

5–54. The girl has a mass of 17 kg and mass center at G_g, and the tricycle has a mass of 10 kg and mass center at G_t. Determine the normal reactions at each wheel for equilibrium.

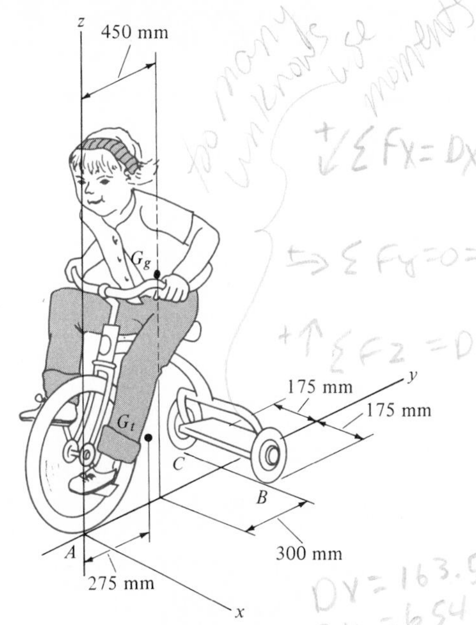

Prob. 5–54

5–55. Determine the force reactions acting along the short links connected at joints A, B, D, and F of the space truss. Note that the links are two-force members.

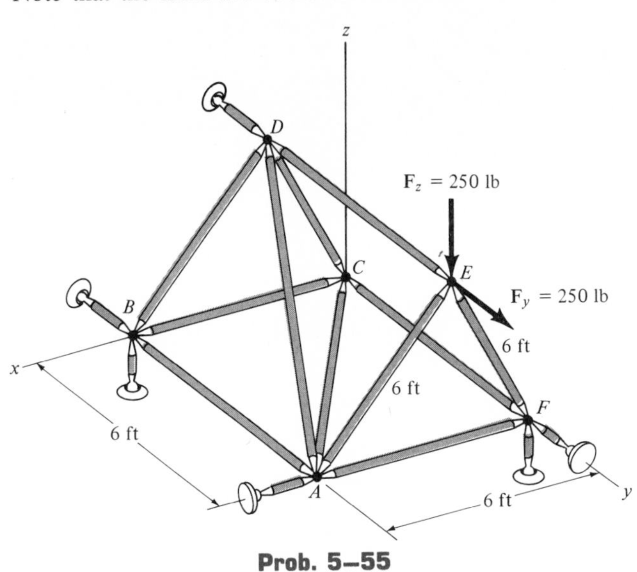

Prob. 5–55

***5–56.** The stiff-leg derrick used on ships is supported by a ball-and-socket joint at D and two cables BA and BC. The cables are attached to a smooth collar ring at B, which allows rotation of the derrick about the z axis. If the derrick supports a crate having a mass of 100 kg, determine the tension in cables BA and BC and the x, y, z components of reaction at D.

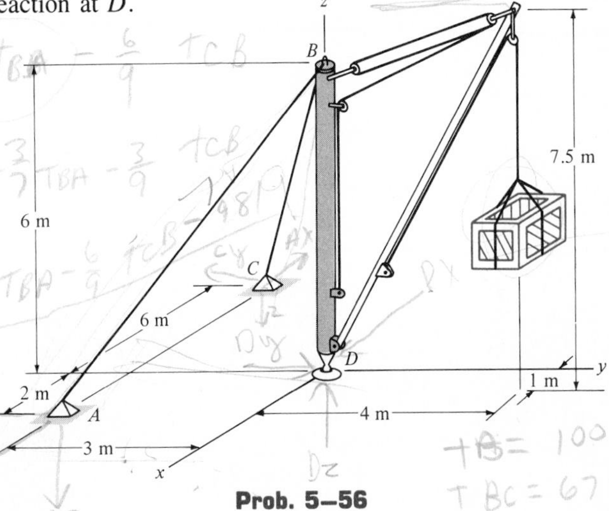

Prob. 5–56

5–57. The pole for a power line is subjected to the two cable forces of 60 lb, each force lying in a plane parallel to the x-y plane. If the tension in the guy wire AB is 80 lb, determine the x, y, z components of reaction at the fixed base of the pole, O.

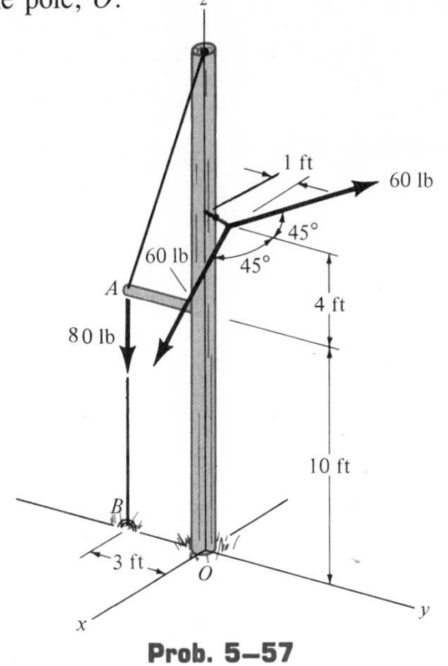

Prob. 5–57

5–58. The nonhomogeneous door of a large pressure vessel has a weight of 125 lb and a center of gravity at G. Determine the magnitudes of the resultant force and resultant couple moment developed at the hinge A, needed to support the door in any open position.

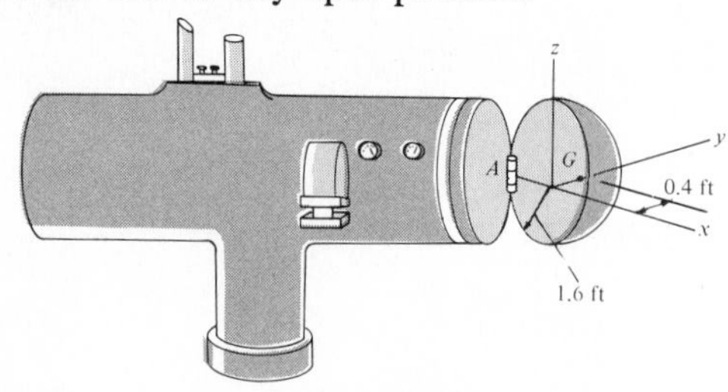

Prob. 5–58

5–59. The wing of the jet aircraft is subjected to a thrust force of $T = 8$ kN from its engine and the resultant lift force $L = 45$ kN. If the mass of the wing is 2.1 Mg and the mass center is at G, determine the x, y, z components of reaction where the wing is fixed to the fuselage at A.

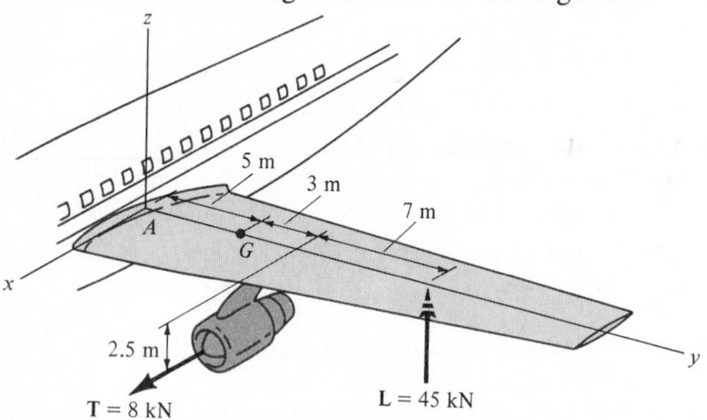

Prob. 5–59

***5–60.** The cart supports the uniform crate having a mass of 60 kg. Determine the vertical reactions on the three casters or rollers at A, B, and C.

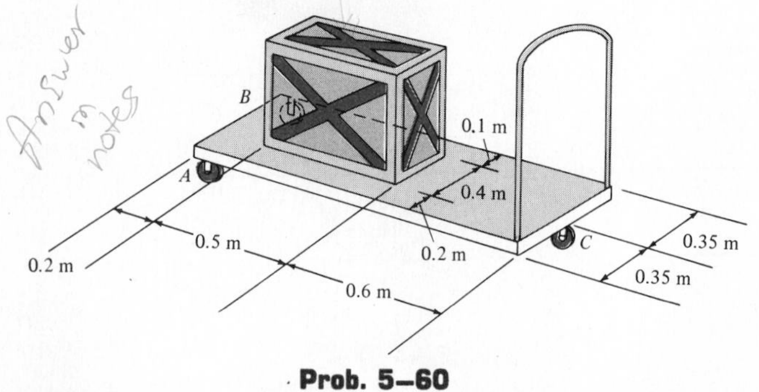

Prob. 5–60

5–61. The boom AC is supported at A by a ball-and-socket joint and by two cables BDC and CE. Cable BDC is continuous and passes over a frictionless pulley at D. Calculate the tension in the cables and the x, y, z components of reaction at A if a crate, having a weight of 80 lb, is suspended from the boom.

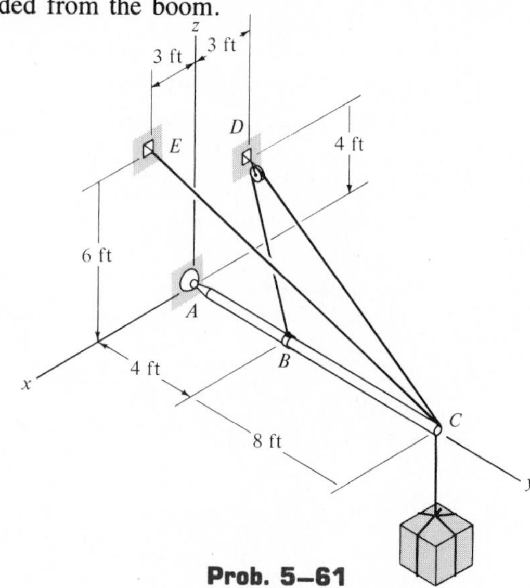

Prob. 5–61

5–62. Determine the tension in the supporting cables BC and BD and the components of reaction at the ball-and-socket joint A of the boom. The boom supports a drum having a weight of 200 lb at F. Points C and D lie in the x-y plane. *Hint:* Analyze the forces on a free-body diagram of the *entire* boom, including the attached drum.

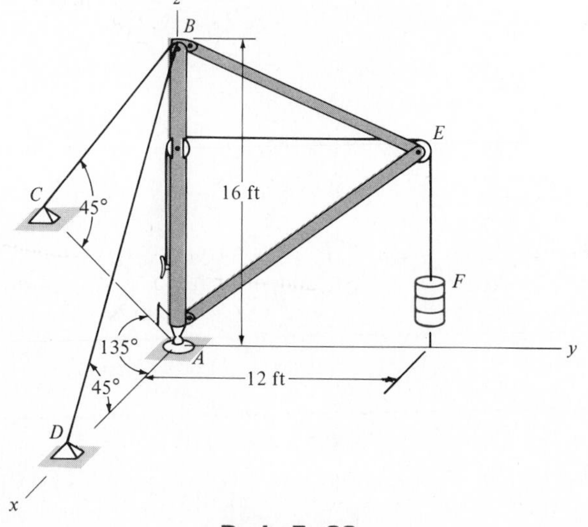

Prob. 5–62

Review Problems

5–63. Determine the components of reaction at the fixed support A for the beam loaded as shown.

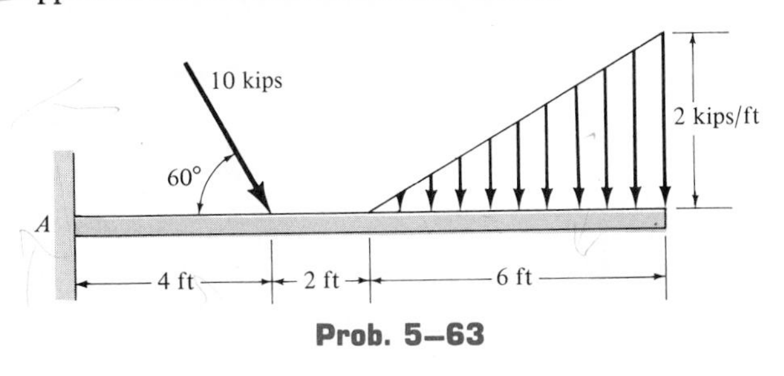

Prob. 5–63

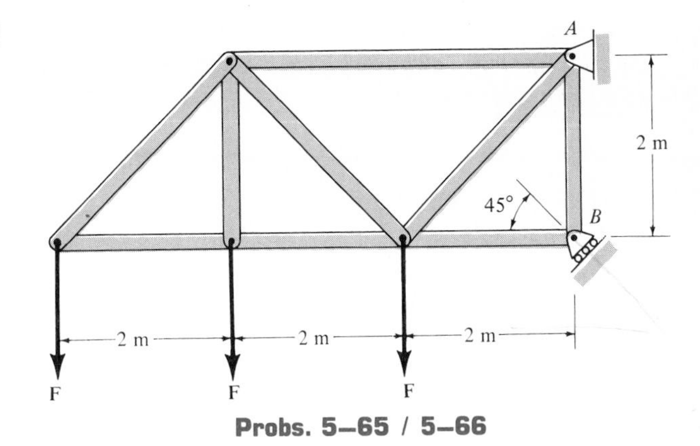

Probs. 5–65 / 5–66

***5–64.** Determine the reactions at the roller A and pin B for equilibrium of the member.

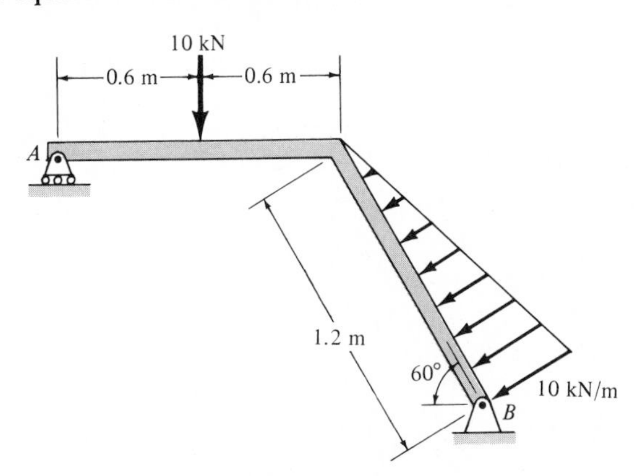

Prob. 5–64

5–65. Determine the horizontal and vertical components of reaction at the pin A and the reaction at the roller B required to support the truss. Set $F = 600$ N.

5–66. If the roller at B can sustain a maximum load of 3 kN, determine the largest magnitude of each of the three forces **F** that can be supported by the truss.

5–67. If the pipe assembly is subjected to a couple moment **M** having components of $M_x = 40\ \text{N}\cdot\text{m}$, $M_y = M_z = 0$ and a force **F** having components of $F_x = 30$ N, $F_y = 60$ N, $F_z = -60$ N, determine the x, y, z components of the supporting force and moment acting at the fixed support C.

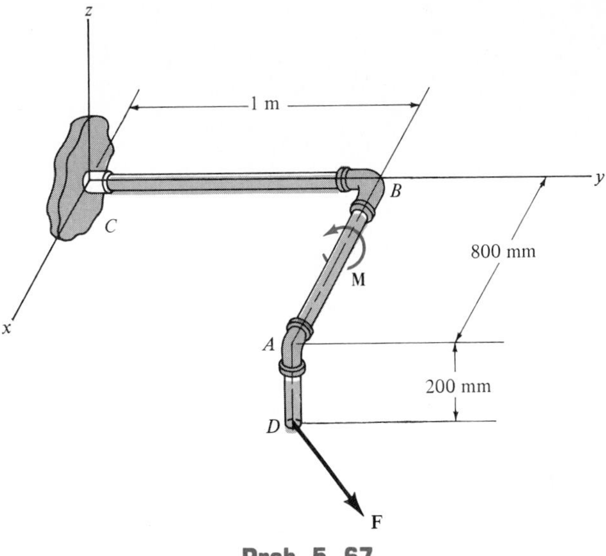

Prob. 5–67

***5–68.** The windlass is subjected to a load of 150 lb. Determine the horizontal force **P** needed to hold the handle in the position shown, and the components of reaction at the ball-and-socket joint A and the smooth bearing B. The bearing at B is in proper alignment and exerts only force reactions on the windlass.

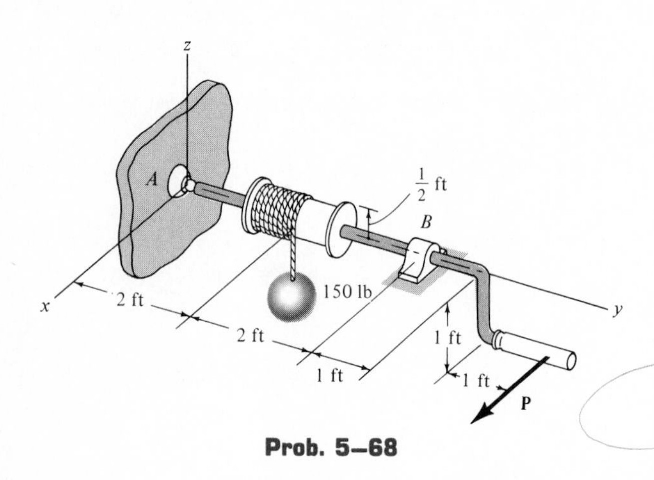

Prob. 5–68

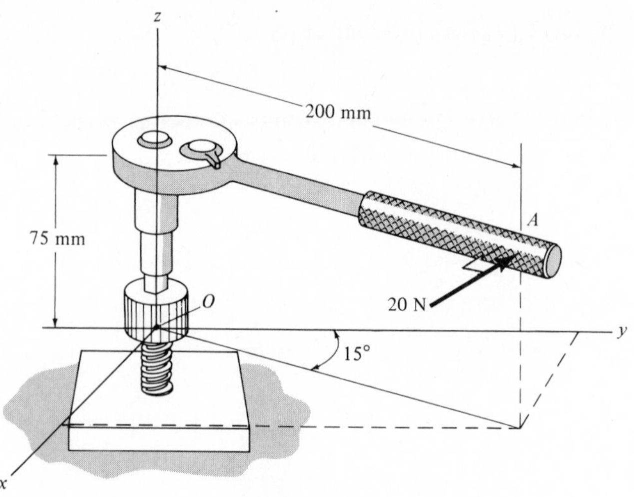

Prob. 5–70

5–69. Determine the reactions at the pin B and roller A for equilibrium.

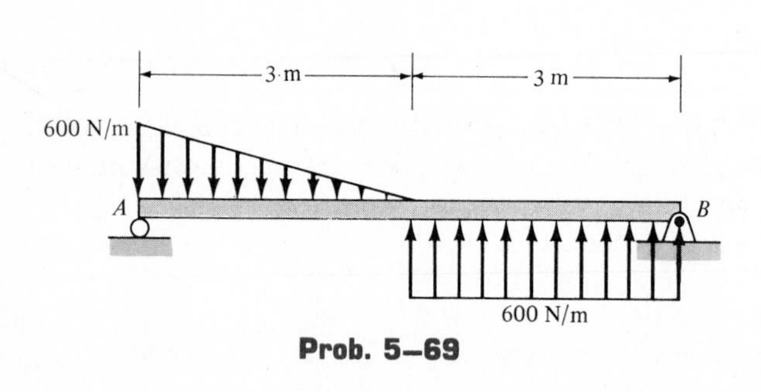

Prob. 5–69

5–70. The 20-N horizontal force is applied perpendicular to the handle of the socket wrench. Determine the x, y, z components of force and moment which the bolt at O exerts on the socket for equilibrium.

5–71. A *Russell's traction* is used for immobilizing femoral fractures C. If the lower leg has a weight of 8 lb, determine the weight W that must be suspended at D in order for the leg to be held in the position shown. Also, what is the tension force **F** in the femur and the distance $\bar{x}$ which locates the center of gravity G of the lower leg? Neglect the size of the pulley at B.

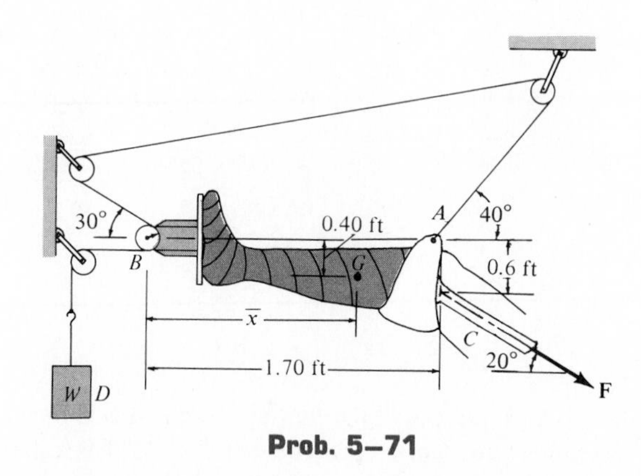

Prob. 5–71

***5–72.** The smooth cylinder has a weight of 20 lb and is suspended from a spring which has a stiffness of $k = 5$ lb/ft. If the spring has an unstretched length of $l_0 = 1$ ft, determine the angle θ for equilibrium.

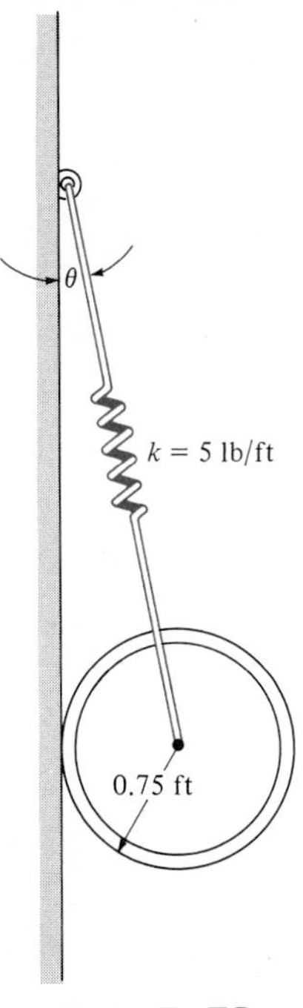

Prob. 5–72

5–74. Determine the magnitude of the force **F** acting on the semicircular plate so the force in each of the cables *BD* and *CD* is 200 lb. What are the components of reaction at the ball-and-socket joint for this loading?

5–75. If the spring is unstretched when $\theta = 0°$, determine the weight of the uniform bar *AB* if it hangs in the equilibrium position at $\theta = 30°$.

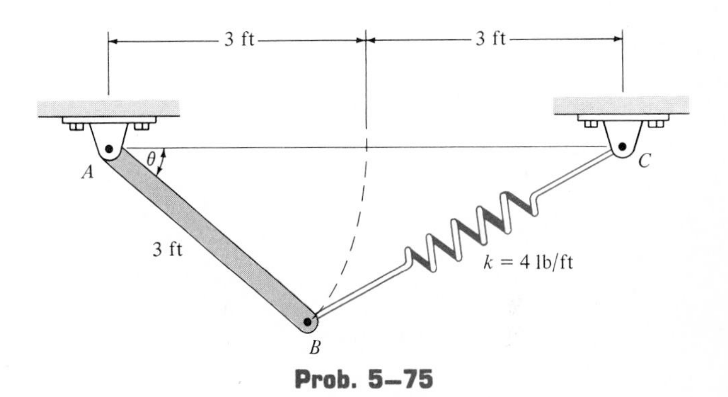

Prob. 5–75

5–73. The semicircular plate supports a load of $F = 150$ lb. Determine the tension in cables *BD* and *CD* and the components of reaction at the ball-and-socket joint at *A* for equilibrium.

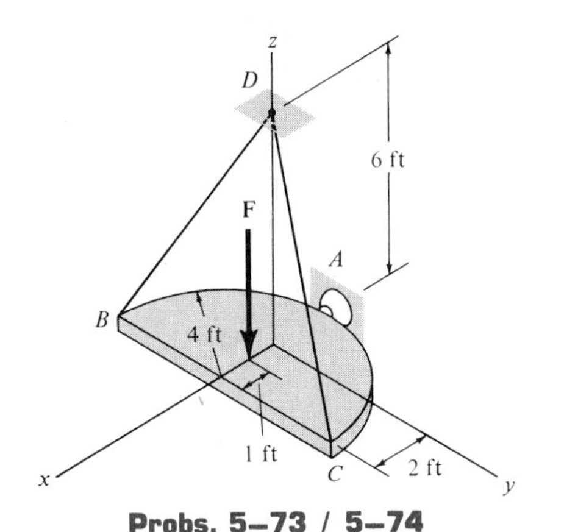

Probs. 5–73 / 5–74

***5–76.** If the maximum intensity of the distributed load acting on the beam is $w = 4$ kN/m, determine the reactions at the pin *A* and roller *B* for equilibrium.

5–77. If either the pin at *A* or the roller at *B* can support a load no greater than 6 kN, determine the maximum intensity of the distributed load w kN/m, so that failure of a support does not occur.

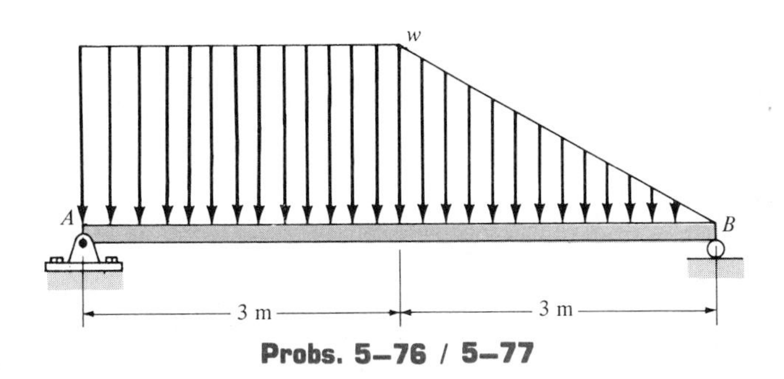

Probs. 5–76 / 5–77

5–78. The bent rod is supported at A, B, and C by smooth bearings. Compute the x, y, z components of reaction at the bearings if the rod is subjected to a 200-lb vertical force and a 30-lb · ft couple as shown. The bearings are in proper alignment and therefore exert only force reactions on the rod.

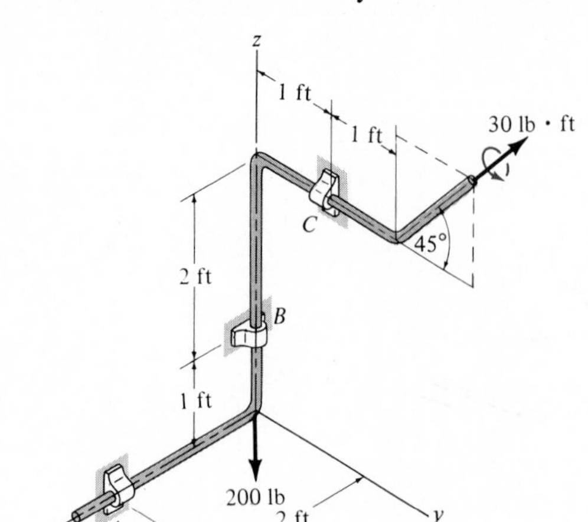

Prob. 5–78

5–79. Two smooth tubes A and B, each having the same weight W, are suspended at their ends by cords of equal length. A third tube C is placed between A and B. Determine the greatest weight of C without upsetting equilibrium.

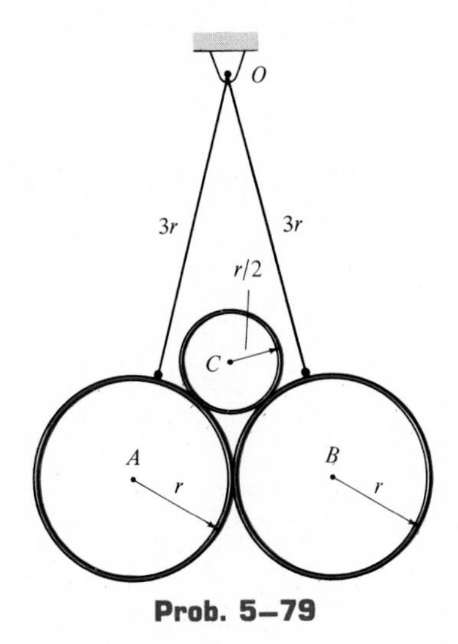

Prob. 5–79

Structural Analysis

In this chapter we will use the equations of equilibrium to analyze structures composed of pin-connected members. The analysis is based upon the principle that if a structure is in equilibrium, then each of its members is also in equilibrium. By applying the equations of equilibrium to the various parts of a simple truss, frame, or machine, we will be able to determine all the forces acting at the connections. Later, in Chapter 7, we extend this analysis and show how to determine the internal forces in a member.

The topics in this chapter are very important since they provide practice in drawing free-body diagrams, using the principle of action, equal but opposite force reaction, and applying the equations of equilibrium.

Simple Trusses 6.1

A *truss* is a structure composed of slender members joined together at their end points. The members commonly used in construction consist of wooden struts, metal bars, angles, or channels. The joint connections are usually formed by bolting or welding the ends of the members to a common plate, called a *gusset plate,* as shown in Fig. 6–1, or by simply passing a large bolt or pin through each of the members.

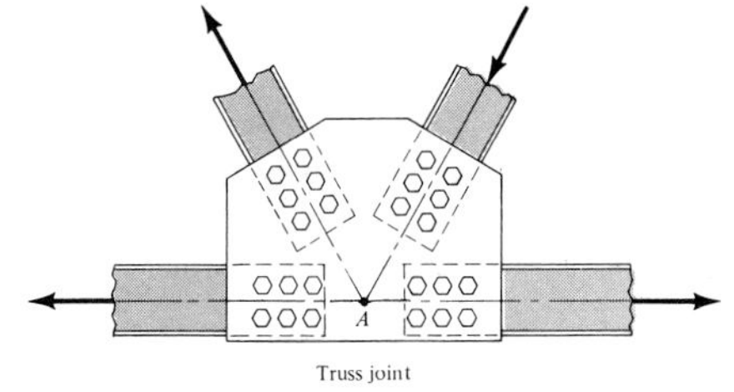

Truss joint

Fig. 6–1

Planar Trusses. *Planar* trusses lie in a single plane and are often used to support roofs and bridges. The truss *ABCDE*, shown in Fig. 6–2*a*, is an

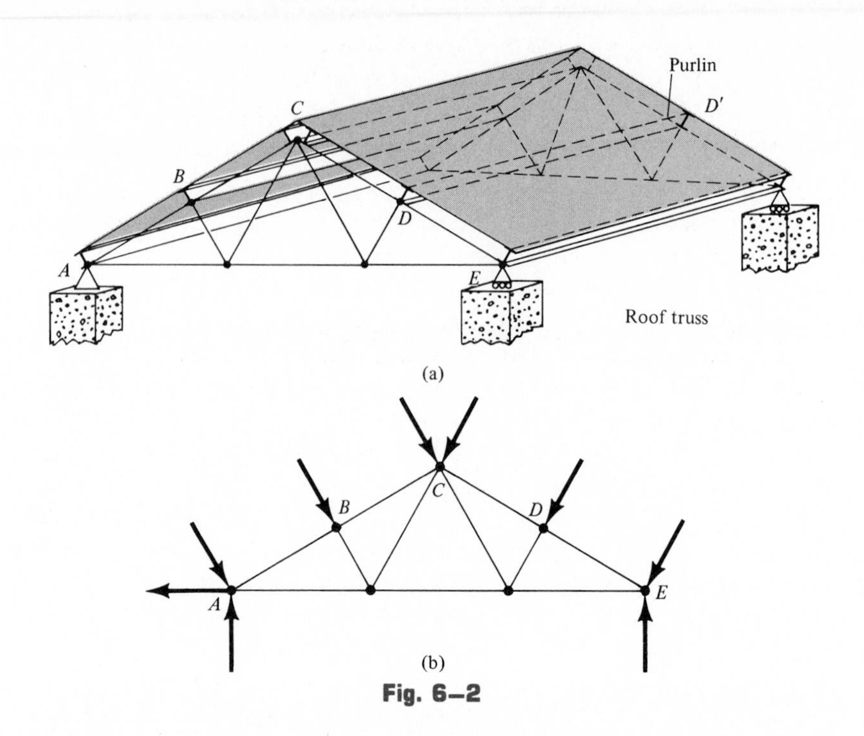

Fig. 6–2

example of a typical roof-supporting truss. In this figure, the roof load is transmitted to the truss *at the joints* by means of a series of *purlins,* such as beam DD'. Since the imposed loading acts in the same plane as the truss, Fig. 6–2*b*, the analysis of the forces developed in the truss members is two-dimensional.

In the case of a bridge, such as shown in Fig. 6–3*a*, the load on the *deck* is first transmitted to *stringers,* then to *floor beams,* and finally to the *joints B, C,* and *D* of the two supporting side trusses. Like the roof truss, the bridge truss loading is coplanar, Fig. 6–3*b*.

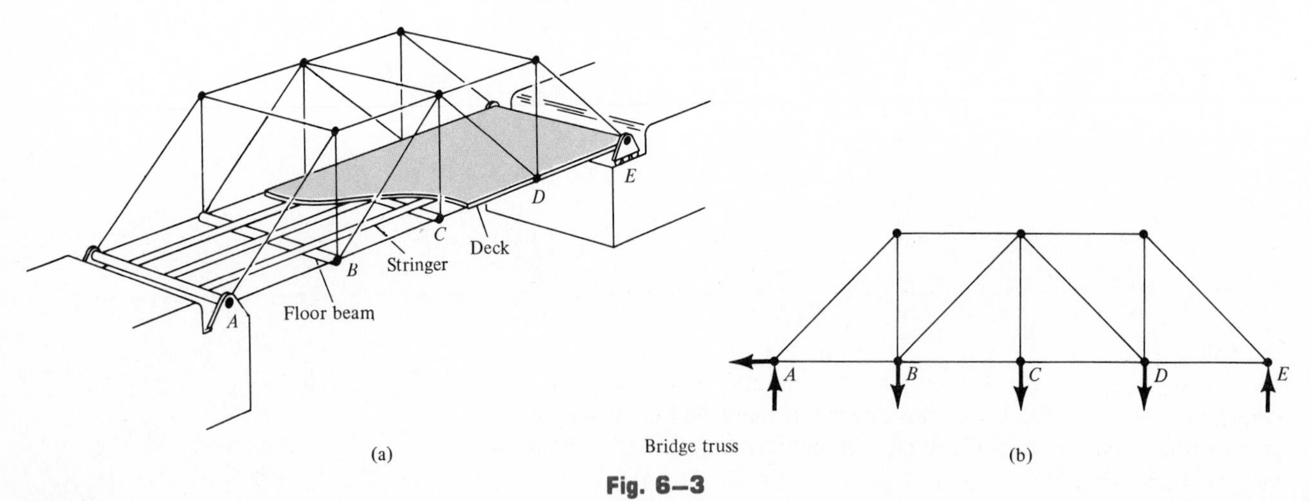

Fig. 6–3

When bridge or roof trusses extend over large distances, a rocker or roller is commonly used for supporting one end, joint E in Figs. 6–2a and 6–3a. This type of support allows freedom for expansion or contraction due to temperature or application of loads.

Assumptions for Design. To design both the members and the connections of a truss, it is first necessary to determine the *force* developed in each member when the truss is subjected to a given loading. In this regard, two important assumptions will be made:

1. *The members are joined together by smooth pins*. In cases where bolted or welded joint connections are used, this assumption is satisfactory provided the center lines of the joining members are concurrent at a point, as in the case of point A in Fig. 6–1.
2. *All loadings are applied at the joints*. In most situations, such as for bridge and roof trusses, this assumption is true. Frequently in the force analysis the weight of the members is neglected, since the force supported by the members is large in comparison with their weight. If the weight is to be included in the analysis, it is generally satisfactory to apply it as a vertical force, half of its magnitude applied at each end of the member.

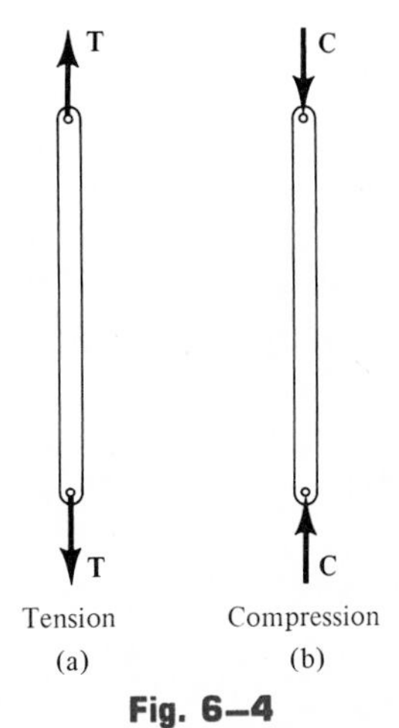

Fig. 6–4

Because of these two assumptions, *each truss member acts as a two-force member,* and therefore the forces acting at the ends of the member must be directed along the axis of the member. If the force tends to *elongate* the member, it is a *tensile force* (**T**), Fig. 6–4a; whereas, if the force tends to *shorten* the member, it is a *compressive force* (**C**), Fig. 6–4b. In the actual design of a truss it is important to state whether the nature of the force is tensile or compressive. Most often, compression members must be made *thicker* than tension members, because of the buckling or column effect that occurs in compression members.

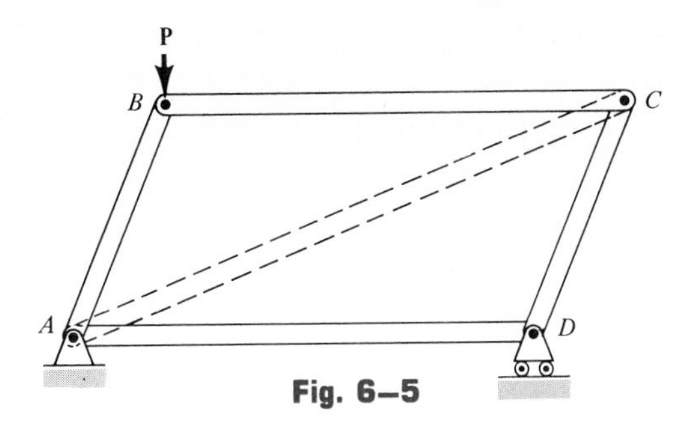

Fig. 6–5

Simple Truss. To prevent collapse, the framework of a truss must be rigid. Obviously, the four-bar frame $ABCD$ in Fig. 6–5 will collapse unless a diagonal, such as AC, is added for support. The simplest framework which is rigid or stable is a *triangle*. Consequently, a *simple truss* is constructed by *starting* with a basic triangular element, such as ABC in Fig. 6–6, and connecting two members (AD and BD) to form an additional element. Thus it is seen that as each additional element of two members is placed on the truss, the number of joints is increased by one.

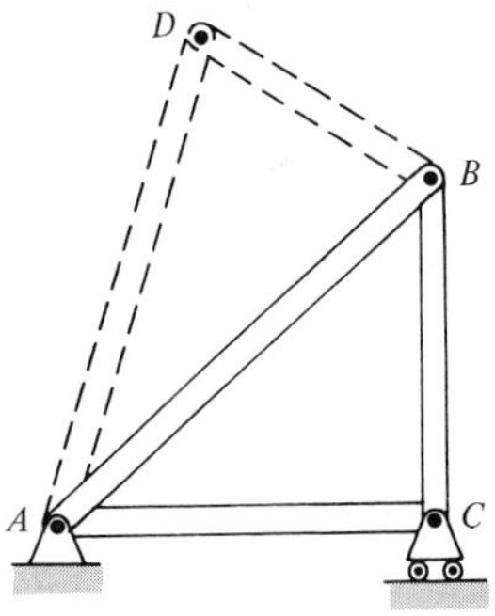

Fig. 6–6

6.2 The Method of Joints

If a truss is in equilibrium, then each of its joints must also be in equilibrium. Hence, the method of joints consists of satisfying the equilibrium conditions for the forces exerted *on the pin* at each joint of the truss. Since the truss members are all straight two-force members lying in the same plane, the force system acting at each pin is *coplanar and concurrent*. Consequently, rotational or moment equilibrium is automatically satisfied at the joint (or pin), and it is only necessary to satisfy $\Sigma F_x = 0$ and $\Sigma F_y = 0$ to ensure translational or force equilibrium.

When using the method of joints, it is *first* necessary to draw the joint's free-body diagram before applying the equilibrium equations. In this regard, recall that the *line of action* of each member force acting on the joint is *specified* from the geometry of the truss, since the force in a member passes along the axis of the member. As an example, consider joint B of the truss in Fig. 6–7*a*. From the free-body diagram, Fig. 6–7*b*, the only unknowns are the *magnitudes* of the forces in members BA and BC. As shown, $\mathbf{F}_{BA}$ is "pulling" on the pin, which means that member BA is in *tension;* whereas $\mathbf{F}_{BC}$ is "pushing" on the pin, and consequently member BC is in *compression*.

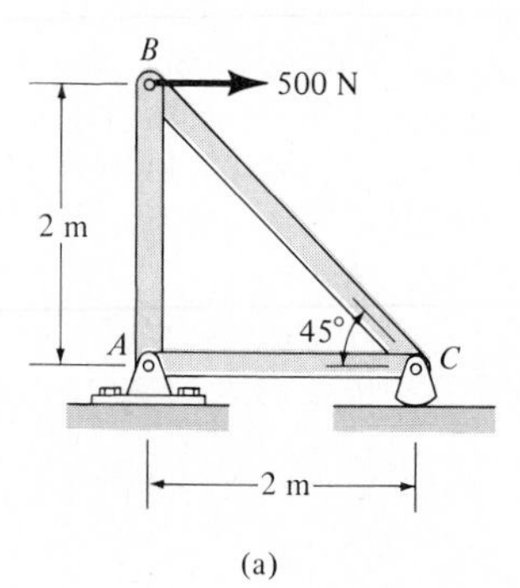

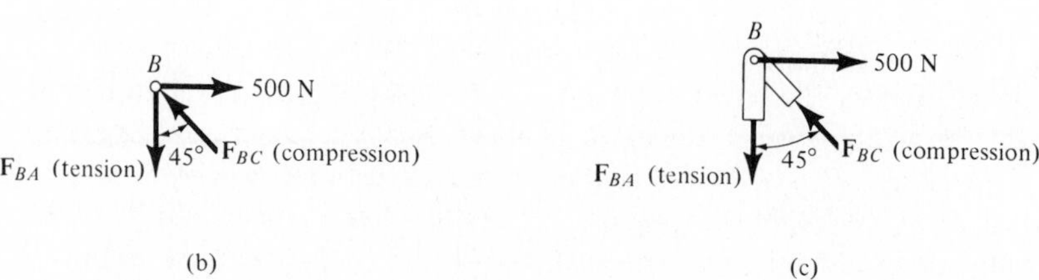

Fig. 6–7

These effects are clearly demonstrated by isolating the joint with small segments of the member connected to the pin, Fig. 6–7*c*. Notice that pushing or pulling on these small segments indicates the effect of the member being either in compression or tension.

In all cases, the joint analysis should start at a joint having at least one known force and at most two unknown forces, as in Fig. 6–7*b*. In this way, application of $\Sigma F_x = 0$ and $\Sigma F_y = 0$ yields two algebraic equations which can be solved for the two unknowns. The correct "arrowhead" sense of direction of an unknown member force can, in many cases, be determined "by inspection." For example, $\mathbf{F}_{BC}$ in Fig. 6–7*b* must push on the pin (compression) since its horizontal component F_{BC} sin 45° must balance the 500-N force ($\Sigma F_x = 0$). Likewise, $\mathbf{F}_{BA}$ is a tensile force since it balances the vertical component F_{BC} cos 45° ($\Sigma F_y = 0$). In more complicated cases, the sense of direction of an unknown member force can be *assumed;* then, after applying the equilibrium equations, the assumed directional sense can be verified from the numerical results. A *positive* answer indicates that the sense of direction is *correct,* whereas a *negative* answer indicates that the sense of direction shown on the free-body diagram must be *reversed.*

PROCEDURE FOR ANALYSIS

The following procedure provides a means for analyzing a truss using the method of joints.

Draw the free-body diagram of a joint having at least one known force and at most two unknown forces. (If this joint is at one of the supports, it will be necessary to know the external reactions at the truss support.) By inspection, attempt to show the unknown forces acting on the joint with the correct "arrowhead" sense of direction. Orient the x and y axes such that the forces on the free-body diagram can be easily resolved into their x and y components and then apply the two force equilibrium equations $\Sigma F_x = 0$ and $\Sigma F_y = 0$. Solve for the two unknown member forces and verify their correct directional sense.

Continue to analyze each of the other joints, where again it is necessary to choose a joint having at most two unknowns and at least one known force. In this regard, realize that once the force in a member is found from the analysis of a joint at one of its ends, the result can be used to analyze the forces acting on the joint at its other end. Strict adherence to the principle of action, equal but opposite reaction must, of course, be observed. Remember, a member in *compression* "pushes" on the joint and a member in *tension* "pulls" on the joint.

Once the force analysis of the truss has been completed, the size of the members and their connections can then be determined using the theory of strength of materials.

Example 6–1

Determine the force in each member of the truss shown in Fig. 6–8*a* and indicate whether the members are in tension or compression.

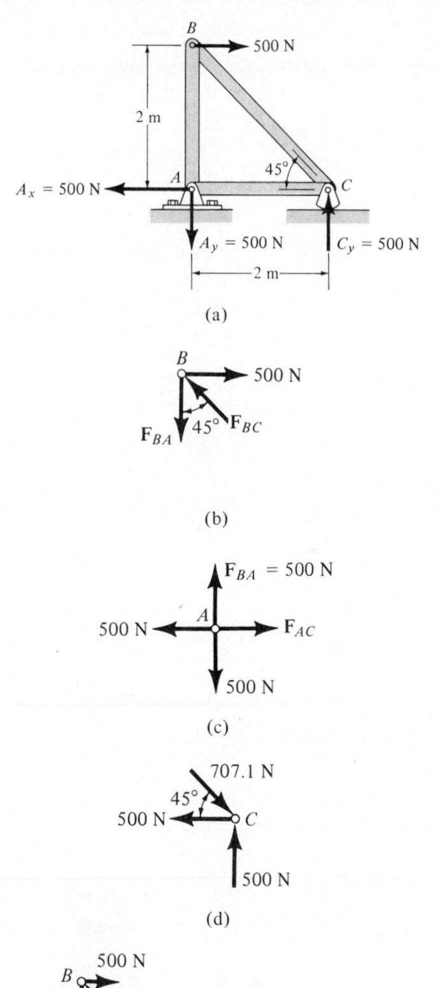

Solution

If the reactions at *A* and *C* had not been computed (although on the figure they are shown), the joint analysis would have to begin at joint *B*. Why?

Joint B. The free-body diagram of the pin at *B* is shown in Fig. 6–8*b*. Three forces act on the pin: the external force of 500 N and the *two* unknown forces developed by members *BA* and *BC*. Applying the equations of joint equilibrium, we have

$$\overset{+}{\rightarrow}\Sigma F_x = 0; \quad 500 - F_{BC} \sin 45° = 0 \qquad F_{BC} = 707.1 \text{ N} \quad \text{(C)}$$

$$+\uparrow \Sigma F_y = 0; \quad F_{BC} \cos 45° - F_{BA} = 0 \qquad F_{BA} = 500 \text{ N} \quad \text{(T)}$$

Since the forces in members *BA* and *BC* have been determined, one can proceed to analyze the forces at joints *A* and *C*.

Joint A. The free-body diagram of the pin at *A* is shown in Fig. 6–8*c*. Here the effect of the *support* on the pin is represented by the reactions A_x = 500 N and A_y = 500 N. Since *BA* was found to be in tension, $\mathbf{F}_{BA}$ (=500 N) pulls on the pin (refer to Fig. 6–8*e*).
Applying the equilibrium equations yields

$$\overset{+}{\rightarrow}\Sigma F_x = 0; \qquad F_{AC} - 500 = 0 \qquad F_{AC} = 500 \text{ N} \quad \text{(T)}$$

$$+\uparrow \Sigma F_y = 0; \qquad 500 - 500 \equiv 0 \quad \text{(check)}$$

Joint C. With the forces in all the members known, the results can be checked, in part, from the free-body diagram of the last joint *C*, Fig. 6–8*d*.

$$\overset{+}{\rightarrow}\Sigma F_x = 0; \qquad -500 + 707.1 \cos 45° \equiv 0 \quad \text{(check)}$$

$$+\uparrow \Sigma F_y = 0; \qquad 500 - 707.1 \sin 45° \equiv 0 \quad \text{(check)}$$

The results of the analysis are summarized in Fig. 6–8*e*.

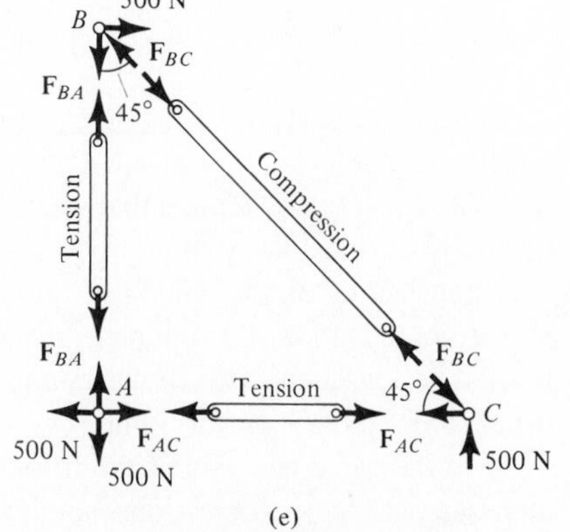

Fig. 6–8

Example 6–2

Determine the forces acting in all the members of the truss shown in Fig. 6–9*a*. The reactions at the supports are shown in the figure.

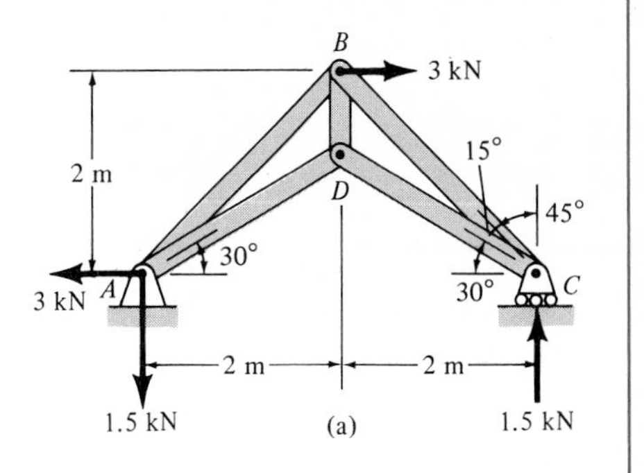

Solution

Since the reactions have been computed, the analysis may begin at either joint *C* or *A*. Why?

Joint C (Fig. 6–9*b*)

$\xrightarrow{+}\Sigma F_x = 0; \qquad -F_{CD}\cos 30° + F_{CB}\sin 45° = 0$

$+\uparrow \Sigma F_y = 0; \quad 1.5 + F_{CD}\sin 30° - F_{CB}\cos 45° = 0$

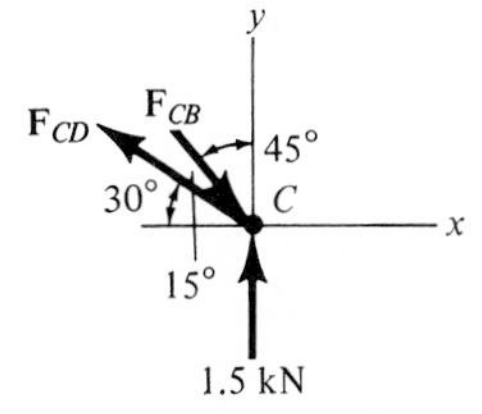

These two equations must be solved *simultaneously* for each of the two unknowns. Note, however, that a *direct solution* for one of the unknown forces may be obtained by applying a force summation along an axis that is *perpendicular* to the direction of the other unknown force. For example, summing forces along the y' axis, which is perpendicular to the direction of $\mathbf{F}_{CD}$, Fig. 6–9*c*, yields a direct solution for F_{CB}.

$+\nearrow\Sigma F_{y'} = 0;$

$1.5\cos 30° - F_{CB}\sin 15° = 0 \quad F_{CB} = 5.02 \text{ kN} \quad (C)$ ***Ans.***

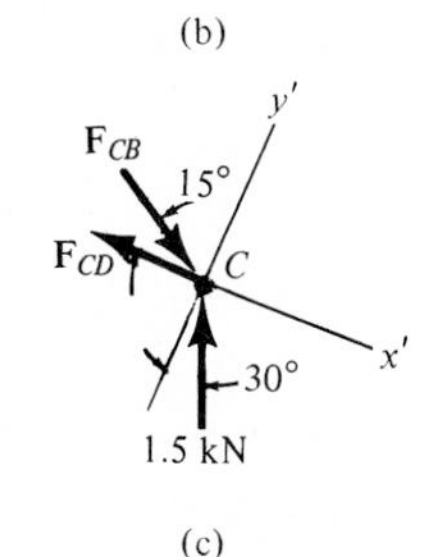

In a similar fashion, summing forces along the y'' axis, Fig. 6–9*d*, yields a direct solution for F_{CD}.

$+\nearrow\Sigma F_{y''} = 0;$

$1.5\cos 45° - F_{CD}\sin 15° = 0 \qquad F_{CD} = 4.10 \text{ kN} \quad (T)$ ***Ans.***

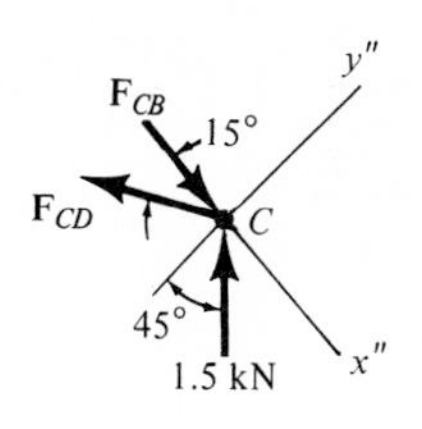

Joint D (Fig. 6–9*e*)

$\xrightarrow{+}\Sigma F_x = 0; \qquad -F_{DA}\cos 30° + 4.10\cos 30° = 0$

$F_{DA} = 4.10 \text{ kN} \quad (T)$ ***Ans.***

$+\uparrow \Sigma F_y = 0; \quad F_{DB} - 2(4.10\sin 30°) = 0 \quad F_{DB} = 4.10 \text{ kN} \quad (T)$ ***Ans.***

Joint B (Fig. 6–9*f*)

$\xrightarrow{+}\Sigma F_x = 0; \qquad F_{BA}\cos 45° - 5.02\cos 45° + 3 = 0$

$F_{BA} = 0.777 \text{ kN} \quad (C)$ ***Ans.***

$+\uparrow \Sigma F_y = 0; \quad 0.777\sin 45° - 4.10 + 5.02\sin 45° \equiv 0 \quad \text{(check)}$

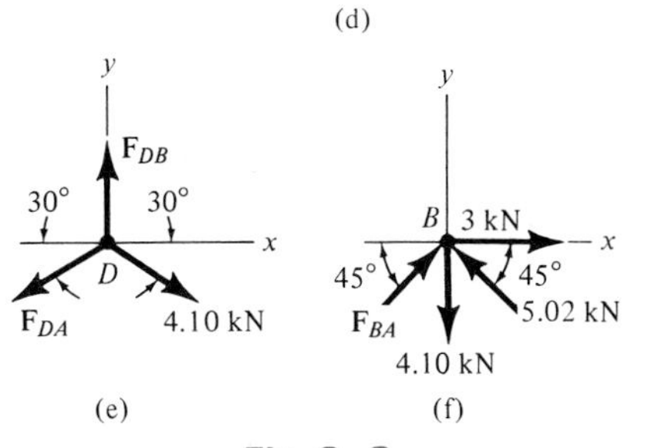

Fig. 6–9

Example 6–3

Determine the force in each member of the truss shown in Fig. 6–10*a*. Indicate whether the members are in tension or compression.

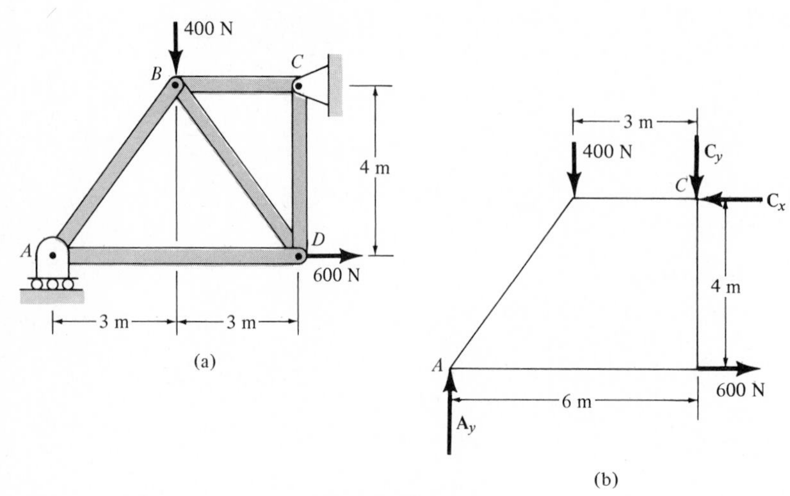

Fig. 6–10a and b

Solution

Support Reactions. No joint can be analyzed until the support reactions are first determined. Why? A free-body diagram of the entire truss is given in Fig. 6–10*b*. Applying the equations of equilibrium, we have

$$\overset{+}{\rightarrow}\Sigma F_x = 0; \qquad 600 - C_x = 0 \qquad C_x = 600 \text{ N}$$

$$\circlearrowleft + \Sigma M_C = 0; \quad -A_y(6) + 400(3) + 600(4) = 0 \qquad A_y = 600 \text{ N}$$

$$+\uparrow \Sigma F_y = 0; \qquad 600 - 400 - C_y = 0 \qquad C_y = 200 \text{ N}$$

The analysis can now start at either joint *A* or *C*. The choice is arbitrary, since there are one known and two unknown member forces acting at each of these joints.

Joint A (Fig. 6–10*c*) As shown on the free-body diagram, there are three forces that act at joint *A*. The inclination of $\mathbf{F}_{AB}$ is determined from the geometry of the truss. By inspection, can you see why this force is assumed to be compressive and $\mathbf{F}_{AD}$ tensile? Applying the equations of equilibrium, we have

$$+\uparrow \Sigma F_y = 0; \quad 600 - \tfrac{4}{5}F_{AB} = 0 \qquad F_{AB} = 750 \text{ N (C)} \qquad \textit{Ans.}$$

$$\overset{+}{\rightarrow}\Sigma F_x = 0; \quad F_{AD} - \tfrac{3}{5}(750) = 0 \qquad F_{AD} = 450 \text{ N (T)} \qquad \textit{Ans.}$$

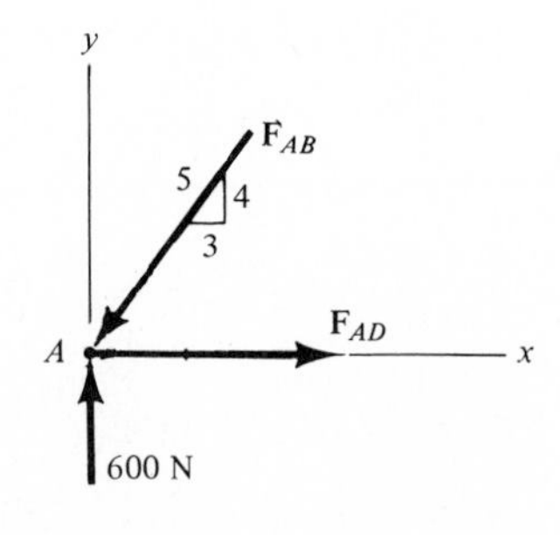

Fig. 6–10c

Joint D (Fig. 6–10*d*) This joint is chosen next since, by inspection of Fig. 6–10*a*, the force in *AD* is known and two unknown forces in *BD* and *CD* can be determined. Applying the equations of equilibrium, Fig. 6–10*d*, we have

$\overset{+}{\rightarrow}\Sigma F_x = 0; \quad -450 + \frac{3}{5}F_{DB} + 600 = 0 \qquad F_{DB} = -250 \text{ N}$

The negative sign indicates $\mathbf{F}_{DB}$ acts in the *opposite direction* to that shown in Fig. 6–10*d*.* Hence,

$$F_{DB} = 250 \text{ N} \quad (\text{T}) \qquad \textit{Ans.}$$

To determine $\mathbf{F}_{DC}$, we can either correct the direction of $\mathbf{F}_{DB}$ and then apply $\Sigma F_y = 0$, or apply the equation to Fig. 6–10*d* and retain the negative sign for F_{DB}, i.e.,

$+\uparrow \Sigma F_y = 0; \quad -F_{DC} - \frac{4}{5}(-250) = 0 \qquad F_{DC} = 200 \text{ N} \quad (\text{C}) \qquad \textit{Ans.}$

Joint C (Fig. 6–10*e*)

$\overset{+}{\rightarrow}\Sigma F_x = 0; \quad F_{CB} - 600 = 0 \qquad F_{CB} = 600 \text{ N} \quad (\text{C}) \qquad \textit{Ans.}$

$+\uparrow \Sigma F_y = 0; \quad 200 - 200 \equiv 0 \quad (\text{check})$

The results may be checked in part by analyzing the "last joint" *B*. The analysis is summarized in Fig. 6–10*f*, which shows the correct free-body diagram for each joint and member.

*The proper direction could have been determined by inspection, prior to applying $\Sigma F_x = 0$.

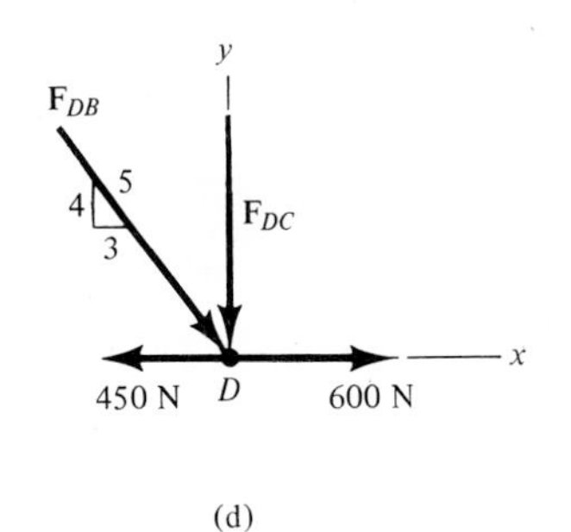

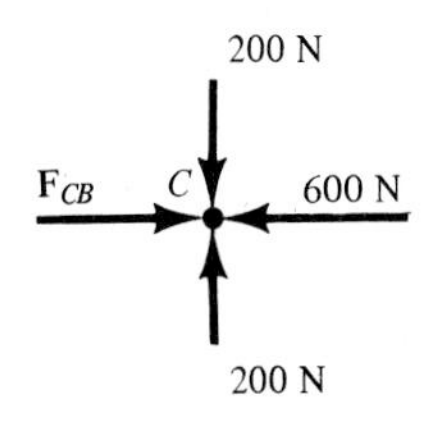

Fig. 6–10d and e

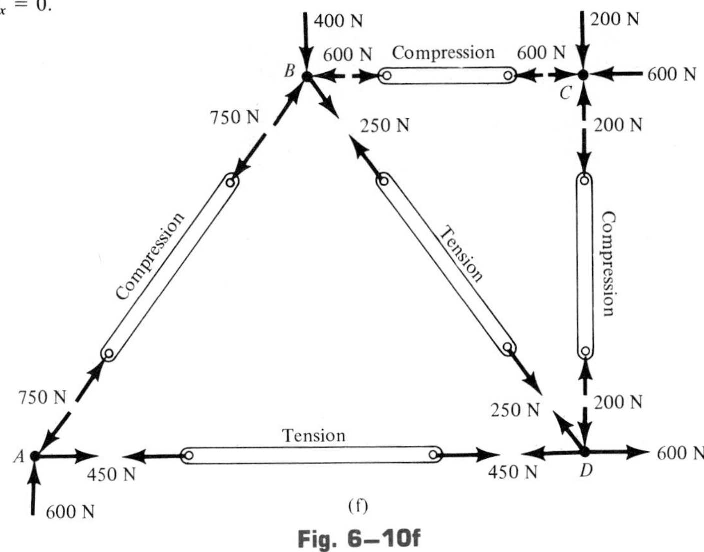

Fig. 6–10f

6.3 Zero-Force Members

Truss analysis using the method of joints is greatly simplified if one is able to first determine those members which support *no loading*. These *zero-force members* are used to increase the stability of the truss during construction and to provide support if the applied loading is changed.

The zero-force members of a truss can generally be determined by inspection of the joints. For example, if two members are connected at a right angle to each other at a joint *that has no external load,* as shown in Fig. 6–11*a*, the force in each member must be zero in order to maintain equilibrium. Furthermore, this is true regardless of the angle, say θ, between the members ($\theta \neq 0°$; $\theta \neq 180°$), Fig. 6–11*b*. Zero-force members also occur at joints having geometries as shown in Fig. 6–12*a* and *b*. In both cases, *no external load acts on the joint,* so that a force summation in the x direction, which is *perpendicular* to the two collinear members, requires that $F_1 = 0$. Particular attention should be directed to these conditions of joint geometry and loading, since the analysis of a truss can be considerably simplified by *first* spotting the zero-force members.

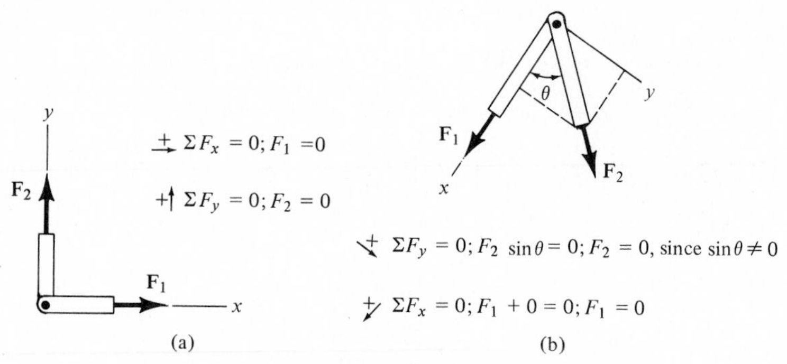

Fig. 6–11

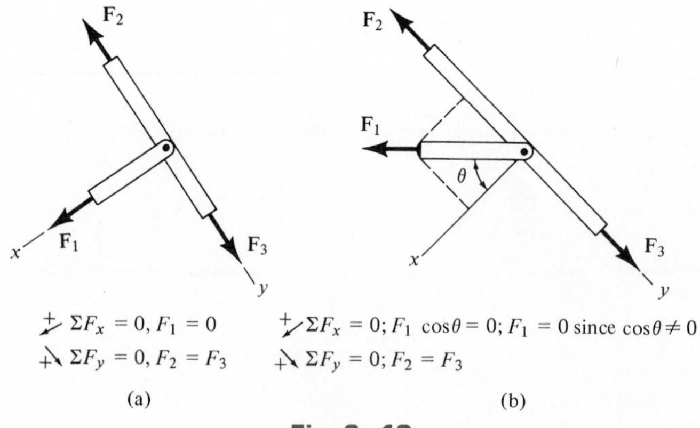

Fig. 6–12

Example 6–4

Using the method of joints, determine all the members of the *Fink truss* shown in Fig. 6–13*a* which are subjected to zero force.

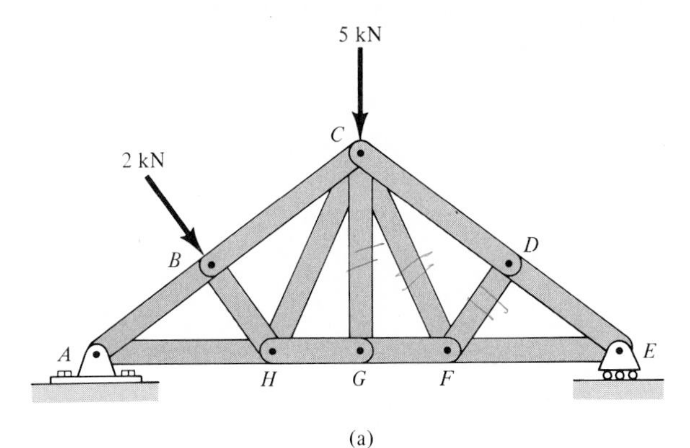

Fig. 6–13a

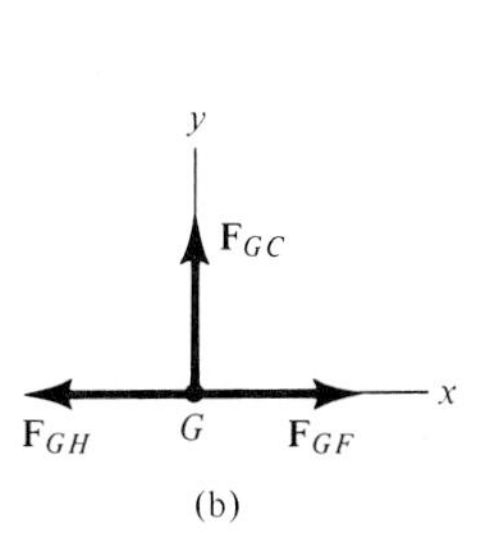

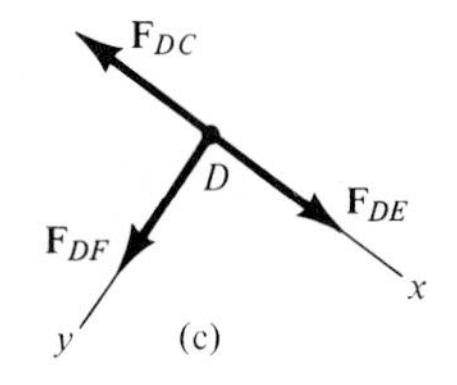

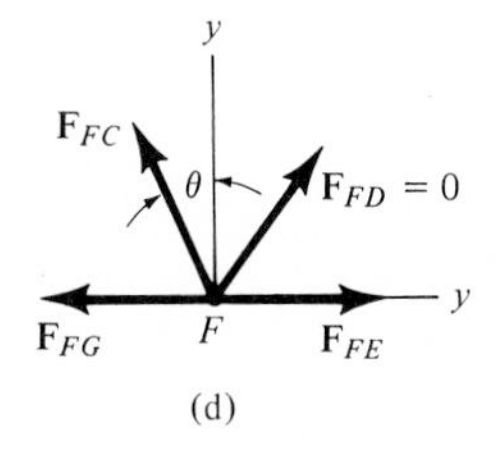

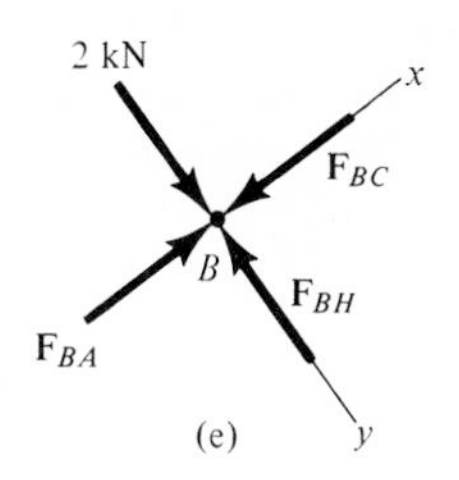

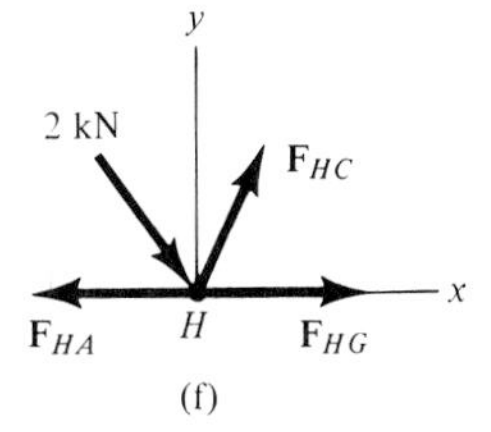

Fig. 6–13b, c, d, e and f

Solution

Looking for joint situations which are similar to those outlined in Figs. 6–11 and 6–12, we have

Joint G, Fig. 6–13*b*, which is like Fig. 6–12*a*

$+\uparrow \Sigma F_y = 0;$ $\quad F_{GC} = 0$ *Ans.*

Joint D, Fig. 6–13*c*, which is like Fig. 6–12*a*

$+\swarrow \Sigma F_y = 0;$ $\quad F_{DF} = 0$ *Ans.*

Joint F, Fig. 6–13*d*, which is like Fig. 6–12*b*

$+\uparrow \Sigma F_y = 0;$ $\quad F_{FC} \cos \theta = 0$, since $\theta \neq 0$; $\quad F_{FC} = 0$ *Ans.*

Note that if joint *B* is analyzed, Fig. 6–13*e*,

$+\searrow \Sigma F_y = 0;$ $\quad 2 - F_{BH} = 0 \quad F_{BH} = 2$ kN (C)

Consequently, the numerical value of F_{HC} must satisfy $\Sigma F_y = 0$, Fig. 6–13*f*, and therefore *HC* is *not* a zero-force member.

Problems

6–1. The truss, used to support a balcony, is subjected to the loading shown. Approximate each joint as a pin and determine the force in each member. State whether the members are in tension or compression. Note that the pin at E exerts *only* a force component $\mathbf{E}_x$ along the two-force member DE. ($\mathbf{E}_y = \mathbf{0}$.)

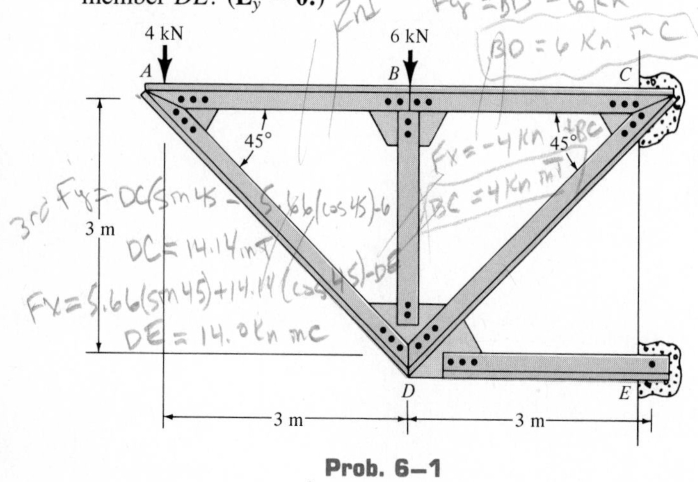

Prob. 6–1

6–2. Determine the force in each member of the truss and indicate whether the members are in tension or compression. Note that the pin at G exerts only a force component $\mathbf{G}_x$ along the two-force member GB. ($\mathbf{G}_y = \mathbf{0}$.)

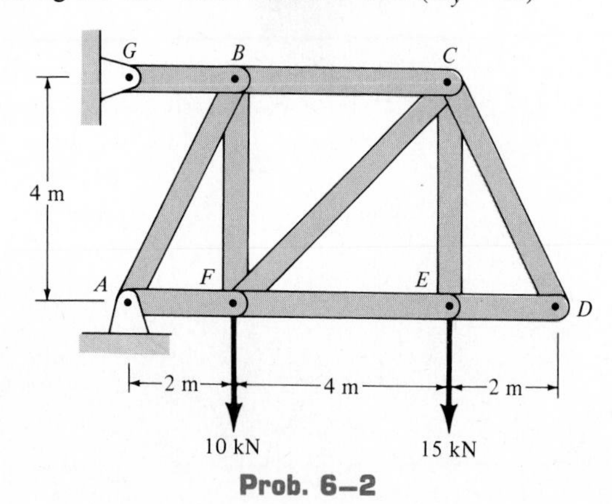

Prob. 6–2

6–3. Compute the force in each member of the *Warren truss* and indicate whether the members are in tension or compression. All the members are 3 m long.

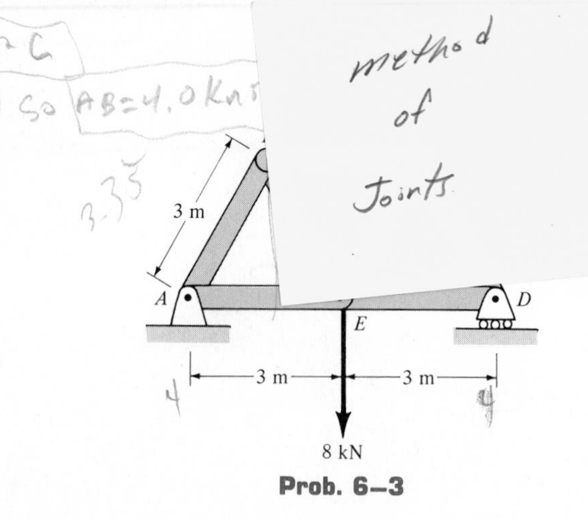

Prob. 6–3

***6–4.** The beam supported by the truss is subjected to a uniform distributed load of 4 kN/m. Assuming the central 2 m of load is supported by joint C and the remaining portions by joints B and D, show that the vertical force at C due to the distributed load is 8 kN and that at B and D the force is 4 kN. Then calculate the force in members BC, GB, GC, GA, and AB, and state whether the members are in tension or compression. *Hint:* Start the analysis at joint G and show that the force in member GB is zero.

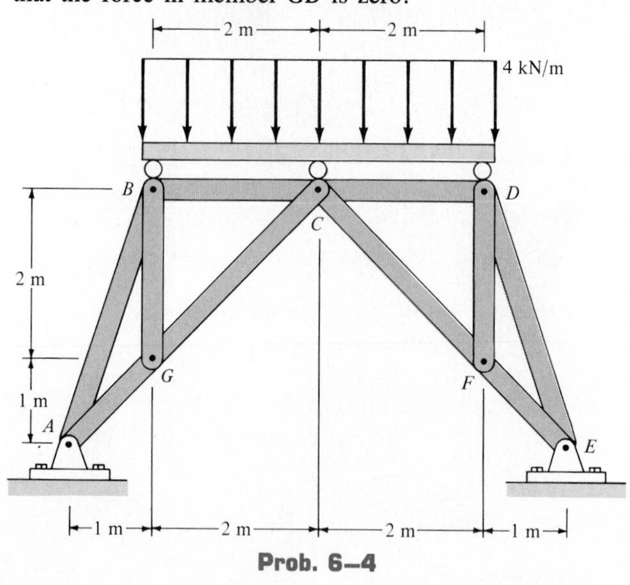

Prob. 6–4

6–5. Determine the force in each member of the truss and indicate whether the members are in tension or compression.

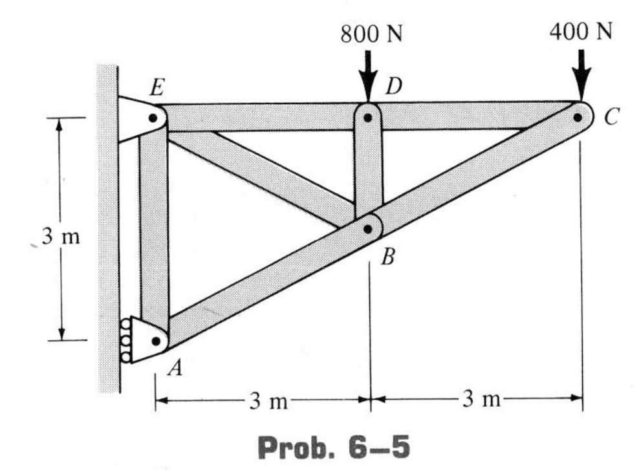

Prob. 6–5

6–6. Determine the force in each member of the truss, and indicate whether the members are in tension or compression.

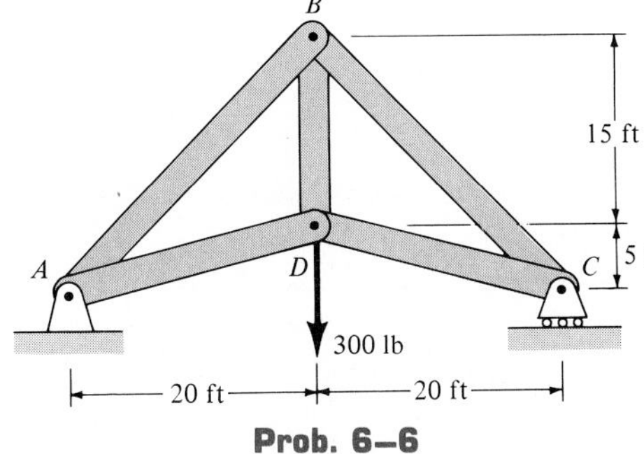

Prob. 6–6

6–7. The trussed building bent is subjected to a loading of 800 lb. Approximate each joint as a pin and determine the force in each member. State whether the members are in tension or compression. *Hint:* Since the load and truss are symmetrical, only half the truss has to be analyzed.

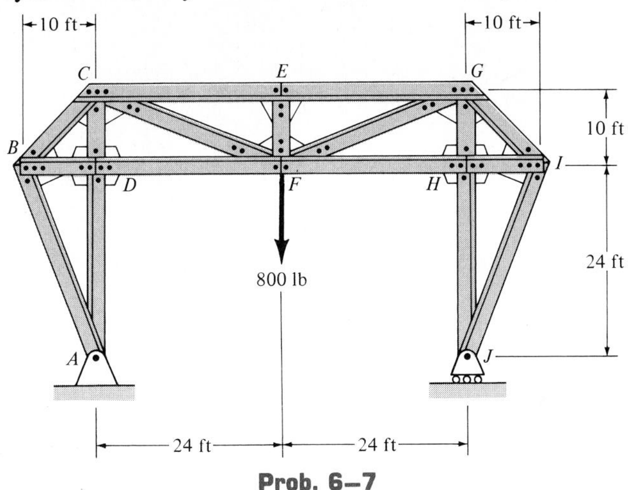

Prob. 6–7

***6–8.** A sign is subjected to a wind loading that exerts horizontal forces of 300 lb on joints *B* and *C* of one of the side supporting trusses. Determine the force in members *BC, CD, DB,* and *DE* of the truss and state whether the members are in tension or compression. *Hint:* First analyze joint *C*, then joint *D*.

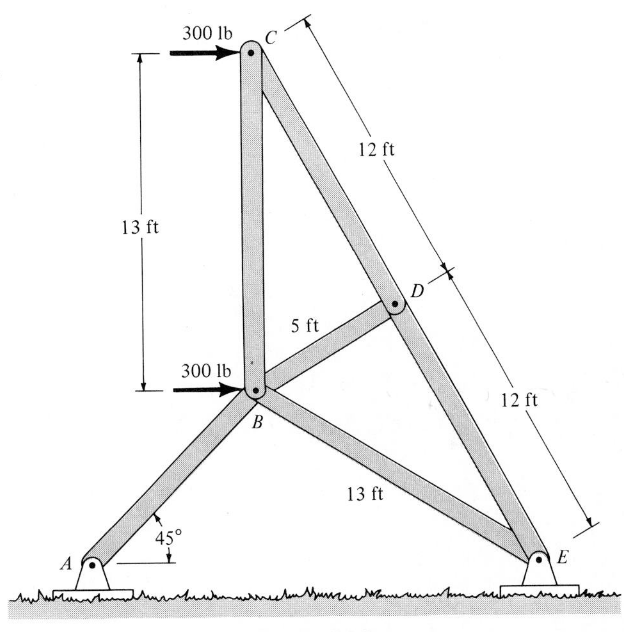

Prob. 6–8

6–9. Determine the force in members *PE* and *KR* of the truss and indicate whether the members are in tension or compression. Also, indicate all zero-force members.

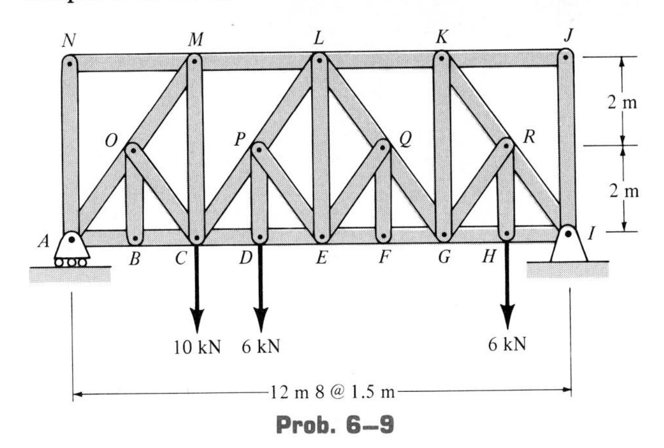

Prob. 6–9

6–10. Determine the force in members LK, AB, and BK of the aircraft frame and indicate whether the members are in tension or compression. All triangles, except ALK, are equilateral. Assume the frame is connected to the fuselage by a short link at A and by a pin at J.

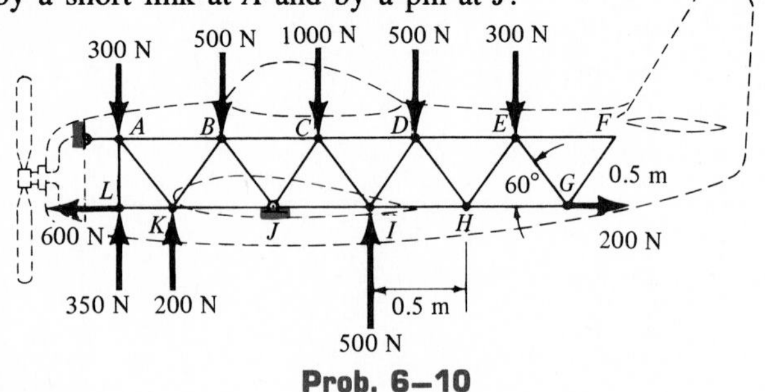

Prob. 6–10

6–11. The *scissors truss* is used to support the roof load. Determine the force in members BF and FD, and indicate whether the members are in tension or compression.

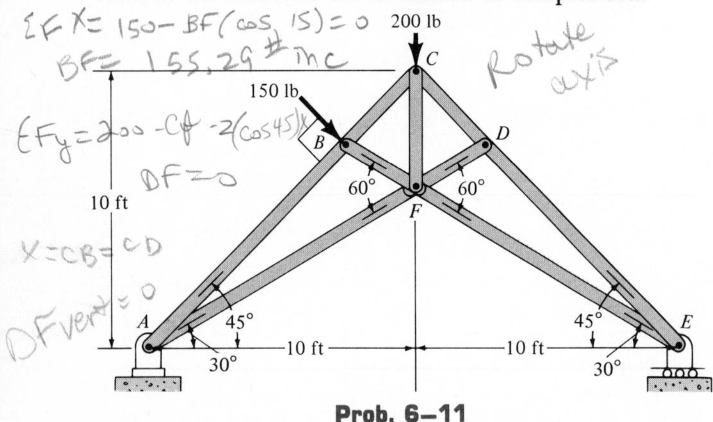

Prob. 6–11

***6–12.** Determine the force in members CD and CM of the *Baltimore bridge truss* and indicate whether the members are in tension or compression. Also, indicate all zero-force members.

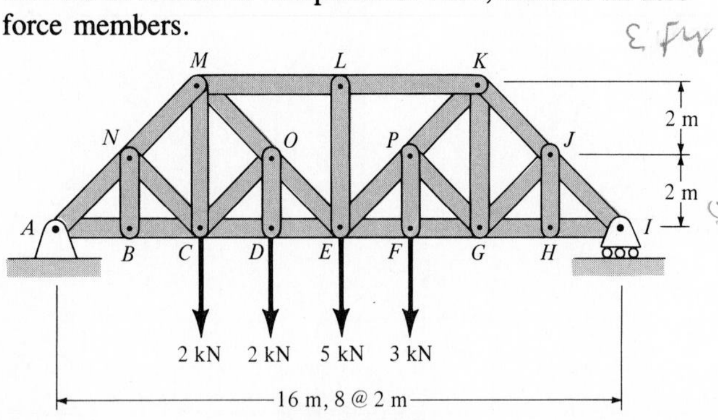

Prob. 6–12

6–13. Determine the force in each member of the truss and indicate whether the members are in tension or compression. *Hint:* Start the analysis at joint B by summing the forces along member BC and show $F_{BC} = 0$. Then analyze joint C, etc.

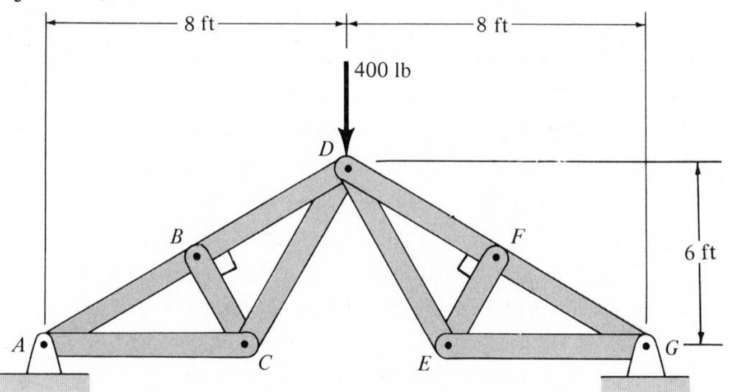

Prob. 6–13

6–14. For the given loading, determine the force in members CD, CJ, and KJ of the *Howe roof truss*. Indicate whether the members are in tension or compression.

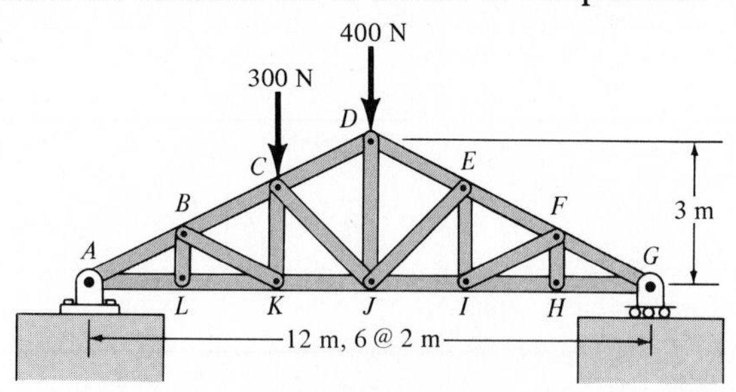

Prob. 6–14

6–15. Determine the force in each member of the truss. State whether the members are in tension or compression.

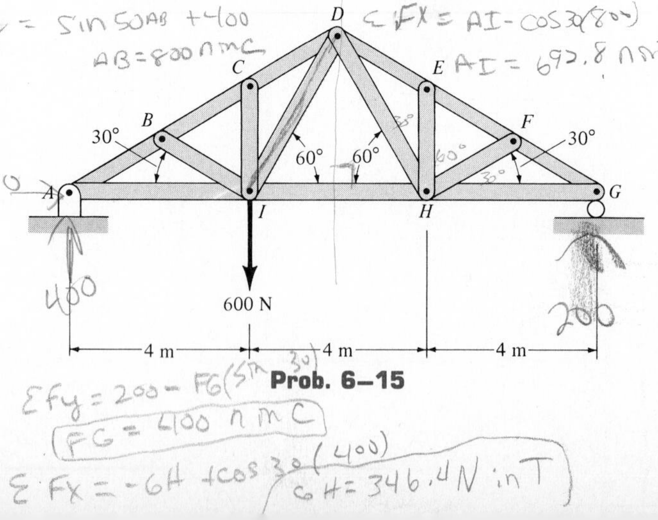

Prob. 6–15

The Method of Sections 6.4

If the forces in only a few members of a truss are to be found, the method of sections generally provides the most direct means of obtaining these forces. The *method of sections* consists of passing an *imaginary section* through the truss, thus cutting it into two parts. Provided the entire truss is in equilibrium, then each of the two parts must also be in equilibrium; and as a result, the three equations of equilibrium may be applied to either one of these two parts to determine the member forces at the "cut section."

When the method of sections is used to determine the force in a particular member, a decision must be made as to how to "cut" or section the truss. Since only *three* independent equilibrium equations ($\Sigma F_x = 0$, $\Sigma F_y = 0$, $\Sigma M_O = 0$) can be applied to the isolated portion of the truss, one should try to select a section that, in general, passes through not more than *three* members in which the forces are unknown. For example, consider the truss in Fig. 6–14*a*. If the force in member *GC* is to be determined, section *aa* would be appropriate. The free-body diagrams of the two parts are shown in Fig. 6–14*b* and *c*. In particular, note that the line of action of each cut member force is specified from the *geometry* of the truss, since the force in a member passes along the axis of the member. Also, the member forces acting on one part of the truss are equal but opposite to those acting on the other part—Newton's third law. As shown, members assumed to be in *tension* (*BC* and *GC*) are subjected to a "pull," whereas the member in *compression* (*GF*) is subjected to a "push."

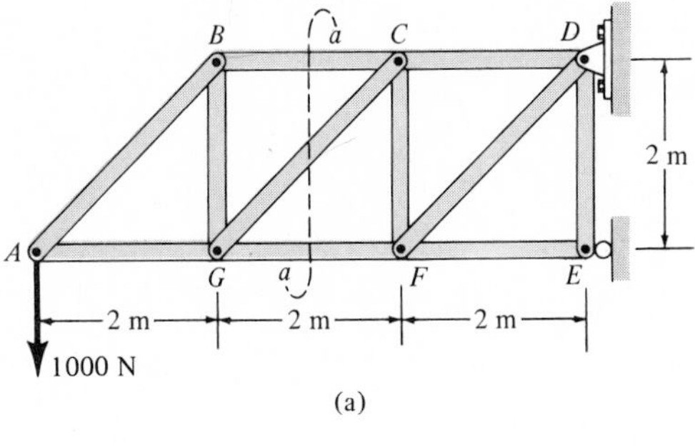

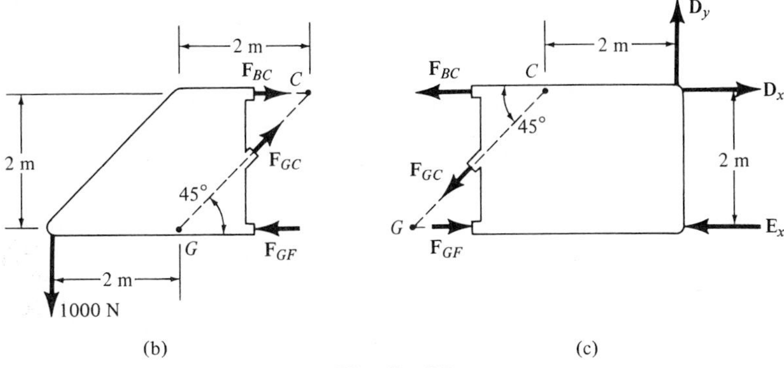

Fig. 6–14

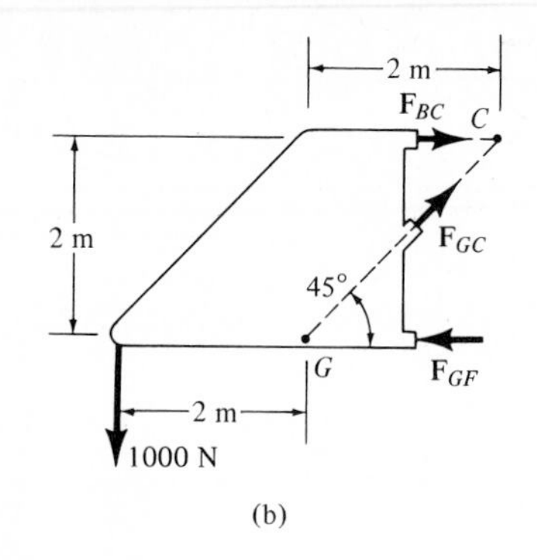

(b)

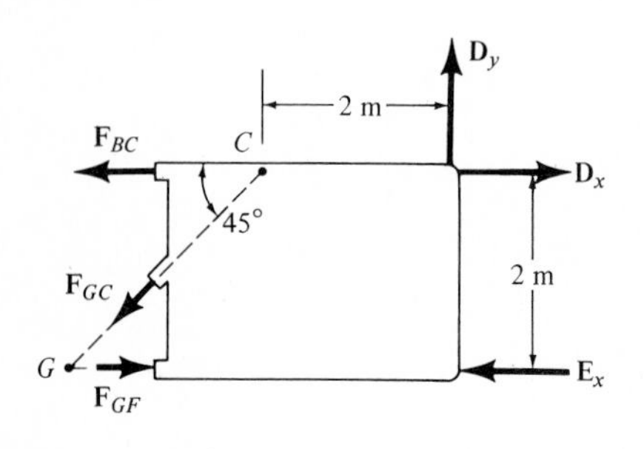

(c)

Fig. 6–14b,c

The three unknown member forces $\mathbf{F}_{BC}$, $\mathbf{F}_{GC}$, and $\mathbf{F}_{GF}$ can be obtained by applying the three equilibrium equations to the free-body diagram in Fig. 6–14*b*. If, however, the free-body diagram in Fig. 6–14*c* is considered, the three support reactions $\mathbf{D}_x$, $\mathbf{D}_y$, and $\mathbf{E}_x$ will have to be determined *first*. Why? (This, of course, is done in the usual manner by considering a free-body diagram of the *entire truss*.) When applying the equilibrium equations, one should consider ways of writing the equations so as to yield a *direct solution* for each of the unknowns, rather than having to solve simultaneous equations. For example, summing moments about C in Fig. 6–14*b* would yield a direct solution for $\mathbf{F}_{GF}$ since $\mathbf{F}_{BC}$ and $\mathbf{F}_{GC}$ create zero moment about C. Likewise, $\mathbf{F}_{BC}$ can be directly obtained by summing moments about G. Finally, $\mathbf{F}_{GC}$ can be found directly from a force summation in the vertical direction since $\mathbf{F}_{GF}$ and $\mathbf{F}_{BC}$ have no vertical components.

The correct "arrowhead" sense of direction of an unknown member force can in many cases be determined "by inspection." For example, $\mathbf{F}_{BC}$ is a tensile force as represented in Fig. 6–14*b*, since moment equilibrium about G requires that $\mathbf{F}_{BC}$ create a moment opposite to that of the 1000-N force. Also, $\mathbf{F}_{GC}$ is tensile since its vertical component must balance the 1000-N force acting downward. In more complicated cases, the sense of an unknown member force may be *assumed*. If the solution yields a *negative* magnitude it indicates that its "arrowhead' direction is *opposite* to that shown on the free-body diagram.

PROCEDURE FOR ANALYSIS

The following procedure provides a means for applying the method of sections to determine the forces in the members of a truss.

Free-Body Diagram. Make a decision as to how to "cut" or section the truss through the members where forces are to be determined. Before isolating the appropriate section, it may first be necessary to determine the truss's *external* reactions, so that the three equilibrium equations are used *only* to solve for member forces at the cut section. Draw the free-body diagram of that part of the sectioned truss which has the least number of forces acting on it. By inspection, attempt to show the unknown member forces acting with the correct "arrowhead" sense of direction.

Equations of Equilibrium. Try to apply the three equations of equilibrium such that simultaneous solution of equations is avoided. In this regard, moments should be summed about a point that lies at the intersection of the lines of action of two unknown forces, so that the third unknown force is determined directly from the moment equation. If two of the unknown forces are *parallel,* forces may be summed *perpendicular* to the direction of these unknowns to determine *directly* the third unknown force.

The following examples numerically illustrate these concepts.

Example 6–5

Determine the force in members *GE*, *GC*, and *BC* of the bridge truss shown in Fig. 6–15*a*. Indicate whether the members are in tension or compression.

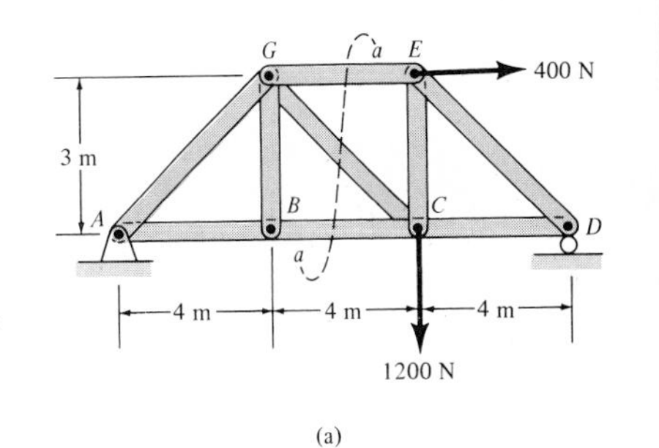

(a)

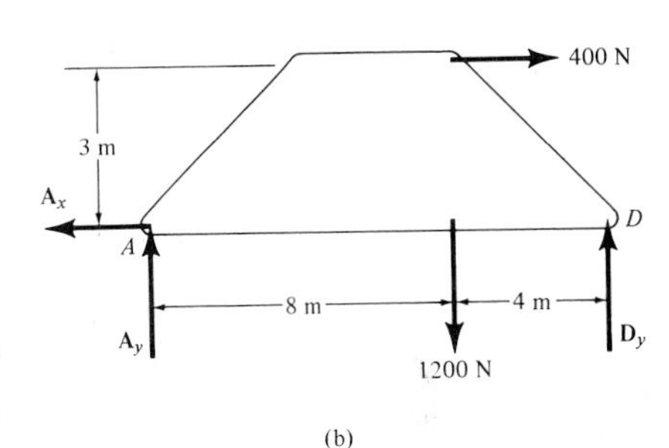

(b)

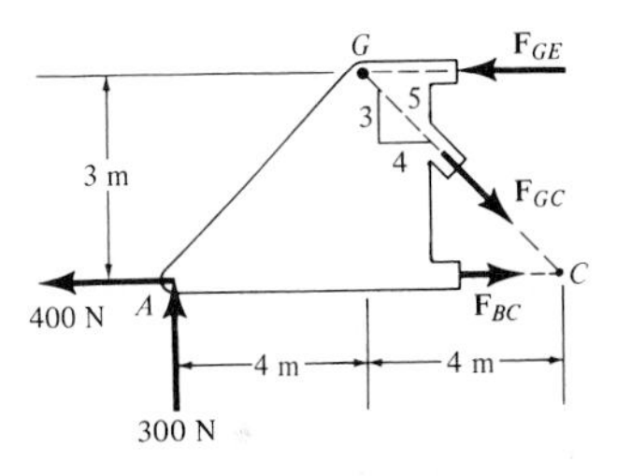

(c)

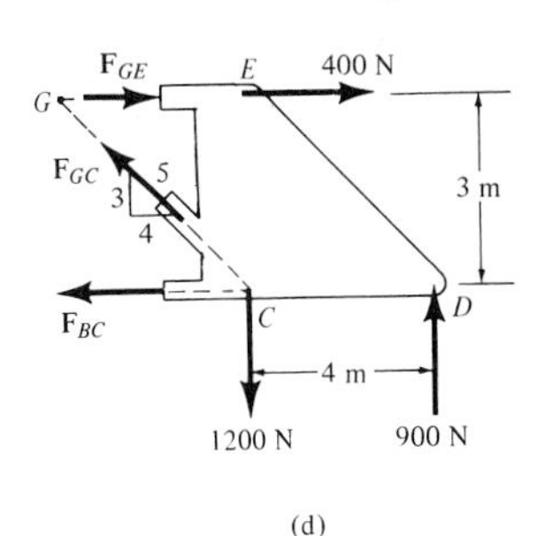

(d)

Fig. 6–15

Solution

Section *aa* in Fig. 6–15*a* has been chosen since it cuts through the *three* members whose forces are to be determined. In order to use the method of sections, however, it is *first* necessary to determine the external reactions at *A* or *D*. Why? A free-body diagram of the entire truss is shown in Fig. 6–15*b*. Applying the equations of equilibrium, we have

$$\overset{+}{\rightarrow}\Sigma F_x = 0; \qquad 400\text{ N} - A_x = 0 \qquad A_x = 400\text{ N}$$

$$\circlearrowleft + \Sigma M_A = 0; \quad -1200\text{ N}(8\text{ m}) - 400\text{ N}(3\text{ m}) + D_y(12\text{ m}) = 0$$

$$D_y = 900\text{ N}$$

$$+\uparrow \Sigma F_y = 0; \quad A_y - 1200\text{ N} + 900\text{ N} = 0 \qquad A_y = 300\text{ N}$$

Free-Body Diagrams. The free-body diagrams of the sectioned truss are shown in Fig. 6–15*c* and *d*. For the analysis the free-body diagram in Fig. 6–15*c* will be used since it involves the least number of forces.

Equations of Equilibrium. Summing moments about point *G* eliminates $\mathbf{F}_{GE}$ and $\mathbf{F}_{GC}$ and yields a direct solution for F_{BC}.

$$\circlearrowleft + \Sigma M_G = 0; \quad -300\text{ N}(4\text{ m}) - 400\text{ N}(3\text{ m}) + F_{BC}(3\text{ m}) = 0$$

$$F_{BC} = 800\text{ N} \quad (\text{T}) \qquad \textit{Ans.}$$

In the same manner, by summing moments about point *C* we obtain a direct solution for F_{GE}.

$$\circlearrowleft + \Sigma M_C = 0; \qquad -300\text{ N}(8\text{ m}) + F_{GE}(3\text{ m}) = 0$$

$$F_{GE} = 800\text{ N} \quad (\text{C}) \qquad \textit{Ans.}$$

Since $\mathbf{F}_{BC}$ and $\mathbf{F}_{GE}$ have no vertical components, summing forces in the *y* direction directly yields F_{GC}, i.e.,

$$+\uparrow \Sigma F_y = 0; \quad 300\text{ N} - \tfrac{3}{5}F_{GC} = 0 \qquad F_{GC} = 500\text{ N} \quad (\text{T}) \qquad \textit{Ans.}$$

Obtain these results by applying the equations of equilibrium to the free-body diagram shown in Fig. 6–15*d*.

Example 6–6

Determine the force in members *FG* and *CF* of the truss shown in Fig. 6–16*a*. Indicate whether the members are in tension or compression. The reactions at the supports are shown in the figure.

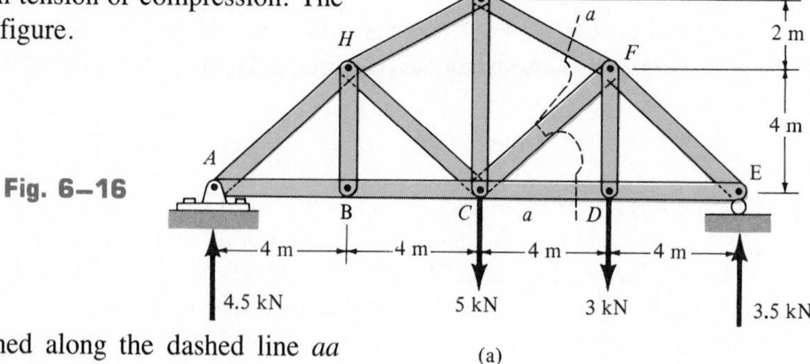

Fig. 6–16

(a)

Solution

Free-Body Diagram. The truss is sectioned along the dashed line *aa* shown in Fig. 6–16*a*. This section will "expose" the internal forces in members *FG* and *CF* as "external" on a free-body diagram of either the right or left portion of the truss. The free-body diagram of the right portion, which is the easiest to analyze, is shown in Fig. 6–16*b*. There are three unknowns, $\mathbf{F}_{FG}$, $\mathbf{F}_{CF}$, and $\mathbf{F}_{CD}$.

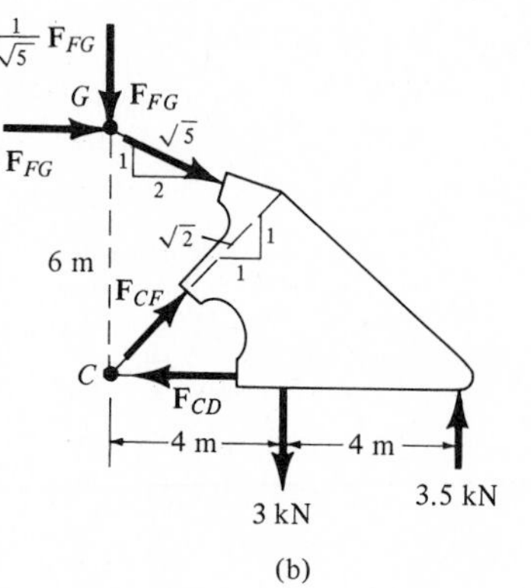

(b)

Equations of Equilibrium. The most direct method for solving this problem requires application of the moment equation applied about a point that eliminates two of the unknown forces. Hence, to obtain $\mathbf{F}_{FG}$ directly, we will take moments about point *C*. If $\mathbf{F}_{FG}$ is applied at *F*, then the moments of *two* components of $\mathbf{F}_{FG}$ must be computed about *C*. Instead, a simpler calculation results if $\mathbf{F}_{FG}$ is resolved into rectangular components and, by the principle of transmissibility, this force is extended to point *G*. (Although *G* is not on the free-body diagram, the moment effect of $\mathbf{F}_{FG}$ about *C* can be computed from *G*.) Here the moment of only the horizontal component must be considered. We have

$$\circlearrowleft+\Sigma M_C = 0; \qquad F_{FG}\left(\frac{2}{\sqrt{5}}\right)(6\text{ m}) - (3\text{ kN})(4\text{ m}) + (3.5\text{ kN})(8\text{ m}) = 0$$

$$F_{FG} = 2.98\text{ kN} \quad (C) \qquad \textit{Ans.}$$

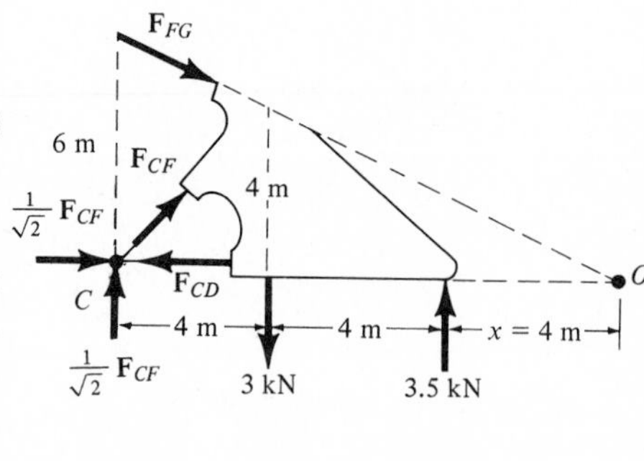

(c)

To obtain $\mathbf{F}_{CF}$ we will eliminate $\mathbf{F}_{FG}$ and $\mathbf{F}_{CD}$ and sum moments about point *O*, Fig. 6–16*c*. Here the principle of transmissibility is used to move $\mathbf{F}_{CF}$ to point *C*. Note that the location of point *O* measured from *E* was determined from proportional triangles, i.e., $4/(4 + x) = 6/(8 + x)$, $x = 4$ m, or in another manner, the slope of member *GF* has a drop of 2 m to a horizontal distance of $CD = 4$ m. Since *FD* is 4 m long, Fig. 6–16*a*, then from *D* to *O*, Fig. 6–16*c*, the distance must be 8 m. Applying the moment equation, we have

$$\circlearrowleft+\Sigma M_O = 0; \quad -F_{CF}\left(\frac{1}{\sqrt{2}}\right)(12\text{ m}) + (3\text{ kN})(8\text{ m}) - (3.5\text{ kN})(4\text{ m}) = 0$$

$$F_{CF} = 1.18\text{ kN} \quad (C) \qquad \textit{Ans.}$$

Try to obtain this same answer by using the result $F_{FG} = 2.98$ kN and summing forces in the *y* direction ($\Sigma F_y = 0$), Fig. 6–16*b*.

Example 6–7

Determine the force in member GC of the truss shown in Fig. 6–17a. Indicate whether the member is in tension or compression. The reactions of the supports are shown in the figure.

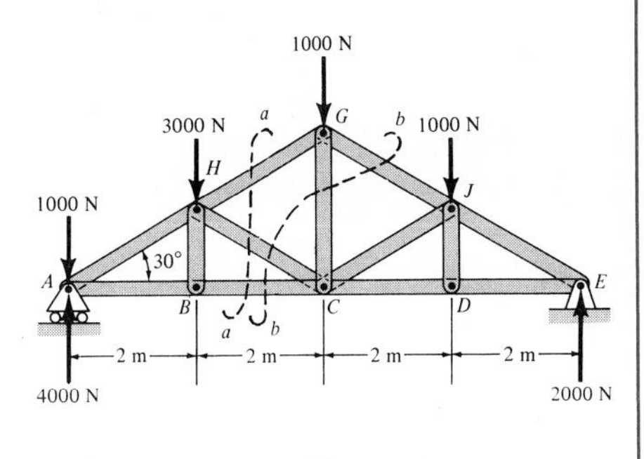

(a)

Solution

Free-Body Diagrams. By the method of sections, any imaginary vertical section that cuts through GC, Fig. 6–17a, will also have to cut through three other members of which the forces are unknown. For example, section bb cuts through GJ, GC, HC, and BC. If a free-body diagram of the left side of this section is considered, Fig. 6–17c, it is possible to determine F_{GJ} by summing moments about C to eliminate the other three unknowns; however, F_{GC} cannot be determined from the remaining two equilibrium equations. It is therefore necessary to determine F_{BC} or F_{HC} *before* using section bb. For example, it is possible to determine F_{HC} by considering the adjacent section aa, Fig. 6–17a. A free-body diagram of the portion of the truss to the left of this section is given in Fig. 6–17b.

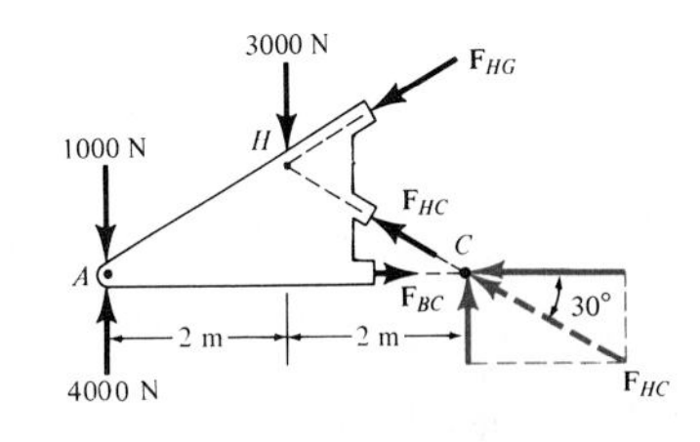

(b)

Equations of Equilibrium. Summing moments about point A one can obtain F_{HC} directly. A simple way to do this is to resolve $\mathbf{F}_{HC}$ into its rectangular components and, by the principle of transmissibility, to extend this force to point C as shown. The moments of $\mathbf{F}_{HC}$ cos 30°, $\mathbf{F}_{BC}$, and $\mathbf{F}_{HG}$ are all zero about point A. Hence,

$$\circlearrowleft +\Sigma M_A = 0; \quad -3000 \text{ N}(2 \text{ m}) + F_{HC} \sin 30^\circ \,(4 \text{ m}) = 0$$

$$F_{HC} = 3000 \text{ N} \quad \text{(C)}$$

The force in GC can now be obtained from Fig. 6–17c by summing moments about E. Again extending $\mathbf{F}_{HC}$ to point C and resolving it into rectangular components as shown, we have

$$\circlearrowleft +\Sigma M_E = 0; \quad 1000 \text{ N}(8 \text{ m}) - 4000 \text{ N}(8 \text{ m}) + 3000 \text{ N}(6 \text{ m})$$
$$+ 1000 \text{ N}(4 \text{ m}) + F_{GC}(4 \text{ m}) - F_{HC} \sin 30^\circ \,(4 \text{ m}) = 0$$

Substituting $F_{HC} = 3000$ N and solving yields

$$F_{GC} = 2000 \text{ N} \quad \text{(T)} \qquad \textit{Ans.}$$

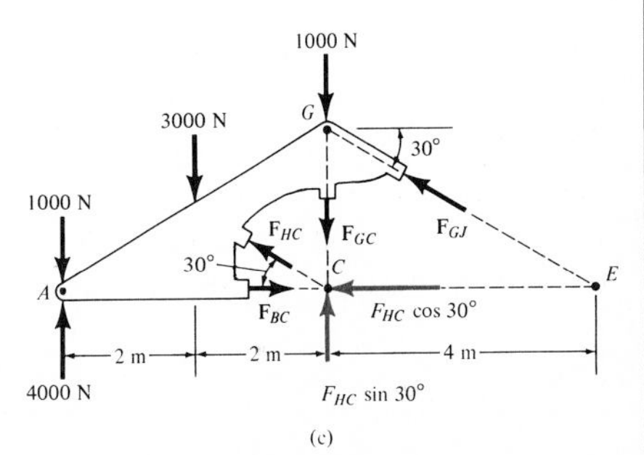

(c)

Fig. 6–17

Problems

***6–16.** The *Howe bridge truss* is subjected to the loading shown. Determine the force in members *HD*, *CD*, and *HG*, and indicate whether the members are in tension or compression.

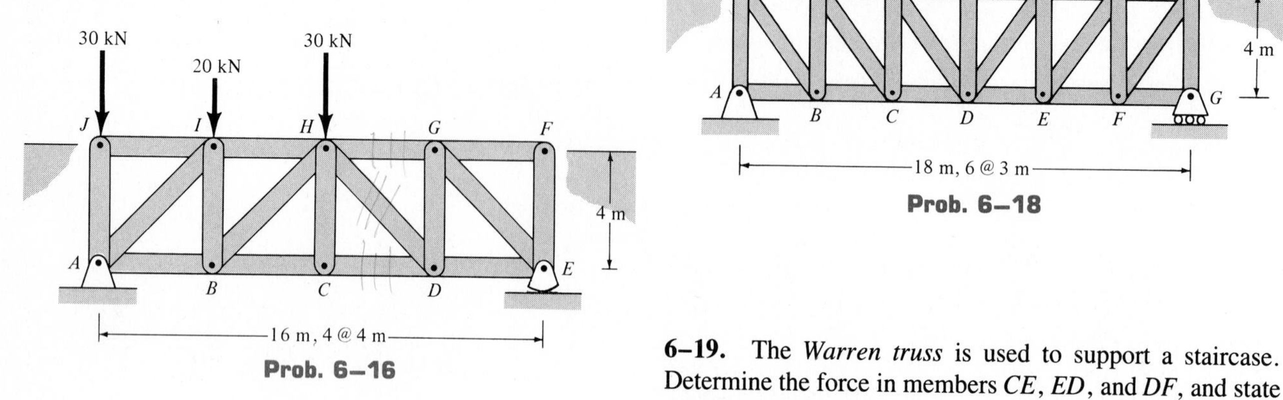

Prob. 6–16

Prob. 6–18

6–17. Determine the force in members *BC*, *HC*, and *HG* of the bridge truss, and indicate whether the members are in tension or compression.

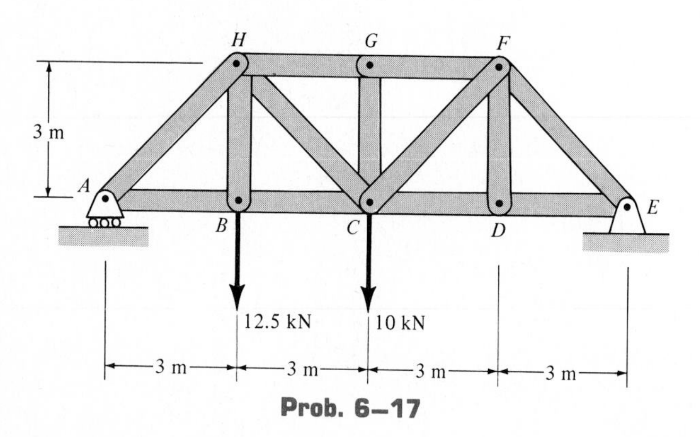

Prob. 6–17

6–18. The *Pratt bridge truss* is subjected to the loading shown. Determine the force in members *CD*, *LD*, and *LK*, and indicate whether these members are in tension or compression.

6–19. The *Warren truss* is used to support a staircase. Determine the force in members *CE*, *ED*, and *DF*, and state whether the members are in tension or compression. Assume all joints are pinned.

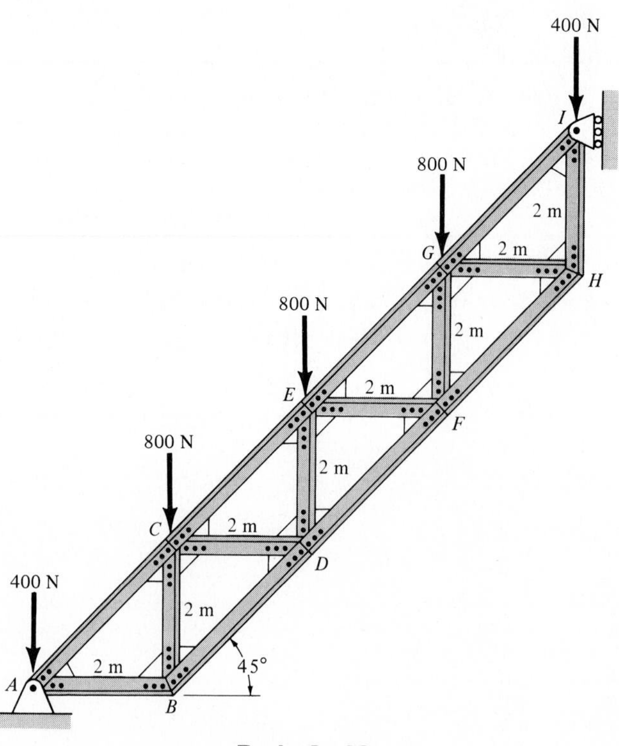

Prob. 6–19

***6–20.** A tower used in an electrical substation supports a power line which exerts a horizontal tension of $T = 500$ lb on each side truss of the tower as shown. Determine the force in members BC, CM, and LM of a side truss and indicate whether the members are in tension or compression.

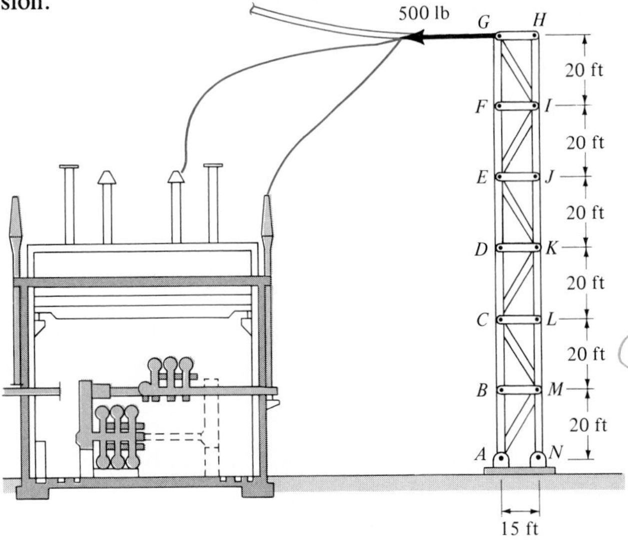

Prob. 6–20

6–21. Determine the force developed in members GB and GF of the bridge truss, and indicate whether the members are in tension or compression.

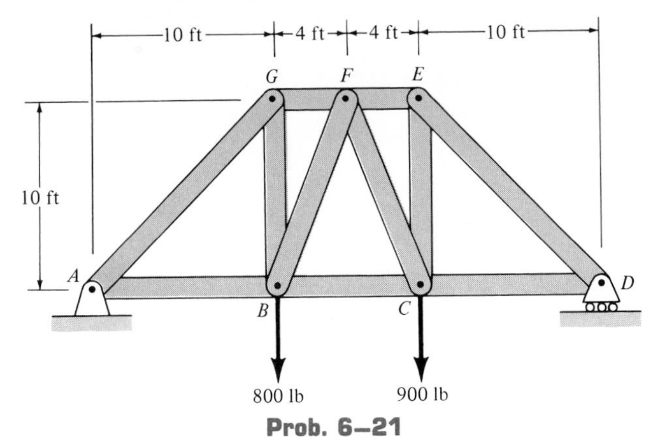

Prob. 6–21

6–22. The *Howe roof truss* supports the vertical loading shown. Determine the force in members KJ, CD, and KD, and indicate whether the members are in tension or compression.

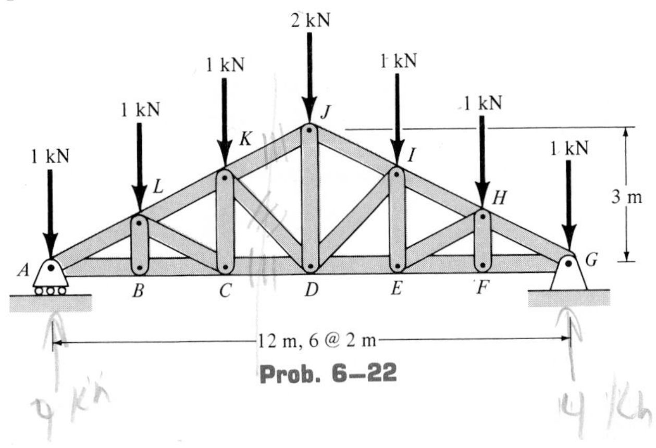

Prob. 6–22

6–23. A crane is constructed from two side trusses. If a load of 4 kN is suspended from one of these trusses as shown, determine the force in members FG, GK, and KJ. State whether the members are in tension or compression. Assume the joints are pin-connected. *Suggestion:* Analyze the forces acting on an appropriate free-body diagram of a sectioned portion of the right side of the truss.

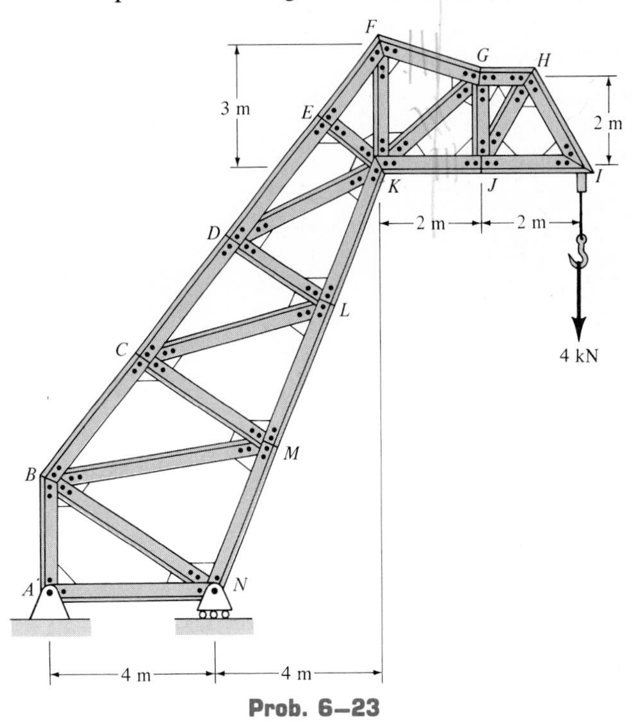

Prob. 6–23

***6–24.** Determine the force in members HC, HG, and BC of the *King-post truss* and indicate whether the members are in tension or compression.

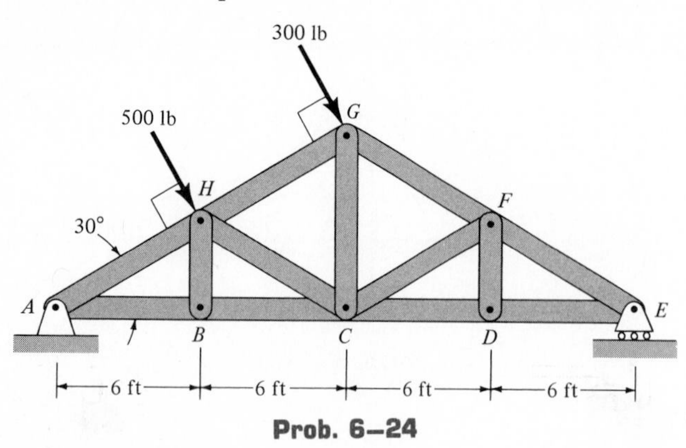

Prob. 6–24

6–25. Determine the force in members GF, FB, and BC of the *Fink truss* and indicate whether the members are in tension or compression.

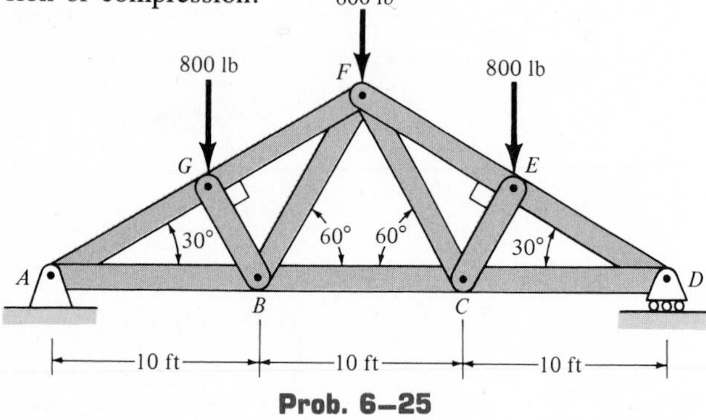

Prob. 6–25

6–26. Compute the force in members BG, HG, and BC of the truss and indicate whether the members are in tension or compression.

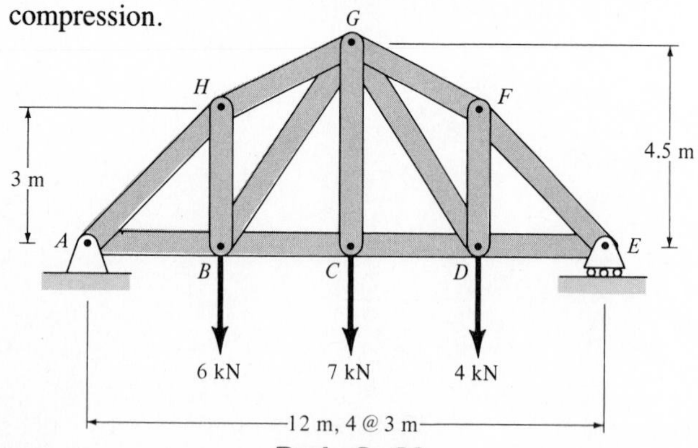

Prob. 6–26

6–27. Determine the force developed in members DE, EQ, and KJ of the side truss of the *hammer-head crane*. Assume that each side truss supports a load of $P = 4000$ lb as shown. Indicate whether the members are in tension or compression. *Suggestion:* Analyze the forces on an appropriate free-body diagram of a section on the *left side* of the truss. Note that QJ is a zero-force member.

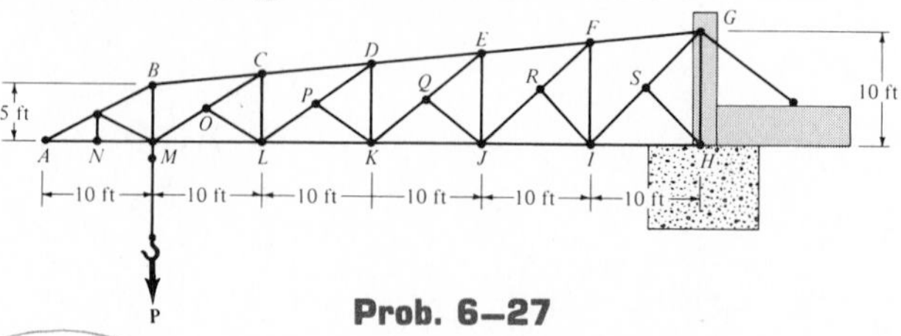

Prob. 6–27

***6–28.** A "K" *truss* used for scaffolding is loaded as shown. Determine the force in members ML and CD using a

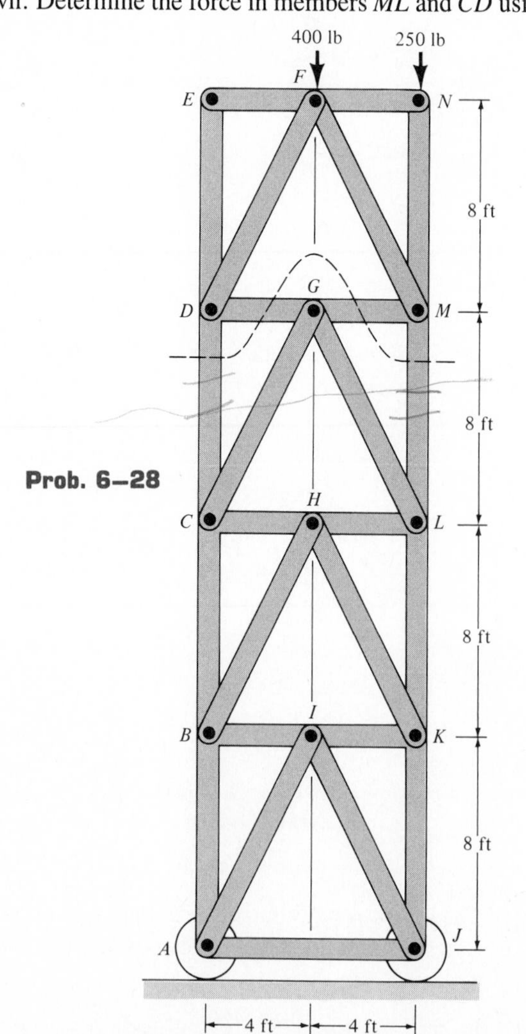

Prob. 6–28

free-body diagram of the upper part of the dashed section shown. Assume that all joints are pin-connected.

6–29. Determine the force in members *LK* and *MK* of the *K truss* and indicate whether the members are in tension or compression. *Hint:* See Prob. 6–28 for an appropriate section to determine the force in *LK*.

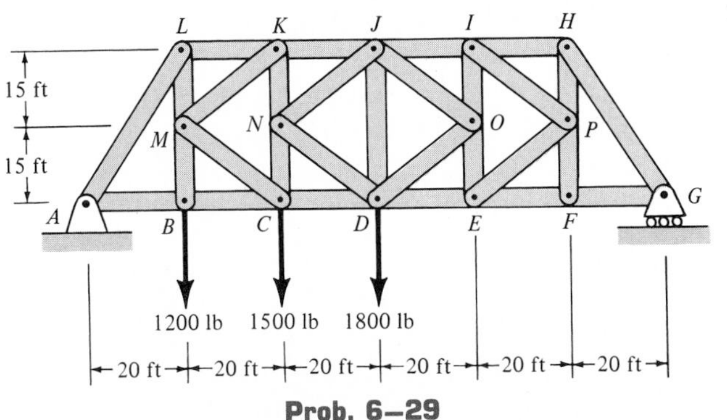

Prob. 6–29

6–30. Determine the force in members *GF*, *CF*, and *CD* of the roof truss and indicate whether the members are in tension or compression.

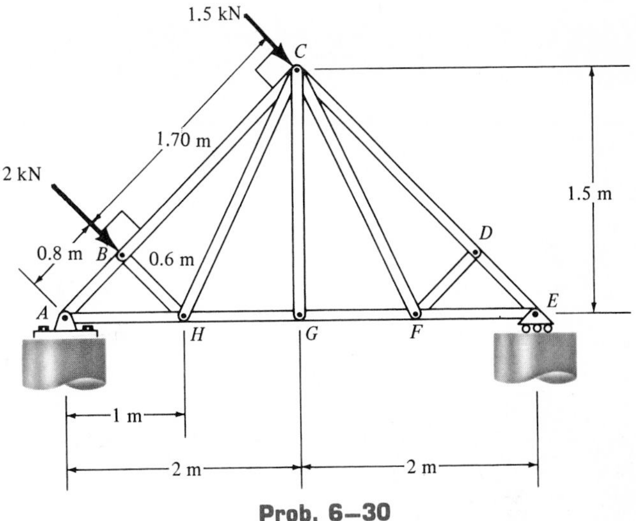

Prob. 6–30

Space Trusses 6.5*

A *space truss* consists of members joined together at their ends to form a stable three-dimensional structure. The simplest element of a space truss is a *tetrahedron,* formed by connecting six members together, as shown in Fig. 6–18. Any additional members added to this basic element would be redundant in supporting the force **P.** A *simple space truss* can be built from this basic tetrahedral element by adding three additional members and a joint forming a system of multiconnected tetrahedrons.

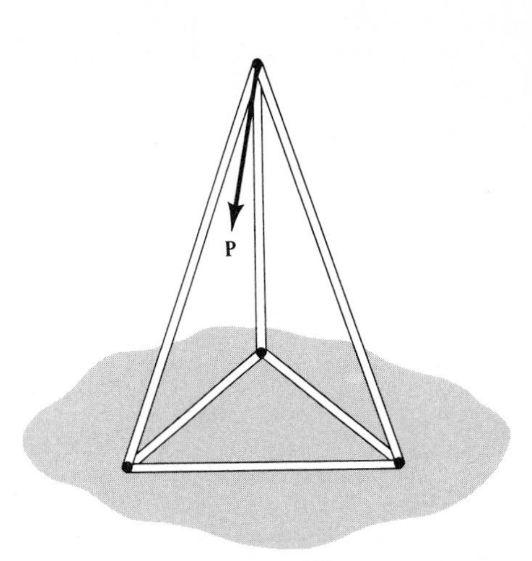

Fig. 6–18

Assumptions for Design. The members of a space truss may be treated as two-force members provided the external loading is applied at the joints and the joints consist of ball-and-socket connections. These assumptions are justified if the welded or riveted connections of the joined members intersect at a common point and the weight of the members can be neglected. In cases where the weight of a member is to be included in the analysis, it is generally satisfactory to apply it as a vertical force, half of its magnitude applied at each end of the member.

PROCEDURE FOR ANALYSIS

Either the method of sections or the method of joints can be used to determine the forces developed in the members of a simple space truss.

Method of Sections. If only a *few* member forces are to be determined, the method of sections may be used. When an imaginary section is passed through a truss, and the truss is separated into two parts, the force system acting on one of the parts must satisfy the *six* scalar equilibrium equations: $\Sigma F_x = 0$, $\Sigma F_y = 0$, $\Sigma F_z = 0$, $\Sigma M_x = 0$, $\Sigma M_y = 0$, $\Sigma M_z = 0$ (Eqs. 5–4 and 5–5). By proper choice of the section and axes for summing forces and moments, many of the unknown member forces in a space truss can be computed *directly,* using a single equilibrium equation.

Method of Joints. Generally, if the forces in *all* the members of the truss must be determined, the method of joints is most suitable for the analysis. When using the method of joints, it is necessary to solve the three scalar equilibrium equations $\Sigma F_x = 0$, $\Sigma F_y = 0$, $\Sigma F_z = 0$ at each joint. The solution of many simultaneous equations can be avoided if the force analysis begins at a joint having at least one known force and at most three unknown forces.

The following example numerically illustrates this procedure.

Example 6–8

Determine the forces acting in the members of the space frame shown in Fig. 6–19*a*. Indicate whether the members are in tension or compression.

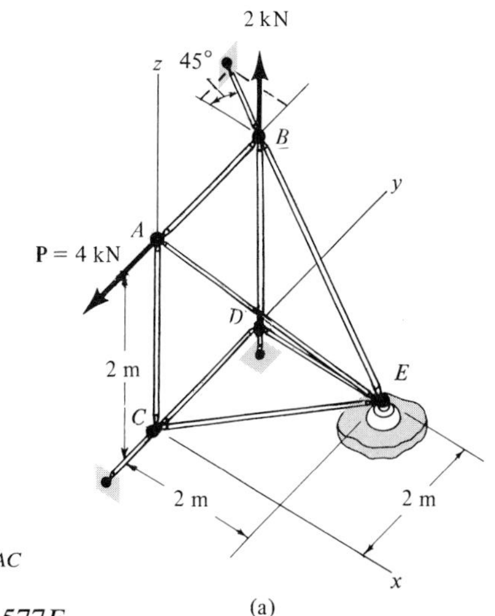

(a)

Solution

Since there are one known force and three unknown forces acting at joint A, the force analysis of the truss will begin at joint A.

Joint A (Fig. 6–19*b*) Each force on the free-body diagram is first expressed in terms of its x, y, z components. Note that a position vector extending from A to E has components of $r_x = 2$ m, $r_y = 2$ m, $r_z = -2$ m. Also, $r_{AE} = \sqrt{12}$ m. Thus,

$$P_x = 0 \qquad P_y = -4 \text{ kN} \qquad P_z = 0$$

$$(F_{AB})_x = 0 \qquad (F_{AB})_y = F_{AB} \qquad (F_{AB})_z = 0$$

$$(F_{AC})_x = 0 \qquad (F_{AC})_y = 0 \qquad (F_{AC})_z = -F_{AC}$$

$$(F_{AE})_x = F_{AE}\left(\frac{2}{\sqrt{2}}\right) = 0.577F_{AE} \qquad (F_{AE})_y = 0.577F_{AE} \qquad (F_{AE})_z = -0.577F_{AE}$$

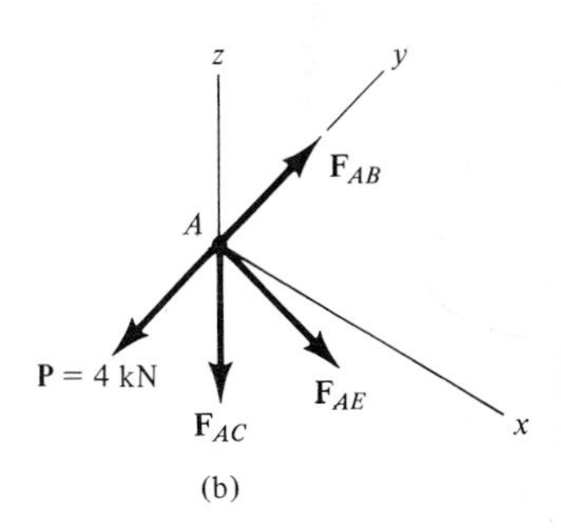

(b)

Applying the equations of equilibrium, we have

$\Sigma F_x = 0;$ $\quad 0.577F_{AE} = 0$

$\Sigma F_y = 0;$ $\quad -4 + F_{AB} + 0.577F_{AE} = 0$

$\Sigma F_z = 0;$ $\quad -F_{AC} - 0.577F_{AE} = 0$

$$F_{AC} = F_{AE} = 0 \qquad \textit{Ans.}$$

$$F_{AB} = 4 \text{ kN} \quad \text{(T)} \qquad \textit{Ans.}$$

Since F_{AB} is known, joint B may be analyzed next.

Joint B (Fig. 6–19*c*)

$\Sigma F_x = 0;$ $\quad -R_B \cos 45° + 0.707F_{BE} = 0$

$\Sigma F_y = 0;$ $\quad -4 + R_B \sin 45° = 0$

$\Sigma F_z = 0;$ $\quad 2 + F_{BD} - 0.707F_{BE} = 0$

$$R_B = F_{BE} = 5.66 \text{ kN} \quad \text{(T)}, \qquad F_{BD} = 2 \text{ kN} \quad \text{(C)} \qquad \textit{Ans.}$$

The three equations of equilibrium may be applied directly to the force systems on the free-body diagrams of joints D and C, since the force components are easily determined. Show that

$$F_{DE} = F_{DC} = F_{CE} = 0 \qquad \textit{Ans.}$$

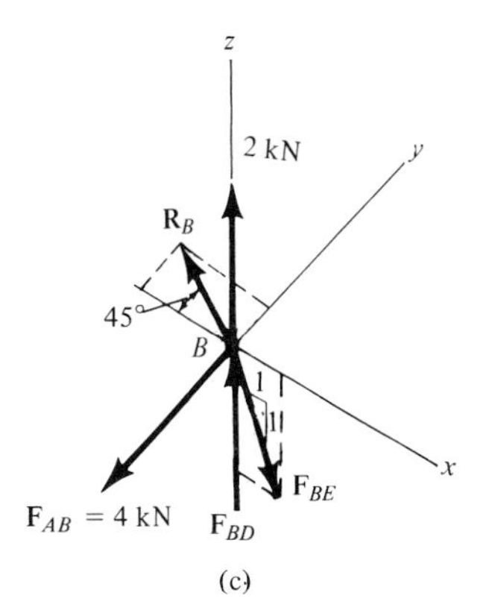

(c)

Fig. 6–19

Problems

6–31. The space truss is used to support vertical forces at joints B, C, and D. Determine the force in each of the members and indicate whether the members are in tension or compression.

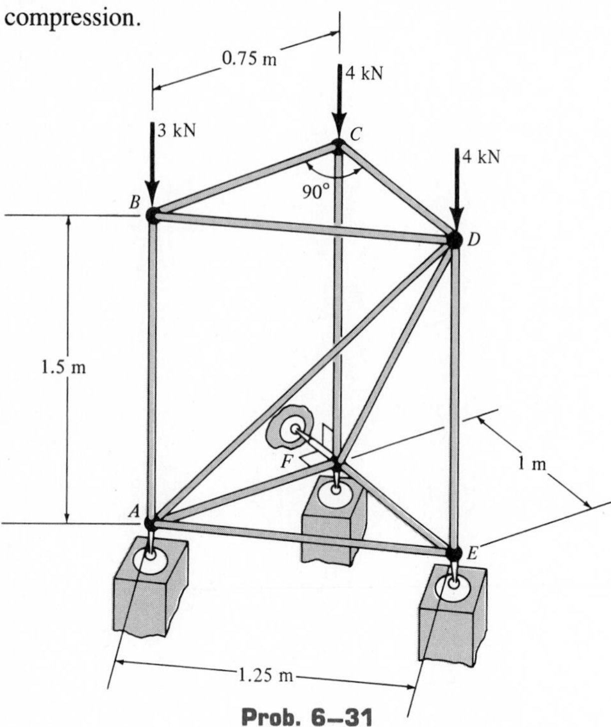

Prob. 6–31

***6–32.** Two space trusses are used to equally support the uniform 150-kg sign. Determine the force developed in members AB, AC, and BC of truss $ABCD$, and indicate whether these members are in tension or compression. Horizontal short links support the truss at joints B, C, and D.

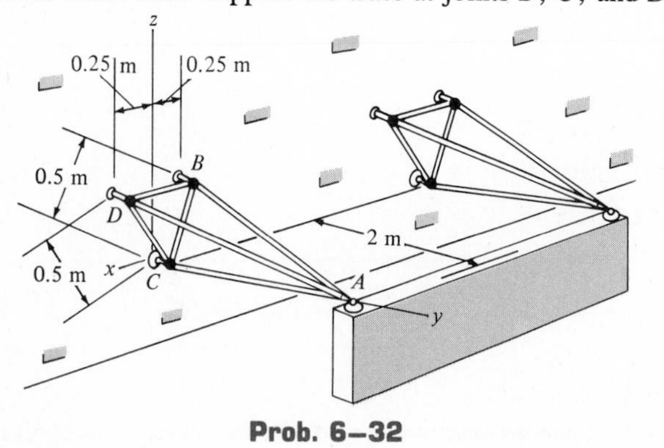

Prob. 6–32

6–33. Determine the force in each of the members of the space truss and indicate whether the members are in tension or compression. The truss is supported by a ball-and-socket joint at A and short links at B and C.

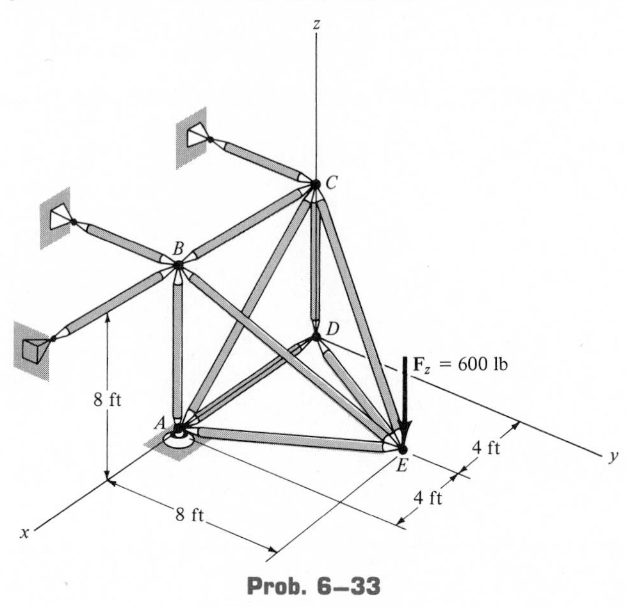

Prob. 6–33

6–34. Determine the force in members BD, AD, and AF of the space truss and state whether the members are in tension

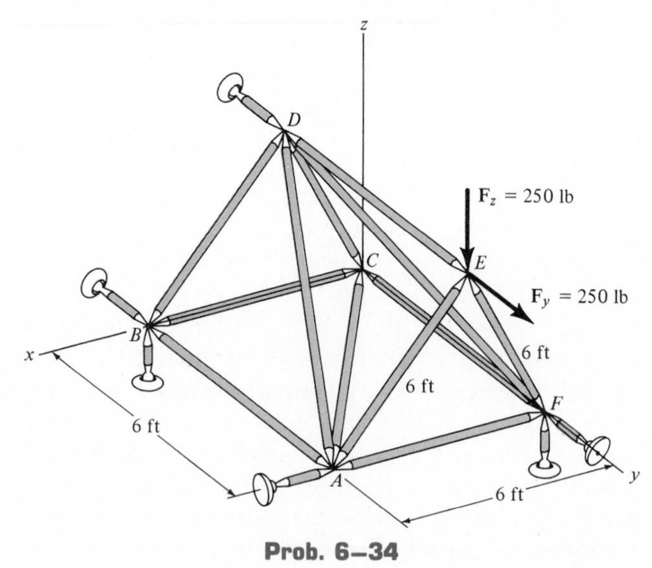

Prob. 6–34

or compression. The truss is supported by short links at A, B, D, and F. The external reactions are listed as the answers to Prob. 5–55.

6–35. Determine the force in each of the members of the space truss when joint A is subjected to the force $\mathbf{P}_y$. Indicate whether the members are in tension or compression. The support reactions at B, C, and D are to be determined.

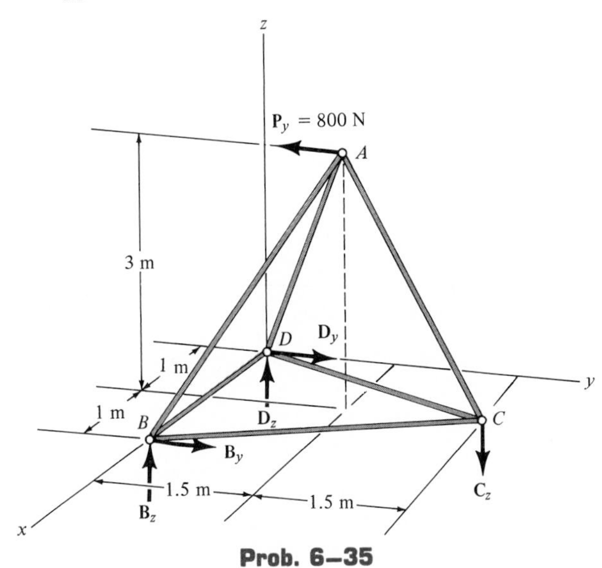

Prob. 6–35

***6–36.** The *tower truss* is subjected to a horizontal force $\mathbf{P}_y$ at its apex. Determine the force in members CD and DB and state whether the members are in tension or compression. The truss is supported by short links at A, G, and F.

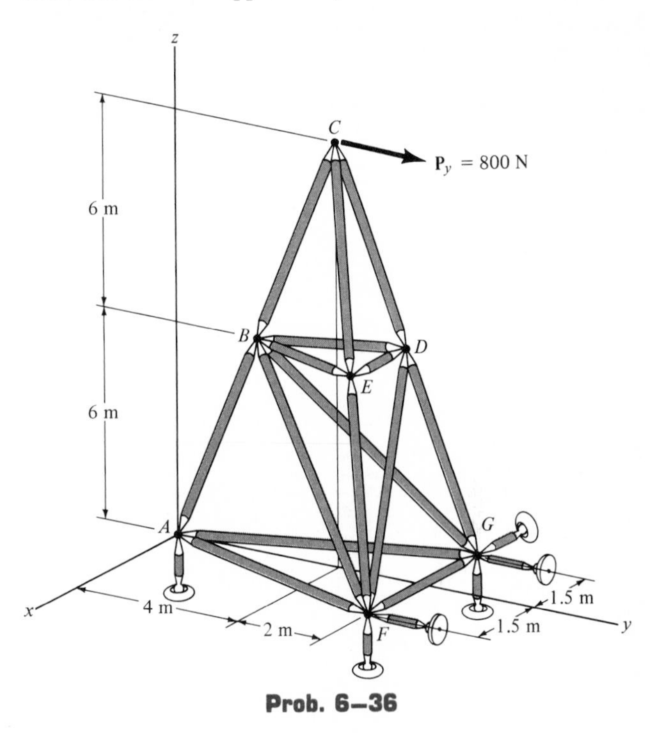

Prob. 6–36

Frames and Machines 6.6

Frames and machines are two common types of structures which are often composed of pin-connected *multiforce members,* i.e., members that are subjected to more than two forces. *Frames* are generally stationary and are used to support loads, whereas *machines* contain moving parts and are designed to transmit and alter the effect of forces. Provided a frame or machine is properly constrained and contains no more supports or members than are necessary to prevent collapse, then the forces acting at the joints and supports can be determined by applying the equations of equilibrium ($\Sigma F_x = 0$, $\Sigma F_y = 0$, $\Sigma M_O = 0$) to each member. Once the forces at the joints are obtained, it is possible to *design* the size of the members, connections, and supports using the theory of strength of materials.

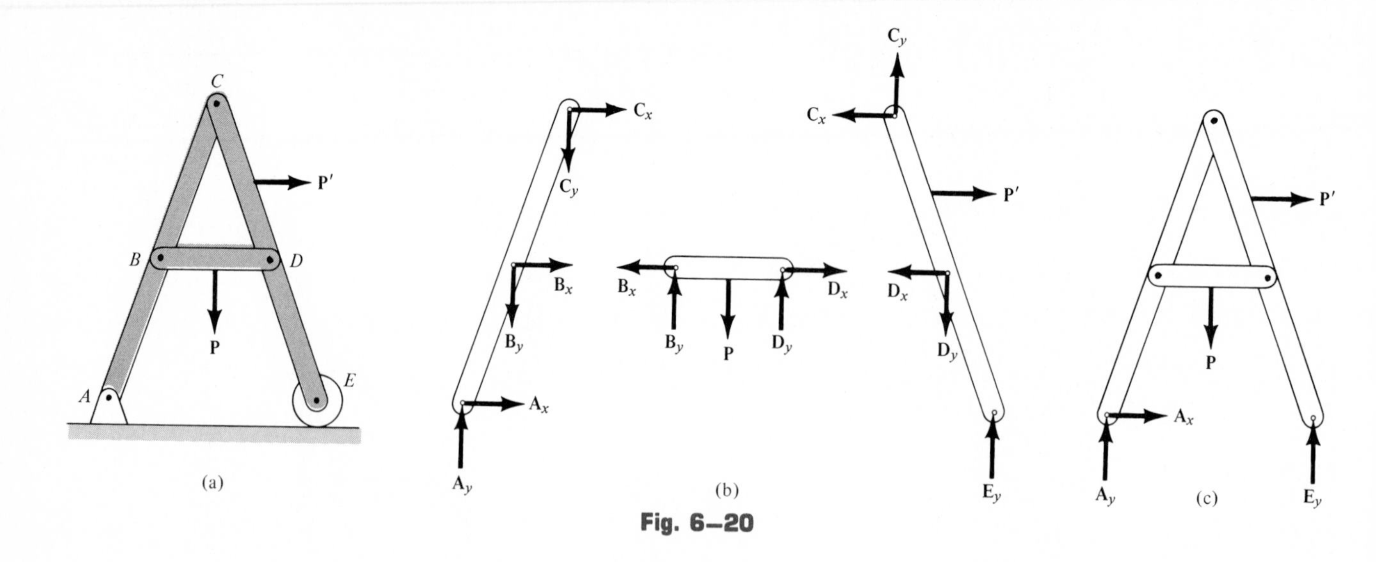

Fig. 6–20

To illustrate the method of force analysis, consider the three-member frame shown in Fig. 6–20*a* which is subjected to loads **P** and **P′**. The free-body diagrams of each member are shown in Fig. 6–20*b*. Notice that two components of reaction exist at each pin connection and that these components are of equal magnitude but opposite direction on corresponding connected members. In total there are nine unknowns; however, nine equations of equilibrium can be written, three for each member, so the problem is *statically determinate,* that is, all the reactions can be determined from the equilibrium equations. For the actual solution it is *also* possible, and sometimes convenient, to consider a portion of the frame or its entirety when applying some of these nine equations. For example, a free-body diagram of the entire frame is shown in Fig. 6–20*c*. By inspection this structure remains rigid without the supports, due to the connecting bar *BD*. Consequently, one can determine the three reactions $\mathbf{A}_x$, $\mathbf{A}_y$, and $\mathbf{E}_y$ on this "rigid" (connected) body, then analyze *any* two members of the frame, Fig. 6–20*b*, to obtain the other six unknowns. Furthermore, the answers can be checked in part by applying the three equations of equilibrium to the remaining "third" member. In general, then, this problem can be solved by writing *at most* nine equilibrium equations using free-body diagrams of any members and/or combinations of connected members. Any more than nine equations written would *not* be unique from the original nine and would only serve to check the results.

Consider now the two-member frame shown in Fig. 6–21*a*. Here the free-body diagrams of the members reveal six unknowns, Fig. 6–21*b*; however, six equilibrium equations, three for each member, can be written, so again the problem is statically determinate. As in the previous case, a free-body diagram of the entire frame can also be used for part of the analysis, Fig. 6–21*c*.

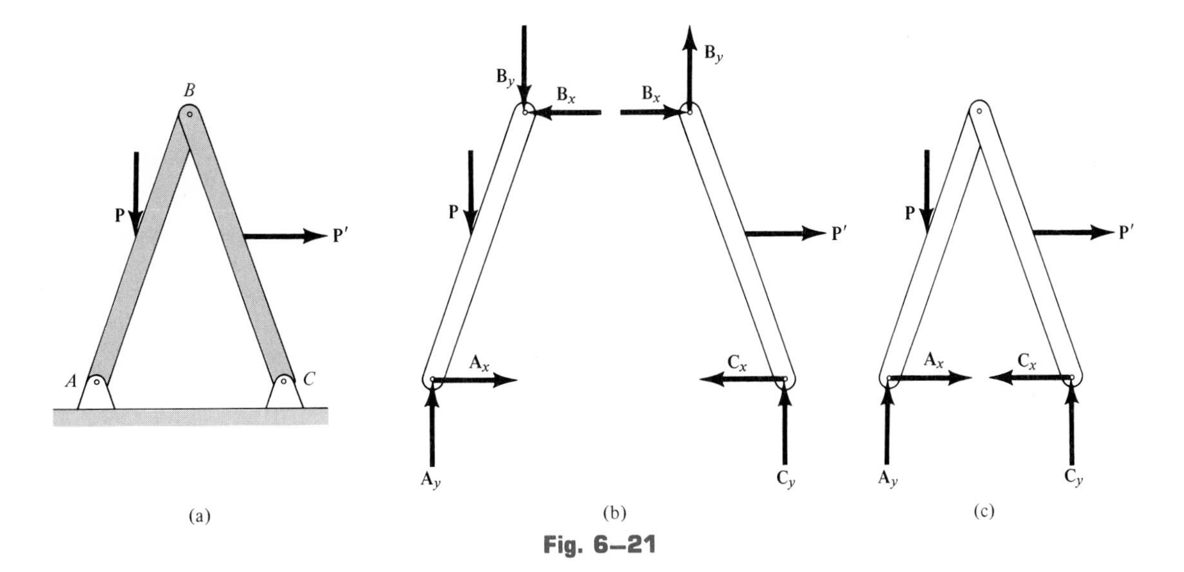

Fig. 6–21

For this case, though, the frame will *collapse* if the pin supports at A and C are removed. Collapsing occurs due to rotation at the pin B. Although this is possible, the force system acting on the entire frame must still hold it in equilibrium. Hence, if so desired, all six unknowns can be determined by applying the three equilibrium equations to the frame, Fig. 6–21c, and to either one of its members, Fig. 6–21b.

The above two examples illustrate a general rule that if a structure (frame or machine) is properly supported and contains no more supports or members than are necessary to prevent collapse, then the unknown forces at the supports and connections can be determined from the equations of equilibrium applied to each member. Also, if the structure remains *rigid* (noncollapsible) when the supports are removed (Fig. 6–20a), then the three support reactions can be determined by applying the three equilibrium equations to the entire structure (Fig. 6–20c). However, if the structure is *nonrigid* (collapsible) after removing the supports (Fig. 6–21a), then it must be dismembered and equilibrium of the individual members must be considered in order to obtain enough equations to determine all the support reactions (Fig. 6–21b). (In Fig. 6–21c there are *four* unknown support reactions.)

PROCEDURE FOR ANALYSIS

The following procedure provides a method for determining the *joint reactions* of frames and machines (structures) composed of multiforce members.

Free-Body Diagrams. Disassemble the structure and draw a free-body diagram of each member. If the structure is *noncollapsible* when the supports are removed, it is often more convenient to supplement a member free-body diagram with a free-body diagram of the *entire structure*. If this is done, simultaneous solution of equilibrium equations is generally avoided in the analysis.

Forces common to two members act with equal magnitudes but opposite directions on the respective free-body diagrams of the members. Recall that all *two-force members,* regardless of their shape, have equal but opposite collinear forces acting at the ends of the member (Sec. 5–3).* The unknown forces acting at the joints of multiforce members should be represented by their rectangular components. In many cases it is possible to tell by inspection the proper ''arrowhead'' sense of direction of the unknown forces; however, if this seems difficult, the directional sense can be assumed.

Equations of Equilibrium. Count the total number of unknowns to make sure that an equivalent number of equilibrium equations can be written for solution. Recall that in general three equilibrium equations can be written for each rigid body. Many times, the solution for the unknowns will be straightforward if moments are summed about a point that lies at the intersection of the lines of action of as many unknown forces as possible. If after obtaining the solution an unknown force magnitude is found to be negative, it means the directional sense of the force is reversed from that shown on the free-body diagrams.

The examples that follow illustrate this procedure. All these examples should be *thoroughly understood* before proceeding to solve the problems.

*Try to form a habit of *immediately spotting* the two-force members in the structure, since this considerably simplifies the force analysis of the structure. Examples include members *BC* in Probs. 6–44 and 6–47, and *BHF* and *BD* in Prob. 6–57.

Example 6–9

Determine the horizontal and vertical components of force which the pin at C exerts on member $ABCD$ of the loaded frame shown in Fig. 6–22a.

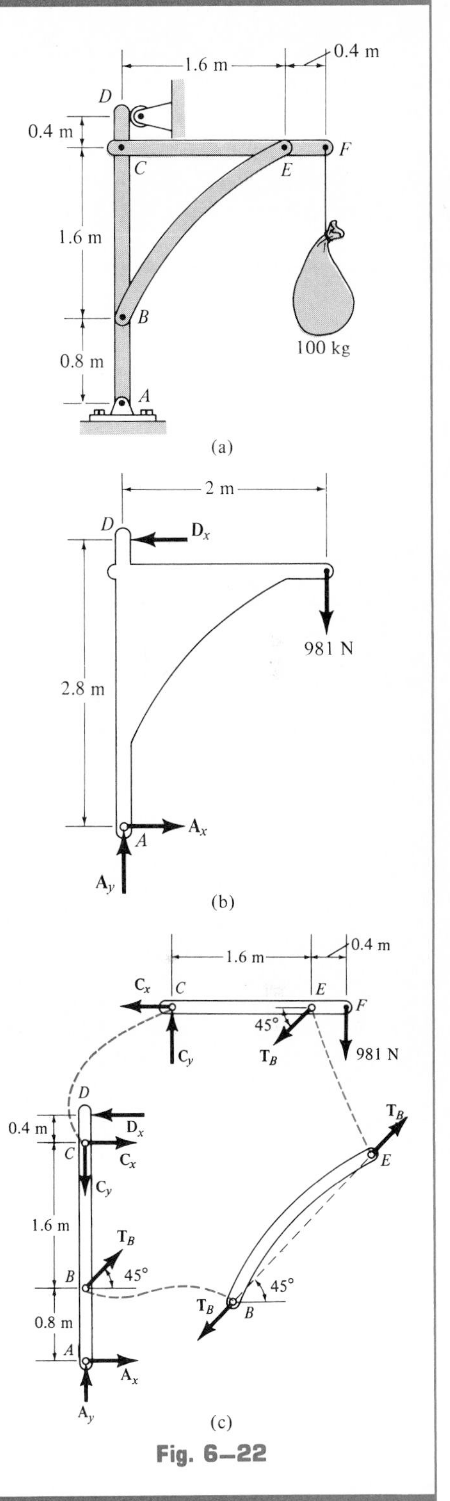

Fig. 6–22

Solution

Free-Body Diagrams. Since the frame is noncollapsible when the supports are removed, a free-body diagram of it will be considered for the analysis, Fig. 6–22b.* Also, the free-body diagram of each frame member is shown in Fig. 6–22c. Notice that member BE is a two-force member. There are two components of force at C, since two connecting members are pinned together there. As shown by the colored dashed lines, the principle of action, equal but opposite reaction of forces must be applied at joints B, C, and E when the separate free-body diagrams are drawn.

Equations of Equilibrium. The six unknowns A_x, A_y, T_B, C_x, C_y, and D_x will be determined from the equations of equilibrium applied to the entire frame and then to member CEF. We have

Entire Frame

$$\circlearrowleft +\Sigma M_A = 0; \quad -981\text{ N}(2\text{ m}) + D_x(2.8\text{ m}) = 0 \qquad D_x = 700.7\text{ N}$$

$$\overset{+}{\rightarrow}\Sigma F_x = 0; \quad A_x - 700.7\text{ N} = 0 \qquad A_x = 700.7\text{ N}$$

$$+\uparrow \Sigma F_y = 0; \quad A_y - 981\text{ N} = 0 \qquad A_y = 981\text{ N}$$

Member CEF

$$\circlearrowleft +\Sigma M_C = 0; \quad -981\text{ N}(2\text{ m}) - (T_B \sin 45°)(1.6\text{ m}) = 0$$

$$T_B = -1734.2\text{ N}$$

$$\overset{+}{\rightarrow}\Sigma F_x = 0; \quad -C_x - (-1734.2 \cos 45°\text{ N}) = 0$$

$$C_x = 1226.2\text{ N} \qquad \textit{Ans.}$$

$$+\uparrow \Sigma F_y = 0; \quad C_y - (-1734.2 \sin 45°\text{ N}) - 981\text{ N} = 0$$

$$C_y = -245.2\text{ N} \qquad \textit{Ans.}$$

Since the magnitudes of forces $\mathbf{T}_B$ and $\mathbf{C}_y$ were calculated as negative quantities, they were assumed to be acting in the wrong sense of direction on the free-body diagrams, Fig. 6–22c. The correct directions of these forces might have been determined "by inspection" *before* applying the equations of equilibrium to member CEF. As shown in Fig. 6–22c, moment equilibrium about point E on member CEF indicates that $\mathbf{C}_y$ must actually act *downward* to counteract the moment created by the 981-N force about point E. Similarly, summing moments about point C, it is seen that the vertical component of force $\mathbf{T}_B$ must actually act *upward*. The above calculations can be checked in part by applying the three equilibrium equations to member $ABCD$, Fig. 6–22c.

*The brace BE prevents the frame from collapsing.

Example 6–10

The piston-and-link mechanism shown in Fig. 6–23*a* is used as a toggle press for the purpose of compressing material contained within the cylinder *C*. If a force of 800 N is applied perpendicular to the handle of the lever, determine the compressive force exerted by the piston and the horizontal and vertical components of force at pin *D*. Assume that the surface of contact between the piston and the cylinder wall is smooth.

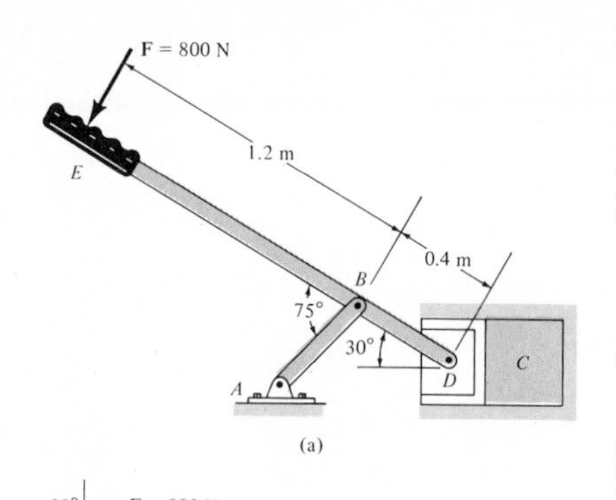

(a)

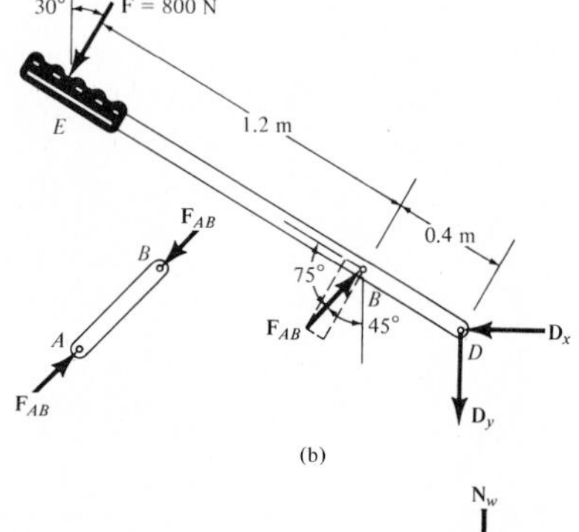

(b)

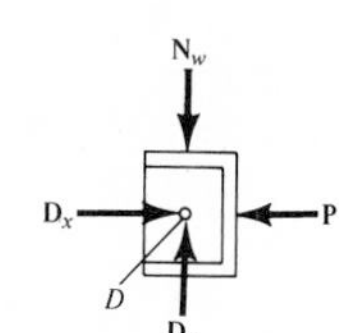

Fig. 6–23

Solution

Free-Body Diagrams. A free-body diagram of each of the three members is shown in Fig. 6–23*b*.* Member *AB* is a two-force member and hence only the magnitude of force acting at its two end points is unknown. Since there is a pin connection at *D*, there are two unknown components $\mathbf{D}_x$ and $\mathbf{D}_y$ acting on both the piston and the lever. As shown, the principle of action, equal but opposite reaction of forces must be applied at joints *B* and *D* when the separate free-body diagrams are drawn. In particular, four components of force act on the piston: $\mathbf{D}_x$ and $\mathbf{D}_y$ represent the effect of the pin (or lever *EBD*), $\mathbf{N}_w$ is the *resultant force* of the wall, and $\mathbf{P}$ is the resultant compressive force of the material within the cylinder.

Equations of Equilibrium. The five unknowns are determined as follows:

Lever EBD

$$\circlearrowright + \Sigma M_D = 0; \quad 800 \text{ N}(1.6 \text{ m}) - F_{AB} \sin 75° \,(0.4 \text{ m}) = 0$$

$$F_{AB} = 3312.9 \text{ N}$$

$$\xrightarrow{+} \Sigma F_x = 0; \quad -800 \sin 30° \text{ N} + (3312.9) \sin 45° \text{ N} - D_x = 0$$

$$D_x = 1942.2 \text{ N} \qquad \textit{Ans.}$$

$$+\uparrow \Sigma F_y = 0; \quad -800 \cos 30° \text{ N} + (3312.9) \cos 45° \text{ N} - D_y = 0$$

$$D_y = 1649.4 \text{ N} \qquad \textit{Ans.}$$

Piston

$$\xrightarrow{+} \Sigma F_x = 0; \quad -P + 1942.2 \text{ N} = 0$$

$$P = 1942.2 \text{ N} \qquad \textit{Ans.}$$

$$+\uparrow \Sigma F_y = 0; \quad -N_w + 1649.4 = 0$$

$$N_w = 1649.4 \text{ N}$$

The analysis indicates that by using this "machine," a small input force (0.800 kN) applied to the handle yields a much larger output force (1.94 kN) at the piston.

*A free-body diagram of the entire mechanism is not considered here, since the mechanism will collapse or rotate about the pin *B* when the supports are removed.

Example 6–11

The beam shown in Fig. 6–24*a* is pin-connected at *B* and supports a triangular distributed load. Determine the reactions at the supports.

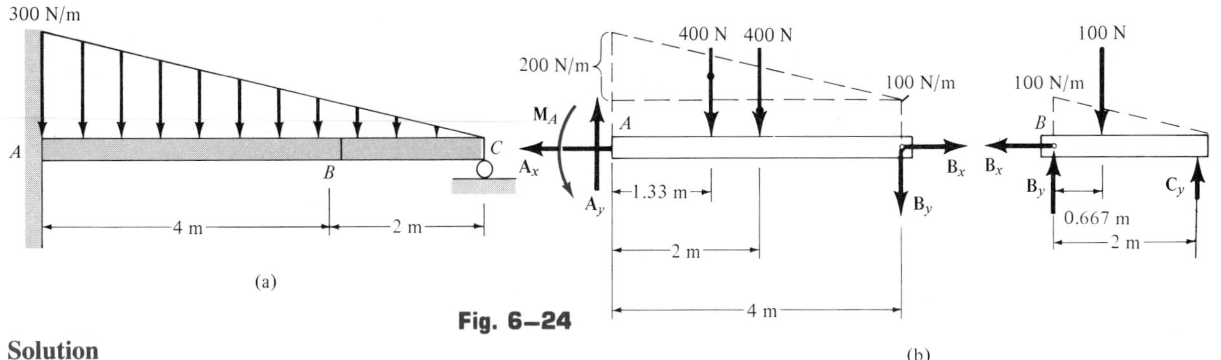

Fig. 6–24

Solution

Free-Body Diagrams. The free-body diagrams of the two beam segments are shown in Fig. 6–24*b*.* Note that it is important to keep the distributed loading in its *exact* position on the beam until *after* the members are separated. This loading is *then* simplified to a series of resultant forces, calculated as follows:

$$F_1 = \tfrac{1}{2}(200 \text{ N/m})(4 \text{ m}) = 400 \text{ N} \qquad \bar{x}_1 = \tfrac{1}{3}(4 \text{ m}) = 1.33 \text{ m}$$

$$F_2 = (100 \text{ N/m})(4 \text{ m}) = 400 \text{ N} \qquad \bar{x}_2 = \tfrac{1}{2}(4 \text{ m}) = 2 \text{ m}$$

$$F_3 = \tfrac{1}{2}(100 \text{ N/m})(2 \text{ m}) = 100 \text{ N} \qquad \bar{x}_3 = \tfrac{1}{3}(2 \text{ m}) = 0.667 \text{ m}$$

Furthermore, the couple $\mathbf{M}_A$ is a free vector and can act *anywhere* on member *AB* (not *BC*).

Equations of Equilibrium.

Segment BC

$$\overset{+}{\leftarrow}\Sigma F_x = 0; \qquad B_x = 0$$

$$\circlearrowleft+\Sigma M_B = 0; \qquad -100 \text{ N}(0.667 \text{ m}) + C_y(2 \text{ m}) = 0$$

$$+\uparrow \Sigma F_y = 0; \qquad B_y - 100 \text{ N} + C_y = 0$$

Segment AB

$$\overset{+}{\leftarrow}\Sigma F_x = 0; \qquad A_x - B_x = 0$$

$$\circlearrowleft+\Sigma M_A = 0; \quad M_A - 400 \text{ N}(1.33 \text{ m}) - 400 \text{ N}(2 \text{ m}) - B_y(4 \text{ m}) = 0$$

$$+\uparrow \Sigma F_y = 0; \qquad A_y - 400 \text{ N} - 400 \text{ N} - B_y = 0$$

Solving each of these equations successively, using previously calculated results, we obtain

$$A_x = 0 \qquad A_y = 866.7 \text{ N} \qquad M_A = 1600 \text{ N}\cdot\text{m} \qquad \textit{Ans.}$$

$$B_x = 0 \qquad B_y = 66.7 \text{ N}$$

$$C_y = 33.3 \text{ N} \qquad \textit{Ans.}$$

*The *four* support reactions cannot *all* be obtained from the *three* equilibrium equations applied to the entire beam because the beam will collapse due to rotation about *B* when the supports are removed.

Example 6–12

The smooth disk shown in Fig. 6–25*a* is pinned at *D* and has a weight of 20 lb. Neglecting the weight of the other members, determine the horizontal and vertical components of reaction at pins *B* and *D*.

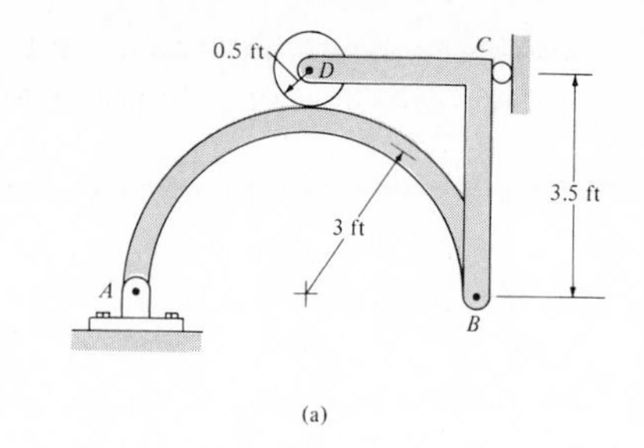

(a)

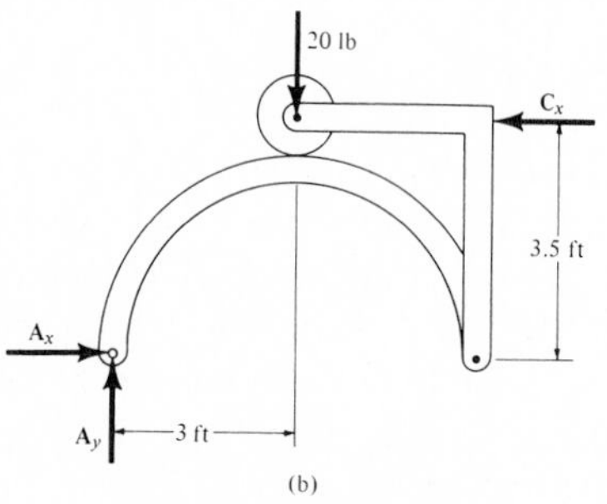

(b)

Fig. 6–24a and b

Solution

Free-Body Diagrams. Since the frame is observed to be noncollapsible when its supports are removed, a free-body diagram of it will be used for the analysis, Fig. 6–25*b*. Also, free-body diagrams of the members are shown in Fig. 6–25*c*.

Equations of Equilibrium. The eight unknowns can of course be obtained by applying the eight equilibrium equations to each member—three to member *AB*, three to member *BCD*, and two to the disk. (Moment equilibrium is automatically satisfied for the disk.) If this is done, however, all the results can be obtained only from a simultaneous solution of some of the equations. (Try it and find out.) To avoid this situation, it is best to first determine the three support reactions on the *entire* frame; then, using these results, the remaining five equilibrium equations can be applied to two other parts in order to solve successively for the other unknowns.

Entire Frame

$\circlearrowright + \Sigma M_A = 0; \quad -20 \text{ lb}(3 \text{ ft}) + C_x(3.5 \text{ ft}) = 0 \qquad C_x = 17.1 \text{ lb}$

$\xrightarrow{+} \Sigma F_x = 0; \qquad A_x - 17.1 \text{ lb} = 0 \qquad A_x = 17.1 \text{ lb}$

$+\uparrow \Sigma F_y = 0; \qquad A_y - 20 \text{ lb} = 0 \qquad A_y = 20 \text{ lb}$

Member AB

$\xrightarrow{+} \Sigma F_x = 0; \qquad 17.1 \text{ lb} - B_x = 0 \qquad B_x = 17.1 \text{ lb}$ ***Ans.***

$\circlearrowright + \Sigma M_B = 0; \quad -20 \text{ lb}(6 \text{ ft}) + N_D(3 \text{ ft}) = 0 \qquad N_D = 40 \text{ lb}$

$+\uparrow \Sigma F_y = 0; \quad 20 \text{ lb} - 40 \text{ lb} + B_y = 0 \qquad B_y = 20 \text{ lb}$ ***Ans.***

Disk

$\xleftarrow{+} \Sigma F_x = 0; \qquad D_x = 0$ ***Ans.***

$+\uparrow \Sigma F_y = 0; \quad 40 \text{ lb} - 20 \text{ lb} - D_y = 0 \qquad D_y = 20 \text{ lb}$ ***Ans.***

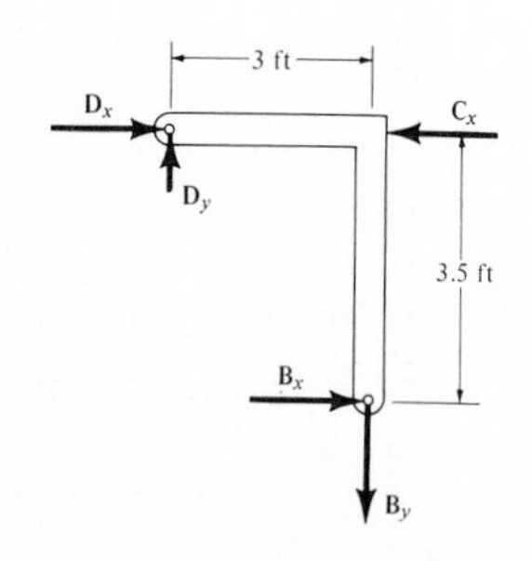

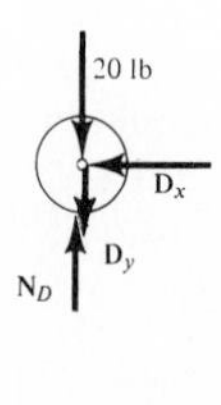

(c)

Fig. 6–25c

Example 6–13

Determine the horizontal and vertical force components acting at the pin connections B and C of the loaded frame shown in Fig. 6–26a.

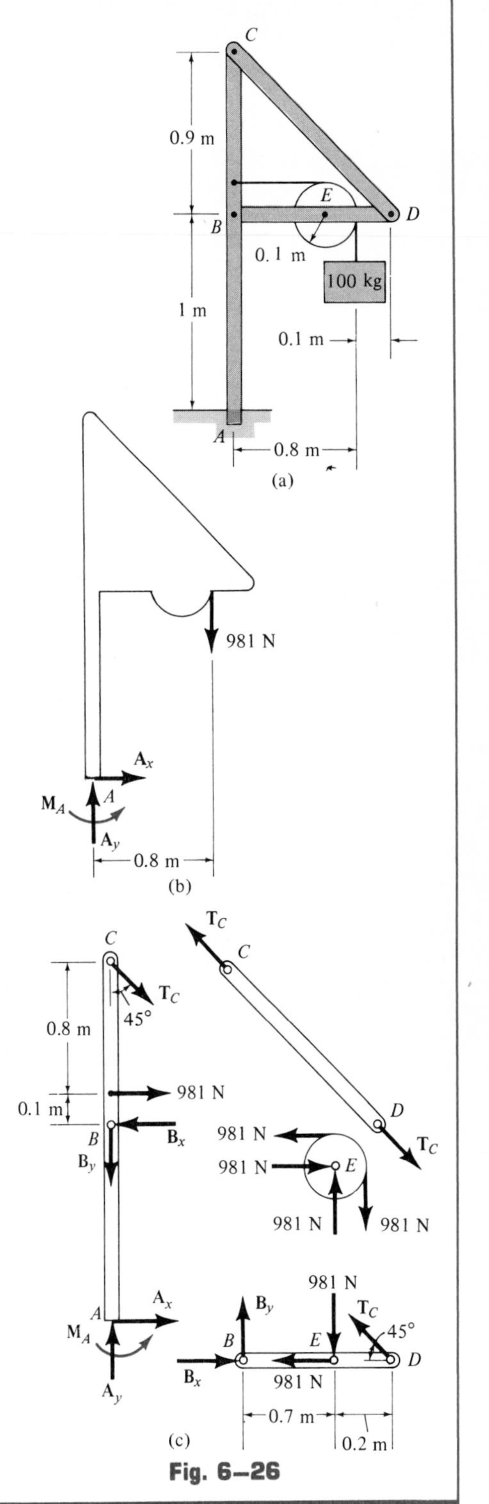

Fig. 6–26

Solution

Free-Body Diagrams. Since the frame is observed to be noncollapsible when the supports are removed, a free-body diagram of it will be used for the analysis, Fig. 6–26b. Also, the free-body diagrams for each of the members are shown in Fig. 6–26c. Notice that member CD is a two-force member. Furthermore, the pulley is held in equilibrium by 981-N force components exerted on it by the pin at point E. By inspection, both force and moment equilibrium of the pulley are satisfied. By Newton's third law, the pulley pin exerts equal but opposite 981-N force components on member BED. Since the cable is *removed* from member ABC, a force of 981 N must pull horizontally to the left on this member.

Equations of Equilibrium. The six unknowns will be determined by applying the equilibrium equations to the entire frame and member BED. We have,

Entire Frame

$$\circlearrowleft +\Sigma M_A = 0; \quad M_A - 981\text{ N}(0.8\text{ m}) = 0 \qquad M_A = 784.8\text{ N}\cdot\text{m}$$

$$\overset{+}{\rightarrow}\Sigma F_x = 0; \qquad A_x = 0$$

$$+\uparrow \Sigma F_y = 0; \quad A_y - 981\text{ N} = 0 \qquad A_y = 981\text{ N}$$

Member BED

$$\circlearrowleft +\Sigma M_B = 0; \quad -981\text{ N}(0.7\text{ m}) + T_C \sin 45^\circ\,(0.9\text{ m}) = 0$$

$$T_C = 1079.2\text{ N} \qquad \textit{Ans.}$$

$$\overset{+}{\rightarrow}\Sigma F_x = 0; \quad B_x - 981\text{ N} - 1079.2\cos 45^\circ\text{ N} = 0$$

$$B_x = 1744.0\text{ N} \qquad \textit{Ans.}$$

$$+\uparrow \Sigma F_y = 0; \quad 1079.2\sin 45^\circ\text{ N} - 981\text{ N} + B_y = 0$$

$$B_y = 218.0\text{ N} \qquad \textit{Ans.}$$

It is suggested that the above calculations be checked by applying the three equations of equilibrium to member ABC.

Example 6–14

Determine the tension in the cables and also the force **P** required to support the 600-N force using the frictionless pulley system shown in Fig. 6–27*a*.

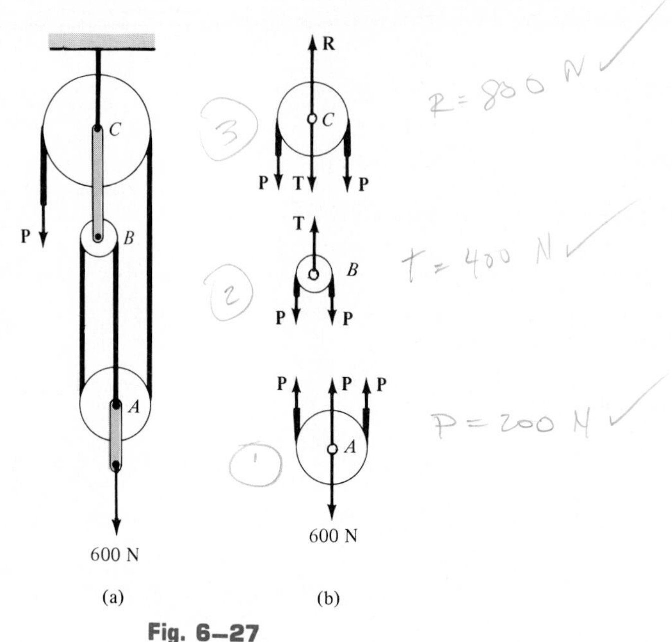

Fig. 6–27

Solution

Free-Body Diagrams. A free-body diagram of each pulley and a portion of the contacting cable is given in Fig. 6–27*b*. Since the cable is *continuous* and the pulleys are frictionless, the cable has a *constant tension* of P N acting throughout its length (see Example 5–6). The link connection between pulleys B and C is a two-force member and therefore it has an unknown tension of T N acting on it. Notice that the *principle of action, equal but opposite reaction* must be carefully observed for forces **P** and **T** when the *separate* free-body diagrams are drawn.

Equations of Equilibrium. The three unknowns are obtained as follows:

Pulley A

$+\uparrow \Sigma F_y = 0; \qquad 3P - 600\text{ N} = 0 \qquad P = 200\text{ N}$ *Ans.*

Pulley B

$+\uparrow \Sigma F_y = 0; \qquad T - 2P = 0 \qquad T = 400\text{ N}$ *Ans.*

Pulley C

$+\uparrow \Sigma F_y = 0; \qquad R - 2P - T = 0 \qquad R = 800\text{ N}$ *Ans.*

Example 6–15

Determine the tension in the cables and also the force **P** required to support the 600-N force using the frictionless pulley system shown in Fig. 6–28*a*.

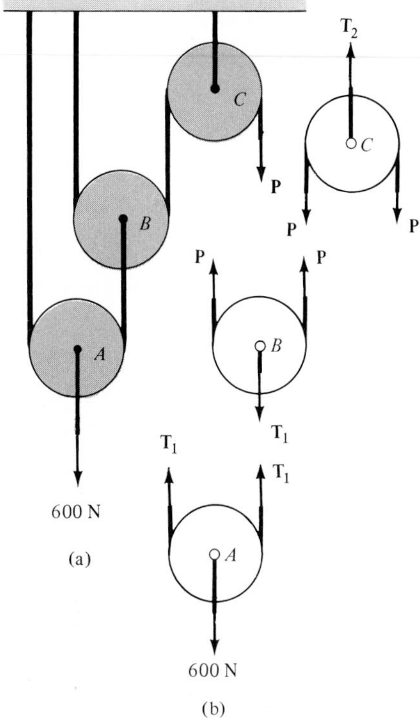

Fig. 6–28

Solution

Free-Body Diagrams. A free-body diagram of each pulley and a portion of its contacting cable is shown in Fig. 6–28*b*. Observe how the principle of action, equal but opposite reaction is applied to forces $\mathbf{T}_1$ and **P** on the separate free-body diagrams.

Equations of Equilibrium. The three unknowns are obtained as follows:

Pulley A

$+\uparrow \Sigma F_y = 0; \qquad 2T_1 - 600 \text{ N} = 0 \qquad T_1 = 300 \text{ N}$ ***Ans.***

Pulley B

$+\uparrow \Sigma F_y = 0; \qquad 2P - T_1 = 0 \qquad P = 150 \text{ N}$ ***Ans.***

Pulley C

$+\uparrow \Sigma F_y = 0; \qquad T_2 - 2P = 0 \qquad T_2 = 300 \text{ N}$ ***Ans.***

Example 6–16

A man having a weight of 150 lb supports himself by means of the cable and pulley system shown in Fig. 6–29*a*. If the seat has a weight of 15 lb, determine the equilibrium force that he must exert on the cable at *A* and the force he exerts on the seat. Neglect the weight of the cables and pulleys.

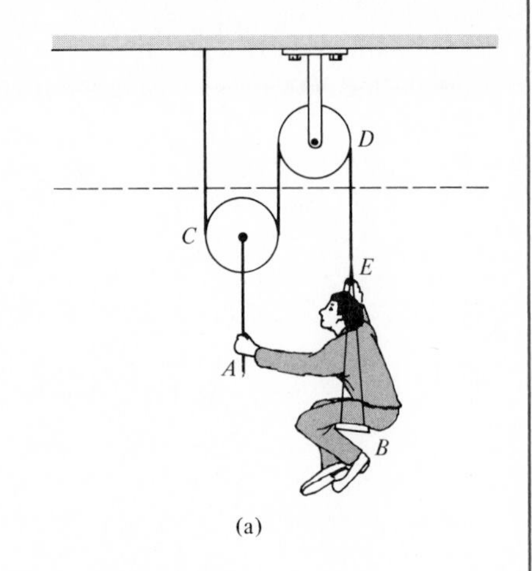

(a)

Solution I

Free-Body Diagrams. The free-body diagrams of the man, seat, and pulley *C* are shown in Fig. 6–29*b*. The *two* cables are subjected to tensions $\mathbf{T}_A$ and $\mathbf{T}_E$, respectively. The man is subjected to three forces: his weight, the tension $\mathbf{T}_A$ of cable *AC*, and the reaction $\mathbf{N}_s$ of the seat.

Equations of Equilibrium. The three unknowns are obtained as follows:

Man

$$+\uparrow \Sigma F_y = 0; \qquad T_A + N_s - 150 \text{ lb} = 0 \qquad (1)$$

Seat

$$+\uparrow \Sigma F_y = 0; \qquad T_E - N_s - 15 \text{ lb} = 0 \qquad (2)$$

Pulley C

$$+\uparrow \Sigma F_y = 0; \qquad 2T_E - T_A = 0 \qquad (3)$$

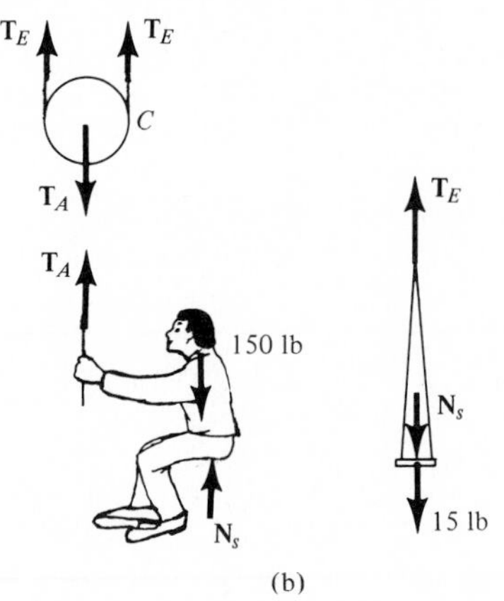

(b)

The magnitude of force $\mathbf{T}_E$ can be determined by adding Eqs. (1) and (2) to eliminate N_s and then using Eq. (3). The other unknowns are then obtained by resubstitution of T_E.

$$T_A = 110 \text{ lb} \qquad \textit{Ans.}$$

$$T_E = 55 \text{ lb}$$

$$N_s = 40 \text{ lb} \qquad \textit{Ans.}$$

Solution II

Free-Body Diagram. By using the dashed section shown in Fig. 6–29*a*, the man, pulley, and seat can be considered as a *single system,* Fig. 6–29*c*. Here $\mathbf{N}_s$ and $\mathbf{T}_A$ are *internal* forces and hence are not included on the "combined" free-body diagram.

Equations of Equilibrium. Applying $\Sigma F_y = 0$ yields a *direct* solution for T_E.

$$+\uparrow \Sigma F_y = 0; \quad 3T_E - 15 \text{ lb} - 150 \text{ lb} = 0 \qquad T_E = 55 \text{ lb}$$

The other unknowns can be obtained from Eqs. (2) and (3).

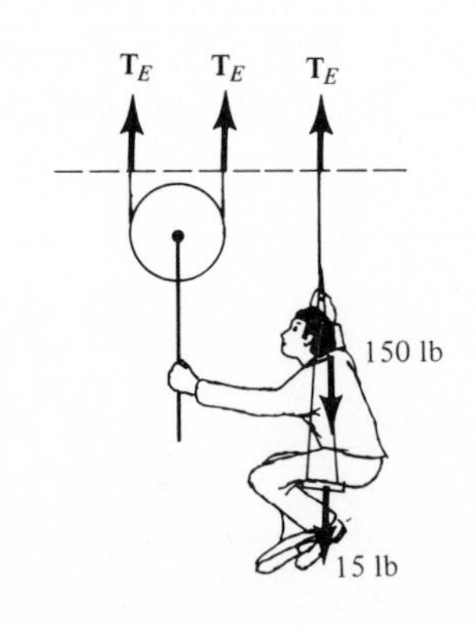

(c)

Fig. 6–29

Example 6–17

The frame shown in Fig. 6–30*a* is supported at A by a pin and at E by a smooth wall. Determine the horizontal and vertical components of force that each pin connection exerts on the frame. The pulley at C is frictionless.

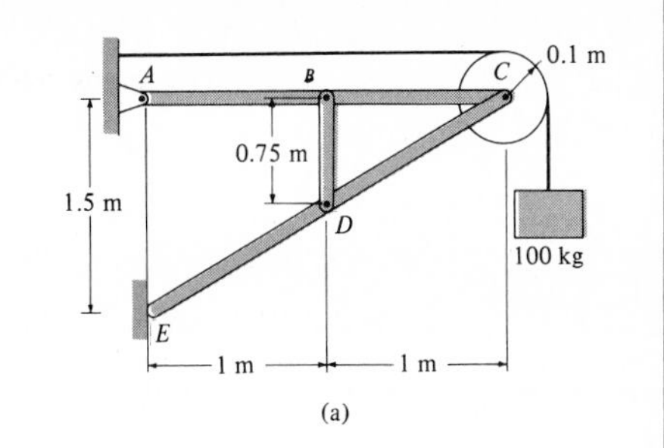

(a)

Solution

Free-Body Diagrams. Since the frame is noncollapsible when the supports are removed, a free-body diagram of the entire frame is considered for the analysis, Fig. 6–30*b*. Free-body diagrams of each of its members, including the pulley and pin at C, are shown in Fig. 6–30*c*. In particular, note that BD is a two-force member and forces of 981 N must be exerted by the pin at C to hold the pulley in equilibrium. Also, note that *three* force interactions occur *on pin C:* force components $\mathbf{C}_x$ and $\mathbf{C}_y$ representing the force exerted by member ABC, the force components $\mathbf{C}_{x'}$ and $\mathbf{C}_{y'}$ representing the force exerted by member EDC, and finally the two 981-N force components exerted by the pulley.*

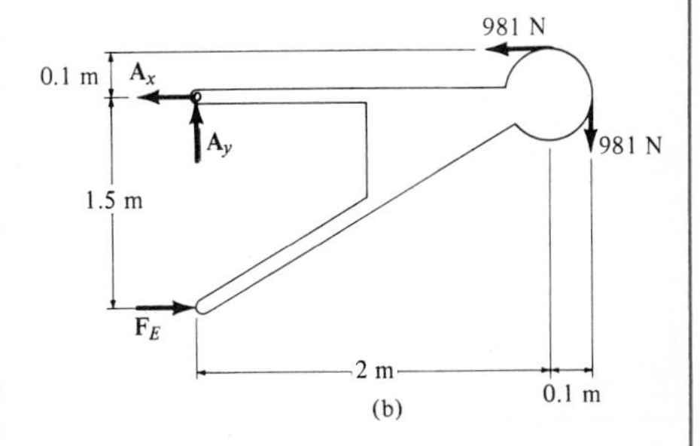

(b)

Equations of Equilibrium. To avoid simultaneous solution of equations, the eight unknowns will be determined by first applying the three equations of equilibrium to the entire frame; then, using these results, the remaining five unknowns will be determined from the equations of equilibrium applied to member ABC and pin C.

Entire Frame

$$\circlearrowleft + \Sigma M_A = 0; \quad -981 \text{ N}(2.1 \text{ m}) + 981 \text{ N}(0.1 \text{ m}) + F_E(1.5 \text{ m}) = 0$$

$$F_E = 1308 \text{ N}$$

$$\overset{+}{\rightarrow}\Sigma F_x = 0; \quad -A_x + 1308 \text{ N} - 981 \text{ N} = 0 \qquad A_x = 327 \text{ N} \qquad \textit{Ans.}$$

$$+\uparrow \Sigma F_y = 0; \quad A_y - 981 \text{ N} = 0 \qquad A_y = 981 \text{ N} \qquad \textit{Ans.}$$

Member ABC

$$\circlearrowleft + \Sigma M_C = 0; \quad -981 \text{ N}(2 \text{ m}) + F(1 \text{ m}) = 0 \quad F = 1962 \text{ N} \qquad \textit{Ans.}$$

$$\overset{+}{\rightarrow}\Sigma F_x = 0; \quad C_x - 327 \text{ N} = 0 \quad C_x = 327 \text{ N} \qquad \textit{Ans.}$$

$$+\uparrow \Sigma F_y = 0; \quad 981 \text{ N} - 1962 \text{ N} + C_y = 0 \qquad C_y = 981 \text{ N} \qquad \textit{Ans.}$$

Pin C

$$\overset{+}{\rightarrow}\Sigma F_x = 0; \quad C_{x'} - 327 \text{ N} - 981 \text{ N} = 0 \qquad C_{x'} = 1308 \text{ N} \qquad \textit{Ans.}$$

$$+\uparrow \Sigma F_y = 0; \quad C_{y'} - 981 \text{ N} - 981 \text{ N} = 0 \qquad C_{y'} = 1962 \text{ N} \qquad \textit{Ans.}$$

These results can be checked in part by applying the equations of equilibrium to member EDC.

Before solving the following problems, it is suggested that a brief review be made of all the previous examples. This may be done by covering over the solutions and trying to locate the two-force members, drawing the free-body diagrams, and conceiving ways of applying the equations of equilibrium to obtain the solution.

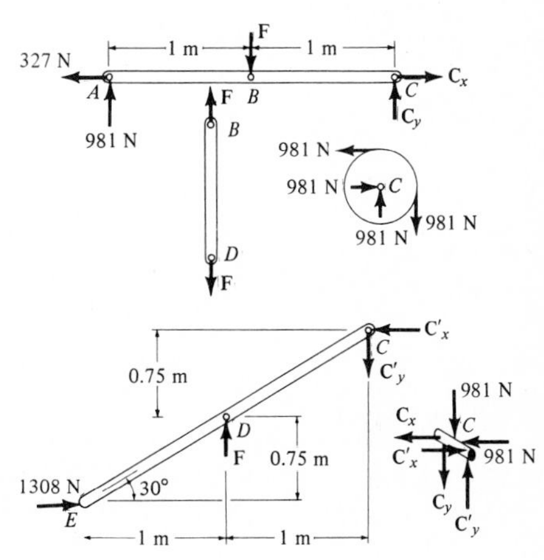

(c)

Fig. 6–30

*Free-body diagrams of pins B and D on the frame are not considered in this analysis since they each connect *only two members*, and consequently, equilibrium simply requires the action of one member on the pin to be equal and opposite to that of the other.

Problems

6–37. In each case, determine the force **P** needed to hold the 20-lb block in equilibrium.

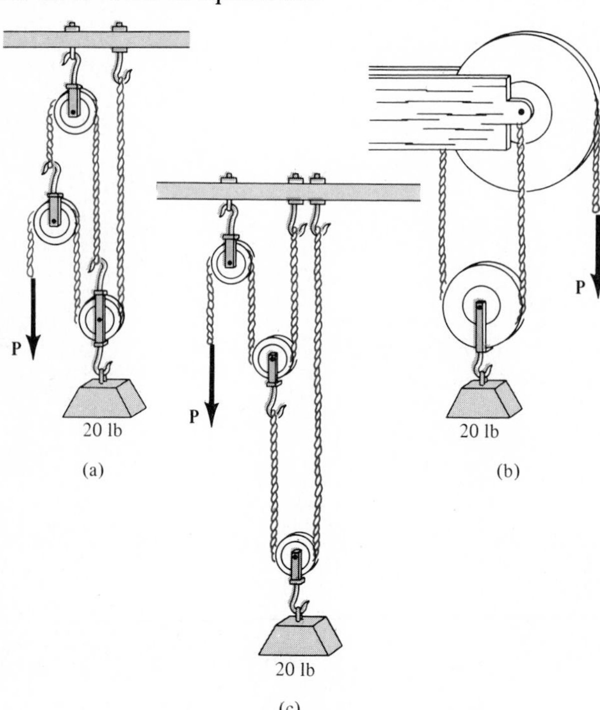

Prob. 6–37

6–38. Determine the force **P** needed to suspend the 60-lb weight. Also, what are the cord reactions at *A* and *B*?

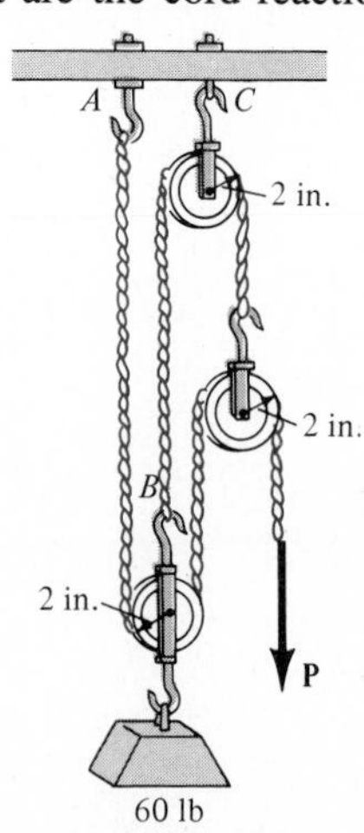

Prob. 6–38

6–39. Determine the force **P** needed to suspend the 80-lb weight. Also, what are the reactions at *A* and *B*?

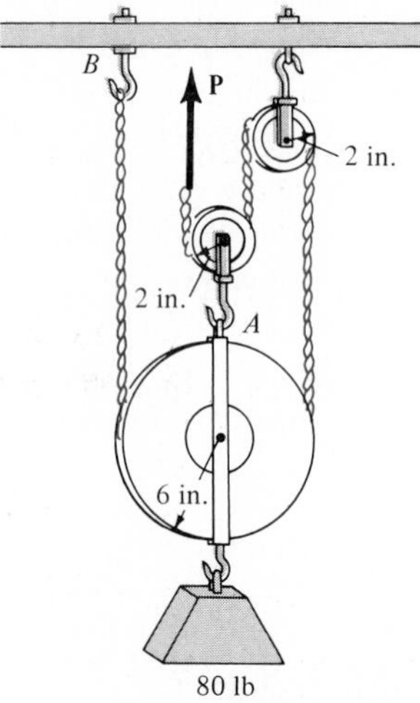

Prob. 6–39

***6–40.** Determine the force **P** needed to support the 10-kg mass using the *Spanish Burton rig*. Also, what are the reactions at the supporting hooks *A*, *B*, and *C*?

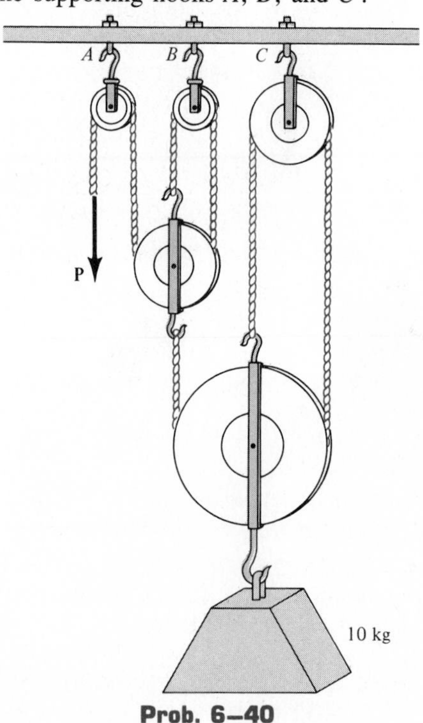

Prob. 6–40

6–41. The principles of a *differential chain block* are indicated schematically in the figure. Determine the magnitude of force **P** needed to support the 800-N force. Also, compute the distance x where the cable must be attached to bar AB so the bar remains horizontal. All pulleys have a radius of 60 mm.

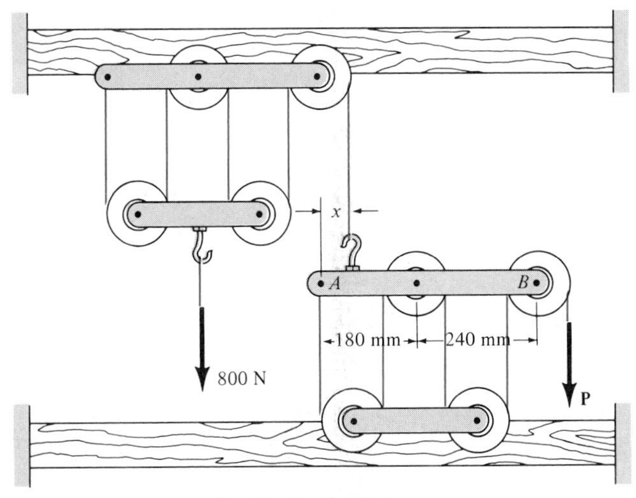

Prob. 6–41

6–42. Compute the tension **T** in the cord, and determine the angle θ that the pulley-supporting link AB makes with the vertical. Neglect the mass of the pulleys and the link. The block has a weight of 100 lb and the cord is attached to the pin at B. The pulleys have radii of $r_1 = 2$ in. and $r_2 = 1$ in.

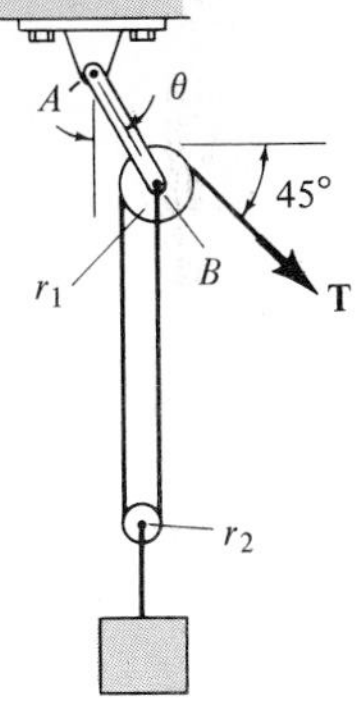

Prob. 6–42

6–43. A force of 4 lb is applied to the handles of the pliers. Determine the force developed on the smooth bolt B and the reaction that the pin A exerts on its attached members.

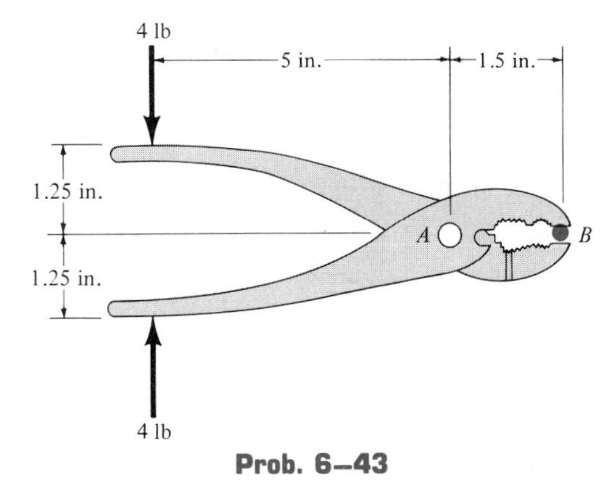

Prob. 6–43

***6–44.** Determine the horizontal and vertical components of force at pins A and C of the two-member frame.

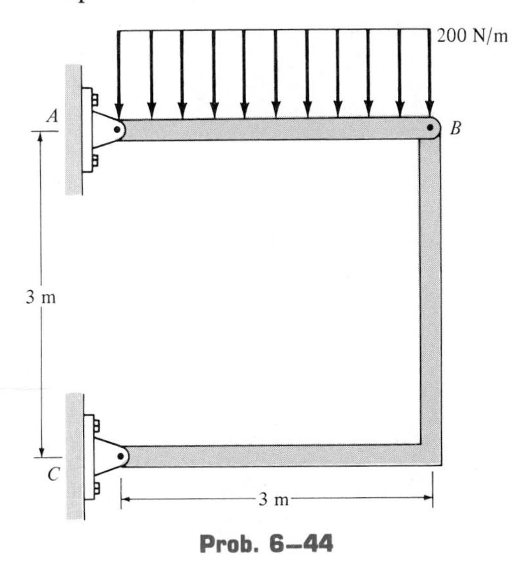

Prob. 6–44

6–45. Determine the horizontal and vertical components of force at pin A.

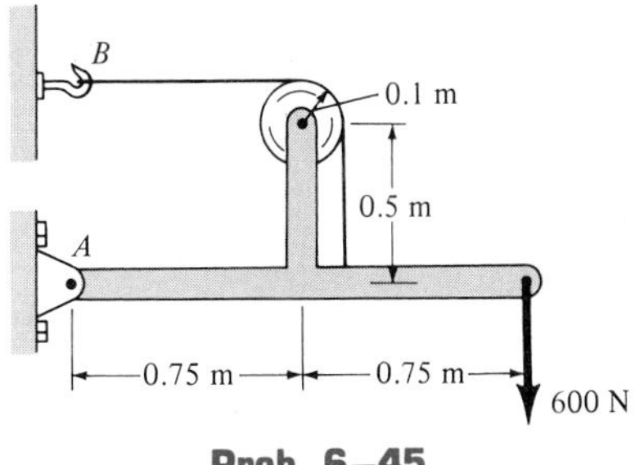

Prob. 6–45

6–46. Determine the horizontal and vertical components of force at pins B and D of the four-member frame.

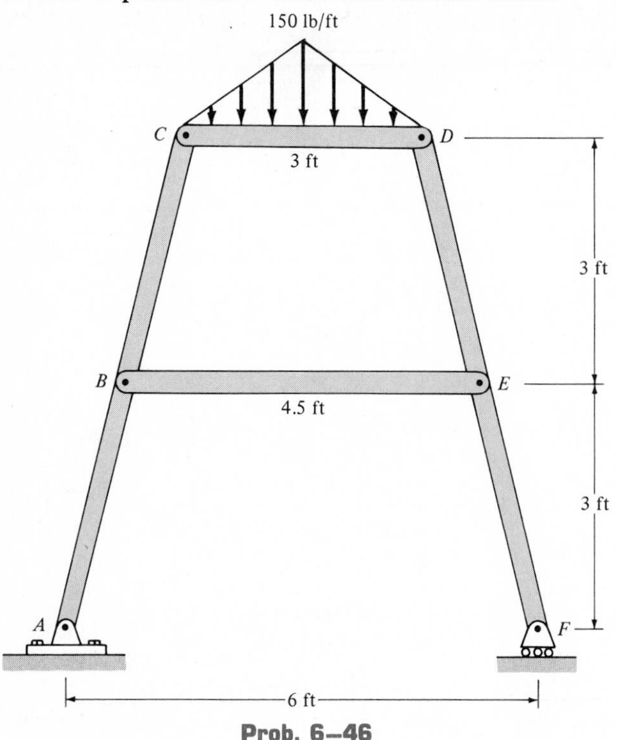

Prob. 6–46

6–47. Determine the horizontal and vertical components of force at pin A.

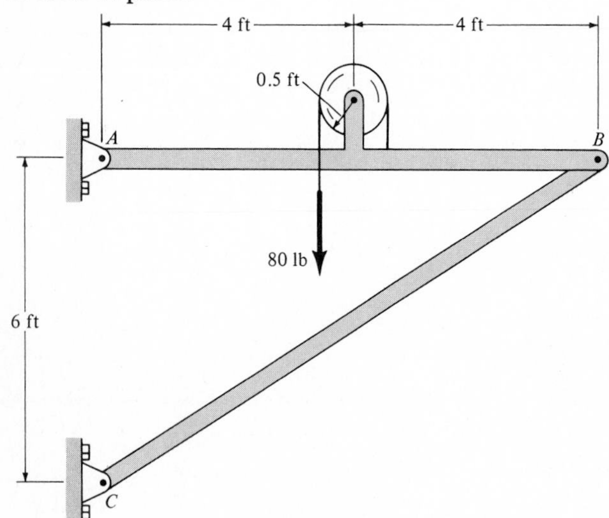

Prob. 6–47

***6–48.** Determine the horizontal and vertical components of force that pins A, B, and C exert on their connecting members.

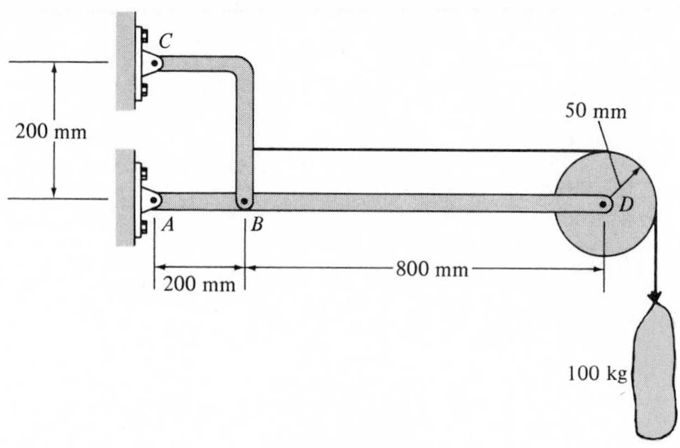

Prob. 6–48

6–49. A man having a weight of 160 lb attempts to lift himself using one of the two methods shown. Determine the total force he must exert on bar AB in each case and the normal reaction he exerts on the platform at C. Neglect the weight of the platform.

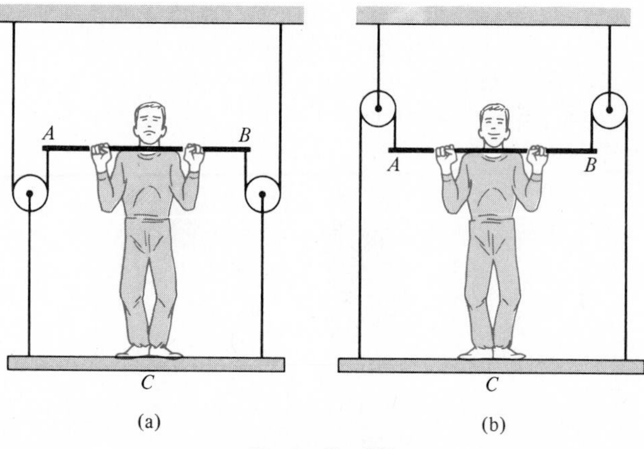

Prob. 6–49

6–50. The bucket of the backhoe and its contents have a weight of 800 lb and a center of gravity at G. Determine the forces in the hydraulic cylinder AB and in links AC and AD in order to hold the load in the position shown. The bucket is pinned at E. *Hint:* AB, AC, and AD are all two-force members. Analyze the forces on the bucket, then analyze joint A.

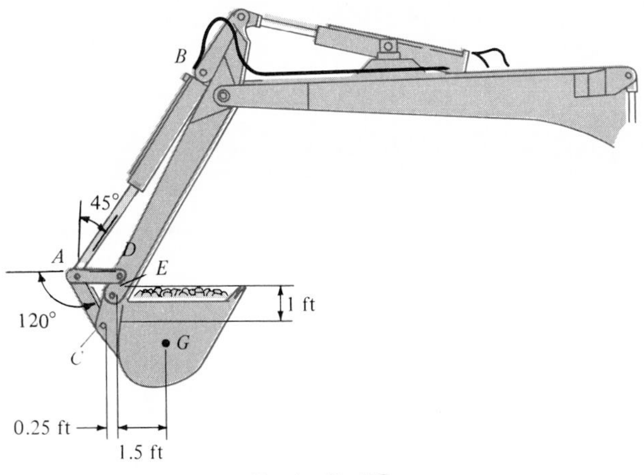

Prob. 6–50

6–51. Determine the force which the smooth roller at C exerts on beam AB. Also, what are the horizontal and vertical components of reaction at pin A? Neglect the weight of the frame.

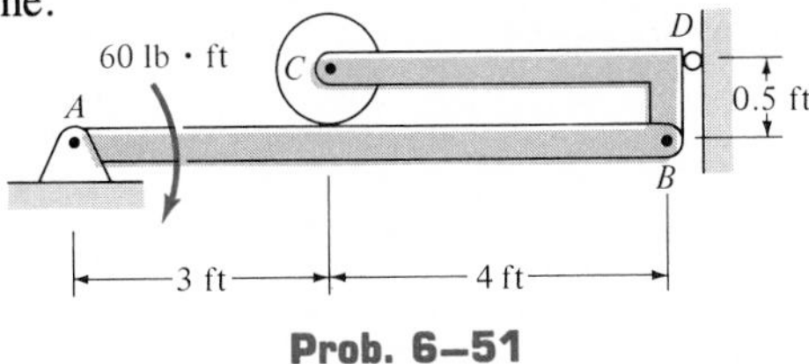

Prob. 6–51

***6–52.** Determine the resultant forces at pins B and C of the four-member frame.

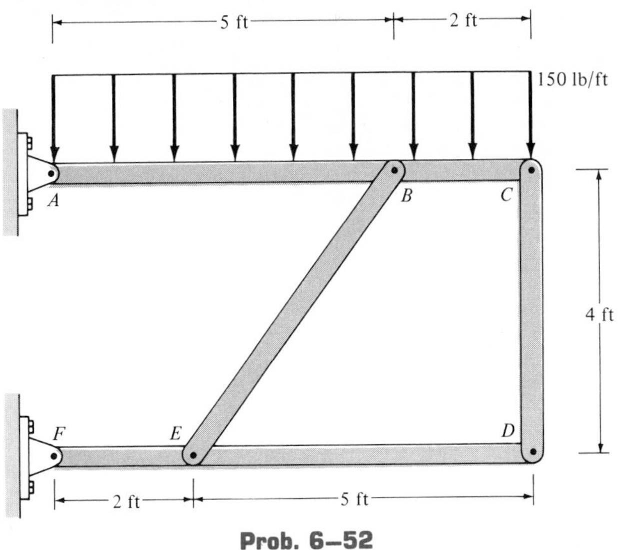

Prob. 6–52

6–53. Determine the force that the jaws J of the metal cutters exert on the smooth cable C if 100-N forces are applied to the handles. The jaws are pinned at E and A, and D and B. There is also a pin at F.

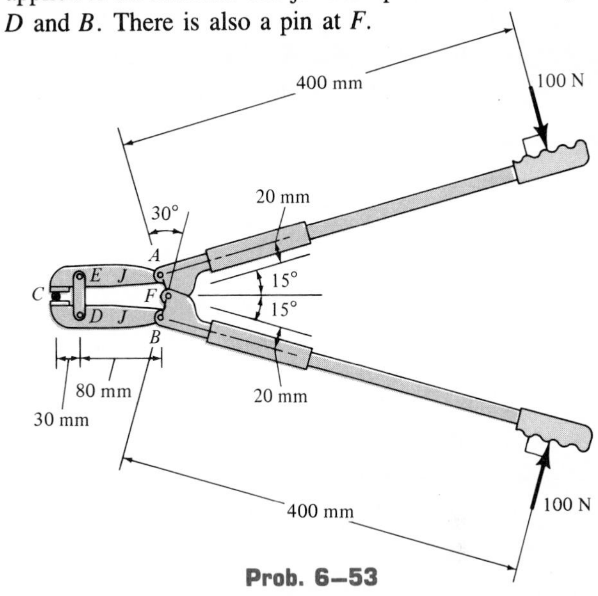

Prob. 6–53

6–54. Using a series of levers is more efficient than using a single lever. For example, the 6-Mg truck is balanced on a scale using either (a) three levers or (b) one lever as shown. In case (a) determine the force **P** required for equilibrium of the truck and in case (b), using the same force **P**, determine the length l of the single lever necessary for balancing the truck.

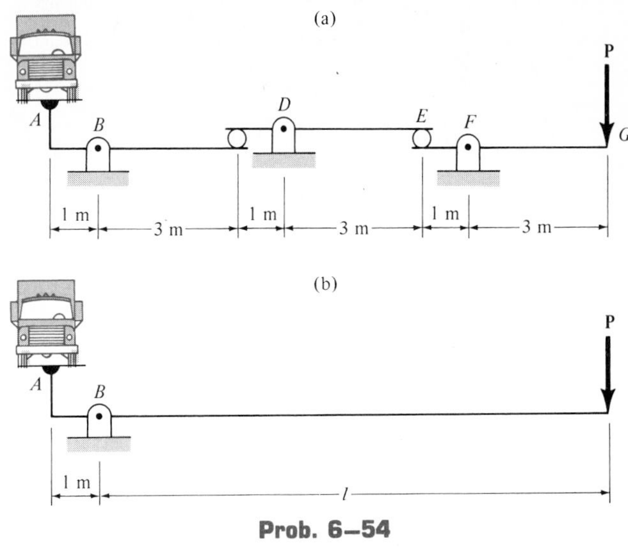

Prob. 6–54

6–55. Determine the reactions at the wall A and the rocker C. The two members are connected by a pin at B.

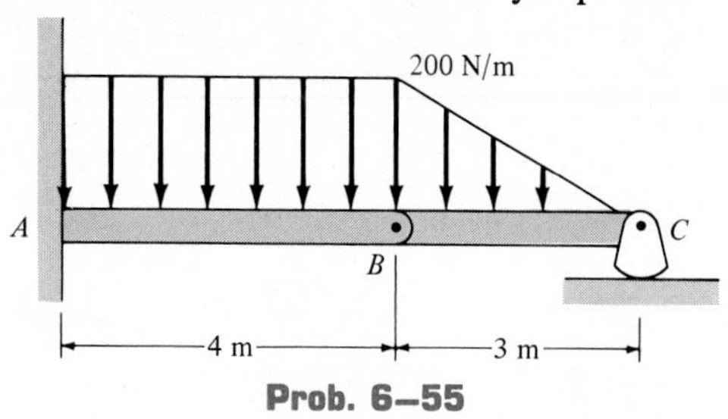

Prob. 6–55

***6–56.** The floor beams AB and BC are stiffened using the two tie rods CD and AD. Determine the force along each rod when the floor beams are subjected to a uniform load of 80 lb/ft. Assume the three contacting members at B are smooth and the joints at A, C, and D are pins. *Hint:* Members AD, CD, and BD are two-force members.

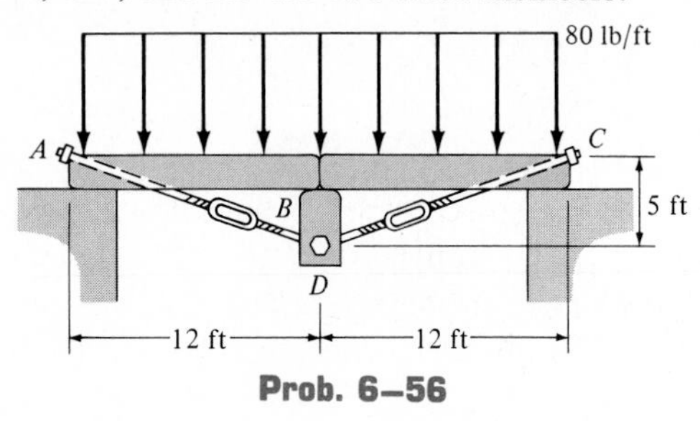

Prob. 6–56

6–57. The hoist supports the 125-kg engine. Determine the force the load creates in member DB and in member FB, which contains the hydraulic cylinder H.

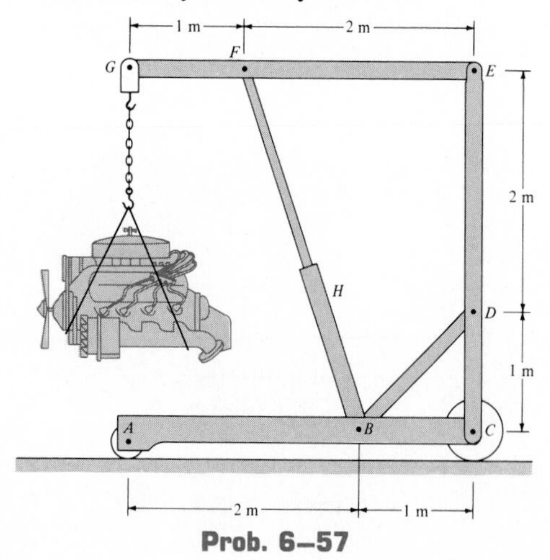

Prob. 6–57

6–58. Multiple levers used in truck scales, discussed in Prob. 6–54, can be made *shorter* and more effective by using a *compound arrangement* such as shown for the pan scale. If the mass on the platform is 10 kg, determine the reactions at pins A, B, and C and the distance x of the 50-g mass to keep the scale in balance.

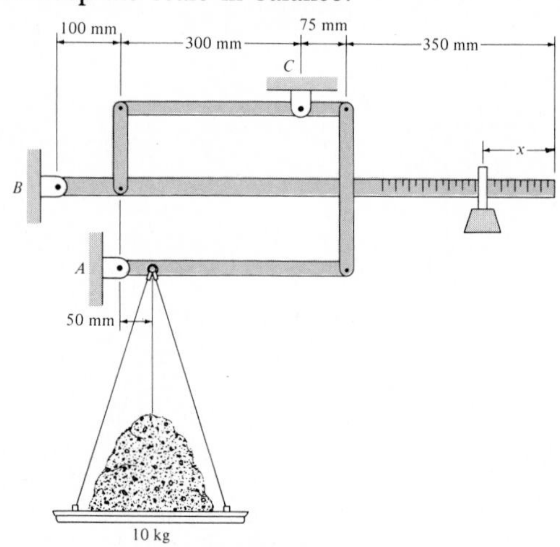

Prob. 6–58

6–59. The scissors lift consists of *two* sets of cross members and *two* hydraulic cylinders, DE, symmetrically located on *each side* of the platform. The platform has a uniform mass of 60 kg, with a center of gravity at G_1. The load of 85 kg, with center of gravity at G_2, is centrally located on each side of the platform. Determine the force in each of the hydraulic cylinders for equilibrium. Rollers are located at B and D.

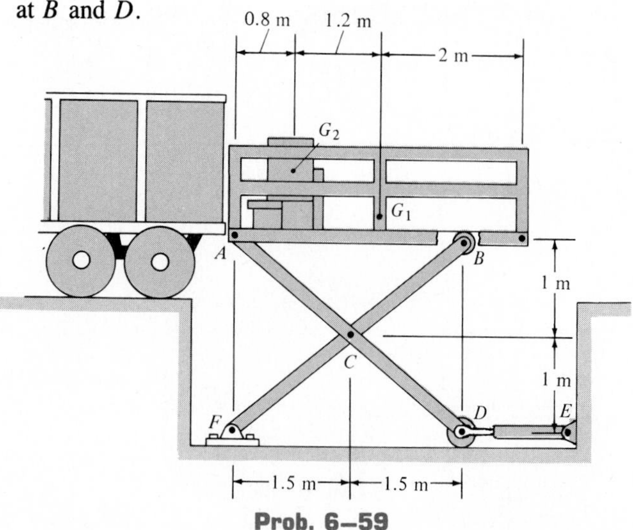

Prob. 6–59

*6–60. The spring mechanism is used as a shock absorber for a load applied to the drawbar AB. Determine the equilibrium length of each spring when the 50-N force is applied. Each spring has an unloaded length of 200 mm, and the drawbar slides along the smooth guide posts CG and EF. The bottom springs pass around the guide posts and the ends of all springs are attached to their respective members.

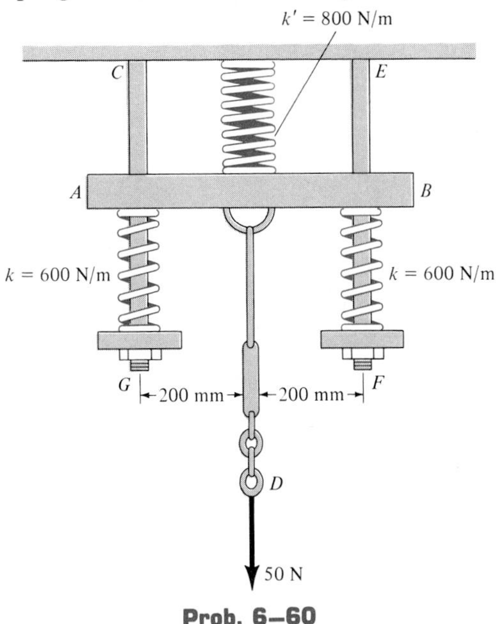

Prob. 6–60

6–61. If a force of $P = 6$ lb is applied perpendicular to the handle of the mechanism, determine the magnitude of force **F** for equilibrium. The members are pin-connected at A, B, C, and D.

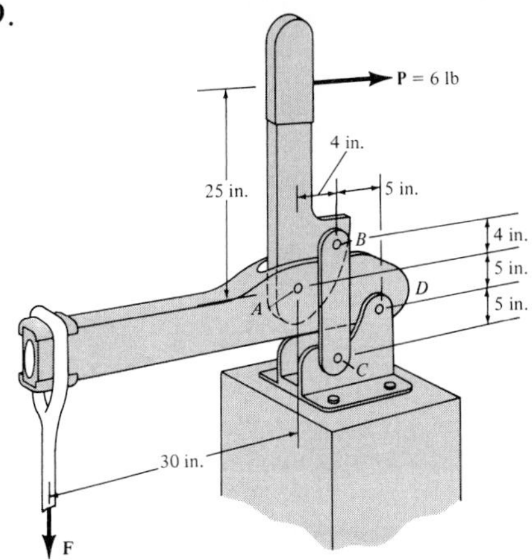

Prob. 6–61

6–62. The three pin-connected members shown in the *top view* support a *downward* force of 60 lb at G. If only vertical forces are supported at the connections B, C, E and pad supports A, D, F, determine the reactions at each pad.

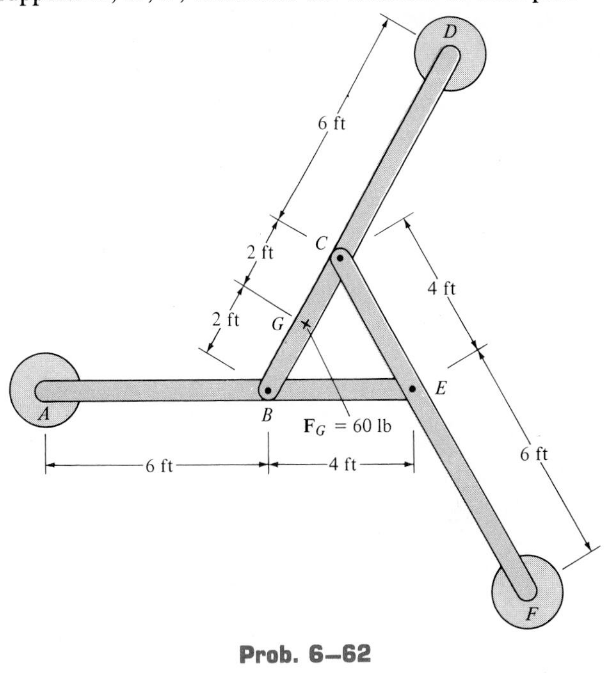

Prob. 6–62

6–63. The four-member "A" frame supports a vertical force of 300 N. If it is assumed that the supports at A, E, and G are smooth collars and all other joints are ball-and-sockets, determine the x, y, z force components which member BD exerts on EDC and FG.

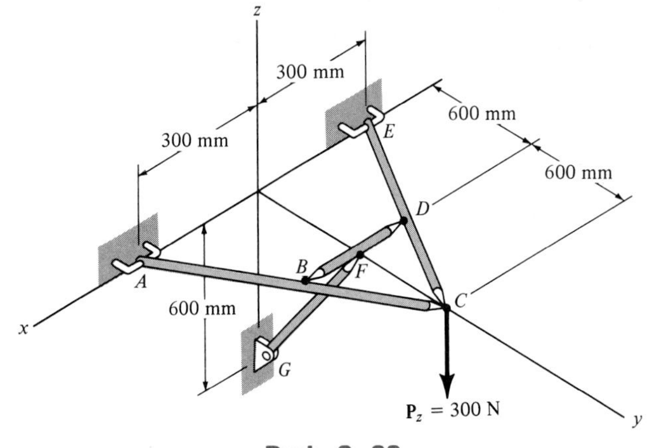

Prob. 6–63

Review Problems

*6–64. Determine the clamping force exerted on the smooth pipe at B if a force of 20 lb is applied to the handles of the pliers. The pliers are pinned together at A.

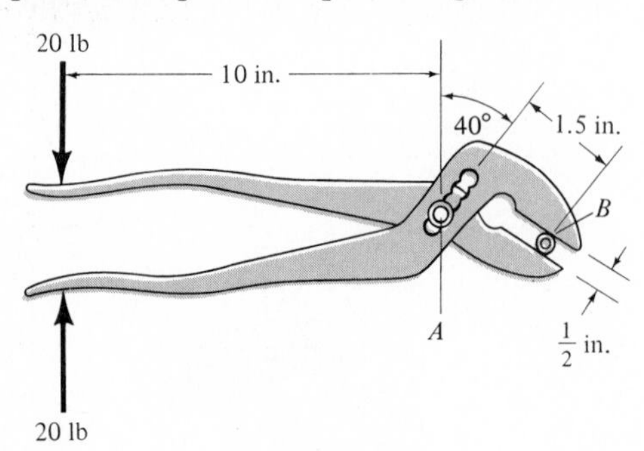

Prob. 6–64

6–65. Determine the horizontal and vertical components of force that the pins at A and C exert on the two-bar mechanism.

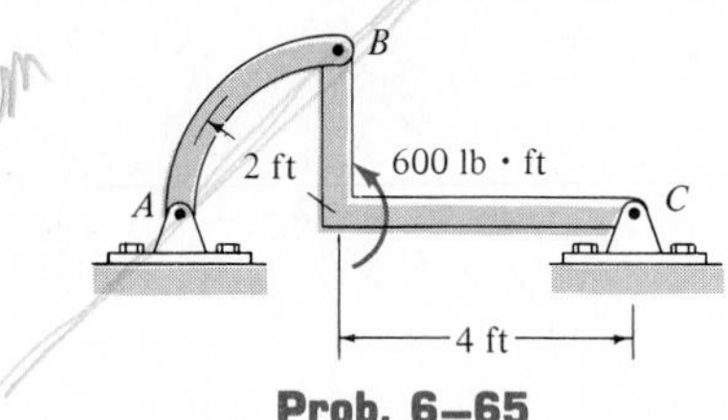

Prob. 6–65

6–66. Determine the force in each member of the truss. Indicate whether the members are in tension or compression.

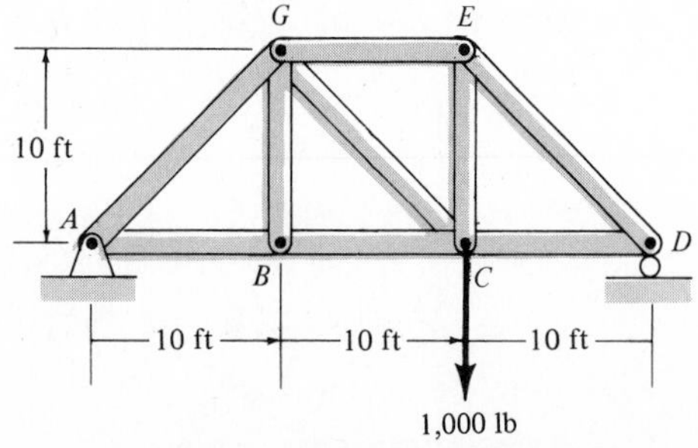

Probs. 6–66 / 6–67

6–67. Determine the force in members GE, GC, and BC of the truss. Indicate whether the members are in tension or compression.

*6–68. The jack shown supports a 350-kg automobile engine. Determine the compression in the hydraulic cylinder C and the magnitude of force that pin B exerts on the horizontal member BDE.

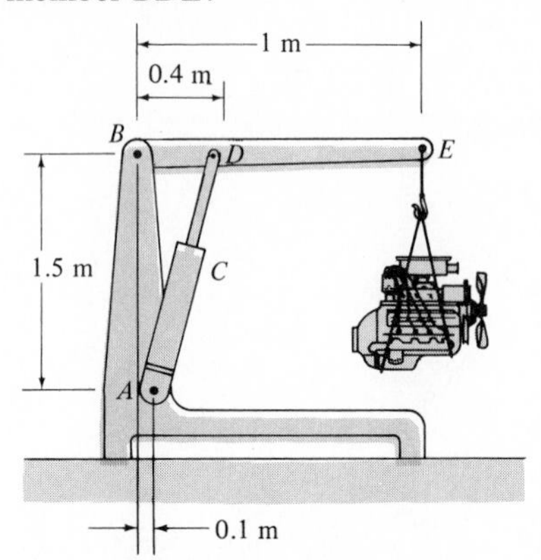

Probs. 6–68 / 6–69

6–69. If the maximum compressive force which the hydraulic cylinder C can sustain is 9 kN, determine the largest mass of the engine at E which can be supported. The arm BDE is horizontal. What are the horizontal and vertical components of force at the pin B for this loading?

6–70. Determine the force in member BC of the truss. Indicate whether the member is in tension or compression.

6–71. Determine the force in member GJ of the truss. Indicate whether the member is in tension or compression.

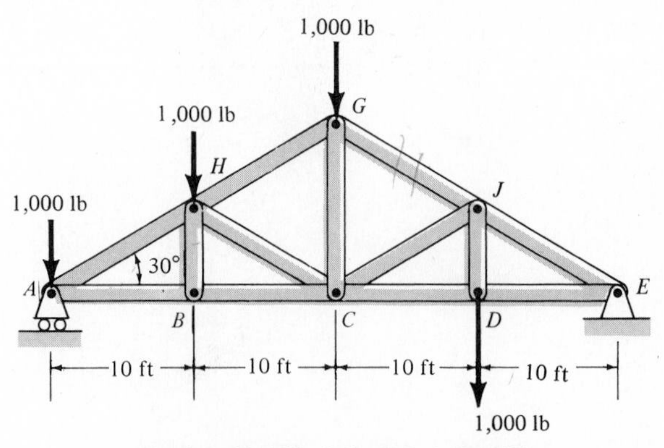

Probs. 6–70 / 6–71 / 6–72

*6–72. Determine the force in member GC of the truss. Indicate whether the member is in tension or compression.

6–73. Determine the forces which the pins at A and B exert on the two-member frame.

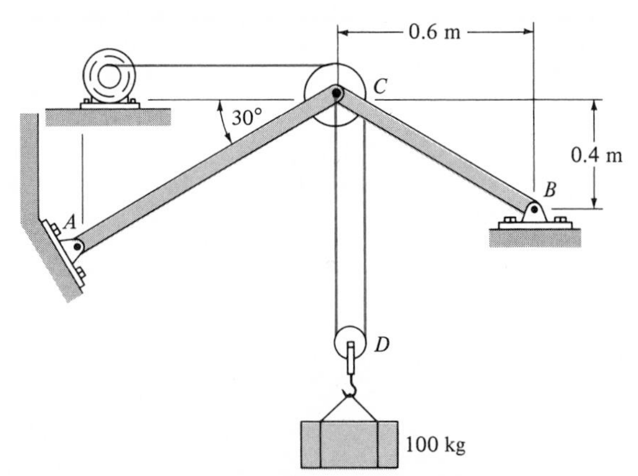

Probs. 6–73 / 6–74

6–74. Determine the horizontal and vertical components of force which the pin at C exerts on member AC and member BC of the two-member frame.

6–75. Determine the horizontal and vertical components of reaction at the pin supports A and E of the compound beam assembly.

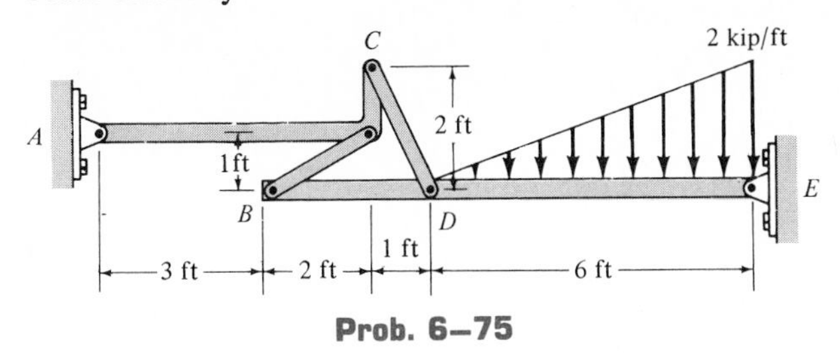

Prob. 6–75

*6–76. Determine the horizontal and vertical components of force that the pins at A and B exert on the two-member frame. Set $F = 0$.

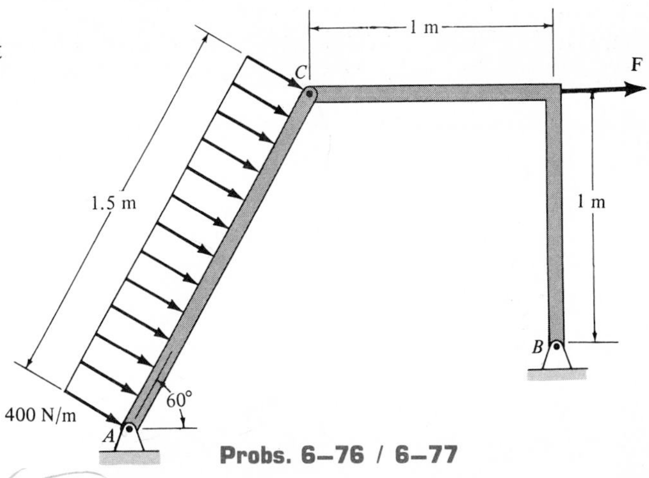

Probs. 6–76 / 6–77

6–77. Determine the horizontal and vertical components of force that pins A and B exert on the two-member frame. Set $F = 500$ N.

6–78. Determine the force developed in each of the members of the space truss. Indicate whether the members are in tension or compression. The crate has a weight of 150 lb.

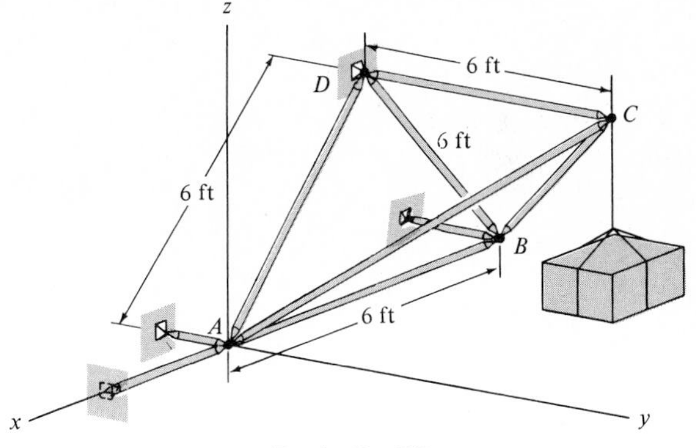

Prob. 6–78

7 Internal Forces

In Chapter 6 we discussed application of the equations of equilibrium to determine the forces acting at the *connections* of a structural member. Once these forces are obtained we can extend the equilibrium analysis to compute the member's *internal loadings*. In this chapter we will develop a technique for finding the internal loading at specific *points* within a structural member, and then we will generalize this method to find the point-to-point variation of loading along the axis of a beam. A graph showing this variation in the internal load will allow us to find the critical point where the *maximum* internal loading occurs. This information is important for the proper design of a structure using the theory of strength of materials.

7.1 Internal Forces Developed in Structural Members

The design of a structural member requires an investigation of the forces acting *within* the member which are necessary to balance the forces acting external to it. The *method of sections* can be used to determine these *internal* forces. This requires that a ''cut'' or section be made perpendicular to the axis of the member at the point where the internal loading is to be determined. A free-body diagram of either segment of the ''cut'' member is isolated and the internal loads are then determined from the equations of equilibrium. The loadings obtained in this manner will actually represent the *resultant* of a distribution of force per unit area, called *stress,* which acts over the member's cross-sectional area at the cut section. Using the theory of strength of materials, one can relate this stress distribution to the resultant loadings and thereby develop a formula which can be used for the design of the member.

Types of Internal Loadings. There are four types of internal loadings that can be resisted by a structural member. Using the method of sections and results from the theory of strength of materials, the effect of each of these loadings will now be illustrated. Throughout this chapter we will assume the member's cross-sectional area is symmetrical about axes which pass through its geometric center or centroid, and the applied external forces pass through the centroid of the cross-sectional area.

1. ***Axial force.*** If a tensile or compressive force, $\mathbf{F}_T$ or $\mathbf{F}_C$, is applied externally along the longitudinal axis of a member, Fig. 7–1*a*, then the internal stress distribution is uniform. The *resultant* of this distribution is called the *axial force* **A.** For equilibrium, **A** must also act along the member's longitudinal axis, and pass through the *centroid* or geometric center of the cross-sectional area.
2. ***Shear force.*** If an external force **P** is applied perpendicular to the axis of a member, Fig. 7–1*b*, it causes an internal stress distribution acting tangent to the member's cross section. The *resultant* of this stress distribution is called the *shear force* **V,** which is also tangent to the cross section.
3. ***Bending moment.*** When an external moment or couple $\mathbf{M}_c$ is applied perpendicular to the axis of a member, Fig. 7–1*c*, the internal distribution of stress is directed perpendicular to the member's cross-sectional area and varies linearly from a "neutral" axis *aa* passing through the member's centroid. The *resultant* of this stress distribution is called the *bending moment* **M.**

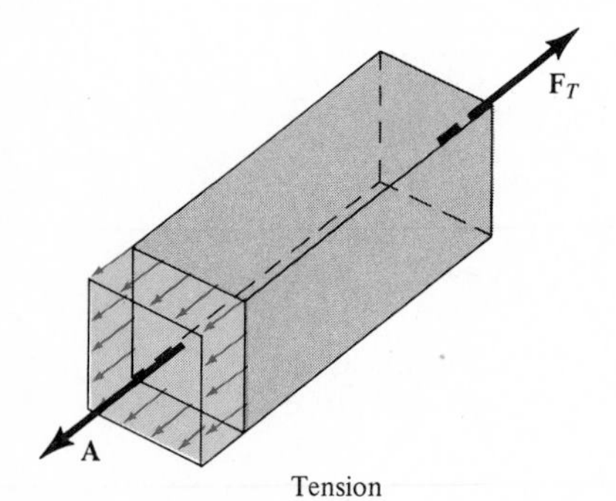

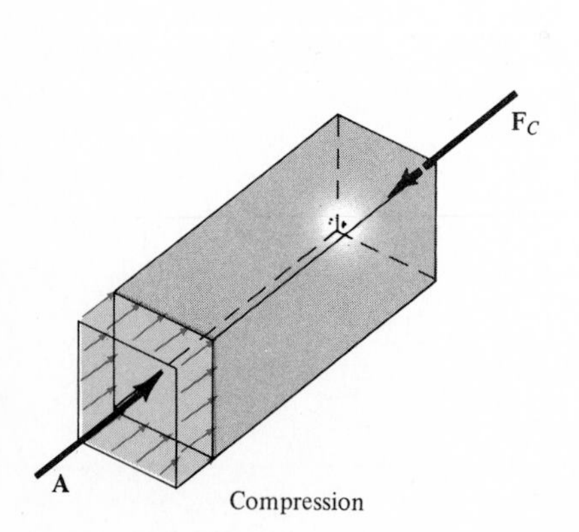

Internal axial force **A**

(a)

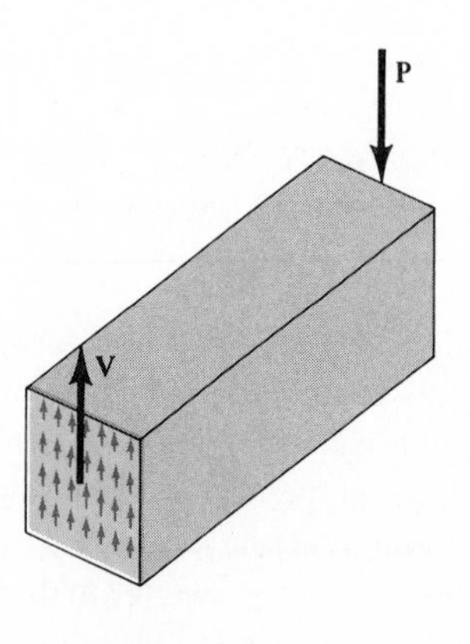

Internal shear force **V**

(b)

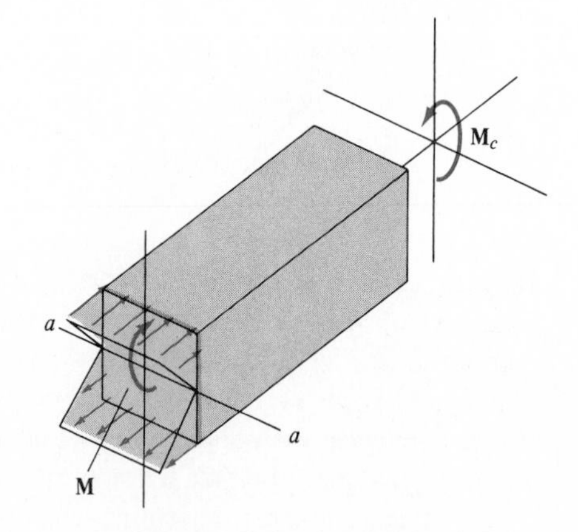

Internal bending moment **M**

(c)

Fig. 7–1a,b,c

4. ***Torsional moment.*** If an applied external moment or torque **T**′ tends to twist a circular member about its longitudinal axis it causes an internal distribution of stress which varies linearly when measured in a radial direction as shown in Fig. 7–1*d*. The *resultant* of this stress distribution is called the *torque* or torsional moment **T.**

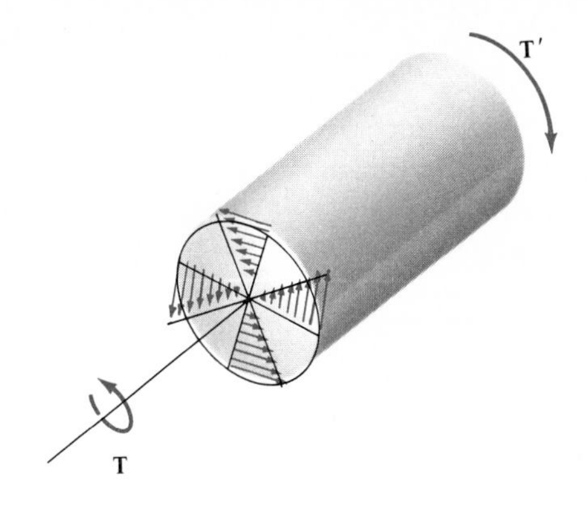

Internal torsional moment **T**

(d)

Fig. 7–1d

In general then, whenever a structural member is subjected to an external force system, the *internal* loading acting at a cut section of the member will consist of either an axial force **A,** shear force **V,** bending moment **M,** torsional moment **T,** or a *combination* of these resultants. Also, each of these loadings will be different at various sections along the axis of the member.

PROCEDURE FOR ANALYSIS

The following procedure provides a means for applying the method of sections to determine the internal axial force, shear force, bending moment, and torsional moment at a specific location in a member.

Support Reactions. Before the member is ''cut'' or sectioned, it may first be necessary to determine the member's support reactions, so that the equilibrium equations are used only to solve for the internal loadings when the member is sectioned. If the member is part of a frame or machine, the pin reactions can be determined using the methods of Sec. 6–6.

Free-Body Diagram. Keep all distributed loadings, couple moments, torques, and forces acting on the member in their *exact location,* then pass an imaginary section through the member, perpendicular to its axis at the point where the internal loading is to be determined. Draw a free-body diagram of one of the ''cut'' segments on either side of the section and indicate the unknown resultants **A, V, M,** and **T** at the section. In particular, if the member is subjected to a *coplanar* system of forces, only **A, V,** and **M** act at the section.

Equations of Equilibrium. Apply the equations of equilibrium to obtain the unknowns **A, V, M,** and **T.** In most cases, moments should be summed at the section about axes passing through the *centroid* of the member's cross-sectional area, in order to eliminate the unknowns **A** and **V** and thereby obtain direct solutions for **M** and **T.** If the solution of the equilibrium equations yields a quantity having a negative magnitude, the assumed directional sense of the quantity is opposite to that shown on the free-body diagram.

The following examples numerically illustrate this procedure.

Example 7–1

The composite bar is fixed at its end and is loaded as shown in Fig. 7–2*a*. Determine the axial force at points *B*, *C*, and *D*.

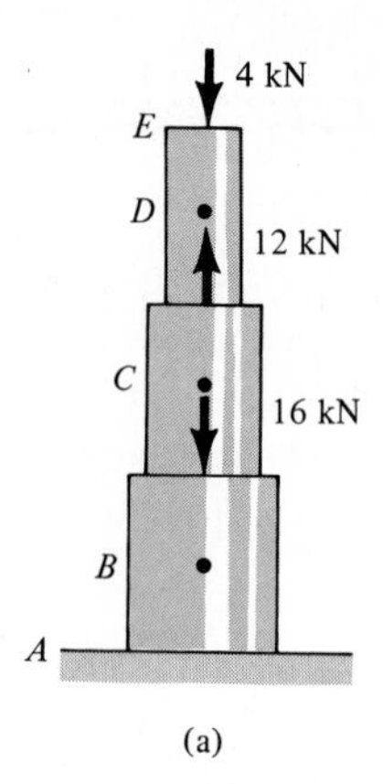

(a)

Solution

Support Reactions. A free-body diagram of the bar is shown in Fig. 7–2*b*. By inspection, only an axial force $\mathbf{F}_A$ acts at the fixed support since the loads are symmetrically applied along the bar's axis.

$$+\uparrow \Sigma F_z = 0; \quad F_A - 16\text{ kN} + 12\text{ kN} - 4\text{ kN} = 0 \qquad F_A = 8\text{ kN}$$

Free-Body Diagrams. The internal forces at *B*, *C*, and *D* will be found using the free-body diagrams of the sectioned bar shown in Fig. 7–2*c*. In particular, section *BA* was chosen here since it contains the *least* number of forces.

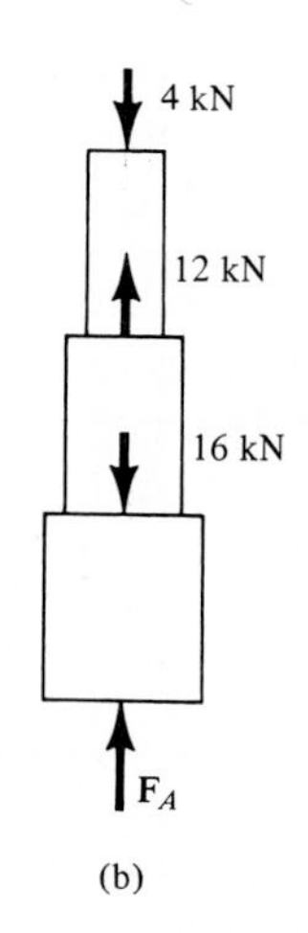

(b)

Equations of Equilibrium.

Segment DE

$$+\uparrow \Sigma F_z = 0; \qquad F_D - 4\text{ kN} = 0 \qquad F_D = 4\text{ kN} \qquad \textit{Ans.}$$

Segment CE

$$+\uparrow \Sigma F_z = 0; \quad -F_C + 12\text{ kN} - 4\text{ kN} = 0 \qquad F_C = 8\text{ kN} \qquad \textit{Ans.}$$

Segment AB

$$+\uparrow \Sigma F_z = 0; \qquad 8\text{ kN} - F_B = 0 \qquad F_B = 8\text{ kN} \qquad \textit{Ans.}$$

Try working this problem in the following manner: Determine $\mathbf{F}_B$ from segment *BE*; using this result, determine $\mathbf{F}_C$ from segment *BC*; finally determine $\mathbf{F}_D$ from segment *CD*. Note that this approach does not require solution for the support reaction at *A*.

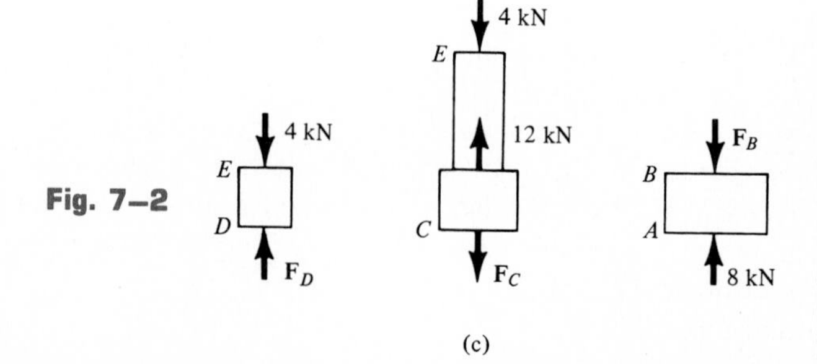

(c)

Fig. 7–2

Example 7–2

A built-up circular shaft is subjected to three concentrated torques as shown in Fig. 7–3*a*. Determine the internal torque at points B, C, and D.

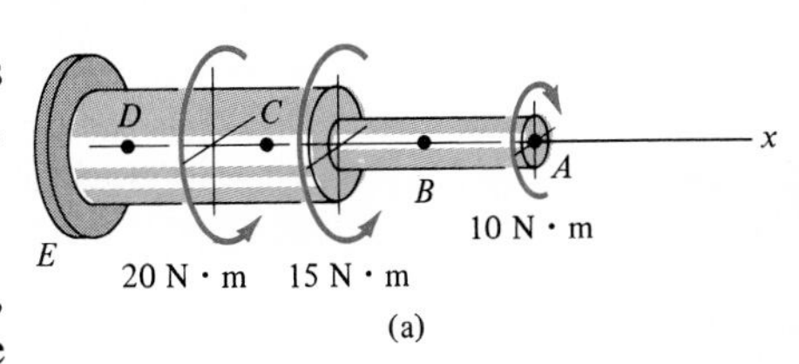

(a)

Solution

Support Reactions. Since the shaft is subjected only to collinear torques, a torque reaction occurs at the support. Using the right-hand rule to define the direction of the torques, we require

$\Sigma M_x = 0;$

$$-10\ \text{N}\cdot\text{m} + 15\ \text{N}\cdot\text{m} + 20\ \text{N}\cdot\text{m} - T_E = 0 \qquad T_E = 25\ \text{N}\cdot\text{m}$$

Free-Body Diagrams. The internal torques at B, C, and D will be found using the free-body diagrams of the sectioned shaft shown in Fig. 7–3*c*. Notice that $\mathbf{T}_D$ acts with equal magnitude and opposite direction on each side of the section at D—Newton's third law.

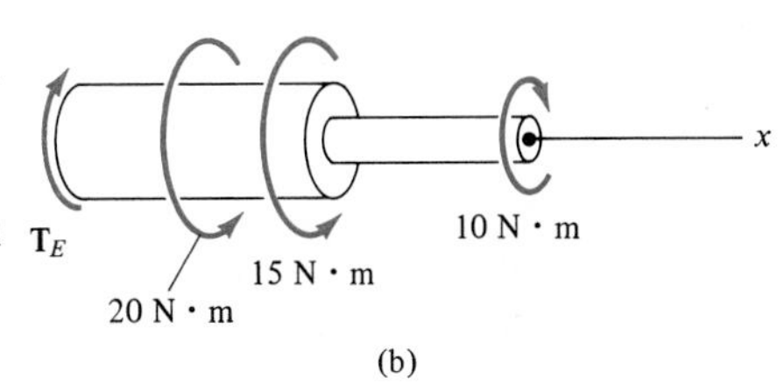

(b)

Equations of Equilibrium. Applying the equation of moment equilibrium along the shaft's axis, we have

Segment AB

$$\Sigma M_x = 0; \qquad -10\ \text{N}\cdot\text{m} + T_B = 0 \qquad T_B = 10\ \text{N}\cdot\text{m} \qquad \textit{Ans.}$$

Segment DE

$$\Sigma M_x = 0 \qquad T_D - 25\ \text{N}\cdot\text{m} = 0 \qquad T_D = 25\ \text{N}\cdot\text{m} \qquad \textit{Ans.}$$

Segment CD Using the above result,

$$\Sigma M_x = 0; \quad T_C + 20\ \text{N}\cdot\text{m} - 25\ \text{N}\cdot\text{m} = 0 \qquad T_C = 5\ \text{N}\cdot\text{m} \qquad \textit{Ans.}$$

Try working the problem in the following manner: Determine $\mathbf{T}_B$ from segment AB, $\mathbf{T}_C$ from segment AC, and $\mathbf{T}_D$ from segment AD. Note that this approach does not require a solution for the support reaction at E.

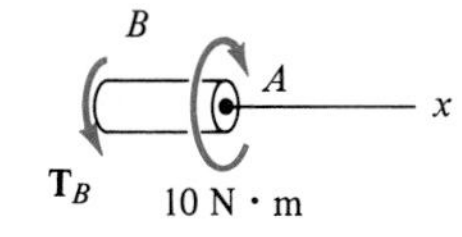

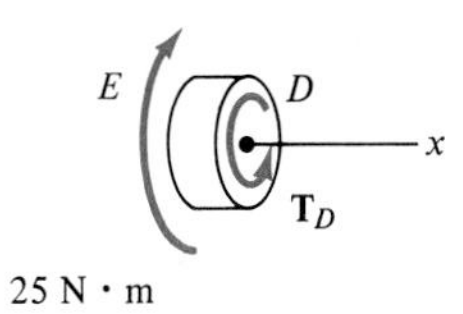

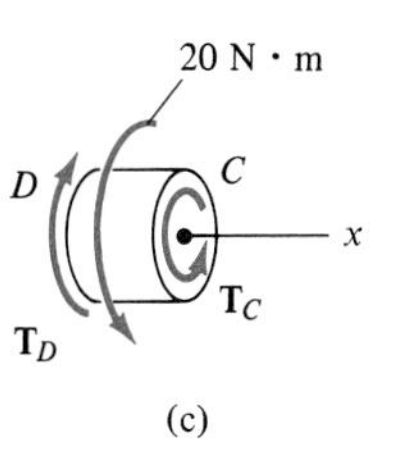

(c)

Fig. 7–3

Example 7–3

The three smooth plates of the joint shown in Fig. 7–4*a* are connected by two bolts. Determine the shear force in each bolt at section *bb* between the plates and the resultant axial force supported by the top plate at section *aa*. Assume that each bolt carries an equal amount of shear force.

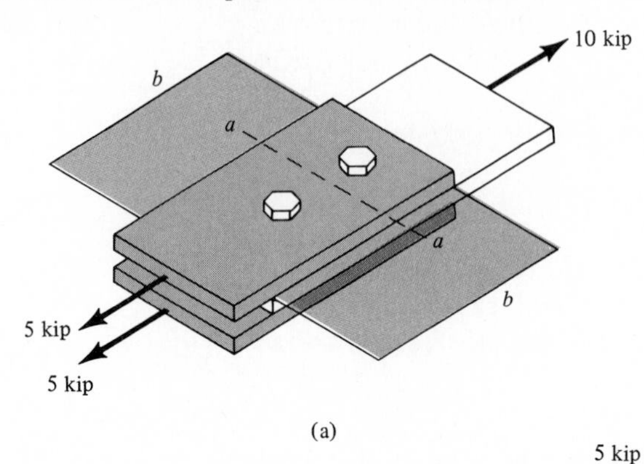

(a)

Fig. 7–4

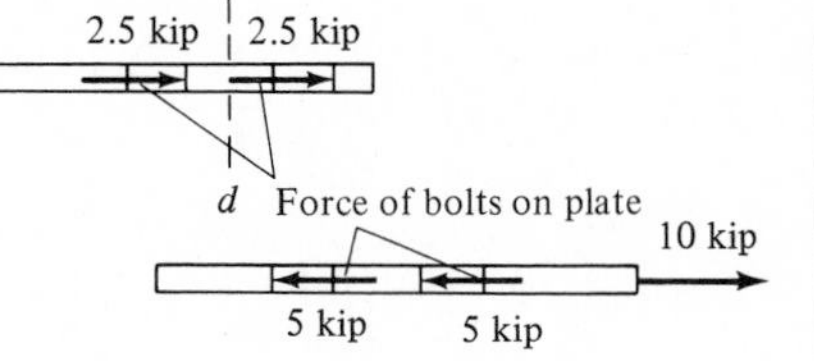

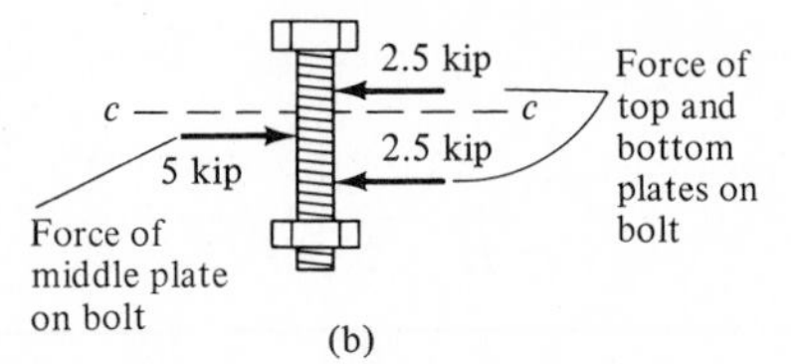

(b)

Solution

Support Reactions. The free-body diagrams and equilibrium requirements for the top *or* bottom plate, the middle plate, and a bolt are shown in Fig. 7–4*b*.

Free-Body Diagrams. The free-body diagrams of sections *bb* and *aa* are shown in Fig. 7–4*c*.

Equations of Equilibrium. Applying the equations of force equilibrium, we have

$$\xleftarrow{+}\Sigma F_x = 0; \qquad 5\text{ kip} - 2V_b = 0 \qquad V_b = 2.5\text{ kip} \qquad \textit{Ans.}$$

$$\xleftarrow{+}\Sigma F_x = 0; \quad 5\text{ kip} - 2.5\text{ kip} - A_p = 0 \qquad A_p = 2.5\text{ kip} \qquad \textit{Ans.}$$

Note: It is also possible to solve this problem in another way. For example, show that sections *cc* through the bolt and *dd* through the plate, Fig. 7–4*b*, yield the results previously obtained.

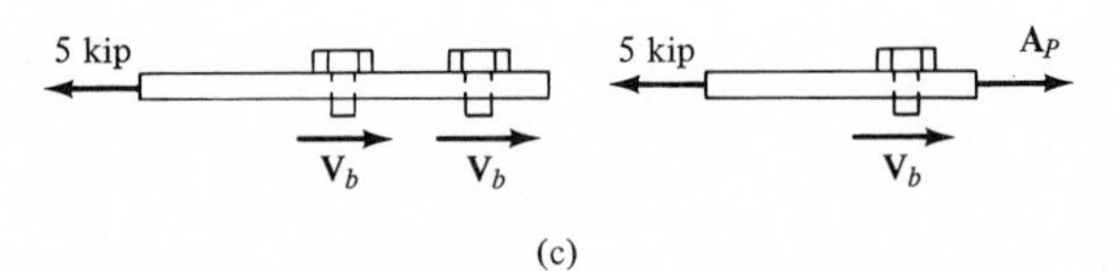

(c)

Example 7–4

Determine the axial force, shear force, and bending moment acting just to the left, point B, and just to the right, point C, of the 6-kN concentrated force shown on the beam in Fig. 7–5a.

Fig. 7–5

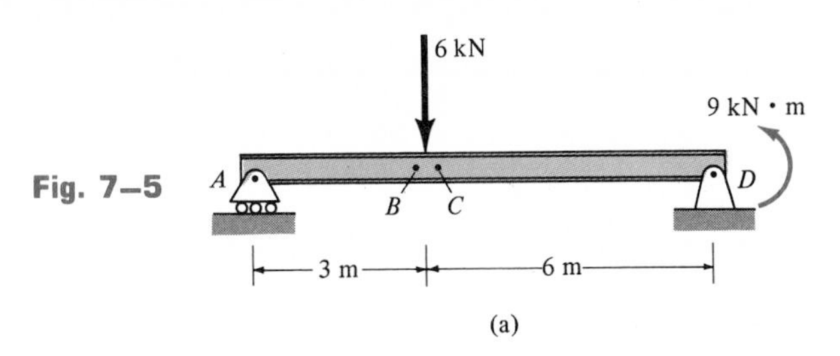

(a)

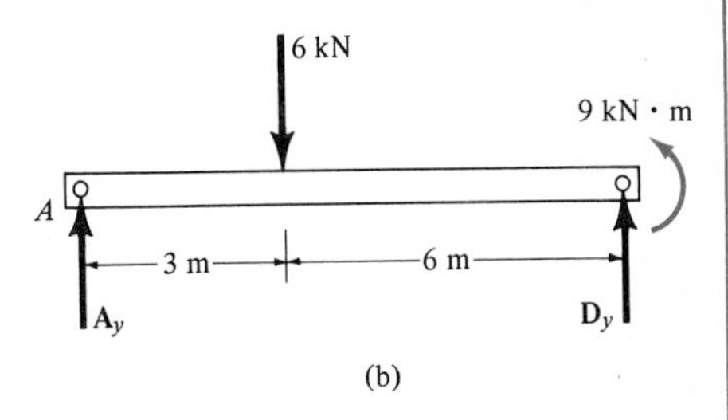

(b)

Solution

Support Reactions. The free-body diagram of the beam is shown in Fig. 7–5b. When computing the reactions, note that the 9-kN · m couple is a free vector and can therefore act anywhere on the beam. We have

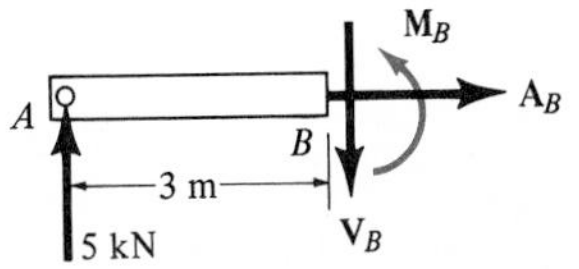

$$\circlearrowleft +\Sigma M_A = 0; \quad -(6\text{ kN})(3\text{ m}) + 9\text{ kN}\cdot\text{m} + D_y(9\text{ m}) = 0$$
$$D_y = 1\text{ kN}$$

$$+\uparrow \Sigma F_y = 0; \quad A_y - 6\text{ kN} + 1\text{ kN} = 0$$
$$A_y = 5\text{ kN}$$

(c)

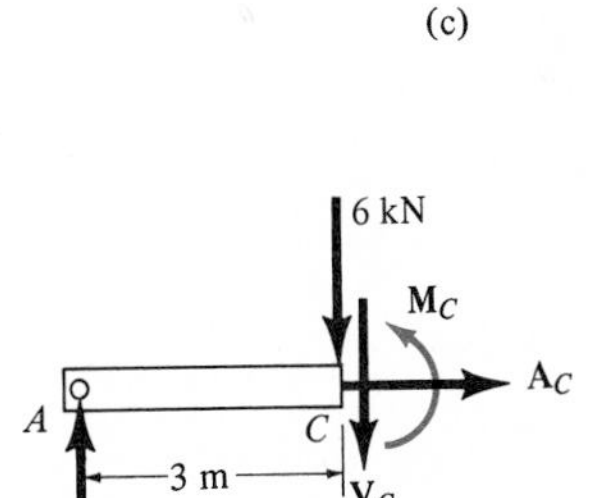

(d)

Free-Body Diagrams. The free-body diagrams of the left segments AB and AC of the beam are shown in Fig. 7–5c and d. Note that the 9-kN · m couple moment is *not included* on these diagrams. This load must be kept in its original position until *after* the section is made and the appropriate body isolated; only then will the true loading situation be represented.

Equations of Equilibrium

Segment AB

$\overset{+}{\rightarrow}\Sigma F_x = 0; \quad A_B = 0$ *Ans.*

$+\uparrow \Sigma F_y = 0; \quad 5\text{ kN} - V_B = 0 \qquad V_B = 5\text{ kN}$ *Ans.*

$\circlearrowleft +\Sigma M_B = 0; \quad -(5\text{ kN})(3\text{ m}) + M_B = 0 \qquad M_B = 15\text{ kN}\cdot\text{m}$ *Ans.*

Segment AC

$\overset{+}{\rightarrow}\Sigma F_x = 0; \quad A_C = 0$ *Ans.*

$+\uparrow \Sigma F_y = 0; \quad 5\text{ kN} - 6\text{ kN} - V_C = 0 \qquad V_C = -1\text{ kN}\cdot\text{m}$ *Ans.*

$\circlearrowleft +\Sigma M_C = 0; \quad -(5\text{ kN})(3\text{ m}) + M_C = 0 \qquad M_C = 15\text{ kN}\cdot\text{m}$ *Ans.*

Note that the moment arm for the 5-kN force in both cases is approximately 3 m, since B and C are almost coincident.

Example 7–5

Determine the axial force, shear force, and bending moment acting at point B of the two-member frame shown in Fig. 7–6a.

Solution

Support Reactions. A free-body diagram of each member is shown in Fig. 7–6b. Since CD is a two-force member, the equations of equilibrium have to be applied only to member AC.

$$\circlearrowleft +\Sigma M_A = 0; \quad -400 \text{ lb}(4 \text{ ft}) + (\tfrac{3}{5})F_{DC}(8 \text{ ft}) = 0 \qquad F_{DC} = 333.3 \text{ lb}$$

$$\overset{+}{\rightarrow}\Sigma F_x = 0; \quad -A_x + (\tfrac{4}{5})(333.3 \text{ lb}) = 0 \qquad A_x = 266.7 \text{ lb}$$

$$+\uparrow \Sigma F_y = 0; \quad A_y - 400 \text{ lb} + \tfrac{3}{5}(333.3 \text{ lb}) = 0 \qquad A_y = 200 \text{ lb}$$

Free-Body Diagrams. Passing an imaginary section perpendicular to the axis of the beam through point B yields the free-body diagrams of segments AB and BC shown in Fig. 7–6c. Although the distributed load can be simplified to a single resultant force, it is important when constructing these diagrams to keep this loading exactly where it is until *after* the section is made. Why? Also, notice that $\mathbf{A}_B$, $\mathbf{V}_B$, and $\mathbf{M}_B$ act with equal magnitude but opposite direction on each segment—Newton's third law.

Equations of Equilibrium. Applying the equations of equilibrium to segment AB, we have

$$\overset{+}{\rightarrow}\Sigma F_x = 0; \quad A_B - 266.7 \text{ lb} = 0 \qquad A_B = 266.7 \text{ lb} \qquad \textit{Ans.}$$

$$+\uparrow \Sigma F_y = 0; \quad 200 \text{ lb} - 200 \text{ lb} - V_B = 0 \qquad V_B = 0 \qquad \textit{Ans.}$$

$$\circlearrowleft +\Sigma M_B = 0; \quad M_B - 200 \text{ lb}(4 \text{ ft}) + 200 \text{ lb}(2 \text{ ft}) = 0$$

$$M_B = 400 \text{ lb} \cdot \text{ft} \qquad \textit{Ans.}$$

As an exercise, try to obtain the foregoing results using segment BC.

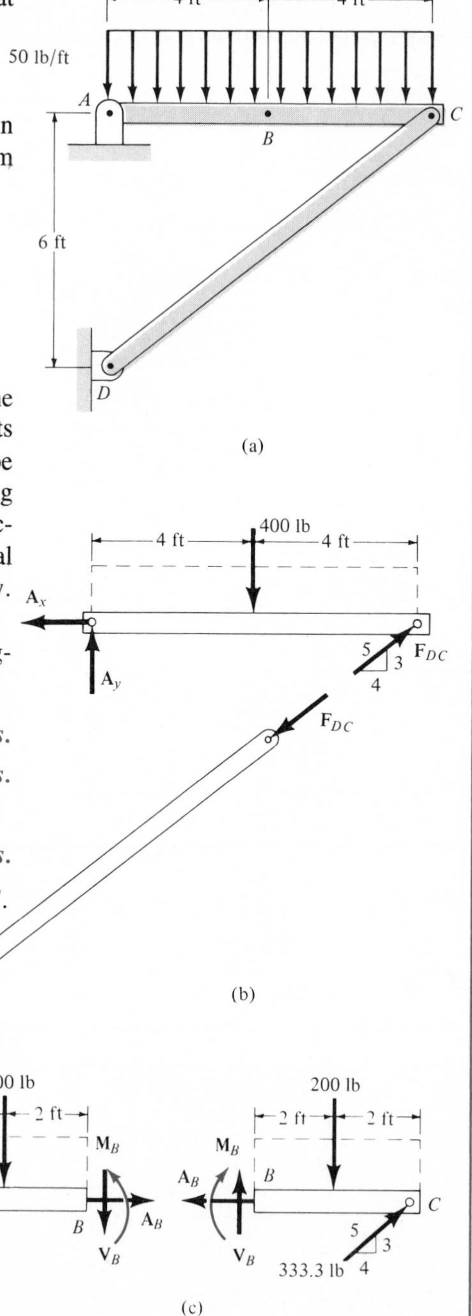

Fig. 7–6

Example 7–6

A force having x, y, z components shown in Fig. 7–7a acts at the corner of a beam extended from a fixed wall. Determine the internal loadings at a section passing through point A.

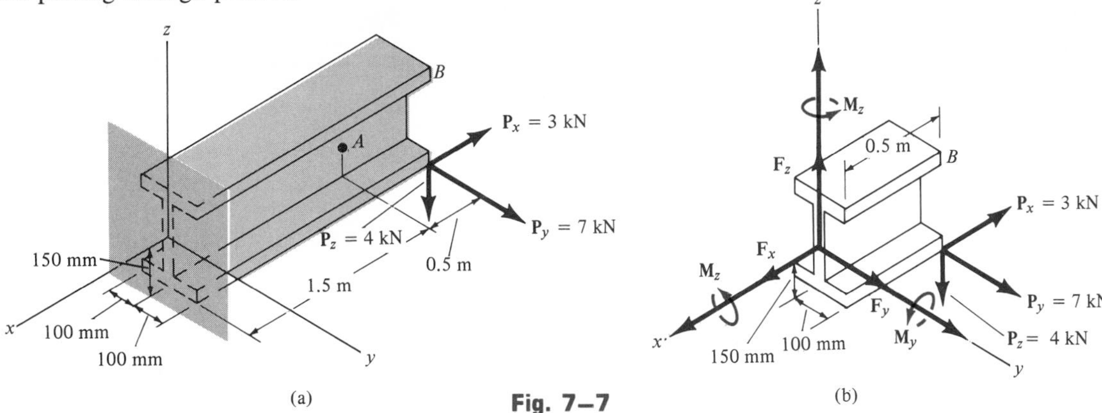

Fig. 7–7

Solution

This problem can be solved by considering a section of the beam that will *not* involve the support reactions.

Free-Body Diagram. The free-body diagram of segment AB is shown in Fig. 7–7b. The components of the resultant force $\mathbf{F}_A$ and moment $\mathbf{M}_A$ pass through the *centroid* of the cross-sectional area at A.

Equations of Equilibrium

$\Sigma F_x = 0;\qquad F_x = 3\text{ kN}$ *Ans.*

$\Sigma F_y = 0;\qquad F_y = -7\text{ kN}$ *Ans.*

$\Sigma F_z = 0;\qquad F_z = 4\text{ kN}$ *Ans.*

$\Sigma M_x = 0;\qquad M_x - 4(0.1) + 7(0.15) = 0$

$M_x = -0.65\text{ kN}\cdot\text{m}$ *Ans.*

$\Sigma M_y = 0;\qquad M_y - 4(0.5) + 3(0.15) = 0$

$M_y = 1.55\text{ kN}\cdot\text{m}$ *Ans.*

$\Sigma M_z = 0;\qquad M_z + 3(0.1) - 7(0.5) = 0$

$M_z = 3.20\text{ kN}\cdot\text{m}$ *Ans.*

Note that $\mathbf{F}_x$ represents the axial force A, whereas $\mathbf{F}_y$ and $\mathbf{F}_z$ are components of the shear force $V = \sqrt{F_y^2 + F_z^2}$. Also, the torsional moment is $\mathbf{T} = \mathbf{M}_x$, and the bending moment is determined from the components $\mathbf{M}_y$ and $\mathbf{M}_z$, i.e., $M = \sqrt{(M_y)^2 + (M_z)^2}$.

Problems

7–1. The axial forces act on the shaft as shown. Determine the internal axial force at points *A* and *B*.

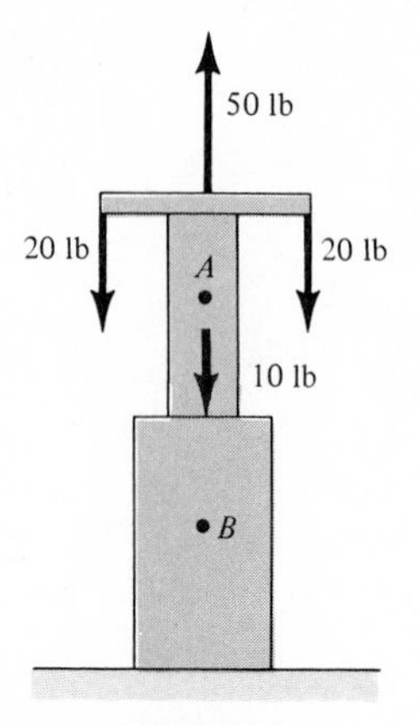

Prob. 7–1

7–2. The three axial forces act on the shaft as shown. Determine the internal axial force at points *A*, *B*, and *C*.

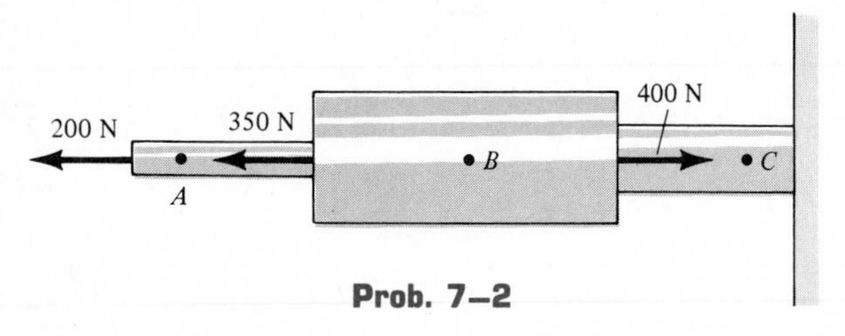

Prob. 7–2

7–3. The lap joint is connected using four bolts as shown. Draw a free-body diagram of bolt *A*, then section the bolt between the plates and determine the shear force necessary to support the 50-lb load. Neglect friction between the plates and assume each bolt carries an equal amount of shear force.

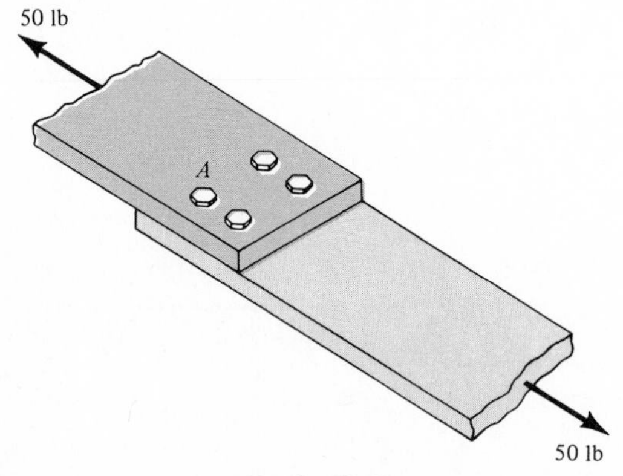

Prob. 7–3

***7–4.** The joint is held together using a single bolt as shown. Draw a free-body diagram of the bolt and then make appropriate sections through the bolt in order to determine the shear force between (a) plates *A* and *B*, and (b) plates *B* and *C*.

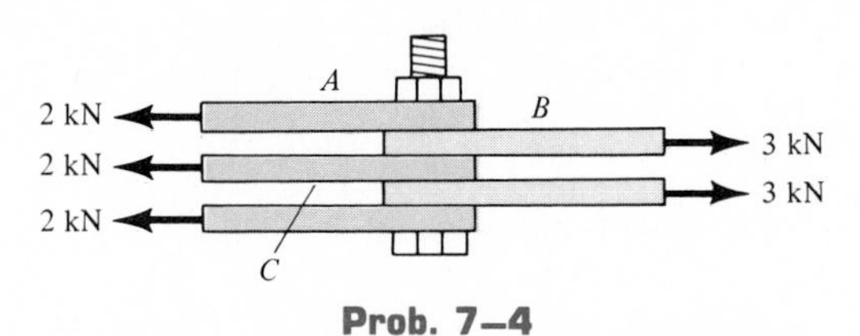

Prob. 7–4

7–5. The shaft is supported by smooth bearings at *A* and *B* and subjected to the torques shown. Determine the internal torques at points *C*, *D*, and *E*.

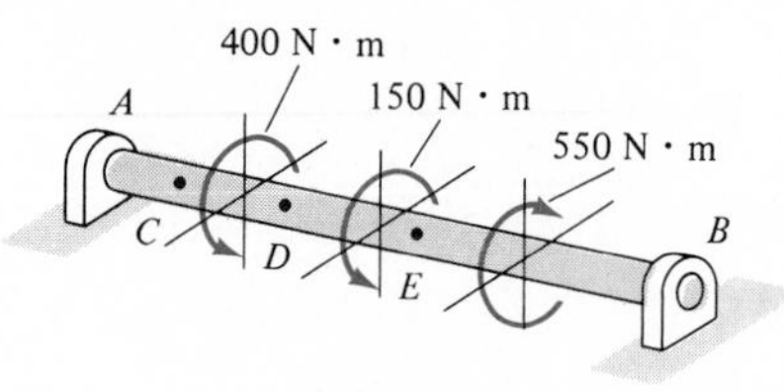

Prob. 7–5

7–6. Three torques act on the shaft as shown. Determine the internal torque at points *A*, *B*, *C*, and *D*.

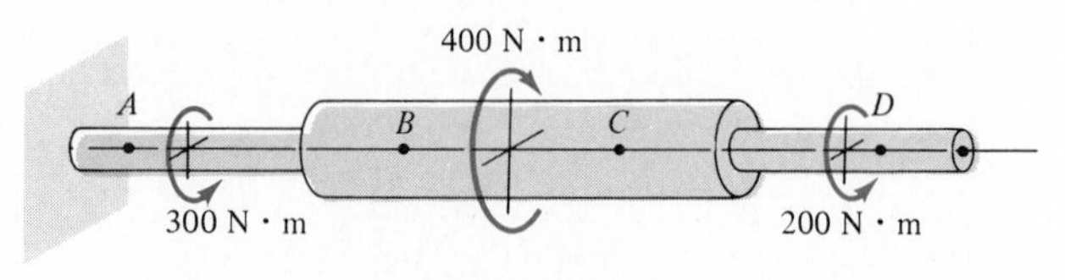

Prob. 7–6

7–7. Determine the axial force, shear force, and moment at point C.

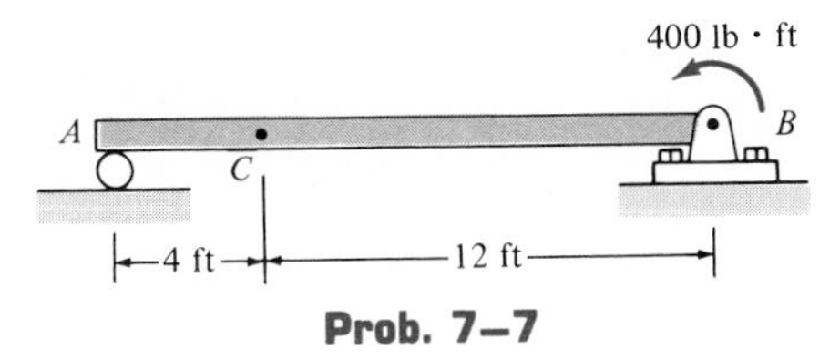

Prob. 7–7

***7–8.** The work platform supports an 80-kg man having a mass center at G. Determine the axial force, shear force, and bending moment at point E of the telescopic column AB due to this loading.

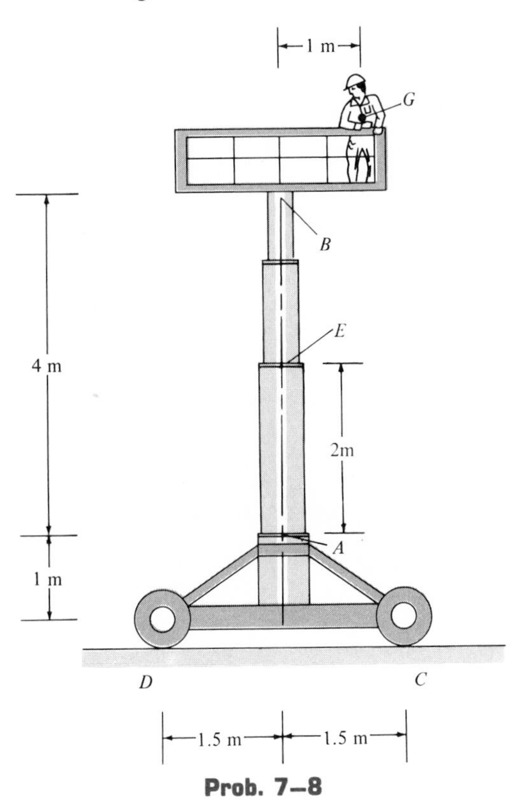

Prob. 7–8

7–9. Determine the axial force, shear force, and moment at point A if the clamp exerts a compressive force of $F = 65$ lb on board B. Force **F** acts along the central axis of the screw.

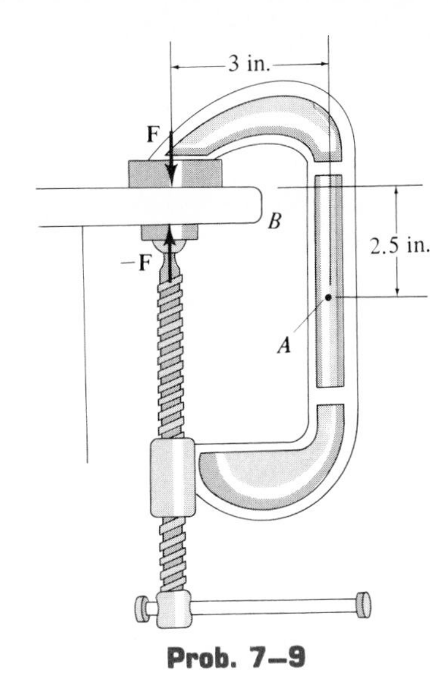

Prob. 7–9

7–10. Determine the internal axial force, shear force, and moment at point D.

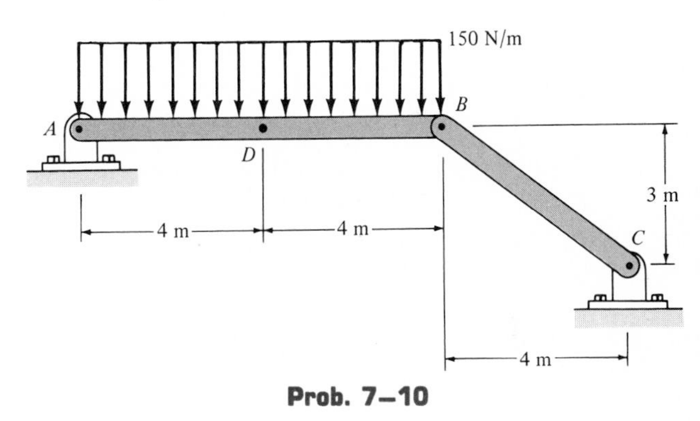

Prob. 7–10

7–11. Determine the internal axial force, shear force, and moment at points E and D. There is a pin or hinge at B.

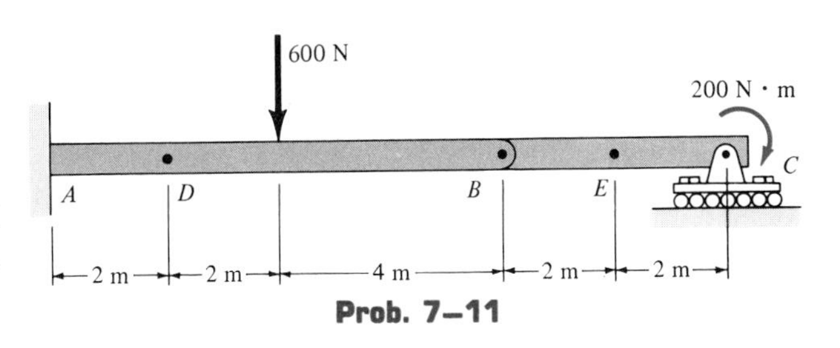

Prob. 7–11

***7–12.** Determine the internal axial force, shear force, and moment at point C.

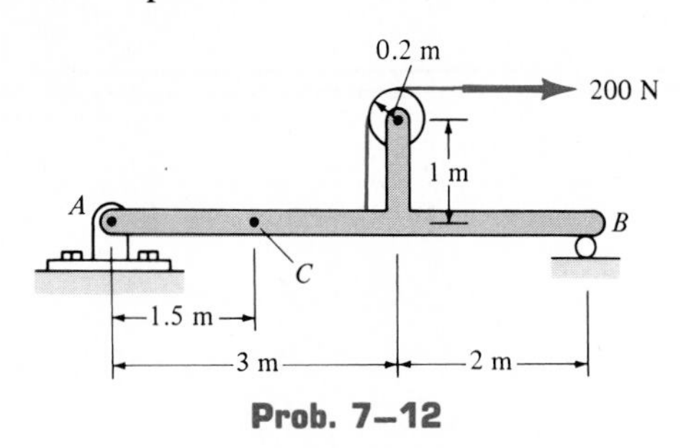

Prob. 7–12

7–13. Determine the internal axial force, shear force, and moment at point C.

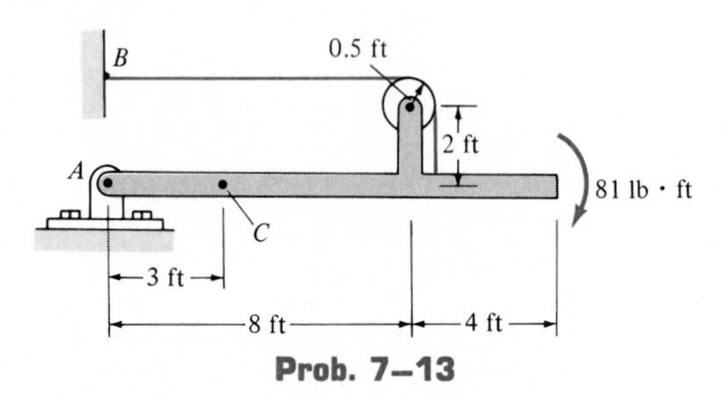

Prob. 7–13

7–14. Determine the internal axial force, shear force, and moment at point D of the two-member frame.

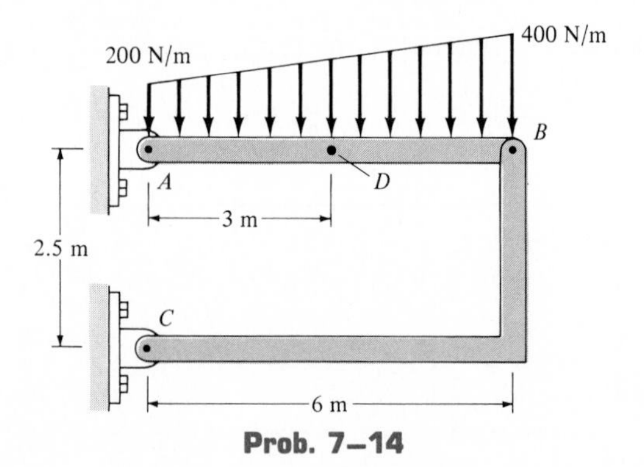

Prob. 7–14

7–15. Determine the internal axial force, shear force, and moment at point C of the beam.

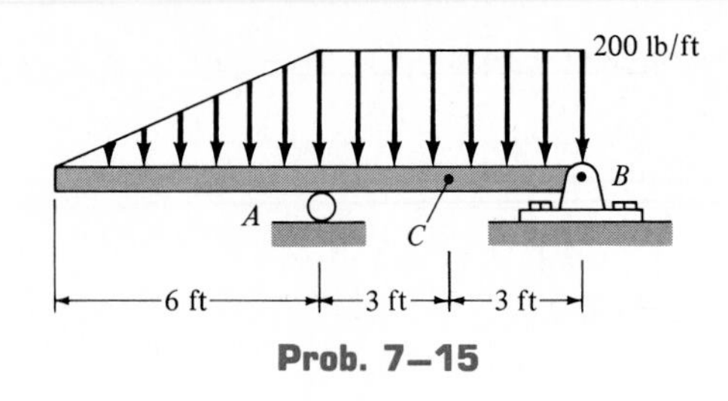

Prob. 7–15

***7–16.** Determine the internal axial force, shear force, and moment at point D of the beam.

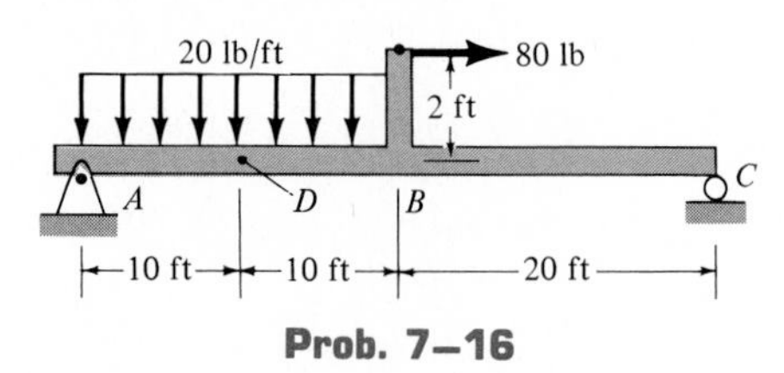

Prob. 7–16

7–17. Determine the internal axial force, shear force, and moment acting at point A of the smooth hook.

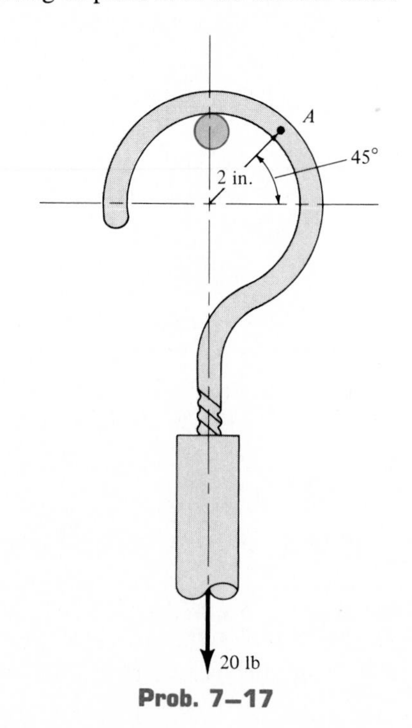

Prob. 7–17

7–18. A vertical force of 80 N is applied to the handle of the pipe wrench as shown. Determine the x, y, z components of internal force and moment at point A.

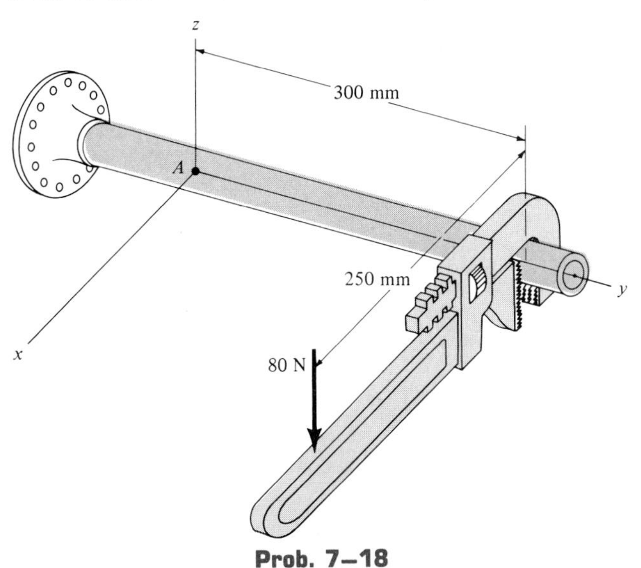

Prob. 7–18

7–19. A 400-lb force acts on the corner of the concrete column. If the column has a uniform density of 150 lb/ft^3, determine the resultant internal force and moment components acting in the x, y, and z directions at a horizontal section taken through point O.

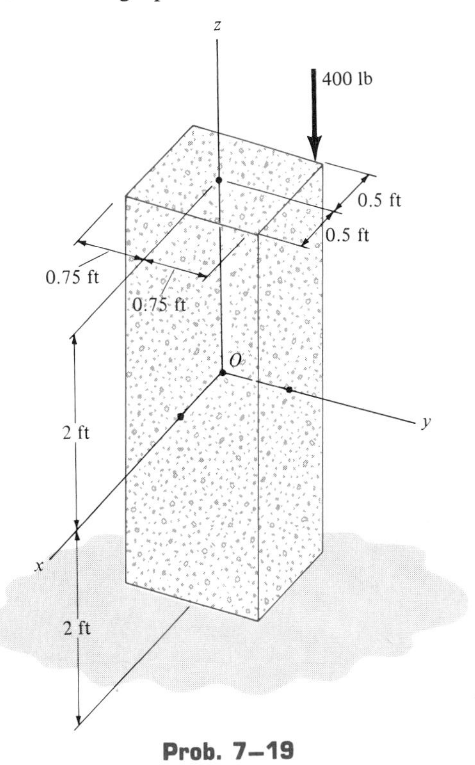

Prob. 7–19

7.2* Shear and Bending-Moment Diagrams for a Beam

Beams are structural members which are designed to support loadings applied perpendicular to their axis. In general, beams are long, straight bars, having a constant cross-sectional area. The actual design of such members requires a detailed knowledge of the *variation* of the internal shear force V and bending moment M acting at *each point* along the axis of the beam. After the force and bending-moment analysis is complete, one can then use the theory of strength of materials to determine the beam's required cross-sectional area.

The *variations* of V and M as a function of the position x of an *arbitrary point* along the beam's axis can be obtained by using the method of sections discussed in Sec. 7–1. Here, however, it is necessary to locate the imaginary section at an arbitrary distance x from the end of the beam rather than at a specified point. If the results are plotted, the graphical variations of V and M as a function of x are termed the *shear diagram* and *bending-moment diagram*, respectively.

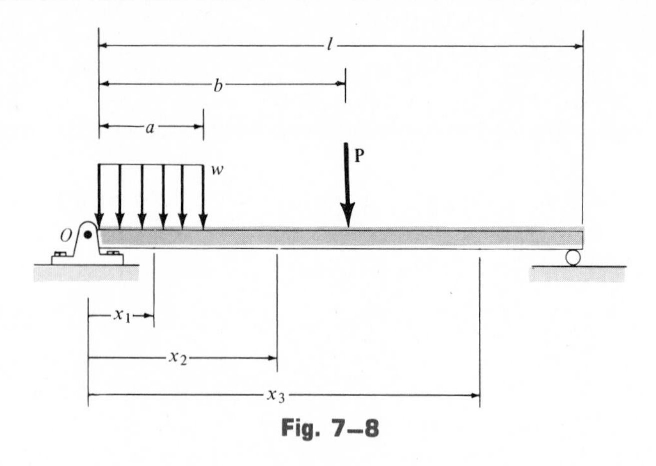

Fig. 7–8

In general, the internal shear and bending-moment functions will be discontinuous, or their slope will be discontinuous at points where a distributed load changes or where concentrated forces or couples are applied. Because of this, shear and bending-moment functions must be determined for *each segment* of the beam located between any two discontinuities of loading. For example, sections located at x_1, x_2, and x_3 will have to be used to describe the variation of V and M throughout the length of the beam in Fig. 7–8. These functions will be valid *only* within regions from O to a for x_1, from a to b for x_2, and from b to l for x_3.

The internal axial force will not be considered in the following discussion for two reasons. In most cases, the loads applied to a beam act perpendicular to the beam's axis and hence produce only an internal shear force and bending moment. For design purposes, the beam's resistance to shear, and particularly to bending, is more important than its ability to resist axial force.

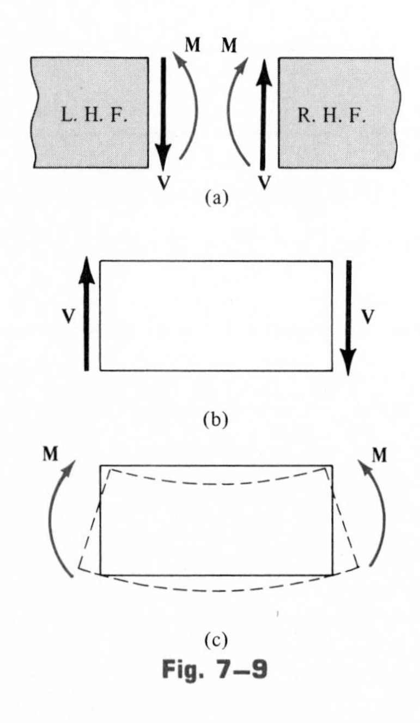

Fig. 7–9

Sign Convention. Before presenting a method for determining the shear and bending moment as functions of x and later plotting these functions (shear and bending-moment diagrams), it is first necessary to establish a *sign convention* so as to define "positive" and "negative" shear force and bending moment acting in the beam. [This is analogous to assigning coordinate directions x positive to the right and y positive upward, when plotting a function $y = f(x)$.] The sign convention to be adopted here is illustrated in Fig. 7–9*a*. On the *left-hand face* (L.H.F.) of a beam segment, the internal shear force **V** acts downward and the internal moment **M** acts counterclockwise. In accordance with Newton's third law, an equal and opposite shear force and bending moment must act on the *right-hand face* (R.H.F.) of a segment. Perhaps an easy way to remember this sign convention is to isolate a small beam segment and note that *positive shear tends to rotate the segment clockwise* (Fig. 7–9*b*) and a *positive bending moment tends to bend the segment, if it were elastic, concave upward* (Fig. 7–8*c*).

PROCEDURE FOR ANALYSIS

The following procedure provides a method for constructing the shear and bending-moment diagrams for a beam.

Support Reactions. Determine all the reactive couples and forces acting on the beam and resolve the forces into components acting perpendicular and parallel to the beam's axis.

Shear and Moment Functions. Specify separate coordinates x having an origin at the beam's *left end* and extending to regions of the beam between concentrated forces and/or couples and where there is no discontinuity of distributed loading. Section the beam perpendicular to its axis at each distance x and from the free-body diagram of one of the segments, determine the unknowns V and M at the cut section as functions of x. On the free-body diagram, **V** and **M** should be shown acting in their *positive directions,* in accordance with the sign conventions given in Fig. 7–9. V is obtained from $\Sigma F_y = 0$ and M is obtained by summing moments about point S located at the cut section, $\Sigma M_S = 0$.

Shear and Moment Diagrams. Plot the shear diagram (V versus x) and the moment diagram (M versus x). If computed values of the functions describing V and M are *positive,* the magnitudes are plotted above the x axis, whereas negative values are plotted below the axis. Generally, it is convenient to plot the shear and bending-moment diagrams directly below the free-body diagram of the beam.

The following examples numerically illustrate this procedure.

Example 7–7

Draw the shear and bending-moment diagrams for the beam shown in Fig. 7–10*a*.

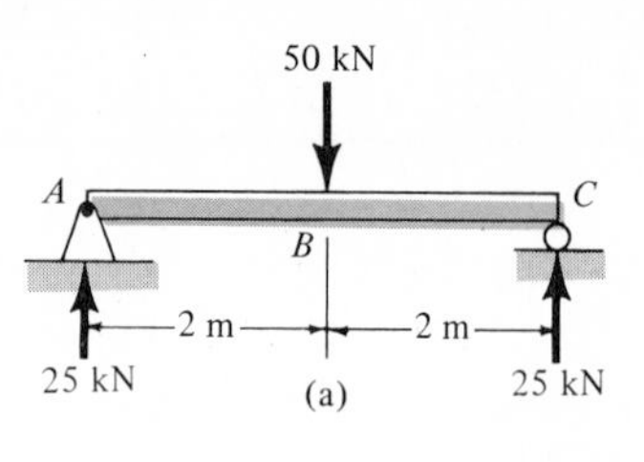

(a)

Solution

Support Reactions. The support reactions have been computed, Fig. 7–10*a*.

Shear and Moment Functions. The beam is sectioned at an arbitrary distance x from point A, extending within the region AB, and the free-body diagram of the left segment is shown in Fig. 7–10*b*. The unknowns **V** and **M** are indicated acting in the *positive direction* on the right-hand face of the segment according to the established sign convention. Why? Applying the equilibrium equations yields.

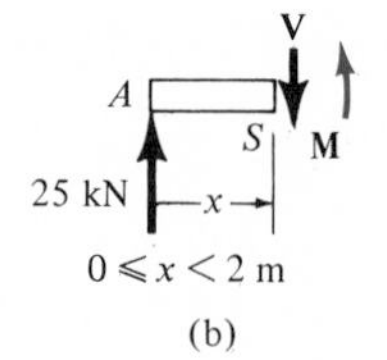

(b)

$$+\uparrow \Sigma F_y = 0; \qquad V = 25 \text{ kN} \qquad (1)$$

$$\circlearrowleft + \Sigma M_S = 0; \qquad M = 25x \text{ kN}\cdot\text{m} \qquad (2)$$

A free-body diagram for a left segment of the beam extending a distance x within the region BC is shown in Fig. 7–10*c*. As always, **V** and **M** are shown acting in the positive sense. Hence,

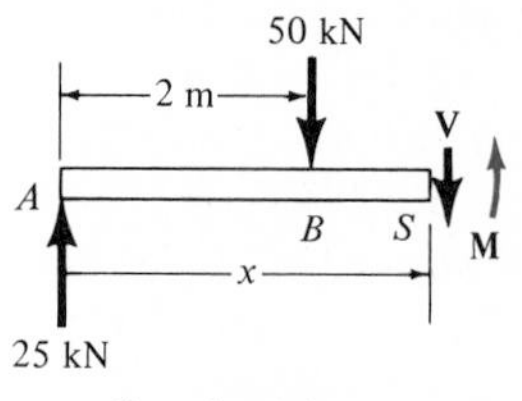

(c)

$$+\uparrow \Sigma F_y = 0; \qquad 25 - 50 - V = 0$$

$$V = -25 \text{ kN} \qquad (3)$$

$$\circlearrowleft + \Sigma M_S = 0; \qquad M + 50(x - 2) - 25(x) = 0$$

$$M = (100 - 25x) \text{ kN}\cdot\text{m} \qquad (4)$$

Shear and Moment Diagrams. When Eqs. (1) to (4) are plotted within the regions in which they are valid, the shear and bending-moment diagrams shown in Fig. 7–10*d* are obtained. The shear diagram indicates that the internal shear force is always 25 kN (positive) along beam segment AB. Just to the right of point B, the shear force changes sign and remains at a constant value of -25 kN for segment BC. The moment diagram starts at zero, increases linearly to point C at $x = 2$ m, where $M_{max} = 25$ kN(2 m) $= 50$ kN · m, and thereafter decreases back to zero.

It is seen in Fig. 7–10*d* that the graph of the shear and moment diagrams is discontinuous at points of concentrated force, i.e., points A, B, and C. For this reason, as stated earlier, it is necessary to express both the shear and bending-moment functions separately for regions between concentrated loads. It should be realized, however, that all loading discontinuities are mathematical, arising from the *idealization of a concentrated force and couple*. Physically, all loads are applied over a finite area, and if this load variation could be accounted for, all shear and bending-moment diagrams would actually be continuous over the beam's entire length.

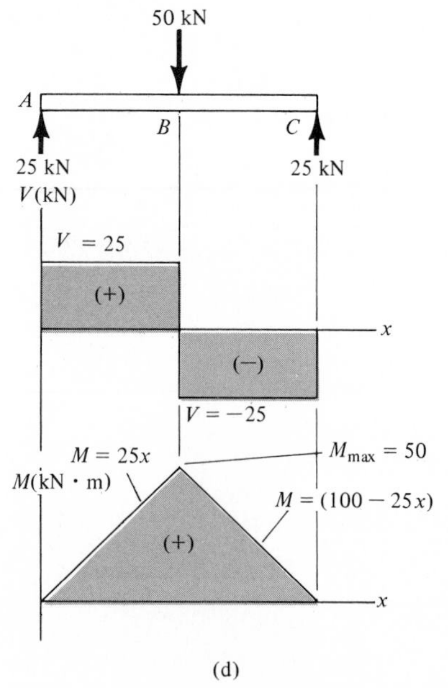

(d)

Fig. 7–10

Example 7–8

Draw the shear and bending-moment diagrams for the beam shown in Fig. 7–11*a*.

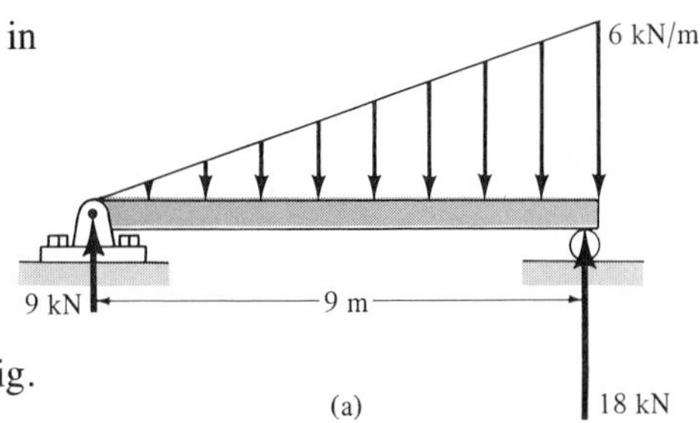

(a)

Solution

Support Reactions. The support reactions have been computed, Fig. 7–11*a*.

Shear and Moment Functions. A free-body diagram for a left segment of the beam is shown in Fig. 7–11*b*. The distributed loading acting on this segment, which has an intensity of $\frac{2}{3}x$ at its end, is replaced by a concentrated force *after* the segment is isolated as a free-body diagram. Since the distributed load now has a length x, the *magnitude* of the *concentrated force* is equal to $\frac{1}{2}(x)(\frac{2}{3}x) = \frac{1}{3}x^2$. This force *acts through the centroid* of the distributed loading area, a distance $\frac{1}{3}x$ from the right end. Applying the two equations of equilibrium yields

(b)

$$+\uparrow \Sigma F_y = 0; \qquad 9 - \frac{1}{3}x^2 - V = 0$$

$$V = \left(9 - \frac{x^2}{3}\right) \text{ kN} \tag{1}$$

$$\circlearrowleft +\Sigma M_S = 0; \qquad M + \frac{1}{3}x^2\left(\frac{x}{3}\right) - 9x = 0$$

$$M = \left(9x - \frac{x^3}{9}\right) \text{ kN}\cdot\text{m} \tag{2}$$

Shear and Moment Diagrams. The shear and bending-moment diagrams shown in Fig. 7–11*c* are obtained by plotting Eqs. (1) and (2).

The point of *zero shear* can be found using Eq. (1):

$$V = 9 - \frac{x^2}{3} = 0$$

$$x = 5.20 \text{ m}$$

This value of x happens to represent the point on the beam where the *maximum moment* occurs (see Sec. 7–3). Using Eq. (2), we have

$$M_{max} = \left(9(5.20) - \frac{(5.20)^3}{9}\right) \text{ kN}\cdot\text{m}$$

$$= 31.18 \text{ kN}\cdot\text{m}$$

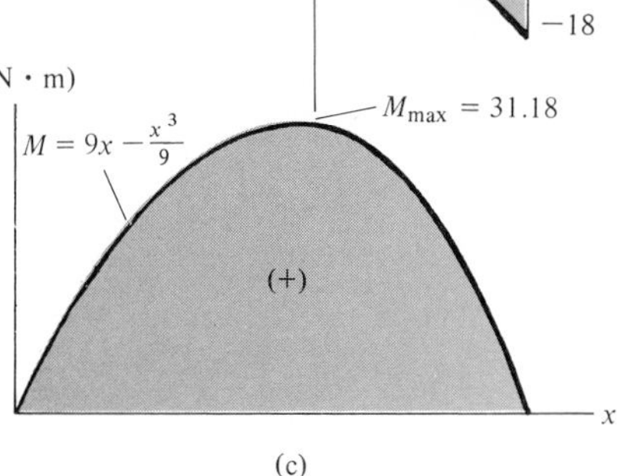

(c)

Fig. 7–11

★7.3 Relations Between Distributed Load, Shear, and Bending Moment

In cases where a beam is subjected to several concentrated forces, couples, and distributed loads, the method of constructing the shear and bending-moment diagrams discussed in Sec. 7–2 may become quite tedious. In this section a simpler method for constructing these diagrams is discussed—a method based upon differential relations that exist between the load, shear, and bending moment.

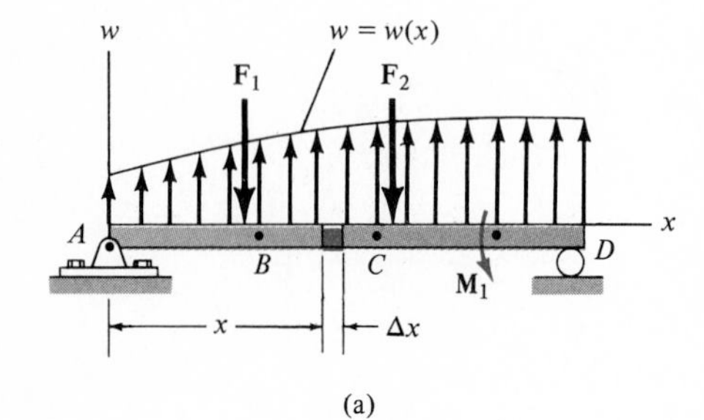

(a)

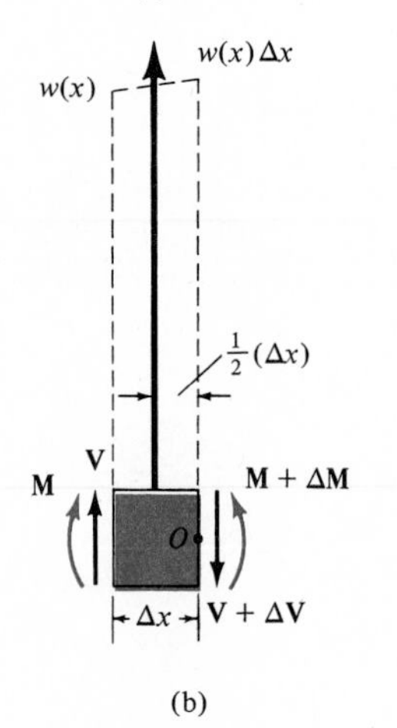

(b)

Fig. 7–12

Consider the beam AD shown in Fig. 7–12a, which is subjected to an arbitrary distributed loading $w = w(x)$ and a series of concentrated loadings. In the following discussion, *the distributed load will be considered positive when the loading acts upward* as shown. The free-body diagram for a small segment of the beam having a length Δx is shown in Fig. 7–12b. Since this segment has been chosen at a point x along the beam which is *not* subjected to a concentrated force or couple, any results obtained will not apply at points of concentrated loading. The internal shear force and bending moment shown on the free-body diagram are assumed to act in the *positive direction* according to the established sign convention, Fig. 7–9. Note that both the shear force and moment acting on the right-hand face must be increased by a small, finite amount in order to keep the segment in equilibrium. The distributed loading has been replaced by a concentrated force $w(x)\,\Delta x$ that acts at a fractional distance $\frac{1}{2}(\Delta x)$ from the right end. Applying the equations of equilibrium, we have

$$+\uparrow \Sigma F_y = 0; \qquad V + w(x)\,\Delta x - (V + \Delta V) = 0$$

$$\Delta V = w(x)\,\Delta x$$

$$\circlearrowleft\!+\Sigma M_O = 0; \quad -V\,\Delta x - M - w(x)\,\Delta x[\tfrac{1}{2}(\Delta x)] + (M + \Delta M) = 0$$

$$\Delta M = V\Delta x + w(x)\tfrac{1}{2}(\Delta x)^2$$

Dividing by Δx and realizing that Δx is small, the term $w(x)\frac{1}{2}(\Delta x)$ approaches zero and can be neglected. Thus,

$$\frac{\Delta V}{\Delta x} = w(x) \tag{7–1}$$

slope of shear diagram = distributed load intensity

and

$$\frac{\Delta M}{\Delta x} = V \tag{7–2}$$

slope of moment diagram = shear

These two equations provide a convenient means for plotting the shear and bending-moment diagrams. For a specific point in the beam, Fig. 7–13*a*, Eq. 7–1 states that the *slope* of the *shear diagram* is equal to the intensity of the *distributed loading* at the point, Fig. 7–13*b*, while Eq. 7–2 states that the *slope* of the *moment diagram* is equal to the intensity of *shear force* at the point, Fig. 7–13*c*. In particular, if the shear is equal to zero, $\Delta M/\Delta x = 0$, and therefore points of zero shear correspond to points of maximum (or possibly minimum) moment.

Equations 7–1 and 7–2 may be rewritten in the form $\Delta V = w(x)\,\Delta x$ and $\Delta M = V\,(\Delta x)$. Noting that $w(x)\,\Delta x$ and $V\,(\Delta x)$ represent the colored segmental areas under the distributed loading curve and the shear diagram, we can sum these areas between two points B and C on the beam and write

$$\Delta V_{BC} = \Sigma w(x)(\Delta x)$$

change in shear = area under loading curve (7–3)

and

$$\Delta M_{BC} = \Sigma V\,(\Delta x)$$

change in moment = area under shear diagram (7–4)

Equation 7–3 states that the *change in shear* between B and C is equal to the *area* under the *distributed loading curve* between these points. Similarly, from Eq. 7–4, the *change in moment* between B and C is equal to the *area* under the *shear diagram* from B to C.

From the above derivation and the discussion in Sec. 7–2, it should be noted that Eqs. 7–1 and 7–3 cannot be used at points where a concentrated force acts, since these equations do not account for the sudden change in shear at these points. Similarly, because of a discontinuity of moment, Eqs. 7–2 and 7–4 cannot be used at points where a couple is applied.

The following two examples illustrate the application of these equations for the construction of the shear and bending-moment diagrams. After working through these examples, it is recommended that Examples 7–7 and 7–8 be solved using this method.

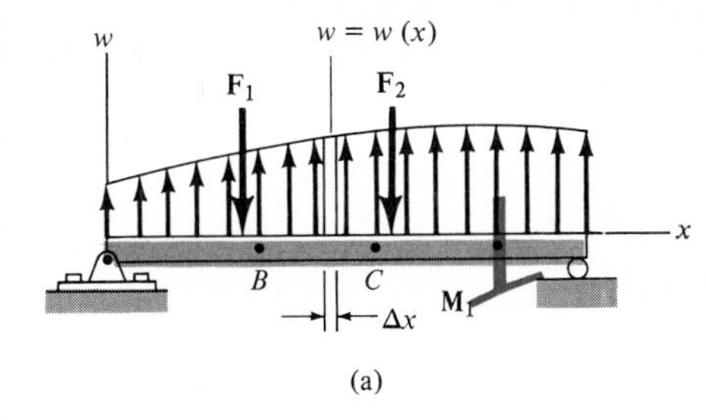

(a)

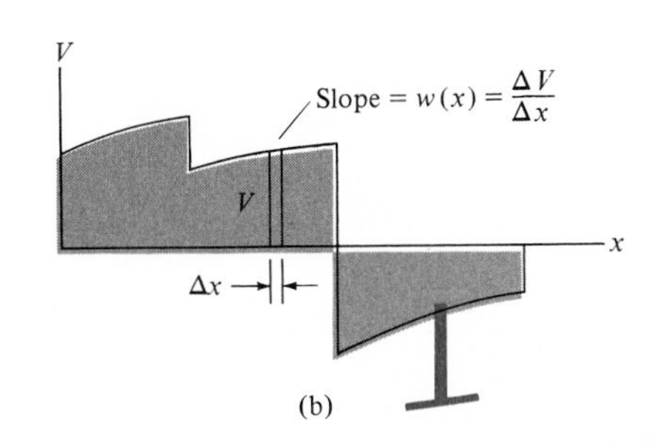

(b)

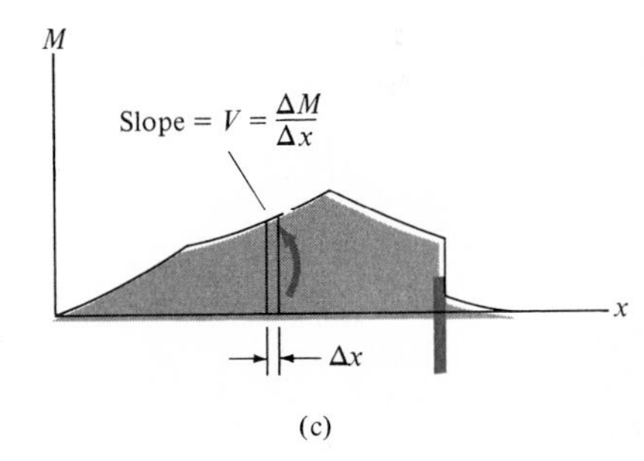

(c)

Fig. 7–13

Example 7–9

Draw the shear and moment diagrams for the beam in Fig. 7–14*a*.

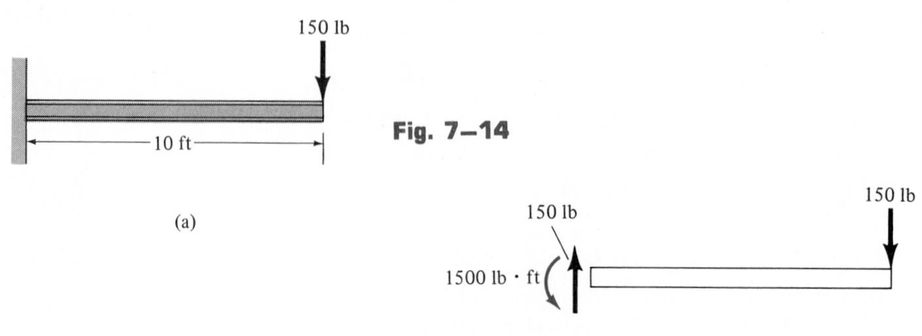

Fig. 7–14

Solution

Support Reactions (Fig. 7–14*b*). The reactions are calculated and shown on a free-body diagram of the beam.

(b)

Shear Diagram (Fig. 7–14*c*). Using the established sign convention, Fig. 7–9, the shear at the ends of the beam is plotted first, i.e., $x = 0$, $V = +150$ lb; $x = 10$ ft, $V = +150$ lb. Since $w = 0$ for $0 < x < 10$ ft, the slope of the shear diagram is zero ($\Delta V/\Delta x = w = 0$), and therefore a horizontal straight line connects the plotted points.

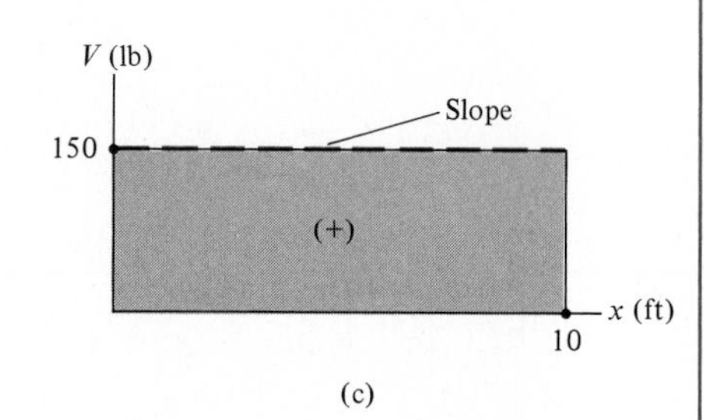

(c)

Moment Diagram (Fig. 7–14*d*). Again, from the established sign convention, Fig. 7–9, the moment at the ends of the beam is plotted first, i.e., $x = 0$, $M = -1500$ lb · ft; $x = 10$ ft, $M = 0$. The shear diagram is constant, $V = +150$ lb, for $0 < x < 10$ ft, so that the slope of the moment diagram will also be constant (positive), ($\Delta M/\Delta x = V = +150$). Hence, the plotted points are connected by a straight positive sloped line as shown in the figure.

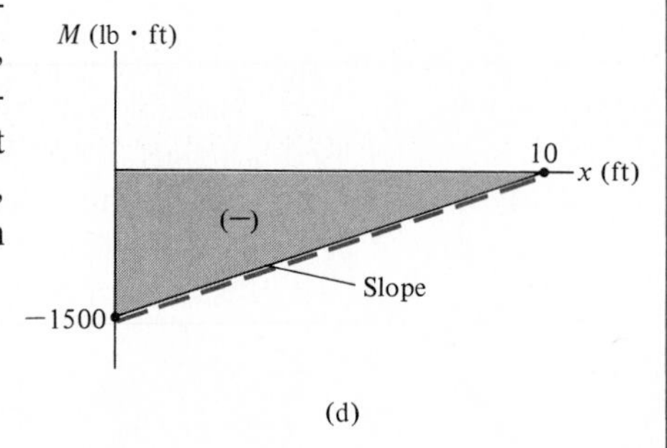

(d)

Example 7–10

Draw the shear and moment diagrams for the beam shown in Fig. 7–15*a*.

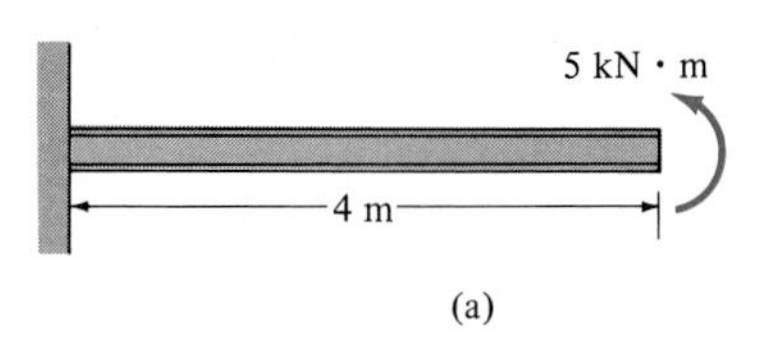

(a)

Fig. 7–15

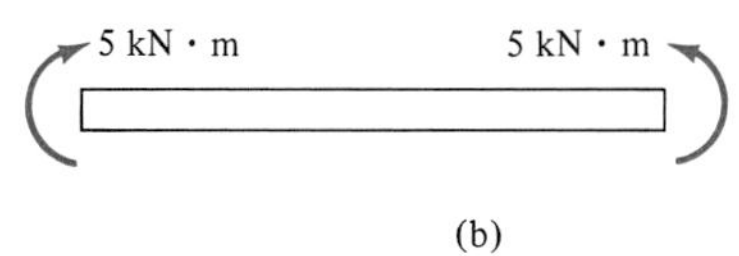

(b)

Solution

Support Reactions (Fig. 7–15*b*). The reactions at the fixed support have been calculated and are shown on the free-body diagram of the beam.

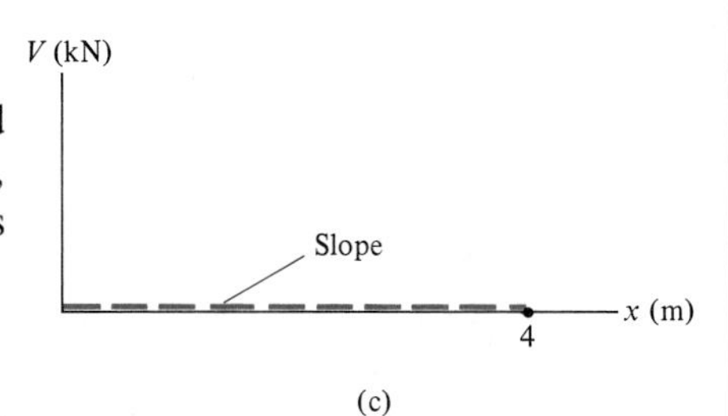

(c)

Shear Diagram (Fig. 7–15*c*). The shear at the end points $x = 0$ and $x = 4$ m is plotted first. Since no load exists on the beam for $0 < x < 4$ m, the slope $\Delta V/\Delta x = 0$. Therefore, the line which connects the end points indicates that the shear is zero throughout the beam.

Moment Diagram (Fig. 7–15*d*). The moment $+5$ kN · m at the beam's end points $x = 0$ and $x = 4$ m is plotted first. The slope $\Delta M/\Delta x = 0$ for the moment diagram since the shear is zero for $0 < x < 4$ m. Therefore, the moment diagram is rectangular.

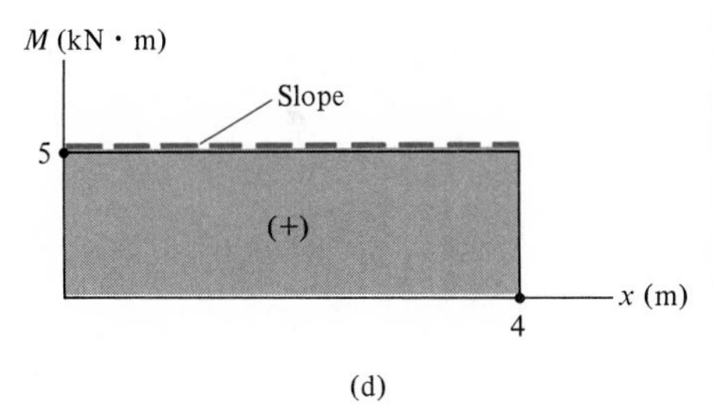

(d)

Example 7–11

Draw the shear and moment diagrams for the beam shown in Fig. 7–16*a*.

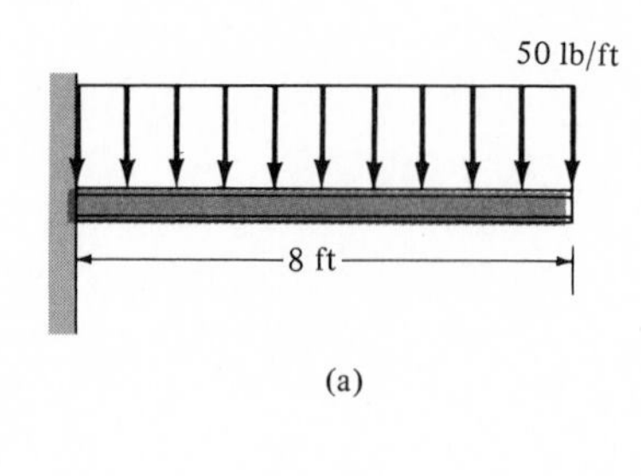

(a)

Fig. 7–16

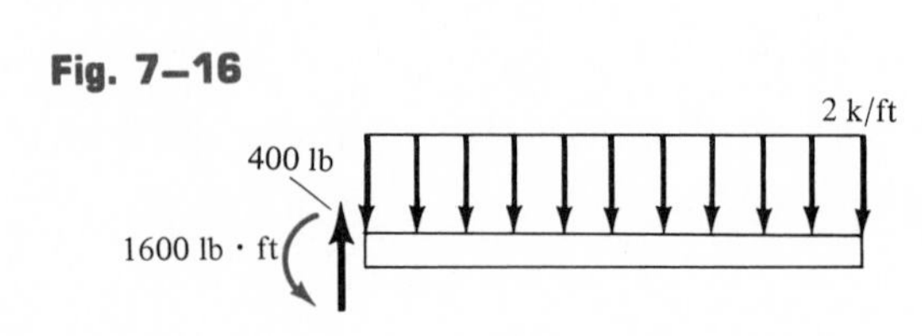

(b)

Solution

Support Reactions (Fig. 7–16*b*). The reactions at the fixed support have been calculated and are shown on the free-body diagram of the beam.

Shear Diagram (Fig. 7–16*c*). The shear at the end points $x = 0$ and $x = 8$ ft is plotted first. Here $\Delta V/\Delta x = -50$, so that a straight *negative* sloping line connects the end points.

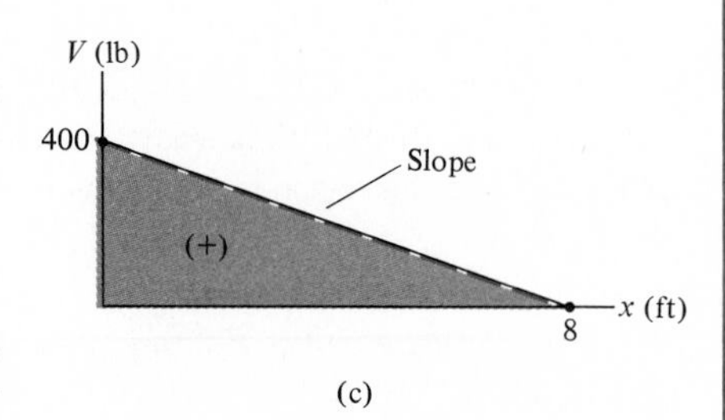

(c)

Moment Diagram (Fig. 7–16*d*). The moment at the beam's end points $x = 0$ and $x = 8$ ft is plotted first. Successive values of shear on the shear diagram indicate the slope $\Delta M/\Delta x = V$ is always positive yet *linearly decreasing* from $\Delta M/\Delta x = 400$ at $x = 0$ to $\Delta M/\Delta x = 0$ at $x = 8$ ft. Thus, a parabola having this characteristic connects the end points.

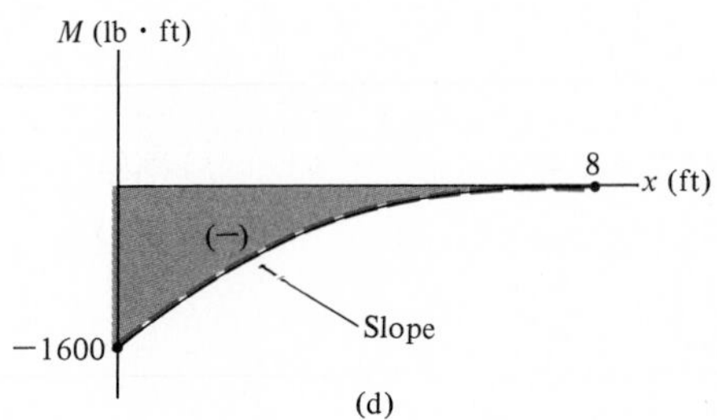

(d)

Example 7–12

Draw the shear and moment diagrams for the beam shown in Fig. 7–17*a*.

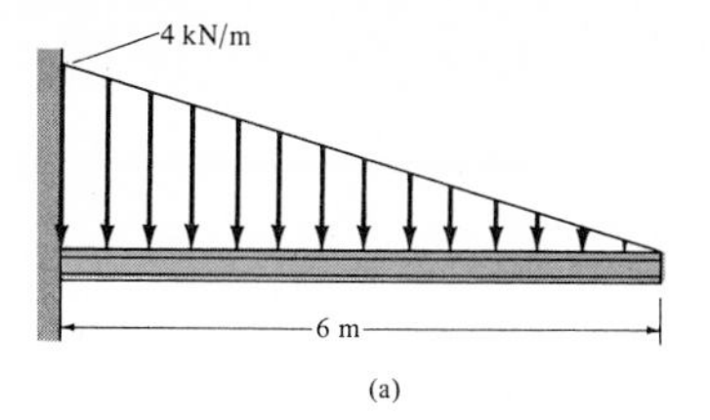

(a)

Fig. 7–17

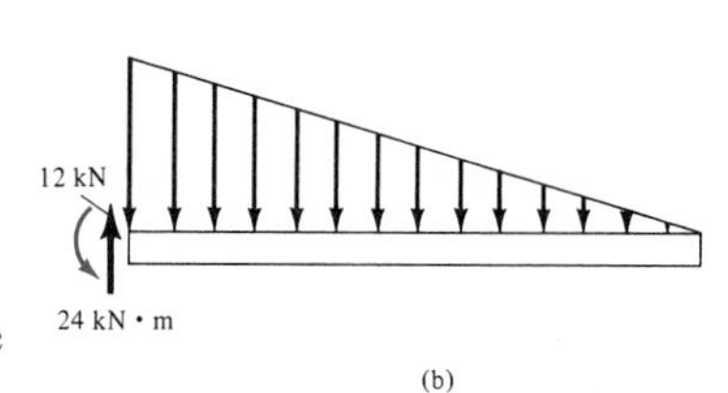

(b)

Solution

Support Reactions (Fig. 7–17*b*). The reactions at the fixed support have been calculated and are shown on the free-body diagram of the beam.

Shear Diagram (Fig. 7–17*c*). The shear at the end points $x = 0$ and $x = 6$ m is plotted first. From the load diagram, the slope $\Delta V/\Delta x = w$ is negatively decreasing from $\Delta V/\Delta x = -4$ at $x = 0$ to $\Delta V/\Delta x = 0$ at $x = 6$ m. A parabola having this characteristic therefore connects the end points of the shear diagram.

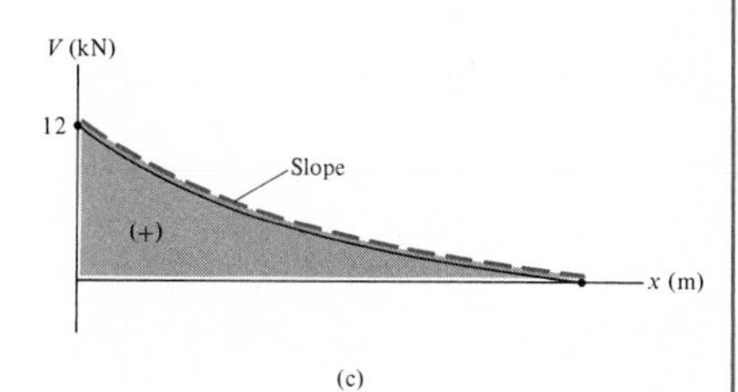

(c)

Moment Diagram (Fig. 7–17*d*). The moment at the beam's end points $x = 0$ and $x = 6$ m is plotted first. From the shear diagram the slope varies parabolically, from $\Delta M/\Delta x = +12$ at $x = 0$ to $\Delta M/\Delta x = 0$ at $x = 6$ m. The curve connecting the plotted end points that has this characteristic is a cubic function of x as shown in the figure.

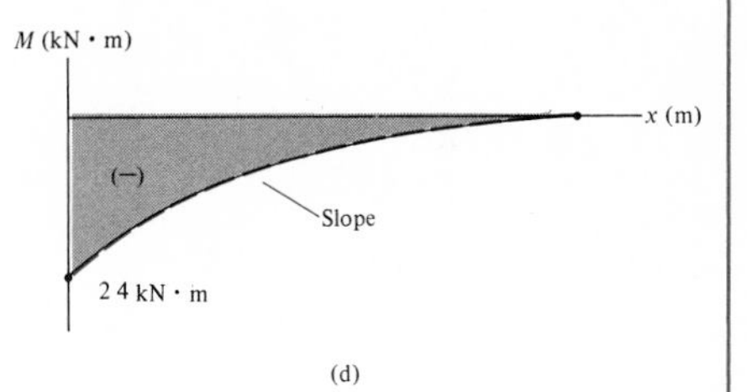

(d)

Example 7–13

Draw the shear and moment diagrams for the beam in Fig. 7–18*a*.

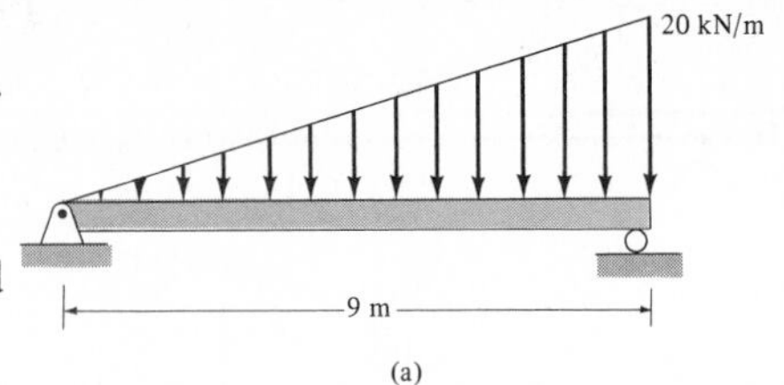

(a)

Solution

Support Reactions (Fig. 7–18*b*). The reactions have been calculated and are shown on the free-body diagram of the beam.

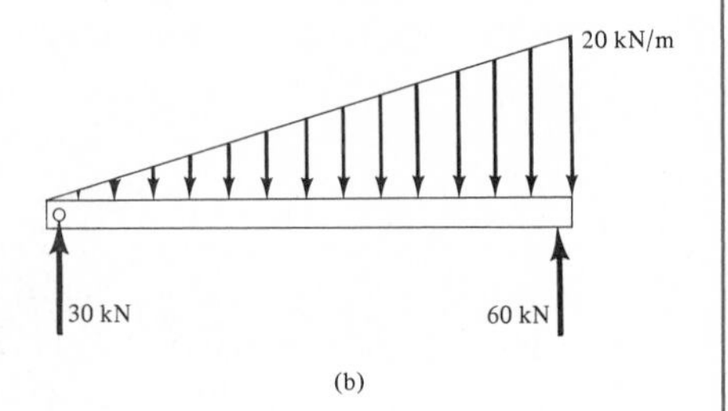

(b)

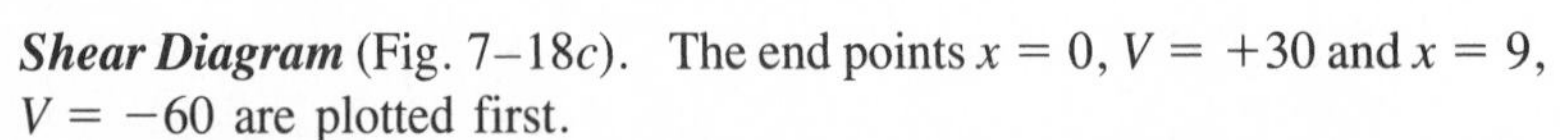

Shear Diagram (Fig. 7–18*c*). The end points $x = 0, V = +30$ and $x = 9$, $V = -60$ are plotted first.

As shown on the load diagram, the slope of the shear diagram varies from $\Delta V/\Delta x = 0$ at $x = 0$ to $\Delta V/\Delta x = -20$ at $x = 9$. For $0 \leqslant x \leqslant 9$ the slope is increasingly negative since the distributed load is increasingly negative ($\Delta V/\Delta x = w$).

The point of zero shear can be found by using the method of sections for a beam segment of length x, Fig. 7–18*c*. We require $V = 0$ so that

$$+\uparrow \Sigma F_y = 0; \quad 30 - \frac{1}{2}\left[20\left(\frac{x}{9}\right)\right]x = 0 \qquad x = 5.20 \text{ m}$$

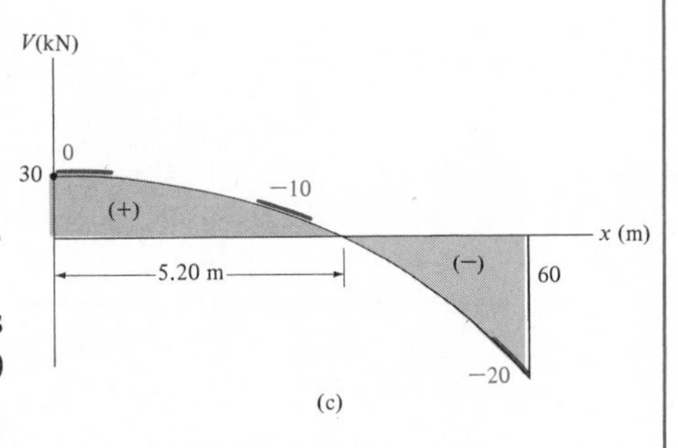

(c)

Moment Diagram (Fig. 7–18*d*). The end points $x = 0, M = 0$ and $x = 0$, $M = 0$ are plotted first.

From the shear diagram the slope of the moment diagram is $\Delta M/\Delta x = +30$ at $x = 0$ and $\Delta M/\Delta x = 0$ at $x = 5.20$. For $0 \leqslant x \leqslant 5.20$ the slope is decreasingly positive since the shear is decreasingly positive. Likewise, for $5.20 \leqslant x \leqslant 9$ the slope is increasingly negative.

The maximum value of moment is at $x = 5.20$ since $\Delta M/\Delta x = V = 0$ at this point. From the free-body diagram in Fig. 7–18*e* we have

$$\circlearrowleft +\Sigma M_S = 0; \quad -30(5.20) + \frac{1}{2}\left[20\left(\frac{5.20}{9}\right)\right](5.20)\left(\frac{5.20}{3}\right) + M = 0$$

$$M = 103.9 \text{ kN} \cdot \text{m}$$

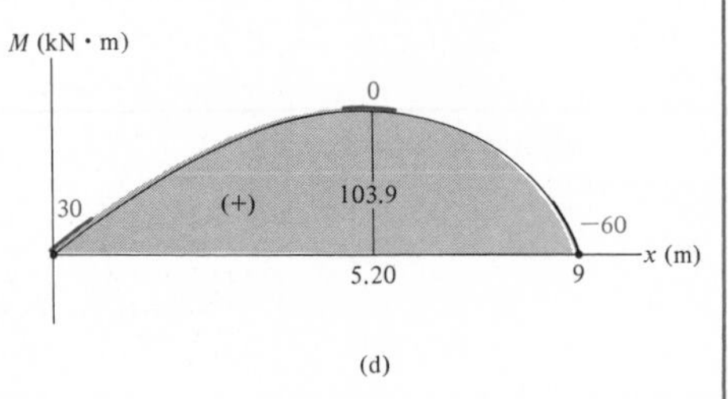

(d)

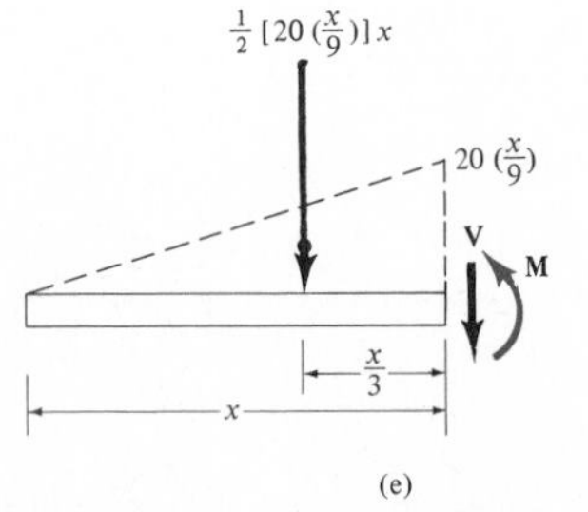

(e)

Fig. 7–18

Example 7–14

Sketch the shear and bending-moment diagrams for the beam shown in Fig. 7–19*a*.

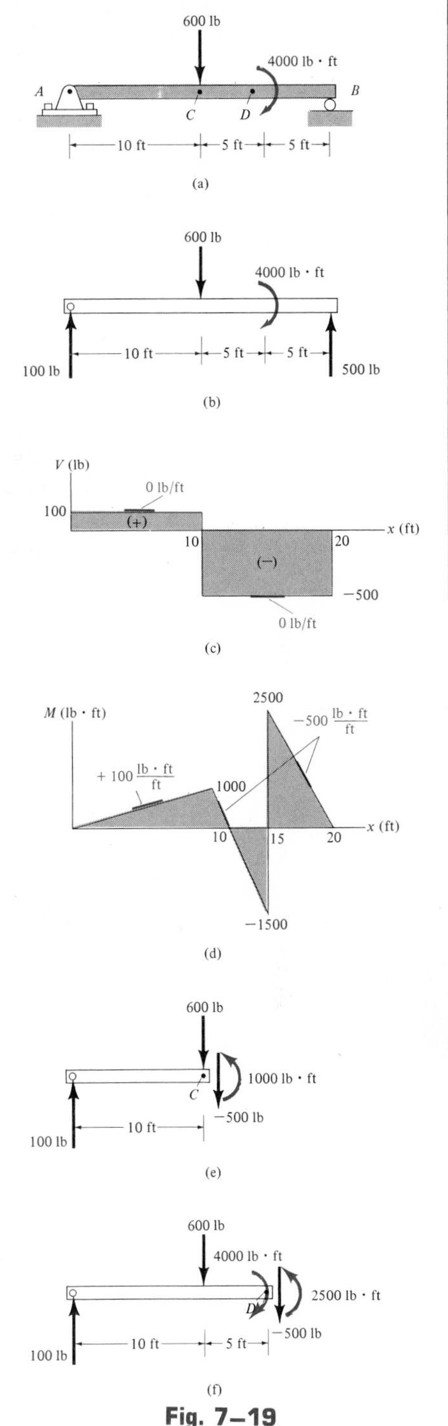

Fig. 7–19

Solution

Support Reactions. The reactions are calculated and indicated on the free-body diagram, Fig. 7–19*b*.

Shear Diagram. The values of the shear at the end points A and B are plotted first. From the sign convention, Fig. 7–9, $V_A = +100$ kN and $V_B = -500$ kN, Fig. 7–19*c*. At an intermediate point between A and C the slope of the shear diagram is zero since $w(x) = \Delta V/\Delta x = 0$. Hence the shear retains its value of $+100$ kN within this region. At C the shear is *discontinuous* since there is a *concentrated force* of 600 lb there. The value of the shear just to the right of C ($+500$ lb) can be found by sectioning the beam at this point. This yields the free-body diagram shown in Fig. 7–19*e*. This point ($V = -500$ lb) is plotted on the shear diagram. As before, $w(x) = 0$ from C to B, so the slope $\Delta V/\Delta x = 0$. The diagram closes to the value of -500 lb at B as shown. One might wonder why no jump or discontinuity in shear occurs at D, the point where the 4000 lb · ft moment is applied, Fig. 7–19*a*. Such is not the case as indicated by the equilibrium conditions on the free-body diagram in Fig. 7–19*f*. It may also be noted that the shear diagram can be constructed quickly by ''following the load'' on the free-body diagram, Fig. 7–19*b*. In this regard, beginning at A the 100-lb force acts upward, so $V_A = +100$ lb. No load acts between A and C, so the shear remains constant. At C the 600-lb force is down, so the shear jumps down 600 lb, from 100 lb to -500 lb. Again the shear is constant (no load) and ends at -500 lb, point B, which closes the diagram back to zero since the 500-lb force on the beam acts upward.

Moment Diagram. Since the moment at each end of the beam is zero, these two points are plotted first, Fig. 7–19*d*. The slope of the moment diagram from A to C is constant since $\Delta M/\Delta x = V = +100$. The value of the moment at C can be determined by the method of sections, Fig. 7–19*e*, or by computing the area under the shear diagram between A and C, i.e., $\Delta M_{AC} = M_C - M_A = (100 \text{ lb})(10 \text{ ft}) = 1000 \text{ lb} \cdot \text{ft}$. Since $M_A = 0$, then $M_C = 0 + 1000 \text{ lb} \cdot \text{ft} = 1000 \text{ lb} \cdot \text{ft}$. From C to D the slope is $\Delta M/\Delta x = V = -500$, Fig. 7–19*c*. The area under the shear diagram between points C and D is $\Delta M_{CD} = M_D - M_C = (-500 \text{ lb})(5 \text{ ft}) = -2500 \text{ lb} \cdot \text{ft}$. Thus $M_D = 1000 - 2500 = -1500 \text{ lb} \cdot \text{ft}$. A jump occurs at point D due to the concentrated couple moment of 4000 lb · ft. The method of sections, Fig. 7–19*f*, gives a value of $+2500$ lb · ft just to the right of D. From this point, the slope of $\Delta M/\Delta x = -500$ is maintained until the diagram closes to zero at B, Fig. 7–19*d*.

Example 7–15

Draw the shear and moment diagrams for the beam in Fig. 7–20*a*.

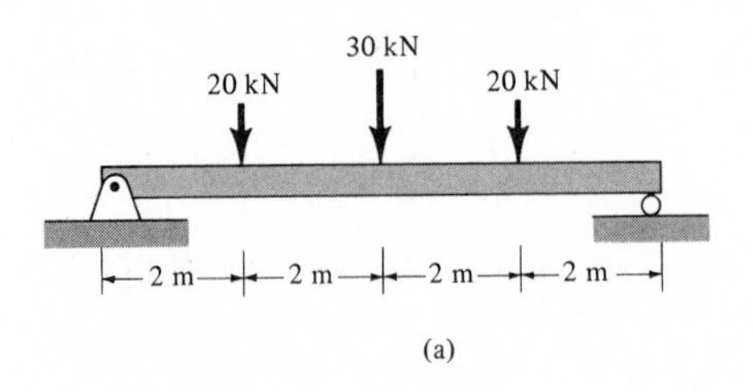

(a)

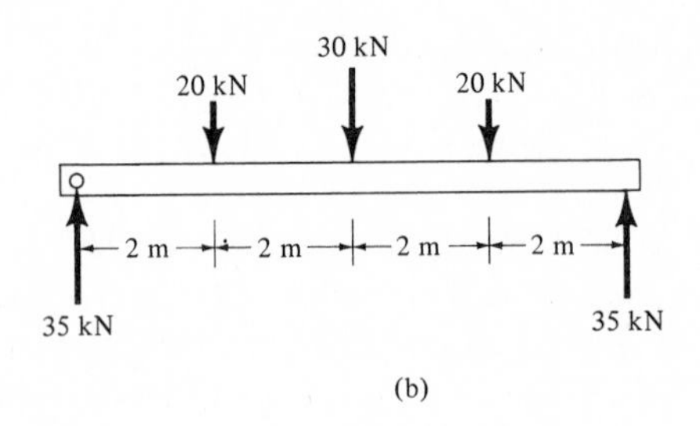

(b)

Solution

Support Reactions (Fig. 7–20*b*). The reactions at the supports are shown on the free-body diagram.

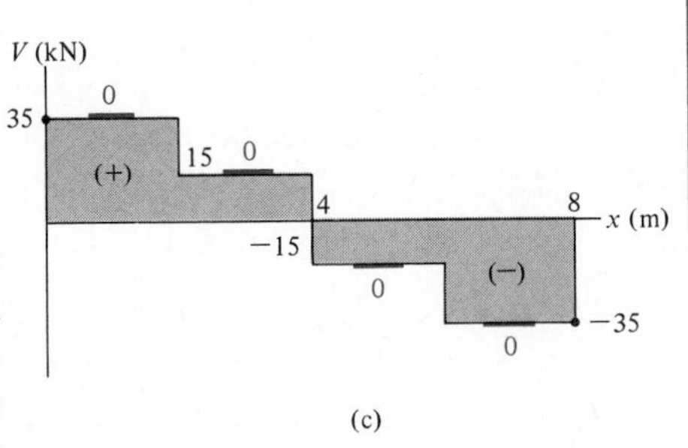

(c)

Shear Diagram (Fig. 7–20*b*). The end points $x = 0$, $V = +35$ and $x = 8$, $V = -35$ are plotted first.

Since there is no distributed load on the beam, the slope of the shear diagram throughout the beam's length is zero, i.e., $\Delta V/\Delta x = 0$. There is a discontinuity or "jump" of the shear diagram, however, at each concentrated force. Using the procedure of the previous example, an easy way to plot these jumps is to note that the "jumps" follow the load; that is, when the force is downward, the jump is downward, and vice versa. Thus, the 20-kN force at $x = 2$ m "pushes" the shear force down from 35 kN to 15 kN; the 30-kN force at $x = 4$ m "pushes" the shear down from 15 kN to -15 kN, etc. We can also obtain numerical values for the shear at each point on the beam by using the method of sections as indicated for $x = 2^+$ m in Fig. 7–20*e*.

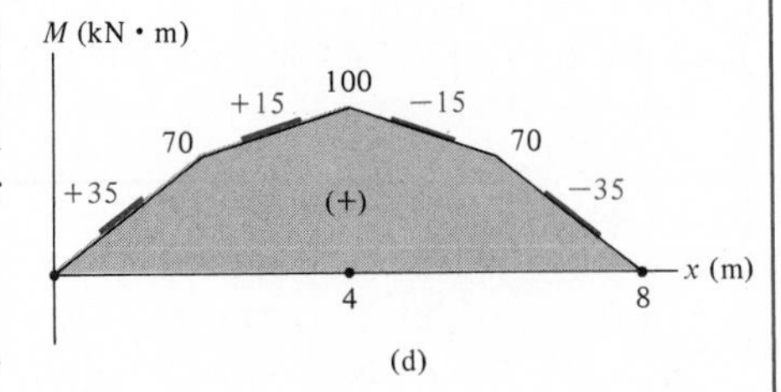

(d)

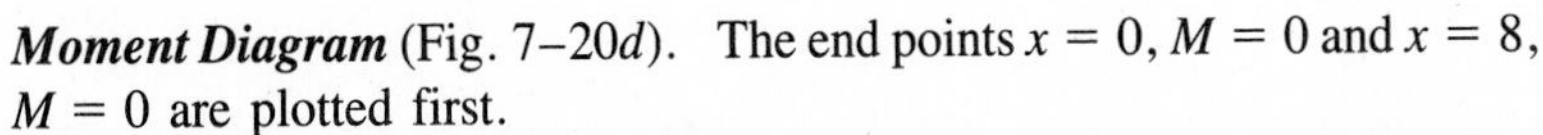

Moment Diagram (Fig. 7–20*d*). The end points $x = 0$, $M = 0$ and $x = 8$, $M = 0$ are plotted first.

Since the shear is constant in each region of the beam, the moment diagram has a corresponding constant positive or negative slope as indicated on the diagram. Numerical values of the moment at any point can be computed from the area under the shear diagram or by the method of sections. For example, at $x = 2$ m, $\Delta M = \Sigma V(\Delta x) = 35(2) = 70$, so $M|_{x=2} = M|_{x=0} + 70 = 0 + 70 = 70$. Also, see Fig. 7–20*e*.

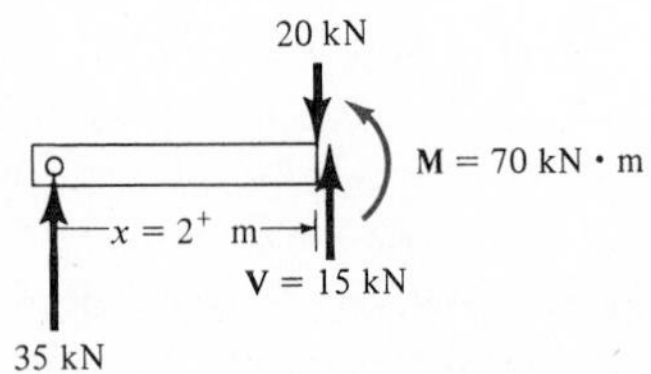

(e)

Fig. 7–20

Example 7–16

Draw the shear and bending-moment diagrams for the beam shown in Fig. 7–21*a*.

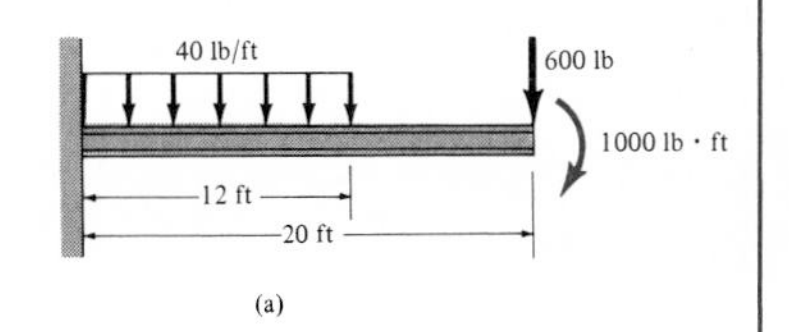

(a)

Solution

Support Reactions (Fig. 7–21*b*). The reactions at the fixed support have been calculated and are shown on the free-body diagram of the beam.

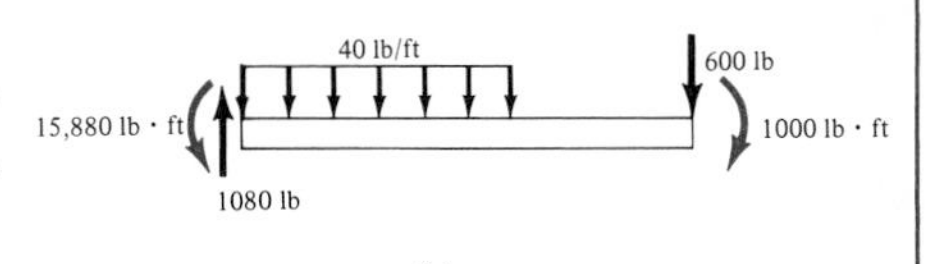

(b)

Shear Diagram (Fig. 7–21*c*). Using the established sign convention, Fig. 7–9, the shear at the ends of the beam is plotted first, i.e., $x = 0$, $V = +1080$; $x = 20$, $V = +600$.

Since the uniform distributed load is *negative* (downward) and *constant*, then the slope of the shear diagram is $\Delta V/\Delta x = w = -40$ for $0 \leq x < 12$ as indicated.

The magnitude of shear at $x = 12$ is $+600$. This can be determined by first finding the area under the load diagram between $x = 0$ and $x = 12$, i.e., $\Delta V = \Sigma w(x)(\Delta x) = (-40)(12) = -480$. Thus $V|_{x=12} = V|_{x=0} + (-480) = 1080 - 480 = 600$. Also, we can obtain this value by using the method of sections, Fig. 7–21*e*, where for equilibrium $V = +600$.

Since the load between $12 < x \leq 20$ is $w = 0$, then $\Delta V/\Delta x = 0$ as indicated. This brings the value of the shear to the required value of $V = 60$ at $x = 20$.

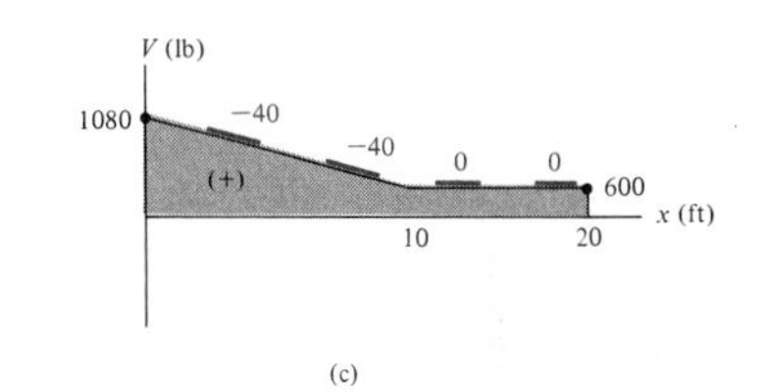

(c)

Moment Diagram (Fig. 7–21*d*). Again, using the established sign convention, Fig. 7–9, the moments at the ends of the beam are plotted first, i.e., $x = 0$, $M = -15{,}880$; $x = 20$, $M = -1000$.

Numerical values of shear change from $+1080$ to $+600$ for $0 \leq x < 12$. Each value of shear gives the slope of the moment diagram since $\Delta M/\Delta x = V$. As indicated, at $x = 0$, $\Delta M/\Delta x = +1080$ and at $x = 12$, $\Delta M/\Delta x = +600$. For $0 < x < 12$ specific values of the shear diagram are positive but linearly decreasing. Hence, the moment diagram is parabolic with a linear decreasing slope.

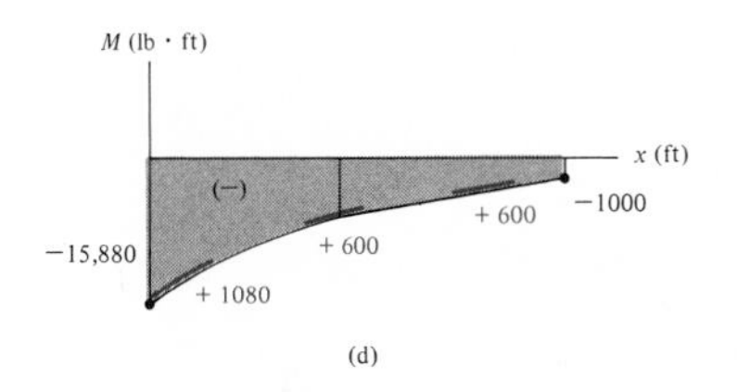

(d)

The magnitude of moment at $x = 12$ ft is -5800. This can be found from the area under the shear diagram, i.e., $\Delta M = \Sigma V(\Delta x) = 600(12) + \frac{1}{2}(1080 - 600)(12) = +10{,}080$, so that $M|_{x=12} = M|_{x=0} + 10{,}080 = -15{,}880 + 10{,}080 = -5800$. The more "basic" method of sections can also be used, where equilibrium at $x = 12$ requires $M = -5800$, Fig. 7–21*e*.

The moment diagram has a constant slope for $12 < x \leq 20$ since $\Delta M/\Delta x = V = +600$. This brings the value of $M = -1000$ at $x = 20$, as required.

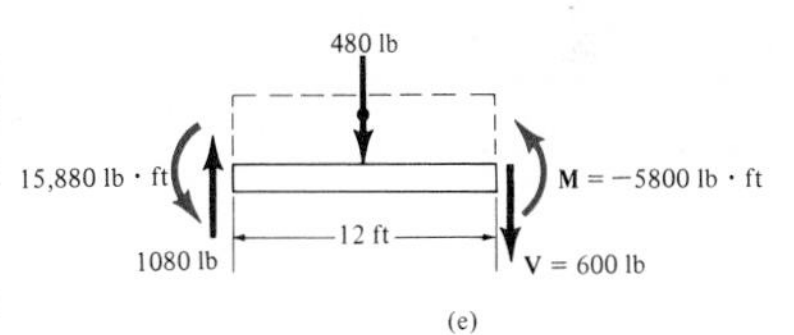

(e)

Fig. 7–21

Problems

*7–20. Draw the shear and bending-moment diagrams for the beam.

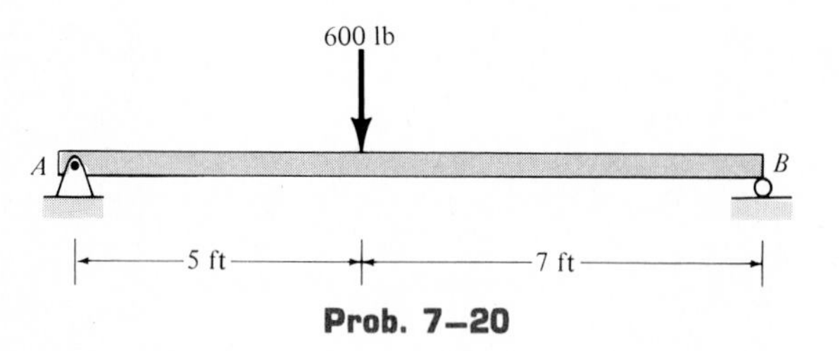

Prob. 7–20

7–21. Draw the shear and bending-moment diagrams for the cantilever beam.

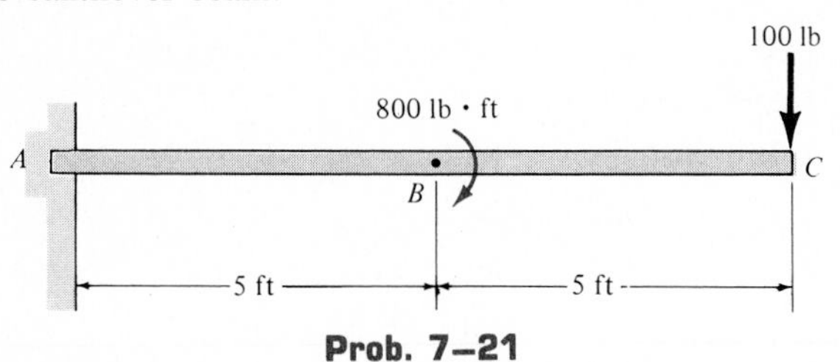

Prob. 7–21

7–22. Draw the shear and bending-moment diagrams for the beam.

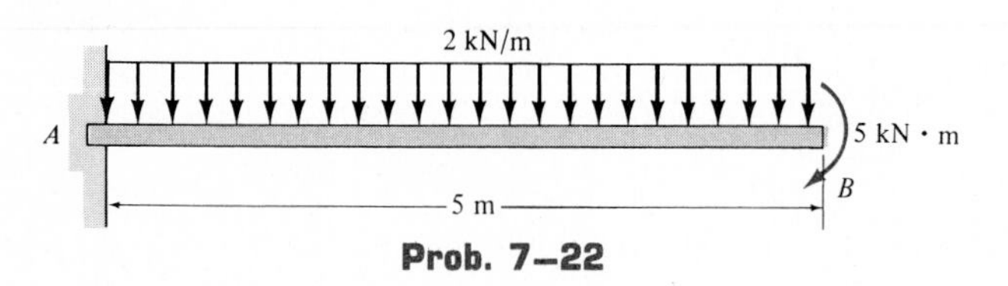

Prob. 7–22

7–23. Draw the shear and bending-moment diagrams for the beam.

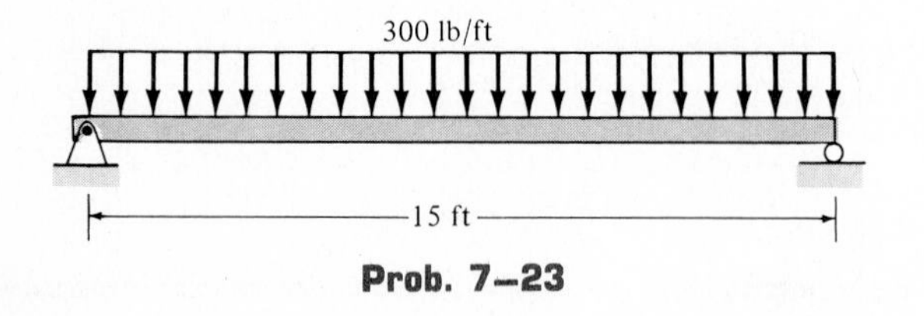

Prob. 7–23

*7–24. Draw the shear and bending-moment diagrams for the beam.

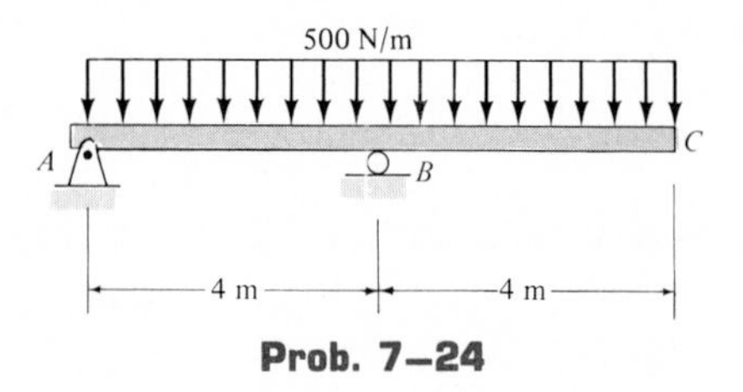

Prob. 7–24

7–25. Draw the shear and bending-moment diagrams for the beam.

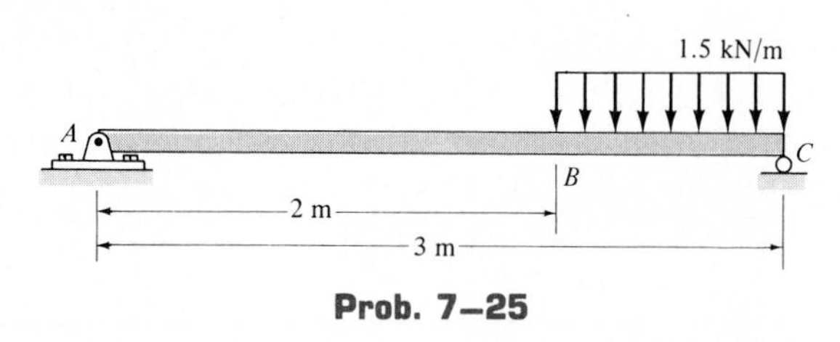

Prob. 7–25

7–26. Draw the shear and bending-moment diagrams for the beam.

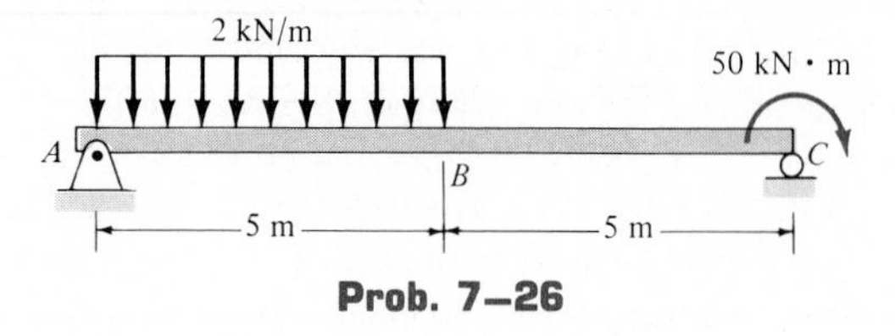

Prob. 7–26

7–27. Draw the shear and bending-moment diagrams for the beam.

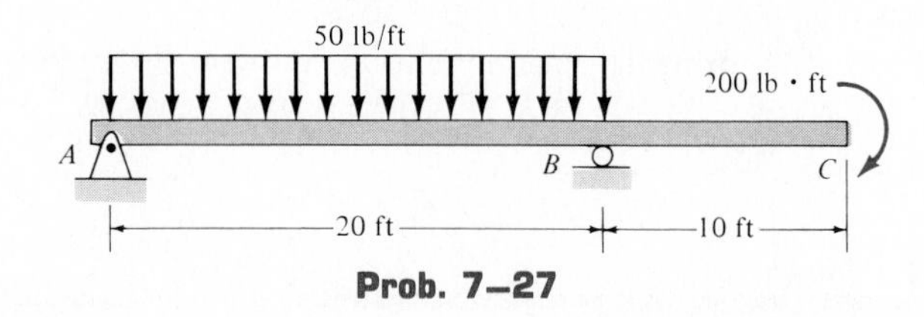

Prob. 7–27

*7–28. Draw the shear and bending-moment diagrams for the beam.

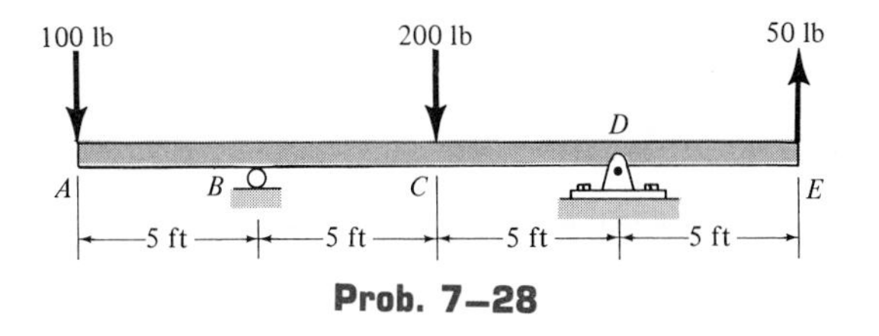

Prob. 7–28

7–29. Draw the shear and bending-moment diagrams for the beam.

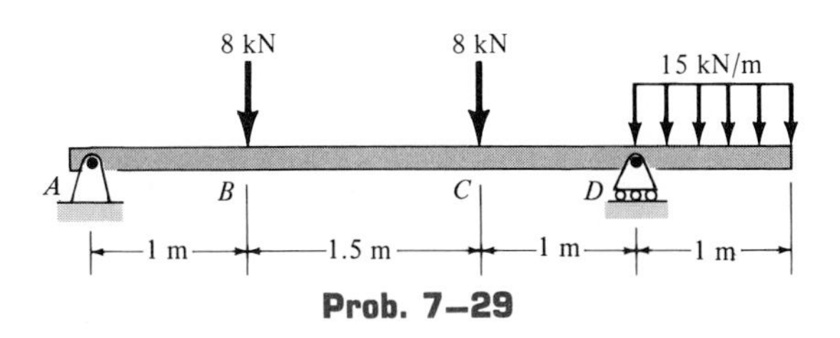

Prob. 7–29

7–30. Draw the shear and bending-moment diagrams for the beam.

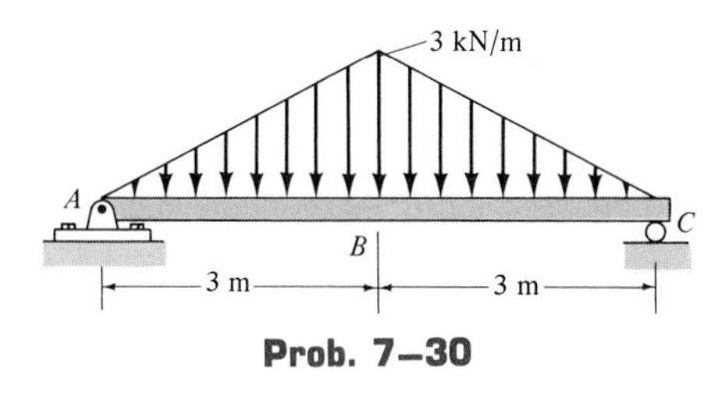

Prob. 7–30

7–31. Draw the shear and bending-moment diagrams for the beam.

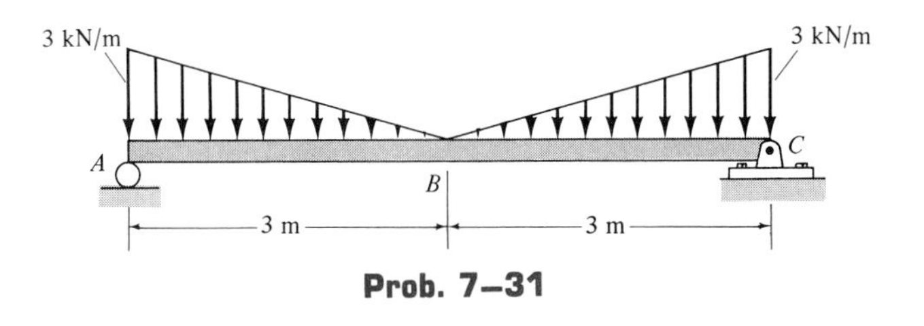

Prob. 7–31

*7–32. Draw the shear and bending-moment diagrams for the beam.

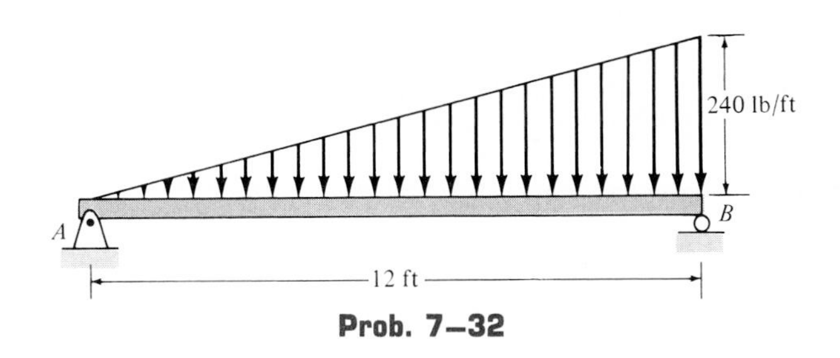

Prob. 7–32

7–33. Draw the shear and bending-moment diagrams for the beam.

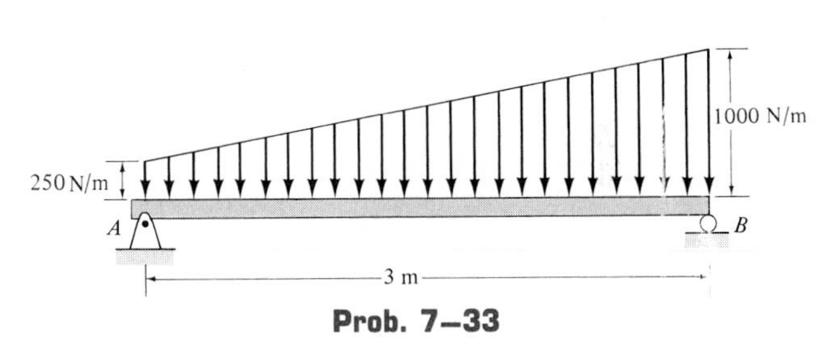

Prob. 7–33

7–34. Draw the shear and bending-moment diagrams for the lathe spindle if it is subjected to the concentrated loads shown. The vertical equilibrium forces at bearings A and B are to be calculated.

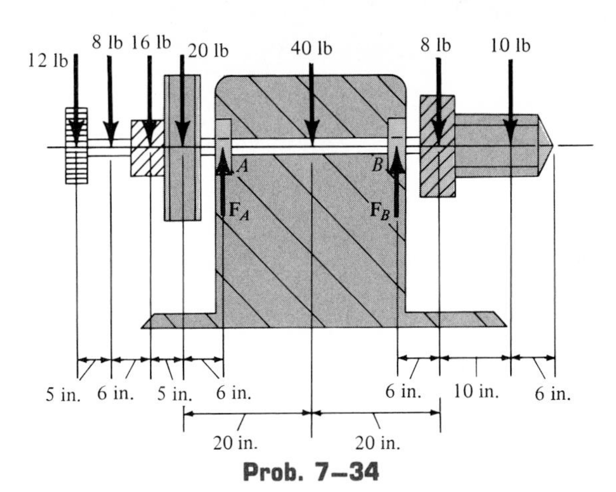

Prob. 7–34

7–35. Draw the shear and bending-moment diagrams for the boom AB of the jib crane. The boom has a mass of 20 kg/m and the trolley supports a force of 600 N in the position shown. Neglect the size of the trolley.

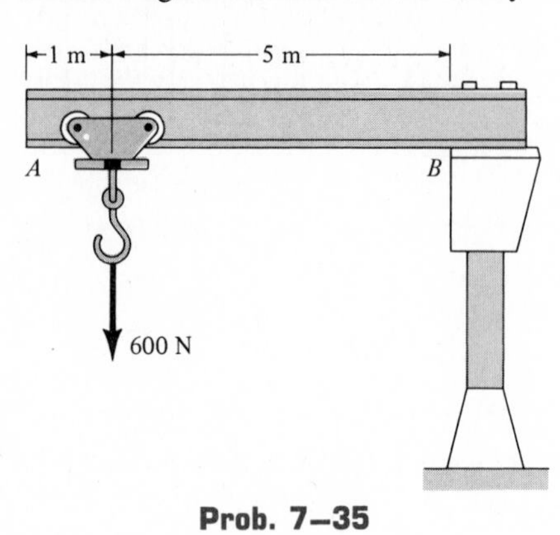

Prob. 7–35

***7–36.** The jib crane carries a load of 8 kN on the trolley T, which rolls along the bottom flange of the beam CD. Determine the position x of the trolley that will create the greatest bending moment in the column at E. What is this moment? As a result of end constraints, $0.3\ \text{m} \leq x \leq 3.9\ \text{m}$.

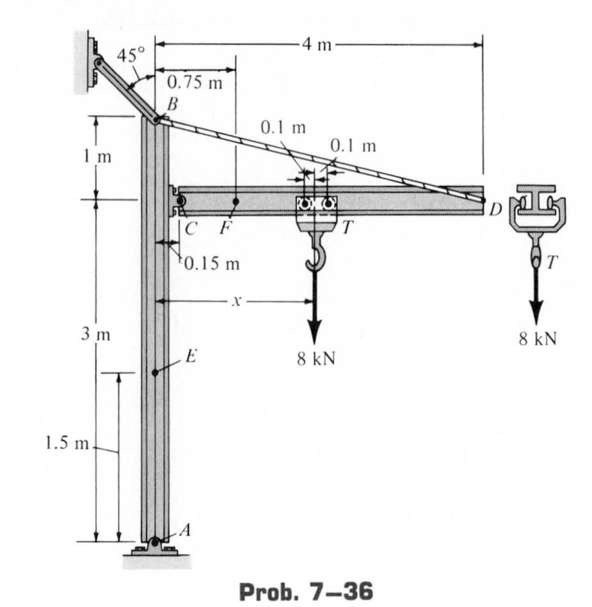

Prob. 7–36

Review Problems

7–37. Determine the x, y, z components of the internal axial force, shear force, and moment at point B of the rod.

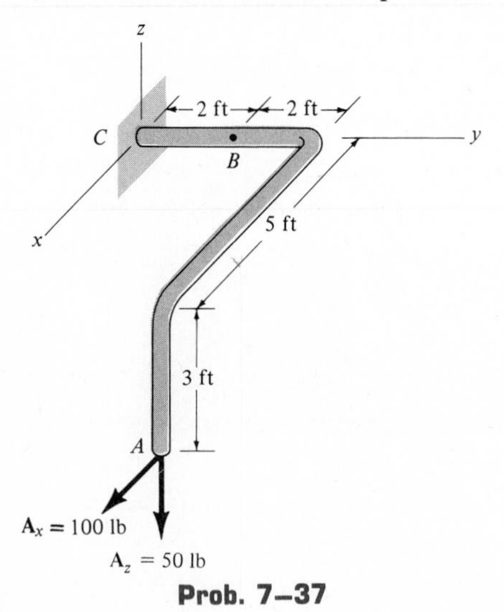

Prob. 7–37

7–38. Draw the shear and bending-moment diagrams for beam CD.

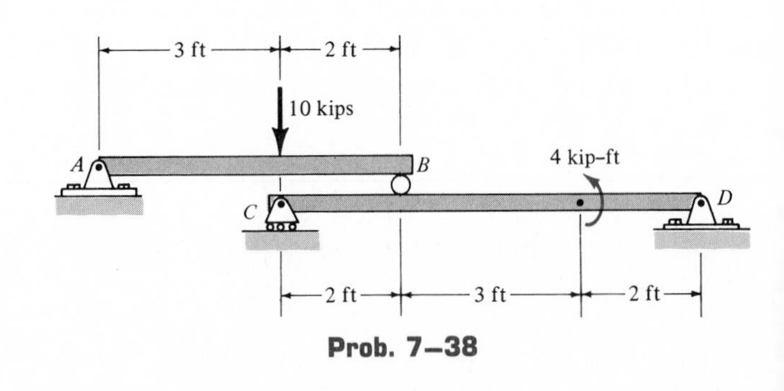

Prob. 7–38

7–39. Determine the internal axial force, shear force, and moment at point E of the frame.

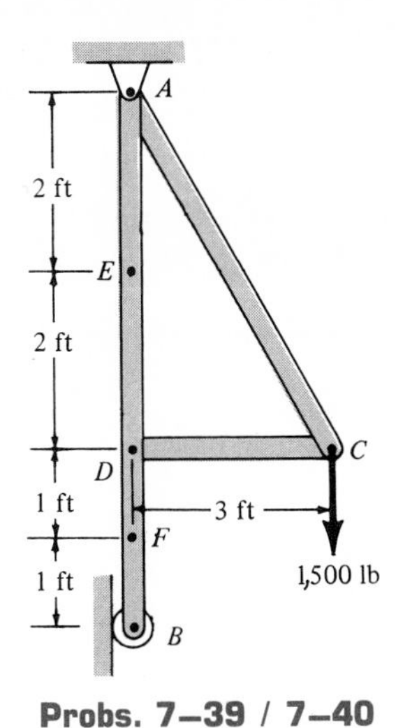

Probs. 7–39 / 7–40

*7–40. Determine the internal axial force, shear force, and moment at point F of the frame.

7–41. Determine the internal axial force, shear force, and moment at point E of the frame.

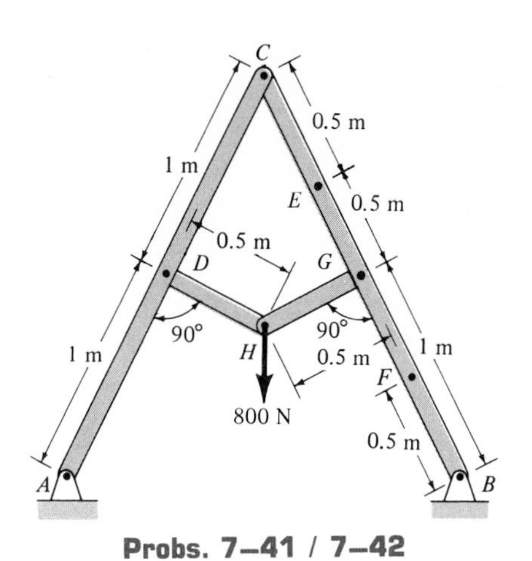

Probs. 7–41 / 7–42

7–42. Determine the internal axial force, shear force, and moment at point F of the frame.

7–43. Draw the shear and bending-moment diagrams for the beam.

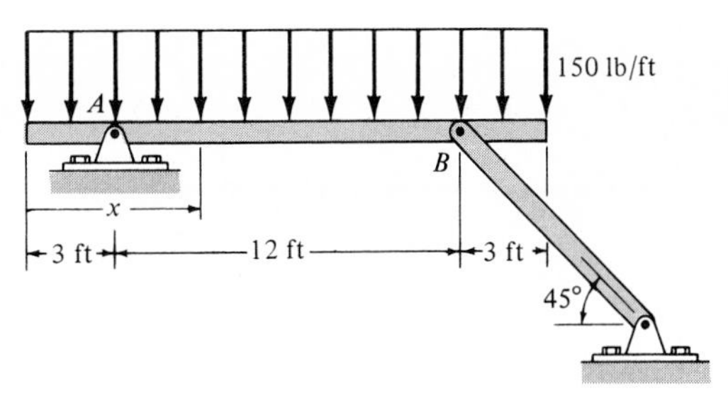

Probs. 7–43 / 7–44

*7–44. Determine the internal shear force and bending moment in the beam for 3 ft $< x <$ 15 ft.

7–45. Draw the shear and bending moment diagrams for the beam.

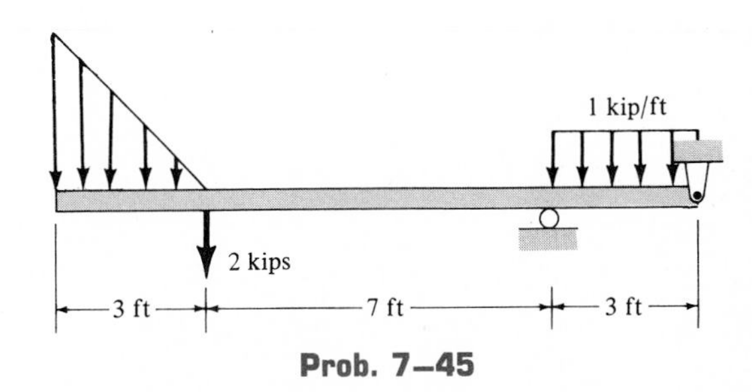

Prob. 7–45

7–46. Determine the internal axial force, shear force, and the bending moment as a function of $0° \leq \theta \leq 180°$ and $0 \leq x \leq 2$ ft for the beam loaded as shown.

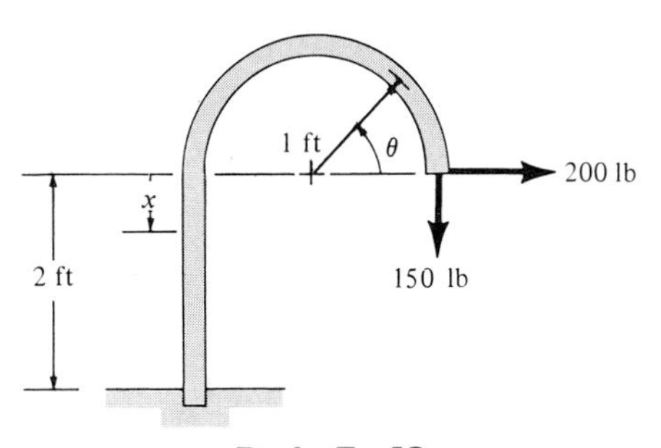

Prob. 7–46

7–47. Determine the internal axial force, shear force, and moment at point E of the oleo strut AB of the aircraft landing gear.

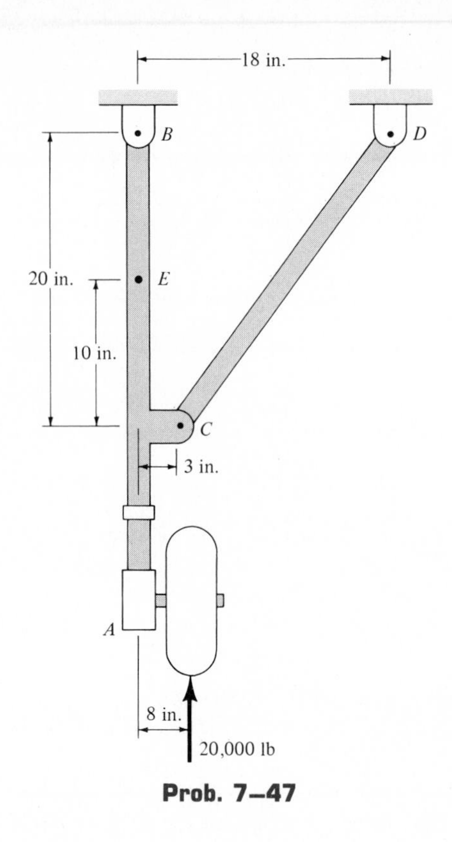

Prob. 7–47

Friction

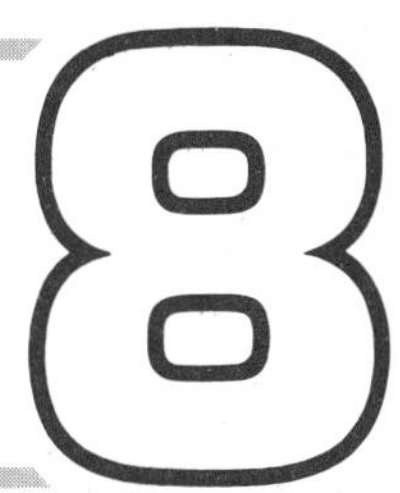

In the previous chapters the surfaces of contact between two bodies were considered to be perfectly *smooth*. Because of this, the force of interaction between the bodies always acts *normal* to the surface at points of contact. In reality, however, all surfaces are *rough,* and depending upon the nature of the problem, the ability of a body to support a *tangential* as well as a *normal* force at its contacting surface must be considered. The tangential force is caused by friction, and in this chapter we will show how to analyze problems involving frictional forces. Specific application will include frictional forces on screws, bearings, disks, and belts. The analysis of rolling resistance is given in the last part of the chapter.

8.1 Characteristics of Dry Friction

Friction may be defined as a force of resistance acting on a body which prevents or inhibits any possible slipping of the body. This force always acts *tangent* to the surface at points of contact with other bodies and is directed so as to oppose the possible or existing motion of the body at these points.

In general, two types of friction can occur between surfaces. *Fluid friction* exists when the contacting surfaces are separated by a film of fluid (gas or liquid). The nature of fluid friction is studied in fluid mechanics since it depends upon knowledge of the velocity of the fluid and the fluid's ability to resist shearing force. In this book only the effects of *dry friction* will be presented. This type of friction is often called *Coulomb friction,* since its characteristics were extensively studied by C. A. Coulomb in 1781. Specifically, dry friction occurs between the contacting surfaces of rigid bodies in the absence of a lubricating fluid.

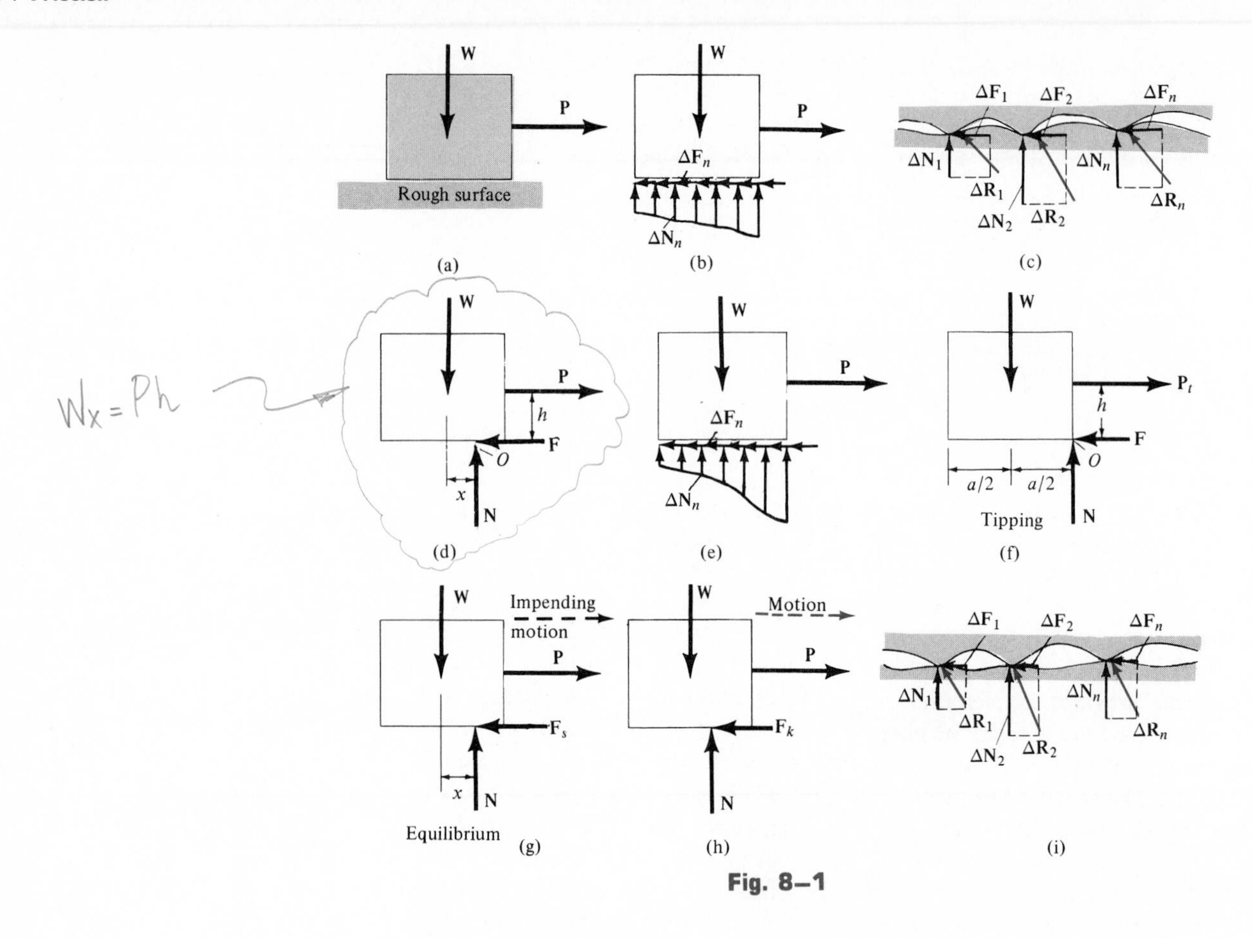

Fig. 8–1

Theory of Dry Friction. The theory of dry friction can best be explained by considering what effects are caused by pulling horizontally on a block of uniform weight **W** which is resting on a rough horizontal floor, Fig. 8–1*a*. As shown on the free-body diagram of the block, Fig. 8–1*b*, the floor exerts a distribution of both *normal force* $\Delta \mathbf{N}_n$ and *frictional force* $\Delta \mathbf{F}_n$ along the contacting surface. For equilibrium, the normal forces must act *upward* to balance the block's weight **W** and the frictional forces act to the left to prevent the applied force **P** from moving the block to the right. Close examination of the contacting surfaces between the floor and block reveals how these frictional and normal forces develop, Fig. 8–1*c*. It can be seen that many microscopic irregularities exist between the two surfaces and, as a result, reactive forces $\Delta \mathbf{R}_n$ are developed at each of the protuberances.* These forces act at all points of contact and, as shown, each reactive force contributes both a frictional component $\Delta \mathbf{F}_n$ and a normal component $\Delta \mathbf{N}_n$.

For simplicity in the foregoing analysis, the effect of these distributed load-

*Besides mechanical interactions as explained here, a detailed treatment of the nature of frictional forces might also include the effects of temperature, density, cleanliness, and electrical attraction between the contacting surfaces.

ings will be indicated by their *resultants* **N** and **F** and then represented on the free-body diagram of the block as shown in Fig. 8–1*d*. Clearly, **F** always acts *tangent to the contacting surface, opposite* to the direction of **P.** The normal force **N** is directed upward to balance the weight **W** and acts on the bottom of the block, a distance x to the right of the line of action of **W.** This location for **N** is necessary to balance the "tipping effect" caused by **P.** For example, if **P** is applied at a height h from the surface, Fig. 8–1*d*, then moment equilibrium about point O is satisfied if $Wx = Ph$ or $x = Ph/W$.

Tipping. Provided the block does *not slip,* any increase in P causes a corresponding increase in x and, as a result, this tends to concentrate the distribution of normal force farther toward the block's right corner, thereby increasing the chance for tipping, Fig. 8–1*e*. Indeed, *tipping* occurs if the contacting surface is "rough" enough to hold the block from slipping and the applied force $P = P_t \geqslant (W/h)(a/2)$, where $x = a/2$, Fig. 8–1*f*.

Impending Motion. In cases where h is small or the surfaces of contact are rather "slippery," the frictional force **F** may *not* be great enough to balance the magnitude of **P,** and consequently the block will tend to slip *before* it can tip. In other words, as the magnitude of **P** is slowly increased, the magnitude of **F** correspondingly increases until it attains a certain *maximum value* F_s, called the *limiting static frictional force,* Fig. 8–1*g*. When this value is reached, the block is in *unstable equilibrium,* since any further increase in P will cause motion. Experimentally, it has been determined that the magnitude of the limiting static frictional force $\mathbf{F}_s$ is *directly proportional* to the magnitude of the resultant normal force **N.** This may be expressed mathematically as

$$F_s = \mu_s N \tag{8–1}$$

where the constant of proportionality, μ_s (mu "sub" s), is called the *coefficient of static friction.*

Typical values for μ_s, found in most engineering handbooks, are given in Table 8–1. It should be noted that μ_s is dimensionless and depends only upon the characteristics of the two surfaces in contact. Furthermore, a wide range of values is given for each value of μ_s, since experimental testing was done under variable conditions of roughness and cleanliness of the contacting surfaces. For applications, therefore, it is important that both caution and judgment be exercised when selecting a coefficient of friction for a given set of conditions. When an exact calculation of F_s is required, the coefficient of friction should be determined directly by an experiment that involves the two materials to be used.

Table 8–1 Typical Values for μ_s

Contact Materials	*Coefficient of Static Friction* (μ_s)
Metal on ice	0.03–0.05
Wood on wood	0.30–0.70
Leather on wood	0.20–0.50
Leather on metal	0.30–0.60
Aluminum on aluminum	1.10–1.70

Motion. If the magnitude of **P** is increased so that it becomes greater than F_s, the frictional force at the contacting surfaces drops slightly to a smaller value F_k, called the *kinetic frictional force*. The block will *not* be held in equilibrium ($P > F_k$); instead, it begins to slide with increasing speed, Fig. 8–1*h*. The drop made in the frictional force magnitude, from F_s (static) to F_k (kinetic), can be explained by again examining the surfaces of contact, Fig. 8–1*i*. Here it is seen that when $P > F_s$, then **P** essentially "lifts" the block out of its settled position and causes it to "ride" on *top* of the peaks at the contacting surface. Consequently, the resultant contact forces $\Delta\mathbf{R}_n$ are aligned slightly more in the vertical direction than before (Fig. 8–1*c*), and hence contribute *smaller* frictional components, $\Delta\mathbf{F}_n$, as when the irregularities are meshed.

Experiments with sliding blocks indicate that the magnitude of the resultant frictional force ($\mathbf{F}_k$) is directly proportional to the magnitude of the resultant normal force $\mathbf{N}$. This may be expressed mathematically as

$$F_k = \mu_k N \tag{8-2}$$

where the constant of proportionality, μ_k, is called the *coefficient of kinetic friction*. Typical values for μ_k are approximately 25 per cent *smaller* than those listed in Table 8–1 for μ_s.

Rules of Dry Friction. As a result of *experiments* that pertain to the foregoing discussion, the following rules which apply to bodies subjected to dry friction may be stated.

1. The frictional force acts *tangent* to the contacting surfaces in a direction *opposed* to the *relative motion* or tendency for motion of one surface against another.
2. The magnitude of the maximum static frictional force $\mathbf{F}_s$ that can be developed is independent of the area of contact provided the normal pressure is not great enough to severely deform or crush the contacting surfaces of the bodies.
3. The magnitude of the maximum static frictional force is generally greater than the magnitude of the kinetic frictional force for any two surfaces of contact. However, if one of the bodies is moving with a *very low velocity* over the surface of another, F_k becomes approximately equal to F_s.
4. When *slipping* at the point of contact is *about to occur,* the magnitude of the maximum static frictional force is proportional to the magnitude of the normal force at the point of contact, such that $F_s = \mu_s N$ (Eq. 8–1).
5. When *slipping* at the point of contact is *occurring,* the magnitude of the kinetic frictional force is proportional to the magnitude of the normal force at the point of contact, such that $F_k = \mu_k N$ (Eq. 8–2).

Angle of Friction. It should be observed that Eqs. 8–1 and 8–2 have a specific, yet *limited,* use in the solution of friction problems. In particular, the frictional force acting at a contacting surface is determined from $F_k = \mu_k N$ *only* if *relative motion* is occurring between the two surfaces. Furthermore, if two bodies are *stationary,* the magnitude of frictional force, F, *does not necessarily* equal $\mu_s N$; instead, F must satisfy the inequality $F \leq \mu_s N$. Only when *impending motion* occurs does F reach its upper limit, $F = F_s = \mu_s N$. This situation may be better understood by considering the block shown in Fig. 8–2*a*, which is acted upon by a force $\mathbf{P}$. In this case consider $P = F_s$, so that the block is on the *verge of sliding*. For equilibrium, the normal force $\mathbf{N}$

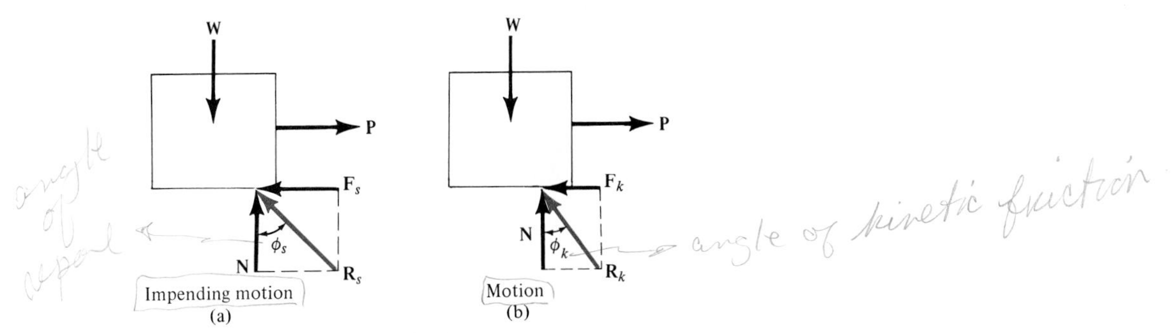

Fig. 8–2

and frictional force $\mathbf{F}_s$ combine to create a resultant $\mathbf{R}_s$. The angle ϕ_s that $\mathbf{R}_s$ makes with $\mathbf{N}$ is called the *angle of static friction* or the *angle of repose*. From the figure,

$$\phi_s = \tan^{-1}\left(\frac{F_s}{N}\right) = \tan^{-1}\left(\frac{\mu_s N}{N}\right) = \tan^{-1}\mu_s$$

Provided the block is *not in motion,* any horizontal force $P < F_s$ causes a resultant $\mathbf{R}$ which has a line of action directed at an angle ϕ from the vertical such that $\phi \leqslant \phi_s$. If $\mathbf{P}$ creates *motion* of the block, then $P = F_k$. In this case, the resultant $\mathbf{R}_k$ has a line of action defined by ϕ_k, Fig. 8–2*b*. This angle is referred to as the *angle of kinetic friction,* where

$$\phi_k = \tan^{-1}\left(\frac{F_k}{N}\right) = \tan^{-1}\left(\frac{\mu_k N}{N}\right) = \tan^{-1}\mu_k$$

By comparison, $\phi_s > \phi_k$.

8.2 Problems Involving Dry Friction

If a rigid body is in equilibrium when it is subjected to a system of forces which includes the effect of friction, the force system must not only satisfy the equations of equilibrium but *also* the laws that govern the frictional forces.

Types of Frictional Problems. In general, there are two types of mechanics problems involving dry friction.

In the "first type" of problem, one has to determine the force needed to either cause impending motion or slipping of a body along *all* its contacting surfaces. If impending motion occurs, the static frictional forces reach their maximum value, i.e., $F_s = \mu_s N$; whereas if the body is slipping or sliding, kinetic frictional forces are developed, $F_k = \mu_k N$ (see Examples 8–1 to 8–3).

In the "second type" of problem, the applied loads acting on the body (or system of bodies) are either known or have to be determined; and for the analysis one has to *check* if the equilibrium frictional force F developed at each surface of contact is less than or equal to the maximum static frictional force that can be developed at these points, $F \leqslant \mu_s N$ (see Examples 8–4 to 8–6).

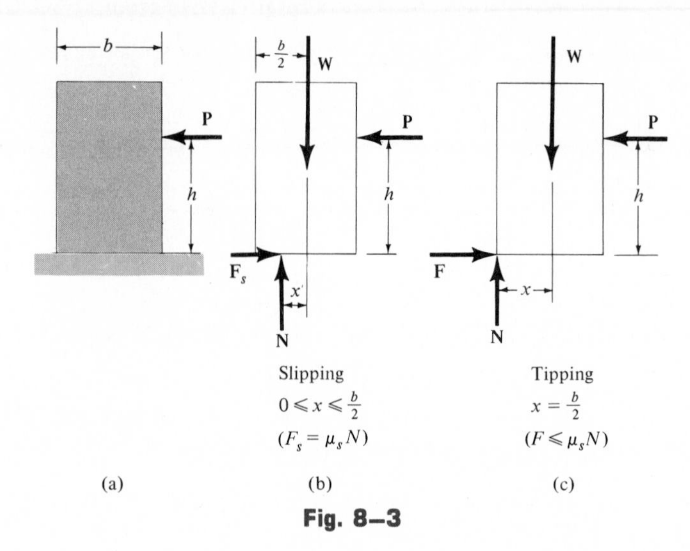

Fig. 8–3

In some cases, this second type of problem may involve several ways in which a body (or system of bodies) may move out of equilibrium and, as a result, each situation will have to be *investigated separately*. For example, a block resting on a rough surface, Fig. 8–3*a*, may either slip, Fig. 8–3*b*, or tip, Fig. 8–3*c*, as the force **P** is increased in magnitude. The actual situation is determined by calculating P for each case, and then choosing the case for which P is the smallest. If in *both cases* the *same value* for P is calculated, then both slipping and tipping occur simultaneously. In cases where a curved body is in contact with another surface, *rolling* rather than tipping may be possible. For example, consider the cylinder wedged between the floor and the lever, Fig. 8–4*a*. As the magnitude of **P** is increased, two possibilities for motion exist: the cylinder can roll along the floor without slipping at A, and as a result it would have to slip at B, Fig. 8–4*b*; or it could roll along the lever and slip at A, Fig. 8–4*c*. If the analysis of each of these cases reveals the *same magnitude* for **P**, then slipping at both surfaces, with no rolling, occurs.

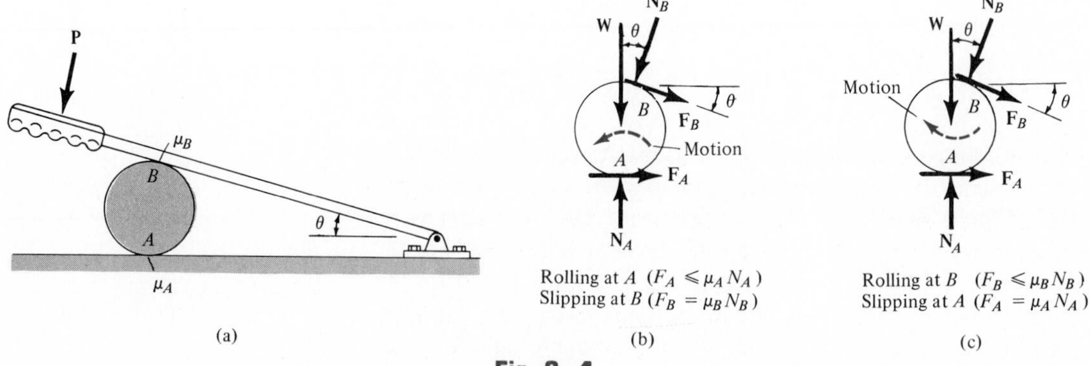

Fig. 8–4

Equilibrium Versus Frictional Equations. As stated above, if a body can be subjected to various types of movement, the frictional forces at some of the points of contact may *not* be equal to their maximum static value; instead, $F < \mu_s N$. In this case, F must be determined from the equations of equilibrium, and since F is an "equilibrium force," the sense of direction of **F** can be *assumed*. The correct direction of **F** is made known *after* the equations of equilibrium are applied and the magnitude of **F** is determined. In particular, if the magnitude is negative, the directional sense of **F** is the reverse of that which was assumed. This convenience of *assuming* the sense of direction of **F** is possible because the equilibrium equations are *vector* equations; i.e., the scalar equilibrium equations equate to zero the *components of vectors* acting in the *same direction*.

In cases where the scalar frictional equation $F_s = \mu_s N$ is used in the solution of a problem, the convenience of *assuming* the directional sense of **F** is *lost*, since the frictional equation relates the magnitudes of two vectors that act in *perpendicular directions* ($\mathbf{F} = \mathbf{F}_s$ is always perpendicular to **N**). Consequently, **F** *must* be shown acting with its *correct direction* on the free-body diagram. In this regard, the direction of the frictional force will *always* be such as to either *oppose the relative motion or impend the motion of the body* over its contacting surface.

PROCEDURE FOR ANALYSIS

The following procedure provides a method for solving equilibrium problems involving dry friction.

Free-Body Diagrams. Draw the necessary free-body diagrams and determine the number of equations required for a complete solution. Recall that only three equations of coplanar equilibrium can be written for each body. Consequently, if there are more unknowns than equations of equilibrium, it will be necessary to apply the frictional equation, $F_s = \mu_s N$, at some, if not all, points of contact to obtain the required number of equations.*

Equations of Friction and Equilibrium. Apply the necessary frictional equations and the equations of equilibrium and solve for the unknowns.

The following example problems numerically illustrate this procedure.

*In cases where the problem is of the "second type," several points of application of the frictional equation may be possible. For example, consider again the cylinder in Fig. 8–4*a*. If the weight **W** is known, the cylinder is subjected to four unknown force magnitudes: N_A, F_A, N_B, and F_B. The solution for these unknowns will be *unique* if it satisfies the *three* equilibrium equations *and only one* frictional equation. Consequently, the possibilities are either $F_B = \mu_B N_B$ (in which case $F_A \leq \mu_A N_A$), Fig. 8–4*b*, or $F_A = \mu_A N_A$ (in which case $F_B \leq \mu_B N_B$), Fig. 8–4*c*.

Example 8–1

The crate shown in Fig. 8–5*a* has a mass of 20 kg. Determine the force **P** so that the crate is on the verge of moving up the plane. The coefficient of static friction is $\mu_s = 0.3$.

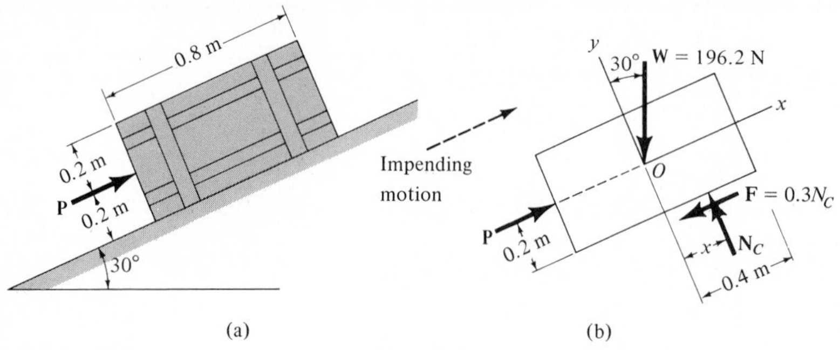

Fig. 8–5

Solution

Free-Body Diagram. As shown in Fig. 8–5*b*, the resultant $\mathbf{N}_C$ must act a distance x from the centerline of the crate in order to counteract the tipping effect caused by **P.** There are *four* unknowns: P, F, N_C, and x. Since *three* equations of equilibrium are available, the frictional equation applies at the contacting surface and provides the necessary fourth equation for the solution. (Use of this equation is also implied by the problem statement, since if the crate is on the verge of slipping up the plane, the maximum static frictional force must be developed on the crate.) Note that **F** must act *down* the plane to prevent upward motion of the crate. It is important that the correct direction of **F** be specified on the free-body diagram since the frictional equation is to be used in the solution.

Equations of Friction and Equilibrium. Writing the frictional equation yields

$F_s = \mu_s N;$ $\qquad F = 0.3N_C$

Using this result and applying the equations of equilibrium, we have

$+\nearrow\Sigma F_x = 0;$ $\qquad P - 0.3N_C - 196.2 \sin 30° = 0$ (1)

$+\nwarrow\Sigma F_y = 0;$ $\qquad N_C - 196.2 \cos 30° = 0$ (2)

$\circlearrowleft +\Sigma M_O = 0;$ $\qquad N_C(x) - 0.3N_C(0.2) = 0$ (3)

Solving for the three unknowns, we obtain

$$N_C = 169.9 \text{ N}$$

$$P = 149.1 \text{ N} \qquad \textit{Ans.}$$

$$x = 0.06 \text{ m}$$

Note that since the answer for x is positive, its direction to the right of the "y" axis was assumed correctly. Furthermore, $x < 0.4$ m, so that the crate will not tip over.

Example 8–2

The uniform plank shown in Fig. 8–6*a* has a mass of 15 kg and rests against a floor and wall for which the coefficients of static friction are $(\mu_s)_A = 0.30$ and $(\mu_s)_B = 0.20$, respectively. Determine the distance s to which a man having a mass of 70 kg can climb without causing the plank to slip.

Solution

Free-Body Diagram. As shown in Fig. 8–6*b*, there are *five* unknowns, N_B, F_B, N_A, F_A, and s, on the free-body diagram. These can be determined from the *three* equations of equilibrium and *two* frictional equations applied at points A and B. The frictional forces $\mathbf{F}_A$ and $\mathbf{F}_B$ must be drawn in their correct direction so that they oppose the tendency for motion of the plank, Fig. 8–6*b*. Why?

Equations of Friction and Equilibrium. Writing the frictional equations,

$$F_s = \mu_s N; \qquad F_A = 0.3N_A$$
$$F_B = 0.2N_B$$

Using these results and applying the equations of equilibrium yields

$$\overset{+}{\rightarrow}\Sigma F_x = 0; \qquad 0.3N_A - N_B = 0$$
$$+\uparrow \Sigma F_y = 0; \qquad N_A - 686.7 - 147.15 + 0.2N_B = 0$$
$$\circlearrowleft + \Sigma M_A = 0; \qquad -686.7(s \cos 40°) - 147.15(2 \cos 40°)$$
$$+ 0.2N_B(4 \cos 40°) + N_B(4 \sin 40°) = 0$$

Solving these equations simultaneously, we get

$$N_A = 786.7 \text{ N}$$
$$N_B = 236.0 \text{ N}$$
$$s = 1.00 \text{ m} \qquad \textit{Ans.}$$

(a)

Fig. 8–6

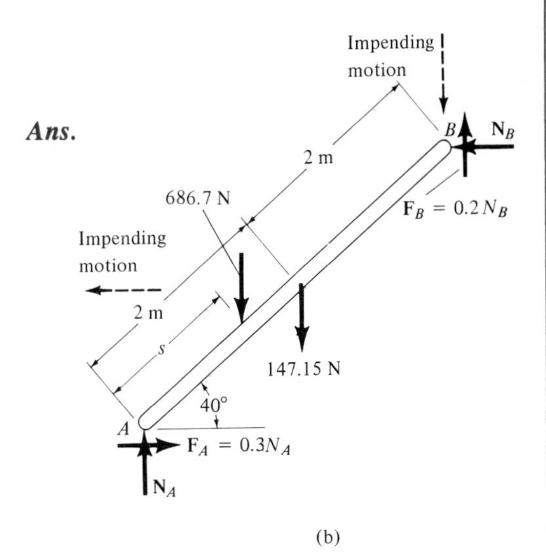

(b)

Example 8–3

The pipe shown in Fig. 8–7*a* is gripped between two levers that are pinned together at C. If the coefficient of friction between the levers and the pipe is $\mu = 0.3$, determine the maximum angle θ at which the pipe can be gripped without slipping. Neglect the weight of the pipe.

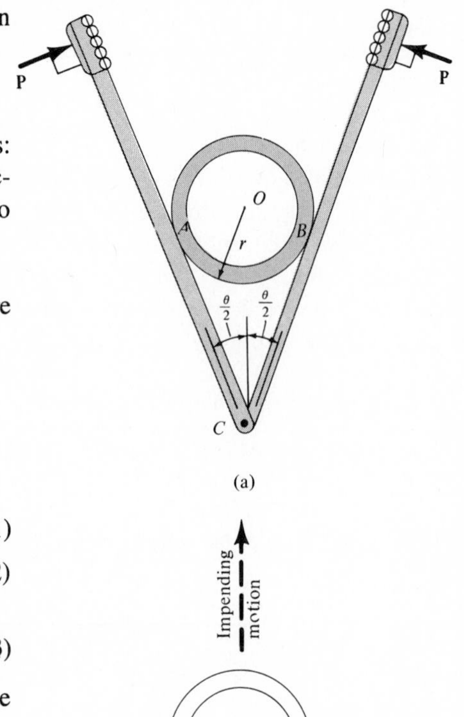

(a)

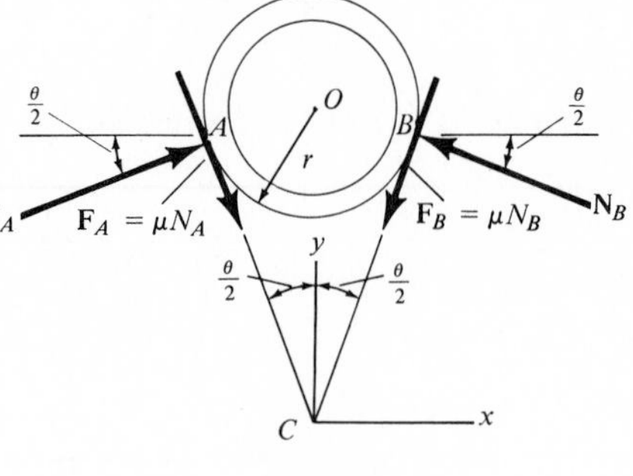

(b)

Fig. 8–7

Solution

Free-Body Diagram. As shown in Fig. 8–7*b*, there are five unknowns: N_A, F_A, N_B, F_B, and θ. The *three* equations of equilibrium and *two* frictional equations at A and B apply. The frictional forces act toward C to prevent upward motion of the pipe.

Equations of Friction and Equilibrium. The frictional equations are

$$F_s = \mu_s N; \qquad F_A = \mu N_A$$
$$F_B = \mu N_B$$

Using these results, and applying the equations of equilibrium, yields

$$\overset{+}{\rightarrow}\Sigma F_x = 0; \quad N_A \cos(\theta/2) + \mu N_A \sin(\theta/2) - N_B \cos(\theta/2) - \mu N_B \sin(\theta/2) = 0 \qquad (1)$$

$$\circlearrowleft +\Sigma M_O = 0; \qquad -\mu N_B(r) + \mu N_A(r) = 0 \qquad (2)$$

$$+\uparrow \Sigma F_y = 0; \quad N_A \sin(\theta/2) - \mu N_A \cos(\theta/2) + N_B \sin(\theta/2) - \mu N_B \cos(\theta/2) = 0 \qquad (3)$$

From either Eq. (1) or (2) it is seen that $N_A = N_B$. This could also have been determined directly from the symmetry of both geometry and loading. Substituting the result into Eq. (3), we obtain

$$\sin(\theta/2) - \mu \cos(\theta/2) = 0$$

so that

$$\tan(\theta/2) = \frac{\sin(\theta/2)}{\cos(\theta/2)} = \mu = 0.3$$

$$\theta = 2 \tan^{-1} 0.3 = 33.4° \qquad \textit{Ans.}$$

Example 8–4

The homogeneous block shown in Fig. 8–8*a* has a weight of 20 lb and rests on the incline for which $\mu_s = 0.55$. Determine the largest angle of tilt, θ, of the plane before the block moves.

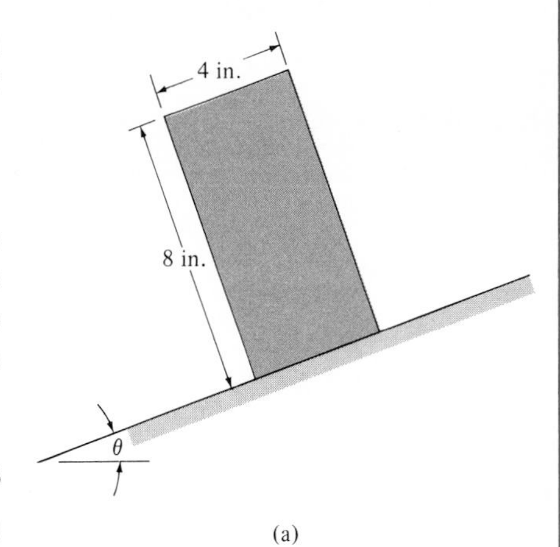

(a)

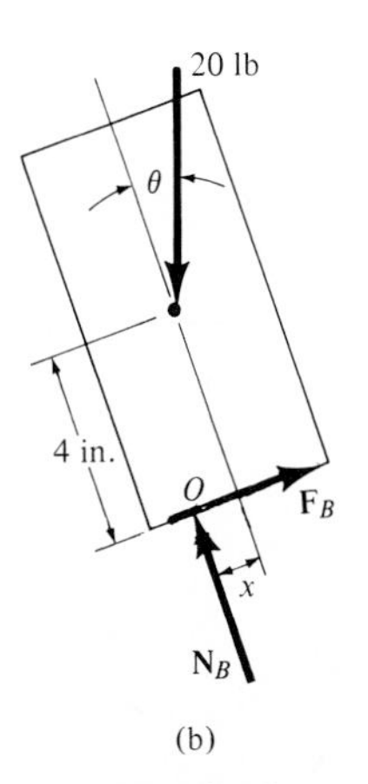

(b)

Fig. 8–8

Solution

Free-Body Diagram. As shown in Fig. 8–8*b*, the dimension x is used to locate the position of the resultant normal force $\mathbf{N}_B$ under the block. There are *four* unknowns, θ, N_B, F_B, and x. *Three* equations of equilibrium are available. The *fourth* equation is obtained by investigating the conditions for tipping or sliding of the block.

Equations of Equilibrium. Applying the equations of equilibrium yields

$+\swarrow\Sigma F_x = 0;$ $\qquad 20 \sin\theta - F_B = 0 \qquad (1)$

$+\nwarrow\Sigma F_y = 0;$ $\qquad N_B - 20\cos\theta = 0 \qquad (2)$

$\circlearrowleft +\Sigma M_O = 0;$ $\qquad 20\sin\theta(4) - 20\cos\theta(x) = 0 \qquad (3)$

(Impending Motion of Block.) This requires use of the frictional equation

$F_s = \mu_s N;$ $\qquad F_B = 0.55N_B \qquad (4)$

Solving Eqs. (1) to (4) yields

$$N_B = 17.5 \text{ lb} \qquad F_B = 9.64 \text{ lb} \qquad \theta = 28.8^\circ \qquad x = 2.2 \text{ in.}$$

Since $x = 2.2$ in. > 2 in., the block will tip *before* sliding.

(Tipping of Block.) This requires

$$x = 2 \text{ in.} \qquad (5)$$

Solving Eqs. (1) to (3) using Eq. (5) yields

$$N_B = 17.9 \text{ lb} \qquad F_B = 8.94 \text{ lb}$$
$$\theta = 26.6^\circ \qquad \textit{Ans.}$$

Example 8–5

Beam AB is subjected to a uniform load of 200 N/m and is supported at B by column BC, Fig. 8–9a. If the static coefficients of friction at B and C are $\mu_B = 0.2$ and $\mu_C = 0.5$, determine the force $\mathbf{P}$ needed to pull the column out from under the beam. Neglect the weight of the members.

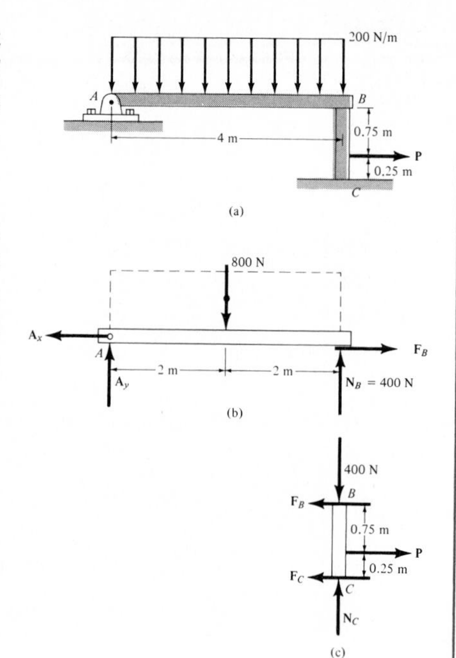

Fig. 8–9

Solution

Free-Body Diagrams. The free-body diagram of beam AB is shown in Fig. 8–9b. Applying $\Sigma M_A = 0$, we obtain $N_B = 400$ N. This result is shown on the free-body diagram of the column, Fig. 8–9c. Referring to this member, the *four* unknowns F_B, P, F_C, and N_C are determined from the *three* equations of equilibrium and *one* frictional equation applied either at B or C.

Equations of Equilibrium and Friction. Applying the equations of equilibrium, we have

$\overset{+}{\rightarrow}\Sigma F_x = 0;$ $\quad P - F_B - F_C = 0 \qquad (1)$

$+\uparrow \Sigma F_y = 0;$ $\quad N_C - 400 = 0 \qquad (2)$

$\circlearrowleft +\Sigma M_C = 0;$ $\quad -P(0.25) + F_B(1) = 0 \qquad (3)$

(Column Slips only at B.) This requires $F_C \leqslant \mu_C N_C$ and

$F_B = \mu_B N_B;$ $\quad F_B = 0.2(400) = 80$ N

Using this result and solving Eqs. (1) to (3), we obtain

$$P = 320 \text{ N}$$
$$F_C = 240 \text{ N}$$
$$N_C = 400 \text{ N}$$

Since $F_C = 240 > \mu_C N_C = 0.5(400) = 200$ N, the other case of movement must be investigated.

(Column Slips only at C.) Here $F_B \leqslant \mu_B N_B$ and

$F_C = \mu_C N_C;$ $\quad F_C = 0.5 N_C \qquad (4)$

Solving Eqs. (1) to (4) yields

$$P = 266.7 \text{ N} \qquad \textit{Ans.}$$
$$N_C = 400 \text{ N}$$
$$F_C = 200 \text{ N}$$
$$F_B = 66.7 \text{ N}$$

Obviously, this case of movement occurs first since it requires a *smaller* value for P.

Example 8–6

Determine the normal force that must be exerted on the 100-kg spool shown in Fig. 8–10a to push it up the 20° incline at constant velocity. The coefficients of static and kinetic friction at the points of contact are $(\mu_s)_A = 0.18$, $(\mu_k)_A = 0.15$ and $(\mu_s)_B = 0.45$, $(\mu_k)_B = 0.4$.

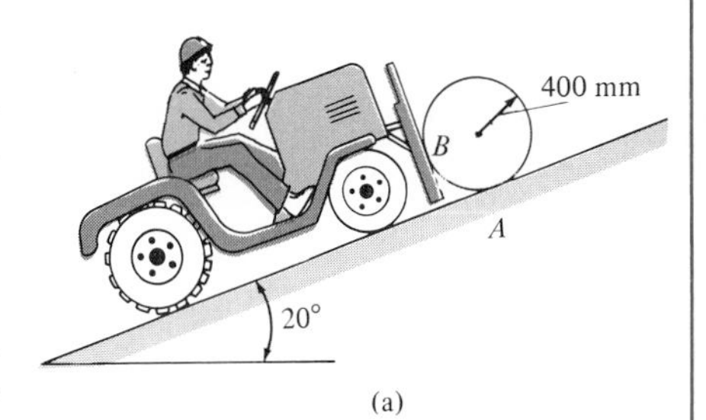

(a)

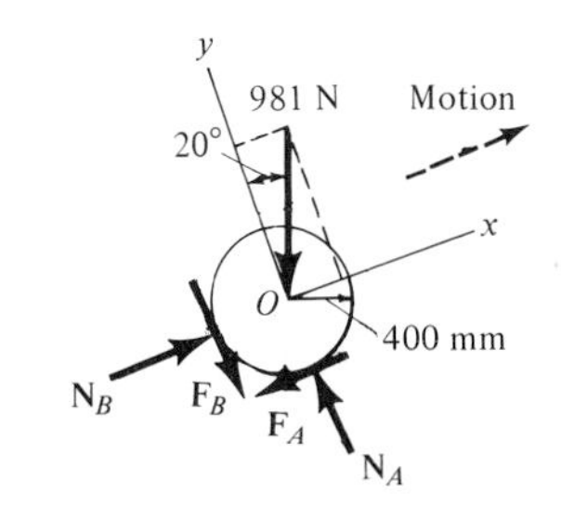

(b)

Fig. 8–10

Solution

Free-Body Diagram. As shown in Fig. 8–10b, there are four unknowns N_A, F_A, N_B, and F_B acting on the spool. These can be determined from the *three* equations of equilibrium and *one* frictional equation, which applies either at A or B. If slipping only occurs at A, the spool will *slide* up the incline; whereas if slipping only occurs at B, the spool *rolls* up the incline. The problem requires determination of N_B.

Equations of Equilibrium and Friction

$$+\nearrow\Sigma F_x = 0; \qquad -F_A + N_B - 981 \sin 20° = 0 \qquad (1)$$

$$+\nwarrow\Sigma F_y = 0; \qquad N_A - F_B - 981 \cos 20° = 0 \qquad (2)$$

$$\circlearrowleft +\Sigma M_O = 0; \qquad F_B(400 \text{ mm}) - F_A(400 \text{ mm}) = 0 \qquad (3)$$

Spool Rolls up Incline. In this case $F_A \leqslant 0.18N_A$ and

$$(F_k)_B = (\mu_k)_B N_B; \qquad F_B = 0.40N_B \qquad (4)$$

The direction of the frictional force at B must be correctly specified. Why? Since the spool is being forced up the plane, $\mathbf{F}_B$ acts downward to prevent the clockwise rolling motion of the spool, Fig. 8–10b. Solving Eqs. (1) to (4), we have

$$N_A = 1145.5 \text{ N} \qquad F_A = 223.7 \text{ N} \qquad N_B = 559.2 \text{ N} \qquad F_B = 223.7 \text{ N}$$

The assumption regarding no slipping at A should be checked.

$$F_A \leqslant (\mu_s)_A N_A; \qquad 223.7 \text{ N} \overset{?}{\leqslant} 0.18(1145.5) = 206.2 \text{ N}$$

The inequality does *not apply*, and therefore slipping occurs at A and not at B. Hence, the other case of motion must be investigated.

Spool Slides up Incline. In this case, $F_B \leqslant 0.45N_B$ and

$$(F_k)_A = (\mu_k)_A N_A; \qquad F_A = 0.15N_A \qquad (5)$$

Solving Eqs. (1) to (3) and (5) yields

$$N_A = 1084.5 \text{ N} \qquad F_A = 162.7 \text{ N} \qquad N_B = 498.2 \text{ N} \qquad F_B = 162.7 \text{ N}$$

The validity of the solution ($N_B = 498.2$ N) can be checked by testing the assumption that indeed no slipping occurs at B.

$$F_B \leqslant 0.45N_B; \quad 162.7 \text{ N} < 0.45(498.2 \text{ N}) = 224.2 \text{ N} \quad \text{(check)}$$

Problems

Assume that $\mu = \mu_s = \mu_k$ in the following problems where specified.

8–1. The crate has a mass of 350 kg and is subjected to a towing force **P** acting at a 20° angle with the horizontal. If $\mu_s = 0.5$, determine the magnitude of **P** to just start the crate moving down the plane.

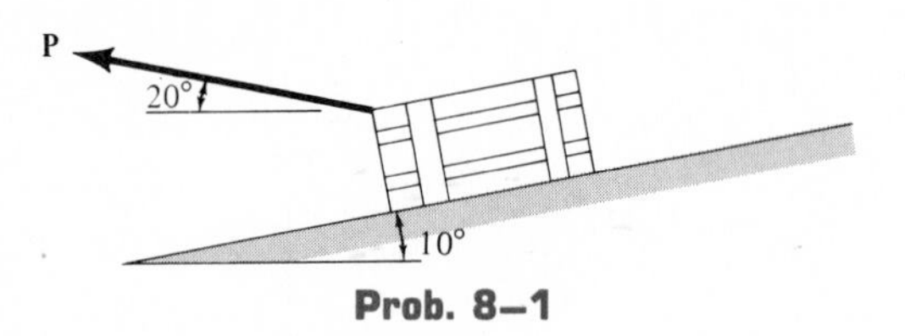

Prob. 8–1

8–2. The uniform dresser has a weight of 80 lb and rests on a tile floor for which $\mu_s = 0.25$. If the man pushes on it in the direction shown, determine the smallest magnitude of force **F** needed to move the dresser. Also, if the man has a weight of 150 lb, determine the smallest coefficient of friction between his shoes and the floor so he doesn't slip.

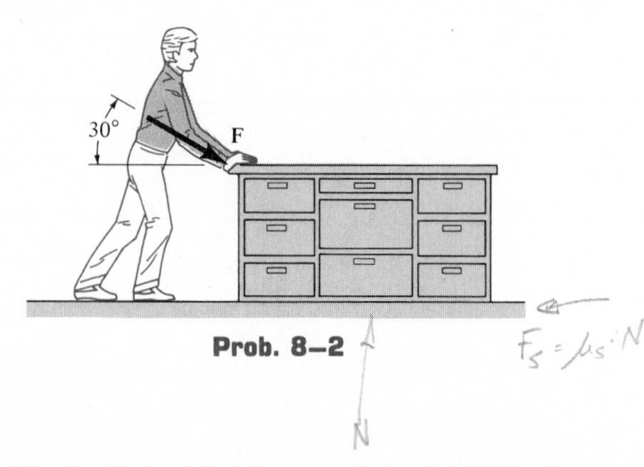

Prob. 8–2

8–3. The jeep and driver have a mass of 1.2 Mg. Determine the steepest slope θ which the jeep can climb at constant speed if (a) driving power is applied only to the rear wheels B, while the front wheels A are free to roll, and (b) driving power is applied to all four wheels. The coefficient of friction between the wheels and the ground is $\mu = 0.6$.

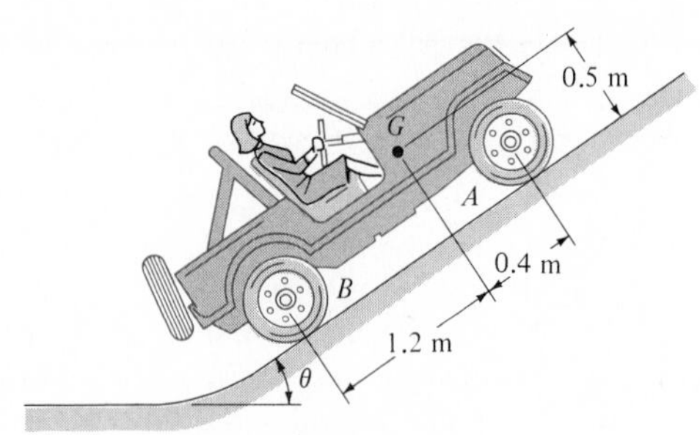

Prob. 8–3

***8–4.** A 17-kg ladder has a center of mass at G. If the coefficients of friction at A and B are $\mu_A = 0.3$ and $\mu_B = 0.2$, respectively, determine the smallest horizontal force that the man must exert on the ladder at point C in order to push the ladder forward.

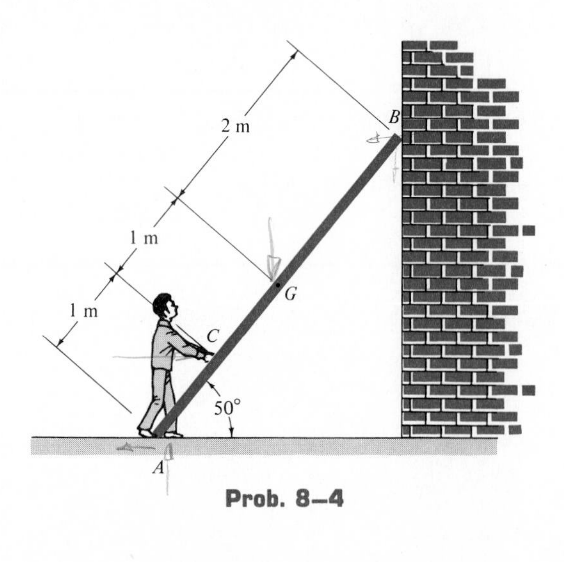

Prob. 8–4

8–5. The motorcyclist travels with constant velocity along a straight, horizontal, banked road. If he aligns his bike so that the tires are perpendicular to the road at A, determine the normal and frictional forces at A. The man has a mass of 60 kg and a mass center at G_c, and the motorcycle has a mass of 120 kg and a mass center at G_m. If the coefficient of friction at A is $\mu_A = 0.4$, will the bike slip?

Prob. 8–5

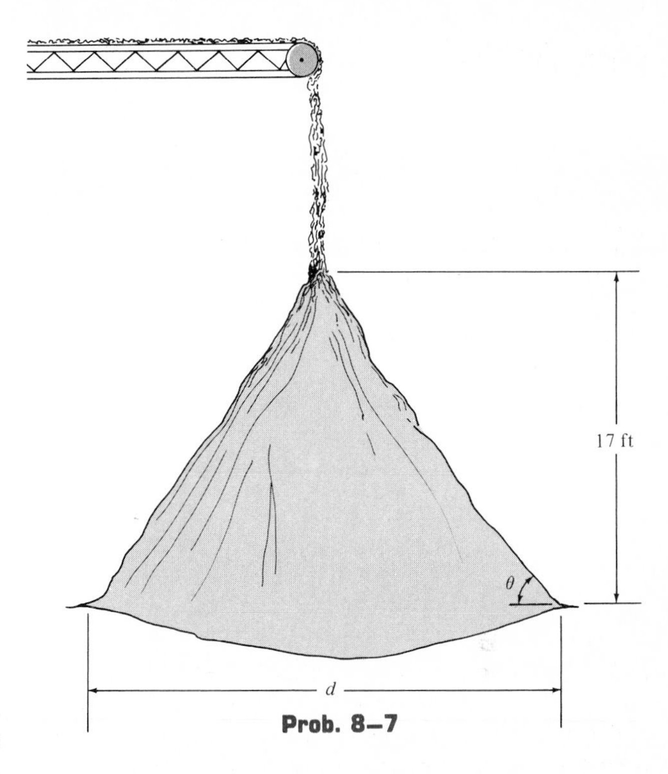

Prob. 8–7

8–6. A cable is wrapped around the inner core of a spool. Determine the magnitude of the vertical force **P** that must be applied to the end of the cable to rotate the spool. The coefficient of friction between the spool and the contacting surfaces at A and B is $\mu = 0.6$. The spool has a mass of 100 kg.

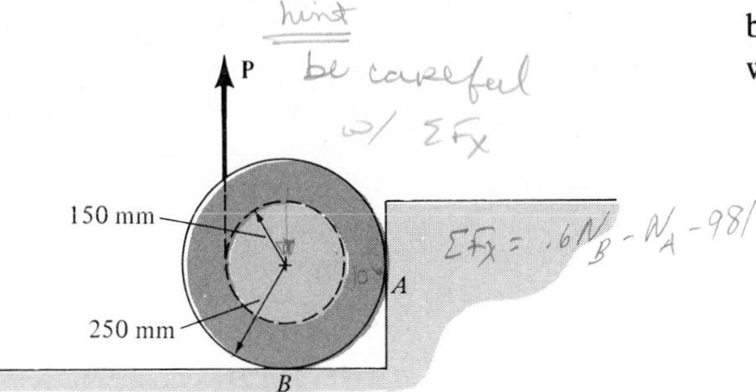

Prob. 8–6

8–7. Gravel is stored in a conical pile at a materials yard. If the height of the pile is 17 ft and the coefficient of static friction between the gravel particles is $\mu_s = 0.4$, determine the approximate diameter d of the pile. For the calculation, neglect the "irregularities" of the particles, and first determine the angle θ of the pile by considering one of the particles to be represented by a *block* resting on an inclined plane of angle θ, where $\mu_s = 0.4$.

***8–8.** A uniform beam has a mass of 18 kg and rests on two surfaces at points A and B. Determine the maximum distance x to which the girl can slowly walk up the beam before it begins to slip. The girl has a mass of 50 kg and walks up the beam with a constant velocity.

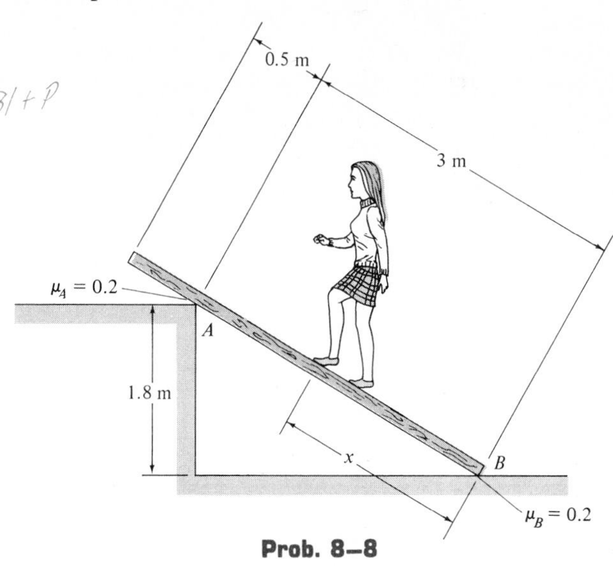

Prob. 8–8

8–9. A 35-kg disk rests on an inclined surface for which $\mu_s = 0.2$. Determine the maximum vertical force **P** that may be applied to link AB without causing the disk to slip at C.

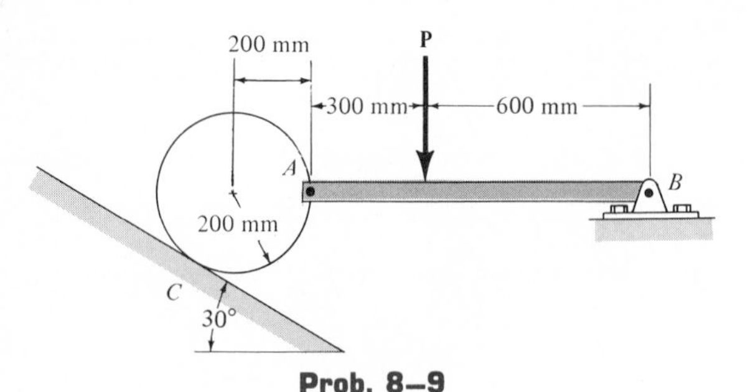

Prob. 8–9

8–10. A man attempts to support a stack of books horizontally by applying a compressive force of $F = 120$ N to the ends of the stack with his hands. If each book has a mass of 0.95 kg, determine the greatest number of books that can be supported in the stack. The coefficient of friction between the man's hands and a book is $\mu_h = 0.6$ and between any two books, $\mu_b = 0.4$.

Prob. 8–10

8–11. The coiled belt spring is used as a feeding device for small boxes of merchandise, each box having a weight of 4 lb. If the coil exerts a *constant* force of $P = 8$ lb on the box at A, determine the smallest number of boxes that can be kept under constant pressure by the device without any movement of the belt. The coefficient of friction between the belt and each box $\mu_b = 0.35$. Neglect tipping of the boxes and any friction that the spring exerts on A. The spring is attached to the ground.

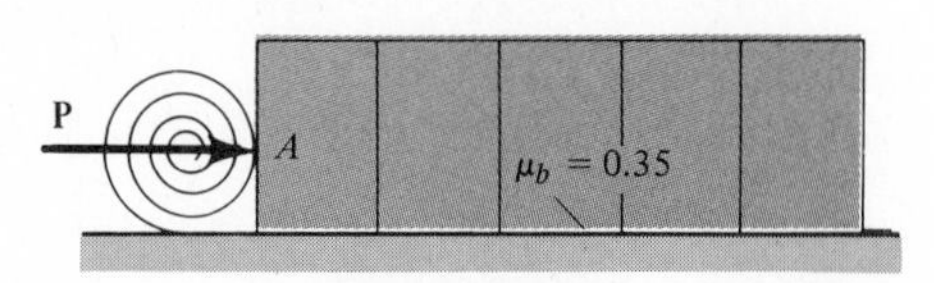

Prob. 8–11

***8–12.** A boy having a mass of 70 kg attempts to walk on a board which is supported by a pin at A and a post BC. If $\mu_B = 0.3$ and $\mu_C = 0.5$, determine the maximum angle θ of alignment for the post so that he can safely reach the other side. Neglect the mass of the board and post and the thickness of the post in the calculation. *Hint:* The post is a two-force member.

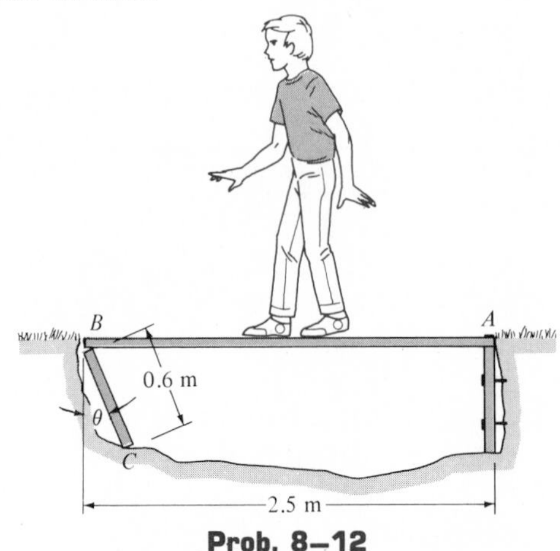

Prob. 8–12

8–13. If a meter stick of uniform mass is placed on your index fingers A and B at the 200-mm and 900-mm position shown, determine what will happen as your hands are slowly drawn close together, i.e., at what mark on the meter stick will they meet? Take $\mu = 0.3$. The mass of the stick is 0.4 kg.

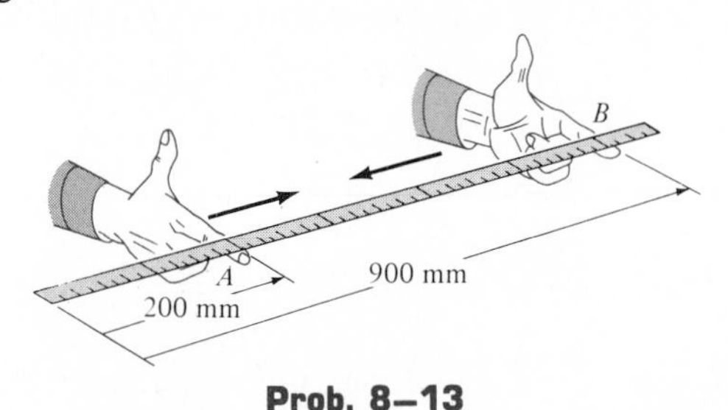

Prob. 8–13

8–14. A man attempts to lift the uniform 40-lb ladder to an upright position by applying a force **F** perpendicular to the ladder at rung R. Determine the coefficient of friction between the ladder and the ground at A if the ladder begins to slip on the ground when his hands reach a height of 6 ft.

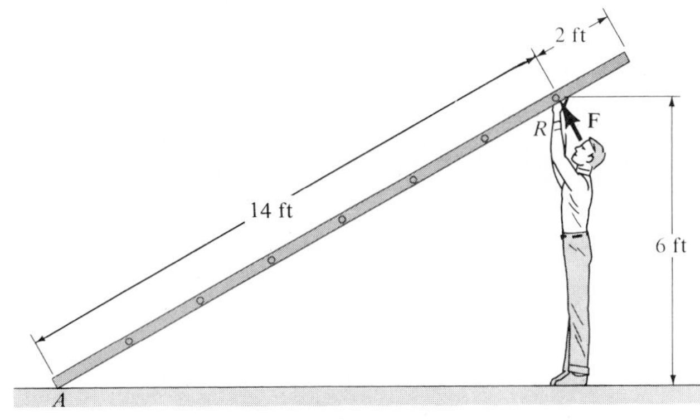

Prob. 8–14

8–15. A winch is mounted at E on the front of a 3-Mg pickup truck. As the cable is drawn in, it loads a crate onto the truck. Determine the largest mass m of the crate that can be loaded without causing the truck to move. The truck is braked only at its rear wheels and has a center of mass at G. The coefficient of static friction between the wheels and the ground is $\mu_w = 0.4$, and between the crate and the ground $\mu_c = 0.5$. The pulley at D is frictionless.

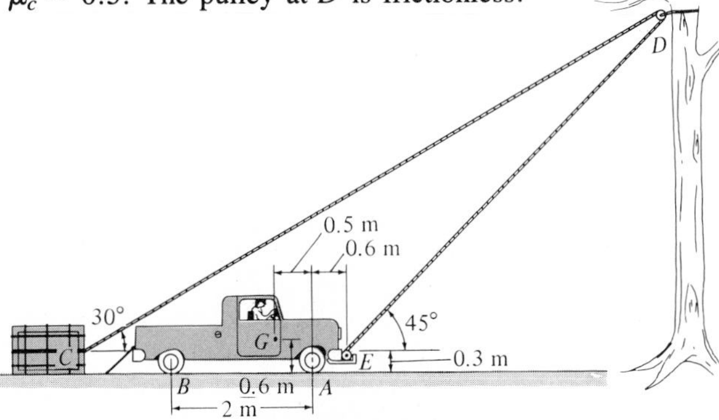

Prob. 8–15

***8–16.** The beam is adjusted to the horizontal position by means of a wedge located at its right support. If the coefficient of friction between the wedge and the two surfaces of contact is $\mu = 0.25$, determine the horizontal force **P** required to push the wedge forward.

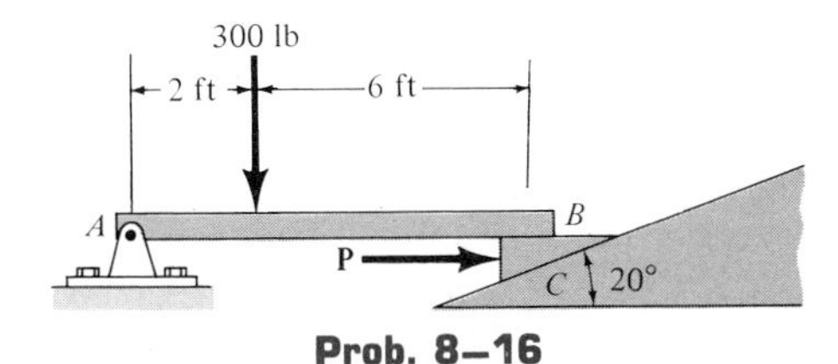

Prob. 8–16

8–17. The two pin-connected rods each have a weight of 15 lb. If the coefficient of static friction at C is $\mu_s = 0.5$, determine the maximum angle θ of spread for equilibrium.

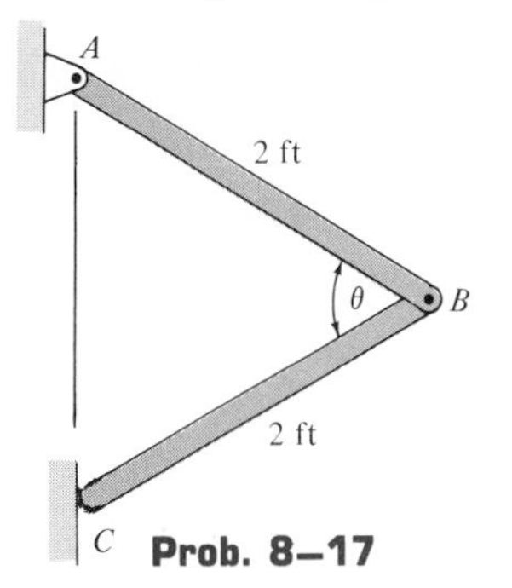

Prob. 8–17

8–18. Determine the minimum applied force **P** required to move wedge A to the right. The spring is compressed a distance of 175 mm. Neglect the weight of A and B. The coefficient of static friction for all contacting surfaces is $\mu_s = 0.35$. Neglect friction at the rollers.

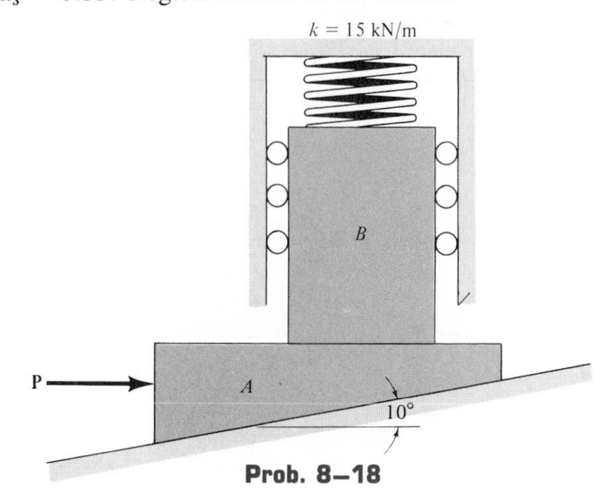

Prob. 8–18

8–19. The beam is supported by a pin at A and a roller at B which has negligible weight and a radius of 15 mm. If the coefficient of static friction is $\mu_B = \mu_C = 0.3$, determine the largest angle θ of the incline so that the roller does not slip for any force **P** applied to the beam.

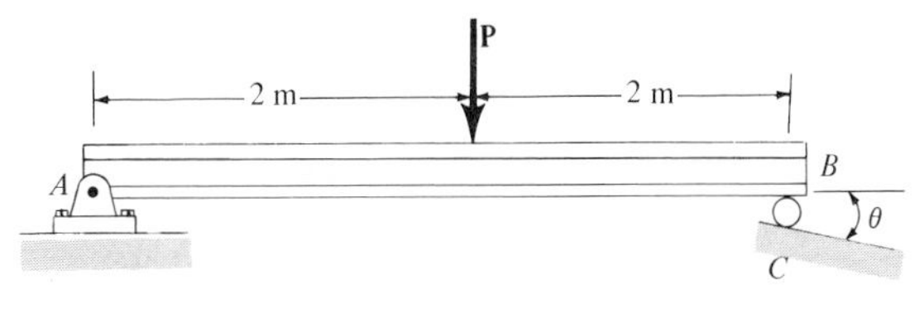

Prob. 8–19

***8–20.** The board can be adjusted vertically by tilting it up and sliding the smooth pin A along the vertical guide G. When placed horizontally, the bottom C bears along the edge of the guide, where $\mu = 0.3$. Determine the smallest dimension d which will support any applied force $\mathbf{F}$ without causing the board to slip downward.

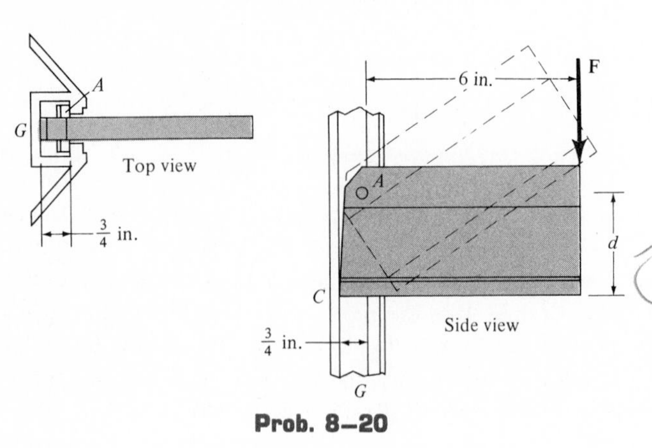

Prob. 8–20

8–21. A 150-lb roofer walks slowly in an upright position down along the surface of a dome that has a radius of curvature of $r = 80$ ft. If the coefficient of friction between his shoes and the dome is $\mu = 0.8$, determine the angle θ at which he first begins to slip.

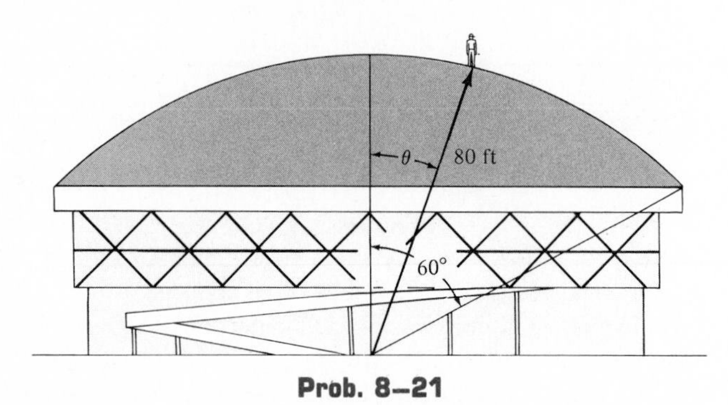

Prob. 8–21

8–22. An ax is driven into the tree stump. If the coefficient of static friction between the ax and the wood is $\mu_s = 0.25$, determine the smallest angle θ of the blade that will cause the ax to be "self-locking," i.e., so it will not slip out. Neglect the weight of the ax.

Prob. 8–22

8–23. If the coefficient of static friction between the drum and brake mechanism is $\mu_s = 0.4$, determine the horizontal and vertical components of reaction at pin O. Does the 160-N force prevent the drum from rotating? Neglect the weight and thickness of the brake. The drum has a mass of 25 kg.

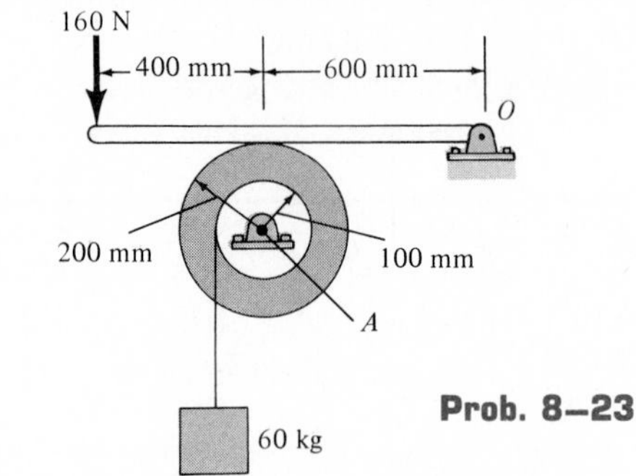

Prob. 8–23

***8–24.** The truck has a weight of 12,000 lb and a center of gravity at G. If it is traveling gradually off a road onto a

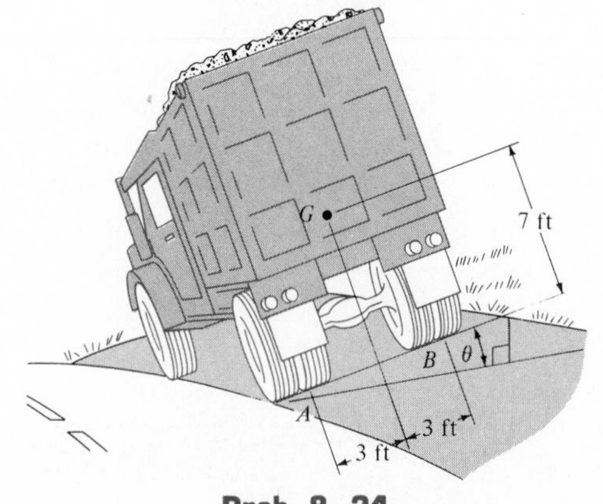

Prob. 8–24

steep shoulder, determine the greatest angle θ at which it can be parked so that its does not slide downward or tip over. Take $\mu_A = \mu_B = 0.4$. *Hint:* Draw a free-body diagram of the truck viewed from the back. For tipping require the normal force at B to be zero.

8–25. The refrigerator has a weight of 200 lb and a center of gravity at G. Determine the force **P** required to move it. Will the refrigerator tip or slip? Take $\mu = 0.4$.

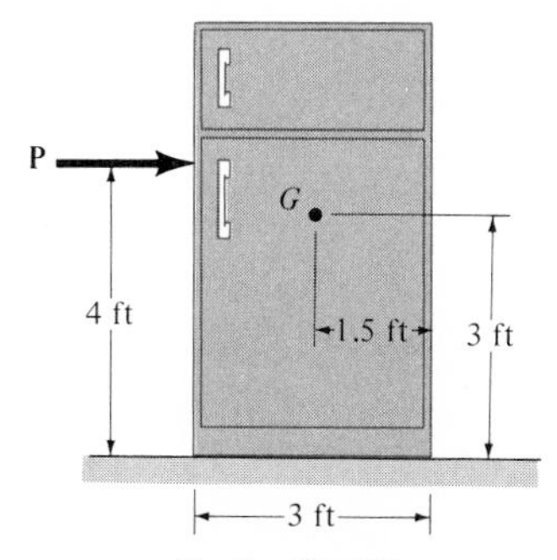

Prob. 8–25

8–26. The boy at D has a mass of 50 kg, a center of mass at G, and stands on a plank at the position shown. The plank is pin-supported at A and rests on a post at B. Neglecting the weight of the plank and post, determine the magnitude of force **P** his friend (?) at E must exert in order to pull out the post. Take $\mu_B = 0.3$ and $\mu_C = 0.8$.

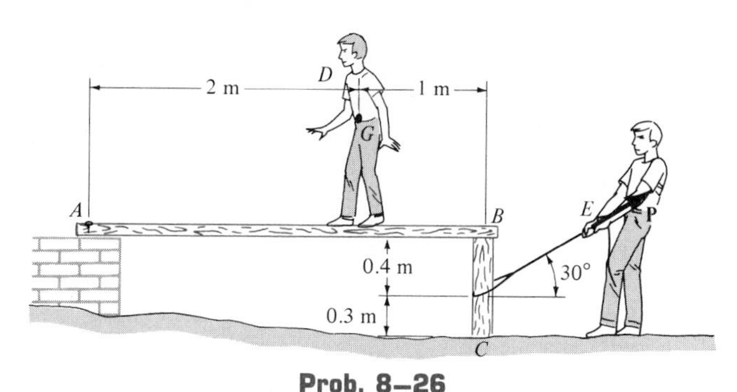

Prob. 8–26

8–27. Determine the minimum force **F** needed to push the two 75-kg cylinders up the incline. The force acts parallel to the plane and the coefficients of friction at the contacting surfaces are $\mu_A = 0.3$, $\mu_B = 0.25$, and $\mu_C = 0.4$. Each cylinder has a radius of 150 mm.

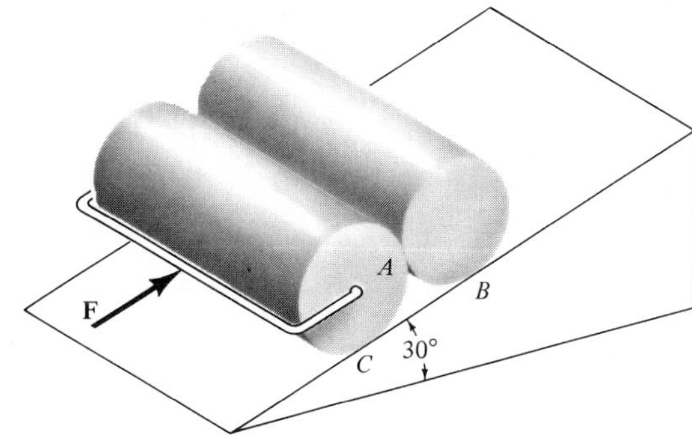

Prob. 8–27

***8–28.** The beam AB has a negligible mass and is subjected to a force of 200 N. It is supported at one end by a pin and at the other end by a spool having a mass of 35 kg. If a cable is wrapped around the inner core of the spool, determine the minimum cable force **P** needed to move the spool from under the beam. The coefficients of friction at B and D are $\mu_B = 0.4$ and $\mu_D = 0.2$, respectively.

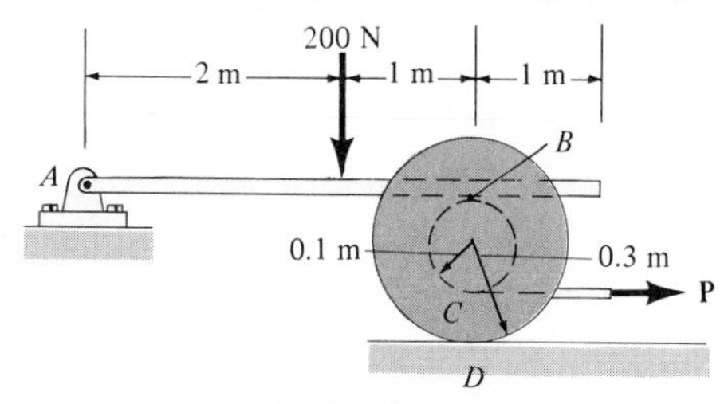

Prob. 8–28

8–29. Block C has a mass of 50 kg and is confined between two walls by smooth rollers. If the block rests on top of a spool that has a mass of 40 kg, determine the minimum cable force **P** needed to move the spool. The cable is wrapped around the spool's inner core. The coefficients of friction at A and B are $\mu_A = 0.3$ and $\mu_B = 0.6$.

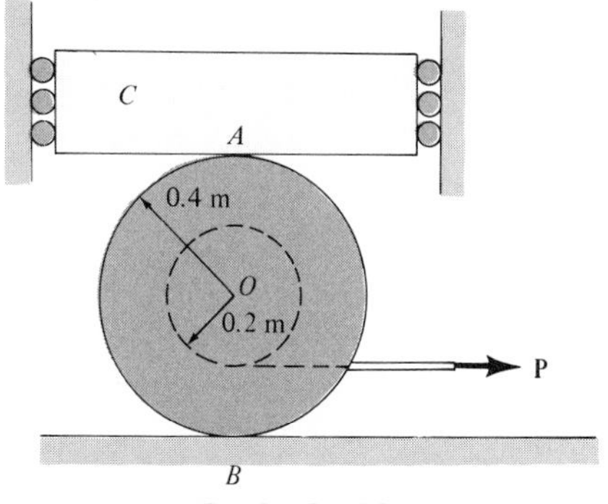

Prob. 8–29

★8.3 Frictional Forces on Screws

A *screw* may be thought of simply as an inclined plane wrapped around a cylinder. A nut initially at position A on the screw shown in Fig. 8–11 will move up to B when rotated 360° around the screw. This rotation is equivalent to translating the nut up an inclined plane of height p and length $l = 2\pi r$, where r is the mean radius of the thread. The rise p is often referred to as the *pitch* of the screw, where the *pitch angle* is given by $\theta_p = \tan^{-1}(p/2\pi r)$.

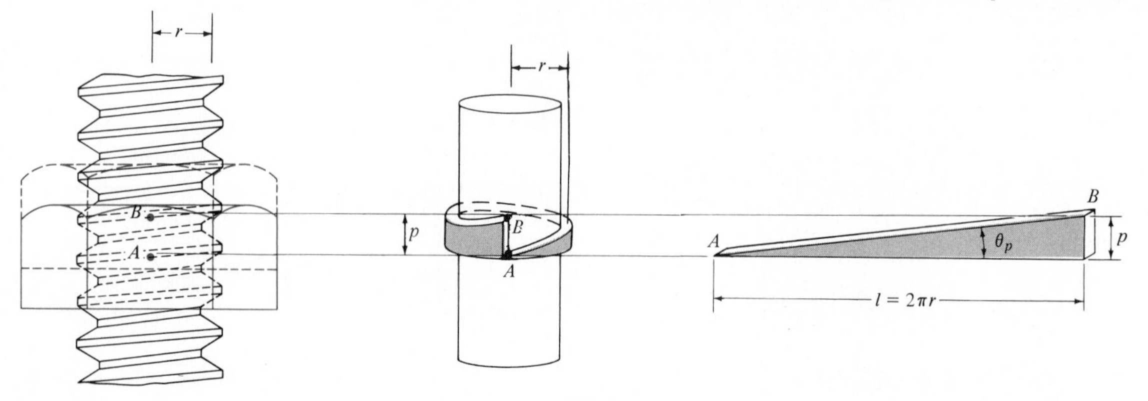

Fig. 8–11

In most cases screws are used as fasteners; however, in many types of machines screws are incorporated to transmit power or motion from one part of the machine to another. A *square-threaded screw* is most commonly used for the latter purpose, especially when large forces are applied along the axis of the screw.

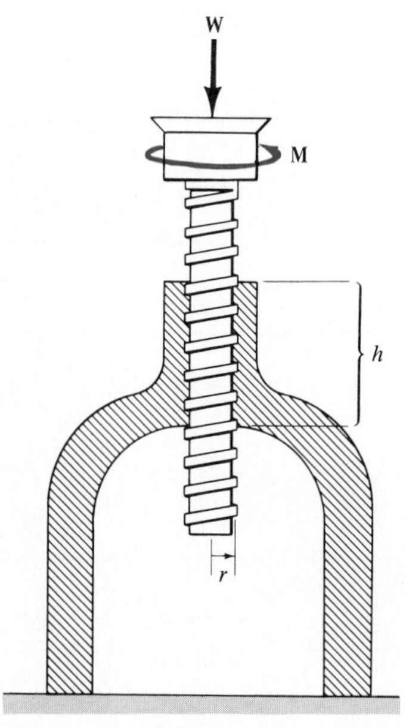

Fig. 8–12

Frictional Analysis. When a screw is subjected to large axial loads, the frictional forces developed at the thread become important in the analysis to determine the force needed to turn the screw. Consider, for example, the square-threaded jack screw shown in Fig. 8–12, which supports the vertical load **W** and twisting moment **M**.* If moments are summed about the axis of the screw, **M** can be thought of as being equivalent to the moment of a horizontal force **S** acting at the mean radius r of the thread, so that $M = Sr$. The reactive forces of the jack to loads S and W are actually distributed over the circumference of the screw thread in contact with the screw hole in the jack, that is, within region h shown in Fig. 8–12. For simplicity, this portion of thread can be imagined as being unwound from the screw and represented as a simple block resting on an inclined plane having the screw's pitch angle

*For applications, **M** is developed by applying a horizontal force **P** at a right angle to the end of a lever that would be fixed to the screw.

θ_p, Fig. 8–13*a*. The inclined plane represents the inside *supporting thread* of the jack base. The block is subjected to the total axial load **W** acting on the jack, the horizontal force **S** (which is related to the applied moment **M**), and the *resultant force* **R** exerted by the plane. As shown, force **R** has components acting normal, **N**, and tangent, **F**, to the contacting surfaces.

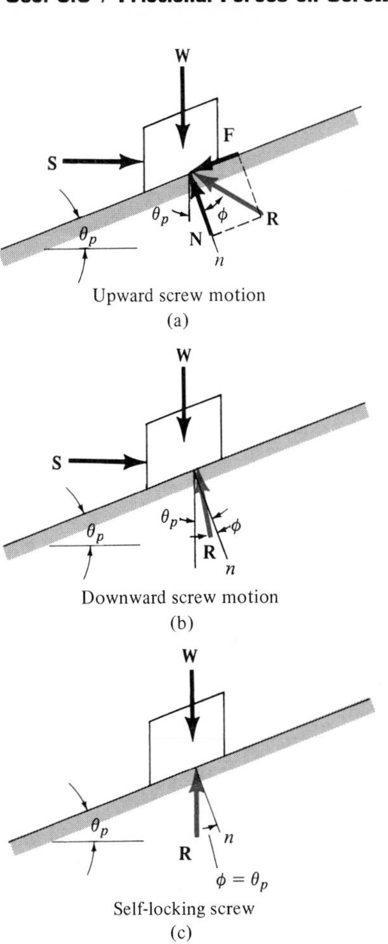

Upward screw motion
(a)
Downward screw motion
(b)
Self-locking screw
(c)
Fig. 8–13

Upward Screw Motion. Provided M is great enough, the screw (and hence the block) is either on the verge of upward impending motion or motion is occurring. Under these conditions, **R** acts at an angle $(\phi + \theta_p)$ from the vertical as shown in the figure, where $\phi = \tan^{-1}(F/N) = \tan^{-1}(\mu N/N) = \tan^{-1}\mu$. Applying the two force equations of equilibrium to the block, we obtain

$$\xrightarrow{+}\Sigma F_x = 0; \qquad S - R\sin(\phi + \theta_p) = 0$$

$$+\uparrow \Sigma F_y = 0; \qquad R\cos(\phi + \theta_p) - W = 0$$

Eliminating R and solving for S, then substituting this value into the equation $M = Sr$, yields

$$M = Wr\tan(\phi + \theta_p) \qquad (8\text{–}3)$$

This equation gives the required value M necessary to cause upward impending motion of the screw when $\phi = \phi_s = \tan^{-1}\mu_s$ (the angle of static friction). If ϕ is replaced by $\phi_k = \tan^{-1}\mu_k$ (the angle of kinetic friction), Eq. 8–3 would give a smaller value M necessary to maintain uniform upward motion of the screw.

Downward Screw Motion. When the load **W** is to be *lowered,* the direction of **M** is reversed, in which case the angle ϕ (ϕ_s or ϕ_k) lies on the opposite side of the normal n to the plane supporting the block. This case is shown in Fig. 8–13*b* for $\phi < \theta_p$. Thus, Eq. 8–3 becomes

$$M = Wr\tan(\phi - \theta_p) \qquad (8\text{–}4)$$

Self-locking Screw. If the moment **M** is *removed,* the screw will remain *self-locking;* i.e., it will support the load **W** *by friction forces alone* provided $\phi > \theta_p$. However, if $\phi = \phi_s = \theta_p$, Fig. 8–13*c*, the screw will be on the verge of rotating downward. When $\theta_p > \phi_s$, a restraining moment is needed to prevent the screw from rotating downward.

Example 8–7

The turnbuckle shown in Fig. 8–14 has a square thread with a mean radius of 5 mm and a pitch of 2 mm. If the coefficient of friction between the screw and the turnbuckle is $\mu_s = 0.25$, determine the moment **M** that must be applied to draw the end screws closer together. Is the turnbuckle "self-locking"?

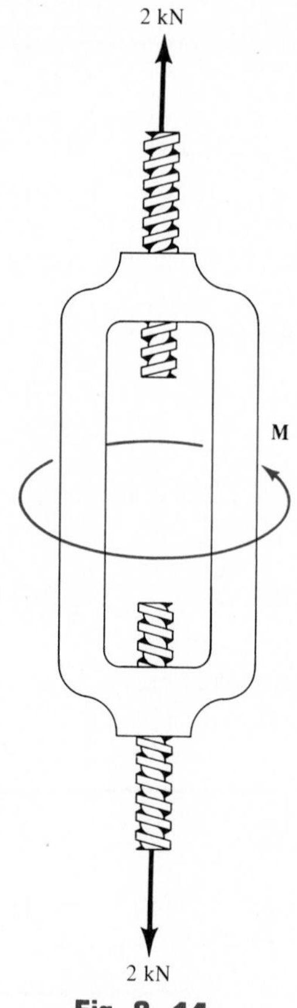

Fig. 8–14

Solution

The moment may be obtained by using Eq. 8–3. Since friction at two screws must be overcome, this requires

$$M = 2[Wr \tan(\phi_s + \theta_p)] \qquad (1)$$

Here, $W = 2000$ N, $r = 5$ mm, $\phi_s = \tan^{-1}\mu_s = \tan^{-1}(0.25) = 14.04°$, and $\theta_p = \tan^{-1}(p/2\pi r) = \tan^{-1}(2\text{ mm}/[2\pi(5\text{ mm})]) = 3.64°$. Substituting these values into Eq. (1) and solving gives

$$M = 2[(2000\text{ N})(5\text{ mm})\tan(14.04° + 3.64°)]$$

$$M = 6375.1\text{ N}\cdot\text{mm} = 6.38\text{ N}\cdot\text{m} \qquad \textit{Ans.}$$

When the moment is *removed*, the turnbuckle will be self-locking; i.e., it will not unscrew, since $\phi_s > \theta_p$.

Frictional Forces on Collar Bearings, Pivot Bearings, and Disks 8.4*

Pivot or *collar bearings* are commonly used to support the *axial* or *normal loads* of a rotating shaft. These two types of support are shown in Fig. 8–15. Provided the bearings are not lubricated, or only partially lubricated, the laws of dry friction may be applied to determine the moment **M** needed to turn the shaft when the shaft supports an axial force **P**.

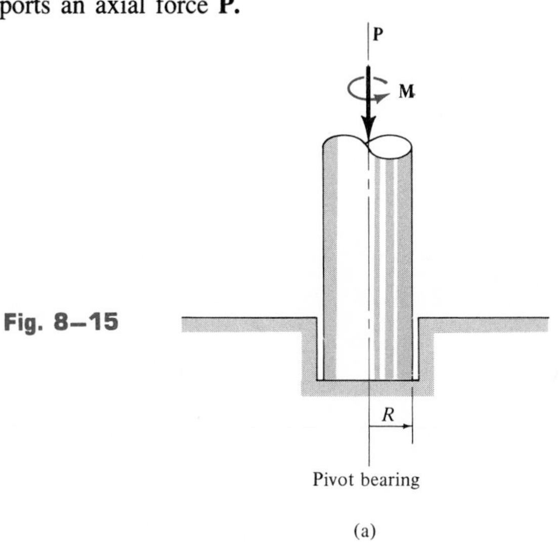

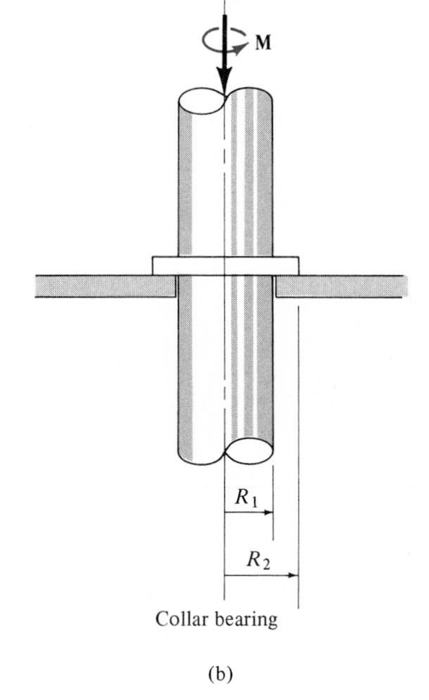

Fig. 8–15

Frictional Analysis. Consider, for example, the end of a collar-bearing shaft shown in Fig. 8–16, which is subjected to an axial force **P** and has a total bearing or contact area $\pi(R_2^2 - R_1^2)$. In the following analysis the normal pressure p is considered to be *uniformly distributed* over this area—a reasonable assumption provided the bearing is new and evenly supported. Since $\Sigma F_z = 0$, p, measured as a force per unit area, is $p = P/\pi(R_2^2 - R_1^2)$.

The moment needed to cause impending rotation of the shaft can be determined from moment equilibrium of the frictional forces $d\mathbf{F}$ developed at the bearing surface by applying $\Sigma M_z = 0$. A small area element $dA = (r\,d\theta)(dr)$, shown in Fig. 8–16, is subjected to both a normal force $dN = p\,dA$ and an associated frictional force

$$dF = \mu_s\,dN = \mu_s p\,dA = \frac{\mu_s P}{\pi(R_2^2 - R_1^2)}\,dA$$

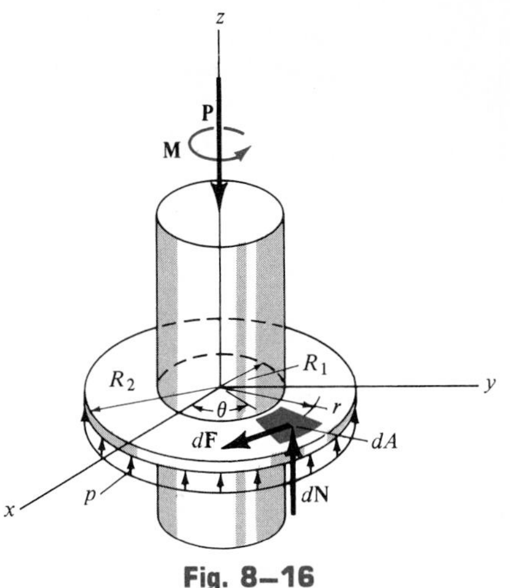

Fig. 8–16

The normal force does not create a moment about the z axis of the shaft; however, the frictional force does, namely $dM = r\,dF$. Integration is needed to compute the total moment created by all the frictional forces acting on differential areas dA. Therefore, for impending rotational motion,

$$\Sigma M_z = 0; \qquad M - \int_A r\, dF = 0$$

Using $dA = (r\, d\theta)(dr)$ and integrating over the entire bearing area yields

$$M = \int_{R_1}^{R_2} \int_0^{2\pi} r \left[\frac{\mu_s P}{\pi(R_2^2 - R_1^2)} \right] (r\, d\theta\, dr) = \frac{\mu_s P}{\pi(R_2^2 - R_1^2)} \int_{R_1}^{R_2} r^2\, dr \int_0^{2\pi} d\theta$$

or

$$M = \tfrac{2}{3}\mu_s P \left(\frac{R_2^3 - R_1^3}{R_2^2 - R_1^2} \right) \qquad (8\text{–}5)$$

This equation gives the magnitude of moment required for impending rotation of the shaft. The frictional moment developed at the end of the shaft, when it is *rotating* at constant speed, can be found by substituting μ_k for μ_s in Eq. 8–5.

When $R_2 = R$ and $R_1 = 0$, as in the case of a pivot bearing, Fig. 8–15*a*, Eq. 8–5 reduces to

$$M = \tfrac{2}{3}\mu_s PR \qquad (8\text{–}6)$$

Recall from the initial assumption that both Eqs. 8–5 and 8–6 apply only for bearing surfaces subjected to *constant pressure*. If the pressure is not uniform, a variation of the pressure as a function of the bearing area must be determined before integrating to obtain the moment. The following example illustrates this concept.

Example 8–8

The pivot bearing shown in Fig. 8–17 is subjected to an axial load of 60 lb. If the coefficient of friction at the contact surface is $\mu = 0.3$, determine the torsional moment **M** which must be applied to turn the shaft.

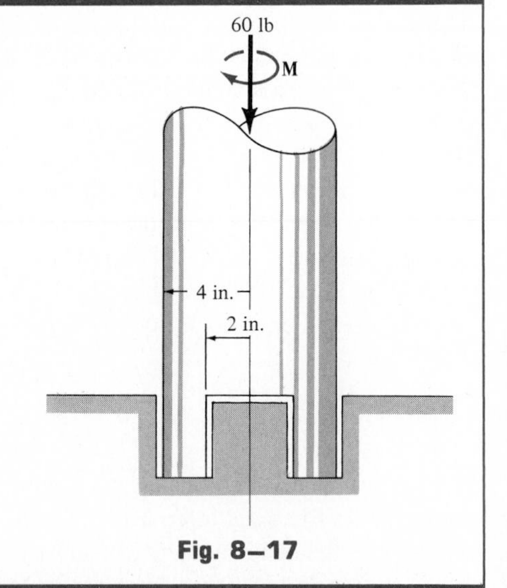

Fig. 8–17

Solution

Assuming the normal pressure is *uniformly distributed* under the bearing, we can apply Eq. 8–5, i.e.,

$$M = \tfrac{2}{3}\mu P \left(\frac{R_2^3 - R_1^3}{R_2^2 - R_1^2} \right)$$

$$= \tfrac{2}{3}(0.3)(60 \text{ lb}) \left(\frac{(4 \text{ in.})^3 - (2 \text{ in.})^3}{(4 \text{ in.})^2 - (2 \text{ in.})^2} \right)$$

$$= 56.0 \text{ lb} \cdot \text{in.} = 4.67 \text{ lb} \cdot \text{ft} \qquad \textit{Ans.}$$

Frictional Forces on Journal Bearings 8.5*

When a shaft or axle is subjected to lateral loads, a *journal bearing* is commonly used for support. Well-lubricated journal bearings are subjected to the laws of fluid mechanics, in which the viscosity of the lubricant, the speed of rotation, and the amount of clearance between the shaft and bearing are needed to determine the frictional resistance of the bearing. When the bearing is not lubricated or only partially lubricated, however, a reasonable analysis of the frictional resistance can be based on the laws of dry friction.

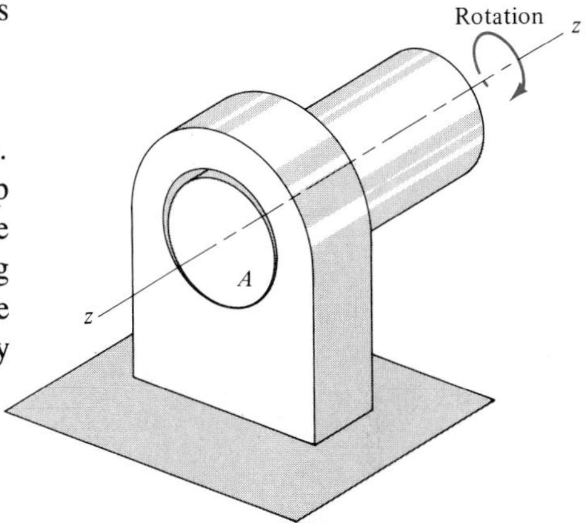

(a)

Frictional Analysis. A typical journal-bearing support is shown in Fig. 8–18*a*. As the shaft rotates in the direction shown in the figure, it rolls up against the wall of the bearing to some point *A* where slipping occurs. If the lateral load acting at the end of the shaft is **W,** it is necessary that the bearing reactive force **R** acting at *A* be equal and opposite to **W,** Fig. 8–18*b*. The moment needed to maintain constant rotation of the shaft can be found by summing moments about the *z* axis of the shaft; i.e.,

$$\Sigma M_z = 0; \qquad -M + (R \sin \phi_k)r = 0$$

or

$$M = Rr \sin \phi_k$$

where ϕ_k is the angle of kinetic friction defined by $\tan \phi_k = F/N = \mu_k N/N = \mu_k$. In Fig. 8–18*c*, it is seen that $r \sin \phi_k = r_f$. The dashed circle with radius r_f is called the *friction circle,* and as the shaft rotates, the reaction **R** will always be tangent to it. If the bearing is partially lubricated, μ_k is small, and therefore $\mu_k = \tan \phi_k \approx \sin \phi_k \approx \phi_k$. Under these conditions, a reasonable approximation to the moment needed to overcome the frictional resistance becomes

$$M \approx Rr\mu_k \tag{8–7}$$

The following example illustrates a common application of this equation.

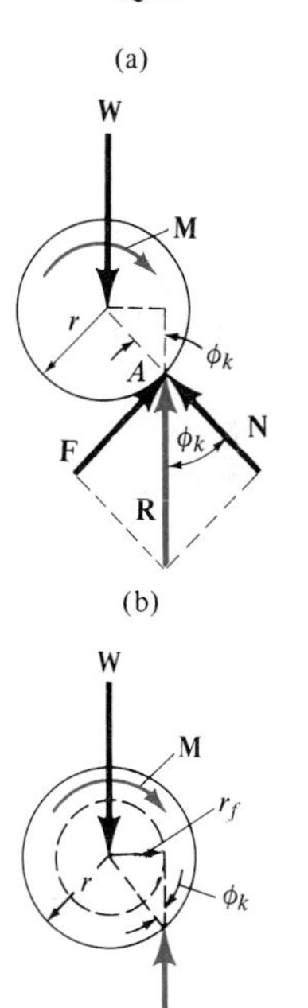

Fig. 8–18

Example 8–9

The 100-mm-diameter pulley shown in Fig. 8–19*a* fits loosely on a 10-mm-diameter shaft for which the coefficients of static and kinetic friction are $\mu = 0.4$. Determine the minimum tension **T** in the belt to (a) raise the 100-kg block at constant velocity, and (b) lower the block at constant velocity. Assume no slipping occurs between the belt and pulley and neglect the weight of the pulley.

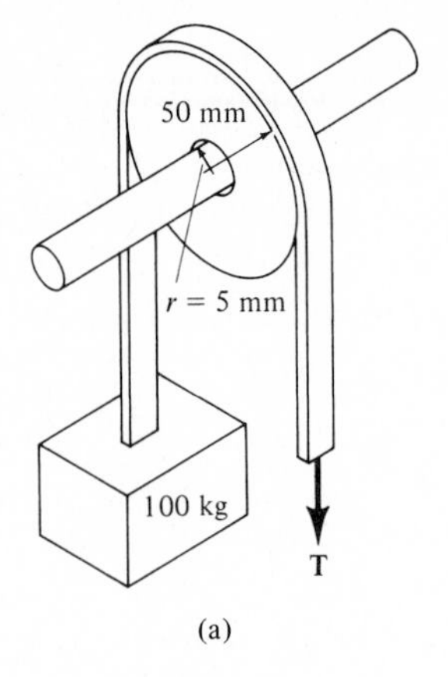

(a)

Solution

Part (a). A free-body diagram of the pulley is shown in Fig. 8–19*b*. When the pulley is subjected to cable tensions of 981 N each, the pulley makes contact with the shaft at point P_1. As the tension **T** is *increased,* the pulley will roll around the shaft to point P_2 before motion impends. From the figure, the friction circle has a radius $r_f = r \sin \phi$. Using the simplification $\sin \phi \approx \mu$, $r_f \approx r\mu = (5 \text{ mm})(0.4) = 2 \text{ mm}$, so that summing moments about P_2 gives

$$\circlearrowleft +\Sigma M_{P_2} = 0; \qquad 981 \text{ N}(52 \text{ mm}) - T(48 \text{ mm}) = 0$$

$$T = 1062.8 \text{ N} \qquad \textit{Ans.}$$

Part (b). When the block is lowered, the resultant force **R** acting on the shaft passes through point P_3, as shown in Fig. 8–19*c*. Summing moments about this point yields

$$\circlearrowleft +\Sigma M_{P_3} = 0; \qquad 981 \text{ N}(48 \text{ mm}) - T(52 \text{ mm}) = 0$$

$$T = 905.5 \text{ N} \qquad \textit{Ans.}$$

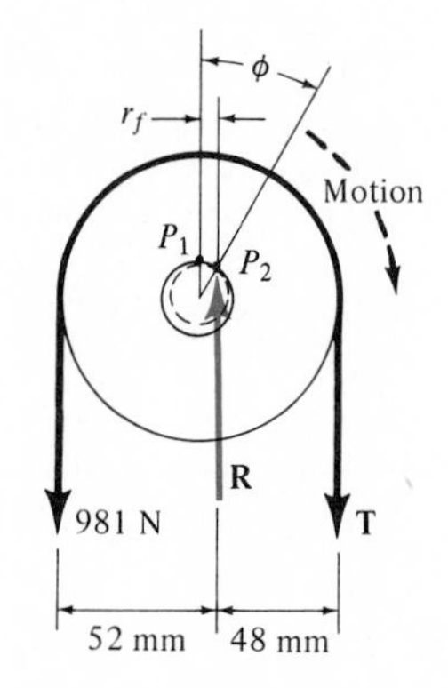

(b)

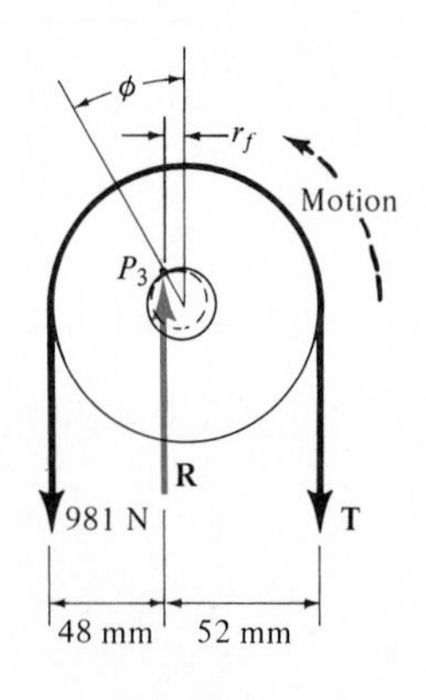

(c)

Fig. 8–19

Problems

8–30. The square-threaded bolt is used to join two plates together. If the bolt has a mean diameter of $d = 0.75$ in. and a pitch of $p = 0.6$ in., determine the smallest torque **M** required to loosen the bolt if the tension in the bolt is $T = 600$ lb. The coefficient of static friction between the threads and the bolt is $\mu_s = 0.4$.

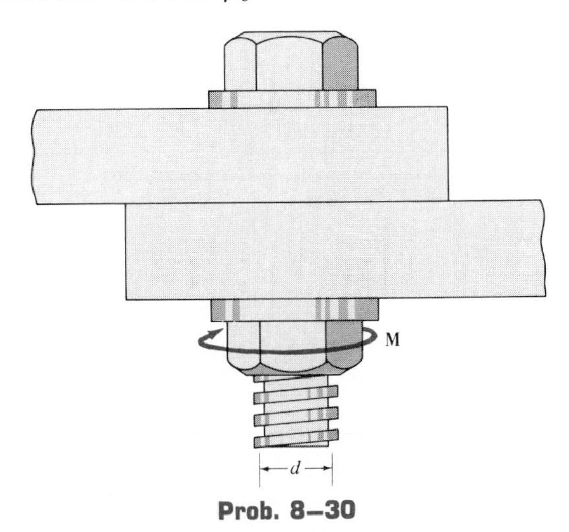

Prob. 8–30

8–31. The square-threaded screw of the clamp has a mean diameter of 14 mm and a pitch of 6 mm. If $\mu_s = 0.2$ for the threads, and the torque applied to the handle is 1.5 N · m, determine the compressive force **F** on the block.

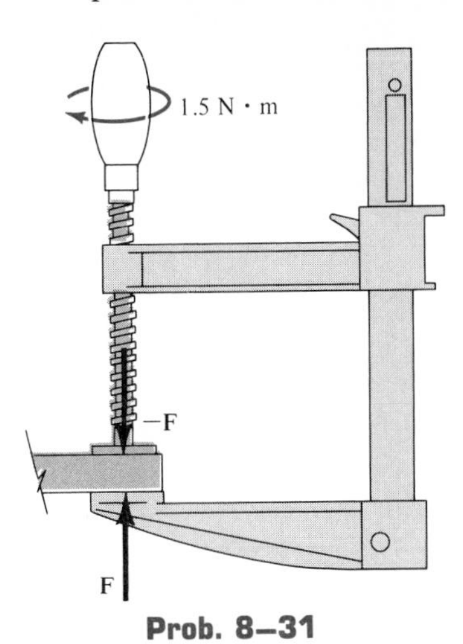

Prob. 8–31

***8–32.** The braking mechanism consists of two pinned arms and a square-threaded screw with left and right-hand threads. Thus, when turned, the screw draws the two arms together. If the pitch of the screw is 4 mm, the mean diameter 12 mm, and the coefficient of friction is $\mu = 0.35$, determine the tension in the screw when a torque of 3 N · m is applied to the screw. If the coefficient of friction between the brake pads A and B and the circular shaft is $\mu' = 0.5$, determine the maximum torque **M** the shaft can resist.

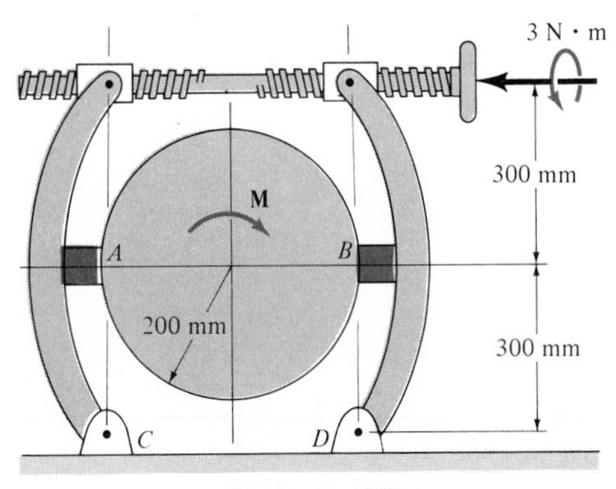

Prob. 8–32

8–33. The plate A is engaged to plate B using three springs. If each spring exerts a force of 120 lb on plate B and $\mu = 0.35$ for the contacting surfaces, determine the maximum torque **M** that may be transmitted across the plates. Assume the pressure created is uniformly distributed between the plates.

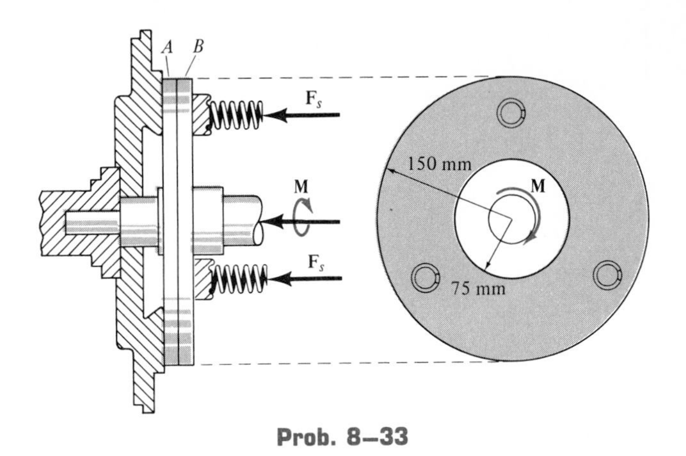

Prob. 8–33

8–34. The square-threaded screw of the vise clamp has a mean diameter of 30 mm and a pitch of 4 mm. If $\mu_s = 0.25$ for the threads, and the force applied perpendicular to the handle is $P = 100$ N, determine the compressive force in the block.

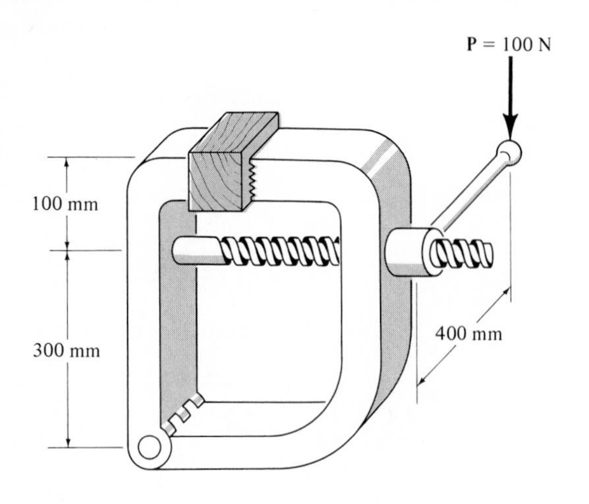

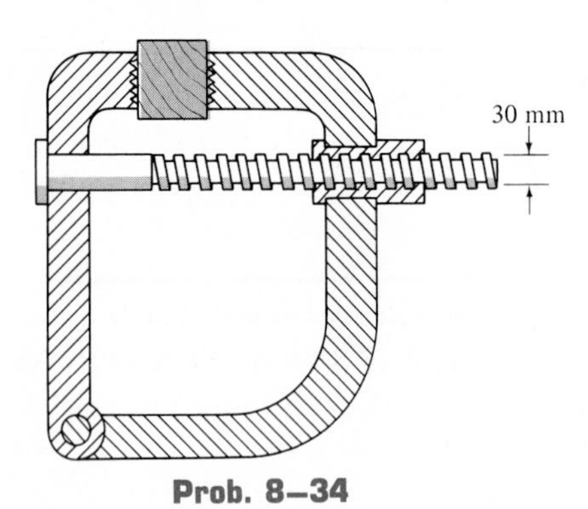

Prob. 8–34

8–35. The shaft has a square-threaded screw with a pitch of $p = 8$ mm and a mean radius of 15 mm. If it is in contact with a plate gear having a mean radius of 30 mm, determine the resisting torque **M** on the plate gear which can be overcome if a torque of 7 N · m is applied to the shaft. The coefficient of friction at the screw is $\mu = 0.2$. Neglect friction of the bearings located at A and B.

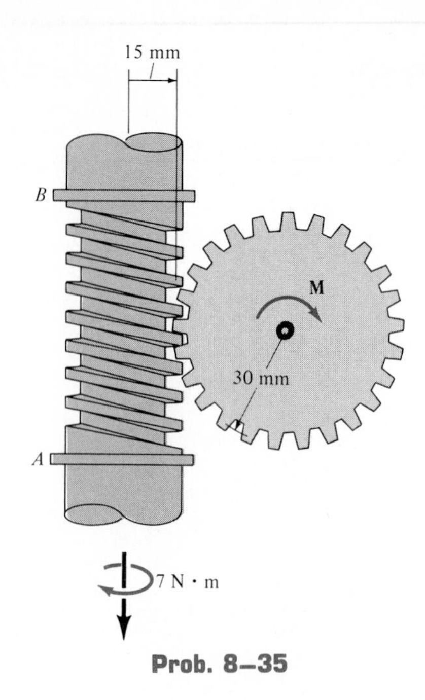

Prob. 8–35

***8–36.** The collar bearing uniformly supports an axial force of $P = 500$ lb. If a torque of $M = 10$ lb · ft is required to overcome friction developed by the bearing, determine the coefficient of friction μ acting at the bearing surface.

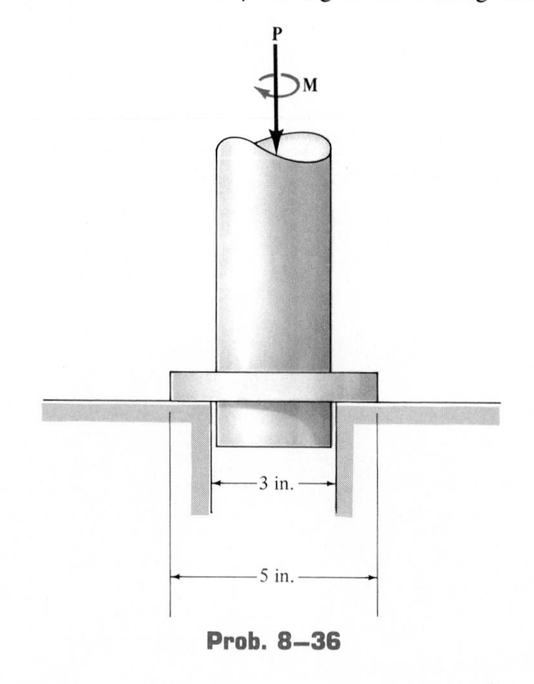

Prob. 8–36

8–37. Determine the torque **M** required to turn the shaft which supports an axial force of 6 kN. The coefficient of friction is $\mu = 0.5$.

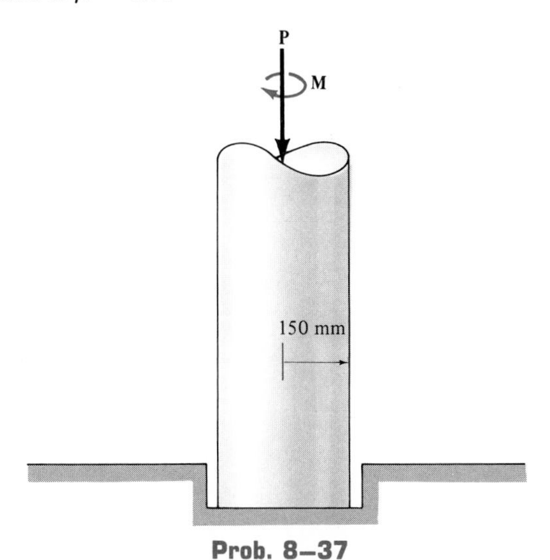

Prob. 8–37

8–38. The *double-collar bearing* is subjected to an axial force of $P = 4$ kN. Assuming that collar A supports $0.6P$ and collar B supports $0.4P$, both with a uniform distribution of pressure, determine the maximum frictional torque **M** that may be resisted by the bearing. $\mu_s = 0.2$ for both collars.

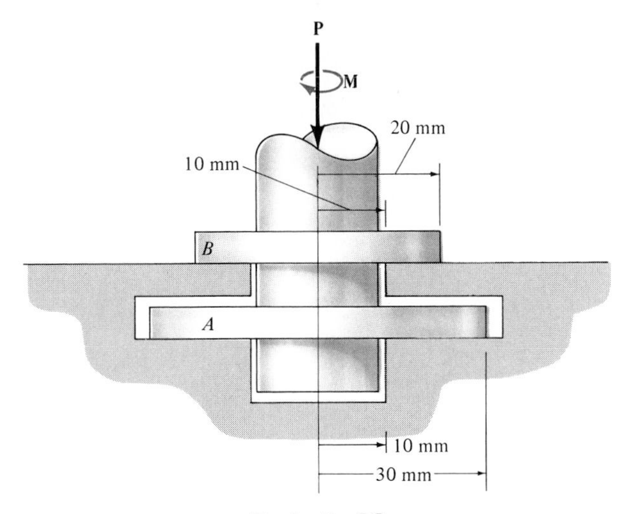

Prob. 8–38

8–39. The collar bushing B fits loosely over the fixed shaft S, which has a radius of 20 mm. Determine the smallest force $\mathbf{T}_A$ needed to pull the belt downward at A. Also determine the resultant normal and frictional components of force developed on the collar bushing if $\mu = 0.3$ between the collar and the shaft. Assume that the belt does not slip on the collar; rather, the collar slips on the shaft.

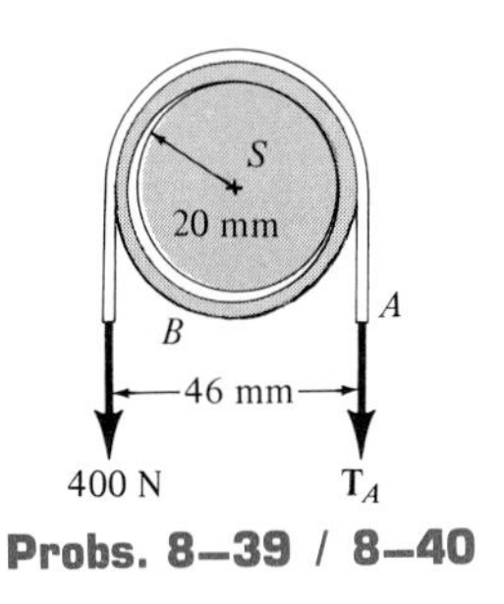

Probs. 8–39 / 8–40

***8–40.** If the smallest tension force $T_A = 500$ N is required to pull the belt downward at A over the shaft S, determine the coefficient of friction between the loosely fitting collar bushing B and the shaft. Assume that the belt does not slip on the collar; rather, the collar slips on the shaft.

8–41. A disk having an outer diameter of 12 in. fits loosely over a fixed shaft having a diameter of 4 in. If the coefficient of friction between the disk and the shaft is $\mu = 0.15$, determine the smallest vertical force **F**, acting on the rim, which must be applied to the disk to cause it to slip over the shaft. The disk weighs 10 lb.

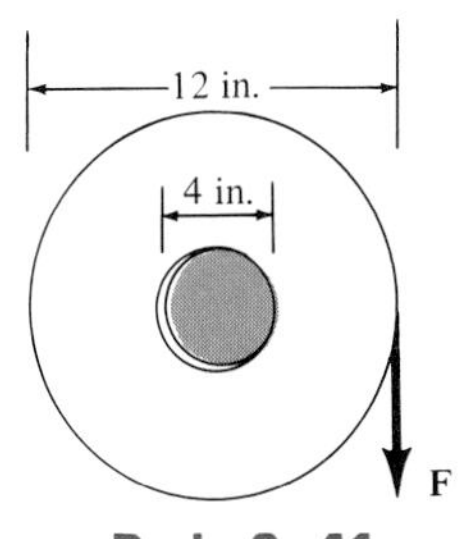

Prob. 8–41

8–42. The collar fits *loosely* around a fixed shaft that has a radius of 2 in. If the coefficient of friction between the shaft and the collar is $\mu = 0.3$, determine the tension **T** in the horizontal segment of the belt so that the belt can be lowered in the direction of the 20-lb force with a constant speed. Assume that the belt does not slip on the collar; rather, the collar slips on the shaft. Neglect the weight and thickness of the belt and collar. The radius, measured from the center of the collar to the mean thickness of the belt, is 2.25 in.

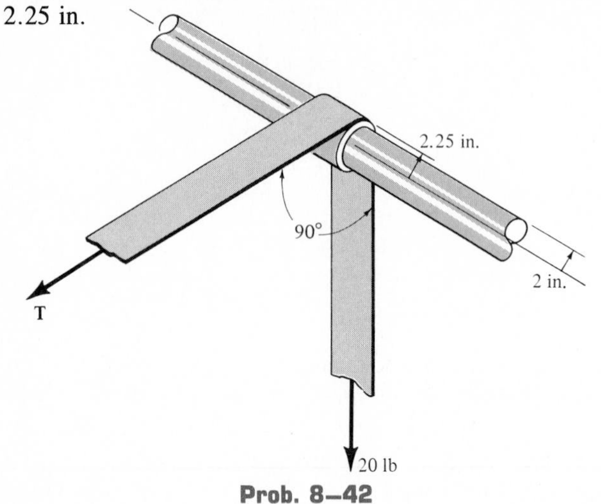

Prob. 8–42

8–43. The axle of the pulley fits loosely in a 50-mm-diameter pin hole. If $\mu = 0.30$, determine the minimum tension **T** required to raise the 50-kg block. Neglect the weight of the pulley and assume that the cord does not slip on the pulley.

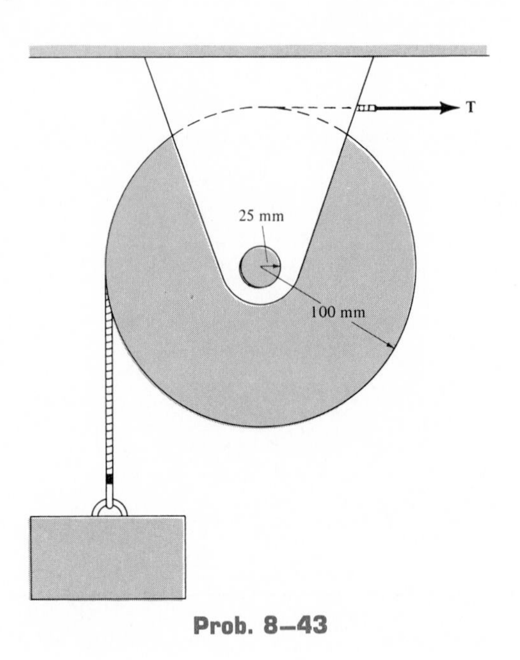

Prob. 8–43

*8.6 Frictional Forces on Flat Belts

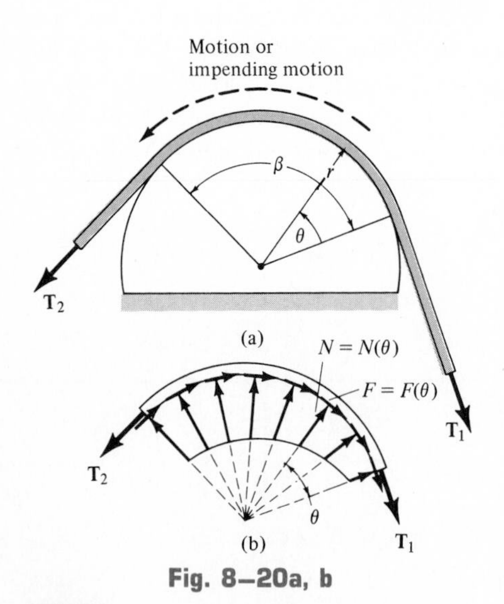

Fig. 8–20a, b

In the design of belt drives or band brakes it is necessary to determine the frictional forces developed between a belt and its contacting surface. Consider the flat belt passing over a fixed drum of radius r, as shown in Fig. 8–20a. The *total* angle of belt contact in radians is β and the coefficient of friction between the two surfaces is μ. If it is known that the tension acting in the belt on the right of the drum is $\mathbf{T}_1$, let it be required to find the tension $\mathbf{T}_2$ needed to pull the belt counterclockwise over the surface of the drum. *Obviously, T_2 must be greater than T_1 since the belt must overcome the resistance of friction at the surface of contact.*

Frictional Analysis. A free-body diagram of the belt is shown in Fig. 8–20b. It is seen that the normal force **N** and the frictional force **F**, acting at different points along the contacting surface of the belt, will vary both in magnitude and direction. Due to this unknown force distribution, the analysis of the problem will proceed on the basis of initially studying the forces acting on a differential element of the belt.

A free-body diagram of an element having a length $ds = r\,d\theta$ is shown in Fig. 8–20*c*. Assuming either impending motion or motion of the belt, the magnitude of frictional force $dF = \mu\,dN$. This force opposes the sliding motion of the belt and thereby increases the magnitude of tensile force acting in the belt by dT. Applying the two force equations of equilibrium, we have

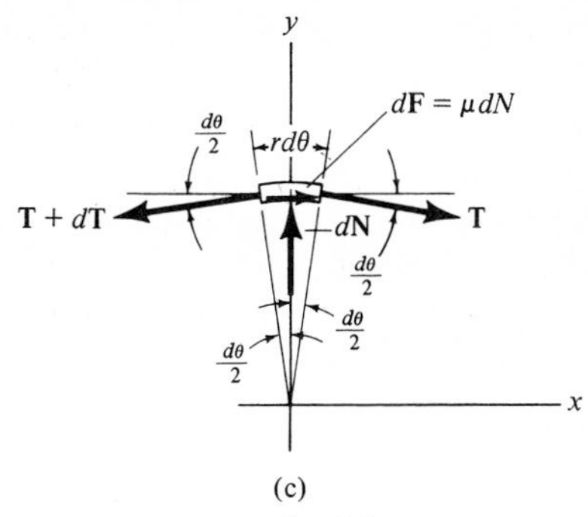

(c)

Fig. 8–20c

$$\overset{+}{\rightarrow}\Sigma F_x = 0; \quad T\cos\left(\frac{d\theta}{2}\right) + \mu\,dN - (T + dT)\cos\left(\frac{d\theta}{2}\right) = 0$$

$$+\uparrow\Sigma F_y = 0; \quad dN - (T + dT)\sin\left(\frac{d\theta}{2}\right) - T\sin\left(\frac{d\theta}{2}\right) = 0$$

Since $d\theta$ is of *infinitesimal size,* $\sin(d\theta/2)$ and $\cos(d\theta/2)$ can be replaced by $d\theta/2$ and 1, respectively. Also, the *product* of two infinitesimals dT and $d\theta/2$ may be neglected when compared to infinitesimals of the first order. The above two equations therefore reduce to

$$\mu\,dN = dT$$

and

$$dN = T\,d\theta$$

Eliminating dN yields

$$\frac{dT}{T} = \mu\,d\theta$$

Integrating this equation between all the points of contact that the belt makes with the drum, and noting that $T = T_1$ at $\theta = 0$ and $T = T_2$ at $\theta = \beta$, yields

$$\int_{T_1}^{T_2}\frac{dT}{T} = \mu\int_0^{\beta} d\theta$$

or

$$\ln\frac{T_2}{T_1} = \mu\beta$$

Solving for T_2, we obtain

$$T_2 = T_1 e^{\mu\beta} \qquad (8\text{–}8)$$

where T_2, T_1 = belt tensions; $\mathbf{T}_1$ opposes the direction of motion (or impending motion) of the belt, while $\mathbf{T}_2$ acts in the direction of belt motion (or impending motion); because of friction, $T_2 > T_1$

μ = coefficient of static or kinetic friction between the belt and the surface of contact

β = angle of belt to surface contact, measured in radians

e = 2.718. . . , base of the natural logarithm

Note that Eq. 8–8 is *independent* of the *radius* of the drum and instead depends upon the angle of belt to surface contact, β. Furthermore, as indicated by the integration, this equation is valid for flat belts placed on *any shape* of contacting surface.

Example 8–10

The maximum tension that can be developed in the belt shown in Fig. 8–21a is 500 N. If the pulley at A is free to rotate and the coefficient of static friction at the fixed drums B and C is $\mu_s = 0.25$, determine the largest mass of the cylinder that can be lifted by the belt. Assume that the force $\mathbf{T}$ applied at the end of the belt is directed vertically downward, as shown.

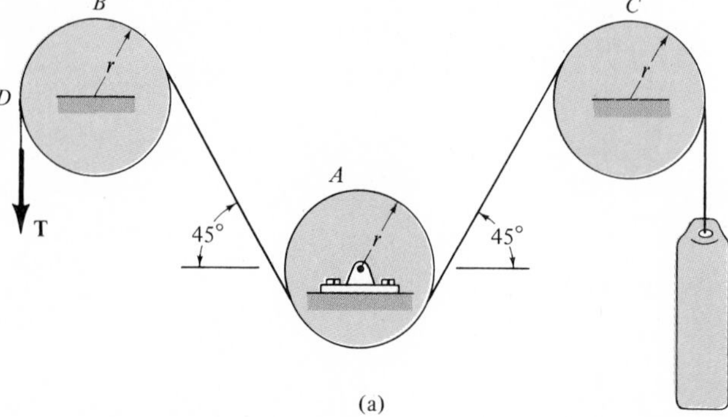

(a)

Solution

Lifting the cylinder, which has a weight $W = mg$, causes the belt to move counterclockwise over the drums at B and C; hence, the maximum tension T_2 in the belt occurs at D. Thus, $T_2 = 500$ N. A section of the belt passing over the drum at B is shown in Fig. 8–21b. Since $180° = \pi$ rad, the angle of contact between the drum and the belt is $\beta = (135°/180°)\pi = 3\pi/4$ rad. Using Eq. 8–8, we have

$$T_2 = T_1 e^{\mu_s \beta}; \qquad 500 \text{ N} = T_1 e^{0.25[(3/4)\pi]}$$

Hence,

$$T_1 = \frac{500 \text{ N}}{e^{0.25[(3/4)\pi]}} = \frac{500}{1.80} = 277.4 \text{ N}$$

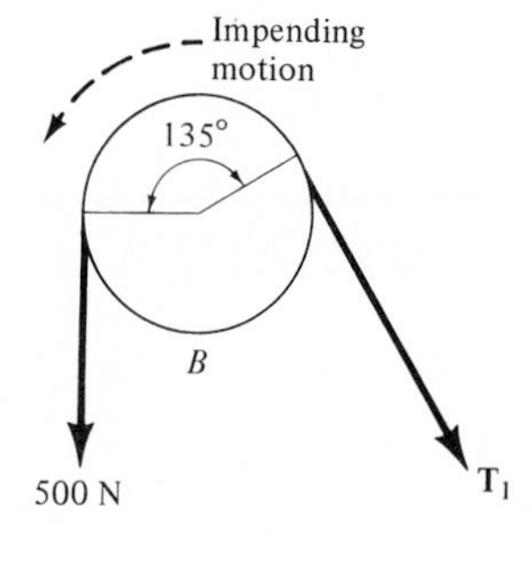

(b)

Since the pulley at A is free to rotate, equilibrium requires that the tension in the belt remains the *same* on both sides of the pulley.

The section of the belt passing over the drum at C is shown in Fig. 8–21c. The load $W < 277.4$ N. Why? Applying Eq. 8–8, we obtain

$$T_2 = T_1 e^{\mu_s \beta}; \qquad 277.4 = W e^{0.25[(3/4)\pi]}$$

$$W = 153.9 \text{ N}$$

so that

$$m = \frac{W}{g} = \frac{153.9}{9.81}$$

$$= 15.7 \text{ kg} \qquad \textit{Ans.}$$

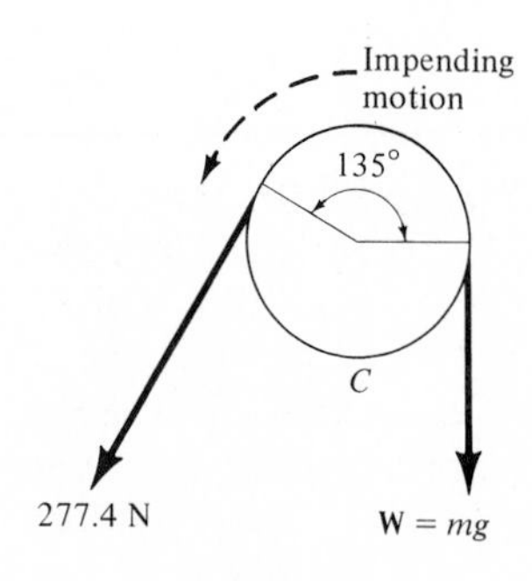

(c)

Fig. 8–21

Rolling Resistance 8.7*

If a *rigid* cylinder of weight **W** rolls at constant velocity along a *rigid* surface, the normal force exerted by the surface on the cylinder acts at the tangent point of contact, as shown in Fig. 8–22*a*. Under these conditions, provided the cylinder does not encounter frictional resistance from the air, motion would continue indefinitely. Actually, however, no materials are perfectly rigid; and therefore the reaction of the surface on the cylinder consists of a *distribution* of normal pressure. If we assume that the cylinder is much harder than the surface, then a small "hill" is formed which is caused by the indentation of the cylinder into the surface, Fig. 8–22*b*. Since the cylinder can never really "climb" over this "hill," the normal pressure is always present, and consequently, to maintain motion, a driving force **P** must be applied to the cylinder. For analysis, the normal pressure distribution can be replaced by its *resultant* force **N,** which acts at an angle θ from the vertical, Fig. 8–22*c*. To keep the cylinder in equilibrium, i.e., moving at constant velocity, it is necessary that **N** be *concurrent* with the driving force **P** and the weight **W.** Summing moments about point A gives $Wa = P(r \cos \theta)$. Since the deformations of the cylinder are generally small in relation to the radius, $\cos \theta \approx 1$; hence,

$$Wa \approx Pr$$

or

$$P \approx \frac{Wa}{r} \qquad (8\text{–}9)$$

The distance a is termed the *coefficient of rolling resistance,* which has the dimension of length. For instance, $a \approx 0.5$ mm for a mild steel wheel rolling on a steel rail. For hardened steel ball bearings on steel $a \approx 0.1$ mm. Experimentally, though, this factor is difficult to measure, since it depends upon such parameters as the rate of rotation of the cylinder, the elastic properties of the contacting surfaces, and the frictional effects of the surfaces. For this reason, little reliance is placed on the data for determining a. The analysis presented here does, however, indicate why a heavy load offers greater resistance to motion than a light load under the same conditions. Furthermore, since the force needed to *roll* the cylinder over the surface is much less than that needed to *slide* the cylinder across the surface, the analysis indicates why roller or ball bearings are often used to minimize the frictional resistance between moving parts.

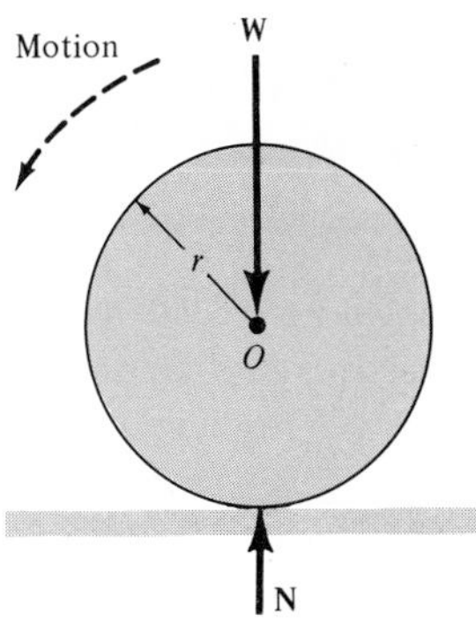

Rigid surface of contact
(a)

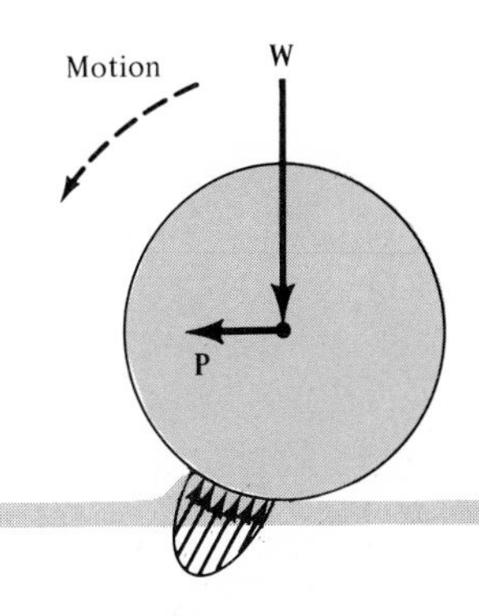

Soft surface of contact
(b)

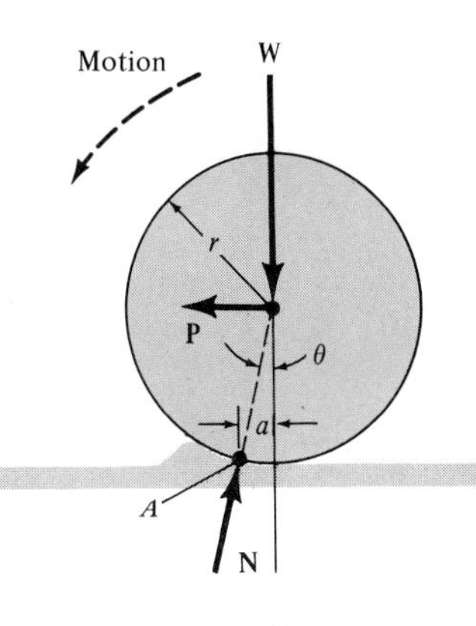

(c)

Fig. 8–22

Example 8–11

A 10-kg steel wheel shown in Fig. 8–23*a* has a radius of 100 mm and rests on an inclined plane made of wood. If θ is increased so that the wheel begins to roll down the incline with constant velocity when $\theta = 1.2°$, determine the coefficient of rolling resistance.

Solution

As shown on the free-body diagram, Fig. 8–23*b*, when the wheel has impending motion, the normal reaction **N** acts at point A defined by the dimension a. Resolving the weight into rectangular components parallel and perpendicular to the incline, and summing moments about point A, yields (approximately)

$$\circlearrowleft +\Sigma M_A = 0; \quad -98.1 \cos 1.2°(a) + 98.1 \sin 1.2°(100) = 0$$

Solving, we obtain

$$a = 2.1 \text{ mm} \qquad \textit{Ans.}$$

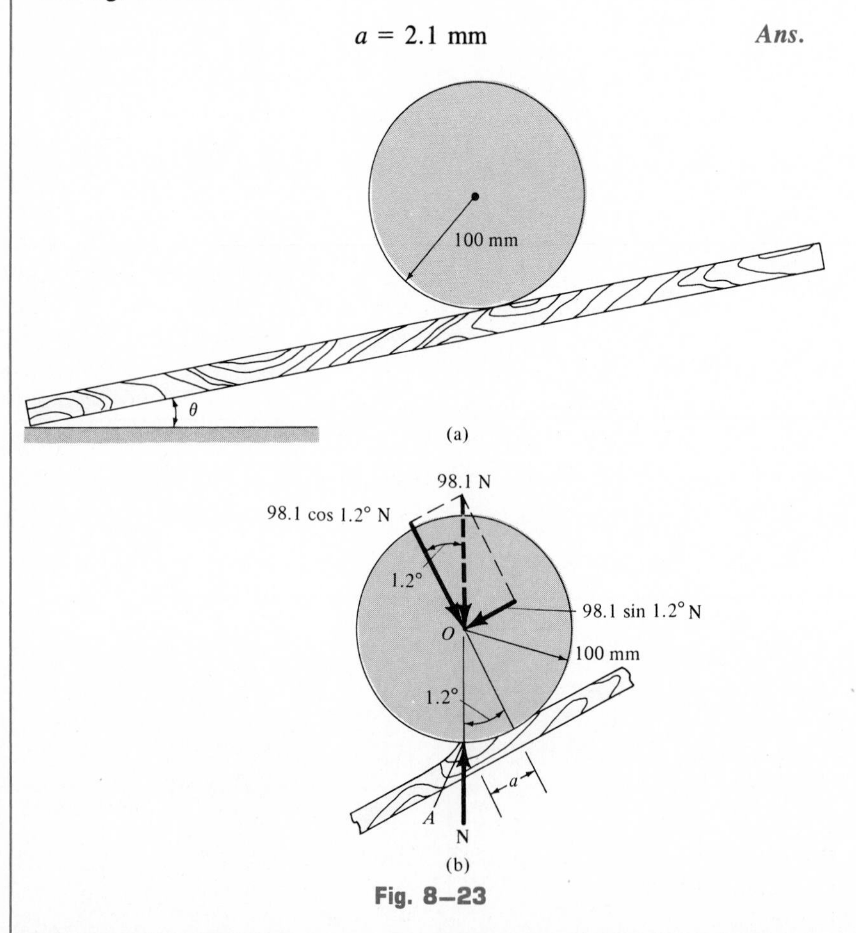

Fig. 8–23

Problems

*8–44. The truck, which has a mass of 3.4 Mg, is to be lowered down the slope by a rope that is wrapped around a tree. If the wheels are free to roll and the man at A can resist a pull of 500 N, determine the minimum number of turns the rope should be wrapped around the tree to lower the truck at a constant speed. The coefficient of friction between the tree and rope is $\mu = 0.4$.

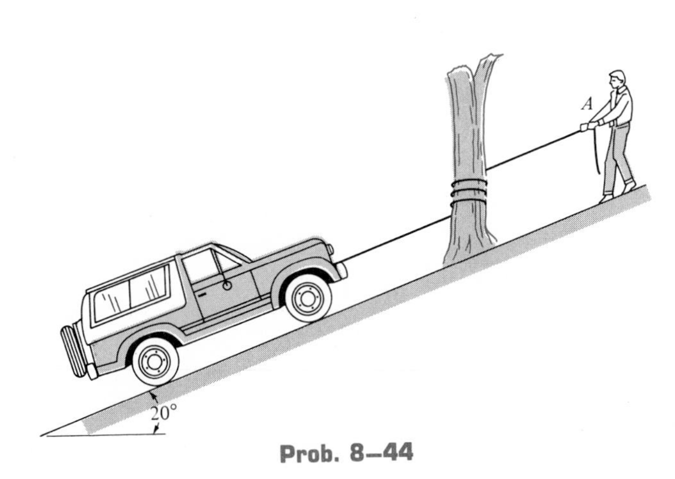

Prob. 8–44

8–45. Determine the *minimum* tension in the rope at points A and B that is necessary to maintain equilibrium. Take $\mu = 0.4$ between the rope and the fixed post D. The rope is wrapped only once around the post.

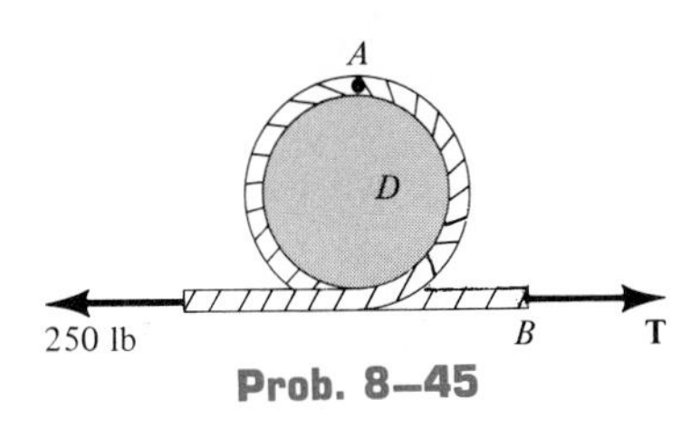

Prob. 8–45

8–46. If a torque of $M = 150$ N · m is applied to the disk, determine the minimum vertical force **P** that must be applied to the band brake to prevent the disk from rotating. The coefficient of friction between the band and the disk is $\mu = 0.45$. The disk has a mass of 60 kg.

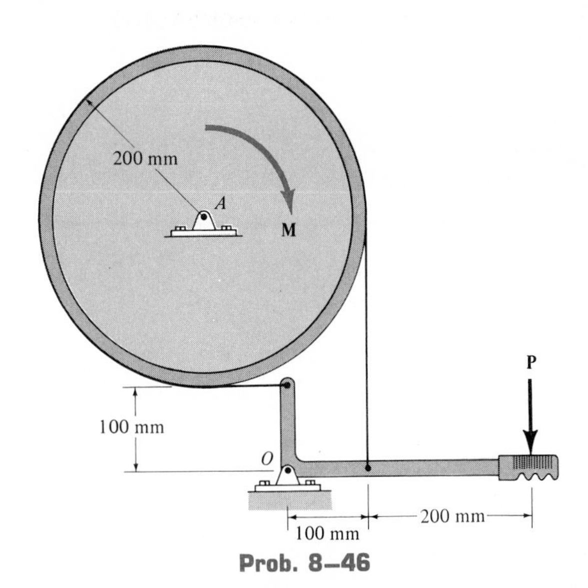

Prob. 8–46

8–47. The boat has a weight of 500 lb and is held in position off the side of a ship by the spars at A and B. A man having a weight of 130 lb gets in the boat, wraps a rope around an overhead boom at C, and ties it to the ends of the boat as shown. If the boat is disconnected from the spars, determine the *minimum number* of *half turns* the rope must make around the boom so that the boat can be safely lowered into the water. The coefficient of friction between the rope and the boom is $\mu = 0.15$. *Hint:* The problem requires that the normal force between the man's feet and the boat be as small as possible.

Prob. 8–47

***8–48.** A cable is attached to a 60-lb plate B and passes over a fixed disk at C. Using the coefficients of friction shown in the figure, determine the smallest force $\mathbf{T}$ needed to hold the block.

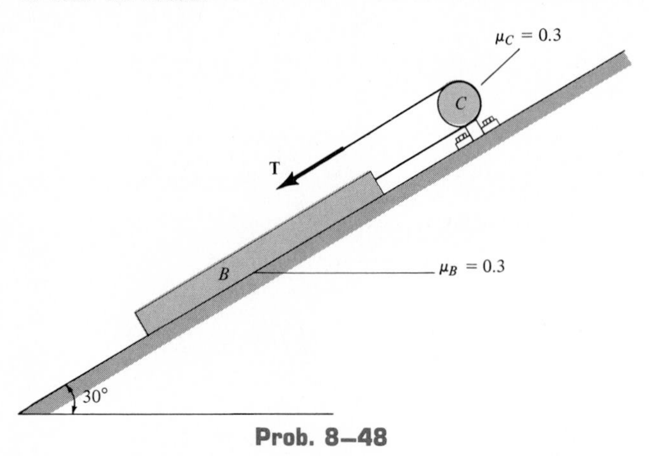

Prob. 8–48

8–49. A ring is tied to the end of a rope at B, and the rope is wrapped around a frictionless pulley A and passes through the ring as shown. If the rope is subjected to a tension $\mathbf{T}$ and the coefficient of friction between the rope and the ring is $\mu = 0.3$, determine the angle θ for equilibrium.

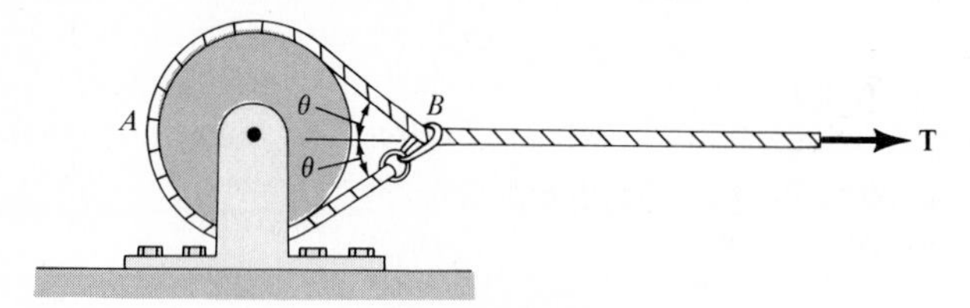

Prob. 8–49

8–50. A tank car has a weight of $60(10^3)$ lb and is supported by eight wheels, each of which has a diameter of 30 in. If the coefficient of rolling resistance is 0.015 in. between the tracks and each wheel, determine the magnitude of horizontal force required to overcome the rolling resistance of the wheels.

8–51. Experimentally, it is found that a cylinder having a diameter of 120 mm rolls with a constant speed down an inclined plane having a slope of 18 mm/m. Determine the coefficient of rolling resistance for the cylinder.

***8–52.** Two 8-kg blocks are attached to a cord that passes over two fixed drums. If $\mu = 0.3$ at the drums, determine the smallest vertical force $\mathbf{P}$ needed to suspend the blocks when $\theta = 30°$. The tension in the cord is zero when the blocks just touch the *ground*, i.e., $\theta = 0°$ and $P = 0$.

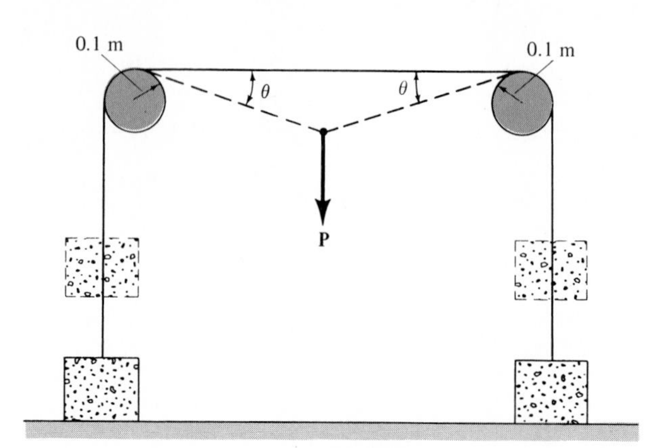

Prob. 8–52

8–53. The lawn roller has a mass of 80 kg. If the arm BA is held at an angle of 30° from the horizontal and the coefficient of rolling resistance for the roller is 25 mm, determine the force $\mathbf{F}$ needed to push the roller at constant speed. Neglect friction developed at the axle, A, and assume that the resultant force $\mathbf{F}$ acting on the handle is applied along arm BA.

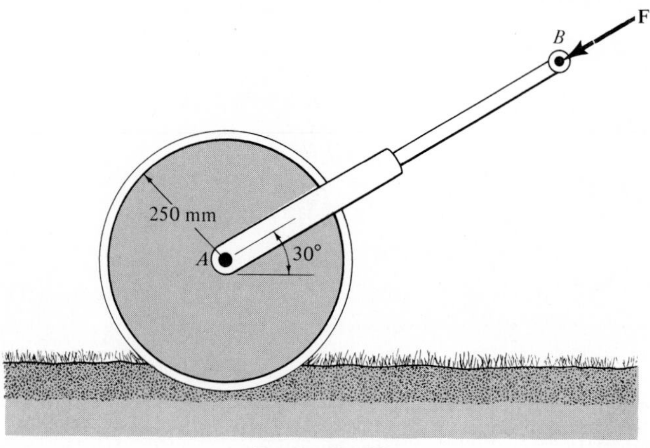

Prob. 8–53

8–54. The pipe is subjected to a load that has a weight **W**. If the coefficients of rolling resistance for the pipe's top and bottom surfaces are a_A and a_B, respectively, show that a force having a magnitude of $P = [W(a_A + a_B)]/2r$ is required to move the load and thereby roll the pipe forward. Neglect the weight of the pipe.

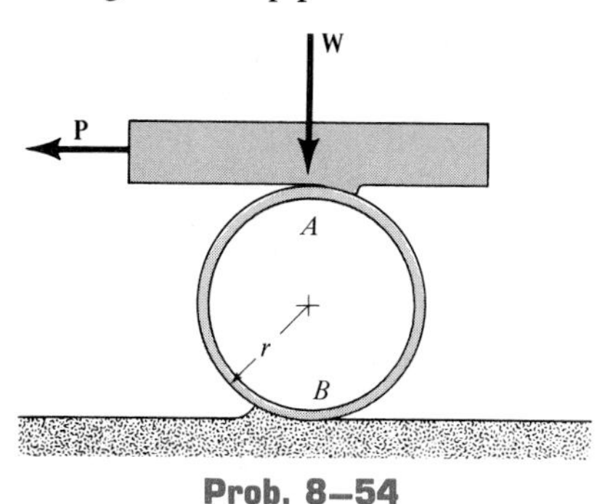

Prob. 8–54

8–55. The 1.2-Mg steel beam is moved over a level surface using a series of 30-mm-diameter rollers for which the coefficient of rolling resistance is 0.4 mm at the ground and 0.2 mm at the bottom surface of the beam. Determine the horizontal force **P** needed to push the beam forward at a constant speed. *Hint:* Use the result of Prob. 8–54.

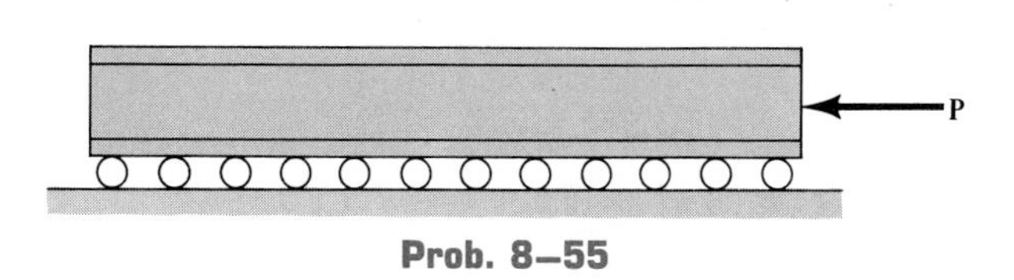

Prob. 8–55

Review Problems

***8–56.** A "hawser" is wrapped around a fixed "capstan" to secure a ship for docking. If the tension in the rope, caused by the ship, is $P = 1{,}500$ lb, determine the least number of complete turns the rope must be wrapped around the capstan in order to prevent slipping of the rope. The greatest horizontal force that a longshoreman can exert on the rope is 90 lb. The coefficient of friction is $\mu = 0.3$.

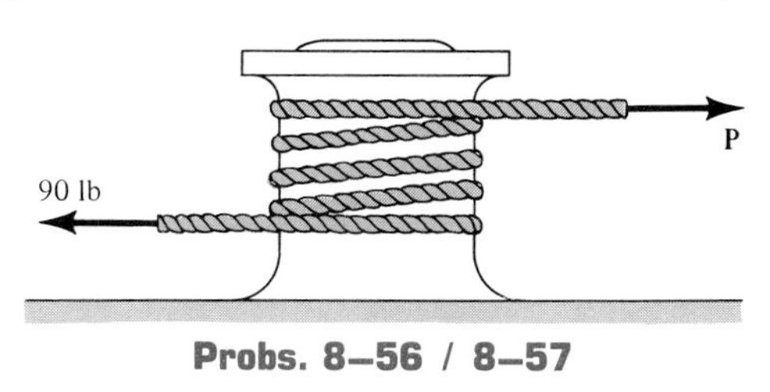

Probs. 8–56 / 8–57

8–57. If the rope is wrapped two times around the capstan, determine the maximum force **P** that can be applied without causing the rope to slip. Take $\mu = 0.3$.

8–58. Blocks A and B have a mass of 1.5 kg and 2 kg, respectively. Determine the smallest coefficient of friction between the inclines and the blocks which will prevent the blocks from moving.

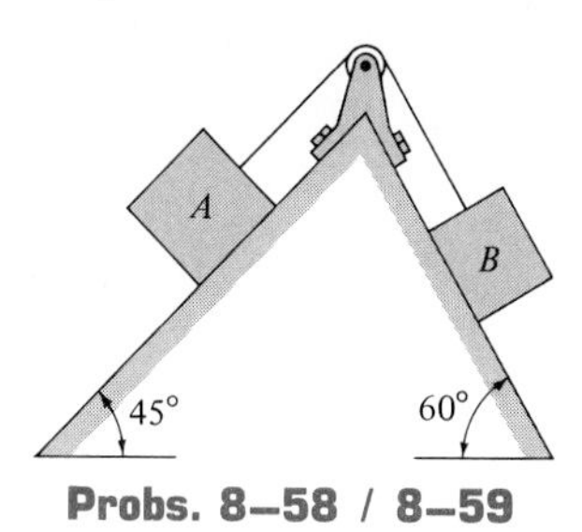

Probs. 8–58 / 8–59

8–59. If block A has a mass of 1.5 kg, determine the largest mass of block B without causing motion of the system. The coefficient of friction between the blocks and inclines is $\mu = 0.2$.

***8–60.** If the coefficient of friction between blocks A and B is $\mu_{AB} = 0.8$ and between B and the floor $\mu_{BC} = 0.1$, determine the minimum force **F** which will cause block A to move.

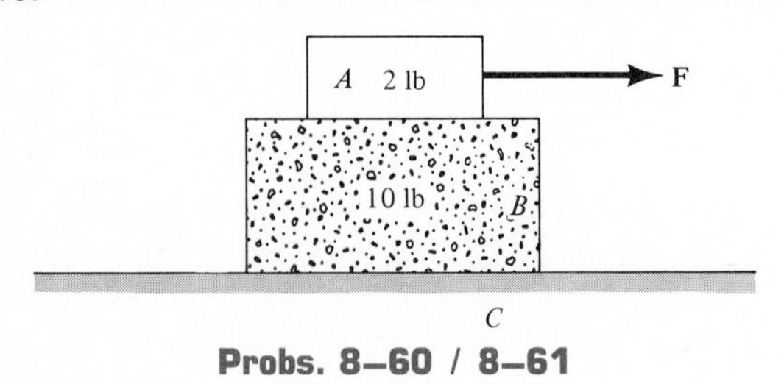

Probs. 8–60 / 8–61

8–61. If the coefficient of friction between blocks A and B is $\mu_{AB} = 0.8$, determine the coefficient of friction between block B and the floor so that when the force **F** is increased the blocks slip at all their contacting surfaces at the same time. What is the magnitude of force **F** needed to move the blocks?

8–62. Blocks A and B weigh 10 lb and 25 lb, respectively. If the coefficient of friction between A and B is $\mu_{AB} = 0.4$ and between B and the floor C, $\mu_{BC} = 0.2$, determine the maximum horizontal force **P** that can be applied without causing motion.

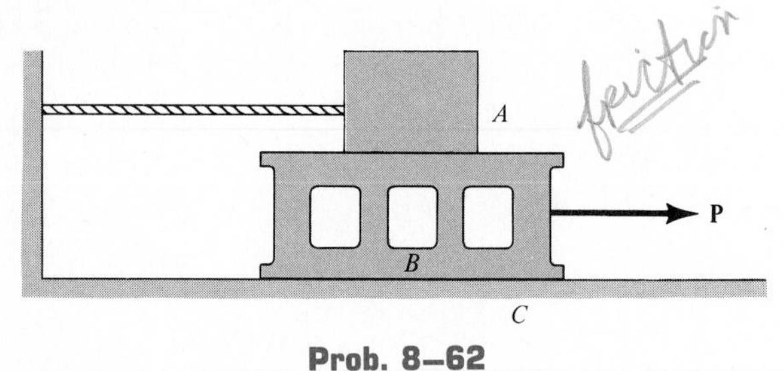

Prob. 8–62

8–63. A 12-in.-diameter disk fits loosely over a 3-in.-diameter shaft for which the coefficient of friction is $\mu = 0.15$. If the disk weighs 100 lb, determine the vertical tangential force which must be applied to the disk to cause it to slip over the shaft.

***8–64.** If the coefficient of friction between the 500-lb drum and the inclined planes is $\mu = 0.3$, determine the couple moment **M** needed to rotate the drum.

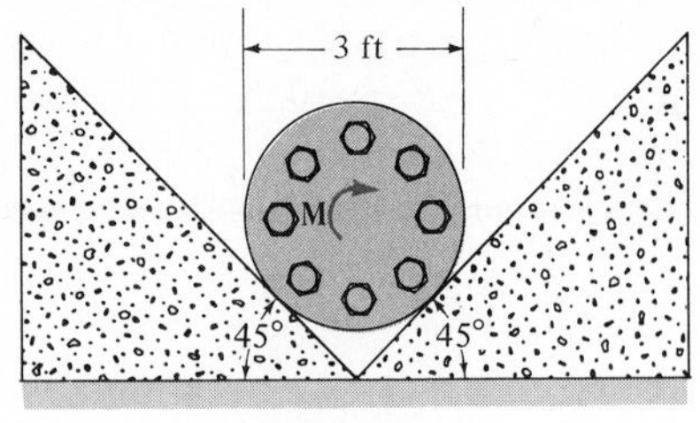

Prob. 8–64

8–65. The four wheels on an automobile each have a diameter of 2.5 ft. If the coefficient of rolling resistance is 0.013 with the road, and the car weighs 4000 lb, determine the horizontal force **P** that is required to overcome the rolling resistance of the wheels.

8–66. The automobile jack is subjected to a vertical load of 600 lb. If the square-threaded screw has a pitch of $\frac{1}{4}$ in. and a mean diameter of $\frac{1}{2}$ in., determine the force **P** which must be applied to the handle to raise the load. Take $\mu = 0.25$. The supporting plate exerts only vertical forces at A and B, and each cross-link has a length of 4 in.

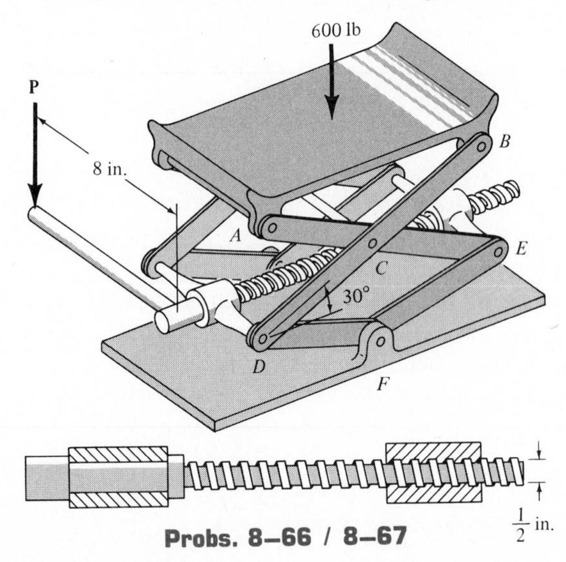

Probs. 8–66 / 8–67

8–67. Determine the minimum force **P** needed to hold the handle of the automobile jack in equilibrium. The jack supports a load of 600 lb, and the square-threaded screw has a pitch of $\frac{1}{4}$ in. and a mean diameter of $\frac{1}{2}$ in. The supporting plate exerts only vertical forces at A and B, and each cross link has a length of 4 in.

Center of Gravity and Centroid

In this chapter we will discuss the method used to determine the location of the center of gravity and center of mass for a system of discrete particles, and then we will expand its application to include a body of arbitrary shape. The same method of analysis will also be used to determine the geometric center, or centroid, of lines, areas, and volumes. Once the centroid has been located, we will then show how to obtain the area and volume of a surface of revolution and determine the effect of fluid pressure and other distributed loadings acting on a surface.

9.1 Center of Gravity and Center of Mass for a System of Particles

Center of Gravity. The *center of gravity* for a system of particles is defined as the point of application for the total weight of all the particles in the system. To determine its location consider the system of n particles shown in Fig. 9–1a which lie along the x axis. The weights of the particles constitute a system of coplanar parallel forces which can be reduced to a single resultant

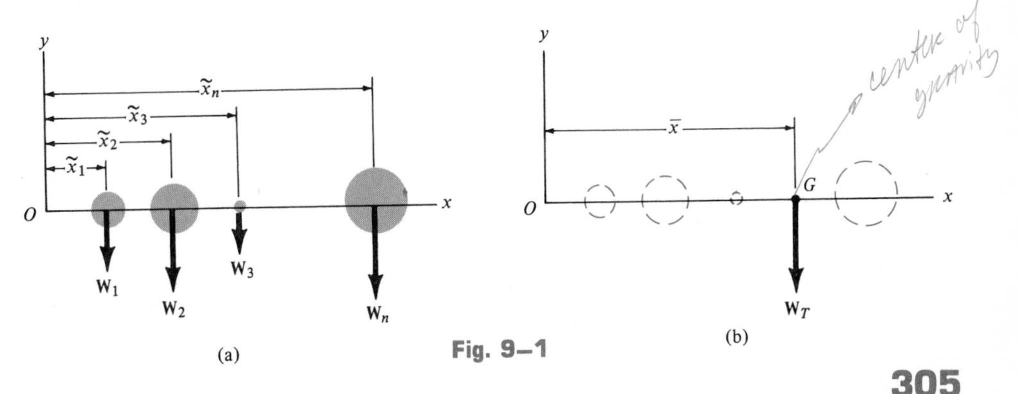

Fig. 9–1

force $\mathbf{W}_T$ acting at a specific location $\bar{x}$, Fig. 9–1b. Here the magnitude of $\mathbf{W}_T$ represents the total weight of all the particles, i.e.,

$$+\downarrow F_R = \Sigma F; \qquad W_T = \Sigma W = W_1 + W_2 + \cdots + W_n$$

The location of $\mathbf{W}_T$ is determined by applying the principle of moments about point O. This requires that the moment of $\mathbf{W}_T$ about O, Fig. 9–1b, be equivalent to the sum of the moments of the weights of all the particles about O, Fig. 9–1a, i.e.,

$$\circlearrowleft + M_{R_O} = \Sigma M_O; \qquad \bar{x}\Sigma W = \Sigma \tilde{x} W = \tilde{x}_1 W_1 + \tilde{x}_2 W_2 + \cdots + \tilde{x}_n W_n$$

where $\tilde{x}$ in the summation represents the *algebraic distance* measured from the origin O to the associated particle of weight W. Solving for $\bar{x}$ yields

$$\bar{x} = \frac{\Sigma \tilde{x} W}{\Sigma W}$$

Point G in Fig. 9–1b is termed the center of gravity. Notice that its location is not necessarily at one of the particles; rather, it is often a point located in space.

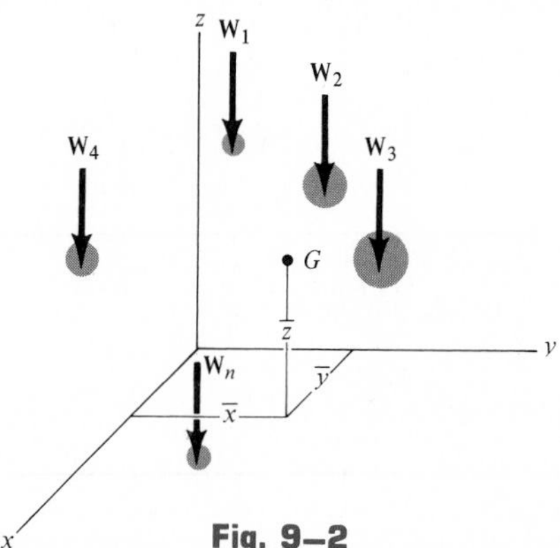

Fig. 9–2

In the most general case, the system of particles is scattered within a region of space, Fig. 9–2. However, if the location $(\tilde{x}, \tilde{y}, \tilde{z})$ of each particle is defined, then we can apply the principle of moments about each of the three axes in order to determine the location of the system's center of gravity $G(\bar{x}, \bar{y}, \bar{z})$.* The equations necessary for doing this are

$$\bar{x} = \frac{\Sigma \tilde{x} W}{\Sigma W} \qquad \bar{y} = \frac{\Sigma \tilde{y} W}{\Sigma W} \qquad \bar{z} = \frac{\Sigma \tilde{z} W}{\Sigma W} \qquad (9\text{–}1)$$

*If the weights of the particles are all parallel to the z axis, as shown in Fig. 9–2, then they will create no moment about this axis. However, to determine $\bar{z}$ it may be helpful to imagine the coordinate system and the particles fixed in it as being rotated 90° about either the x or y axis. This way one can "visualize" how the z-axis moments are developed.

Center of Mass. To study problems concerning the motion of *matter* under the influence of force, i.e., dynamics, it is necessary to locate a point called the *center of mass*. Provided the acceleration of gravity g for every particle is constant, then $W = mg$. Substituting into Eqs. 9–1 and canceling g from both the numerator and denominator yields

$$\bar{x} = \frac{\Sigma \tilde{x} m}{\Sigma m} \qquad \bar{y} = \frac{\Sigma \tilde{y} m}{\Sigma m} \qquad \bar{z} = \frac{\Sigma \tilde{z} m}{\Sigma m} \tag{9–2}$$

By comparison then, the location of the center of gravity coincides with that of the center of mass.* Recall, however, that particles have "weight" only when under the influence of a gravitational attraction, whereas the center of mass is independent of gravity. For example, it would be meaningless to define the center of gravity of a system of particles representing the planets of our solar system, while the center of mass of this system is important.

9.2 Center of Gravity and Centroids of Composite Bodies

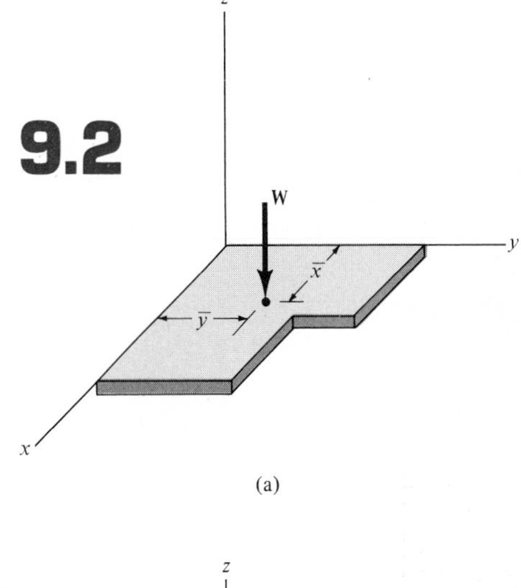

(a)

Center of Gravity. In many cases a body can be sectioned or divided into several parts having simpler shapes. Provided the weight and location of the center of gravity of each of these "composite parts" are known, one can then determine the center of gravity for the entire body. For example, the plate in Fig. 9–3*a* consists of two rectangular "composite" parts, Fig. 9–3*b*. Each part has a specified weight, $\mathbf{W}_1$ and $\mathbf{W}_2$, which passes through its center of gravity as shown. If each of these weights is treated in the same way as a system of particles, then application of the principle of moments to each of the composite parts yields formulas analogous to Eqs. 9–1. Thus, in three dimensions, the necessary formulas for finding the center of gravity for a composite body become

$$\bar{x} = \frac{\Sigma \tilde{x} W}{\Sigma W} \qquad \bar{y} = \frac{\Sigma \tilde{y} W}{\Sigma W} \qquad \bar{z} = \frac{\Sigma \tilde{z} W}{\Sigma W} \tag{9–3}$$

where $\tilde{x}$, $\tilde{y}$, and $\tilde{z}$ represent the *algebraic distances* from the center of gravity of each composite part to the origin of the coordinates, and ΣW represents the sum of the weights of the composite parts or simply the *total weight* of the body.

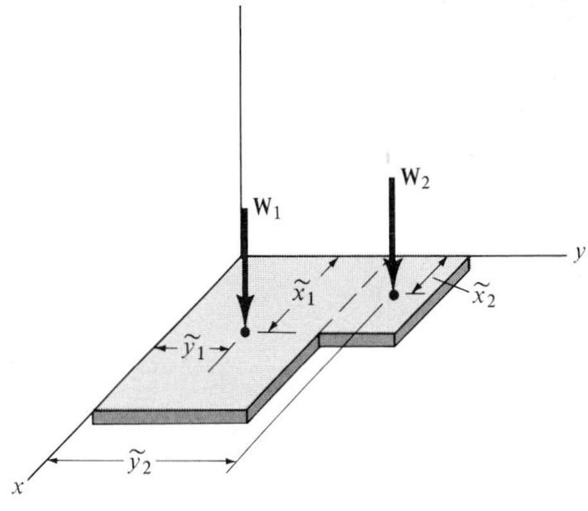

(b)

Fig. 9–3

*This is not true in the exact sense since the weights are not parallel to each other, rather they are all *concurrent* at the earth's center. Furthermore, the acceleration of gravity g is actually different for each particle since it depends upon the distance from the earth's center to the particle. For all practical purposes, however, both of these effects can be neglected.

Centroid. The *centroid* is a point which defines the *geometrical center* of an object. It can be obtained from formulas similar to those used to determine a body's center of gravity. In particular, if the material composing a body is uniform or *homogeneous,* the density ρ or weight/volume will be *constant* throughout the body. In this case, the body's weight becomes $W = \rho V$. Substituting this relation into Eqs. 9–3, ρ will factor out of the summations and *cancel* from both the numerator and denominator. The resulting formulas define the centroid of the body, since they are independent of the body's weight and instead depend only on the body's *geometrical properties*.

Volume. The resulting equations defining the centroidal location of the volume of a body are

$$\bar{x} = \frac{\Sigma \tilde{x} V}{\Sigma V} \qquad \bar{y} = \frac{\Sigma \tilde{y} V}{\Sigma V} \qquad \bar{z} = \frac{\Sigma \tilde{z} V}{\Sigma V} \tag{9–4}$$

Area. If the "body" is represented as a plate of constant thickness t and density ρ, then the weight of its composite parts is $W = \rho(t)A$, where A is the area of each part. Substitution into Eqs. 9–3 leads to

$$\bar{x} = \frac{\Sigma \tilde{x} A}{\Sigma A} \qquad \bar{y} = \frac{\Sigma \tilde{y} A}{\Sigma A} \qquad \bar{z} = \frac{\Sigma \tilde{z} A}{\Sigma A} \tag{9–5}$$

Line. If the "body" is represented as a rod of constant cross-sectional area A and density ρ, then substitution of $W = \rho(A)L$, where L is the length of each composite part, into Eqs. 9–3 yields

$$\bar{x} = \frac{\Sigma \tilde{x} L}{\Sigma L} \qquad \bar{y} = \frac{\Sigma \tilde{y} L}{\Sigma L} \qquad \bar{z} = \frac{\Sigma \tilde{z} L}{\Sigma L} \tag{9–6}$$

The centroids for common shapes representing lines, areas, shells, and volumes are given in the table on the inside back cover. It should be noted that the centroids of some shapes may be partially or completely specified by using *symmetry conditions*. For example, the centroid C for the rod shown in Fig. 9–4 must lie along the y axis since for every segment a distance $+\tilde{x}$ to the right of the y axis there is an identical segment at a distance $-\tilde{x}$ to the left. The total moment for all the segments will therefore cancel, i.e., $\Sigma \tilde{x} L = 0$ (Eqs. 9–6) so that $\bar{x} = 0$. In cases where a shape has two or three axes of symmetry, it follows that the centroid lies at the intersection of these axes, Fig. 9–5.

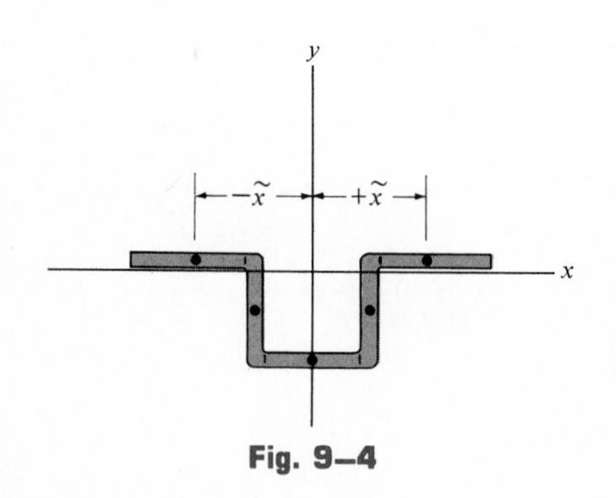

Fig. 9–4

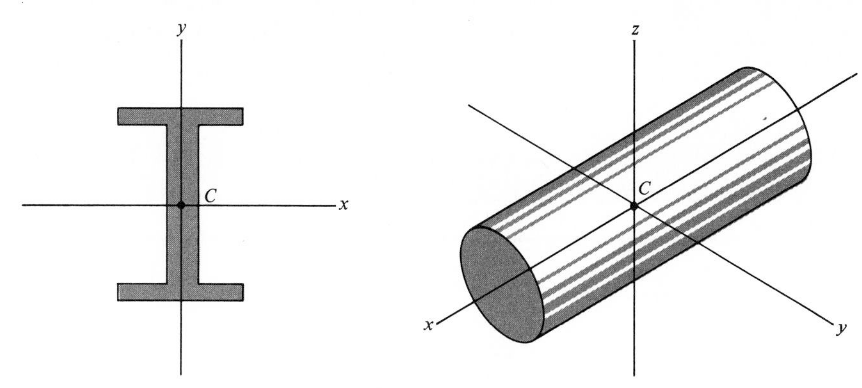

Fig. 9–5

Besides locating the center of gravity of homogeneous bodies, computations for the centroids of volume and area shapes are important in other fields of mechanics. For example, in structural mechanics, the centroid of the cross-sectional area of a beam or column must be located in order to properly design these members. Furthermore, it was shown in Sec. 4–8 that a distributed surface loading may be reduced to a single resultant force having a line of action that passes through the *centroid* of the *area* described by the *loading diagram*.

PROCEDURE FOR ANALYSIS

The following procedure provides a method for determining the center of gravity of a body or the centroid of a composite geometrical object represented by a line, area, or volume.

Composite Parts. Using a sketch, divide the body or object into a finite number of composite parts. If a *hole,* or geometric region having no material, represents one of the parts, the weight or "size" of the hole is considered a *negative quantity*.

Moment Arms. Establish the coordinate axes on the sketch and determine the location $\tilde{x}$, $\tilde{y}$, $\tilde{z}$ of the center of gravity or centroid of each part.

Summations. Determine $\bar{x}$, $\bar{y}$, $\bar{z}$ by applying the center of gravity equations (Eqs. 9–3) or the analogous centroid equations. If an object is *symmetrical* about an axis, recall that the centroid of the object lies on the axis.

If desired, the calculations can be arranged in tabular form, as indicated in the following examples.

Example 9–1

Locate the centroid of the wire shown in Fig. 9–6*a*.

Solution

Composite Parts. The wire is divided into three segments as shown in Fig. 9–6*b*.

Moment Arms. The location of the centroid for each piece is determined and indicated on the sketch. In particular, the centroid of segment 1 is determined by using the table on the inside back cover.

Summations. Taking the data from Fig. 9–6*b*, the calculations are tabulated as follows:

Segment	L (mm)	$\tilde{x}$ (mm)	$\tilde{y}$ (mm)	$\tilde{z}$ (mm)	$\tilde{x}L$ (mm^2)	$\tilde{y}L$ (mm^2)	$\tilde{z}L$ (mm^2)
1	$\pi(60) = 188.5$	60	−38.2	0	11,310	−7200	0
2	40	0	20	0	0	800	0
3	20	0	40	−10	0	800	−200
	$\Sigma L = 248.5$				$\Sigma\tilde{x}L = 11{,}310$	$\Sigma\tilde{y}L = -5600$	$\Sigma\tilde{z}L = -200$

Thus,

$$\bar{x} = \frac{\Sigma\tilde{x}L}{\Sigma L} = \frac{11{,}310}{248.5} = 45.5 \text{ mm} \qquad \textit{Ans.}$$

$$\bar{y} = \frac{\Sigma\tilde{y}L}{\Sigma L} = \frac{-5600}{248.5} = -22.5 \text{ mm} \qquad \textit{Ans.}$$

$$\bar{z} = \frac{\Sigma\tilde{z}L}{\Sigma L} = \frac{-200}{248.5} = -0.8 \text{ mm} \qquad \textit{Ans.}$$

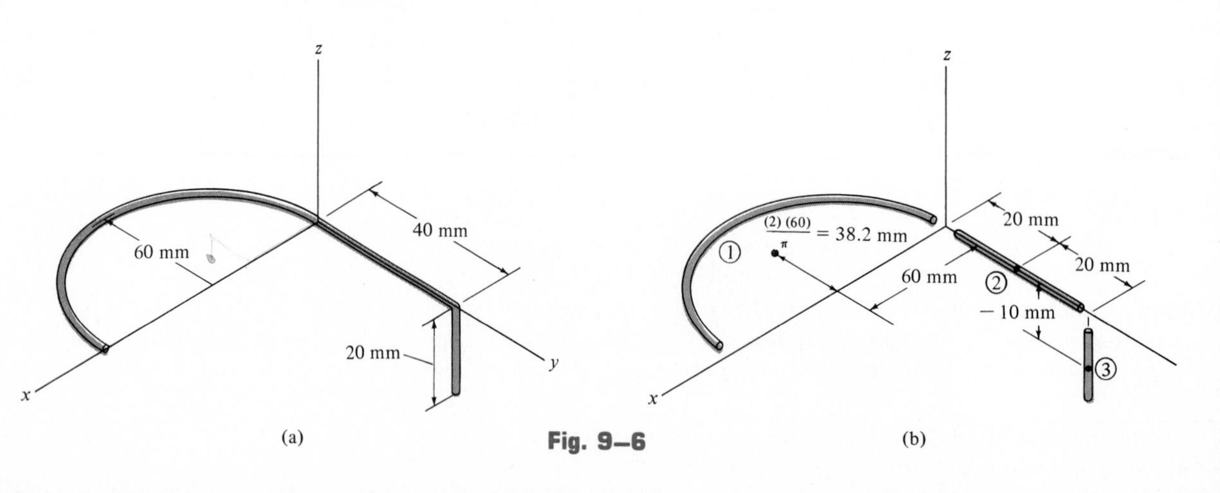

Fig. 9–6

Example 9–2

Locate the centroid of the plate area shown in Fig. 9–7*a*.

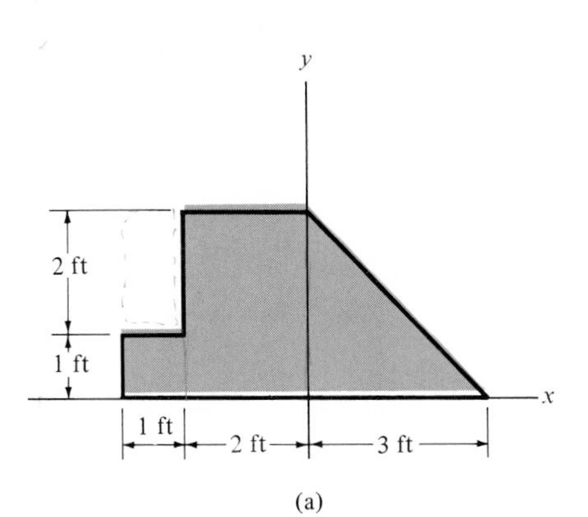

(a)

Fig. 9–7

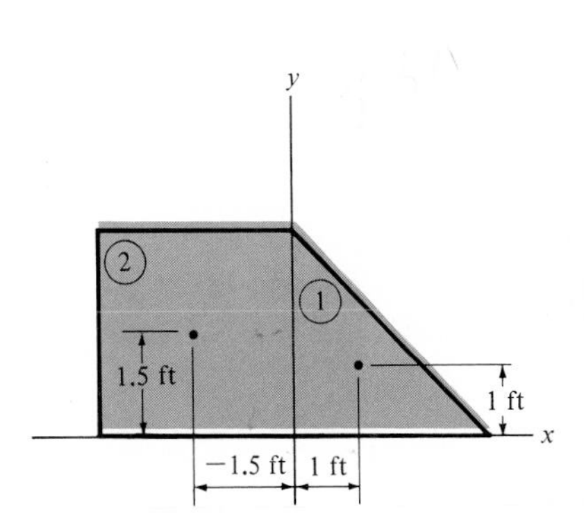

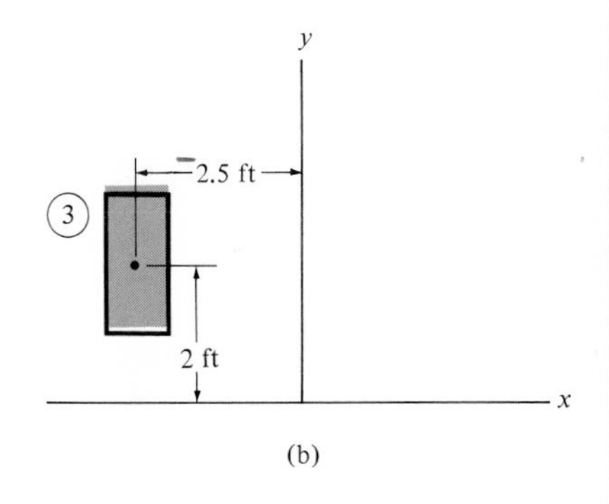

(b)

Solution

Composite Parts. The plate is divided into three segments as shown in Fig. 9–7*b*. Here the area of the small rectangle (3) is considered "negative" since it must be subtracted from the larger one (2).

Moment Arms. The centroid of each segment is located as indicated in the figure. Note that the $\tilde{x}$ coordinates of (2) and (3) are *negative*.

Summations. Taking the data from Fig. 9–7*b*, the calculations are tabulated as follows:

Segment	A (ft²)	$\tilde{x}$ (ft)	$\tilde{y}$ (ft)	$\tilde{x}A$ (ft³)	$\tilde{y}A$ (ft³)
1	$\frac{1}{2}(3)(3) = 4.5$	1	1	4.5	4.5
2	$(3)(3) = 9$	−1.5	1.5	−13.5	13.5
3	$-(2)(1) = -2$	−2.5	2	5	−4
	$\Sigma A = 11.5$			$\Sigma\tilde{x}A = -4$	$\Sigma\tilde{y}A = 14$

Thus,

$$\bar{x} = \frac{\Sigma\tilde{x}A}{\Sigma A} = \frac{-4}{11.5} = -0.348 \text{ ft} \qquad \textit{Ans.}$$

$$\bar{y} = \frac{\Sigma\tilde{y}A}{\Sigma A} = \frac{14}{11.5} = 1.22 \text{ ft} \qquad \textit{Ans.}$$

Example 9–3

Locate the center of mass of the composite assembly shown in Fig. 9–8*a*. The conical frustum has a density of $\rho_c = 8\ \text{Mg/m}^3$ and the hemisphere has a density of $\rho_h = 4\ \text{Mg/m}^3$.

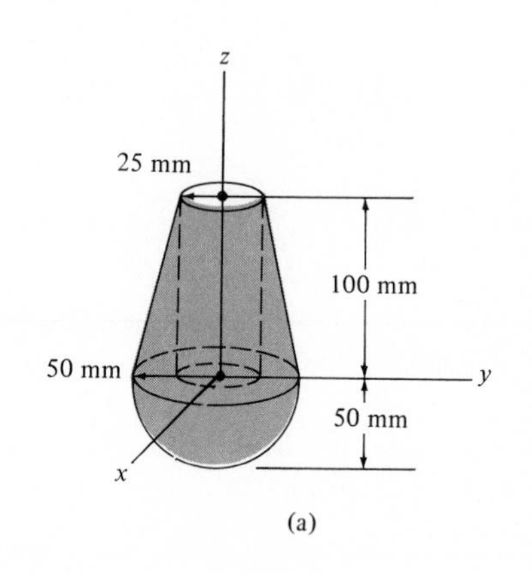

Fig. 9–8

Solution

Composite Parts. The assembly can be thought of as consisting of four segments as shown in Fig. 9–8*b*. For the calculations, segments 3 and 4 must be considered as "negative" volumes in order that the four pieces, when added together, yield the total composite shape shown in Fig. 9–8*a*.

Moment Arm. Using the table on the inside back cover, the computations for the centroid $\tilde{z}$ of each piece are shown in the figure.

Summations. Because of *symmetry,* note that

$$\bar{x} = \bar{y} = 0 \qquad \textit{Ans.}$$

Since $W = mg$ and g is constant, the third of Eqs. 9–3 becomes $\bar{z} = \Sigma\tilde{z}m/\Sigma m$. The mass of each piece can be computed from $m = \rho V$ and used for the calculations. Also, $1\ \text{Mg/m}^3 = 10^{-6}\ \text{kg/mm}^3$, so that

Segment	m (kg)	$\tilde{z}$ (mm)	$\tilde{z}m$ (kg · mm)
1	$8(10^{-6})(\frac{1}{3})\pi(50)^2(200) = 4.189$	50	209.440
2	$4(10^{-6})(\frac{2}{3})\pi(50)^3 = 1.047$	−18.75	−19.635
3	$-8(10^{-6})(\frac{1}{3})\pi(25)^2(100) = -0.524$	100 + 25 = 125	−65.450
4	$-8(10^{-6})\pi(25)^2(100) = -1.571$	50	−78.540
	$\Sigma m = 3.141$		$\Sigma\tilde{z}m = 45.815$

Thus,

$$\bar{z} = \frac{\Sigma\tilde{z}W}{\Sigma W} = \frac{\Sigma\tilde{z}m}{\Sigma m} = \frac{45.815}{3.141} = 14.6\ \text{mm} \qquad \textit{Ans.}$$

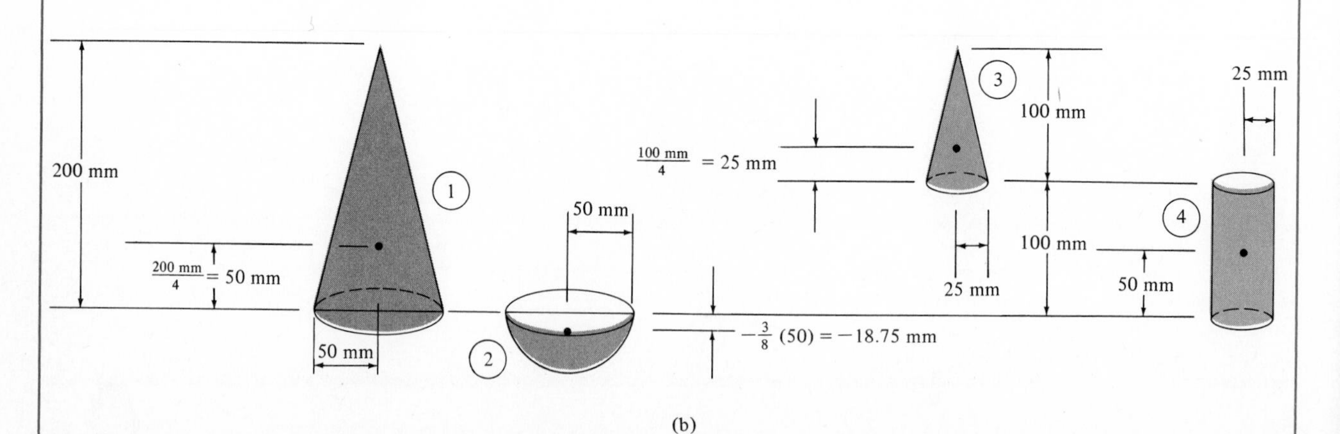

(b)

Problems

9–1. The truss is made from seven members, each having a mass of 6 kg/m. Locate the position $(\bar{x}, \bar{y})$ of the center of mass. Neglect the mass of the gusset plates at the joints.

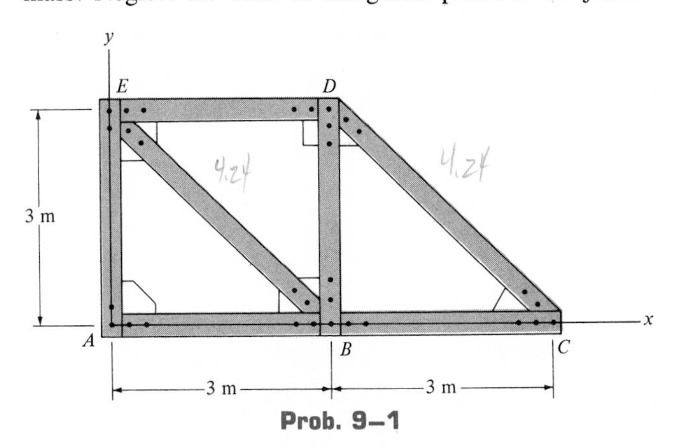

Prob. 9–1

9–2. Locate the center of gravity $G(\bar{x}, \bar{y})$ of the street light. Neglect the thickness of each segment. The material density per unit length of each segment is as follows: $\rho_{AB} = 12$ kg/m, $\rho_{BC} = 8$ kg/m, $\rho_{CD} = 5$ kg/m, and $\rho_{DE} = 2$ kg/m.

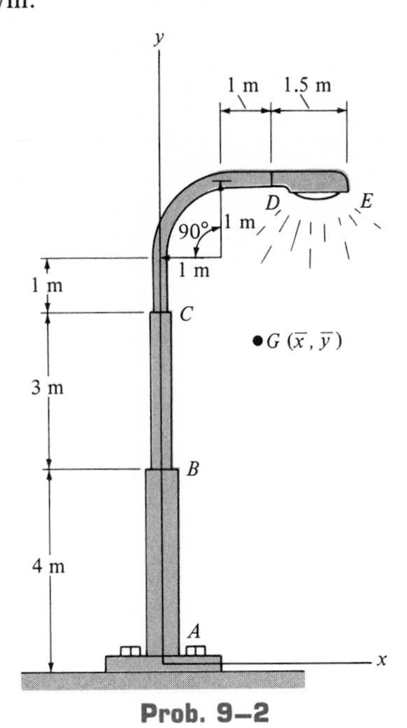

Prob. 9–2

9–3. Locate the center of gravity of the homogeneous wire.

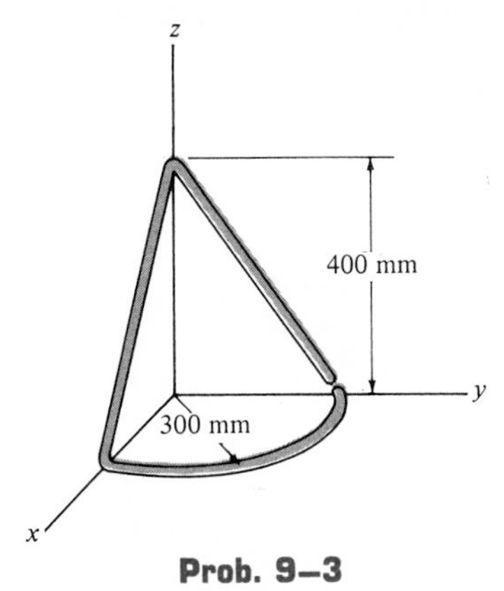

Prob. 9–3

***9–4.** Determine the distance $\bar{y}$ to the centroidal axis $\bar{x}\bar{x}$ of the beam's cross-sectional area. Neglect the size of the corner welds at A and B for the calculation.

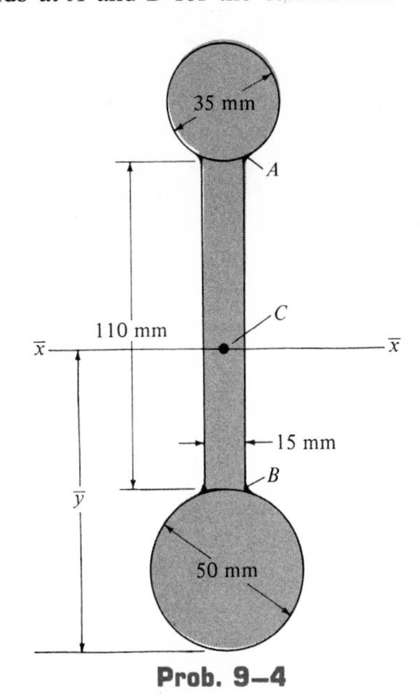

Prob. 9–4

9–5. Determine the location of the centroid for the structural shape used in the construction of an airplane wing. Neglect the thickness of the member.

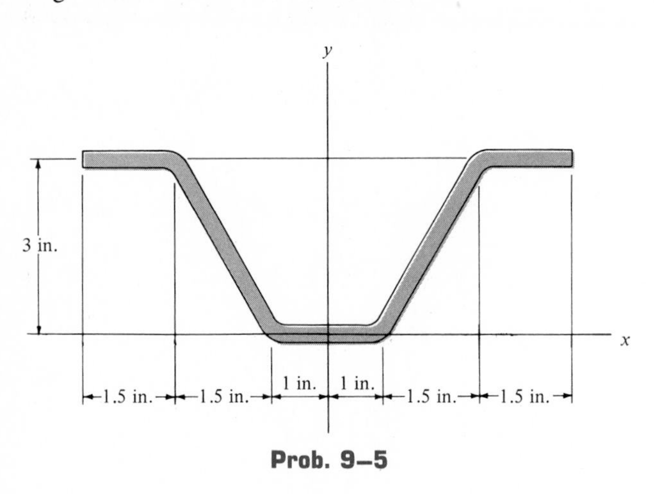

Prob. 9–5

9–6. Locate the centroid $C(\overline{x}, \overline{y})$ of the channel's cross-sectional area.

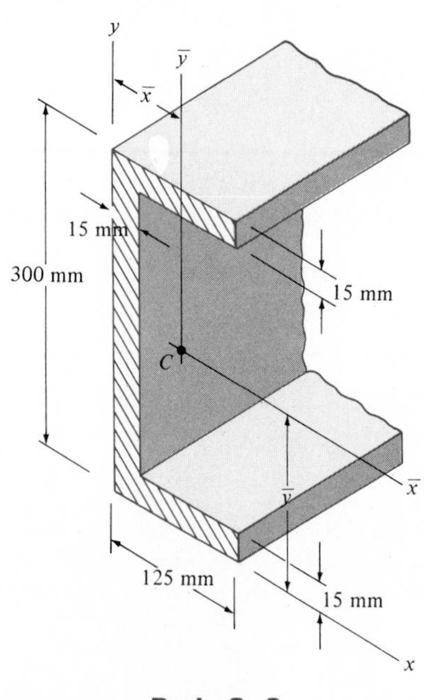

Prob. 9–6

9–7. Determine the distance $\overline{y}$ to the centroid of the shaded area, measured up from the x axis.

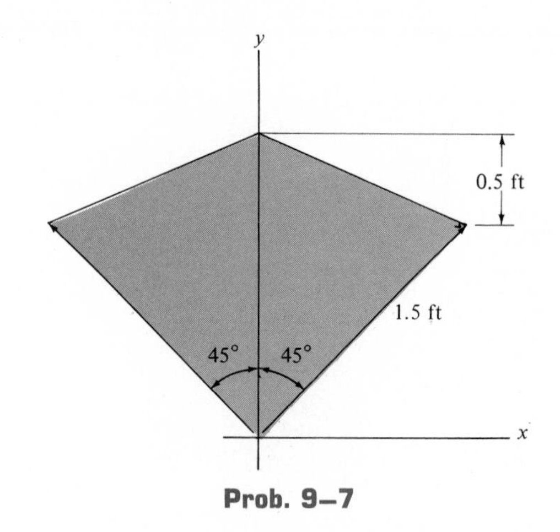

Prob. 9–7

***9–8.** Determine the distance $\overline{y}$ to the centroidal axis $\overline{x}\overline{x}$ of the beam's cross-sectional area.

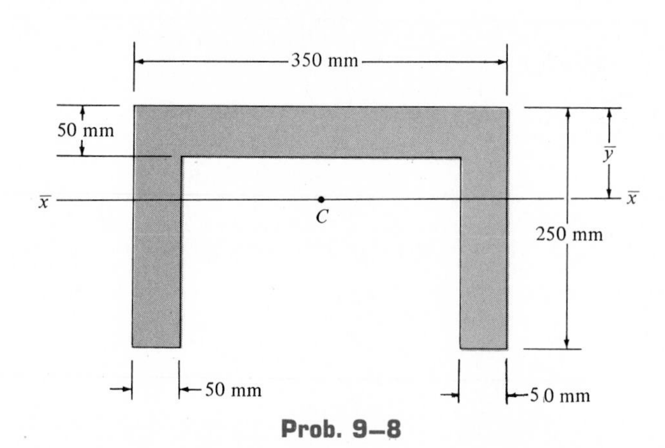

Prob. 9–8

9–9. Determine the distance $\overline{y}$ to the centroidal axis $\overline{x}\overline{x}$ of the beam's cross-sectional area. Neglect the size of the corner welds at A and B for the calculation.

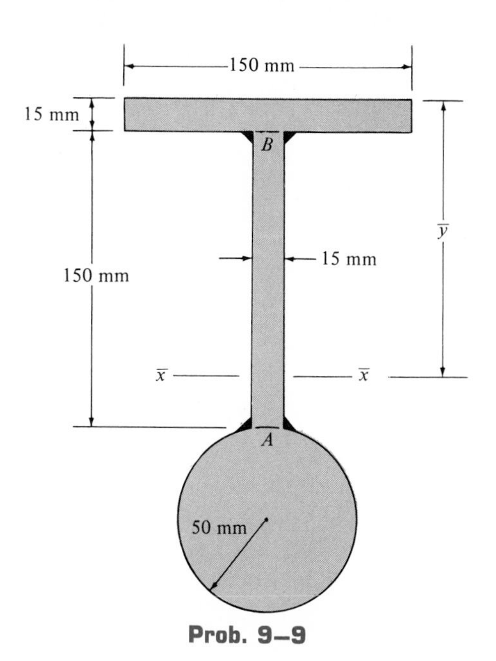

Prob. 9–9

9–11. Determine the distance $\bar{y}$ to the centroidal axis $\bar{x}\bar{x}$ of the beam's cross-sectional area.

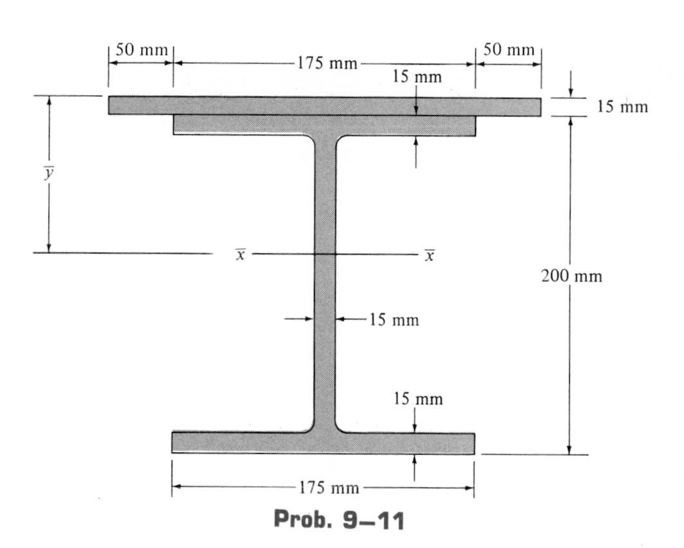

Prob. 9–11

9–10. Locate the centroid $\bar{y}$ of the cross-sectional area of the beam constructed from a plate, channel, and four angles. Handbook values for the areas and centroids C_c and C_a of the channel and one of the angles are listed. Neglect the size of all the rivet heads, R, for the calculation.

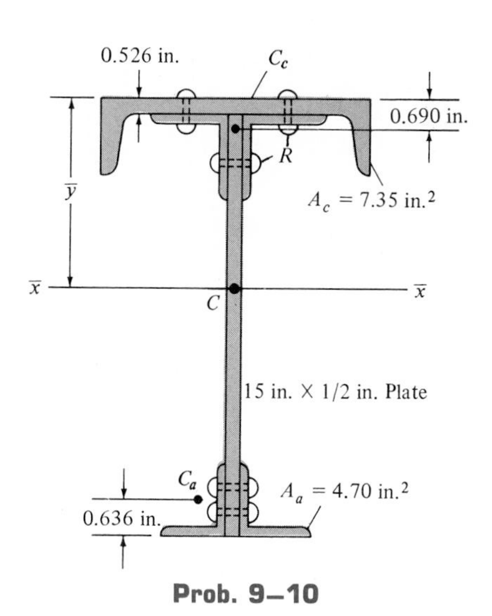

Prob. 9–10

***9–12.** Determine the center of gravity of the airplane. The location of the various items and their weight are tabulated in the figure.

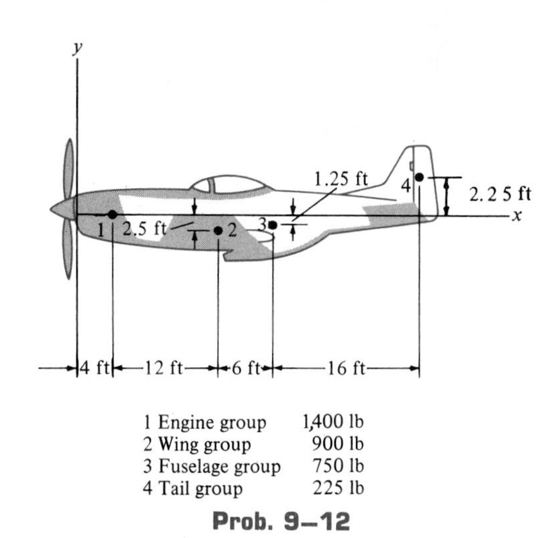

Prob. 9–12

9–13. Locate the centroid of the cross-sectional area of the beam.

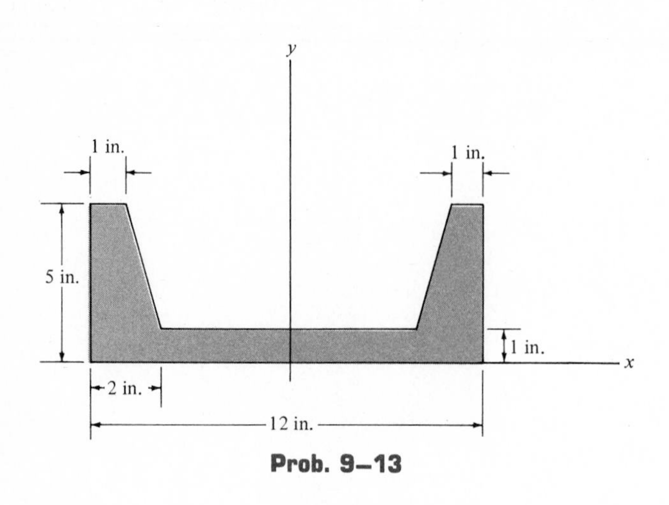

Prob. 9–13

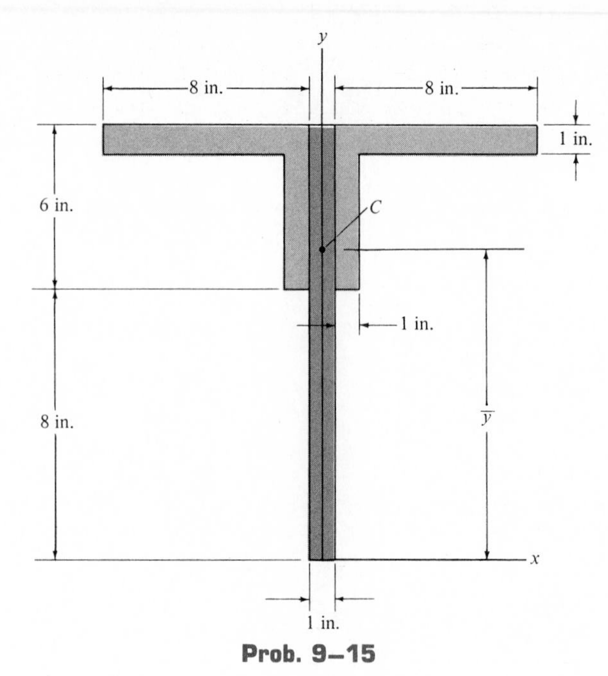

Prob. 9–15

9–14. A toy skyrocket consists of a solid conical top, $\rho_t = 500\ \text{kg/m}^3$, a hollow cylinder, $\rho_c = 400\ \text{kg/m}^3$, and a stick having a circular cross section, $\rho_s = 300\ \text{kg/m}^3$. Determine the length of the stick, x, so that the center of mass G of the skyrocket is located along line aa.

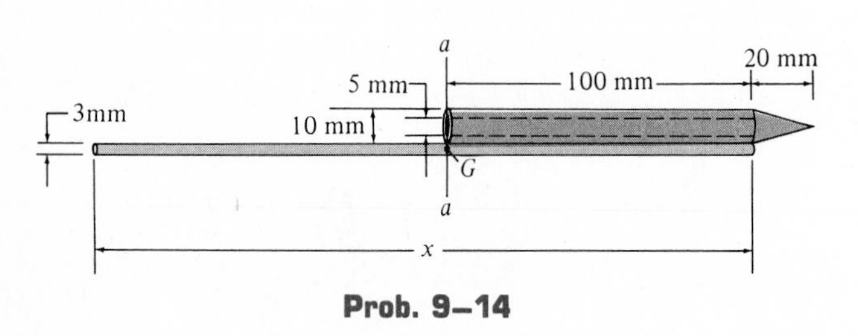

Prob. 9–14

9–15. Determine the distance $\overline{y}$ to the centroid of the cross-sectional area of the beam.

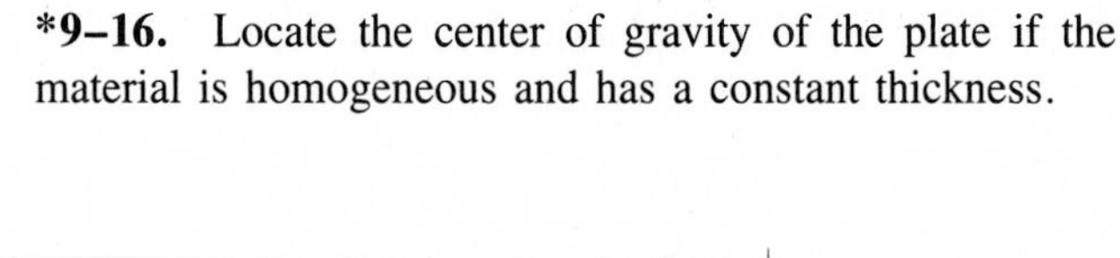

***9–16.** Locate the center of gravity of the plate if the material is homogeneous and has a constant thickness.

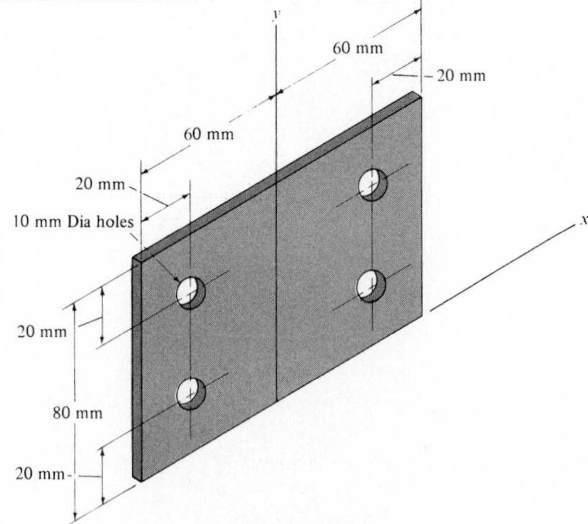

Prob. 9–16

9–17. Divide the plate into parts, and using the grid for measurement, determine the location of the centroid of the plate. Each part is parabolic in shape as defined by the boundary lines.

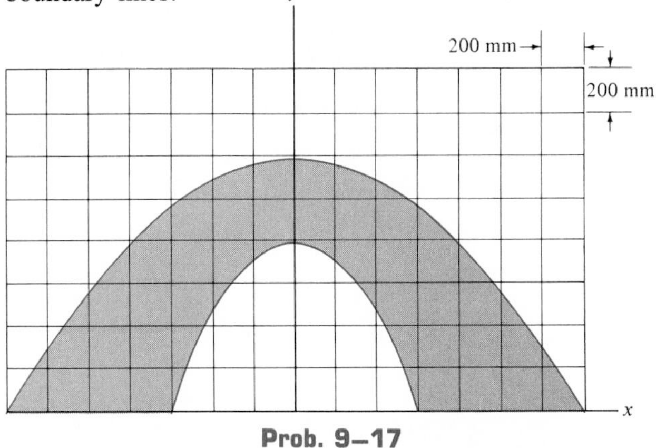

Prob. 9–17

9–18. Locate the center of mass of the casting that is formed from a hollow cylinder having a density of 8 Mg/m^3 and a hemisphere having a density of 3 Mg/m^3.

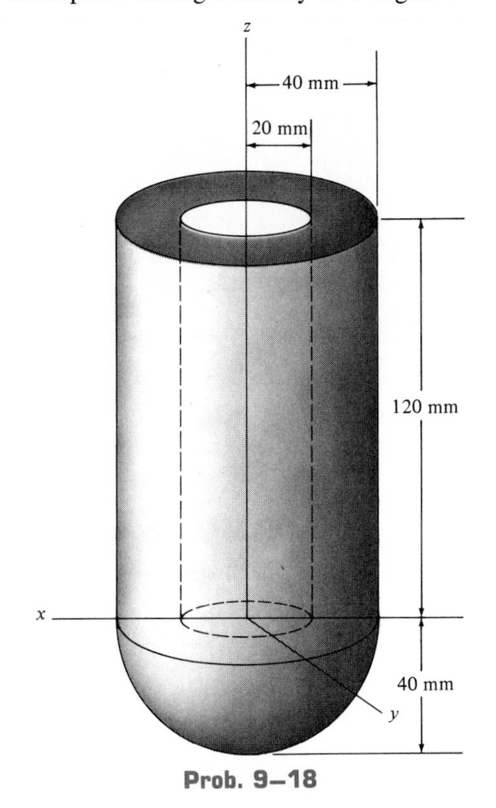

Prob. 9–18

9–19. Determine the location $\bar{z}$ of the center of mass of the assembly. The material has a mass density of $\rho = 3\ Mg/m^3$. There is a hole bored through the assembly's center.

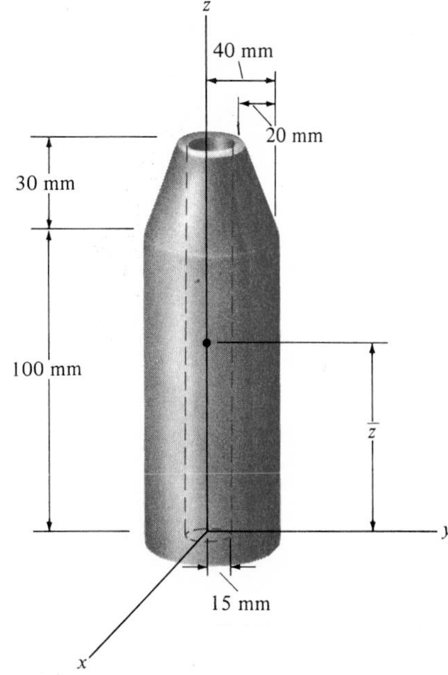

Prob. 9–19

***9–20.** Locate the center of gravity of the two-block assembly. The densities of materials A and B are $\rho_A = 150\ lb/ft^3$ and $\rho_B = 400\ lb/ft^3$, respectively.

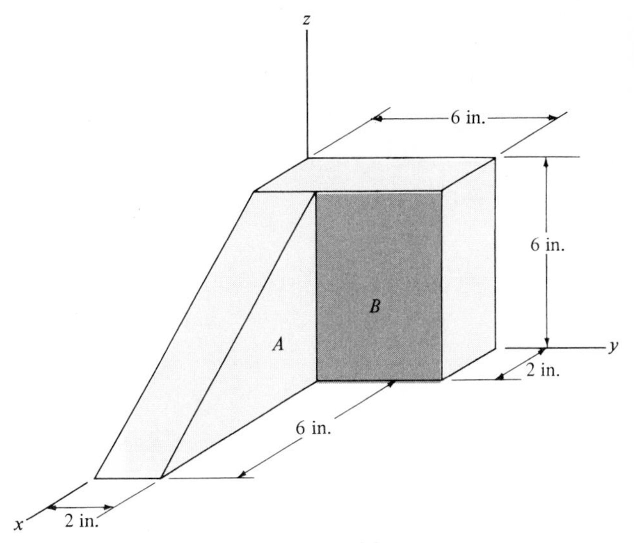

Prob. 9–20

9.3 Center of Gravity and Centroids Using Integration

Center of Gravity. The equations used to determine the location of the center of gravity for a system of discrete particles or a composite body may also be used to determine the location of this point for a body of any arbitrary shape. If the body is divided into an infinite number of segments or particles, then a typical segment located at the point $(\tilde{x}, \tilde{y}, \tilde{z})$ will have a *differential* weight d**W**. Consequently, application of Eqs. 9–1 requires *integration* rather than a discrete summation of the terms. We therefore have

$$\bar{x} = \frac{\int \tilde{x}\, dW}{\int dW} \qquad \bar{y} = \frac{\int \tilde{y}\, dW}{\int dW} \qquad \bar{z} = \frac{\int \tilde{z}\, dW}{\int dW} \tag{9–7}$$

In order to properly use these equations, the differential weight dW must be expressed in terms of the associated volume dV. If ρ represents the density of the body, measured as a weight/volume, then $dW = \rho\, dV$ and therefore

$$\bar{x} = \frac{\int \tilde{x}\rho\, dV}{\int \rho\, dV} \qquad \bar{y} = \frac{\int \tilde{y}\rho\, dV}{\int \rho\, dV} \qquad \bar{z} = \frac{\int \tilde{z}\rho\, dV}{\int \rho\, dV} \tag{9–8}$$

Here integration must be performed throughout the entire volume of the body. Since the gravitational force field for the body is assumed to be both parallel and uniform, these equations can also be used to determine the body's *center of mass,* provided the density ρ is expressed as a mass/volume.

Centroid. If the material composing the body is uniform or homogeneous, then ρ will factor out of the integrals in Eqs. 9–8, and therefore cancel from both the numerator and denominator. The resulting formulas depend only on the body's *geometry* and thereby locate the centroid of the body.

Volume. Using a differential element dV, the coordinates for the centroid C of the volume of space occupied by an object are

$$\bar{x} = \frac{\int \tilde{x}\, dV}{\int dV} \qquad \bar{y} = \frac{\int \tilde{y}\, dV}{\int dV} \qquad \bar{z} = \frac{\int \tilde{z}\, dV}{\int dV} \tag{9–9}$$

Area. Using a differential element dA, the coordinates for the centroid C of an area are

$$\bar{x} = \frac{\int \tilde{x}\, dA}{\int dA} \qquad \bar{y} = \frac{\int \tilde{y}\, dA}{\int dA} \qquad \bar{z} = \frac{\int \tilde{z}\, dA}{\int dA} \tag{9–10}$$

Line. Using a differential element dL, the coordinates for the centroid C of a line are

$$\bar{x} = \frac{\int \tilde{x}\, dL}{\int dL} \qquad \bar{y} = \frac{\int \tilde{y}\, dL}{\int dL} \qquad \bar{z} = \frac{\int \tilde{z}\, dL}{\int dL} \tag{9–11}$$

It is important to remember that Eqs. 9–8 to 9–11 were all formulated using the *principle of moments*. Hence, the terms $\tilde{x}$, $\tilde{y}$, $\tilde{z}$ in these equations refer to the "moment arms" or perpendicular distances from the coordinate planes to the *center of gravity or centroid of the differential element* used in the equations. If possible, the differential element should be chosen such that it is of differential size or thickness in only *one direction*. When this is done only a single integration is required to cover the entire region.

PROCEDURE FOR ANALYSIS

The following procedure provides a method for determining the center of gravity or centroid of an object using a single integration.

Differential Element. Specify the coordinate axes and choose an appropriate differential element for integration. For lines this element dL is represented as a differential line segment; for areas the element dA is generally a rectangle, having a finite height and differential width; and for volumes the element dV can be a circular disk, having a finite radius and differential thickness. Locate the element so that it intersects the boundary of the line, area, or volume at an *arbitrary point* (x, y, z).

Size and Moment Arms. Express the length dL, area dA, or volume dV of the element in terms of the coordinates used to define the boundary of the object. Determine the perpendicular distances (moment arms) from the coordinate planes to the centroid or center of gravity of the element. For an x, y, z coordinate system, these dimensions are represented by the coordinates $\tilde{x}$, $\tilde{y}$, $\tilde{z}$.

Integrations. Substitute the data computed above into the appropriate equations (Eqs. 9–8 to 9–11) and perform the integrations. Note that integration can be accomplished when the function in the integrand is expressed in terms of the *same variable as the differential thickness of the element*. The limits of the integral are then defined from the two extreme locations of the element's differential thickness, so that when the elements are "summed" or the integration performed, the entire region is covered.

The following examples numerically illustrate this procedure.

Example 9–4

Locate the centroid of the thin rod bent into the shape of a quarter circle having a radius r as shown in Fig. 9–9.

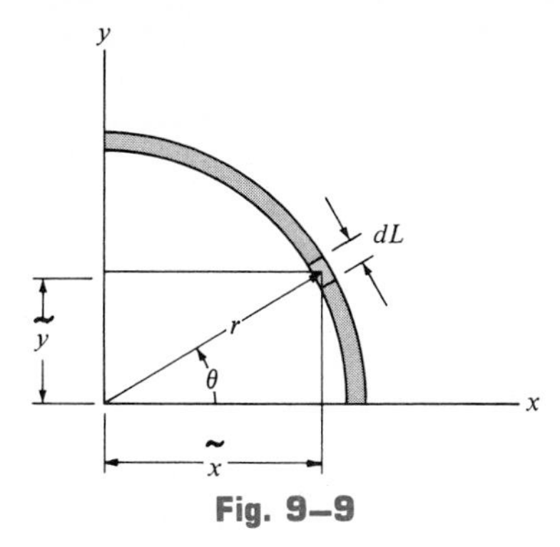

Fig. 9–9

Solution

Differential Element. The boundary of the rod can easily be defined using polar coordinates. The "arc" element intersects the curve at the arbitrary point (r, θ), Fig. 9–9.

Length and Moment Arms. The length of the element is $dL = r\,d\theta$, and the centroid is located at $\tilde{x} = r\cos\theta$, $\tilde{y} = r\sin\theta$.

Integrations. Using Eqs. 9–11 and integrating with respect to θ yields

$$\bar{x} = \frac{\int \tilde{x}\,dL}{\int dL} = \frac{\int_0^{\pi/2} r\cos\theta(r\,d\theta)}{\int_0^{\pi/2} r\,d\theta} = \frac{r^2}{\frac{\pi}{2}r} = \frac{2r}{\pi} \qquad \textit{Ans.}$$

$$\bar{y} = \frac{\int \tilde{y}\,dL}{\int dL} = \frac{\int_0^{\pi/2} r\sin\theta(r\,d\theta)}{\int_0^{\pi/2} r\,d\theta} = \frac{r^2}{\frac{\pi}{2}r} = \frac{2r}{\pi} \qquad \textit{Ans.}$$

Actually, by inspection only $\bar{x}$ has to be calculated, since $\bar{x} = \bar{y}$ due to symmetry.

Example 9–5

Locate the centroid of the triangular area shown in Fig. 9–10*a*.

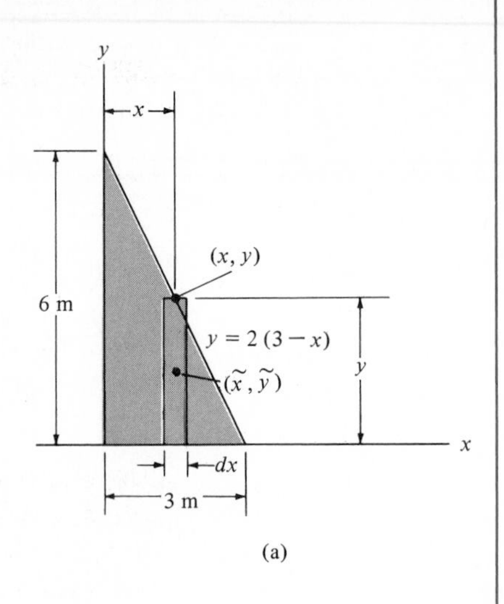

Fig. 9–10

Solution I

Differential Element. A rectangular differential element of thickness dx is shown in Fig. 9–10*a*. The element intersects the curve at the arbitrary point (x, y).

Area and Moment Arms. The area of the element is $dA = y\,dx$, and the centroid is located at $\tilde{x} = x$, $\tilde{y} = \frac{1}{2}y$.

Integrations. Using Eqs. 9–10 and integrating with respect to x yields

$$\bar{x} = \frac{\int_A \tilde{x}\,dA}{\int_A dA} = \frac{\int_0^3 xy\,dx}{\int_0^3 y\,dx} = \frac{\int_0^3 x\,2(3-x)\,dx}{\int_0^3 2(3-x)\,dx} = \frac{\frac{6}{2}x^2 - \frac{2}{3}x^3 \Big|_0^3}{6x - \frac{2}{2}x^2 \Big|_0^3} = 1\text{ m}$$

Ans.

$$\bar{y} = \frac{\int_A \tilde{y}\,dA}{\int_A dA} = \frac{\int_0^3 (\frac{1}{2}y)\,y\,dx}{\int_0^3 y\,dx} = \frac{\int_0^3 \frac{4}{2}(3-x)^2\,dx}{\int_0^3 2(3-x)\,dx} = \frac{2(9x - \frac{6}{2}x^2 + \frac{1}{3}x^3)\Big|_0^3}{6x - \frac{2}{2}x^2 \Big|_0^3} = 2\text{ m}$$

Ans.

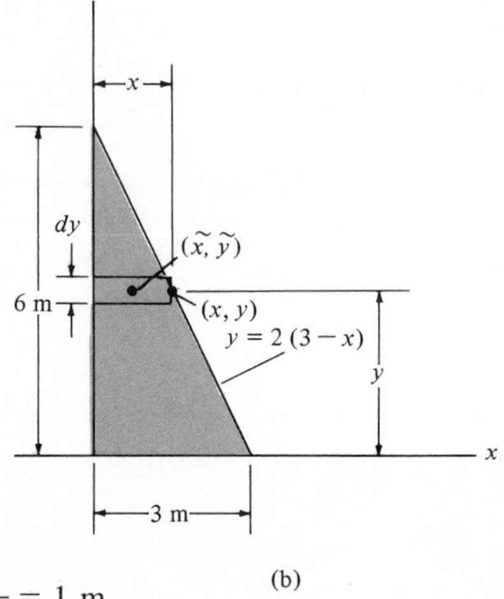

Solution II

Differential Element. A rectangular differential element of thickness dy is shown in Fig. 9–10*b*. As before, the element intersects the curve at the arbitrary point (x, y).

Area and Moment Arms. The area of the element is $dA = x\,dy$, and the centroid is located at $\tilde{x} = \frac{1}{2}x$, $\tilde{y} = y$.

Integrations. Using Eqs. 9–10, and integrating with respect to y yields

$$\bar{x} = \frac{\int_A \tilde{x}\,dA}{\int_A dA} = \frac{\int_0^6 (\frac{1}{2}x)\,x\,dy}{\int_0^6 x\,dy} = \frac{\int_0^6 \frac{1}{2}(3 - \frac{1}{2}y)^2\,dy}{\int_0^6 (3 - \frac{1}{2}y)\,dy} = \frac{\frac{1}{2}(9y - \frac{3}{2}y^2 + \frac{1}{12}y^3)\Big|_0^6}{3y - \frac{1}{4}y^2 \Big|_0^6} = 1\text{ m}$$

Ans.

$$\bar{y} = \frac{\int_A \tilde{y}\,dA}{\int_A dA} = \frac{\int_0^6 y\,x\,dy}{\int_0^6 x\,dy} = \frac{\int_0^6 y(3 - \frac{1}{2}y)\,dy}{\int_0^6 (3 - \frac{1}{2}y)\,dy} = \frac{\frac{3}{2}y^2 - \frac{1}{6}y^3 \Big|_0^6}{3y - \frac{1}{4}y^2 \Big|_0^6} = 2\text{ m}$$

Ans.

Example 9–6

Locate the centroid of the area of the plate shown in Fig. 9–11a.

Solution I

Differential Element. A differential element of thickness dx is shown in Fig. 9–11a. The element intersects the curve at the arbitrary point (x, y).

Area and Moment Arms. The area of the element is $dA = y\,dx$, and the centroid is located at $\tilde{x} = x$, $\tilde{y} = y/2$.

Integrations. Using Eqs. 9–10 and integrating with respect to x yields

$$\bar{x} = \frac{\int_A \tilde{x}\,dA}{\int_A dA} = \frac{\int_0^1 xy\,dx}{\int_0^1 y\,dx} = \frac{\int_0^1 x^3\,dx}{\int_0^1 x^2\,dx} = \frac{0.250}{0.333} = 0.75 \text{ m} \qquad \textbf{\textit{Ans.}}$$

$$\bar{y} = \frac{\int_A \tilde{y}\,dA}{\int_A dA} = \frac{\int_0^1 \left(\frac{y}{2}\right) y\,dx}{\int_0^1 y\,dx} = \frac{\int_0^1 \left(\frac{x^2}{2}\right) x^2\,dx}{\int_0^1 x^2\,dx} = \frac{0.100}{0.333} = 0.3 \text{ m} \qquad \textbf{\textit{Ans.}}$$

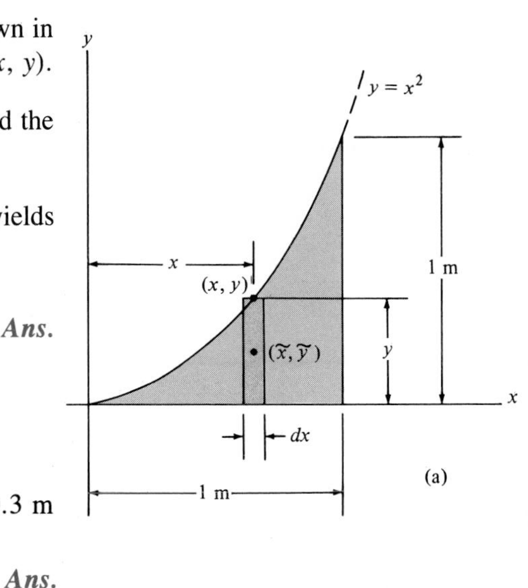

Solution II

Differential Element. The differential element of thickness dy is shown in Fig. 9–11b. Again the element intersects the curve at the arbitrary point (x, y).

Area and Moment Arms. The area of the element is $dA = (1 - x)\,dy$, and the centroid is located at

$$\tilde{x} = x + \left(\frac{1-x}{2}\right) = \frac{1+x}{2}, \qquad \tilde{y} = y$$

Integrations. Using Eqs. 9–10, and integrating with respect to y, we obtain

$$\bar{x} = \frac{\int_A \tilde{x}\,dA}{\int_A dA} = \frac{\int_0^1 \left(\frac{1+x}{2}\right)(1-x)\,dy}{\int_0^1 (1-x)\,dy} = \frac{\frac{1}{2}\int_0^1 (1-y)\,dy}{\int_0^1 (1-\sqrt{y})\,dy}$$

$$= \frac{0.250}{0.333} = 0.75 \text{ m} \qquad \textbf{\textit{Ans.}}$$

$$\bar{y} = \frac{\int_A \tilde{y}\,dA}{\int_A dA} = \frac{\int_0^1 y(1-x)\,dy}{\int_0^1 (1-x)\,dy} = \frac{\int_0^1 (y - y^{3/2})\,dy}{\int_0^1 (1-\sqrt{y})\,dy} = \frac{0.100}{0.333} = 0.3 \text{ m} \qquad \textbf{\textit{Ans.}}$$

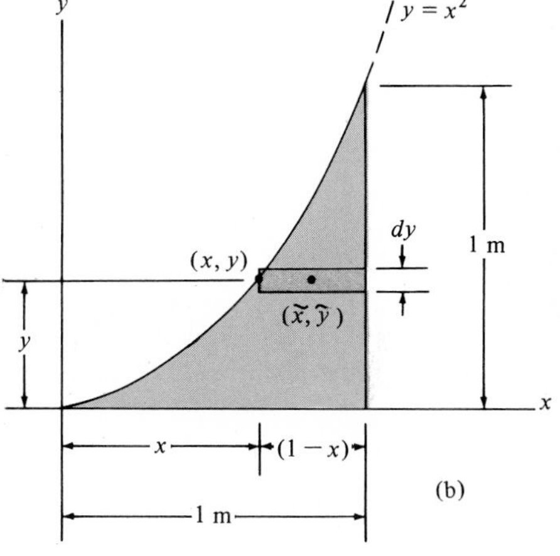

Fig. 9–11

Example 9–7

Locate the $\bar{y}$ centroid for the paraboloid of revolution, which is generated by revolving the shaded area shown in Fig. 9–12 about the y axis.

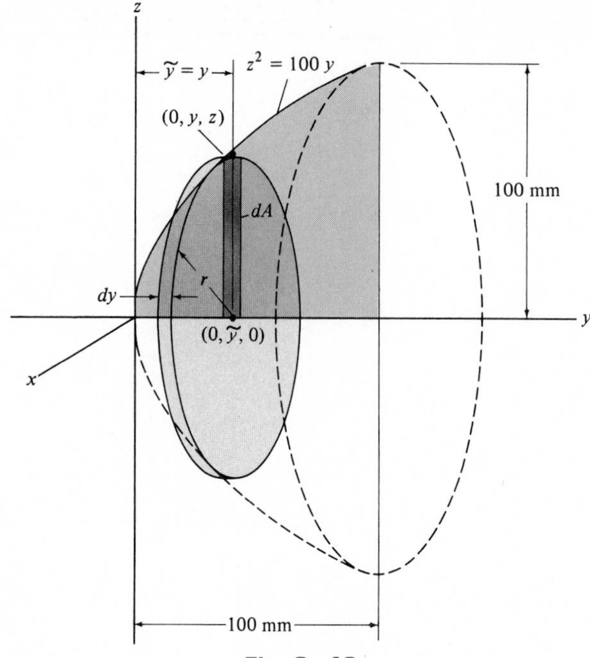

Fig. 9–12

Solution

Differential Element. An element having the shape of a *thin disk* is chosen, Fig. 9–12. This element has a radius of $r = z$ and a thickness of dy. In this "disk" method of analysis, the element of planar area, dA, is always taken *perpendicular* to the axis of revolution. Here the element intersects the generating curve at the arbitrary point $(0, y, z)$.

Volume and Moment Arm. The volume of the element is $dV = (\pi z^2)\,dy$, and the centroid is located at $\tilde{y} = y$.

Integrations. Using the second of Eqs. 9–9 and integrating with respect to y yields

$$\bar{y} = \frac{\int_V \tilde{y}\,dV}{\int_V dV} = \frac{\int_0^{100} y(\pi z^2)\,dy}{\int_0^{100} (\pi z^2)\,dy} = \frac{100\pi \int_0^{100} y^2\,dy}{100\pi \int_0^{100} y\,dy}$$

$$= 66.7 \text{ mm} \qquad \textit{Ans.}$$

Problems

9–21. Locate the centroid $\overline{x}$ of the circular rod. Express the answer in terms of the radius r and semiarc angle α.

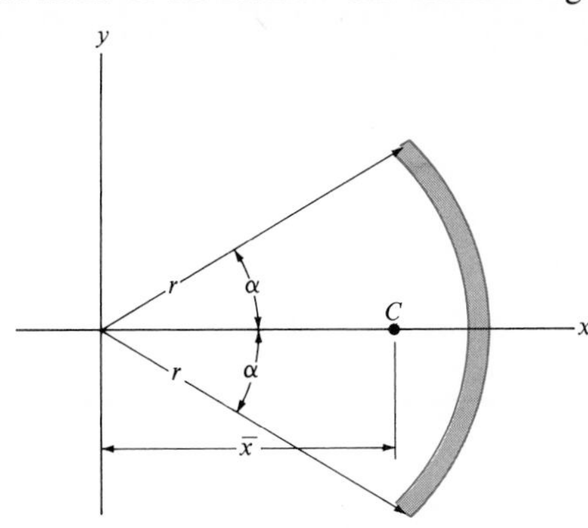

Prob. 9–21

9–22. Locate the centroid of the parabolic area.

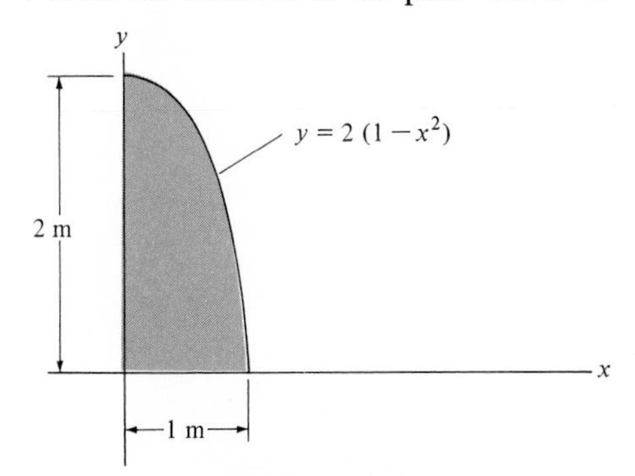

Prob. 9–22

9–23. Locate the centroid of the exparabolic segment of area.

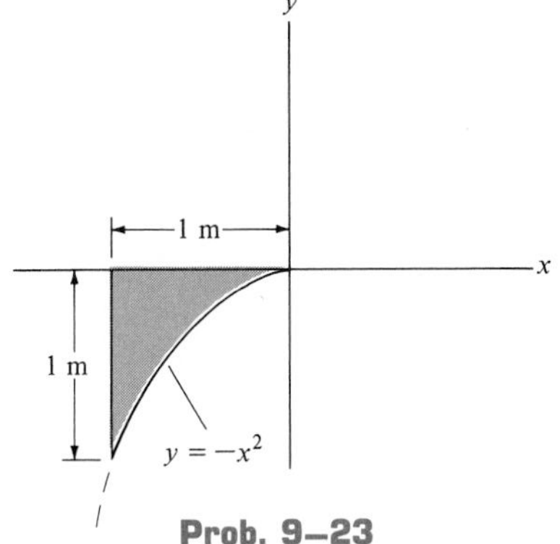

Prob. 9–23

***9–24.** Locate the centroid $\overline{y}$ of the shaded area.

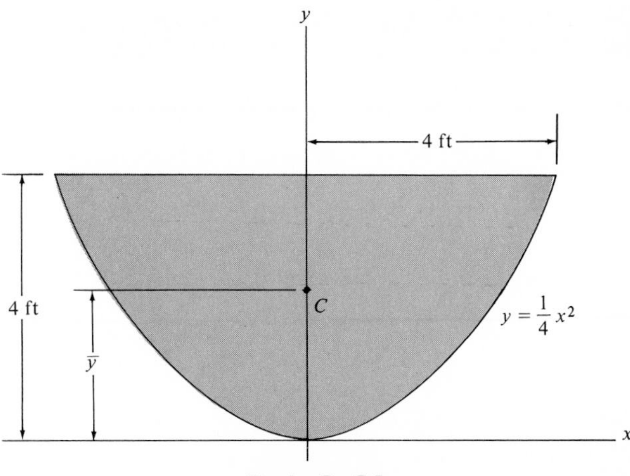

Prob. 9–24

9–25. Locate the centroid of the shaded area.

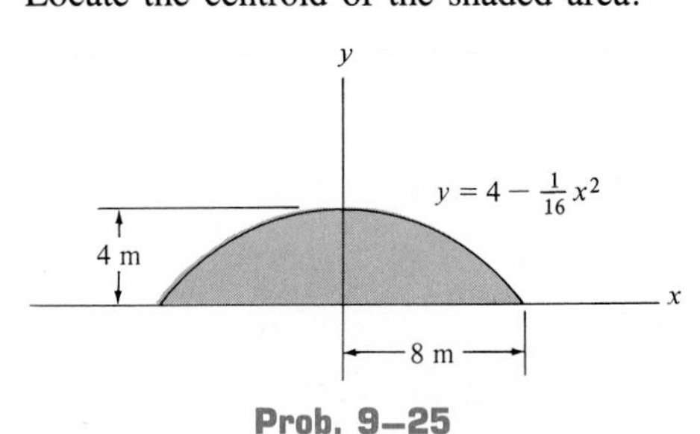

Prob. 9–25

9–26. Locate the centroid of the shaded area.

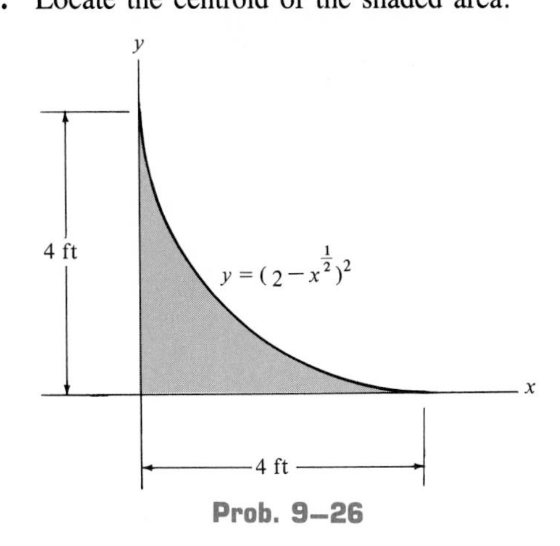

Prob. 9–26

9–27. Locate the centroid of the shaded elliptical sector.

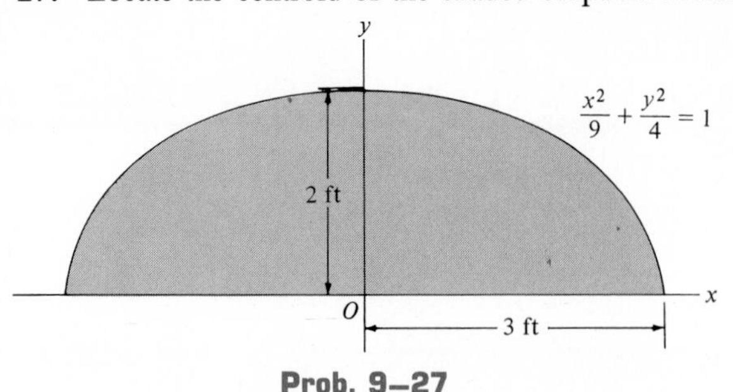

Prob. 9–27

***9–28.** Locate the centroid for the triangular area.

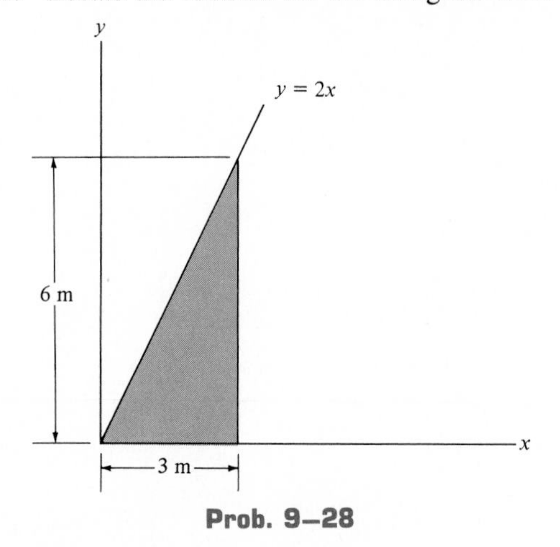

Prob. 9–28

9–29. Locate the centroid of the shaded area.

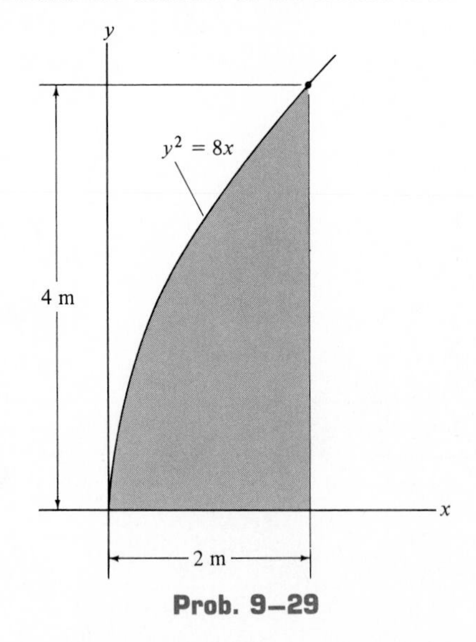

Prob. 9–29

9–30. Locate the center of gravity of the homogeneous "bell-shaped" volume formed by revolving the shaded area about the y axis.

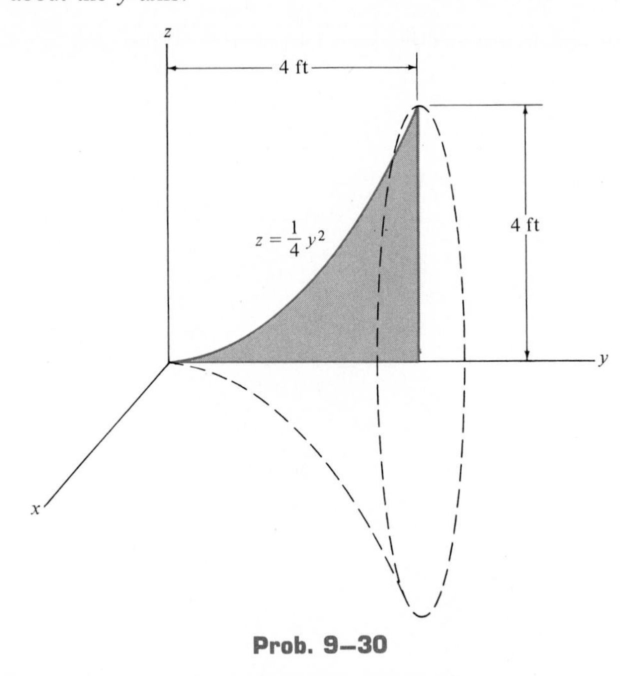

Prob. 9–30

9–31. Locate the center of gravity of the frustum of the paraboloid. The material is homogeneous.

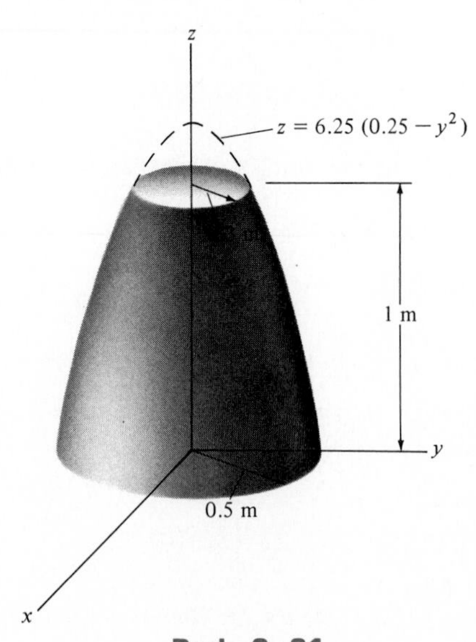

Prob. 9–31

***9–32.** Locate the center of gravity of the volume generated by revolving the shaded area about the z axis. The material is homogeneous.

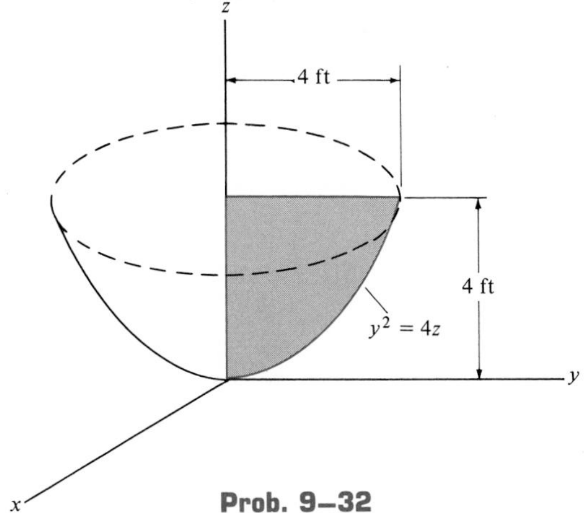
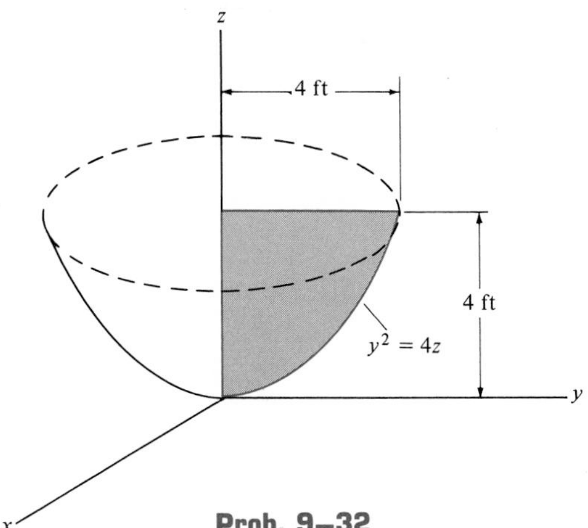

Prob. 9–32

9–33. Locate the center of gravity of the homogeneous cone formed by rotating the shaded area about the y axis.

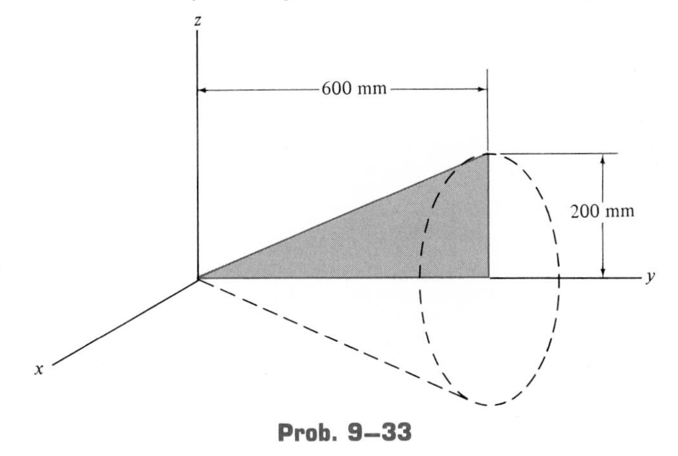

Prob. 9–33

9–34. Locate the center of gravity of the volume generated by revolving the shaded area about the y axis. The material is homogeneous.

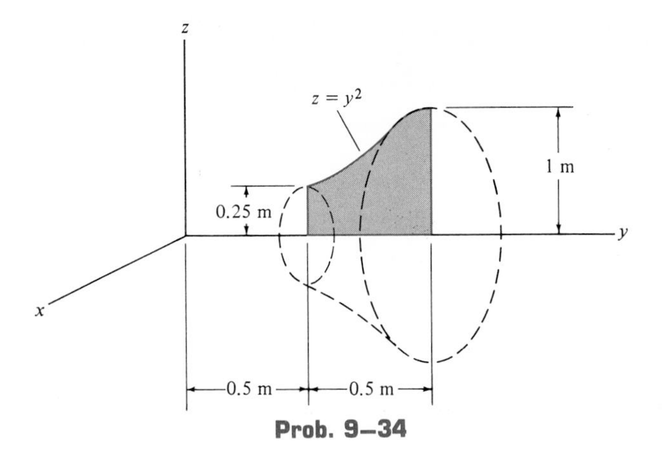

Prob. 9–34

9.4* Theorems of Pappus and Guldinus

The *theorems of Pappus and Guldinus*, which were first developed by Pappus of Alexandria during the third century A.D. and again at a later time by the Swiss mathematician Paul Guldin or Guldinus (1577–1643), are used to find the surface area or volume of any solid of revolution.

A *surface area of revolution* is generated by revolving a *plane curve* about a nonintersecting fixed axis in the plane of the curve; whereas a *volume of revolution* is generated by revolving a *plane area* about a nonintersecting fixed axis in the plane of the area. For example, if the line *AB* shown in Fig. 9–13 is rotated about a fixed axis, it generates the surface area of a cone; if the triangular area *ABC* shown in Fig. 9–14 is rotated about the axis, it generates the volume of a cone.

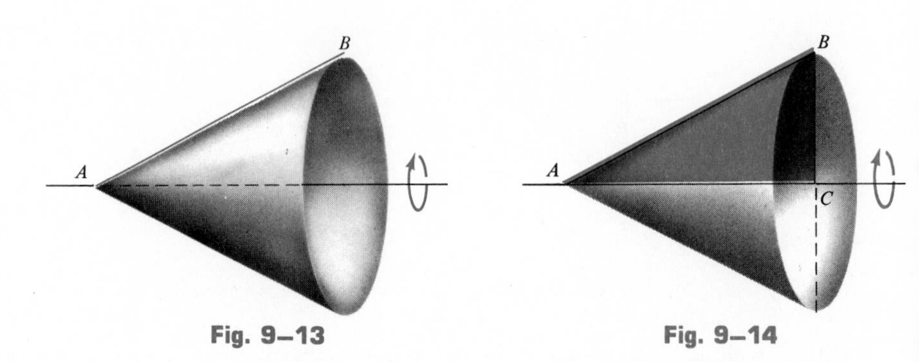

Fig. 9–13

Fig. 9–14

The statements and proofs of the theorems of Pappus and Guldinus follow. The proofs require that the generating curves and areas do *not* cross the axis about which they are rotated; otherwise, two sections on either side of the axis would generate areas or volumes having opposite signs and hence cancel each other.

Surface Area. *The area of a surface of revolution equals the product of the length of the generating curve and the distance traveled by the centroid of the curve in generating the surface area.*

Proof. When a differential length dL of the curve shown in Fig. 9–15 is revolved about an axis through a distance $2\pi r$, it generates a ring having a surface area $dA = 2\pi r\, dL$. The entire surface area, generated by revolving the entire curve about the axis, is therefore $A = 2\pi \int_L r\, dL$. Since $\int_L r\, dL = \bar{r}L$, where $\bar{r}$ locates the centroid C of the generating curve L, the area becomes $A = 2\pi\bar{r}L$. In general, though,

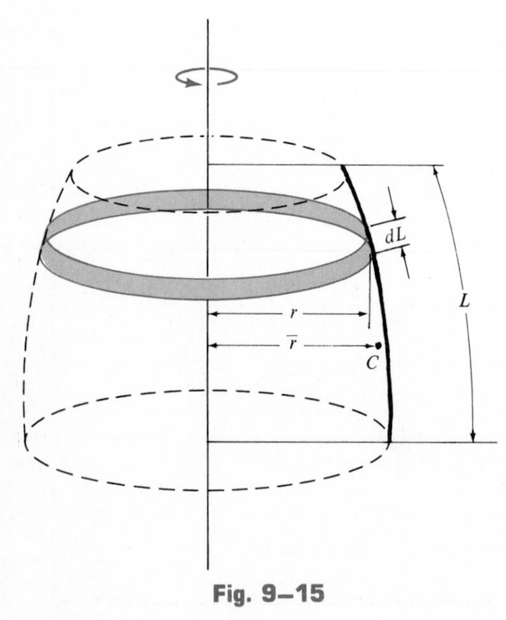

Fig. 9–15

$$A = \theta\bar{r}L \tag{9–12}$$

where A = surface area of revolution
θ = angle of revolution measured in radians, $\theta \leq 2\pi$
$\bar{r}$ = distance from the axis of revolution to the centroid of the generating curve
L = length of the generating curve

Volume. *The volume of a surface of revolution equals the product of the generating area and the distance traveled by the centroid of the area in generating the volume.*

Proof. When the differential area dA shown in Fig. 9–16 is revolved about an axis through a distance $2\pi r$, it generates a ring having a volume $dV = 2\pi r\, dA$. The entire volume, generated by revolving A about the axis, is therefore $V = 2\pi \int_A r\, dA$. Since $\int_A r\, dA = \bar{r}A$, where $\bar{r}$ locates the centroid C of the generating area A, the volume becomes $V = 2\pi \bar{r} A$. In general, though,

$$V = \theta \bar{r} A \qquad (9\text{–}13)$$

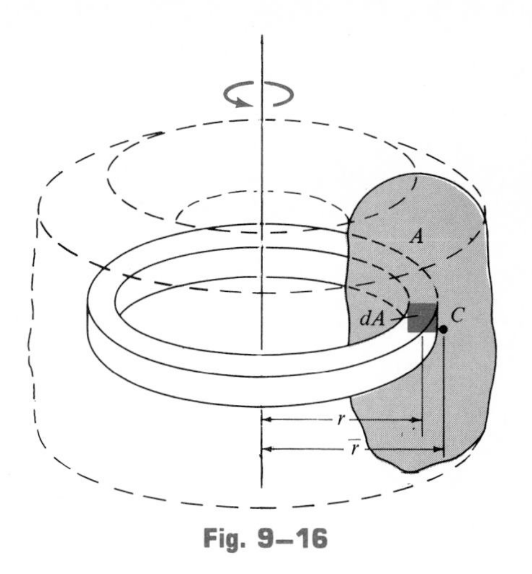

Fig. 9–16

where V = volume of revolution
θ = angle of revolution measured in radians, $\theta \leqslant 2\pi$
$\bar{r}$ = distance from the axis of revolution to the centroid of the generating area
A = generating area

The following examples illustrate application of the above two theorems.

Example 9–8

Show that the surface area of a sphere is $4\pi R^2$.

Solution

The sphere in Fig. 9–17 is generated by rotating a semicircular *arc* about the x axis. Using the table listed on the inside back cover, it is seen that the centroid of this arc is located at a distance $\bar{y} = 2R/\pi$ from the x axis of rotation. Since the centroid moves through an angle of $\theta = 2\pi$ rad in generating the sphere, applying Eq. 9–12 gives the surface area of the sphere as:

$$A = \theta \bar{y} L; \qquad A = 2\pi\left(\frac{2R}{\pi}\right)\pi R = 4\pi R^2 \qquad \textit{Ans.}$$

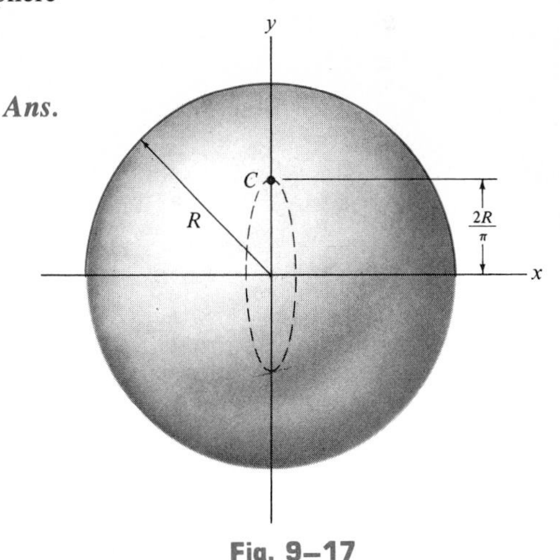

Fig. 9–17

Example 9–9

Determine the mass of concrete needed to construct the arched beam shown in Fig. 9–18. The density of concrete is $\rho_c = 2.1\ \text{Mg/m}^3$.

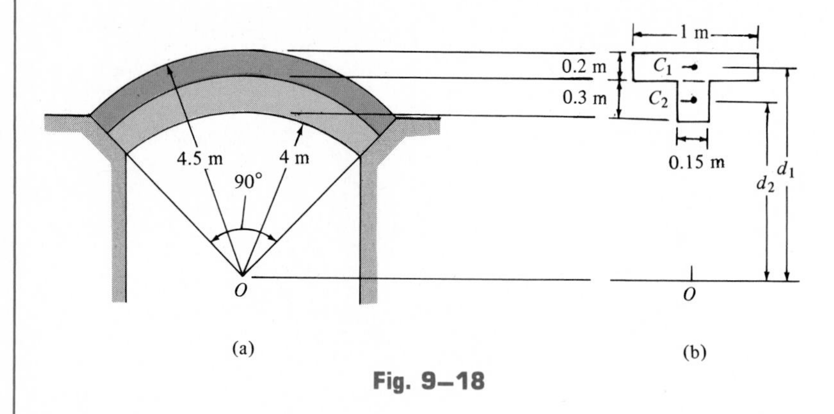

Fig. 9–18

Solution

The mass of the arch may be determined by using the density ρ_c provided the volume of concrete is known. The cross-sectional area of the beam is composed of two rectangles having centroids at points C_1 and C_2 as shown in Fig. 9–18*b*. The volume generated by each of these areas is equal to the product of the cross-sectional area and the distance traveled by its centroid. In generating the arch, the centroids move through an angle of $\theta = \pi(90°/180°) = 1.57$ radians. Point C_1 acts at a distance of $d_1 = 4.5 - 0.1 = 4.4$ m from the center of rotation, point O; whereas C_2 acts at $d_2 = 4 + 0.15 = 4.15$ m from O. Applying Eq. 9–13, we obtain the total volume:

$$
\begin{aligned}
V &= \Sigma \theta \bar{x} A \\
&= (1.57\ \text{rad})(4.4\ \text{m})(1\ \text{m})(0.2\ \text{m}) + (1.57\ \text{rad})(4.15)(0.15\ \text{m})(0.3\ \text{m}) \\
&= 1.676\ \text{m}^3
\end{aligned}
$$

The required mass of concrete is then

$$m = \rho_c V = (2.1\ \text{Mg/m}^3)(1.676\ \text{m}^3) = 3.52\ \text{Mg} \qquad \textit{Ans.}$$

Problems

9–35. Compute both the area and the centroidal distance $\bar{x}$ of the parabolic shaded region using the table on the inside back cover. Then, using the second theorem of Pappus–Guldinus, compute the volume of the solid generated by revolving the shaded area about the *aa* axis.

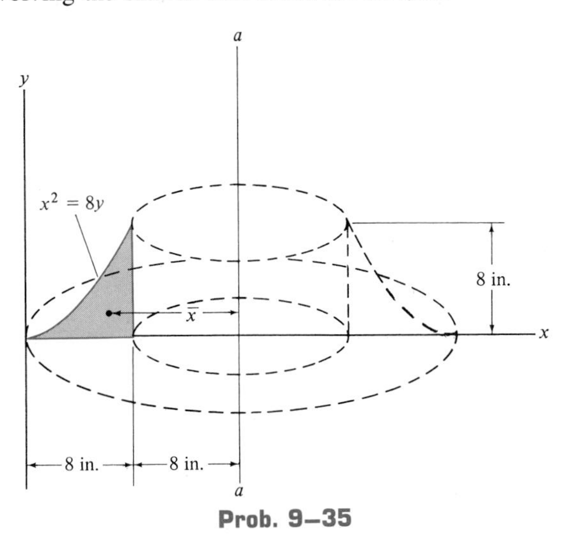

Prob. 9–35

***9–36.** Sand is piled between two walls as shown. Assume the pile to be a quarter section of a cone and use the second theorem of Pappus–Guldinus to determine its volume.

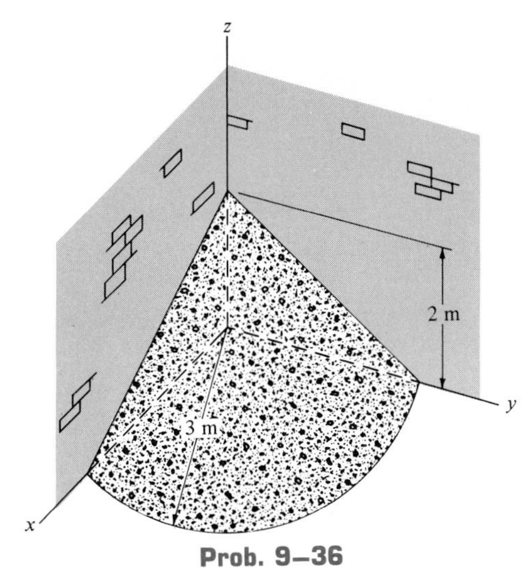

Prob. 9–36

9–37. Determine the area and the centroidal distance $\bar{y}$ of the parabolic shaded area using the table on the inside back cover. Then, using the second theorem of Pappus–Guldinus, determine the volume of a paraboloid formed by revolving the area about the x axis.

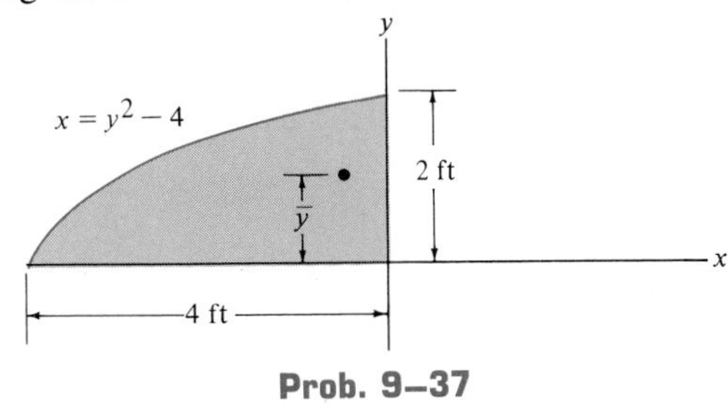

Prob. 9–37

9–38. A circular sea wall is made of concrete. Determine the total weight of the wall if the concrete has a density of $\rho = 150\ \text{lb/ft}^3$.

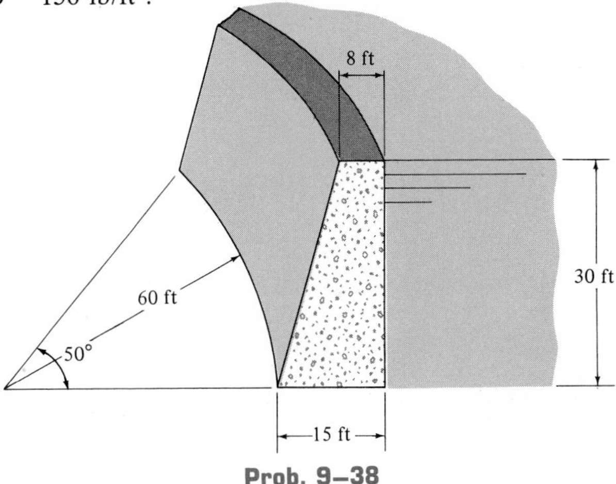

Prob. 9–38

9–39. Determine the volume of concrete needed to construct the circular curb.

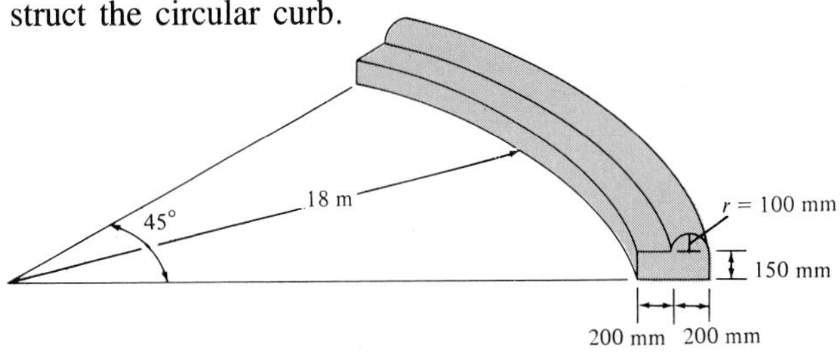

Prob. 9–39

***9–40.** Determine the approximate amount of paint needed to cover the surface of the water storage tank. Assume that a liter of paint covers $2.5\ m^2$. Also, what is the total inside volume of the tank?

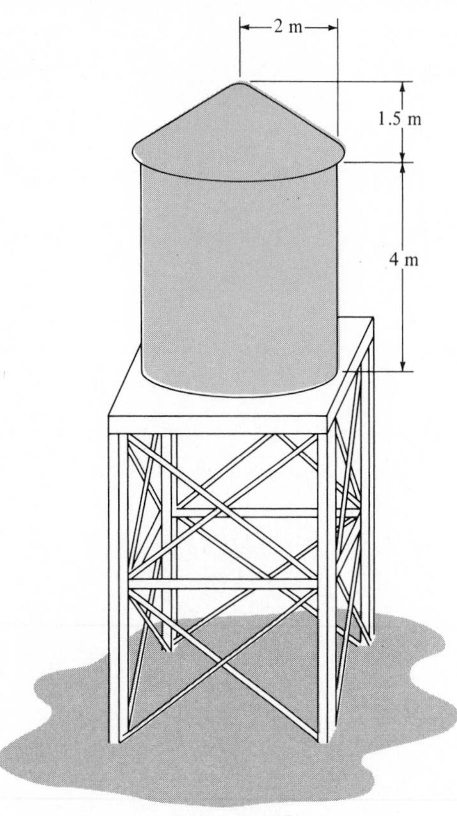

Prob. 9–40

9–41. Determine the surface area and volume of the torus.

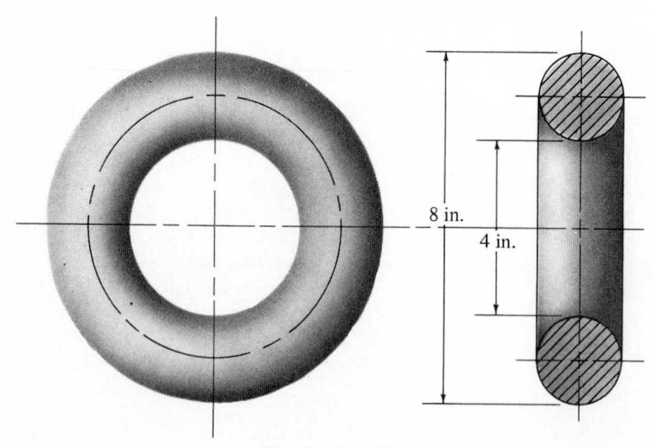

Prob. 9–41

9–42. A steel wheel has a diameter of 840 mm and a cross section as shown in the figure. Determine the total mass of the wheel if $\rho = 5\ Mg/m^3$.

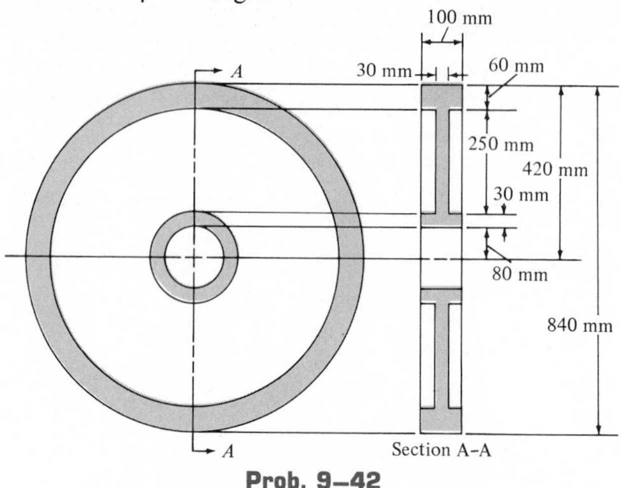

Prob. 9–42

9–43. Determine the height h to which liquid should be poured into the conical cup so that it contacts half the surface area on the inside of the cup.

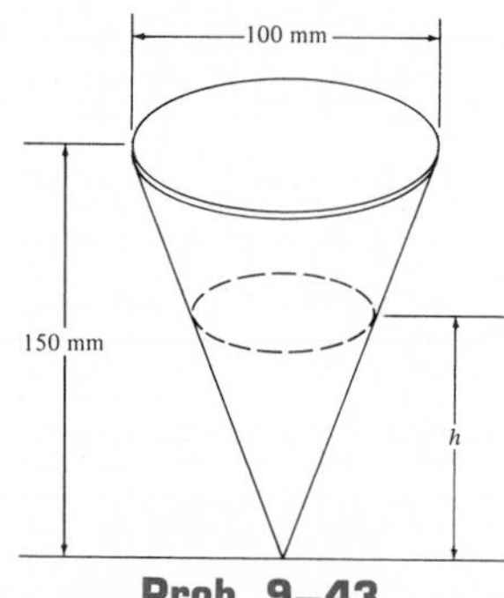

Prob. 9–43

***9–44.** Determine the surface area and the volume of material required to make the casting.

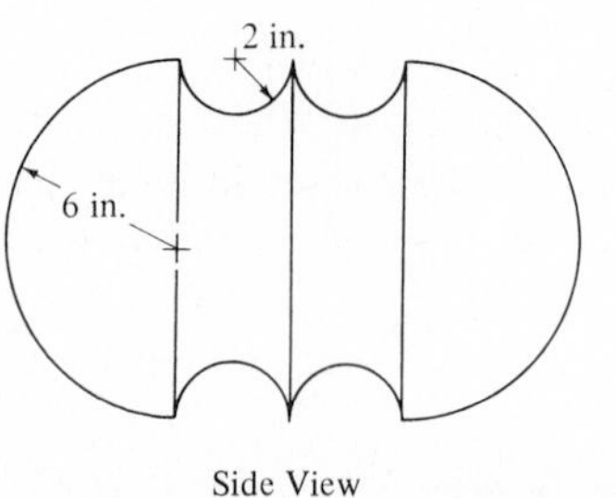

Prob. 9–44

Distributed Loadings 9.5*

In Sec. 4–8 we discussed the method for both reducing a simple distributed loading to a single resultant force and specifying the location of the resultant. In this section and the next we will consider the most common case of a distributed pressure loading which is uniform along an axis of a flat body upon which the loading is applied. An example of such a loading is shown in Fig. 9–19*a*. The intensity of this loading at each point is measured as a force per unit area, defined as the *pressure p*, which can be measured in units of lb/ft^2 or pascals (Pa), where 1 Pa = 1 N/m^2. The entire loading on the plate is therefore a system of parallel forces, infinite in number and each acting on a separate differential area of the plate. Here the loading function $p = p(x)$ is only a function of x since the pressure is uniform along the y axis. If we multiply $p = p(x)$ by the width a of the plate, we obtain $w = [p(x)]a = w(x)$. The loading function, $w = w(x)$, is a measure of force per unit length rather than force per unit area, and consequently the load-intensity diagram for $w = w(x)$ represents an infinite number of coplanar parallel forces, distributed along the edge of the plate, Fig. 9–19*b*.

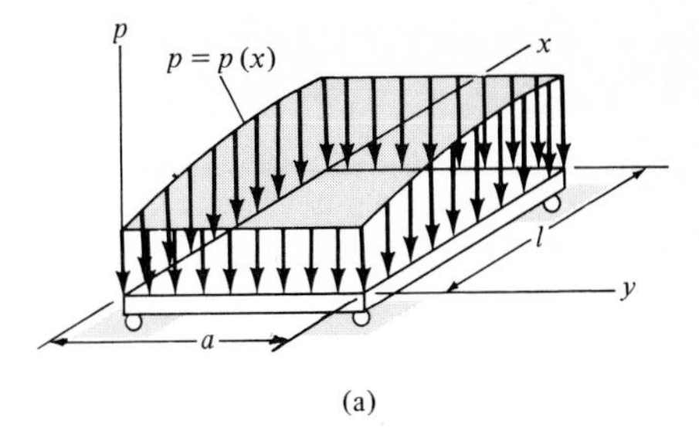

(a)

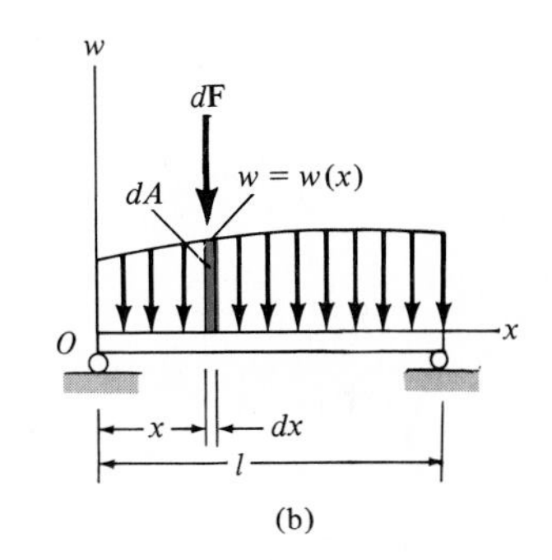

(b)

Magnitude of Resultant Force. As discussed in Sec. 4–8, the magnitude of $\mathbf{F}_R$, Fig. 9–19*c*, is determined by summing each force $dF = w(x)\,dx = dA$ over the entire length l. Fig. 9–19*b*, i.e.,

$$+\downarrow F_R = \Sigma F; \qquad F_R = \int_l w(x)\,dx = \int_A dA = A \qquad (9\text{–}14)$$

Here F_R equals the area under the loading curve.

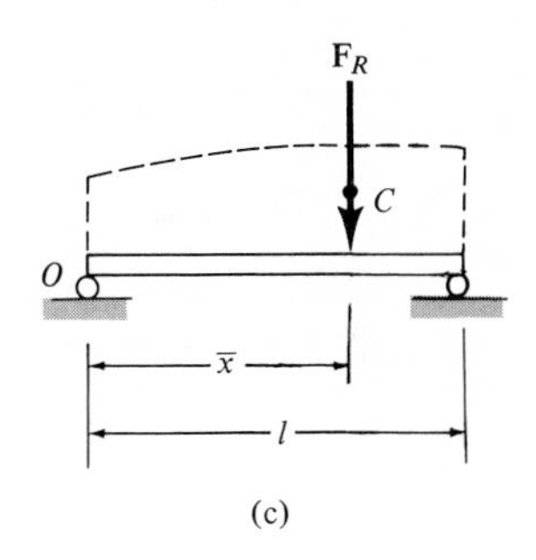

(c)

Location of Resultant Force. Applying the principle of moments about point O, Figs. 9–19*b* and 9–19*c*, we have

$$\circlearrowleft + M_{R_O} = \Sigma M_O; \qquad \bar{x}F_R = \int_A x\,dA$$

or

$$\bar{x} = \frac{\displaystyle\int_A x\,dA}{\displaystyle\int_A dA} \qquad (9\text{–}15)$$

Here $\bar{x}$ locates the centroid of the loading curve area.

Once $\bar{x}$ is determined, $\mathbf{F}_R$ by symmetry passes through point $(\bar{x}, a/2)$ on the surface of the plate, Fig. 9–19*d*. Regarding the three-dimensional pressure loading $p(x)$, Fig. 9–19*a*, we can therefore conclude that *the resultant force has a magnitude equal to the volume under the distributed loading curve $p = p(x)$ and a line of action which passes through the centroid (geometric center) of this volume.*

The following example numerically illustrates an application of Eqs. 9–14 and 9–15.

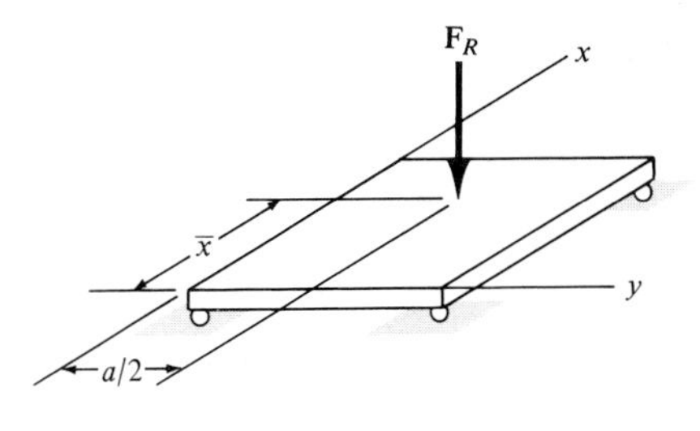

(d)

Fig. 9–19

Example 9–10

Determine the magnitude and location of the resultant force acting on the beam in Fig. 9–20*a*.

Solution

As shown, the colored differential area element $dA = w\,dx = 60x^2\,dx$ will be used for integration. Applying Eq. 9–14, by summing these elements from $x = 0$ to $x = 2$ m, we obtain the resultant force $\mathbf{F}_R$,

$F_R = \Sigma F;$

$$F_R = \int_A dA = \int_0^2 60x^2\,dx = 60\left[\frac{x^3}{3}\right]_0^2 = 60\left[\frac{2^3}{3} - \frac{0^3}{3}\right]$$

$$= 160 \text{ N} \qquad \textit{Ans.}$$

Since the element of area dA is located an arbitary distance x from O, the location $\bar{x}$ of $\mathbf{F}_R$ *measured from O*, Fig. 9–20*b*, is determined from Eq. 9–15:

$$\bar{x} = \frac{\int_A x\,dA}{\int_A dA} = \frac{\int_0^2 x(60\,x^2)\,dx}{160} = \frac{60\left[\frac{x^4}{4}\right]_0^2}{160} = \frac{60\left[\frac{2^4}{4} - \frac{0^4}{4}\right]}{160}$$

$$= 1.5 \text{ m} \qquad \textit{Ans.}$$

These results may be checked by using the table on the inside back cover, where it is shown that for a semiparabolic area of height a, length b, and shape shown in Fig. 9–20*a*,

$$A = \frac{ab}{3} = \frac{240(2)}{3} = 160 \text{ N} \quad \text{and} \quad \bar{x} = \frac{3}{4}b = \frac{3}{4}(2) = 1.5 \text{ m}$$

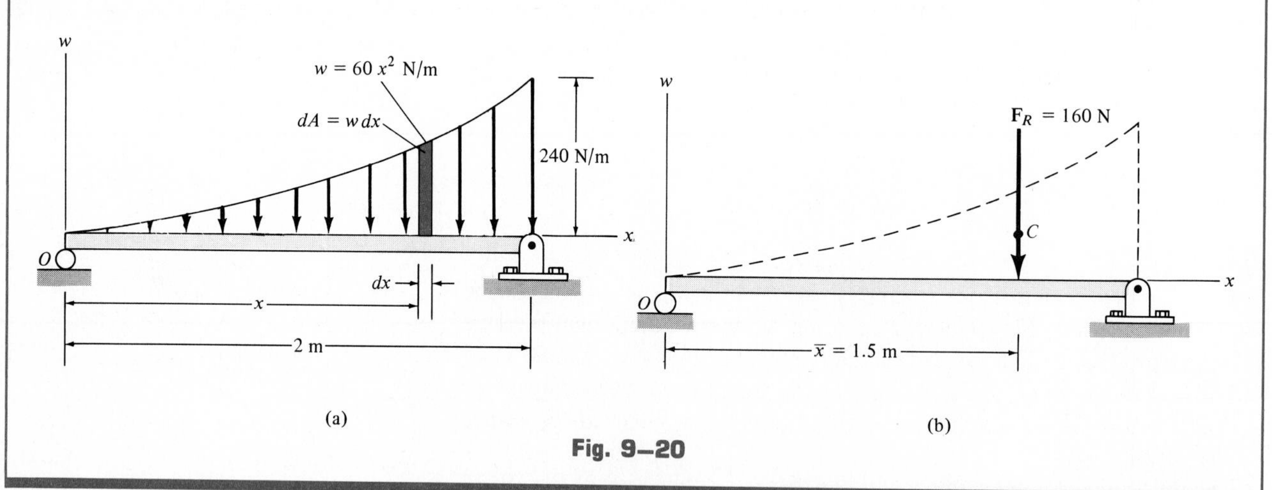

Fig. 9–20

Problems

9–45. Replace the loading by an equivalent force and couple moment acting at point O.

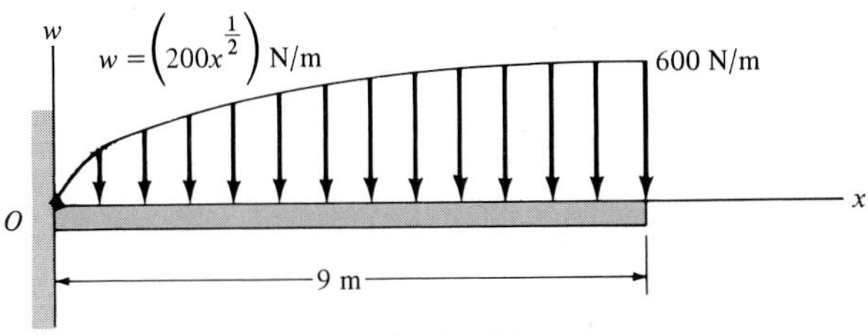

Prob. 9–45

9–46. The lifting force along the wing of a jet aircraft consists of a uniform distribution along AB, and a semiparabolic distribution along BC with origin at B. Replace this loading by a single resultant force and specify its location measured from point A.

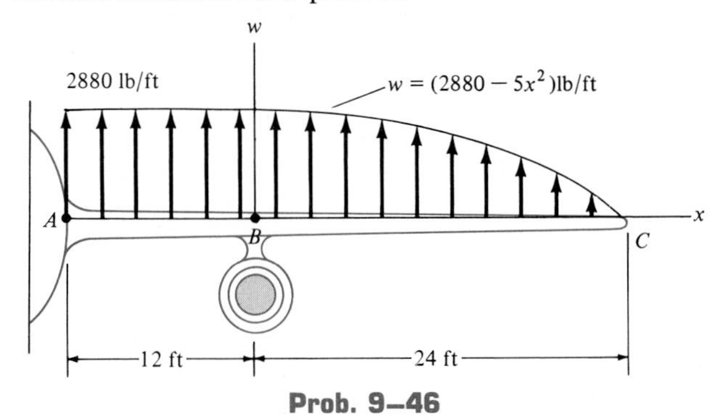

Prob. 9–46

9–47. The beam is subjected to the parabolic loading. Determine an equivalent force and couple system at point A.

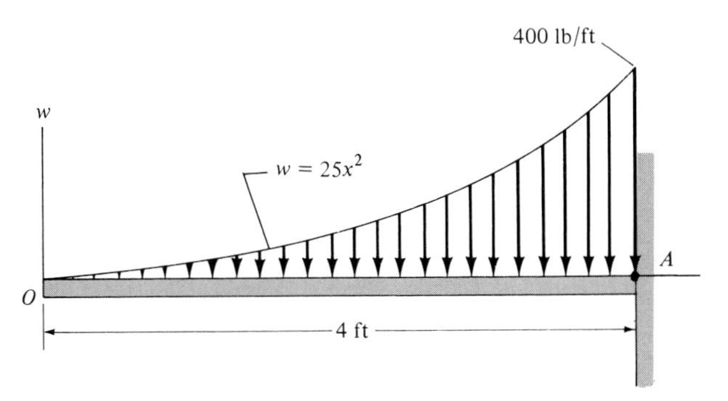

Prob. 9–47

***9–48.** The wind loading on the chemical distillation tower varies in the manner shown. Determine the resultant force of the wind and specify its location on the tower, measured from its base, $y = 0$.

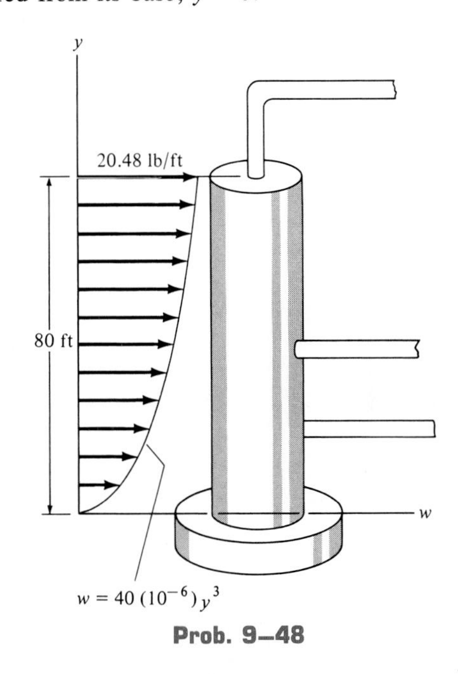

Prob. 9–48

9–49. The wind has blown sand over a platform such that the intensity of load can be approximated by the function $w = (\frac{1}{2}x^3)$ N/m. Simplify this distributed loading to a single concentrated force and specify the magnitude and location of the force measured from A.

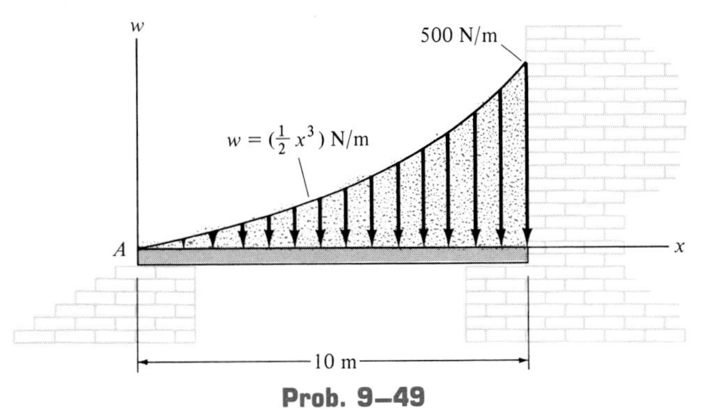

Prob. 9–49

9–50. The form is used to cast concrete columns. Determine the resultant force that wet concrete exerts along the plate A, $0.5\text{ m} \leq z \leq 3\text{ m}$, if the pressure due to the concrete varies as shown. Specify the location of the resultant force, measured from the top of the column.

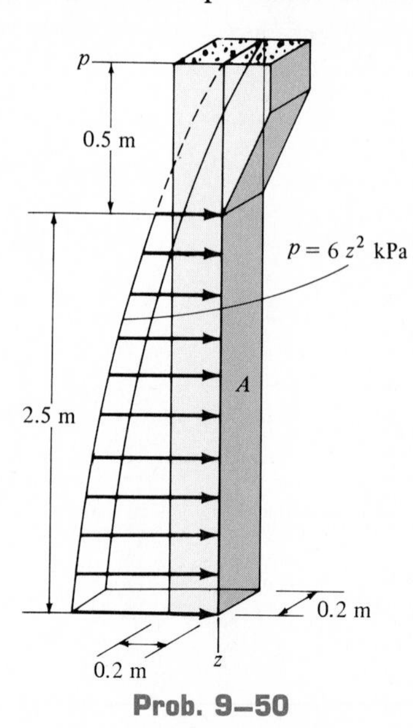

Prob. 9–50

9–51. The rectangular bin is filled with coal, which creates a pressure distribution along wall A that varies as shown, i.e., $p = (4z^3)\text{ lb/ft}^2$, where z is measured in feet. Compute the resultant force created by the coal, and specify its location measured from the top surface of the coal.

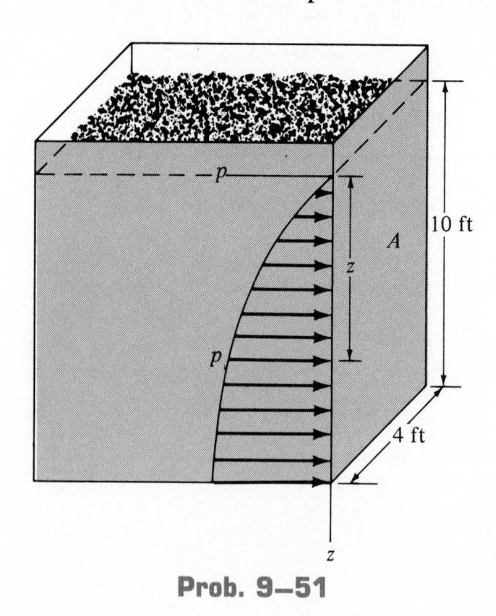

Prob. 9–51

***9–52.** The wind, acting on a square plate, generates a pressure distribution that is parabolic. Determine the magnitude and location of the resultant force.

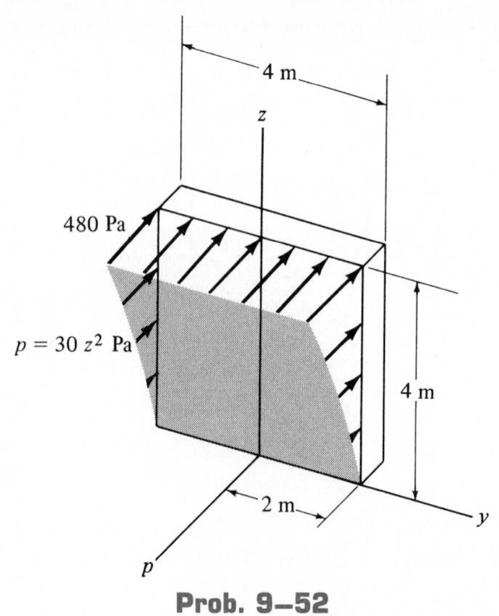

Prob. 9–52

9–53. A wind loading creates a positive pressure on one side of the chimney and a negative (suction) pressure on the other side, as shown. If this pressure loading acts uniformly along the chimney's length, determine the magnitude of the resultant force created by the wind.

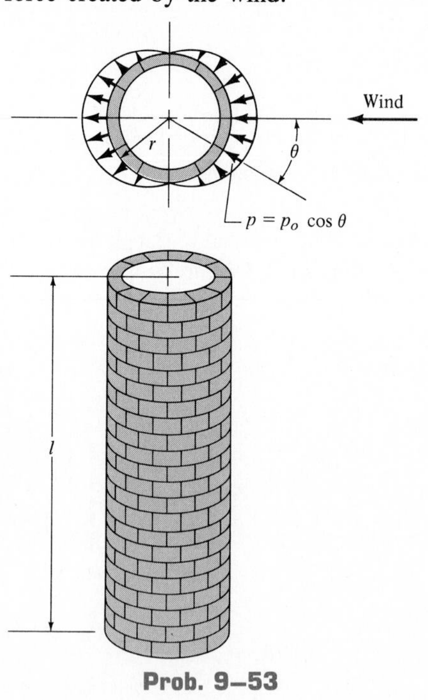

Prob. 9–53

9.6* Fluid Pressure Acting on a Submerged Surface

In the preceding section we discussed the method used to simplify a distributed loading which is uniform along an axis of a flat surface. In this section we will apply this method to the surface of a body which is submerged in a fluid.

Pressure Loading. According to Pascal's law, a fluid at rest creates a pressure p at a point that is the *same* in *all* directions. The magnitude of p, measured as a force per unit area, depends upon the mass density ρ of the fluid and the depth z of the point from the fluid surface. The relationship can be expressed mathematically as

$$p = \rho g z \tag{9-16}$$

where g is the acceleration of gravity. Equation 9–16 is only valid for fluids that are *incompressible,* as in the case of most liquids. Gases are compressible fluids, and since their density changes significantly with both altitude and temperature, Eq. 9–16 cannot be used.

Consider now what effect the pressure of a liquid has at points A, B, and C, located on the top surface of the submerged plate shown in Fig. 9–21. Since points A and B are both at depth z_2 from the liquid surface, the *pressure* at these points has a magnitude of $p_2 = \rho g z_2$. Likewise, point C is at depth z_1; hence, $p_1 = \rho g z_1$. In all cases, the pressure acts *normal* to the surface area dA located at the point, Fig. 9–21.

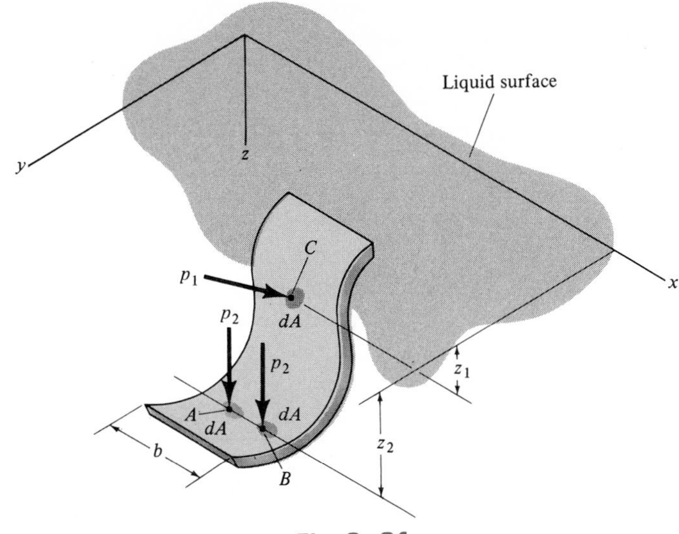

Fig. 9–21

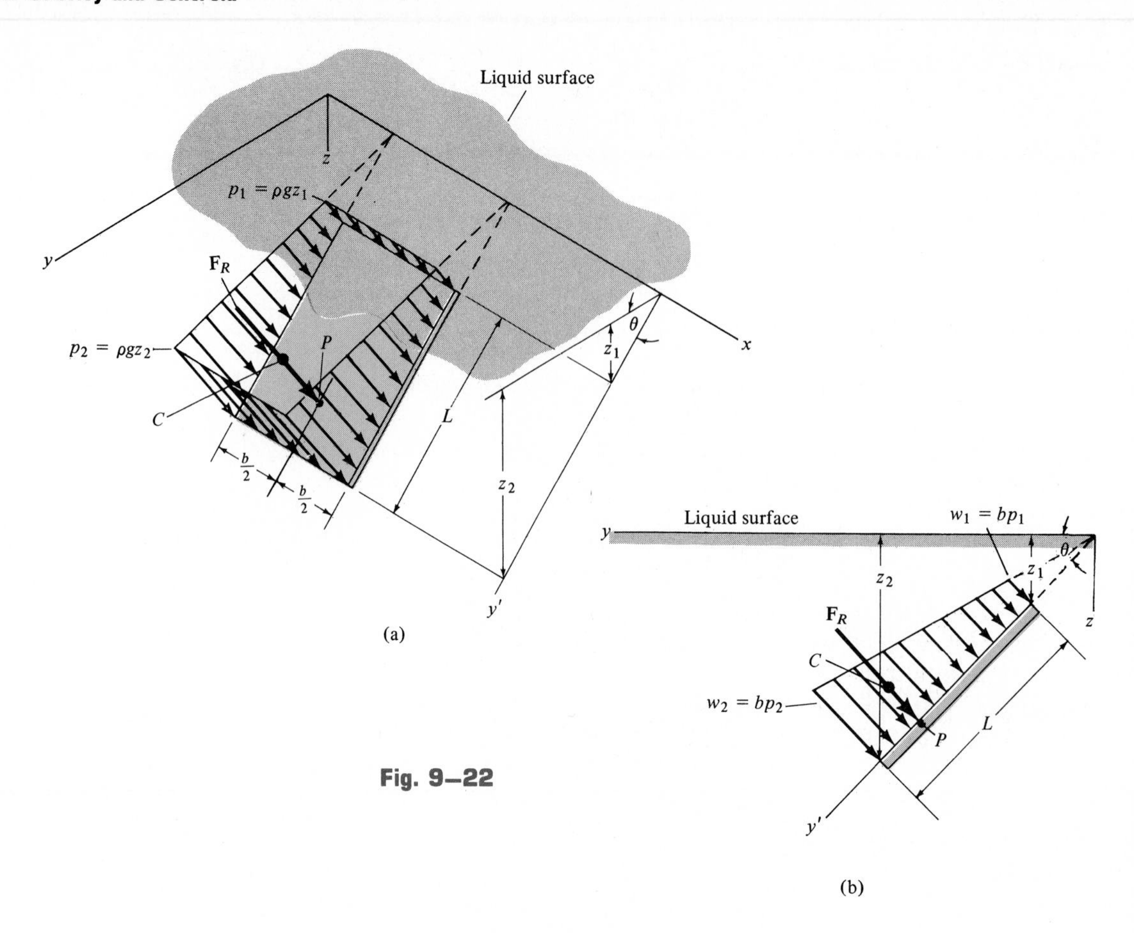

Fig. 9–22

Flat Plate of Constant Width. A flat rectangular plate of constant width, which is submerged in a liquid having a mass density ρ, is shown in Fig. 9–22*a*. The plane of the plate makes an angle θ with the horizontal, such that its top edge is located at a depth z_1 from the liquid surface and its bottom edge is located at a depth z_2. Since pressure varies linearly with depth (Eq. 9–16), the distribution of pressure over the plate's surface is represented by a trapezoidal volume of loading. The magnitude of the *resultant force* $\mathbf{F}_R$ is equal to the *volume* of this loading diagram and has a *line of action* that passes through the volume's centroid C. Note that $\mathbf{F}_R$ does *not* act at the center of the plate; rather, it acts at point P, called the *center of pressure*.

Since the plate has a *uniform width*, the loading distribution may also be viewed in two dimensions, Fig. 9–22*b*. Here the loading intensity is measured as force/length and varies linearly from $w_1 = bp_1 = b\rho g z_1$ to $w_2 = bp_2 = b\rho g z_2$. Consequently, the magnitude of $\mathbf{F}_R$ equals the trapezoidal *area* and $\mathbf{F}_R$ has a *line of action* that passes through the area's *centroid* C. For numerical applications, the area and location of the centroid for a trapezoid are tabulated on the inside back cover.

Curved Plate of Constant Width. When the submerged plate is curved, the pressure acting normal to the plate continually changes direction, and therefore calculation of the magnitude of $\mathbf{F}_R$ and its location P is more difficult than for a flat plate. Three- and two-dimensional views of the loading distribution are shown in Fig. 9–23*a* and *b*, respectively. Here integration can be used to determine both F_R (the volume or area under the loading diagrams) and the location of the centroid C or center of pressure P.

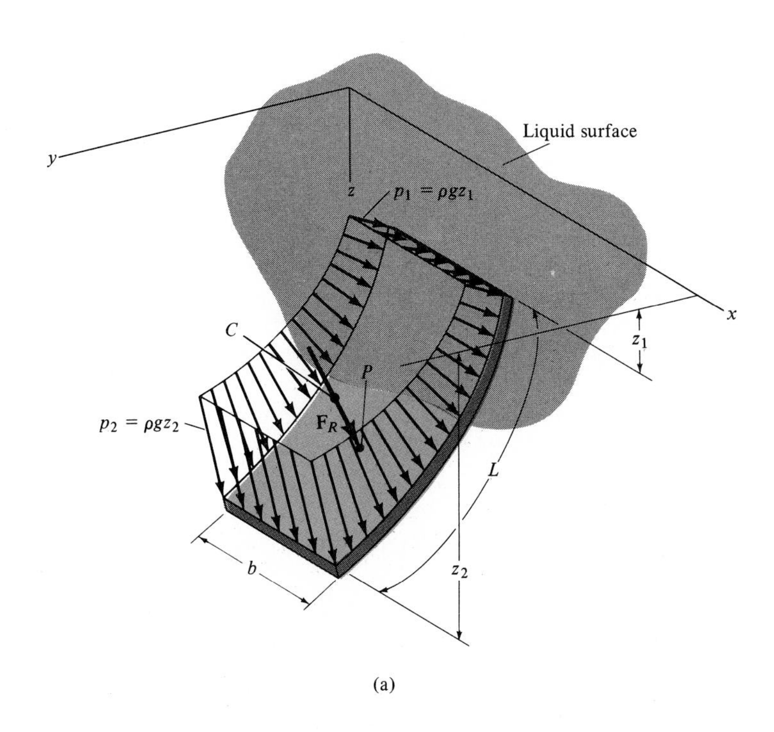

(a)

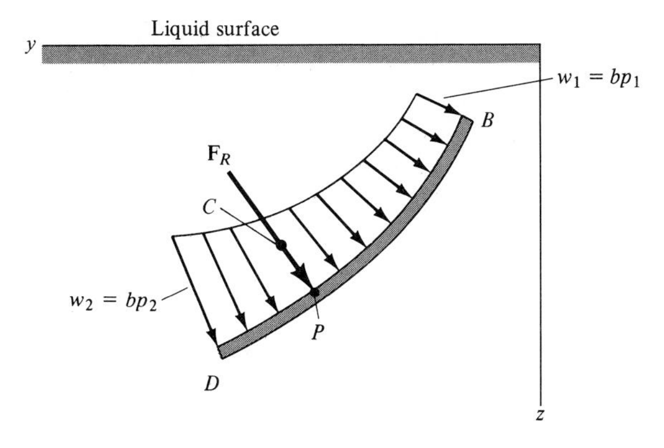

(b)

Fig. 9–23

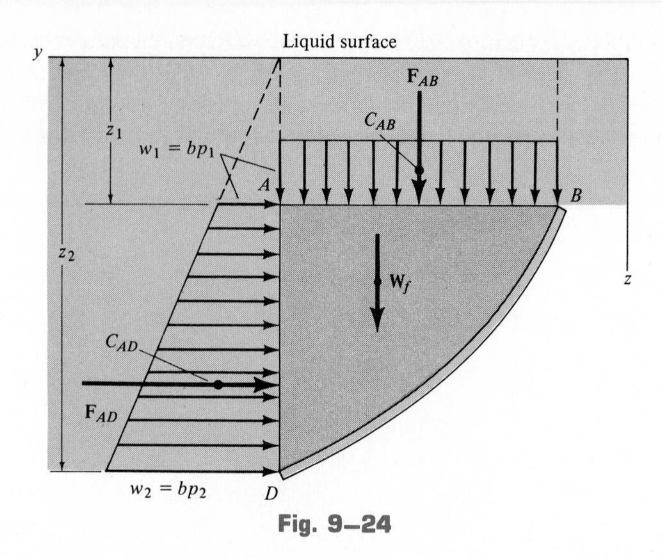

Fig. 9–24

A simpler method exists, however, for calculating the magnitude of $\mathbf{F}_R$ and its location along a curved (or flat) plate having a *constant width*. This method requires separate calculations for the horizontal and vertical *components* of $\mathbf{F}_R$. For example, the distributed loading acting on the top surface of the curved plate *DB* in Fig. 9–23*b* can be represented by the *equivalent loading* shown in Fig. 9–24. Here it is seen that *DB* supports a horizontal force along its *vertical projection AD*. This force, $\mathbf{F}_{AD}$, has a magnitude that equals the area under the trapezoid and acts through the centroid C_{AD} of this area. The distributed loading along the *horizontal projection AB* is constant, since all points lying in this plane are at the same depth from the surface of the liquid. The magnitude of $\mathbf{F}_{AB}$ is simply the area of the rectangle. This force acts through the centroid C_{AB} (or midpoint) of *AB*. In addition to $\mathbf{F}_{AB}$, the curved surface *DB* must also support the downward *weight of fluid* $\mathbf{W}_f$ contained within the block *BDA*. This force has a magnitude of $W_f = (\rho g b)(Area_{BDA})$ and acts through the centroid of *BDA*. Summing the three coplanar forces shown in Fig. 9–24 yields $\mathbf{F}_R = \Sigma \mathbf{F} = \mathbf{F}_{AD} + \mathbf{F}_{AB} + \mathbf{W}_f$. The center of pressure *P* is determined by applying the principle of moments to the force system and its resultant about a convenient reference point. The final results for $\mathbf{F}_R$ and its location *P* will be equivalent to those shown in Fig. 9–23*a*.

Example 9–11

Determine the magnitude and location of the resultant hydrostatic force acting on the submerged plate AB shown in Fig. 9–25a. The plate has a width of 1.5 m; $\rho_w = 1000\ \text{kg/m}^3$.

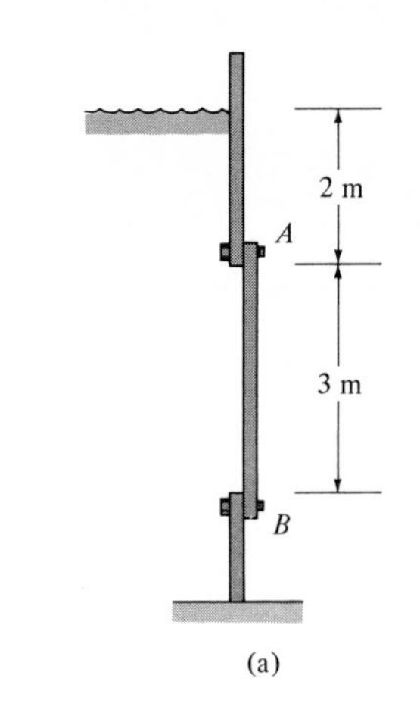

(a)

Solution

Since the plate has a constant width, the distributed loading can be viewed in two dimensions as shown in Fig. 9–25b. The intensity of the load at A and B is computed as

$$w_A = b\rho_w g z_A = (1.5\ \text{m})(1000\ \text{kg/m}^3)(9.81\ \text{m/s}^2)(2\ \text{m})$$
$$= 29.4\ \text{kN/m}$$
$$w_B = b\rho_w g z_B = (1.5\ \text{m})(1000\ \text{kg/ m}^3)(9.81\ \text{m/s}^2)(5\ \text{m})$$
$$= 73.6\ \text{kN/m}$$

The magnitude of the resultant force $\mathbf{F}_R$ created by the distributed load is

$$F_R = (\text{area of trapezoid})$$
$$= \tfrac{1}{2}(3)(29.4 + 73.6) = 154.5\ \text{kN} \qquad \textit{Ans.}$$

This force acts through the centroid of the area,

$$h = \frac{1}{3}\left(\frac{2(29.4) + 73.6}{29.4 + 73.6}\right)(3) = 1.29\ \text{m} \qquad \textit{Ans.}$$

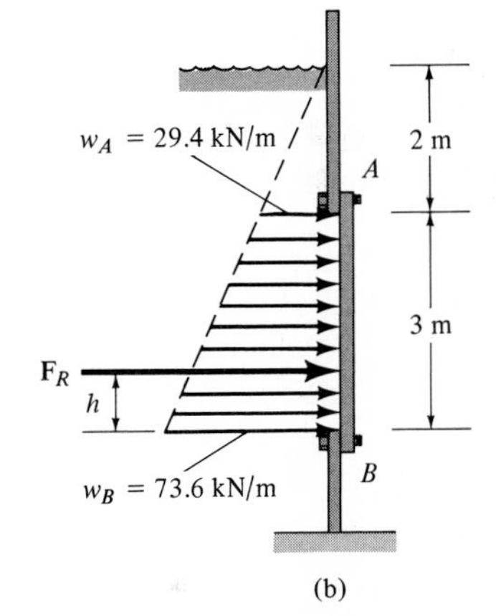

(b)

measured upward from B, Fig. 9–25b.

The same results can be obtained by considering two components of $\mathbf{F}_R$ defined by the triangle and rectangle shown in Fig. 9–25c. Each force acts through its associated centroid and has a magnitude of

$$F_{Re} = (29.4\ \text{kN/m})(3\ \text{m}) = 88.2\ \text{kN}$$
$$F_t = \tfrac{1}{2}(44.2\ \text{kN/m})(3\ \text{m}) = 66.3\ \text{kN}$$

Hence,

$$F_R = F_{Re} + F_t = 88.2 + 66.3 = 154.5\ \text{kN} \qquad \textit{Ans.}$$

The location of $\mathbf{F}_R$ is determined by the principle of moments applied at B, Fig. 9–25b and c, i.e.,

$$\circlearrowleft+(M_R)_B = \Sigma M_B; \quad -(154.5)h = -88.2(1.5) - 66.3(1)$$
$$h = 1.29\ \text{m} \qquad \textit{Ans.}$$

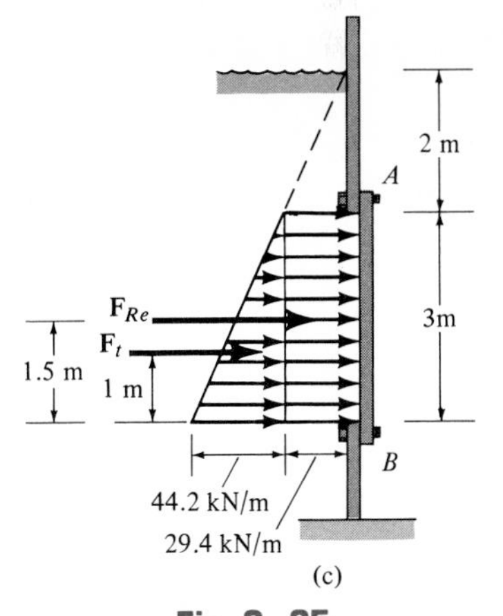

(c)

Fig. 9–25

Example 9–12

Determine the magnitude of the resultant hydrostatic force acting on the surface of a sea wall shaped in the form of a parabola as shown in Fig. 9–26*a*. The wall is 5 m long; $\rho_w = 1020 \text{ kg/m}^3$.

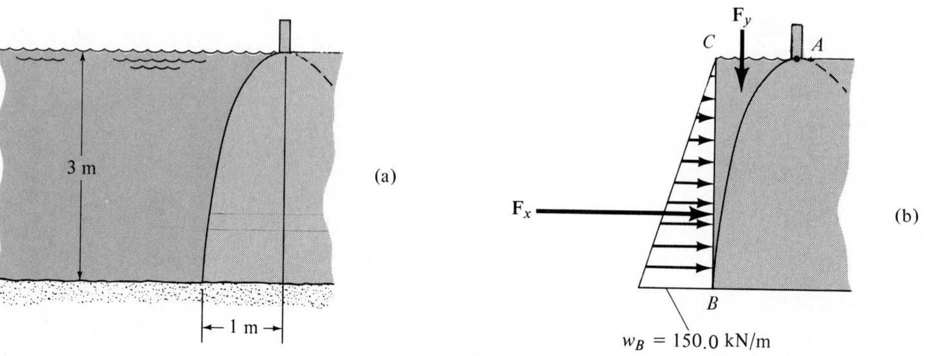

Fig. 9–26

Solution

The horizontal and vertical components of the resultant force will be calculated, Fig. 9–26*b*. Since

$$w_B = b\rho_w g z_B = 5 \text{ m}(1020 \text{ kg/m}^3)(9.81 \text{ m/s}^2)(3 \text{ m}) = 150.0 \text{ kN/m}$$

then

$$F_x = \tfrac{1}{2}(3 \text{ m})(150.0 \text{ kN/m}) = 225.0 \text{ kN}$$

The area of the parabolic sector *ABC* can be determined using the table on the inside back cover. Hence, the weight of water within this region is

$$F_y = (\rho_w g b)(Area_{ABC})$$
$$= (1020 \text{ kg/m}^3)(9.81 \text{ m/s}^2)5 \text{ m}[\tfrac{1}{3}(1 \text{ m})(3 \text{ m})] = 50.0 \text{ kN}$$

The resultant force is therefore

$$F_R = \sqrt{F_x^2 + F_y^2} = \sqrt{(225.0)^2 + (50.0)^2}$$
$$= 230.5 \text{ kN} \qquad \textit{Ans.}$$

Problems

9–54. Determine the magnitude of the resultant hydrostatic force acting on the dam and its location measured from the top surface of the water. The width of the dam is 8 m; $\rho_w = 1.0\ \text{Mg/m}^3$.

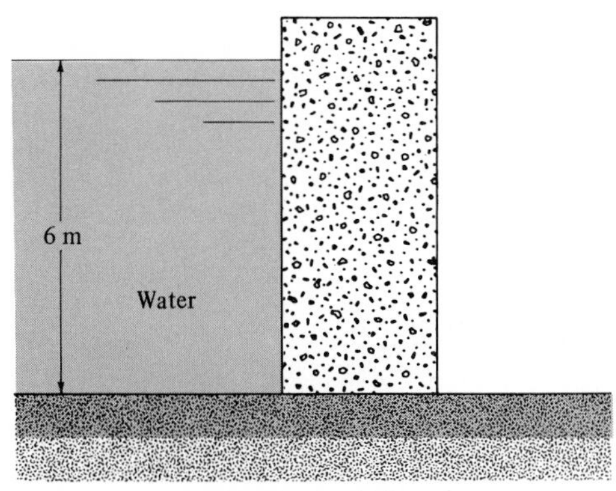

Prob. 9–54

9–55. When the tide water A subsides, the tide gate automatically swings open to drain the marsh B. For the condition of high tide shown, determine the horizontal resultant forces developed at the hinge C and stop block D. The length of the gate is 6 m and its height is 4 m. $\rho_w = 1\ \text{Mg/m}^3$.

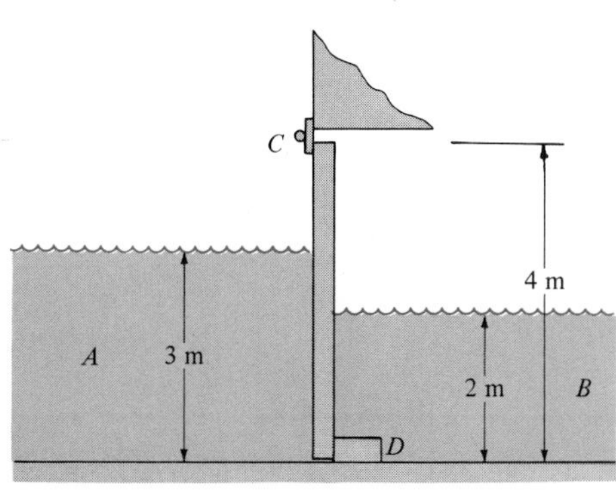

Prob. 9–55

***9–56.** The concrete "gravity" dam is held in place by its own weight. If the density of concrete is $\rho_c = 2.5\ \text{Mg/m}^3$, and water has a density of $\rho_w = 1.0\ \text{Mg/m}^3$, determine the smallest width d that will prevent the dam from overturning about its end A.

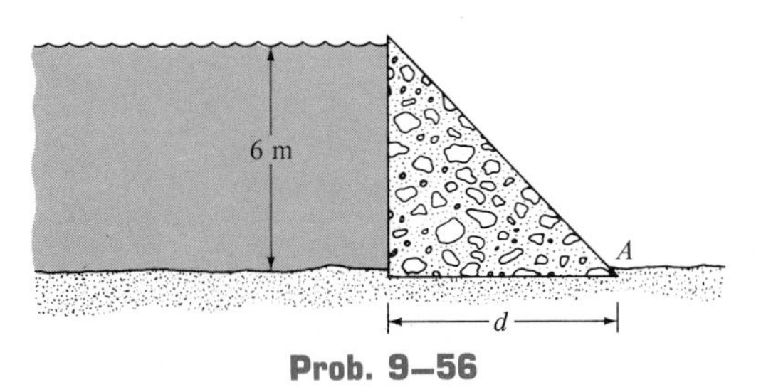

Prob. 9–56

9–57. The tank is 1.25 m wide on each side and 4 m high. If it is filled to a depth of 1 m with water and 3 m with oil, determine the resultant force created by both of these fluids along side AB of the tank and its location measured from the top of the tank. $\rho_o = 0.90\ \text{Mg/m}^3$ and $\rho_w = 1.0\ \text{Mg/m}^3$.

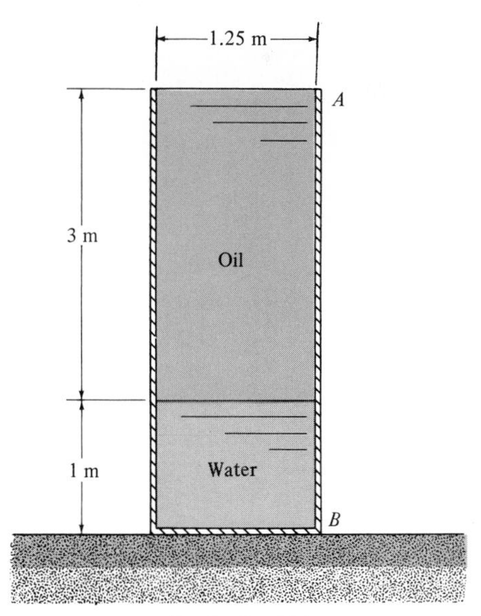

Prob. 9–57

9–58. The storage tank contains oil having a density of $\rho_o = 0.90\ \text{Mg/m}^3$. If the tank is 1.5 m wide, calculate the resultant force acting on the inclined side AB of the tank, caused by the oil, and specify its location along AB, measured from A.

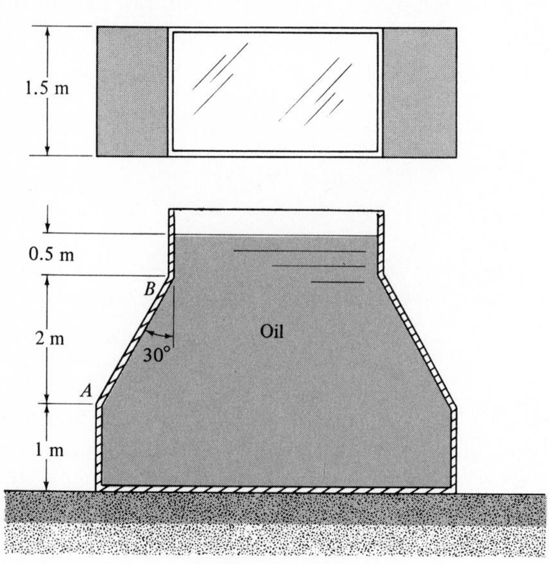

Prob. 9–58

***9–60.** The semicircular tunnel passes under a river which is 9 m deep. Determine the resultant hydrostatic force acting per meter of length along the length of the tunnel. The tunnel is 6 m wide; $\rho_w = 1.0\ \text{Mg/m}^3$.

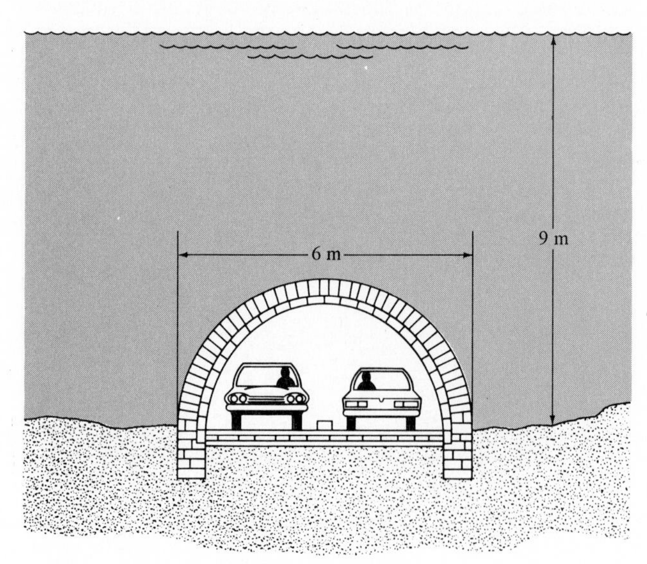

Prob. 9–60

9–59. Determine the magnitude and location of the resultant hydrostatic force acting on each of the cover plates A and B. The density of water is $\rho_w = 1.0\ \text{Mg/m}^3$.

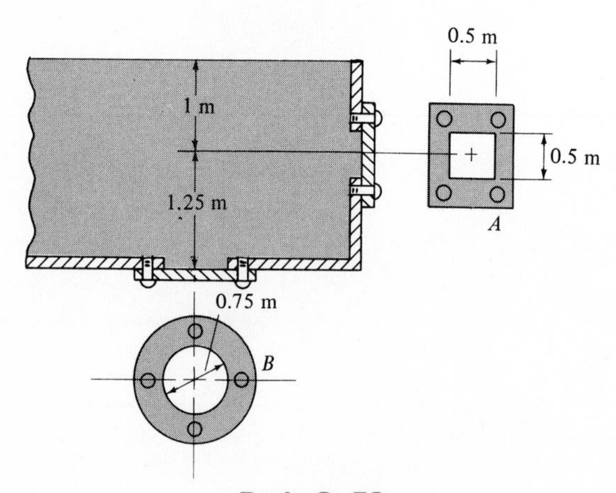

Prob. 9–59

9–61. Determine the magnitude of the resultant hydrostatic force acting per meter of length on the sea wall; $\rho_w = 1.0\ \text{Mg/m}^3$.

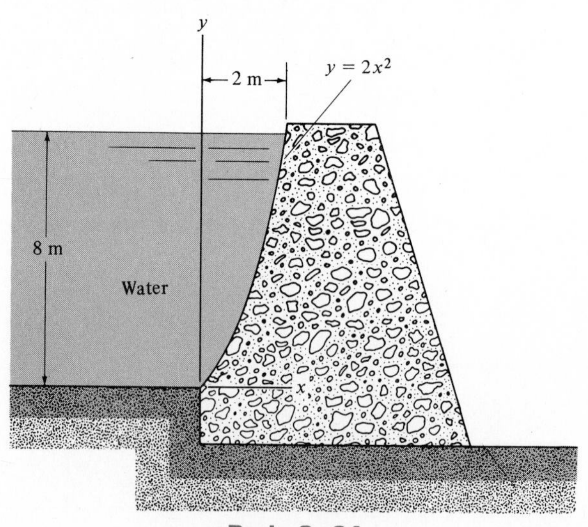

Prob. 9–61

Review Problems

9–62. Locate the center of gravity of the homogeneous rod.

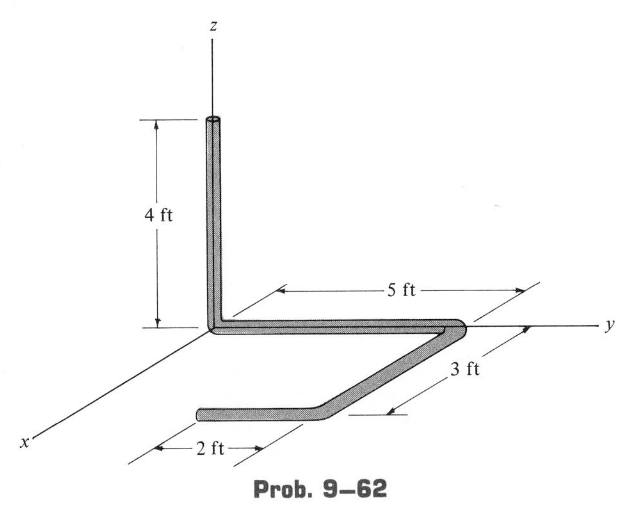

Prob. 9–62

9–63. Determine the distance $\overline{y}$ to the centroidal axis $\overline{x}\,\overline{x}$ of the beam's cross-sectional area. Neglect the size of the corner welds at B for the calculation.

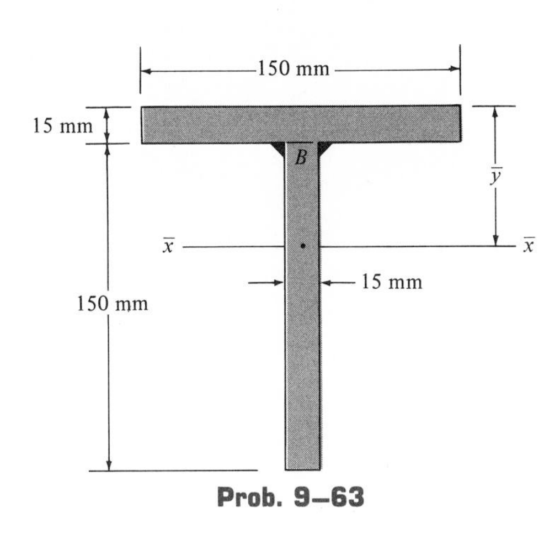

Prob. 9–63

***9–64.** Locate the center of gravity $\overline{y}$ of the volume generated by revolving the shaded area about the y axis. The material is homogeneous.

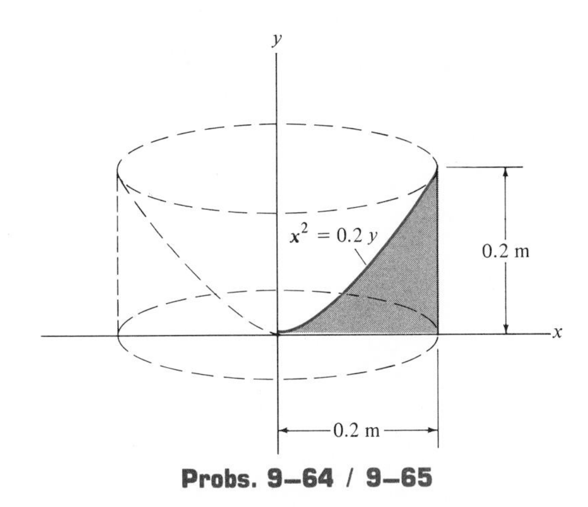

Probs. 9–64 / 9–65

9–65. Using integration, compute the area and the centroidal distance $\overline{x}$ for the shaded area. Then using the second theorem of Pappus–Guldinus, compute the volume of the solid generated by revolving the shaded area about the y axis.

9–66. Locate the distance $\overline{x}$ to the centroid C of the shaded area.

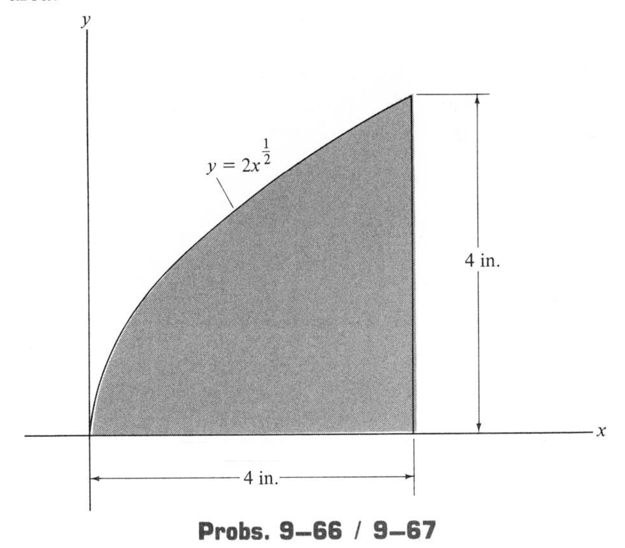

Probs. 9–66 / 9–67

9–67. Locate the distance $\overline{y}$ to the centroid C of the shaded area.

***9–68.** Locate the distance $\overline{y}$ to the centroid C of the shaded area.

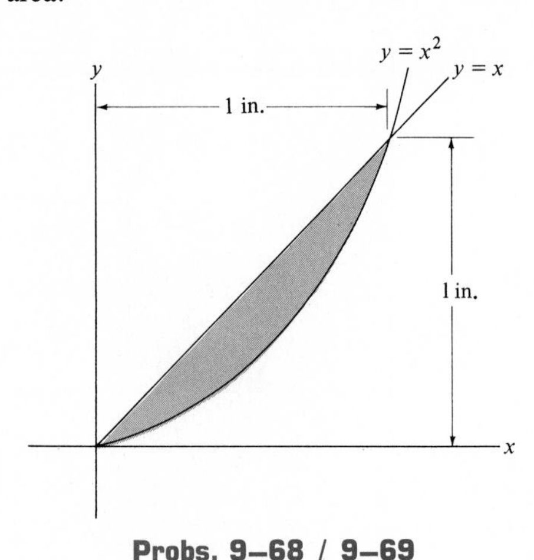

Probs. 9–68 / 9–69

9–69. Locate the distance $\overline{x}$ to the centroid C of the shaded area.

9–70. Locate the center of gravity of the volume generated by revolving the shaded area about the x axis. The material is homogeneous.

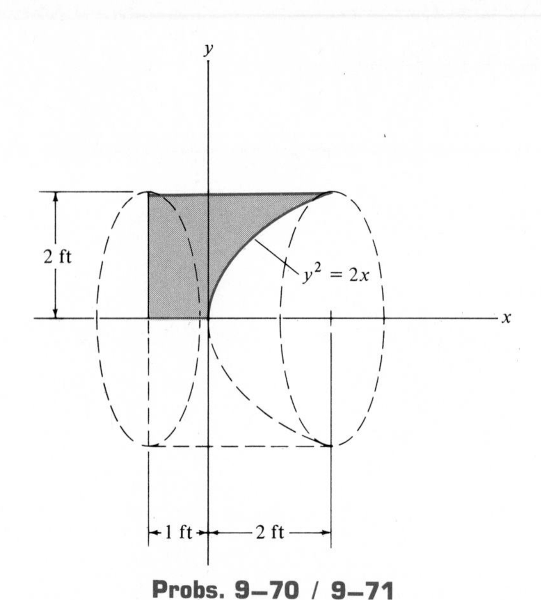

Probs. 9–70 / 9–71

9–71. Using integration, compute the area and the centroidal distance $\overline{y}$ for the shaded area. Then using the second theorem of Pappus–Guldinus, compute the volume of the solid generated by revolving the shaded area about the x axis.

Moments of Inertia

In Chapter 9 the *centroid* of an area has been defined by considering the *first moment* of the area about an axis, i.e., for the computation it is necessary to evaluate integrals of the form $\int x\,dA$. There are many important topics in mechanics, however, which require evaluation of an integral of the *second moment* of an area. These integrals are of the form $\int x^2\,dA$, where x is the "moment arm" measured from the element dA to an axis that is either perpendicular to, or lying in the plane of, the area. In engineering practice such integrals are referred to as the moments of inertia for an area.* The terminology "moment of inertia" as used here is actually a misnomer; however, it has been adopted because of the similarity with integrals of the same form related to mass.

Methods used to determine both the area and mass moments of inertia will be discussed in this chapter. Computation of the area moment of inertia has important applications in structual design, and a body's mass moment of inertia is frequently used to solve problems in dynamics.

10.1 Definition of Moments of Inertia for Areas

The moment of inertia of an area originates whenever one computes the moment of a distributed load that varies linearly from the moment axis. A common example of this type of loading occurs in the study of fluid mechanics. It was pointed out in Sec. 9–6 that the pressure, or force per unit area, exerted at a point located a distance z below the surface of a liquid is $p = \rho g z$, Eq. 9–16, where ρ is the mass density of the liquid. Thus, the magnitude of force

*The discussion of area moments of inertia is extended in Appendix B to include product and principal moments of inertia.

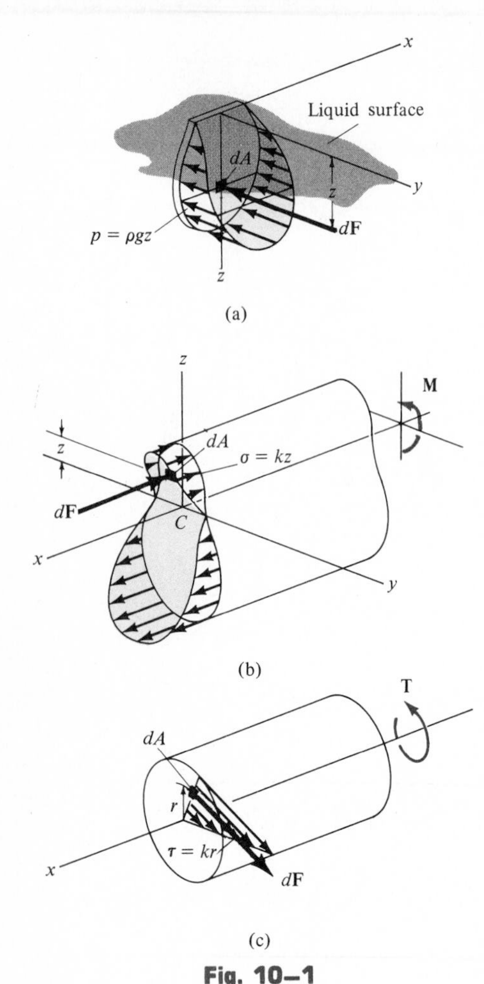

Fig. 10–1

exerted by a liquid on the area dA of the submerged plate shown in Fig. 10–1a is $dF = p\,dA = \rho gz\,dA$. The moment of this force about the x axis of the plate is $dM = dFz = \rho gz^2\,dA$ and, therefore, the moment created by the entire distributed (pressure) loading is $M = \rho g\int z^2\,dA$. Here the integral represents the moment of inertia of the area of the plate about the x axis.

The area moment of inertia also appears when one relates the *normal stress* σ (sigma), or force per unit area, acting on a transverse cross section of an elastic beam, to the applied external moment **M** which causes bending of the beam. From the theory of strength of materials, it can be shown that the stress within the beam varies linearly with its distance from an axis passing through the centroid C of the beam's cross-sectional area; i.e., $\sigma = kz$, Fig. 10–1b. The magnitude of force acting on the area element dA, shown in the figure, is therefore $dF = \sigma\,dA = kz\,dA$. Since this force is located a distance z from the y axis, the moment of $d\mathbf{F}$ about the y axis is $dM = dFz = kz^2\,dA$. The resulting moment of the entire stress distribution is equal to the applied moment **M**; hence, $M = k\int z^2\,dA$.

As a third example to illustrate the formulation of area moments of inertia, consider the elastic twisting of a circular shaft about its longitudinal axis, as shown in Fig. 10–1c. The applied external torsional moment **T** causes a *shearing stress* τ (tau) to be distributed over the shaft's cross section which varies linearly along any radial line segment; i.e., $\tau = kr$. The magnitude of **T** may be related to the internal stress distribution by summing moments about the x axis of the shaft, which results in $T = \int r\,dF = \int r\tau\,dA = k\int r^2\,dA$. Note that in this case the moment of inertia (or integral) differs from that of the other two examples, because the "moment" axis is *perpendicular* to the plane of the area and not lying within it.

Since the moments of inertia for areas frequently appear in design formulas used in fluid mechanics, strength of materials, and structural mechanics, it is important that the engineer become familiar with the methods used to compute these quantities.

Moment of Inertia. Consider the area A, shown in Fig. 10–2, which lies in the x-y plane. By definition, the moments of inertia of the differential planar area dA about the x and y axes are $dI_x = y^2\,dA$ and $dI_y = x^2\,dA$, respectively. For the entire area the *moment of inertia* is determined by integration, i.e.,

$$I_x = \int y^2\,dA$$

$$I_y = \int x^2\,dA \qquad (10\text{–}1)$$

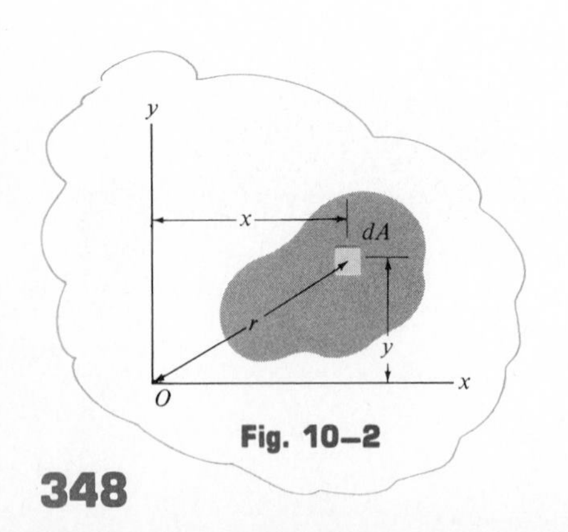

Fig. 10–2

Integrals of this type must be evaluated in the solution of the fluid-pressure and elastic-bending problems mentioned in reference to Fig. 10–1a and b.

We can also formulate the second moment of the differential area dA about the pole O or z axis, Fig. 10–2. This is referred to as the polar moment of inertia, $dJ_O = r^2\,dA$. Here r is the perpendicular distance from the pole (z

axis) to the element dA. For the entire area the *polar moment of inertia* is

$$J_O = \int r^2\, dA = I_x + I_y \tag{10–2}$$

The relationship between J_O and I_x, I_y is possible since $r^2 = x^2 + y^2$, Fig. 10–2. In the previous discussion, computation for J_O would have been necessary for solving the elastic-torsion problem of Fig. 10–1*c*.

From the above formulations it is seen that I_x, I_y, and J_O will *always* be *positive,* since they involve the product of distance squared and area. Furthermore, the units for moment of inertia involve length raised to the fourth power, e.g., m^4, mm^4, or ft^4, in^4. UNITS

Parallel-Axis Theorem for an Area 10.2

If the moment of inertia for an area is known about an axis passing through its centroid, we can determine the moment of inertia of the area about a corresponding parallel axis using the *parallel-axis theorem*. To derive this theorem, consider finding the moment of inertia of the shaded area shown in Fig. 10–3 about the x axis. In this case, a differential element dA is located at an arbitrary distance y from the centroidal $\bar{x}$ axis, whereas the *fixed distance* between the parallel x and $\bar{x}$ axes is defined as d_y. Since the moment of inertia of dA about the x axis is $dI_x = (y + d_y)^2\, dA$, then for the entire area,

$$I_x = \int (y + d_y)^2\, dA$$

$$= \int y^2\, dA + 2d_y \int y\, dA + d_y^2 \int dA$$

The first term on the right represents the moment of inertia of the area about the $\bar{x}$ axis, $\bar{I}_{\bar{x}}$. The second term is zero since the $\bar{x}$ axis passes through the area's centroid C, i.e., $\int y\, dA = \bar{y}A = 0$. The final result is therefore

$$I_x = \bar{I}_{\bar{x}} + Ad_y^2 \tag{10–3}$$

A similar expression can be written for I_y, i.e.,

$$I_y = \bar{I}_{\bar{y}} + Ad_x^2 \tag{10–4}$$

And finally, for the polar moment of inertia about an axis perpendicular to the x-y plane and passing through the pole O (z axis), Fig. 10–3, we have

$$J_O = \bar{J}_C + Ad^2 \tag{10–5}$$

The form of each of these equations states that *the moment of inertia of an area about an axis is equal to the area's moment of inertia about a parallel axis passing through the centroid plus the product of the area and the square of the perpendicular distance between the axes.*

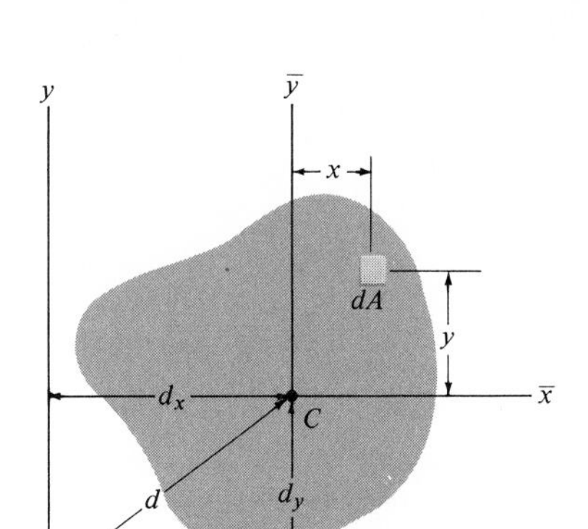

Fig. 10–3

10.3 Radius of Gyration of an Area

The *radius of gyration* of a planar area is often used for column design in structural mechanics. Provided the area and moments of inertia are *known,* the radii of gyration are determined from the formulas

$$k_x = \sqrt{\frac{I_x}{A}} \qquad k_y = \sqrt{\frac{I_y}{A}} \qquad k_O = \sqrt{\frac{J_O}{A}} \tag{10–6}$$

Note that the form of these equations is easily remembered since it is similar to that for finding the moment of inertia of a differential area about an axis. For example, $I_x = k_x^2 A$; whereas for a differential area, $dI_x = y^2\, dA$.

10.4 Moments of Inertia for an Area by Integration

When the boundaries for a planar area can be expressed by mathematical functions, Eqs. 10–1 may be integrated to determine the moments of inertia for the area. If the element of area chosen for integration has a differential size in two directions as shown in Fig. 10–2, a double integration must be performed to evaluate the moment of inertia. Most often, however, it is easier to choose an element having a differential size or thickness in only one direction, because then the evaluation requires only a single integration.

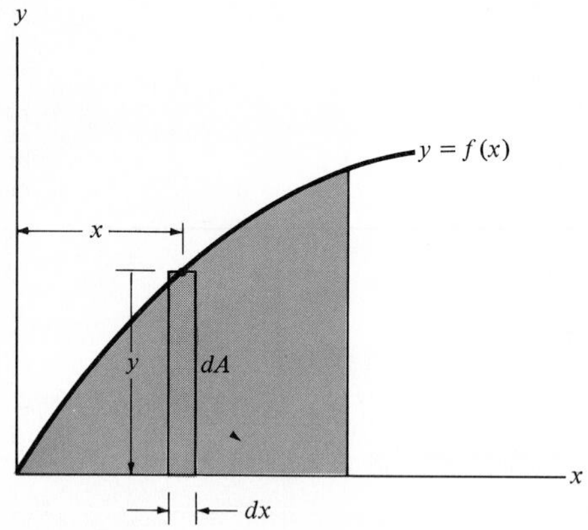

(a)

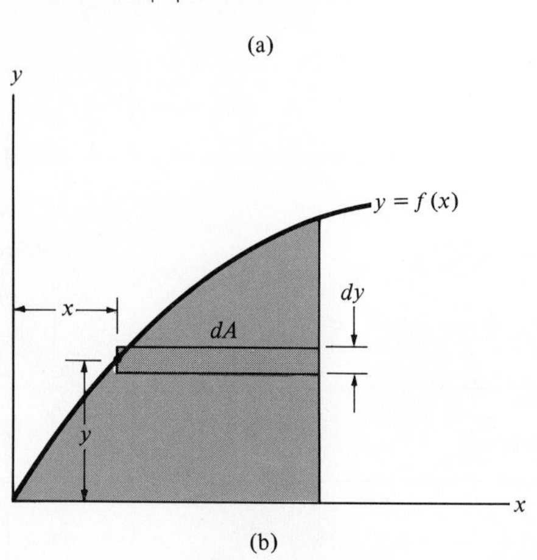

(b)

Fig. 10–4

PROCEDURE FOR ANALYSIS

If a single integration is performed to determine the moment of inertia of an area about an axis, then Eqs. 10–1 can be applied only if the *length* of the element is *parallel* to the axis. This situation occurs when using the rectangular element shown in Fig. 10–4*a* for computing I_y of the area. *Direct application* of Eq. 10–1, $I_y = \int x^2\, dA$, can be made in this case since *all parts* of the element lie at the *same* moment-arm distance x from the y axis.* Notice that I_x cannot be computed using Eq. 10–1, $I_x = \int y^2\, dA$, since all parts of the element do not lie at the same moment-arm distance from the x axis. Instead, an element of thickness dy must be chosen, Fig. 10–4*b*.

The following examples illustrate the above procedure.

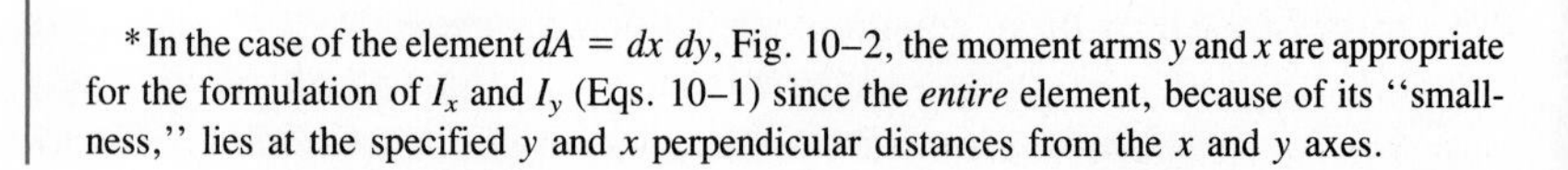

*In the case of the element $dA = dx\, dy$, Fig. 10–2, the moment arms y and x are appropriate for the formulation of I_x and I_y (Eqs. 10–1) since the *entire* element, because of its "smallness," lies at the specified y and x perpendicular distances from the x and y axes.

Example 10–1

Determine the moment of inertia for the rectangular area shown in Fig. 10–5 with respect to (a) the centroidal $\bar{x}$ axis, (b) the axis x_b passing through the base of the rectangle, and (c) the pole or z axis perpendicular to the $\bar{x}$-$\bar{y}$ plane and passing through the centroid C.

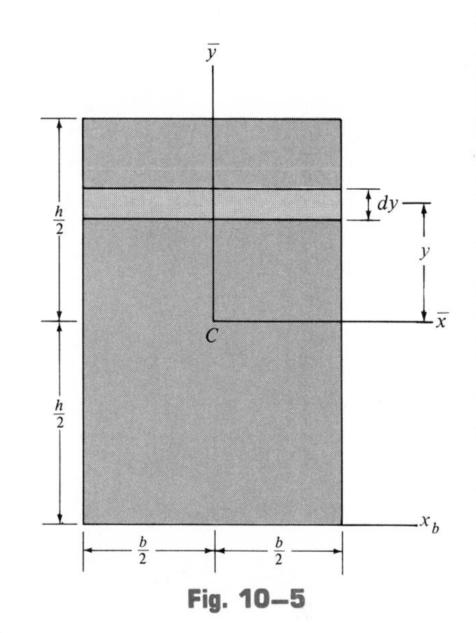

Fig. 10–5

Solution

Part (a). The differential element shown in Fig. 10–5 is chosen for integration. Because of its location and orientation, the *entire element* is at a distance y from the $\bar{x}$ axis. Here it is necessary to integrate from $y = -h/2$ to $y = h/2$. Since $dA = b\,dy$, then

$$\bar{I}_{\bar{x}} = \int y^2\,dA = \int_{-h/2}^{h/2} y^2(b\,dy) = b\int_{-h/2}^{h/2} y^2\,dy$$

$$= \frac{1}{12}bh^3 \qquad \textit{Ans.}$$

Part (b). The moment of inertia about an axis passing through the base of the rectangle can be obtained by using the result of part (a) and applying the parallel-axis theorem, Eq. 10–3.

$$I_{x_b} = \bar{I}_{\bar{x}} + Ad_y^2$$

$$= \frac{1}{12}bh^3 + bh\left(\frac{h}{2}\right)^2 = \frac{1}{3}bh^3 \qquad \textit{Ans.}$$

Part (c). The moment of inertia $\bar{I}_{\bar{y}}$ may be found by interchanging the dimensions b and h in the result of part (a), in which case

$$\bar{I}_{\bar{y}} = \frac{1}{12}hb^3$$

Using Eq. 10–2, the polar moment of inertia about C is

$$\bar{J}_C = \bar{I}_{\bar{x}} + \bar{I}_{\bar{y}} = \frac{1}{12}bh(h^2 + b^2) \qquad \textit{Ans.}$$

Example 10–2

Compute the moments of inertia of the shaded area shown in Fig. 10–6*a* about the x and y axes.

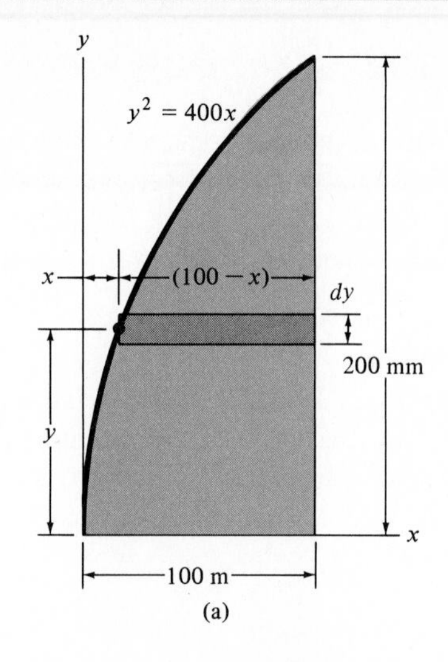

Solution

For I_x a differential element of area that is *parallel* to the x axis, as shown in Fig. 10–6*a*, is chosen for integration. Since the element has a thickness dy and intersects the curve at the arbitrary point (x, y), the area is $dA = (100 - x)\,dy$. Furthermore, all parts of the element lie at the same distance y from the x axis. Hence, integrating with respect to y, from $y = 0$ to $y = 200$ mm, yields

$$I_x = \int y^2\,dA = \int y^2(100 - x)\,dy$$

$$= \int_0^{200} y^2\left(100 - \frac{y^2}{400}\right)dy = 100\int_0^{200} y^2\,dy - \frac{1}{400}\int_0^{200} y^4\,dy$$

$$= 106.7(10^6)\text{ mm}^4 \qquad \textit{Ans.}$$

For I_y a differential element that is parallel to the y axis is chosen for integration, Fig. 10–6*b*. Why is this element appropriate? Here the element intersects the curve at the arbitrary point (x, y), so that the area is $dA = y\,dx$. Integrating with respect to x, from $x = 0$ to $x = 100$ mm, yields

$$I_y = \int x^2\,dA = \int x^2 y\,dx$$

$$= \int_0^{100} x^2(20x^{1/2})\,dx = 20\int_0^{100} x^{5/2}\,dx$$

$$= 28.6(10^6)\text{ mm}^4 \qquad \textit{Ans.}$$

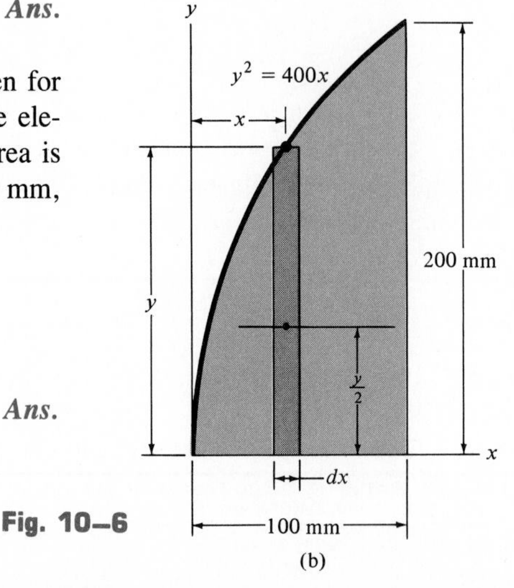

Fig. 10–6

Example 10–3

Determine the moment of inertia with respect to the x axis of the circular area shown in Fig. 10–7.

Solution

Using the differential element shown in Fig. 10–7, since $dA = 2x\,dy$, we have

$$I_x = \int y^2\,dA = \int y^2(2x)\,dy$$

$$= \int_{-a}^{a} y^2(2\sqrt{a^2 - y^2})\,dy = \frac{\pi a^4}{4} \qquad \textit{Ans.}$$

Fig. 10–7

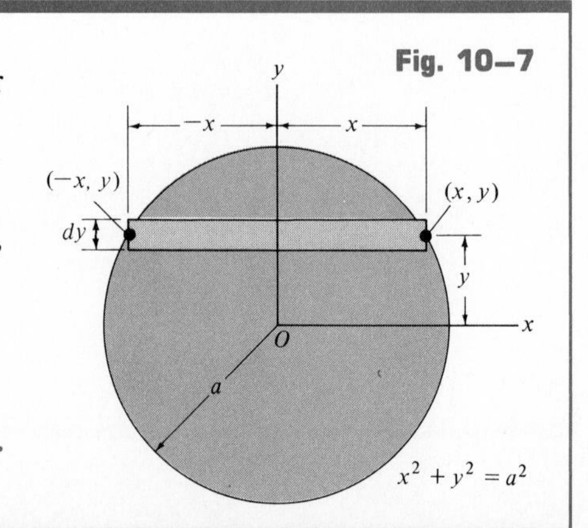

Problems

10–1. The irregular area has a moment of inertia about the AA axis of $35(10^6)$ mm^4. If the total area is $1.2(10^4)$ mm^2, determine the moment of inertia of the area about the BB axis. The DD axis passes through the centroid C of the area.

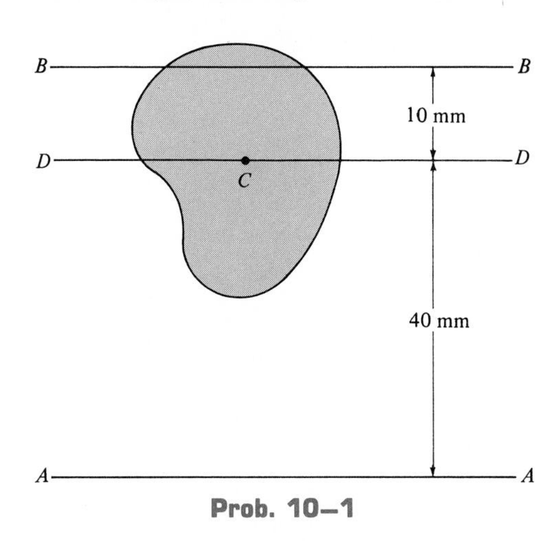

Prob. 10–1

10–2. The polar moment of inertia of the area is $\bar{J}_C = 23$ in.^{4}, computed about the z axis passing through the centroid C. If the moment of inertia about the y axis is 5 in.^{4} and the moment of inertia about the x' axis is 40 in.^{4}, determine the area A.

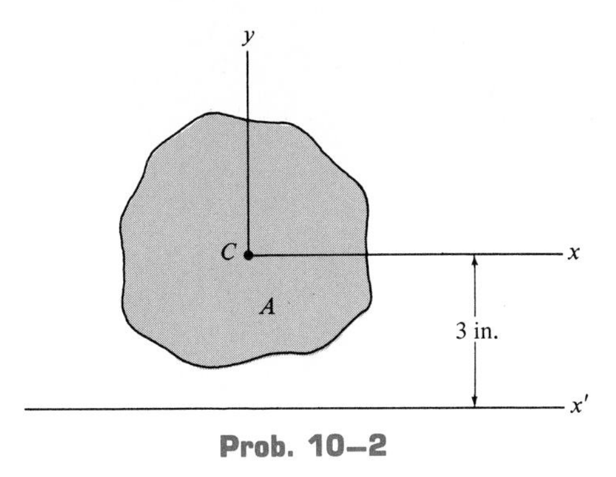

Prob. 10–2

10–3. Determine the moment of inertia of the area about the x axis.

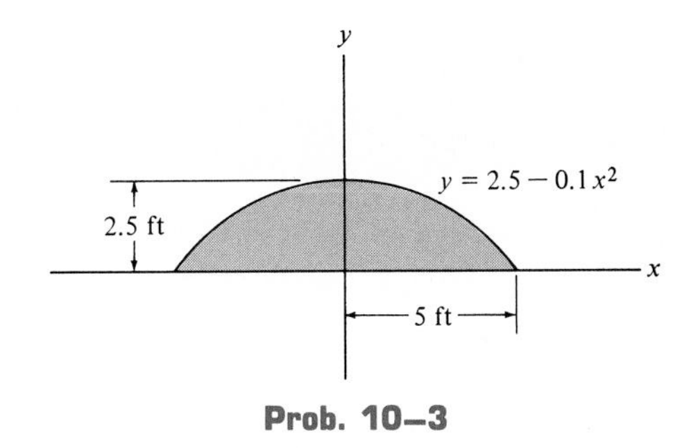

Prob. 10–3

***10–4.** Determine the moments of inertia I_x and I_y of the shaded elliptical area. What is the polar moment of inertia about the origin O?

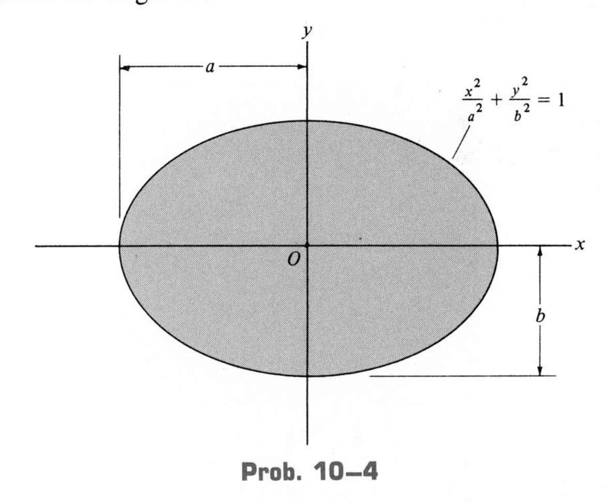

Prob. 10–4

10–5. Determine the moment of inertia of the semicircular area about the x axis. Then, using the parallel-axis theorem, compute the moment of inertia about the $\bar{x}$ axis that passes through the centroid C.

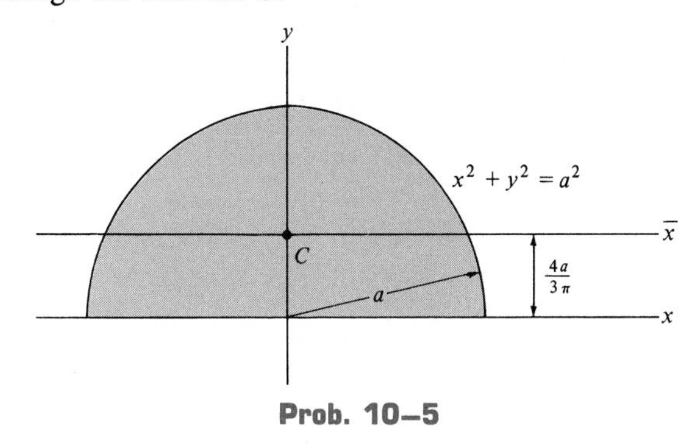

Prob. 10–5

10–6. Determine the moment of inertia of the parabolic area about the x axis.

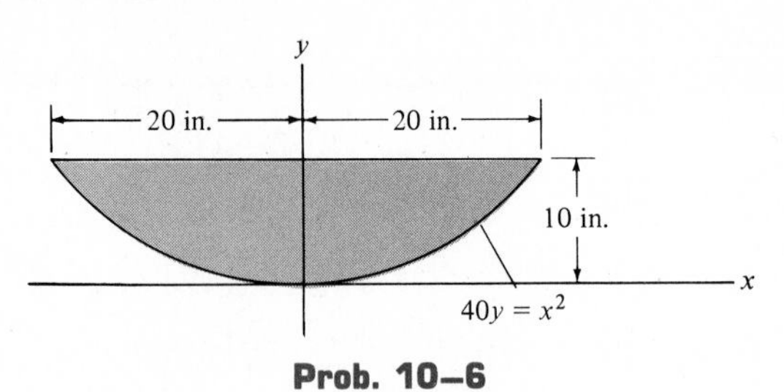

Prob. 10–6

10–7. Determine the radius of gyration k_y of the parabolic area.

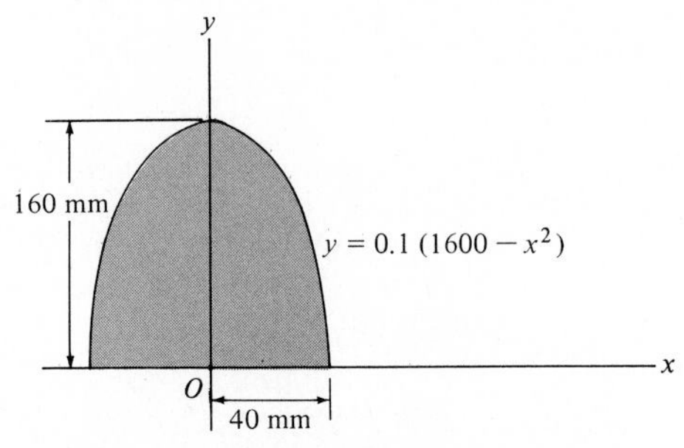

Prob. 10–7

***10–8.** Determine the moment of inertia of the area about the x axis. Then, using the parallel-axis theorem, compute the moment of inertia about the $\bar{x}$ axis that passes through the centroid C of the area. $\bar{y} = 120$ mm.

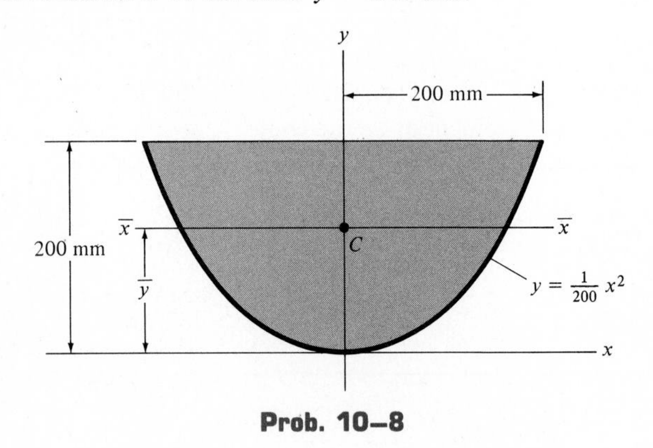

Prob. 10–8

10–9. Determine the radius of gyration k_O of the shaded area about an axis perpendicular to the x-y plane and passing through the origin O.

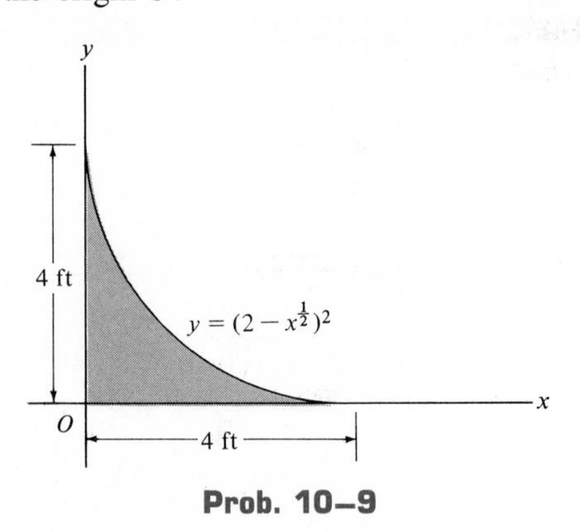

Prob. 10–9

10–10. Determine the moment of inertia of the quarter circular area about the x axis.

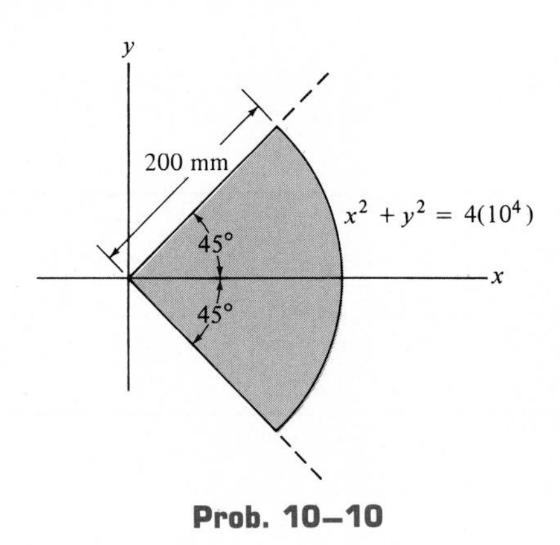

Prob. 10–10

Moments of Inertia for Composite Areas 10.5

A composite area consists of a series of connected "simpler" area shapes, such as semicircles, rectangles, and triangles. Provided the moment of inertia of each of these shapes is known or can be computed about a common axis, then the moment of inertia of the composite area equals the *algebraic sum* of the moments of inertia of all its composite parts.

PROCEDURE FOR ANALYSIS

The following procedure provides a method for determining the moment of inertia of a composite area about a reference axis.

Composite Parts. Using a sketch, divide the area into its composite parts and indicate the perpendicular distance from the *centroid* of each part to the reference axis.

Parallel-Axis Theorem. The moment of inertia of each part should be computed about its centroidal axis, which is parallel to the reference axis. (See the table given on the inside back cover.) If the centroidal axis does not coincide with the reference axis, the parallel-axis theorem, $I = \bar{I} + Ad^2$, should be used to determine the moment of inertia of the part about the reference axis.

Summation. The moment of inertia of the entire area about the reference axis is determined by summing the results of its composite parts. In particular, if a composite area has a "hole," the moment of inertia for the composite is found by "subtracting" the moment of inertia for the hole from the moment of inertia of the entire area including the hole.

Example 10–4

Compute the moment of inertia of the composite area shown in Fig. 10–8*a* about the x axis.

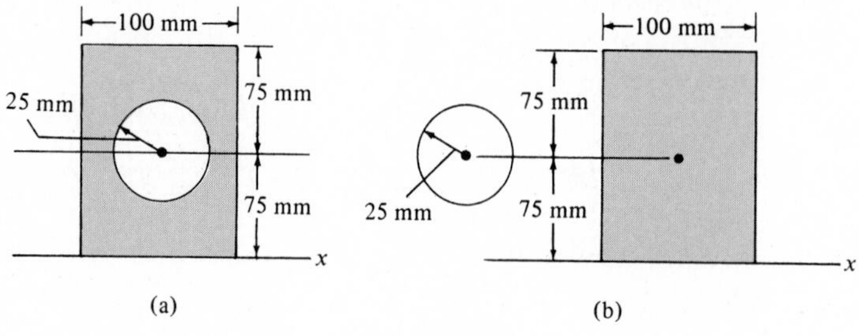

Fig. 10–8

Solution

Composite Parts. The composite area is obtained by *subtracting* the circle from the rectangle as shown in Fig. 10–8*b*. The centroid of each area is located in the figure.

Parallel-Axis Theorem. The moments of inertia about the x axis are computed using the parallel-axis theorem and the data contained in the table on the inside back cover.

Circle

$$I_x = \bar{I}_{\bar{x}} + Ad_y^2$$
$$= \frac{1}{4}\pi(25)^4 + \pi(25)^2(75)^2 = 11.4(10^6)\text{ mm}^4$$

Rectangle

$$I_x = \bar{I}_{\bar{x}} + Ad_y^2$$
$$= \frac{1}{12}(100)(150)^3 + (100)(150)(75)^2 = 112.5(10^6)\text{ mm}^4$$

Summation. The moment of inertia for the composite area is thus

$$I_x = -11.4(10^6) + 112.5(10^6)$$
$$= 101.1(10^6)\text{ mm}^4 \qquad \textit{Ans.}$$

Example 10–5

Compute the moments of inertia of the beam's cross-sectional area shown in Fig. 10–9*a* about the *x* and *y* centroidal axes.

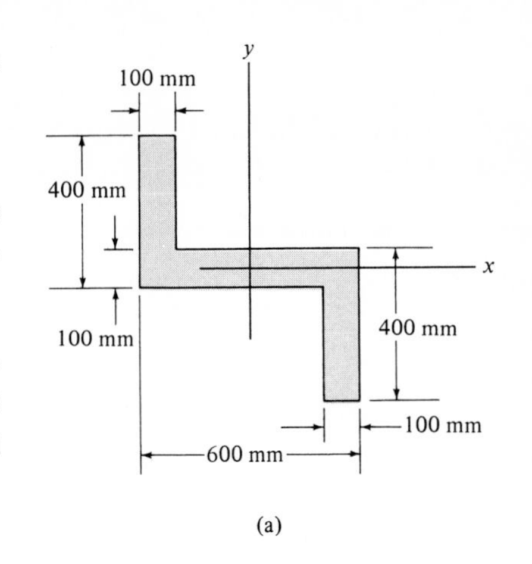

(a)

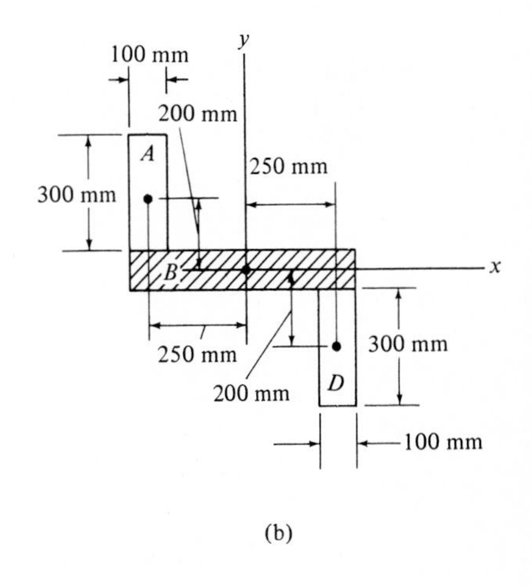

(b)

Fig. 10–9

Solution

Composite Parts. The cross section can be considered as three composite rectangular areas *A*, *B*, and *D* shown in Fig. 10–9*b*. For the calculation, the centroid of each of these rectangles is located in the figure.

Parallel-Axis Theorem. From the table on the inside back cover, or Example 10–1, the moment of inertia of a rectangle about its centroidal axis is $\bar{I} = \frac{1}{12}bh^3$. Hence, using the parallel-axis theorem for rectangles *A* and *D*, the computations are as follows:

Rectangle A

$$I_x = \bar{I}_{\bar{x}} + Ad_y^2 = \tfrac{1}{12}(100)(300)^3 + (100)(300)(200)^2$$
$$= 14.25(10^8)\text{ mm}^4$$
$$I_y = \bar{I}_{\bar{y}} + Ad_x^2 = \tfrac{1}{12}(300)(100)^3 + (100)(300)(250)^2$$
$$= 19(10^8)\text{ mm}^4$$

Rectangle B

$$I_x = \tfrac{1}{12}(600)(100)^3 = 0.50(10^8)\text{ mm}^4$$
$$I_y = \tfrac{1}{12}(100)(600)^3 = 18(10^8)\text{ mm}^4$$

Rectangle D

$$I_x = \bar{I}_{\bar{x}} + Ad_y^2 = \tfrac{1}{12}(100)(300)^3 + (100)(300)(200)^2$$
$$= 14.25(10^8)\text{ mm}^4$$
$$I_y = \bar{I}_{\bar{y}} + Ad_x^2 = \tfrac{1}{12}(300)(100)^3 + (100)(300)(250)^2$$
$$= 19(10^8)\text{ mm}^4$$

Summation. The moments of inertia for the entire cross section are thus

$$I_x = 14.25(10^8) + 0.50(10^8) + 14.25(10^8)$$
$$= 29(10^8)\text{ mm}^4 \qquad \textit{Ans.}$$
$$I_y = 19(10^8) + 18(10^8) + 19(10^8) = 56(10^8)\text{ mm}^4 \qquad \textit{Ans.}$$

Problems

10–11. Determine the moments of inertia I_x and I_y of the beam's cross-sectional area.

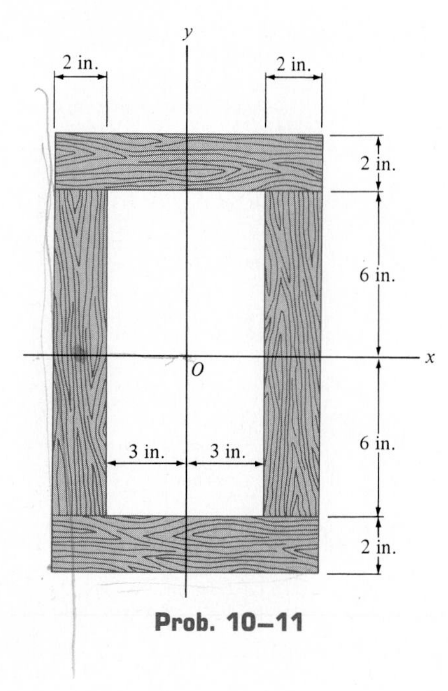

Prob. 10–11

***10–12.** The fuselage of an airplane is made up of stringers covered with sheet metal as shown in (a). If the combined areas of the angles and the sheet are tabulated and assumed to be concentrated at points 1 through 10 as shown in (b), determine the centroidal distance $\bar{y}$ and the moment of inertia about a horizontal $\overline{x}\overline{x}$ axis passing through the centroid C of the total cross section. *Hint:* Using the area elements, note that $\bar{y} = \Sigma\tilde{y}A/\Sigma A$ and $I_{\overline{xx}} = \Sigma\tilde{y}^2A$.

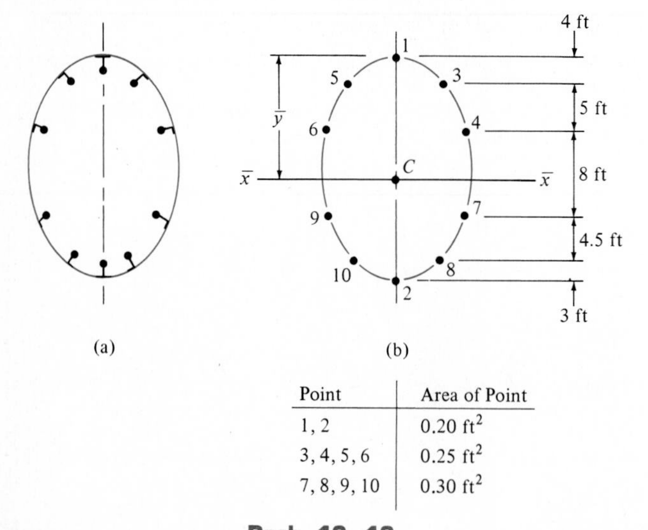

Point	Area of Point
1, 2	0.20 ft^2
3, 4, 5, 6	0.25 ft^2
7, 8, 9, 10	0.30 ft^2

Prob. 10–12

10–13. Determine the moment of inertia of the beam's cross-sectional area with respect to the $\overline{x}\overline{x}$ axis passing through the centroid C of the cross section. Neglect the size of the corner welds at A and B for the calculation. $\bar{y} = 104.3$ mm.

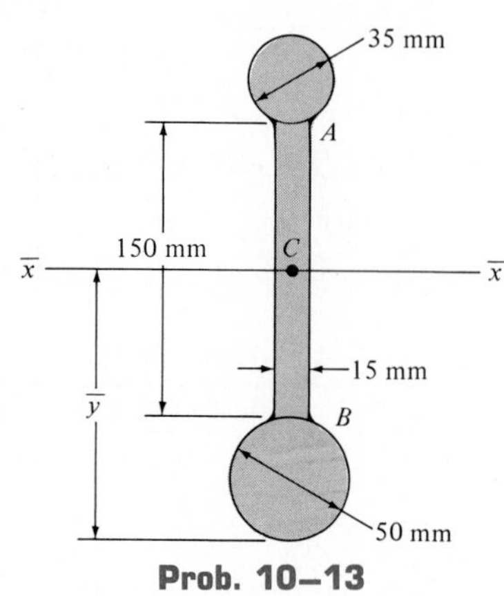

Prob. 10–13

10–14. Compute the polar moments of inertia J_O for the cross-sectional area of the solid shaft and tube. What percentage of J_O is contributed by the tube to that of the solid shaft?

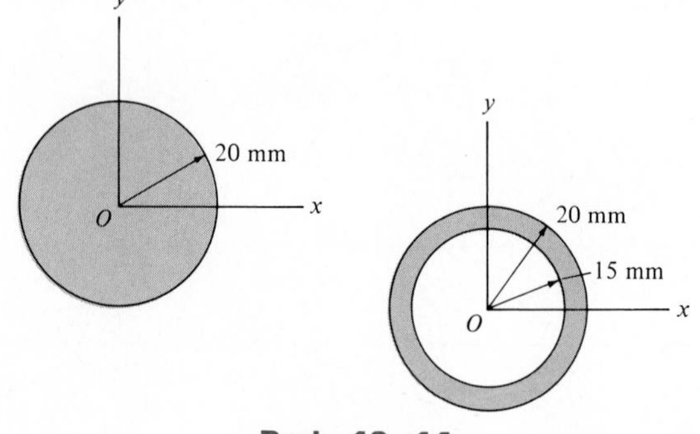

Prob. 10–14

10–15. Determine the moments of inertia $I_{\overline{x}}$ and $I_{\overline{y}}$ for the channel section. $\overline{x} = 33.9$ mm, $\overline{y} = 150$ mm.

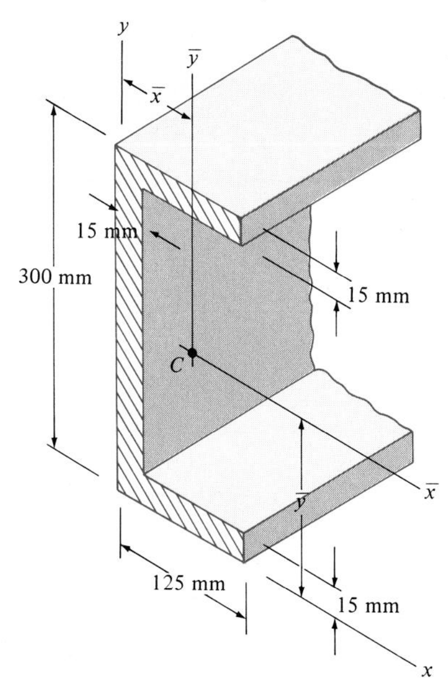

Prob. 10–15

***10–16.** Determine the location $\overline{y}$ of the centroid C of the beam's cross-sectional area. Then compute the moment of inertia of the area about the $\overline{x}\,\overline{x}$ axis.

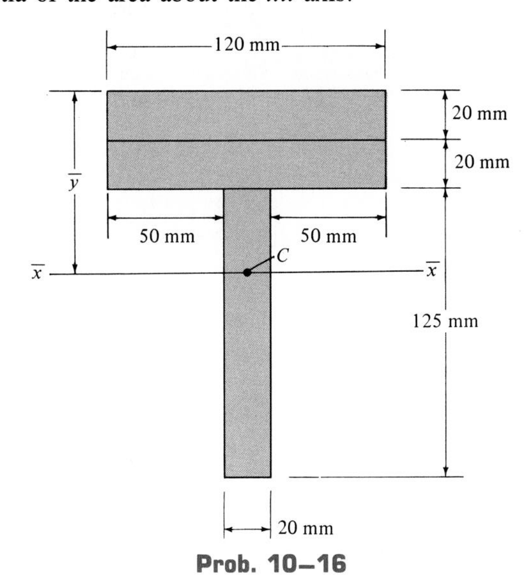

Prob. 10–16

10–17. Determine the moments of inertia I_x and I_y of the shaded area.

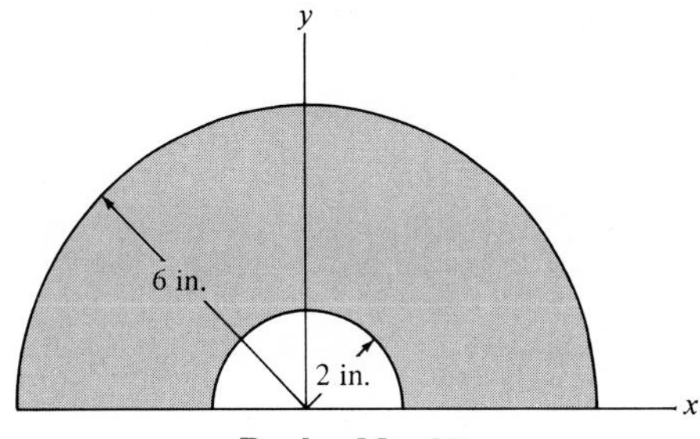

Prob. 10–17

10–18. Determine the moment of inertia of the beam's cross-sectional area with respect to the $\overline{x}\,\overline{x}$ axis passing through the centroid C. $\overline{y} = 0.875$ in.

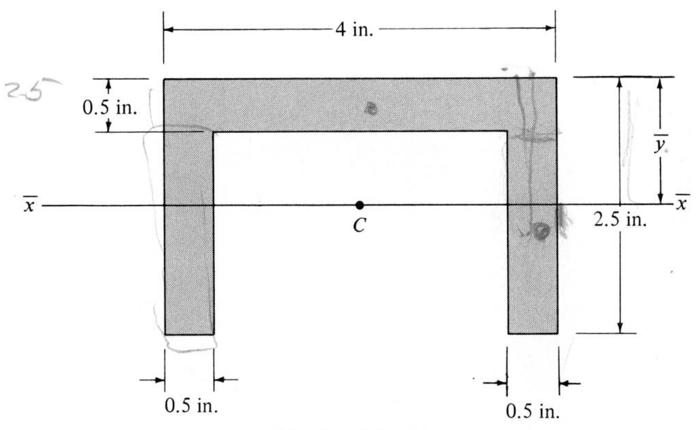

Prob. 10–18

10–19. The composite beam consists of a wide-flange beam and a cover plate welded together as shown. Determine the moment of inertia of the cross-sectional area with respect to the $\overline{x}\,\overline{x}$ centroidal axis. $\overline{y} = 77.8$ mm.

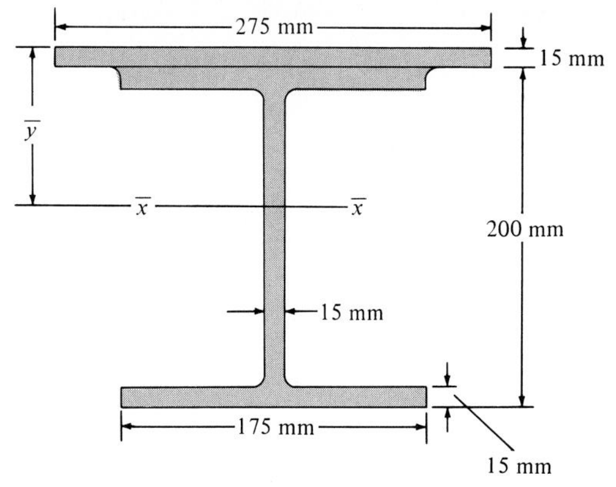

Prob. 10–19

***10–20.** Determine the moment of inertia of the beam's cross-sectional area about the xx axis.

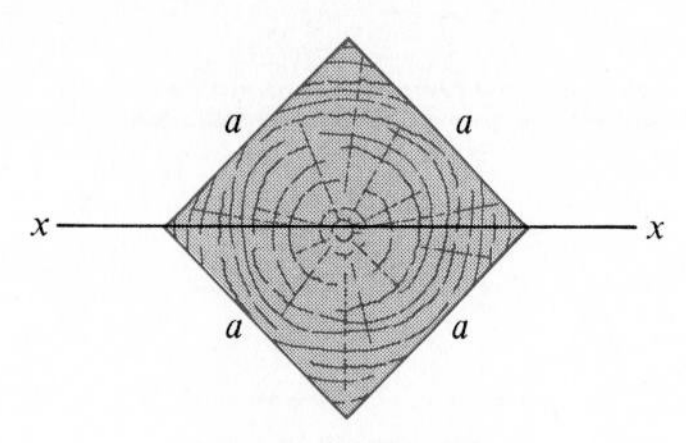

Prob. 10–20

10–21. The composite cross section for the column consists of two cover plates riveted to two channels. Determine the radius of gyration $k_{\bar{x}}$ with respect to the centroidal $\overline{xx}$ axis. Each channel has a cross-sectional area of $A_c = 11.8$ in.^{2} and a moment of inertia $(I_{\bar{x}})_c = 349$ in.^{4}.

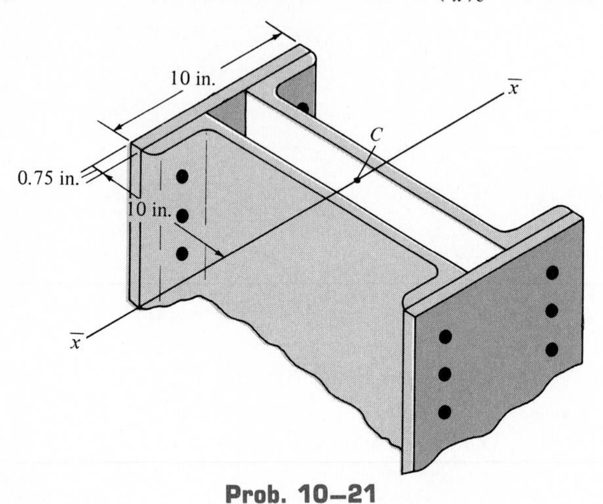

Prob. 10–21

10–22. Compute the polar radius of gyration, k_O, for the ring and show that for small thicknesses, $t = r_2 - r_1$, k_O is equal to the mean radius r_m of the ring.

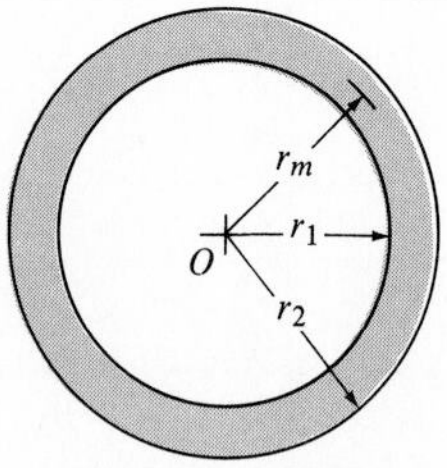

Prob. 10–22

10–23. Determine the moments of inertia I_x and I_y of the Z-section. The origin of coordinates is at the centroid C.

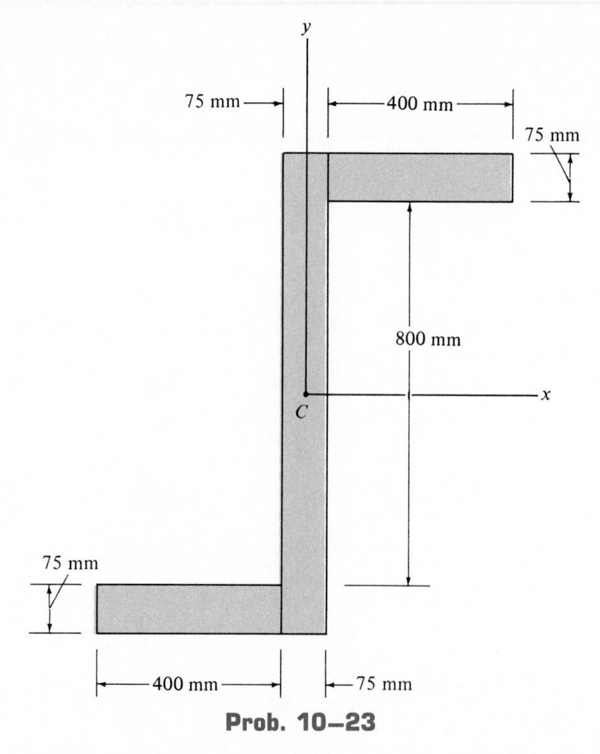

Prob. 10–23

***10–24.** Determine the moment of inertia of the beam's cross-sectional area with respect to the $\bar{x}$ centroidal axis. Neglect the size of all the rivet heads, R, for the calculation. Handbook values for the area, moment of inertia, and centroids of the channel and one angle located at the top and one at the bottom of the beam are listed in the figure. $\bar{y} = 8.46$ in.

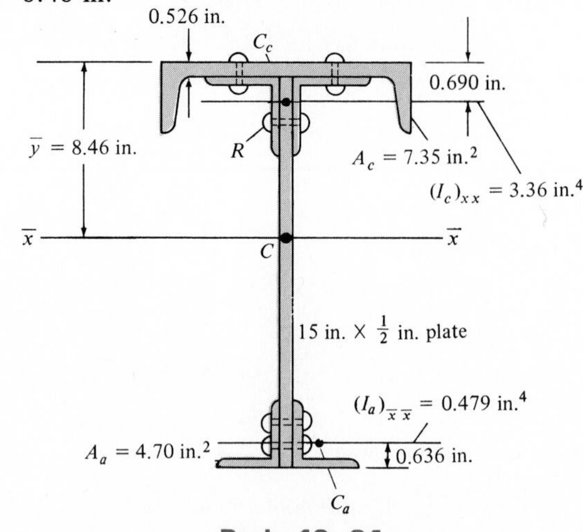

Prob. 10–24

Mass Moment of Inertia 10.6

The *mass moment of inertia* of a body is a property which measures the resistance of the body to angular motion. For example, consider the rigid body shown in Fig. 10–10. Here the applied external forces cause an *unbalanced moment* about the *aa* axis which gives the body an angular acceleration $\boldsymbol{\alpha}$. The analysis of the motion is treated in the study of dynamics. An important equation used in this regard is $\Sigma M = I\alpha$, where ΣM defines the moment of the external forces about the *aa* axis and I is a property of the body called the *mass moment of inertia*. This term is computed by integrating over the entire body the "second moment" about the *aa* axis of all the differential-sized elements of mass *dm*, Fig. 10–10. If the moment arm is r, extending from the axis *aa* to the arbitrary element *dm*, then

$$I = \int_m r^2\, dm \qquad (10\text{–}7)$$

Since the formulation involves the distance r, the value of I is *unique* for each axis *aa* about which it is computed. The axis which is generally chosen for analysis, however, passes through the body's mass center G. The moment of inertia computed about this axis will be defined as I_G. Since the moment of inertia is an important property, used throughout the study of dynamics, methods used for its calculation will now be discussed.

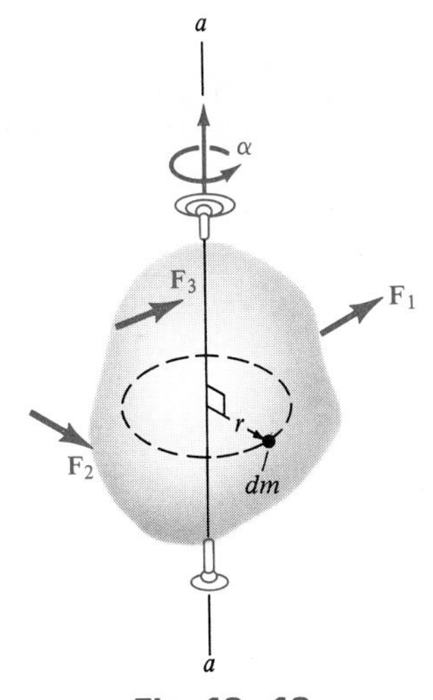

Fig. 10–10

PROCEDURE FOR ANALYSIS

In this treatment, we will only consider bodies having surfaces which are generated by revolving a curve about an axis. An example of such a body which is generated about the z axis is shown in Fig. 10–11.

If the body consists of material having a variable mass density, $\rho = \rho(x, y, z)$, the elemental mass dm of the body may be expressed in terms of its density and volume as $dm = \rho\, dV$. Substituting in Eq. 10–7, the body's moment of inertia is then computed using *volume elements* for integration, i.e.,

$$I = \int_V r^2 \rho \, dV \tag{10–8}$$

In the special case of ρ being a *constant*, this term may be factored out of the integral and the integration is then purely a function of geometry,

$$I = \rho \int_V r^2 \, dV \tag{10–9}$$

When the elemental volume chosen for integration has differential sizes in all three directions, e.g., $dV = dx\,dy\,dz$, Fig. 10–11a, the moment of inertia of the body must be computed using "triple integration." The integration process can, however, be simplified to a *single integration* provided the chosen elemental volume has a differential size or thickness in only *one direction*. Shell or disk elements are often used for this purpose.

Shell Element. If a *shell element* having a height z, radius $r = y$, and thickness dy is chosen for integration, Fig. 10–11b, then the volume is $dV = (2\pi y)(z)\,dy$. This element may be used in Eq. 10–8 or 10–9 for computing the moment of inertia I_z of the body about the z axis, since the *entire element,* due to its "thinness," lies at the *same* perpendicular distance $r = y$ from the z axis (see Example 10–6).

Disk Element. If a disk element having a radius $r = y$ and a thickness dz is chosen for integration, Fig. 10–11c, then the volume is $dV = (\pi y^2)\,dz$. In this case, however, the element is *finite* in the radial direction, and consequently parts of it *do not* all lie at the *same radial distance* r from the z axis. As a result, Eq. 10–8 or 10–9 *cannot* be used to determine I_z. Instead, to perform the integration using this element, it is first necessary to determine the moment of inertia *of the element* about the z axis and then integrate this result (see Example 10–7).

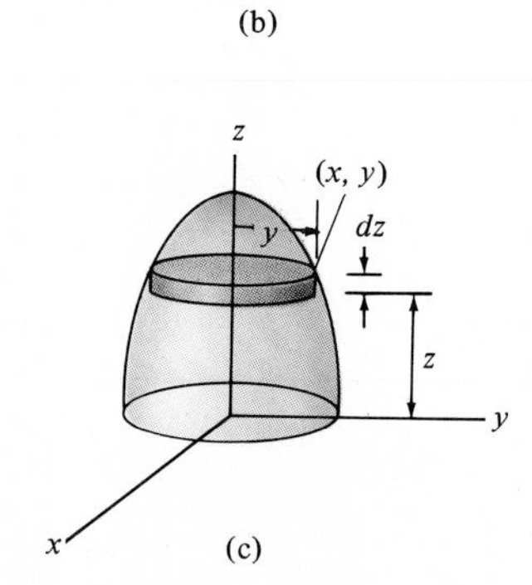

Fig. 10–11

Example 10–6

Determine the moment of inertia of the right circular cylinder shown in Fig. 10–12*a* about the *z* axis. The mass density ρ of the material is constant.

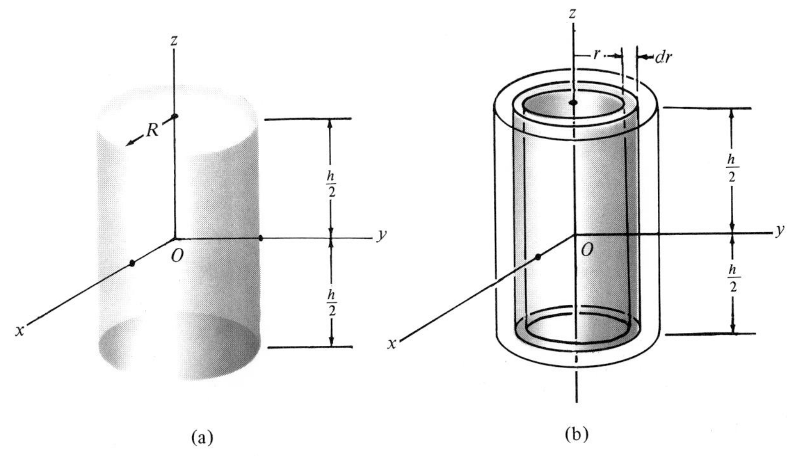

Fig. 10–12

Solution

Shell Element. This problem may be solved using the *shell element* in Fig. 10–12*b* and single integration. The volume of the element is $dV = (2\pi r)(h)\,dr$, so that the mass is $dm = \rho\,dV = \rho(2\pi hr\,dr)$. Since the *entire element* lies at the same distance *r* from the *z* axis, the moment of inertia *of the element* is

$$dI_z = r^2\,dm = \rho 2\pi hr^3\,dr$$

Integrating over the entire region of the cylinder yields

$$I_z = \int_m r^2\,dm = \rho 2\pi h \int_0^R r^3\,dr = \frac{\rho\pi}{2}R^4 h$$

The mass of the cylinder is

$$m = \int_m dm = \rho 2\pi h \int_0^R r\,dr = \rho\pi h R^2$$

so that

$$I_z = \frac{1}{2}mR^2 \qquad \textit{Ans.}$$

Example 10–7

A solid is formed by revolving the shaded area shown in Fig. 10–13*a* about the y axis. If the mass density of the material is 5 slug/ft³, determine the moment of inertia about the y axis.

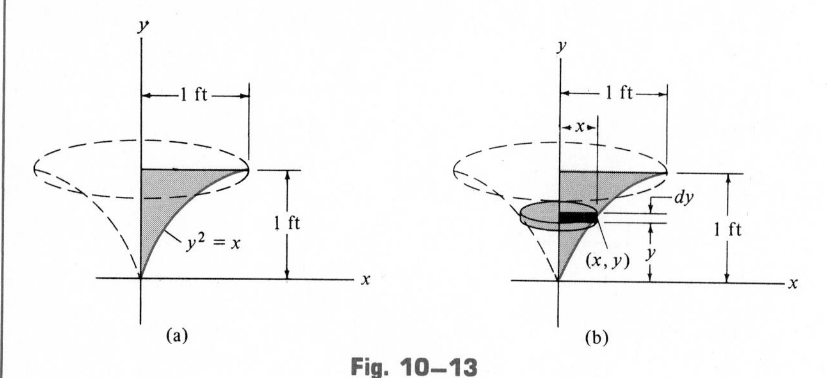

Fig. 10–13

Solution

Disk Element. The moment of inertia will be computed using a *disk element,* as shown in Fig. 10–13*b*. Here the element intersects the curve at the arbitrary point (x, y) and has a mass

$$dm = \rho\, dV = \rho(\pi x^2)\, dy$$

Although all portions of the element are *not* located at the same distance from the y axis, it is still possible to determine the moment of inertia dI_y *of the element* about the y axis. In the preceding example it was shown that the moment of inertia of a cylinder about its longitudinal axis is $I = \frac{1}{2}mR^2$, where m and R are the mass and radius of the cylinder. Since the height of the cylinder is not involved in this formula, the moment of inertia of the disk element in Fig. 10–13*b* is

$$dI_y = \tfrac{1}{2}(dm)x^2 = \tfrac{1}{2}[\rho(\pi x^2)\, dy]x^2$$

Substituting $x = y^2$, $\rho = 5$ slug/ft³, and integrating with respect to y, from $y = 0$ to $y = 1$ ft, yields the moment of inertia for the entire solid.

$$I_y = \frac{\pi 5}{2}\int_0^1 x^4\, dy = \frac{\pi 5}{2}\int_0^1 y^8\, dy = 0.873 \text{ slug}\cdot\text{ft}^2 \qquad \textit{Ans.}$$

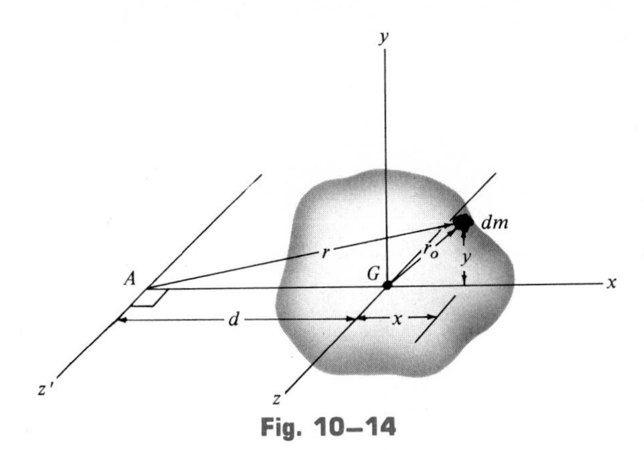

Fig. 10–14

Parallel-Axis Theorem. If the moment of inertia of the body about an axis passing through the body's mass center is known, then the body's moment of inertia may be determined about any other *parallel axis* by using the *parallel-axis theorem*. This theorem can be derived by considering the body shown in Fig. 10–14. The z axis passes through the body's mass center G, whereas the corresponding *parallel axis* z' lies at a constant distance d away. Selecting the differential element of mass dm which is located at point (x, y) and using the Pythagorean theorem, $r^2 = (d + x)^2 + y^2$, we can express the moment of inertia of the body computed about the z' axis as

$$I = \int_m r^2\,dm = \int_m [(d + x)^2 + y^2]\,dm$$

$$= \int_m (x^2 + y^2)\,dm + 2d\int_m x\,dm + d^2\int_m dm$$

Since $r_o^2 = x^2 + y^2$, the first integral represents I_G. The second integral equals *zero,* since the z axis passes through the body's mass center, so that $\int x\,dm = \bar{x}\int dm$ since $\bar{x} = 0$. Finally, the third integral represents the total mass m of the body. Hence, the moment of inertia about the z' axis can be written as

$$I = I_G + md^2 \qquad (10\text{–}10)$$

where

I_G = moment of inertia about the z axis passing through the mass center G
m = mass of the body
d = perpendicular distance between the parallel axes

Radius of Gyration. Occasionally, the moment of inertia of a body about a specified axis is reported in handbooks using the *radius of gyration*. When this length, k, and the body's mass m are known, the body's moment of inertia is determined from the equation

$$I = mk^2 \quad \text{or} \quad k = \sqrt{\frac{I}{m}} \tag{10–11}$$

Note the *similarity* between the definition of k in this formula and r in the equation $dI = r^2\, dm$, which defines the moment of inertia of an elemental mass dm of the body about an axis.

Composite Bodies. The parallel-axis theorem is often used to determine the moment of inertia of composite shapes when the moment of inertia I_G of each of the composite parts is either known or can be computed by integration.* For example, if the body is constructed of a number of simple shapes such as disks, spheres, and rods, the moment of inertia of the body about any axis z' can be determined by adding algebraically the moments of inertia of all the composite shapes computed about the z' axis. Algebraic addition is necessary since a composite part must be considered as a negative quantity if it has already been counted as part of another part—for example a "hole" subtracted from a solid plate. The parallel-axis theorem is needed for the calculations if the center of gravity of each composite part does not lie on the z' axis.

*See the table given on the inside back cover of this book.

Example 10–8

If the plate shown in Fig. 10–15*a* has a density of 8000 kg/m^3 and a thickness of 10 mm, compute its moment of inertia about an axis directed perpendicular to the page and passing through point O.

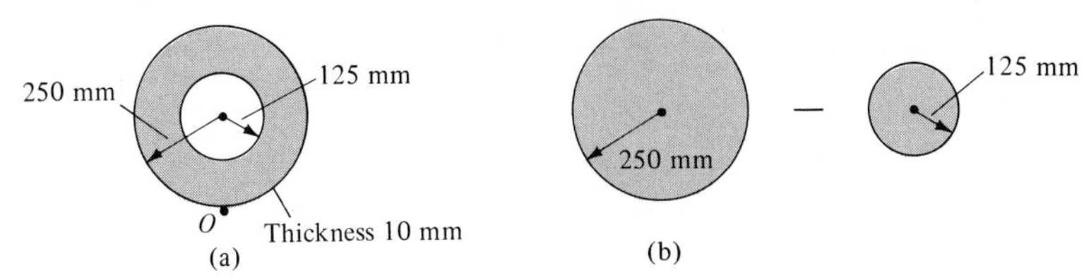

Fig. 10–15

Solution

The plate consists of two composite parts, Fig. 10–15*b*. The 250-mm-radius disk *minus* a 125-mm-radius disk. The moment of inertia about O can be determined by computing the moment of inertia of each of these parts about O and then *algebraically* adding the results. The computations are performed by using the parallel-axis theorem in conjunction with the data listed in the table on the inside back cover.

Disk. The moment of inertia of a thin disk about an axis perpendicular to the plane of the disk is $I_G = \frac{1}{2}mr^2$. The mass centers of *both* the 250-mm-radius disk and the 125-mm-radius disk (hole) are located at a distance of 0.25 m from point O. For the 250-mm-radius disk, we have

$$m_d = \rho_d V_d = 8000 \text{ kg/m}^3[\pi(0.25 \text{ m})^2(0.01 \text{ m})] = 15.71 \text{ kg}$$

$$\begin{aligned}(I_O)_d &= \tfrac{1}{2}m_d r_d^2 + m_d d^2 \\ &= \tfrac{1}{2}(15.71 \text{ kg})(0.25 \text{ m})^2 + (15.71 \text{ kg})(0.25 \text{ m})^2 \\ &= 1.473 \text{ kg}\cdot\text{m}^2\end{aligned}$$

Hole. For the 125-mm-radius disk (hole), we have

$$m_h = \rho_h V_h = 8000 \text{ kg/m}^3[\pi(0.125 \text{ m})^2(0.01 \text{ m})] = 3.93 \text{ kg}$$

$$\begin{aligned}(I_O)_h &= \tfrac{1}{2}m_h r_h^2 + m_h d^2 \\ &= \tfrac{1}{2}(3.93 \text{ kg})(0.125 \text{ m})^2 + (3.93 \text{ kg})(0.25 \text{ m})^2 \\ &= 0.276 \text{ kg}\cdot\text{m}^2\end{aligned}$$

The moment of inertia of the plate about point O is therefore

$$\begin{aligned}I_O &= (I_O)_d - (I_O)_h \\ &= 1.473 - 0.276 \\ &= 1.197 \text{ kg}\cdot\text{m}^2\end{aligned}$$ *Ans.*

Problems

10–25. Determine the moment of inertia of the homogeneous sphere having a mass density of $\rho = 6$ slug/ft^3 and radius of 2 ft with respect to the y axis.

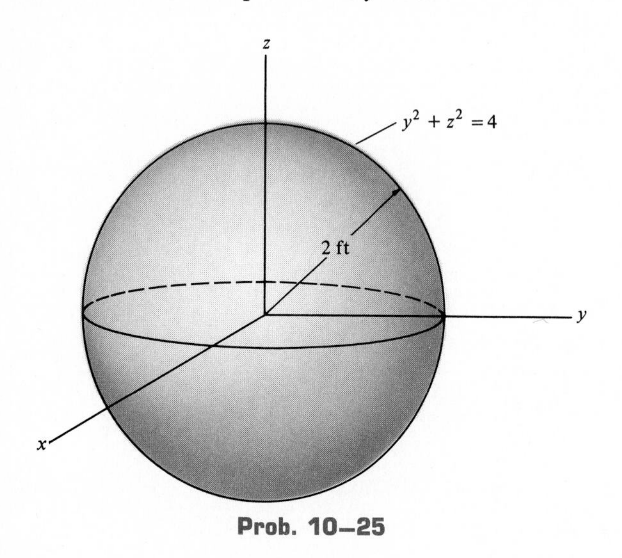

Prob. 10–25

10–26. The right circular cone is formed by revolving the shaded area around the x axis. Determine the moment of inertia I_x. The cone has a density of 4 Mg/m^3.

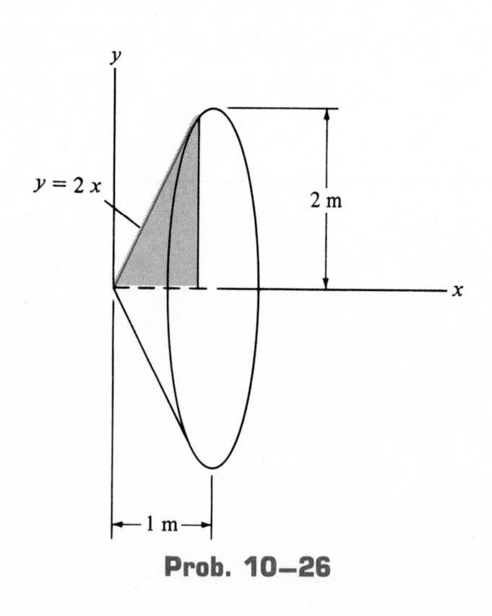

Prob. 10–26

10–27. The solid is formed by revolving the shaded area around the x axis. Determine the moment of inertia about the x axis. The mass density of the material is $\rho =$ 5 Mg/m^3.

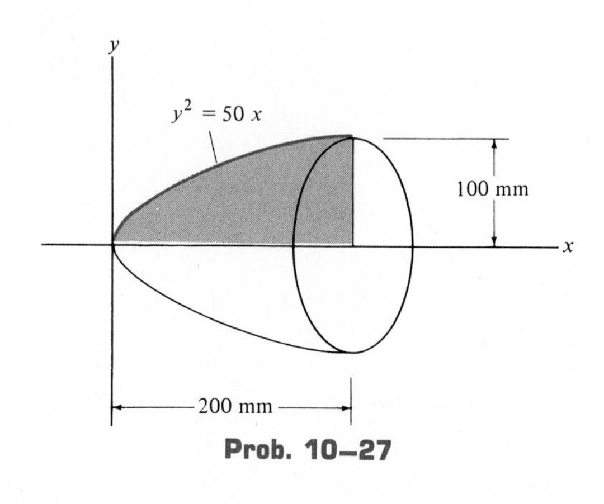

Prob. 10–27

***10–28.** An ellipsoid is formed by rotating the shaded area about the x axis. Determine the moment of inertia of this body with respect to the x axis. The density of the material is $\rho = 4$ Mg/m^3.

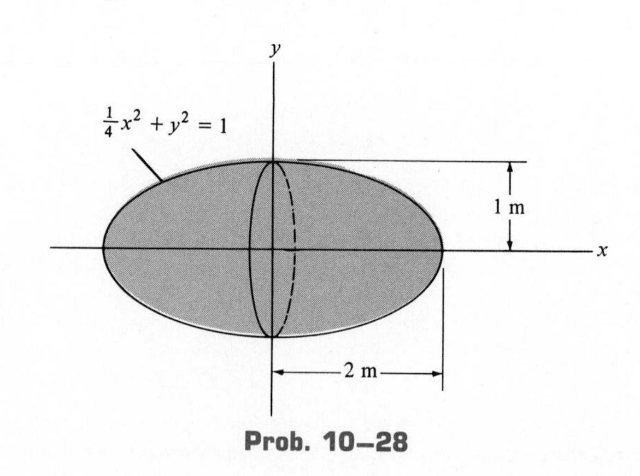

Prob. 10–28

10–29. Determine the moment of inertia of the hemispherical solid about the y axis. The density of the material is $\rho = 4\ \text{slug/ft}^3$.

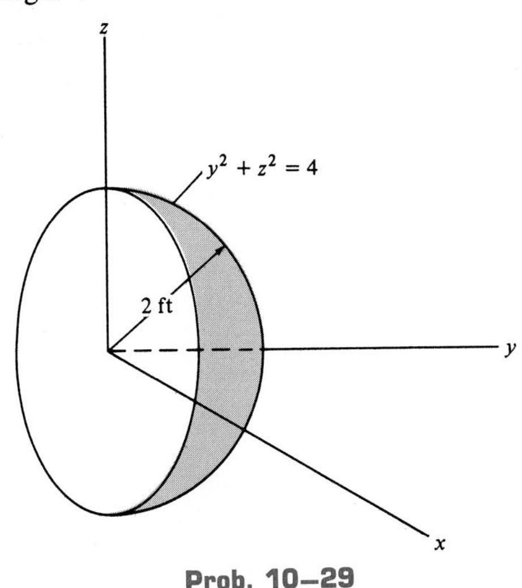

Prob. 10–29

10–30. A concrete solid is formed by rotating the shaded area about the y axis. Determine the moment of inertia I_y. The density of material is $\rho = 150\ \text{lb/ft}^3$.

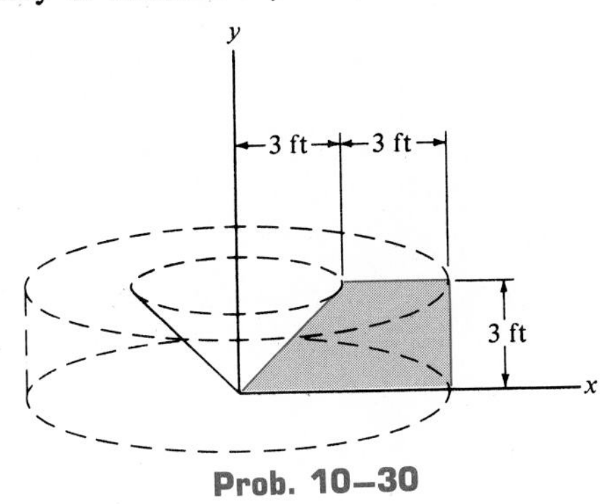

Prob. 10–30

10–31. The assembly consists of a uniform rod AB having a mass of 0.2 kg and two solid spheres C and D having a mass of 0.4 kg and 0.6 kg, respectively. Determine $\bar{x}$, which locates the center of mass G, and then calculate the moment of inertia about an axis perpendicular to the page and passing through G.

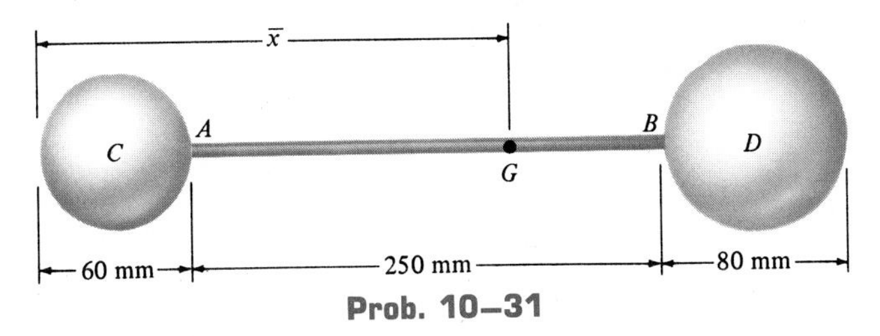

Prob. 10–31

***10–32.** The pulley has the cross section shown. If the density of the material is $\rho = 400\ \text{lb/ft}^3$, determine the pulley's moment of inertia about the xx axis.

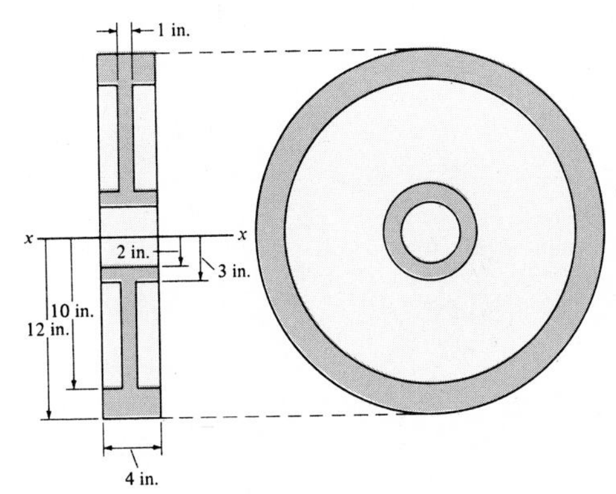

Prob. 10–32

10–33. Determine the moment of inertia of the wheel with respect to the z axis. The rim R and plate P both have a density of $\rho = 600\ \text{lb/ft}^3$.

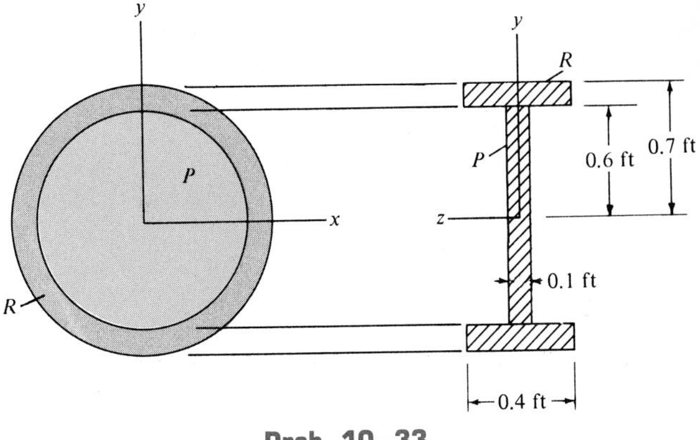

Prob. 10–33

Review Problems

10–34. A column has a built-up cross section of four equal-sized angles as shown. Neglecting the inertia of the lacing and rivets, determine the radius of gyration of the column about the $\overline{x}\overline{x}$ centroidal axis. Handbook values for the area, moment of inertia, and the location of the centroid for one of the angles are listed in the figure.

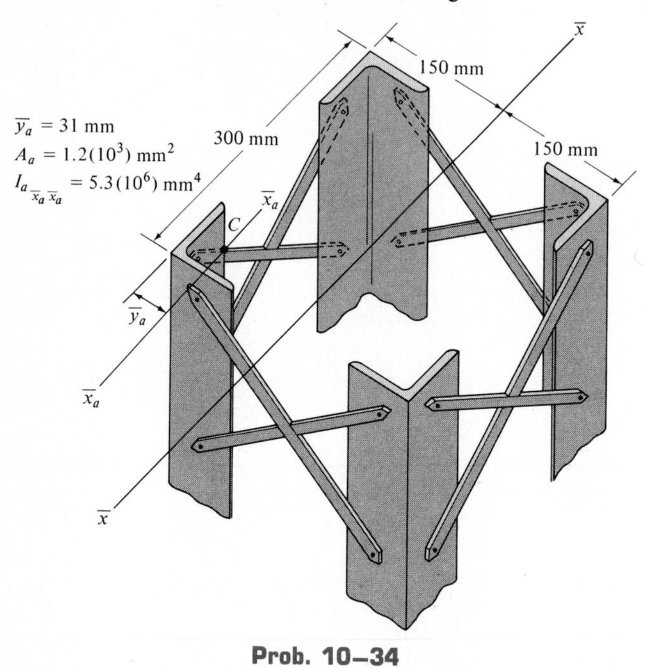

Prob. 10–34

10–35. Determine the moments of inertia I_x and I_y of the shaded area.

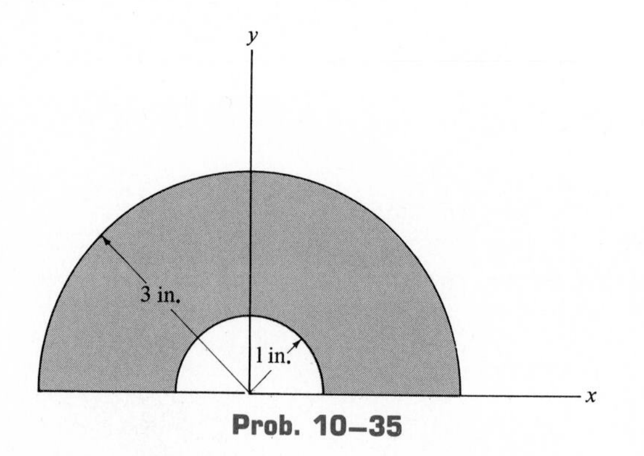

Prob. 10–35

***10–36.** Determine the moments of inertia I_x and I_y of the shaded area.

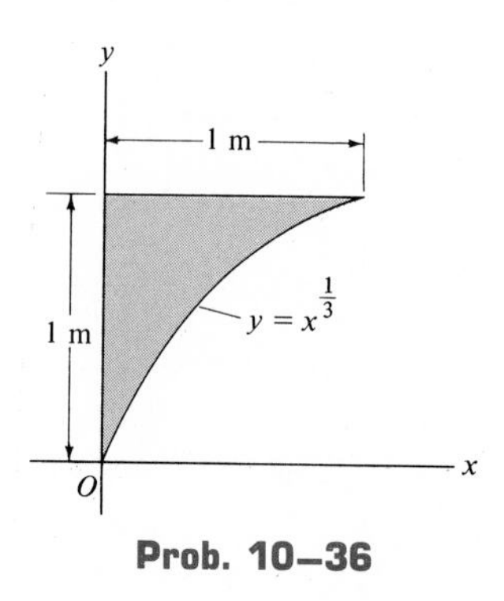

Prob. 10–36

10–37. Determine the moment of inertia I_x of the shaded area about the x axis.

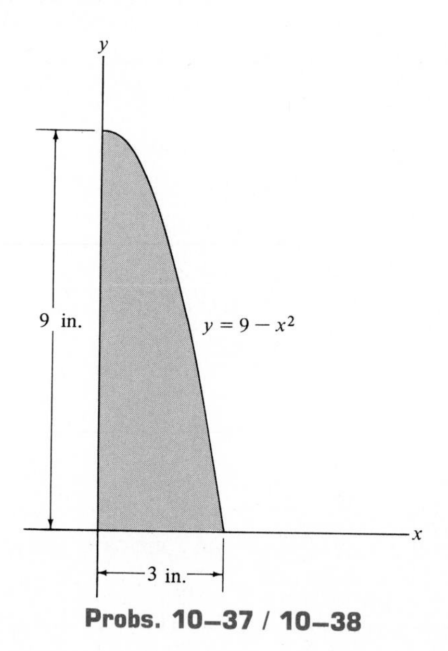

Probs. 10–37 / 10–38

10–38. Determine the moment of inertia I_y of the shaded area about the y axis.

10–39. Determine the moment of inertia I_x of the solid. The density of the material is $\rho = 6\ \mathrm{Mg/m^3}$.

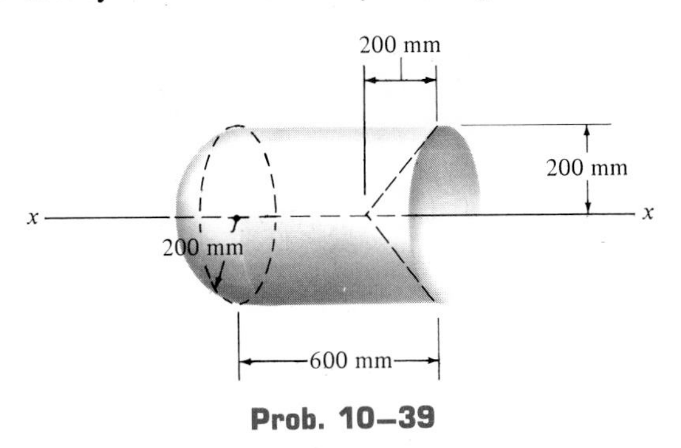

Prob. 10–39

***10–40.** Determine the moment of inertia of the homogeneous pyramid of mass m with respect to the z axis. The density of the material is ρ. *Suggestion:* Use a rectangular plate element having a volume of $dV = (2x)(2y)\,dz$.

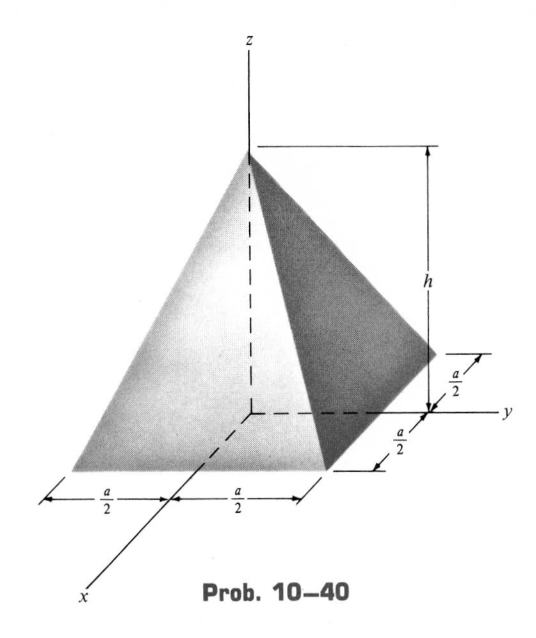

Prob. 10–40

11 Virtual Work

In the previous chapters, statics problems have been solved using the equations of equilibrium. It is also possible to solve these problems using the principle of virtual work. This method has been found to be very useful for determining the equilibrium configuration or position of a series of connected rigid bodies. Although application requires more mathematical sophistication, using calculus, than the conventional vector approach, using the equations of equilibrium, it will be shown that once the equation of virtual work is established the solution may be obtained *directly,* without having to dismember the system to obtain relationships between forces occurring at the connections.

11.1 Definition of Work and Virtual Work

Work of a Force. To understand the principle of virtual work, it is first necessary to define the *work done by a force* when the force undergoes a differential displacement along its path s, Fig. 11–1a. If θ is the angle made between the tails of the force and displacement vectors, the *differential* amount of work done by $\mathbf{F}$ is a *scalar quantity* defined as

$$dU = F\,ds \cos\theta$$

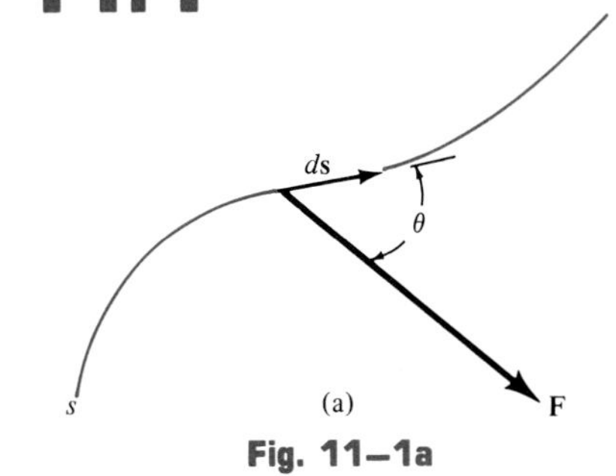

Fig. 11–1a

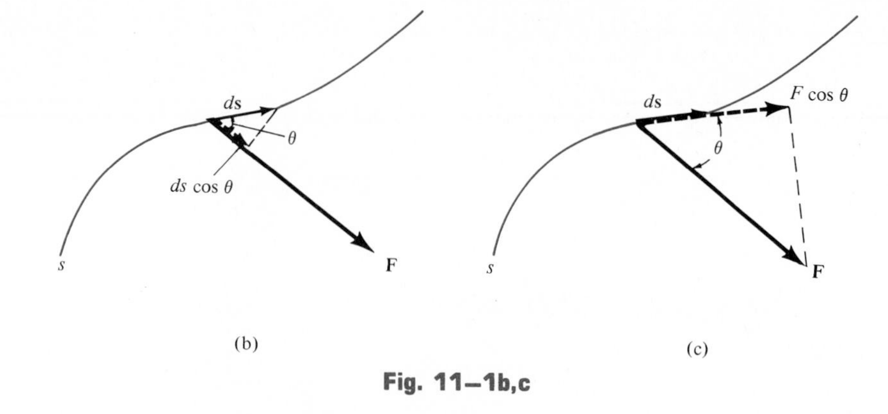

Fig. 11–1b,c

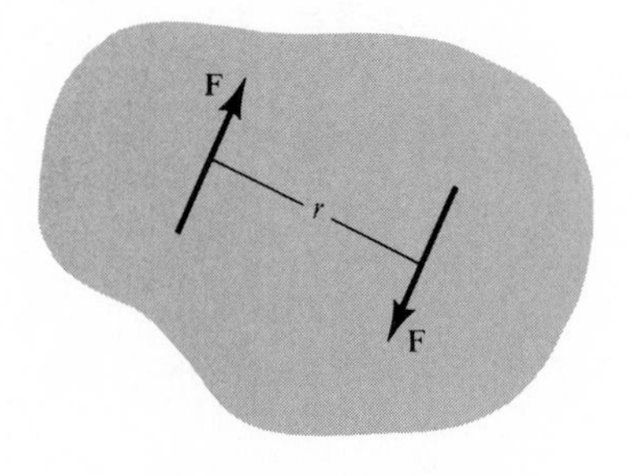

(a)

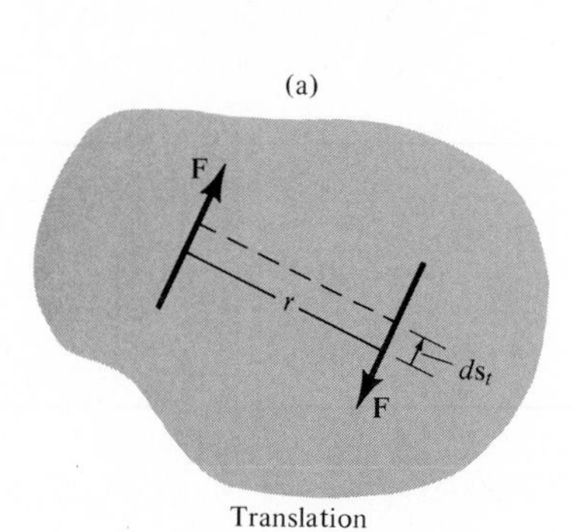

Translation

(b)

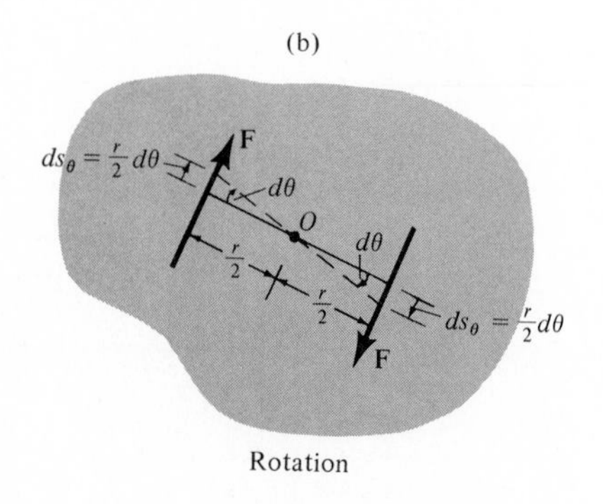

Rotation

(c)

Fig. 11–2

Hence, the work may be interpreted in one of two ways, either as the product of F and the component of the displacement in the direction of **F**, i.e., $ds \cos \theta$, Fig. 11–1*b*, or as the product of ds and the component of force in the direction of $d\mathbf{s}$, i.e., $F \cos \theta$, Fig. 11–1*c*. If $F \cos \theta$ and $d\mathbf{s}$ are in the *same direction,* the work is *positive;* if they are in *opposite directions,* the work is *negative*. Furthermore, if **F** is *perpendicular* to $d\mathbf{s}$, $\theta = 90°$ and the work done by **F** is *zero*.

The basic unit for work combines the units of force and displacement. In the SI system a *joule* (J) is equivalent to the work done by a force of 1 newton which moves 1 meter in the direction of the force (1 J = 1 N · m). In the FPS system work is defined in units of ft · lb. The moment of a force has the same combination of units; however, the concepts of moment and work are in no way related. A moment is a vector quantity, whereas work is a scalar.

Work of a Couple. The two forces of a couple do work when the couple *rotates* about an axis perpendicular to the plane of the couple. To show this, consider the body in Fig. 11–2*a*, which is subjected to a couple having a magnitude of $M = Fr$. Any general differential displacement of the body can be considered as a combination of a translation and rotation. When the body *translates* such that the *component of displacement* along the line of action of each force is $d\mathbf{s}_t$, clearly the "positive" work of one force ($F ds_t$) *cancels* the "negative" work of the other ($-F ds_t$), Fig. 11–2*b*. Consider now a differential *rotation* $d\boldsymbol{\theta}$ of the body about an axis perpendicular to the plane of the couple, which intersects the plane at the midpoint O, Fig. 11–2*c*. (For the derivation, any other point in the plane may also be considered.) As shown, each force undergoes a displacement $ds_\theta = (r/2)\, d\theta$ in the direction of the force; hence, the total work done is

$$dU = F\left(\frac{r}{2}\, d\theta\right) + F\left(\frac{r}{2}\, d\theta\right) = (Fr)\, d\theta$$

or

$$dU = M\, d\theta$$

The resultant work is *positive* when the rotational sense of **M** is in the *same* direction as that of $d\boldsymbol{\theta}$, and negative when they are in opposite directions. As in the case of the moment vector, the *direction* of $d\boldsymbol{\theta}$ is defined by the right-hand rule, where the fingers of the right hand follow the rotation or "curl" and the thumb indicates the direction of $d\boldsymbol{\theta}$. Hence, the line of action of $d\boldsymbol{\theta}$ must be *parallel* to the line of action of **M**. This is always the case if movement of the body occurs in the *same plane*. If the body rotates in space, however, the *component of* $d\boldsymbol{\theta}$ in the direction of **M** is required.

Virtual Work. The definitions of the work of a force and a couple have been presented in terms of *actual movements* expressed by the differential displacement $d\mathbf{s}$ and rotation $d\boldsymbol{\theta}$. Consider now an *imaginary* or *virtual movement*, which indicates a displacement or rotation that is *assumed* and *does not actually exist*. These movements are first-order differential quantities and will be denoted by the symbols $\delta\mathbf{s}$ and $\delta\boldsymbol{\theta}$ (delta **s** and delta $\boldsymbol{\theta}$), respectively. The *virtual work* done by a force undergoing a virtual displacement is

$$\delta U = F \cos \theta \; \delta s \qquad (11\text{–}1)$$

Similarly, when a couple undergoes a virtual rotation in the plane of the couple forces, the *virtual work* is

$$\delta U = M \; \delta\theta \qquad (11\text{–}2)$$

11.2 Principle of Virtual Work for a Particle and a Rigid Body

If a particle is in equilibrium, the resultant of a force system acting on it must be equal to zero. Hence, if the particle undergoes an imaginary or virtual displacement in the x, y, or z direction, the virtual work (δU) done by the force system must be equal to zero since the components $\Sigma F_x = 0$, $\Sigma F_y = 0$, $\Sigma F_z = 0$. Mathematically, this may be expressed as

$$\delta U = 0$$

For example, if the particle in Fig. 11–3 is given a virtual displacement $\delta\mathbf{x}$, only the x components of the forces acting on the particle do work. (No work is done by the y and z components since they are perpendicular to the displacement.) The virtual work equation is therefore

$$\delta U = 0; \qquad F_{1x}\,\delta x + F_{2x}\,\delta x + F_{3x}\,\delta x = 0$$

Factoring out δx, which is common to every term, yields

$$(F_{1x} + F_{2x} + F_{3x})\,\delta x = 0$$

Since $\delta x \neq 0$, this equation is satisfied only if the sum of the force components in the x direction is equal to zero, i.e., $\Sigma F_x = 0$. Two other virtual work

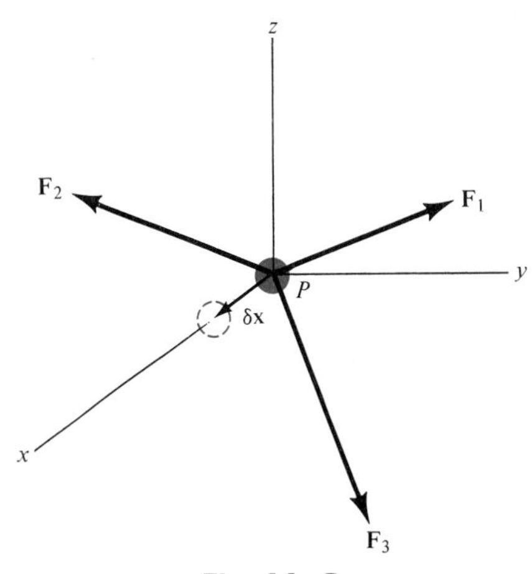

Fig. 11–3

equations can be written by assuming virtual displacements $\delta\mathbf{y}$ and $\delta\mathbf{z}$ in the y and z directions, respectively. Doing this, however, amounts to satisfying the equilibrium equations $\Sigma F_y = 0$ and $\Sigma F_z = 0$ for the particle.

In a similar manner, a rigid body that is subjected to a coplanar force system will be in equilibrium provided $\Sigma F_x = 0$, $\Sigma F_y = 0$, and $\Sigma M_O = 0$. We can also write a set of three virtual work equations for the body, each of which requires $\delta U = 0$. If these equations involve separate virtual translations in the x and y directions and a virtual rotation about an axis perpendicular to the x-y plane and passing through point O, then it can be shown that they will correspond to the above mentioned three equilibrium equations. When writing these equations, it is *not necessary* to include the work done by the *internal forces* acting on the body since the body *does not deform* when subjected to an external loading, and furthermore, when the body moves through a virtual displacement, the internal forces occur in equal but opposite collinear pairs, so that the corresponding work done by each pair of forces *cancels*.

As in the case of a particle, however, no added advantage would be gained by solving rigid-body equilibrium problems using the principle of virtual work. This is because for each application of the virtual-work equation the virtual displacement, common to every term, factors out, leaving an equation that could have been obtained by *direct application* of the equilibrium equations.

11.3 Principle of Virtual Work for a System of Connected Rigid Bodies

The method of virtual work is most suitable for solving equilibrium problems that involve a system of several *connected* rigid bodies such as shown in Fig. 11–4. For each of these systems, the number of *independent virtual-work equations* that can be written depends upon the number of *independent virtual displacements* that can be made by the system.

Degrees of Freedom. In general, *for a system of connected bodies, the number of independent virtual displacements equals the minimum number of independent coordinates q needed to specify completely the location of all members of the system.* Each independent coordinate gives the system a *degree of freedom* that must be consistent with the constraining action of the supports. Thus, an n-degree-of-freedom system requires n independent coordinates q_n to specify the location of all its members with respect to a fixed reference point. For example, the link and sliding-block arrangement shown in Fig. 11–4*a* is an example of a one-degree-of-freedom system. The independent coordinate $q = \theta$ may be used to specify the location of the two connecting links and the block. The coordinate x could also be used as the independent coordinate. However, since the block is constrained to move within the slot, x is not independent of θ; rather it can be related to θ using the cosine law, $b^2 = a^2 + x^2 - 2ax \cos \theta$. The double-link arrangement,

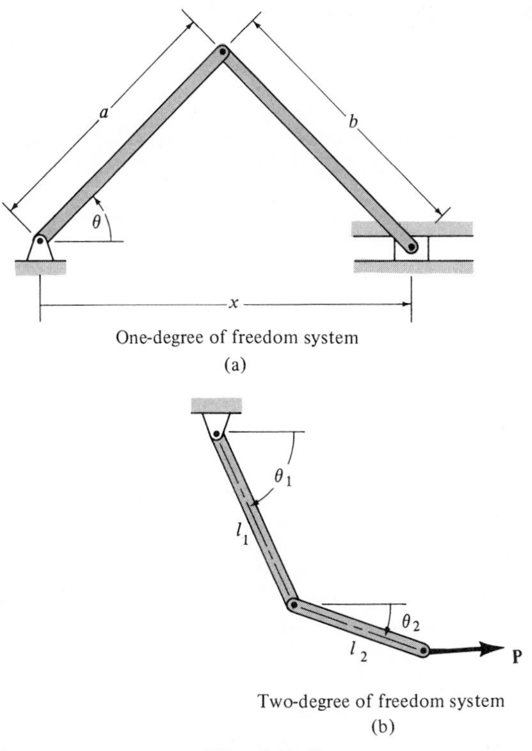

One-degree of freedom system
(a)

Two-degree of freedom system
(b)

Fig. 11–4

shown in Fig. 11–4*b*, is an example of a two-degree-of-freedom system. To specify the location of each link, the coordinate angles θ_1 and θ_2 must be known, since a rotation of one link is independent of a rotation of the other.

Principle of Virtual Work. The principle of virtual work for a system of rigid bodies whose connections are *frictionless* may be stated as follows: *A system of connected rigid bodies is in equilibrium provided the virtual work done by all the external forces and couples acting on the system is zero for each independent virtual displacement of the system.* Mathematically, this may be expresssed as

$$\delta U = 0 \tag{11-3}$$

where δU represents the virtual work of all the external forces (and couples) acting on the system during any independent virtual displacement.

It has already been pointed out that if a system has n degrees of freedom it takes n independent coordinates q_n to completely specify the location of the system. Hence, for the system it is possible to write n independent virtual-work equations; one for each independent coordinate, while the remaining $n - 1$ coordinates are held *fixed*.

PROCEDURE FOR ANALYSIS

The following procedure provides a method for applying the equation of virtual work to solve problems involving a system of frictionless connected rigid bodies having a single degree of freedom.

Free-Body Diagram. Draw the free-body diagram of the entire system of connected bodies and define the *independent coordinate* q. Sketch the "deflected position" of the system on the free-body diagram when the system undergoes a *positive* virtual displacement δq. From this, specify the "active" forces and couples, that is, those that do work.

Virtual Displacements. Indicate *position coordinates* s_i, measured from a *fixed point* on the free-body diagram to each of the i number of "active" forces and couples. Each coordinate axis should be in the *same direction* as the line of action of the "active" force to which it is directed.

Relate each of the position coordinates s_i to the independent coordinate q; then *differentiate* these expressions in order to express the virtual displacements δs_i in terms of δq.

Virtual-Work Equation. Write the *virtual-work equation* for the system assuming that, whether possible or not, all the position coordinates s_i undergo *positive* virtual displacements δs_i. Using the relations for δs_i, express the work of *each* "active" force and couple in the equation in terms of the single independent virtual displacement δq. By factoring out this common displacement, one is left with an equation that generally can be solved for an unknown force, couple, or equilibrium position.

If the system contains n degrees of freedom, n independent coordinates q_n must be specified. In this case, follow the above procedure and let *only one* of the independent coordinates undergo a virtual displacement δq_n, while the remaining $n - 1$ coordinates are held fixed. In this way, n virtual-work equations can be written, one for each independent coordinate.

The following examples should help to clarify application of this procedure.

Example 11–1

Determine the angle θ for equilibrium of the two-member linkage shown in Fig. 11–5*a*. Each member has a mass of 10 kg.

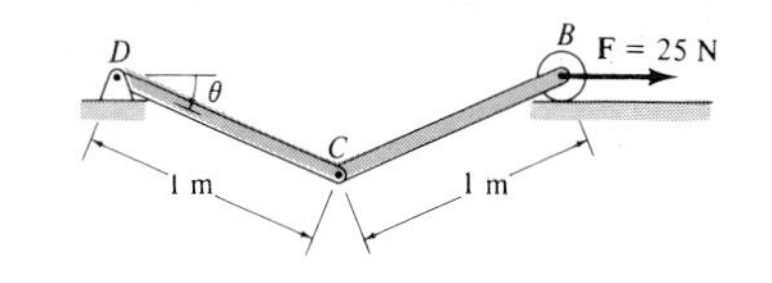

(a)

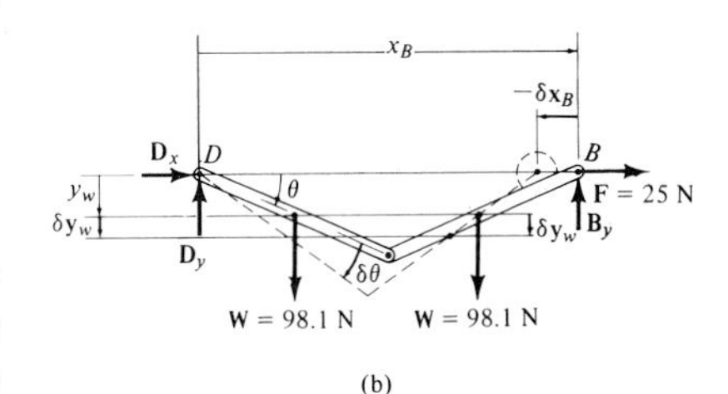

(b)

Fig. 11–5

Solution

Free-Body Diagram. The system has only one degree of freedom, since the location of both links may be specified by the single independent coordinate $(q =)\ \theta$. As shown on the free-body diagram in Fig. 11–5*b*, when θ undergoes a *positive* (clockwise) virtual rotation $\delta\theta$, only the active forces, **F** and the two 98.1-N weights, do work. (The reactive forces $\mathbf{D}_x$ and $\mathbf{D}_y$ are fixed, and $\mathbf{B}_y$ does not move along its line of action.)

Virtual Displacements. If the origin of coordinates is established at the *fixed* pin support D, the location of **F** and **W** may be specified by the *position coordinates* x_B and y_w, as shown in the figure. In order to compute the work, note that these coordinates are in the *same direction* as the lines of action of their associated forces.

Expressing the position coordinates in terms of the independent coordinate θ and taking the derivatives yields

$$x_B = 2(1\ \cos\theta)\ \text{m} \qquad \delta x_B = -2\sin\theta\ \delta\theta\ \text{m} \qquad (1)$$

$$y_w = \tfrac{1}{2}(1\ \sin\theta)\ \text{m} \qquad \delta y_w = 0.5\cos\theta\ \delta\theta\ \text{m} \qquad (2)$$

It is seen by the *signs* of these equations, and indicated in Fig. 11–5*b*, that an *increase* in θ (i.e., $\delta\theta$) causes a *decrease* in x_B and an *increase* in y_w.

Virtual-Work Equation. For *positive* virtual displacements δx_B and δy_w, the forces **W** and **F** do positive work since the forces and their corresponding displacements would be in the same direction. Hence, the virtual-work equation for the displacement $\delta\theta$ is

$$\delta U = 0; \qquad W\,\delta y_w + W\,\delta y_w + F\,\delta x_B = 0 \qquad (3)$$

Substituting Eqs. (1) and (2) into Eq. (3) in order to relate the virtual displacements to the common virtual displacement $\delta\theta$ yields

$$98.1(0.5\cos\theta\ \delta\theta) + 98.1(0.5\cos\theta\ \delta\theta) + 25(-2\sin\theta\ \delta\theta) = 0$$

Notice that the "negative work" done by **F** (force in the opposite direction to displacement) has been *accounted for* in the above equation by the "negative sign" of Eq. (1). Factoring out the *common displacement* $\delta\theta$ and solving for θ, noting that $\delta\theta \neq 0$, yields

$$(98.1\cos\theta - 50\sin\theta)\ \delta\theta = 0$$

$$\theta = \tan^{-1}\frac{98.1}{50} = 63.0^\circ \qquad \textit{Ans.}$$

If this problem had been solved using the equations of equilibrium, it would have been necessary to dismember the links and apply three scalar equations to *each* link. The principle of virtual work, by means of calculus, has eliminated this task so that the answer is obtained in a very direct manner.

Example 11-2

Using the principle of virtual work, determine the angle θ required to maintain equilibrium of the mechanism shown in Fig. 11–6*a*. Neglect the weight of the links. The spring is unstretched when $\theta = 0°$ and it maintains a horizontal position due to the roller.

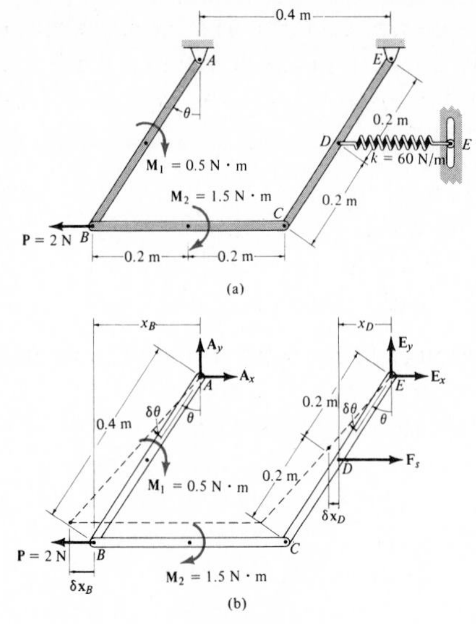

Fig. 11–6

Solution

Free-Body Diagram. The mechanism has one degree of freedom, and therefore the location of each member may be specified using the independent coordinate θ. When θ undergoes a *positive* virtual displacement $\delta\theta$, as shown on the free-body diagram in Fig. 11–6*b*, links *AB* and *EC* rotate by the same amount since they have the same length, and link *BC* only translates. Since a couple does work *only* when it rotates, the work done by $\mathbf{M}_2$ is zero. The reactive forces at *A* and *E* do no work since the supports do not translate.

Virtual Displacements. The position coordinates x_B and x_D are *parallel* to the lines of action of **P** and **F** and locate these forces with respect to the *fixed points A* and *E*. From Fig. 11–6*b*,

$$x_B = 0.4 \sin\theta \text{ m}$$
$$x_D = 0.2 \sin\theta \text{ m}$$

Thus,

$$\delta x_B = 0.4 \cos\theta\, \delta\theta \text{ m}$$
$$\delta x_D = 0.2 \cos\theta\, \delta\theta \text{ m}$$

Virtual-Work Equation. Applying the equation of virtual work, noting that $\mathbf{F}_s$ is opposite to *positive* $\delta\mathbf{x}_D$ displacement and hence does negative work, we obtain

$$\delta U = 0; \qquad M_1\,\delta\theta + P\,\delta x_B - F_s\,\delta x_D = 0$$

Relating each of the virtual displacements to the *common* virtual displacement $\delta\theta$ yields

$$0.5\,\delta\theta + 2(0.4\cos\theta\,\delta\theta) - F_s(0.2\cos\theta\,\delta\theta) = 0$$
$$(0.5 + 0.8\cos\theta - 0.2F_s\cos\theta)\,\delta\theta = 0 \qquad (1)$$

For the arbitrary angle θ, the spring is stretched a distance of $x = x_D = (0.2 \sin\theta)$ m; and therefore, $F_s = kx = 60$ N/m$(0.2\sin\theta)$ m $= (12\sin\theta)$ N. Substituting into Eq. (1) and noting that $\delta\theta \neq 0$, we have

$$0.5 + 0.8\cos\theta - 0.2(12\sin\theta)\cos\theta = 0$$

Since $\sin 2\theta = 2\sin\theta\cos\theta$, then

$$1 = 2.4\sin 2\theta - 1.6\cos\theta$$

Solving for θ by trial and error yields

$$\theta = 36.3° \qquad \textit{Ans.}$$

Example 11–3

Using the principle of virtual work, determine the force that the spring must exert in order to hold the mechanism shown in Fig. 11–7*a* in equilibrium when $\theta = 45°$. Neglect the weight of the members.

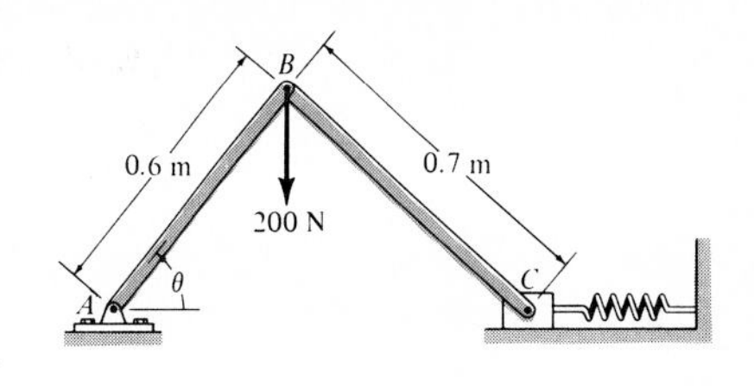

(a)

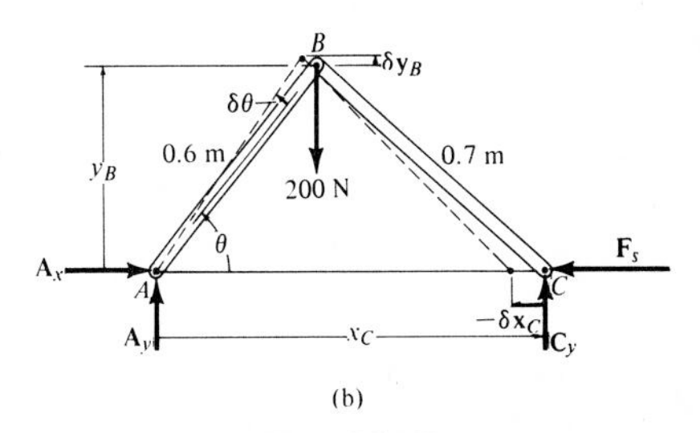

(b)

Fig. 11–7

Solution

Free-Body Diagram. As shown on the free-body diagram, Fig. 11–7*b*, the system has one degree of freedom, defined by the independent coordinate θ. When θ undergoes a *positive* virtual displacement $\delta\theta$, only $\mathbf{F}_s$ and the 200-N force do work.

Virtual Displacements. Forces $\mathbf{F}_s$ and 200 N are located from the fixed origin *A* using position coordinates y_B and x_C. From Fig. 11–7*b*, x_C can be related to θ by the "law of cosines." Hence,

$$(0.7)^2 = (0.6)^2 + x_C^2 - 2(0.6)x_C \cos\theta \qquad (1)$$

$$0 = 0 + 2x_C\,\delta x_C - 1.2\,\delta x_C \cos\theta + 1.2x_C \sin\theta\,\delta\theta$$

$$\delta x_C = \frac{1.2x_C \sin\theta}{1.2\cos\theta - 2x_C}\,\delta\theta \qquad (2)$$

Also,

$$y_B = 0.6\sin\theta$$

$$\delta y_B = 0.6\cos\theta\,\delta\theta \qquad (3)$$

Virtual-Work Equation. When y_B and x_C undergo *positive* virtual displacements δy_B and δx_C, $\mathbf{F}_s$ and 200 N do *negative work,* since they both act in the opposite direction to $\delta\mathbf{y}_B$ and $\delta\mathbf{x}_C$. Hence,

$$\delta U = 0; \qquad -200\,\delta y_B - F_s\,\delta x_C = 0$$

Substituting Eqs. (2) and (3) into this equation, factoring out $\delta\theta$, and solving for F_s yields

$$-200(0.6\cos\theta\,\delta\theta) - F_s\frac{1.2x_C\sin\theta}{1.2\cos\theta - 2x_C}\,\delta\theta = 0$$

$$F_s = \frac{-120\cos\theta(1.2\cos\theta - 2x_C)}{1.2x_C\sin\theta}$$

At the required equilibrium position $\theta = 45°$, the corresponding value of x_C can be found by using Eq. (1), in which case

$$x_C^2 - 1.2\cos 45°\,x_C - 0.13 = 0$$

Solving for the positive root yields

$$x_C = 0.981 \text{ m}$$

Thus,

$$F_s = \frac{-120\cos 45°\,[1.2\cos 45° - 2(0.981)]}{1.2(0.981)\sin 45°} = 113.5 \text{ N} \quad \textit{Ans.}$$

Example 11–4

Using the principle of virtual work, determine the equilibrium position of the two-bar linkage shown in Fig. 11–8*a*. Neglect the weight of the links.

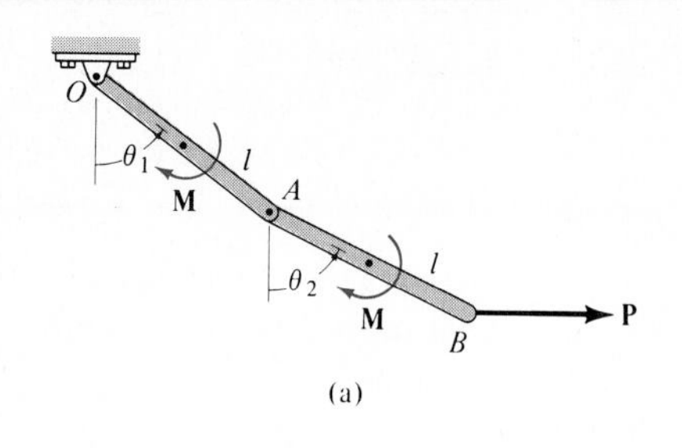

Solution

The system has two degrees of freedom, since the *independent coordinates* θ_1 and θ_2 must be known to locate the position of both links. The position coordinate x_B, measured from the fixed point O, is used to specify the location of **P**, Fig. 11–8*b* and *c*.

If θ_1 is held *fixed* and θ_2 varies by an amount $\delta\theta_2$, as shown in Fig. 11–8*b*, the virtual-work equation becomes

$$[\delta U = 0]_{\theta_2}; \qquad P(\delta x_B)_{\theta_2} - M\,\delta\theta_2 = 0 \qquad (1)$$

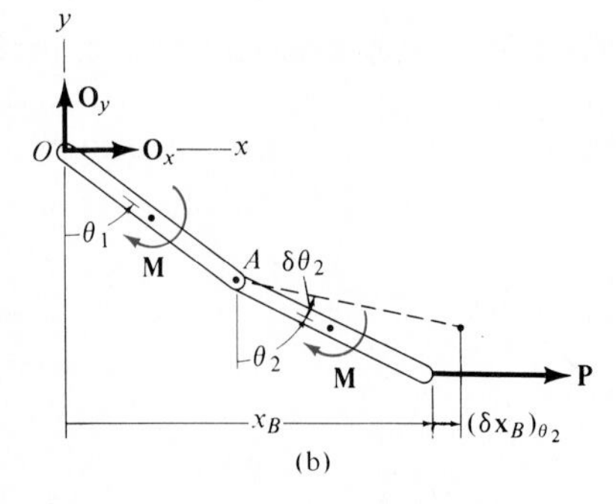

where P and M represent the magnitudes of the applied force and couple acting on link AB.

When θ_2 is held *fixed* and θ_1 varies by an amount $\delta\theta_1$, as shown in Fig. 11–8*c*, the virtual-work equation becomes

$$[\delta U = 0]_{\theta_1}; \qquad P(\delta x_B)_{\theta_1} - M\,\delta\theta_1 - M\,\delta\theta_1 = 0 \qquad (2)$$

The *position coordinate* x_B may be related to the independent coordinates θ_1 and θ_2 by the equation

$$x_B = l\sin\theta_1 + l\sin\theta_2 \qquad (3)$$

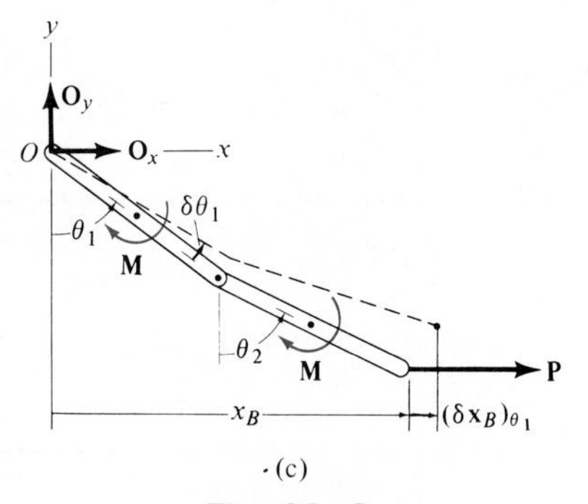

Fig. 11–8

To obtain the variation of δx_B in terms of $\delta\theta_2$, it is necessary to take the *partial derivative* of x_B with respect to θ_2 since x_B is a function of both θ_1 and θ_2. Hence,

$$\frac{\partial x_B}{\partial \theta_2} = l\cos\theta_2 \qquad (\delta x_B)_{\theta_2} = l\cos\theta_2\,\delta\theta_2$$

Substituting into Eq. (1), we have

$$(Pl\cos\theta_2 - M)\,\delta\theta_2 = 0$$

Since $\delta\theta_2 \neq 0$, then

$$\theta_2 = \cos^{-1}\left(\frac{M}{Pl}\right) \qquad \textit{Ans.}$$

Using Eq. (3) to obtain the variation of x_B with θ_1 yields

$$\frac{\partial x_B}{\partial \theta_1} = l\cos\theta_1 \qquad (\delta x_B)_{\theta_1} = l\cos\theta_1\,\delta\theta_1$$

Substituting into Eq. (2), we have

$$(Pl\cos\theta_1 - 2M)\,\delta\theta_1 = 0$$

Since $\delta\theta_1 \neq 0$, then

$$\theta_1 = \cos^{-1}\left(\frac{2M}{Pl}\right) \qquad \textit{Ans.}$$

Problems

11–1. Compute the force developed in the spring required to keep the 6-kg rod in equilibrium when $\theta = 30°$.

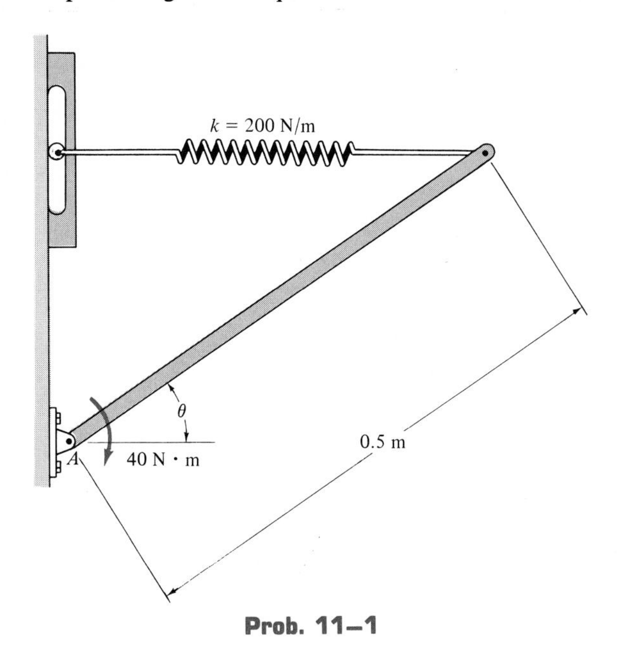

Prob. 11–1

11–2. The thin rod of weight **W** rests against the smooth wall and floor. Determine the magnitude of force **P** needed to hold it in equilibrium.

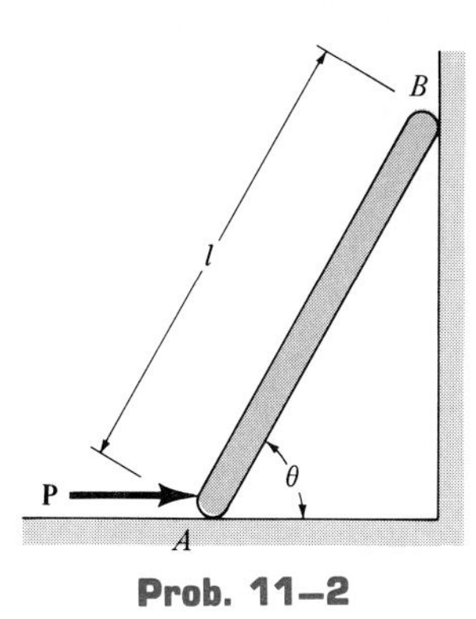

Prob. 11–2

11–3. The members of the mechanism are pin-connected. If a horizontal force of 400 N acts at A, determine the angle θ for equilibrium. The spring is unstretched when $\theta = 90°$.

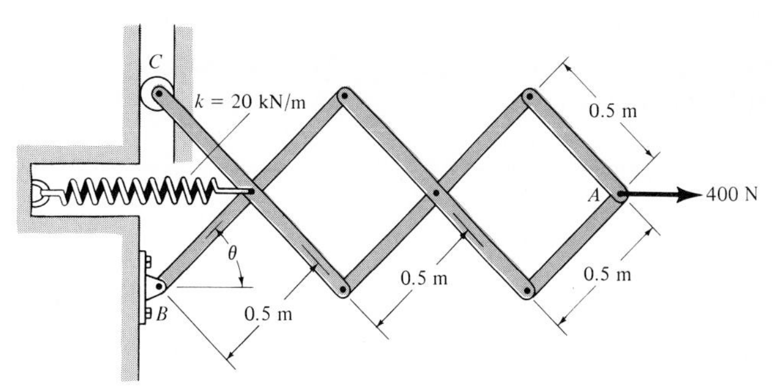

Prob. 11–3

***11–4.** The spring has an unstretched length of 0.3 m. Determine the angle θ for equilibrium if the uniform links each have a mass of 5 kg.

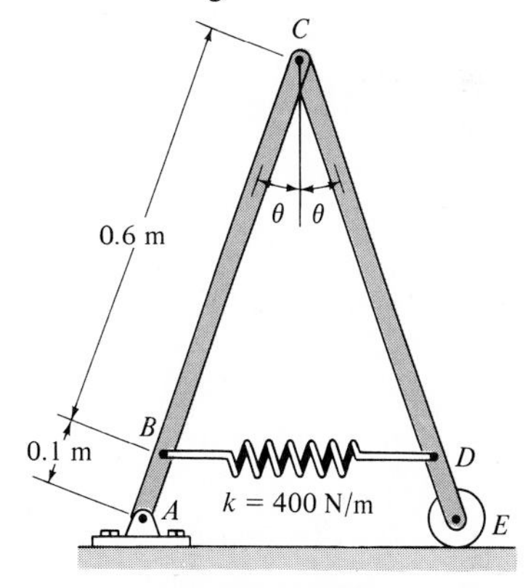

Prob. 11–4

11–5. If each of the three links of the mechanism has a weight of 20 lb, determine the angle θ for equilibrium. The spring, which always remains horizontal, is unstretched when $\theta = 0°$.

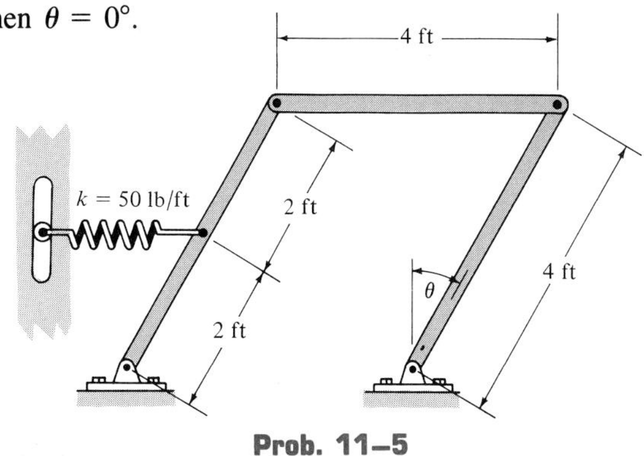

Prob. 11–5

11–6. The uniform rod OA has a weight of 10 lb. When the rod is in the vertical position, $\theta = 0°$, the spring is unstretched. Determine the angle θ for equilibrium if the end of the spring wraps around the periphery of the disk as the disk turns.

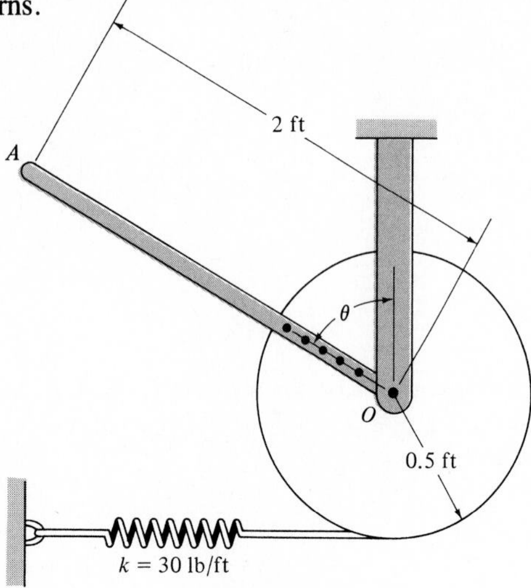

Prob. 11–6

11–7. If a force $P = 30$ lb is applied perpendicular to the handle of the toggle press, determine the compressive force developed at C; $\theta = 30°$.

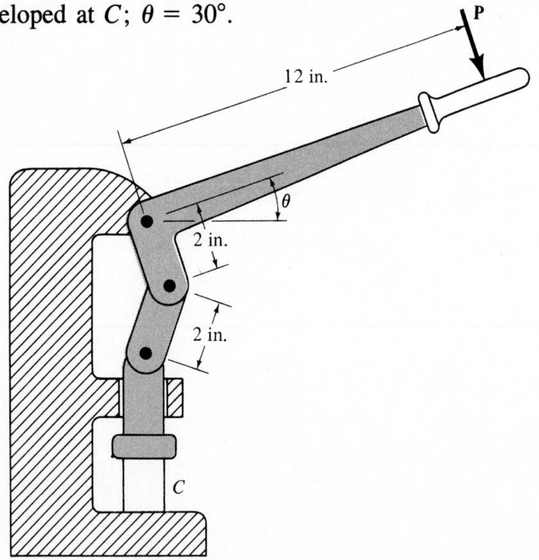

Prob. 11–7

***11–8.** The punch press consists of the ram R, connecting rod AB, and a flywheel. If a torque of $M = 50$ N · m is applied to the flywheel as shown, determine the force **F** applied at the ram to hold the rod in the position $\theta = 60°$.

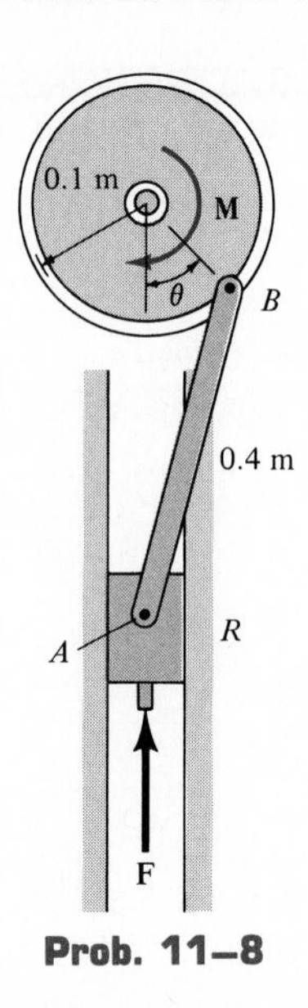

Prob. 11–8

11–9. Determine the force **P** required to lift the 30-lb block using the differential hoist. The lever arm is fixed to the upper pulley and turns with it.

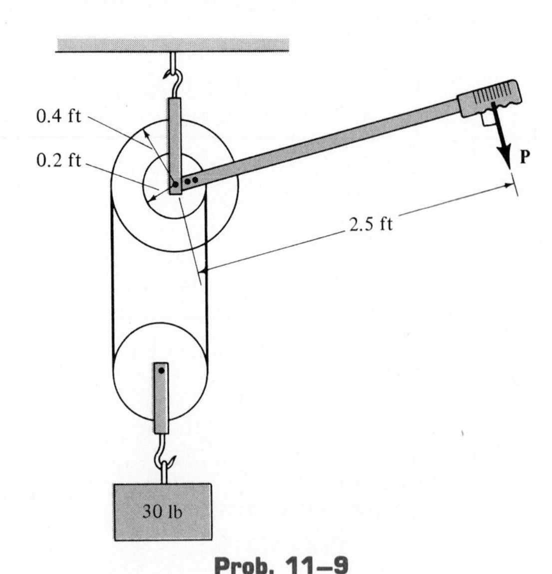

Prob. 11–9

11–10. The three-bar mechanism of negligible weight is subjected to a couple $M_A = 8$ lb · ft. Determine the magnitude of the couple $\mathbf{M}_D$ needed to maintain the equilibrium position $\theta = 30°$.

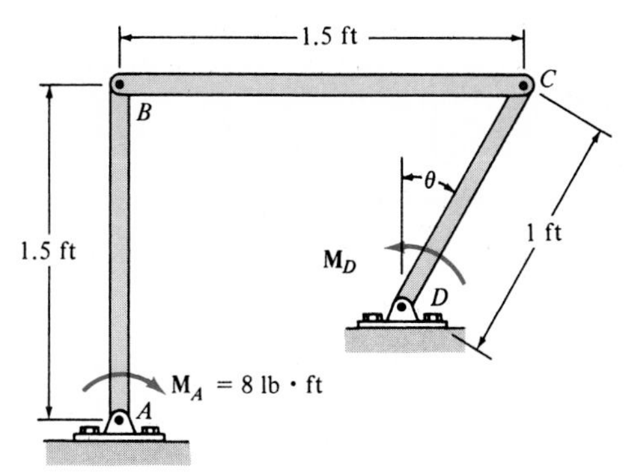

Prob. 11–10

11–11. Determine the force **F** needed to lift the block having a weight of 100 lb. *Hint:* First show that the coordinates s_A and s_B are related to the *constant* vertical length l of the cord by the equation $l = s_A + 2s_B$.

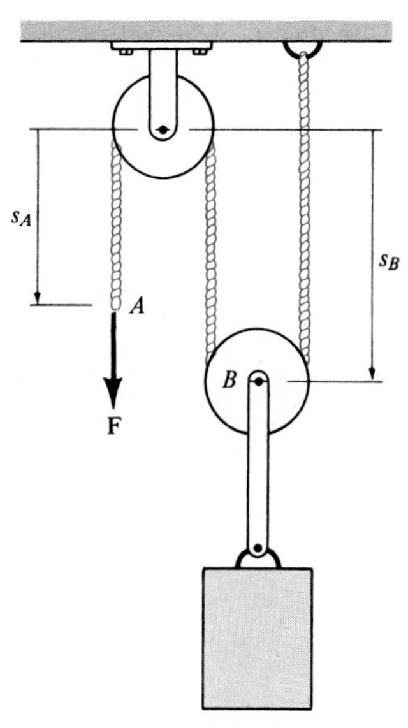

Prob. 11–11

***11–12.** Determine the force **F** acting on the cord which is required to maintain equilibrium of the horizontal 20-kg bar CB. *Hint:* First show that the coordinates s_A and s_B are related to the *constant* vertical length l of the cord by the equation $5s_B - s_A = l$.

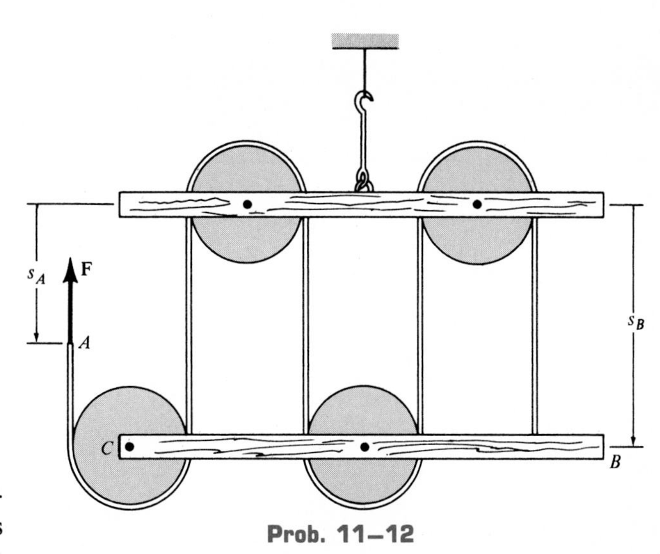

Prob. 11–12

11–13. The piston C moves vertically between the two smooth walls. If the spring has a stiffness of $k = 15$ lb/in. and is unstretched when $\theta = 0°$, determine the couple **M** that must be applied to AB to hold the mechanism in equilibrium when $\theta = 30°$.

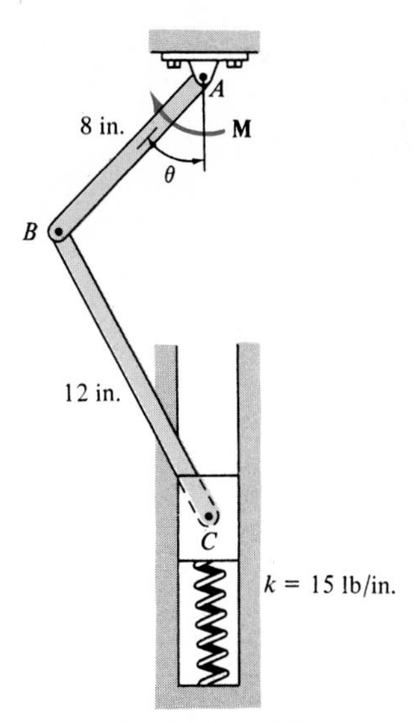

Prob. 11–13

11–14. Determine the vertical force **P** needed to hold the 7-kg block in equilibrium. The pulley has a mass of 2 kg. Neglect the weight of the cords. *Hint:* First show that the position coordinates s_A and s_B are related to the *constant* vertical length l of the cord by the equation $l = 2s_B - s_A$.

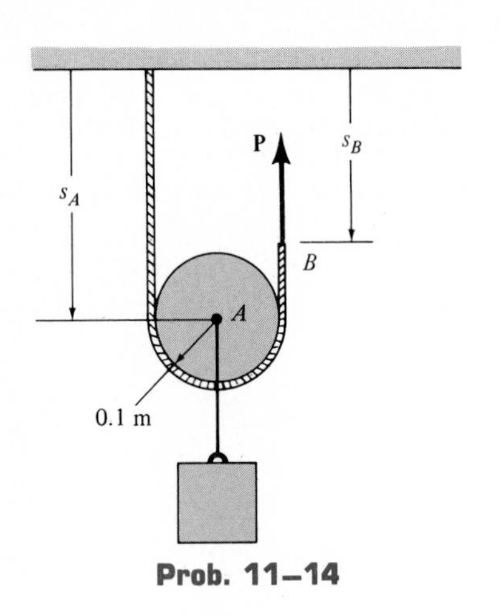

Prob. 11–14

***11–16.** The *Roberval balance* is in equilibrium when no weights are placed on the pans A and B. If two masses m_A and m_B are placed at *any* location a and b on the pans, show that equilibrium is maintained if $m_A d_A = m_B d_B$.

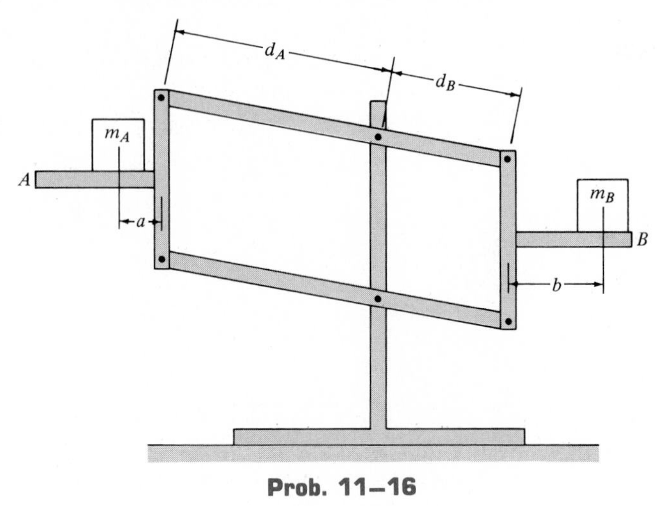

Prob. 11–16

11–15. The machine shown is used for forming metal plates. It consists of two toggles ABC and DEF, which are operated by the hydraulic cylinder H. The toggles push the movable bar G forward, pressing the plate p into the cavity. If the force which the plate exerts on the head is $P = 8$ kN, determine the force **F** in the hydraulic cylinder when $\theta = 30°$.

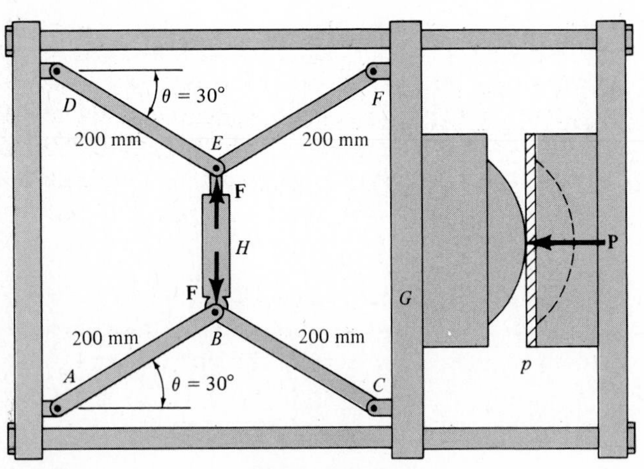

Prob. 11–15

11–17. The truck is weighed on the highway inspection scale. If a known mass m is placed a distance s from the fulcrum B of the scale, determine the mass of the truck m_t if its center of gravity is located at a distance d from point C. When the scale is empty, the weight of the lever ABC balances the scale CDE.

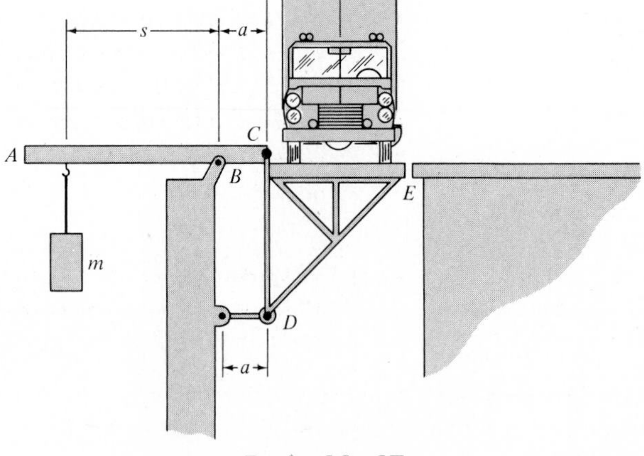

Prob. 11–17

11–18. Determine the force **P** that must be applied to the cord wrapped around the shaft at C, which is necessary to lift the mass m. As the mass is lifted, the pulley rolls on a cord that winds up on shaft B and unwinds from shaft A.

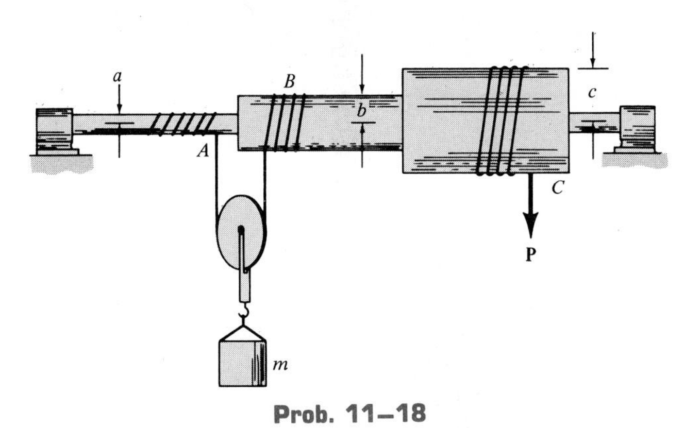

Prob. 11–18

11–19. The chain puller is used to draw two ends of a chain together in order to attach the "master link." The device is operated by turning the screw S, which pushes the bar AB downward, thereby drawing the tips C and D towards one another. If the sliding contacts at A and B are smooth, determine the force **F** maintained by the screw at E which is required to develop a drawing tension of 120 lb in the chains.

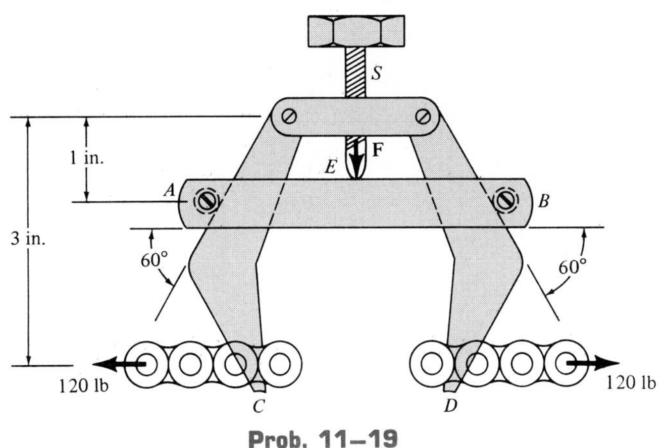

Prob. 11–19

***11–20.** A 3-lb disk is attached to the end of rod ABC. If the rod is supported by a smooth slider block at C and rod BD, determine the angle θ for equilibrium. Neglect the weight of the rods and the slider.

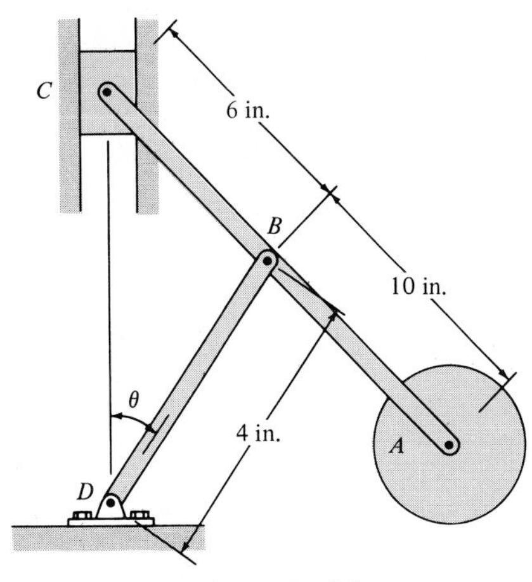

Prob. 11–20

11–21. A horizontal force acts on the end of the link as shown. Determine the angles θ_1 and θ_2 for equilibrium of the two links. Each link is uniform and has a mass m.

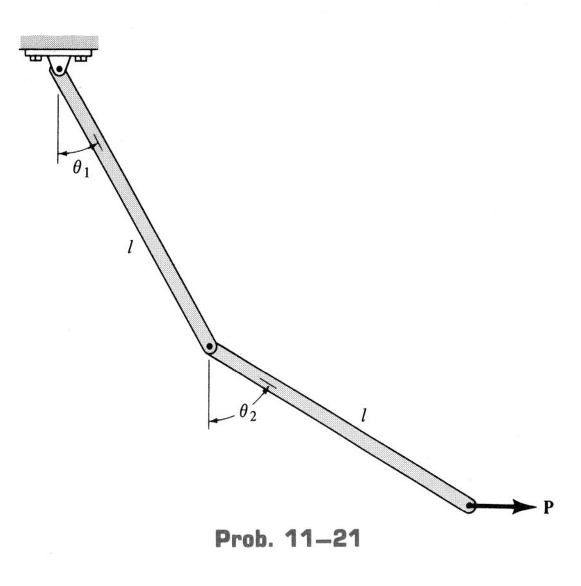

Prob. 11–21

11–22. The loading arm is used to transfer oil from a ship to land. When the device is in use, the oil flows through the nozzle at C and exits at D as shown by the arrows. If each of the arms AB and BC has a total mass m and they are pin-connected at A, B, and I, determine the mass of blocks J and F that will keep the system in balance for any angle θ or ϕ. J is attached to the extended portion of arm AB and has a center of gravity at G_1; whereas F has a center of gravity at G_2 and is attached to a link that is pin-connected at A and E. Neglect the weight of links AEH, EI, and the extended portions IB of BC and KA of AB. Assume the weight of AB and BC acts through their centers.

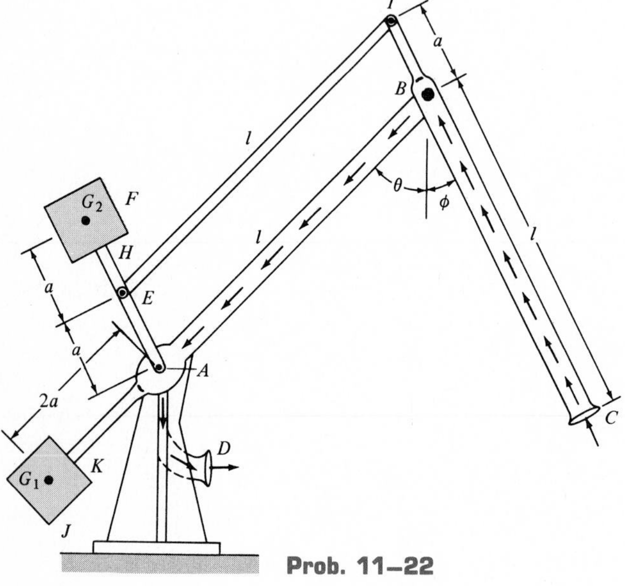

Prob. 11–22

11–23. Rods AB and BC have a center of gravity located at their midpoints. If all contacting surfaces are smooth and BC has a mass of 100 kg, determine the appropriate mass of AB required for equilibrium.

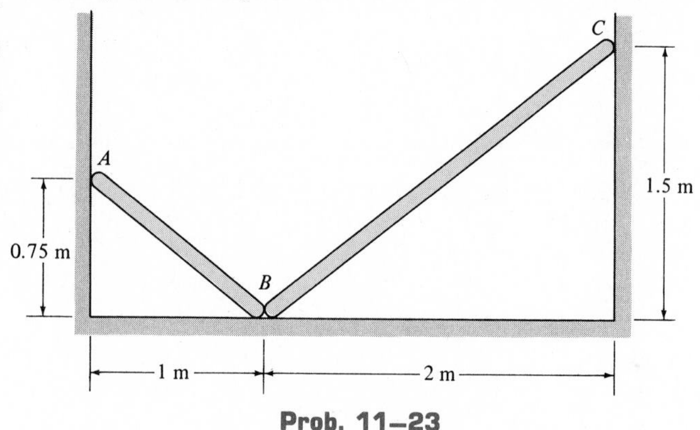

Prob. 11–23

***11–24.** Determine the mass of A and B required to hold the 400-g desk lamp in balance for any angle θ or ϕ. Neglect the weight of the mechanism and the size of the lamp.

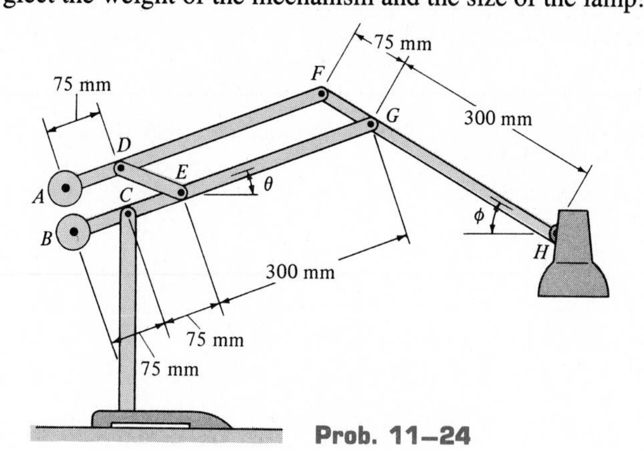

Prob. 11–24

Mathematical Expressions

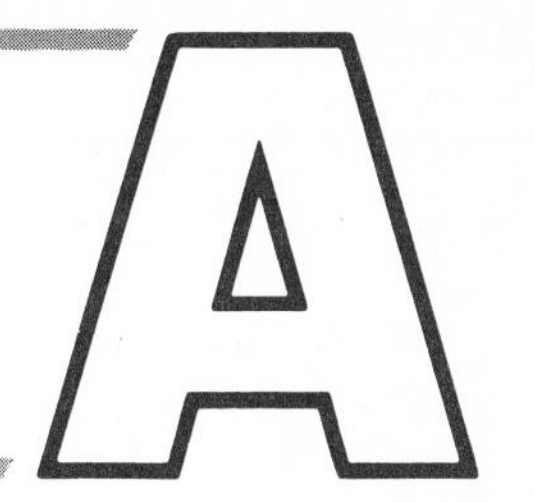

Quadratic Formula:

If $ax^2 + bx + c = 0$, then $x = \dfrac{-b \pm \sqrt{b^2 - 4ac}}{2a}$

Trigonometric Identities:

$\sin^2 \theta + \cos^2 \theta = 1$

$\sin 2\theta = 2 \sin \theta \cos \theta$

$\cos 2\theta = \cos^2 \theta - \sin^2 \theta$

$\tan \theta = \dfrac{\sin \theta}{\cos \theta}$

$1 + \tan^2 \theta = \sec^2 \theta$

$1 + \cot^2 \theta = \csc^2 \theta$

Integrals:

$\int x^n \, dx = \dfrac{x^{n+1}}{n+1}, \; n \neq -1$

$\int \sin x \, dx = -\cos x$

$\int \cos x \, dx = \sin x$

Derivatives:

$\dfrac{d}{dx}(u^n) = nu^{n-1}\dfrac{du}{dx}$

$\dfrac{d}{dx}(uv) = u\dfrac{dv}{dx} + v\dfrac{du}{dx}$

$\dfrac{d}{dx}\left(\dfrac{u}{v}\right) = \dfrac{v\dfrac{du}{dx} - u\dfrac{dv}{dx}}{v^2}$

$\dfrac{d}{dx}(\sin u) = \cos u\dfrac{du}{dx}$

$\dfrac{d}{dx}(\cos u) = -\sin u\dfrac{du}{dx}$

$\dfrac{d}{dx}(\tan u) = \sec^2 u\dfrac{du}{dx}$

$\dfrac{d}{dx}(\cot u) = -\csc^2 u\dfrac{du}{dx}$

$\dfrac{d}{dx}(\sec u) = \tan u \sec u\dfrac{du}{dx}$

$\dfrac{d}{dx}(\csc u) = -\csc u \cot u\dfrac{du}{dx}$

Product and Principal Moments of Inertia for an Area

B

Product of Inertia for an Area B.1

In general, the moment of inertia for an area is different for every axis about which it is computed. In some applications of structural design it is necessary to know the orientation of those axes which give, respectively, the maximum and minimum moment of inertia for the area. The method for determining this is discussed in the next section. To use this method, however, one must first compute the product of inertia for the area as well as its moments of inertia for given x, y axes.

The product of inertia for an element of area located at point (x, y) as shown in Fig. B–1, is defined as $dI_{xy} = xy\ dA$. Thus, for the entire area A, the *product of inertia* is

$$I_{xy} = \int xy\ dA \tag{B–1}$$

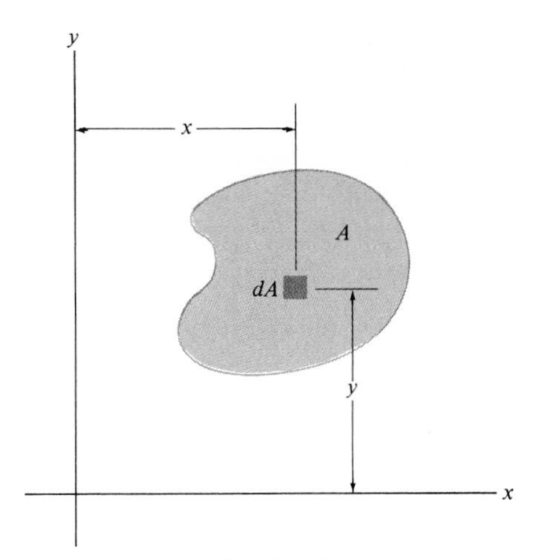

Fig. B–1

If the element of area chosen has a differential size in two directions, as shown in Fig. B–1, a double integration must be performed to evaluate I_{xy}. Most often, however, it is easier to choose an element having a differential size or thickness in only one direction, in which case the evaluation requires only a single integration (see Example B–1).

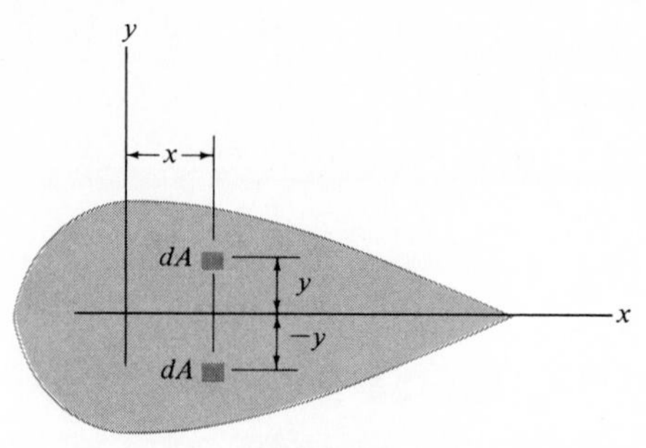

Fig. B–2

Like the moment of inertia, the product of inertia has units of length raised to the fourth power, e.g., m^4, mm^4 or ft^4, $in.^4$. However, since x or y may be a negative quantity, while the element of area is always positive, the product of inertia may be positive, negative, or zero, depending upon the location and orientation of the coordinate axes. For example, the product of inertia I_{xy} for an area will be *zero* if either the x or y axis is an axis of *symmetry* for the area. To show this, consider the shaded area in Fig. B–2, where for every element dA located at point (x, y) there is a corresponding element dA located at $(x, -y)$. Since the products of inertia for these elements are respectively $xy\,dA$ and $-xy\,dA$, the algebraic sum or integration of all the elements of area that are chosen in this way will cancel each other. Consequently, the product of inertia for the total area becomes zero. It also follows from the definition of I_{xy} that the "sign" of this quantity depends upon the quadrant where the area is located, Fig. B–3.

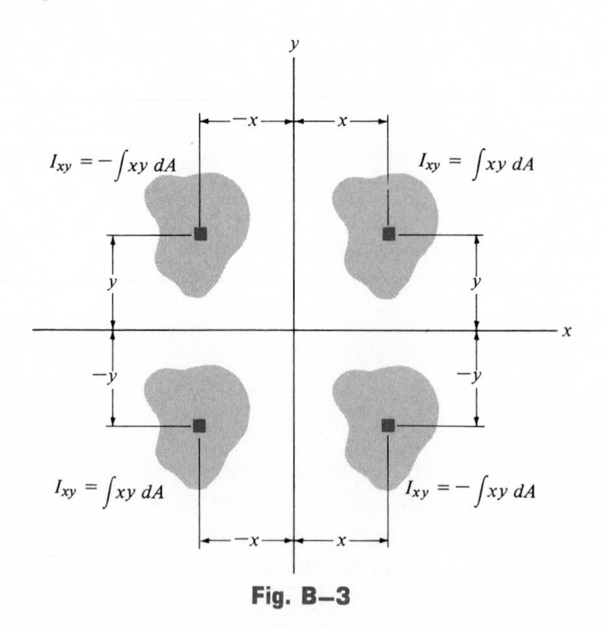

Fig. B–3

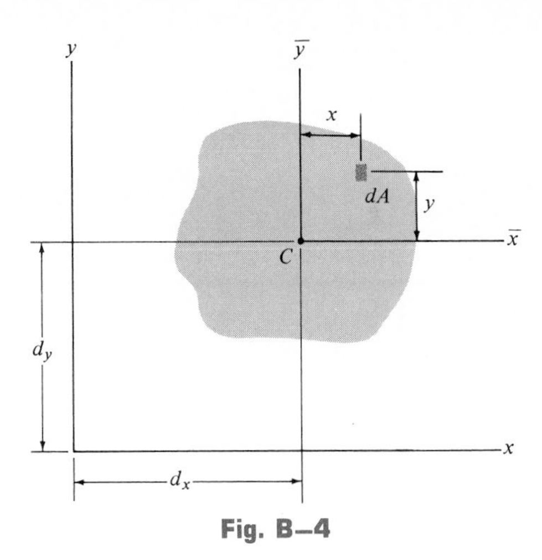

Fig. B–4

Parallel-Axis Theorem. Consider the shaded area shown in Fig. B–4, where $\bar{x}$ and $\bar{y}$ represent a set of axes passing through the *centroid* of the area, and x and y represent a corresponding set of parallel axes. Since the product of inertia of dA with respect to the x and y axes is $dI_{xy} = (x + d_x)(y + d_y)\, dA$, then for the entire area,

$$I_{xy} = \int (x + d_x)(y + d_y)\, dA$$

$$= \int xy\, dA + d_x \int y\, dA + d_y \int x\, dA + d_x d_y \int dA$$

The first term on the right represents the product of inertia of the area with respect to the centroidal axis, $\bar{I}_{\bar{x}\bar{y}}$. The integrals in the second and third terms are zero since the moments of the area are taken about the centroidal axis. Realizing that the fourth integral represents the total area A, we therefore have as the final result,

$$I_{xy} = \bar{I}_{\bar{x}\bar{y}} + A d_x d_y \tag{B–2}$$

The similarity between this equation and the parallel-axis theorem for moments of inertia should be noted. In particular, it is important that the *algebraic signs* for d_x and d_y be maintained when applying Eq. B–2. As illustrated in Example B–2, the parallel-axis theorem finds important application in determining the product of inertia of a *composite area* with respect to a set of x, y axes.

Example B–1

Determine the product of inertia I_{xy} of the triangle shown in Fig. B–5*a*.

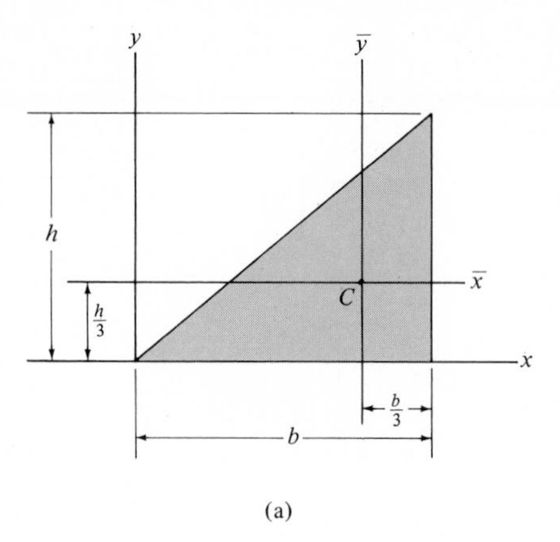

(a)

Solution I

Consider the differential element that has a thickness dx and area $dA = y\,dx$, Fig. B–5*b*. The product of inertia of the element about the x, y axes is determined using the parallel-axis theorem.

$$dI_{xy} = d\bar{I}_{\tilde{x}\tilde{y}} + dA\,\tilde{x}\tilde{y}$$

where $(\tilde{x}, \tilde{y})$ locates the *centroid* of the element. Since $dI_{\tilde{x}\tilde{y}} = 0$, due to symmetry, and $\tilde{x} = x$, $\tilde{y} = y/2$, then

$$dI_{xy} = 0 + (y\,dx)x\left(\frac{y}{2}\right) = \left(\frac{h}{b}x\,dx\right)x\left(\frac{h}{2b}x\right)$$

$$= \frac{h^2}{2b^2}x^3\,dx$$

Integrating with respect to x from $x = 0$ to $x = b$ yields

$$I_{xy} = \frac{h^2}{2b^2}\int_0^b x^3\,dx = \frac{b^2h^2}{8} \qquad \textit{Ans.}$$

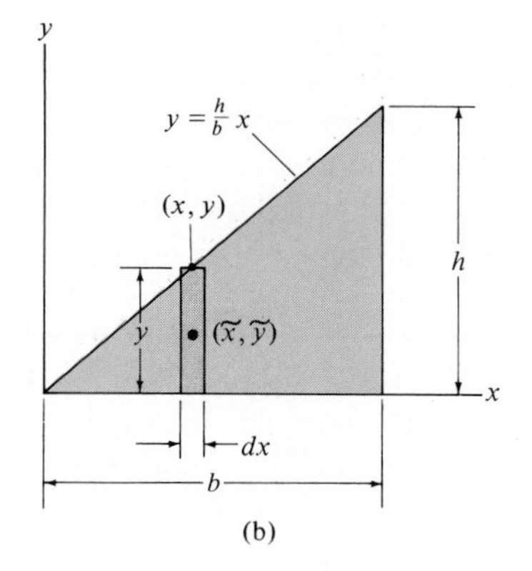

(b)

Solution II

Consider a differential element that has a thickness dy and area $dA = (b - x)\,dy$, as shown in Fig. B–5*c*. The *centroid* is located at point $\tilde{x} = x + (b - x)/2 = (b + x)/2$, $\tilde{y} = y$, so that the product of inertia of the element becomes

$$dI_{xy} = d\bar{I}_{\tilde{x}\tilde{y}} + dA\,\tilde{x}\tilde{y}$$

$$= 0 + (b - x)\,dy\left(\frac{b + x}{2}\right)y$$

$$= \left(b - \frac{b}{h}y\right)dy\left(\frac{b + \frac{b}{h}y}{2}\right)y = \frac{1}{2}y\left(b^2 - \frac{b^2}{h^2}y^2\right)dy$$

Integrating with respect to y from $y = 0$ to $y = h$ yields

$$I_{xy} = \frac{1}{2}\int_0^h y\left(b^2 - \frac{b^2}{h^2}y^2\right)dy = \frac{b^2h^2}{8} \qquad \textit{Ans.}$$

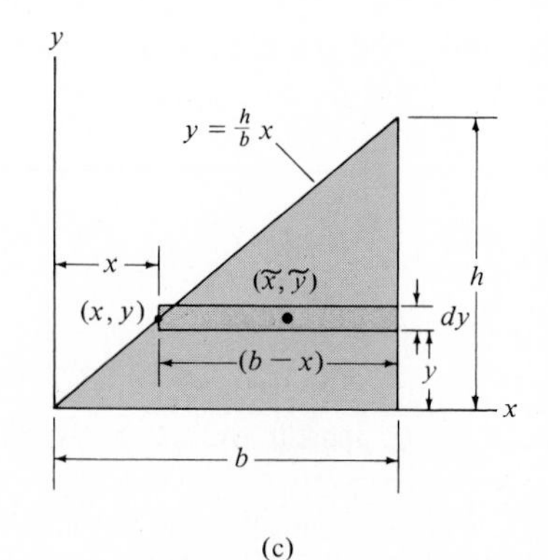

(c)

Fig. B–5

Example B–2

Compute the product of inertia of the beam's cross-sectional area, shown in Fig. B–6*a*, about the *x* and *y* centroidal axes.

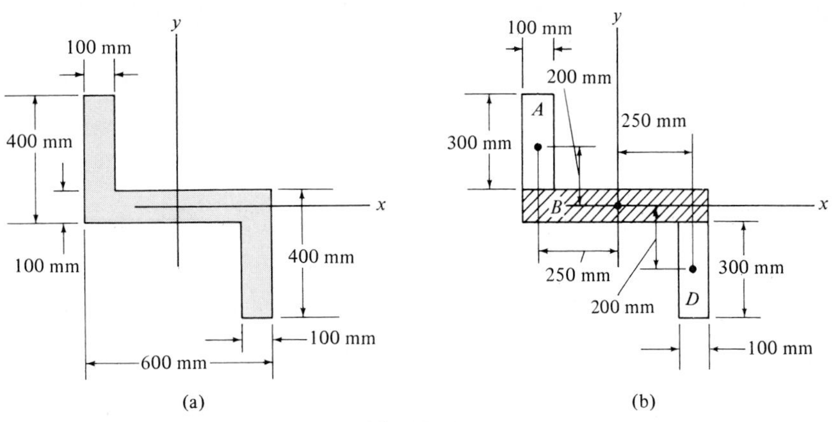

Fig. B–6

Solution

As in Example 10–5, the cross section can be considered as three composite rectangular areas *A*, *B*, and *D*, Fig. B–6*b*. The coordinates for the centroid of each of these rectangles are shown in the figure. Due to symmetry, the product of inertia of *each rectangle* is *zero* about a set of $\bar{x}$, $\bar{y}$ axes that pass through the rectangle's centroid. Hence, application of the parallel-axis theorem to each of the rectangles yields

Rectangle A

$$
\begin{aligned}
I_{xy} &= \bar{I}_{\bar{x}\bar{y}} + A d_x d_y \\
&= 0 + (300)(100)(-250)(200) \\
&= -15(10^8)\ \text{mm}^4
\end{aligned}
$$

Rectangle B

$$
\begin{aligned}
I_{xy} &= \bar{I}_{\bar{x}\bar{y}} + A d_x d_y \\
&= 0 + 0 \\
&= 0
\end{aligned}
$$

Rectangle D

$$
\begin{aligned}
I_{xy} &= \bar{I}_{\bar{x}\bar{y}} + A d_x d_y \\
&= 0 + (300)(100)(250)(-200) \\
&= -15(10^8)\ \text{mm}^4
\end{aligned}
$$

The product of inertia for the entire cross section is thus

$$I_{xy} = [-15(10^8)] + 0 + [-15(10^8)] = -30(10^8)\ \text{mm}^4 \qquad \textit{Ans.}$$

B.2 Moments of Inertia for an Area About Inclined Axes

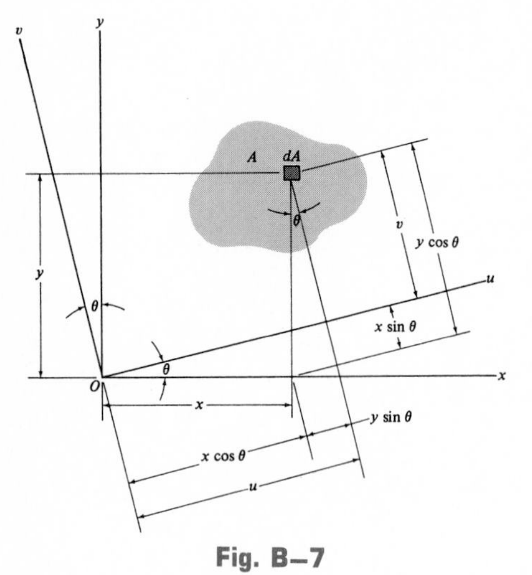

Fig. B–7

In structural mechanics, it is sometimes necessary to calculate the moments and product of inertia I_u, I_v, and I_{uv} for an area with respect to a set of inclined u and v axes when the values for θ, I_x, I_y, and I_{xy} are *known*. From Fig. B–7 the perpendicular distance from the area element dA may be related to the axes of the two coordinate systems by using the *transformation equations*

$$u = x\cos\theta + y\sin\theta \qquad v = y\cos\theta - x\sin\theta$$

Using these equations, the moments and product of inertia of dA about the u and v axes become

$$dI_u = v^2\,dA = (y\cos\theta - x\sin\theta)^2\,dA$$
$$dI_v = u^2\,dA = (x\cos\theta + y\sin\theta)^2\,dA$$
$$dI_{uv} = uv\,dA = (x\cos\theta + y\sin\theta)(y\cos\theta - x\sin\theta)\,dA$$

Expanding each expression and integrating, realizing that $I_x = \int y^2\,dA$, $I_y = \int x^2\,dA$, and $I_{xy} = \int xy\,dA$, we obtain

$$I_u = I_x\cos^2\theta + I_y\sin^2\theta - 2I_{xy}\sin\theta\cos\theta$$
$$I_v = I_x\sin^2\theta + I_y\cos^2\theta + 2I_{xy}\sin\theta\cos\theta$$
$$I_{uv} = I_x\sin\theta\cos\theta - I_y\sin\theta\cos\theta + I_{xy}(\cos^2\theta - \sin^2\theta)$$

These equations may be simplified by using the trigonometric identities $\sin 2\theta = 2\sin\theta\cos\theta$ and $\cos 2\theta = \cos^2\theta - \sin^2\theta$, in which case

$$I_u = \frac{I_x + I_y}{2} + \frac{I_x - I_y}{2}\cos 2\theta - I_{xy}\sin 2\theta$$
$$I_v = \frac{I_x + I_y}{2} - \frac{I_x - I_y}{2}\cos 2\theta + I_{xy}\sin 2\theta \qquad \text{(B–3)}$$
$$I_{uv} = \frac{I_x - I_y}{2}\sin 2\theta + I_{xy}\cos 2\theta$$

Note that if the first and second equations are added together, it is seen that the polar moment of inertia about the z axis passing through point O is *independent* of the orientation of the u and v axes, i.e.,

$$J_O = I_u + I_v = I_x + I_y$$

Principal Moments of Inertia. From Eqs. B–3, it may be seen that I_u, I_v, and I_{uv} depend upon the angle of inclination, θ, of the u, v axes. We will now determine the orientation of the u, v axes about which the moments of inertia for the area, I_u and I_v, are maximum and minimum. This particular set of axes is called the *principal axes* of the area, and the corresponding moments of inertia with respect to these axes are called the *principal moments of inertia*. In general, there is a set of principal axes for every chosen origin O, although in structural mechanics the area's centroid is an important location for O.

The angle $\theta = \theta_p$, which defines the orientation of the principal axes for the area, may be found by differentiating the first of Eqs. B–3 with respect to θ and setting the result equal to zero. Thus,

$$\frac{dI_u}{d\theta} = -2\left(\frac{I_x - I_y}{2}\right)\sin 2\theta - 2I_{xy}\cos 2\theta = 0$$

Therefore, at $\theta = \theta_p$,

$$\tan 2\theta_p = \frac{-I_{xy}}{\dfrac{I_x - I_y}{2}} \tag{B–4}$$

This equation has two roots, θ_{p_1} and θ_{p_2}, which specify the inclination of the principal axes. Because of the nature of the tangent, the values of $2\theta_{p_1}$ and $2\theta_{p_2}$ are 180° apart, so that θ_{p_1} and θ_{p_2} are 90° apart. Assuming that I_{xy} and $(I_x - I_y)$ are both positive quantities, the sine and cosine of $2\theta_{p_1}$ and $2\theta_{p_2}$ can be obtained from the triangles shown in Fig. B–8, which are based upon Eq. B–4.

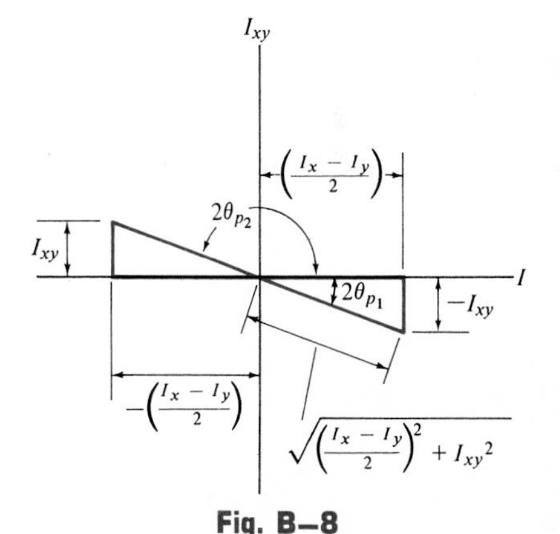

Fig. B–8

For θ_{p_1}

$$\sin 2\theta_{p_1} = -I_{xy}\Big/\sqrt{\left(\frac{I_x - I_y}{2}\right)^2 + I_{xy}^2}$$

$$\cos 2\theta_{p_1} = \left(\frac{I_x - I_y}{2}\right)\Big/\sqrt{\left(\frac{I_x - I_y}{2}\right)^2 + I_{xy}^2}$$

For θ_{p_2}

$$\sin 2\theta_{p_2} = I_{xy}\Big/\sqrt{\left(\frac{I_x - I_y}{2}\right)^2 + I_{xy}^2}$$

$$\cos 2\theta_{p_2} = -\left(\frac{I_x - I_y}{2}\right)\Big/\sqrt{\left(\frac{I_x - I_y}{2}\right)^2 + I_{xy}^2}$$

If these two sets of trigonometric relations are substituted into the first or second of Eqs. B–3 and simplified, the result is

$$I_{\substack{\max \\ \min}} = \frac{I_x + I_y}{2} \pm \sqrt{\left(\frac{I_x - I_y}{2}\right)^2 + I_{xy}^2} \tag{B–5}$$

Depending upon the sign chosen, this result gives the maximum or minimum moment of inertia for the area. Furthermore, if the above trigonometric relations for θ_{p_1} and θ_{p_2} are substituted into the third of Eqs. B–3, it may be seen that $I_{uv} = 0$; that is, the *product of inertia with respect to the principal axes is zero*. Since it was indicated in Sec. B–1 that the product of inertia is zero with respect to any symmetrical axis, it therefore follows that *any symmetrical axis represents a principal axis of inertia for the area*.

Example B–3

Determine the principal moments of inertia for the beam's cross-sectional area shown in Fig. B–9*a* with respect to an axis passing through the centroid.

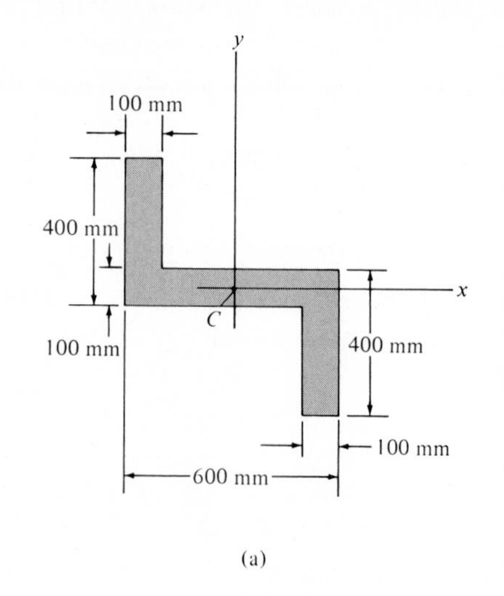

(a)

Solution

The moments and product of inertia of the cross section with respect to the x, y axes have been computed in Examples 10–5 and B–2. The results are

$$I_x = 29(10^8)\ \text{mm}^4 \qquad I_y = 56(10^8)\ \text{mm}^4 \qquad I_{xy} = -30(10^8)\ \text{mm}^4$$

Using Eq. B–4, the angles of inclination of the principal axes u and v are and v are

$$\tan 2\theta_p = \frac{-I_{xy}}{\dfrac{I_x - I_y}{2}} = \frac{30(10^8)}{\dfrac{29(10^8) - 56(10^8)}{2}} = -2.22$$

$$2\theta_{p_1} = -65.8° \qquad \text{and} \qquad 2\theta_{p_2} = 114.2°$$

Thus, as shown in Fig. B–9*b*,

$$\theta_{p_1} = -32.9° \qquad \text{and} \qquad \theta_{p_2} = 57.1°$$

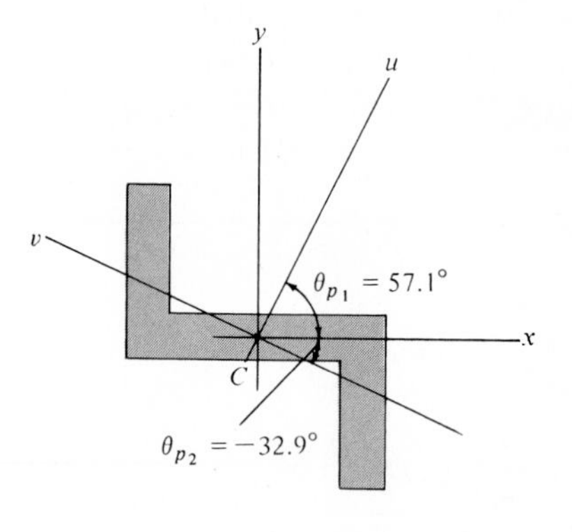

(b)

Fig. B–9

The principal moments of inertia with respect to the u and v axes are determined by using Eq. B–5. Hence,

$$I_{\substack{\max \\ \min}} = \frac{I_x + I_y}{2} \pm \sqrt{\left(\frac{I_x - I_y}{2}\right)^2 + I_{xy}^2}$$

$$= \frac{29(10^8) + 56(10^8)}{2} \pm \sqrt{\left[\frac{29(10^8) - 56(10^8)}{2}\right]^2 + [-30(10^8)]^2}$$

$$= 42.5(10^8) \pm 32.9(10^8)$$

or

$$(I_u)_{\max} = 75.4(10^8)\ \text{mm}^4 \qquad (I_v)_{\min} = 9.6(10^8)\ \text{mm}^4 \qquad \textit{Ans.}$$

Specifically, the maximum moment of inertia, $(I_u)_{\max} = 75.4(10^8)$ mm^4, occurs with respect to the u axis, since *by inspection* most of the cross-sectional area is farthest away from this axis. Stated in a general manner, $(I_u)_{\max}$ occurs about an axis located within $\pm 45°$ of the axis (x or y) which has the largest I (I_x or I_y). (To conclude this mathematically, substitute the data with $\theta = 57.1°$ into the first of Eqs. B–3.)

Mohr's Circle for Moments of Inertia B.3

Equations B–3 to B–5 have a graphical solution that is convenient to use and generally easy to remember. Squaring the first and third of Eqs. B–3 and adding, it is found that

$$\left(I_u - \frac{I_x + I_y}{2}\right)^2 + I_{uv}^2 = \left(\frac{I_x - I_y}{2}\right)^2 + I_{xy}^2 \qquad \text{(B–6)}$$

In a given problem, I_u and I_{uv} are *variables,* and I_x, I_y, and I_{xy} are *known constants*. Thus, Eq. B–6 may be written in compact form as

$$(I_u - a)^2 + I_{uv}^2 = R^2$$

When this equation is plotted, the resulting graph represents a *circle* of radius

$$R = \sqrt{\left(\frac{I_x - I_y}{2}\right)^2 + I_{xy}^2}$$

having its center located at point $(a, 0)$, where $a = (I_x + I_y)/2$. The circle so constructed is called *Mohr's circle,* named after the German engineer Otto Mohr (1835–1918).

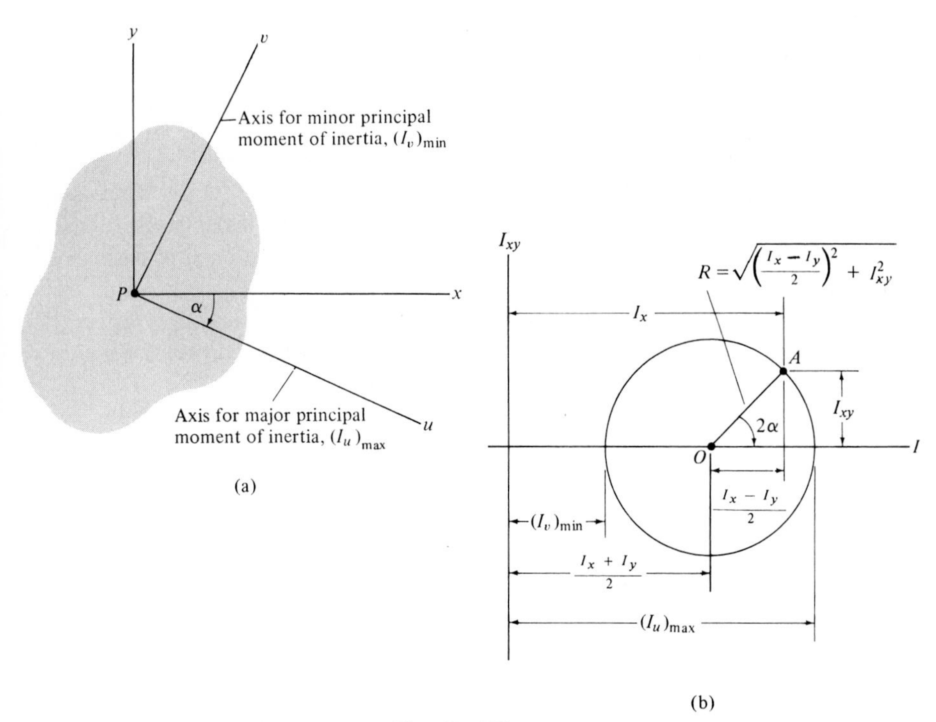

Fig. B–10

PROCEDURE FOR ANALYSIS

There are several methods for plotting Mohr's circle as defined by Eq. B–6. The main purpose in using the circle here is to have a convenient means for transforming I_x, I_y, and I_{xy} into the principal moments of inertia. The following procedure provides a method for doing this.

Compute I_x, I_y, I_{xy}. Establish the x, y axes for the area, with the origin located at the point P of interest, and determine I_x, I_y, and I_{xy} (Fig. B–10*a*).

Construct the Circle. Construct a rectangular coordinate system such that the abscissa represents the moment of inertia I, and the ordinate represents the product of inertia I_{xy}, Fig. B–10*b*. Determine the center of the circle, O, which is located at a distance $(I_x + I_y)/2$ from the origin, and plot the "controlling point" A having coordinates (I_x, I_{xy}). By definition, I_x is always positive, whereas I_{xy} will be either positive or negative. Connect the controlling point A with the center of the circle, and determine the distance OA by trigonometry. This distance represents the radius of the circle, Fig. B–10*b*. Finally, draw the circle.

Principal Moments of Inertia. The points where the circle intersects the abscissa give the values of the principal moments of inertia $(I_v)_{min}$ and $(I_u)_{max}$. Notice that the *product of inertia will be zero at these points,* Fig. B–10*b*.

Principal Axes. To find the direction of the major principal axis, determine by trigonometry the angle 2α, *measured from the radius OA to the direction line of the positive abscissa,* Fig. B–10*b*. This angle represents twice the angle from the x axis of the area in question to the axis of maximum moment of inertia $(I_u)_{max}$, Fig. B–10*a*. Both the angle on the circle, 2α, and the angle on the area, α, *must be measured in the same sense,* as shown in Fig. B–10. The axis for minimum moment of inertia $(I_u)_{min}$ is perpendicular to the axis for $(I_u)_{max}$.

Using trigonometry, the above procedure may be verified to be in accordance with the equations developed in Sec. B–2.

Example B–4

Using Mohr's circle, determine the principal moments of inertia for the beam's cross-sectional area, shown in Fig. B–11*a*, with respect to an axis passing through the centroid.

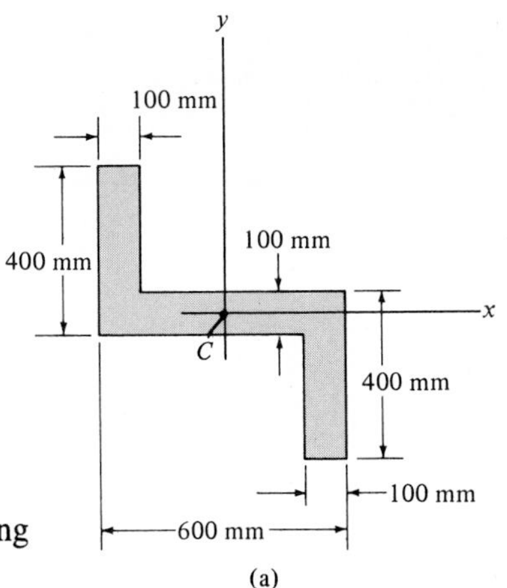

(a)

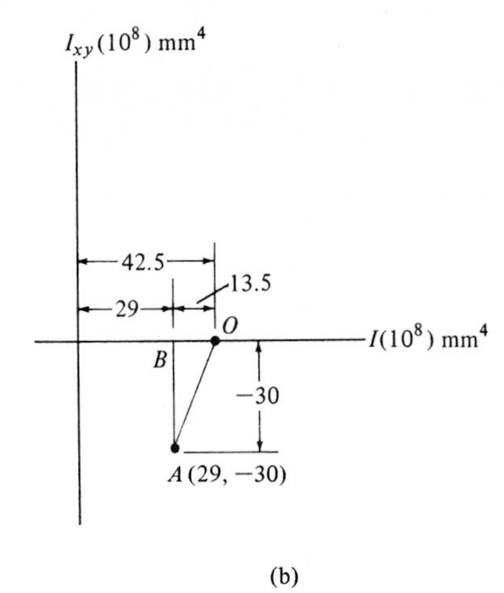

(b)

Solution

The problem will be solved using the procedure outlined above.

Compute I_x, I_y, I_{xy}. The moments of inertia and the product of inertia have been determined in Examples 10–5 and B–2 with respect to the x, y axes shown in Fig. B–11*a*. The results are $I_x = 29(10^8)$ mm^4, $I_y = 56(10^8)$ mm^4, and $I_{xy} = -30(10^8)$ mm^4.

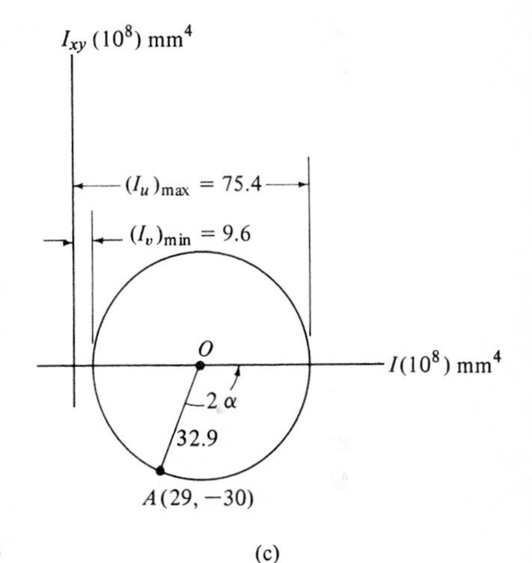

(c)

Construct the Circle. The I and I_{xy} axes are shown in Fig. B–11*b*. The center of the circle, O, lies at a distance $(I_x + I_y)/2 = (29 + 56)/2 = 42.5$ from the origin. When the controlling point $A(29, -30)$ is connected to point O, the radius OA is determined from the triangle OBA using the Pythagorean theorem.

$$OA = \sqrt{(13.5)^2 + (-30)^2} = 32.9$$

The circle is constructed in Fig. B–11*c*.

Principal Moments of Inertia. The circle intersects the I axis at points (75.4, 0) and (9.6, 0). Hence,

$$(I_u)_{max} = 75.4(10^8)\ \text{mm}^4 \qquad \textit{Ans.}$$

$$(I_v)_{min} = 9.6(10^8)\ \text{mm}^4 \qquad \textit{Ans.}$$

Principal Axes. As shown in Fig. B–11*c*, the angle 2α is determined from the circle by measuring counterclockwise from OA to the direction of the *positive I* axis. Hence,

$$2\alpha = 180° - \sin^{-1}\left(\frac{|BA|}{|OA|}\right) = 180° - \sin^{-1}\left(\frac{30}{32.9}\right) = 114.2°$$

The principal axis for $(I_u)_{max} = 75.4(10^8)$ mm^4 is therefore oriented at an angle $\alpha = 57.1°$, measured *counterclockwise*, from the *positive x* axis to the *positive u* axis. The v axis is perpendicular to this axis. The results are shown in Fig. B–11*d*.

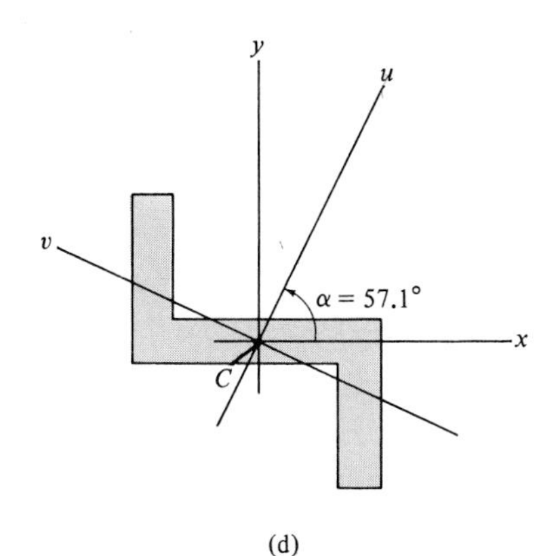

(d)

Fig. B–11

Problems

B–1. Determine the product of inertia of the shaded area with respect to the x and y axes.

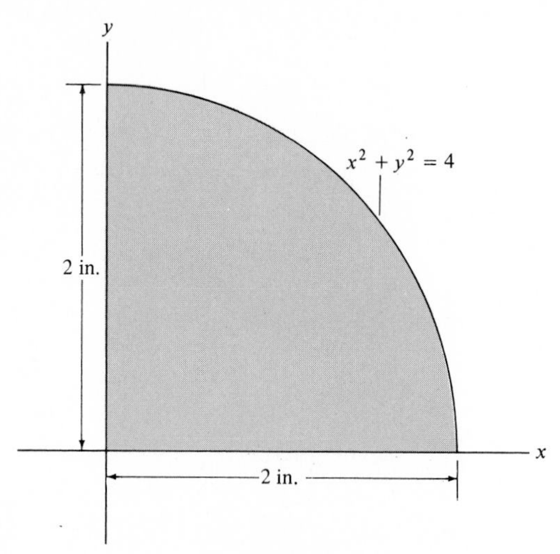

Prob. B–1

B–2. Determine the product of inertia of the parabolic area with respect to the x and y axes.

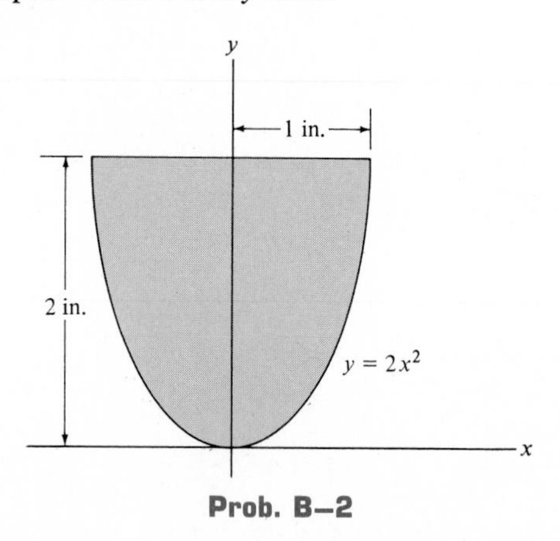

Prob. B–2

B–3. Determine the product of inertia I_{xy} of the right half of the parabolic area in Prob. B–2, bounded by the lines $y = 2$ in. and $x = 0$.

***B–4.** Determine the product of inertia of the shaded area with respect to the x and y axes.

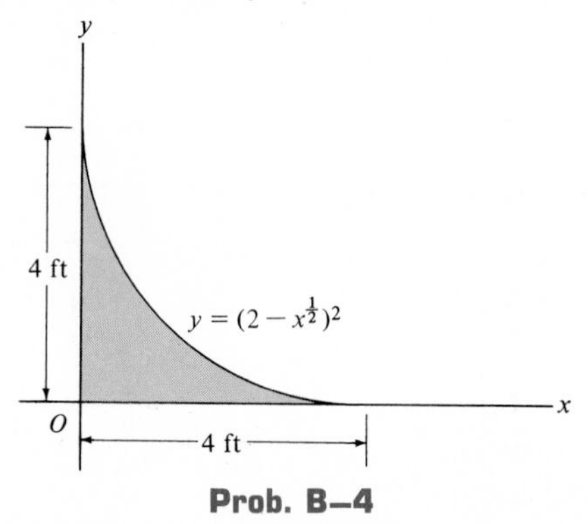

Prob. B–4

B–5. Determine the product of inertia of the shaded area of the ellipse with respect to the x and y axes.

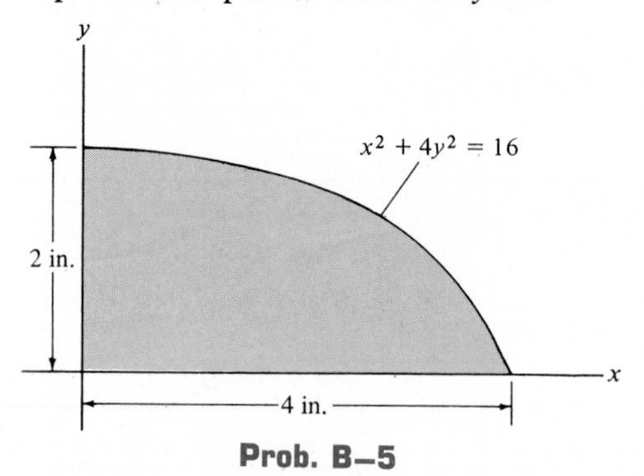

Prob. B–5

B–6. Determine the product of inertia for the angle's cross-sectional area with respect to the x and y axes having their origin located at the centroid C. Assume all corners to be square.

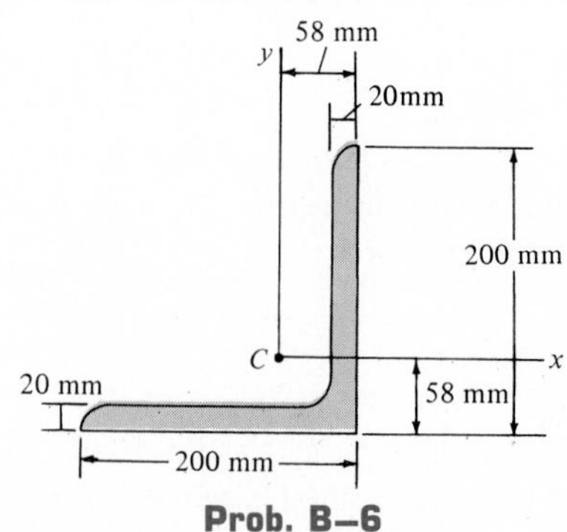

Prob. B–6

B–7. Determine the product of inertia of the beam's cross-sectional area with respect to the x and y axes that have their origin located at the centroid C.

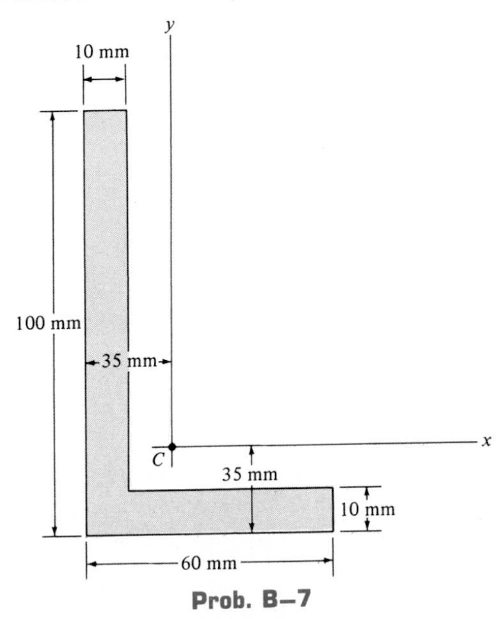

Prob. B–7

***B–8.** Determine the product of inertia of the angle's cross-sectional area with respect to the x and y axes that have their origin located at the centroid C.

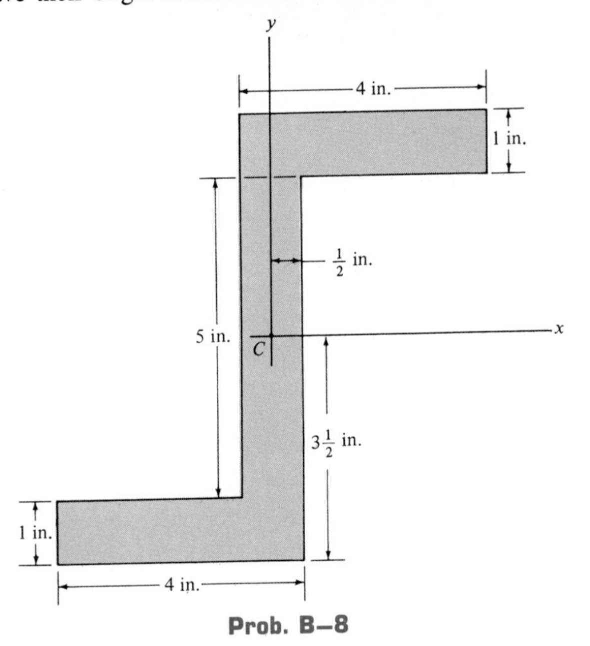

Prob. B–8

B–9. Determine the product of inertia $\bar{I}_{xy}$ of the cross-sectional area of the channel with respect to the $\bar{x}$ and $\bar{y}$ axes. $\bar{x} = 33.9$ mm, $\bar{y} = 150$ mm.

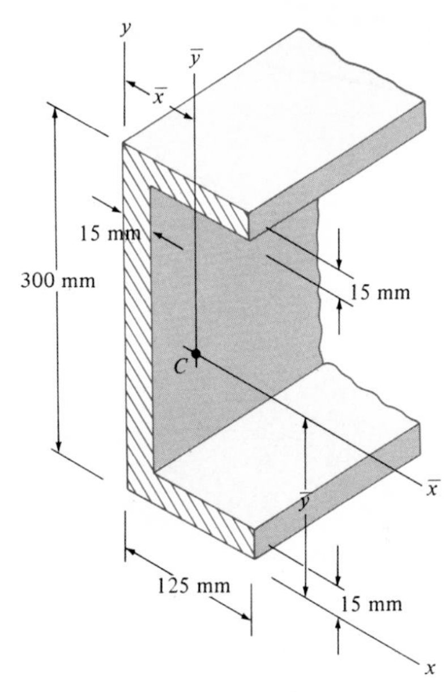

Prob. B–9

B–10. The area of the cross section of an airplane wing has the following properties about the $\bar{x}$ and $\bar{y}$ axes passing through the centroid C: $I_{\bar{x}} = 450$ in.^{4}, $I_{\bar{y}} = 1{,}730$ in.^{4}, $I_{\bar{x}\bar{y}} = 138$ in.^{4}. Determine the orientation of the principal axes and the principal moments of inertia.

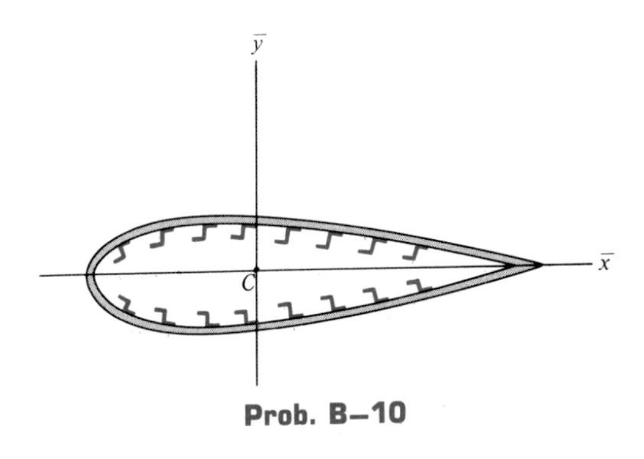

Prob. B–10

B–11. Determine the moments of inertia I_u and I_v and the product of inertia I_{uv} for the rectangular area. The u and v axes pass through the centroid C.

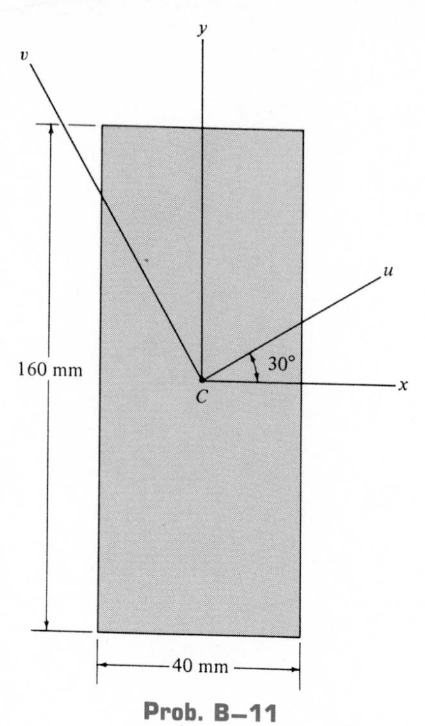

Prob. B–11

***B–12.** Determine the directions of the principal axes with origin located at point O, and the principal moments of inertia for the rectangular area about these axes.

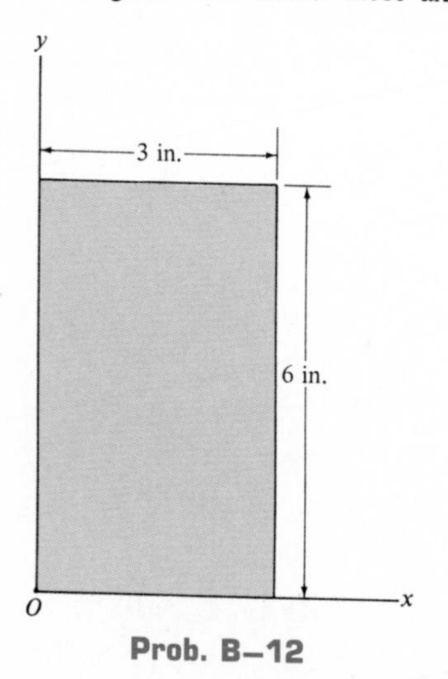

Prob. B–12

B–13. Locate the centroid, $\bar{y}$, and compute the moments of inertia I_u and I_v of the channel section. The u and v axes have their origin at the centroid C. For the calculation, assume all corners to be square.

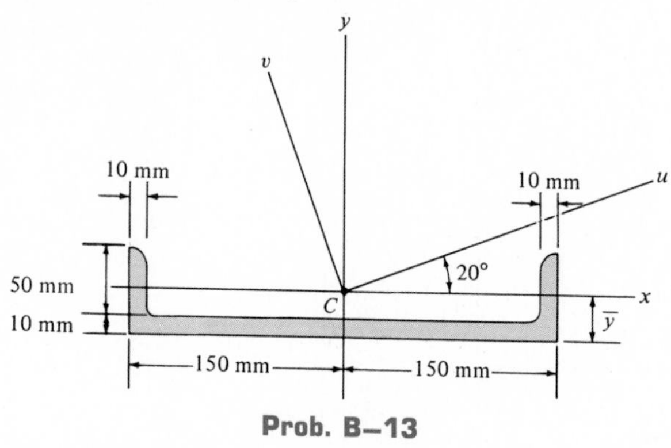

Prob. B–13

B–14. Determine the moments of inertia I_u and I_v and the product of inertia I_{uv} for the semicircular area.

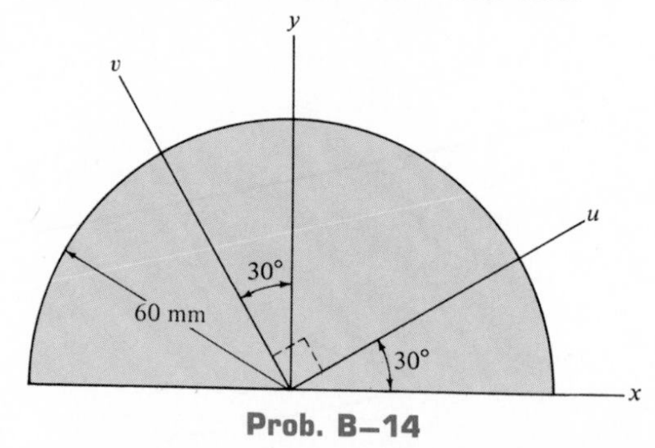

Prob. B–14

B–15. Locate the centroid, $\bar{y}$, and determine the orientation of the principal centroidal axes for the composite area. What are the moments of inertia with respect to these axes?

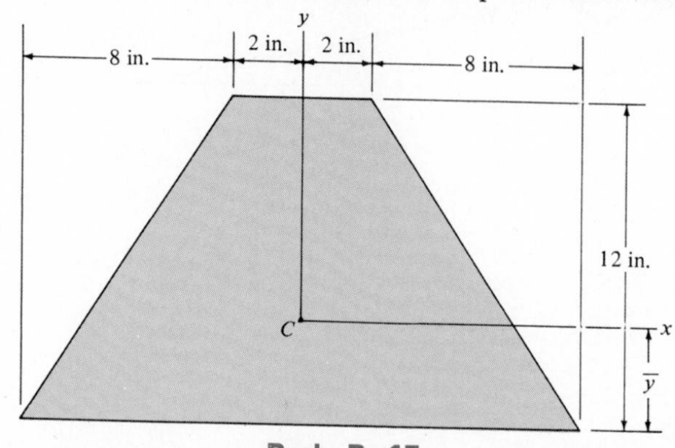

Prob. B–15

*B–16. Determine the principal moments of inertia of the composite area with respect to a set of principal axes that have their origin located at the centroid C. Use the equations developed in Sec. B–2. $I_{xy} = -1296(10^4)\ \text{mm}^4$.

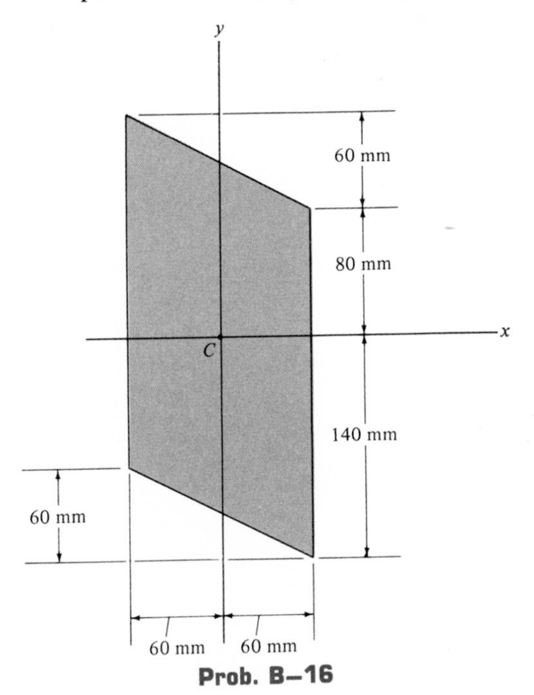

Prob. B–16

B–17. Compute the moments of inertia I_u and I_v of the shaded area.

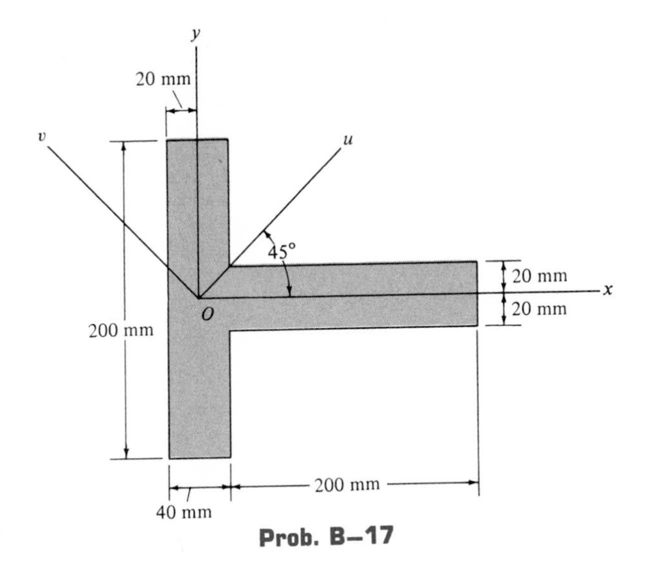

Prob. B–17

B–18. Determine the principal moments of inertia for the angle's cross-sectional area with respect to a set of principal axes that have their origin located at the centroid C. Use the equations developed in Sec. B–2. For the calculation, assume all corners to be square.

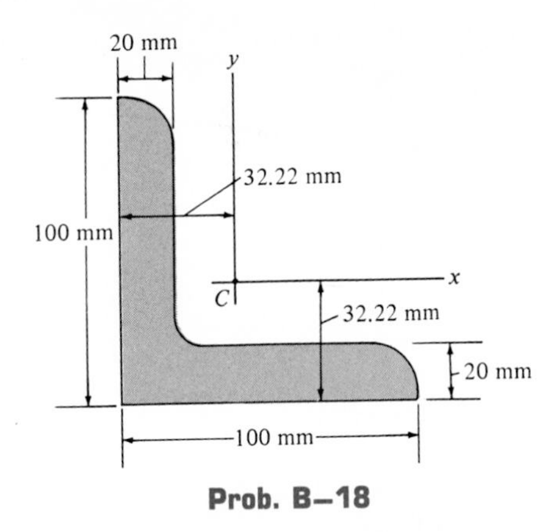

Prob. B–18

B–19. Construct Mohr's circle for the shaded area in Prob. B–12.

*B–20. Solve Prob. B–15 using Mohr's circle.

B–21. Solve Prob. B–16 using Mohr's circle.

Answers*

Chapter 1

1–1. **(a)** 98.1 N, **(b)** 4.905 mN, **(c)** 44.14 kN.
1–2. 1 Mg/m^3.
1–5. **(a)** $40(10^3)$ kN^2, **(b)** $25(10^{-6})$ mm^2, **(c)** 0.064 km^3.
1–6. **(a)** N, **(b)** MN/m, **(c)** N/s^2, **(d)** MN/s.
1–7. **(a)** Gg/s, **(b)** kN/m, **(c)** kN/(kg · s).
1–9. 0.0209 lb/ft^2, 101.4 kPa.
1–10. 7.41 μN.
1–11. **(a)** 18.6 slug, **(b)** 272.2 kg, **(c)** 2.67 kN, **(d)** 98.8 lb, **(e)** 272.2 kg.

Chapter 2

2–1. 500 N, $\measuredangle$ 36.9°.
2–2. 12.49 kN, $\measuredangle$ 43.9°.
2–3. 72.11 lb, $\measuredangle$ 73.9°.
2–5. 1.0 kN, 1.73 kN.
2–6. 606.2 lb, 350.0 lb.
2–7. 73.2 N, 141.4 N.
2–9. 12.9 lb, −35.4 lb.
2–10. 61.4 N, −43.0 N.
2–11. $F_{\parallel} = 45.9$ lb, $F_{\perp} = 65.5$ lb.
2–13. 5.45 mN, 3.55 mN.
2–14. $F_R = 3.44$ kN, In same direction, $F_{R_{max}} = 4.50$ kN.
2–15. 80.3 lb, $\theta = 46.2°$.
2–17. 214.9 lb, $\theta = 52.7°$.
2–18. $F_{AB} = 734.8$ N, $F_{CA} = 819.6$ N.
2–19. $F_1 = 14.0$ lb, $F_2 = 8.35$ lb.
2–21. 252.2 N, 82.5° $\measuredangle$.
2–22. 300 N, 36.9° $\measuredangle$.
2–23. 3.57 kN, 5.22 kN.
2–25. 88.3 lb, 43.8°.
2–26. 29.7 N, $\measuredangle$ 70.9°.
2–27. −47.5 N, 82.3 N.
2–29. 200 lb, −150 lb.
2–30. 52.5 lb, −460 lb, 463.0 lb
2–31. $F_1 = 14.0$ lb, $F_2 = 8.35$ lb.
2–33. $F_A = 3.57$ kN, $F_B = 5.22$ kN.
2–34. 29.7 N, $\measuredangle$ 70.9°.
2–35. −200 lb, 0, −125 lb, 216.5 lb, 194.9 lb, 112.5 lb.
2–37. $\mathbf{F}_1 \cos\theta$, $F_1 \sin\theta$, 350 N, 0, 0, −100 N, $\theta = 67.0°$, $F_1 = 434.5$ N.

*Note: Answers to every fourth problem are omitted.

2–38. 259.8 N, 150.0 N, −240 N, 100 N.
2–39. 2.50 kN, 3.54 kN, 2.50 kN, 0, −2 kN, 0.
2–41. −1.25 kN, 0, 0, −2.5 kN, 4.33 kN, 0, 1.20 kN, 1.60 kN, 6.20 kN, 101.6°, 102.1°, 16.9°.
2–42. 150.7 lb, 53.3°, 138.8°, 73.8°.
2–43. $F = 63.2$ N, $F_y = 44.7$ N.
2–45. 40.0 lb, 56.6 lb, 40.0 lb.
2–46. 428.9 lb, 62.2°, 110.0°, 144.7°.
2–47. 43.3 lb, −25.0 lb, 28.87 lb, 57.7 lb.
2–49. 12 ft, 70.5°, 48.2°, 131.8°.
2–50. 9 m, 131.8°, 48.2°, 70.5°.
2–51. 376.06 mm.
2–53. 7.68 m.
2–54. 6.72 km.
2–55. −1200 N, −900 N.
2–57. 404.4 lb, 276.2 lb, −101.1 lb.
2–58. 17 ft, −160 lb, −180 lb, 240 lb.
2–59. 1477.3 N, 86.9°, 77.5°, 167.1°.
2–61. 20 N, 40 N, −40 N.
2–62. 96 kN.
2–63. 24 ft, 18 ft, 16 ft.
2–65. 782.6 mm, −9.96 N, −13.42 N, 24.92 N.
2–66. 14.91 m, 7.45 m, −18.63 m.
2–67. 82.7 lb, 9.24°.
2–69. 80.3 lb, 73.8°.
2–70. 80.3 lb, 106.2°.
2–71. −21.8 N, −109.0 N, −436.1 N, −111.3 N, −13.9 N, −278.3 N, 736.92 N.
2–73. 754.3 N, 9.81°.
2–74. 407.5 N, 26.5°.
2–75. 614.9 N, 26.6°, 85.1°, 64.0°.
2–77. 79.5°, 90.8 lb.
2–78. 5.54 ft.

Chapter 3

3–1. 258.8 N, 366.0 N.
3–2. 3.23 kN, 21.8°.
3–3. 180.3 lb, 73.9°.
3–5. 4.45 lb, 6.82 lb.
3–6. 40.8 lb, 48.9 lb.
3–7. 1471.5 N, 882.9 N.
3–9. 126.9 lb, 78.6°.
3–10. 6.02 kN, 41.6°.
3–11. 4.09 N.
3–13. 2.34 ft.
3–14. 18.33 lb.
3–15. $\frac{1}{k_T} = \frac{1}{k_1} + \frac{1}{k_2}$.
3–17. 3.55 mN, 5.03 mN, 2.67 mN, 1.60 mN.
3–18. 7.76 lb, 10.98 lb, 10.97 lb, 4.02 lb, 45.0°, 13.45 lb.
3–19. 39.24 N, 67.97 N, 39.24 N, 39.24 N.
3–21. 2.35 m.
3–22. 292.8 N, 200.0 N, 358.6 N.
3–23. 2.83 kN, 1.10 kN, 3.46 kN.
3–25. 108.7 lb, 47.4 lb, 87.8 lb.
3–26. 11.47 lb, 25.00 lb, 14.97 lb.
3–27. 1320.2 N, 1420.5 N, 1420.5 N.
3–29. $F_{AB} = F_{CB} = 2523.3$ N, $F_{BD} = 3646.8$ N.
3–30. $F_{AC} = F_{AD} = 243.7$ lb, $F_{AB} = 540.8$ lb.
3–31. 18.8 lb, 26.2 lb, 1.80 lb.
3–33. $F_B = F_D = 858.4$ N, $F_C = 0$.
3–34. 319.1 N, 110.5 N, 85.9 N.
3–35. 343.0 N, 294.1 N, 146.9 N.
3–37. 5.33 ft.
3–38. 6 lb.
3–39. 1,000 lb, 424.2 lb, 566.0 lb.
3–41. 106.1 lb.
3–42. 86.2 lb, 175.0 lb, 506.3 lb.
3–43. 1098.1 N, 503.0 N, 407.0 N.
3–45. 18.6 lb.
3–46. 190 mm, 12.3 mm.
3–47. 15.6 kg.

Chapter 4

4–1. **(a)** 400 N·m ↓, **(b)** 600.0 N·m ↓, **(c)** 9.86 kN·m ↓.
4–2. **(a)** 1,103.5 lb·ft ↰, **(b)** 5656.8 N·m ↰.
4–3. 24.1 N·m ↰, 14.5 N·m ↰.
4–5. **(a)** 214.7 N·m ↰, **(b)** 238.6 N →.
4–6. 18 N·m ↓, 120°.
4–7. 81.9 lb·ft ↰, 84.4 lb·ft ↰.
4–9. 500 lb·ft, 2,950 lb·ft.

4–10. 18.39 kN·m.

4–11. 30.06 kN·m ↻, 8.66 kN.

4–13. **(a)** 36.9°, 150 lb·ft, **(b)** 126.9°, 0.

4–14. **(a)** 1200 N·m ↻, **(b)** 21.9 kN·m ↺.

4–15. 20 kip, 24.4 kip.

4–17. 800 N, 133.3 N.

4–18. $\mathbf{M}_C$ depends only on $(a + b)$, changing c has no effect.

4–19. 25,972.7 lb·ft, 15,473.2 lb·ft.

4–21. 450 N ∠ 30°, 2458.8 N·m ↻.

4–22. 15 kN 36.9° ∠, 30 kN·m ↻.

4–23. 390 lb ∠ 22.6°, 204 lb·ft ↻.

4–25. 50 N 60° ∠, 14.24 N·m ↻.

4–26. 17.3 N ∠ 60°, 42.0 N·m ↺.

4–27. 1.18 kN 85.8° ∠, 21.6 kN·m ↺.

4–29. 294.3 N ∠ 40.1°, 39.6 N·m ↻.

4–30. 450.5 N ∠ 27.4°, 1714.3 N·m ↻.

4–31. 78.1 lb ∠ 87.1°, 100 lb·ft ↺.

4–33. 12,100 lb ↓, 10.04 ft.

4–34. 9 kN →, $y = 1.89$ m.

4–35. 600 lb, 130.4 ft.

4–37. 903.8 lb ∠ 23.6°, 6.10 ft.

4–38. 1100 N ↓, 3100 N·m ↻.

4–39. 90 kN ↓, 337.5 kN·m ↻.

4–41. 0.525 kN, 0.171 m.

4–42. 176,000 lb, 150.5 ft.

4–43. 144.3 N ∠ 60.1°, 89.6 N·m ↺.

4–45. 1,128.5 lb 7.05° ∠, 8,480 lb·ft ↺.

4–46. 1.5 m, 175.0 N/m.

4–47. 40 kN·m, 0, 80 kN·m.

4–49. 400 lb·ft, −200 lb·ft, 1200 lb·ft.

4–50. −1765.8 N·m, −1103.6 N·m.

4–51. 805.0 N·m.

4–53. 264.0 lb·ft.

4–54. 621.0 lb·ft.

4–55. 2412.2 N·m.

4–57. 39.2 lb·ft.

4–58. 463.6 lb·ft.

4–59. $0.866\ F_z + 0.5\ F_y$.

4–61. 14.85 N·m.

4–62. 20.2 N.

4–63. −140 N·m, 150 N·m, −180 N·m.

4–65. 4000 lb·ft, −2000 lb·ft, 4000 lb·ft.

4–66. 61.2 N·m, 49.2°, 67.5°, 49.2°..

4–67. 12.2 N·m, 130.9°, 75.8°, 135.7°.

4–69. 50 N, −20 N, −30 N, −50 N·m, 130 N·m, −170 N·m.

4–70. 63.6 N·m, −169.7 N·m, 263.6 N·m.

4–71. 60 lb, −80 lb, 20 lb·ft, 160 lb·ft, 120 lb·ft.

4–73. −120 lb, 0, 0, 0, −3,000 lb·ft, 1,800 lb·ft.

4–74. **(a)** 0, 0, −1500 lb, −1500 lb·ft, 0, 0, **(b)** 0, 0, −1500 lb, −2,375 lb·ft, −750 lb·ft, 0.

4–75. 20 N, 60 N, −30 N, **(a)** −180 N·m, 70 N·m, 20 N·m, **(b)** −180 N·m, 70 N·m, 20 N·m.

4–77. 20 N, 30 N, −20 N, −150 N·m, 80 N·m, −70 N·m.

4–78. 9 kN, 12 kN, 0, −90 kN·m, 60 kN·m, −234 kN·m.

4–79. −100 lb, 200 lb, −200 lb, 750 lb·ft, −1513.4 lb·ft, −2000 lb·ft.

4–81. 0, 20 lb, 60 lb, 90 lb·ft, 0, 50 lb·ft.

4–82. 1.5 m.

4–83. 25 lb, −0.9 ft.

4–85. 31 kN ↓, 0.593 m, −0.267 m.

4–86. 0.2 m, −40 N, −30 N·m.

4–87. 360.6 lb, 56.3°, 860.6 lb.

4–89. 16 kN ↓, 1.14 m.

4–90. 111.5 lb, 22.4° ∠, 300.2 lb·ft ↺.

4–91. 27.7 lb.

4–93. −21.3 lb·in. ↺.

4–94. 35.5 lb·in.

4–95. 1,949.2 lb·ft.

4–97. −2.40 lb, 3.20 lb, −3.60 lb, 9.60 lb·in., 7.20 lb·in., 0.

4–98. −840 N·m, 360 N·m, −660 N·m.

4–99. 4.30 N·m, 75.7°, 75.7°, 159.6°.

4–101. 70.8°, 39.8°, 56.7°.

4–102. 298.0 lb·in., 15.2 lb·in., −200 lb·in.

Chapter 5

5–1. $\mathbf{A}_x$, $\mathbf{A}_y$, $\mathbf{T}_{BC}$.

5–2. $\mathbf{N}_C$, $\mathbf{N}_D$.

5–3. $\mathbf{N}_C$, $\mathbf{N}_D$, $\mathbf{N}_P$.

5–5. $\mathbf{C}_x$, $\mathbf{C}_y$, $\mathbf{T}_{AB}$.

5–6. $\mathbf{B}_x$, $\mathbf{B}_y$, $\mathbf{F}_s$.

5–7. $\mathbf{A}_x$, $\mathbf{A}_y$, $\mathbf{N}_B$.

5–9. $\mathbf{A}_x$, $\mathbf{A}_y$, $\mathbf{T}$.

5–10. $\mathbf{A}_x$, $\mathbf{A}_y$, $\mathbf{B}_x$.

5–11. $\mathbf{A}_x$, $\mathbf{A}_y$, $\mathbf{T}_{BC}$.

5–13. $\mathbf{A}_x$, $\mathbf{A}_y$, $\mathbf{B}_x$.

5–14. $\mathbf{M}_A$, $\mathbf{A}_x$, $\mathbf{A}_y$, $\mathbf{N}_B$.

5–15. $\mathbf{A}_x$, $\mathbf{A}_y$, $\mathbf{T}_{BD}$.

5–17. $A_x = 0$, $A_y = 2$ kN, $B_y = 8$ kN.

5–18. $A_x = 0$, $A_y = 200$ N, $B_y = 400$ N.

5–19. $A_x = 0$, $A_y = 2{,}750$ lb, $B_y = 1{,}000$ lb.

5–21. $B_x = 15.9$ N, $B_y = 120.7$ N.

5–22. $A_x = 154.0$ lb, $A_y = 333.3$ lb, $N_B = 307.9$ lb.

5–23. $F_A = 30$ lb, $F_B = 36.2$ lb, $F_C = 9.38$ lb.

5–25. $F_{CD} = 447.3$ kip, $A_x = 342.6$ kip, $A_y = 172.5$ kip.

5–26. $C_x = 1.58$ kN, $C_y = 5.65$ kN.

5–27. $F_B = 675$ lb, $F_A = 525$ lb.

5–29. $n = 5$, $x = 8$ in.

5–30. **(a)** 305.5 lb, **(b)** 120.9 lb.

5–31. $F_s = 1.5$ N, $P = 1.5$ N, $F = 0.3$ N.

5–33. $w_B = 2{,}190.5$ lb/ft, $w_A = 10{,}714.3$ lb/ft.

5–34. 392.4 kN.

5–35. 30.8°.

5–37. $w_1 = 413.3$ kN/m, $w_2 = 406.7$ kN/m.

5–38. 0.34 m.

5–39. $w_B = 1113.6$ N/m, $w_A = 1437.6$ N/m.

5–41. $F = 5196.2$ N, $N_A = 17\,335.9$ N, $N_B = 24\,904.1$ N.

5–42. $T_1 = 1891.7$ N, $T_2 = 2254.5$ N, $\bar{x} = 2.30$ m.

5–43. 10.4°, $\Delta_A = 0.0785$ m $\downarrow$, $\Delta_B = 0.1046$ m $\uparrow$.

5–45. $A_x = A_y = 0$, $A_z = 200$ N, $B_z = 200$ N, $F_{DC} = 300$ N.

5–46. $A_x = 1.5$ kN, $A_y = 1$ kN, $A_z = 1.25$ kN, $B_x = 3.5$ kN, $B_z = 0.75$ kN, $F_{DC} = 2$ kN.

5–47. $A_x = A_y = 0$, $A_z = 120$ lb, $B_z = 80$ lb, $F_{DC} = 0$.

5–49. $T_A = 208.7$ lb, $T_B = 260.9$ lb, $T_C = 330.4$ lb.

5–50. $A_x = 135.7$ lb, $A_z = 40$ lb, $B_x = 35.7$ lb, $B_z = 40$ lb, $P = 100$ lb.

5–51. $A_x = -1866.7$ N, $A_y = 1866.7$ N, $A_z = 800$ N, $C_y = -1866.7$ N, $C_z = 600$ N, $D_x = 1866.7$ N.

5–53. $A_x = -240$ N, $A_y = -160$ N, $A_z = 480$ N, $(M_A)_x = 4800$ N·m, $(M_A)_y = -7200$ N·m, $(M_A)_z = 0$.

5–54. $F_B = F_C = 68.0$ N, $F_A = 128.8$ N.

5–55. $A_x = 0$, $B_y = 72.17$ lb, $B_z = 125$ lb, $F_y = 72.17$ lb, $F_z = 125$ lb, $D_y = 394.33$ lb.

5–57. $A_x = 0$, $A_y = 84.9$ lb, $A_z = 80$ lb, $M_x = 947.9$ lb·ft, $M_y = M_z = 0$.

5–58. $F_A = 125$ lb, $M_R = 206.2$ lb·ft.

5–59. $A_x = 8.0$ kN, $A_y = 0$, $A_z = 24.4$ kN, $M_x = -572$ kN·m, $M_y = 20$ kN·m, $M_z = 64$ kN·m.

5–61. $A_x = 19.4$ lb, $A_y = 192.0$ lb, $A_z = -25.8$ lb, $T = 62.0$ lb, $F = 110.1$ lb.

5–62. $A_x = 0.0$, $A_y = 150$ lb, $A_z = 562.1$ lb, $T_{BC} = 300$ lb, $T_{BD} = 212.2$ lb.

5–63. $A_x = 5.0$ kips, $A_y = 14.7$ kips, $M_A = 94.6$ kip·ft.

5–65. $A_x = 3.60$ kN, $A_y = 1.80$ kN, $N_B = 5.09$ kN.

5–66. 353.6 N.

5–67. $F_x = -30$ N, $F_y = -60$ N, $F_z = 60$ N, $(M_C)_x = 8$ N·m, $(M_C)_y = -42$ N·m, $(M_C)_z = -18$ N·m.

5–69. $A_y = 300$ N, $B_x = 0$, $B_y = 1200$ N.

5–70. $F_x = 19.3$ N, $F_y = -5.18$ N, $F_z = 0$, $M_x = 0.388$ N·m, $M_y = 1.45$ N·m, $M_z = -4.00$ N·m.

5–71. $\bar{x} = 1.45$ ft, $F = 12.6$ lb. $T = 10.8$ lb.

5–73. $T_{BD} = T_{CD} = 70.1$ lb, $A_x = 37.5$ lb, $A_y = 0$, $A_z = 37.5$ lb.

5–74. $F = 427.6$ lb, $A_x = 106.9$ lb, $A_y = 0$, $A_z = 320.7$ lb.

5–75. 5.35 lb.

5–77. 2.18 kN/m.

5–78. $A_y = 168.2$ lb, $A_z = 368.2$ lb, $B_x = -357.6$ lb, $B_y = -168.2$ lb, $C_x = 357.6$ lb, $C_z = -168.2$ lb.

5–79. $W_C = 0.776W$.

Chapter 6

6–1. $F_{AD} = 5.66$ kN (C), $F_{AB} = 4$ kN (T), $F_{BC} = 4$ kN (T), $F_{BD} = 6$ kN (C), $F_{DC} = 14.14$ kN (T), $F_{DE} = 14$ kN (C).

6–2. $F_{CD} = F_{ED} = F_{FE} = 0$, $F_{EC} = 15$ kN (T), $F_{CF} = 21.2$ kN (C), $F_{CB} = 15$ kN (T), $F_{AF} = 15$ kN (C), $F_{FB} = 25$ kN (T), $F_{AB} = 28.0$ kN (C).

6–3. $F_{AB} = 4.62$ kN (C), $F_{AE} = 2.31$ kN (T), $F_{BE} = 4.62$ kN (T), $F_{BC} = 4.62$ kN (C), $F_{DE} = 2.31$ kN (T), $F_{DC} = 4.62$ kN (C), $F_{EC} = 4.62$ kN (T).

6–5. $F_{CB} = 894.4$ N (C), $F_{CD} = 800$ N (T), $F_{ED} = 800$ N (T), $F_{DB} = 800$ N (C),

$F_{EB} = 894.4$ N (T), $F_{AB} = 1788.9$ N (C), $F_{AE} = 800$ N (T).

6–6. $F_{AB} = 282.8$ lb (C), $F_{AD} = 206.2$ lb (T), $F_{CB} = 282.8$ lb (C), $F_{CD} = 206.2$ lb (T), $F_{BD} = 400$ lb (T).

6–7. $F_{HJ} = 400$ lb (C), $F_{HG} = 400$ lb (C), $F_{GF} = 1{,}040$ lb (T), $F_{GE} = 960$ lb (C), $F_{JI} = F_{EF} = F_{IG} = F_{HI} = F_{HF} = 0$.

6–9. $F_{PE} = 3.75$ kN (C), $F_{KJ} = 0$, $F_{IJ} = 0$, $F_{RK} = 8.75$ kN (C), $F_{NM} = F_{AN} = F_{BO} = F_{OC} = F_{FQ} = F_{EQ} = 0$.

6–10. $F_{LK} = 600$ N (T), $F_{AB} = 1732.0$ N (T), $F_{BK} = 288.7$ N (C).

6–11. $F_{BF} = 155.3$ lb (C), $F_{FD} = 0$.

6–13. $F_{FE} = F_{ED} = F_{EG} = F_{AC} = F_{CD} = F_{BC} = 0$, $F_{AB} = F_{BD} = F_{GF} = F_{FD} = 333.3$ lb (C).

6–14. $F_{KJ} = 800$ N (T), $F_{CJ} = 282.9$ N (C), $F_{CD} = 671.0$ N (C).

6–15. $F_{FG} = 400$ N (C), $F_{GH} = F_{IH} = 346.4$ N (T), $F_{FH} = F_{EH} = F_{DH} = F_{BI} = F_{CI} = 0$, $F_{FE} = F_{DE} = 400$ N (C), $F_{AB} = F_{BC} = F_{CD} = 800$ N (C), $F_{AI} = F_{ID} = 692.8$ N (T).

6–17. $F_{BC} = 14.4$ kN (T), $F_{HC} = 2.65$ kN (T), $F_{HG} = 5.42$ kN (C).

6–18. $F_{LD} = 0$, $F_{LK} = 112.5$ kN (C), $F_{CD} = 112.5$ kN (T).

6–19. $F_{DF} = 2263.1$ N (T), $F_{CE} = 4526.2$ N (C), $F_{ED} = 400$ N (C).

6–21. $F_{GB} = 835.7$ lb (T), $F_{GF} = 835.7$ lb (C).

6–22. $F_{CD} = 5$ kN (T), $F_{KJ} = 4.47$ kN (C), $F_{KD} = 1.41$ kN (C).

6–23. $F_{KJ} = 4000$ N (C), $F_{FG} = 5962.8$ N (T), $F_{GK} = 1885.9$ N (C).

6–25. $F_{BF} = 692.8$ lb (T), $F_{BC} = 1{,}212.5$ lb (T), $F_{GF} = 1{,}800$ lb (C).

6–26. $F_{HG} = 10.1$ kN (C), $F_{BC} = 8.0$ kN (T), $F_{BG} = 1.83$ kN (T).

6–27. $F_{DE} = 11{,}485.6$ lb (T), $F_{KJ} = 15{,}000$ lb (C), $F_{QE} = 4{,}569.6$ lb (T).

6–29. $F_{LK} = 1{,}933.3$ lb (C), $F_{MK} = 1{,}416.7$ lb (C).

6–30. $F_{CF} = 0$, $F_{CD} = 2.23$ kN (C), $F_{GF} = 1.78$ kN (T).

6–31. $F_{CF} = F_{DE} = 4$ kN (C), $F_{BA} = 3$ kN (C), $F_{CB} = F_{CD} = F_{BD} = F_{DF} = F_{DA} = F_{EF} = F_{EA} = F_{FA} = 0$.

6–33. $F_{BE} = 900$ lb (T), $F_{BA} = 600$ lb (C), $F_{AE} = 670.8$ lb (C), $F_{BC} = F_{AC} = F_{AD} = F_{DE} = F_{CD} = F_{CE} = 0$.

6–34. $F_{BD} = 144.3$ lb (C), $F_{AD} = 239.3$ lb (T), $F_{AF} = 0$.

6–35. $F_{CA} = 894.4$ N (T), $F_{CD} = F_{CB} = 210.8$ N (C), $F_{AB} = F_{AD} = 466.7$ N (C).

6–37. **(a)** 5 lb, **(b)** 10 lb, **(c)** 5 lb.

6–38. $P = 15$ lb, $T_A = 15$ lb, $T_B = 30$ lb.

6–39. $P = 20$ lb, $T_A = 40$ lb, $T_B = 20$ lb.

6–41. 40 N, 240 mm.

6–42. 14.6°.

6–43. $F_B = 13.3$ lb, $A_x = 0$, $A_y = 17.3$ lb.

6–45. $A_x = 1500$ N, $A_y = 600$ N.

6–46. $D_y = 112.5$ lb, $F_{BE} = B_x = D_x = 56.25$ lb, $B_y = 0$.

6–47. $A_x = 46.7$ lb, $A_y = 45.0$ lb.

6–49. $F = 160$ lb, $N_P = 320$ lb, $F = 80$ lb, $N_P = 80$ lb.

6–50. $F_{CA} = 1{,}674.8$ lb, $F_{BA} = 2{,}051.2$ lb, $F_{AD} = 2{,}287.8$ lb.

6–51. $A_y = 0$, $A_x = 120$ lb, $C_y = 15$ lb.

6–53. 20.6 kN.

6–54. **(a)** 2180 N, **(b)** 27 m.

6–55. $C_y = 100$ N, $B_y = 200$ N, $A_x = 0$, $M_A = 2400$ N·m, $A_y = 1000$ N.

6–57. $F_{FB} = 1938.9$ N, $F_{DB} = 2601.3$ N.

6–58. $A_x = 0$, $A_y = 85.02$ N, $C_x = 0$, $C_y = 16.35$ N, $x = 158$ mm, $B_x = 0$, $B_y = 2.78$ N.

6–59. 1066.9 N.

6–61. 9.42 lb.

6–62. $A_y = 24.4$ N, $D_y = 20.8$ N, $F_y = 14.7$ N.

6–63. $F_{GF} = 848.5$ N, $D_y = D_z = 300$ N, $D_x = 225$ N.

6–65. $A_x = A_y = C_x = C_y = 100$ lb.

6–66. $F_{AG} = 471.4$ lb (C), $F_{AB} = F_{BC} = 333.3$ lb (T), $F_{DE} = 942.9$ lb (C), $F_{EC} = F_{DC} = 666.7$ lb (T), $F_{EG} = 666.7$ lb (C), $F_{GC} = 471.4$ lb (T).

6–67. $F_{GC} = 471.4$ lb (T), $F_{BC} = 333.3$ lb (T), $F_{GE} = 666.7$ lb (C).

6–69. 359.8 kg, $B_x = 1.77$ kN, $B_y = 5.30$ kN.

6–70. 2,598.1 lb (T).

6–71. 2,000 lb (C).

6–73. $F_{AC} = 758.8$ N, $F_{BC} = 200.3$ N.

6–74. $(F_{AC})_x = (F_{AC})_y = 490.5$ N, $(F_{BC})_x = (F_{BC})_y = 0$.

6–75. $A_x = 8.31$ kip, $A_y = 0.308$ kip, $E_x = 8.31$ kip, $E_y = 5.69$ kip.

6–77. $A_x = 483.0$ N, $A_y = 236.6$ N, $B_x = B_y = 536.6$ N.

6–78. $F_{CA} = F_{CB} = 122.5$ lb (C), $F_{CD} = 173.2$ lb (T), $F_{BD} = F_{AD} = 86.6$ lb (T), $F_{BA} = 0$.

Chapter 7

7–1. $A_A = 10$ lb, $A_B = 0$.

7–2. $T_A = 200$ N, $T_B = 550$ N, $T_C = 150$ N.

7–3. 12.5 lb.

7–5. $T_C = 0$, $T_D = 400$ N·m, $T_E = 550$ N·m.

7–6. $T_A = 100$ N·m, $T_B = 200$ N·m, $T_C = 200$ N·m, $T_D = 0$.

7–7. $A_C = 0$, $V_C = 25$ lb, $M_C = 100$ lb·ft.

7–9. $A_A = 65$ lb, $V_A = 0$, $M_A = 16.25$ lb·ft.

7–10. $A_D = 800$ N, $V_D = 0$, $M_D = 1200$ N·m.

7–11. $A_E = 0$, $V_E = 50$ N, $M_E = 100$ N·m, $A_D = 0$, $V_D = 550$ N, $M_D = 900$ N·m.

7–13. $A_C = 32.4$ lb, $V_C = M_C = 0$.

7–14. $A_D = 2400$ N, $V_D = 50$ N, $M_D = 1350$ N·m.

7–15. $A_C = 0$, $V_C = 200$ lb, $M_C = 300$ lb·ft.

7–17. $A_A = 14.14$ lb, $V_A = 14.14$ lb, $M_A = 28.28$ lb·in.

7–18. $V_x = 0$, $A_y = 0$, $V_z = 80$ N, $M_x = 24$ N·m, $M_y = -20$ N·m, $M_z = 0$.

7–19. $A_O = 850$ lb, $(V_O)_x = 0$, $(V_O)_y = 0$, $M_x = 300$ lb·ft, $M_y = 200$ lb·ft, $M_z = 0$.

7–21. $0 \le x < 5$, $V = 100$, $M = 100x - 1800$; $5 < x \le 10$, $V = 100$, $M = 100x - 1000$.

7–22. $0 \le x \le 5$, $V = 10 - 2x$, $M = -30 + 10x - x^2$.

7–23. $0 \le x \le 15$, $V = 2250 - 300x$, $M = 2250x - 150x^2$.

7–25. $0 \le x < 2$, $V = 0.25$, $M = 0.25x$; $2 < x \le 3$, $V = 3.25 - 1.5x$, $M = -3 + 3.25x - 0.75x^2$.

7–26. $0 \le x < 5$, $V = 2.5 - 2x$, $M = 2.5x - x^2$; $5 < x < 10$, $V = -7.5$, $M = -7.5x + 25$.

7–27. $0 < x \le 20$, $V = 490 - 50x$, $M = 490x - 25x^2$; $20 < x \le 30$, $V = 0$, $M = -200$.

7–29. $x = 1^+$, $V = -2.14$, $M = 5.86$; $x = 3.5^+$, $V = 15$, $M = -7.5$.

7–30. $x = 3$; $V = 0$, $M = 9$

7–31. $x = 0$ and $x = 6$, $V = 3$, $M = 0$; $x = 3$, $V = 0$, $M = 4.5$.

7–33. $V = 750 - 250x - 125x^2$ $M = 750x - 125x^2 - 41.67x^3$.

7–34. $x = 22^-$, $V = -56$, $M = -696$; $\bar{x} = 62^-$, $V = 18$, $M = -208$.

7–35. $x = 1^+$, $V = -796.2$, $M = -98.1$; $x = 6$, $V = -1777.2$, $M = -6531.6$.

7–37. $B_x = -100$ lb, $B_y = 0$, $B_z = 50$ lb, $(M_B)_x = 100$ lb·ft, $(M_B)_y = 50$ lb·ft, $(M_B)_z = 200$ lb·ft.

7–38. $x = 2^-$, $V = -1.14$, $M = 9.71$, $x = 5^+$, $V = -1.14$, $M = 2.29$.

7–39. $A_E = 0$, $V_E = 375$ lb, $M_E = 750$ lb·ft.

7–41. $A_E = 223.6$ N, $V_E = 447.2$ N, $M_E = 223.6$ N.

7–42. $A_F = 223.6$ N, $V_F = 447.2$ N, $M_F = 223.6$ N·m.

7–43. $x = 3^-$, $V = -450$, $M = -675$; $x = 15^-$, $V = -900$, $M = -675$.

7–45. $x = 3^-$, $V = -4.5$, $M = -9$; $x = 10^+$, $V = 19.7$, $M = -54.5$.

7–46. $0 \leqslant x \leqslant \pi$, $A = 150 \cos\theta + 200 \sin\theta$, $V = 150 \sin\theta - 200 \cos\theta$, $M = 150 \cos\theta + 200 \sin\theta - 150$ $0 \leqslant x \leqslant 1$, $V = 200$, $A = -150$, $M = 150 + 200x$.

7–47. $A_E = 11\ 111.1$ lb, $V_E = 6666.7$ lb, $M_E = 5555.6$ lb·ft.

Chapter 8

8–1. 980.7 N.

8–2. 27 lb, 0.17.

8–3. **(a)** 10.5°, **(b)** 31.0°.

8–5. 1659.3 N, 603.9 N.

8–6. 530.7 N.

8–7. 85 ft.

8–9. 182.0 N.

8–10. 12.

8–11. 6.

8–13. 500 mm.

8–14. 0.415.

8–15. 1.01 Mg.

8–17. 28.1°.

8–18. 2391.2 N.

8–19. 33.4°.

8–21. 38.7°.

8–22. 28.1°.

8–23. Drum slips, $O_x = O_y = 106.7$ N.

8–25. 75 lb, tipping occurs.

8–26. 264.3 N.

8–27. 1051.1 N.

8–29. 588.8 N.

8–30. 29.7 lb·ft.

8–31. 619.6 N.
8–33. 14.7 N·m.
8–34. 6.77 kN.
8–35. 48.3 N·m.
8–37. 300 N·m
8–38. 15.4 N·m.
8–39. 1036.7 N, 311.0 N, 682.4 N.
8–41. 0.53 lb.
8–42. 13.6 lb.
8–43. 545.5 N.
8–45. $T_A = 71.1$ lb, $T_B = 20.3$ lb.
8–46. 318.2 N.
8–47. 540°, 8.4 lb.
8–49. 49.6°.
8–50. 60 lb.
8–51. 1.08 mm.
8–53. 96.7 N.
8–54. $W(a_A + a_B)/2r$.
8–55. 235.4 N.
8–57. 3904 lb.
8–58. 0.325.
8–59. 1.66 kg.
8–61. 1.6 lb, 0.13.
8–62. 11 lb.
8–79. 41.6 lb.
8–81. 3.66 lb.

Chapter 9

9–1. 2.42 m, 1.31 m.
9–2. 0.2 m, 4.37 m.
9–3. 112.2 mm, 112.2 mm, 136.0 mm.
9–5. 1.63 m
9–6. 33.9 mm
9–7. 0.874 ft.
9–9. 154.4 in.
9–10. 6.42 in.
9–11. 77.8 mm.
9–13. 1.64 in.
9–14. 483 mm.
9–15. 10.48 in.
9–17. 544 mm.
9–18. 0, 0, 52.5 mm.
9–19. 58.1 mm.
9–21. $\bar{x} = \dfrac{r \sin \alpha}{\alpha}$.
9–22. 0.375 m, 0.8 m.
9–23. −0.75 m, −0.3 m.
9–25. 0, 1.6 ft.
9–26. 0.8 ft, 0.8 ft.
9–27. 0, 0.849 ft.
9–29. 1.2 m, 1.5 m.
9–30. 0, 3.33 ft, 0.
9–31. 0, 0, 0.422 m.
9–33. 0, 0, 450 mm.
9–34. 0, 0.847 mm, 0.
9–35. 21.3 in.^2, 6.0 in., 1340.4 in.^3
9–37. 5.33 ft^2, 0.75 ft, 25.1 ft^3.
9–38. $3.12(10^6)$ lb.
9–39. 1.08 m^3.
9–41. 118.4 in.^2, 59.2 in.^3
9–42. 0.0276 m^3, 138 kg.
9–43. 106.1 mm.
9–45. 3600 N ↓, 19.44 kN·m ↻.
9–46. $80.64(10^3)$ lb, 14.6 ft.
9–47. 533.3 lb ↓, 533.3 lb·ft ↻.
9–49. 1250 N, 8 m.
9–50. 10.75 kN, 2.26 m.
9–51. $40(10)^3$ lb, 8 ft.
9–53. $4rlp_o$.
9–54. 1.41 MN, 4 m.
9–55. $F_D = 100.6$ kN, $C_x = 46.6$ kN.
9–57. 88.90 kN, 2.67 m.
9–58. 45.87 kN, 0.898 m.
9–59. 2452.5 N →, 1.02 m; 9.75 kN ↓, center of plate B.
9–61. 321.2 kN/m.
9–62. 0.75 ft, 2.54 ft, 0.57 ft.
9–63. 487.5 mm.
9–65. $12.6(10^{-3})$ m^3.
9–66. 2.4 in.
9–67. 1.5 in.
9–69. 0.5 in.
9–70. −0.11 ft.
9–71. 25.13 ft^3.

Chapter 10

10–1. $17(10^6)\ \text{mm}^4$.
10–2. $2.44\ \text{in.}^2$
10–3. $47.2(10^3)\ \text{ft}^4$.
10–5. $\dfrac{\pi a^4}{8}$, $0.1098a^4$.
10–6. $11{,}429\ \text{in.}^4$
10–7. 17.89 mm.
10–9. 1.51 ft.
10–10. $114.28(10^6)\ \text{mm}^4$.
10–11. $2549.3\ \text{in.}^4$, $1117.3\ \text{in.}^4$
10–13. $30.24(10^6)\ \text{mm}^4$.
10–14. $251.3(10^3)\ \text{mm}^4$, 68.4%.
10–15. $100.8(10^6)\ \text{mm}^4$, $10.8(10^6)\ \text{mm}^4$.
10–17. $502.7\ \text{in.}^4$
10–18. $2.27\ \text{in.}^4$
10–19. $76.3(10^6)\ \text{mm}^4$.
10–21. $7.74\ \text{in.}^4$
10–23. $16.9(10^9)\ \text{mm}^4$, $4.22(10^9)\ \text{mm}^4$.
10–25. $321.7\ \text{slug}\cdot\text{ft.}^2$
10–26. $20.1\ \text{Mg}\cdot\text{m}^2$.
10–27. $52.34(10^6)\ \text{kg}\cdot\text{m}^2$.
10–29. $107.2\ \text{slug}\cdot\text{ft}^2$.
10–30. $28.09(10^3)\ \text{slug}\cdot\text{ft}^2$.
10–31. 215.8 mm, $0.0264\ \text{kg}\cdot\text{m}^2$.
10–33. $1.67\ \text{slug}\cdot\text{ft}^2$.
10–34. $89.2(10^6)\ \text{mm}^4$.
10–35. $31.42\ \text{in.}^4$, $31.42\ \text{in.}^4$
10–37. $333.3\ \text{in.}^4$
10–38. $32.4\ \text{in.}^4$
10–39. $10.05\ \text{kg}\cdot\text{m}^2$.

Chapter 11

11–1. 211.0 N.
11–2. $P = \dfrac{1}{2}\dfrac{W}{\tan\theta}$.
11–3. 78.5°.
11–5. 0°, 36.9°.
11–6. 0°, 73.1°.
11–7. 180 lb.
11–9. 1.20 lb.
11–10. $4.62\ \text{lb}\cdot\text{ft}$.
11–11. 50 lb.
11–13. $42.52\ \text{lb}\cdot\text{in}$.
11–14. 44.1 N.
11–15. 4.62 kN.
11–17. $m_t = m\left(\dfrac{s}{a}\right)$.
11–18. $P = \left(\dfrac{b-a}{2c}\right)mg$.
11–19. 311.8 lb.
11–21. $\theta_2 = \tan^{-1}\left(\dfrac{2P}{mg}\right)$, $\theta_1 = \tan^{-1}\left(\dfrac{2P}{3mg}\right)$.
11–22. $W_J = W\left(\dfrac{3l}{4a}\right)$, $W_F = W\left(\dfrac{l}{4a}\right)$.
11–23. 100 kg.

Appendix B

B–1. $2\ \text{in.}^4$
B–2. 0.
B–3. $0.667\ \text{in.}^4$
B–5. $8\ \text{in.}^4$
B–6. $17.05(10^6)\ \text{mm}^4$.
B–7. $-0.45(10^6)\ \text{mm}^4$.
B–9. 0.
B–10. $\theta = 6.08°$, $2006.6\ \text{in.}^4$, $175.4\ \text{in.}^4$
B–11. $I_u = 1.045(10^7)\ \text{mm}^4$, $I_v = 4.05(10^6)\ \text{mm}^4$, $I_{uv} = 5.54(10^6)\ \text{mm}^4$.
B–13. $\bar{y} = 12.5\ \text{mm}$, $I_u = 5.8(10^6)\ \text{mm}^4$, $I_v = 38.5(10^6)\ \text{mm}^4$.
B–14. $5.09(10^6)\ \text{mm}^4$, $5.09(10^6)\ \text{mm}^4$.
B–15. x and y are principal axes, 4.667 in., $1472\ \text{in.}^4$, $2496\ \text{in.}^4$
B–17. $85.3(10^6)\ \text{mm}^4$, $85.3(10^6)\ \text{mm}^4$.
B–18. $4.92(10^6)\ \text{mm}^4$, $1.36(10^6)\ \text{mm}^4$.
B–19. $249.6\ \text{in.}^4$, $20.4\ \text{in.}^4$
B–21. $116.4(10^6)\ \text{mm}^4$, $29.7(10^6)\ \text{mm}^4$.

Index

Mechanics for Engineers
DYNAMICS

MECHANICS FOR ENGINEERS

DYNAMICS

Russell C. Hibbeler

Macmillan Publishing Company
New York
Collier Macmillan Publishers
London

Printed in the United States of America.

Macmillan Publishing Company
866 Third Avenue, New York, New York 10022

Collier Macmillan Canada, Inc.

Library of Congress Cataloging in Publication Data

Hibbeler, R. C.
Mechanics for engineers—dynamics.

Rev. ed. of: Engineering mechanics—dynamics. 3rd ed. 1983. Includes index.
1. Dynamics. I. Hibbeler, R. C. Engineering mechanics—dynamics. II. Title.
TA352.H54 1985 531'.3'02462 84-3859
ISBN 0-02-354260-8

Printing: 1 2 3 4 5 6 7 8 Year: 5 6 7 8 9 0 1 2 3

ISBN 0-02-354260-8

Preface

The purpose of this book is to provide the student with a clear and thorough presentation of the theory and application of the principles of engineering mechanics. Emphasis is placed on developing the student's ability to analyze problems—a most important skill for any engineer. In order to achieve this objective the contents of each chapter are organized into well defined sections. Selected groups of sections contain an explanation of specific topics, a "procedure for analysis" which provides a systematic approach for applying the theory, illustrative example problems, and a set of homework problems.

Numerous problems in the book depict realistic situations encountered in engineering practice. It is hoped that this realism will both stimulate the student's interest in engineering mechanics and provide a means for developing the skill to reduce any such problem from its physical description to a model or symbolic representation to which the principles of mechanics may be applied. Both SI and FPS units are used equally throughout the book. Furthermore, in any set, the problems are arranged in order of increasing difficulty* and the answers to all but every fourth problem, which is indicated by an asterisk, are listed in the back of the book.

Most of the text material has been organized so that topics within each section are placed into subgroups defined by boldface titles. The purpose of this is to present a structured method for introducing each new definition or concept, and to provide a convenient means for later reference or review. As stated above, a "procedure for analysis" is used in many sections of the book. These guides provide the student with a logical and orderly method to follow when applying the theory. Most often the first step in the procedure

*Review problems, at the end of each chapter, are presented in random order.

will require drawing a diagram. By doing so, the student forms the habit of tabulating the necessary data while focusing on the physical aspects of the problem and its associated geometry. If this step is correctly performed, applying the relevant equations of mechanics becomes somewhat methodical, since the data can be taken directly from the diagram. The example problems are solved using this outlined method in order to clarify its numerical application.

Since mathematics provides a systematic means of applying the principles of mechanics, the student is expected to have prior knowledge of algebra, geometry, trigonometry, and some calculus. Occasionally, the example problems are solved using several different methods of analysis so that the student develops the ability to use mathematics as a tool, whereby the solution of any problem may be carried out in the most direct and effective manner.

Contents. The contents of this book are presented in 10 chapters. In particular, the kinematics of a particle is discussed in Chapter 12,* followed by a discussion of particle kinetics in Chapter 13 (equation of motion), Chapter 14 (work and energy), and Chapter 15 (impulse and momentum). A similar sequence of presentation is given for the planar motion of a rigid body: Chapter 16 (planar kinematics), Chapter 17 (equation of motion), Chapter 18 (work and energy), and Chapter 19 (impulse and momentum). If desired, it is possible to cover Chapters 12 through 19 in the following order with no loss in continuity: Chapters 12 and 16 (kinematics), Chapters 13 and 17 (equation of motion), Chapters 14 and 18 (work and energy), and Chapters 15 and 19 (impulse and momentum).

Time permitting, some of the material included in Chapter 20, (special applications) may be covered in the course, at points where the instructor finds it beneficial. Chapter 21 (vibrations) may be included if the student has the necessary mathematical background. Sections of the book which are considered to be beyond the scope of the basic dynamics course are indicated by a star and may be omitted. Note, however, that this material provides a suitable reference for basic principles when it is covered in more advanced courses.

Acknowledgements. I have endeavored to write this book so that it will appeal to both the student and the instructor. Many people helped in its development. I wish to acknowledge the valuable suggestions and comments made by T. G. Carley, University of Tennessee; M. K. Kurtz, Florida Institute of Technology; W. Liddell, Jr., Auburn University; L. R. Mack, University of Texas; V. C. Matzen, North Carolina State University; J. P. Uldrick, United States Naval Academy; and William H. Walston, Jr., University of Maryland. Many thanks are also extended to all of my students and to those in the teaching profession who have taken the time to send me their suggestions and comments. Although the list is too long to mention, it is hoped that those who

*The first 11 chapters of this sequence form the contents of *Mechanics for Engineers: Statics*.

have given help will accept this anonymous recognition. Lastly, I should like to acknowledge the assistance of my wife, Conny, who has once again been quite helpful in preparing the manuscript for publication.

Russell Charles Hibbeler

Contents

Kinematics of a Particle

12

Kinetics of a Particle: Force and Acceleration

Kinetics of a Particle: Work and Energy

93

Kinetics of a Particle: Impulse and Momentum

129

Planar Kinematics of a Rigid Body

173

Planar Kinetics of a Rigid Body: Force and Acceleration

17

Planar Kinetics of a Rigid Body: Work and Energy

Planar Kinetics of a Rigid Body: Impulse and Momentum

19

Special Applications

Vibrations

APPENDIXES

Mathematical Expressions

Answers

Index

Mechanics for Engineers
DYNAMICS

60 mph = 88 ft/sec

12 Kinematics of a Particle

geometry of motion

Engineering mechanics consists of the study of both statics and dynamics. *Statics* deals with the equilibrium of bodies at rest or moving with constant velocity, whereas *dynamics* deals with bodies having accelerated motion. In general, dynamics is divided into two parts: *kinematics,* which is concerned with the geometrical aspects of motion, and *kinetics,* which is concerned with the analysis of the forces causing the motion. For simplicity in presenting the theory of both kinematics and kinetics, in this book particle dynamics will be discussed first, followed by topics in rigid-body dynamics.

We will begin our study of dynamics by discussing particle kinematics, and therefore we must limit application to those objects that have dimensions that are of no consequence in the analysis of the problem. In most problems encountered, one is interested in bodies of finite size, such as rockets, projectiles, or vehicles. Such objects may be considered as particles, provided motion of the body is characterized by motion of its mass center and any rotation of the body is neglected.

This chapter begins with a study of the absolute motion of a particle, which is motion measured with respect to a fixed coordinate system. In this regard, motion along a straight line will be studied before the more general motion along a curved path. Afterward, the relative motion between two particles will be considered, using a translating coordinate system.

12.1 Rectilinear Kinematics

Rectilinear motion occurs when a particle moves along a straight-line path. The kinematics of the motion is characterized by specifying the particle's position, velocity, and acceleration.

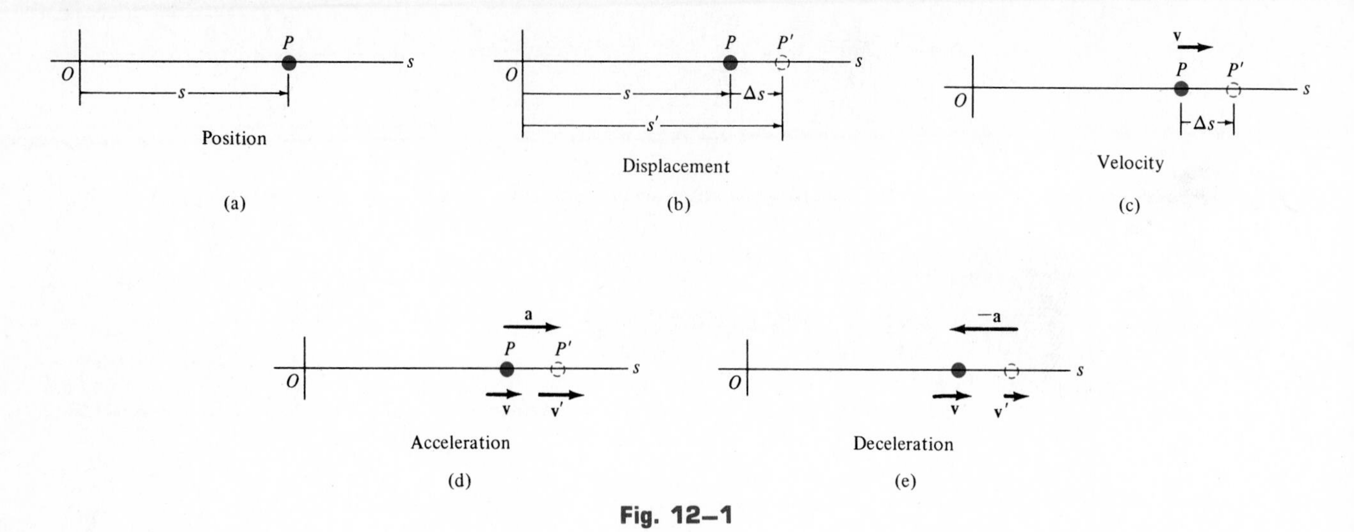

Fig. 12–1

Position. Consider the particle at point P shown in Fig. 12–1*a*. The coordinate s, which is measured from the fixed origin O, is used to define the *position* of the particle at any given instant. Since motion is along a straight-line path, the position coordinate has the properties of a vector. We can regard its *magnitude* as the *distance* from O to P, usually measured in meters (m) or feet (ft), and its *direction* by the algebraic sign of s. Although the choice is arbitrary, in this case the direction of s is positive when the particle is located to the right of the origin and negative when it is located to the left.

Displacement. The *displacement* of the particle is defined as the *change* in its *position*. For example, if the particle moves from P to P', Fig. 12–1*b*, the displacement is $\Delta s = s' - s$. Here Δs is *positive* since the particle's final position is to the *right* of its initial position, i.e., $s' > s$. Likewise, if the final position is to the *left*, Δs is *negative*.

The displacement of a particle, which is a vector, should be distinguished from the distance the particle travels. Specifically, the *distance traveled* is a scalar quantity and represents the total length of path traversed by the particle. This length is *always positive*.

Velocity. If the particle moves through a displacement Δs from P to P' during the time interval Δt, Fig. 12–1*c*, the *average velocity* of the particle during this time interval is

$$v_{\text{avg}} = \frac{\Delta s}{\Delta t}$$

If we take smaller and smaller values of Δt, the magnitude of Δs becomes smaller and smaller. Consequently, the *instantaneous velocity* is defined as $v = \lim_{\Delta t \to 0} (\Delta s/\Delta t)$ or

$$v = \frac{ds}{dt} \tag{12–1}$$

The sign used to define the *direction* of the velocity is the same as that of Δs (or ds). For example, if the particle is moving to the *right,* Fig. 12–1*c*, the velocity is *positive;* whereas if it is moving to the *left*, the velocity is *negative*. The *magnitude* of the velocity is known as the *speed* and is generally expressed in units of m/s or ft/s.

Occasionally, the term "average speed" is used. The *average speed* is a positive scalar and is defined as the total distance traveled by a particle, s_T, divided by the elapsed time Δt, i.e.,

$$(v_{sp})_{avg} = \frac{s_T}{\Delta t}$$

Acceleration. Provided the instantaneous velocities for the particle are known at the two points P and P', the *average acceleration* for the particle during the time interval Δt is defined as

$$a_{avg} = \frac{\Delta v}{\Delta t}$$

Here Δv represents the difference in the velocities during the time interval Δt, i.e., $\Delta v = v' - v$, Fig. 12–1*d*.

The *instantaneous acceleration* at time t is found by taking smaller and smaller values of Δt and corresponding smaller and smaller values of Δv, so that $a = \lim_{\Delta t \to 0} (\Delta v/\Delta t)$ or

$$a = \frac{dv}{dt} \tag{12–2}$$

Both the average and instantaneous acceleration can be either positive or negative. In particular, when the particle is *slowing down* it is said to be *decelerating*. In this case, v' in Fig. 12–1*e* is *less* than v and so $\Delta v = v' - v$ will be negative. Consequently, a will also be negative and therefore it will act to the *left,* in the opposite direction to v. Also, note that when the *velocity* is *constant,* the *acceleration* is *zero* since $\Delta v = v - v = 0$. Units commonly used to express the magnitude of acceleration are m/s^2 or ft/s^2.

A differential relation involving the displacement, velocity, and acceleration along the path may be obtained by eliminating the time differential dt between Eqs. 12–1 and 12–2 and equating. This yields

$$a\,ds = v\,dv \tag{12–3}$$

Constant Acceleration, $a = a_c$. When the acceleration is constant, each of the three kinematic equations $a_c = dv/dt$, $v = ds/dt$, and $a_c\, ds = v\, dv$ may be integrated to obtain formulas that relate a_c, v, s, and t.

Velocity as a Function of Time. Integrate $a_c = dv/dt$, assuming that initially $v = v_0$ at $t = 0$.

$$\int_{v_0}^{v} dv = \int_0^t a_c\, dt$$

$$v - v_0 = a_c(t - 0)$$

$$v = v_0 + a_c t \qquad (12\text{–}4)$$

Position as a Function of Time. Integrate $v = ds/dt = v_0 + a_c t$, assuming that initially $s = s_0$ at $t = 0$.

$$\int_{s_0}^{s} ds = \int_0^t (v_0 + a_c t)\, dt$$

$$s - s_0 = v_0(t - 0) + a_c(\tfrac{1}{2}t^2 - 0)$$

$$s = s_0 + v_0 t + \tfrac{1}{2}a_c t^2 \qquad (12\text{–}5)$$

Velocity as a Function of Position. Either solve for t in Eq. 12–4 and substitute into Eq. 12–5, or integrate $v\, dv = a_c\, ds$, assuming that initially $v = v_0$ when $s = s_0$.

$$\int_{v_0}^{v} v\, dv = \int_{s_0}^{s} a_c\, ds$$

$$\tfrac{1}{2}v^2 - \tfrac{1}{2}v_0^2 = a_c(s - s_0)$$

$$v^2 = v_0^2 + 2a_c(s - s_0) \qquad (12\text{–}6)$$

The magnitudes and signs of s_0, v_0, and a_c, used in these equations, are determined from the chosen origin and positive direction of the s axis.

It is important to remember that the above equations are useful *only when the acceleration is constant.* A common example of this motion occurs when a body falls freely toward the earth. If air resistance is neglected and the distance of fall is short, then the constant *downward* acceleration of the body when it is close to the earth is approximately 9.81 m/s^2 or 32.2 ft/s^2.*

*The proof is given in Example 13–2.

PROCEDURE FOR ANALYSIS

Often a mathematical relationship between *any two* of the four variables a, v, s, and t can be established by observation or experiment. When this is the case, the relationships between the remaining variables can be obtained by either differentiation or integration, using the kinematic equations $a = dv/dt$, $v = ds/dt$, or $a\,ds = v\,dv$.* Since each of these equations relates *three* variables, then, if a variable is *known* as a function of another variable, a third variable can be determined by *choosing the kinematic equation which relates all three*. For example, suppose that the *acceleration* is known as a function of *position*, $a = f(s)$. The *velocity* can be determined from $a\,ds = v\,dv$ since $f(s)$ can be substituted for a to yield $f(s)\,ds = v\,dv$. Solution for v requires integration.† Note that the velocity *cannot* be obtained by using $a = dv/dt$, since $f(s)\,dt = dv$ *cannot* be integrated.

Whenever the kinematic equations are applied, one should specify the positive direction of the position coordinate. By doing so, the *directions* of s, v, and a can then be determined from the *algebraic signs* of their numerical quantities. In the following examples we will indicate the positive coordinate direction alongside each kinematic equation as it is applied.

*Some standard differentiation and integration formulas are given in Appendix A.

†The position and velocity must be known at a given instant in order to evaluate either the constant of integration if an indefinite integral is used, or the limits of integration if a definite integral is used.

Example 12–1

The bicyclist in Fig. 12–2 has an acceleration of 2 ft/s². If he starts from rest, determine his velocity and position at $t = 5$ s.

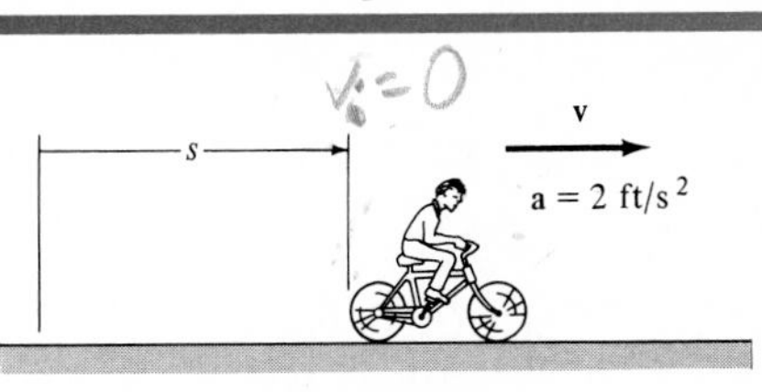

Fig. 12–2

Solution

Velocity. This problem can be solved using the formulas of *constant acceleration* with $a_c = 2$ ft/s². Since the time limit is specified, and initially $v_0 = 0$, the velocity is determined from

$$(\overset{+}{\rightarrow}) \qquad v = v_0 + a_c t$$

$$v = 0 + 2(5) = 10 \text{ ft/s} \qquad \textit{Ans.}$$

Position. In a similar manner, the bicyclist's position is determined from

$$(\overset{+}{\rightarrow}) \qquad s = s_0 + v_0 t + \tfrac{1}{2} a_c t^2$$

$$s = 0 + 0(5) + \tfrac{1}{2}(2)(5)^2 = 25 \text{ ft} \qquad \textit{Ans.}$$

Since the final velocity has been computed, it is also possible to obtain the position as follows:

$$(\overset{+}{\rightarrow}) \qquad v^2 = v_0^2 + 2a_c(s - s_0)$$

$$(10)^2 = (0)^2 + 2(2)(s - 0)$$

$$s = 25 \text{ ft} \qquad \textit{Ans.}$$

Example 12–2

The car in Fig. 12–3 moves in a straight line such that for a short time its velocity is defined by $v = (9t^2 + 2t)$ ft/s, where t is measured in seconds. Determine its position and acceleration when $t = 3$ s.

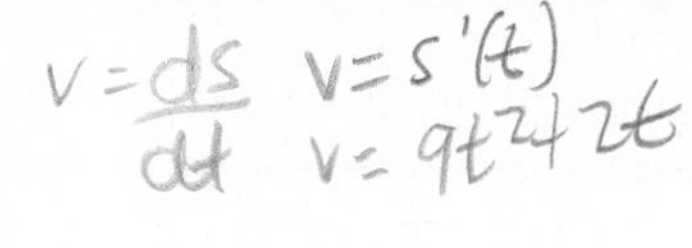

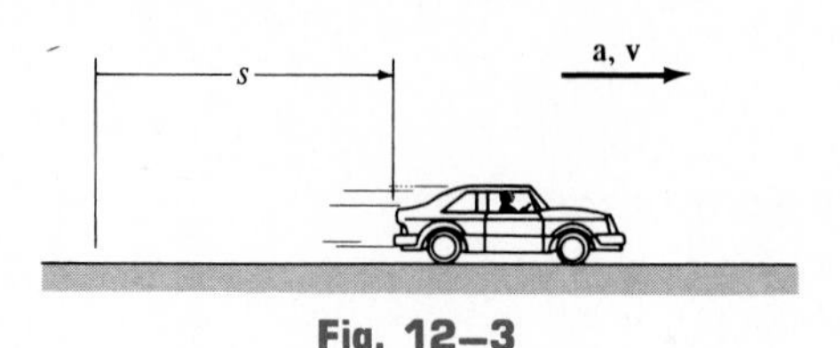

Fig. 12–3

Solution

Position. The car's velocity is given as a function of time so that its position can be determined from $v = ds/dt$, since this equation relates v, s, and t. Noting that $s_0 = 0$ at $t_0 = 0$, we have*

$$(\overset{+}{\rightarrow}) \qquad v = \frac{ds}{dt} = (9t^2 + 2t)$$

$$\int_0^s ds = \int_0^t (9t^2 + 2t)\, dt$$

$$s\Big|_0^s = \frac{9}{3}t^3 + \frac{2}{2}t^2\Big|_0^t$$

$$s = 3t^3 + t^2$$

When $t = 3$ s,

$$s = 3(3)^3 + (3)^2 = 90 \text{ ft} \qquad \textit{Ans.}$$

Acceleration. The acceleration is determined from $a = dv/dt$, since this equation relates a, v, and t.

$$(\overset{+}{\rightarrow}) \qquad a = \frac{dv}{dt} = \frac{d}{dt}(9t^2 + 2t)$$

$$= 18t + 2$$

When $t = 3$ s,

$$a = 18(3) + 2 = 56 \text{ ft/s}^2 \qquad \textit{Ans.}$$

*The *same result* can be obtained by evaluating a constant of integration C rather than using definite limits on the integral. For example, integrating $ds = (9t^2 + 2t)\, dt$ yields $s = 3t^3 + t^2 + C$. Using the condition that at $t_0 = 0$, $s_0 = 0$, then $C = 0$.

Example 12–3

A metallic particle is subjected to the influence of a magnetic field such that it travels vertically through a fluid that extends from plate A to plate B, Fig. 12–4. If the particle is released from rest at C, $s = 100$ mm, and the acceleration is measured as $a = (4s)$ m/s^2, where s is in meters, determine the velocity of the particle when it reaches plate B; $s = 200$ mm.

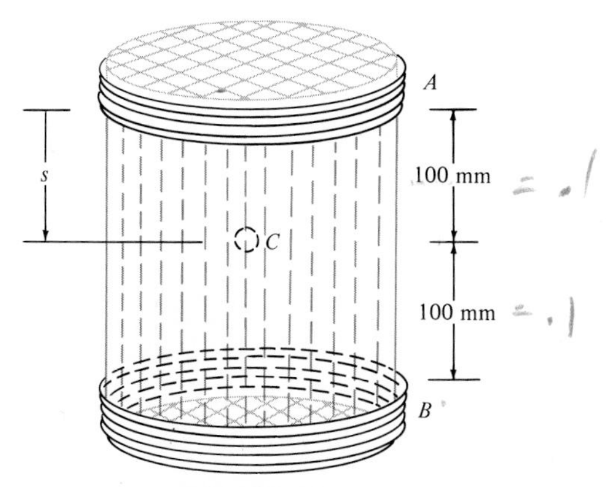

Fig. 12–4

Solution

Velocity. Knowing the acceleration as a function of position, the velocity as a function of position can be obtained by using $v\,dv = a\,ds$. Why? Realizing that $v_0 = 0$ at $s_0 = 100$ mm $= 0.1$ m, we have

$(+\downarrow)$
$$v\,dv = a\,ds$$
$$\int_0^v v\,dv = \int_{0.1}^s 4s\,ds$$
$$\frac{1}{2}v^2\Big|_0^v = \frac{4}{2}s^2\Big|_{0.1}^s$$
$$v = 2(s^2 - 0.01)^{1/2} \qquad (1)$$

When $s = 200$ mm $= 0.2$ m,

$$v_B = 0.3464 \text{ m/s} = 346.4 \text{ mm/s} \qquad \textit{Ans.}$$

The positive root is chosen since the particle is traveling downward, i.e., in the $+s$ direction.

Note: Since the velocity is known as a function of position, Eq. (1), the *time* for the particle to travel from C to B can be obtained by integrating $v = ds/dt$, where $s_0 = 0.1$ m at $t_0 = 0$.

Example 12–4

A small projectile is fired vertically downward into a fluid medium with an initial velocity of 60 m/s. If fluid resistance causes a deceleration of the projectile which is equal to $a = (-0.4v^3)$ m/s^2, where v is measured in m/s, determine the velocity v in 4 s after the projectile is fired.

Solution

Velocity. The acceleration is given as a function of velocity so that the velocity as a function of time can be obtained from $a = dv/dt$, since this equation relates v, a, and t. (Why not use $v = v_0 + a_c t$?) Integrating, with the initial condition that $v_0 = 60$ m/s at $t_0 = 0$, yields

$$(+\downarrow) \qquad a = \frac{dv}{dt} = -0.4v^3$$

$$\left(\frac{-1}{0.4}\right)\int_{60}^{v} v^{-3}\,dv = \int_0^t dt$$

$$\left(\frac{-1}{0.4}\right)\left(\frac{1}{-2}\right) v^{-2}\Big|_{60}^{v} = t$$

$$\frac{1}{0.8}\left[\frac{1}{v^2} - \frac{1}{(60)^2}\right] = t$$

$$v = \left\{\left[\frac{1}{(60)^2} + 0.8t\right]^{-1/2}\right\} \text{ m/s} \qquad (1)$$

Here the positive root is taken, since the projectile is moving downward. When $t = 4$ s,

$$v = 0.559 \text{ m/s} \qquad \textit{Ans.}$$

Note: Since the velocity is known as a function of time, Eq. (1), the position as a function of time can be obtained by integrating $v = ds/dt$, using the initial condition $s_0 = 0$ at $t_0 = 0$.

Example 12–5

A boy tosses a ball in the vertical direction off the side of a cliff, as shown in Fig. 12–5. If the initial velocity of the ball is 15 m/s upward, and the ball is released 40 m from the bottom of the cliff, determine the maximum height s_B reached by the ball and the speed of the ball just before it hits the ground. During the entire time the ball is in motion, it is subjected to a constant downward acceleration of 9.81 m/s^2 due to gravity. Neglect the effect of air resistance.

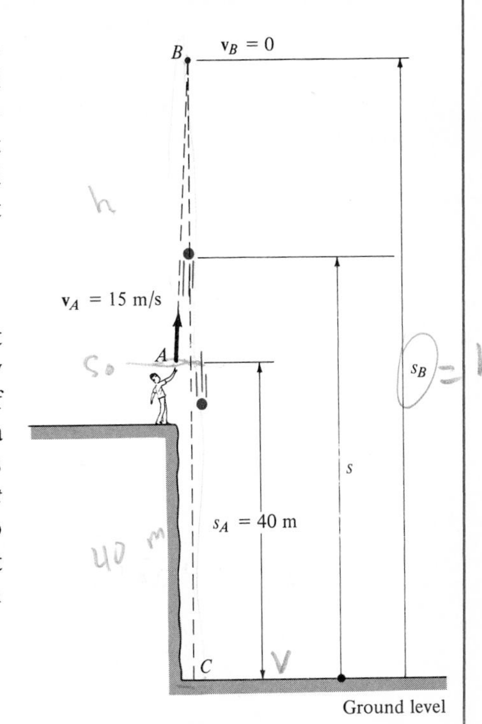

Fig. 12–5

Solution

Maximum Height. The origin for the position coordinate is taken at ground level as shown in the figure. At the maximum height s_B the velocity $v_B = 0$. Furthermore, the ball is thrown from an initial height of $s_A = +40$ m. Since the ball is thrown *upward* at $t = 0$, it is subjected to a velocity of $v_A = +15$ m/s (positive since it is in the same direction as positive displacement). For the entire motion, the acceleration is *constant* such that $a_c = -9.81$ m/s^2 (negative since it acts in a direction *opposite* to positive velocity or positive displacement). Since a_c is *constant* throughout the entire motion, the position may be related to the velocity at points A and B as follows:

$$(+\uparrow) \qquad v_B^2 = v_A^2 + 2a_c(s_B - s_A)$$
$$0 = (15)^2 + 2(-9.81)(s_B - 40)$$

so that

$$s_B = 51.5 \text{ m} \qquad \textit{Ans.}$$

Velocity. To obtain the velocity of the ball just before it hits the ground, Eq. 12–6 can be applied between points B and C, Fig. 12–5,

$$(+\uparrow) \qquad v_C^2 = v_B^2 + 2a_c(s_C - s_B)$$
$$= 0 + 2(-9.81)(0 - 51.5)$$
$$v_C = -31.8 \text{ m/s} \qquad \textit{Ans.}$$

The negative root was chosen since the ball is moving *downward*.

Similarly, Eq. 12–6 may also be applied between points A and C, i.e.,

$$(+\uparrow) \qquad v_C^2 = v_A^2 + 2a_c(s_C - s_A)$$
$$= 15^2 + 2(-9.81)(0 - 40)$$
$$v_C = -31.8 \text{ m/s} \qquad \textit{Ans.}$$

Note: It should be realized that the ball is subjected to a *deceleration* from A to B of 9.81 m/s^2, and then from B to C it is *accelerated* at this rate. Furthermore, even though the ball momentarily comes to *rest* at B ($v_B = 0$) the acceleration is 9.81 m/s^2 downward!

Example 12–6

A particle moves along a horizontal straight line such that its velocity is given by $v = (3t^2 - 6t)$ m/s, where t is the time in seconds. If it is initially located at the origin O, determine the distance traveled during the time interval $t = 0$ to $t = 3.5$ s, the average velocity, and the average speed of the particle during this time interval.

Solution

Distance Traveled. Since the velocity is related to time, a function that relates position to time may be found by integrating $v = ds/dt$ with the condition that at $t_0 = 0$, $s_0 = 0$.

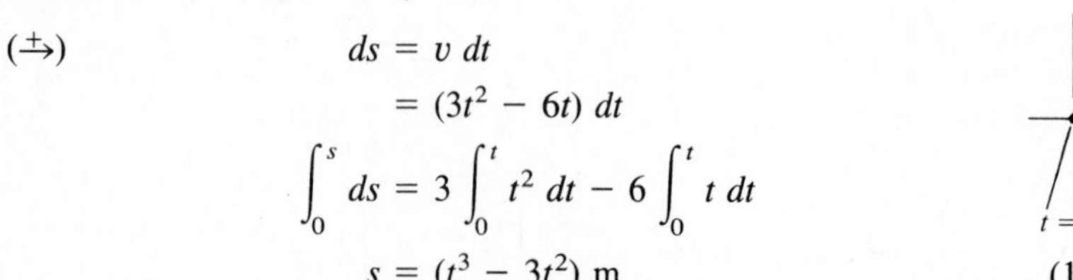

$(\overset{+}{\rightarrow})$

$$ds = v\,dt$$

$$= (3t^2 - 6t)\,dt$$

$$\int_0^s ds = 3\int_0^t t^2\,dt - 6\int_0^t t\,dt$$

$$s = (t^3 - 3t^2)\text{ m} \qquad (1)$$

(a)

(b)

Fig. 12–6

In order to determine the distance traveled in 3.5 s, it is necessary to investigate the path of motion during this time. The graph of the velocity function, Fig. 12–6*a*, reveals that for $0 \leq t < 2$ s the velocity is *negative*, which means the particle is traveling to the *left*, and for $t > 2$ s the velocity is *positive* and hence the particle is traveling to the *right*. Also, since the velocity changes sign when $t = 2$ s, the particle reverses its direction at this instant. The particle's position when $t = 0$, $t = 2$ s, and $t = 3.5$ s can be computed from Eq. (1). This yields

$$s|_{t=0} = 0, \qquad s|_{t=2\text{ s}} = -4\text{ m}, \qquad s|_{t=3.5\text{ s}} = 6.12\text{ m}$$

The path is shown in Fig. 12–6*b*. Hence, the distance traveled in 3.5 s is

$$s_T = 4 + 4 + 6.12 = 14.12\text{ m} \qquad \textit{Ans.}$$

Velocity. The *displacement* from $t = 0$ to $t = 3.5$ s is

$$\Delta s = s|_{t=3.5\text{ s}} - s|_{t=0} = 6.12 - 0 = 6.12\text{ m}$$

so that the average velocity is

$$v_{avg} = \frac{\Delta s}{\Delta t} = \frac{6.12}{3.5 - 0} = 1.75\text{ m/s} \qquad \textit{Ans.}$$

The average speed is defined in terms of the *distance traveled* s_T. Hence,

$$(v_{sp})_{avg} = \frac{s_T}{\Delta t} = \frac{14.12}{3.5 - 0} = 4.03\text{ m/s} \qquad \textit{Ans.}$$

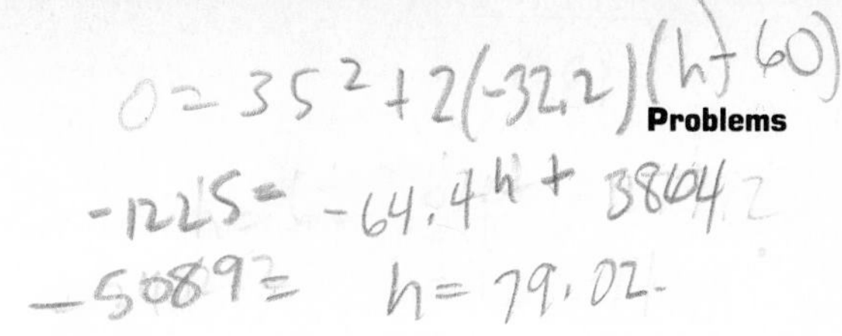

Problems

12–1. After traveling a distance of 150 m, a car reaches a speed of 85 km/h, starting from rest. Determine the car's *constant* acceleration.

12–2. A car is traveling at a speed of 80 ft/s when the brakes are suddenly applied, causing a constant deceleration of 10 ft/s^2. Determine the time required to stop the car and the distance traveled before stopping.

12–3. A missile is fired vertically such that its altitude is defined by $s = (2t^3 + 5t^2 + 14t)$ ft, where t is measured in seconds. Determine the missile's position, velocity, and acceleration when $t = 5$ s.

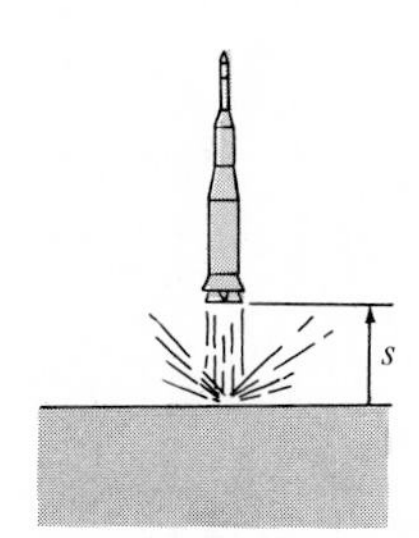

Prob. 12–3

***12–4.** A particle is moving along a straight line through a fluid medium such that its speed is measured as $v = (2t)$ m/s, where t is in seconds. If it is released from rest at $s = 0$, determine its position when $t = 3$ s.

12–5. If a particle has an initial velocity of $v_0 = 12$ ft/s to the right, determine its position when $t = 10$ s, if $a = 2$ ft/s^2 to the left. Originally $s_0 = 0$.

12–6. A particle is moving along a straight-line path such that its position is defined by $s = (9t^2 + 15)$ ft, where t is measured in seconds. Determine (a) the displacement of the particle during the time interval from $t = 1$ s to $t = 4$ s, (b) the average velocity of the particle during this time interval, and (c) the acceleration when $t = 1$ s.

12–7. A ball is thrown vertically upward from the top of a building with an initial velocity of $v_A = 35$ ft/s. Determine (a) how high above the top of the building the ball will go before it stops at B, (b) the time t_{AB} it takes to reach its maximum height, and (c) the total time t_{AC} needed for it to reach the ground at C from the instant it is released.

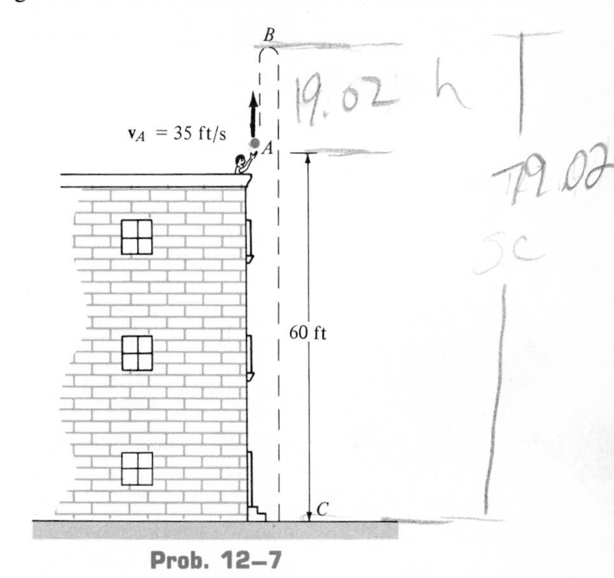

Prob. 12–7

***12–8.** A car is initially traveling to the right with a speed of 25 ft/s. If it is subjected to a constant deceleration of 5 ft/s^2 (directed to the left), determine its velocity when $t = 4$ s. What is the displacement during this time interval?

12–9. The cork from a champagne bottle is fired vertically upward. If it takes a *total* of 6 s to rise from the bottle top and then fall to its initial elevation, determine the velocity in m/s at which it left the bottle. What is the total distance the cork travels during this time? *Note:* In reality, air resistance on a light object, such as a cork, has an appreciable effect on the motion and should be accounted for in the analysis. This is discussed in Chapter 13.

12–10. Starting from rest, a particle moving in a straight line has an acceleration of $a = (2t - 6)$ m/s^2, where t is in seconds. What is the particle's velocity when $t = 6$ s, and what is its position when $t = 11$ s?

12–11. Initially an object is moving to the right at the rate of 8 m/s when it passes a given point O. If it then receives a constant deceleration (to the left) of 0.5 m/s^2, determine its position from O and the speed it will have when $t = 20$ s.

***12–12.** A ball is thrown vertically upward from the top of a 30-m-high building with an initial velocity of 5 m/s. At the same instant another ball is thrown upward from the ground with an initial velocity of 20 m/s. Determine the height from the ground and the time at which they pass.

12–13. A particle moves along a straight-line path such that in 2 s it moves from an initial position $s_A = +2$ ft to a position $s_B = -3$ ft. Then in another 3 s, it moves from s_B to $s_C = +5$ ft. Determine the particle's average velocity and average speed during the 5-s time interval.

12–14. A small metal particle passes downward through a fluid medium while being subjected to the attraction of a magnetic field such that its position is observed to be $s = (15t^3 - 3t)$ mm, where t is measured in seconds. Determine (a) the particle's displacement from $t = 2$ s to $t = 4$ s, and (b) the velocity and acceleration of the particle when $t = 5$ s.

12–15. A train is initially traveling along a straight track at a speed of 90 km/h. For 6 s it is subjected to a constant deceleration of 0.5 m/s^2, and then for the next 5 s it has a constant deceleration a_c. Determine the magnitude of a_c so that the train stops at the end of the 11-s time period.

***12–16.** A particle has an initial velocity of $v_0 = 12$ ft/s to the right. Determine the total distance it travels in 10 s if $a = 2$ ft/s^2 to the left.

12–17. A sandbag is dropped from a balloon which is ascending vertically at a constant speed of 6 m/s. If the bag is released with the same upward velocity of 6 m/s at $t = 0$ and hits the ground in $t = 8$ s, determine the altitude of the balloon at the instant the bag hits the ground and the speed at which the bag hits the ground.

12–18. A car, initially at rest, moves along a straight road with constant acceleration such that it attains a velocity of 60 ft/s when $s = 150$ ft. Then after being subjected to *another* constant acceleration, it attains a final velocity of 100 ft/s when $s = 325$ ft. Determine the average velocity and average acceleration of the car for the entire 325-ft displacement.

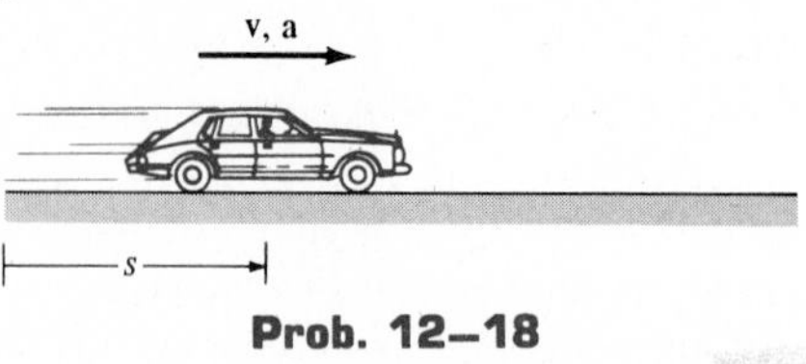

Prob. 12–18

12–19. A particle moves with accelerated motion such that $a = -ks$, where s is the distance from the starting point and k is a proportionality constant which is to be determined. When $s = 2$ ft the velocity is 4 ft/s, and when $s = 3.5$ ft the velocity is 8 ft/s. What is s when $v = 0$?

***12–20.** A projectile, initially at the origin, moves along a straight-line path through a fluid medium such that its velocity is defined as $v = 1800(1 - e^{-0.3t})$ mm/s, where t is measured in seconds. Determine the displacement of the projectile during the first 3 s.

12–21. Determine the time required for a car to travel 1 km along a road if it starts from rest, reaches a maximum speed at some intermediate point, and then stops at the end of the road. The car can accelerate at 2 m/s^2 and decelerate at 3 m/s^2.

12–22. At the same instant, two cars A and B start from rest at a stop light. Car A has a constant acceleration of $a_A = 6$ m/s^2, while car B has an acceleration of $a_B = (3t^{3/2})$ m/s^2, where t is measured in seconds. Determine the distance between the cars when A reaches a speed of $v_A = 90$ km/h.

12–23. A particle moves along a straight-line path with an acceleration of $a = (15/s)$ ft/s^2, where s is measured in feet. Determine the particle's velocity when $s = 8$ ft, if it starts from rest when $s = 1$ ft.

***12–24.** The speed of a particle traveling along a straight-line path within a liquid is measured as a function of its position as $v = (100 - s)$ mm/s, where s is given in millimeters. Determine (a) the particle's deceleration when it is located at point A, where $s_A = 75$ mm, and (b) the distance the particle travels before it stops.

12–25. As a body is projected to a high altitude above the earth's *surface,* the variation of the acceleration of gravity with respect to altitude y must be taken into account. Neglecting air resistance, this acceleration is determined from the formula $a = -g_o[R^2/(R + y)^2]$, where g_o is the constant gravitational acceleration at sea level, R is the radius of the earth, and the positive direction is measured upward. If $g_o = 9.81$ m/s^2 and $R = 6356$ km, determine the minimum initial velocity (escape velocity) at which a projectile should be shot vertically from the earth's surface so that it does not fall back to the earth. *Hint:* This requires that $v = 0$ as $y \rightarrow \infty$.

12–26. Accounting for the variation of gravitational acceleration a with respect to altitude y (see Prob. 12–25), derive an equation that relates the velocity of a freely falling particle to its altitude. Assume that the particle is released from rest at an altitude y_o from the earth's surface. With what velocity does the particle strike the earth if it is released from rest at an altitude of $y_o = 700$ km? Use the numerical data in Prob. 12–25.

Graphical Solutions 12.2

When a particle's motion during a time period is erratic, it may be difficult to obtain a continuous mathematical function to describe its position, velocity, or acceleration. Instead, the motion may best be described graphically using a series of curves. If this graph describes the behavior of any two of the variables a, v, s, t, a graph describing the behavior of the other variables can be established by using the kinematic equations $a = dv/dt$, $v = ds/dt$, $a\,ds = v\,dv$. The following situations frequently occur.

Given the s-t Graph, Construct the v-t Graph. If the position of a particle can be *experimentally determined* during a time period t, the s-t graph for the particle can be plotted, Fig. 12–7a. Since $v = ds/dt$, *the v-t graph can be established by measuring the slope (ds/dt) of the s-t graph at various times and plotting the results.* (In a graphical sense, the slope is measured with a ruler and protractor.) For example, measurement of the slopes v_0, v_1, v_2, and v_3 at the intermediate points (t_0, s_0), (t_1, s_1), (t_2, s_2), and (t_3, s_3) on the s-t graph, Fig. 12–7a, gives corresponding points on the v-t graph shown in Fig. 12–7b.

It may also be possible to establish the v-t graph *mathematically,* provided the curves of the s-t graph can be expressed in the form of equations $s = f(t)$. Corresponding equations describing curves of the v-t graph are then determined by *differentiation* since $v = ds/dt = d(f(t))/dt$.

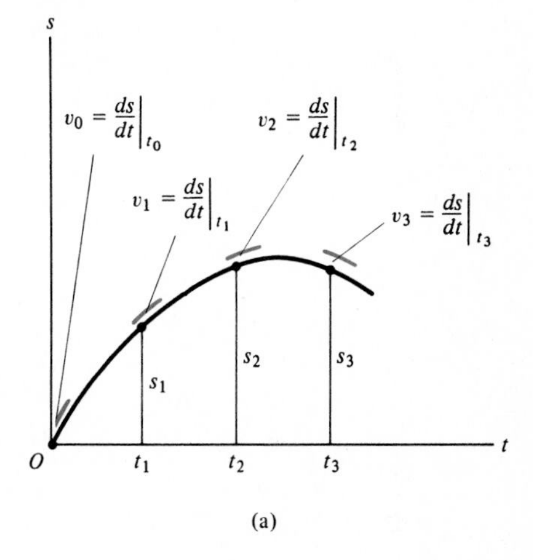

(a)

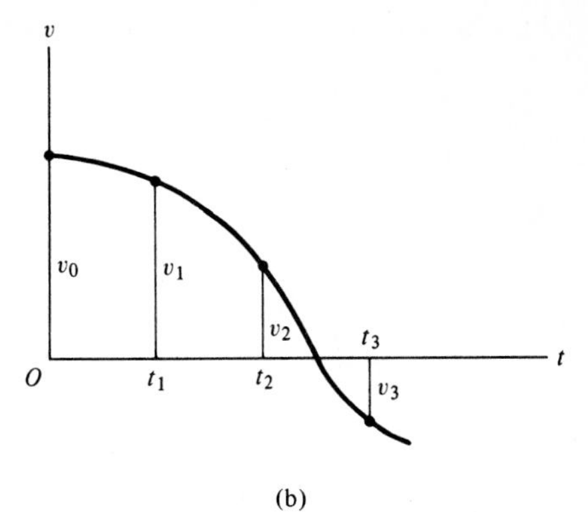

(b)

Fig. 12–7

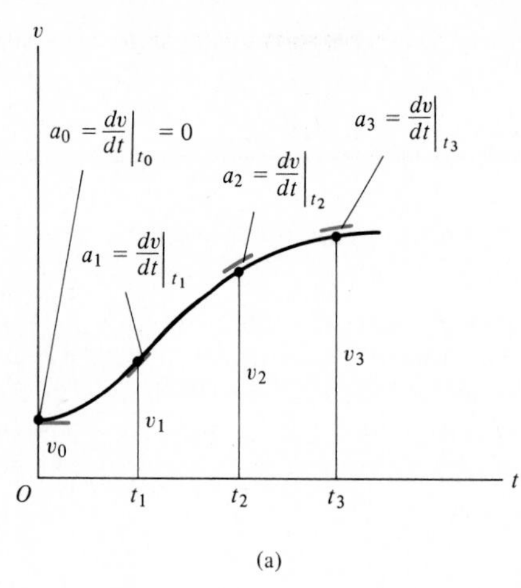

(a)

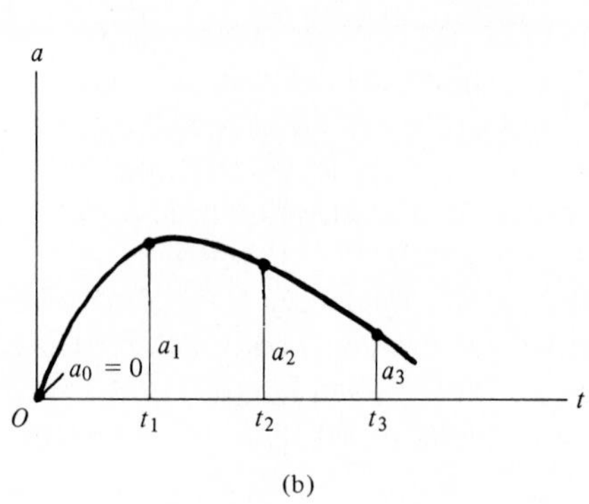

(b)

Fig. 12–8

Slope

Given the *v-t* Graph, Construct the *a-t* Graph. When the particle's *v-t* graph is known, as in Fig. 12-8*a*, the acceleration can be determined at any instant using $a = dv/dt$. Hence, *the a-t graph can be established by measuring the slope (dv/dt) of the v-t graph at various times and plotting the results.* For example, measurement of the slopes a_0, a_1, a_2, and a_3 at the intermediate points (t_0, v_0), (t_1, v_1), (t_2, v_2), and (t_3, v_3) on the *v-t* graph, Fig. 12–8*a*, yields corresponding points on the *a-t* graph shown in Fig. 12–8*b*.

The segmented curves of the *a-t* graph can also be determined *mathematically* provided the equations of the corresponding curves of the *v-t* graph are known, $v = g(t)$. This is done by simply taking the *derivative* since $a = dv/dt = d(g(t))/dt$.

Since differentiation reduces a polynomial of degree *n* to that of degree *n*-1, then from the above explanation, if the *s-t* graph is parabolic (second-degree curve), the *v-t* graph will be a sloping line (first-degree curve), and the *a-t* graph will be constant or a horizontal line (zero-degree curve).

Example 12–7

A car moves along a straight road such that its position is described by the graph shown in Fig. 12–9*a*. Construct the *v-t* and *a-t* graphs for the time period $0 \leqslant t \leqslant 30$ s.

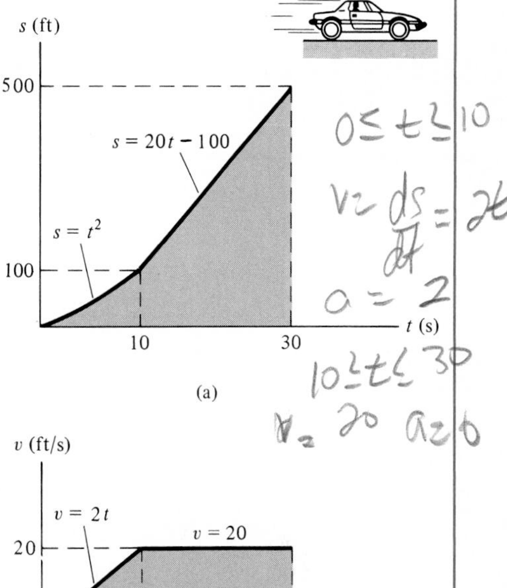

(a)

Solution

v-t Graph. Since $v = ds/dt$, the *v-t* graph can be determined by differentiating the equations defining the *s-t* graph, Fig. 12–9*a*. We have

$$0 \leqslant t < 10 \text{ s}; \qquad s = t^2 \qquad v = \frac{ds}{dt} = 2t$$

$$10 \text{ s} < t \leqslant 30 \text{ s}; \quad s = 20t - 100 \qquad v = \frac{ds}{dt} = 20$$

The results are plotted in Fig. 12–9*b*. We can also obtain specific values of v by measuring the *slope* of the *s-t* graph at a given instant. For example, at $t = 20$ s, the slope of the *s-t* graph is determined from the straight line from 10 s to 30 s, i.e.,

$$t = 20 \text{ s}; \qquad v = \frac{\Delta s}{\Delta t} = \frac{500 - 100}{30 - 10} = 20 \text{ ft/s}$$

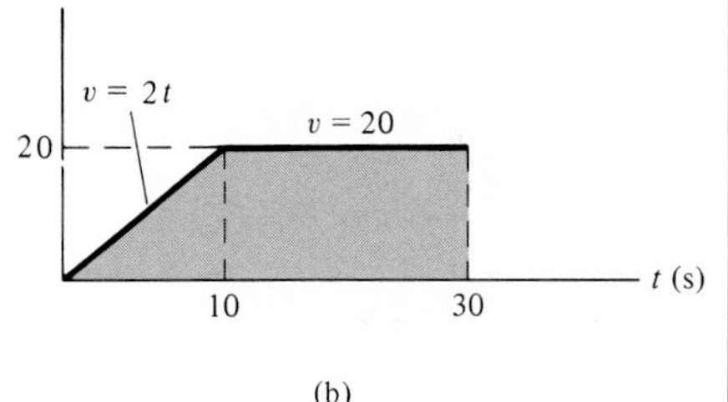

(b)

a-t Graph. Since $a = dv/dt$, the *a-t* graph can be determined by differentiating the equations defining the lines of the *v-t* graph. This yields

$$0 \leqslant t < 10 \text{ s}; \qquad v = 2t \qquad a = \frac{dv}{dt} = 2$$

$$10 < t \leqslant 30 \text{ s}; \qquad v = 20 \qquad a = \frac{dv}{dt} = 0$$

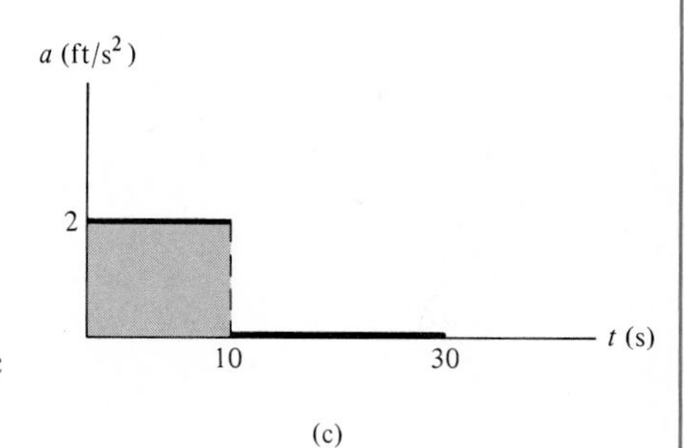

(c)

Fig. 12–9

The results are plotted in Fig. 12–9*c*. How can you determine the specific value of the car's acceleration at $t = 5$ s?

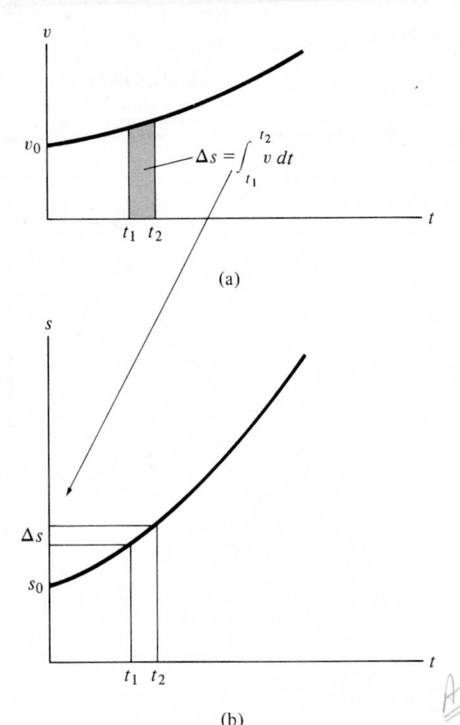

Fig. 12–10

Given the *a-t* Graph, Construct the *v-t* Graph. If the *a-t* graph is given, Fig. 12–10*a*, the *v-t* graph may be constructed using the equation $a = dv/dt$, written in integrated form as $\Delta v = \int a\,dt$. In this case, *the change in the particle's speed during a period of time is equal to the area under the a-t graph during the same time period,* Fig. 12–10*b*. (In a graphical sense, any *small* area may be approximated as a trapezoid or rectangle.) Using this method, one begins with knowing the particle's initial velocity v_0 and adding (algebraically) to this small area increments Δv determined from the *a-t* graph. In this manner, one determines successive points, $v_1 = v_0 + \Delta v$, etc., for the *v-t* graph. Notice that an algebraic addition of area is necessary, since areas lying above the *t* axis correspond to an increase in *v* (''positive'' area), whereas those lying below the *t* axis indicate a decrease in *v* (''negative'' area).

If the curves of the *a-t* graph can be described by a series of equations, then each of these equations may be *integrated* to yield equations describing the corresponding curves of the *v-t* graph. Hence, if the *a-t* graph is linear (first-degree curve), the integration will yield a *v-t* graph that is parabolic (second-degree curve), etc.

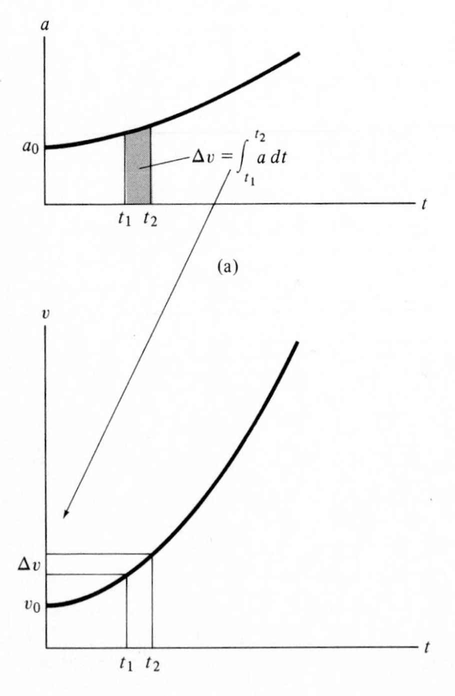

Fig. 12–11

Given the *v-t* Graph, Construct the *s-t* Graph. When the *v-t* graph is given, Fig. 12–11*a*, it is possible to determine the *s-t* graph using $v = ds/dt$, written in integrated form as $\Delta s = \int v\,dt$. In this case *the particle's displacement during a period of time is equal to the area under the v-t graph during the same time period,* Fig. 12–11*b*. In the same manner as stated above, one begins by knowing the particle's initial position s_0 and adding (algebraically) to this small area increments Δs determined from the *v-t* graph.

If it is possible to describe the segments of the *v-t* graph by a series of equations, then each of these equations may be *integrated* to yield equations describing the *s-t* graph.

Example 12–8

The rocket sled in Fig. 12–12*a* starts from rest and travels along a straight track such that it accelerates at a constant rate for $t = 10$ s and then decelerates at a constant rate. Draw the v-t and s-t graphs and determine the time t' needed to stop the sled. How far has the sled traveled?

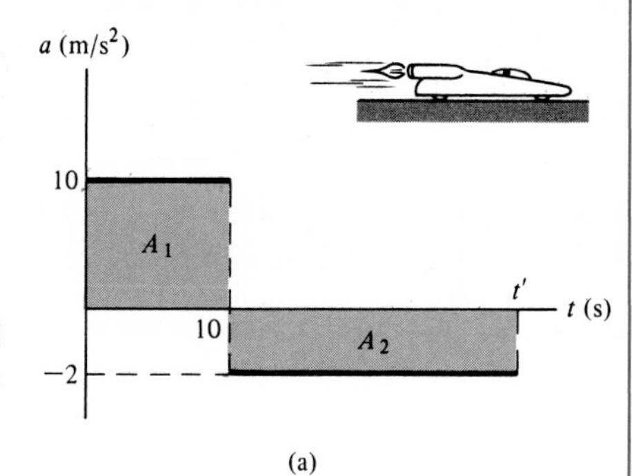

(a)

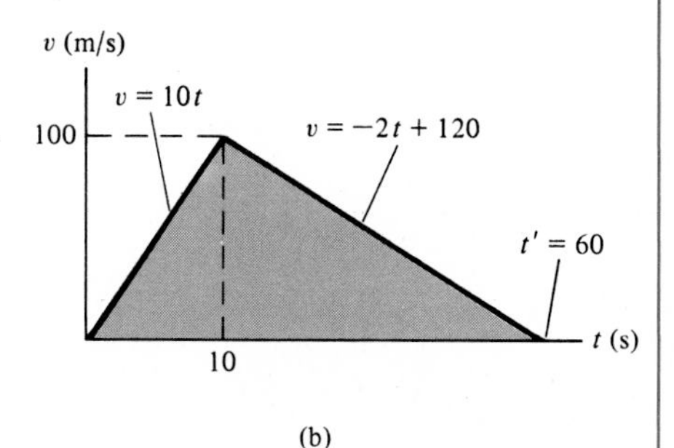

(b)

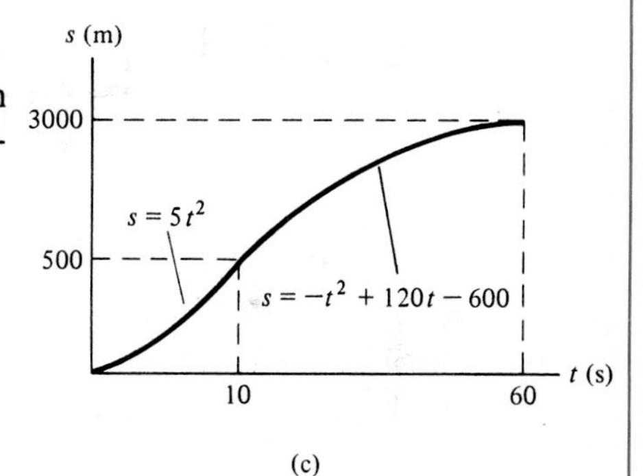

(c)

Fig. 12–12

Solution

v-t Graph. Since $dv = a\,dt$, the v-t graph is determined by integrating the straight-line segments of the a-t graph. Using the initial condition $v = 0$ at $t = 0$, we have

$$0 \leqslant t < 10 \text{ s}; \quad a = 10; \qquad \int_0^v dv = \int_0^t 10\,dt$$

$$v = 10t$$

When $t = 10$ s, $v = 10(10) = 100$ m/s. Using this as the initial condition for the next time period, we have

$$10 \text{ s} < t \leqslant t'; \quad a = -2; \qquad \int_{100}^v dv = \int_{10}^t -2\,dt$$

$$v = -2t + 120$$

At $t = t'$ we require $v = 0$. This yields

$$t' = 60 \text{ s} \qquad \textit{Ans.}$$

The results are shown in Fig. 12–12*b*.

s-t Graph. Since $ds = v\,dt$, integrating the equations of the v-t graph yields the corresponding equations of the s-t graph. Using the initial condition $s = 0$ at $t = 0$, we have

$$0 \leqslant t < 10 \text{ s}; \quad v = 10t; \qquad \int_0^s ds = \int_0^t 10t\,dt$$

$$s = 5t^2$$

When $t = 10$ s, $s = 5(10)^2 = 500$ m. Using this initial condition,

$$10 < t \leqslant 60 \text{ s}; \quad v = -2t + 120;$$

$$\int_{500}^s ds = \int_{10}^t (-2t + 120)\,dt$$

$$s - 500 = -t^2 + 120t - (-(10)^2 + 120(10))$$

$$s = -t^2 + 120t - 600$$

When $t' = 60$ s, the position is

$$s = -(60)^2 + 120(60) - 600 = 3000 \text{ m} \qquad \textit{Ans.}$$

The s-t graph is shown in Fig. 12–12*c*. Note that a direct solution for s at $t' = 60$ s is possible, since the *triangular area* under the v-t graph would yield the displacement $\Delta s = s - 0$ from $t = 0$ to $t' = 60$ s. Hence,

$$\Delta s = \tfrac{1}{2}(60)(100) = 3000 \text{ m} \qquad \textit{Ans.}$$

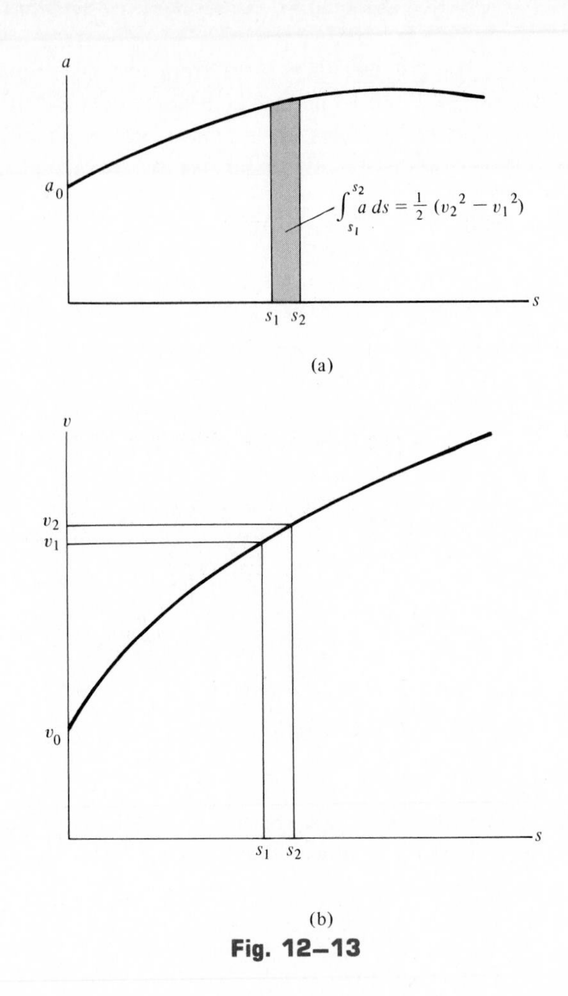

Fig. 12–13

Given the *a-s* Graph, Construct the *v-s* Graph. In some cases an *a-s* graph is known, so that points on the *v-s* graph can be determined by using $v\,dv = a\,ds$. Integrating this equation between the limits $v = v_1$ at $s = s_1$ and $v = v_2$ at $s = s_2$, we have $\frac{1}{2}(v_2^2 - v_1^2) = \int_{s_1}^{s_2} a\,ds$. Thus, small segments of area under the *a-s* graph, $\int_{s_1}^{s_2} a\,ds$, shown colored in Fig. 12–13*a*, equal one-half the difference in the squares of the speed, $\frac{1}{2}(v_2^2 - v_1^2)$. By approximation of the area, $\int_{s_1}^{s_2} a\,ds$, it is possible to compute the value of v_2 at s_2 if an initial value of v_1 at s_1 is known, i.e., $v_2 = (2\int_{s_1}^{s_2} a\,ds + v_1^2)^{1/2}$, Fig. 12–13*b*. The *v-s* graph can be constructed in this manner starting from the initial velocity v_0.

Perhaps a more direct way to construct the *v-s* graph is to first determine the equations which define the various curves of the *a-s* graph. Then the corresponding equations defining the curves of the *v-s* graph can be obtained directly from integration, using $v\,dv = a\,ds$.

Given the v-s Graph, Construct the a-s Graph. If the v-s graph is known, the acceleration a at any position s can be determined using the following graphical procedure. At any point $C(v, s)$, Fig. 12–14a, the slope dv/ds of the v-s graph is determined. The perpendicular line drawn from this slope intersects the s axis at point B. The distance AB represents the magnitude of the particle's acceleration a, Fig. 12–14b, since the two colored triangles shown in Fig. 12–14a are similar. (Recall that $a\,ds = v\,dv$, so that $\tan\theta = dv/ds = a/v$.) Of course, one must work with a consistent set of units when using this procedure; for example, if v is measured in m/s and s in meters, then a will be measured in m/s^2.

We can also determine the curves describing the a-s graph analytically, provided the equations of the corresponding curves of the v-s graph are known. As above, this requires application of $a\,ds = v\,dv$.

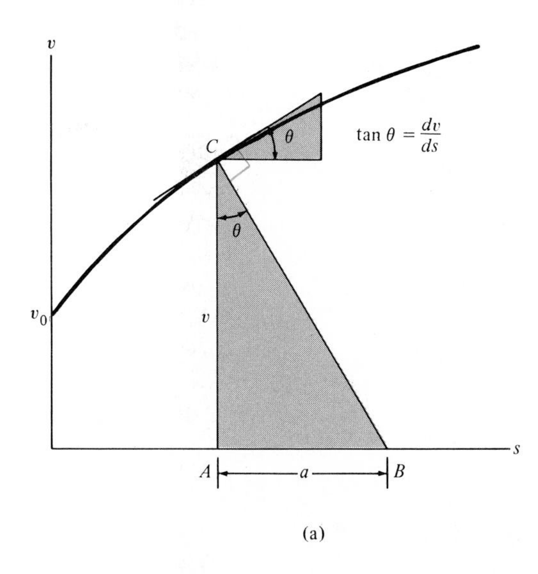

(a)

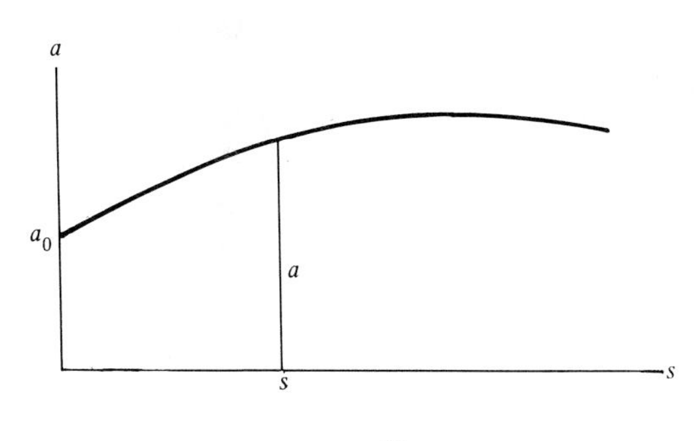

(b)

Fig. 12–14

also: $\Delta S = V_1 (t_2 - t_1) + (\text{Area}_{a\text{-}t})t_2$

Example 12–9

The v-s graph describing the motion of a motorcycle is shown in Fig. 12–15a. Construct the a-s graph of the motion.

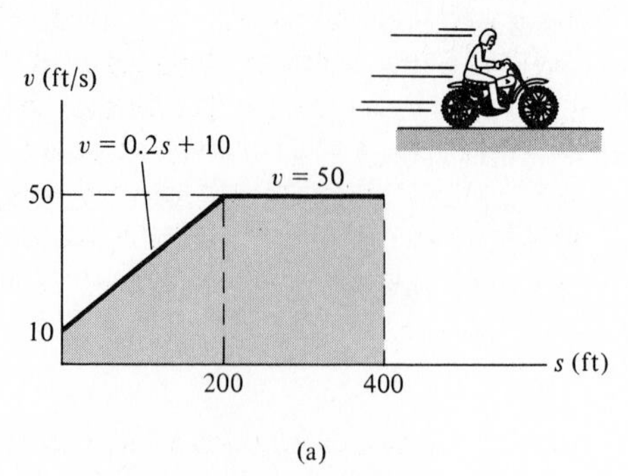

(a)

Fig. 12–15a

Solution

a-s Graph. Since the equations of the v-s graph are given, the a-s graph can be determined using the equation $a\,ds = v\,dv$, which yields

$$0 \leq s < 200 \text{ ft}; \quad v = 0.2s + 10; \qquad a = v\frac{dv}{ds} = (0.2s + 10)\frac{d}{ds}(0.2s + 10)$$

$$= 0.04s + 2$$

$$200 \text{ ft} < s \leq 400 \text{ ft}; \quad v = 50; \qquad a = v\frac{dv}{ds} = (50)\frac{d}{ds}(50)$$

$$= 0$$

The results are plotted in Fig. 12–15b.

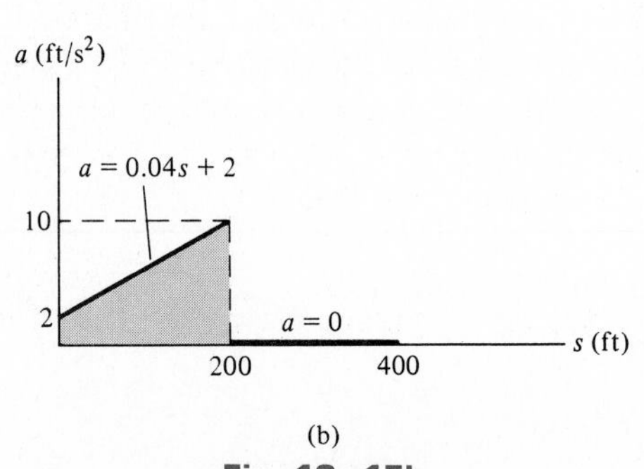

(b)

Fig. 12–15b

Problems

12–27. A car travels up a hill with the speed shown. Compute the total distance the car moves until it stops ($t = 60$ s). Plot the a-t graph.

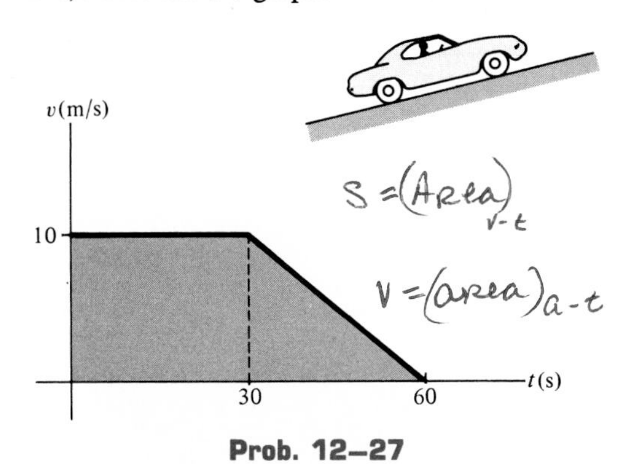

Prob. 12–27

***12–28.** The speed of a train during the first minute of its motion has been recorded as follows:

t (s)	0	20	40	60
v (m/s)	0	16	21	24

Plot the v-t graph, approximating the curve as straight line segments between the given points. Determine the total distance traveled.

12–29. A race car starting from rest moves along a straight track with an acceleration as shown, where $t \geq 10$ s, $a = 8\ \text{m/s}^2$. Determine the time t for the car to reach a speed of 50 m/s and construct the v-t graph that describes the motion until the time t.

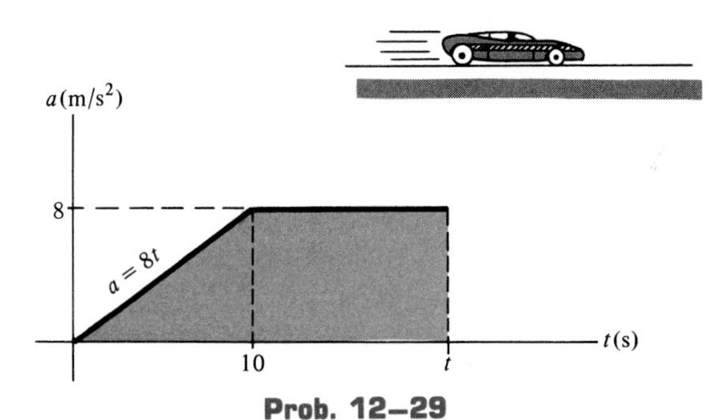

Prob. 12–29

12–30. A man riding upward in a freight elevator accidentally drops a package off the elevator when it is 100 ft from the ground. If the elevator maintains a constant upward speed of 4 ft/s, determine how high the elevator is from the ground the instant the package hits the ground. Draw the v-t curve for the package during the time it is in motion. Assume that the package was released with the same upward speed as the elevator, and that the acceleration of the package is constant, acting downward with a magnitude of $32.2\ \text{ft/s}^2$.

12–31. A two-stage missile is fired vertically from rest with an acceleration as shown. In 15 s the first stage A burns out and the second stage B ignites. Plot the v-t and s-t graphs which describe the motion of the second stage for $0 \leq t \leq 20$ s.

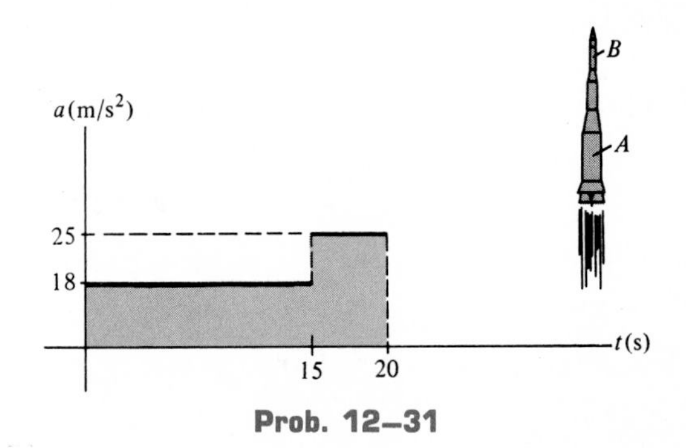

Prob. 12–31

***12–32.** From experimental data, the motion of a jet plane while traveling along a runway is defined by the v-t graph shown. Construct the s-t and a-t graphs for the motion.

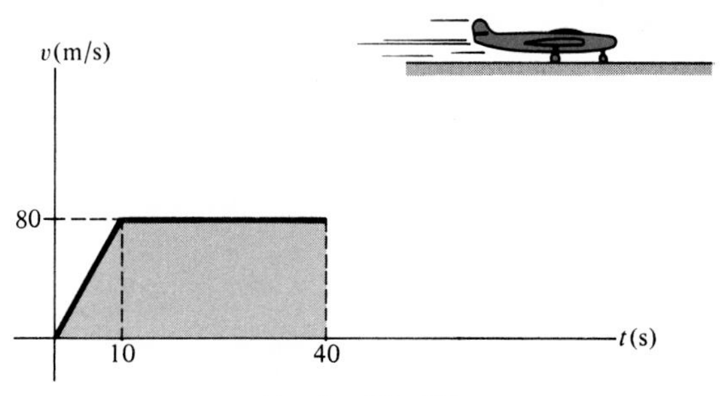

Prob. 12–32

12–33. The *s-t* graph for a train has been experimentally determined. From the data, construct the *v-t* and *a-t* graphs for the motion; $0 \leqslant t \leqslant 40$ s. For $0 \leqslant t \leqslant 20$ s, the curve is $s = (3.75t^2)$ ft, where t is in seconds.

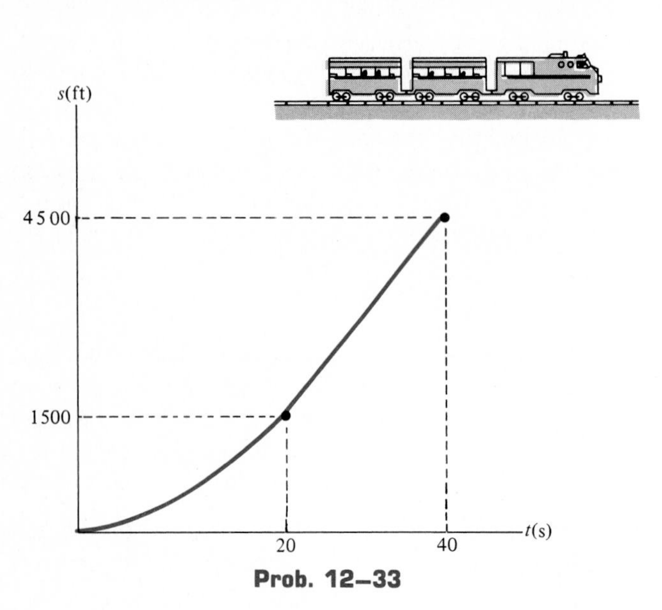

Prob. 12–33

12–34. The *v-s* graph for a rocket sled is shown. Determine the acceleration of the sled when $s = 100$ m and $s = 175$ m.

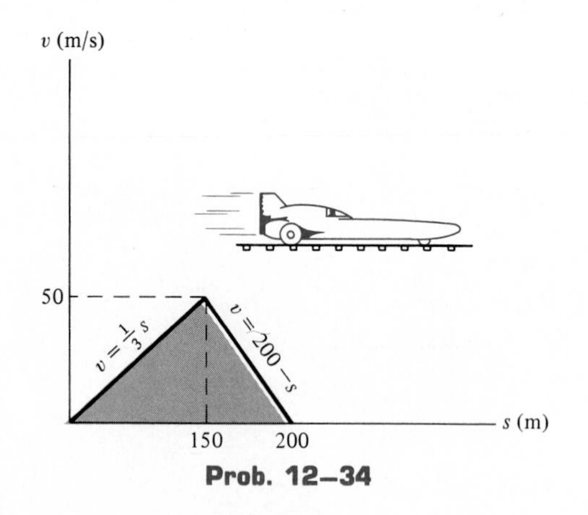

Prob. 12–34

12–35. The *a-t* graph for an automobile is shown. Plot the *v-t* and *s-t* graphs if the automobile starts from rest at $t = 0$. At what time t' does the car stop?

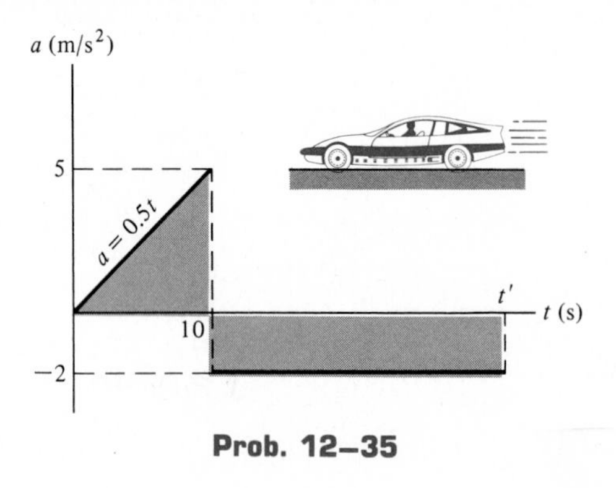

Prob. 12–35

***12–36.** The *a-s* graph for a train is given for the first 200 m of its motion. Plot the *v-s* graph. The train starts from rest.

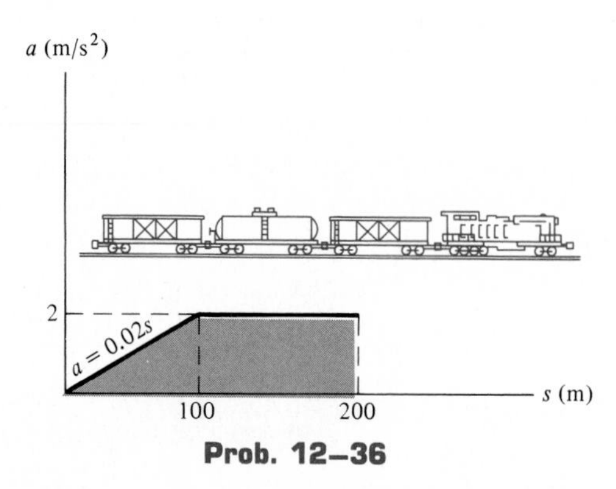

Prob. 12–36

12–37. The rocket car starts from rest and is subjected to a constant acceleration of $a_c = 15$ ft/s² for $0 \leqslant t < 10$ s.

The brakes are then applied, which causes a deceleration at the rate shown until the car stops. Determine the car's maximum speed and the time t when it stops.

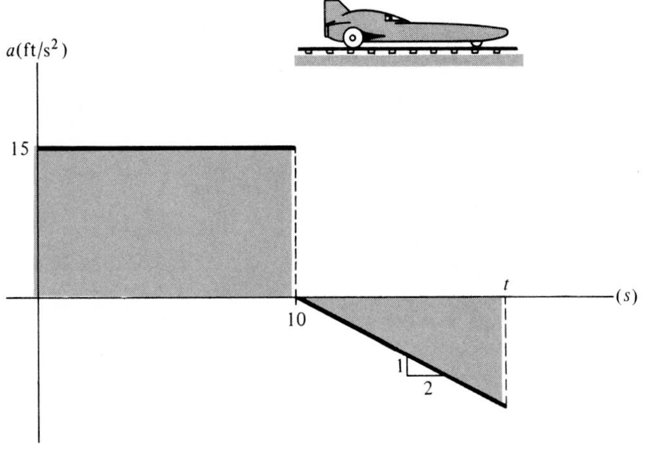

Prob. 12–37

12–38. Two cars start from rest side by side and race along a straight track. Car A accelerates at 5 m/s^2 for 10 s and then maintains a constant speed. Car B accelerates at 6 m/s^2 until reaching a constant speed of 30 m/s and then maintains this speed. Construct the a-t, v-t, and s-t graphs for each car until t = 15 s. What is the distance between the two cars when t = 15 s?

12–39. The a-s graph for a rocket moving along a straight horizontal track has been experimentally determined. If the rocket starts from rest, determine its speed at the instants s = 50 ft, 150 ft, and 200 ft, respectively.

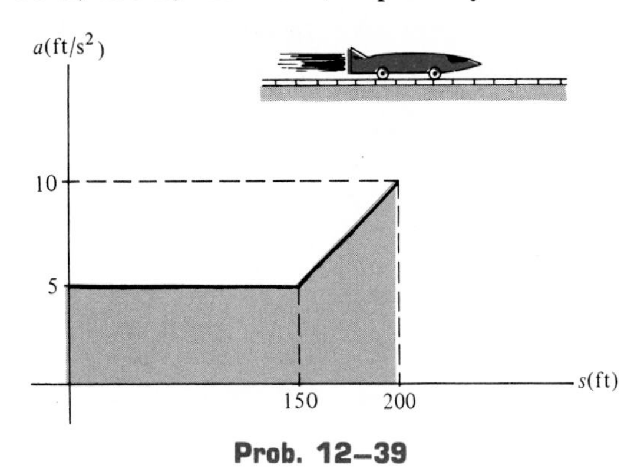

Prob. 12–39

***12–40.** A freight train starts from rest at t = 0 and travels with a constant acceleration of 0.5 ft/s^2. After a time t' it maintains a constant speed so that when t = 160 s the train has traveled 2000 ft. Determine the time t' and draw the v-t graph for the motion.

motion along curved path

General Curvilinear Motion 12.3

When a particle moves along a curved path, the motion is called *curvilinear motion*. Because the path is often represented in three dimensions, *vector analysis* will be used to formulate the particle's position, velocity, and acceleration. In this section the general aspects of curvilinear motion are discussed, and in subsequent sections rectangular, and normal and tangential coordinate systems often used to describe this motion will be introduced.*

*In Sec. 20–1 cylindrical coordinates are considered.

Position. The position of a particle, measured from a fixed point O and located at point P on a space curve, will be designated by the *position vector* $\mathbf{r} = \mathbf{r}(t)$ shown in Fig. 12–16*a*. This vector is a function of time since, in general, both its magnitude and direction change as the particle moves along the curve.

Displacement. Suppose that during a small time interval Δt the particle moves a distance Δs along the curve to a new position P', defined by $\mathbf{r}' = \mathbf{r} + \Delta\mathbf{r}$, Fig. 12–16*b*. The *displacement* $\Delta\mathbf{r}$ of the particle is determined by vector subtraction, i.e., $\Delta\mathbf{r} = \mathbf{r}' - \mathbf{r}$.

Velocity. During the time Δt, the *average velocity* of the particle is defined as

$$\mathbf{v}_{\text{avg}} = \frac{\Delta\mathbf{r}}{\Delta t}$$

The *instantaneous velocity* is determined from this equation by letting $\Delta t \to 0$ so that $\Delta\mathbf{r}$ approaches the *tangent* to the curve at point P. Hence, $\mathbf{v} = \lim_{\Delta t \to 0} (\Delta\mathbf{r}/\Delta t)$ or

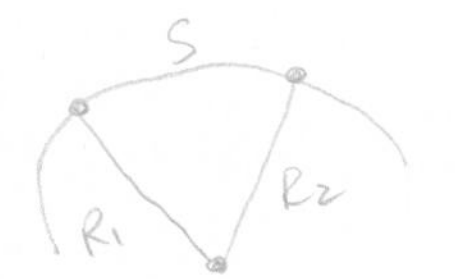

$$\mathbf{v} = \frac{d\mathbf{r}}{dt} \tag{12–7}$$

Since $d\mathbf{r}$ will be tangent to the path at P, the *direction* of $\mathbf{v}$ is *also tangent to the path,* Fig. 12–16*c*. The *magnitude* of $\mathbf{v}$, which is called the *speed,* may be obtained by noting that the magnitude of the displacement $\Delta\mathbf{r}$ is the length of the straight line segment from P to P', Fig. 12–16*b*. Realizing that this length, Δr, approaches the arc length Δs as $\Delta t \to 0$, we have $v = \lim_{\Delta t \to 0} (\Delta r/\Delta t) = \lim_{\Delta t \to 0} (\Delta s/\Delta t)$, or

$$v = \frac{ds}{dt} \tag{12–8}$$

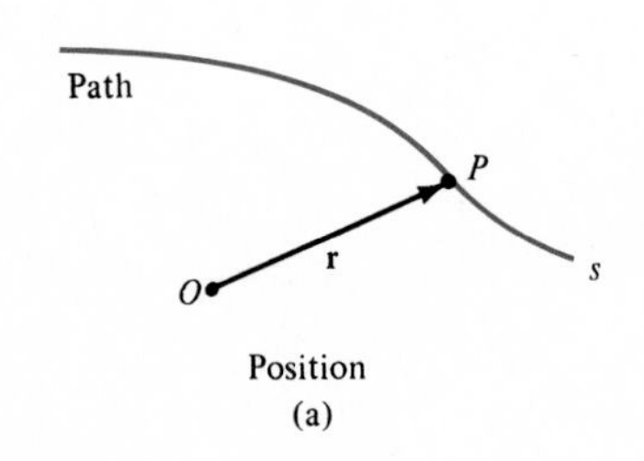

Position
(a)

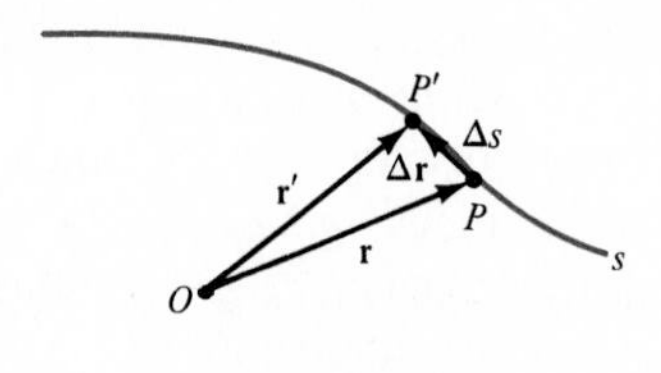

Displacement
(b)

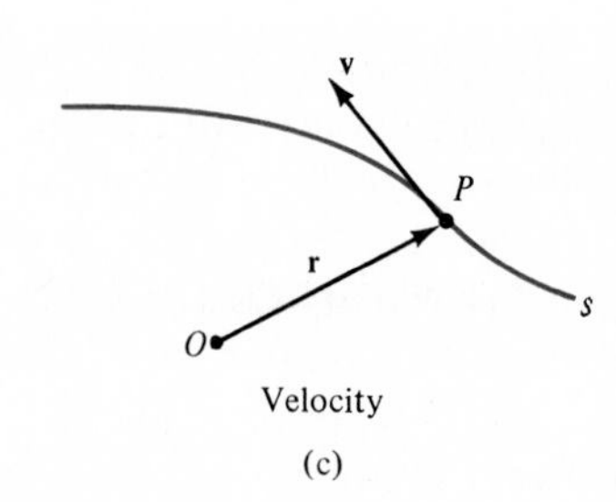

Velocity
(c)

Fig. 12–16a–c

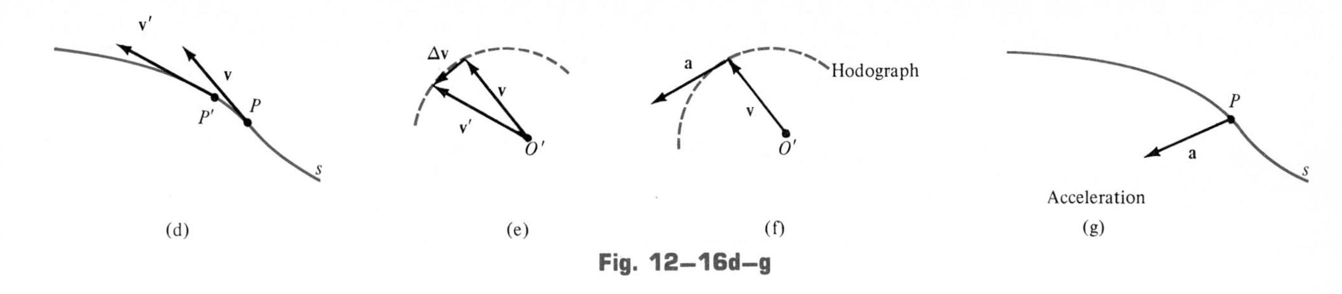

Fig. 12–16d–g

Thus, the *speed* may be obtained by differentiating the path function s with respect to time.

Acceleration. If the particle has a velocity $\mathbf{v}$ at time t and a velocity $\mathbf{v}' = \mathbf{v} + \Delta\mathbf{v}$ at $t + \Delta t$, Fig. 12–16*d*, then the *average acceleration* of the particle during the time interval Δt is

$$\mathbf{a}_{\text{avg}} = \frac{\Delta \mathbf{v}}{\Delta t}$$

where $\Delta\mathbf{v} = \mathbf{v}' - \mathbf{v}$. To study this time rate of change, the two velocity vectors in Fig. 12–16*d* are plotted in Fig. 12–16*e* such that their tails are located at the fixed point O' and their tips reach points on the dashed curve. This curve is called a *hodograph,* and when constructed, it describes the locus of points for the tip of the velocity vector in the same manner as the *path s* describes the locus of points for the tip of the position vector, Fig. 12–16*a*.

To obtain the *instantaneous acceleration,* let $\Delta t \to 0$ so that in the limit $\Delta\mathbf{v}$ becomes *tangent to the hodograph* and we have $\mathbf{a} = \lim_{\Delta t \to 0} (\Delta\mathbf{v}/\Delta t)$, or

$$\mathbf{a} = \frac{d\mathbf{v}}{dt} \tag{12–9}$$

Using Eq. 12–7, we can also write

$$\mathbf{a} = \frac{d^2\mathbf{r}}{dt^2}$$

By definition of the derivative, $\mathbf{a}$ acts *tangent to the hodograph,* Fig. 12–16*f*, and therefore, *in general,* **a** *is not tangent to the path of motion,* Fig. 12–16*g*.

12.4 Curvilinear Motion: Rectangular Components

If the path of motion of a particle lies in the x, y plane, a *fixed* set of x, y axes can be established and the motion of the particle expressed in terms of its component motions along these axes.

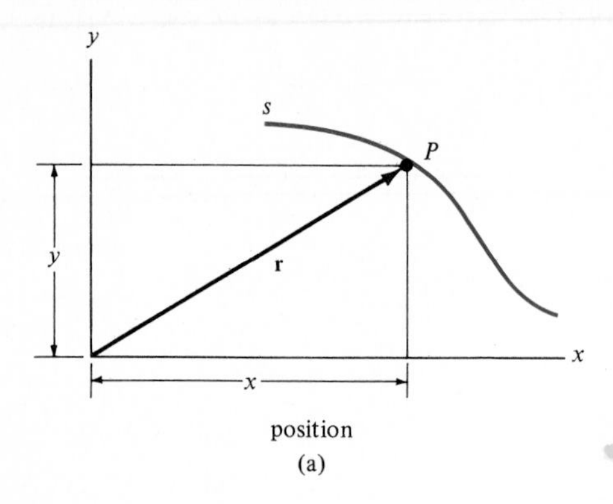

position
(a)

y
s
P
$\mathbf{v}_x$
θ_v
$\mathbf{v}_y$
$\mathbf{v}$
x

velocity
(b)

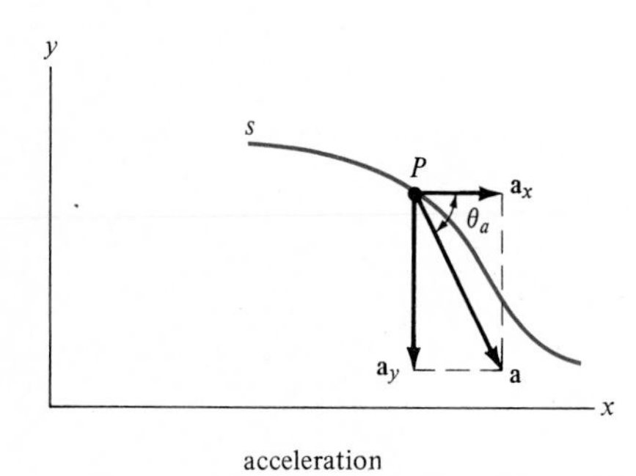

acceleration
(c)

Fig. 12–17

Position. If at a given instant the particle is located at point $P(x, y)$ along the fixed curved path s, Fig. 12–17a, its location is defined by the *position vector* $\mathbf{r}$ having components

$$r_x = x \qquad r_y = y \tag{12–10}$$

Because of the motion and the geometry of the path, the x, y components of $\mathbf{r}$ are generally functions of time, i.e., $x = x(t)$ and $y = y(t)$, so that $\mathbf{r} = \mathbf{r}(t)$.

Provided the components of $\mathbf{r}$ have been defined, Eqs. 12–10, then the *magnitude* of $\mathbf{r}$ is determined from

$$r = \sqrt{x^2 + y^2} \tag{12–11}$$

and its direction θ measured from the horizontal, Fig. 12–17a, is $\theta_r = \tan^{-1}(y/x)$.

Velocity. The first time derivative of $\mathbf{r}$ yields the velocity $\mathbf{v}$ of the particle. When taking the derivative, it is necessary to account for changes in *both* the magnitudes and directions of the vector components. However, since the x, y reference frame is *fixed,* the directions of the components do not change with time. Consequently, the time derivatives of Eqs. 12–10 yield

$$v_x = \frac{dx}{dt} \qquad v_y = \frac{dy}{dt} \tag{12–12}$$

The velocity has a *magnitude* defined by

$$v = \sqrt{v_x^2 + v_y^2} \tag{12–13}$$

and a *direction* that is specified by $\theta_v = \tan^{-1}(v_y/v_x)$. This direction is *always tangent* to the path, as shown in Fig. 12–17b.

Acceleration. The acceleration of the particle is obtained by taking the time derivative of Eqs. 12–12. We have

$$a_x = \frac{dv_x}{dt} \qquad a_y = \frac{dv_y}{dt} \tag{12–14}$$

The acceleration has a *magnitude* defined by

$$a = \sqrt{a_x^2 + a_y^2} \tag{12–15}$$

and a *direction* specified by $\theta_a = \tan^{-1}(a_y/a_x)$. Since $\mathbf{a}$ represents the time rate of *change* in velocity, in general $\mathbf{a}$ will *not* be tangent to the path traveled by the particle, Fig. 12–17c.

For Curvilinear Motion
i.e. motion along a curved path

Example 12–10

At any instant the position of the kite in Fig. 12–18*a* is defined by the coordinates $x = (30t)$ ft and $y = (9t^2)$ ft, where t is given in seconds. Determine (a) the equation which describes the path and the distance of the kite from the boy when $t = 2$ s, (b) the magnitude and direction of the velocity when $t = 2$ s, and (c) the magnitude and direction of the acceleration when $t = 2$ s.

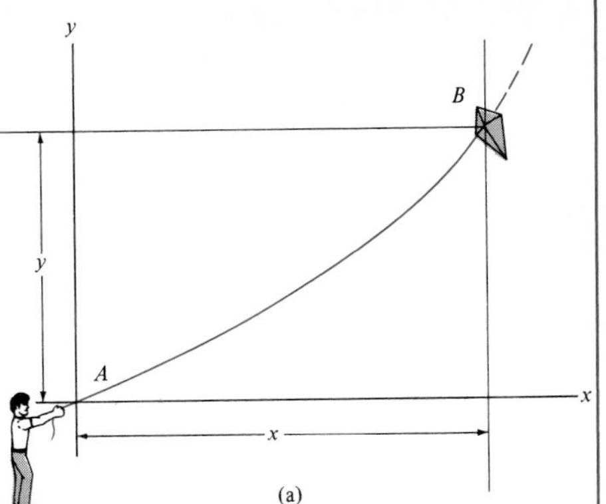

(a)

Solution

Position. The equation of the path is determined by eliminating t from the expressions for x and y, i.e., $t = x/30$ so that $y = 9(x/30)^2$ or

$$y = \frac{1}{100}x^2 \qquad \textit{Ans.}$$

This is the equation of a parabola, Fig. 12–18*a*. At $t = 2$ s, the kite is located at

$$x = 30(2) = 60 \text{ ft}, \qquad y = 9(2)^2 = 36 \text{ ft}$$

Using Eq. 12–11, the distance AB is therefore

$$r = \sqrt{(60)^2 + (36)^2} = 70.0 \text{ ft} \qquad \textit{Ans.}$$

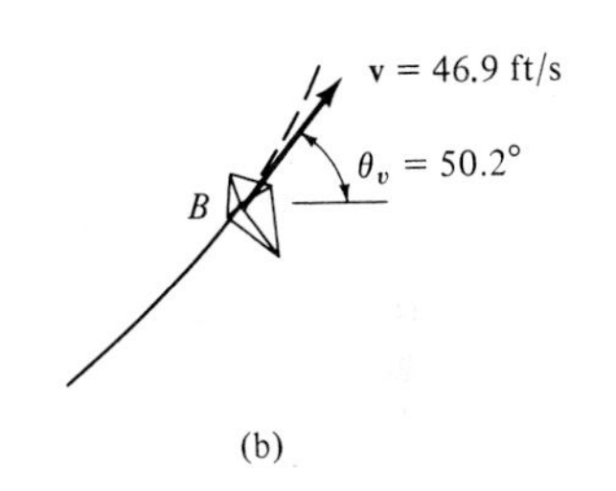

(b)

Velocity. Using Eqs. 12–12, the components of velocity at $t = 2$ s are

$$v_x = \frac{d}{dt}(30t) = 30 \text{ ft/s}$$

$$v_y = \frac{d}{dt}(9t^2) = 18t\Big|_{t=2\text{ s}} = 36 \text{ ft/s}$$

At $t = 2$ s, the magnitude of velocity is

$$v = \sqrt{(30)^2 + (36)^2} = 46.9 \text{ ft/s} \qquad \textit{Ans.}$$

The direction is tangent to the path, Fig. 12–18*b*, where

$$\theta_v = \tan^{-1}\frac{36}{30} = 50.2° \qquad \textit{Ans.}$$

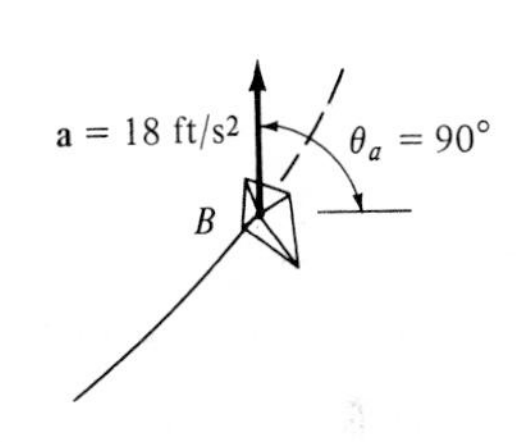

(c)

Fig. 12–18

Acceleration. The components of acceleration are determined from Eqs. 12–14,

$$a_x = \frac{d}{dt}(30) = 0$$

$$a_y = \frac{d}{dt}(18t) = 18 \text{ ft/s}^2$$

Here the acceleration is constant for all time and thus

$$a = \sqrt{(0)^2 + (18)^2} = 18 \text{ ft/s}^2 \qquad \textit{Ans.}$$

The direction, as shown in Fig. 12–18*c*, is

$$\theta_a = \tan^{-1}\frac{18}{0} = 90° \qquad \textit{Ans.}$$

12.5 Motion of a Projectile

Rectangular components are often used to study the free-flight motion of a projectile. To illustrate the concepts involved in the kinematic analysis, consider a projectile fired from a gun located at point $((s_x)_0, (s_y)_0)$, as shown in Fig. 12–19. The path is defined in the x-y plane such that the initial velocity is $\mathbf{v}_0$, having components $(\mathbf{v}_x)_0$ and $(\mathbf{v}_y)_0$. When air resistance is neglected, the only force acting on the projectile is its weight, which creates a *constant downward acceleration* of approximately $a_c = g = 9.81 \text{ m/s}^2$ or $g = 32.2 \text{ ft/s}^2$.*

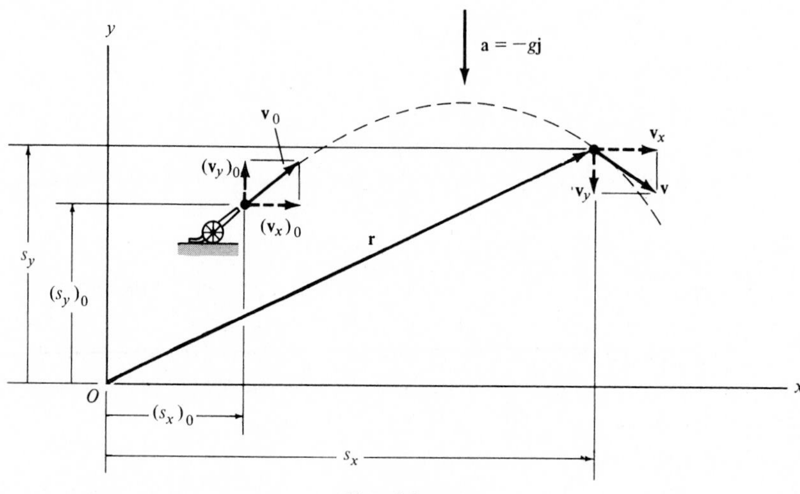

Fig. 12–19

Horizontal Motion. Since the component of acceleration in the x direction is $a_x = 0$, application of the constant acceleration equations, 12–4 to 12–6, yields

$$(\overset{+}{\rightarrow})\, v = v_0 + a_c t; \qquad v_x = (v_x)_0$$

$$(\overset{+}{\rightarrow})\, s = s_0 + v_0 t + \tfrac{1}{2} a_c t^2; \qquad s_x = (s_x)_0 + (v_x)_0 t$$

$$(\overset{+}{\rightarrow})\, v^2 = v_0^2 + 2a_c(s - s_0); \qquad v_x = (v_x)_0$$

The first and last equations indicate that *the horizontal component of velocity remains constant throughout the motion.*

Vertical Motion. Since the positive y axis is directed upward, the component of acceleration in the y direction is $a_y = -g$. Applying Eqs. 12–4 to 12–6, we get

*This assumes that the earth's gravitational field does not vary with altitude (see Example 13–2).

$$(+\uparrow)\,v = v_0 + a_c t; \qquad v_y = (v_y)_0 - gt$$

$$(+\uparrow)\,s = s_0 + v_0 t + \tfrac{1}{2}a_c t^2; \qquad s_y = (s_y)_0 + (v_y)_0 t - \tfrac{1}{2}gt^2$$

$$(+\uparrow)\,v^2 = v_0^2 + 2a_c(s - s_0); \qquad v_y^2 = (v_y)_0^2 - 2g(s_y - (s_y)_0)$$

Recall that the last equation can be formulated on the basis of eliminating the time t between the first two equations, and therefore *only two of the above three equations are independent of one another*.

At any instant, the resultant position **r** and velocity **v** are defined as the *vector sum* of their horizontal and vertical components, which are determined from the above equations and shown in Fig. 12–19.

PROCEDURE FOR ANALYSIS

Using the above results, the following procedure provides a method for solving problems concerning free-flight projectile motion.

Kinematic Diagram. Sketch the trajectory of the particle, establish the x, y coordinate axes, and between any *two points* on the path specify the given problem data and the three unknowns. In all cases the acceleration of gravity acts downward. The particle's initial and final velocities should be represented in terms of their x and y components.

Kinematic Equations. Depending upon the known data and what is to be determined, a choice should be made as to which three of the following four equations should be applied between the two points on the path to obtain the most direct solution to the problem.

Horizontal Motion. The *velocity* in the horizontal or x direction is *constant*, i.e., $(v_x) = (v_x)_0$, and

$$s_x = (s_x)_0 + (v_x)_0 t$$

Vertical Motion. In the vertical or y direction only two of the following three equations can be used for solution.

$$v_y = (v_y)_0 + a_c t$$

$$v_y^2 = (v_y)_0^2 + 2a_c(s_y - (s_y)_0)$$

$$s_y = (s_y)_0 + (v_y)_0 t + \tfrac{1}{2}a_c t^2$$

The following examples numerically illustrate application of this procedure.

Example 12–11

A cannon ball is fired from point A with a horizontal muzzle velocity of 120 m/s, as shown in Fig. 12–20. If the cannon is located at an elevation of 60 m above the ground, determine the time for the cannon ball to strike the ground and the range R.

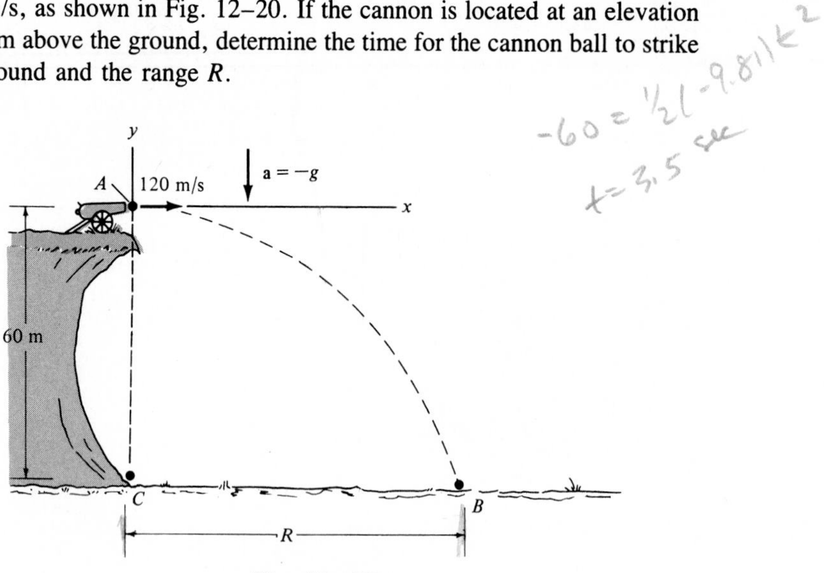

Fig. 12–20

Solution

Kinematic Diagram. The origin of coordinates is established at the initial point A, Fig. 12–20. The initial velocity of the cannon ball has components of $(v_A)_x = 120$ m/s and $(v_A)_y = 0$. Also, between points A and B the acceleration is $a_y = -9.81$ m/s^2. Since $(v_B)_x = (v_A)_x = 120$ m/s, the three unknowns are $(v_B)_y$, R, and the time of flight t.

Vertical Motion. The vertical distance from A to B is known, and therefore we can obtain a direct solution for t by using the equation

$(+\uparrow)$

$$s_y = (s_y)_0 + (v_y)_0 t + \tfrac{1}{2} a_c t^2$$

$$-60 = 0 + 0 + \tfrac{1}{2}(-9.81)t^2$$

$$t = 3.50 \text{ s}$$ ***Ans.***

[This calculation also indicates that if an object is released from rest at A at the instant the cannon is fired (horizontally), it will strike the ground at C the same instant the cannon ball strikes the ground at B.]

Horizontal Motion. Since t has been calculated, R is determined as follows:

$(\overset{+}{\rightarrow})$

$$s_x = (s_x)_0 + (v_x)_0 t$$

$$R = 0 + 120(3.50)$$

$$R = 419.7 \text{ m}$$ ***Ans.***

Example 12–12

A ball is thrown from a position 5 ft above the ground to the roof of a 40-ft-high building, as shown in Fig. 12–21. If the initial velocity of the ball is 70 ft/s, inclined at an angle of 60° from the horizontal, determine the range or horizontal distance R from the point where the ball is thrown to where it strikes the roof.

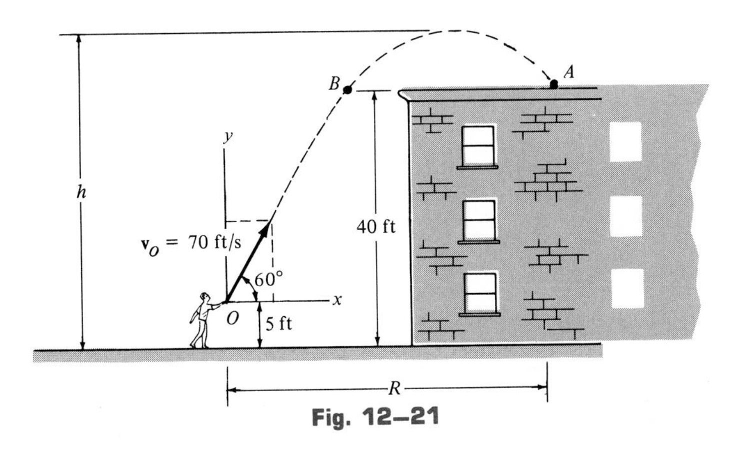

Fig. 12–21

Solution

Kinematic Diagram. When the motion is analyzed between points O and A, the three unknowns are represented as the horizontal distance R, time of flight t_{OA}, and vertical component of velocity $(v_y)_A$. With the origin of coordinates at O, Fig. 12–21, the ball's initial velocity has components of

$$(v_x)_O = 70 \cos 60° = 35 \text{ ft/s} \rightarrow$$
$$(v_y)_O = 70 \sin 60° = 60.62 \text{ ft/s} \uparrow$$

Also, $(v_x)_A = (v_x)_O = 35$ ft/s and $a_y = -32.2$ ft/s^2.

Horizontal Motion

$(\overset{+}{\rightarrow})$
$$(s_x)_A = (s_x)_O + (v_x)_O t_{OA}$$
$$R = 0 + 35 t_{OA} \tag{1}$$

Vertical Motion. Relating t_{OA} to the initial and final elevations of the ball,

$(+\uparrow)$
$$(s_y)_A = (s_y)_O + (v_y)_O t_{OA} + \tfrac{1}{2} a_c t_{OA}^2$$
$$(40 - 5) = 0 + 60.62 t_{OA} + \tfrac{1}{2}(-32.2) t_{OA}^2$$

Solving for the two roots, using the quadratic formula, we have $t_{OB} = 0.712$ s and $t_{OA} = 3.05$ s. The first root (shortest time) is designated as t_{OB}, since it represents the time needed for the ball to reach point B, which has the same elevation as point A, Fig. 12–21. Substituting t_{OA} into Eq. (1) and solving for R yields

$$R = 35(3.05) = 106.8 \text{ ft} \qquad \textit{Ans.}$$

Using only one equation, can you show that the maximum height reached by the ball, Fig. 12–21, is $h = 62.1$ ft?

Example 12–13

When a ball is kicked from A as shown in Fig. 12–22, it just clears the top of a wall at B as it reaches its maximum height. Knowing that the distance from A to the wall is 20 m and the wall is 4 m high, determine the initial speed at which the ball was kicked. Neglect the size of the ball.

Solution

Kinematic Diagram. The unknowns are represented by the initial speed v_A, angle of inclination θ, and the time t_{AB} to travel from A to B, Fig. 12–22. At the highest point, B, the velocity $(v_y)_B = 0$, and $(v_x)_B = (v_x)_A = v_A \cos\theta$.

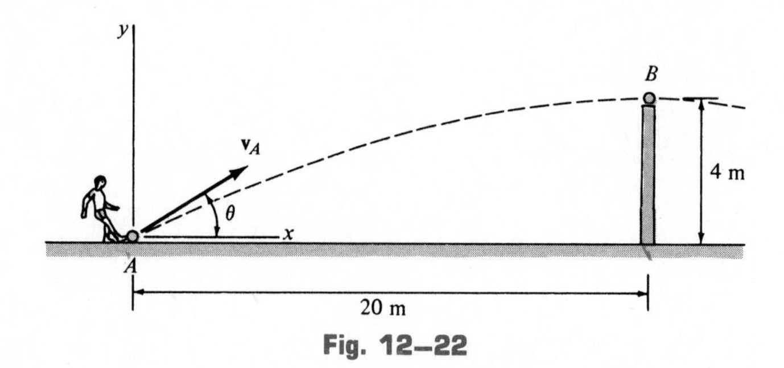

Fig. 12–22

Horizontal Motion

$(\overset{+}{\rightarrow})$

$$(s_x)_B = (s_x)_A + (v_x)_A t_{AB}$$

$$20 = 0 + (v_A \cos\theta) t_{AB} \tag{1}$$

Vertical Motion

$(+\uparrow)$

$$(v_y)_B = (v_y)_A + a_c t_{AB}$$

$$0 = v_A \sin\theta - 9.81 t_{AB} \tag{2}$$

$(+\uparrow)$

$$(v_y)_B^2 = (v_y)_A^2 + 2a_c[(s_y)_B - (s_y)_A]$$

$$0 = v_A^2 \sin^2\theta + 2(-9.81)(4 - 0) \tag{3}$$

To obtain v_A, eliminate t_{AB} from Eqs. (1) and (2), which yields

$$v_A^2 \sin\theta \cos\theta = 196.2 \tag{4}$$

Solve for v_A^2 in Eq. (4) and substitute into Eq. (3), so that

$$\frac{\sin\theta}{\cos\theta} = \tan\theta = \frac{2(9.81)(4)}{196.2} = 0.4$$

$$\theta = \tan^{-1}(0.4) = 21.8^\circ$$

Then using Eq. (4), the required initial speed is

$$v_A = \sqrt{\frac{196.2}{(\sin 21.8^\circ)(\cos 21.8^\circ)}} = 23.9 \text{ m/s}$$ ***Ans.***

Problems

12–41. A particle, originally at rest and located at point (3 ft, 2 ft), is subjected to an acceleration having components of $a_x = 6t$ ft/s^2, $a_y = 12t^2$ ft/s^2. Determine the x, y coordinates of the particle's position at $t = 1$ s.

12–42. The curvilinear motion of a particle is defined by $r_x = 4t^2$ and $r_y = 4t + 6$, where the r_x, r_y position is given in meters and the time in seconds. Determine the magnitudes and directions of the particle's velocity and acceleration when $t = 3$ s.

12–43. If the velocity of a particle has components of $v_x = 2t^2$ ft/s, $v_y = 10t^{1/2}$ ft/s, determine the magnitude and direction of the particle's acceleration when $t = 3$ s.

***12–44.** A particle moves in the x-y plane such that its position is defined by the components $r_x = (\sin^2 \theta)$ ft, $r_y = (\cos 2\theta)$ ft, where θ is in radians. If $\theta = (4t^2)$ degrees, where t is measured in seconds, determine the x, y components of the particle's velocity and acceleration when $t = 1$ s.

12–45. The flight path of a jet aircraft as it takes off is defined by the parametric equations $x = 1.25t^2$ and $y = 0.03t^3$, where t is the time after take-off, measured in seconds, and x and y are given in meters. If the plane starts to level off at $t = 40$ s, determine at this instant (a) the horizontal distance it is from the airport, (b) its altitude, (c) its speed, and (d) the magnitude of its acceleration.

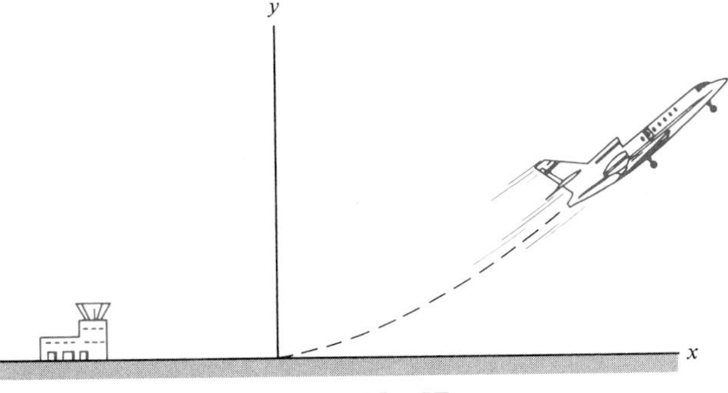

Prob. 12–45

12–46. A particle moves with curvilinear motion in the x-y plane such that the y component of motion is described by the equation $y = 4t^3$, where y is in meters and t is in seconds. If the particle starts from rest at the origin when $t = 0$, and maintains a *constant* acceleration in the x direction of 10 m/s^2, determine the particle's speed when $t = 2$ s.

12–47. For a short time the missile moves along the parabolic path $y = (18 - 2x^2)$ km. If motion along the ground is measured as $x = (4t - 3)$ km, where t is in seconds, determine the magnitudes of the missile's velocity and acceleration when $t = 1$s.

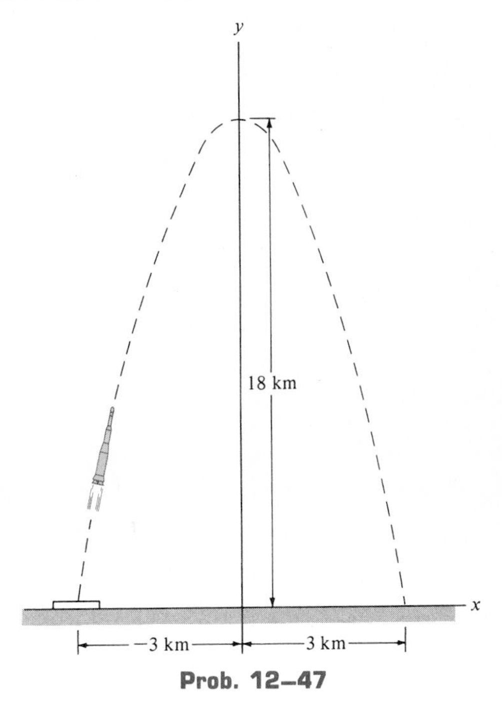

Prob. 12–47

***12–48.** A particle has curvilinear planar motion described by the parametric equations $x = 4\cos(\pi/2)t$ and $y = 2\sin(\pi/2)t$, where t is in seconds and the coordinates are in meters. Determine the equation of the path by squaring each equation and eliminating the parameter t. Then compute the particle's position, velocity, and acceleration when $t = 1$ s.

12–49. The car travels from A to B, then from B to C as shown in the figure. Compute the displacement of the car and the distance traveled.

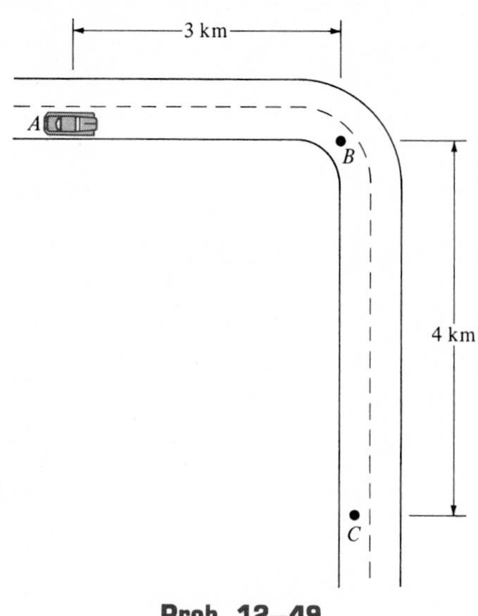

Prob. 12–49

12–50. A particle travels along the curve from A to B in 1 s. If it takes 3 s for it to go from A to C, determine its *average velocity* when it goes from B to C.

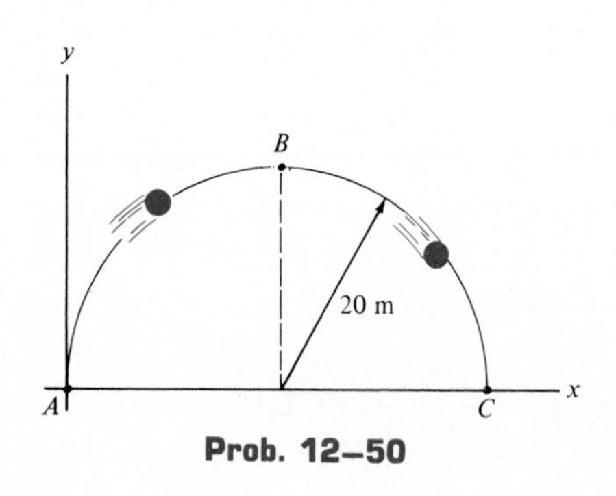

Prob. 12–50

12–51. A car traveling along a curved road has the velocities indicated on the figure when it arrives at points A, B, and C. If it takes 3 s to go from A to B, and then 5 s to go from B to C, determine the average acceleration between points A and B and between points A and C. Point B is an inflection point on the curve.

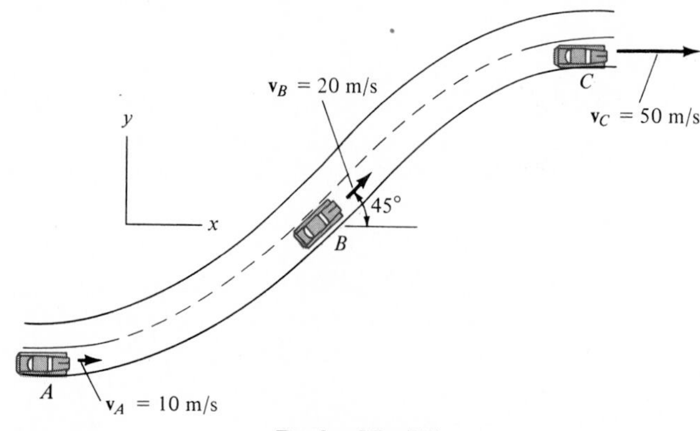

Prob. 12–51

***12–52.** The balloon A is ascending at the rate of $v_A = 10$ km/h and is being carried horizontally by the wind at $v_w = 20$ km/h. If a ballast bag is dropped from the balloon such that it takes 8 s to reach the ground, determine the balloon's altitude h at the instant the bag was released. Assume that the bag was released from the balloon with the same velocity as the balloon.

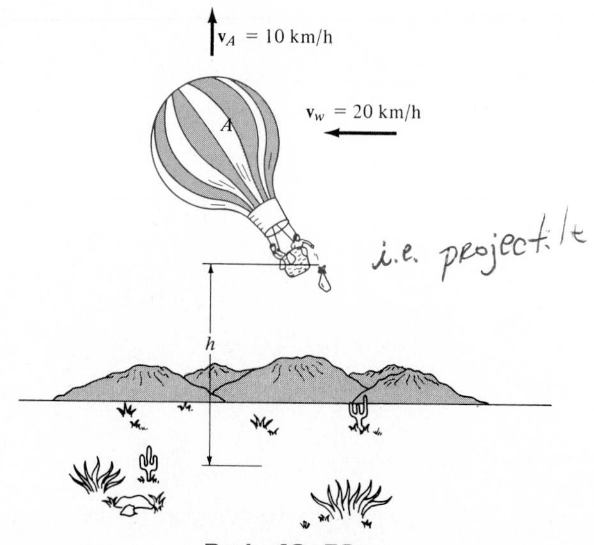

Prob. 12–52

h = 291.7 meters

12–53. A basketball is tossed from A at an angle of 30° from the horizontal. Determine the speed v_A at which the ball is released in order to make the basket B. With what speed does the ball pass through the hoop? Neglect the size of the ball in the calculation.

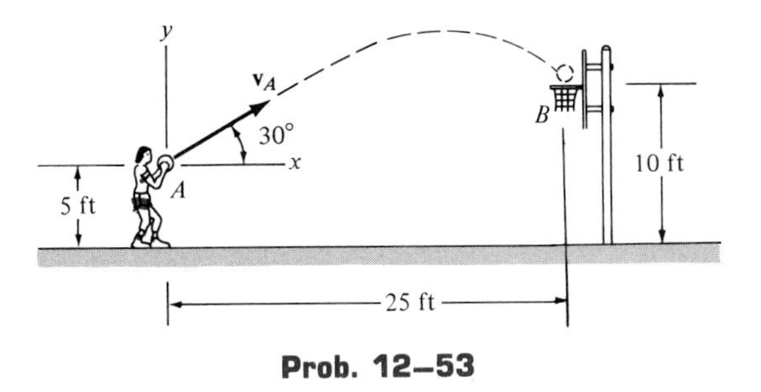

Prob. 12–53

12–54. The plane is flying horizontally with a constant speed of 200 ft/s at an altitude of 3,000 ft. If the pilot drops a package with the same horizontal speed of 200 ft/s, determine the angle θ at which he must sight the target B so that when the package is released it falls and strikes the target. Air resistance neglected, explain why the package appears to remain directly beneath the plane as it falls.

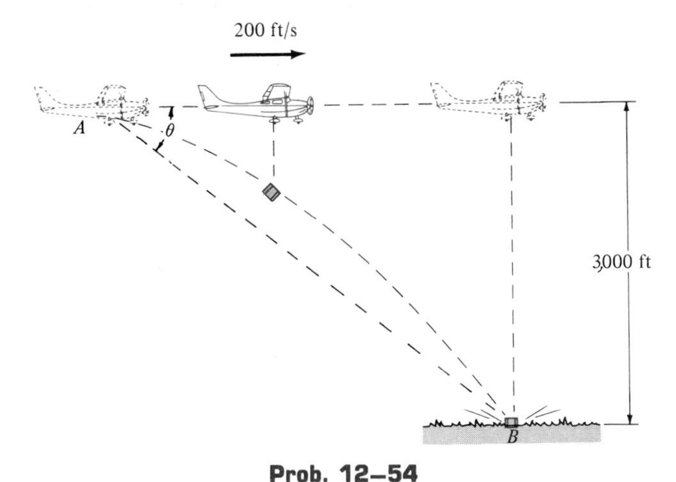

Prob. 12–54

12–55. The motorcyclist attempts to jump over a series of cars and trucks and land smoothly on the other ramp, i.e., such that his velocity is tangent to the ramp at B. Determine the launch speed v_A necessary to make the jump.

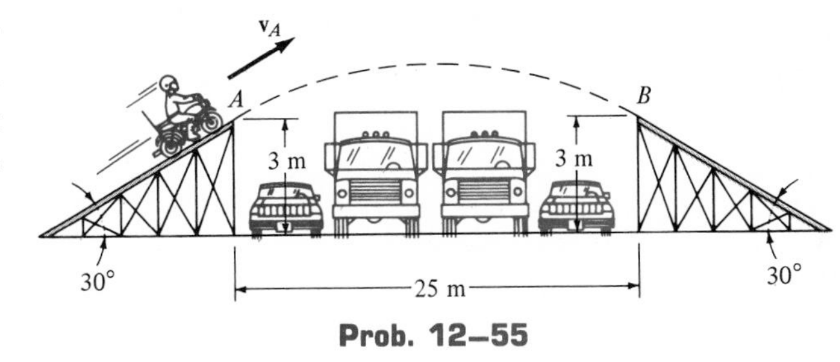

Prob. 12–55

***12–56.** The boy throws a snowball such that it strikes the wall of the building at the maximum height of its trajectory. If it takes $t = 1.5$ s to travel from A to B, determine the velocity $\mathbf{v}_A$ at which it was thrown, the angle of release θ, and the height h.

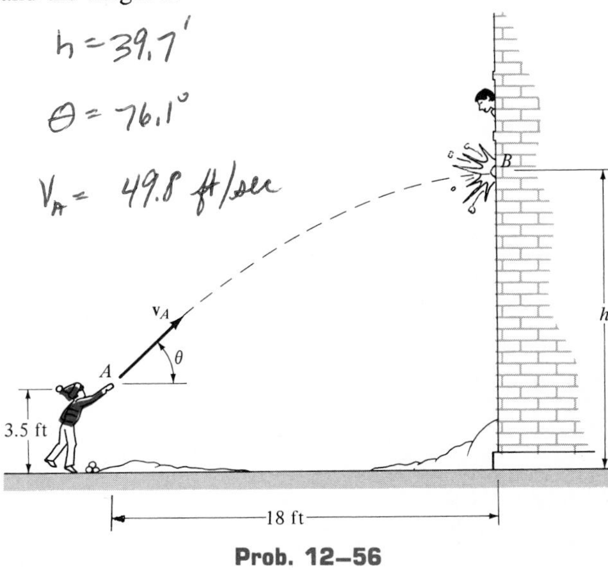

Prob. 12–56

12–57. A boy at A throws a ball 45° from the horizontal such that it strikes the slope at B. Determine the speed at which the ball is thrown and the time of flight.

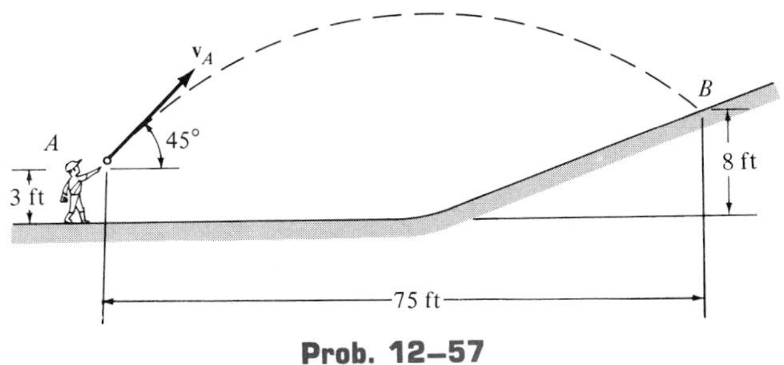

Prob. 12–57

12–58. The water sprinkler, positioned at the base of a hill, releases a stream of water with a velocity of 20 ft/s at the instant shown. Determine the point $B(x, y)$ where the water strikes the ground on the hill. Assume that the hill is defined by the equation $y = (0.05x^2)$ ft, and neglect the size of the sprinkler.

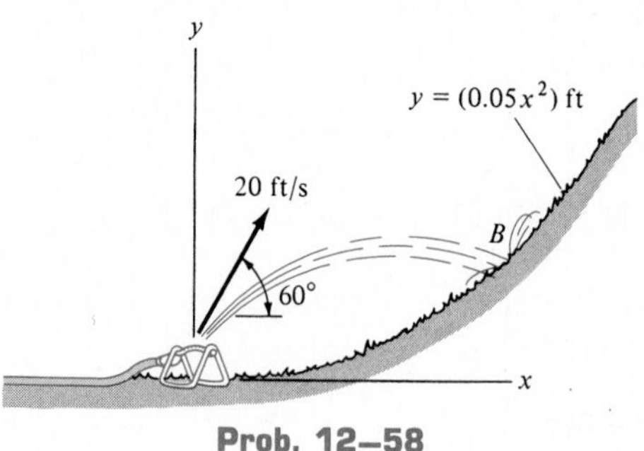

Prob. 12–58

12–59. A girl at A throws an acorn directly at a squirrel perched on a branch at B. If the initial speed of the acorn is 16 m/s and the squirrel, out of fright, happens to fall from rest off the branch at the instant the acorn is released at A, show that the squirrel can still catch the acorn, and determine how far h the squirrel falls before the catch is made.

Prob. 12–59

***12–60.** The missile at A takes off from rest and rises vertically to B, where its fuel runs out in $t = 8$ s. If the acceleration varies with time as shown, determine its height h_B and speed v_B. If by internal controls the missile is then suddenly pointed 45° as shown, and allowed to travel in free flight, determine the maximum height attained, h_C, and the range R to where it crashes at D.

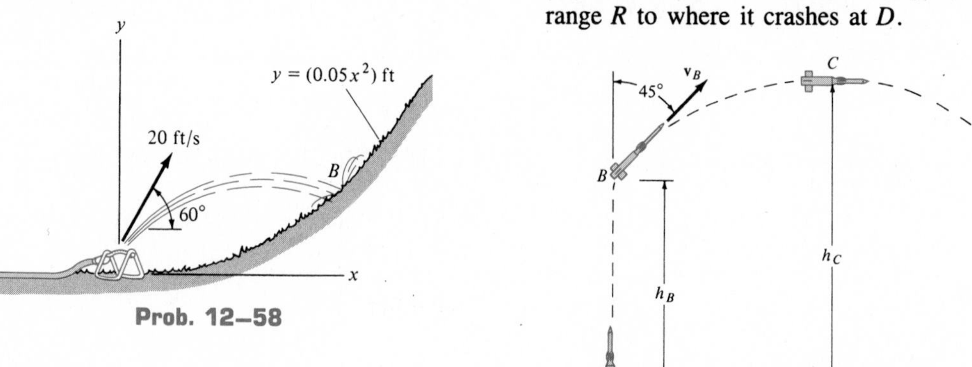

Prob. 12–60

12–61. A ball is thrown downward on the 30° inclined plane so that when it rebounds perpendicular to the incline it has a velocity of $v_A = 40$ ft/s. Determine the distance R where it strikes the plane at B.

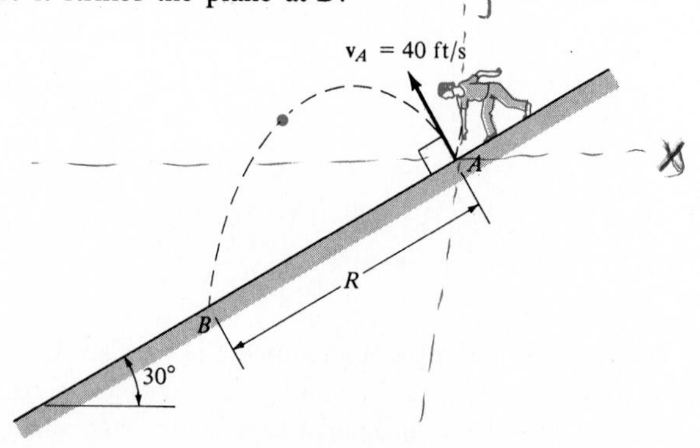

Prob. 12–61

12.6 Curvilinear Motion: Normal and Tangential Components

When the path along which a particle is moving is *known,* it is often convenient to describe the motion using n and t coordinates which act normal and tangent to the path, respectively, and at the instant considered have their *origin located at the particle.*

Planar Motion. Consider the particle shown in Fig. 12–23a, which is moving in a plane such that at a given instant it is located at point P, a position s on the curve. The t axis is *tangent* to the curve at O and is positive in the direction of *increasing* s. A unique choice for the *normal axis* can be made by considering the fact that geometrically the curve consists of a series of differential arc segments ds. As shown in Fig. 12–23b, each segment is constructed from the arc of an associated circle having a *radius of curvature* ρ (rho) and *center of curvature* O'. The normal axis n which will be chosen is perpendicular to the t axis and is directed from P *toward* the center of curvature O', Fig. 12–23a. This positive direction is *always* on the concave side of the curve. The plane which contains the n and t axes is referred to as the *osculating plane,* and in this case it is fixed in the plane of motion.

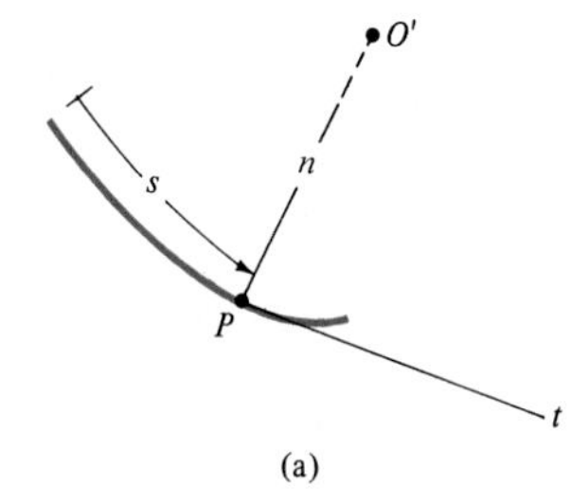

Velocity. As indicated in Sec. 12–3, **v** has a *direction* that is always tangent to the curve, Fig. 12–23b, and a *magnitude* that is determined by taking the time derivative of the path function s, i.e., $v = ds/dt$ (Eq. 12–8). Hence

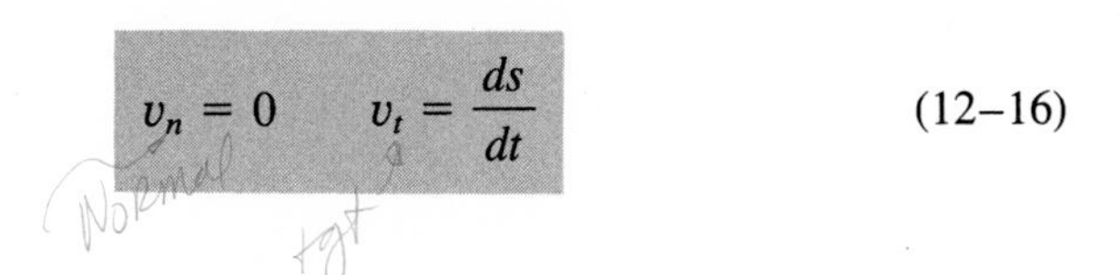

$$v_n = 0 \qquad v_t = \frac{ds}{dt} \tag{12–16}$$

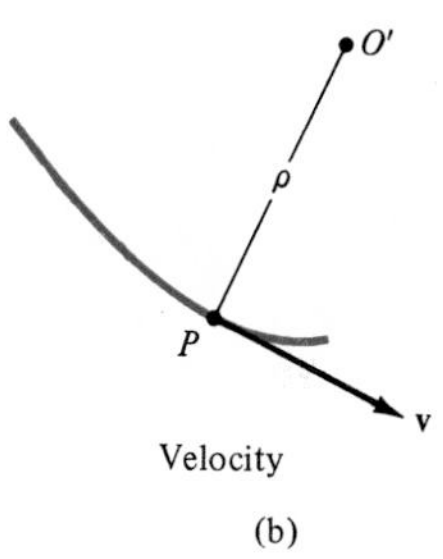

Fig. 12–23a,b

or simply

$$v = \frac{ds}{dt} \tag{12–17}$$

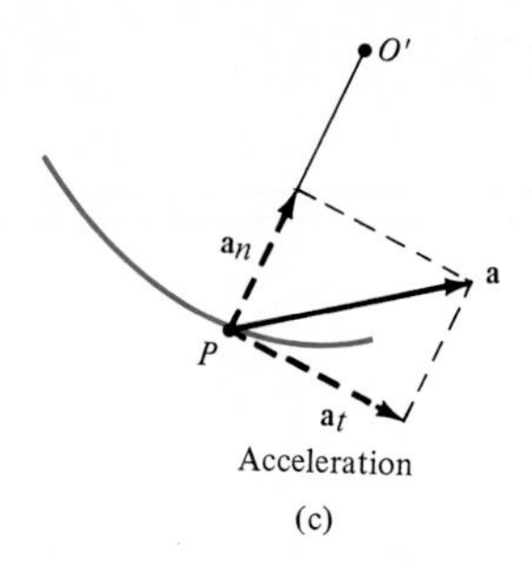

Acceleration

(c)

Acceleration. The acceleration of the particle can be resolved into its normal and tangential components as shown in Fig. 12–23*c*, where

$$\mathbf{a} = \mathbf{a}_n + \mathbf{a}_t \tag{12–18}$$

In order to determine these components, we must consider the change in the magnitude and direction of the particle's velocity when the particle moves from point P to P' during the time Δt, Fig. 12–23*d*. The change made in the velocity can be determined if we superimpose the velocities $\mathbf{v}$ and $\mathbf{v}'$, shown exaggerated in Fig. 12–23*e*. Here it is seen that the *normal component* $\Delta \mathbf{v}_n$ represents the *change in the direction of* $\mathbf{v}$. For small angles $\Delta\theta$, $\Delta v_n \approx v\Delta\theta$. Also,* $\Delta\theta = \Delta s/\rho$, Fig. 12–23*d*. Hence, the normal component of acceleration becomes

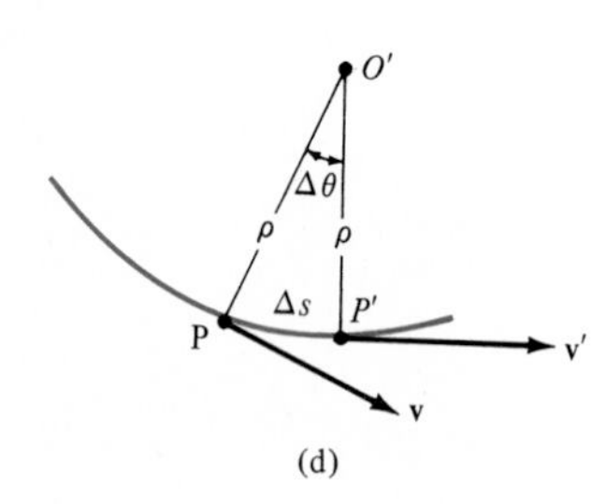

(d)

$$a_n = \lim_{\Delta t \to 0} \frac{\Delta v_n}{\Delta t} = \frac{v}{\rho} \lim_{\Delta t \to 0} \frac{\Delta s}{\Delta t} = \frac{v^2}{\rho}$$

The *tangential component* $\Delta \mathbf{v}_t$ represents the *change in the magnitude of* $\mathbf{v}$, Fig. 12–23*e*. Since $v = v_t$, Eqs. 12–16 and 12–17, we have

$\Delta \mathbf{v}_n$ (change in direction of $\mathbf{v}$)

$\mathbf{v}'$ $\Delta\theta$ $\Delta \mathbf{v}$ $\mathbf{v}$

$\Delta \mathbf{v}_t$ (change in magnitude of $\mathbf{v}$)

(e)

Fig. 12–23c,d,e

$$a_t = \lim_{\Delta t \to 0} \frac{\Delta v_t}{\Delta t} = \lim_{\Delta t \to 0} \frac{\Delta v}{\Delta t} = \frac{dv}{dt}$$

In summary, then, the two components of acceleration are

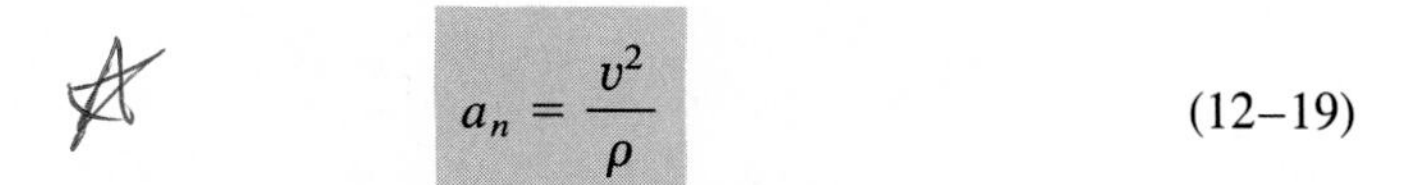

$$a_n = \frac{v^2}{\rho} \tag{12–19}$$

and

$$a_t = \frac{dv}{dt} \quad \text{or} \quad a_t = v\frac{dv}{ds} \tag{12–20}$$

Since these components are perpendicular, Fig. 12–23*c*, the *magnitude* of acceleration is

$$a = \sqrt{a_n^2 + a_t^2} \tag{12–21}$$

*Recall the arc s of a circle of radius r is determined from $s = \theta r$, where θ is in radians.

PROCEDURE FOR ANALYSIS

Motion of a particle along a *known* path can be expressed in terms of its normal and tangential components. The following is a summary of the equations necessary for doing this.

Velocity. The particle's *velocity* always acts tangent to the path in the direction of motion. Its magnitude is found from the time derivative of the path function.

$$v = \frac{ds}{dt}$$

Tangential Acceleration. The *tangential component of acceleration creates a change in the magnitude of velocity*. This component acts in the direction of motion if the particle's speed is increasing or in the opposite direction if the speed is decreasing. The magnitudes of $\mathbf{v}$ and $\mathbf{a}_t$ are related to the path position s and time t by the equations of rectilinear motion: namely,

$$a_t = \frac{ds}{dt} \qquad a_t\,ds = v\,dv$$

If a_t is *constant*, $a_t = (a_t)_c$, the above equations, when integrated, yield

$$s = s_0 + v_0 t + \tfrac{1}{2}(a_t)_c t^2$$
$$v = v_0 + (a_t)_c t$$
$$v^2 = v_0^2 + 2(a_t)_c(s - s_0)$$

Normal Acceleration. The *normal component of acceleration creates a change in the direction of the particle's velocity*. This component is *always* directed toward the center of curvature of the path, i.e., along the positive n axis. Its magnitude is determined from

$$a_n = \frac{v^2}{\rho}$$

If the path is expressed as $y = f(x)$, the radius of curvature ρ is computed from the equation*

$$\rho = \left| \frac{[1 + (dy/dx)^2]^{3/2}}{d^2y/dx^2} \right|$$

The following examples numerically illustrate application of these equations.

*The derivation of this result is given in any standard calculus text.

Example 12–14

A skier travels with a constant speed of 6 m/s along the parabolic path $y = \frac{1}{20}x^2$ shown in Fig. 12–24. Determine the magnitude of the velocity and acceleration at the instant he arrives at A. Neglect the size of the skier in the calculation.

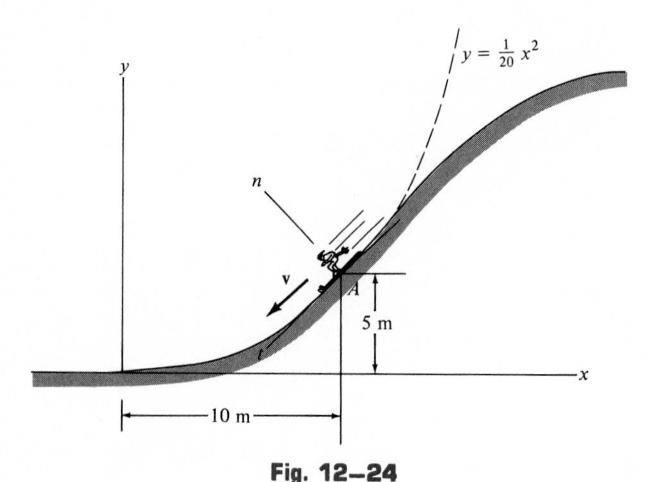

Fig. 12–24

Solution

Velocity. By definition, the velocity is always directed tangent to the path, Fig. 12–24. From the problem statement,

$$v = 6 \text{ m/s} \qquad \textit{Ans.}$$

Acceleration. The components of acceleration are found from $a_t = dv/dt$ and $a_n = v^2/\rho$. The radius of curvature of the path at A (10 m, 5 m) must first be determined. Since $y = \frac{1}{20}x^2$, then $dy/dx = \frac{1}{10}x$ and $d^2y/dx^2 = \frac{1}{10}$, so that

$$\rho = \left|\frac{[1 + (dy/dx)^2]^{3/2}}{d^2y/dx^2}\right| = \left|\frac{[1 + (\frac{1}{10}x)^2]^{3/2}}{\frac{1}{10}}\right|_{x=10\text{ m}} = 28.3 \text{ m}$$

Thus

$$a_n = \frac{v^2}{\rho} = \frac{(6)^2}{28.3} = 1.27 \text{ m/s}^2$$

$$a_t = \frac{dv}{dt} = \frac{d}{dt}(6) = 0$$

Since $\mathbf{a}_A$ acts in the direction of the positive n axis,

$$a_A = 1.27 \text{ m/s}^2 \qquad \textit{Ans.}$$

Example 12–15

A race car C travels around the horizontal circular track that has a radius of 300 ft, Fig. 12–25. If the car increases its speed at a constant rate of 7 ft/s^2 starting from rest, determine the time needed for it to reach an acceleration of 8 ft/s^2. What is its speed at this instant?

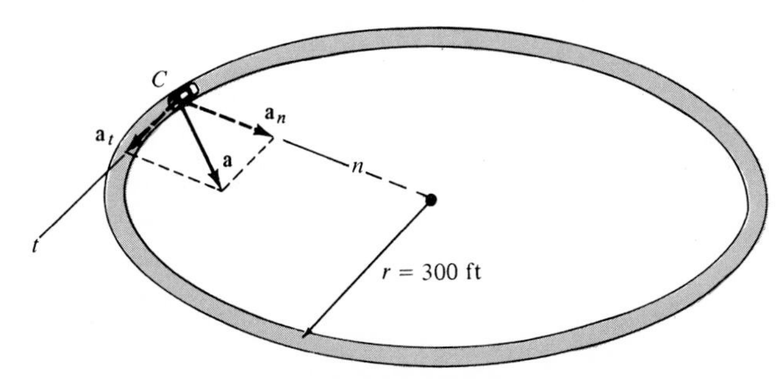

Fig. 12–25

Solution

Acceleration. In general, $\mathbf{a} = \mathbf{a}_t + \mathbf{a}_n$. Since $\mathbf{a}_t$ represents the change in the velocity's magnitude, then $a_t = 7$ ft/s^2. The normal component, representing the change in the velocity's direction, is $a_n = v^2/\rho$. Since $\rho = 300$ ft and

$$v = v_0 + (a_t)_c t$$
$$v = 0 + 7t$$

then

$$a_n = \frac{v^2}{\rho} = \frac{(7t)^2}{300} = 0.163t^2 \text{ ft/s}^2$$

Adding $\mathbf{a}_t$ and $\mathbf{a}_n$ vectorially, Fig. 12–25, the magnitude of acceleration can be expressed as

$$a = \sqrt{a_t^2 + a_n^2}$$
$$8 = \sqrt{(7)^2 + (0.163t^2)^2}$$

Solving for t yields

$$0.163t^2 = \sqrt{(8)^2 - (7)^2}$$
$$t = 4.87 \text{ s} \qquad \textit{Ans.}$$

Velocity. The speed at time $t = 4.87$ s is

$$v = 7t = 7(4.87) = 34.1 \text{ ft/s} \qquad \textit{Ans.}$$

Example 12–16

A car starts from rest at point A and travels along the horizontal track shown in Fig. 12–26a. During the motion, the increase in speed is $a_t = (0.2t)\ \text{m/s}^2$, where t is in seconds. If it takes 5 seconds to reach point B, determine the magnitude of the car's acceleration when it arrives at point B.

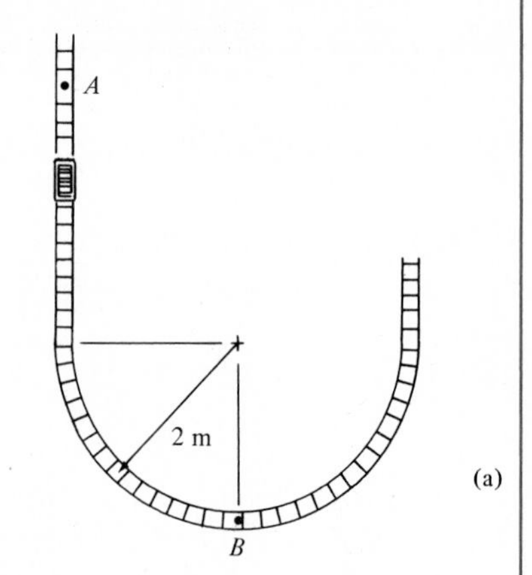

Solution

The acceleration components are determined from $a_t = dv/dt$ and $a_n = v^2/\rho$. To use these equations, however, it is first necessary to formulate v and dv/dt so that they may be evaluated at B. Since $v_A = 0$ at $t = 0$, then

$$a_t = \frac{dv}{dt} = 0.2t$$

$$\int_0^v dv = \int_0^t 0.2t\,dt$$

$$v = 0.1t^2$$

When $t = 5$ s, we have

$$(a_t)_B = 0.2(5) = 1.0\ \text{m/s}^2$$

$$v_B = 0.1(5)^2 = 2.5\ \text{m/s}$$

At B, $\rho_B = 2$ m, so that

$$(a_n)_B = \frac{v_B^2}{\rho_B} = \frac{(2.5)^2}{2} = 3.13\ \text{m/s}^2$$

The magnitude of $\mathbf{a}_B$, Fig. 12–26b, is therefore

$$a_B = \sqrt{(1.0)^2 + (3.13)^2} = 3.28\ \text{m/s}^2 \qquad \textit{Ans.}$$

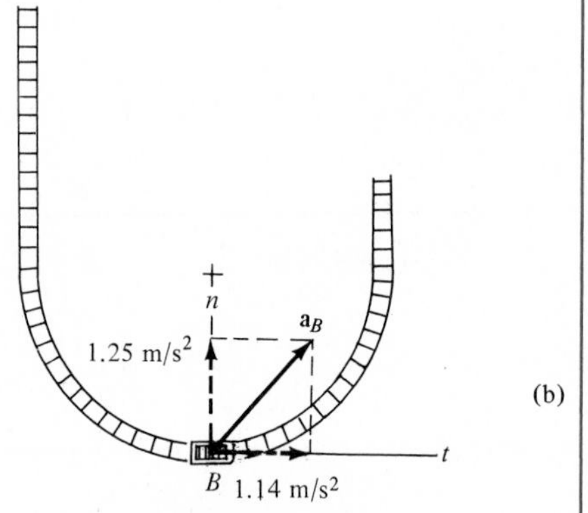

Fig. 12–26

Problems

12–62. A boat is traveling along a circular path having a radius of 20 m. Determine the magnitude of the boat's acceleration if at a given instant the boat's speed is $v = 5$ m/s and the rate of increase in speed is $\dot{v} = 2$ m/s^2.

12–63. A train travels along a horizontal circular curve that has a radius of 200 m. If the speed of the train is uniformly increased from 30 km/h to 45 km/h in 5 s, determine the magnitude of the acceleration at the instant the speed of the train is 40 km/h.

***12–64.** An automobile is traveling with a *constant speed* along a horizontal circular curve that has a radius of $\rho = 750$ ft. If the magnitude of acceleration is $a = 8$ ft/s^2, determine the speed at which the automobile is traveling.

12–65. At a given instant the train engine at E has a speed of 30 m/s and an acceleration of 14 m/s^2 acting in the direction shown. Determine the rate of increase in the train's speed and the radius of curvature of the path.

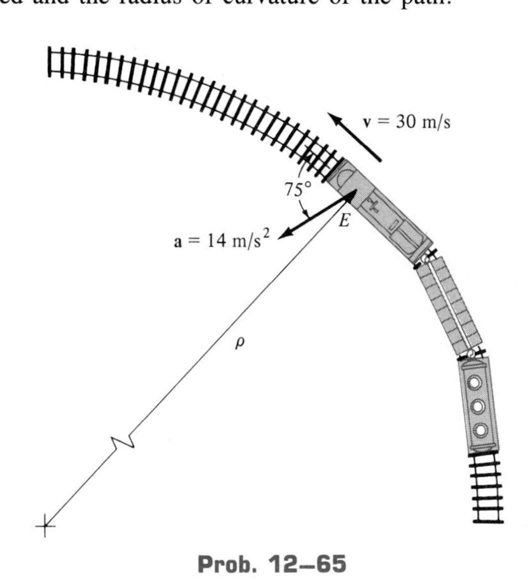

Prob. 12–65

12–66. A sled is traveling down along a curve which can be approximated by the parabola $y = \frac{1}{4}x^2$. When point B on the runner is coincident with point A on the curve ($x_A = 2$ m, $y_A = 1$ m), the speed of B is measured as $v_B = 8$ m/s and the increase in speed is $dv_B/dt = 4$ m/s^2. Determine the magnitude of the acceleration of point B at this instant.

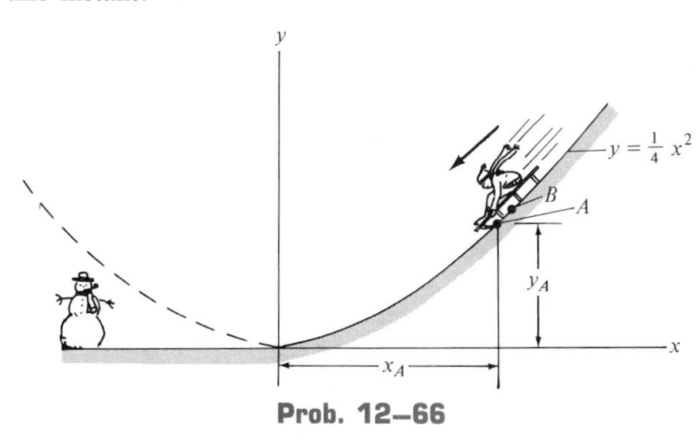

Prob. 12–66

12–67. A spiral transition curve is used on railroads to connect a straight portion of the track with a curved portion. If the spiral is defined by the equation $y = (10^{-6})x^3$, where x and y are in feet, determine the magnitude of the acceleration of a train engine moving with a constant speed of 40 ft/s when it is at point $x = 600$ ft.

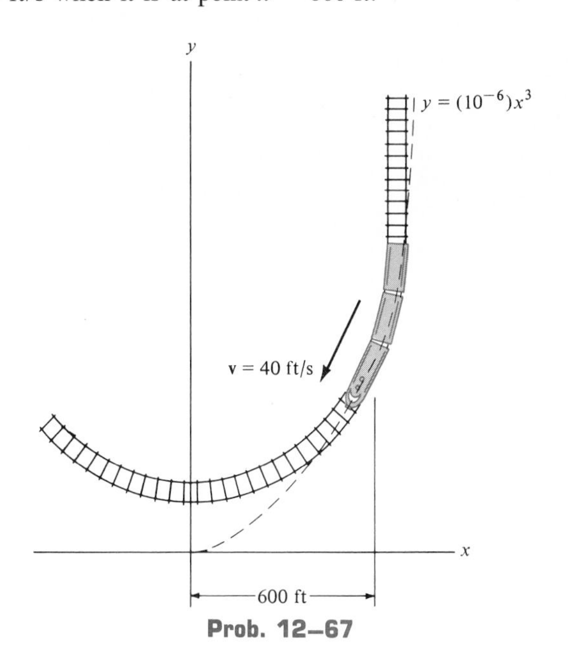

Prob. 12–67

***12–68.** A particle P moves along the curve $y = (x^2 - 4)$ ft with a constant speed of 6 ft/s. Determine the magnitude of acceleration when the particle reaches the point (0, − 4 ft).

12–69. Starting from rest, the motorboat travels around the circular path, $\rho = 60$ m, at a speed of $v = (0.8t)$ m/s, where t is measured in seconds. Determine the magnitudes of the boat's velocity and acceleration just after it has traveled 20 m.

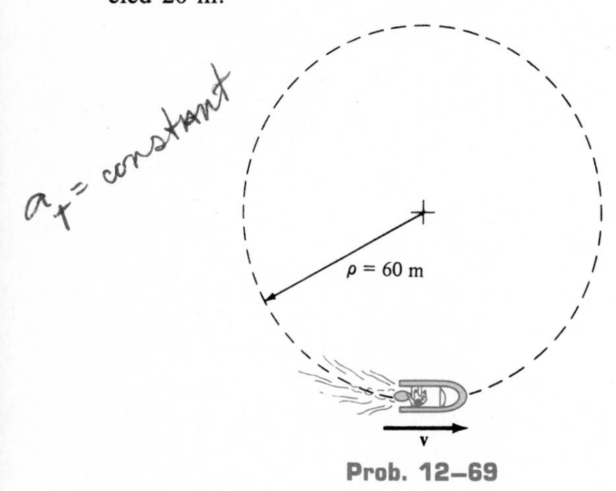

Prob. 12–69

12–70. A race car has an initial speed of $v_A = 15$ m/s when it is at A. If it increases its speed along the circular track, $\rho = 200$ m, at the rate of $a_t = (0.4s)$ m/s², where s is measured in meters, determine the normal and tangential components of the car's acceleration when $s = 100$ m.

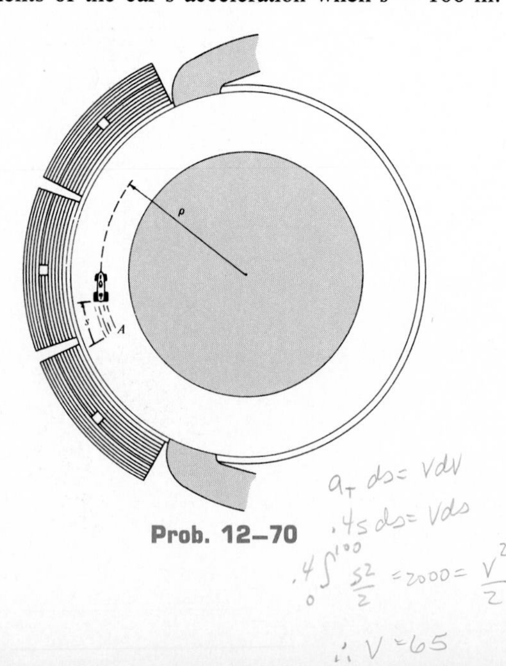

Prob. 12–70

12–71. If the race car in Prob. 12–70 has an initial speed of $v_A = 15$ m/s at A, compute the velocity when the car has traveled 50 m.

***12–72.** When the motorcyclist is at A he increases his speed along the vertical circular path at the rate of $\dot{v} = (0.3t)$ ft/s², where t is in seconds. If he starts from rest when he is at A, determine his velocity and acceleration when he reaches B, where $s_{AB} = 314.16$ ft.

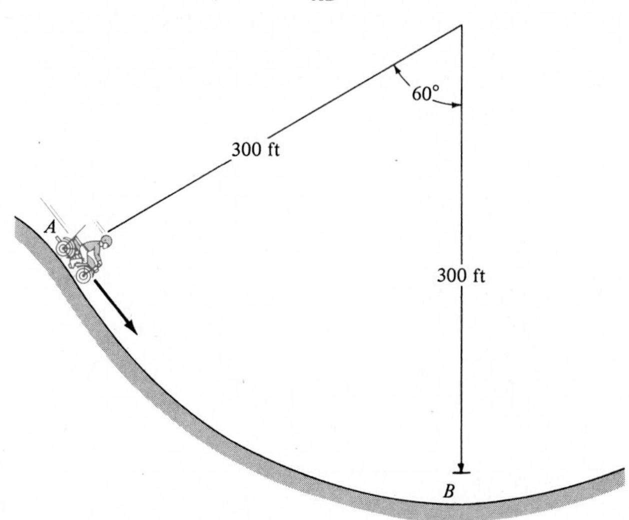

Prob. 12–72

12–73. A package is dropped from the plane which is flying with a constant horizontal velocity of $v_A = 150$ ft/s. Determine the tangential and normal components of acceleration and the radius of curvature of the path of motion (a) at the moment the package is released at A, where it has a horizontal velocity of $v_A = 150$ ft/s and $h = 1500$ ft, and (b) *just before* it strikes the ground at B. *Hint:* The acceleration of the package is $g = 32.2$ ft/s², downward.

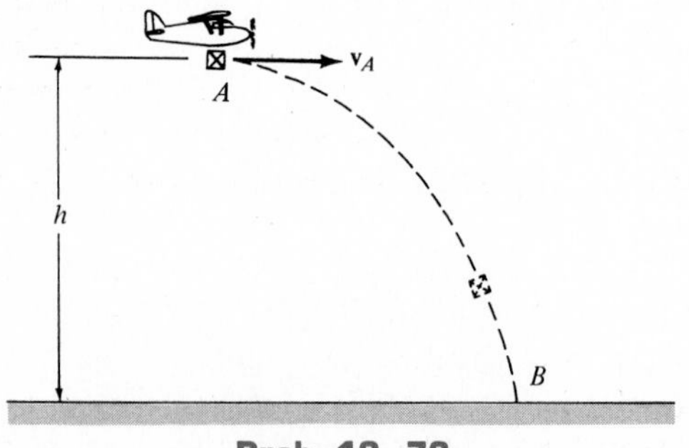

Prob. 12–73

12.7 Absolute-Dependent-Motion Analysis of Two Particles

In some types of problems the motion of one particle will *depend* upon the corresponding motion of another particle. This dependency commonly occurs if the particles are interconnected by inextensible cords which are wrapped around pulleys. For example, the movement of block A downward along the inclined plane in Fig. 12–27 will cause a corresponding movement of block B up the other incline. We can show this mathematically by first specifying the positions of the blocks using the coordinates s_A and s_B, each measured from the *fixed point* O. Since the total cord length l_T is constant, the coordinates are related by the equation

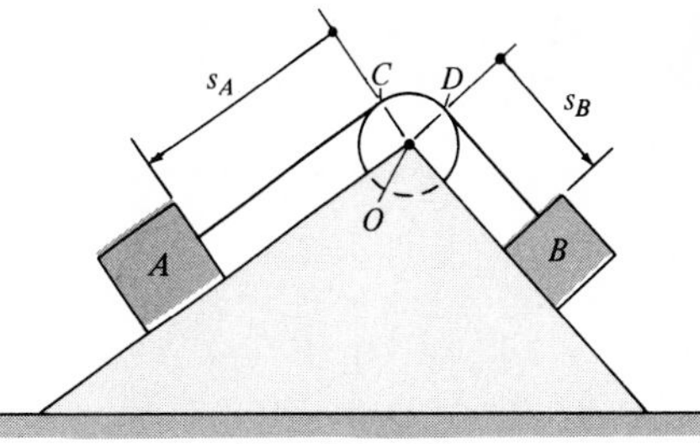

Fig. 12–27

$$s_A + s_B = l_T - l_{CD} = l'$$

where l_{CD} is the length of cord passing over arc CD.* Taking the time derivative of this expression yields a relation between the velocities of the blocks, i.e.,

$$\frac{ds_B}{dt} + \frac{ds_A}{dt} = 0$$

or

$$v_B = -v_A$$

The negative sign indicates that positive motion of block A (downward in the direction of positive s_A) causes a corresponding negative motion (upward) of block B.

In a similar manner, time differentiation of the velocities yields the relation between the accelerations, i.e.,

$$a_B = -a_A$$

A more complicated example involving dependent motion of two blocks is shown in Fig. 12–28*a*. In this case, the position of block A is specified by s_A and the position of the *end* of the cord from which block B is suspended is defined by s_B. Notice that as before each coordinate is measured from a *fixed point* and is directed along the block's path of motion. The *constant* cord length l, which represents the total length of cord *minus* the constant cord segments which are colored, can be related to the position coordinates by the equation

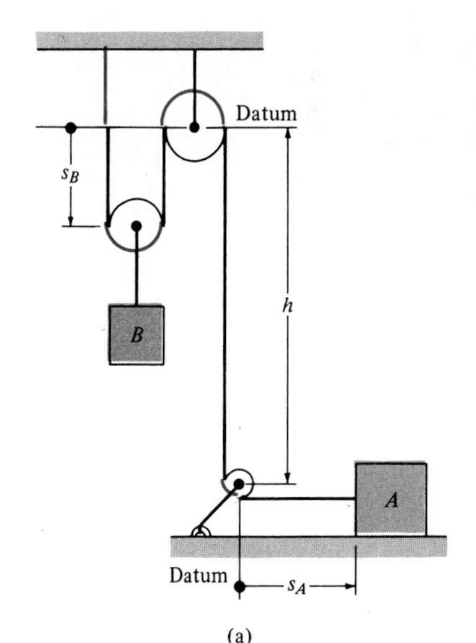

(a)
Fig. 12–28a

$$2s_B + h + s_A = l$$

*Note that when the blocks move, the coordinates s_A and s_B measure the lengths of the *changing segments* of the cord. The cord length l_{CD} always remains *constant*.

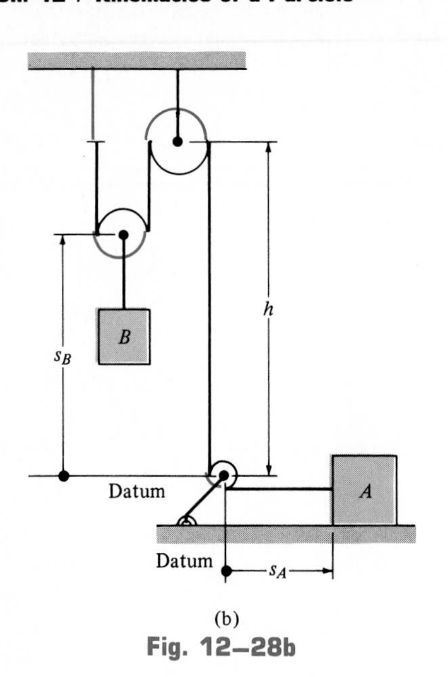

(b)
Fig. 12–28b

Since l and h are constants, the two time derivatives yield

$$2v_B = -v_A \qquad 2a_B = -a_A$$

Hence, when B moves downward ($+s_B$), A moves to the left ($-s_A$) at two times the motion.

This example can also be worked by defining the position of block B from the center of the bottom pulley (a fixed point), Fig. 12–28b. In this case

$$2(h - s_B) + h + s_A = l$$

Time differentiation yields

$$2v_B = v_A \qquad 2a_B = a_A$$

Here the signs are the same. Why?

PROCEDURE FOR ANALYSIS

The dependent motion of particle A located on a cord may be related to the motion of another particle B located on the *same cord* by using the following procedure.

Position-Coordinate Equation. Establish position coordinates which extend from a fixed point (or datum) to each of the particles. It is *not necessary* that the *fixed point* (or datum) be the *same* for both of these coordinates; however, it is *important* that each coordinate axis selected be directed along the *path of motion* of the particle.

Using geometry, relate the coordinates to the total length of the cord, l_T, or to that portion of cord, l, which *excludes* the segments that do not change length as the particles move—such as the segments wrapped over pulleys.

Time Derivatives. The two successive time derivatives of the position-coordinate equation yield the required velocity and acceleration equations which relate the motions of the two particles. The signs of the terms in these equations will be consistent with those that specify the positive and negative directions of the position coordinates.

If a problem involves a *system* of two or more cords wrapped around pulleys, then the motion of a point on one cord must be related to the motion of a point on another cord using the above procedure. Separate equations are written for corresponding cords of the system and the positions of the two particles are then related by these equations (see Examples 12–18 and 12–19).

Example 12–17

Determine the speed of block A in Fig. 12–29. Block B has an upward speed of 6 ft/s.

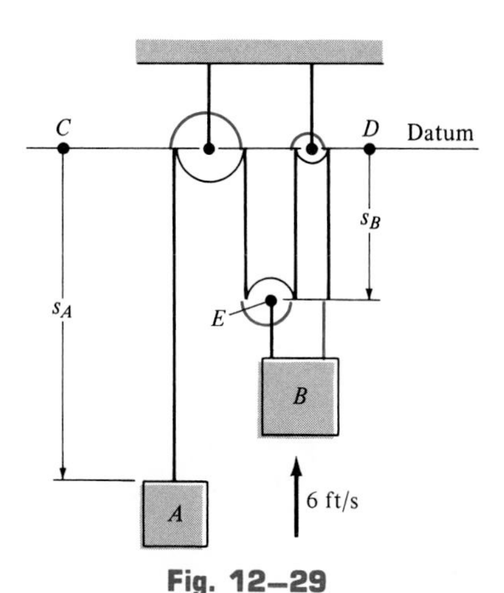

Fig. 12–29

Solution

Position-Coordinate Equation. As shown in Fig. 12–29, there is *one* cord in this system having segments which are changing length. Position coordinates s_A and s_B will be used since each is measured from a fixed point (C or D) and extends along the block's *path of motion*. In particular, s_B is directed to point E since motion of B and E is the *same*.

The colored segments of the cord in Fig. 12–29 remain at a constant length and do not have to be considered as the blocks move. The remaining length of cord, l, is also constant and is related to the changing position coordinates s_A and s_B by the equation

$$s_A + 3s_B = l$$

Time Derivative. Taking the time derivative yields

$$v_A + 3v_B = 0$$

so that when $v_B = -6$ ft/s (upward), then

$$v_A = 18 \text{ ft/s} \quad \text{(downward)} \qquad \textit{Ans.}$$

Example 12–18

Determine the speed of block A in Fig. 12–30. Block B has an upward speed of 6 ft/s.

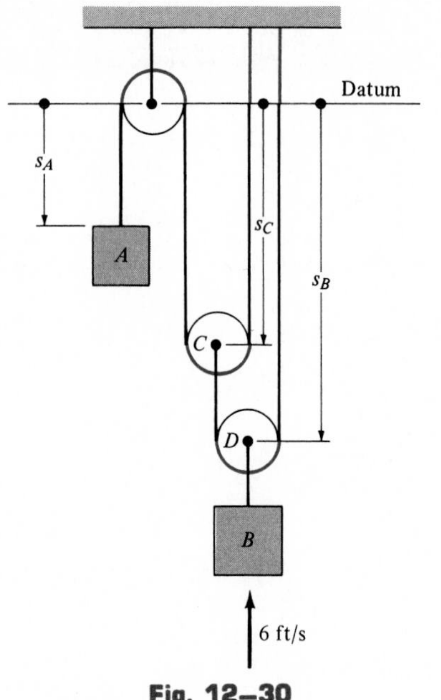

Fig. 12–30

Solution

Position-Coordinate Equation. As shown in Fig. 12–30, this system has two cords which have segments that change length. Motion of the end of the cord attached to A can be related to the motion of the end of the cord attached to the pulley at C using coordinates s_A and s_C. Likewise, the motion of the end of the cord at C can be related to the motion of the end of the cord at D using position coordinates s_C and s_B.

The colored segments of the cords in Fig. 12–30 do not have to be considered in the analysis. Why? For the remaining cord lengths, say l_1 and l_2, we have

$$s_A + 2s_C = l_1 \qquad s_B + (s_B - s_C) = l_2$$

Eliminating s_C yields an equation defining the positions of both blocks, i.e.,

$$s_A + 4s_B = 2l_2 + l_1$$

Time Derivative. The time derivative gives

$$v_A + 4v_B = 0$$

so that when $v_B = -6$ ft/s (upward), then

$$v_A = 24 \text{ ft/s} \quad \text{(downward)} \qquad \textbf{\textit{Ans.}}$$

Example 12–19

Determine the speed with which block B rises in Fig. 12–31 if the end of the cable A is pulled down with a speed of 2 m/s.

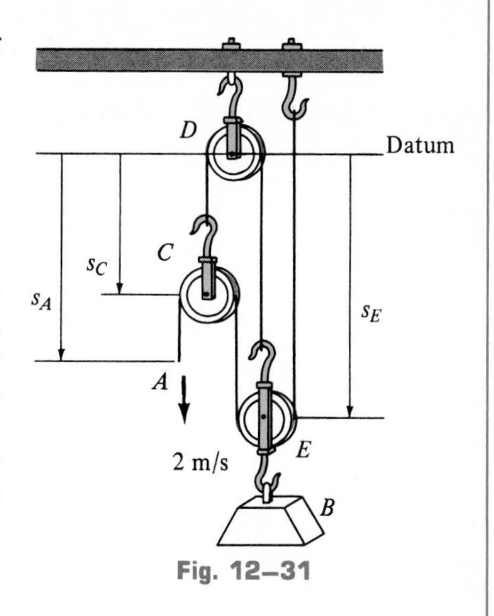

Fig. 12–31

Solution

Position-Coordinate Equation. By inspection this system has two cables. Three position coordinates have been selected. These are all measured from the horizontal datum passing through the fixed point D. The position of the end of the cable at A is defined by s_A, the position of the block B (or pulley E) is defined by s_E, and the position of the pulley at C is defined by s_C.

Excluding the colored segments of the cables in Fig. 12–31, the remaining constant cable lengths l_1 and l_2 (with the hook dimensions) can be expressed in terms of the three position coordinates. We have

$$(s_A - s_C) + (s_E - s_C) + s_E = l_1$$
$$s_C + s_E = l_2$$

Eliminating s_C yields

$$s_A + 4s_E = l_1 + 2l_2$$

This equation relates the position s_E of the block to the position s_A of point A.

Time Derivative. The time derivative gives

$$v_A + 4v_E = 0$$

so that when $v_A = 2$ m/s (downward), then

$$v_E = -0.5 \text{ m/s} \quad \text{or} \quad v_E = 0.5 \text{ m/s (upward)} \qquad \textit{Ans.}$$

12.8 Relative-Motion Analysis of Two Particles Using Translating Axes

Throughout this chapter the absolute motion of a particle has been determined using a single fixed reference frame for measurement. There are many cases, however, where the path of motion for a particle is complicated, so that it may be feasible to analyze the motion in parts by using two or more frames of reference. For example, the motion of a particle located at the tip of an airplane propeller, while the plane is in flight, is more easily described if one observes first the motion of the airplane from a fixed reference and then superimposes (vectorially) the circular motion of the particle measured from a reference attached to the airplane. Any type of coordinates—rectangular, cylindrical, etc.—may be chosen to describe these two different motions.

In this section only *translating frames of reference* will be considered for the analysis. Relative-motion analysis of particles using rotating frames of reference will be treated in Sec. 20–3, since such an analysis depends upon prior knowledge of the kinematics of line segments.

Position. Consider particles A and B, which move along the arbitrary paths aa and bb, respectively, as shown in Fig. 12–32a. The *absolute position* of each particle, $\mathbf{r}_A$ and $\mathbf{r}_B$, is measured from the common origin O of the *fixed* x, y, z reference frame. The origin of a second frame of reference x', y', z' is attached to and moves with particle A. The axes of this frame do not rotate; rather they are *only permitted to translate* relative to the fixed frame. The *relative position* of "B with respect to A" is observed from this moving frame and is designated by $\mathbf{r}_{B/A}$, called a *relative-position vector*. Using vector addition, the three vectors shown in Fig. 12–32a can be related by the equation*

$$\mathbf{r}_B = \mathbf{r}_A + \mathbf{r}_{B/A} \tag{12–22}$$

Velocity. An equation that relates the velocities of the particles can be determined by taking the time derivative of Eq. 12–22, i.e.,

$$\mathbf{v}_B = \mathbf{v}_A + \mathbf{v}_{B/A} \tag{12–23}$$

Here $\mathbf{v}_B = d\mathbf{r}_B/dt$ and $\mathbf{v}_A = d\mathbf{r}_A/dt$ refer to *absolute velocities,* since they are measured from the fixed frame of reference; whereas the *relative velocity* $\mathbf{v}_{B/A} = d\mathbf{r}_{B/A}/dt$ is observed from the translating reference frame. The above equation therefore states that the absolute velocity of B is equal to the absolute velocity of A plus (vectorially) the relative velocity of "B with respect to A," as measured by a *translating observer,* Fig. 12–32b.

*An easy way to remember the setup of this equation, and others like it, is to note the "cancellation" of the subscript A between the two terms, i.e., $\mathbf{r}_B = \mathbf{r}_{\not{A}} + \mathbf{r}_{B/\not{A}}$.

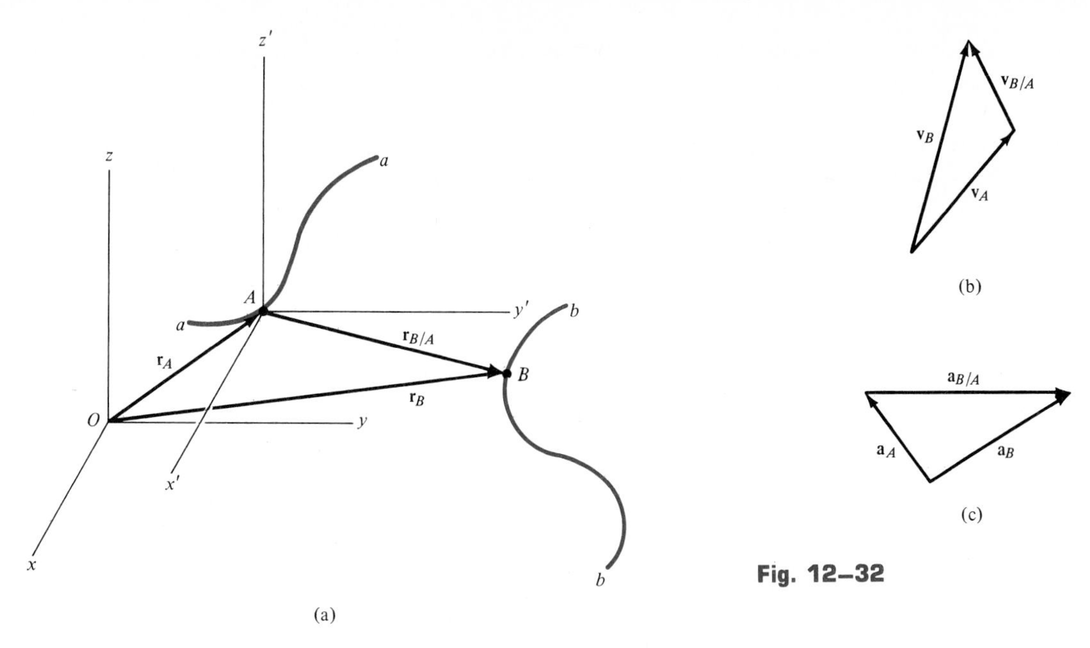

Fig. 12–32

Acceleration. The time derivative of Eq. 12–23 yields a similar vector relationship between the *absolute* and *relative accelerations* of particles A and B,

$$\mathbf{a}_B = \mathbf{a}_A + \mathbf{a}_{B/A} \tag{12–24}$$

Here $\mathbf{a}_{B/A}$ is the acceleration of B as seen by an observer located at A and translating with the x', y', z' reference frame. The vector addition is shown in Fig. 12–32*c*.

PROCEDURE FOR ANALYSIS

When applying the relative-position equation, $\mathbf{r}_B = \mathbf{r}_A + \mathbf{r}_{B/A}$, it is first necessary to specify the locations of the fixed x, y, z and translating x', y', z' axes. Usually, the origin A of the translating axes is located at a point having a *known position,* $\mathbf{r}_A$, Fig. 12–32*a*. A graphical representation of the vector addition $\mathbf{r}_B = \mathbf{r}_A + \mathbf{r}_{B/A}$ can be shown, and both the known and unknown quantities labeled on this sketch. Since vector addition forms a triangle there can be at most *two unknowns,* represented by the magnitudes and/or directions of the vector quantities. These unknowns can be solved for either graphically, using trigonometry (law of sines, law of cosines), or by resolving each of the three vectors $\mathbf{r}_B$, $\mathbf{r}_A$, and $\mathbf{r}_{B/A}$ into rectangular components, thereby generating two scalar equations. The latter method is illustrated in the example problems which follow.

The relative-motion equations $\mathbf{v}_B = \mathbf{v}_A + \mathbf{v}_{B/A}$ and $\mathbf{a}_B = \mathbf{a}_A + \mathbf{a}_{B/A}$ are applied in the same manner as explained above, except in this case the origin O of the fixed x, y, z axes does not have to be specified, Fig. 12–32*b* and *c*.

Example 12–20

Water drips from a faucet at the rate of five drops per second as shown in Fig. 12–33. Determine the vertical separation between two consecutive drops after the lower drop has attained a velocity of 3 m/s.

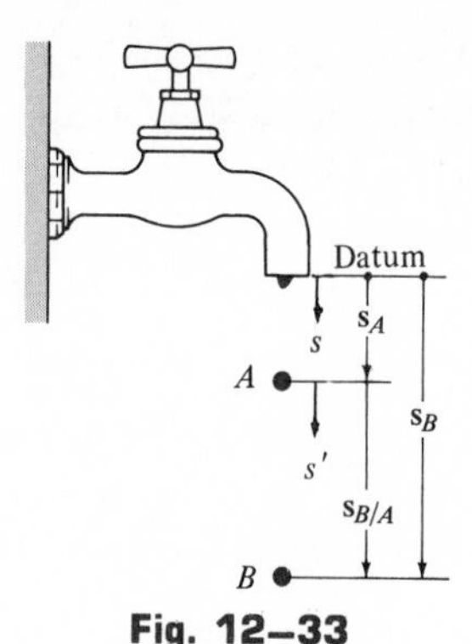

Fig. 12–33

Solution

If the first and second drops of water are denoted as B and A, respectively, Fig. 12–33, the separation $s_{B/A}$ (position of B with respect to A) can be determined by the equation

$$s_{B/A} = s_B - s_A \tag{1}$$

Here the origin of the fixed frame of reference is located at the head of the faucet (datum) with positive s downward. The origin of the moving frame s' is at drop A. Since each drop is subjected to rectilinear motion, having a constant downward acceleration of $a_c = g = 9.81 \text{ m/s}^2$, Eq. 12–4 may be applied with the initial condition $v_0 = 0$ to determine the time needed for drop B to attain a velocity of 3 m/s.

$(+\downarrow)$
$$v = v_0 + a_c t$$
$$3 = 0 + 9.81 t_B$$
$$t_B = 0.306 \text{ s}$$

Since one drop falls every fifth of a second (0.2 s), drop A falls for a time

$$t_A = 0.306 \text{ s} - 0.2 \text{ s} = 0.106 \text{ s}$$

before drop B attains a velocity of 3 m/s.

The position of each drop can be found by means of Eq. 12–5, with the condition $s_0 = 0$, $v_0 = 0$.

$(+\downarrow)$
$$s = s_0 + v_0 t + \tfrac{1}{2} a_c t^2$$
$$s_B = 0 + 0 + \tfrac{1}{2}(9.81)(0.306)^2$$
$$= 0.459 \text{ m}$$
$$s_A = 0 + 0 + \tfrac{1}{2}(9.81)(0.106)^2$$
$$= 0.055 \text{ m}$$

Applying Eq. (1) yields

$$s_{B/A} = s_B - s_A = 0.459 - 0.055 = 0.404 \text{ m}$$
$$= 404 \text{ mm}$$
Ans.

Example 12–21

A train, traveling at a constant speed of 60 mi/h, crosses over a road as shown in Fig. 12–34*a*. If an automobile *A* is traveling at 45 mi/h along the road, determine the relative velocity of the train with respect to the automobile.

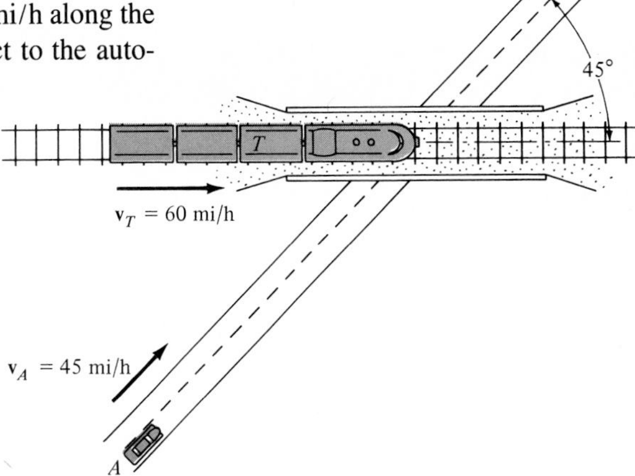

Solution

The relative velocity $\mathbf{v}_{T/A}$ is determined from $\mathbf{v}_T = \mathbf{v}_A + \mathbf{v}_{T/A}$. Here $\mathbf{v}_T$ and $\mathbf{v}_A$ are known in *both* magnitude and direction. The unknowns will be represented by the components of $\mathbf{v}_{T/A}$. Using the *x*, *y* axes in Fig. 12–34*b* we have

$$\mathbf{v}_T = \mathbf{v}_A + \mathbf{v}_{T/A}$$

$$\underset{\rightarrow}{60 \text{ mi/h}} = \underset{\measuredangle 45^\circ}{45 \text{ mi/h}} + \underset{\rightarrow}{(v_{T/A})_x} + \underset{\downarrow}{(v_{T/A})_y}$$

Resolving each vector into its *x* and *y* components yields

$$(\overset{+}{\rightarrow}) \qquad 60 = 45 \cos 45^\circ + (v_{T/A})_x + 0$$

$$(+\uparrow) \qquad 0 = 45 \sin 45^\circ + 0 - (v_{T/A})_y$$

Solving, we obtain the previous results,

$$(v_{T/A})_x = 28.2 \text{ mi/h} \rightarrow$$

$$(v_{T/A})_y = 31.8 \text{ mi/h} \downarrow$$

The magnitude of $\mathbf{v}_{T/A}$ is thus

$$v_{T/A} = \sqrt{(28.2)^2 + (31.8)^2} = 42.5 \text{ mi/h} \qquad \textit{Ans.}$$

The direction, defined from the *x* axis, is

$$\tan \theta = \frac{(v_{T/A})_y}{(v_{T/A})_x} = \frac{31.8}{28.2}$$

$$\theta = 48.4^\circ \qquad \textit{Ans.}$$

Note that the vector addition shown in Fig. 12–34*b* indicates the correct sense for $\mathbf{v}_{T/A}$. This figure anticipates the answer and can be used to check it.

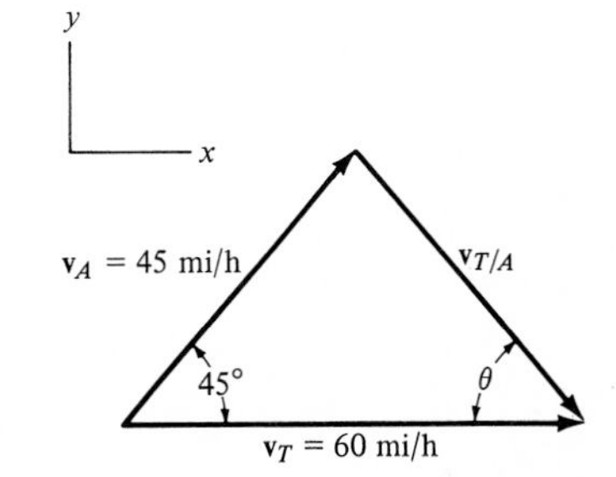

Fig. 12–34a,b

Example 12–22

Two jet planes are flying horizontally at the same elevation, as shown in Fig. 12–35*a*. Plane *A* is flying along a straight-line path, and at the instant shown it has a speed of 700 km/h and an acceleration of 50 km/h^2. Plane *B* is flying along a circular path at 600 km/h and decreasing its speed at the rate of 100 km/h^2. Determine the velocity and acceleration of *B* as measured by the pilot in *A*.

Solution

Velocity. Plane *A* is traveling with rectilinear motion and a *translating frame of reference, x′, y′*, will be attached to it. Applying the relative-velocity equation in scalar form since the velocity vectors of both planes are parallel at the instant shown, Fig. 12–35*b*, we have

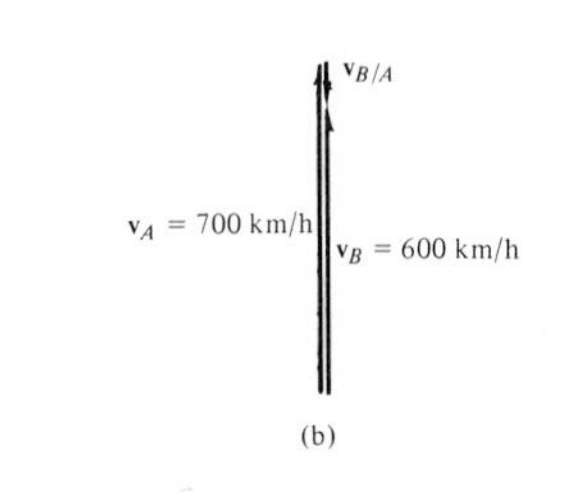

$(+\uparrow)$

$$v_B = v_A + v_{B/A}$$
$$600 = 700 - v_{B/A}$$
$$v_{B/A} = 100 \text{ km/h} \downarrow \qquad \textbf{\textit{Ans.}}$$

Acceleration. Plane *B* has both tangential and normal components of acceleration, since it is flying along a *curved path*. From Eq. 12–19, the magnitude of the normal component is

$$(a_B)_n = \frac{v_B^2}{\rho} = \frac{(600)^2}{400} = 900 \text{ km/h}^2$$

Applying the relative-acceleration equation, Fig. 12–35*c*, we have

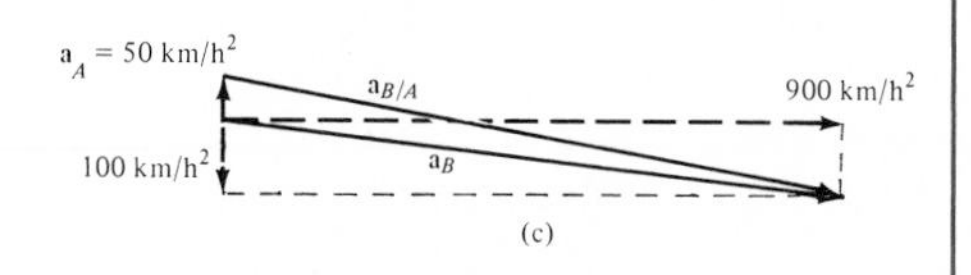

Fig. 12–35

$$\mathbf{a}_B = \mathbf{a}_A + \mathbf{a}_{B/A}$$
$$\underset{\downarrow}{100} + \underset{\rightarrow}{900} = \underset{\uparrow}{50} + \underset{\rightarrow}{(a_{B/A})_x} + \underset{\downarrow}{(a_{B/A})_y}$$

Thus,

$(\overset{+}{\rightarrow})$
$$0 + 900 = 0 + (a_{B/A})_x$$
$(+\uparrow)$
$$-100 + 0 = 50 - (a_{B/A})_y$$

Solving, we obtain

$$(a_{B/A})_x = 900 \text{ km/h}^2 \rightarrow \qquad (a_{B/A})_y = 150 \text{ km/h}^2 \downarrow$$

The magnitude and direction of $\mathbf{a}_{B/A}$ are therefore

$$a_{B/A} = 912.4 \text{ km/h}^2 \qquad \theta = 9.46° \qquad \textbf{\textit{Ans.}}$$

Notice that the solution to this problem is possible using a translating frame of reference, since the pilot in plane *A* is indeed "translating."

Problems

12–74. In each case, if the end of the cable at A is pulled down with a speed of 2 m/s, determine the speed at which block B rises.

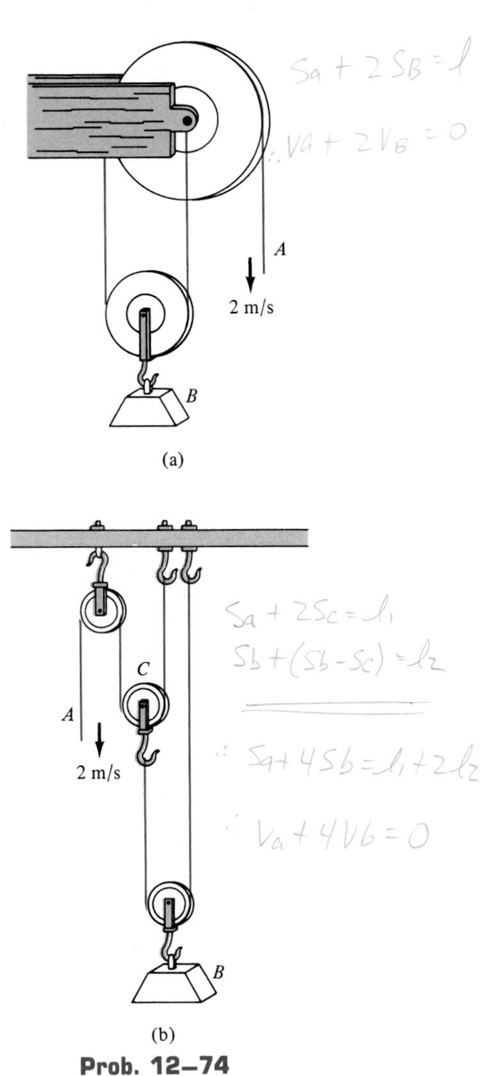

Prob. 12–74

12–75. The mine car is being pulled up the inclined plane using the motor M and the rope-and-pulley arrangement shown. Determine the speed v_P at which a point P on the cable must be traveling toward the motor to move the car up the plane with a constant speed of $v = 5$ m/s. *Hint:* Establish the datum through the top pulley, and locate the position of the center of the car and the position of point P.

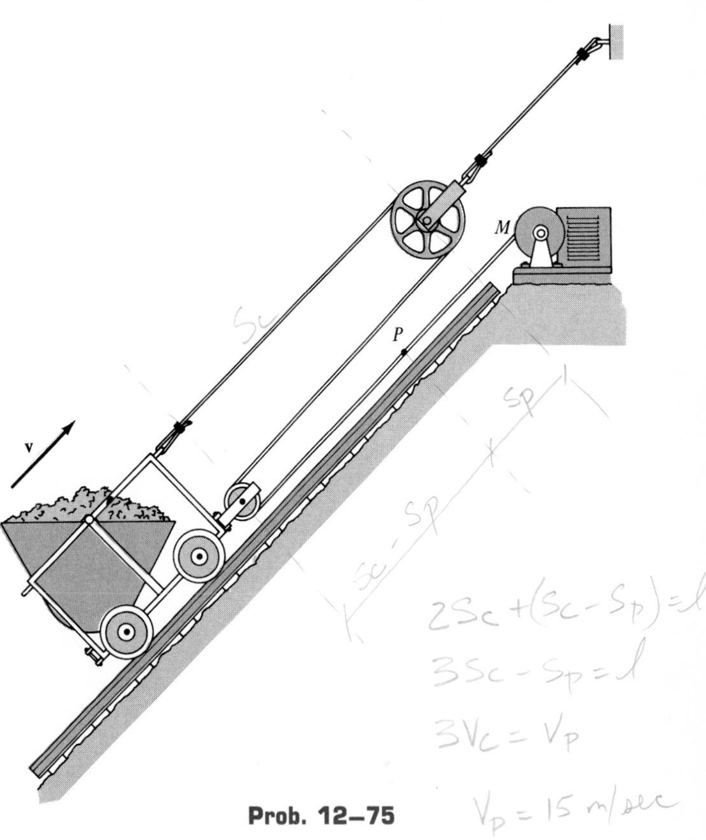

Prob. 12–75

***12–76.** The device shown is used for catapulting the slider A. If this is done by using the hydraulic cylinder H to draw rod BC in at a rate of $v_{BC} = 8$ m/s, determine the velocity of the slider.

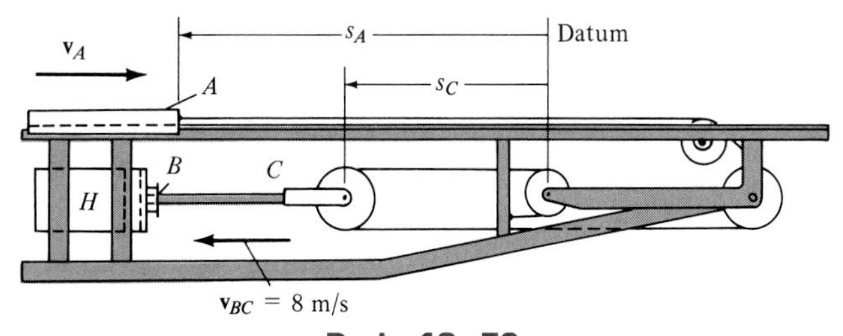

Prob. 12–76

12–77. The hook B on the oil rig is supported by a cable which is connected to the drum at A, passes over a pulley at E, down around the hook's pulley at B, up and around another pulley at E, and is then attached to the drum at C. Place the datum through E, and determine the time needed to hoist the hook from $h = 0$ to $h = 120$ ft if (a) drum A draws the cable in at a speed of 4 ft/s and drum C is stationary, and (b) drum A draws in the cable at 3 ft/s and drum C draws in the cable at 2 ft/s.

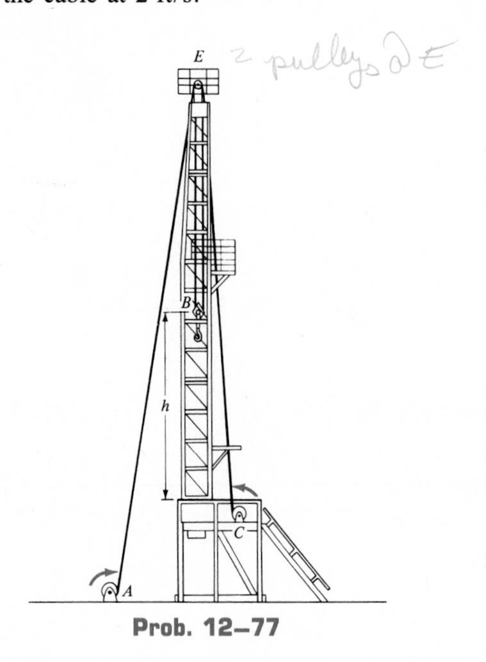

Prob. 12–77

12–78. The pulley arrangement shown is used to obtain a high mechanical advantage. If the cylinder C remains *fixed* and hydraulically pushes the plunger P downward with a speed of 2 ft/s, determine the speed of point A.

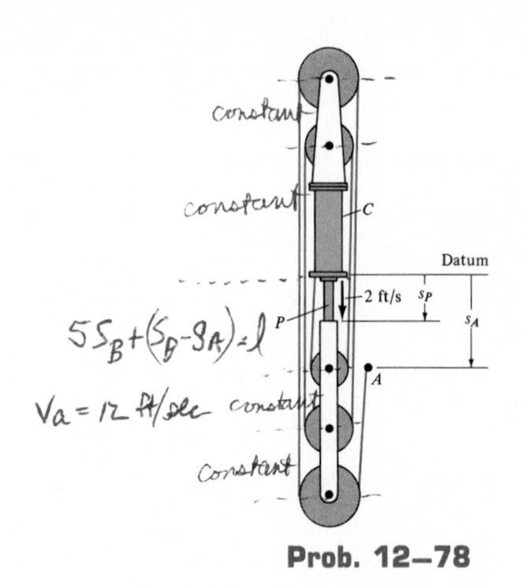

Prob. 12–78

12–79. Three blocks, A, B, and C, move along a straight-line path with constant velocities. If the relative velocity of A with respect to B is 6 m/s (moving to the right), and the relative velocity of B with respect to C is -2 m/s (moving to the left), determine the absolute velocities of A and B. The velocity of C is 4 m/s to the right.

i.e. $V_B = V_C + V_{B/C}$ → see bottom of pg 50

***12–80.** Two planes, A and B, are flying at the same altitude. If their velocities are $v_A = 450$ mi/h and $v_B = 400$ mi/h such that the angle between their straight-line courses is $\theta = 75°$, determine the velocity of plane B with respect to plane A.

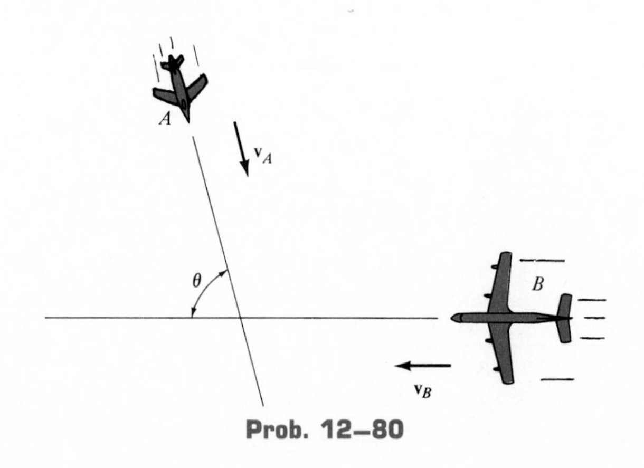

Prob. 12–80

12–81. The airplane can fly at a speed of $v_{P/w} =$ 700 km/h in still air. The point of destination is located at a bearing of north 60° east. If the wind is blowing from north to south at $v_w = 50$ km/h, determine the bearing angle θ at which the pilot must direct the plane to stay on course. What is the velocity of the plane, v_P?

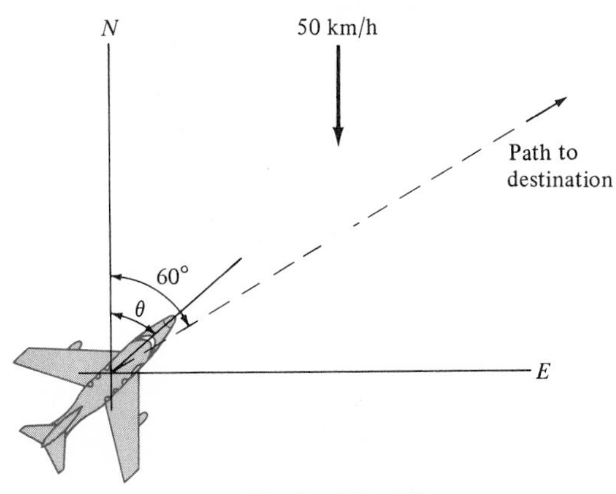

Prob. 12–81

12–82. Two boats leave the pier P at the same time and travel in the directions shown. If $v_A = 40$ ft/s and $v_B = 30$ ft/s, determine the speed of boat A relative to boat B. How long after leaving the pier will the boats be 1000 ft apart?

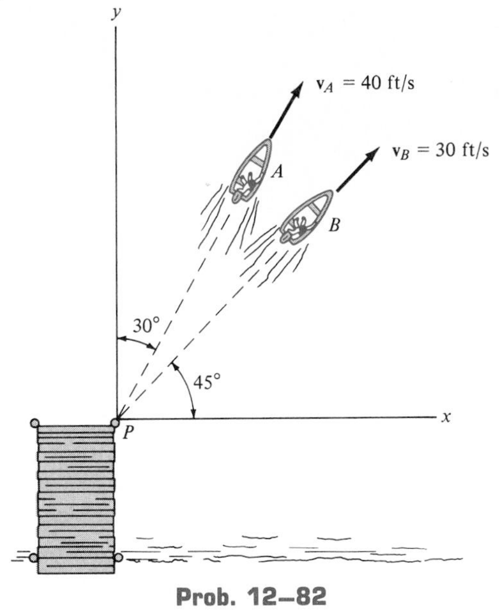

Prob. 12–82

12–83. The motorboat can travel with a speed of $v_B = 12$ ft/s in still water. If it heads for C at the opposite side of the river, and the river flows with a speed of $v_R = 5$ ft/s, determine the resultant velocity of the boat. How far downstream, d, is the boat carried when it reaches the other side at D?

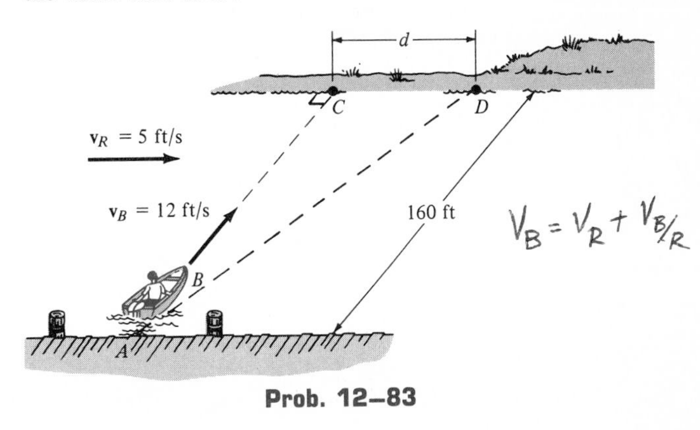

Prob. 12–83

***12–84.** A passenger in an automobile observes that raindrops make an angle of 30° with the horizontal as the auto travels forward with a speed of 50 km/h. Compute the terminal (constant) velocity $\mathbf{v}_r$ of the rain if it is assumed to fall vertically.

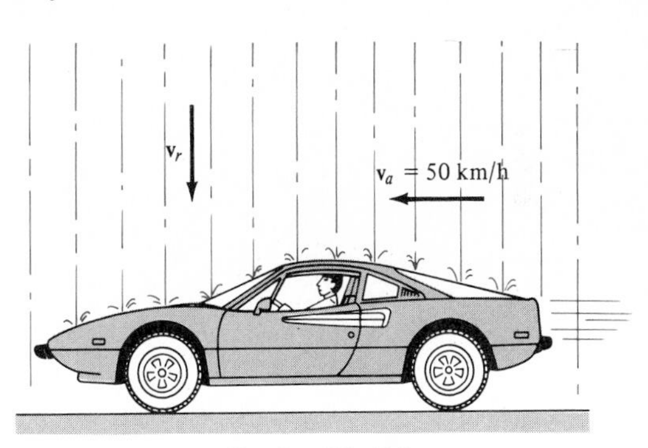

Prob. 12–84

12–85. A fly traveling horizontally at a constant speed enters the open window of a train and leaves through the opposite window 3 m away 0.75 s later. If the fly travels perpendicular to the train's motion and the train is traveling at 3 m/s, determine the speed and direction of flight of the fly observed by a passenger on the train.

12–86. A man is riding on a mail car traveling forward at 15 mi/h. Determine the direction θ, measured in the horizontal plane, in which he must throw a package P at 20 mi/h so that it has a velocity $\mathbf{v}_P$ which is perpendicular to his direction of travel.

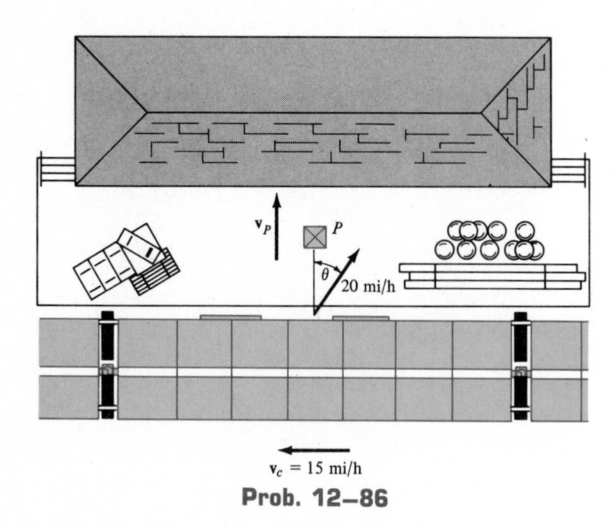

Prob. 12–86

12–87. A man walks at 5 km/h in the direction of a 20-km/h wind. If raindrops fall vertically at 7 km/h in *still air,* determine the direction in which the drops appear to fall with respect to the man. *Hint:* First determine the velocity of the rain, then determine the velocity of the rain with respect to the man.

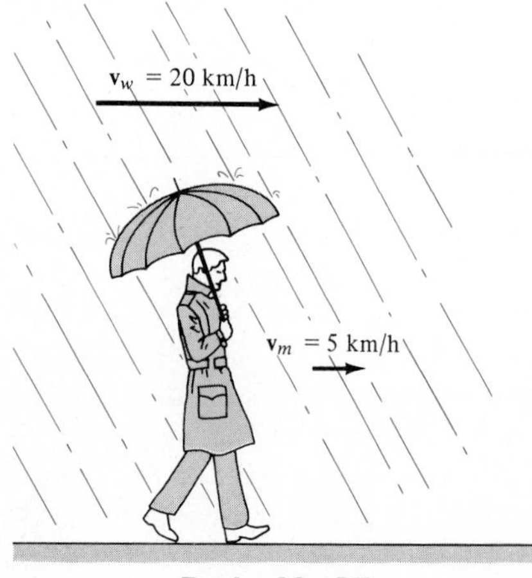

Prob. 12–87

***12–88.** An aircraft carrier is traveling forward with a velocity of 30 km/h. At the instant shown, the plane at A has just taken off and has attained a forward horizontal air speed of 250 km/h, measured from still water. If the plane at B is traveling along the runway of the carrier at 200 km/h in the direction shown, determine the velocity of A with respect to B. *Hint:* First determine the velocity of B measured from still water, then determine $v_{A/B}$.

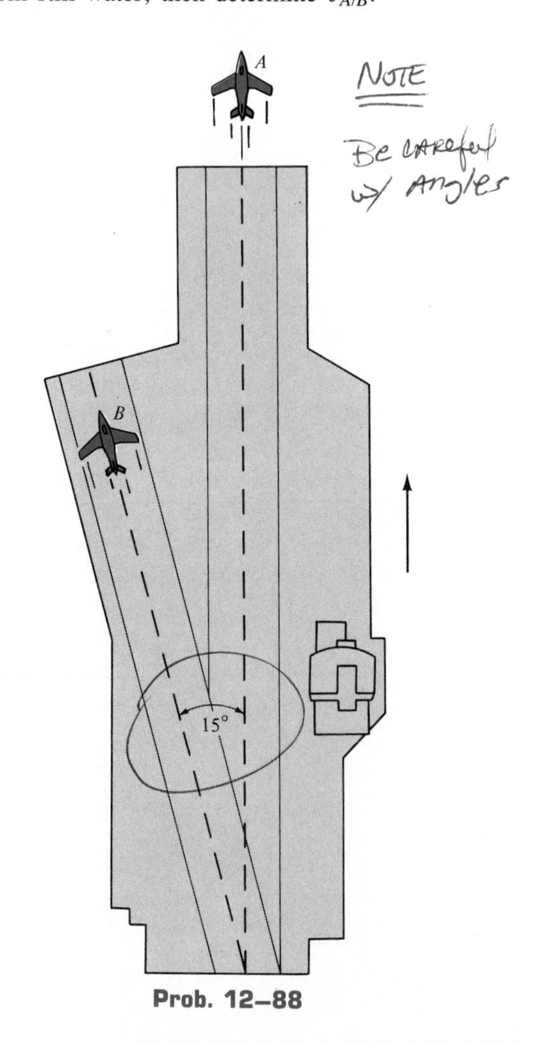

Prob. 12–88

NOTE
Be careful w/ Angles

12–89. At the instant shown, cars A and B are traveling at speeds of 20 mi/h and 45 mi/h, respectively. If B is accelerating at 1600 mi/h^2 while A maintains a constant speed, determine the velocity and acceleration of A with respect to B. *Hint:* A has n and t components of acceleration.

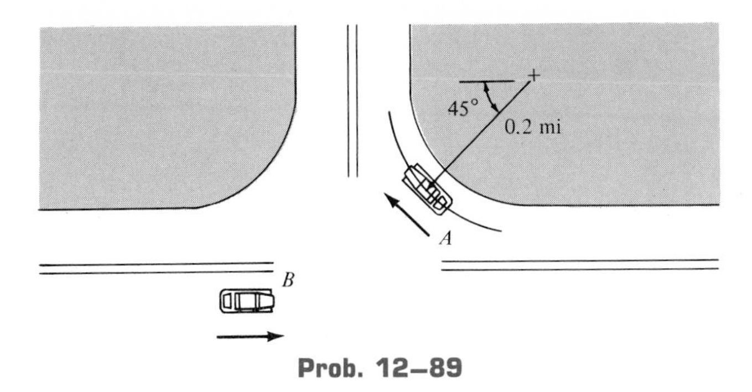

Prob. 12–89

12–90. A passenger in the automobile B observes the motion of the train car A. At the instant shown, the train has a speed of 18 m/s and is reducing its speed at a rate of 1.5 m/s^2. The automobile is accelerating at 2 m/s^2 and has a speed of 25 m/s. Determine the velocity and acceleration of A with respect to B. The train is moving along a curve of radius $r = 300$ m. *Hint:* A has n and t components of acceleration.

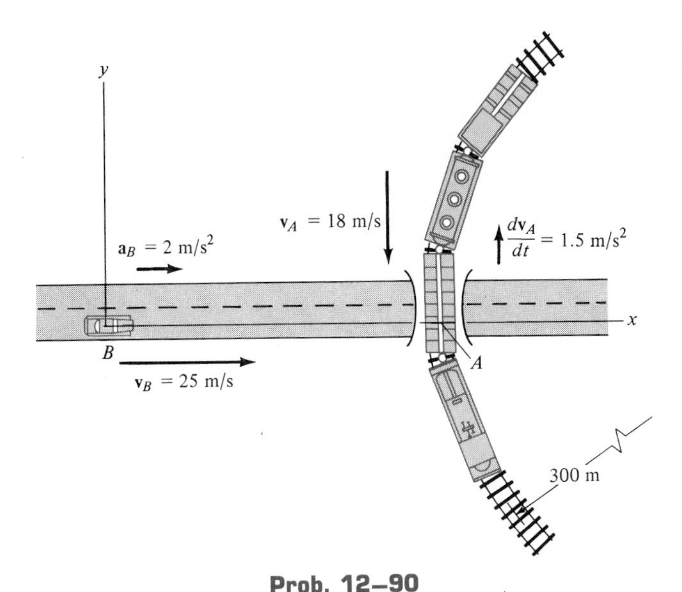

Prob. 12–90

12–91. At the instant shown, car A has a speed of 80 km/h, which is being increased at the rate of 4000 km/h^2 as the car enters an expressway. At the same instant, car B is decelerating at 3000 km/h^2 while traveling forward at 120 km/h. Determine the velocity and acceleration of A with respect to B. *Hint:* A has n and t components of acceleration.

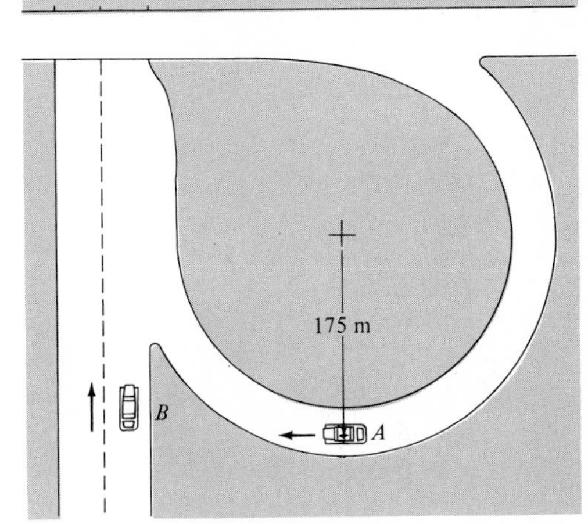

Prob. 12–91

***12–92.** The boy B is running down the slope at a constant speed of 15 ft/s. At the instant he reaches A, his friend (?) throws an egg horizontally off the side of the slope, which later strikes him on the head. Determine the time t of flight, the launch velocity $\mathbf{v}_0$, and the speed with which the egg scrambles on B. Experimental verification is not recommended.

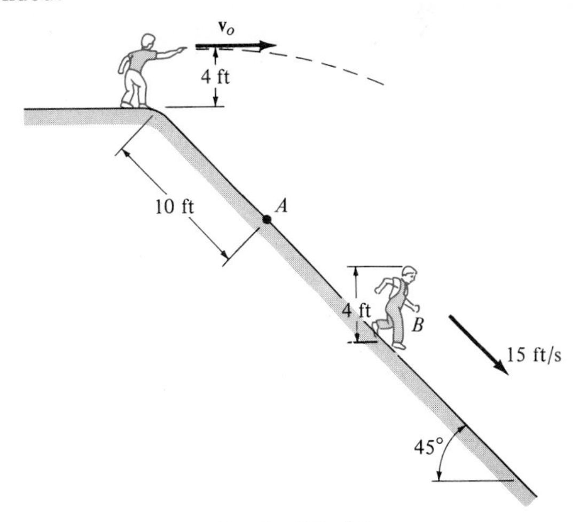

Prob. 12–92

Review Problems

12–93. A particle moves along a straight-line path such that the displacement is defined by the equation $s = 4t^3 - 16t^2 + 3$, where s is expressed in feet and t in seconds. From time $t = 0$, determine (a) the total distance traveled in order to reduce the particle's velocity to zero and (b) the time at which the acceleration is zero.

12–94. A particle moves with a constant speed of $v = 5$ ft/s along a curve $y = 5x^2 - 2$, where the x, y coordinates are measured in feet. Determine the point on the curve where the maximum acceleration occurs and compute its value.

12–95. The projectile is fired horizontally from a cannon at A with a velocity of 600 ft/s. Determine the range R to where it strikes the ground at B and the time of flight.

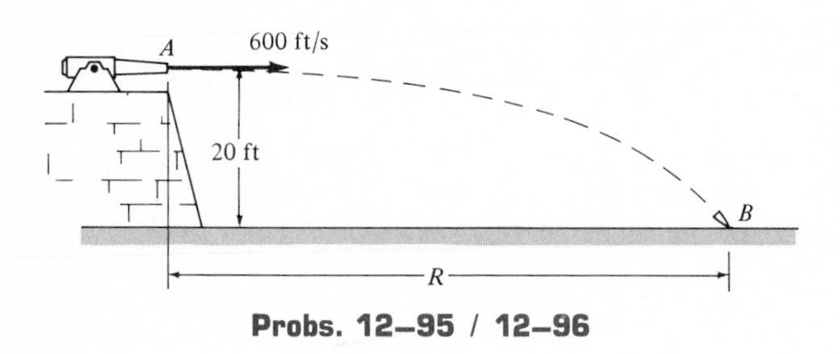

Probs. 12–95 / 12–96

***12–96.** The projectile is fired horizontally from a cannon at A with a velocity of 600 ft/s. At what angle, measured from the horizontal, and with what velocity does it strike the ground at B?

12–97. A particle moves along a straight-line path with an acceleration of $a = (3t^2 - 2)$ mm/s^2, where t is measured in seconds. Determine the velocity and position of the particle when $t = 2$ s. When $t = 0$, $v = 0$, $s = 0$.

12–98. The v-t graph for a car is shown. From the data, construct the s-t and a-t graphs.

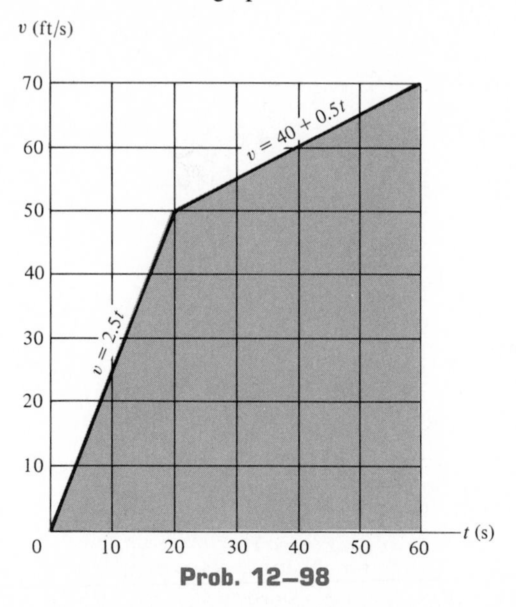

Prob. 12–98

12–99. Determine the speed of the automobile if it has the acceleration shown and is traveling on a road which has a radius of curvature of $\rho = 50$ m. Is this speed constant? Why or why not?

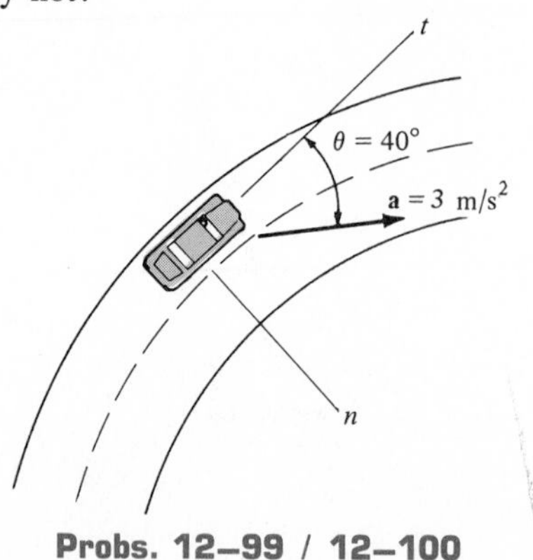

Probs. 12–99 / 12–100

***12–100.** At a given instant, the automobile has a speed of 16 m/s and an acceleration of 3 m/s^2 acting in the direction shown. Determine the radius of curvature of the road and the rate of increase of the automobile's speed.

12–101. Measurements of the velocity of a particle moving along a straight line have been recorded in the table. Determine the acceleration of the particle when $t = 7$ s. The equation that fits this data is $v = (4 + t^2)$ ft/s, where t is measured in seconds.

v (ft/s)	t (s)
4	0
5	1
20	4
40	6
68	8
104	10
148	12

Probs. 12–101 / 12–102

12–102. Using the v vs. t data, determine the position of the particle when $t = 6$ s if $s = 0$ when $t = 0$.

12–103. If the toboggan has a speed of $v_A = 8$ m/s when it reaches point A, determine if the riders reach the other side B of the gorge, and if so, at what angle, measured from the horizontal, does the toboggan strike the snow?

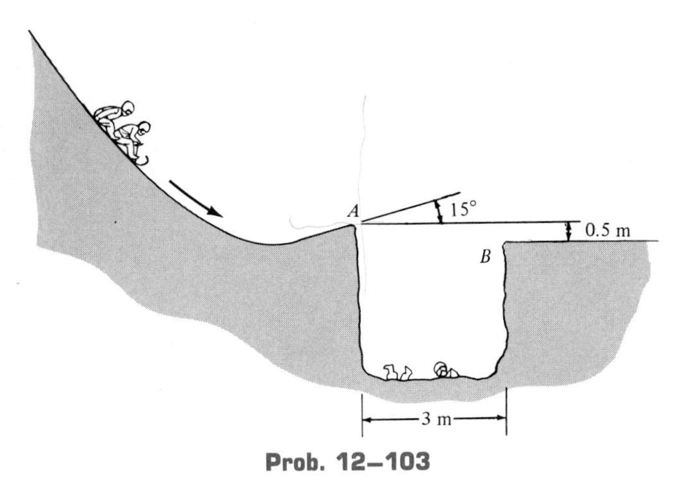

Prob. 12–103

***12–104.** The rocket is given an acceleration which varies with time as shown. If it is launched vertically from rest, compute the height h to which it travels when $t = 20$ s and its velocity at the instant $t = 15$ s.

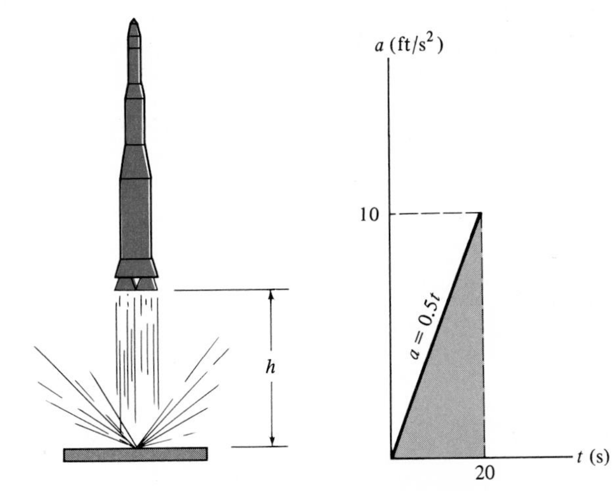

Probs. 12–104 / 12–105

12–105. Construct the v-t and s-t graphs which describe the motion of the rocket. The rocket starts from rest.

12–106. The cable is drawn into the motor M. If the hoist H is moving upward at 6 ft/s, determine the speed of point P.

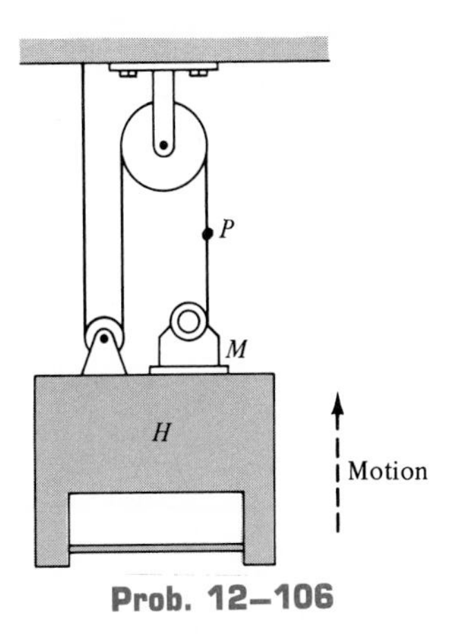

Prob. 12–106

12–107. The sports car can accelerate at 10 m/s^2 and decelerate at 3 m/s^2. If the maximum speed it can attain is 60 m/s, determine the shortest time it takes to travel 900 m starting from rest and then stopping when s = 900 m.

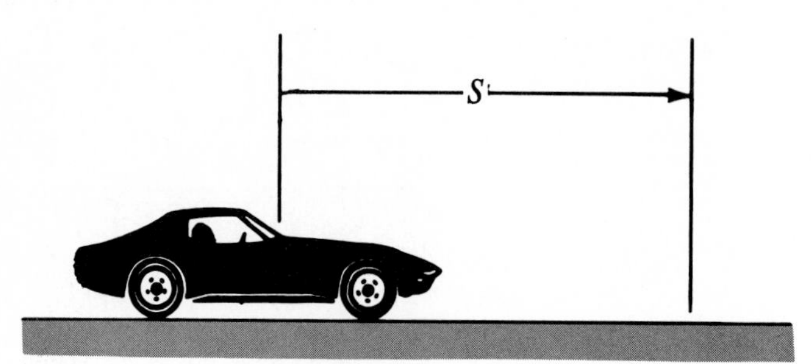

Prob. 12–107

***12–108.** The pilot of fighter plane F is following 1.5 km behind the pilot of bomber B. Both planes are originally traveling at 120 m/s. In an effort to pass the bomber, the pilot in F gives his plane a constant acceleration of 12 m/s^2. Determine the speed at which the pilot in the bomber sees the pilot of the fighter plane pass if at the start of the passing operation the bomber is decelerating at 3 m/s^2. Neglect the effect of any turning.

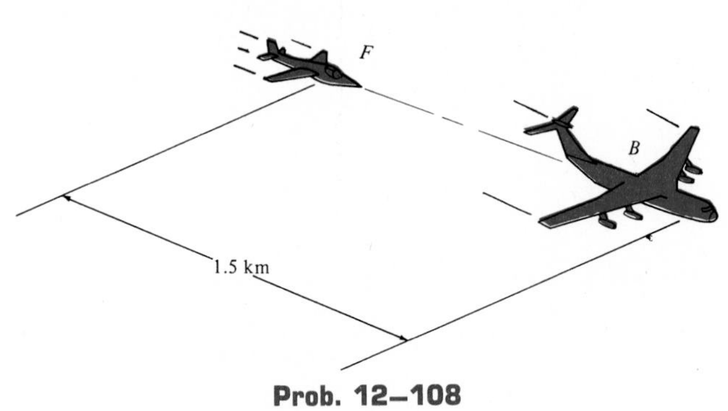

Prob. 12–108

Kinetics of a Particle: Force and Acceleration

In Chapter 12 we developed the methods needed to formulate the acceleration of a particle in terms of its velocity and position. In this chapter we will use these concepts when applying Newton's second law of motion, $\mathbf{F} = m\mathbf{a}$, to study the effects caused by an unbalanced force acting on a particle. Depending upon the geometry of the path, the analysis of problems will be performed using rectangular, or normal and tangential coordinates.*

13.1 Newton's Laws of Motion

Many of the earlier notions about dynamics were dispelled after 1590 when Galileo performed experiments to study the motions of pendulums and falling bodies. The conclusions drawn from these experiments gave some insight as to the effects of forces acting on bodies in motion. The general motion of a body subjected to forces was not known, however, until 1687, when Isaac Newton first presented three basic laws governing the motion of a particle. In a slightly reworded form, Newton's three laws of motion can be stated as follows:

First Law. A particle originally at rest, or moving in a straight line with a constant velocity, will remain in this state provided the particle is not subjected to an unbalanced force.

Second Law. A particle acted upon by an unbalanced force **F** experiences an acceleration **a** that has the same direction as the force and a magnitude that is directly proportional to the force.†

Third Law. For every force acting on a particle, the particle exerts an equal, opposite, and collinear reactive force.

* Applications using cylindrical coordinates are discussed in Chapter 20.

† Stated another way, the unbalanced force acting on the particle is proportional to the time rate of change of the particle's linear momentum.

The first and third laws were used extensively in developing the concepts of statics. Although these laws are also considered in dynamics, Newton's second law of motion forms the basis for most of this study, since this law relates the accelerated motion of a particle to the forces that act on it. Measurements of force and acceleration can be recorded in a laboratory so that in accordance with the second law, if a known unbalanced force $\mathbf{F}_1$ is applied to a particle, the acceleration $\mathbf{a}_1$ of the particle may be measured. Since the force and acceleration are directly proportional, the constant of proportionality, m, may be determined from the ratio $m = F_1/a_1$. Provided the units of measurement are consistent, a different unbalanced force $\mathbf{F}_2$ applied to the particle will create an acceleration $\mathbf{a}_2$, such that $F_2/a_2 = m$. In both cases the ratio will be the same and the acceleration and the force, both being vector quantities, will have the same direction. The scalar m is called the *mass* of the particle. Being constant during any acceleration, m provides a quantitative measure of the resistance of the particle to a change in its velocity.

If the mass of the particle is m, Newton's second law of motion may be written in mathematical form as

$$\mathbf{F} = m\mathbf{a}$$

This equation, which is referred to as the *equation of motion,* is one of the most important formulations in mechanics. In 1905, however, Albert Einstein placed limitations on its use for describing general particle motion. In developing the theory of relativity, he discovered that *time* is not an absolute quantity as assumed by Newton; as a result, it has been shown that the equation of motion fails to *accurately* predict the behavior of a particle, especially when the particle's speed approaches the speed of light ($3.0(10^8)$ m/s). Developments of the theory of quantum mechanics by Schrödinger and his colleagues indicate further that conclusions drawn from using this equation are also invalid when particles move within an atomic distance of one another. For the most part, however, these requirements regarding particle speed and size are not encountered in engineering problems, so that their effects will not be considered in this book.

Einstein

Newton's Law of Gravitational Attraction. Shortly after formulating his three laws of motion, Newton postulated a law governing the mutual attraction between any two particles. In mathematical form this law can be expressed as

$$F = G\frac{m_1 m_2}{r^2} \qquad (13\text{–}1)$$

where F = force of attraction between the two particles

G = universal constant of gravitation; according to experimental evidence $G = (6.673(10^{-11})\ \text{m}^3)/(\text{kg}\cdot\text{s}^2)$

m_1, m_2 = mass of each of the two particles

r = distance between the centers of the two particles

Any two particles or bodies have a mutually attractive (gravitational) force acting between them. In the case of a particle located at or near the surface of the earth, however, the only attractive force having any sizable magnitude is that of the earth's gravitation. This force is termed the "weight" and, for our purpose, it will be the only gravitational force considered.

Mass and Weight. We have defined the *mass* of a body as an *absolute* quantity since the measurement of mass can be made at any location. The *weight* of a body, however, is *not absolute* since it is measured in a gravitational field, and hence its magnitude depends upon where the measurement is made.

very important

The mass and weight of a body are measured differently in the SI and FPS systems of units, and the method of defining the units should be thoroughly understood.

SI System of Units. In the SI system the mass of the body is specified in kilograms and the weight must be calculated using the equation of motion, $F = ma$. Hence, if a body has a mass of m (kg) and is located at a point where the acceleration due to gravity is g (m/s^2), then the weight is expressed in *newtons* as $W = mg$ (N), Fig. 13–1*a*. In particular, if the body is located on the earth at sea level and at a latitude of 45° (considered the "standard location"), the acceleration due to gravity is $g = 9.806\ 65$ m/s^2. For calculations, the value $g = 9.81$ m/s^2 will be used, so that

$$W = mg \text{ (N)} \quad (g = 9.81 \text{ m/s}^2) \tag{13–2}$$

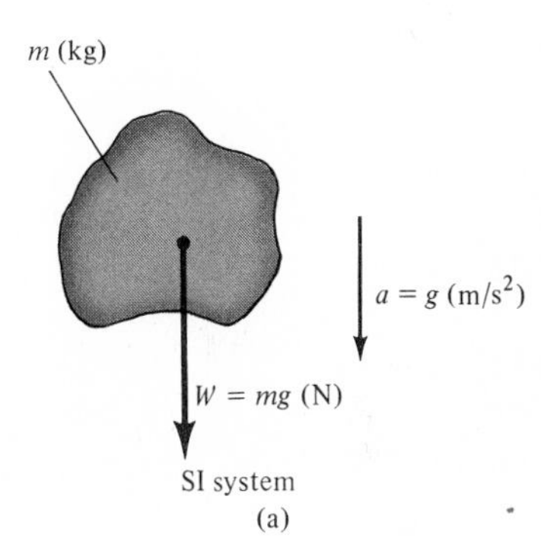

SI system
(a)

Therefore, a body of mass 1 kg has a weight of 9.81 N; a 2-kg body weighs 19.62 N; and so on.

FPS System of Units. In the FPS system the weight of the body is specified in pounds and the mass must be calculated from $F = ma$. Hence, if a body has a weight of W (lb) and is located at a point where the acceleration due to gravity is g (ft/s^2), then the mass is expressed in *slugs* as $m = W/g$ (slug), Fig. 13–1*b*. Since the acceleration of gravity at the standard location is approximately 32.2 ft/s^2 (= 9.81 m/s^2), the mass of the body measured in slugs is

$$m = \frac{W}{g} \text{(slug)} \quad (g = 32.2 \text{ ft/s}^2) \tag{13–3}$$

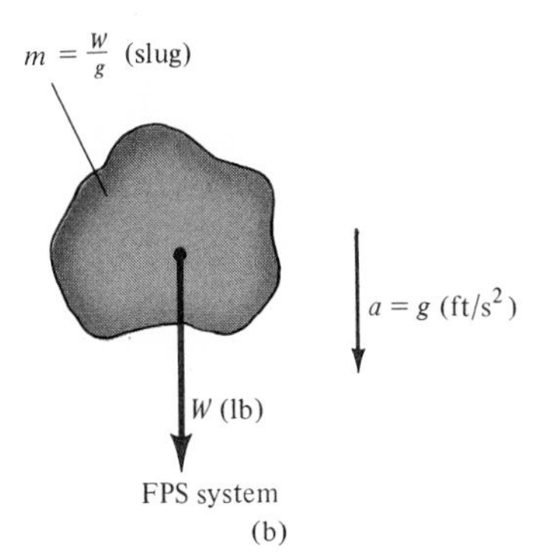

FPS system
(b)

Fig. 13–1

Therefore, a body weighing 32.2 lb has a mass of 1 slug; a 64.4-lb body has a mass of 2 slugs; and so on.

1 slug = 1 lb sec²/ft

13.2 The Equation of Motion

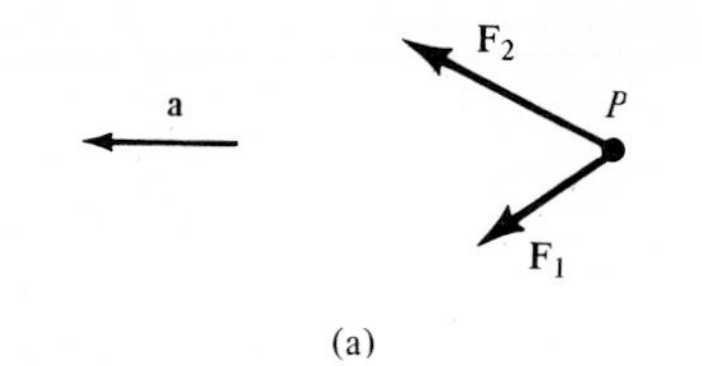

(a)

When more than one force acts on a particle, the resultant force is determined by a vector summation of all the forces; i.e., $\mathbf{F}_R = \Sigma\mathbf{F}$. For this more general case, the equation of motion may be written as

$$\Sigma\mathbf{F} = m\mathbf{a} \tag{13–4}$$

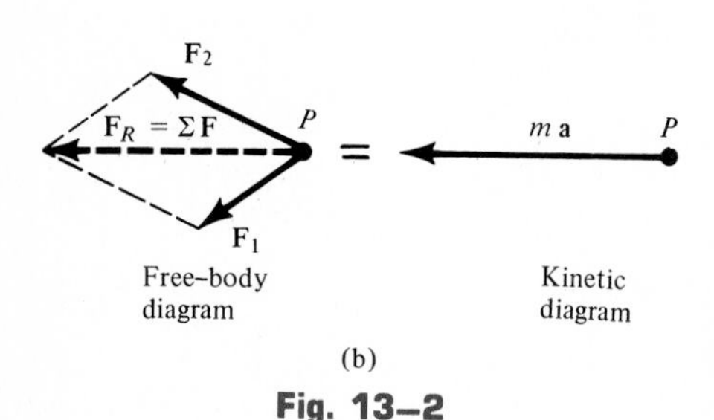

(b)

Fig. 13–2

To illustrate the use of this equation, consider the particle P, shown in Fig. 13–2*a*, which has a mass m and is subjected to the action of two forces, $\mathbf{F}_1$ and $\mathbf{F}_2$. The *free-body diagram,* constructed in Fig. 13–2*b*, graphically accounts for all the forces acting on the particle, $\Sigma\mathbf{F} = \mathbf{F}_1 + \mathbf{F}_2$, whereas the *kinetic diagram* graphically accounts for the vector $m\mathbf{a}$, Fig. 13–2*b*. As shown, the particle accelerates in the direction of $\mathbf{F}_R$, such that the magnitude of $a = F_R/m$. In particular, if $\mathbf{F}_R = \Sigma\mathbf{F} = \mathbf{0}$, the acceleration is also zero, in which case the particle will either remain at *rest* or move along a straight-line path with *constant velocity*. Such are the conditions of *static equilibrium,* Newton's first law of motion.

It is very important to realize that the $m\mathbf{a}$ vector, which is shown on the kinetic diagram, is *not* the same as a force. Instead, it is the result of an unbalanced force which gives the body's mass an acceleration or change in velocity. For example, consider the passenger riding in a train that is accelerating, Fig. 13–3*a*. The forward motion of the train creates a horizontal force $\mathbf{F}$ which the seat exerts on her back, Fig. 13–3*b*. By the equation of motion, it is this unbalanced *force* which gives her a forward acceleration ($\mathbf{F} = m\mathbf{a}$). No force exists which pushes her back toward the seat, although this is the sensation she receives.

Inertial Frame of Reference. Whenever the equation of motion is applied, it is required that measurements of the acceleration be made from a *Newtonian* or *inertial frame of reference*. Such a *coordinate system does not rotate and is either fixed or translates in a given direction with a constant velocity (zero acceleration)*. This definition ensures that the particle's *acceleration* meas-

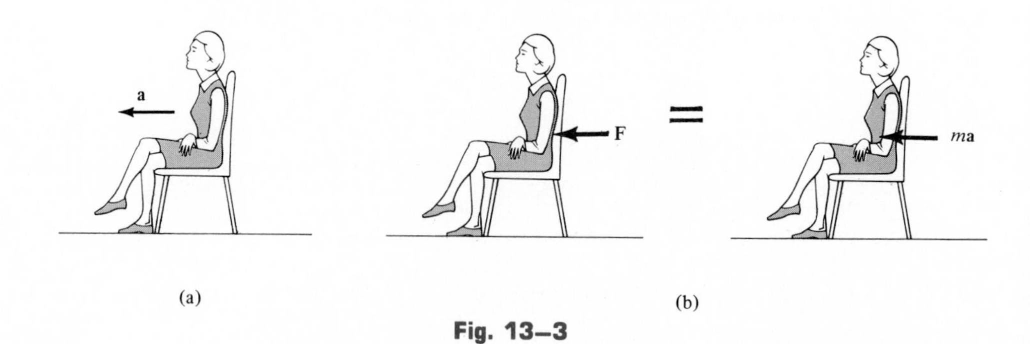

(a) (b)

Fig. 13–3

ured by observers in two different inertial frames of reference will always be the *same*. When studying the motions of rockets and satellites it is justifiable to consider the inertial reference frame as fixed to the stars, whereas dynamics problems concerned with motions on or near the surface of the earth may be solved by using an inertial frame which is assumed fixed to the earth. Even though the earth *rotates* both about its own axis and about the sun, the acceleration created by these rotations can be neglected in most computations.

Equation of Motion for a System of Particles 13.3

The equation of motion will now be extended to include a system of particles isolated within an enclosed region in space, as shown in Fig. 13–4*a*. In particular, there is no restriction in the way the particles are connected, and as a result the following analysis will apply equally well to the motion of a solid, liquid, or gas system. At the instant considered, the arbitrary *i*th particle, having a mass m_i, is subjected to a set of internal forces and a resultant external force. The *internal forces*, represented symbolically as $\Sigma \mathbf{f}_i$, are reactive forces which the other particles each exert on the *i*th particle. The *resultant external force* $\mathbf{F}_i$ represents the effect of gravitational, electrical, magnetic, or contact forces between adjacent bodies or particles not included within the system.

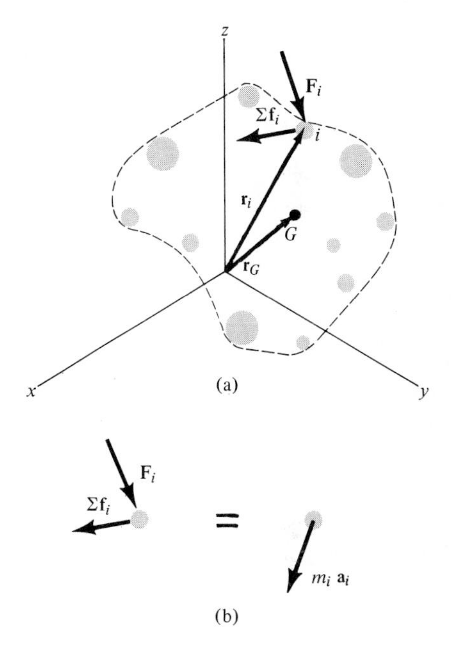

Fig. 13–4

The free-body and kinetic diagrams for the *i*th particle are shown in Fig. 13–4*b*. Applying the equation of motion yields

$$\Sigma \mathbf{F} = m\mathbf{a}; \qquad \mathbf{F}_i + \Sigma \mathbf{f}_i = m_i \mathbf{a}_i$$

When the equation of motion is applied to each of the other particles of the system, similar equations will result. However, when all these equations are added together *vectorially*, only the sum of the external forces, $\Sigma \mathbf{F} = \Sigma \mathbf{F}_i$, acting on the particles, will remain. The vector sum of the internal forces acting within the system is zero, since these forces occur in equal but opposite collinear pairs and therefore cancel each other. Hence, the equation of motion, written for the *entire system of particles*, becomes

$$\Sigma \mathbf{F} = \Sigma m_i \mathbf{a}_i$$

If $\mathbf{r}_G$ is a position vector which locates the *center of mass G* of the particles, Fig. 13–4*a*, then by definition of the center of mass $m\mathbf{r}_G = \Sigma m_i \mathbf{r}_i$, where $m = \Sigma m_i$ is the total mass of all the particles. Differentiating this equation twice with respect to time, assuming no mass is entering or leaving the system, yields

$$m\mathbf{a}_G = \Sigma m_i \mathbf{a}_i$$

Substituting this result into the above equation, we obtain

$$\Sigma \mathbf{F} = m\mathbf{a}_G \qquad (13\text{–}5)$$

This equation states that the sum of the external forces acting on the system of particles is equal to the mass $m = \Sigma m_i$ of a single "fictitious" particle times its acceleration. This fictitious particle is located at the center of mass G of all the particles.

Since, in reality, all particles must have finite size to possess mass, Eq. 13–5 justifies application of the equation of motion to a *body* that is represented as a single particle.

13.4 Equations of Motion: Rectangular Coordinates

When a particle is moving relative to an inertial x, y, z frame of reference, the forces acting on the particle, as well as its acceleration, may be expressed in terms of their x, y, z components, Fig. 13–5. The equations of motion for the particle can then be represented as

$$\Sigma \mathbf{F} = m\mathbf{a}$$
$$\Sigma F_x \mathbf{i} + \Sigma F_y \mathbf{j} + \Sigma F_z \mathbf{k} = m(a_x \mathbf{i} + a_y \mathbf{j} + a_z \mathbf{k})$$

Since the respective **i, j,** and **k** components must be equivalent, the solution of the above equation can be represented in terms of the following three scalar equations:

$$\begin{aligned} \Sigma F_x &= m a_x \\ \Sigma F_y &= m a_y \\ \Sigma F_z &= m a_z \end{aligned} \qquad (13\text{–}6)$$

If all the forces acting on the particle lie in the x-y plane, the particle will only have motion in this plane, and therefore only the first two of Eqs. 13–6 may be used to specify the motion.

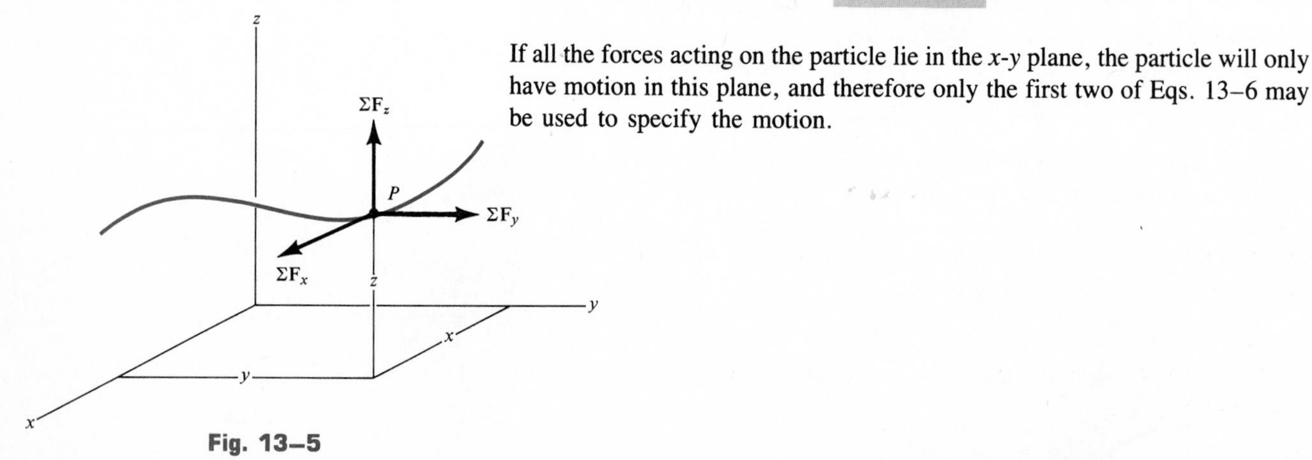

Fig. 13–5

PROCEDURE FOR ANALYSIS

The following procedure provides a method for solving problems using the equations of motion:

Free-Body and Kinetic Diagrams. Establish the x, y, z inertial coordinate system and draw the particle's free-body and kinetic diagrams. As stated previously, the free-body diagram is a graphical representation of all the forces ($\Sigma\mathbf{F}$) which act on the particle; whereas the kinetic diagram graphically accounts for $m\mathbf{a}$, Fig. 13–2*b*. If the particle's acceleration is *unknown,* assume that $m\mathbf{a}$ acts in the *direction* of the *positive* inertial coordinate(s).*

Equations of Motion. Resolve the forces and $m\mathbf{a}$ into their components directly from the free-body and kinetic diagrams and apply the equations of motion.

(Friction.) If the particle contacts a rough surface, it may be necessary to use the *frictional equation,* which relates the coefficient of kinetic friction μ_k to the magnitudes of the frictional and normal forces $\mathbf{F}_f$ and $\mathbf{N}$ acting at the surfaces of contact,† i.e., $F_f = \mu_k N$.

(Spring.) If the particle is connected to an *elastic spring* having negligible mass, the magnitude of spring force $\mathbf{F}_s$ can be related to the stretch or compression x of the spring by the equation $F_s = kx$, where k is the spring's stiffness measured as a force per unit length.

Kinematics. Use the equations of kinematics if a complete solution cannot be obtained strictly from the equation of motion. In this regard, if the velocity or position of the particle is to be found, it will be necessary to apply the proper kinematic equations once the particle's acceleration is determined from $\Sigma\mathbf{F} = m\mathbf{a}$.

If the *acceleration is a function of time,* use $a = dv/dt$ and $v = ds/dt$ which, when integrated, yield the particle's velocity and position.

If the *acceleration is a function of displacement,* integrate $a\,ds = v\,dv$ to obtain the velocity as a function of position.

If the *acceleration is constant,* use

$$v = v_0 + a_c t \qquad s = s_0 + v_0 t + \tfrac{1}{2}a_c t^2 \qquad v^2 = v_0^2 + 2a_c(s - s_0)$$

to determine the position or velocity of the particle.

*This provides a mathematical convenience to account for the proper sign of $\mathbf{a}$ if the equations of kinematics are used in the solution.

†A review of friction in Sec. 8–1, *Mechanics for Engineers: Statics,* is suggested *before* solving the problems.

Example 13–1

The 50-kg crate shown in Fig. 13–6*a* rests on a horizontal plane for which the coefficient of kinetic friction is $\mu_k = 0.3$. If the crate is subjected to a 400-N towing force as shown, determine the velocity of the crate in 5 s starting from rest.

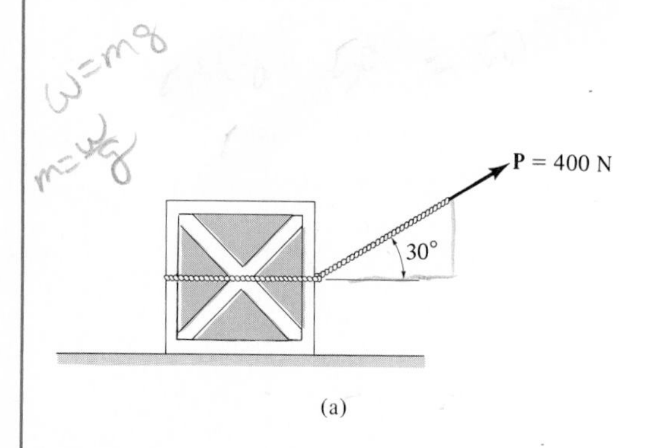

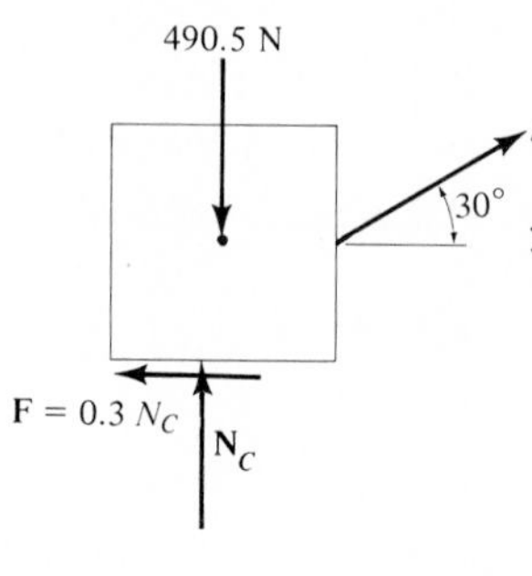

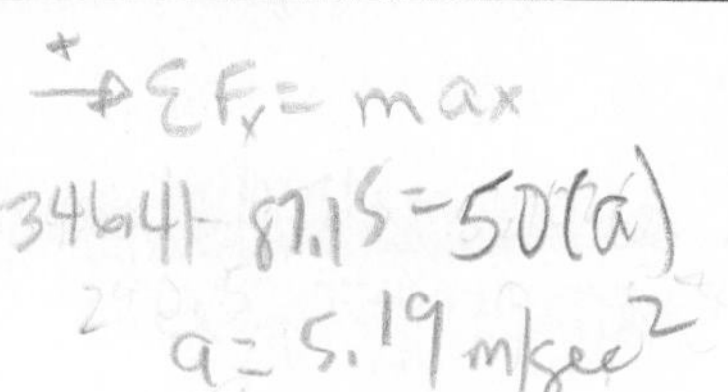

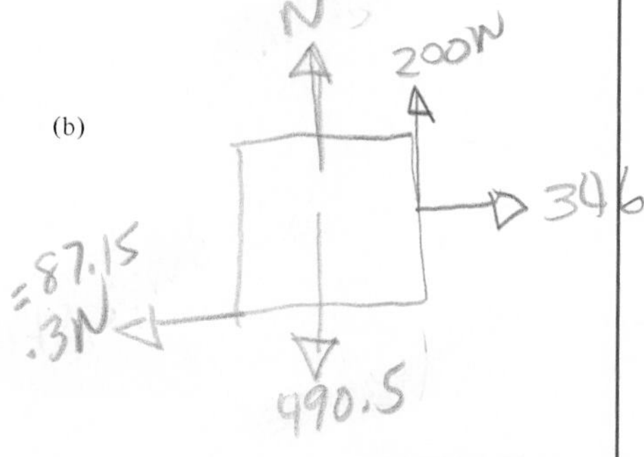

50 a

(b)

Fig. 13–6

Solution

Free-Body and Kinetic Diagrams. From Eq. 13–2, the weight of the crate is $W = mg = 50\text{ kg}(9.81\text{ m/s}^2) = 490.5\text{ N}$. As shown in Fig. 13–6*b*, the frictional force has a magnitude of $F = \mu_k N_C$ and acts to the left, since it opposes the motion of the crate. Note that $m\mathbf{a}$ acts in the direction of increasing displacement.

Equations of Motion. Using the data shown on the free-body and kinetic diagrams, we have

$$\overset{+}{\rightarrow}\Sigma F_x = ma_x; \qquad 400\cos 30^\circ - 0.3\,N_C = 50a \qquad (1)$$

$$+\uparrow \Sigma F_y = ma_y; \quad N_C - 490.5 + 400\sin 30^\circ = 0 \qquad (2)$$

Solving Eq. (2) for N_C, substituting the result into Eq. (1), and solving for a yields

$$N_C = 290.5\text{ N}$$

$$a = 5.19\text{ m/s}^2$$

Kinematics. Since the acceleration is *constant,* and the initial velocity is zero, the velocity of the crate in 5 s is

$$(\overset{+}{\rightarrow}) \qquad \begin{aligned} v &= v_0 + a_c t \\ &= 0 + 5.19(5) \\ &= 26.0\text{ m/s} \end{aligned} \qquad \textit{Ans.}$$

Example 13–2

A 10-kg projectile is fired vertically upward from the ground, with an initial velocity of 50 m/s, Fig. 13–7*a*. Determine the maximum height to which it will travel if atmospheric resistance is neglected.

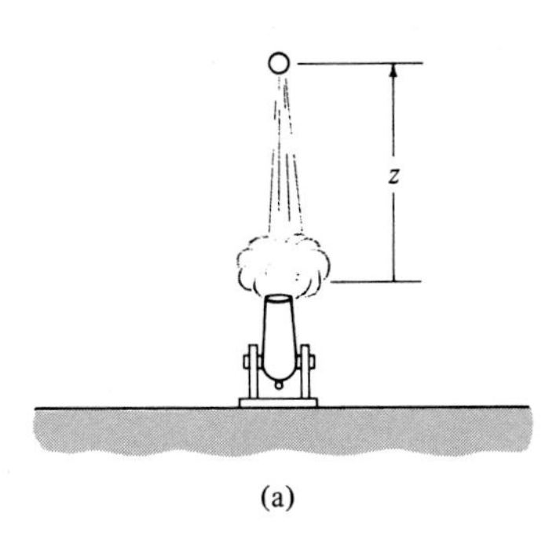

(a)

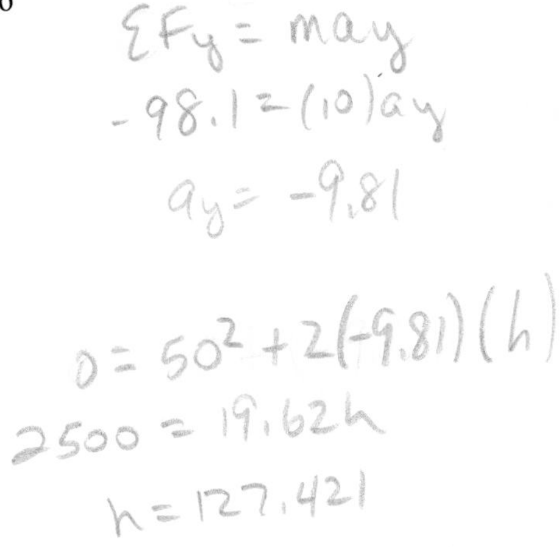

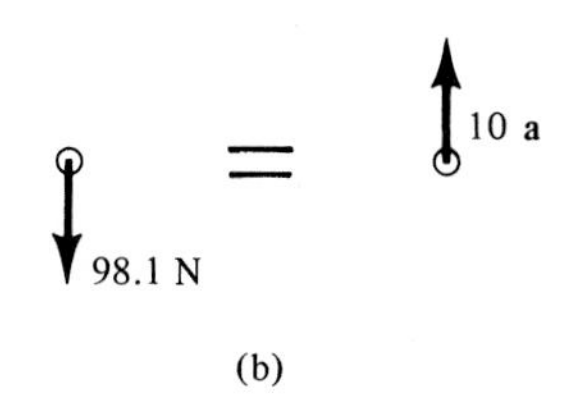

(b)

Fig. 13–7

Solution

Part (a)

Free-Body and Kinetic Diagrams. As shown in Fig. 13–7*b*, the projectile's weight is $W = mg = 10(9.81) = 98.1$ N. We assume that $m\mathbf{a}$ acts upward in the *positive* z direction.

Equation of Motion

$+\uparrow \Sigma F_z = ma_z;$ $\qquad -98.1 = 10a$

$$a = -9.81 \text{ m/s}^2$$

The result indicates that the projectile, like every object having free-flight motion near the earth's surface, is subjected to a constant downward acceleration of 9.81 m/s^2.

Kinematics. Since the acceleration is *constant*, the maximum height h can be obtained using Eq. 12–6. Initially, $s_0 = 0$ and $v_0 = 50$ m/s, and at the maximum height, $v = 0$; therefore,

$(+\uparrow)$ $\qquad v^2 = v_0^2 + 2a_c(s - s_0)$

$$0 = (50)^2 + 2(-9.81)(h - 0)$$

$$h = 127.4 \text{ m} \qquad \textit{Ans.}$$

Example 13–3

The crate shown in Fig. 13–8*a* has a weight of 50 lb and is acted upon by a force having a variable magnitude of $P = 20t$, where P is in pounds and t is in seconds. Compute the crate's velocity 2 s after **P** has been applied. The crate's initial velocity is $v_0 = 3$ ft/s down the plane, and the coefficient of kinetic friction between the crate and the plane is $\mu_k = 0.3$.

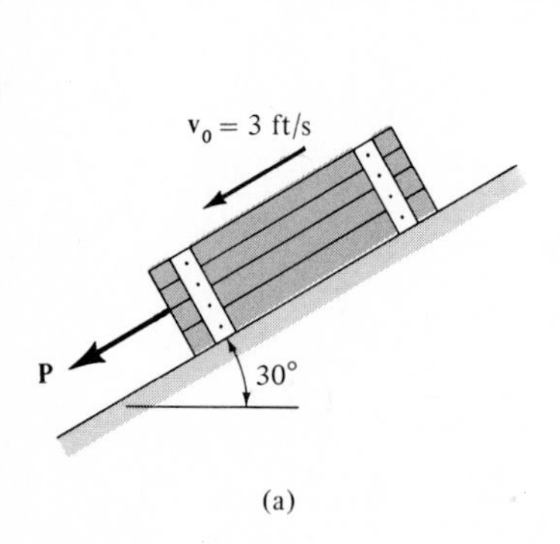

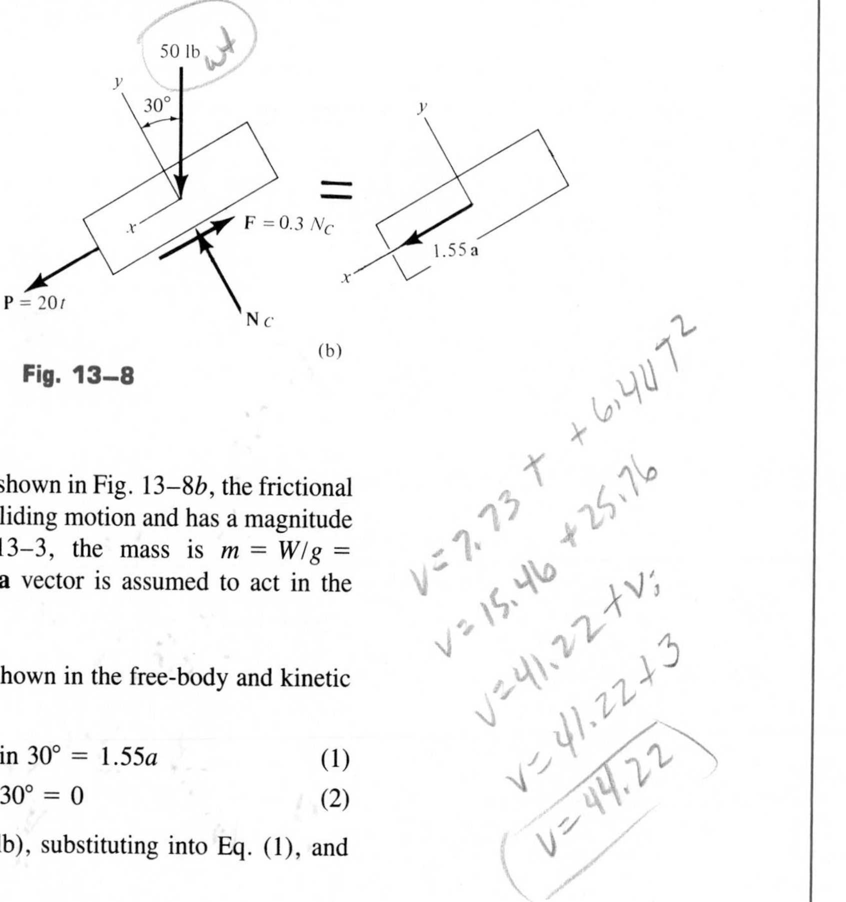

Fig. 13–8

Solution

Free-Body and Kinetic Diagrams. As shown in Fig. 13–8*b*, the frictional force is directed opposite to the crate's sliding motion and has a magnitude of $F = \mu_k N_C = 0.3N_C$. Using Eq. 13–3, the mass is $m = W/g = 50/32.2 = 1.55$ slug. As usual, the $m\mathbf{a}$ vector is assumed to act in the *positive x* direction.

Equations of Motion. Using the data shown in the free-body and kinetic diagrams, we have

$$+\swarrow \Sigma F_x = ma_x; \quad 20t - 0.3N_C + 50 \sin 30° = 1.55a \qquad (1)$$

$$+\nwarrow \Sigma F_y = ma_y; \quad N_C - 50 \cos 30° = 0 \qquad (2)$$

Solving for N_C in Eq. (2) ($N_C = 43.3$ lb), substituting into Eq. (1), and simplifying yields

$$a = 12.88t + 7.73 \qquad (3)$$

Kinematics. Since the acceleration is a function of time, the velocity of the crate is obtained by using $a = dv/dt$ with the initial condition that $v_0 = 3$ ft/s at $t = 0$. We have

$$(+\swarrow) \qquad dv = a\,dt$$

$$\int_3^v dv = \int_0^t (12.88t + 7.73)dt$$

$$v = 6.44t^2 + 7.73t + 3$$

When $t = 2$ s,

$$v = 44.2 \text{ ft/s} \qquad \textit{Ans.}$$

Example 13–4

A smooth 2-kg collar, shown in Fig. 13–9a, is attached to a spring having a stiffness of $k = 30$ N/m and an unstretched length of 1 m. If the collar is released from rest at A, determine its acceleration and the normal force of the rod on the collar at the instant $y = 0.75$ m.

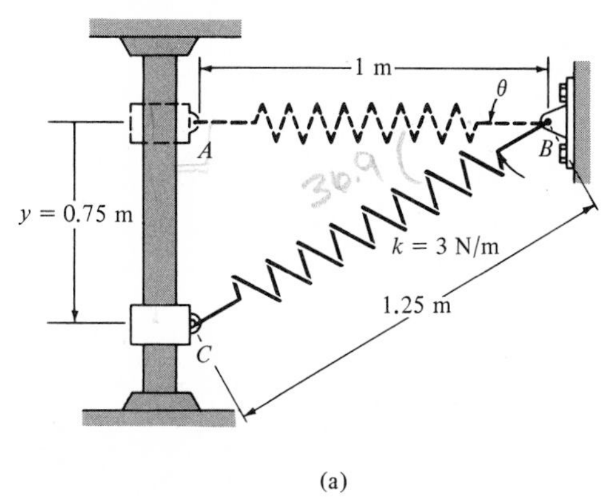

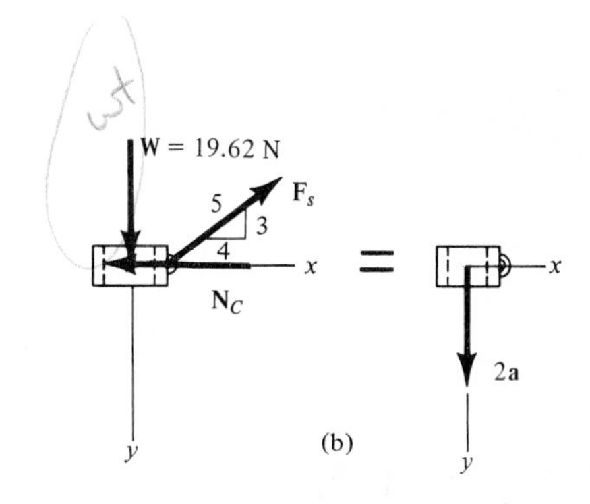

Fig. 13–9

Solution

Free-Body and Kinetic Diagrams. The free-body and kinetic diagrams of the collar when it is located at the position $y = 0.75$ m are shown in Fig. 13–9b. Note that the weight is $W = 2(9.81) = 19.62$ N and $\mathbf{F}_s$ acts on the slope of a 3-4-5 triangle. Furthermore, the collar is assumed to be accelerating so that "**2a**" acts downward in the *positive* y direction.

Equations of Motion. From the data in Fig. 13–9b,

$$\xrightarrow{+}\Sigma F_x = ma_x; \qquad -N_C + \left(\frac{4}{5}\right)F_s = 0 \qquad (1)$$

$$+\downarrow \Sigma F_y = ma_y; \qquad 19.62 - \left(\frac{3}{5}\right)F_s = 2a \qquad (2)$$

The magnitude of the spring force is a function of the stretch s of the spring, i.e., $F_s = ks$. Here the *unstretched length* is $AB = 1$ m, Fig. 13–9a; therefore, $s = CB - AB = 1.25\text{ m} - 1\text{ m} = 0.25$ m. Since $k = 3$ N/m, then

$$F_s = ks = (30\text{ N/m})(0.25\text{ m}) = 7.5\text{ N} \qquad (3)$$

Substituting this value into Eqs. (1) and (2), we obtain

$$N_C = 6.0\text{ N} \qquad \textit{Ans.}$$

$$a = 7.56\text{ m/s}^2 \qquad \textit{Ans.}$$

Example 13–5

The 100-kg block A shown in Fig. 13–10a is released from rest. If the mass of the pulleys and the cord is neglected, determine the speed of the 20-kg block B in 2 s.

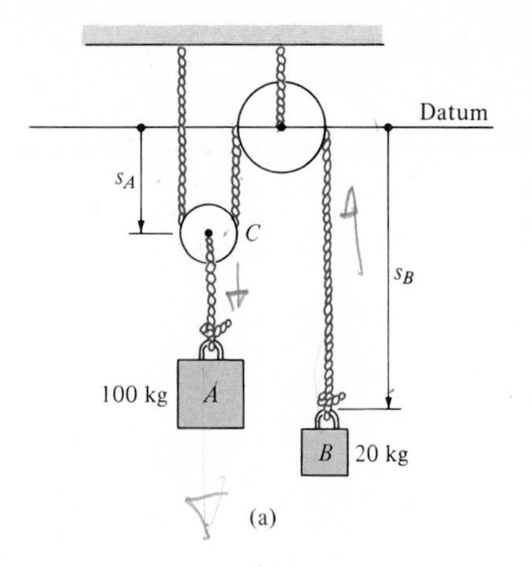

(a)

Solution

Motion of blocks A and B will be analyzed separately.

Free-Body and Kinetic Diagrams. Since the mass of the pulleys is neglected, the equilibrium condition for pulley C is shown in Fig. 13–10b. The free-body and kinetic diagrams for blocks A and B are shown in Fig. 13–10c and d, respectively. Here we *assume* both blocks accelerate downward, in the direction of $+s_A$ and $+s_B$. Why?

(b)

Equations of Motion

Block A, Fig. 13–10c:

$$+\downarrow \Sigma F_y = ma_y; \qquad 981 - 2T = 100a_A \qquad (1)$$

Block B, Fig. 13–10d:

$$+\downarrow \Sigma F_y = ma_y; \qquad 196.2 - T = 20a_B \qquad (2)$$

Kinematics. The necessary third equation is obtained by studying the kinematics of the pulley arrangement in order to relate a_A to a_B. Using the technique developed in Sec. 12–7, the coordinates s_A and s_B measure the positions of A and B from the fixed datum, Fig. 13–10a. It is seen that

$$2s_A + s_B = l$$

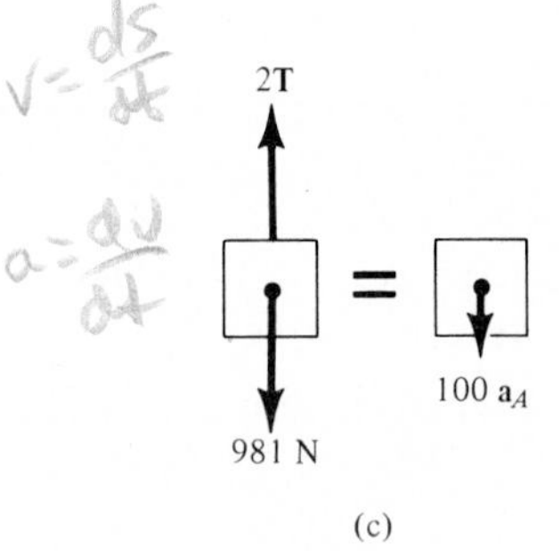

(c)

where l is constant and represents the total vertical length of cable. Differentiating this expression twice with respect to time yields

$$2a_A = -a_B \qquad (3)$$

Hence when block A accelerates *downward*, block B accelerates *upward* at twice the amount. Notice, however, that in writing Eqs. (1) to (3), the *positive direction was always assumed downward.* It is important to be *consistent* in this assumption since we are seeking a simultaneous solution of equations. The solution yields

$$T = 327.0 \text{ N}$$
$$a_A = 3.27 \text{ m/s}^2$$
$$a_B = -6.54 \text{ m/s}^2$$

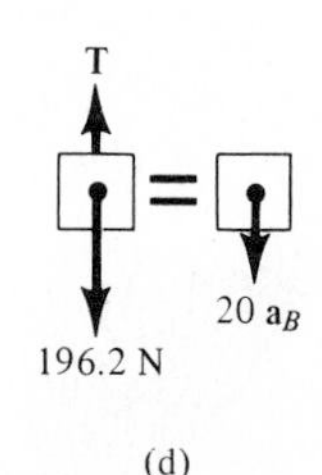

(d)

Fig. 13–10

Since a_B is constant, the velocity of block B in 2 s is thus

$$(+\downarrow) \qquad v = v_0 + a_B t$$
$$= 0 + (-6.54)(2)$$
$$= -13.08 \text{ m/s} \qquad \textit{Ans.}$$

The negative sign indicates that block B is moving upward.

Problems

Except when stated otherwise, throughout this chapter assume that the coefficients of static and kinetic friction are equal, i.e., $\mu = \mu_s = \mu_k$.

13–1. Determine the gravitational attraction between two spheres which are just touching each other. Each sphere has a mass of 10 kg and a radius of 200 mm.

13–2. The moon has a mass of $7.3(10^{22})$ kg and the earth's mass is $6.0(10^{24})$ kg. If their centers are $3.9(10^8)$ m apart, determine the gravitational attractive force between the two bodies.

13–3. The planets Mars and Earth have diameters of 6775 and 12 755 km, respectively. The mass of Mars is 0.107 times that of the earth. If a body weighs 400 N on the surface of the earth, what would its weight be on Mars? Also, what is the mass of the body and the acceleration of gravity on Mars?

***13–4.** Determine the force which a boy having a weight of 75 lb exerts on the floor of an elevator when the elevator is (a) at rest, (b) descending with a constant velocity of 5 ft/s, and (c) ascending with a constant acceleration of 6 ft/s^2. $\Sigma F_y = ma_y$

13–5. Each of the three barges has a mass of 30 Mg, whereas the tugboat has a mass of 12 Mg. As the barges are being pulled forward with a constant velocity of 4 m/s, the tugboat must overcome the frictional resistance of the water, which is 2 kN for each barge and 1.5 kN for the tugboat. If the cable between *A* and *B* breaks, determine the acceleration of the tugboat. watch the units

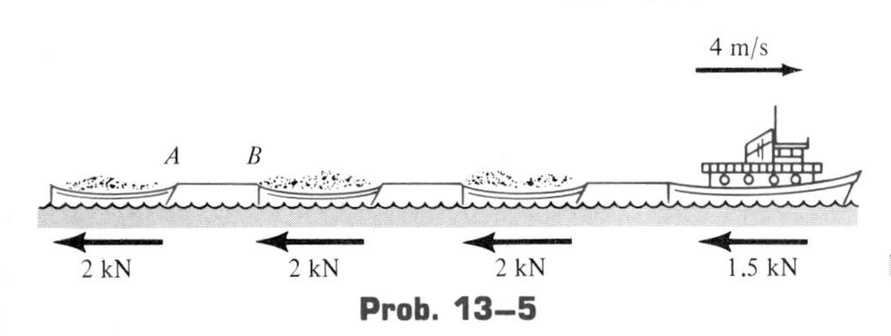

Prob. 13–5

13–6. A block having a mass of 2 kg is placed on a spring scale located in an elevator that is moving downward. If the scale reading, which measures the force in the spring, is 20 N, determine the acceleration of the elevator. Neglect the mass of the scale.

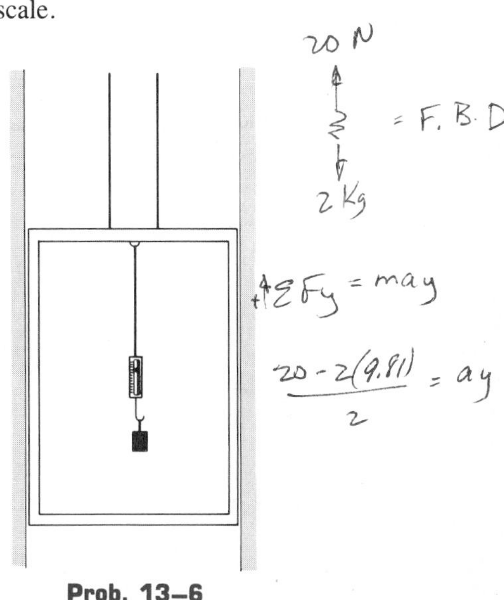

Prob. 13–6

13–7. The 300-kg bar *B*, originally at rest, is being towed over a series of small rollers. Compute the force in the cable when $t = 5$ s, if the motor *M* is drawing in the cable for a short time at a rate of $v = (0.4t^2)$ m/s, where *t* is in seconds ($0 \leqslant t \leqslant 6$ s). How far does the bar move in 5 s? Neglect the mass of the cable, pulley *P*, and the rollers.

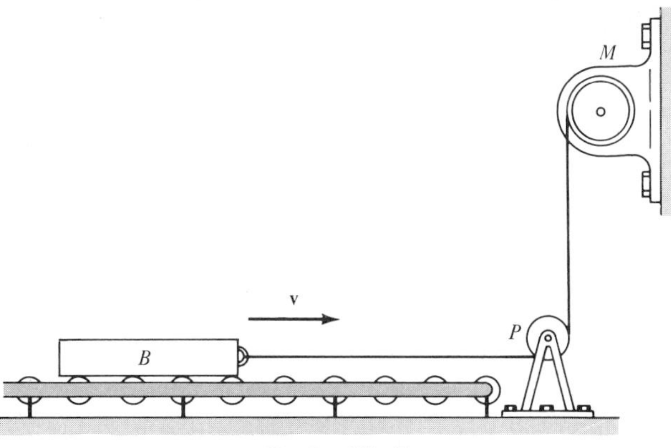

Prob. 13–7

***13–8.** A crate having a mass of 60 kg falls horizontally off the back of a truck which is traveling at 80 km/h. Compute the coefficient of kinetic friction between the road and the crate if the crate slides 45 m on the ground with no tumbling along the road before coming to rest. Assume that the initial speed of the crate along the road is 80 km/h.

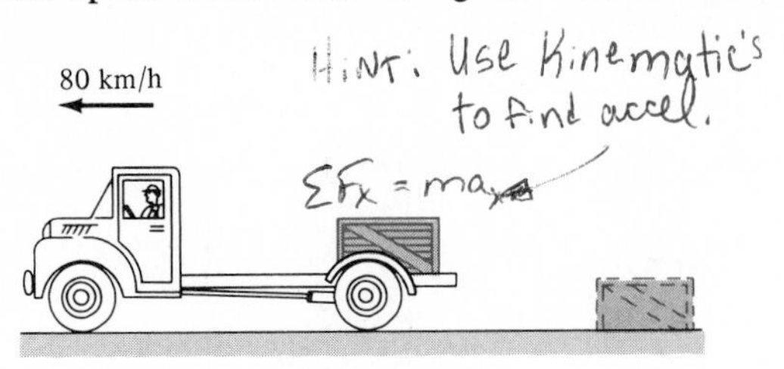

Prob. 13–8

13–9. A 1.5-lb brick is released from rest at A and slides down the inclined roof. If the coefficient of friction between the roof and the brick is $\mu = 0.3$, determine the speed at which the brick strikes the gutter G.

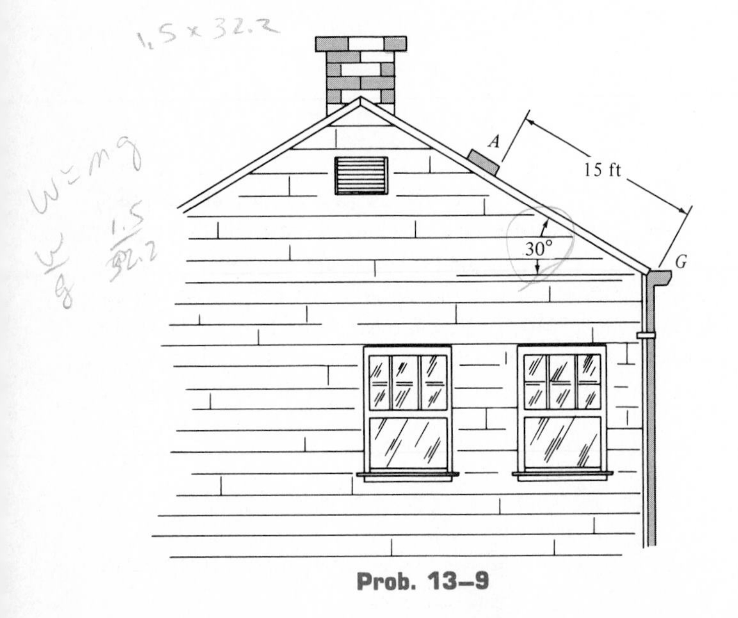

Prob. 13–9

13–10. The conveyor belt C is moving at 6 m/s. If the static coefficient of friction between the conveyor and the 10-kg box B is $\mu = 0.2$, determine the shortest time the conveyor can stop so that the box does not shift or move on the belt.

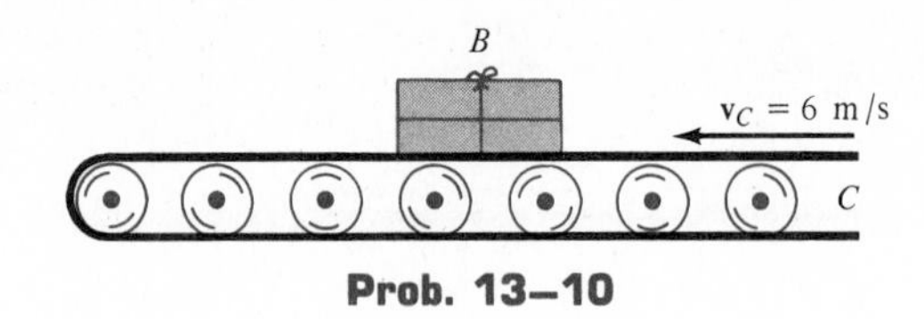

Prob. 13–10

13–11. The conveyor delivers each 15-lb crate to the ramp at A such that at A the crate's velocity is $v_A = 4$ ft/s, directed down *along* the ramp. If the kinetic coefficient of friction between each crate and the ramp is $\mu_k = 0.3$, determine the speed at which each crate slides off the ramp at B. Assume that no tipping occurs.

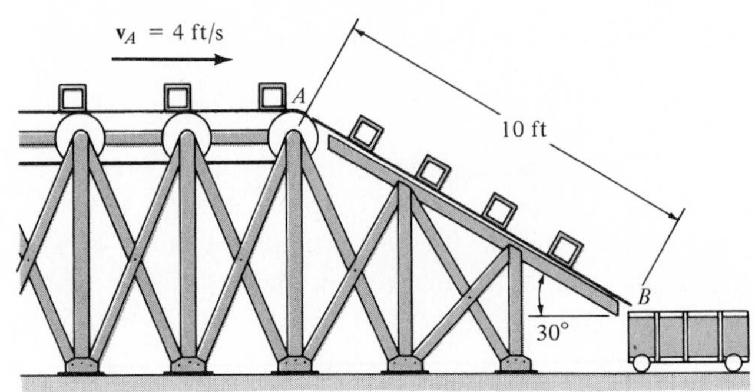

Prob. 13–11

***13–12.** The two boxcars A and B have a weight of 20,000 lb and 30,000 lb, respectively. If they are freely coasting down the incline when the brakes are applied to the wheels of car A, determine the force in the coupling C between the two cars. The coefficient of friction between the wheels of A and the tracks is $\mu = 0.3$. The wheels of car B are free to roll. *Suggestion:* Solve the problem by representing single resultant normal forces acting on A and B, respectively.

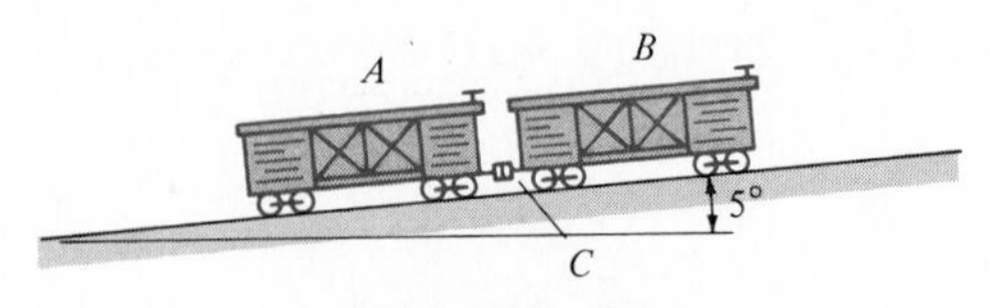

Prob. 13–12

13–13. Solve Prob. 13–12 if $\mu = 0.5$.

13–14. A parachutist having a mass of 70 kg is falling at 10 m/s when he opens his parachute at a very high altitude. If the atmospheric drag resistance is $F_D = 350$ N, determine his velocity when he has fallen 20 m.

Prob. 13–14

13–15. A car entered in a soap-box derby starts from rest at A. If the resistance caused by the air is $F_D = 125$ N, determine the car's speed in $t = 2$ s.

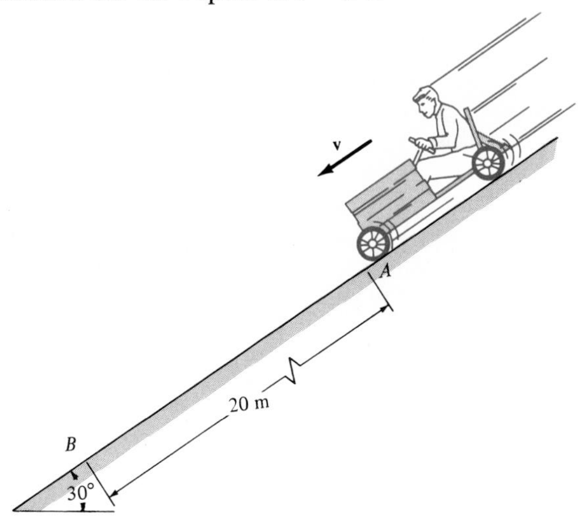

Prob. 13–15

***13–16.** In the event of an emergency, a nuclear reactor is shut down by dropping the control rod R, having a weight of 20 lb, into the reactor core C. If the rod is immersed in "heavy" water, which offers a drag resistance to downward motion of $F_D = 5$ lb, determine the distance s it must descend into the reactor core before it attains a speed of 15 ft/s. The rod is released from rest when $s = 0$. Neglect the effects of buoyancy.

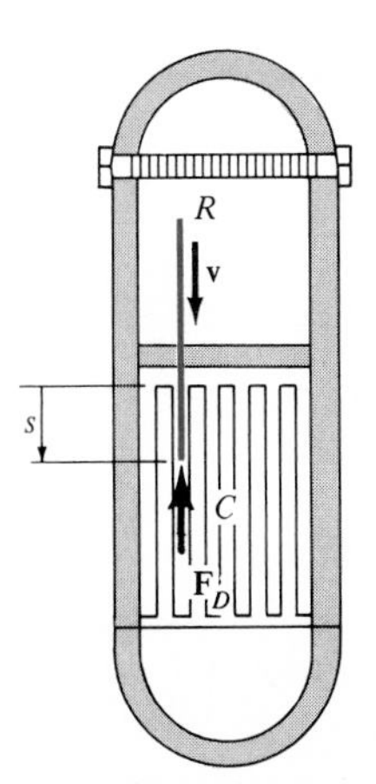

Prob. 13–16

13–17. The tanker has a weight of $8(10^8)$ lb and is traveling forward at $v_0 = 3$ ft/s in still water when the engines are shut off. If the drag resistance of the water is proportional to the speed of the tanker at any instant and can be approximated by $F_D = (4(10^5)v)$ lb, where v is measured in ft/s, determine the time needed for the tanker's speed to become 1.5 ft/s. Given the initial velocity of $v_0 = 3$ ft/s, through what distance must the tanker travel before it stops?

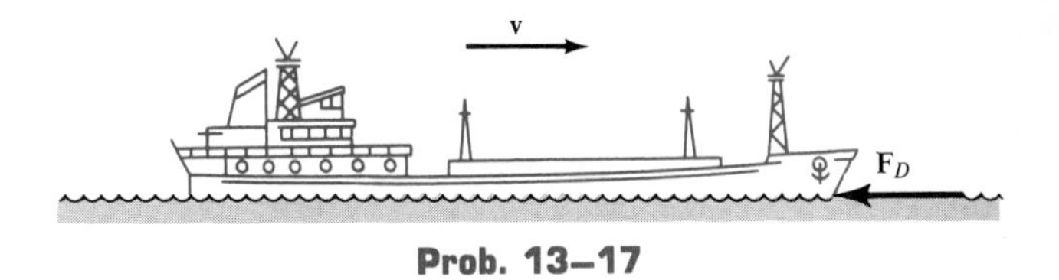

Prob. 13–17

13–18. The safe S has a weight of 200 lb and is supported by the rope and pulley arrangement shown. If the end of the rope is given to a boy B of weight 80 lb, determine his acceleration if in the confusion he doesn't let go of the rope.

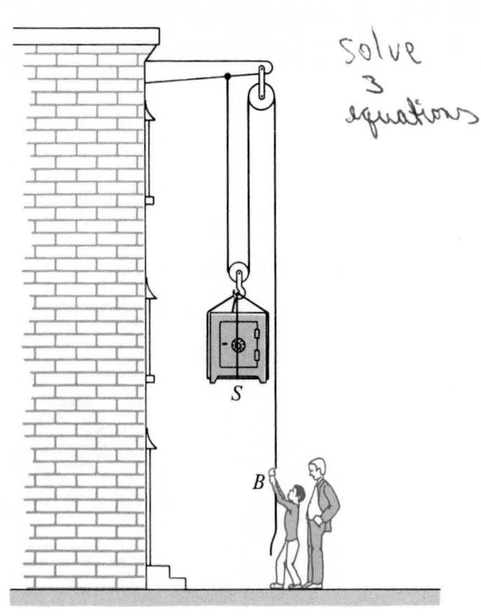

Prob. 13–18

13–19. If cylinders B and C have a mass of 15 kg and 10 kg, respectively, determine the required mass of A so that it does not move when all the cylinders are released. Neglect the mass of the pulleys and the cord.

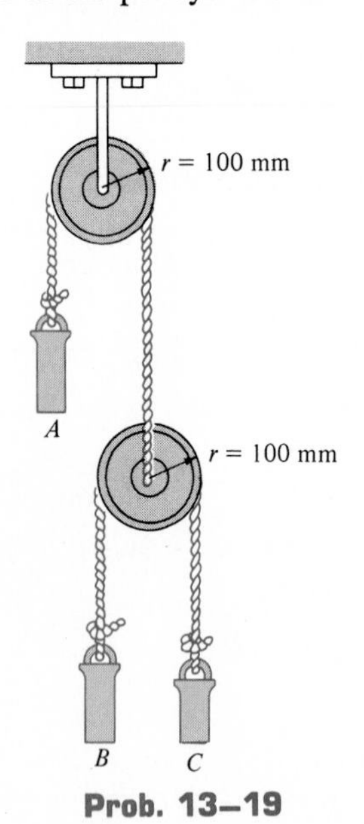

Prob. 13–19

***13–20.** The 2-kg shaft CA passes through a smooth journal bearing at B. Initially, the springs, which are coiled loosely around the shaft, are unstretched when no force is applied to the shaft. In this position $s = s' = 250$ mm and the shaft is originally at rest. If a horizontal force of $F = 5$ kN is applied, determine the speed of the shaft at the instant $s = 50$ mm, $s' = 450$ mm. The ends of the springs are attached to the bearing at B and the caps at C and A.

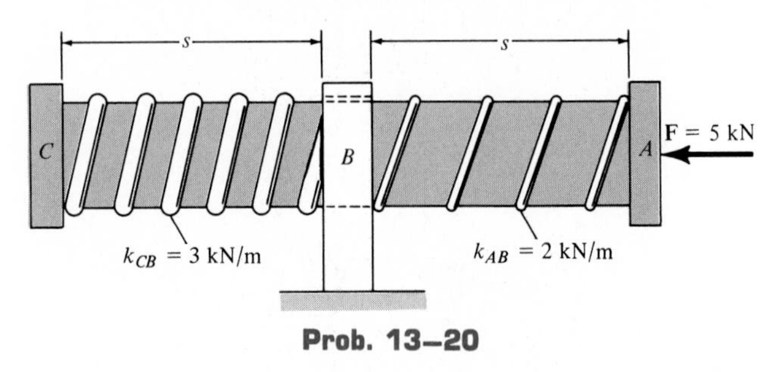

Prob. 13–20

13–21. The spring mechanism is used as a shock absorber for railroad cars. Determine the maximum compression of spring HI if the fixed bumper R of a 5-Mg railroad car, rolling freely at 2 m/s, strikes the plate P. Bar AB slides along the guide paths CE and DF. The ends of all springs are attached to their respective members and are originally unstretched.

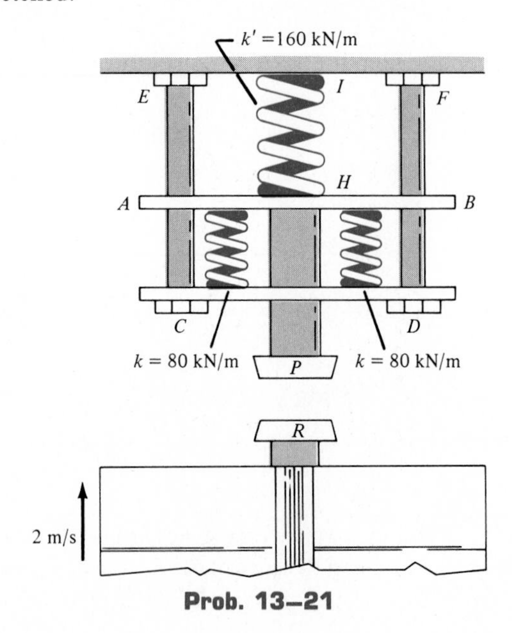

Prob. 13–21

13–22. Determine the acceleration of block A when the system is released. The coefficient of friction and the weight of each block are indicated in the figure. Neglect the mass of the pulleys and cords.

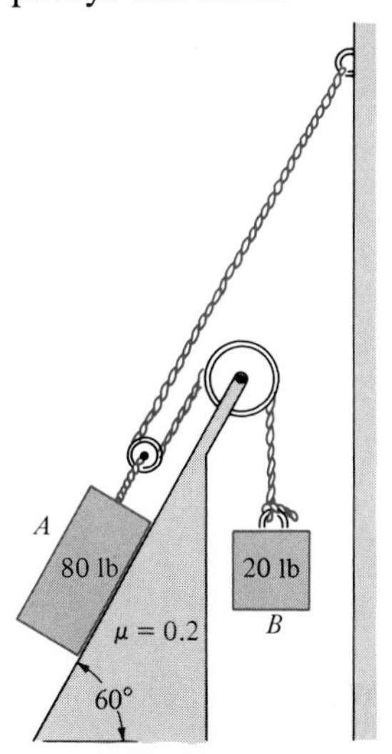

Prob. 13–22

13–23. Determine the tension developed in the two cords and the acceleration of each block. Neglect the mass of the pulleys and cords. *Hint:* Since the system consists of *two* cords, relate the motion of block A to C, and of block B to C. Then, by elimination, relate the motion of A to B.

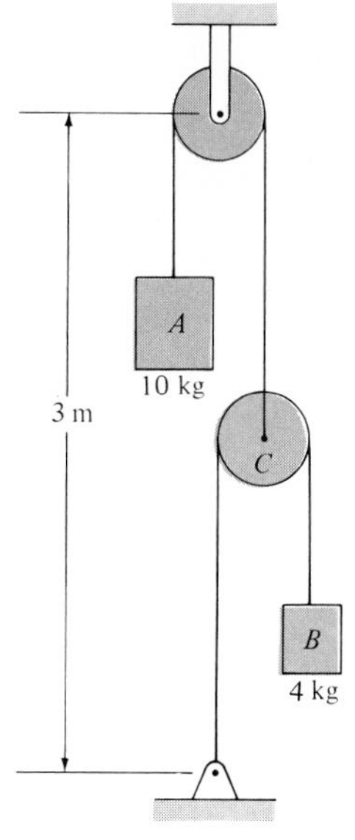

Prob. 13–23

***13–24.** Block A has a mass of 30 kg and block B has a mass of 100 kg. Determine their velocities in $t = 3$ s after they are released from rest. Neglect the mass of the pulleys and cords.

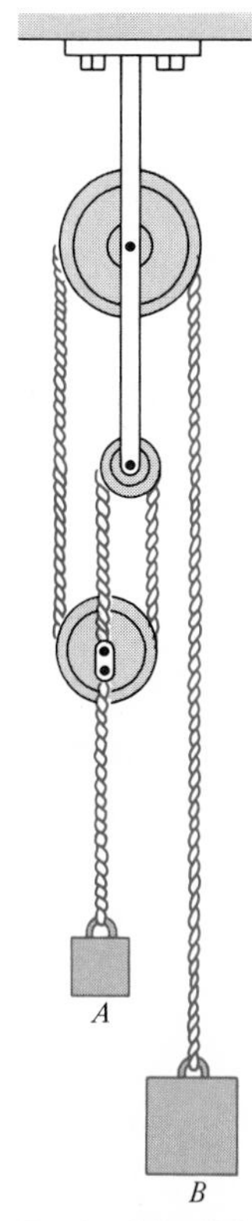

Prob. 13–24

13–25. The 2-kg collar C is free to slide along the smooth shaft AB. Determine the acceleration of collar C if (a) the shaft is fixed from moving, and (b) collar A, which is fixed to shaft AB, moves to the left at constant velocity along the horizontal guide. In both cases, the shaft moves in the vertical plane.

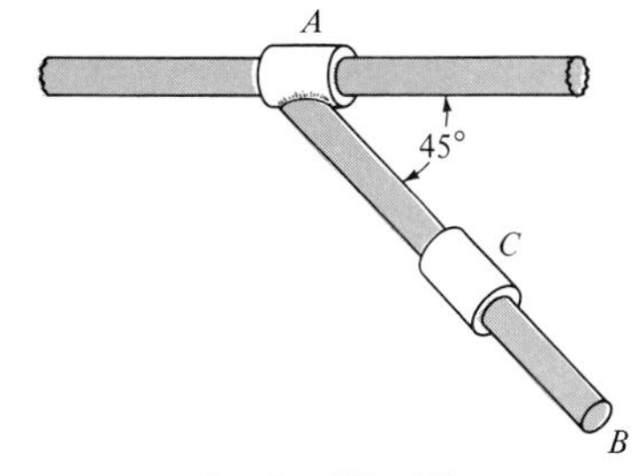

Prob. 13–25

13–26. The 30-lb crate is being hoisted upward with a constant acceleration of 6 ft/s^2. If the uniform beam AB has a weight of 200 lb, determine the components of reaction at A. Neglect the size and mass of the pulley at B. *Hint:* First find the tension in the cable, then analyze the forces on the beam using statics.

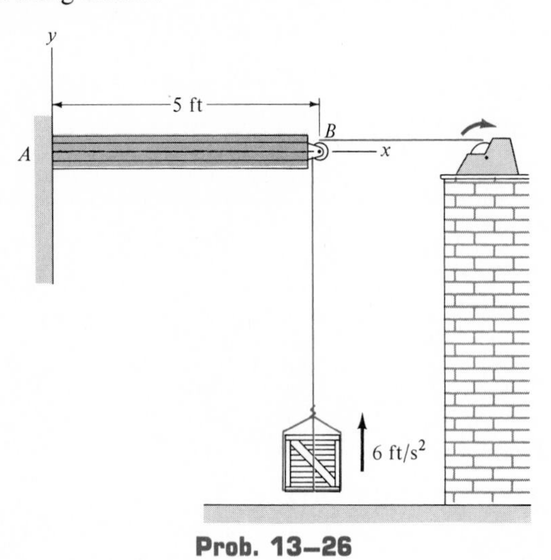

Prob. 13–26

13–27. The 10-kg block A rests on the 50-kg plate B in the position shown. Neglecting the mass of the rope and pulley, and friction, determine the time needed for block A to slide 0.5 m *on the plate* when the system is released from rest.

Prob. 13–27

13.5 Equations of Motion: Normal and Tangential Coordinates

When a particle moves over a curved path which is known, the equation of motion for the particle may be written in the normal and tangential directions. We have

$$\Sigma \mathbf{F} = m\mathbf{a}$$

$$\Sigma F_t \mathbf{u}_t + \Sigma F_n \mathbf{u}_n = m\mathbf{a}_t + m\mathbf{a}_n$$

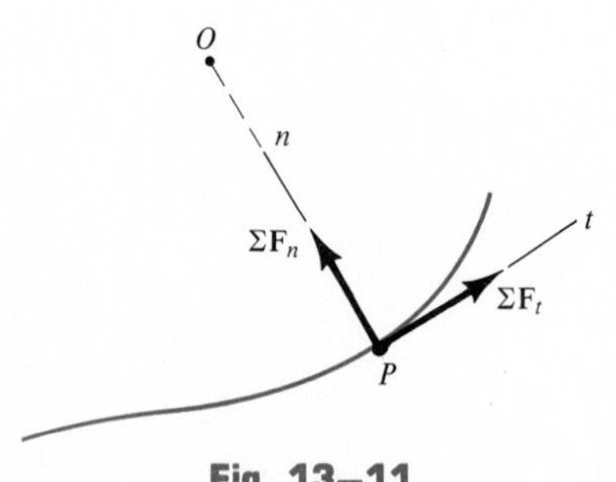

Fig. 13–11

Here ΣF_n and ΣF_t represent the sums of all the force components acting on the particle in the normal and tangential directions, respectively, Fig. 13–11. Since the respective $\mathbf{u}_t$ and $\mathbf{u}_n$ components must be equivalent, the following two scalar equations result:

$$\Sigma F_t = ma_t \qquad \Sigma F_n = ma_n \tag{13–7}$$

The first equation indicates that the change in the magnitude of velocity ($a_t = dv/dt$) is caused by the resultant tangential force acting on the particle. Hence, if $\Sigma \mathbf{F}_t$ acts in the direction of motion, the particle's speed will increase; whereas if it acts in the opposite direction, the particle will slow down.

The second of Eqs. 13–7 states that the change in the direction of velocity ($a_n = v^2/\rho$) is caused by the resultant normal force acting on the particle. This resultant, $\Sigma \mathbf{F}_n$, will always act toward the center of curvature, O, of the path, Fig. 13–11, in the same direction as $\mathbf{a}_n$. In particular, when the particle is traveling on a constrained circular path with a constant speed, there is a normal force exerted on the particle by the constraint. This force is termed a *centripetal force*. The equal but opposite force exerted by the particle on the constraint is called a *centrifugal force*.

PROCEDURE FOR ANALYSIS

The procedure given in Sec. 13–4 may be stated as follows when applied to problems involving n and t coordinates:

Free-Body and Kinetic Diagrams. Establish the inertial n-t coordinate system at the particle and draw the particle's free-body and kinetic diagrams. When drawing the kinetic diagram, assume the vectors $m\mathbf{a}_n$ and $m\mathbf{a}_t$ act in the positive direction of n and t if they are unknown.

Equations of Motion. Apply the equations of motion, Eqs. 13–7.

Kinematics. Formulate the tangential and normal components of acceleration, i.e., $a_t = dv/dt$ or $a_t = v\,dv/ds$ and $a_n = v^2/\rho$. If the path is defined as $y = f(x)$, the radius of curvature can be obtained from $\rho = |[1 + (dy/dx)^2]^{3/2}/(d^2y/dx^2)|$.

The following examples numerically illustrate application of this procedure.

Example 13–6

The ball B is fastened to the end of a 1-m-long string, Fig. 13–12*a*. If air resistance is neglected, then the ball, as it moves with constant speed, describes a horizontal circular path in which the string OB generates a locus of points defining the surface of a cone. Determine the speed of the ball along its circular path if $\theta = 45°$.

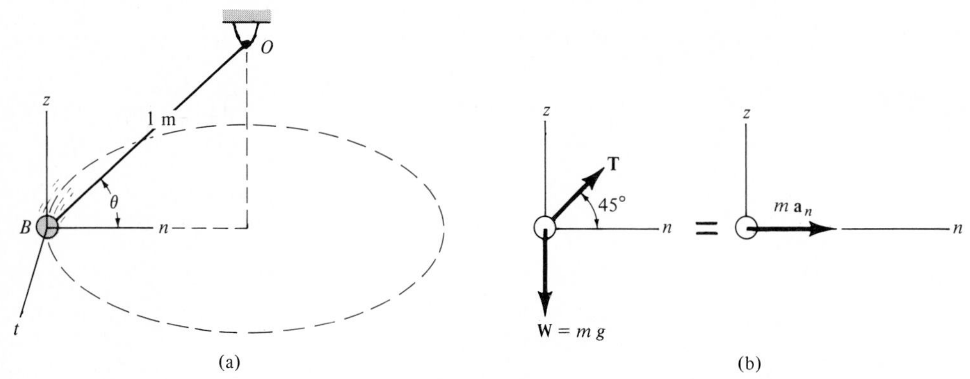

Fig. 13–12

Solution

Free-Body and Kinetic Diagrams. As shown in Fig. 13–12*b*, the weight of the ball is represented as $W = mg$, where $g = 9.81 \text{ m/s}^2$. Since the ball moves around the circle with constant speed, only a normal component of acceleration $\mathbf{a}_n$ (directed toward the center of curvature) is present. Why? This component has a magnitude of $a_n = v^2/\rho = v^2/(1 \cos 45°)$.

Equations of Motion. Using the n-z axes shown, we have

$$+\uparrow \Sigma F_z = ma_z; \qquad T \sin 45° - m(9.81) = 0 \qquad (1)$$

$$\overset{+}{\rightarrow}\Sigma F_n = ma_n; \qquad T \cos 45° = m\frac{v^2}{1 \cos 45°} \qquad (2)$$

Application of the equation of motion in the tangential direction is of no consequence to the solution, since there are no forces in this direction. Solving Eq. (1) for T and substituting into Eq. (2), it is found that m cancels and the speed v of the ball is then

$$v = 2.63 \text{ m/s} \qquad \textit{Ans.}$$

Example 13–7

Determine the banking angle θ of the circular track so that the wheels of the sports car shown in Fig. 13–13*a* will not have to depend upon friction to prevent the car from sliding either up or down the curve. The car travels at a constant speed of 100 ft/s. The radius of the track is 600 ft.

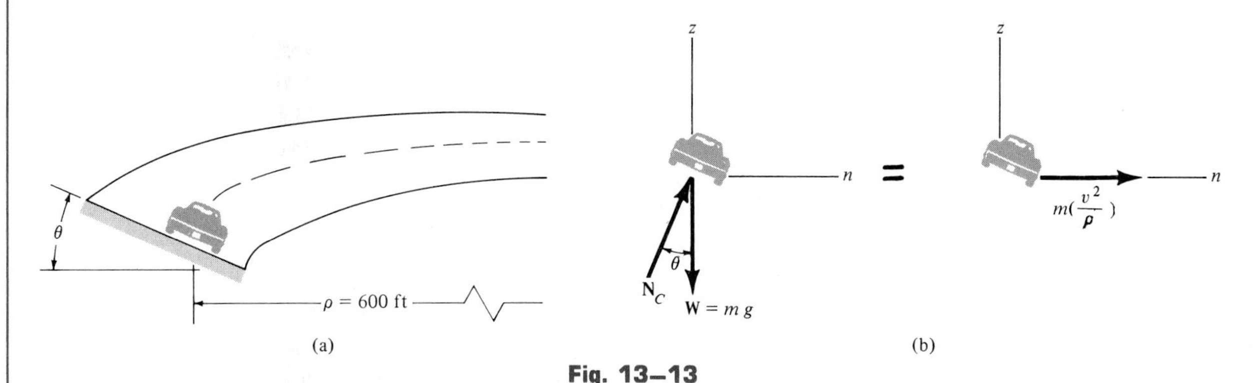

Fig. 13–13

Solution

Free-Body and Kinetic Diagrams. As shown in Fig. 13–13*b*, the car is assumed to have a mass *m*. As stated in the problem, no frictional force acts on the car.

Equations of Motion. Using the *n*-*z* axes shown, realizing that $a_n = v^2/\rho$, we have

$$+\uparrow \Sigma F_z = ma_z; \qquad N_C \cos\theta - mg = 0 \qquad (1)$$

$$\overset{+}{\rightarrow}\Sigma F_n = ma_n; \qquad N_C \sin\theta = m\frac{v^2}{\rho} \qquad (2)$$

Eliminating N_C and m from these equations by dividing Eq. (2) by Eq. (1), we obtain

$$\tan\theta = \frac{v^2}{g\rho} = \frac{(100)^2}{32.2(600)}$$

$$\theta = \tan^{-1}(0.518)$$

$$= 27.4^\circ \qquad \textit{Ans.}$$

A force summation in the tangential direction of motion is of no consequence to the solution. If it were considered, note that $a_t = dv/dt = 0$, since the car moves with *constant speed*.

Example 13–8

The 3-kg disk D is attached to the end of a cord as shown in Fig. 13–14*a*. The other end of the cord is attached to a ball-and-socket joint located at the center of a platform. If the platform is rotating rapidly, and the disk is placed on it and released from rest as shown, determine the time it takes for the disk to reach a speed great enough to break the cord. The maximum tension the cord can sustain is 100 N, and the coefficient of kinetic friction between the disk and the platform is $\mu_k = 0.1$.

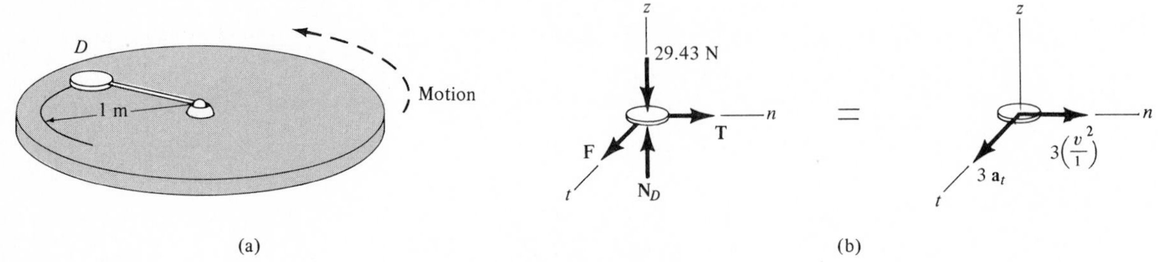

Fig. 13–14

Solution

Free-Body and Kinetic Diagrams. As shown in Fig. 13–14*b*, the disk has *both* normal and tangential components of acceleration as a result of the unbalanced forces **T** and **F**. Since sliding occurs, the frictional force has a magnitude of $F = \mu_k N_D = 0.1N_D$ and a direction that opposes the *relative motion* of the disk with respect to the platform. The weight of the disk is $W = 3(9.81) = 29.43$ N.

Equations of Motion

$\Sigma F_z = ma_z;$ $$N_D - 29.43 = 0 \quad (1)$$

$\Sigma F_t = ma_t;$ $$0.1N_D = 3a_t \quad (2)$$

$\Sigma F_n = ma_n;$ $$T = 3\left(\frac{v^2}{1}\right) \quad (3)$$

Since the maximum tension sustained by the cord is $T = 100$ N, Eq. (3) can be solved for the critical speed v_{cr} of the disk needed to break the cord. Solving all the equations, we obtain

$$N_D = 29.43 \text{ N}$$
$$a_t = 0.981 \text{ m/s}^2$$
$$v_{cr} = 5.77 \text{ m/s}$$

Kinematics. Since a_t is *constant*, the time needed to break the cord is

$$v_{cr} = v_0 + a_t t$$
$$5.77 = 0 + (0.981)t$$
$$t = 5.89 \text{ s}$$

Ans.

Example 13–9

The skier in Fig. 13–15*a* descends the smooth slope, which may be approximated by a parabola. If she has a weight of 120 lb, determine the normal force she exerts on the ground at the instant she arrives at point A, where her velocity is 30 ft/s. Also compute her acceleration at A.

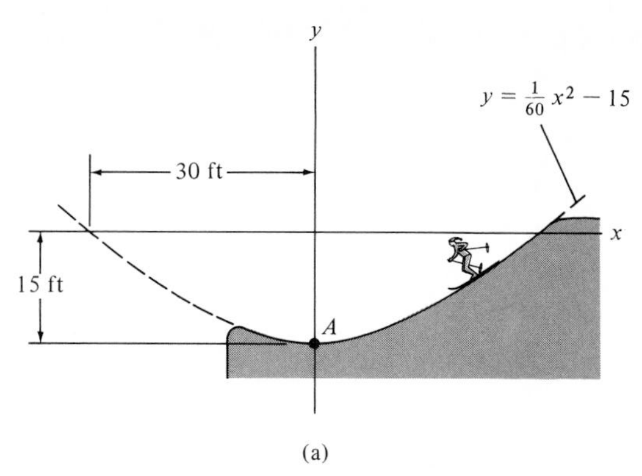

(a)

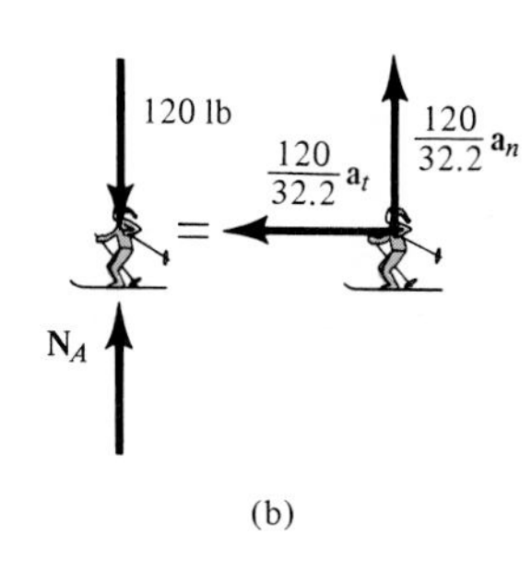

(b)

Fig. 13–15

Solution

Free-Body and Kinetic Diagrams. The free-body and kinetic diagrams for the skier when she is at A are shown in Fig. 13–15*b*. Since the path is *curved*, there are two components of acceleration and hence $m\mathbf{a}_n$ and $m\mathbf{a}_t$ are shown on the kinetic diagram.

Equations of Motion

$$+\uparrow \Sigma F_n = ma_n; \qquad N_A - 120 = \frac{120}{32.2}\left(\frac{(30)^2}{\rho}\right) \tag{1}$$

$$\overset{+}{\leftarrow} \Sigma F_t = ma_t; \qquad 0 = \frac{120}{32.2}a_t \tag{2}$$

The radius of curvature ρ for the path must be computed at point $A(0, -15 \text{ ft})$. Since $y = \frac{1}{60}x^2 - 15$, $dy/dx = \frac{1}{30}x$, $d^2y/dx^2 = \frac{1}{30}$, then at $x = 0$,

$$\rho = \left|\frac{[1 + (dy/dx)^2]^{3/2}}{d^2y/dx^2}\right|_{x=0} = \left|\frac{[1 + (0)^2]^{3/2}}{\frac{1}{30}}\right| = 30 \text{ ft}$$

Substituting into Eq. (1) and solving for N_A, we have

$$N_A = 231.8 \text{ lb} \qquad \textit{Ans.}$$

Kinematics. From Eq. (2),

Since

$$a_t = 0$$

then

$$a_n = \frac{v^2}{\rho} = \frac{(30)^2}{30} = 30 \text{ ft/s}^2$$

$$a = a_n = 30 \text{ ft/s}^2 \uparrow \qquad \textit{Ans.}$$

Problems

***13–28.** Compute the mass of the sun, knowing that the distance from the earth to the sun is $149.6(10^6)$ km. *Hint:* Represent the force of gravitation acting on the earth by using Eq. 13–1.

13–29. When crossing an intersection, a motorcyclist encounters the slight bump or crown caused by the intersecting road. If the crest of the bump has a radius of curvature of $\rho = 50$ ft, determine the maximum constant speed at which he can travel without leaving the surface of the road. Neglect the size of the motorcycle and rider in the calculation. The rider and his motorcycle have a total weight of 350 lb.

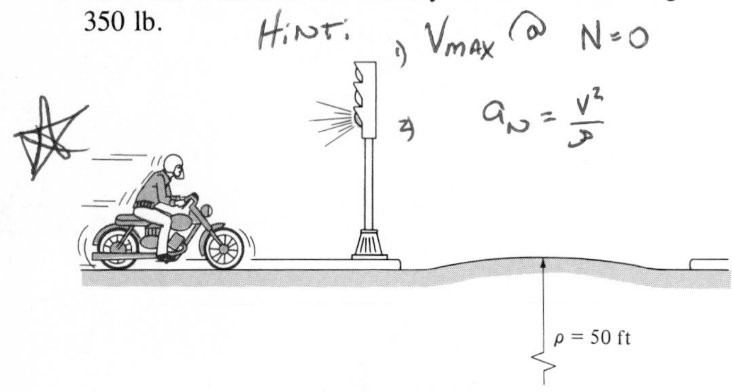

Prob. 13–29

13–30. A boy twirls a 15-lb bucket of water in a vertical circle. If the radius of curvature of the path is 4 ft, determine the minimum speed the bucket must have when it is overhead at A so no water spills out. Neglect the size of the bucket in the calculation. If the bucket were moving at a slightly slower rate than that calculated, would the water fall on the boy when it starts to spill out at A? Explain.

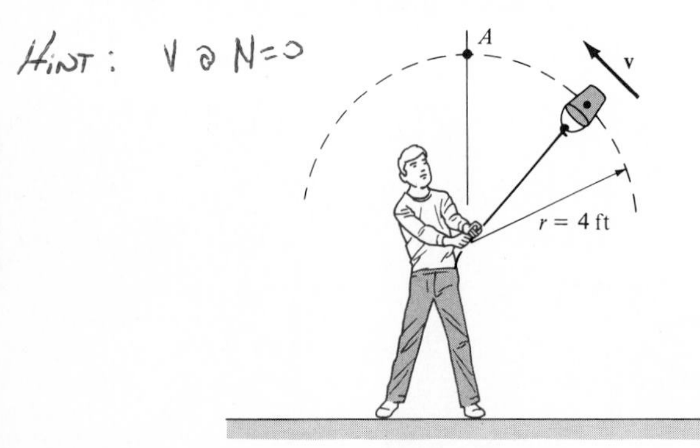

Prob. 13–30

13–31. Compute the constant speed of the cars on the amusement-park ride if it is observed that the cables are directed at 30° from the vertical. Each car including its passengers has a mass of 550 kg. Also, what are the components of force in the n, t, and z directions which a 60-kg passenger exerts on the car during the motion?

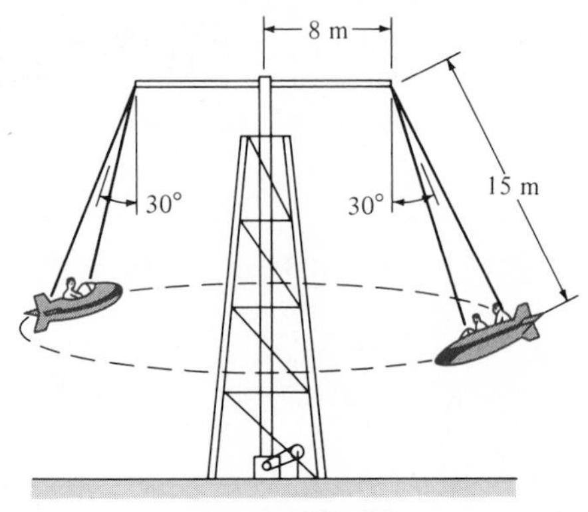

Prob. 13–31

***13–32.** The pendulum bob B has a weight of 5 lb. If it has a speed of 12 ft/s when it reaches point D, determine the tension in string BC at this instant.

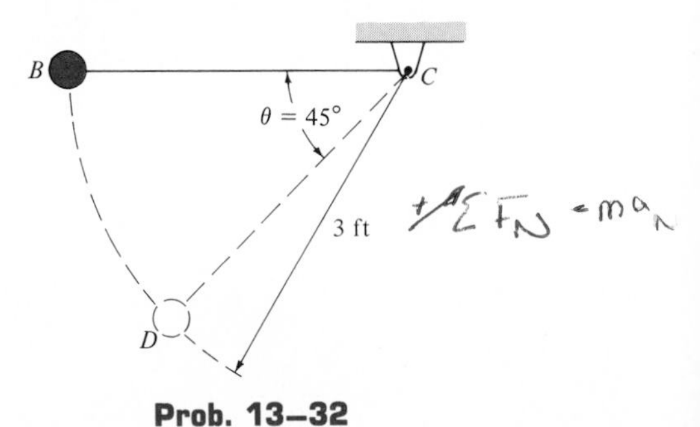

Prob. 13–32

13–33. A toboggan and rider have a total mass of 100 kg and travel down along the (smooth) slope defined by the equation $y = 0.2x^2$. At the instant $x = 8$ m, the toboggan's speed is 4 m/s. At this point, determine the rate of increase in speed and the normal force which the toboggan exerts on the slope. Neglect the size of the toboggan and rider for the calculation. *Hint:* Compute the slope of the path to be $\theta = 72.6°$ using $\tan\theta = dy/dx$.

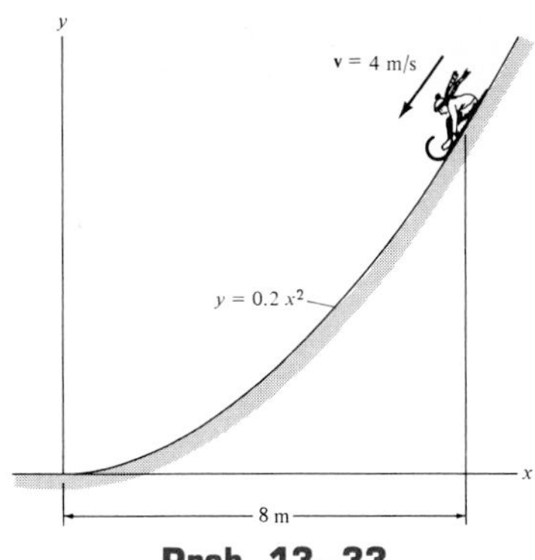

Prob. 13–33

13–34. A sled and rider, weighing 30 lb and 150 lb respectively, travel down a smooth slope that has the shape shown. If the rider does not hold on to the sides of the sled, determine the normal force which he exerts on the sled when the sled is (a) at point B, where it has a velocity of 15 ft/s, and (b) at some point C in mid-air.

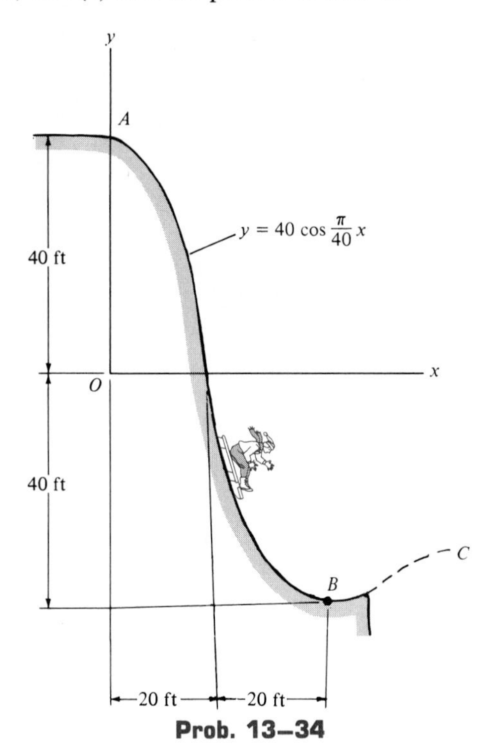

Prob. 13–34

13–35. Packages having a weight of 5 lb ride on the surface of the conveyor belt. If the belt travels at a constant speed of 2 ft/s and the coefficient of friction between the belt and each package is $\mu = 0.4$, determine the angle θ at which the packages first begin to slip off the surface of the belt.

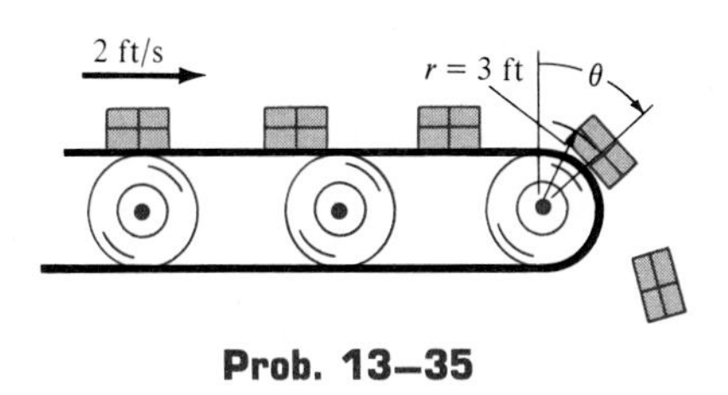

Prob. 13–35

***13–36.** Determine the constant speed of the satellite S so that it circles the earth with an orbit of radius $r = 15$ Mm. The mass of the earth is $5.976(10^{24})$ kg. *Hint:* Use Eq. 13–1.

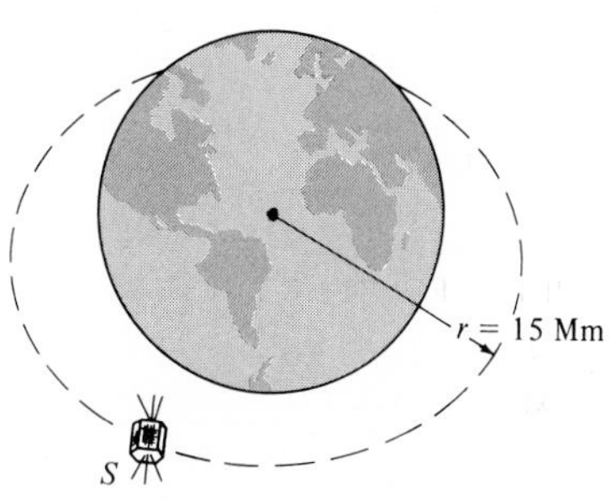

Prob. 13–36

13–37. The smooth block B, having a mass of 0.2 kg, is attached to the vertex A of the right circular cone using a light cord. The cone is rotating at a constant angular rate about the z axis such that the block attains a speed of 0.5 m/s. At this speed, determine the tension in the cord and the reaction which the cone exerts on the block. Neglect the size of the block in the computation. *Hint:* ρ for the block is measured as the perpendicular distance from the z axis to B.

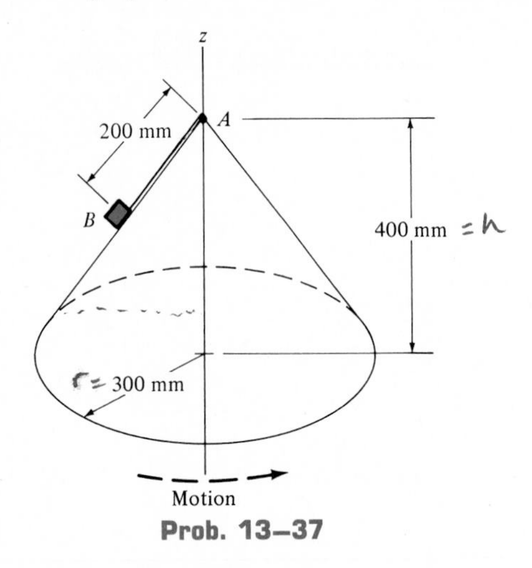

Prob. 13–37

Right Circular Cone

$$V = \frac{1}{3} \cdot \pi \cdot r^2 \cdot h$$

$$S = \pi r \sqrt{r^2 + h^2}$$

13–38. The rotational speed of the disk is controlled by a 30-g smooth contact arm AB which is spring-mounted on the disk. When the disk is *at rest,* the center of mass G of the arm is located 150 mm from the center O, and the preset compression in the spring is 20 mm. If the initial gap between B and the contact at C is 10 mm, determine the (controlling) speed of the arm's mass center, v_G, which will close the gap. The disk rotates in the horizontal plane. The spring has a stiffness of $k = 50$ N/m, and its ends are attached to the contact arm at D and to the disk at E.

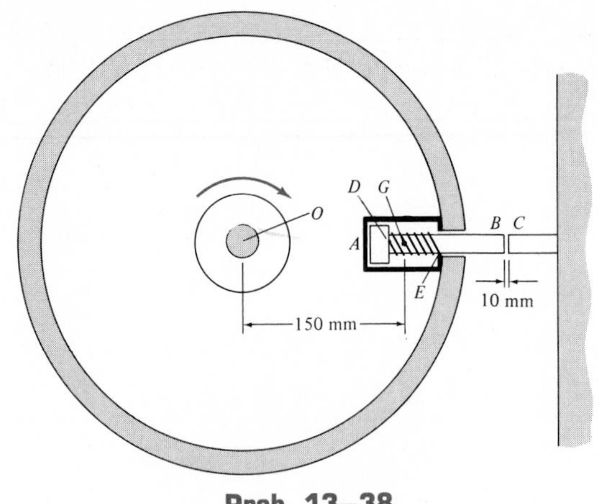

Prob. 13–38

13–39. The collar A, having a mass of 0.75 kg, is attached to a spring having a stiffness of $k = 200$ N/m. When rod BC rotates about the vertical axis, the collar slides outward along the smooth rod DE. If the spring is unstretched when $s = 0$, determine the constant speed of the collar in order that $s = 100$ mm.

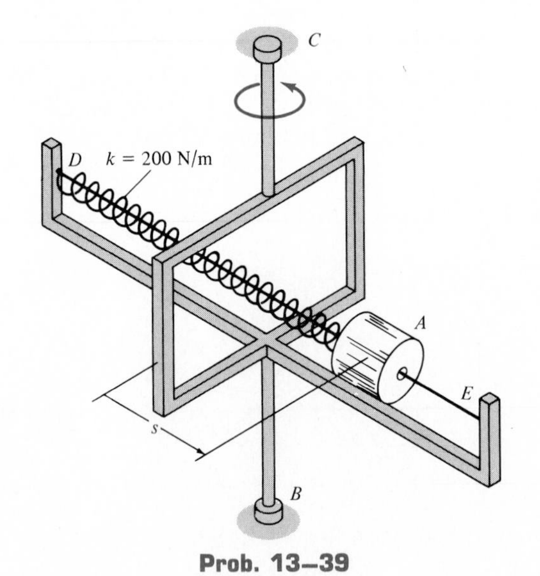

Prob. 13–39

*13–40. An acrobat has a weight of 150 lb and is sitting on a chair which is perched on top of a pole as shown. If by a mechanical drive the pole rotates downward at a constant rate from $\theta = 0°$, such that the acrobat's center of mass G maintains a *constant speed* of $v_a = 10$ ft/s, determine the normal forces he exerts on the back and bottom of the chair at the instant $\theta = 30°$. Neglect friction and assume that the distance from the pivot O to G is $\rho = 15$ ft.

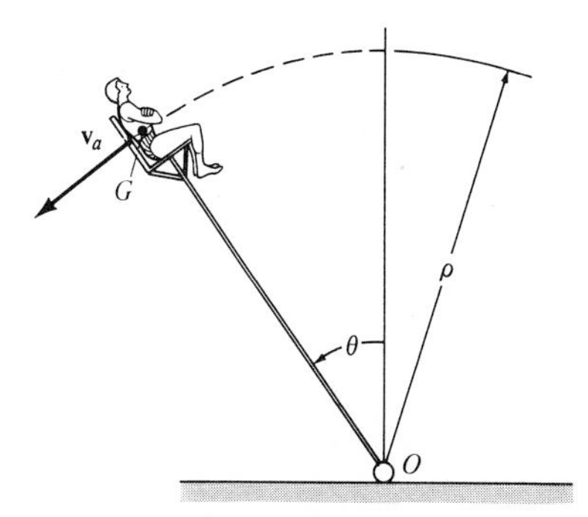

Prob. 13–40

13–41. Solve Prob. 13–40 if the speed of the acrobat's center of mass is increased from $(v_a)_1 = 10$ ft/s at $\theta = 0°$ by a constant rate of $\dot{v}_a = 0.5$ ft/s^2.

13–42. A girl, having a mass of 15 kg, sits motionless relative to the surface of a horizontal platform at a distance of $r = 5$ m from the platform's center. If the angular motion of the platform is *slowly* increased so that the girl's tangential component of acceleration can be neglected, determine the maximum speed which the girl will have before she begins to slip off the platform. The coefficient of friction between the girl and the platform is $\mu = 0.2$.

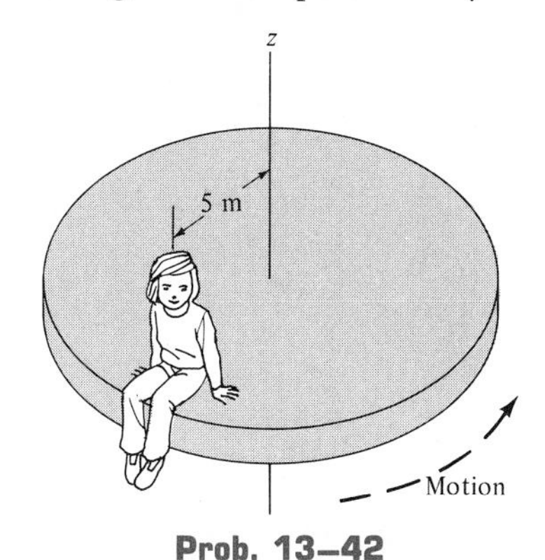

Prob. 13–42

13–43. A collar having a mass of 0.75 kg slides over the surface of a horizontal circular rod for which the coefficient of friction is $\mu = 0.3$. If the collar has a speed of 4 m/s, determine the rate of decrease in the speed (a_t) and the resultant normal force the rod exerts on the collar.

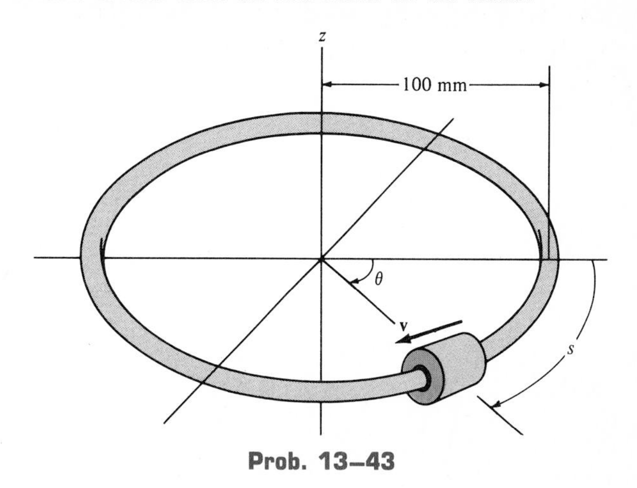

Prob. 13–43

*13–44. A ball having a mass of 2 kg rolls within a vertical circular slot. Determine the force it exerts on the slot when it arrives at points A and B if it has speeds of $v_A = 4$ m/s and $v_B = 6$ m/s at these points. Neglect the rolling motion of the ball in the calculation.

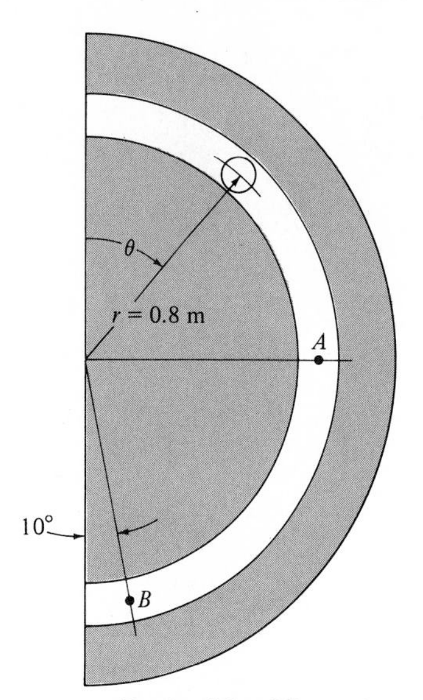

Prob. 13–44

Review Problems

13–45. The uniform block has a weight of 1500 lb and is originally at rest. Determine the constant vertical force **F** which must be applied to give it a speed of 4 ft/s when it rises 20 ft. Determine the tension in each of the cables during the motion. Assume each cable supports an equal force.

13–46. If the uniform 1500-lb block is subjected to a vertical force of $F = 2000$ lb, determine its speed and its elevation three seconds after the force is applied. Originally the block is at rest. Determine the tension in each of the cables during the motion. Assume each cable supports an equal force.

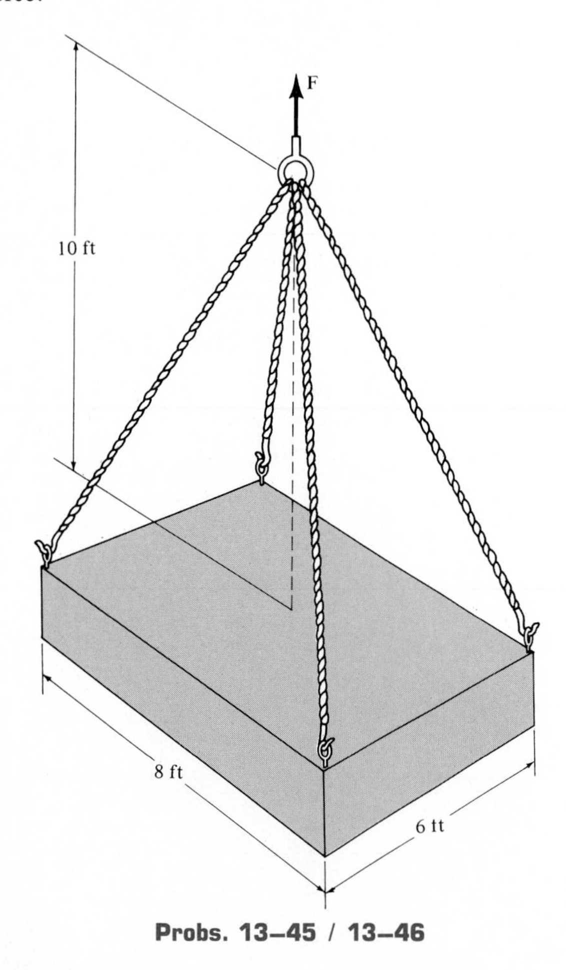

Probs. 13–45 / 13–46

13–47. The sports car, having a mass of 1700 kg, is traveling in the horizontal plane along a 20° banked track which is circular and has a radius of curvature of $\rho = 100$ m. If the coefficient of friction between the tires and the road is $\mu = 0.2$, determine the *maximum constant speed* at which the car can travel without sliding up the track.

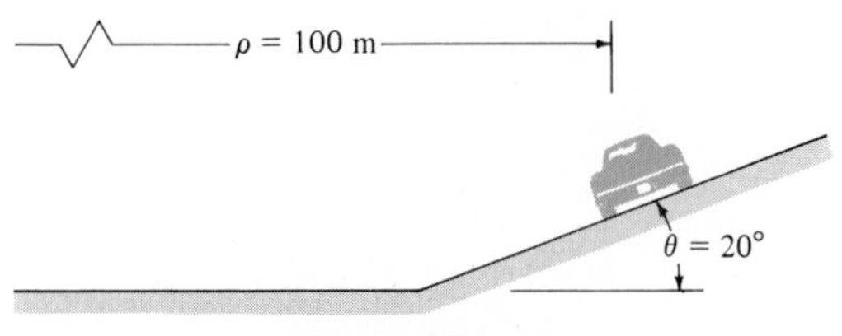

Prob. 13–47

***13–48.** Determine the *minimum constant speed* of the 1700-kg sports car in Prob. 13–47 so that it does not slide down the track. Set $\mu = 0.2$.

13–49. If the motor M exerts a force of 80 lb on the cable, determine the acceleration of the 50-lb crate. Neglect friction.

13–50. If the motor M exerts a force of 80 lb on the cable, determine the acceleration of the 50-lb crate. The coefficient of kinetic friction between the crate and the incline is $\mu = 0.3$.

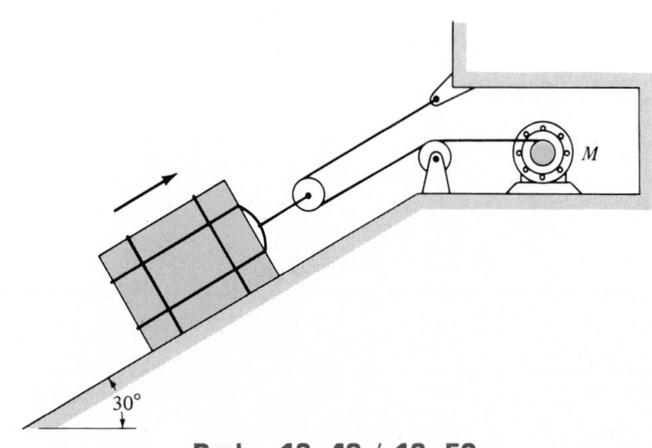

Probs. 13–49 / 13–50

13–51. Blocks A and B have a weight of 15 lb and 10 lb, respectively. Determine the force between blocks A and B if CD is accelerating downward at $a = 3$ ft/s^2. What is the tension in the supporting cords during the motion?

***13–52.** Blocks A and B have a weight of 15 lb and 10 lb, respectively. Determine the maximum downward acceleration of the support CD so that separation between blocks A and B does not occur.

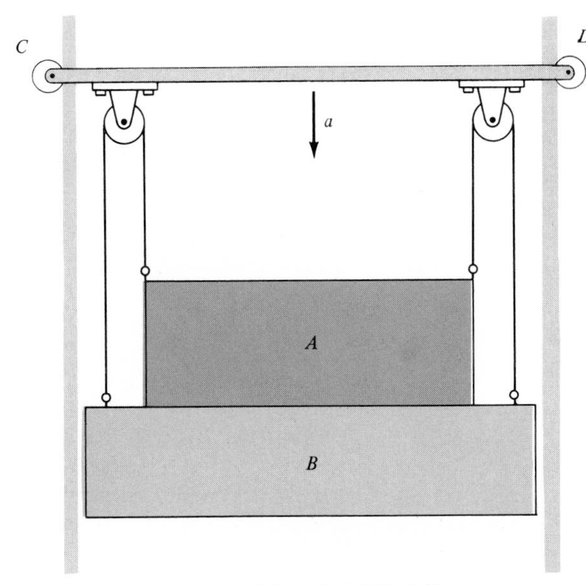

Probs. 13–51 / 13–52

13–53. The amusement park ride called a "rotor" consists of a circular room which rotates about the vertical z axis at a constant rate. Due to this rotation, the passengers are forced up against the wall because of centrifugal force. If the coefficient of friction between each passenger and the wall is $\mu = 0.4$, determine the smallest rate of rotation $\dot{\theta}$ at which time the floor can be dropped without causing the passengers to slide down the wall. Neglect the size of the passengers for the calculation.

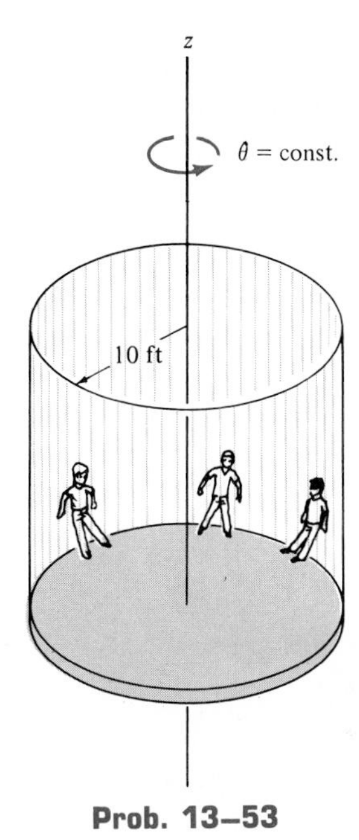

Prob. 13–53

13–54. If the motor draws in the cable with a constant acceleration of 2 m/s^2, determine the compression in members AC and BC.

13–55. Determine the compression in members AC and BC if the motor is unwinding the cable with a constant acceleration of 1 m/s^2.

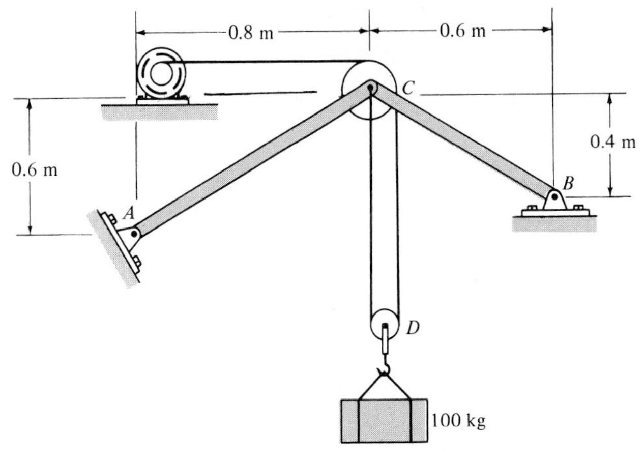

Probs. 13–54 / 13–55

***13–56.** If a horizontal force of $P = 10$ lb is applied to block A, determine the acceleration of block B. Neglect friction. *Hint:* Show that $a_B = a_A \tan 15°$.

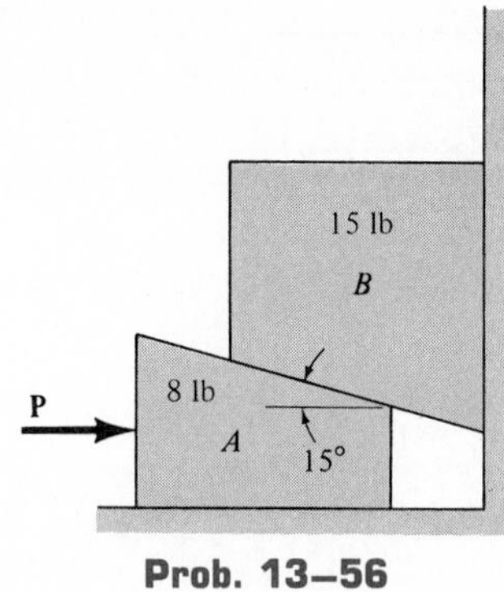

Prob. 13–56

14 Kinetics of a Particle: Work and Energy

In this chapter we will develop an integrated formulation of the equation of motion, referred to as the principle of work and energy. This equation is useful for solving problems which involve force, velocity, and displacement. Later in the chapter the concept of power is discussed and a method is presented for solving kinetics problems using the theorem of conservation of energy. Before presenting these topics, however, it is first necessary to define the work done by various types of forces.

14.1 The Work of a Force

In mechanics, a force **F** does work only when it undergoes a displacement in the direction of the force. If the displacement is $d\mathbf{s}$ and θ is the angle formed between the tails of **F** and $d\mathbf{s}$, Fig. 14–1a, the *work* dU *is a scalar quantity,* defined as

$$dU = F\,ds\cos\theta$$

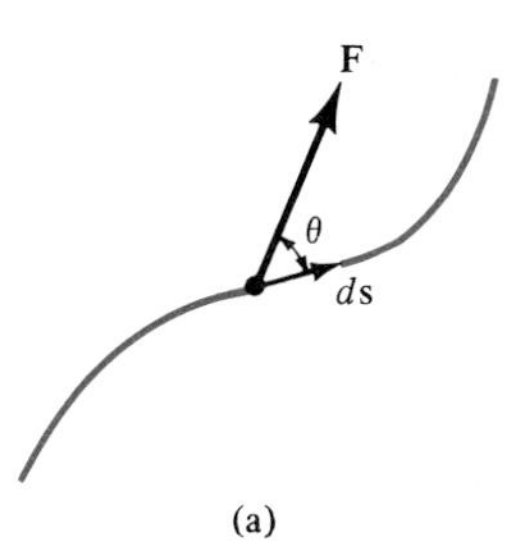

(a)

Fig. 14–1a

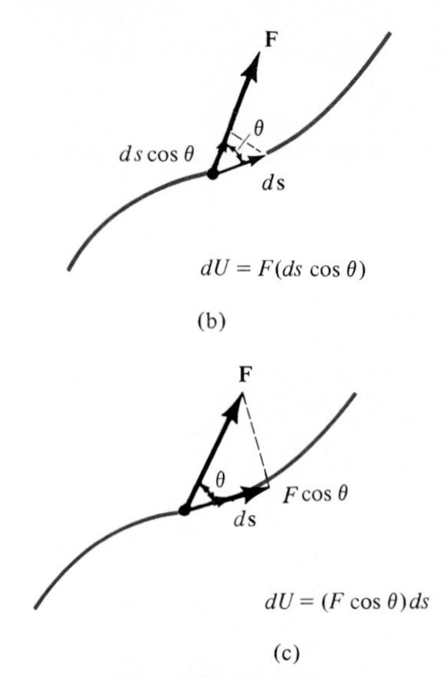

Fig. 14–1b and c

Work, as expressed by this equation, may be interpreted in one of two ways: either as the product of F and the component of displacement in the direction of **F**, i.e., $ds \cos \theta$, Fig. 14–1*b*; or as the product of ds and the component of force in the direction of $d\mathbf{s}$, i.e., $F \cos \theta$, Fig. 14–1*c*. If $F \cos \theta$ and $d\mathbf{s}$ have the *same sense of direction,* the work is *positive;* whereas, if these vectors have an *opposite sense of direction,* the work is *negative*. Note that $dU = 0$ if **F** is *perpendicular* to $d\mathbf{s}$ since $\cos 90° = 0$, or if **F** is applied at a *fixed point,* in which case $d\mathbf{s} = \mathbf{0}$.

The basic unit for work in the SI system is called a joule (J). This unit combines the units of force and displacement. Specifically, 1 *joule* of work is done when a force of 1 newton moves 1 meter along its line of action (1 J = 1 N · m). The moment of a force has this same combination of units (N · m); however, the concepts of moment and work are in no way related. A moment is a vector quantity, whereas work is a scalar. In the FPS system work is defined in ft · lb, which is distinguished from the units for a moment written as lb · ft.

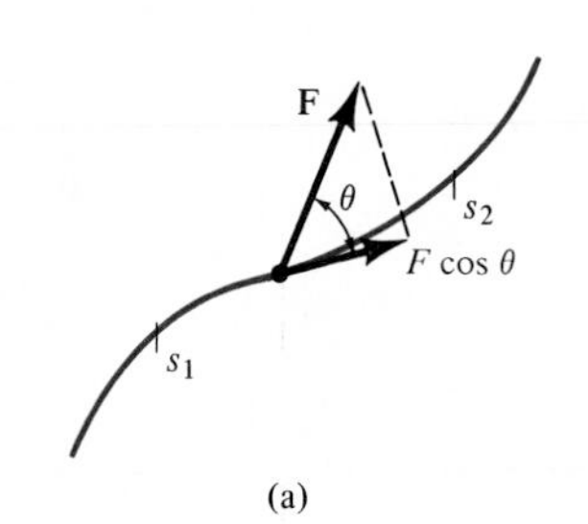

(a)

Work of a Variable Force. If a force undergoes a finite displacement along its path from s_1 to s_2, Fig. 14–2*a*, the work is determined by integration. Provided **F** is a function of position, $\mathbf{F} = \mathbf{F}(s)$, we have

$$U_{1-2} = \int_{s_1}^{s_2} F \cos \theta \, ds \qquad (14\text{–}1)$$

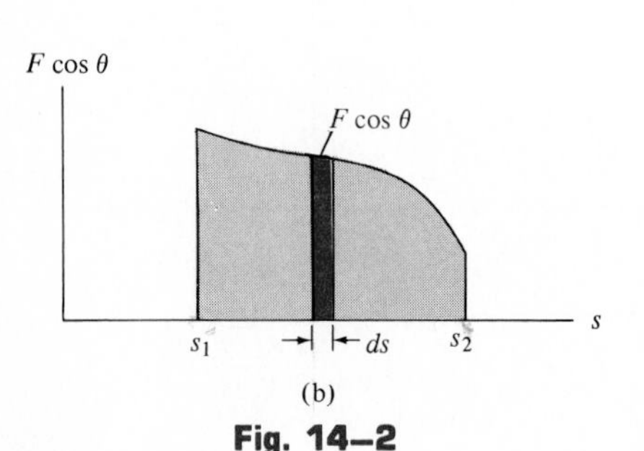

(b)

Fig. 14–2

If the working component of the force, $F \cos \theta$, is plotted versus s, Fig. 14–2*b*, the integral represented in this equation can be interpreted as the *area under the curve* between the points s_1 and s_2.

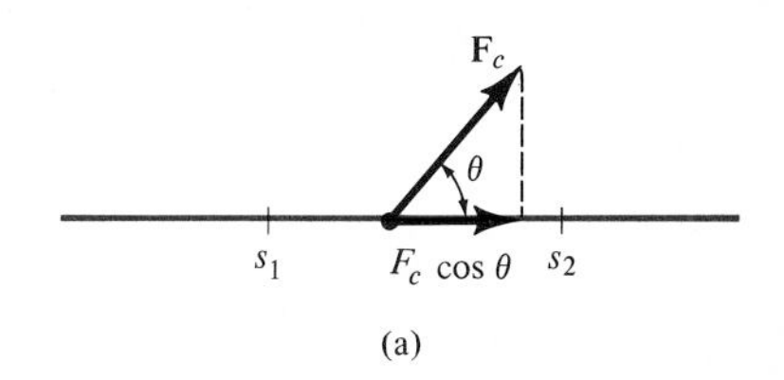

(a)

Work of a Constant Force Moving Along a Straight Line. If the force $\mathbf{F}_c$ has a constant magnitude and acts at a constant angle θ from its straight-line path, Fig. 14–3*a*, then the component of $\mathbf{F}_c$ in the direction of displacement is $F_c \cos\theta$. The work done by $\mathbf{F}_c$ when it is displaced from s_1 to s_2 is determined by Eq. 14–1, in which case

$$U_{1-2} = F_c \cos\theta \int_{s_1}^{s_2} ds$$

or

$$U_{1-2} = F_c \cos\theta(s_2 - s_1) \tag{14–2}$$

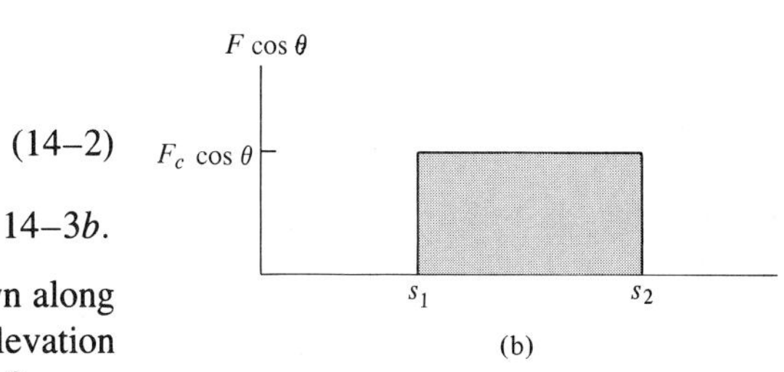

(b)

Fig. 14–3

Here the work of $\mathbf{F}_c$ represents the *area under the rectangle* in Fig. 14–3*b*.

Work of a Weight. Consider a particle (or block) which moves down along the path shown in Fig. 14–4*a* from an initial elevation y_1 to a final elevation y_2, where both elevations are measured from a fixed horizontal reference plane or *datum*.* As the particle *descends* along the path by an amount $d\mathbf{s}$, the displacement component in the direction of its weight $\mathbf{W}$ is $d\mathbf{y}$. Hence, from Fig. 14–4*b*, we have $dy = ds \cos\theta$, so that applying Eq. 14–1, realizing that $\mathbf{W}$ is constant, we obtain

$$U_{1-2} = W \int_{y_1}^{y_2} dy$$

or

$$U_{1-2} = W(y_2 - y_1) \tag{14–3}$$

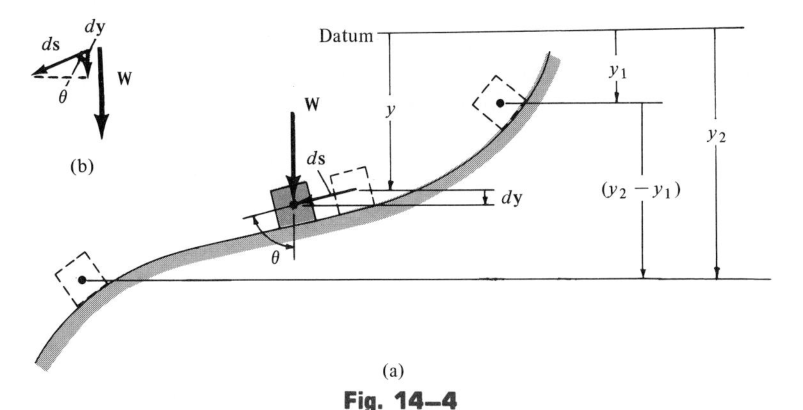

(a)

Fig. 14–4

*Here the weight (force) is assumed to be *constant*. This assumption is suitable for small differences in elevation $(y_2 - y_1)$. If the elevation change is significant, however, a variation of weight with elevation must be taken into account (see Prob. 14–19).

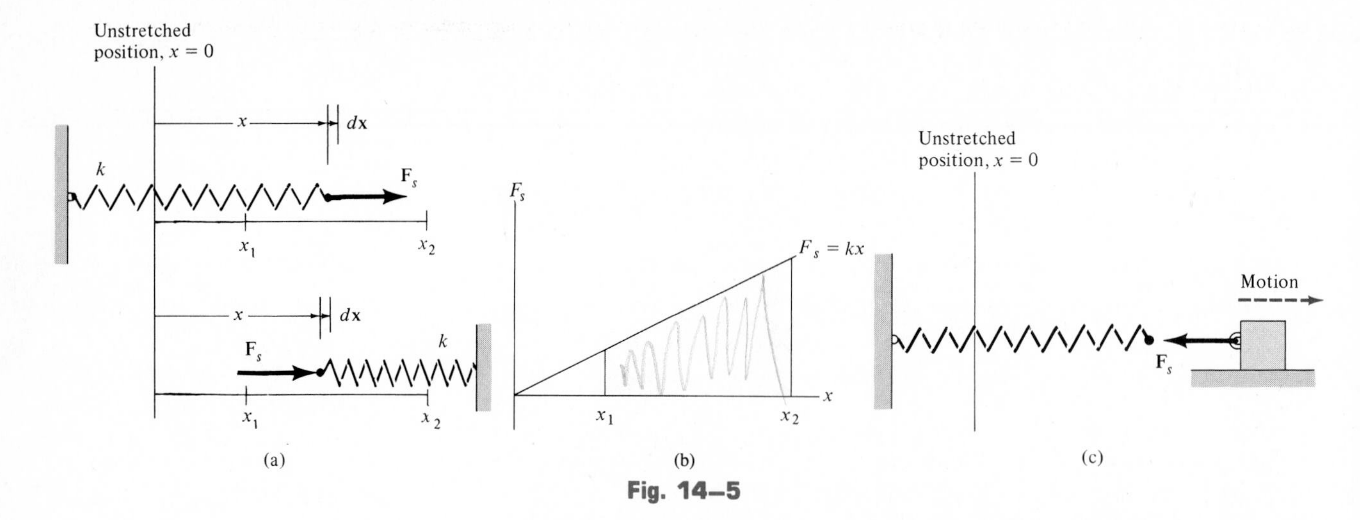

Fig. 14–5

Thus, the work done is equal to the magnitude of the particle's weight times its vertical displacement.* In this case the total work is *positive,* since **W** and $(y_2 - y_1)$ are both *downward,* that is, in the same direction. If the particle is displaced *upward* the work of the weight is *negative.* Why?

Work of a Spring Force. The magnitude of force developed by a linear elastic spring when the spring is displaced a distance x from its unstretched position is $F_s = kx$, where k is the spring stiffness. If the spring is elongated or compressed from a position x_1 to a further position x_2, Fig. 14–5*a*, the work done *on the spring* by $\mathbf{F}_s$ is *positive,* since in each case the force and displacement are in the *same direction.* We require

$$U_{1-2} = \int_{x_1}^{x_2} kx\,dx$$

$$= \left(\tfrac{1}{2}kx_2^2 - \tfrac{1}{2}kx_1^2\right)$$

This equation represents the trapezoidal area under the line $F_s = kx$ versus x, Fig. 14–5*b*.

If a particle (block) is attached to a spring, then the force $\mathbf{F}_s$ exerted on the block is *opposite* to that exerted on the spring, Fig. 14–5*c*. Consequently, the force will do *negative work* on the block when the block moves so as to further elongate (or compress) the spring. Hence, the above equation becomes

$$U_{1-2} = -\left(\tfrac{1}{2}kx_2^2 - \tfrac{1}{2}kx_1^2\right) \qquad (14\text{–}4)$$

When the above equation is used, a mistake in sign will be eliminated if one simply notes the direction of the spring force acting on the body and compares it with the direction of displacement of the body—if both are in the *same direction, positive work* results; if they are *opposite* to one another, the *work is negative.*

*Note that the location of the datum plane is *arbitrary* since the results indicate that the work done depends *only upon the difference in elevation.*

Example 14–1

The 10-kg block shown in Fig. 14–6*a* rests on the rough incline for which the kinetic coefficient of friction is $\mu_k = 0.3$. If the spring is originally unstretched, determine the total work done by all the forces acting on the block when a horizontal force of $P = 400$ N pushes the block up the plane $s = 2$ m.

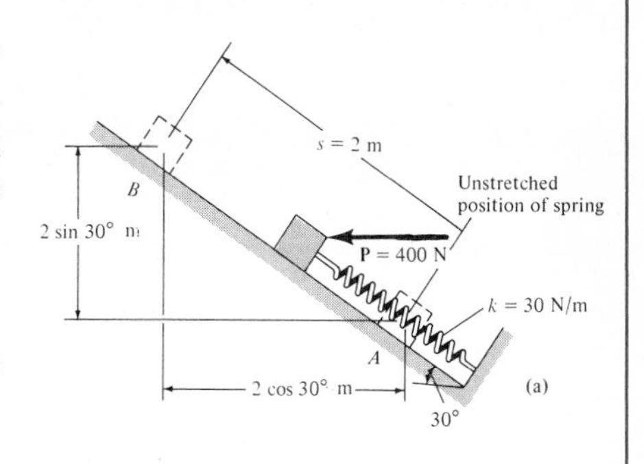

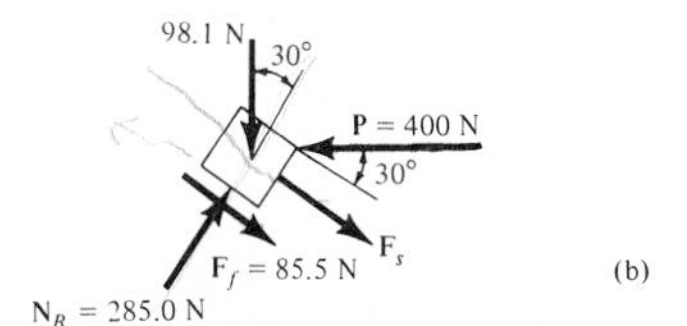

Fig. 14–6

Solution

The free-body diagram of the block is shown in Fig. 14–6*b*. By statics ($\Sigma F_y = 0$), $N_B = 285.0$ N, and $F_f = \mu_k N_B = 0.3(285.0) = 85.5$ N.

***Horizontal Force* P.** Since the force is *constant,* the work is computed using Eq. 14–2. The result can be calculated as the force times the component of displacement in the direction of the force, i.e.,

$$U_P = 400(2 \cos 30°) = 692.8 \text{ J}$$

or, the displacement times the component of force in the direction to displacement, i.e.,

$$U_P = 400 \cos 30°(2) = 692.8 \text{ J}$$

***Friction Force* $\mathbf{F}_f$.** This force is constant and acts in the *opposite* direction to displacement, hence *negative* work is done. Using Eq. 14–2, we have

$$U_f = -(85.5)(2) = -171.0 \text{ J}$$

***Spring Force* $\mathbf{F}_s$.** Since the spring is originally unstretched and in the final position is stretched 2 m, the work of $\mathbf{F}_s$ is

$$U_s = -\tfrac{1}{2}(30)(2)^2 = -60 \text{ J}$$

Why is the work negative?

***Weight* W.** Since the weight acts in the opposite direction to its vertical displacement the work is negative, i.e.,

$$U_W = -98.1(2 \sin 30°) = -98.1 \text{ J}$$

It is also possible to consider the component of weight in the direction of displacement, i.e.,

$$U_W = -(98.1 \sin 30°)2 = -98.1 \text{ J}$$

***Normal Force* $\mathbf{N}_B$.** This force does *no work* since it is *always* perpendicular to the displacement.

The total work performed when the block is displaced 2 m is thus

$$U_T = 692.8 - 171.0 - 60 - 98.1 = 363.7 \text{ J} \qquad \textit{Ans.}$$

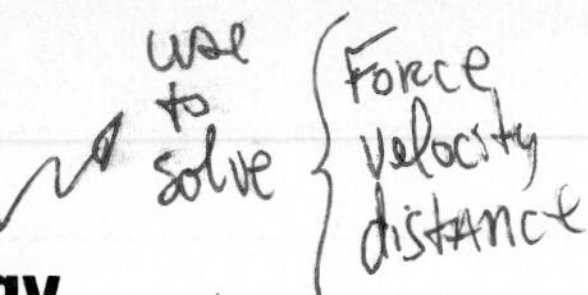

14.2 Principle of Work and Energy

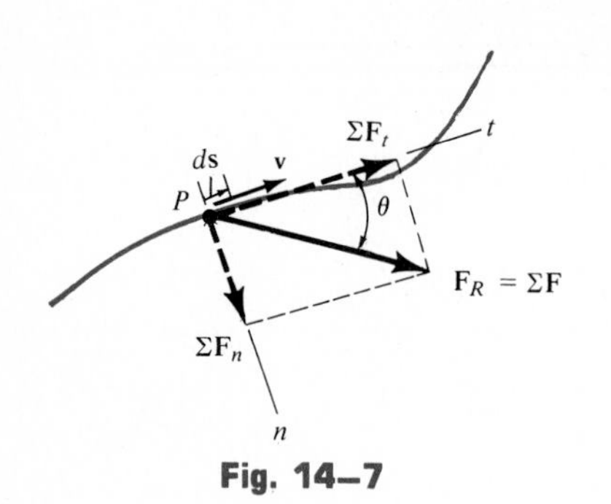

Fig. 14–7

Consider a particle P which is located at some arbitrary point on its path as shown in Fig. 14–7. At the instant considered, P is subjected to a system of external forces, represented by the resultant $\mathbf{F}_R = \Sigma\mathbf{F}$. If the forces are resolved into their normal and tangential components, then during an infinitesimal displacement $d\mathbf{s}$, the normal components, $\Sigma\mathbf{F}_n$, *do no work,* since they do not move in the normal direction; instead, only the tangential components, $\Sigma\mathbf{F}_t$, do work. As a result, consider writing the equation of motion for the particle in the tangential direction.

$$\Sigma F_t = ma_t; \qquad \Sigma F \cos\theta = ma_t$$

During the displacement $d\mathbf{s}$, $a_t = v\,dv/ds$ (Eq. 12–3). Hence,

$$\Sigma F \cos\theta\, ds = mv\, dv$$

Integrating both sides, assuming that initially the particle has a position $s = s_1$ and a speed $v = v_1$, and later $s = s_2$, $v = v_2$, we have

$$\Sigma \int_{s_1}^{s_2} F\cos\theta\, ds = \int_{v_1}^{v_2} mv\, dv$$

or

$$\Sigma \int_{s_1}^{s_2} F\cos\theta\, ds = \tfrac{1}{2}mv_2^2 - \tfrac{1}{2}mv_1^2$$

Using Eq. 14–1, the final result may be written as

$$\Sigma U_{1-2} = \tfrac{1}{2}mv_2^2 - \tfrac{1}{2}mv_1^2 \qquad (14\text{–}5)$$

This equation represents the *principle of work and energy* for the particle. The term on the left is the sum of the work done by *all* the forces acting on the particle as the particle moves from point 1 to point 2. The two terms on the right side, which are of the form $T = \frac{1}{2}mv^2$, define the particle's final and initial *kinetic energy,* respectively. These terms are positive scalar quantities since they do not depend on the direction of the particle's velocity. Furthermore, Eq. 14–5 is dimensionally homogeneous so that the kinetic energy has the same units as work, e.g., joules (J) or ft · lb.

As noted from the derivation, the principle of work and energy represents an integrated form of $\Sigma F_t = ma_t$, acquired by using the kinematic equation $a_t = v\,dv/ds$. Hence, if the particle's initial speed is known, and the work of all the forces acting on the particle can be computed, then Eq. 14–5 provides a *direct means* of obtaining the final speed v_2 of a particle after it undergoes a specified displacement. If instead v_2 is determined by means of the equation of motion, a two-step process is necessary; i.e., apply $\Sigma F_t = ma_t$ to obtain a_t, then integrate $a_t = v\,dv/ds$ to obtain v_2.

When applying Eq. 14–5 it is convenient to rewrite it in the form

$$T_1 + \Sigma U_{1-2} = T_2 \qquad (14\text{–}6)$$

which states that the particle's initial kinetic energy plus the work done by all the forces acting on the particle as it moves from its initial to its final position is equal to the particle's final kinetic energy. It should be noted that Eq. 14–6 is a *scalar equation;* and therefore, only *one unknown* can be obtained by using this equation when it is applied to a single particle. The principle of work and energy cannot be used, for example, to determine forces directed *normal* to the path of motion, since these forces do no work on the particle. For curved paths, however, the magnitude of the normal force is a function of velocity. Hence, it is generally easier to obtain the velocity using the principle of work and energy, and then to substitute this quantity into the equation of motion $\Sigma F_n = mv^2/\rho$ to obtain the normal force.

PROCEDURE FOR ANALYSIS

The principle of work and energy is used to solve kinetics problems that involve *velocity, force,* and *displacement,* since these terms are involved in the formulation. For application it is suggested that the following procedure be used.

Work (Free-Body Diagram). Draw a free-body diagram of the particle when it is located at an intermediate point along its path in order to account for all the forces that do work on the particle.

Principle of Work and Energy. Apply the principle of work and energy, $T_1 + \Sigma U_{1-2} = T_2$. The kinetic energy at the initial and final points is always positive, since it involves the speed squared ($T = \frac{1}{2}mv^2$). For the calculation v must be measured from an inertial reference frame. The work done by each force shown on the free-body diagram is computed by using the appropriate equations developed in Sec. 14–1. Since *algebraic addition* of the work terms is required, it is important that the proper sign of each term be specified. Specifically, work is positive when the force is in the same direction as its displacement, otherwise it is negative.

Numerical application of this procedure is illustrated in the examples following Sec. 14–3.

14.3 Principle of Work and Energy for a System of Particles

The principle of work and energy may be applied to each particle of a system of particles and the results added algebraically. We can then write

$$\Sigma T_1 + \Sigma U_{1-2} = \Sigma T_2 \tag{14–7}$$

Here the system's initial kinetic energy plus the total work done by all the external and internal forces acting on the particles of the system is equal to the system's final kinetic energy. In cases where all the particles are connected by inextensible links or cables, or the particles are contained within a translating rigid body, the work created by the internal forces is zero. This is because these forces occur in equal but opposite pairs, and each pair of forces acting on adjacent particles is displaced by an equal amount.

The procedure for analysis outlined above provides a method for applying Eq. 14–7. However, only one equation applies for the entire system, and therefore if the particles are connected by cords, other equations can generally be obtained by using the kinematic principles outlined in Sec. 12–7 in order to *relate* the particles' speeds (see Example 14–6).

Example 14–2

The 3,200-lb automobile shown in Fig. 14–8*a* is traveling up the 20° incline at a speed of 20 ft/s. If the driver wishes to stop his car in a distance of 15 ft, determine the frictional force at the pavement which must be supplied by the rear wheels.

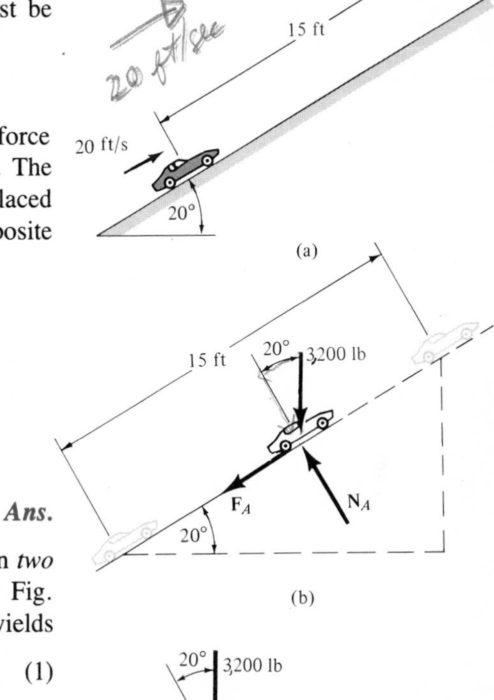

Fig. 14–8

Solution

Work (Free-Body Diagram). As shown in Fig. 14–8*b*, the normal force $\mathbf{N}_A$ does no work since it is never displaced along its line of action. The friction force $\mathbf{F}_A$ is displaced 15 ft, and the 3,200-lb force is displaced 15 sin 20° ft. Both forces do negative work since they act in the opposite direction to their displacement.

Principle of Work and Energy

$$\{T_1\} + \{\Sigma U_{1-2}\} = \{T_2\}$$

$$\{\tfrac{1}{2}(\tfrac{3,200}{32.2})(20)^2\} + \{-3,200(15 \sin 20°) - F_A(15)\} = \{0\}$$

Solving for F_A yields

$$F_A = 230.6 \text{ lb} \qquad \textit{Ans.}$$

Note that if this problem is solved by using the equation of motion *two steps* are involved. First, from free-body and kinetic diagrams, Fig. 14–8*c*, the equation of motion is applied along the incline. This yields

$$+\swarrow \Sigma F_x = ma_x; \qquad F_A + 3,200 \sin 20° = \frac{3,200}{32.2}a \qquad (1)$$

Then, using the integrated form of $a\,ds = v\,dv$ (kinematics),

$$+\swarrow v^2 = v_0^2 + 2a(s - s_0) \qquad (0)^2 = (-20)^2 + 2a(-15 - 0) \qquad (2)$$

Hence,

$$a = 13.3 \text{ ft/s}^2 \qquad F_A = 230.6 \text{ lb} \qquad \textit{Ans.}$$

Example 14–3

A 10-kg block rests on the horizontal surface shown in Fig. 14–9*a*. The spring, which is not attached to the block, has a stiffness of $k = 500$ N/m and is initially compressed 0.2 m from C to A. After the block is released from rest at A, determine its velocity when it passes point D. The coefficient of kinetic friction between the block and the plane is $\mu_k = 0.2$.

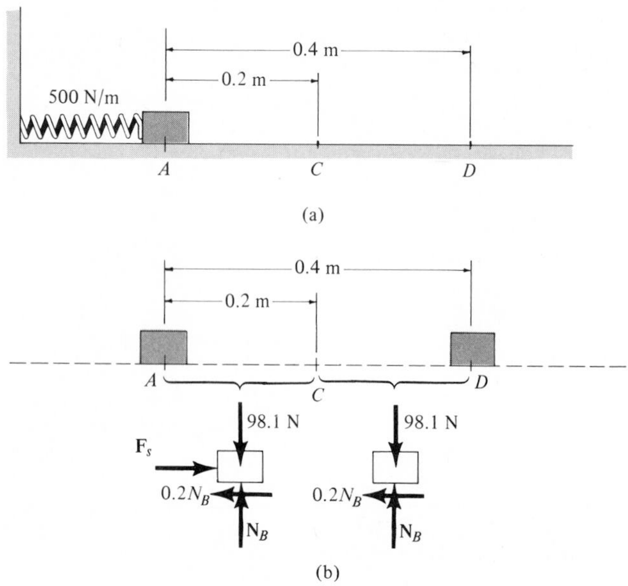

Fig. 14–9

Solution

Work (Free-Body Diagrams). Two free-body diagrams for the block are shown in Fig. 14–9*b*. The block moves under the influence of the spring force $\mathbf{F}_s$ along the 0.2-m-long path AC, after which it continues to slide along the plane to point D. With reference to either free-body diagram, $\Sigma F_y = 0$; hence, $N_B = 98.1$ N. Only the spring and friction forces do work during the displacement—the spring force does positive work from A to C, whereas the frictional force does negative work from A to D. Why?

Principle of Work and Energy

$$\{T_A\} + \{\Sigma U_{A-D}\} = \{T_D\}$$

$$\{\tfrac{1}{2}m(v_A)^2\} + \{\tfrac{1}{2}ks_{AC}^2 - (0.2N_B)s_{AD}\} = \{\tfrac{1}{2}m(v_D)^2\}$$

$$\{0\} + \{\tfrac{1}{2}(500)(0.2)^2 - 0.2(98.1)(0.4)\} = \{\tfrac{1}{2}(10)(v_D)^2\}$$

Solving for v_D, we get

$$v_D = 0.656 \text{ m/s} \qquad \textit{Ans.}$$

Example 14–4

The platform P, shown in Fig. 14–10a, has negligible mass and is tied down so that the 0.4-m-long cords keep the spring compressed 0.6 m when *nothing* is on the platform. If a 2-kg block is placed on the platform and released when the platform is pushed down 0.1 m, determine the maximum height h the block rises in the air, measured from the ground.

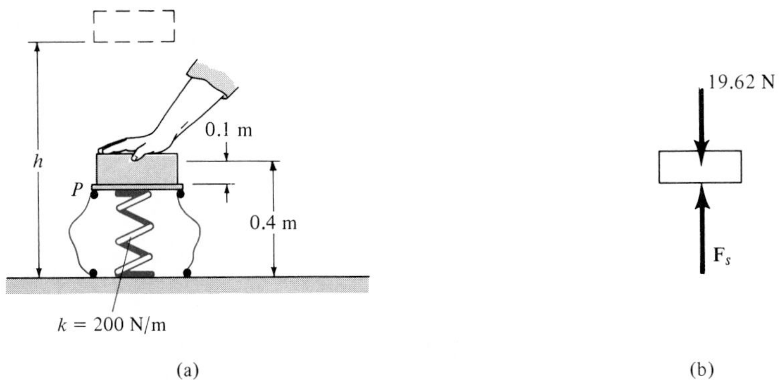

Fig. 14–10

Solution

Work (Free-Body Diagram). Since the block is released from rest and later reaches its maximum height, the initial and final velocities are zero. The free-body diagram of the block when it is still in contact with the platform is shown in Fig. 14–10b. The weight does negative work and the spring force does positive work. Why? In particular, the *initial compression* in the spring is 0.6 m + 0.1 m = 0.7 m. Due to the cords, the spring's *final compression* is 0.6 m (after the block leaves the platform).

Principle of Work and Energy

$$\{T_1\} + \{\Sigma U_{1-2}\} = \{T_2\}$$

$$\{0\} + \{\tfrac{1}{2}(200)(0.7)^2 - \tfrac{1}{2}(200)(0.6)^2 - (19.62)[h - (0.4 - 0.1)]\} = \{0\}$$

Solving yields

$$h = 0.963 \text{ m} \qquad \textit{Ans.}$$

Example 14–5

A block having a mass of 2 kg is given an initial velocity of $v_0 = 1$ m/s when it is at the top surface of the smooth cylinder shown in Fig. 14–11*a*. If the block moves along a path of 0.5-m radius, determine its velocity at the instant $\theta = 30°$. Also, compute the normal force the cylinder exerts on the block.

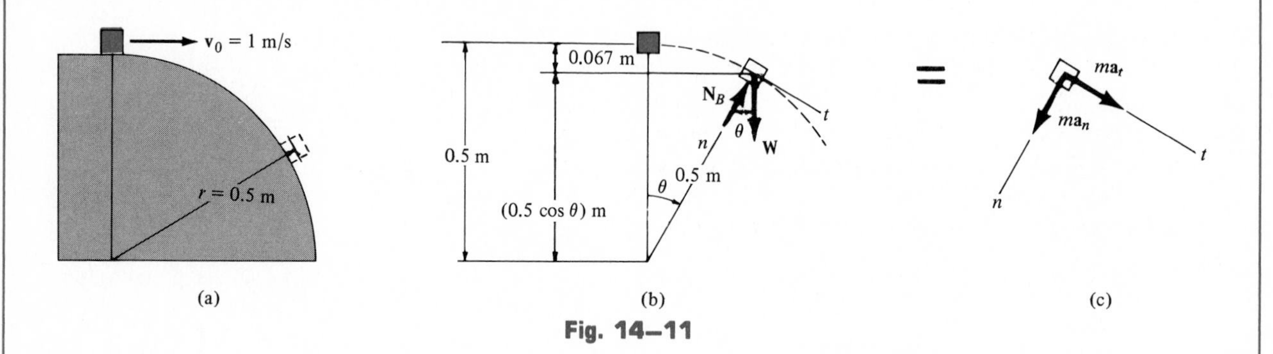

Fig. 14–11

Solution

Work (Free-Body Diagram). Inspection of Fig. 14–11*b* reveals that only the weight $W = 2(9.81) = 19.62$ N does work during the displacement. Here the weight moves through a vertical displacement of $0.5 - 0.5\cos 30° = 0.067$ m, as shown in the figure.

Principle of Work and Energy

$$\{T_1\} + \{\Sigma U_{1-2}\} = \{T_2\}$$

$$\{\tfrac{1}{2}(2)(1)^2\} + \{19.62(0.067)\} = \{\tfrac{1}{2}(2)v_2^2\}$$

or

$$v_2 = 1.52 \text{ m/s} \qquad \textit{Ans.}$$

Equation of Motion. The equation of motion must be applied in the normal direction to obtain N_B. The kinetic diagram is shown in Fig. 14–11*c*. Thus,

$$+\swarrow \Sigma F_n = ma_n; \qquad -N_B + 19.62\cos 30° = 2\left(\frac{(1.52)^2}{0.5}\right)$$

Solving, we obtain

$$N_B = 7.73 \text{ N} \qquad \textit{Ans.}$$

Example 14–6

The blocks A and B shown in Fig. 14–12a have a mass of 10 and 100 kg, respectively. Determine the distance B travels from the point where it is released from rest to the point where its speed becomes 2 m/s.

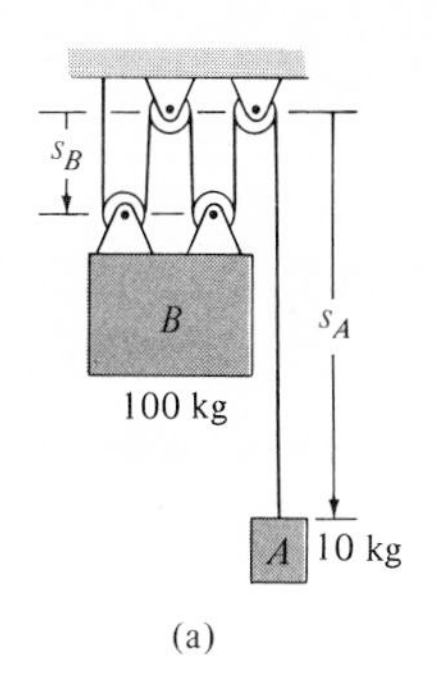

(a)

Solution

This problem may be solved by considering the blocks separately and applying the principle of work and energy to each block. However, the work of the (unknown) cable tension can be eliminated from the analysis by considering blocks A and B together as a *system*.

Work (Free-Body Diagram). If a free-body diagram of B were drawn and $\Sigma F_y = 0$ applied, it would be found that a cable tension of $\frac{1}{4}(100)(9.81) = 245.3$ N is required for *equilibrium*. Since A weighs $10(9.81) = 98.1$ N, which requires a cable tension of only 98.1 N for *equilibrium*, B will move downward while A moves *upward*. As shown on the free-body diagram of the system, Fig. 14–12b, the cable force **T** and reactions $\mathbf{R}_1$ and $\mathbf{R}_2$ do *no work* since these forces represent the reactions at the supports and consequently do not move while the blocks are being displaced.

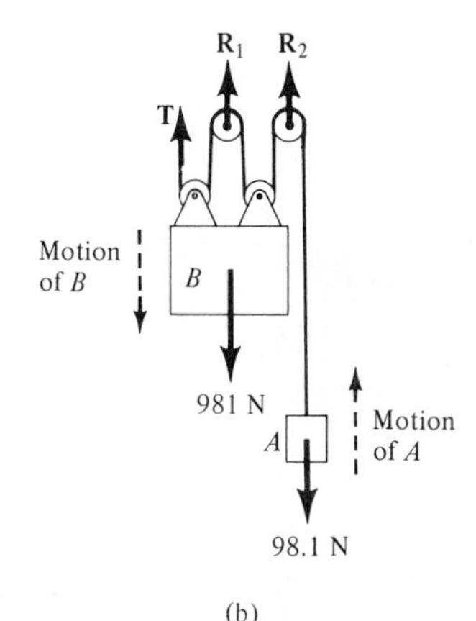

(b)

Fig. 14–12

Principle of Work and Energy. Realizing the blocks are released from rest, we have

$$\{\Sigma T_1\} + \{\Sigma U_{1-2}\} = \{\Sigma T_2\}$$

$$\{\tfrac{1}{2}m_A(v_A)_1^2 + \tfrac{1}{2}m_B(v_B)_1^2\} + \{W_B\Delta s_B - W_A\Delta s_A\} = \{\tfrac{1}{2}m_A(v_A)_2^2 + \tfrac{1}{2}m_B(v_B)_2^2\}$$

$$\{0 + 0\} + \{981\ \Delta s_B - 98.1(\Delta s_A)\} = \{\tfrac{1}{2}(10)(v_A)_2^2 + \tfrac{1}{2}(100)(2)^2\} \qquad (1)$$

Kinematics. Using the methods of kinematics discussed in Sec. 12–7, it may be seen from Fig. 14–12a that at any given instant the total length l of all the vertical segments of cable may be expressed in terms of the position coordinates s_A and s_B as

$$s_A + 4s_B = l$$

Hence, a displacement Δs_B (downward) of block B will cause a corresponding displacement $\Delta s_A = -4\Delta s_B$ (upward); i.e.,

$$|\Delta s_A| = |4\Delta s_B|$$

Taking the time derivative yields

$$|v_A| = |4v_B| = |4(2)| = 8 \text{ m/s}$$

Substituting into Eq. (1) and solving for Δs_B yields

$$\Delta s_B = 0.883 \text{ m} \qquad \textit{Ans.}$$

Problems

Except when stated otherwise, throughout this chapter assume that the coefficients of static and kinetic friction are equal, i.e., $\mu = \mu_s = \mu_k$.

14–1. A freight car having a mass of 15 Mg is towed along a horizontal track. If the car starts from rest and attains a speed of 8 m/s after traveling a distance of 150 m, determine the constant horizontal towing force applied to the car. Neglect friction and the mass of the wheels.

14–2. A crate has a weight of 1,500 lb. If it is pulled along the ground at a constant speed for a distance of 20 ft, and the towing cable makes an angle of 15° with the horizontal, determine the tension in the cable and the work done by the towing force. The coefficient of friction between the ground and the crate is $\mu = 0.55$.

14–3. A car having a mass of 2 Mg strikes a smooth, rigid sign post with an initial speed of 30 km/h. To stop the car, the front end horizontally deforms 0.2 m. If the car is free to roll during the collision, determine the *average* horizontal collision force causing the deformation.

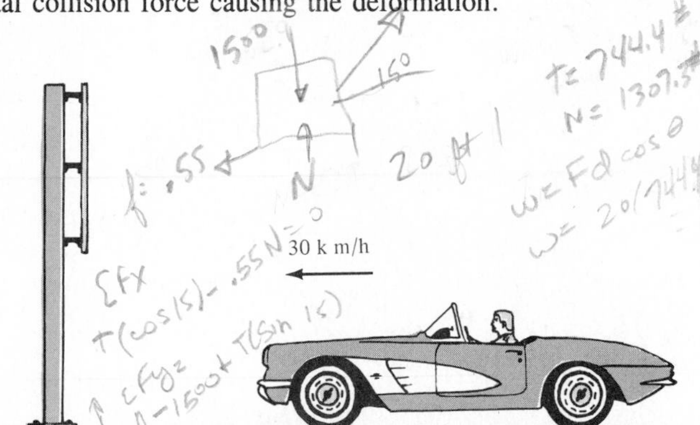

Prob. 14–3

*14–4. A woman having a mass of 70 kg stands in an elevator which has a downward acceleration of 4 m/s² starting from rest. Determine the work done by her weight and the work of the normal force which the floor exerts on her after the elevator descends 8 m. Explain why the work of these forces is different.

14–5. If a 150-lb crate is released from rest at A, determine its speed after it slides 30 ft down the plane. The coefficient of friction between the crate and plane is $\mu = 0.3$.

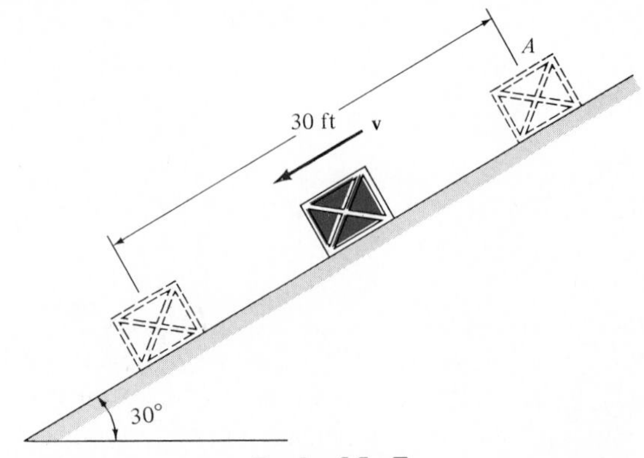

Prob. 14–5

14–6. The 1,500-lb "wrecking" ball, suspended from a 50-ft cable, swings from rest at point A, $\theta = 30°$, toward a wall. After striking the wall it rises to point B, $\phi = 20°$. Determine the amount of energy which is lost during the collision and compute the tension in the cable just before the collision, $\theta = 0°$. Neglect the size of the ball for the calculation. Point C doesn't move, and when $\theta = 0°$ the ball is at its lowest point.

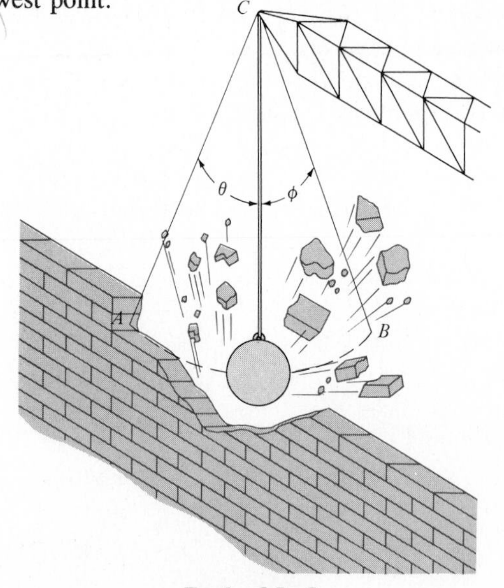

Prob. 14–6

14–7. A car is equipped with a bumper B designed to absorb collisions. The bumper is mounted to the car using pieces of flexible tubing T. Upon collision with a rigid barrier A, a constant horizontal force $\mathbf{F}$ is developed which causes a car deceleration of $3g = 29.43\ \text{m/s}^2$ (the highest safe deceleration for a passenger without a seat belt). If the car and passenger have a total mass of 1.5 Mg and the car is initially coasting with a speed of 1.5 m/s, compute the magnitude of $\mathbf{F}$ needed to stop the car and the deformation x of the bumper tubing.

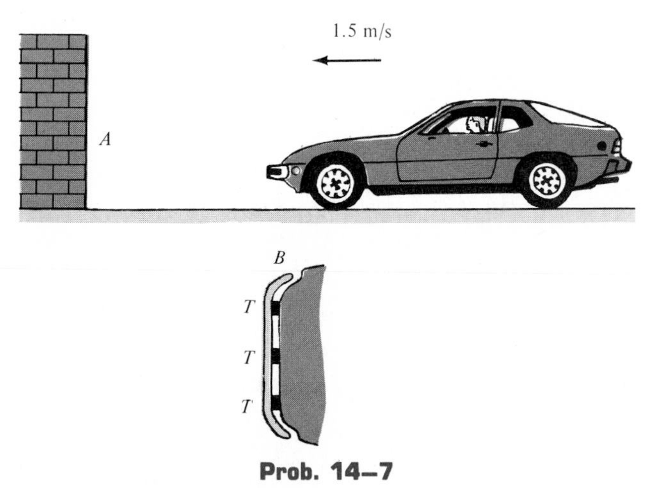

Prob. 14–7

14–9. A car, assumed to be rigid and having a mass of 800 kg, strikes a barrel-barrier installation without the driver applying the brakes. From experiments, the magnitude of the force of resistance $\mathbf{F}_r$, created by deforming the barrels successively, is shown as a function of vehicle penetration. If the car strikes the barrier traveling at $v_c = 70$ km/h, determine approximately the distance s to which the car penetrates the barrier.

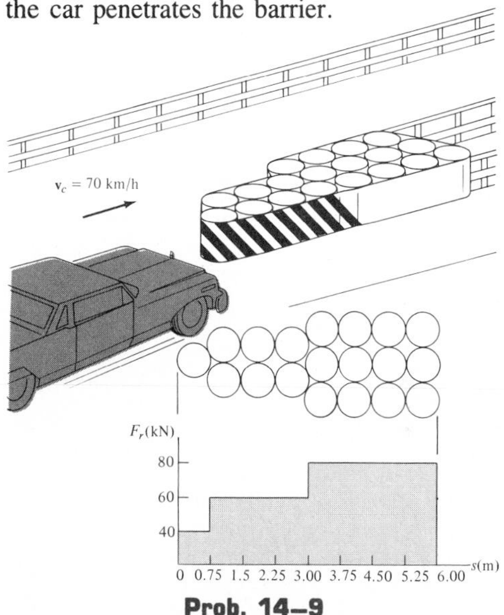

Prob. 14–9

***14–8.** The driving frictional force $\mathbf{F}$ for a car having four-wheel drive has been measured versus its position and is shown in the figure. Determine the maximum frictional force $F_{max} = \mu N$ and the speed of the car at the instant all its wheels begin to slip on the pavement. The car starts from rest and has a mass of 1800 kg. Neglect wind resistance. The coefficient of friction between the tires and pavement is $\mu = 0.5$.

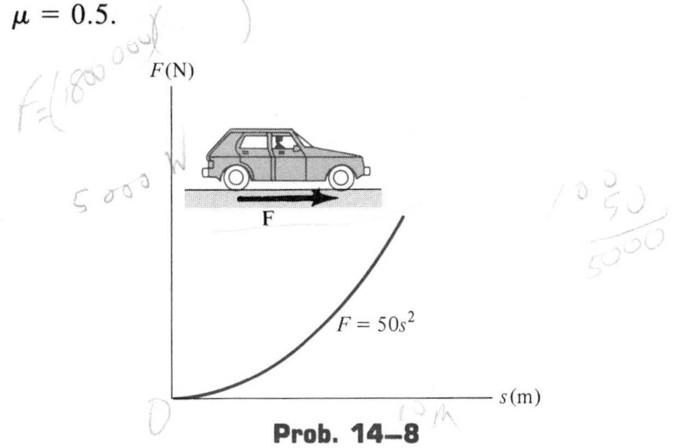

Prob. 14–8

14–10. When at A the bicyclist has a speed of $v_A = 4$ ft/s. If he coasts without pedaling from the top of the hill at A to the shore of B and then leaps off the shore, determine his speed at B and the distance x where he strikes the water at C. The rider and his bicycle have a total weight of 150 lb. Neglect the size of the bicycle and wind resistance.

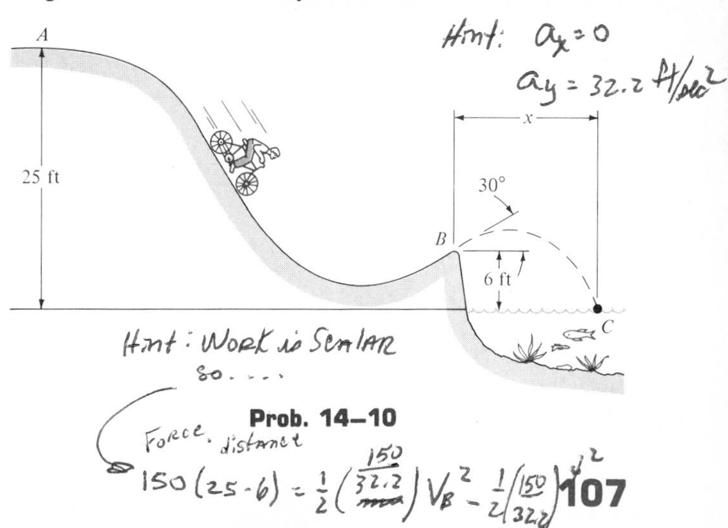

Prob. 14–10

14–11. Coins are placed in a small container C. Using an elastic spring gun, a clerk "fires" the container along the smooth wire from A to a cashier located at B. If the spring in the gun has a stiffness of $k = 2.5$ kN/m and the spring is compressed 100 mm when the gun is fired, determine the greatest mass of coins which can be placed in the container and still allow it to reach the cashier. Neglect the effect of friction and assume the spring becomes completely unstretched when the gun is fired. The empty container has a mass of 75 g. *Hint:* The weight of the container travels through a vertical distance of 2.5 m from A to B.

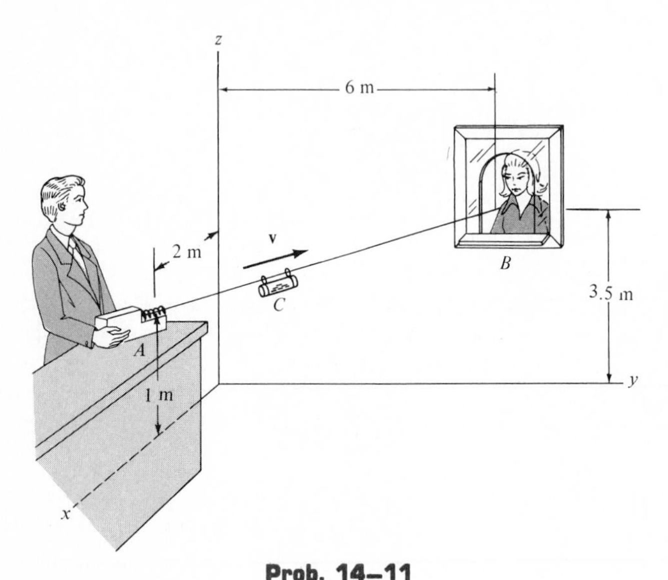

Prob. 14–11

***14–12.** If the cashier in Prob. 14–11 releases the *empty* 75-g container from rest at B, determine the speed at which it strikes the gun at A.

14–13. Determine the speed of the bicycle at B in Prob. 14–10 if $v_A = 8$ ft/s.

14–14. The coefficient of friction between the 2-lb block and the surface is $\mu = 0.2$. The block is acted upon by a horizontal force of $P = 15$ lb and has a speed of 6 ft/s when it is at point A. Determine the maximum deformation of the outer spring B at the instant the block comes to rest. Spring B has a stiffness of $k_B = 20$ lb/ft and the "nested" spring C has a stiffness of $k_C = 40$ lb/ft.

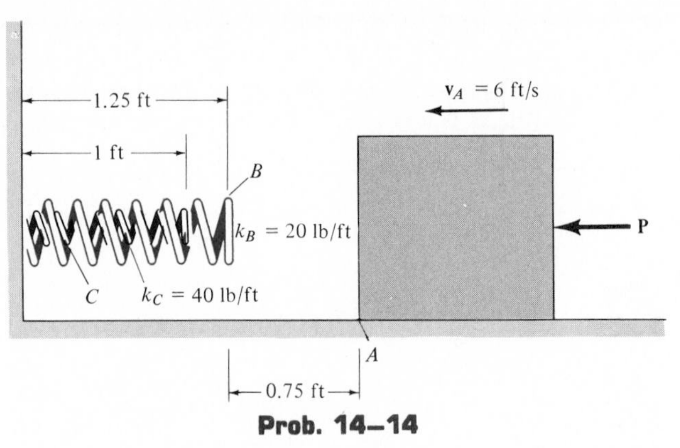

Prob. 14–14

14–15. If the 2-lb block in Prob. 14–14 is pressed against the two springs so that it is 0.75 ft from the wall and then released from rest, with $\mathbf{P} = \mathbf{0}$, determine how far from the wall the block slides before coming to rest. The coefficient of friction is $\mu = 0.2$.

***14–16.** The catapulting mechanism is used to propel the 15-kg slider A to the right along the smooth track. The propelling action is obtained by drawing the pulley attached to rod BC rapidly to the left by means of a piston P. If the piston applies a constant force of $F = 8$ kN to rod BC such that it moves it 0.2 m, determine the speed attained by the slider if it was originally at rest. Neglect the mass of the pulleys, cable, piston, and rod BC.

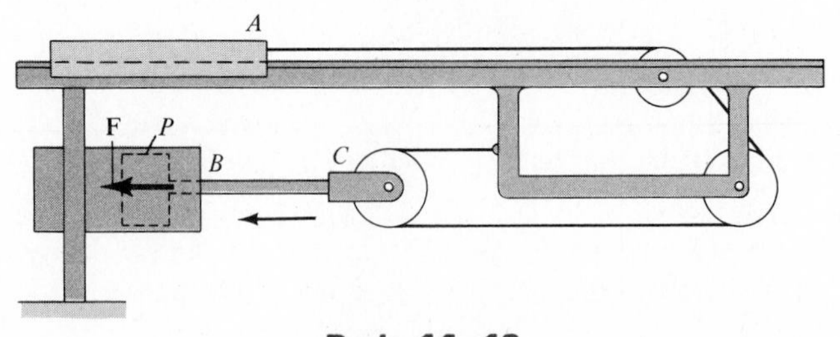

Prob. 14–16

14–17. The block has a mass of $m = 0.5$ kg and moves within the smooth vertical slot. If it starts from rest when the *attached* spring is in the unstretched position at A, determine the *constant* vertical force **F** which must be applied to the cord so that the block attains a speed of $v_B = 2.5$ m/s when it reaches B; $s_B = 0.15$ m. *Hint:* The work of **F** can be determined by finding the difference Δl in cord lengths AC and BC and using $U_F = F\,\Delta l$.

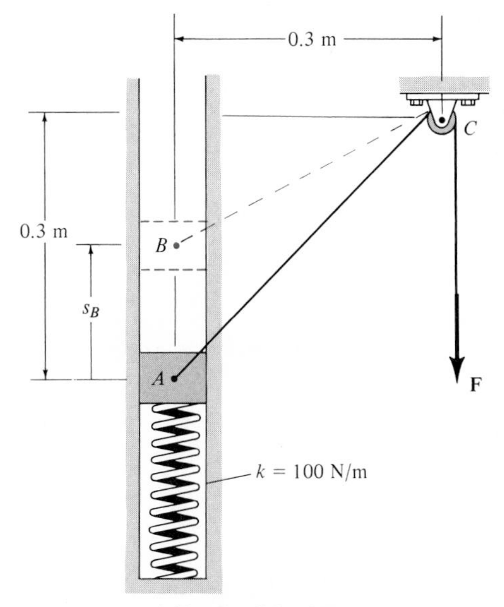

Prob. 14–17

14–18. The "flying car" is a ride at an amusement park, which consists of a car having wheels that roll along a track mounted on a rotating drum. Motion of the car is created by applying the car's brake, thereby gripping the car to the track and allowing it to move with a speed of $v_t = 3$ m/s. If the rider applies the brake when going from B to A and then releases it at the top of the drum, A, so that the car coasts freely down along the track to B ($\theta = \pi$ rad), determine the speed of the car at B and the normal reaction which the drum exerts on the car at B. Neglect friction during the motion from A to B. The rider and car have a total mass of $m = 250$ kg and the center of mass of the car and rider moves along a circular path of radius $r = 8$ m.

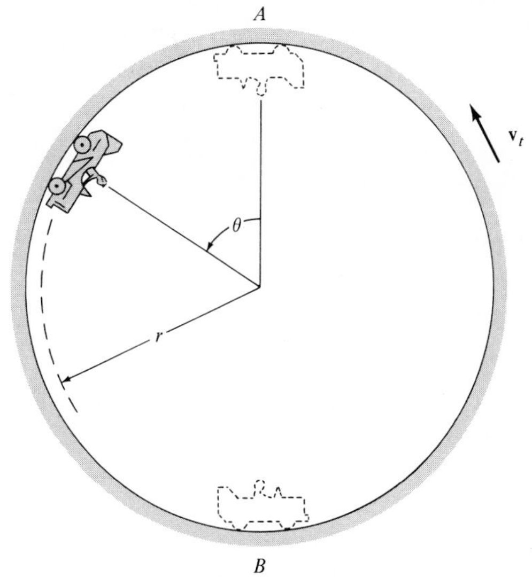

Prob. 14–18

14–19. A rocket of mass m is fired vertically from the surface of the earth, i.e., at $r = r_1$. Assuming that no mass is lost as it travels upward, determine the work it must do against gravity to reach a distance r_2. The force of gravity is $F = GM_e m/r^2$ (Eq. 13–1), where M_e is the mass of the earth and r the distance between the rocket and the center of the earth.

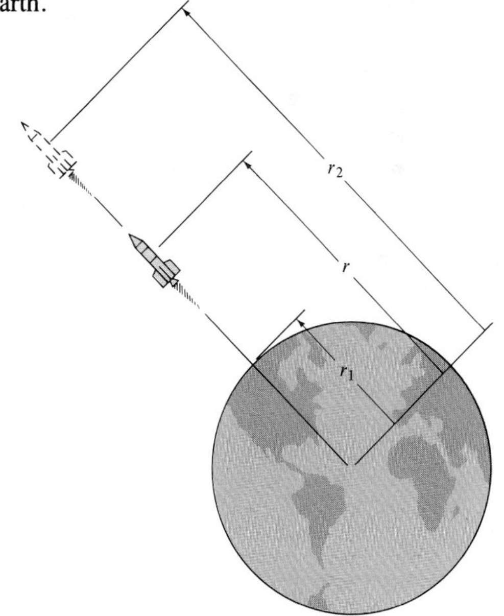

Prob. 14–19

14.4 Power and Efficiency

Power. *Power* is defined as the amount of work performed per unit of time. Hence, the *average power* generated by a machine or engine that performs an amount of work ΔU within the time interval Δt is

$$P_{\text{avg}} = \frac{\Delta U}{\Delta t} \tag{14-8}$$

If the time $\Delta t \rightarrow dt$, and consequently $\Delta U \rightarrow dU$, then the *instantaneous power* is defined as

$$P = \frac{dU}{dt} \tag{14-9}$$

Provided that the work dU is expressed by $dU = F \cos \theta \, ds$, then it is also possible to write

$$P = \frac{dU}{dt} = \frac{F \cos \theta \, ds}{dt}$$

or

$$P = (F \cos \theta)v \tag{14-10}$$

where θ is the angle between F and v. Hence, power is a *scalar*, where in the formulation F represents the component of unbalanced force acting in the direction of the instantaneous velocity v of the object.

The basic units of power used in the SI and FPS systems are the watt (W) and horsepower (hp), respectively. These units are defined as

$$1 \text{ W} = 1 \text{ J/s} = 1 \text{ N} \cdot \text{m/s}$$

$$1 \text{ hp} = 550 \text{ ft} \cdot \text{lb/s}$$

For conversion between the two systems of units, 1 hp = 746 W.

The term "power" provides a useful basis for determining the type of motor or machine which is required to do a certain amount of work in a given time. For example, two pumps may each be able to empty a reservoir if given enough time; however, the pump having the larger power will complete the job sooner.

Efficiency. The *mechanical efficiency* of a machine is defined as the ratio of the output of useful power created by the machine to the input of power supplied to the machine. Hence,

$$\eta = \varepsilon = \frac{\text{power output}}{\text{power input}} \tag{14–11}$$

If work is being done by the machine at a *constant rate,* then the efficiency may be expressed in terms of the ratio of output energy to input energy; i.e.,

$$\varepsilon = \frac{\text{energy output}}{\text{energy input}} \tag{14–12}$$

Since machines consist of a series of moving parts, frictional forces will always be developed within the machine, and as a result, extra work or power is needed to overcome these forces. Consequently, *the efficiency of a machine is always less than 1.*

PROCEDURE FOR ANALYSIS

When computing the power supplied to a body, one must first determine the unbalanced external force **F** acting on the body which causes the motion. This force is usually developed by a machine or engine placed either within or external to the body. If the body is accelerating, it may be necessary to draw the proper free-body and kinetic diagrams and apply the equation of motion ($\Sigma \mathbf{F} = m\mathbf{a}$) to determine **F.** Once **F** has been computed, the power is determined by multiplying the instantaneous velocity by the component of force acting in the direction of **v**, i.e., $P = (F \cos \theta)v$.

In some problems the power may also be computed by calculating the work done per unit of time ($P_{avg} = \Delta U/\Delta t$, or $P = dU/dt$). Depending upon the problem, this work is done either by the external or internal force of a machine or engine, by the weight of the body, or by an elastic spring force acting on the body.

Example 14–7

The motor M of the hoist shown in Fig. 14–13a operates with an efficiency of 0.85. Determine the power that must be supplied to the motor to lift the crate C having a weight of 75 lb. The cable is being drawn in with an acceleration of 4 ft/s^2, and at the instant shown its speed is 2 ft/s. Neglect the mass of the pulley and cable.

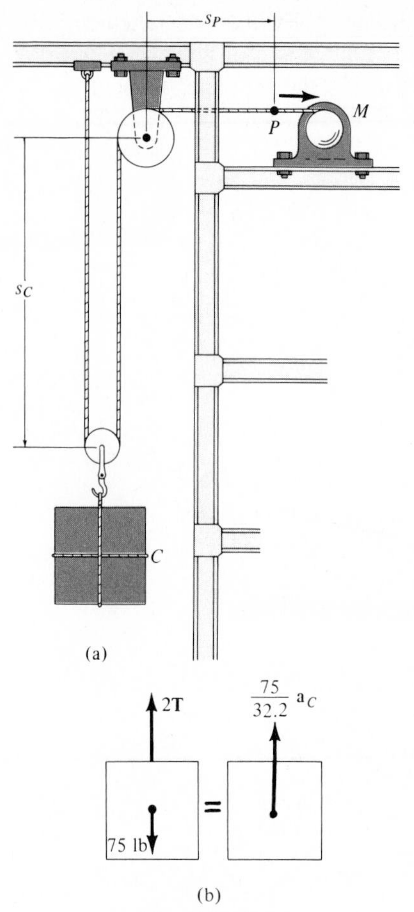

Fig. 14–13

Solution

In order to compute the power output of the motor it is first necessary to determine the tension force in the cable. Since the crate is subjected to an acceleration, this requires application of the equation of motion.

Using the data shown on the free-body and kinetic diagrams, Fig. 14–13b, we have

$$+\uparrow \Sigma F_y = ma_y; \qquad 2T - 75 = \frac{75}{32.2}a_C \qquad (1)$$

The acceleration of the crate can be obtained by using kinematics to relate the motion of the crate to the known motion of a point P located on the cable, Fig. 14–3a. Hence, by the methods of Sec. 12–7, the coordinates s_C and s_P in Fig. 14–13a can be related to a constant portion of cable length l which is changing in the vertical and horizontal directions. We have $2s_C + s_P = l$. Taking the second time derivative of this equation yields

$$2a_C = -a_P$$

From the problem data, $a_P = 4$ ft/s^2; thus, $a_C = (-4 \text{ ft/s}^2)/2 = -2$ ft/s^2. Substituting this result into Eq. (1), *neglecting* the negative sign since it indicates that the acceleration is upward in accordance with the direction of $m\mathbf{a}_C$ in Fig. 14–13b, we have

$$2T - 75 = \frac{75}{32.2}(2)$$

$$T = 39.8 \text{ lb}$$

The power output, measured in units of horsepower, required to draw the cable in at a rate of 2 ft/s is therefore

$$P = Tv = (39.8 \text{ lb})(2 \text{ ft/s})(1 \text{ hp}/550 \text{ lb}\cdot\text{ft/s})$$
$$= 0.145 \text{ hp}$$

This *power output* requires that the motor provide a *power input* of

$$\text{power input} = \frac{1}{\varepsilon}(\text{power output})$$
$$= \frac{1}{0.85}(0.145) = 0.170 \text{ hp} \qquad \textit{Ans.}$$

Since the velocity of the crate is constantly changing, notice that this power requirement is *instantaneous*.

Example 14–8

The car shown in Fig. 14–14*a* has a mass of 1000 kg and an engine running efficiency of $\varepsilon = 0.63$. As it moves forward at a constant speed of $v = 60$ m/s, the wind creates a constant drag resistance on the car of $F_D = 80$ N. Assuming that power is delivered to *all the wheels,* determine the power supplied by the engine. The coefficient of friction between the wheels and the pavement is $\mu = 0.25$.

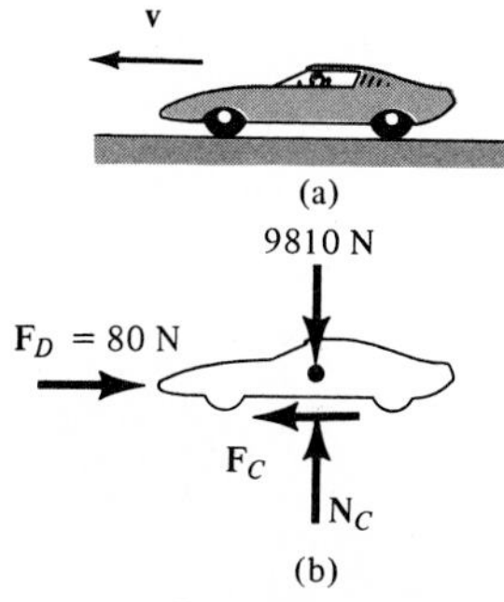

Fig. 14–14

Solution

As shown on the free-body diagram, Fig. 14–14*b*, the normal force $\mathbf{N}_C$ and frictional force $\mathbf{F}_C$ represent the *resultant forces* of all four wheels. In particular, the unbalanced frictional force drives or pushes the car *forward.* This effect is, of course, created by the rotating motion of the wheels on the pavement.

Applying the equations of equilibrium since the car's acceleration is zero, we have

$$\overset{+}{\leftarrow}\Sigma F_x = ma_x; \qquad F_C - 80 = 0; \qquad F_C = 80 \text{ N}$$

$$+\uparrow \Sigma F_y = ma_y; \quad N_C - 9810 = 0 \qquad N_C = 9810 \text{ N}$$

The frictional force reaches its maximum value when $F_{C_{max}} = 0.25N_C = 0.25(9810) = 2452.5$ N. Since $80 \text{ N} < 2452.5 \text{ N}$, the car does not slip on the pavement. The power output of the car is created by the driving (frictional) force $\mathbf{F}_C$. Thus,

$$P = F_C v = 80(60) = 4.8 \text{ kW}$$

The power supplied by the engine (power input) is therefore

$$\text{power input} = \frac{1}{\varepsilon}(\text{power output}) = \frac{1}{0.63}(4.8) = 7.62 \text{ kW} \qquad \textit{Ans.}$$

Problems

***14–20.** The diesel engine of a 400-Mg train increases the train's speed from rest to 10 m/s in 80 s along a horizontal track. Determine the average power developed.

14–21. A man having a weight of 150 lb is able to run up a 15-ft-high flight of stairs in 4 s. Determine the power generated. How long would a 100-W light bulb have to burn to expend the same amount of energy? *Conclusion:* Turn the lights off when not in use!

Prob. 14–21

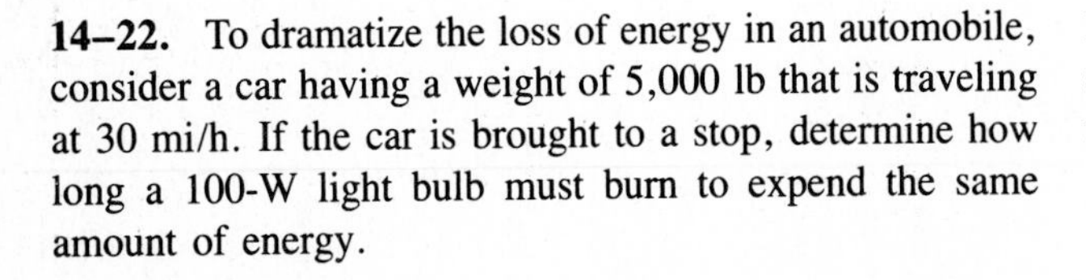

14–22. To dramatize the loss of energy in an automobile, consider a car having a weight of 5,000 lb that is traveling at 30 mi/h. If the car is brought to a stop, determine how long a 100-W light bulb must burn to expend the same amount of energy.

14–23. An electric train car, having a mass of 25 Mg, travels up a 10° incline with a constant speed of 80 km/h. Determine the power required to overcome the force of gravity.

***14–24.** A spring having a stiffness of 5 kN/m is compressed 400 mm. The stored energy in the spring is used to drive a machine which requires 80 W of power. Determine how long the spring can supply energy at the required rate.

14–25. The block and tackle arrangement for the oil rig is shown schematically in the figure. Determine the power output of the draw-work motor M necessary to lift the 800-lb drill pipe upward with a constant speed of 4 ft/s.

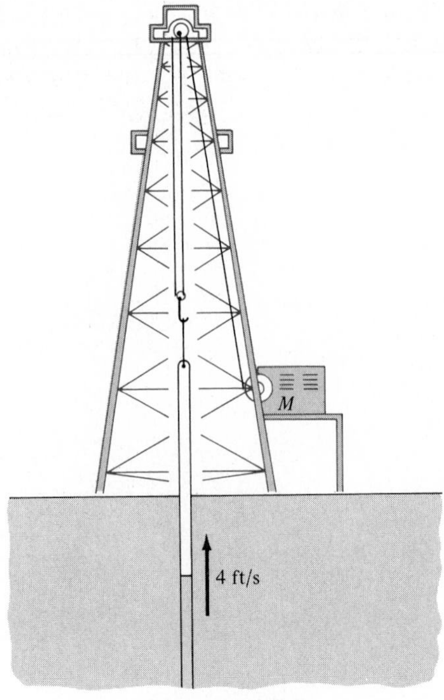

Prob. 14–25

14–26. A motor hoists a crate that has a mass of 50 kg to a height of $h = 6$ m in 3 s. If the indicated power of the motor is 4 kW, determine the motor's efficiency.

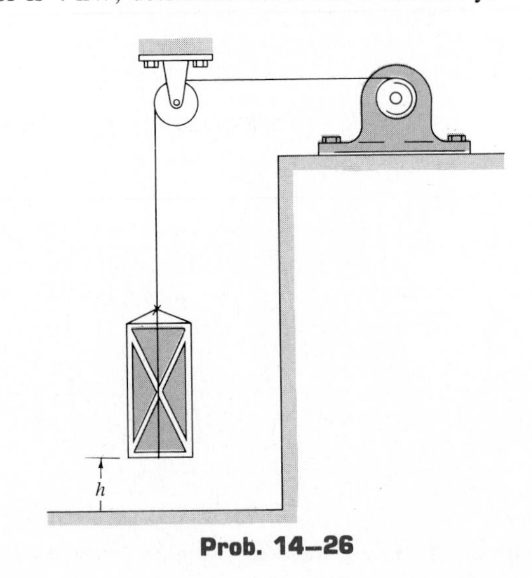

Prob. 14–26

14–27. A truck has a weight of 25,000 lb and an engine which transmits a power of 350 hp to *all* the wheels. Assuming that the wheels do not slip on the ground, determine the angle θ of the largest incline the truck can climb at a constant speed of $v = 50$ ft/s.

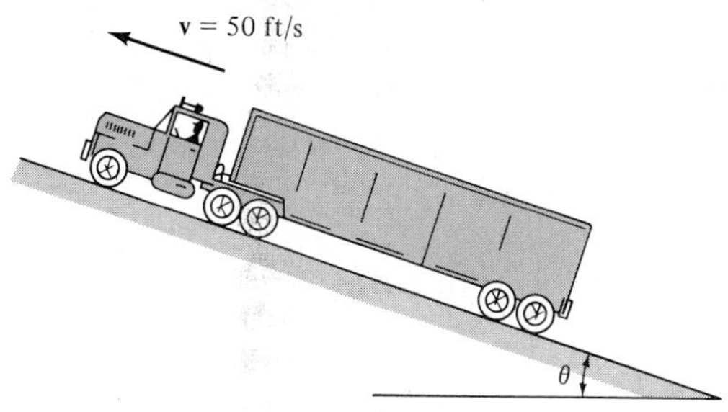

Prob. 14–27

***14–28.** The crate, having a weight of 50 lb, is hoisted by the pulley system and motor M. If the crate starts from rest and, by constant acceleration, attains a speed of 12 ft/s after rising $s = 10$ ft, determine the power that must be supplied to the motor at the instant $s = 10$ ft. The motor has an efficiency of $\varepsilon = 0.74$.

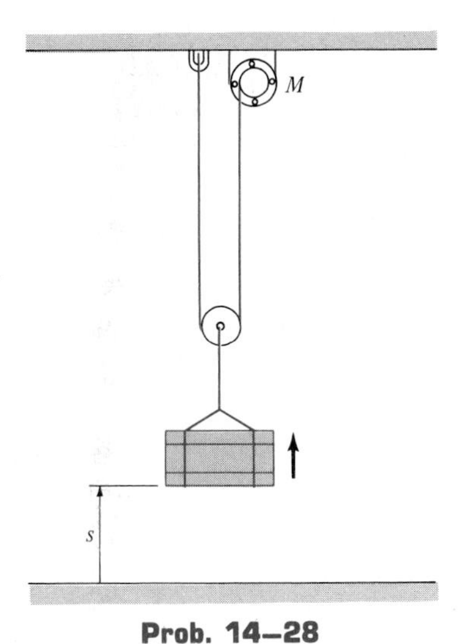

Prob. 14–28

14–29. An automobile having a weight of 3,500 lb travels up a 7° slope at a constant speed of $v = 40$ mi/h. If friction and wind resistance are neglected, determine the power developed by the engine if the automobile has a mechanical efficiency of $\varepsilon = 0.65$.

14–30. A rocket having a total mass of 8 Mg is fired vertically from rest. If the engines provide a constant thrust of $T = 300$ kN, determine the power output of the engines as a function of time. Neglect the effect of drag resistance and the loss of fuel mass and weight.

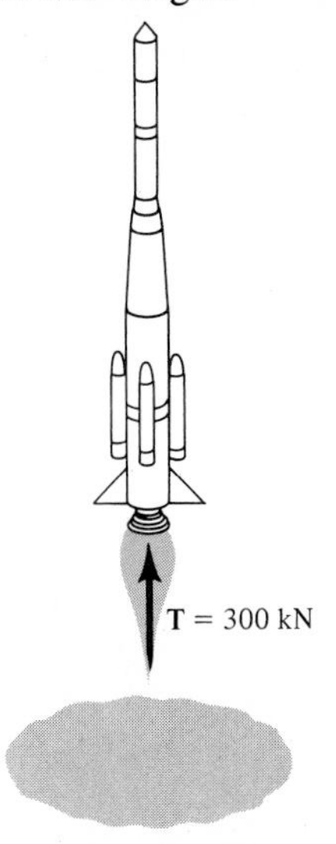

Prob. 14–30

14–31. The elevator E and its freight have a total mass of 400 kg. Hoisting is provided by the motor M and the 60-kg block C. If the motor has an efficiency of $\varepsilon = 0.6$, determine the power that must be supplied to the motor when the elevator is hoisted upward at a constant speed of $v_E = 4$ m/s.

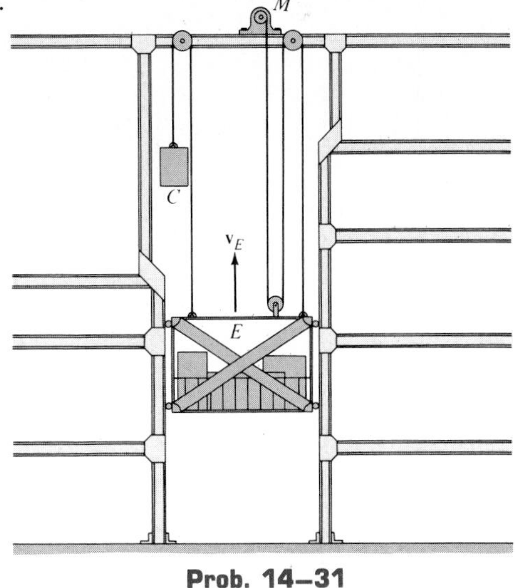

Prob. 14–31

14.5 Conservative Forces and Potential Energy

Conservative Force. A particularly simple type of force acting on a particle is one that depends *only* on the particle's position and is independent of the particle's velocity and acceleration. Furthermore, if the work done by this force in moving the particle from one point to another is *independent of the path* followed by the particle, this force is called a *conservative force*. The weight of a particle and the force of an elastic spring are two examples of conservative forces often encountered in mechanics.

Weight. The work done by the weight of a particle is independent of the path; rather, it depends only on the particle's *vertical displacement*. If this displacement is y (downward), then from Eq. 14–3,

$$U = Wy$$

Elastic Spring. The work done by a spring force *acting on a particle* is independent of the path of the particle, but depends only on the extension or compression x of the spring. If the spring is originally *unstretched,* then from Eq. 14–4,

$$U = -\tfrac{1}{2}kx^2$$

Friction. In contrast to a conservative force, consider the force of friction exerted *on a moving object* by a fixed surface. The work done by the frictional force *depends upon the path*—the longer the path, the greater the work. Consequently, *frictional forces are nonconservative*. The work is dissipated from the body in the form of heat.

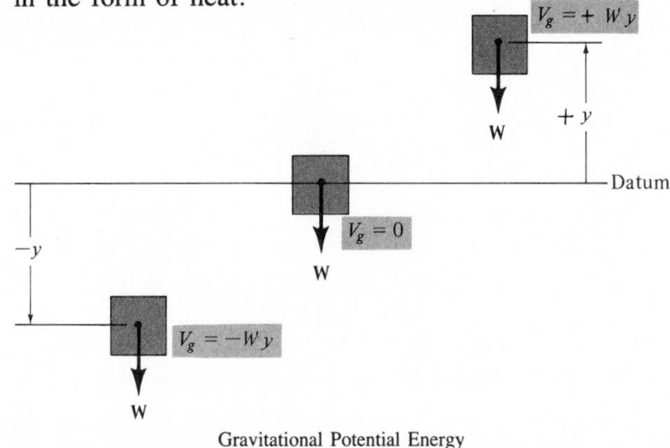

Gravitational Potential Energy

Fig. 14–15

Potential Energy. When a conservative force acts on a particle, the force has the capacity to do work. This capacity, measured as *potential energy, V,* depends only upon the *location* of the particle when acted upon by the force.

Gravitational Potential Energy. If a particle (or block) is located a distance y *above* a datum, as shown in Fig. 14–15, the particle's weight has positive *gravitational potential energy,* V_g, since **W** has the capacity of doing positive work when the particle is moved back down to the datum. This energy can be expressed mathematically as*

$$V_g = +Wy \qquad (14\text{–}13)$$

If the particle is located a distance y *below* the datum, V_g is negative,

$$V_g = -Wy \qquad (14\text{–}14)$$

since the weight does negative work when the particle is moved back up to the datum. At the datum $V_g = 0$.

Elastic Potential Energy. When an elastic spring is elongated or compressed a distance x from its unstretched position, the elastic potential energy V_e which the spring imparts to an attached particle (or block) can be expressed as

$$V_e = +\tfrac{1}{2}kx^2 \qquad (14\text{–}15)$$

Here V_e is *always positive* since, in the deformed position, the force of the spring has the *capacity* for doing positive work on the particle when the spring is returned to its unstretched position, Fig. 14–16.

Potential-Energy Function. In the general case, if a particle is subjected to both gravitational and elastic forces, the particle's potential energy can be expressed as a *potential-energy function,* which is the algebraic sum

$$V = V_g + V_e \qquad (14\text{–}16)$$

Measurement of V depends upon the location of the particle with respect to a selected datum in accordance with Eqs. 14–13 to 14–15.

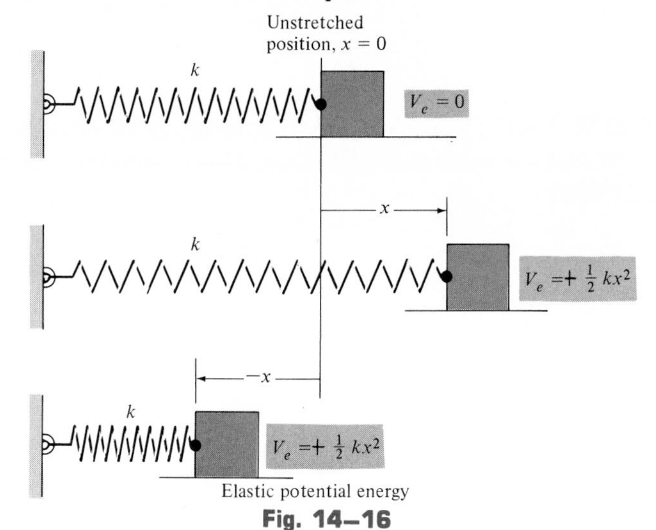

Elastic potential energy
Fig. 14–16

In general, if a particle is located at an arbitrary point (x, y, z) in space, its potential energy or capacity to do work can be defined using a potential-

*Here the weight is assumed to be *constant*. This assumption is suitable for small differences in elevation y. If the elevation change is significant, however, a variation of weight with elevation must be taken into account (see Prob. 14–44).

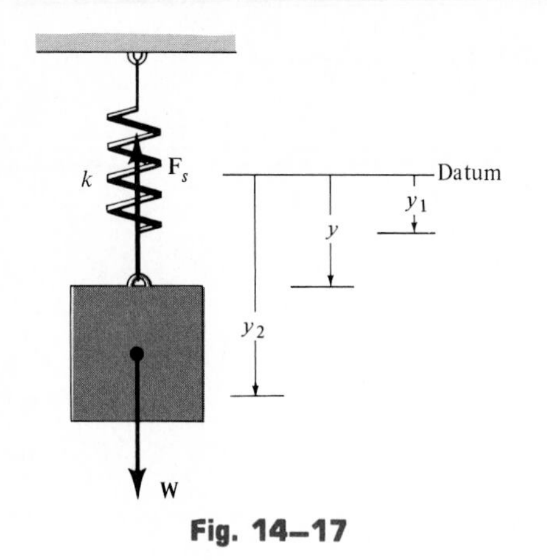

Fig. 14–17

energy function $V = V(x, y, z)$. The work done by a conservative force in moving the particle from point (x_1, y_1, z_1) to point (x_2, y_2, z_2) is then measured by the *difference* of this function; i.e.,

$$U_{1-2} = V_1 - V_2 \tag{14–17}$$

For example, the potential-energy function for a block of weight **W** suspended from a spring can be expressed in terms of its position, y, measured from a datum located at the unstretched length of the spring, Fig. 14–17. We have

$$\begin{aligned} V &= V_g + V_e \\ &= -Wy + \tfrac{1}{2}ky^2 \end{aligned}$$

If the block moves from y_1 to a further downward position y_2, then applying Eq. 14–17 it can be seen that the work of **W** and $\mathbf{F}_s$ is

$$\begin{aligned} U_{1-2} = V_1 - V_2 &= (-Wy_1 + \tfrac{1}{2}ky_1^2) - (-Wy_2 + \tfrac{1}{2}ky_2^2) \\ &= W(y_2 - y_1) - (\tfrac{1}{2}ky_2^2 - \tfrac{1}{2}ky_1^2) \end{aligned}$$

14.6 Conservation of Energy Theorem

When a particle is acted upon by a *system of conservative forces,* the work done by all the forces is $\Sigma U_{1-2} = V_1 - V_2$ (Eq. 14–17). Using the principle of work and energy we can also write $\Sigma U_{1-2} = T_2 - T_1$. Eliminating ΣU_{1-2} from these equations yields

$$T_1 + V_1 = T_2 + V_2 \tag{14–18}$$

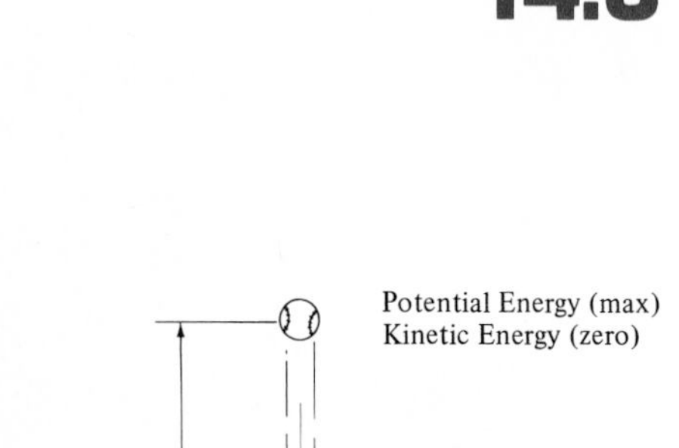

Fig. 14–18

This equation is referred to as the *conservation of energy theorem.* It states that during the motion the sum of the particle's kinetic and potential energy remains *constant.* For example, if a ball of weight **W** is dropped from a height h above the ground (datum), Fig. 14–18, the potential energy of the ball is maximum before it is dropped, at which time its kinetic energy is zero. The total energy of the ball in its initial position is thus

$$E = T_1 + V_1 = 0 + Wh = Wh$$

When the ball has fallen a distance $h/2$, its speed can be determined by using $v^2 = v_0^2 + 2a_c(y - y_0)$, which yields $v = \sqrt{2g(h/2)} = \sqrt{gh}$. The energy of the ball at the mid-height position is therefore

$$E = V_2 + T_2 = W\frac{h}{2} + \frac{1}{2}\frac{W}{g}(\sqrt{gh})^2 = Wh$$

Just before the ball strikes the ground, its potential enegy is zero, and its speed is $v = \sqrt{2gh}$. Here, again, the total energy of the ball is

$$E = V_3 + T_3 = 0 + \frac{1}{2}\frac{W}{g}(\sqrt{2gh})^2 = Wh$$

When the ball comes in contact with the ground, it deforms somewhat, and provided the ground is hard enough, the ball will rebound off the surface, reaching a new height h', which will be less than the height h from which it was first released. The difference in height accounts for an energy loss, $E_l = W(h - h')$, occurring at the moment of collision. Portions of this loss produce noise, deformation of the ball and ground, vibrations, and heat.

System of Particles. An equation similar to Eq. 14–18 can be written for a system of particles, which is based on Eq. 14–7 ($\Sigma T_1 + \Sigma U_{1-2} = \Sigma T_2$). We have

$$\Sigma T_1 + \Sigma V_1 = \Sigma T_2 + \Sigma V_2 \qquad (14\text{–}19)$$

Here, the sum of the system's initial kinetic and potential energies is equal to the sum of the system's final kinetic and potential energies.

PROCEDURE FOR ANALYSIS

The conservation of energy theorem is used to solve problems involving *velocity, displacement,* and *conservative force systems*. For application it is suggested that the following procedure be used.

Potential Energy. Draw two diagrams showing the particle located at its initial and final points along the path. If the particle is subjected to a vertical displacement, establish the fixed horizontal datum from which to measure the particle's gravitational potential energy V_g. Although this position can be selected arbitrarily, it is best to locate the datum either at the initial or final point of the path, since at the datum $V_g = 0$. Data pertaining to the elevation y of the particle from the datum and the extension or compression x of any connecting springs can be determined from the geometry associated with the two diagrams. Recall that the potential energy $V = V_g + V_e$, where $V_g = \pm Wy$ and $V_e = +\frac{1}{2}kx^2$.

Conservation of Energy Theorem. Apply the conservation of energy theorem, $T_1 + V_1 = T_2 + V_2$. When computing the kinetic energy, $T = \frac{1}{2}mv^2$, the particle's speed v must be measured from an inertial reference frame.

It is important to remember that only problems involving conservative force systems may be solved by using the conservation of energy theorem. As stated previously, friction or other drag-resistant forces, which depend upon velocity or acceleration, are nonconservative. The work done by such forces is transformed into thermal energy used to heat up the surfaces of contact, and consequently this energy dissipates into the surroundings and may not be recovered. Therefore, problems involving frictional forces should either be solved by using the principle of work and energy, if it applies, or the equation of motion.

The following example problems numerically illustrate application of the procedure described above.

Example 14–9

The boy and his bicycle shown in Fig. 14–19*a* have a total weight of 125 lb and a center of mass at *G*. If he is coasting, i.e., not pedaling, with a speed of 10 ft/s at the top of the hill *A*, determine the normal force exerted on the wheels of the bicycle when he arrives at *B*, where the radius of curvature of the road is $\rho = 50$ ft. Neglect friction.

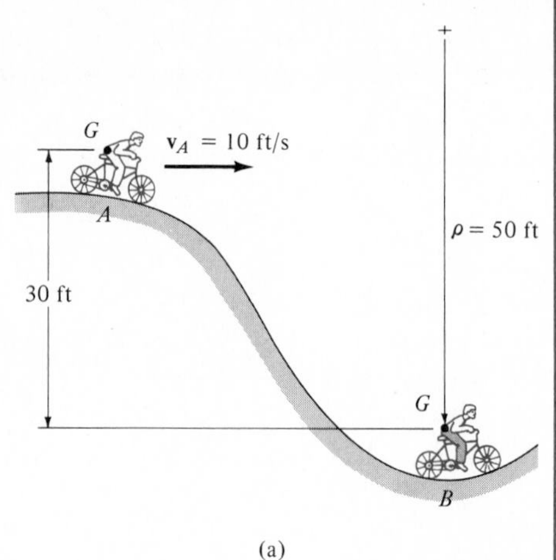

(a)

Solution

Since the normal force does *no work,* it must be obtained using the equation of motion, $\Sigma F_n = m(v^2/\rho)$. We can, however, determine the bicycle's speed at *B* using the conservation of energy theorem.

Potential Energy. Figure 14–19*b* shows the bicycle at points *A* and *B*. For convenience, the potential-energy datum has been established through the center of mass when the bicycle is located at *B*.

Conservation of Energy Theorem

$$\{T_A\} + \{V_A\} = \{T_B\} + \{V_B\}$$

$$\left\{\frac{1}{2}\left(\frac{125}{32.2}\right)(10)^2\right\} + \{(125)(30)\} = \left\{\frac{1}{2}\left(\frac{125}{32.2}\right)(v_B)^2\right\} + \{0\}$$

$$v_B = 45.1 \text{ ft/s}$$

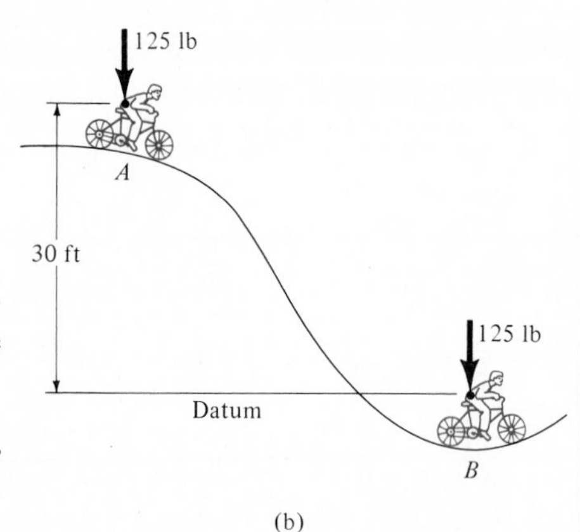

(b)

Equation of Motion. Using the data tabulated on the free-body and kinetic diagrams when the bicycle and rider are at *B*, Fig. 14–19*c*, we have

$$+\uparrow \Sigma F_n = ma_n; \qquad N_B - 125 = \frac{125}{32.2}\frac{(45.1)^2}{50}$$

$$N_B = 282.8 \text{ lb} \qquad \textit{Ans.}$$

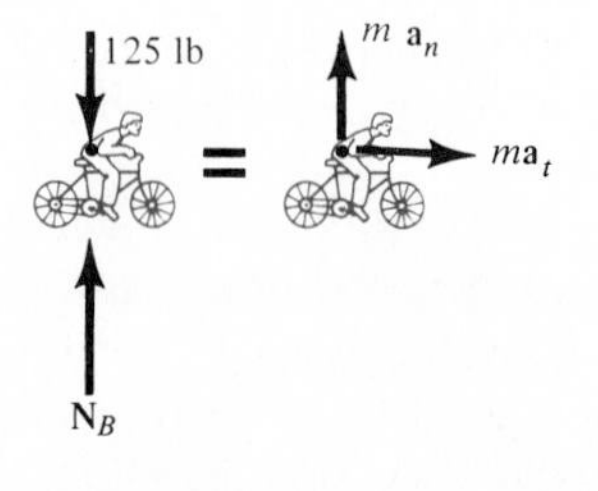

(c)

Fig. 14–19

Example 14–10

The ram R shown in Fig. 14–20a has a mass of 100 kg and is released from rest 0.75 m from the top of a spring A that has a stiffness of k_A = 12 kN/m. If a second spring B, having a stiffness of k_B = 15 kN/m, is "nested" in A, determine the maximum deflection of A needed to stop the downward motion of the ram. The unstretched length of each spring is indicated in the figure.

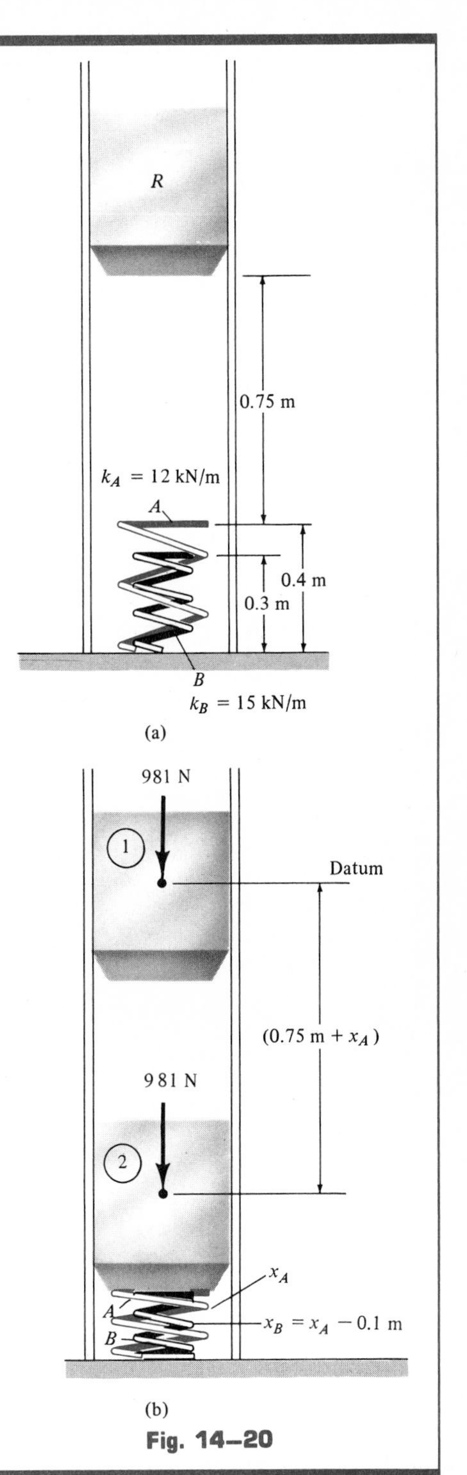

Fig. 14–20

Solution

Potential Energy. We will *assume* that the ram compresses *both* springs at the instant it comes to rest. The datum is located through the center of gravity of the ram at its initial position, Fig. 14–20b. When the kinetic energy is reduced to zero ($v_2 = 0$), A is compressed a distance x_A so that B compresses $x_B = x_A - 0.1$ m.

Conservation of Energy Theorem

$$T_1 + V_1 = T_2 + V_2$$

$$\{0\} + \{0\} = \{0\} + \{\tfrac{1}{2}k_A x_A^2 + \tfrac{1}{2}k_B(x_A - 0.1)^2 - Wh\}$$

$$\{0\} + \{0\} = \{0\} + \{\tfrac{1}{2}(12\ 000)x_A^2 + \tfrac{1}{2}(15\ 000)(x_A - 0.1)^2 - 981(0.75 + x_A)\}$$

Rearranging the terms,

$$13\ 500x_A^2 - 2481x_A - 660.75 = 0$$

Using the quadratic formula, and solving for the positive root,* we have

$$x_A = 0.331 \text{ m} \qquad \textit{Ans.}$$

Since $x_B = 0.331 - 0.1 = 0.231$ m, which is positive, indeed the assumption that *both* springs are compressed by the ram is correct.

*The second root, $x_A = -0.148$ m, does not represent the physical situation. Since positive x is measured downward, the negative sign indicates that spring A would have to be "extended" by an amount of 0.148 m to stop the ram.

Example 14–11

A smooth 2-kg collar C, shown in Fig. 14–21a, fits loosely on the vertical shaft. If the spring is unextended when the collar is in the dashed position A, determine the speed at which the collar is moving when $s = 1$ m, if (a) it is released from rest at A and (b) it is released at A with an *upward* velocity of $v_A = 2$ m/s.

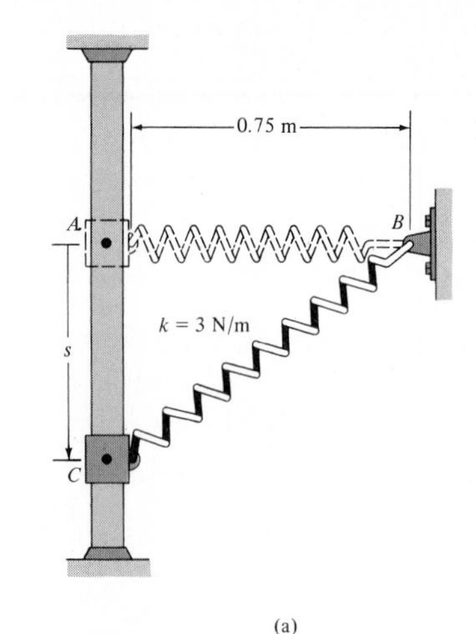

(a)

Solution

Part (a)

Potential Energy. For convenience, the datum is established through AB, Fig. 14–21b. When the collar is at C, the gravitational potential energy is $-(mg)s$, since the collar is *below* the datum, and the elastic potential energy is $\frac{1}{2}kx_{CB}^2$. Here $x_{CB} = 0.5$ m, which represents the *stretch* in the spring as computed in the figure.

Conservation of Energy Theorem.

$$\{T_A\} + \{V_A\} = \{T_C\} + \{V_C\}$$
$$\{0\} + \{0\} = \{\tfrac{1}{2}mv_C^2\} + \{\tfrac{1}{2}kx_{CB}^2 - mgs\}$$
$$\{0\} + \{0\} = \{\tfrac{1}{2}(2)v_C^2\} + \{\tfrac{1}{2}(3)(0.5)^2 - 2(9.81)(1)\}$$
$$v_C = 4.39 \text{ m/s} \qquad \textit{Ans.}$$

Part (b)

Conservation of Energy Theorem. For the case with $v_A = 2$ m/s, using the data in Fig. 14–21b, we have

$$\{T_A\} + \{V_A\} = \{T_C\} + \{V_C\}$$
$$\{\tfrac{1}{2}mv_A^2\} + \{0\} = \{\tfrac{1}{2}mv_C^2\} + \{\tfrac{1}{2}kx_{CB}^2 - mgs\}$$
$$\{\tfrac{1}{2}(2)(2)^2\} + \{0\} = \{\tfrac{1}{2}(2)v_C^2\} + \{\tfrac{1}{2}(3)(0.5)^2 - 2(9.81)(1)\}$$
$$v_C = 4.82 \text{ m/s} \qquad \textit{Ans.}$$

Note that the kinetic energy of the collar depends only on the *magnitude* of velocity, and therefore it is immaterial if the collar is moving up or down at 2 m/s when released at A.

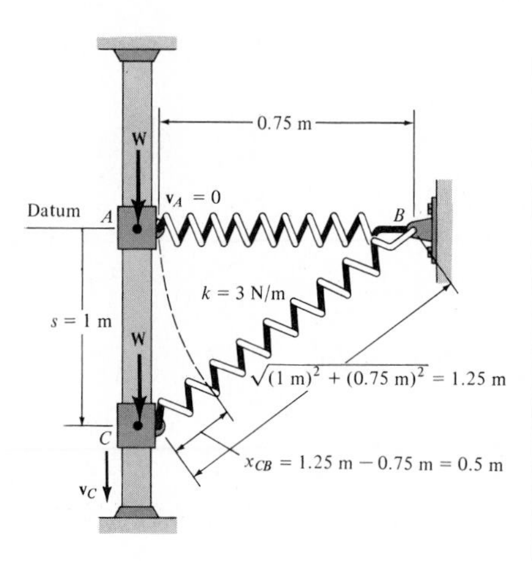

(b)

Fig. 14–21

Problems

***14–32.** Solve Prob. 14–10 using the conservation of energy theorem.

14–33. Solve Prob. 14–11 using the conservation of energy theorem.

14–34. Using the conservation of energy theorem, determine the speed of the car in Prob. 14–18 when it reaches point B.

14–35. The block has a weight of 1.5 lb and slides along the smooth chute AB. It is released from rest at A, which has coordinates of A(5 ft, 0, 10 ft). Determine the speed at which it slides off at B, which has coordinates of B(0, 8 ft, 0).

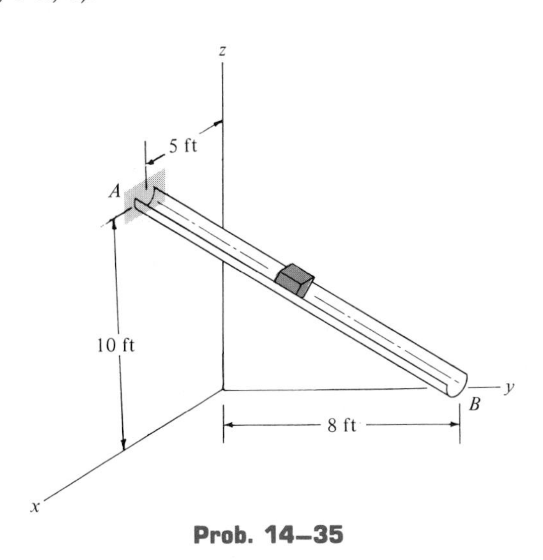

Prob. 14–35

***14–36.** Tarzan has a mass of 100 kg and from rest leaps from the cliff by rigidly holding on to the 10-m-long tree vine, measured from the supporting limb A to his center of gravity. Determine his speed just after the vine strikes the lower limb at B. Also, with what force must he hold on to the vine just before and just after the vine contacts the limb at B? *Hint:* Note that the radius of curvature of his path changes from 10 m to 4 m at this point.

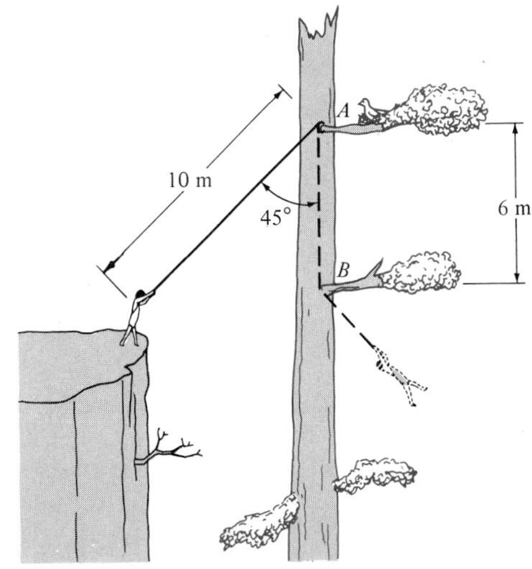

Prob. 14–36

14–37. The roller-coaster car has a mass of 800 kg, including its passenger, and moves from the top of the hill A with a speed of $v_A = 3$ m/s. Determine the minimum height h of the hill crest so that the car travels around both inside loops without leaving the track. Neglect friction and the size of the car. What is the normal reaction on the car when the car is at B and when it is at C?

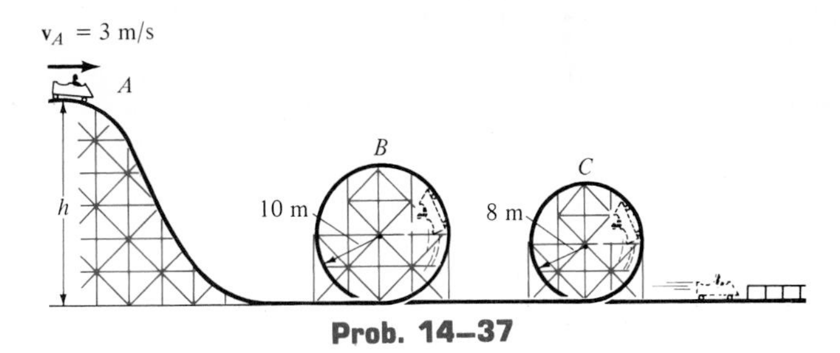

Prob. 14–37

14–38. The firing mechanism of a pinball machine consists of a plunger P having a mass of 0.25 kg and a spring of stiffness $k = 300$ N/m. When $s = 0$, the spring is not compressed. If the arm is pulled back such that $s = 100$ mm and released, determine the speed of the 0.3-kg pinball B *just before* the plunger strikes the stop, i.e., $s = 0$. Assume

Potential Energy of Spring.... $\frac{1}{2}kx^2$

all surfaces of contact to be smooth. The ball moves in the horizontal plane. Neglect friction and the rolling motion of the ball.

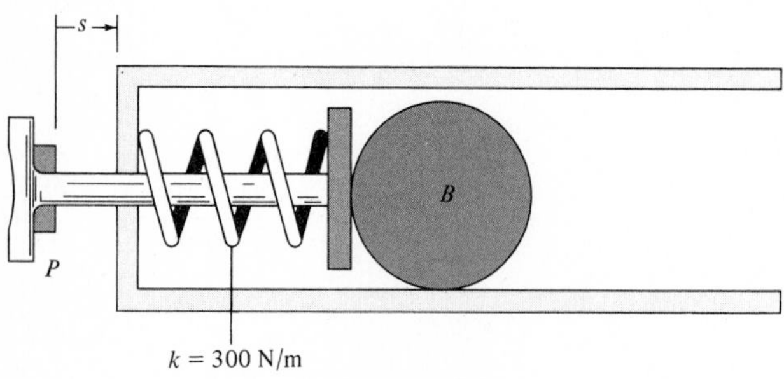

Prob. 14–38

14–39. A block having a mass of 20 kg is attached to four springs. If each spring has a stiffness of $k = 2$ kN/m and an unstretched length of 100 mm, determine the *maximum* downward vertical displacement s_{max} of the block if it is released from rest when $s = 0$.

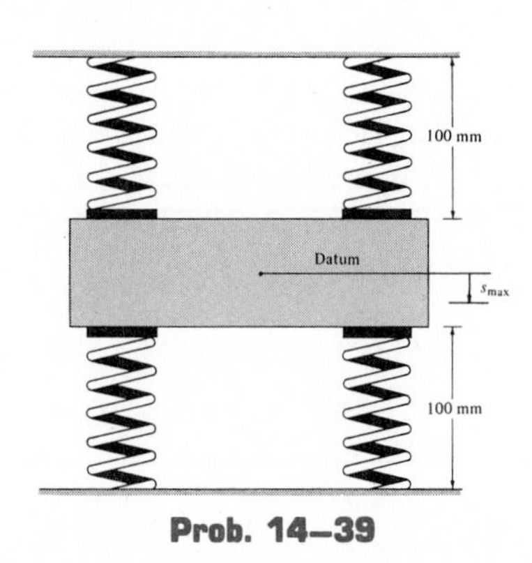

Prob. 14–39

***14–40.** Four inelastic cables C are attached to a plate P and hold the spring 6 in. in compression when *no force* acts on the plate. If a block B, having a weight of 5 lb, is placed on the plate and the plate is pushed down 8 in. and released from rest, determine how high the block rises from the point where it was released. Neglect the mass of the plate. *Suggestion:* See Example 14–4.

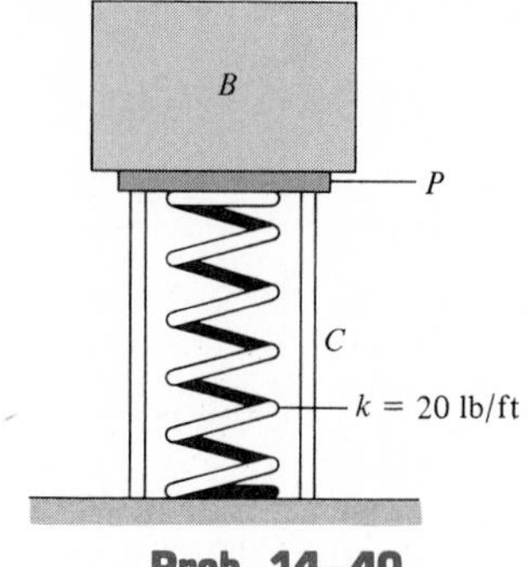

Prob. 14–40

14–41. The block A having a weight of 1.5 lb slides in the smooth horizontal slot. If the block is drawn back so that $s = 1.5$ ft and released from rest, determine its speed at the instant $s = 0$. Each of the two springs has a stiffness of $k = 150$ lb/ft and an unstretched length of 2 ft.

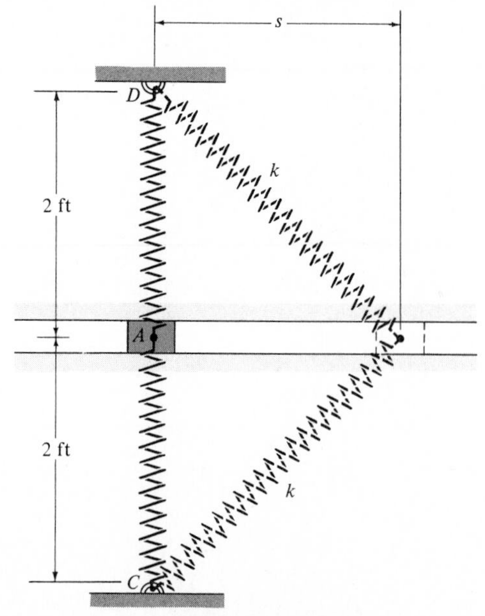

Prob. 14–41

14–42. The roller-coaster car has a speed of 15 ft/s when it is at the crest of a vertical parabolic track. Compute the velocity and the normal force it exerts on the track when it reaches point B. Neglect friction and the mass of the wheels. The total weight of the car and the passengers is 350 lb.

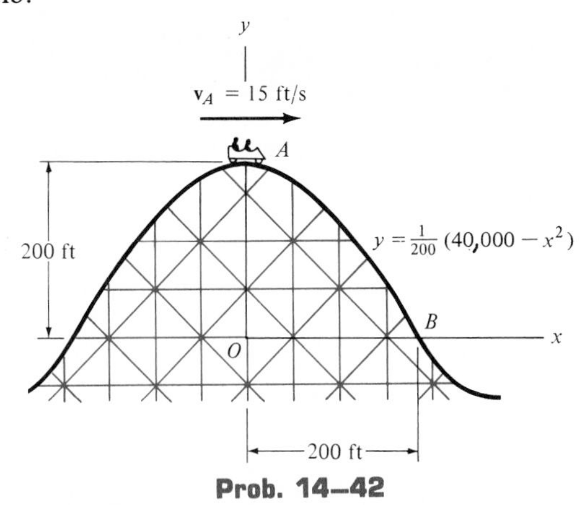

Prob. 14–42

14–43. The car C and its contents have a weight of 600 lb, whereas block B has a weight of 200 lb. If the car is released from rest, determine its speed when it travels 30 ft down the 20° incline. *Suggestion:* To measure the gravitational potential energy, establish separate datums at the initial elevations of B and C.

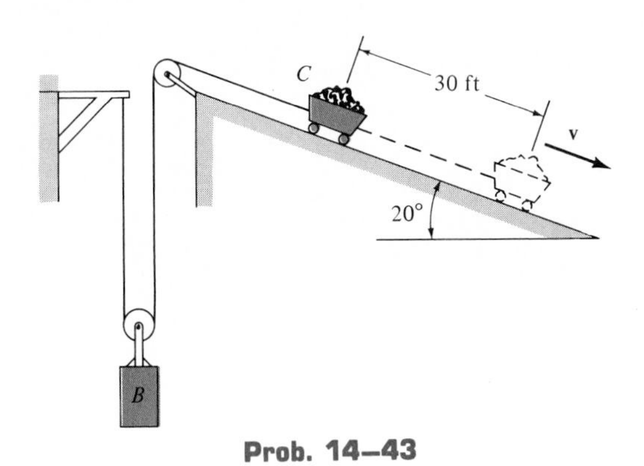

Prob. 14–43

***14–44.** If the mass of the earth is M_e, show that the gravitational potential energy of a body of mass m located a distance r from the center of the earth is $V_g = -GM_em/r$. Recall that the gravitational force acting between the earth and the body is $F = G(M_em/r^2)$ (Eq. 13–1). For the calculation, locate the datum at $r \to \infty$. Also, prove that $\mathbf{F}$ is a conservative force.

14–45. A 70-kg satellite is traveling in free flight along an elliptical orbit such that at A, where $r_A = 20$ Mm, it has a speed of $v_A = 60$ Mm/h. What is the speed of the satellite when it reaches point B, where $r_B = 80$ Mm? *Hint:* See Prob. 14–44, where $M_e = 5.976(10^{24})$ kg and $G = 6.673(10^{-11})\ \text{m}^3/(\text{kg} \cdot \text{s}^2)$.

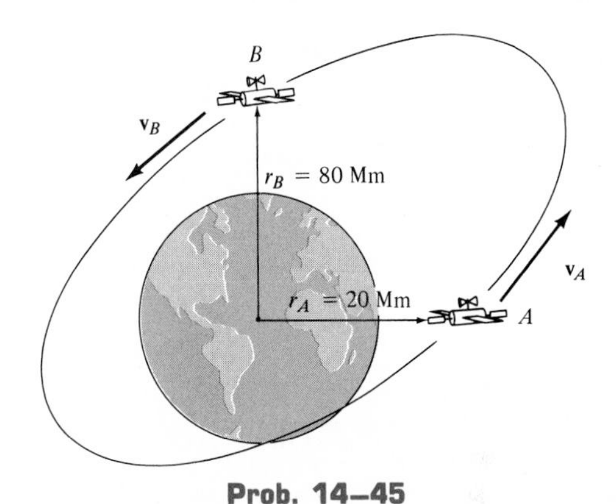

Prob. 14–45

14–46. A toboggan and its two riders have a total mass of 225 kg. Determine the greatest initial speed v_A the toboggan can have at A so that it arrives at C in the shortest time without leaping off the path. The path consists of a 30° sector of a circular arc which has an inflection point at B. Neglect friction. *Hint:* The greatest speed is reached when the toboggan tends to leave the path at B, where the radius of curvature is still 10 m.

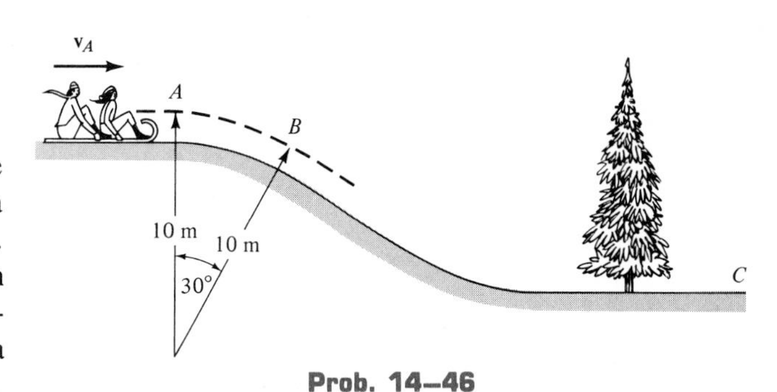

Prob. 14–46

14–47. A tank car is stopped by two spring bumpers A and B, having a stiffness of $k_A = 1.5(10^4)$ lb/ft and $k_B = 2.0(10^4)$ lb/ft, respectively. Bumper A is attached to the car, whereas bumper B is attached to the wall. If the car has a weight of $2.5(10^4)$ lb and is freely coasting at $v_c = 3$ ft/s, compute the maximum deflection of each spring at the instant the bumpers stop the car.

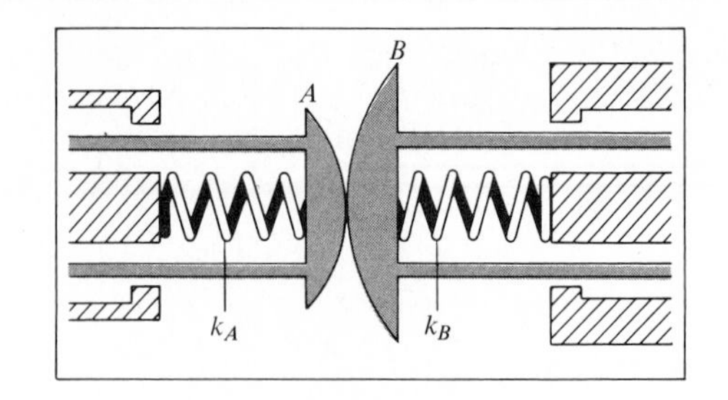

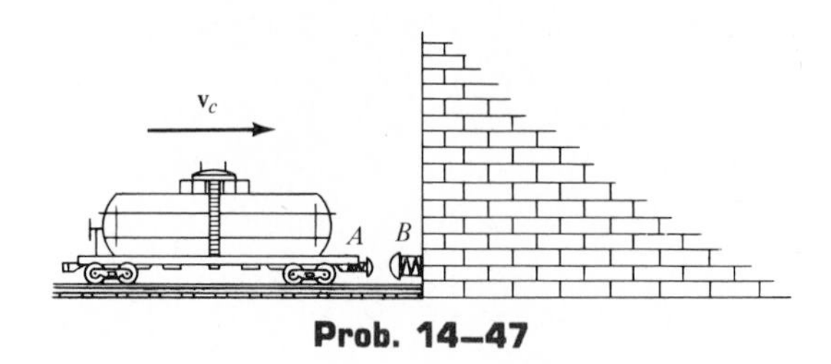

Prob. 14–47

Review Problems

***14–48.** An automobile and driver have a total mass of 1500 Mg. When the automobile is traveling at 20 m/s, the brakes are applied and it is observed that the car slides 30 m before stopping. How far will the car slide if it is traveling at 35 m/s?

14–49. A block having a weight of 20 lb rests on a smooth horizontal surface. If it starts from rest and is acted upon by a horizontal force which varies in magnitude as shown in the graph, determine its approximate speed at the instant it moves 10 ft.

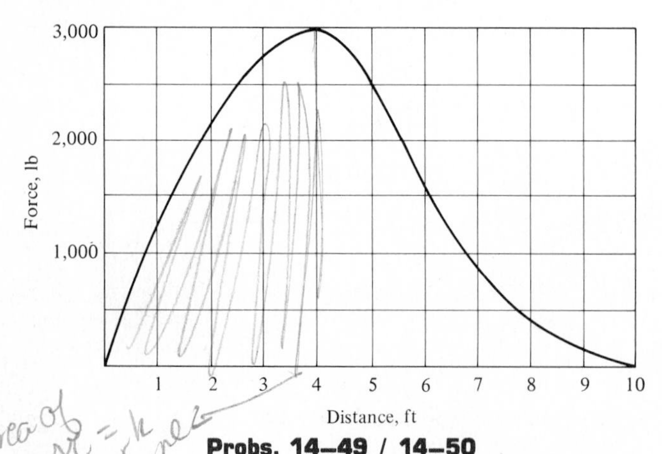

Probs. 14–49 / 14–50

14–50. A block having a weight of 20 lb rests on a smooth horizontal surface. If it starts from rest and is acted upon by a horizontal force which varies in magnitude as shown in the graph, determine its approximate speed when it moves 4 ft.

14–51. An electrically powered train engine draws 30 kW of power from the tracks. If the engine weighs 40 kips, determine the maximum speed it attains in 30 s starting from rest. The mechanical efficiency is $\varepsilon = 0.8$.

***14–52.** An electrically powered train engine draws power from the tracks. If the engine has a mass of 40 Mg, determine the power required if the wind resistance is $F_w = 800$ N when the train is traveling at a constant speed of 20 m/s. The mechanical efficiency is $\varepsilon = 0.8$.

14–53. Determine the smallest amount the spring at B must be compressed against the ball so that when it is released from B it reaches point A. The ball weighs 0.5 lb and its size can be neglected.

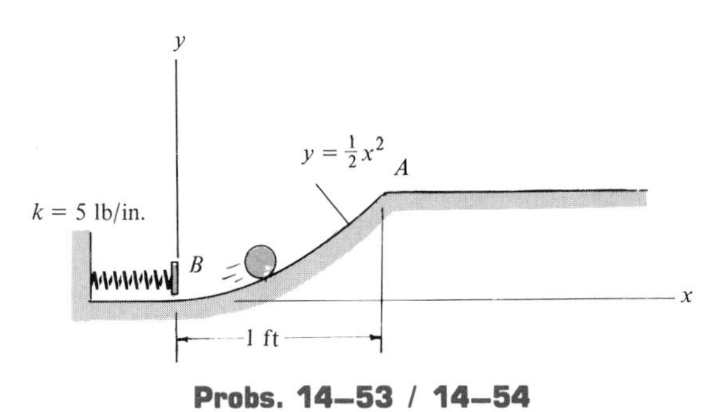

Probs. 14–53 / 14–54

14–54. If the spring is compressed 3 in. against the ball and it is released from rest, determine the approximate horizontal distance measured from A to where it strikes the horizontal plane. The ball weighs 0.5 lb and its size can be neglected.

14–55. The 2-lb smooth collar C is given a velocity of 6 ft/s to the right when $s = 3$ ft. Determine its velocity at the instant $s = 0$. The spring has an unstretched length of 2 ft.

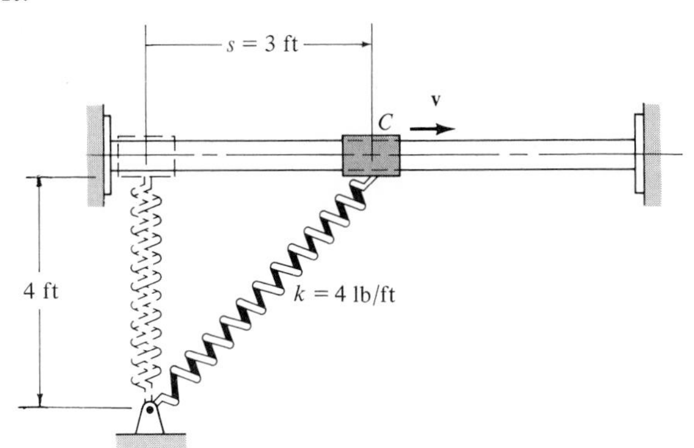

Probs. 14–55 / 14–56

***14–56.** If the 2-lb smooth collar is given a velocity of 6 ft/s to the right when $s = 3$ ft, determine the maximum distance, s_{max}, the collar travels before stopping. The spring has an unstretched length of 2 ft.

14–57. The 30-lb box A is released from rest and slides down along the *smooth* ramp and onto the surface of a cart. If the cart is *fixed from moving,* determine the distance s from the end of the cart to where the box stops. The coefficient of friction between the cart and the box is $\mu = 0.6$.

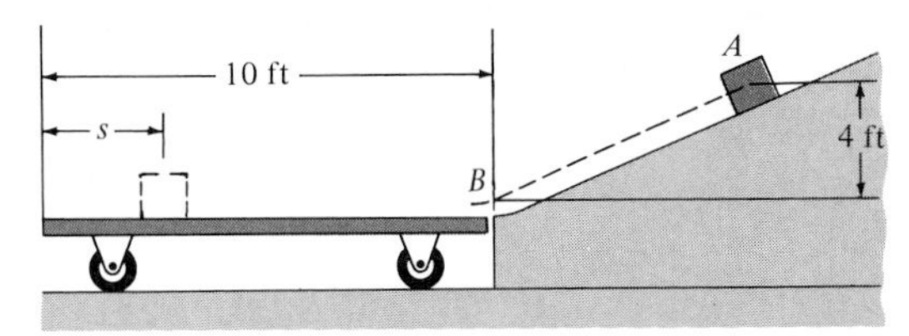

Prob. 14–57

14–58. If the escalator in Prob. 14–59 is *not moving,* determine the required amount of time in which a man having a weight of 150 lb must walk up the steps to generate 100 W of power—the same amount that is needed to power a standard light bulb.

14–59. The escalator steps move with a constant speed of $v = 2$ ft/s. If the steps are 8 in. high and 15 in. in length, determine the horsepower output of a motor needed to lift an average weight of 180 lb per step. There are 30 steps.

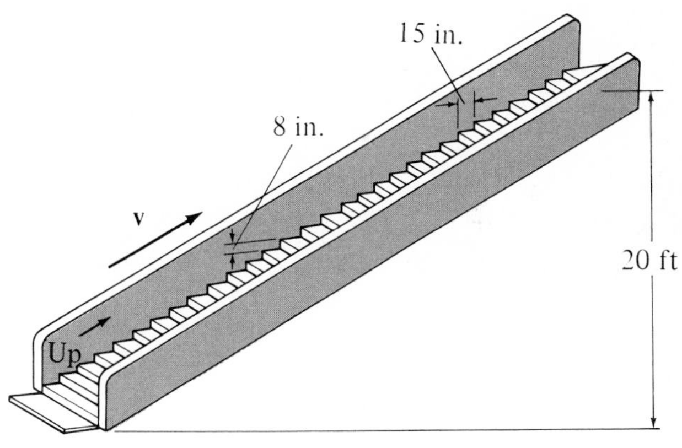

Prob. 14–59

***14–60.** Determine the theoretical escape velocity **v** of a 2-kg particle projected from the earth's surface. Such a velocity would allow the particle to travel an infinite distance before its velocity is decreased to zero. Assume the earth has a radius of 6378 km and neglect air resistance. *Hint:* Use the result of Prob. 14–44.

Kinetics of a Particle: Impulse and Momentum

In this chapter we will develop an integrated formulation of the equation of motion which is useful for solving problems involving force, velocity, and time. The resulting equation is called the principle of impulse and momentum. We will first apply this principle to problems involving linear or straight-line motion, then use it to study motion along a curved path. It will also be shown that impulse and momentum principles provide an important means for analyzing problems involving impact.

15.1 Principle of Linear Impulse and Momentum

Consider a particle of mass m which is subjected to several forces. The equation of motion for the particle can be written as

$$\Sigma \mathbf{F} = m\mathbf{a} = m\frac{d\mathbf{v}}{dt} \tag{15–1}$$

where **a** and **v** indicate the particle's instantaneous acceleration and velocity, respectively. Rearranging the terms and integrating between the limits $\mathbf{v} = \mathbf{v}_1$ at $t = t_1$ and $\mathbf{v} = \mathbf{v}_2$ at $t = t_2$, we have

$$\Sigma \int_{t_1}^{t_2} \mathbf{F}\, dt = m \int_{\mathbf{v}_1}^{\mathbf{v}_2} d\mathbf{v}$$

or

$$\Sigma \int_{t_1}^{t_2} \mathbf{F}\, dt = m\mathbf{v}_2 - m\mathbf{v}_1 \tag{15–2}$$

This equation, which is referred to as the *principle of linear impulse and momentum,* provides a *direct means* of obtaining the particle's final velocity $\mathbf{v}_2$ after a specified time period when the particle's initial velocity is known and the forces acting on the particle are either constant or can be expressed as functions of time. Notice from the derivation that if $\mathbf{v}_2$ is determined using the equation of motion, a two-step process is necessary; i.e., apply $\Sigma\mathbf{F} = m\mathbf{a}$ to obtain $\mathbf{a}$, then integrate $\mathbf{a} = d\mathbf{v}/dt$ to obtain $\mathbf{v}_2$.

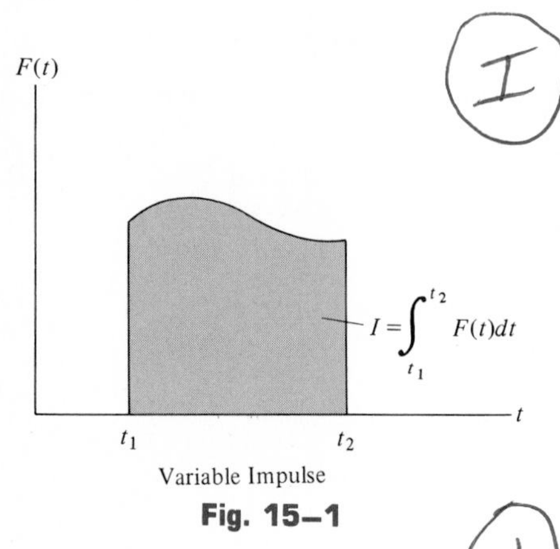

Variable Impulse

Fig. 15–1

Linear Impulse. The integral $\mathbf{I} = \int \mathbf{F}\,dt$ in Eq. 15–2 is defined as the *linear impulse*. This term is a vector quantity which measures the effect of a force during the time the force acts. The impulse vector acts in the same direction as the force, and its magnitude has units of force–time, e.g., N · s or lb · s. If the force is expressed as a function of time, the impulse may be determined by direct evaluation of the integral. In particular, if $\mathbf{F}$ acts in a *constant direction* during the time period t_1 to t_2, the magnitude of the impulse $\mathbf{I} = \int_{t_1}^{t_2} \mathbf{F}\,dt$ can be represented experimentally by the shaded area under the curve of force versus time, Fig. 15–1. If the force is constant in magnitude and direction, the resulting impulse becomes $\mathbf{I} = \int_{t_1}^{t_2} \mathbf{F}_c\,dt = \mathbf{F}_c(t_2 - t_1)$, which represents the shaded rectangular area shown in Fig. 15–2.

Linear Momentum. Each of the two vectors of the form $\mathbf{L} = m\mathbf{v}$ in Eq. 15–2 is defined as the *linear momentum* of the particle. Since m is a scalar, the linear-momentum vector has the same direction as $\mathbf{v}$, and its magnitude mv has units of mass–velocity, e.g., kg · m/s, slug · ft/s.

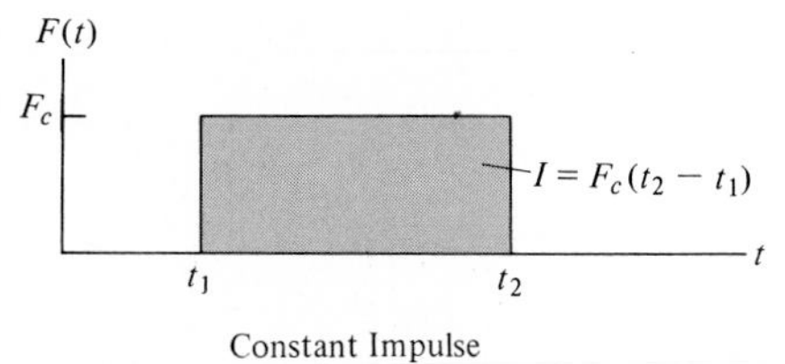

Constant Impulse

Fig. 15–2

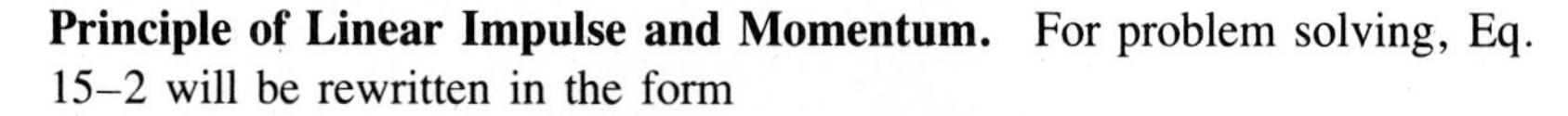

Principle of Linear Impulse and Momentum. For problem solving, Eq. 15–2 will be rewritten in the form

$$m\mathbf{v}_1 + \Sigma\int_{t_1}^{t_2} \mathbf{F}\,dt = m\mathbf{v}_2 \qquad (15\text{–}3)$$

which states that the initial momentum of the particle at t_1 plus the vector sum of all the impulses applied to the particle during the time interval t_1 to t_2 is equivalent to the final momentum of the particle at t_2. These three terms are illustrated graphically on the *impulse and momentum diagrams* shown in Fig. 15–3.

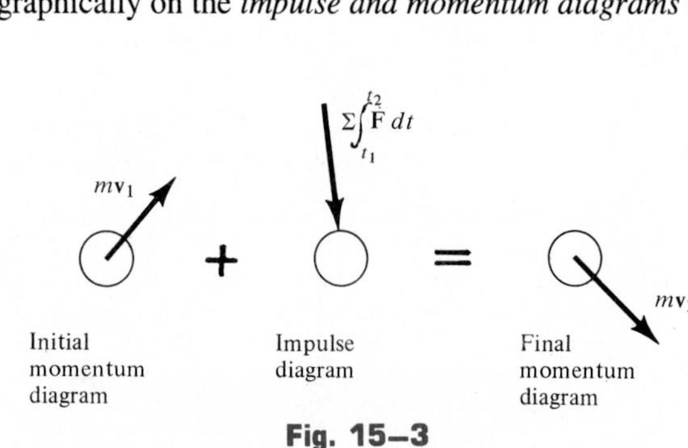

Fig. 15–3

Scalar Equations. For motion in the x-y plane each of the vectors in Eq. 15–3 can be resolved into its x and y components. Hence we can write the following two scalar equations:

$$m(v_x)_1 + \Sigma \int_{t_1}^{t_2} F_x \, dt = m(v_x)_2$$
$$m(v_y)_1 + \Sigma \int_{t_1}^{t_2} F_y \, dt = m(v_y)_2 \qquad (15\text{–}4)$$

These equations represent the principle of linear impulse and momentum for the particle in the x and y directions, respectively.

PROCEDURE FOR ANALYSIS

The principle of linear impulse and momentum is used to solve problems involving *force, time,* and *velocity,* since these terms are involved in the formulation. For application it is suggested that the following procedure be used.

Impulse and Momentum Diagrams. Establish the x, y, z axes and draw the impulse and momentum diagrams for the particle. Each of these diagrams graphically accounts for all the vectors in the equation $m\mathbf{v}_1 + \Sigma \int_{t_1}^{t_2} \mathbf{F} \, dt = m\mathbf{v}_2$. The two *momentum diagrams* are simply outlined shapes of the particle which indicate the direction and magnitude of the particle's initial and final momentum, $m\mathbf{v}_1$ and $m\mathbf{v}_2$, respectively, Fig. 15–3. Similar to the free-body diagram, the *impulse diagram* is an outlined shape of the particle showing all the impulses that act on the particle when it is located at some intermediate point along its path. In general, whenever the magnitude or direction of a force *varies,* the impulse of the force is determined by integration and represented on the impulse diagram as $\mathbf{I} = \int_{t_1}^{t_2} \mathbf{F} \, dt$. If the force is *constant* for the time interval $(t_2 - t_1)$, the impulse applied to the particle is $\mathbf{I} = \mathbf{F}_c(t_2 - t_1)$, acting in the same direction as $\mathbf{F}_c$.

Principle of Impulse and Momentum. Apply the principle of linear impulse and momentum, $m\mathbf{v}_1 + \Sigma \int_{t_1}^{t_2} \mathbf{F} \, dt = m\mathbf{v}_2$. If motion occurs in the x-y plane, the two scalar component equations can be formulated by resolving the vector components *directly* from the impulse and momentum diagrams.

The following examples numerically illustrate application of this procedure.

Example 15–1

The 100-kg crate shown in Fig. 15–4*a* is originally at rest on the smooth horizontal surface. If a force $F = 200$ N, acting at an angle of $\theta = 45°$, is applied to the crate for 10 s, determine the final velocity of the crate and the normal force which the surface exerts on the crate during the time interval.

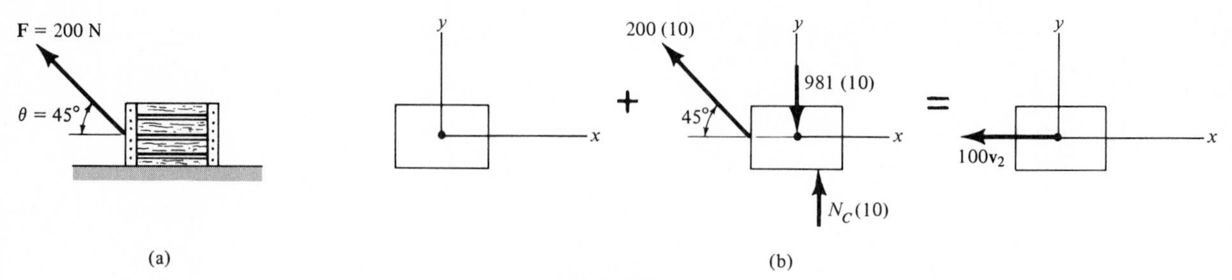

Fig. 15–4

Solution

Impulse and Momentum Diagrams. These three diagrams are shown in Fig. 15–4*b*. Here it has been assumed that during the motion the crate remains on the surface and after 10 s the crate moves to the left with a velocity $\mathbf{v}_2$. Since all the forces acting on the crate are *constant*, the respective impulses are simply the product of the force magnitude and 10 s $(\mathbf{I} = \mathbf{F}_c(t_2 - t_1))$.

Principle of Impulse and Momentum. Resolving the vectors in Fig. 15–4*b* along the *x*,*y* axes, we have

$(\overset{+}{\leftarrow})$

$$m(v_x)_1 + \Sigma \int_{t_1}^{t_2} F_x\, dt = m(v_x)_2$$

$$0 + 200(10)\cos 45° = 100v_2$$

$$v_2 = 14.1 \text{ m/s} \qquad \textit{Ans.}$$

$(+\uparrow)$

$$m(v_y)_1 + \Sigma \int_{t_1}^{t_2} F_y\, dt = m(v_y)_2$$

$$0 + N_C(10) - 981(10) + 200(10)\sin 45° = 0$$

$$N_C = 839.6 \text{ N} \qquad \textit{Ans.}$$

Since no motion occurs in the *y* direction, direct application of $\Sigma F_y = 0$ gives the same result for N_C.

Example 15–2

The crate shown in Fig. 15–5*a* has a weight of 50 lb and is acted upon by a force having a variable magnitude of $P = (20t)$ lb, where t is in seconds. Compute the crate's velocity 2 s after **P** has been applied. The crate has an initial velocity of $v_1 = 3$ ft/s down the plane and the coefficient of kinetic friction between the crate and the plane is $\mu_k = 0.3$.

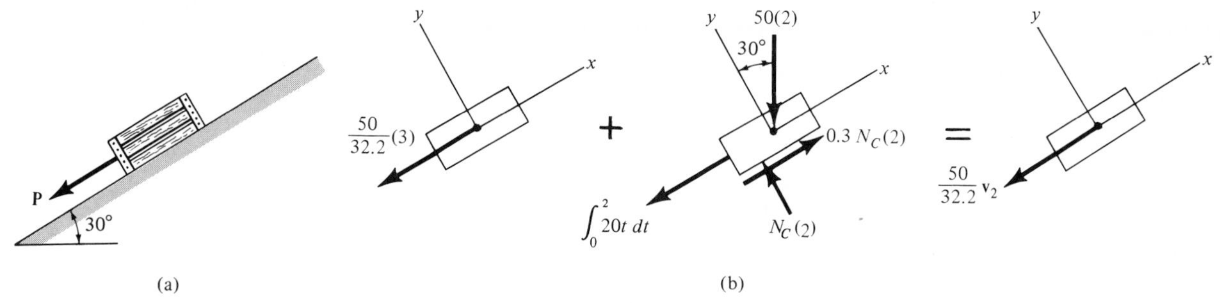

Fig. 15–5

Solution

Impulse and Momentum Diagrams. Since the magnitude of force $P = 20t$ *varies* with time, the impulse, Fig. 15–5*b*, must be determined by *integrating* over the 2-s time interval. The weight, normal force, and frictional force (which acts opposite to the direction of motion) are all *constant,* so that the impulse created by each of these forces is simply the magnitude of the force times 2 s.

Principle of Impulse and Momentum. Summing the vectors shown in Fig. 15–5*b* in the x direction, we have

$$(+\nearrow) \qquad m(v_x)_1 + \Sigma \int_{t_1}^{t_2} F_x \, dt = m(v_x)_2$$

$$-\frac{50}{32.2}(3) - \int_0^2 20t \, dt + 0.3N_C(2) - (50)(2) \sin 30° = -\frac{50}{32.2} v_2$$

$$-4.66 - 40 + 0.6N_C - 50 = -1.55v_2$$

The equation of equilibrium can be applied in the y direction. Why?

$$+\nwarrow \Sigma F_y = 0; \qquad N_C - 50 \cos 30° = 0$$

Solving,

$$N_C = 43.3 \text{ lb} \qquad v_2 = 44.2 \text{ ft/s} \qquad \textit{Ans.}$$

This problem has also been solved using the equation of motion in Example 13–3. The two methods of solution should be compared. Since *force, velocity,* and *time* are involved in the problem, application of the principle of impulse and momentum eliminates the need for using kinematics ($a = dv/dt$) and thereby yields an easier method for solution.

Example 15–3

Blocks A and B shown in Fig. 15–6a have a mass of 3 kg and 5 kg, respectively. If the system is released from rest, determine the velocity of block B in 6 s. Neglect the mass of the pulleys and cord.

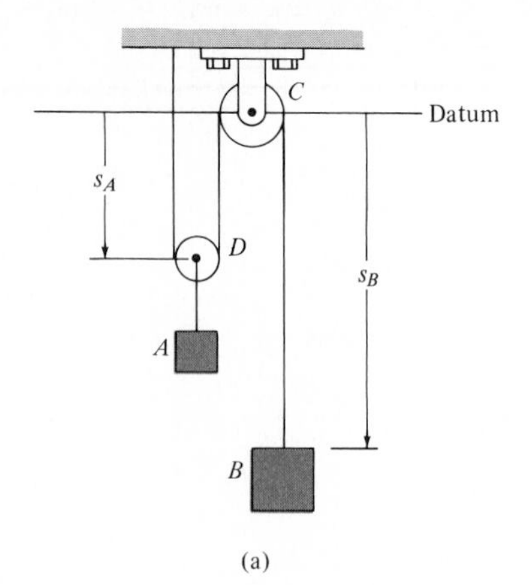

(a)

Solution

Impulse and Momentum Diagrams. The diagrams for each block are shown in Fig. 15–6b. Since the impulses of the blocks' weights are constant, the impulses of the cord tensions are also constant. Furthermore, since the mass of pulley D is neglected, the cord tension $T_A = 2T_B$, Fig. 15–6c.

Principle of Impulse and Momentum

Block A:

$$(+\uparrow) \qquad m(v_A)_1 + \Sigma \int_{t_1}^{t_2} F_y\, dt = m(v_A)_2$$

$$0 + 2T_B(6) - 3(9.81)(6) = -3(v_A)_2 \qquad (1)$$

Block B:

$$(+\downarrow) \qquad m(v_B)_1 + \Sigma \int_{t_1}^{t_2} F_y\, dt = m(v_B)_2$$

$$0 + 5(9.81)(6) - T_B(6) = 5(v_B)_2 \qquad (2)$$

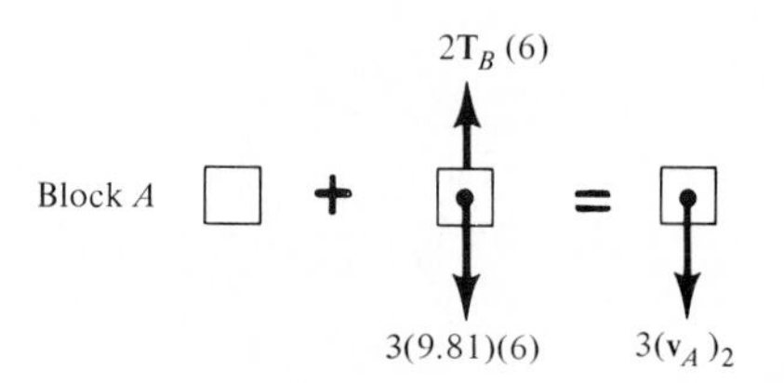

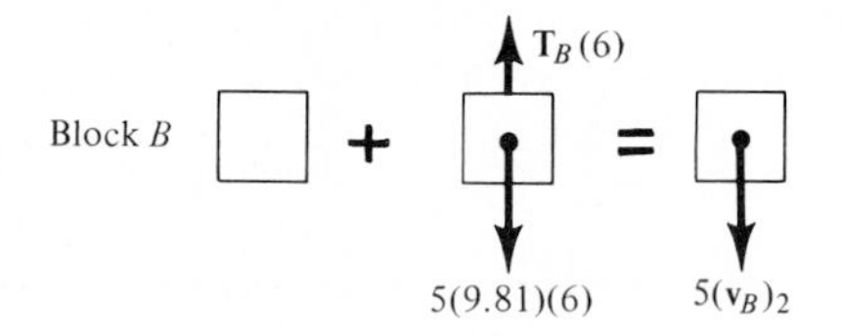

(b)

Kinematics. Since the blocks are subjected to dependent motion, the velocity of A may be related to that of B by using the kinematic analysis discussed in Sec. 12–7. A horizontal datum is established through the fixed point at C, Fig. 15–6a, and the changing positions of the blocks, s_A and s_B, are related to the constant total length l of the vertical segments of the cord by the equation

$$2s_A + s_B = l$$

Taking the time derivative yields

$$2v_A = -v_B \qquad (3)$$

As indicated by the negative sign, when B moves downward A moves upward.* Substituting this result into Eq. (1) and solving Eqs. (1) and (2) yields

$$(v_B)_2 = 35.8 \text{ m/s} \qquad \textit{Ans.}$$

$$T_B = 19.2 \text{ N}$$

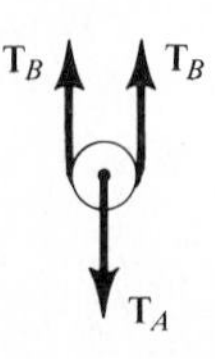

(c)

Fig. 15–6

*Note that the *positive* (downward) directions for $\mathbf{v}_A$ and $\mathbf{v}_B$ are *consistent* in Fig. 15–6a and b and in Eqs. (1) to (3).

HINT: WATCH THE UNITS

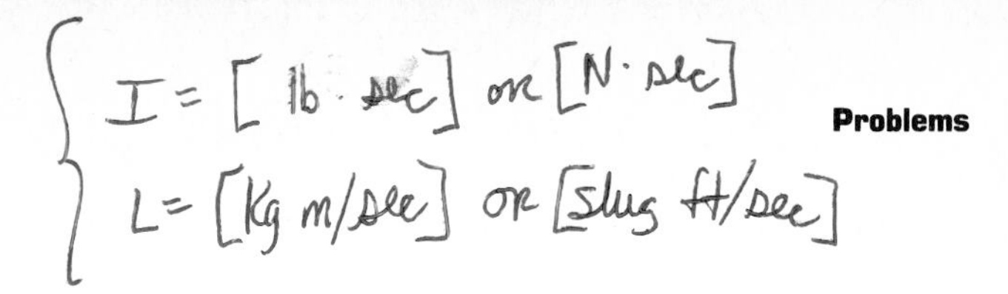

Problems

Except when stated otherwise, throughout this chapter assume that the coefficients of static and kinetic friction are equal, i.e., $\mu = \mu_s = \mu_k$.

15–1. A hammer head H having a weight of 0.3 lb is moving vertically downward at 40 ft/s when it strikes the head of a nail of negligible mass and drives it into a block of wood. Find the impulse on the nail if it is assumed that the grip at A is loose, the handle has a negligible mass, and the hammer stays in contact with the nail while it comes to rest.

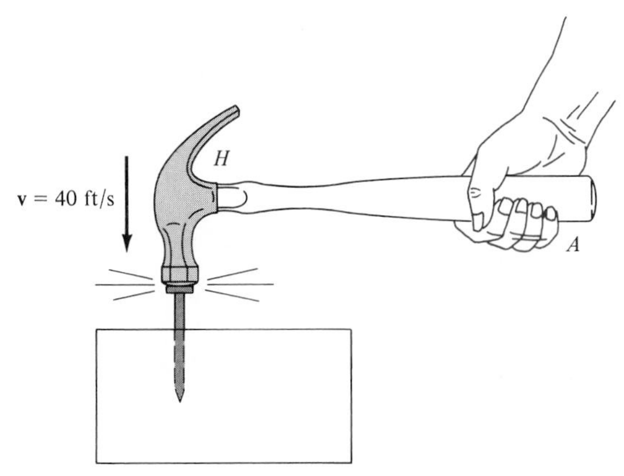

Prob. 15–1

15–2. A cannon ball having a weight of 20 lb is fired upward, in the vertical direction, with a muzzle velocity of 1,500 ft/s. Determine how long it takes before its velocity is reduced to zero, which occurs when it reaches its maximum height. Use the principle of impulse and momentum.

15–3. A car has a weight of 2,500 lb and is traveling forward horizontally at 60 ft/s. If the coefficient of friction between the tires and the pavement is $\mu = 0.6$, determine the time needed to stop the car. The brakes are applied to all four wheels.

***15–4.** A golf ball having a mass of 40 g is struck such that it has an initial velocity of 200 m/s as shown. Determine the horizontal and vertical components of the impulse given to the ball.

Prob. 15–4

15–5. When the 0.4-lb football is kicked, it leaves the ground at an angle of 40° from the horizontal and strikes the ground at the same elevation a distance of 130 ft away. Determine the impulse given to the ball.

Prob. 15–5

15–6. A solid-fueled rocket can be made using a fuel grain with either a hole (a) or starred cavity (b) in the cross section. From experiment the engine thrust-time curves (T vs. t) for the same amount of propellent using these geometries is shown. Determine the total impulse in both cases. Realizing that the mass of the fuel is expelled as it is burned, explain which geometry for the propellent is best for obtaining the maximum range when used in a rocket and why.

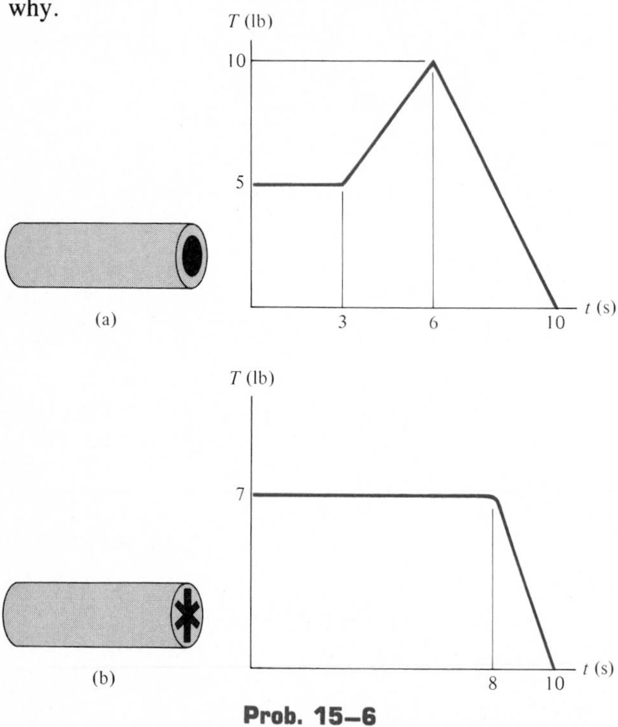

Prob. 15–6

15–7. Packages having a mass of 6 kg slide down a smooth chute and land horizontally with a speed of 3 m/s on the surface of a conveyor belt. If the coefficient of friction between the belt and a package is $\mu = 0.2$, determine the time needed to bring the package to rest on the belt if the belt is moving in the same direction as the package with a speed of $v = 1$ m/s.

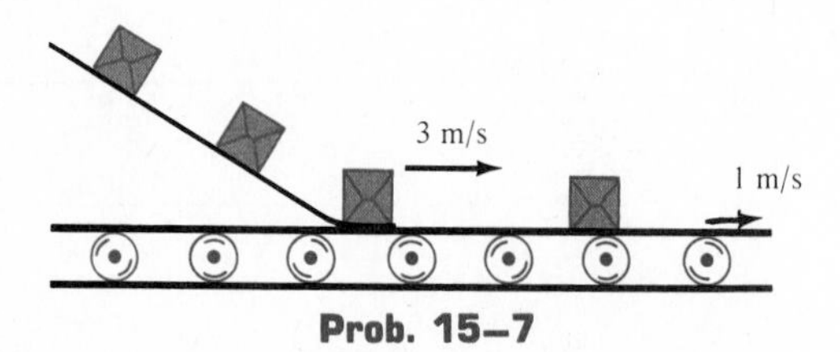

Prob. 15–7

***15–8.** A train consists of a 45-Mg engine and three cars, each having a mass of 30 Mg. If it takes 60 s for the train to increase its speed uniformly to 40 km/h, starting from rest, determine the force **T** developed at the coupling between the engine E and the first car A. The wheels of the engine provide the resultant frictional tractive force **F** which gives the train forward motion, whereas the car wheels roll freely. Determine **F** acting on the engine wheels.

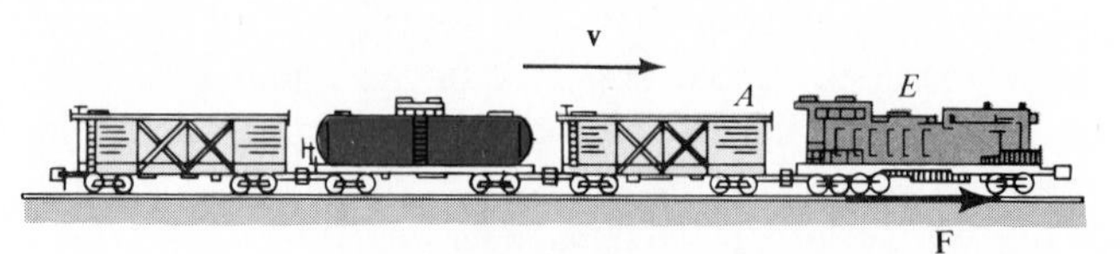

Prob. 15–8

15–9. Assuming that the impulse acting on a 2.5-g bullet, as it passes horizontally through the barrel of a rifle, is constant, determine this force $\mathbf{F}_O$ applied to the bullet when it is fired. The muzzle velocity is 450 m/s when $t = 0.75$ ms. Neglect friction between the bullet and the rifle barrel.

15–10. A hockey puck is traveling to the left with a velocity of $v_1 = 10$ m/s when it is struck by a hockey stick and given a velocity of $v_2 = 20$ m/s as shown. Determine the magnitude of the net impulse exerted by the hockey stick on the puck. The puck has a mass of 0.2 kg.

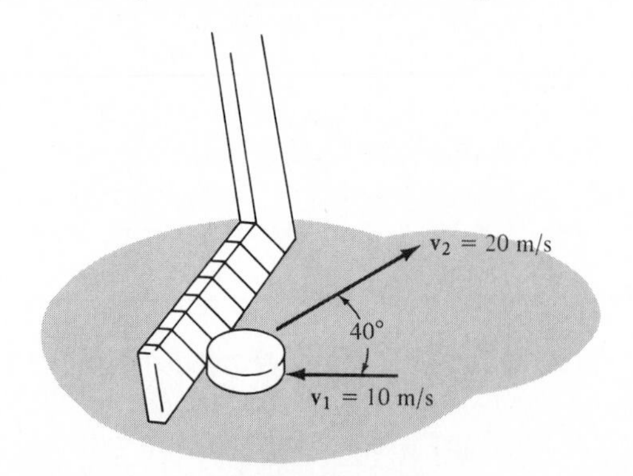

Prob. 15–10

15–11. The fuel-element assembly of a nuclear reactor has a weight of 600 lb. Suspended in a vertical position and initially at rest, it is given an upward speed of 5 ft/s in 0.3 s using a crane hook H. Determine the average tension in cables AC and AB during this time interval.

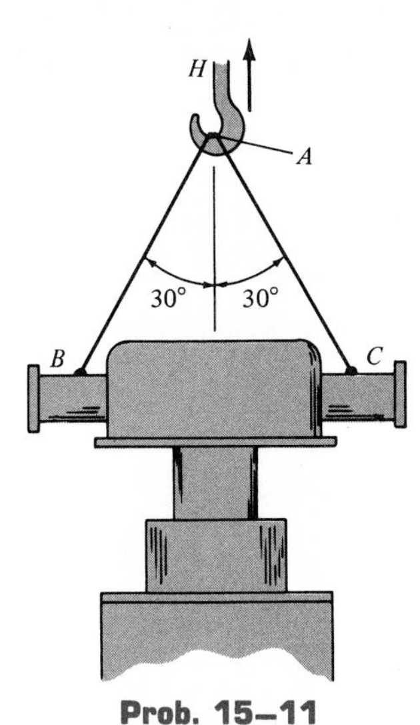

Prob. 15–11

***15–12.** As a 4-lb sphere falls vertically from rest through a liquid, the drag force exerted on it is $F_D = 2$ lb. Determine the time required for the sphere to attain a velocity of 8 m/s.

15–13. Determine the velocities of blocks A and B 2 s after they are released from rest. Neglect the mass of the pulleys and cables.

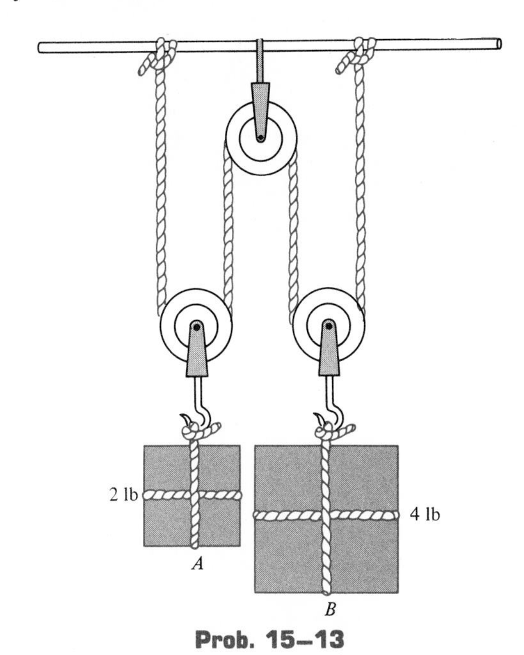

Prob. 15–13

15–14. In cases of emergency, the gas actuator can be used to move a 75-kg block B by exploding a charge C near a pressurized cylinder of negligible mass. As a result of the explosion, the cylinder fractures and the released gas forces the front part of the cylinder, A, to move B forward, giving it a speed of 200 mm/s in 0.4 s. If the coefficient of friction between B and the floor is $\mu = 0.5$, determine the impulse that the actuator must impart to B.

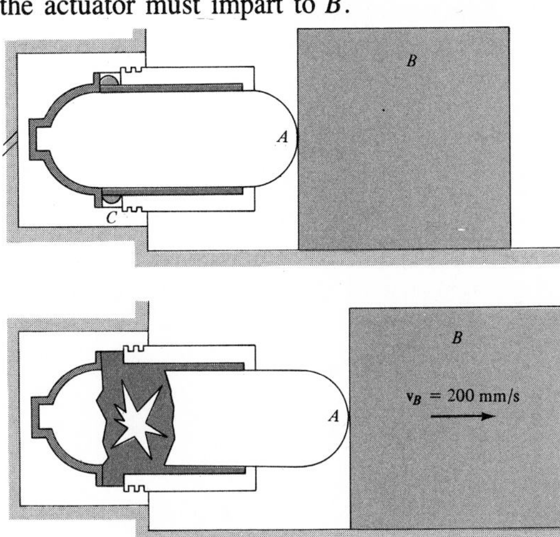

Prob. 15–14

15–15. Solve Prob. 13–24 using the principle of impulse and momentum.

***15–16.** A 30-lb block is initially moving along a smooth horizontal surface with a speed of $v_1 = 6$ ft/s to the left. If it is acted upon by a force **F**, which varies in the manner shown, determine the velocity of the block in 15 s. The argument for the cosine is in radians. *Hint:* The impulse of **F** is the area under the curve, $I = -(250/\pi)$ lb.

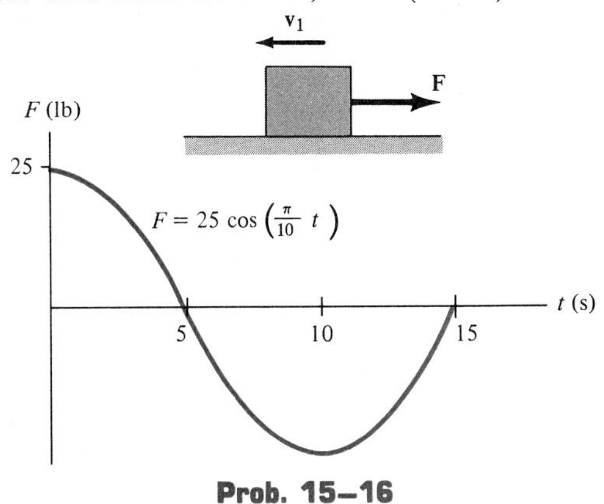

Prob. 15–16

15–17. A jet plane having a mass of 7 Mg is to be launched forward from a stationary position on an aircraft carrier using a catapult that exerts a horizontal force on the plane which varies as shown by the graph. If the carrier is traveling forward with a speed of 40 km/h and the plane is to achieve an air speed of 200 km/h after 5 s, determine the peak force $\mathbf{F}_O$ which must be exerted on the plane. While the catapult is in operation, the jet on the plane exerts a constant horizontal thrust of 70 kN.

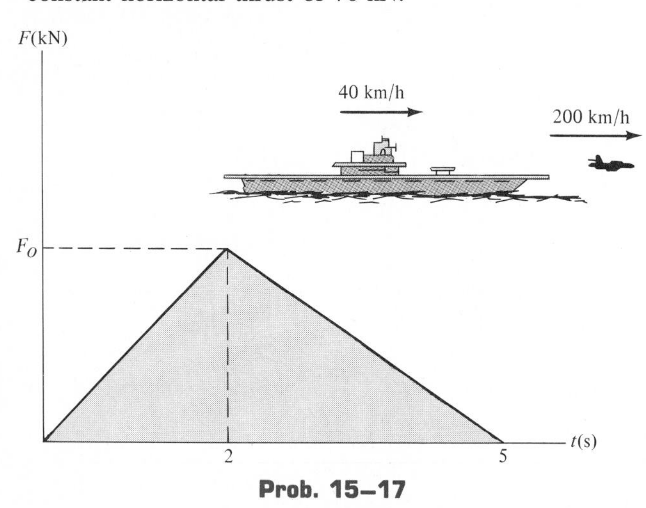

Prob. 15–17

15–18. A tank car has a mass of 30 Mg and is freely rolling to the right with a speed of 0.75 m/s. If it strikes the barrier, determine the horizontal impulse needed to stop the car if the spring in the bumper B has a stiffness of (a) $k \rightarrow \infty$ (bumper is rigid) and (b) $k = 15$ kN/m.

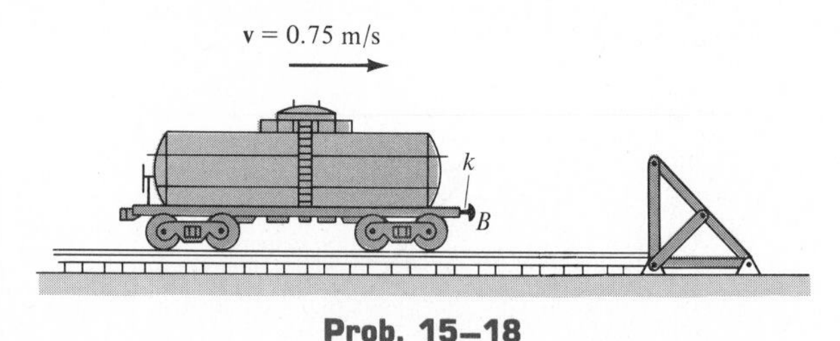

Prob. 15–18

15–19. A cake and plate weighing 1.5 lb rest at the center of a circular table. Without touching the cake a boy attempts to remove the 2-ft-radius tablecloth by quickly pulling on it horizontally. If slipping between the cake plate and tablecloth occurs at all times, determine the longest time the tablecloth can remain in contact with the plate without having the cake wind up on the floor. The coefficient of friction between the cake plate and the tablecloth is $\mu = 0.3$ and between the cake plate and table surface $\mu' = 0.4$. *Hint:* Apply the principles of impulse and momentum and work and energy to the cake for x ft, the distance the cake plate remains on the tablecloth, and for $(2 - x)$ ft when it remains on the table.

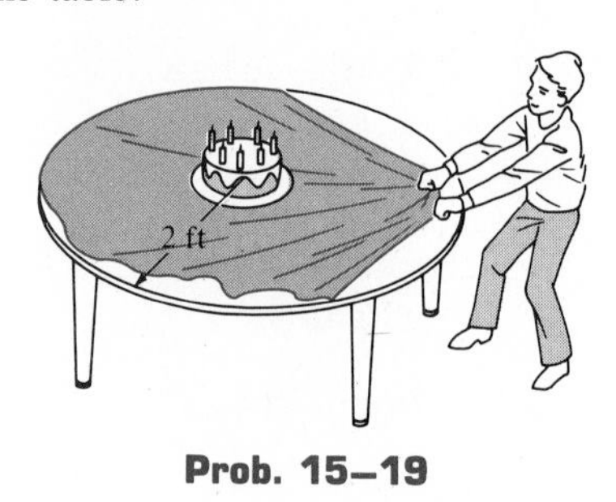

Prob. 15–19

***15–20.** The motor M pulls on the cable with a force $\mathbf{F}$ that has a magnitude which varies as shown on the graph. If the 15-kg crate is originally resting on the floor such that the cable tension is zero when the motor is turned on, determine the speed of the crate when $t = 6$ s. *Hint:* First show that the time needed to *begin* lifting the crate is $t = 2.45$ s.

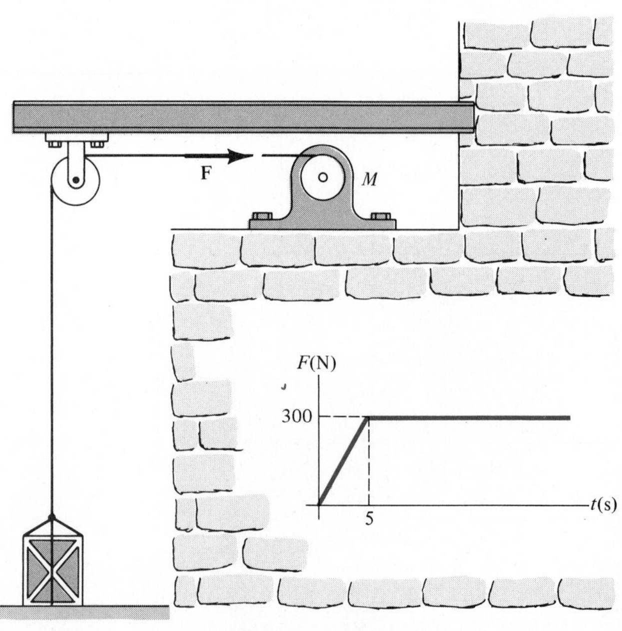

Prob. 15–20

Impulse & Momentum

use to solve ..... { Force, velocity, Time } problems

1) **Impulse** $I = \int_{t_1}^{t_2} F \cdot dt$ [N·s], [lb·s]

2) **Principle of Linear Impulse & Momentum**

$$\Sigma \int_{t_1}^{t_2} F\,dt = \Sigma m V_2 - \Sigma m V_1$$

3) **Momentum** $= L = mV$ [kg m/sec]

4) $m(V_x)_1 + \Sigma \int_{t_1}^{t_2} F_x\,dt = m(V_x)_2$

$m(V_y)_1 + \Sigma \int_{t_1}^{t_2} F_y\,dt = m(V_y)_2$

5) **Conservation of Linear Momentum**

-when impulses are zero.....

$$\Sigma m V_1 = \Sigma m V_2$$

i.e. to find rifle recoil velocity

15–21. If the coefficient of friction between the plane and the 40-kg crate is $\mu = 0.3$, determine the time needed for the force **F** to give the crate a speed of 2 m/s. **F** is always horizontal and has a magnitude of $F = (50t)$ N, where t is measured in seconds. *Hint:* First show that the time needed to overcome friction and *start* the crate moving is $t = 2.35$ s.

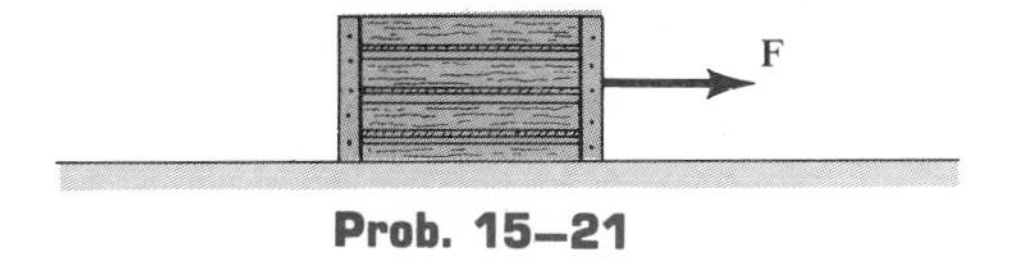

Prob. 15–21

Principle of Linear Impulse and Momentum for a System of Particles

The principle of linear impulse and momentum for a system of particles, Fig. 15–7, may be obtained from $\Sigma \mathbf{F} = \Sigma m_i \mathbf{a}_i$ (see Sec. 13–3), which may be rewritten as

$$\Sigma \mathbf{F} = \Sigma m_i \frac{d\mathbf{v}_i}{dt} \qquad (15\text{–}5)$$

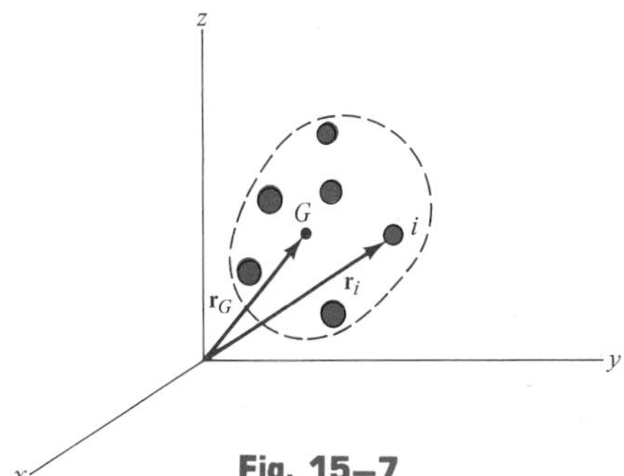

Fig. 15–7

The term on the left side represents only the sum of the *external forces* acting on the system of particles. The internal forces between particles do not appear with this summation, since they occur in equal but opposite collinear pairs and therefore cancel out. Multiplying both sides of Eq. 15–5 by dt and integrating between the limits $t = t_1$, $\mathbf{v}_i = (\mathbf{v}_i)_1$ and $t = t_2$, $\mathbf{v}_i = (\mathbf{v}_i)_2$ yields

$$\Sigma m_i(\mathbf{v}_i)_1 + \Sigma \int_{t_1}^{t_2} \mathbf{F}\, dt = \Sigma m_i(\mathbf{v}_i)_2 \qquad (15\text{–}6)$$

(NOTE: Σ)

This equation states that the initial linear momenta of the system added vectorially to the impulses of all the *external forces* acting on the system during the time period t_1 to t_2 are equal to the system's final linear momenta.

By definition, the location of the mass center G of the system is determined from $m\mathbf{r}_G = \Sigma m_i \mathbf{r}_i$, where $m = \Sigma m_i$ is the total mass of all the particles, and $\mathbf{r}_G$ and $\mathbf{r}_i$ are defined in Fig. 15–7. Taking the time derivatives, we have

$$m\mathbf{v}_G = \Sigma m_i \mathbf{v}_i$$

which states that the total linear momentum of the system of particles is equivalent to the linear momentum of a "fictitious" aggregate particle of mass $m = \Sigma m_i$ moving with the velocity of the mass center G of the system. Substituting into Eq. 15–6 yields

$$m(\mathbf{v}_G)_1 + \Sigma \int_{t_1}^{t_2} \mathbf{F}\, dt = m(\mathbf{v}_G)_2 \qquad (15\text{–}7)$$

This equation states that the initial linear momentum of the aggregate particle plus (vectorially) the external impulses acting on the system of particles during the time interval t_1 to t_2 is equal to the aggregate particle's final linear momentum. Since in reality all particles must have finite size to possess mass, the above equation justifies application of the principle of linear impulse and momentum to a rigid body represented as a single particle.

15.3 Conservation of Linear Momentum for a System of Particles

When the sum of the external impulses acting on a system of particles is zero, Eq. 15–6 reduces to a simplified form,

$$\Sigma m_i(\mathbf{v}_i)_1 = \Sigma m_i(\mathbf{v}_i)_2 \qquad (15\text{–}8)$$

This equation is referred to as the *conservation of linear momentum*. It states that the linear momenta for a system of particles remain constant throughout the time period t_1 to t_2. Since $m\mathbf{v}_G = \Sigma m_i \mathbf{v}_i$, we can also write

$$(\mathbf{v}_G)_1 = (\mathbf{v}_G)_2 \qquad (15\text{–}9)$$

which indicates that the velocity $\mathbf{v}_G$ of the mass center for the system of particles does not change.

To illustrate a situation for which the conservation of linear momentum applies, consider the missile shown in Fig. 15–8*a* which has an intended

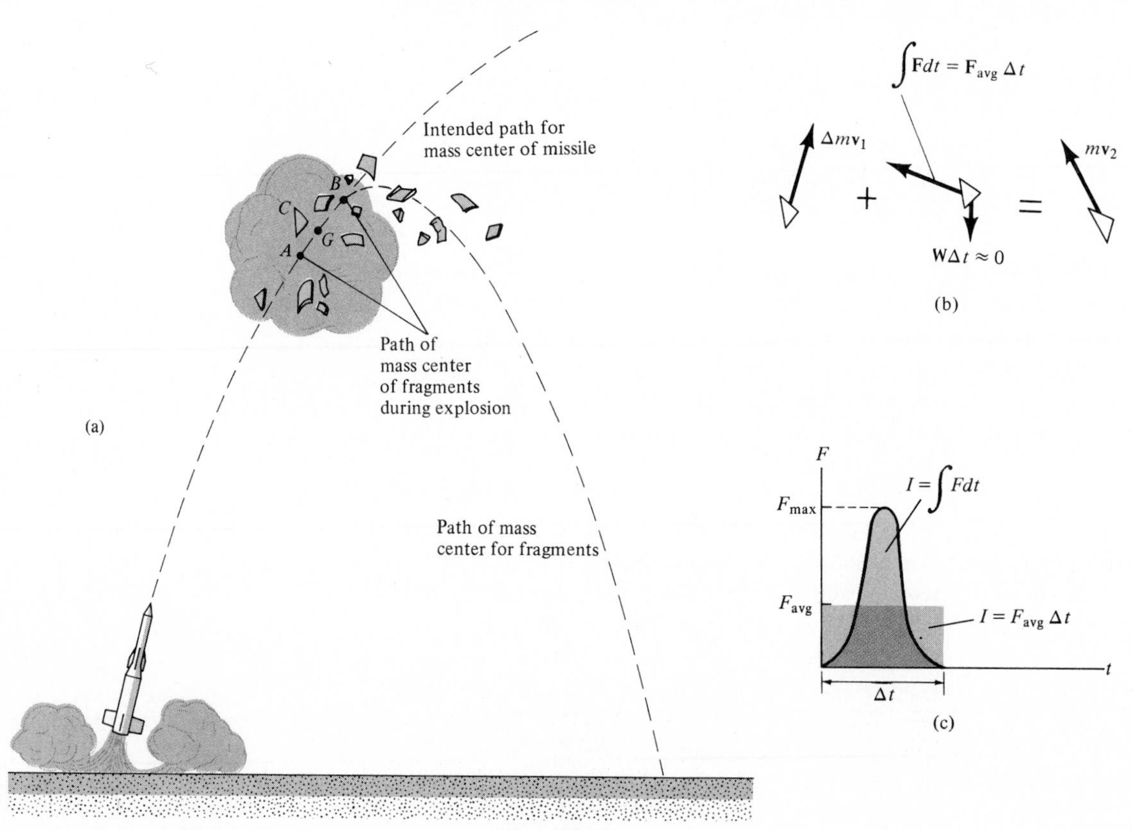

Fig. 15–8

trajectory shown by the dashed path. Suppose that when the missile reaches point A it suddenly blows up. Just before the explosion the velocity of the fragments (or all the particles Σm_i) is the same, since the missile is intact. Just after the explosion, however, the velocity of each fragment changes. Even so, the *total momentum of the system is conserved during the time of the explosion*. There are two reasons for this. First, the fragments are blown apart only by impulses which occur from equal but opposite *internal forces* acting on the system; and, second, the time for explosion is very *short,* so that the *external impulse* created by gravity is negligible in comparison to the large internal explosive impulses given to the system. Furthermore, neglecting atmospheric drag, the mass center G of the fragments will follow path AB, Fig. 15–8a, and this will continue as the time *after the explosion* becomes longer.

If a study is made of the motion of just *one* of the missile fragments instead of the entire system of fragments, the explosive impulse *on the fragment* is considered to be "external" and hence it must be included in the analysis involving impulse and momentum of the fragment. For example, consider fragment C of mass m in Fig. 15–8a. The impulse and momentum diagrams are shown in Fig. 15–8b. Initially, the fragment has a momentum $m\mathbf{v}_1$ *just before* the explosion, where $\mathbf{v}_1$ is the velocity of the missile. During the time Δt of the explosion, both the explosive impulse of $\int \mathbf{F}\,dt$ and the weight impulse $\mathbf{W}(\Delta t)$ act on the fragment. However, as stated above, since the weight is *much smaller* than the explosive force and Δt is *very small,* $\mathbf{W}(\Delta t) \approx \mathbf{0}$ and it can be neglected. Assuming the final momentum of the fragment to be $m\mathbf{v}_2$, then applying the principle of impulse and momentum, we have

$$m\mathbf{v}_1 + \int \mathbf{F}\,dt = m\mathbf{v}_2$$

If the final velocity of the fragment is known, the explosive impulse can be determined, i.e.,

$$\mathbf{I} = \int \mathbf{F}\,dt = m(\mathbf{v}_2 - \mathbf{v}_1)$$

Generally, the variation of the explosive force with time is *not known*. However, it is suspected that during the time of explosion the magnitude of $\mathbf{F}$ will increase sharply to some value $F_{\max}$, as shown in Fig. 15–8c, then decrease sharply to zero. Consequently, the impulse is represented as the *colored area* under the graph. If the magnitude of the impulse can be determined from the above equation, the *average impulsive force* can be calculated for the same time period Δt by noting that the average impulsive loading is equivalent to the rectangular shaded area shown in Fig. 15–8c. In this case, $I = \int F\,dt = F_{\text{avg}}\,\Delta t$, or using the above equation,

$$\mathbf{F}_{\text{avg}} = \frac{m}{\Delta t}(\mathbf{v}_2 - \mathbf{v}_1)$$

Impulsive and Nonimpulsive Forces. A force that is very large and acts for a very short period of time such that it produces a significant change in momentum is called an *impulsive force*. This force normally occurs due to an explosion or the striking of one body against another. In the above example, the force of the explosion acting on the missile fragment may be classified as an impulsive force. By comparison, the weight **W** of the fragment is *nonimpulsive* since it creates *no significant change* in the momentum of the fragment during the time Δt. Consequently, the analysis of a problem may be simplified by *neglecting* the effects of nonimpulsive forces when the principle of impulse and momentum is applied during the time period Δt. In general, *nonimpulsive forces* include the weight of a body, the force imparted by a slightly deformed spring having a relatively small stiffness, or for that matter, any force that is small compared to other larger (impulsive) forces.

PROCEDURE FOR ANALYSIS

Generally, the principle of linear impulse and momentum or the conservation of linear momentum is applied to a *system of particles* in order to determine the final velocities of the particles *just after* the time period considered. By applying these equations to the entire system, the internal impulses acting within the system, which may be unknown, are *eliminated* from the analysis, since they occur in equal but opposite collinear pairs. For application it is suggested that the following procedure be used.

Impulse and Momentum Diagrams. Draw the impulse and momentum diagrams for the system of particles just before, during, and just after the impulsive forces are applied. The conservation of linear momentum applies to the system in a given direction when *no external impulsive forces* act on the system in that direction. To determine if this condition applies, one should investigate the impulse diagram in order to clearly distinguish the external impulsive and nonimpulsive forces from the system's internal impulses.

Momentum Equations. Apply the principle of linear impulse and momentum or the conservation of linear momentum in the appropriate directions. Most often the scalar component equations can be formulated by resolving the vector components *directly* from each of the impulse and momentum diagrams. If the particles are subjected to dependent motion using cables and pulleys, kinematics as discussed in Sec. 12–7 can be used to relate the velocities.

The following examples numerically illustrate application of this procedure.

Example 15–4

The 15-Mg boxcar A is coasting freely at 1.5 m/s on the horizontal track when it encounters a tank car B having a mass of 12 Mg and coasting at 0.75 m/s toward it as shown in Fig. 15–9a. If the cars meet and couple together, determine (a) the speed of both cars just after the coupling and (b) the average force between them if the coupling takes place at 0.8 s.

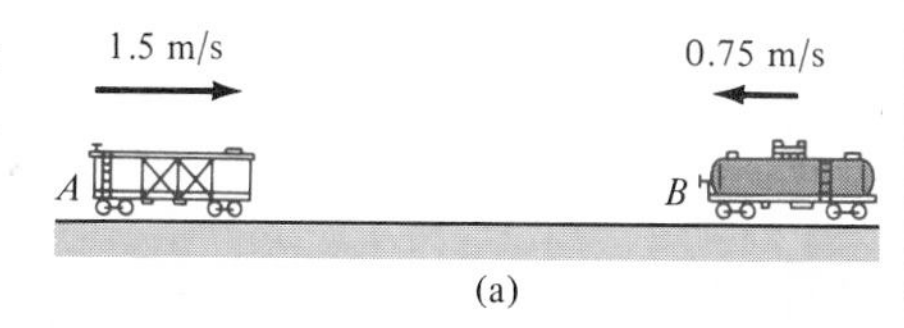

(a)

15 000 (1.5)   12 000 (0.75) $\int$ Fdt 27 000 (v_2)

(b)

15 000 (1.5) $\int$ Fdt 15 000 (0.50)

(c)

Fig. 15–9

Solution

Part (a)

Impulse and Momentum Diagrams. As shown in Fig. 15–9b, we have considered *both* cars as a single system. Hence the coupling impulse, $\int \mathbf{F}\,dt$, is *internal* to the system and therefore will *not* be included in the momentum-impulse analysis. Also, it is assumed both cars, when coupled, move to the right. By inspection, momentum is conserved in the horizontal direction.

Conservation of Linear Momentum

$$\Sigma m v_1 = \Sigma m v_2$$

$$(\overset{+}{\rightarrow}) \qquad m_A(v_A)_1 + m_B(v_B)_1 = (m_A + m_B)v_2$$

$$15\,000(1.5) - 12\,000(0.75) = 27\,000 v_2$$

$$v_2 = 0.50 \text{ m/s} \qquad \textit{Ans.}$$

Part (b). The average (impulsive) coupling force, $\mathbf{F}_{avg}$, can be determined by applying the principle of linear momentum to *either one* of the cars.

Impulse and Momentum Diagrams. As shown in Fig. 15–9c, by isolating the boxcar, the coupling impulse is *external* to the car.

Principle of Impulse and Momentum. Since $\int F\,dt = F_{avg}\,\Delta t = F_{avg}(0.8)$, we have

$$(\overset{+}{\rightarrow}) \qquad m_A(v_A)_1 + \Sigma \int F\,dt = m_A v_2$$

$$15\,000(1.5) - F_{avg}(0.8) = 15\,000(0.50)$$

$$F_{avg} = 18.75 \text{ kN} \qquad \textit{Ans.}$$

Solution was possible since the boxcar's final velocity was obtained in Part (a). Try solving for F_{avg} by applying the principle of impulse and momentum to the tank car.

Example 15–5

The 1,200-lb cannon shown in Fig. 15–10*a* fires an 8-lb projectile with a muzzle velocity of 1,500 ft/s. If firing takes place in 0.03 s, determine (a) the recoil velocity of the cannon just after firing and (b) the average impulsive force acting on the projectile. The cannon support is firmly fixed to the ground and the horizontal recoil of the cannon is absorbed by two springs.

Recoil spring

A

(a)

Solution

Part (a)

Impulse and Momentum Diagrams. As shown in Fig. 15–10*b*, we have considered the projectile and cannon as a single system, since then the impulsive forces, $\int \mathbf{F}\,dt$, between the cannon and projectile are *internal* to the system and are therefore not included in the momentum-impulse analysis of the system. Furthermore, during the time $\Delta t = 0.03$ s, the two recoil springs which are attached to the support each exert a *nonimpulsive force* $\mathbf{F}_s$ on the cannon. This is because Δt is very short, so that during this time the cannon only moves through a very small distance* x. Consequently, $F_s = kx \approx 0$. Hence it may be concluded that momentum for the system is conserved in the *horizontal direction*.

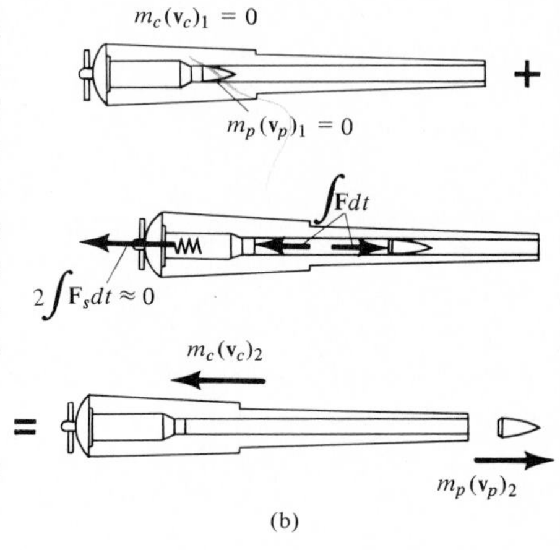

(b)

Conservation of Linear Momentum. Using the data on the momentum diagrams, we have

$$\Sigma m\mathbf{v}_1 = \Sigma m\mathbf{v}_2$$

$$(\overset{+}{\rightarrow}) \qquad m_c(v_c)_1 + m_p(v_p)_1 = -m_c(v_c)_2 + m_p(v_p)_2$$

$$0 + 0 = -\frac{1{,}200}{32.2}(v_c)_2 + \frac{8}{32.2}(1{,}500)$$

$$(v_c)_2 = 10 \text{ ft/s} \qquad \textit{Ans.}$$

$\int \mathbf{F}dt$ $m_p(\mathbf{v}_p)_2$

(c)

Fig. 15–10

Part (b). The average impulsive force exerted by the cannon on the projectile can be determined by applying the principle of linear impulse and momentum to the projectile.

Principle of Impulse and Momentum. Using the data on the impulse and momentum diagrams, Fig. 15–10*c*, noting that $\int F\,dt = F_{avg}\,\Delta t = F_{avg}(0.03)$, we have

$$(\overset{+}{\rightarrow}) \qquad m(v_p)_1 + \Sigma \int F\,dt = m(v_p)_2$$

$$0 + F_{avg}(0.03) = \frac{8}{32.2}(1{,}500)$$

$$F_{avg} = 12{,}422 \text{ lb} \qquad \textit{Ans.}$$

Obtain this same answer by applying the principle of linear impulse and momentum to the cannon.

*If the cannon is firmly fixed to its support (no springs), the reactive force of the support on the cannon must be considered as an external impulse to the system, since the support would allow no movement of the cannon.

Example 15–6

The 350-Mg tugboat T shown in Fig. 15–11a is used to pull the 50-Mg barge B with a rope R. If the initial velocity of the tugboat is $(v_T)_1 = 3$ m/s while the rope is slack, determine the velocity of the tugboat *directly after* towing starts to occur. Assume the rope does not stretch. Neglect the frictional effects of the water.

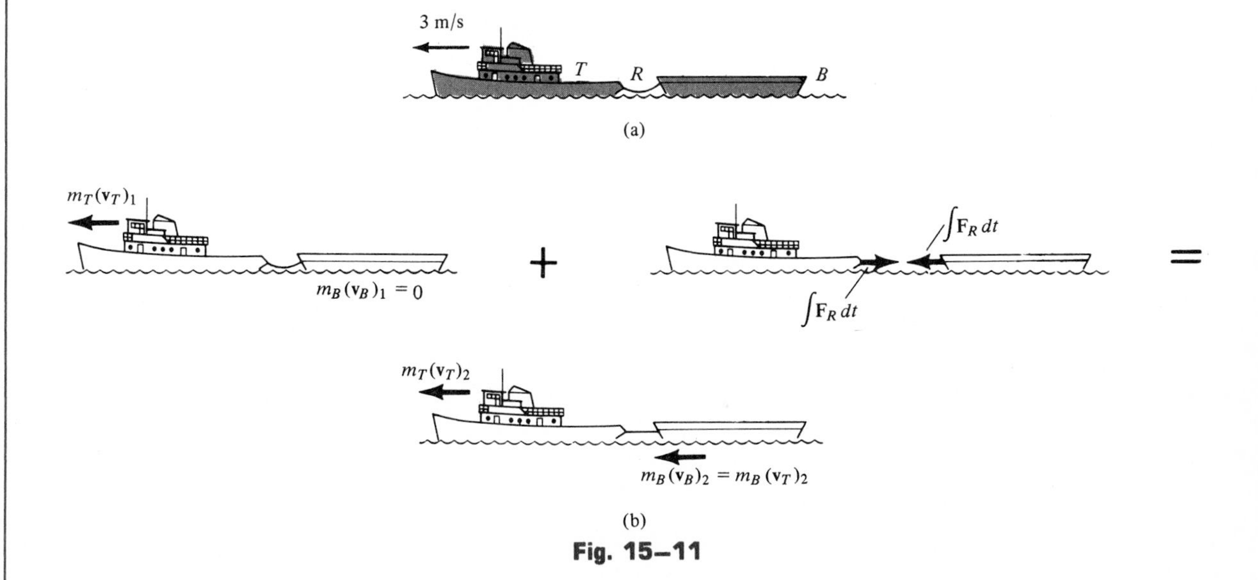

Fig. 15–11

Solution

Impulse and Momentum Diagrams. As shown in Fig. 15–11b, we have considered the entire system (tugboat and barge). Hence, the impulse created by the rope and the barge is *internal* to the system, and therefore momentum of the system is conserved during the instant of towing.

Conservation of Momentum. Noting that $(v_B)_2 = (v_T)_2$, we have

$$(\overset{+}{\leftarrow}) \qquad m_T(v_T)_1 + m_B(v_B)_1 = m_T(v_T)_2 + m_B(v_B)_2$$

$$350(10^3)(3) + 0 = 350(10^3)(v_T)_2 + 50(10^3)(v_T)_2$$

Solving,

$$(v_T)_2 = 2.63 \text{ m/s} \qquad \textit{Ans.}$$

This value represents the tugboat's velocity *just after* the towing impulse. How would you find the towing impulse?

Example 15–7

A rigid pile P shown in Fig. 15–12a has a mass of 800 kg and is driven into the ground using a hammer H that has a mass of 300 kg. The hammer falls from rest from a height of $y_0 = 0.5$ m and strikes the top of the pile. Determine the impulse which the hammer imparts on the pile if the pile is surrounded entirely by loose sand so that after striking the hammer does *not* rebound off the pile.

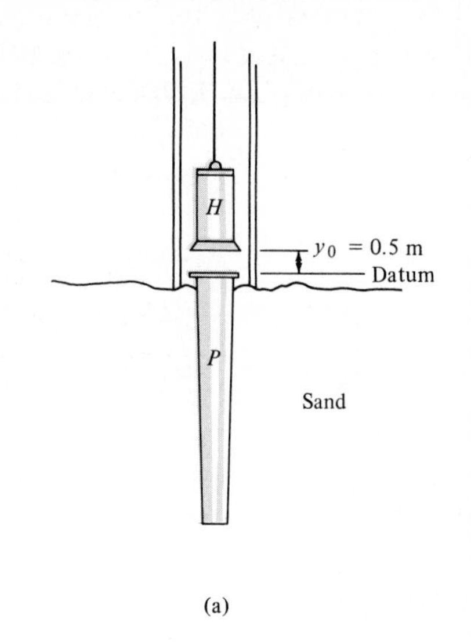

(a)

Solution

Conservation of Energy Theorem. The velocity at which the hammer strikes the pile can be determined using the conservation of energy theorem applied to the hammer. With the datum at the top of the pile, Fig. 15–12a, we have

$$T_0 + V_0 = T_1 + V_1$$
$$\tfrac{1}{2}m_H(v_H)_0^2 + W_H y_0 = \tfrac{1}{2}m_H(v_H)_1^2 + W_H y_1$$
$$0 + 300(9.81)(0.5) = \tfrac{1}{2}(300)(v_H)_1^2 + 0$$
$$(v_H)_1 = 3.13 \text{ m/s}$$

Momentum and Impulse Diagrams. From the physical aspects of the problem, the impulse diagram, Fig. 15–12b, indicates that during the short time occurring just before to just after the *collision* the weights of the hammer and pile and the resistance force $\mathbf{F}_s$ of the soil are all *nonimpulsive*. Furthermore, the impulse $\int \mathbf{R}\, dt$ is internal to the system and therefore cancels. Consequently, momentum is conserved in the vertical direction.

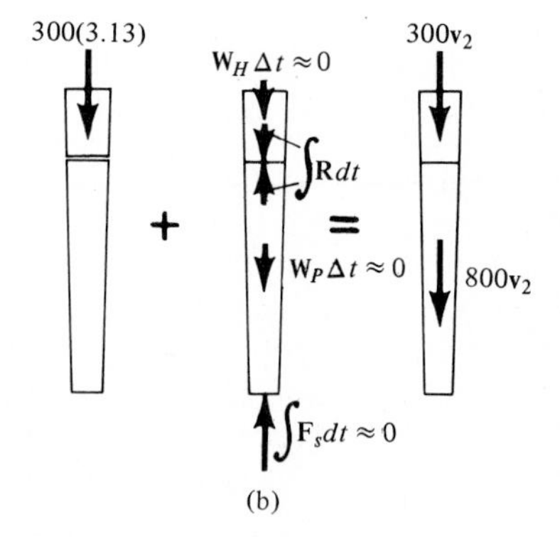

(b)

Conservation of Momentum

$$(+\downarrow) \qquad m_H(v_H)_1 + m_P(v_P)_1 = m_H v_2 + m_P v_2$$
$$300(3.13) + 0 = 300v_2 + 800v_2$$
$$v_2 = 0.854 \text{ m/s}$$

Principle of Impulse and Momentum. The impulse which the hammer imparts to the pile can now be determined since $\mathbf{v}_2$ is known. From the momentum and impulse diagrams for the hammer, Fig. 15–12c, we have

$$(+\downarrow) \qquad m_H(v_H)_1 + \Sigma\int_{t_1}^{t_2} F_y\, dt = m_H v_2$$
$$300(3.13) - \int R\, dt = 300(0.854)$$
$$\int R\, dt = 682.9 \text{ N}\cdot\text{s} \qquad \textit{Ans.}$$

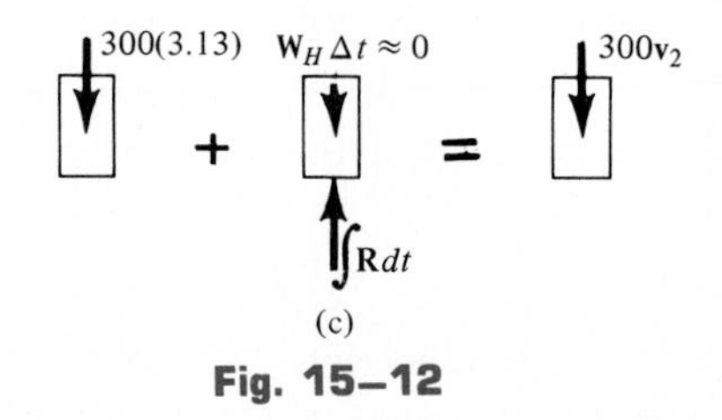

(c)

Fig. 15–12

Try finding the impulse by applying the principle of impulse and momentum to the pile.

Example 15–8

A boy having a mass of 40 kg stands on the 15-kg toboggan, Fig. 15–13*a*. If he walks forward at a constant speed of 0.75 m/s relative to the toboggan, determine the speed of the toboggan measured by an observer standing on the ground. Neglect friction between the bottom of the toboggan and the ground (ice).

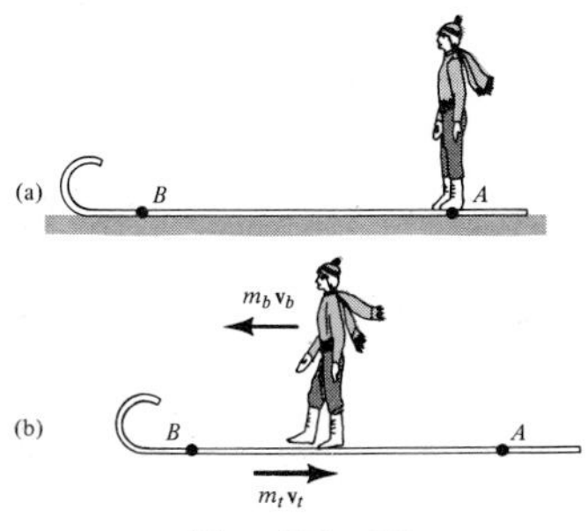

Fig. 15–13

Solution

Momentum Diagram. The unknown frictional force of the boy's shoes on the bottom of the toboggan can be *excluded* from the analysis if the toboggan and boy on it are considered as a single system. In this way the frictional force becomes internal and the conservation of momentum applies. The system's momentum must be zero when the boy is at some intermediate point between A and B, Fig. 15–13*b*.

Conservation of Momentum

$$(\overset{+}{\rightarrow}) \qquad -m_b v_b + m_t v_t = 0 \qquad (1)$$

Here the two unknowns v_b and v_t represent the velocities of the boy and the toboggan measured from a *fixed inertial reference* on the ground.

The problem statement reports the speed of the boy relative to the toboggan, i.e., $v_{b/t} = 0.75$ m/s. This velocity is related to the velocities of the boy and toboggan by the equation $\mathbf{v}_b = \mathbf{v}_t + \mathbf{v}_{b/t}$, Eq. 12–23. Since positive motion is assumed to be to the right in Eq. (1), $\mathbf{v}_b$ and $\mathbf{v}_{b/t}$ are negative, since the boy's motion is to the left. Hence, in scalar form, $-v_b = v_t - v_{b/t}$ and Eq. (1) then becomes

$$m_b(v_t - v_{b/t}) + m_t v_t = 0$$

Substituting in the data and solving for v_t, we have

$$40(v_t - 0.75) + 15v_t = 0$$

$$v_t = 0.545 \text{ m/s} \qquad \textit{Ans.}$$

Problems

15–22. A rifle has a mass of 2.5 kg. If it is loosely gripped and a 1.5-g bullet is fired from it with a muzzle velocity of 1400 m/s, determine the recoil velocity of the rifle just after firing.

15–23. A 0.6-kg brick is thrown into a 25-kg wagon which is initially at rest. If, upon entering, the brick has a velocity of 10 m/s as shown, determine the final velocity of the wagon.

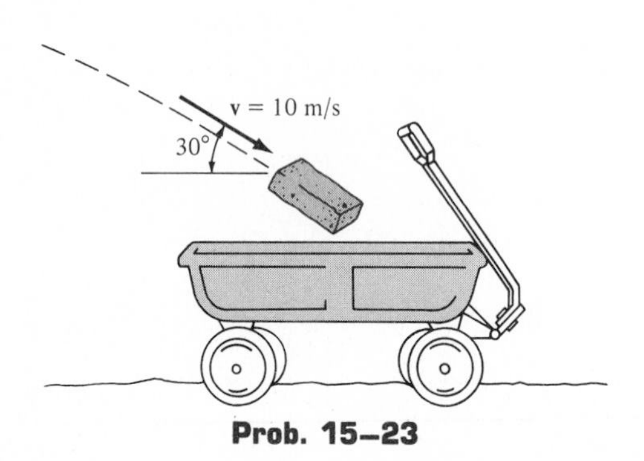

Prob. 15–23

***15–24.** A railroad car having a mass of 15 Mg is coasting at 2 m/s on a horizontal track. At the same time another car having a mass of 10 Mg is coasting at 0.75 m/s in the opposite direction. If the cars meet and couple together, determine the speed of both cars just after the coupling. Compute the difference between the total kinetic energy before and after coupling has occurred, and explain qualitatively what happened to this energy.

15–25. A girl having a weight of 40 lb slides down the smooth slide onto the surface of a 20-lb wagon. Determine the speed of the wagon at the instant the girl stops sliding on it. If someone ties the wagon to the slide at B, determine the horizontal impulse the girl will exert at C in order to stop her motion. Neglect friction and assume that the girl starts from rest at the top of the slide, A. *Hint:* First apply the conservation of energy theorem to obtain the speed of the girl at B.

Prob. 15–25

15–26. A man wearing ice skates throws an 8-kg block with an initial velocity of $v_{b/m} = 2$ m/s, measured relative to himself, in the direction shown. If he is originally at rest and completes the throw in 1.5 s while keeping his legs rigid, determine the horizontal velocity of the man just after releasing the block. What is the vertical reaction of both his skates on the ice during the throw? The man has a mass of 70 kg. Neglect friction and the motion of his arms.

Prob. 15–26

15–27. A boy, having a weight of 90 lb, jumps off a wagon with a relative velocity of $v_{b/w} = 6$ ft/s. If the angle of jump is 30°, determine the horizontal velocity $(\mathbf{v}_w)_2$ of the wagon just after the jump. Originally both the wagon and the boy are at rest. Also, compute the total average impulsive force that all four wheels of the wagon exert on the ground if the boy jumps off in $\Delta t = 0.8$ s. The wagon has a weight of 20 lb.

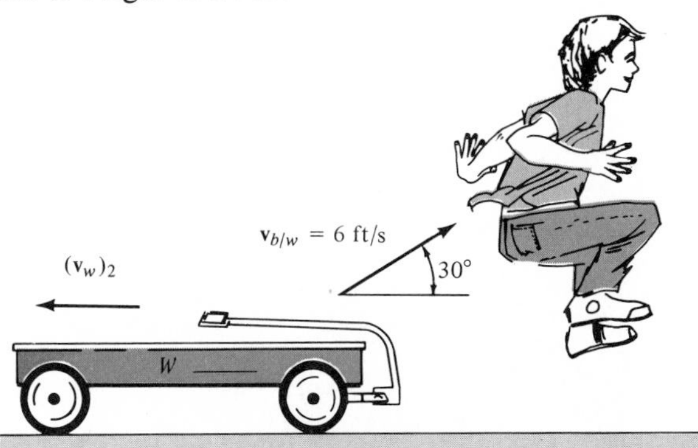

Prob. 15–27

***15–28.** A boy having a weight of 80 lb stands at the end of the 40-lb canoe. If the canoe is not tied to the pier P and the boy runs at 4 ft/s relative to the canoe, determine how far d the canoe moves away from the pier at the instant he leaps off at A. Neglect water resistance. Originally $d = 0$.

Prob. 15–28

15–29. A tugboat T having a mass of 19 Mg is tied to a barge B having a mass of 75 Mg. Determine the common velocity of the barge and tugboat when the rope between them becomes completely stretched. Originally both the tugboat and barge are moving in the same direction with speeds of $(v_T)_1 = 15$ km/h and $(v_B)_1 = 10$ km/h, respectively. Neglect the resistance of the water.

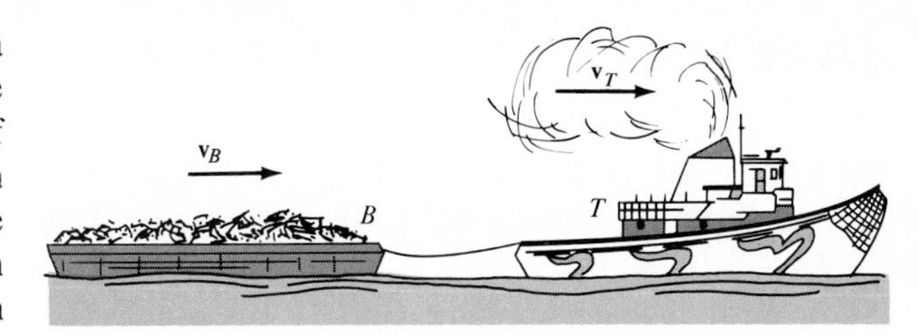

Prob. 15–29

15–30. The two handcars A and B each have a mass of 80 kg. If the man C has a mass of 70 kg and jumps from A with a horizontal *relative* velocity of $v_{C/A} = 2$ m/s and lands on B, determine the velocity of each car after the jump. Neglect the effects of rolling resistance. *Hint:* First study the momentum between A and C, then C and B.

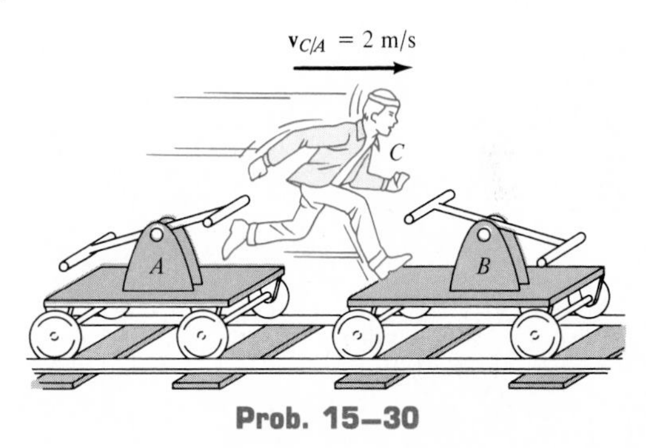

Prob. 15–30

15–31. Two men A and B, each having a weight of 150 lb, stand on the 500-lb raft. Determine the speed imparted to the raft if they run on the raft at a relative speed of 3 ft/s and jump off (a) one at a time and (b) both at the same time. Neglect turning of the raft and water resistance.

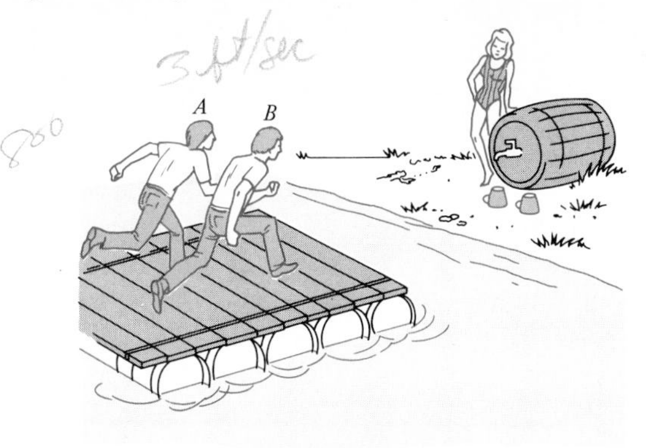

Prob. 15–31

***15–32.** The two toy cars A and B have a weight of 0.4 lb and 0.6 lb, respectively. A spring having a stiffness of 30 lb/ft is attached to one of them and the cars are pressed together such that the spring is compressed 0.2 ft. Determine the speed of each car after they are released from rest.

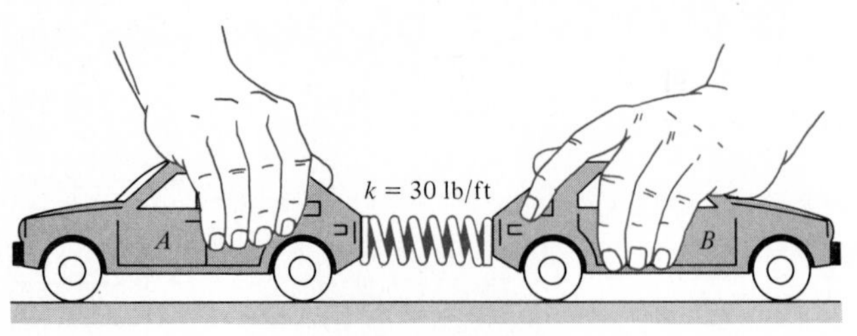

Prob. 15–32

15–33. A toboggan having a weight of 20 lb starts from rest at A and carries a girl and boy having a weight of 100 lb and 70 lb, respectively. When the toboggan reaches the bottom of the slope at B, the boy is pushed off from the back with a horizontal velocity of $v_{b/t} = 4$ ft/s, measured relative to the toboggan. Determine the velocity of the toboggan afterward. Neglect friction in the calculation.

Prob. 15–33

15.4 Impact

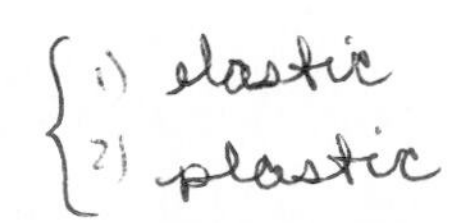

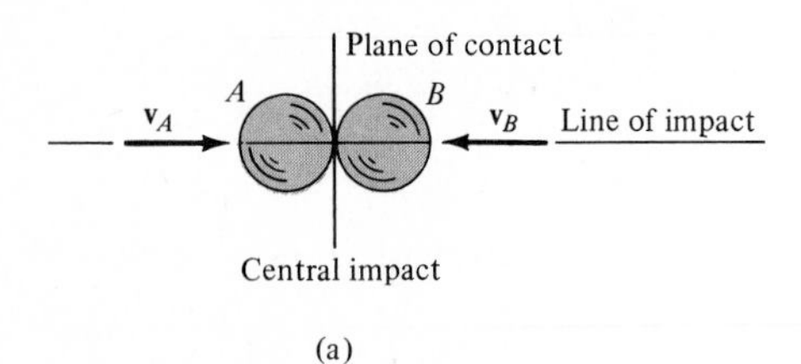

(a)

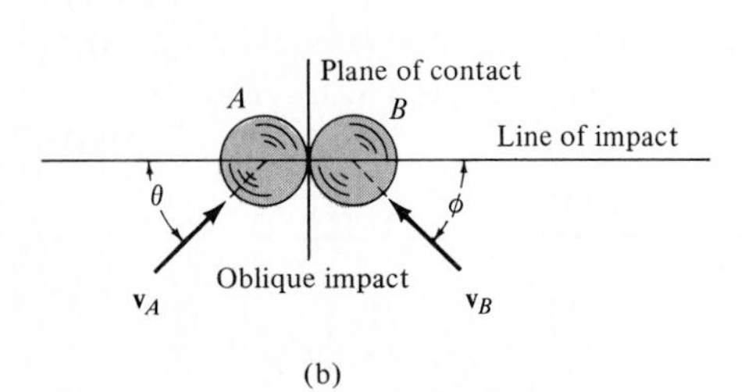

(b)

Fig. 15–14

Impact occurs when two bodies collide with each other during a very *short* interval of time, causing relatively large (impulsive) forces to be exerted between the bodies. The striking of a hammer and nail, or a golf club and ball, are common examples of impact loadings.

In general, there are two types of impact. *Central impact* occurs when the direction of motion of the mass centers of the two colliding particles is along the *line of impact,* Fig. 15–14*a*. When the motion of one or both of the particles is at an angle with the line of impact, Fig. 15–14*b*, the impact is said to be *oblique impact*.

Central Impact. To illustrate the method for analyzing the mechanics of impact, consider the case involving the central impact of particles A and B shown in Fig. 15–15.

1. The particles have the initial momenta shown in Fig. 15–15*a*. Provided that $(v_A)_1 > (v_B)_1$, collision will eventually occur.
2. During the collision the particles will undergo a *period of deformation* such that they exert an equal but opposite deformation impulse $\int \mathbf{P}\, dt$ on each other, Fig. 15–15*b*.
3. Only at the instant of *maximum deformation* will both particles move with a common velocity $\mathbf{v}$, Fig. 15–15*c*.
4. Afterward a *period of restitution* occurs, in which case the particles will either return to their original shape or remain permanently deformed. The

equal but opposite *restitution impulse* $\int \mathbf{R}\,dt$ pushes the particles apart from one another, Fig. 15–15*d*. In reality, the physical properties of any two bodies are such that the deformation impulse is *always greater* than that of restitution, i.e., $\int P\,dt > \int R\,dt$.

5. Just after separation the particles will have the final momenta shown in Fig. 15–15*e*, where $(v_B)_2 > (v_A)_2$.

In most problems the initial velocities of the particles will be *known* and it will be necessary to determine their final velocities $(v_A)_2$ and $(v_B)_2$. In this regard, *momentum* for the *system of particles* is *conserved* since during collision the internal impulses of deformation and restitution *cancel*. Hence, referring to Fig. 15–15*a* and *e*,

$$(\overset{+}{\rightarrow}) \qquad m_A(v_A)_1 + m_B(v_B)_1 = m_A(v_A)_2 + m_B(v_B)_2 \qquad (15\text{–}10)$$

In order to obtain a second equation, necessary to solve for $(v_A)_2$ and $(v_B)_2$, we must apply the principle of impulse and momentum to *each particle*. For example, during the deformation phase for particle A, Fig. 15–15*a* to *c*, we have

$$(\overset{+}{\rightarrow}) \qquad m_A(v_A)_1 - \int P\,dt = m_A(v)$$

For the restitution phase, Fig. 15–15*c* to *e*,

$$(\overset{+}{\rightarrow}) \qquad m_A v - \int R\,dt = m_A(v_A)_2$$

The ratio of the restitution impulse to the deformation impulse is called the *coefficient of restitution, e*. From the above equations, this value for particle A is

$$e = \frac{\int R\,dt}{\int P\,dt} = \frac{v - (v_A)_2}{(v_A)_1 - v}$$

In a similar manner, we can establish e by considering particle B, Fig. 15–15. This yields

$$e = \frac{\int R\,dt}{\int P\,dt} = \frac{(v_B)_2 - v}{v - (v_B)_1}$$

If the unknown v is eliminated from the above two equations, the coefficient of restitution can be expressed in terms of the particles' initial and final velocities as

$$e = \frac{(v_B)_2 - (v_A)_2}{(v_A)_1 - (v_B)_1} \qquad (15\text{–}11)$$

Provided a value for e is specified, Eqs. 15–10 and 15–11 may be solved simultaneously to obtain $(v_A)_2$ and $(v_B)_2$.

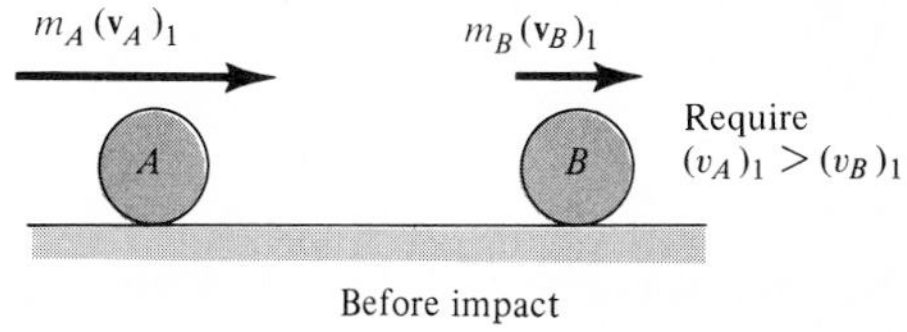

(a)

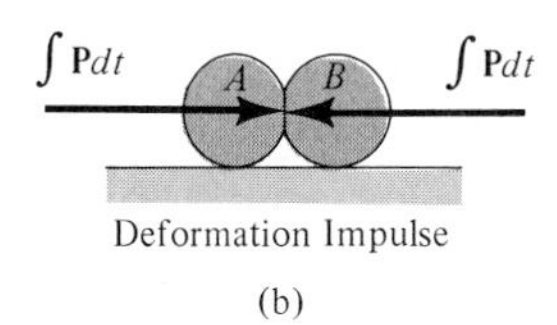

(b)

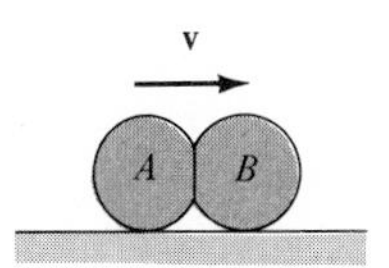

(c)

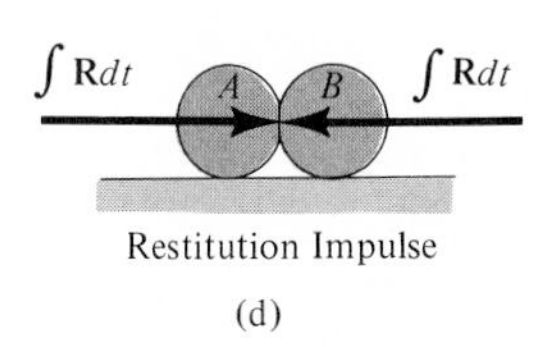

(d)

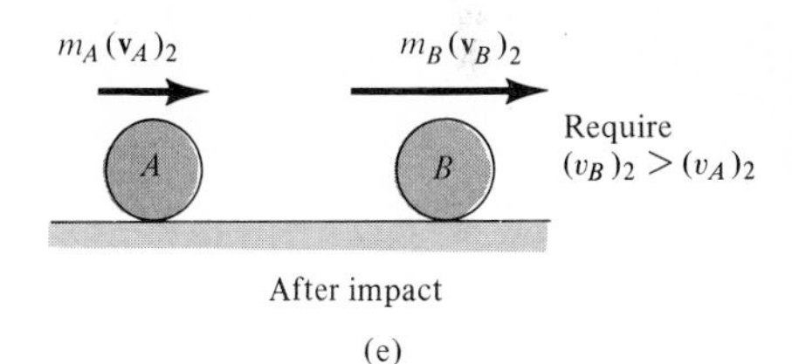

(e)

Fig. 15–15

Coefficient of Restitution. With reference to Fig. 15–15*a* and *e*, it is seen that Eq. 15–11 states that *the coefficient of restitution is equal to the ratio of the relative velocity of the particle's separation just after impact,* $(v_B)_2 - (v_A)_2$, *to the relative velocity of the particle's approach just before impact,* $(v_A)_1 - (v_B)_1$. Specifically,

$$e = \frac{(v_{\text{rel}})_2}{(v_{\text{rel}})_1} = \frac{(v_B)_2 - (v_A)_2}{(v_A)_1 - (v_B)_1} \quad \text{(along line of impact)} \tag{15–12}$$

As noted, this equation applies only along the line of impact. Also, in the derivation, the *positive direction* for the velocities was taken to the right, Fig. 15–15*a* and *e*. The equation is, of course, just as valid if velocities to the left are considered positive. One just has to be consistent once a sign convention is established. If motion of the particles after collision is unknown, the directional sense of the velocities can be assumed. For example, if positive velocities are directed to the right, and both particles are moving *toward* one another before collision, Fig. 15–16*a*, their relative velocity is $(v_A)_1 + (v_B)_1$. After collision, if it is *assumed* that both particles move to the *left*, the relative velocity of *separation* is $-(v_B)_2 + (v_A)_2$, Fig. 15–16*a*. Hence, for this case

$$e = \frac{-(v_B)_2 + (v_A)_2}{(v_A)_1 + (v_B)_1}$$

In a similar manner, particle *B* collides with particle *A* in Fig. 15–16*b* only if $(v_B)_1 > (v_A)_1$. Defining motion to the right as positive, the relative velocity is $-(v_A)_1 + (v_B)_1$. If it is *assumed* that after collision the particles move in opposite directions, Fig. 15–16*b*, the relative velocity is $(v_B)_2 + (v_A)_2$. Hence,

$$e = \frac{(v_B)_2 + (v_A)_2}{-(v_A)_1 + (v_B)_1}$$

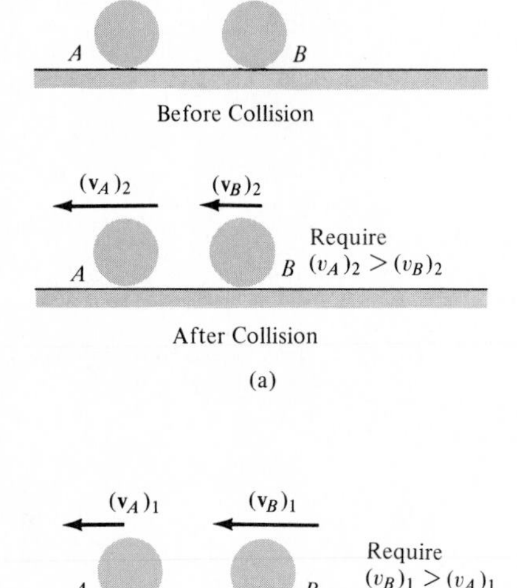

Fig. 15–16

Equation 15–12 provides a relatively simple means for an experimental determination of *e*. By measuring the relative velocities, it has been found that *e* varies appreciably with impact velocity as well as with the size and shape of the colliding bodies. Differences in measurement occur because some of the initial kinetic energy of the bodies is transformed into heat energy as well as creating sound and elastic shock waves when the collision occurs. For these reasons the coefficient of restitution is reliable only when used under conditions which closely approximate those which were known to exist when measurements were made.

Elastic Impact (e = 1). If the collision between the two particles is *perfectly elastic,* the deformation impulse ($\int \mathbf{P}\, dt$) is equal and opposite to the restitution impulse ($\int \mathbf{R}\, dt$). Although in reality this can never be achieved, $e = 1$ for an elastic collision. Under these conditions one can show that no energy is lost in the collision (see Prob. 15–41).

Plastic Impact (e = 0). The impact is said to be *inelastic or plastic* when $e = 0$. In this case there is no restitution impulse given to the particles ($\int \mathbf{R}\, dt = \mathbf{0}$), so that after collision both particles couple or stick *together* and move with a common velocity. In this case the energy lost during collision is a maximum.

PROCEDURE FOR ANALYSIS
(Central Impact)

In most cases the *final velocities* of the two colliding particles are to be determined *just after* they are subjected to direct central impact. Provided the coefficient of restitution, the mass of each particle, and each particle's initial velocity *just before* impact are known, the two equations available for solution are:

1. The conservation of momentum applies to the system of particles, $\Sigma m v_1 = \Sigma m v_2$.
2. The coefficient of restitution, $e = (v_{rel})_2/(v_{rel})_1$, relates the relative velocities of the particles from just before to just after impact.

When applying these two equations, the sense of direction of an unknown velocity can be assumed. If the solution yields a negative magnitude, the velocity acts in the opposite sense of direction.

Oblique Impact. When oblique impact occurs, the particles move away from each other with velocities having unknown directions as well as unknown magnitudes. Provided the initial velocities are known, four unknowns are present in the problem. As shown in Fig. 15–17*a*, these unknowns may be represented as $(v_A)_2$, $(v_B)_2$, θ_2, and ϕ_2. Here we will consider the particles to be smooth.

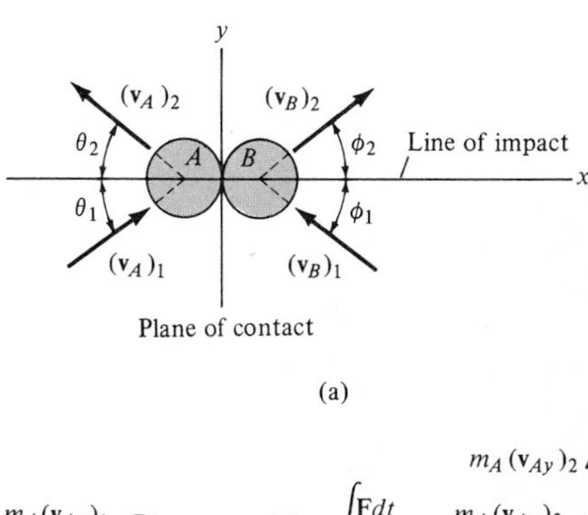

(a)

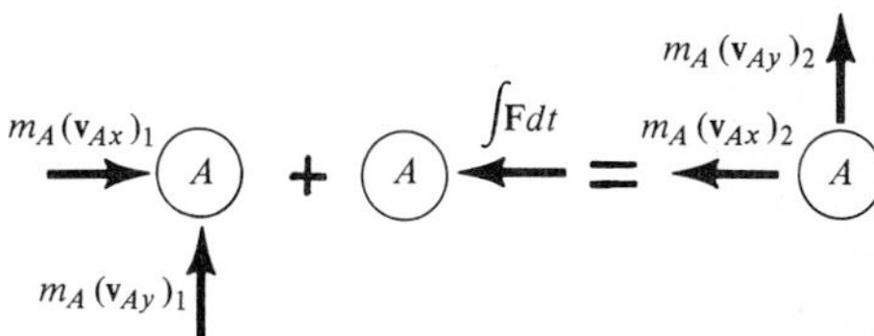

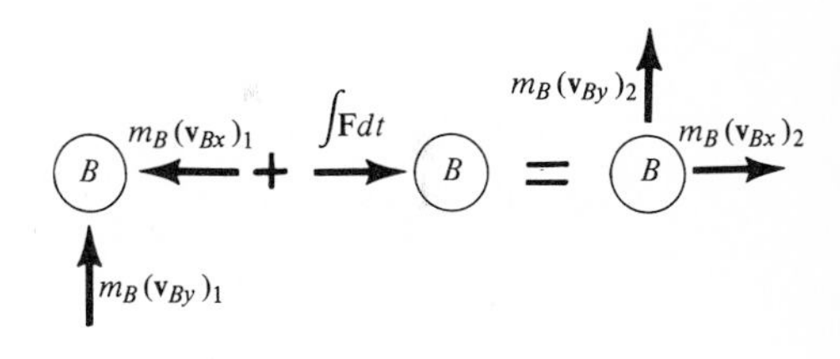

(b)

Fig. 15–17

PROCEDURE FOR ANALYSIS
(Oblique Impact)

If the y axis is established within the plane of contact and the x axis along the line of impact, the impulsive forces of deformation and restitution act *only in the x direction,* Fig. 15–17*b*. Resolving the velocity or momentum vectors into components along the x and y axes, Fig. 15–17*b*, it is possible to write four independent scalar equations in order to determine $(v_{Ax})_2$, $(v_{Ay})_2$, $(v_{Bx})_2$, and $(v_{By})_2$.

1. Momentum of the system is conserved *along the line of impact, x* axis, so that $\Sigma m(v_x)_1 = \Sigma m(v_x)_2$.
2. The coefficient of restitution, $e = (v_{rel})_2/(v_{rel})_1$, relates the relative-velocity *components* of the particles *along the line of impact* (x axis).
3. Momentum of particle A is conserved along the y axis, perpendicular to the line of impact, since no impulse acts on the particle in this direction.
4. Momentum of particle B is conserved along the y axis, perpendicular to the line of impact, since no impulse acts on the particle in this direction.

Application of these four equations is illustrated numerically in Example 15–11.

Example 15–9

The bag A, having a weight of 6 lb, is released from rest at the position $\theta = 0°$, as shown in Fig. 15–18*a*. It strikes an 18-lb box B when $\theta = 90°$. If the coefficient of restitution between the bag and box is $e = 0.5$, determine the velocities of the bag and box just after impact.

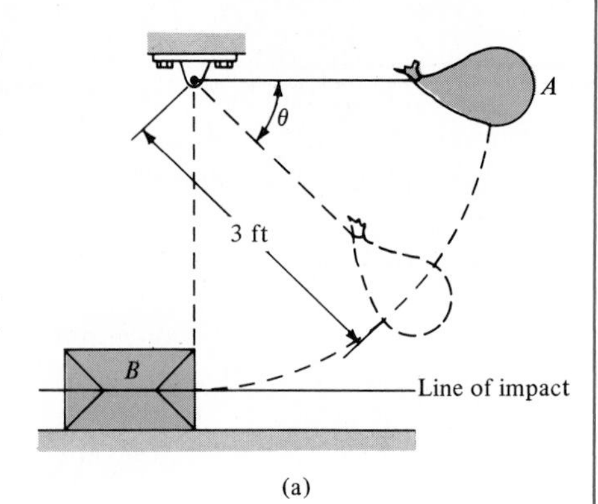

(a)

Solution

This problem involves central impact. Why? Before analyzing the mechanics of the impact, however, it is first necessary to obtain the velocity of the bag *just before* it strikes the box.

Conservation of Energy Theorem. With the datum at $\theta = 0°$, Fig. 15–18*b*, we have

$$T_0 + V_0 = T_1 + V_1$$

$$0 + 0 = \frac{1}{2}\left(\frac{6}{32.2}\right)(v_A)_1^2 - 6(3)$$

$$(v_A)_1 = 13.90 \text{ ft/s}$$

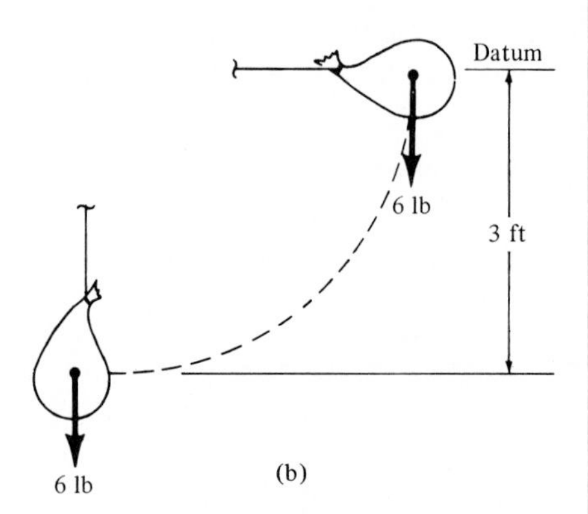

(b)

Conservation of Momentum. The momentum diagrams of A and B just before and just after impact are shown in Fig. 15–18*c*. Here it is assumed that just after collision both bodies continue to travel to the left. Applying the conservation of momentum to the system, we have

$$(\overset{+}{\leftarrow}) \qquad m_B(v_B)_1 + m_A(v_A)_1 = m_B(v_B)_2 + m_A(v_A)_2$$

$$0 + \frac{6}{32.2}(13.90) = \frac{18}{32.2}(v_B)_2 + \frac{6}{32.2}(v_A)_2$$

$$(v_A)_2 = 13.90 - 3(v_B)_2 \qquad (1)$$

Coefficient of Restitution. Applying Eq. 15–12, with *positive* motion to the *left* as in Eq. (1), we have

$$e = \frac{(v_B)_2 - (v_A)_2}{(v_A)_1 - (v_B)_1}$$

$$0.5 = \frac{(v_B)_2 - (v_A)_2}{13.90 - 0}$$

$$(v_A)_2 = (v_B)_2 - 6.95 \qquad (2)$$

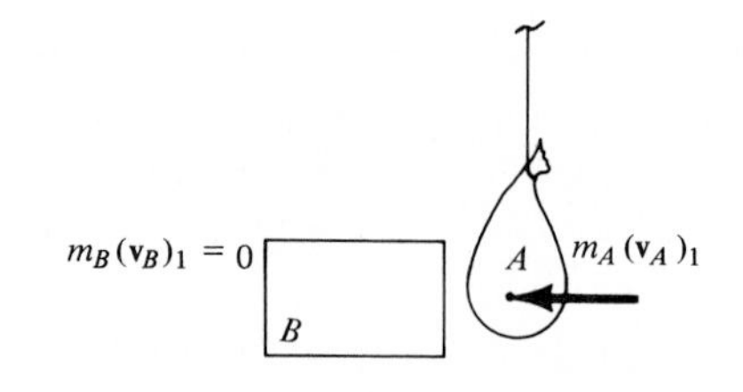

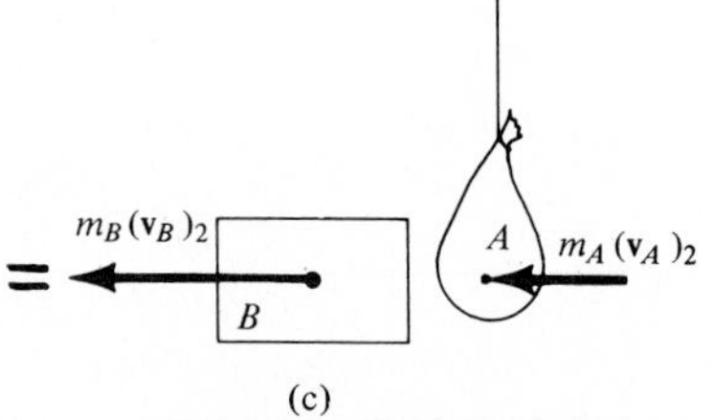

(c)

Fig. 15–18

Solving Eqs. (1) and (2) simultaneously yields

$$(v_A)_2 = -1.74 \text{ ft/s} \quad \text{and} \quad (v_B)_2 = 5.21 \text{ ft/s} \qquad \textit{Ans.}$$

The negative sign for $(v_A)_2$ indicates that the bag moves to the *right* after impact instead of to the left as shown in Fig. 15–18*c*.

Example 15–10

The ball B shown in Fig. 15–19a has a mass of 1.5 kg and is suspended from the ceiling by a 1-m-long elastic cord. If the cord is *stretched* downward 250 mm and the ball is released from rest, determine how far the cord stretches after the ball rebounds from the ceiling. The stiffness of the cord is $k = 800$ N/m and the coefficient of restitution between the ball and ceiling is $e = 0.8$. The ball makes a central impact with the ceiling.

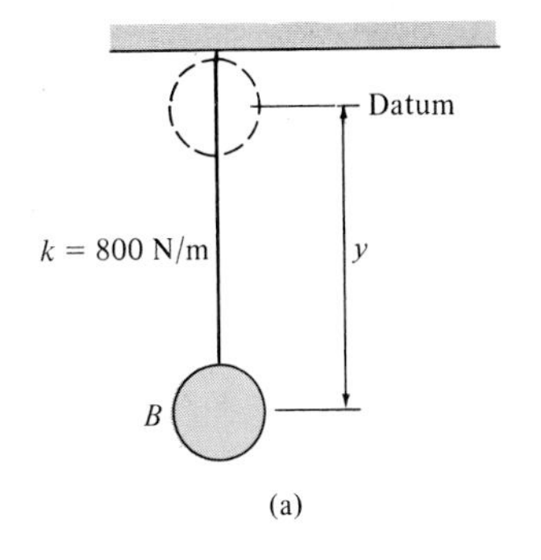

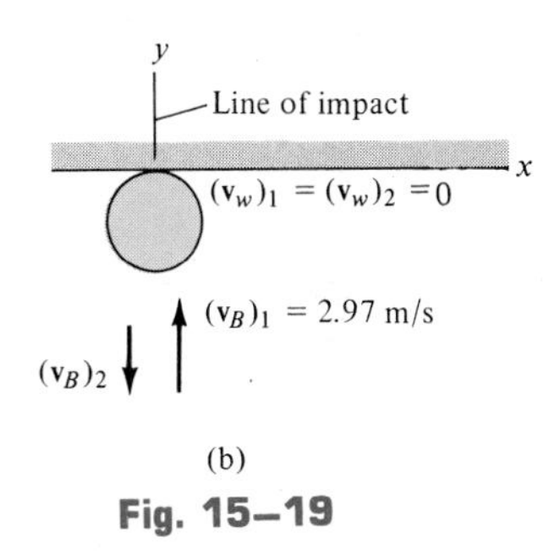

Fig. 15–19

Solution

It is necessary to first obtain the initial velocity of the ball *just before* it strikes the ceiling.

Conservation of Energy Theorem. With the datum located as shown in Fig. 15–19a, realizing that initially $y = y_0 = (1 + 0.25)$ m $= 1.25$ m, we have

$$T_0 + V_0 = T_1 + V_1$$

$$\tfrac{1}{2}m(v_B)_0^2 - W_B y_0 + \tfrac{1}{2}kx^2 = \tfrac{1}{2}m(v_B)_1^2 + 0$$

$$0 - 1.5(9.81)(1.25) + \tfrac{1}{2}(800)(0.25)^2 = \tfrac{1}{2}(1.5)(v_B)_1^2$$

$$(v_B)_1 = 2.97 \text{ m/s}$$

The interaction of the ball with the ceiling will now be considered using the principles of impact. Note that the y axis, Fig. 15–19b, represents the line of impact for the ball. Since an unknown portion of the mass of the ceiling is involved in the impact, the conservation of momentum for the ball–ceiling system will not be written. The "velocity" of this portion of ceiling remains at rest *both* before and after impact.

Coefficient of Restitution. Defining $+y$ upward, we have

$$e = \frac{(v_B)_2 - (v_A)_2}{(v_A)_1 - (v_B)_1}; \qquad 0.8 = \frac{-(v_B)_2 - 0}{0 - 2.97}$$

$$(v_B)_2 = 2.37 \text{ m/s}$$

Conservation of Energy Theorem. The maximum stretch x in the cord may be determined by again applying the conservation of energy theorem to the ball just after collision. Assuming that $y = y_3 = (1 + x_3)$ m, Fig. 15–19a, then

$$T_2 + V_2 = T_3 + V_3$$

$$\tfrac{1}{2}m(v_B)_2^2 + 0 = \tfrac{1}{2}m(v_B)_3^2 - W_B y_3 + \tfrac{1}{2}kx_3^2$$

$$\tfrac{1}{2}(1.5)(2.37)^2 = 0 - 9.81(1.5)(1 + x_3) + \tfrac{1}{2}(800)x_3^2$$

$$400x_3^2 - 14.72x_3 - 18.94 = 0$$

Solving this quadratic equation for the positive root yields

$$x_3 = 0.237 \text{ m} = 237 \text{ mm} \qquad \textit{Ans.}$$

Example 15–11

Two disks A and B, having a mass of 1 and 2 kg, respectively, collide with initial velocities as shown in Fig. 15–20*a*. If the coefficient of restitution for the disks is $e = 0.75$, determine the x and y components of the final velocity of each disk after collision. Neglect friction.

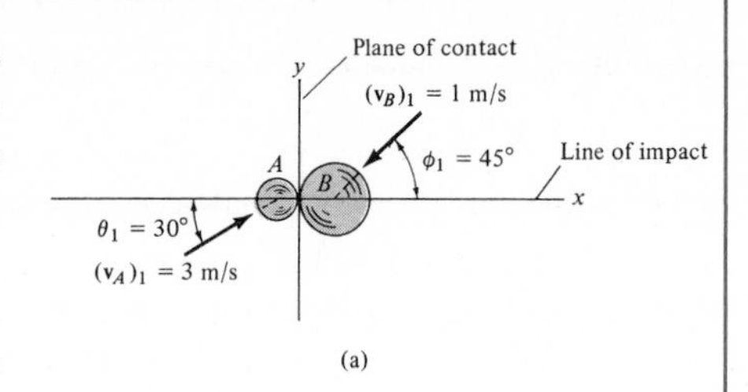

(a)

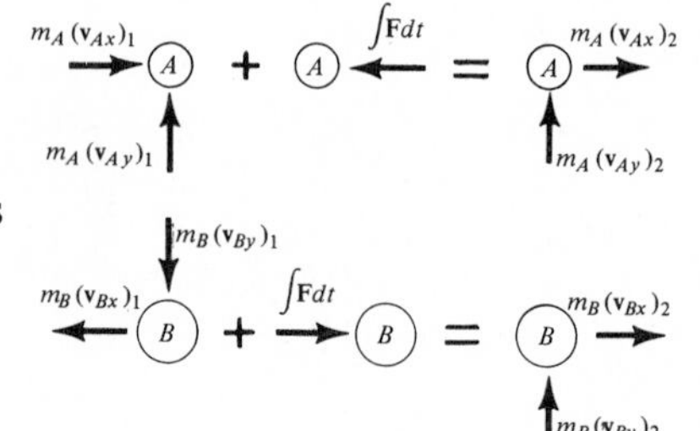

(b)

(c)

Fig. 15–20

Solution

The problem involves *oblique impact*. Why? In order to seek a solution, we have established the x and y axes along the line of impact and the plane of contact, respectively, Fig. 15–20*a*.

Resolving each of the initial velocities into x and y components, we have

$$(v_{Ax})_1 = 3 \cos 30° = 2.60 \text{ m/s}, \qquad (v_{Ay})_1 = 3 \sin 30° = 1.50 \text{ m/s}$$
$$(v_{Bx})_1 = -1 \cos 45° = -0.707 \text{ m/s}, \ (v_{By})_1 = -1 \sin 45° = -0.707 \text{ m/s}$$

Since the impact occurs only in the x direction (line of impact), Fig. 15–20*b*, the conservation of momentum for *both* disks can be applied in this direction. Why?

Conservation of "x" Momentum. In reference to the momentum diagrams, we have

$$(\overset{+}{\rightarrow}) \qquad m_A(v_{Ax})_1 + m_B(v_{Bx})_1 = m_A(v_{Ax})_2 + m_B(v_{Bx})_2$$
$$1(2.60) + 2(-0.707) = 1(v_{Ax})_2 + 2(v_{Bx})_2$$
$$(v_{Ax})_2 + 2(v_{Bx})_2 = 1.18 \qquad (1)$$

Coefficient of Restitution (x). Equation 15–12 is applied along the x axis (line of impact). Since both disks are *assumed* to have components of velocity in the $+x$ direction after collision, Fig. 15–20*b*, we have

$$e = \frac{(v_B)_2 - (v_A)_2}{(v_A)_1 - (v_B)_1}; \qquad 0.75 = \frac{(v_{Bx})_2 - (v_{Ax})_2}{2.60 + 0.707}$$
$$(v_{Bx})_2 - (v_{Ax})_2 = 2.48 \qquad (2)$$

Solving Eqs. (1) and (2) for $(v_{Ax})_2$ and $(v_{Bx})_2$ yields

$$(v_{Ax})_2 = -1.26 \text{ m/s} \ (\leftarrow) \qquad (v_{Bx})_2 = 1.22 \text{ m/s} \ (\rightarrow) \qquad \textit{Ans.}$$

Conservation of "y" Momentum. The momentum of *each disk* is *conserved* in the y direction (plane of contact), since *no impact* occurs in this direction. Hence, in reference to Fig. 15–20*b*.

$$(+\uparrow) \quad m_A(v_{Ay})_1 = m_A(v_{Ay})_2 \qquad (v_{Ay})_2 = 1.50 \text{ m/s} \ (\uparrow) \qquad \textit{Ans.}$$

and

$$(+\uparrow) \quad m_B(v_{By})_1 = m_B(v_{By})_2 \qquad (v_{By})_2 = -0.707 \text{ m/s} \ (\downarrow) \qquad \textit{Ans.}$$

Show that when the velocity components are summed, one obtains the results shown in Fig. 15–20*c*.

Problems

15–34. Ball A has a mass of 250 g and an initial velocity of $(v_A)_1 = 2$ m/s. As it rolls on a horizontal plane, it makes a direct collision with ball B, which has a mass of 200 g and is originally at rest. If both balls are of the same size and the collision is perfectly elastic ($e = 1$), determine the velocity of each ball after the collision. Show that the kinetic energy of the balls before and after collision is the same.

15–35. An ivory ball having a mass of 200 g is released from rest at a height of 500 mm above a very large fixed metal surface. If the ball rebounds to a height of 300 mm above the surface, determine the coefficient of restitution between the ball and the surface.

***15–36.** Ball B has a mass of 0.75 kg and is moving forward with a velocity of $(v_B)_1 = 4$ m/s when it strikes the 2-kg block A, which is originally at rest. If the coefficient of restitution between the ball and the block is $e = 0.6$, compute (a) the velocity of A and B just after collision and (b) the distance block A slides before coming to rest. The coefficient of friction between the block and the surface is $\mu_A = 0.4$.

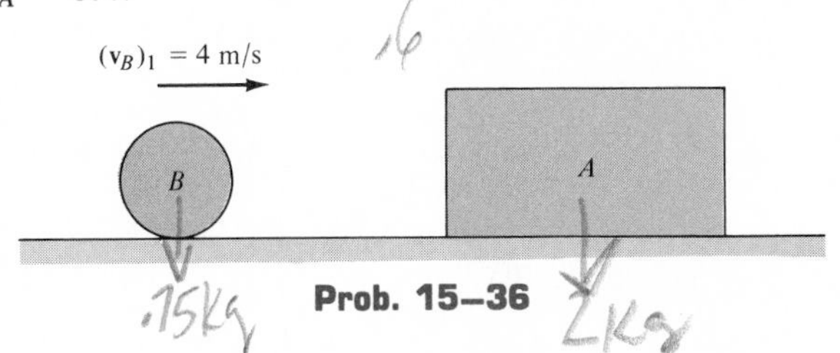

Prob. 15–36

15–37. The girl throws the ball with a vertical velocity of $v_1 = 8$ ft/s as shown. If the coefficient of restitution between the ball and the ground is $e = 0.8$, determine (a) the velocity of the ball just before striking the ground, and just after it rebounds from the ground; and (b) the maximum height to which the ball rises after the first bounce.

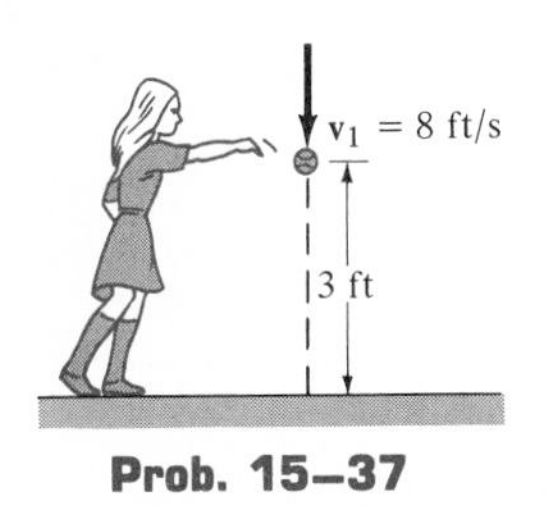

Prob. 15–37

15–38. The drop hammer H has a weight of 900 lb and falls from rest $h = 3$ ft onto a forged anvil plate P that has a weight of 500 lb. The plate is mounted on a set of springs which have a combined stiffness of $k_T = 500$ lb/ft. Determine (a) the velocity of H just before collision, and the velocity of P and H just after collision, and (b) the maximum compression in the springs caused by the impact. The coefficient of restitution between the hammer and the plate is $e = 0.6$. Neglect friction along the vertical guide posts A and B.

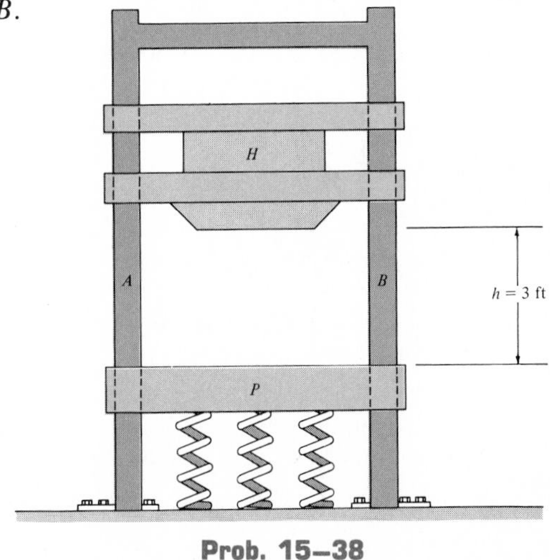

Prob. 15–38

15–39. The 15-lb suitcase A is released from rest at C. After it slides down the smooth ramp it strikes the 10-lb suitcase B, which is originally at rest. If the coefficient of restitution between the suitcases is $e = 0.3$ and the coefficient of kinetic friction between the floor DE and each suitcase is $\mu = 0.4$, determine (a) the velocity of A just before impact, (b) the velocities of A and B just after impact, and (c) the distance B slides before coming to rest.

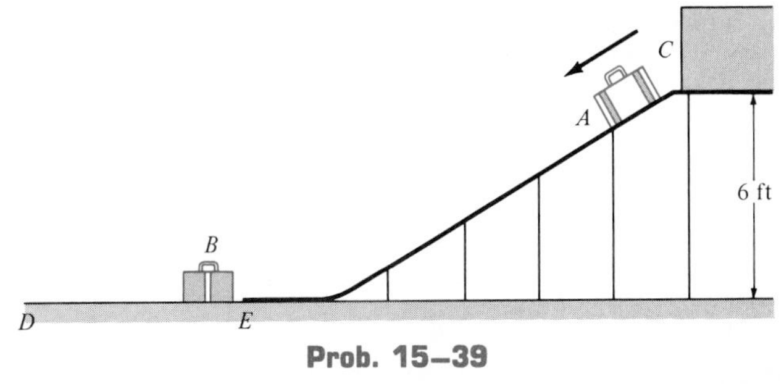

Prob. 15–39

***15–40.** Plates A and B each have a mass of 4 kg and are restricted to move along the frictionless guides. If the coefficient of restitution between the plates is $e = 0.7$, determine (a) the speed of both plates just after collision and (b) the maximum deflection of the spring. Plate A has a velocity of 4 m/s just before striking B. Plate B is originally at rest.

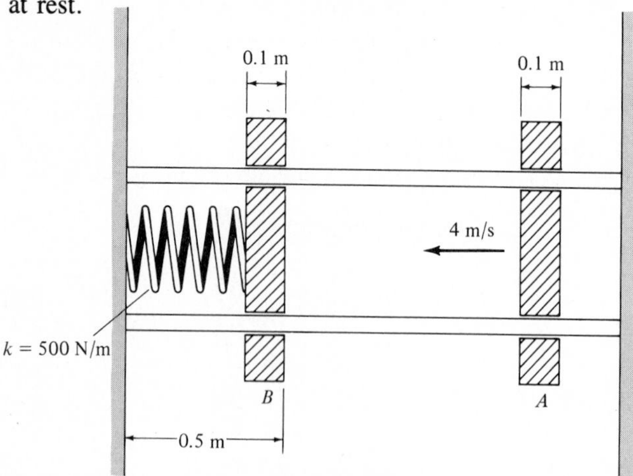

Prob. 15–40

15–41. If two balls A and B have the same mass and are subjected to direct central impact such that the collision is perfectly elastic ($e = 1$), prove that the kinetic energy before collision equals the kinetic energy after collision.

15–42. A stunt driver in car A travels in free flight off the edge of a ramp at C. At the point of maximum height he strikes car B. If the direct horizontal collision is perfectly plastic ($e = 0$), determine the required ramp speed v_C at the end of the ramp C, and the approximate distance s where both cars strike the ground. Each car has a mass of 3.5 Mg. Neglect the size of the cars in the calculation.

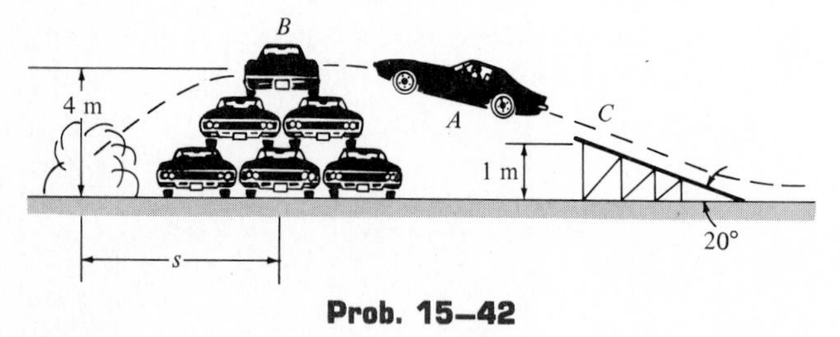

Prob. 15–42

15–43. The four balls shown each have the same mass m. If A and B are rolling forward with velocity **v** and strike C, explain why after collision C and D each move off with velocity **v**. Why doesn't D move off with velocity 2**v**? The collision is elastic, $e = 1$.

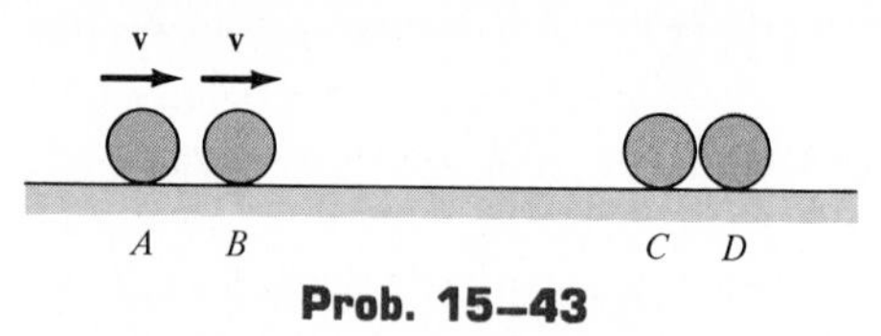

Prob. 15–43

***15–44.** Two smooth billiard balls A and B have an equal mass of $m = 200$ g. If A strikes B with a velocity of $(v_A)_1 = 2$ m/s as shown, determine their final velocities just after collision. Ball B is originally at rest and the coefficient of restitution is $e = 0.75$.

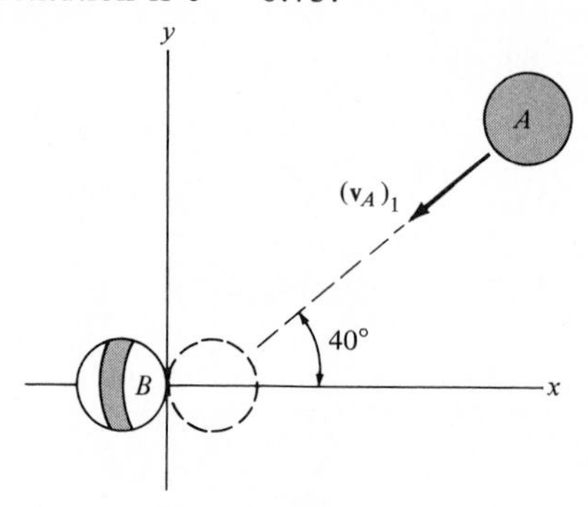

Prob. 15–44

15–45. The "stone" A used in the sport of curling is thrown over the ice track and strikes another "stone" B as shown. If each "stone" has a weight of 47 lb and the coef-

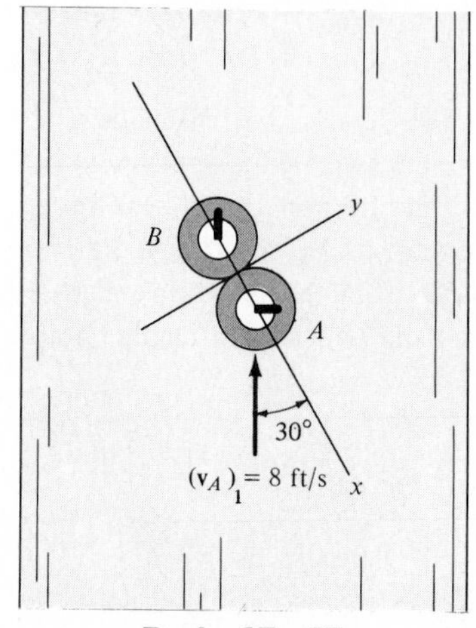

Prob. 15–45

ficient of restitution between the ''stones'' is $e = 0.7$, determine their final speed after collision. Initially A has a velocity of 8 ft/s and B is at rest. Neglect friction.

15–46. The two hockey pucks A and B each have a mass of 200 g. If they collide at O and are deflected along the dashed paths, determine their speed just after impact. Assume that the icy surface over which they slide is smooth. *Hint:* Since the y' axis is *not* along the line of impact, apply the conservation of momentum along the x' and y' axes.

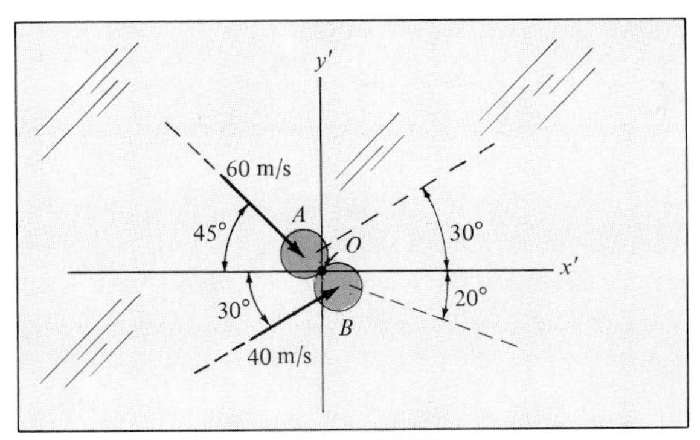

Prob. 15–46

15–47. Two coins A and B have the initial velocities shown just before they collide at point O. If they have weights of $W_A = 13.2(10^{-3})$ lb and $W_B = 6.6(10^{-3})$ lb and the surface upon which they slide is smooth, determine their speed just after impact. The coefficient of restitution is $e = 0.65$.

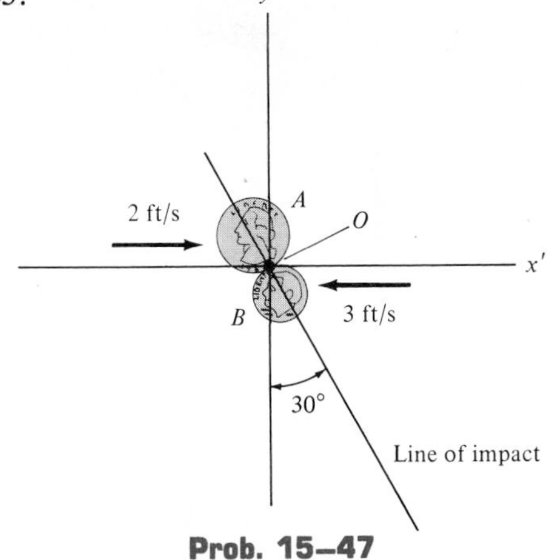

Prob. 15–47

***15–48.** Two cars A and B, each having a weight of 4,000 lb, collide on the icy pavement of an intersection. The direction of motion of each car after collision is measured from snow tracks as shown. If the driver in car A states that he was going 30 mi/h just before collision and that after collision he applied the brakes so that his car skidded 10 ft before stopping, determine the approximate speed of car B just before the collision. Assume that the coefficient of friction between the car wheels and the pavement is $\mu = 0.15$. *Note:* The line of impact has not been defined; furthermore, this information is not needed for the solution.

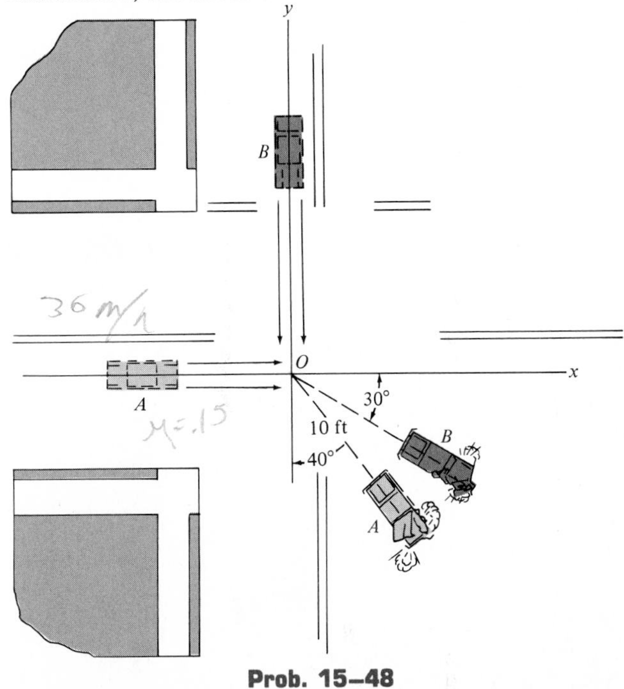

Prob. 15–48

15–49. Ball A strikes ball B with an initial velocity of $(v_A)_1$ as shown. If both balls have the same mass and the collision is perfectly elastic, determine the angle θ after collision. Ball B is originally at rest.

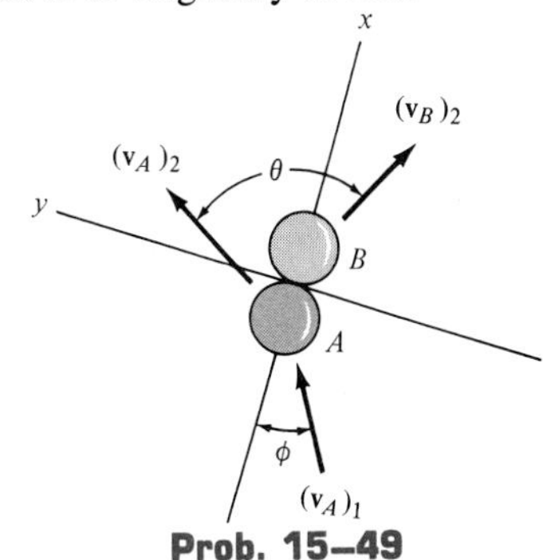

Prob. 15–49

15.5 Angular Momentum

The *angular momentum* of a particle about point O is defined as the "moment" of the particle's linear momentum about O. Since this concept is analogous to finding the moment of a force about a point, the angular momentum, $\mathbf{H}_O$, is sometimes referred to as the *moment of momentum*.

Definition. Consider a particle which is moving along a curve lying in the x-y plane, Fig. 15–21. The *magnitude* of the angular momentum can be computed about point O (actually the z axis) by using the formula

$$H_O = (d)(mv) \tag{15–13}$$

Here d is the moment arm or perpendicular distance from O to the line of action of $m\mathbf{v}$. Common units for this magnitude are kg · m^2/s or slug · ft^2/s. The *direction* of $\mathbf{H}_O$ is defined by the right-hand rule. As shown in Fig. 15–21, the curl of the fingers of the right hand indicates the sense of rotation of $m\mathbf{v}$ about O, so that in this case the thumb (or $\mathbf{H}_O$) is directed perpendicular to the x-y plane along the $+z$ axis.

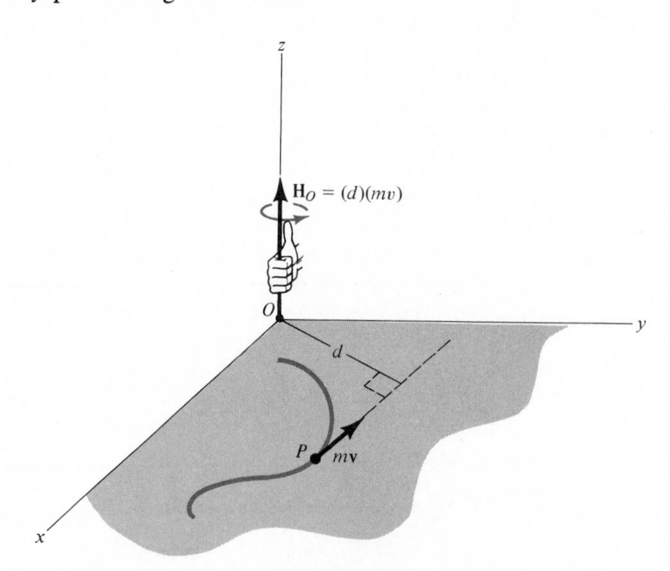

Fig. 15–21

Vector Formulation. If the particle is moving along a space curve, Fig. 15–22, the vector cross product can be used to determine the *angular momentum* about O. In this case

$$\mathbf{H}_O = \mathbf{r} \times m\mathbf{v} \tag{15–14}$$

Angular Momentum of a System of Particles 15.6

Consider the system of particles shown in Fig. 15–22 which are all moving in the x-y plane. The forces acting on the arbitrary ith particle of the system consist of a resultant external force $\mathbf{F}_i$ and the internal forces $\Sigma\mathbf{f}_i$ which act between the ith particle and all other particles within the system. At the instant considered the particle has a velocity $\mathbf{v}_i$. If the x and y force and velocity components are considered, we can write the equations of motion for the ith particle, having a mass m_i, as

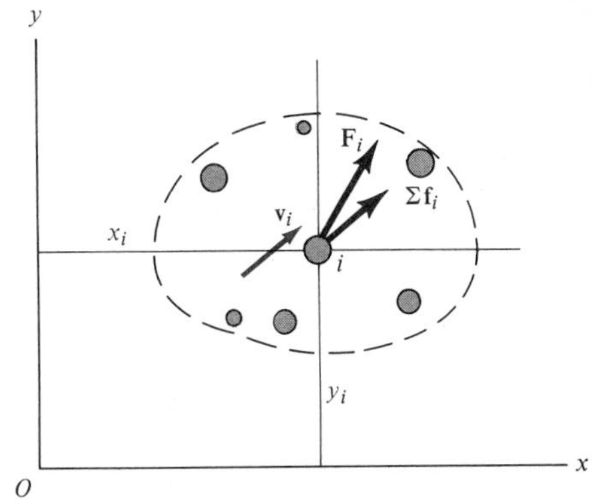

Fig. 15–22

$$F_{ix} + \Sigma f_{ix} = \frac{d}{dt}(m_i v_{ix})$$

$$F_{iy} + \Sigma f_{iy} = \frac{d}{dt}(m_i v_{iy})$$

Summing moments about the fixed point O ($+z$ axis) and using the above two equations, we have

$$\circlearrowleft +\Sigma M_{Oi} = (F_{iy} + \Sigma f_{iy})x_i - (F_{ix} + \Sigma f_{ix})y_i = x_i \frac{d}{dt}(m_i v_{iy}) - y \frac{d}{dt}(m_i v_{ix})$$

By differentiation, we can also write

$$\Sigma M_{Oi} = \frac{d}{dt}(x_i m_i v_{iy} - y_i m_i v_{ix})$$

$$\Sigma M_{Oi} = \frac{d}{dt}(x_i m_i v_{iy}) - \frac{d}{dt}(y_i m_i v_{ix}) = \frac{dH_{Oi}}{dt}$$

or, neglecting the subscript i, for a *single particle* we have

$$\Sigma M_O = \frac{dH_O}{dt} \qquad (15\text{–}14)$$

where ΣM_O is the moment of both the external and internal forces about the fixed point O, and dH_O/dt is the time derivative of the angular momentum taken about O.

If the entire *system of particles* is considered, then the moments created by the internal forces would cancel out. This is because corresponding pairs of internal forces are equal in magnitude, opposite in direction, and collinear. Using a similar equation to represent this condition, we have

$$\Sigma M_O = \frac{dH_O}{dt} \qquad (15\text{–}15)$$

Here ΣM_O represents the sum of the moments of all the external forces acting on the particles about the fixed point O, and dH_O/dt is the time derivative of the angular momentum of the system of particles about O.

A similar result can also be developed for moments summed about the *mass center* G of the system of particles, which may be moving.

$$\Sigma M_G = \frac{dH_G}{dt} \tag{15–15}$$

15.7 Angular Impulse and Momentum Principles

Principle of Angular Impulse and Momentum. If Eq. 15–14 is rewritten in the form $\Sigma \mathbf{M}_O \, dt = d\mathbf{H}_O$ and integrated, we have, assuming that at time $t = t_1$, $\mathbf{H}_O = (\mathbf{H}_O)_1$ and at time $t = t_2$, $\mathbf{H}_O = (\mathbf{H}_O)_2$,

$$\Sigma \int_{t_1}^{t_2} \mathbf{M}_O \, dt = (\mathbf{H}_O)_2 - (\mathbf{H}_O)_1$$

or

$$(\mathbf{H}_O)_1 + \Sigma \int_{t_1}^{t_2} \mathbf{M}_O \, dt = (\mathbf{H}_O)_2 \tag{15–16}$$

This equation is referred to as the *principle of angular impulse and momentum*. The initial and final angular momenta $(\mathbf{H}_O)_1$ and $(\mathbf{H}_O)_2$ are defined as the moment of the linear momentum of the particle at the instants t_1 and t_2, respectively. The second term on the left side, $\Sigma \int_{t_1}^{t_2} \mathbf{M}_O \, dt$, is called the *angular impulse*. It is computed on the basis of integrating, with respect to time, the moments of all the forces acting on the particle over the time interval t_1 to t_2.

In a similar manner, using Eq. 15–15, the principle of angular impulse and momentum for a system of particles may be written as

$$\Sigma(\mathbf{H}_O)_1 + \Sigma \int_{t_1}^{t_2} \mathbf{M}_O \, dt = \Sigma(\mathbf{H}_O)_2 \tag{15–17}$$

Here the first and third terms represent the angular momenta of the system of particles at the instants t_1 and t_2. The second term is the vector sum of the angular impulses given to all the particles during the time period t_1 to t_2. These impulses are created only by the moments of the external forces acting on the system. Hence, if the particle is confined to move in the x-y plane, three independent scalar equations may be written to express the motion; namely,

$$\begin{aligned} m(v_x)_1 + \Sigma \int_{t_1}^{t_2} F_x \, dt &= m(v_x)_2 \\ m(v_y)_1 + \Sigma \int_{t_1}^{t_2} F_y \, dt &= m(v_y)_2 \\ (H_O)_1 + \Sigma \int_{t_1}^{t_2} M_O \, dt &= (H_O)_2 \end{aligned} \tag{15–18}$$

The first two of these equations represent the principle of linear impulse and momentum in the x and y directions, and the third equation represents the principle of angular impulse and momentum about the z axis.

Conservation of Angular Momentum. When the angular impulses acting on a particle are all zero during the time t_1 to t_2, Eq. 15–16 reduces to the following simplified form,

$$(\mathbf{H}_O)_1 = (\mathbf{H}_O)_2 \tag{15–19}$$

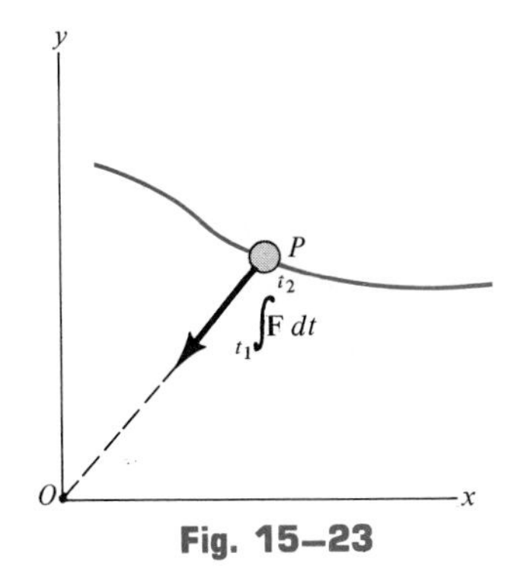

Fig. 15–23

This equation is known as the *conservation of angular momentum*. It states that from t_1 to t_2 the particle's angular momentum remains constant. Obviously, if no external impulse is applied to the particle, both linear and angular momentum will be conserved. In some cases, however, the particle's angular momentum will be conserved and linear momentum may not. An example of this occurs when the particle is subjected *only* to a *central force*. For example, in Fig. 15–23, the impulsive central force **F** is always directed toward point O as the particle moves along the path. Hence, the angular impulse (moment) created by **F** about the z axis passing through point O is always zero, and therefore angular momentum of the particle is conserved about this axis.

Using Eq. 15–17, we can also write the conservation of angular momentum for a system of particles, namely,

$$\Sigma(\mathbf{H}_O)_1 = \Sigma(\mathbf{H}_O)_2 \tag{15–20}$$

In this case the summation must include the angular momenta of all the particles in the system.

PROCEDURE FOR ANALYSIS

When applying the principles of angular impulse and momentum, or the conservation of angular momentum, it is suggested that the following procedure be used.

Impulse and Momentum Diagrams. Draw the impulse and momentum diagrams for the particle. In particular, the impulse diagram provides a convenient means for determining any axis about which angular momentum may be conserved. For this to occur, the moments of the linear impulses about the axis must be zero throughout the time period t_1 to t_2.

Momentum Equations. Apply the principle of angular impulse and momentum, $(\mathbf{H}_O)_1 + \Sigma \int_{t_1}^{t_2} \mathbf{M}_O\, dt = (\mathbf{H}_O)_2$, or if appropriate, the conservation of angular momentum, $(\mathbf{H}_O)_1 = (\mathbf{H}_O)_2$. Each of the terms in these equations can be formulated directly from the data shown on the impulse and momentum diagrams.

If other equations are needed for the problem solution, when appropriate, use the principle of linear impulse and momentum, the principle of work and energy, the equations of motion, or kinematics.

The following examples numerically illustrate application of the above procedure.

Example 15–12

The ball B, shown in Fig. 15–24a, has a weight of 0.8 lb and is attached to a cord which passes through a hole at A in a smooth table. When the ball is $r_1 = 1.75$ ft from the hole it is rotating around in a circle such that its speed is $v_1 = 4$ ft/s. If by applying a force **F** the cord is pulled downward through the hole with a constant speed of $v_c = 6$ ft/s, determine (a) the speed of the ball at the instant it is $r_2 = 0.6$ ft from the hole and (b) the amount of work done by the force **F** in shortening the cord.

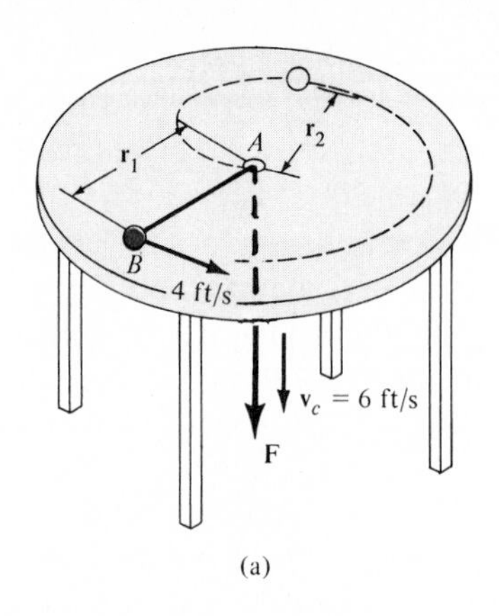

(a)

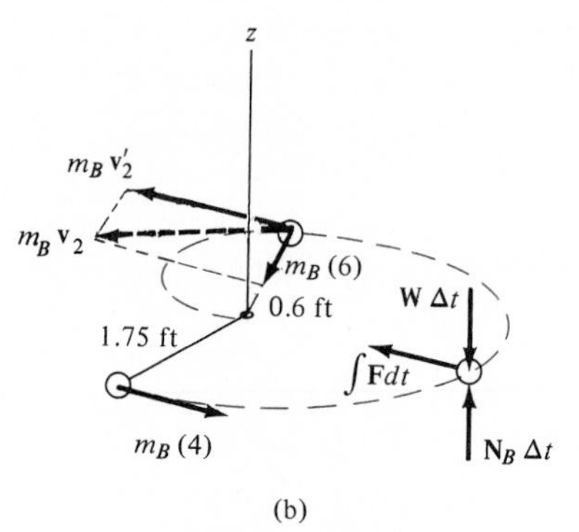

(b)

Fig. 15–24

Solution

Part (a)

Impulse and Momentum Diagrams. As the ball moves from radial position r_1 to r_2, Fig. 15–24b, the three impulses acting on it are all unknown; however, since the cord tension **F** passes through the z axis and **W** and $\mathbf{N}_B$ are parallel to it, the moments, or angular impulses created by the forces, are all *zero* about this axis. Hence, the conservation of angular momentum applies about the z axis.

Conservation of Angular Momentum. As indicated on the second momentum diagram, $m_B\mathbf{v}_2$ is resolved into two components. The radial component, $m_B(6)$, is known; however, it produces zero angular momentum about the z axis. Thus,

$$\mathbf{H}_1 = \mathbf{H}_2$$

$$r_1 m_B v_1 = r_2 m_B v_2'$$

$$1.75\left(\frac{0.8}{32.2}\right)4 = 0.6\left(\frac{0.8}{32.2}\right)v_2'$$

$$v_2' = 11.67 \text{ ft/s}$$

The speed of the ball is thus

$$v_2 = \sqrt{(11.67)^2 + (6)^2}$$
$$= 13.1 \text{ ft/s} \qquad \textit{Ans.}$$

Part (b). The only force that does work on the ball is **F.** (The normal force and weight do not move vertically.) The initial and final kinetic energy of the ball can be determined so that from the principle of work and energy we have

$$T_1 + \Sigma U_{1-2} = T_2$$

$$\frac{1}{2}\left(\frac{0.8}{32.2}\right)(4)^2 + U_{1-2} = \frac{1}{2}\left(\frac{0.8}{32.2}\right)(13.1)^2$$

$$U_{1-2} = 1.94 \text{ ft} \cdot \text{lb} \qquad \textit{Ans.}$$

Example 15–13

The 2-kg block shown in Fig. 15–25*a* rests on a smooth horizontal surface and is attached to a cord which is fixed at point O. If the block is acted upon by a force of $F = 6$ N which is always directed perpendicular to the cord, determine the speed of the block when $t = 3$ s. Originally the block is at rest.

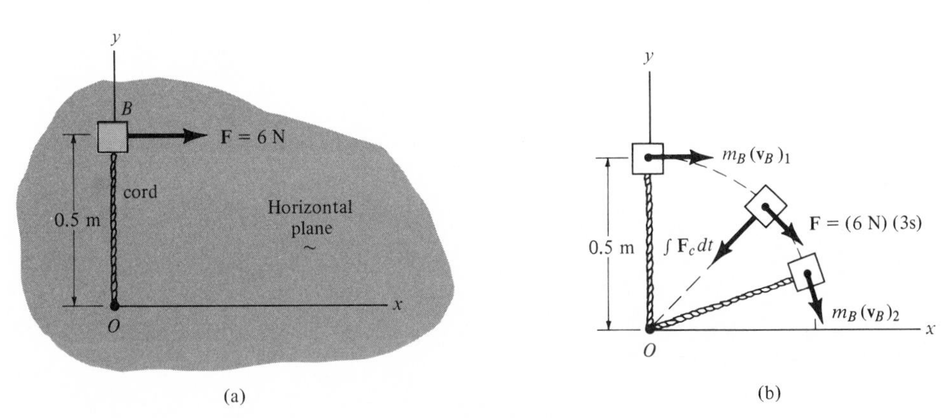

Fig. 15–25

Solution

Impulse and Momentum Diagrams. After the block has been launched, it slides along the dashed circular path shown in Fig. 15–25*b*. By inspection of the impulse diagram, $\mathbf{F}_c$ creates no angular momentum about point O (or the z axis) since the moment of the linear impulse $\int \mathbf{F}_c\, dt$ about O is always zero. ($\mathbf{F}_c$ is a central force.)

Conservation of Angular Momentum. The speed $(\mathbf{v}_B)_2$ can be obtained by applying the principle of angular momentum about O (the z axis), i.e.,

$$(H_z)_1 + \Sigma \int M_z\, dt = (H_z)_2$$

$$0 + (0.5 \text{ m})(6 \text{ N})(3 \text{ s}) = (2 \text{ kg})(v_B)_2$$

$$v_B = 4.5 \text{ m/s} \qquad \textit{Ans.}$$

Example 15–14

The frame shown in Fig. 15–26*a* is originally at rest and supports the 5-kg sphere *A*. A moment of $M = (3t)$ N·m, where *t* is measured in seconds, is applied to shaft *CD* and a force $P = 10$ N is applied perpendicular to arm *AB*. If the mass of the frame is neglected, determine the speed of the sphere in 4 s.

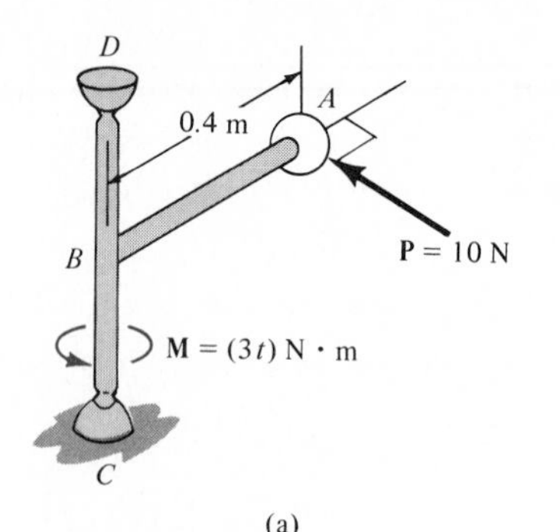

Fig. 15–26

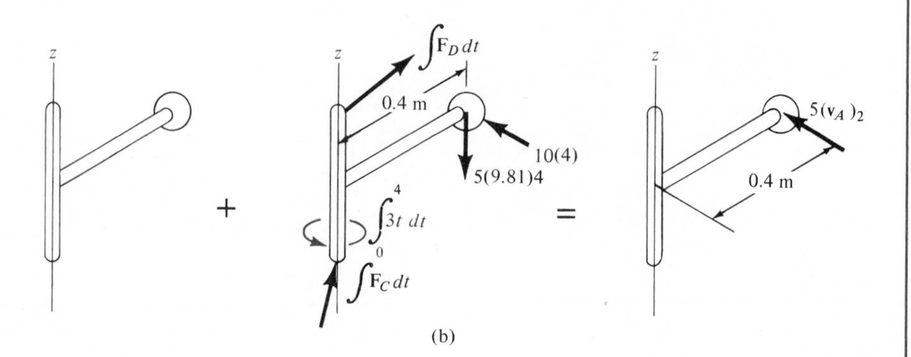

Solution

Impulse and Momentum Diagrams. As shown in Fig. 15–26*b*, the impulses $\int \mathbf{F}_D\, dt$ and $\int \mathbf{F}_C\, dt$ are unknown in both magnitude and direction and represent the effect of the supports on the frame. These impulses can be eliminated from the analysis by applying the principle of angular impulse and momentum about the *z* axis. Why? If this is done, the angular impulse created by the weight of sphere *A* is also eliminated, since it acts parallel to the *z* axis and therefore creates zero moment about this axis.

Principle of Angular Impulse and Momentum

$$(H_z)_1 + \Sigma \int M_z\, dt = (H_z)_2$$

$$(H_{Az})_1 + \int M\, dt + r_{AB}P(\Delta t) = (H_{Az})_2$$

$$0 + \int_0^4 3t\, dt + (0.4)(10)(4) = 5(v_A)_2(0.4)$$

$$24 + 16 = 2(v_A)_2$$

$$(v_A)_2 = 20.0 \text{ m/s} \qquad \textbf{\textit{Ans.}}$$

Hint: Cross Product A×B

Magnitude.... |AB sin θ|

Problems

15–50. Determine the angular momentum $\mathbf{H}_O$ for each of the particles about point O.

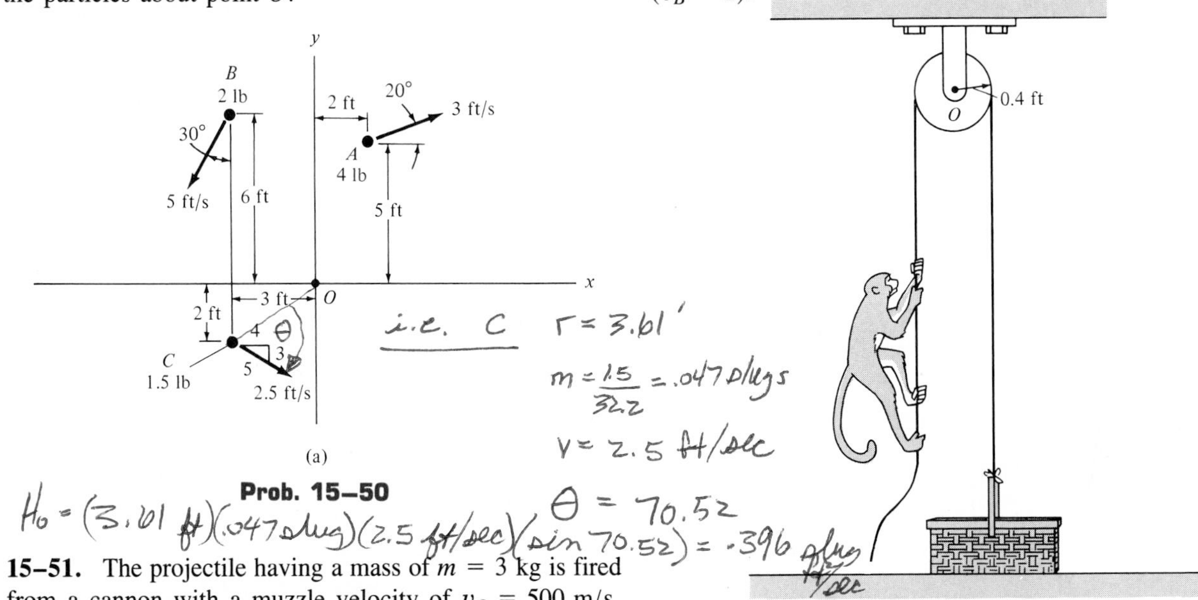

Prob. 15–50

$H_o = (3.61\ ft)(.047\ slug)(2.5\ ft/sec)(\sin 70.52) = .396$ slug ft²/sec

15–51. The projectile having a mass of $m = 3$ kg is fired from a cannon with a muzzle velocity of $v_O = 500$ m/s. Determine the projectile's angular momentum about point O at the instant it is at the maximum height of its trajectory, i.e., when its velocity is horizontal.

Prob. 15–51

***15–52.** A basket and its contents have a weight of 10 lb. Determine the speed at which the basket rises when $t = 3$ s, if initially a monkey having a weight of 20 lb begins to climb upward along the other end of the rope with a constant speed of $v_{m/r} = 2$ ft/s, measured relative to the rope. Neglect the mass of the pulley and rope. *Hint:* Angular momentum is conserved about point O. Assume that the basket moves upward at v_B, then the speed of the monkey is $(v_B - 2)$.

Prob. 15–52

15–53. The two blocks A and B each have a mass of 500 g. The blocks are fixed to the horizontal rods and their initial velocity is 2 m/s in the direction shown. If a couple moment of $M = 0.8$ N·m is applied about CD of the frame, determine the speed of the blocks in 4 s. The mass of the supporting frame is negligible and it is free to rotate about CD.

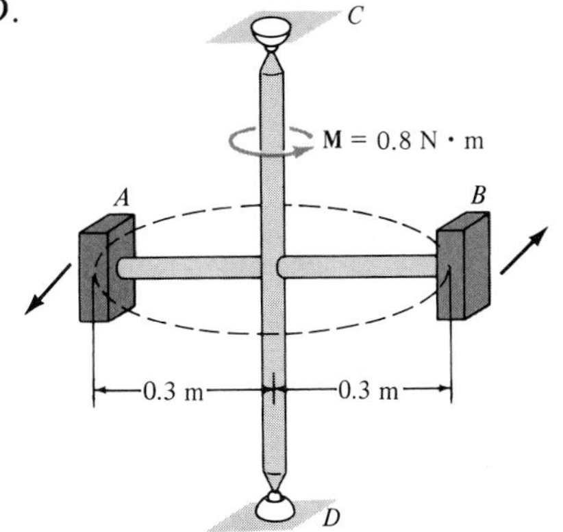

Prob. 15–53

15–54. The four spheres each have a weight of 5 lb and are rigidly attached to a crossbar frame of negligible mass. If a couple $M = (5t + 2)$ lb · ft, where t is in seconds, is applied as shown, determine the speed of each of the spheres in 3 s, starting from rest.

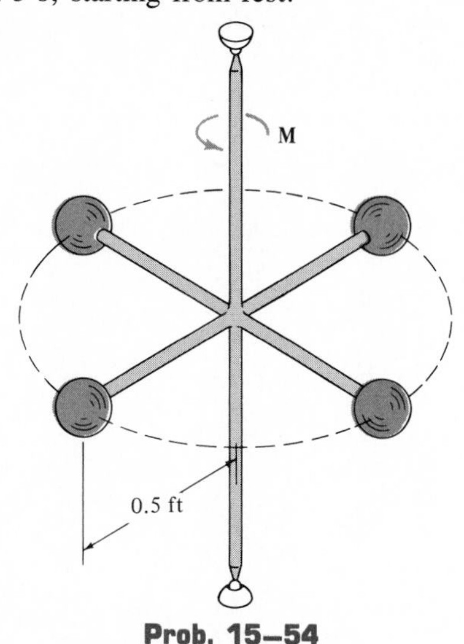

Prob. 15–54

15–55. A girl having a weight of 80 lb is skating around in a circle of radius $r_A = 20$ ft with a speed of $(v_A)_1 = 5$ ft/s while holding on to the end of a rope. If her partner starts to pull the rope inward with a constant speed of $v_r = 2$ ft/s, determine the girl's speed at the instant $r_B = 10$ ft. How much work is done by her partner after pulling in the rope? Neglect friction and assume the girl remains in a rigid position.

Prob. 15–55

***15–56.** A cord of length l is attached to a ball of mass m and a peg having a radius $a \ll l$. If the ball is given an initial velocity $\mathbf{v}_1$, perpendicular to the cord, determine the ball's velocity just after it has wrapped the cord four times around the peg.

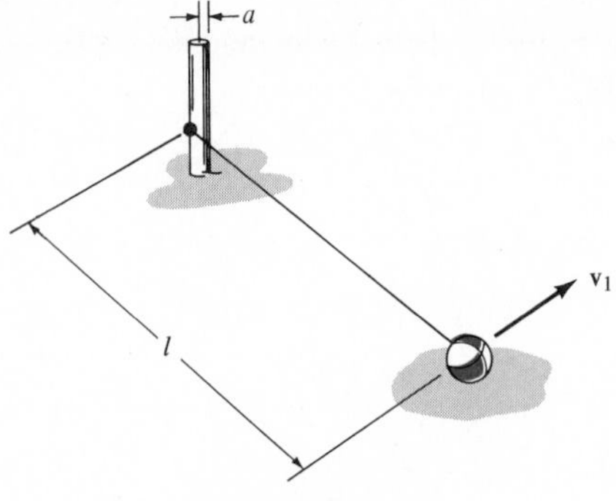

Prob. 15–56

15–57. The boy has a weight of 80 lb. While holding on to a ring he runs in a circle and then lifts his feet off the ground, holding himself in the crouched position shown. *Initially* his center of gravity G is $r_A = 10$ ft from the pole and his velocity is *horizontal* such that $v_A = 8$ ft/s. (a) Use the conservation of energy principle to determine his velocity when he is at B, where $r_B = 7$ ft, $\Delta z = 2$ ft. (b) Use the conservation of angular momentum about the z axis to find $(v_B)_{\text{horiz}}$, then determine the vertical component of his velocity, $(\mathbf{v}_B)_z$, which is causing him to fall downward.

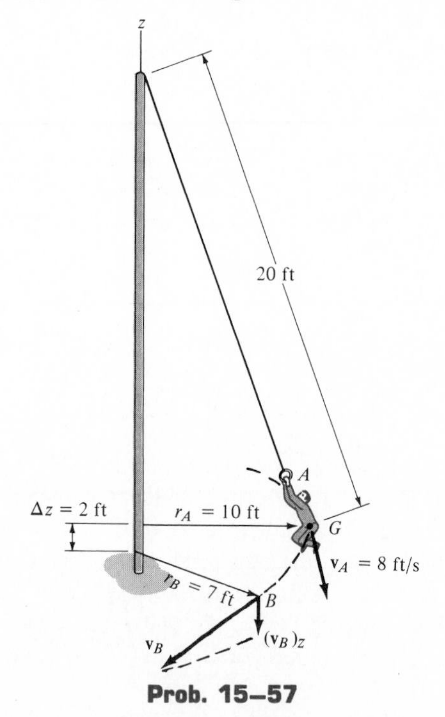

Prob. 15–57

15–58. The elastic cord, having an unstretched length of 300 mm and a stiffness of $k = 600$ N/m, is attached to a fixed point at A and a block at B which has a mass of 2 kg. If the block is given an initial velocity of $v_B = 2$ m/s along the x axis, from the position shown, determine the speed of the block at the instant the cord becomes slack at C. How close, d, does the block come within approaching A? The block slides on the smooth horizontal plane.

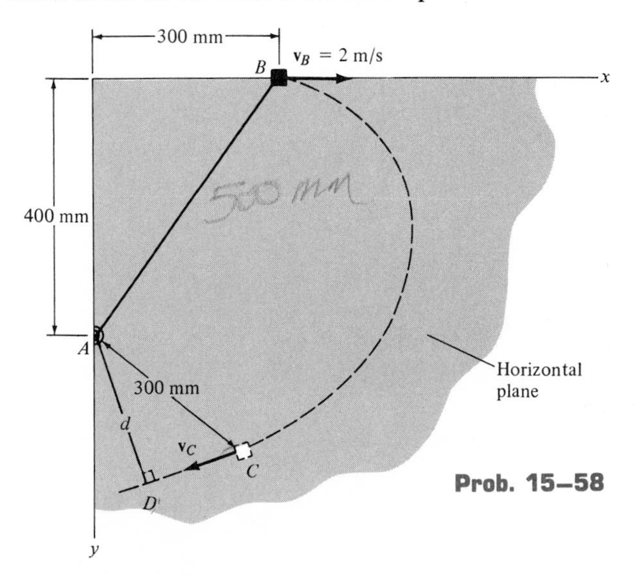

Prob. 15–58

15–59. A toboggan and rider, having a total mass of 150 kg, enter horizontally tangent to a 90° circular curve with a velocity of $v_A = 70$ km/h. If the track is flat and banked at an angle of 60°, determine the velocity $\mathbf{v}_B$ and the angle θ of "descent," measured from the horizontal in a vertical $x - z$ plane, at which the toboggan exits at B. Neglect friction in the calculation. The radius r_B equals 57 m.

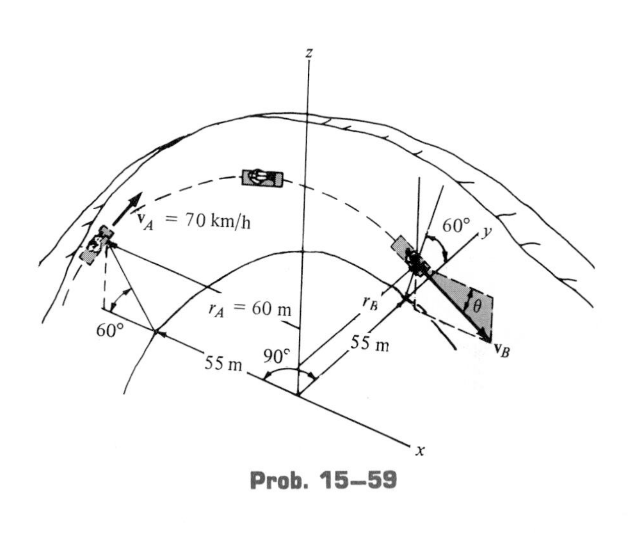

Prob. 15–59

Review Problems

***15–60.** When the 4-kg ballistic pendulum is at rest, $\theta = 0°$, a 2-g bullet strikes and becomes embedded in it. If it is observed that the pendulum swings upward to a maximum angle of $\theta = 6°$, estimate the speed of the bullet just before it strikes the pendulum.

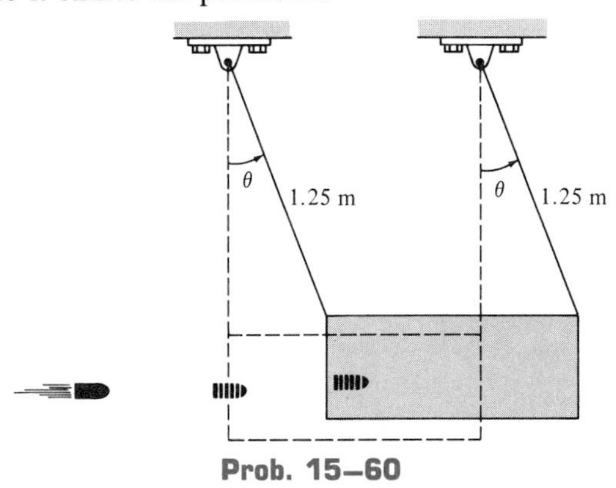

Prob. 15–60

15–61. Determine the impulse given to the 1.5-lb soccer ball when it is at A, and compute its momentum when it reaches its highest point B and just before it strikes the ground at C.

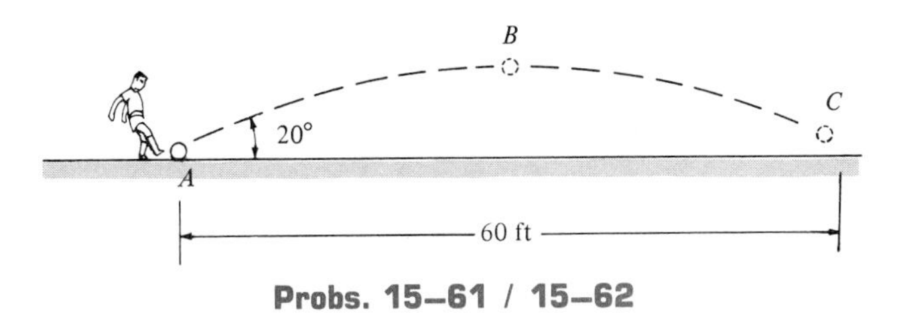

Probs. 15–61 / 15–62

15–62. If the coefficient of restitution between the 1.5-lb soccer ball and the ground is $e = 0.6$, determine the momentum of the ball just after it rebounds from the ground at C.

15–63. The barge has a mass of 20 Mg and supports the automobile which has a mass of 1.5 Mg. If the barge is not tied to the pier P and the automobile is driven forward at a constant speed of 5 m/s, measured relative to the barge, determine the speed at which the barge moves away from the pier. Neglect the resistance of the water.

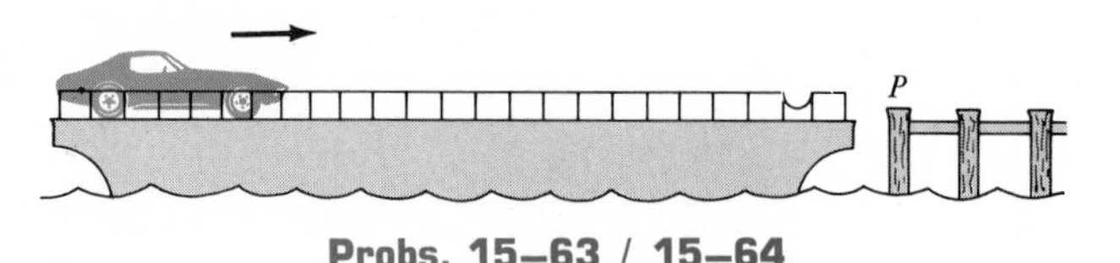

Probs. 15–63 / 15–64

***15–64.** Determine how far the 20-Mg barge moves from the pier if the 1.5-Mg automobile is driven forward 40 m and stops. Neglect the resistance of the water. Both the barge and the automobile are originally at rest.

15–65. The ball A has a weight of 0.5 lb and is released from rest 8 in. above a spring-supported plate P which weighs 0.25 lb. If the coefficient of restitution between the plate and the ball is $e = 0.6$, determine (a) the speed of the plate just after impact and (b) the maximum compression of the spring.

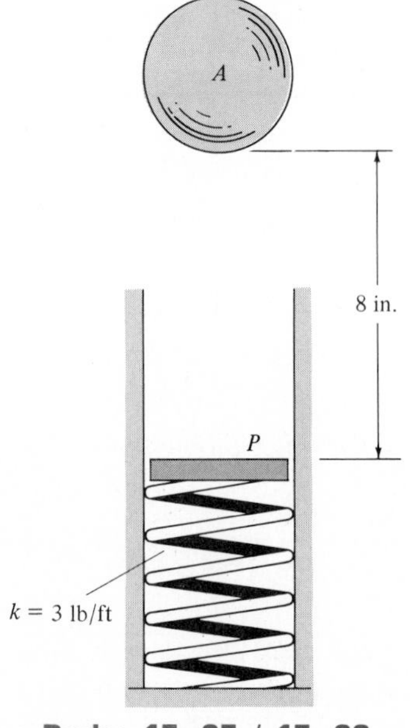

Probs. 15–65 / 15–66

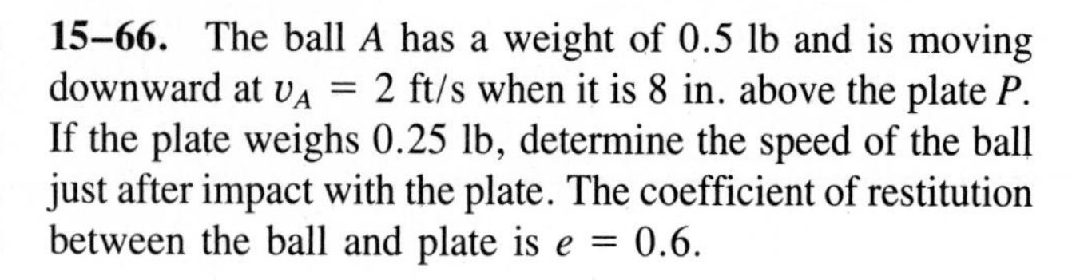

15–66. The ball A has a weight of 0.5 lb and is moving downward at $v_A = 2$ ft/s when it is 8 in. above the plate P. If the plate weighs 0.25 lb, determine the speed of the ball just after impact with the plate. The coefficient of restitution between the ball and plate is $e = 0.6$.

15–67. The two disks A and B have a mass of 3 and 5 kg, respectively. If they collide with the initial velocities shown, determine their velocities just after impact. The coefficient of restitution is $e = 0.65$.

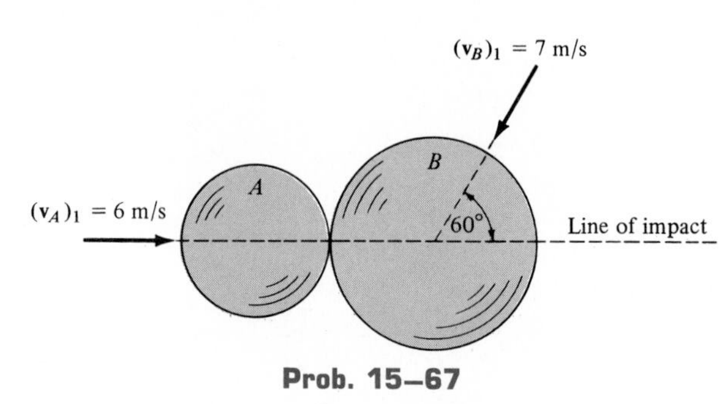

Prob. 15–67

***15–68.** The ball has a weight of 0.5 lb and is dropped from rest at a height of $h = 3$ ft. If the coefficient of restitution between the ball and the board at B is $e = 0.6$, determine the angle ϕ of rebound.

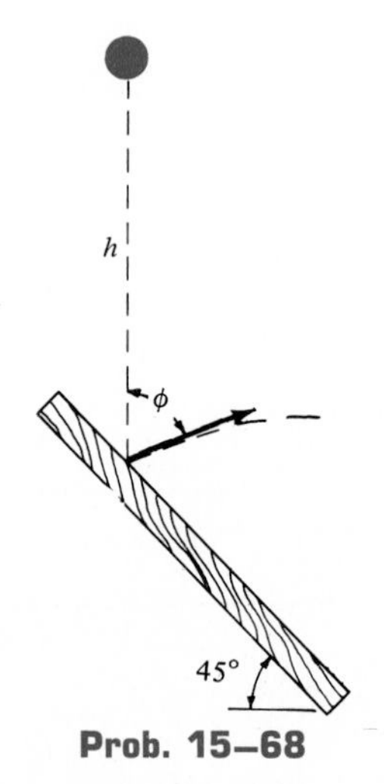

Prob. 15–68

15–69. The billiard ball has a mass m and is moving with a speed of 2.5 m/s when it strikes the side of the pool table at A. If the coefficient of restitution between the ball and the table is $e = 0.6$, determine the speed of the ball just after it strikes the table twice, i.e., at A, then at B.

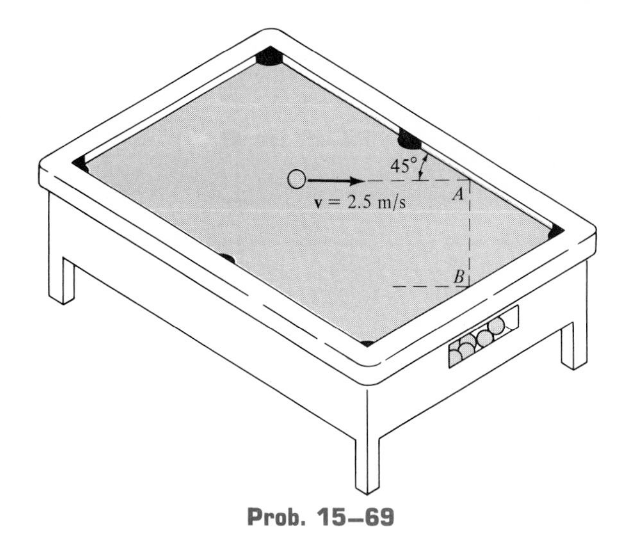

Prob. 15–69

Planar Kinematics of a Rigid Body

In this chapter the kinematics or the geometry of the planar motion for a rigid body will be discussed. This study is important since the design of gears, cams, and mechanisms used for machine operations depends upon the geometry of their motions. Furthermore, once the kinematics of a rigid body is thoroughly understood, it will be possible to apply the equations of motion, which relate the forces on the body to the body's motion.

A rigid body can be subjected to three types of planar motion: namely, translation, rotation about a fixed axis, and general plane motion. We will first define each of these motions, then discuss the analysis of each separately. In all cases it will be shown that rigid-body planar motion is completely specified provided the motions of any two points on the body are known.

Rigid-Body Motion 16.1

When each of the particles of a rigid body moves along a path which is equidistant from a fixed plane, the body is said to undergo *planar motion*. As previously stated, there are three types of planar motion; in order of increasing complexity, they are:

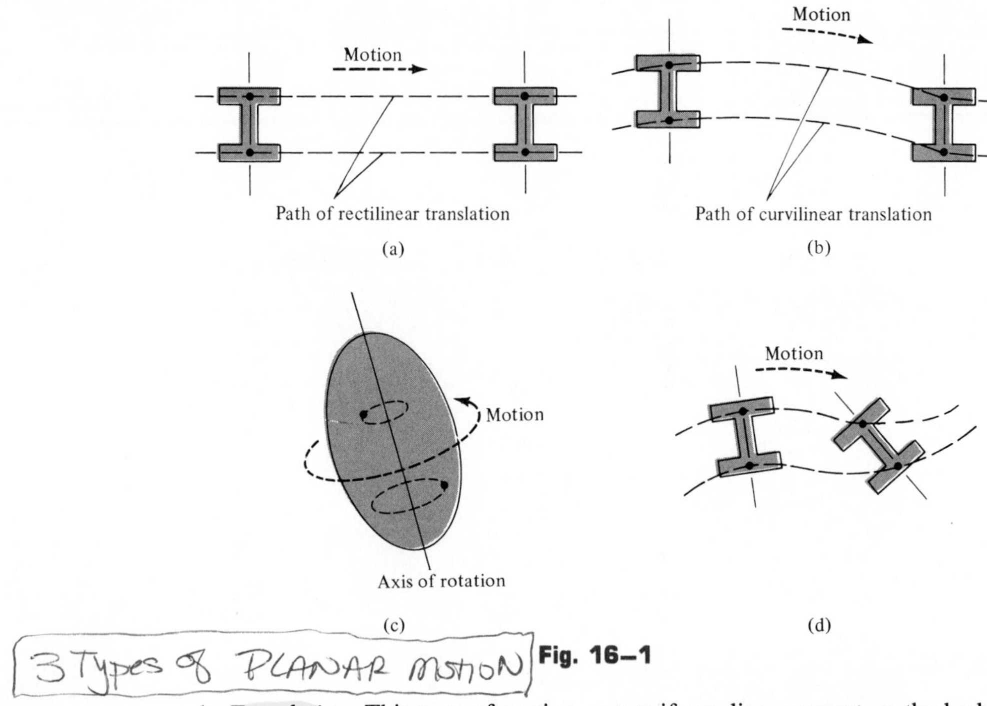

Fig. 16–1

1. *Translation.* This type of motion occurs if any line segment on the body remains parallel to its original direction during the motion. When the paths of motion for all the particles of the body are along parallel straight lines, Fig. 16–1*a*, the motion is called *rectilinear translation*. However, if the paths of motion are along curved lines which are all parallel, Fig. 16–1*b*, the motion is called *curvilinear translation*.
2. *Rotation about a fixed axis*. When a rigid body rotates about a fixed axis, all the particles of the body, except those which lie on the axis of rotation, move along circular paths, Fig. 16–1*c*.
3. *General plane motion*. When a body is subjected to general plane motion, it undergoes a combination of translation *and* rotation, Fig. 16–1*d*. The translation occurs within a reference plane, and the rotation occurs about an axis perpendicular to the reference plane.

The above planar motions are exemplified by the moving parts of the crank mechanism shown in Fig. 16–2.

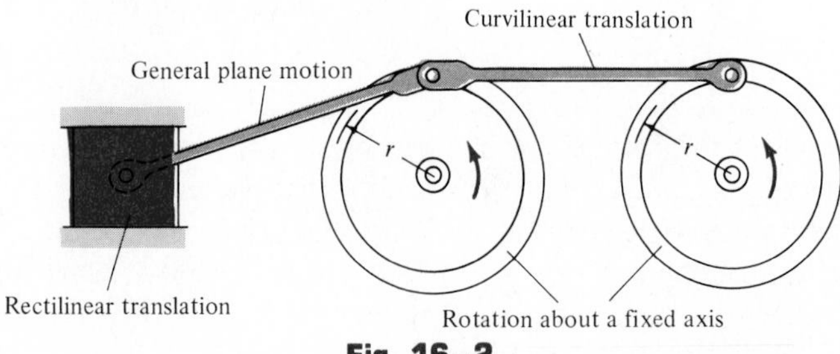

Fig. 16–2

Translation 16.2

Consider a rigid body which is subjected to either rectilinear or curvilinear translation in the x-y plane, Fig. 16–3.

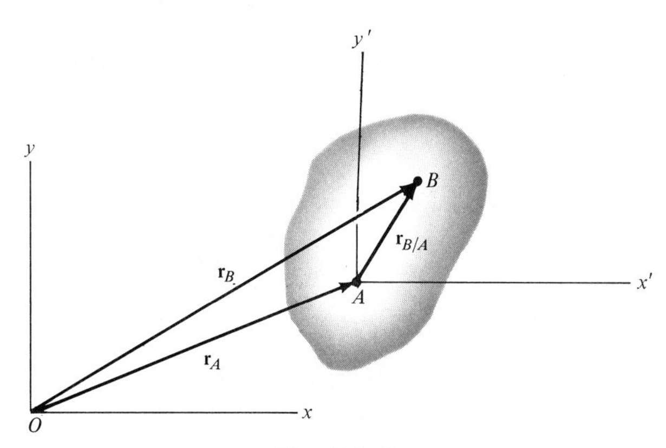

Fig. 16–3

Position. The locations of points A and B in the body are defined from the fixed x, y reference frame by using *position vectors* $\mathbf{r}_A$ and $\mathbf{r}_B$. The translating x', y' coordinate system is *fixed in the body* and has its origin located at A, hereafter referred to as the *base point*. The position of B with respect to A is denoted by the *relative-position vector* $\mathbf{r}_{B/A}$ ("$\mathbf{r}$ of B with respect to A"). By vector addition,

$$\mathbf{r}_B = \mathbf{r}_A + \mathbf{r}_{B/A}$$

Velocity. A relationship between the instantaneous velocities of A and B is obtained by taking the time derivative of the position equation, which yields $\mathbf{v}_B = \mathbf{v}_A + d\mathbf{r}_{B/A}/dt$. Here $\mathbf{v}_A$ and $\mathbf{v}_B$ denote *absolute velocities* since these vectors are measured from the x, y axes. The term $d\mathbf{r}_{B/A}/dt = \mathbf{0}$, since the *magnitude* of $\mathbf{r}_{B/A}$ is constant by definition of a rigid body; and because the body is translating, the *direction* of $\mathbf{r}_{B/A}$ is constant. Therefore,

$$\mathbf{v}_B = \mathbf{v}_A$$

Acceleration. Taking the time derivative of the velocity equation yields a similar relationship between the instantaneous accelerations of A and B,

$$\mathbf{a}_B = \mathbf{a}_A$$

The above two equations indicate that *all points in a rigid body subjected to either curvilinear or rectilinear translation move with the same velocity and acceleration*. As a result, the kinematics of particle motion, discussed in Chapter 12, may also be applied to specify the kinematics of the particles located in a translating rigid body.

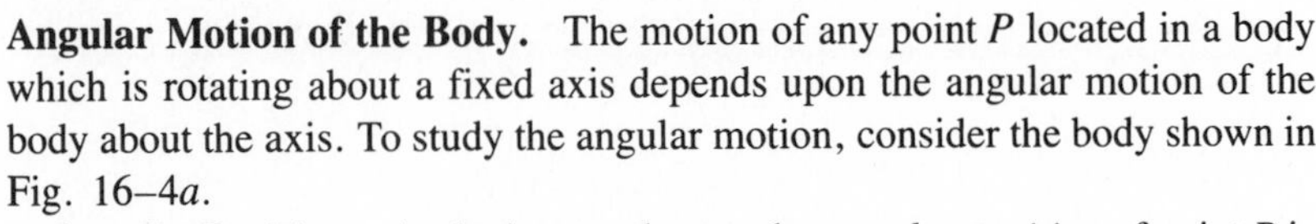

16.3 Rotation About a Fixed Axis

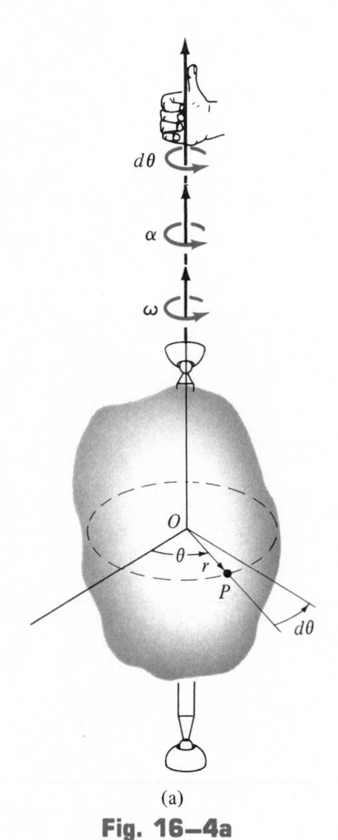

(a)

Fig. 16–4a

Angular Motion of the Body. The motion of any point P located in a body which is rotating about a fixed axis depends upon the angular motion of the body about the axis. To study the angular motion, consider the body shown in Fig. 16–4*a*.

Angular Position. At the instant shown, the *angular position* of point P is defined by the angle θ, measured between a fixed line and the radius r extending from O to P.

Angular Displacement. The change in the angular position, often measured as a differential $d\boldsymbol{\theta}$, is called the *angular displacement*. This vector has a *magnitude* $d\theta$, which is measured in degrees, radians, or revolutions, where 1 rev = 2π rad. The *direction* of $d\boldsymbol{\theta}$ is determined by the right-hand rule; that is, the fingers of the right hand are curled with the sense of rotation, so that in this case the thumb, or $d\boldsymbol{\theta}$, points upward, Fig. 16–4*a*.

Angular Velocity. The time rate of change in the angular position is called the *angular velocity* $\boldsymbol{\omega}$ (omega). The *magnitude* of this vector is

$$\omega = \frac{d\theta}{dt} \qquad (16\text{–}1)$$

which is often measured in rad/s. The *direction* of $\boldsymbol{\omega}$ is the same as that of the angular displacement, Fig. 16–4*a*.

Angular Acceleration. The *angular acceleration* $\boldsymbol{\alpha}$ (alpha) measures the time rate of change of the angular velocity. Hence, the *magnitude* of this vector may be written as

$$\alpha = \frac{d\omega}{dt} \qquad (16\text{–}2)$$

The line of action of $\boldsymbol{\alpha}$ is the same as that for $\boldsymbol{\omega}$, Fig. 16–4*a*; however, its sense of *direction* depends upon whether $\boldsymbol{\omega}$ is increasing or decreasing with time. In particular, if $\boldsymbol{\omega}$ is decreasing $\boldsymbol{\alpha}$ is called an *angular deceleration* and therefore has a sense of direction which is opposite to $\boldsymbol{\omega}$.

By eliminating dt from Eqs. 16–1 and 16–2, it is possible to obtain a differential relation between the angular acceleration, angular velocity, and angular displacement: namely,

$$\alpha\, d\theta = \omega\, d\omega \qquad (16\text{–}3)$$

The similarity between the differential relations for angular motion and those developed for rectilinear motion of a particle ($v = ds/dt$, $a = dv/dt$, and $a\,ds = v\,dv$) should be apparent.

Constant Angular Acceleration. If the angular acceleration of the body is constant, $\boldsymbol{\alpha} = \boldsymbol{\alpha}_c$, Eqs. 16–1, 16–2, and 16–3, when integrated, yield a set of formulas which relate the body's angular velocity, its angular position, and time. These equations are similar to Eqs. 12–4 to 12–6 used for rectilinear motion. The results are

$$\omega = \omega_0 + \alpha_c t \qquad (16\text{–}4)$$

$$\theta = \theta_0 + \omega_0 t + \tfrac{1}{2}\alpha_c t^2 \qquad (16\text{–}5)$$

$$\omega^2 = \omega_0^2 + 2\alpha_c(\theta - \theta_0) \qquad (16\text{–}6)$$

Here θ_0 and ω_0 are the initial values of the body's angular position and angular velocity, respectively.

Motion of Point *P*. As the rigid body in Fig. 16–4*b* rotates, point *P* travels along a *circular path* of radius *r* and with center at point *O*.

Position. The position of *P* is defined by the position vector **r**, which extends from *O* toward *P*.

Velocity. The velocity of *P* has a magnitude which can be determined from the circular motion of *P*. In general, $v = ds/dt$; however, $s = \theta r$. Since *r* is constant, then $v = d(\theta r)/dt = r\,d\theta/dt$. Using Eq. 16–2, we can therefore write the result as

$$v = \omega r \qquad (16\text{–}7)$$

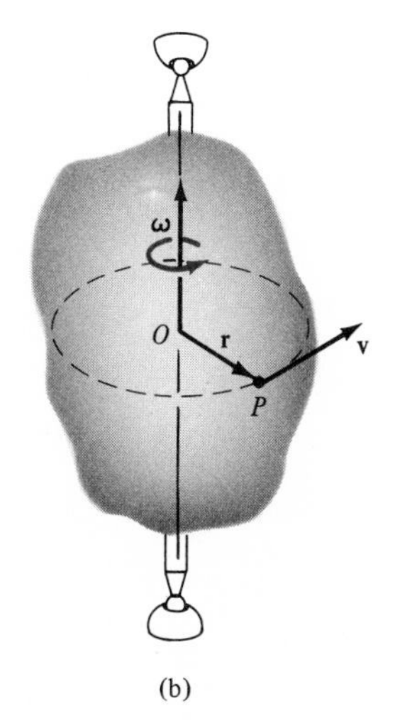

(b)

As shown in Fig. 16–4*b*, the *direction* of **v** is *tangent* to the circular path.

Acceleration. For convenience, the acceleration of *P* will be expressed in terms of its normal and tangential components. Using Eqs. 12–20 and 12–19, $a_t = dv/dt$ and $a_n = v^2/\rho$, noting that $\rho = r$, $v = \omega r$, and $\alpha = d\omega/dt$, we have

$$a_t = \alpha r \qquad (16\text{–}8)$$

$$a_n = \omega^2 r \qquad (16\text{–}9)$$

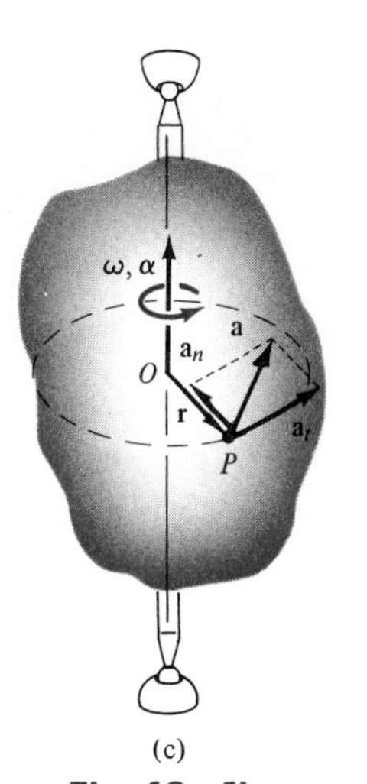

(c)

Fig. 16–4b,c

The *tangential component of acceleration,* Fig. 16–4*c*, represents the change in the magnitude of velocity. If the speed of *P* is increasing, then $\mathbf{a}_t$ acts in the same *direction* as **v**; however, if the speed is decreasing, $\mathbf{a}_t$ acts in the opposite direction of **v**.

The *normal component of acceleration* represents the change in the direction of velocity. The *direction* of $\mathbf{a}_n$ is always toward *O*, the center of the circular path, Fig. 16–4*c*.

PROCEDURE FOR ANALYSIS

In order to determine the velocity and acceleration of a point located in a rigid body that is rotating about a fixed axis, it is first necessary to know the body's angular velocity and angular acceleration.

Angular Motion. If $\boldsymbol{\alpha}$ or $\boldsymbol{\omega}$ are unknown, and $\boldsymbol{\alpha}$ is *not constant,* then the relationships between the angular motions are defined by the differential equations:

$$\omega = \frac{d\theta}{dt} \qquad \alpha = \frac{d\omega}{dt} \qquad \alpha\, d\theta = \omega\, d\omega$$

If the body's angular acceleration is *constant,* the following equations are used for this purpose:

$$\theta = \theta_0 + \omega_0 t + \tfrac{1}{2}\alpha_c t^2$$
$$\omega = \omega_0 + \alpha_c t$$
$$\omega^2 = \omega_0^2 + 2\alpha_c(\theta - \theta_0)$$

Motion of P. When the motion of a point P in the body is to be determined, it is suggested that a *kinematic diagram* accompany the problem solution. This diagram is simply a graphical representation showing the motion of the point (Fig. 16–4*b* or *c*).

The velocity and the two components of acceleration can be determined from the equations

$$v = \omega r$$
$$a_t = \alpha r$$
$$a_n = \omega^2 r$$

The following examples numerically illustrate application of these principles.

Example 16–1

A cord is wrapped around a wheel which is initially at rest as shown in Fig. 16–5. If a force is applied to the cord and gives it an acceleration of $a = (4t)\text{m/s}^2$, where t is in seconds, determine as a function of time (a) the angular velocity of the wheel and (b) the angular position in radians.

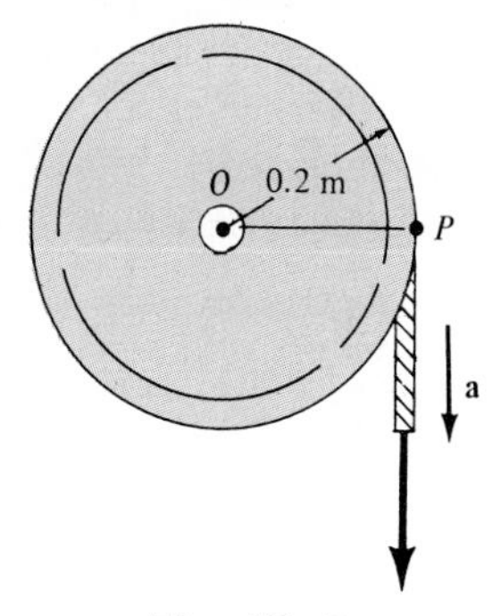

Fig. 16–5

Solution

Part (a). The wheel is subjected to rotation about a fixed axis passing through point O, Fig. 16–5. Thus, point P on the wheel has motion about a circular path, and therefore the acceleration of this point has both tangential and normal components. In particular, $(a_P)_t = (4t)\ \text{m/s}^2$, since the cord is connected to the wheel and *tangent* to it at P. Hence, the angular acceleration of the wheel is

$$\circlearrowleft + \qquad (a_P)_t = \alpha r$$

$$4t = \alpha(0.2)$$

$$\alpha = 20t\ \text{rad/s}^2$$

The wheel's angular velocity $\boldsymbol{\omega}$ can be determined by using $\alpha = d\omega/dt$, since this equation relates α, t, and ω. Integrating, with the initial condition that $\omega_0 = 0$ at $t_0 = 0$, yields

$$\circlearrowleft + \qquad \alpha = \frac{d\omega}{dt} = 20t$$

$$\int_0^\omega d\omega = \int_0^t 20t\ dt$$

$$\omega = 10t^2\ \text{rad/s} \qquad \textit{Ans.}$$

Why is it not possible to use Eq. 16–4 ($\omega = \omega_0 + \alpha_c t$) to obtain this result?

Part (b). The angular position can be computed using $\omega = d\theta/dt$, since this equation relates θ, ω, and t. Using the initial condition $\theta_0 = 0$ at $t_0 = 0$, we have

$$\circlearrowleft + \qquad \frac{d\theta}{dt} = \omega = 10t^2$$

$$\int_0^\theta d\theta = \int_0^t 10t^2\ dt$$

$$\theta = \tfrac{10}{3} t^3\ \text{rad} \qquad \textit{Ans.}$$

Example 16–2

Disk A, shown in Fig. 16–6a, starts from rest and rotates with a constant angular acceleration of $\alpha_A = 2$ rad/s². If disk A is in contact with disk B and no slipping occurs between the disks, determine the angular velocity and angular acceleration of B just after A turns 10 revolutions.

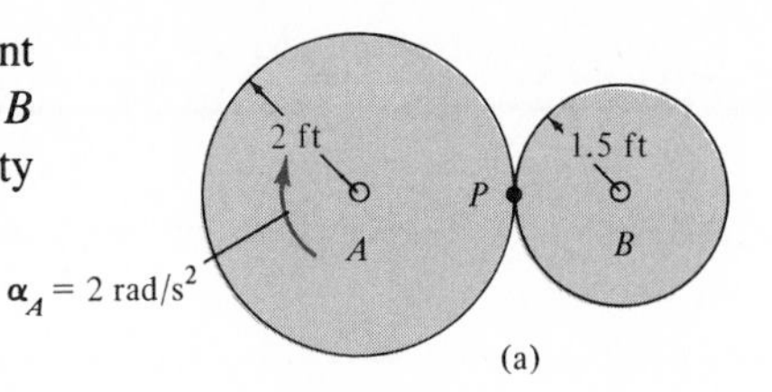

(a)

Solution

The angular velocity of A can be determined using the equations of *constant* angular acceleration since $\alpha_A = 2$ rad/s². There are 2π radians to one revolution, so that

$$\theta_A = 10 \text{ rev}\left(\frac{2\pi \text{ rad}}{1 \text{ rev}}\right) = 62.83 \text{ rad}$$

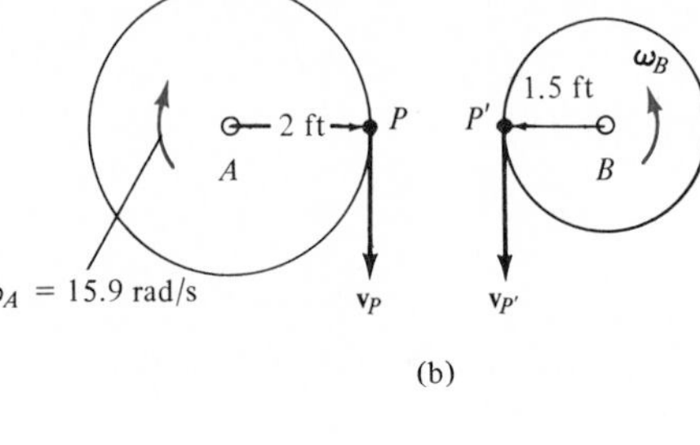

(b)

Thus,

$\circlearrowleft +$
$$\omega^2 = \omega_0^2 + 2\alpha_c(\theta - \theta_0)$$
$$\omega_A^2 = 0 + 2(2)(62.83 - 0)$$
$$\omega_A = 15.9 \text{ rad/s} \circlearrowright$$

As shown in Fig. 16–6b, the speed of the contacting point P on the rim of A is

$+\downarrow$
$$v_P = \omega_A r_A = (15.9)(2) = 31.8 \text{ ft/s} \quad \downarrow$$

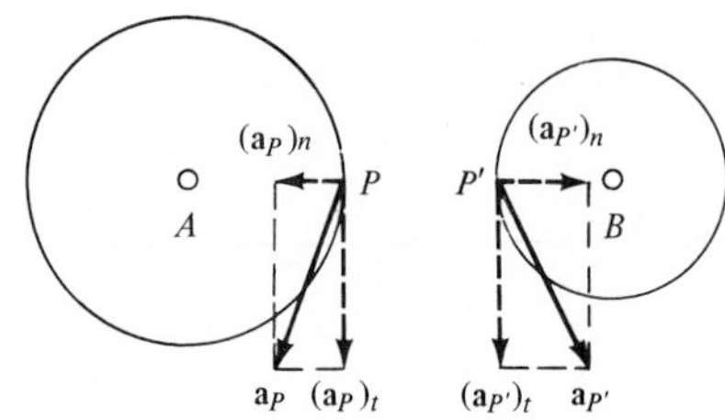

(c)

Fig. 16–6

The velocity is always tangent to the path of motion, so that the speed of point P' on B is the *same* as the speed of P on A. The angular velocity of B is therefore

$\circlearrowleft +$
$$\omega_B = \frac{v_{P'}}{r_B} = \frac{31.8}{1.5} = 21.2 \text{ rad/s} \circlearrowleft \qquad \textit{Ans.}$$

As shown in Fig. 16–6c, the *tangential components* of acceleration of both disks are equal, since the disks are in contact with one another. Hence,

$$(a_P)_t = (a_{P'})_t \quad \text{or} \quad \alpha_A r_A = \alpha_B r_B$$
$$\alpha_B = \alpha_A\left(\frac{r_A}{r_B}\right) = 2\left(\frac{2}{1.5}\right) = 2.67 \text{ rad/s}^2 \circlearrowleft \qquad \textit{Ans.}$$

Notice that the normal components of acceleration $(a_P)_n$ and $(a_{P'})_n$ act in *opposite directions,* since the paths of motion for both points are *different*. Furthermore, $(a_P)_n \neq (a_{P'})_n$, since the *magnitudes* of these components depend upon the radius and angular velocity of each disk, i.e., $(a_P)_n = \omega_A^2 r_A$ and $(a_{P'})_n = \omega_B^2 r_B$.

Problems

16–1. The tub of a washing machine is rotating at 50 rad/s when the power is turned off. If it takes 20 s for the tub to come to rest, determine (a) the constant angular deceleration and (b) the total number of revolutions the tub makes.

16–2. A flywheel has its angular speed increased uniformly from 15 rad/s to 60 rad/s in 120 s. If the diameter of the wheel is 2 ft, determine the normal and tangential components of acceleration of a point on the rim of the wheel when $t = 120$ s, and the total distance the point travels during the time period.

16–3. A wheel has an initial clockwise angular velocity of 8 rad/s and a constant angular acceleration of 2 rad/s^2. Determine the number of revolutions it must undergo to acquire a clockwise angular velocity of 15 rad/s. What time is required?

***16–4.** Gear A is in mesh with gear B as shown. If A starts from rest and has a constant angular acceleration of $\alpha_A = 2$ rad/s^2, determine the time needed for B to attain an angular velocity of $\omega_B = 50$ rad/s.

Prob. 16–4

16–5. During a gust of wind, the blades of the windmill are given an angular acceleration of $\alpha = (0.2\ \theta)$ rad/s^2, where θ is measured in radians. If initially the blades have an angular velocity of 5 rad/s, determine the speed of point P located at the tip of one of the blades just after the blade has turned two revolutions.

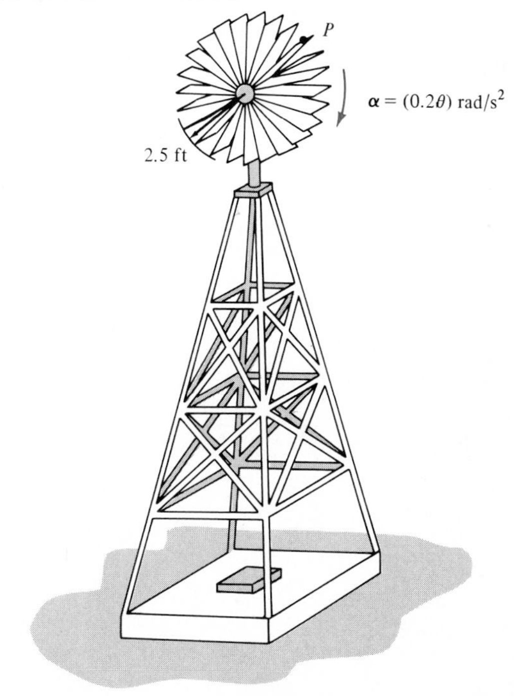

Prob. 16–5

16–6. The phonograph turntable T is driven by the frictional idler wheel A, which simultaneously bears against the inner rim of the turntable and the motor-shaft spindle B. Determine the required diameter d of the spindle if the motor turns it at 30 rad/s and it is required that the turntable rotate at 33 revolutions per minute.

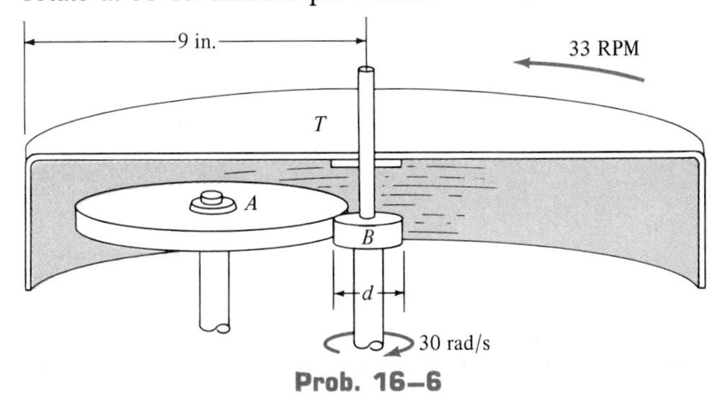

Prob. 16–6

16–7. Arm $ABCD$ is pinned at B and undergoes motion such that $\theta = (0.3t^2)$ rad, where t is measured in seconds. Determine the speed of point A in $t = 1.5$ s and the magnitude of the acceleration of point D at this instant.

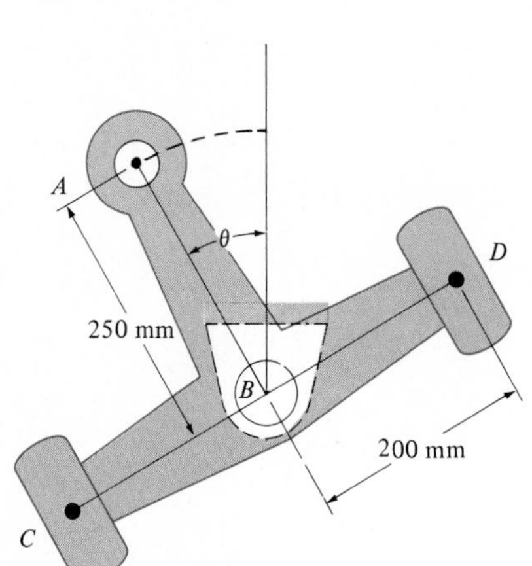

Prob. 16–7

***16–8.** The power of a bus engine is transmitted using the belt-and-pulley arrangement shown. If the engine turns pulley A at 50 rad/s, compute the angular velocities of the generator pulley B and the air-conditioning pulley C. The hub at D is rigidly *connected* to B and turns with it.

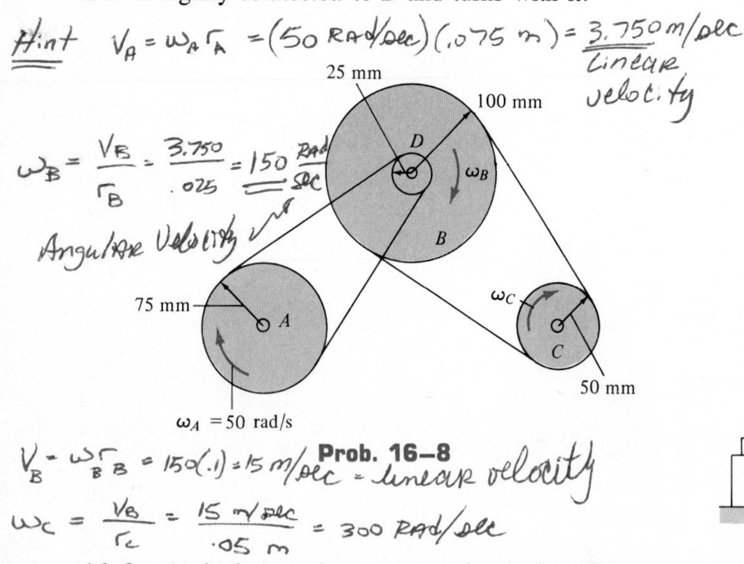

Prob. 16–8

16–9. At the instant shown, gear A is rotating with a constant angular velocity of $\omega_A = 6$ rad/s. Determine the largest angular velocity of gear B and the maximum speed of point C.

Prob. 16–9

16–10. The construction boom AB is used to place the core C of a nuclear reactor into its pressure vessel P. When the top of the boom B is positioned directly over the top of the pressure vessel and held in that position, the core is observed to have a swinging motion such that $\alpha = (-0.30 \sin \theta)$ rad/s². If $\omega = 0$ at $\theta = 10°$, determine the velocity and acceleration of the center of mass G of the core when $\theta = 0°$.

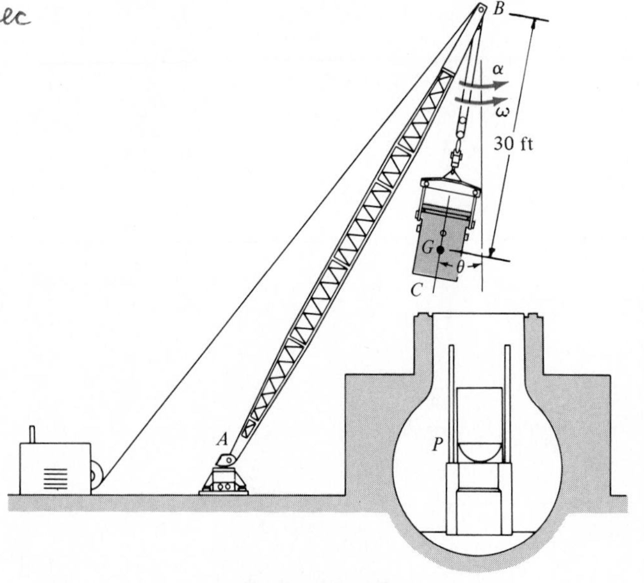

Prob. 16–10

16–11. The sphere starts from rest at $\theta = 0$ and rotates with an angular acceleration of $\alpha = (4\,\theta)$ rad/s^2, where θ is measured in radians. Determine ω and α, and the magnitudes of the velocity and acceleration of point P on the sphere, at the instant $\theta = 6$ rad.

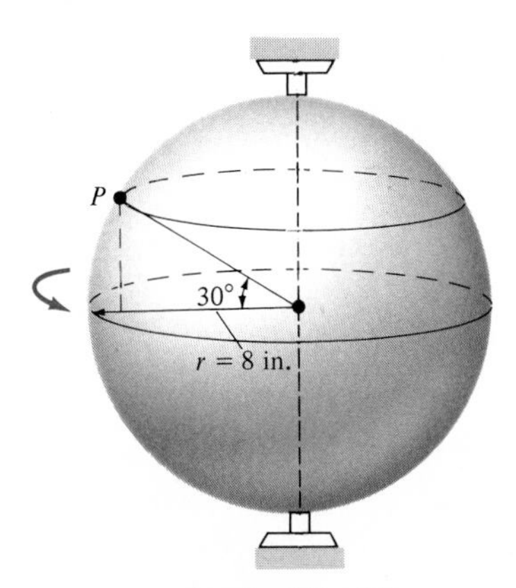

Prob. 16–11

***16–12.** If the hoisting gear A has an initial angular velocity of $\omega_A = 8$ rad/s and an angular deceleration of $\alpha_A = -1.5$ rad/s^2, determine the velocity and acceleration of block C in 2 s.

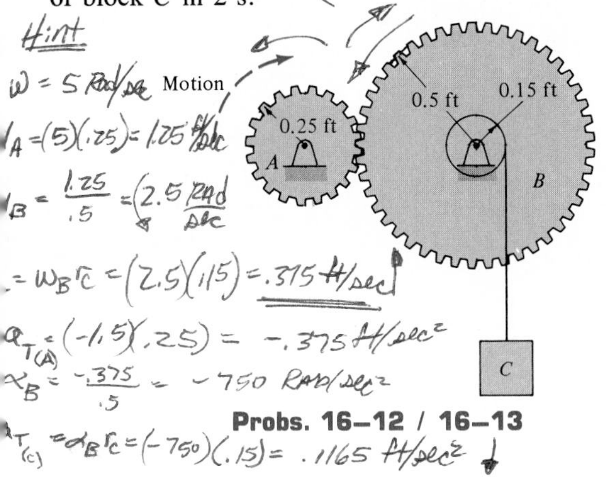

Probs. 16–12 / 16–13

16–13. Solve Prob. 16–12 assuming that the angular deceleration of gear A is $\alpha_A = -1$ rad/s^2, where t is measured in seconds.

16–14. At the instant shown, the horizontal portion of the belt has an acceleration of 3 m/s^2, while points in contact with the outer edge of the pulleys have an acceleration with a magnitude of 5 m/s^2. If the belt does not slip on the pulleys, determine the belt's speed due to the motion.

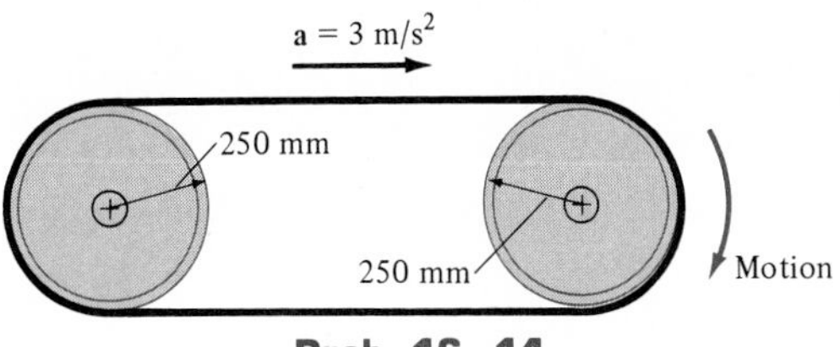

Prob. 16–14

16–15. Just after the fan is turned on, the motor gives the blade an angular acceleration of $\alpha = (20t)$ rad/s^2, where t is in seconds. Determine the speed of the tip P of one of the blades when $t = 2$ s. How many revolutions has the blade turned in 2 s?

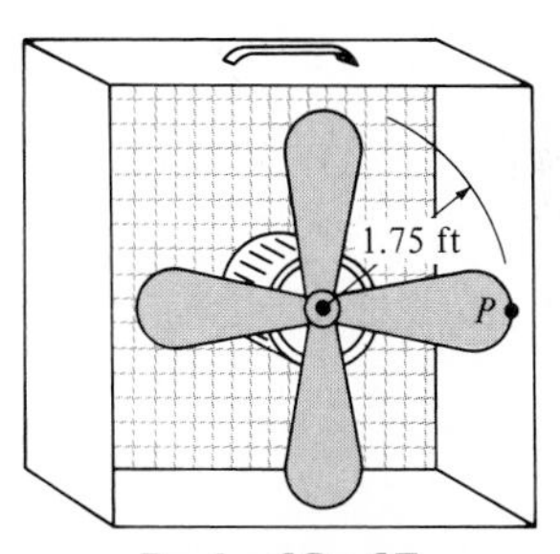

Prob. 16–15

***16–16.** A tape wraps around the wheel which is turning at a constant rate of $\omega = 4$ rad/s. Assuming the unwrapped portion of tape remains horizontal, determine the acceleration of point P on the tape just before and just after it is wrapped on the wheel. Take $r = 75$ mm.

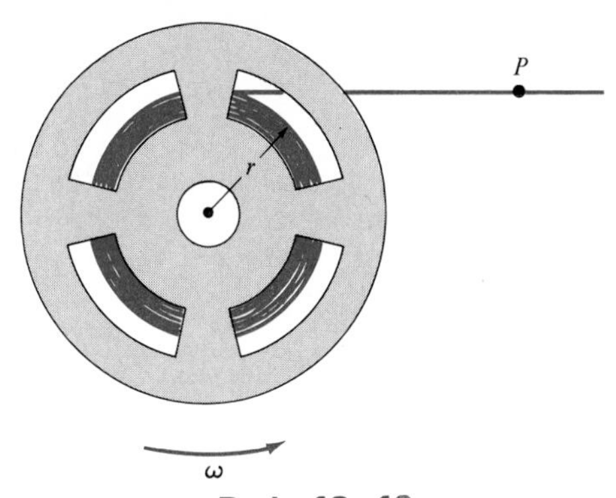

Prob. 16–16

16–17. The rod assembly is supported by ball-and-socket joints at A and B. If it is rotating about the y axis with an angular velocity of $\omega = 6$ rad/s and it has an angular acceleration of $\alpha = 8$ rad/s^2, determine the magnitudes of the velocity and acceleration of point C at the instant shown. Bars AD and BD lie in the x-y plane. *Hint:* Point C is moving in a circular path of radius $AC = 1$ ft.

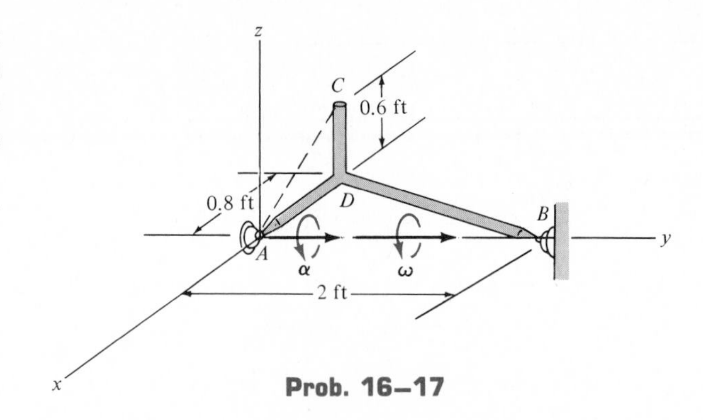

Prob. 16–17

*16.4 Absolute General Plane Motion Analysis

A body subjected to *general plane motion* undergoes a *simultaneous* translation and rotation. If the body is represented by a thin slab, the slab translates in the plane and rotates about an axis perpendicular to the plane. Occasionally, the *position* of a *point* located in a body may be directly related to the *angular position* of a line contained in the body. Then, by *direct application* of the time-differential equations $v = ds/dt$, $a = dv/dt$, $\omega = d\theta/dt$, and $\alpha = d\omega/dt$, the *motion* of the point and the angular motion of the line can be related. In some cases, this procedure may also be used to relate the motions of one body to those of an adjacent connected body.

PROCEDURE FOR ANALYSIS

The following procedure provides a method for relating the absolute motion of a point P in a body to the angular motion of a line contained in the same body.

Position Coordinate Equation. Locate point P lying in the body using a rectilinear position coordinate s, which is measured from a *fixed origin* and is directed along the path of motion of point P. Measure the angular position θ of a line extending from the same origin and lying in the body. From the dimensions of the body, relate s to θ, $s = f(\theta)$, using geometry and/or trigonometry.

Time Derivatives. Take the first derivative of $s = f(\theta)$ with respect to time to get a relationship between v and ω. Take the second time derivative to get a relationship between a and α.

This procedure is illustrated in the following example problems.

Example 16–3

At a given instant, the cylinder of radius r, shown in Fig. 16–7, has an angular velocity $\boldsymbol{\omega}$ and angular acceleration $\boldsymbol{\alpha}$. Determine the velocity and acceleration of its center G if it rolls without slipping.

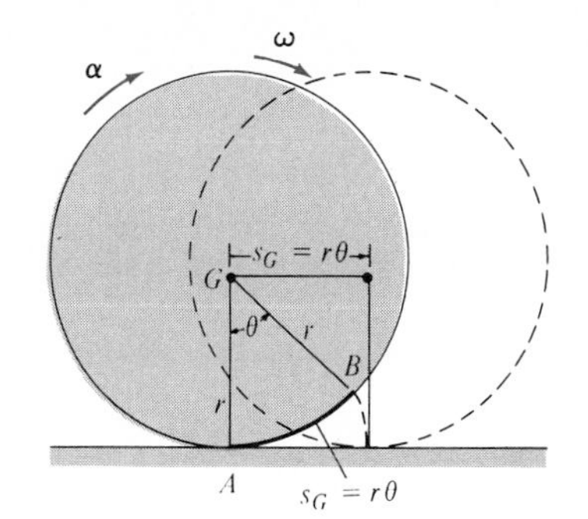

Fig. 16–7

Solution

Position Coordinate Equation. Coordinates s_G and θ are chosen for the analysis since the rectilinear motion of G is defined by $v_G = ds_G/dt$ and the angular motion of the cylinder is defined by $\omega = d\theta/dt$. The origin of the s_G axis is located at a fixed point coincident with G at some instant, Fig. 16–7; then, as the cylinder rotates through an angle θ, succeeding points on the arc length $AB = r\theta$ contact the ground, such that G travels a distance $s_G = r\theta$.

Time Derivatives. Taking successive time derivatives of this equation, realizing that r is constant, $\omega = d\theta/dt$, and $\alpha = d\omega/dt$, gives the necessary relationships

$$s_G = r\theta$$

$$v_G = r\omega \qquad \textit{Ans.}$$

$$a_G = r\alpha \qquad \textit{Ans.}$$

Example 16–4

The end of rod R shown in Fig. 16–8 maintains contact with the cam by means of a spring. If the cam rotates about an axis through point O with an angular acceleration $\boldsymbol{\alpha}$ and angular velocity $\boldsymbol{\omega}$, compute the velocity and acceleration of R when the cam is in the arbitrary position θ.

Solution

Position Coordinate Equation. Coordinates x and θ are chosen for the analysis since the angular motion of the cam is defined by $\omega = d\theta/dt$ and the rectilinear motion of the rod (or horizontal component of the motion of point B) is $v = dx/dt$. These coordinates are measured from the fixed point O and may be related to each other using trigonometry. Since $OC = CB = r\cos\theta$, Fig. 16–8, then

$$x = 2r\cos\theta$$

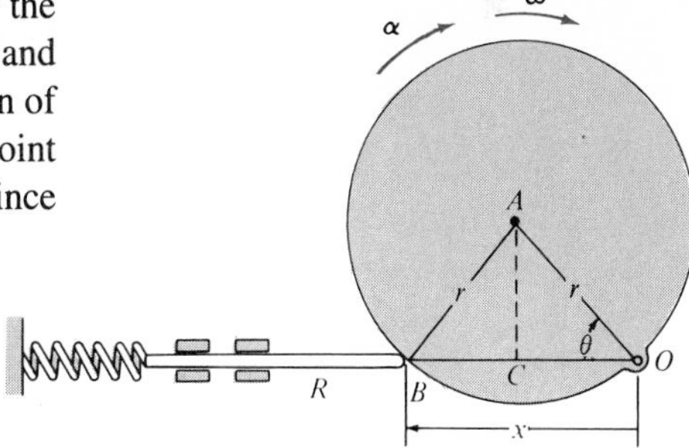

Fig. 16–8

Time Derivatives. Using the chain rule of calculus, we have

$$\frac{dx}{dt} = -2r\sin\theta\frac{d\theta}{dt}$$

$$v = -2r\,\omega\sin\theta \qquad \textit{Ans.}$$

$$\frac{dv}{dt} = -2r\left(\frac{d\omega}{dt}\right)\sin\theta - 2r\omega\left(\cos\theta\frac{d\theta}{dt}\right)$$

$$a = -2r\,\alpha\sin\theta - 2r\omega^2\cos\theta \qquad \textit{Ans.}$$

Problems

16–18. The mechanism is used to convert the *constant* circular motion of rod *AB* into translating motion of rod *CD*. Compute the velocity and acceleration of *CD* for any angle θ of *AB*.

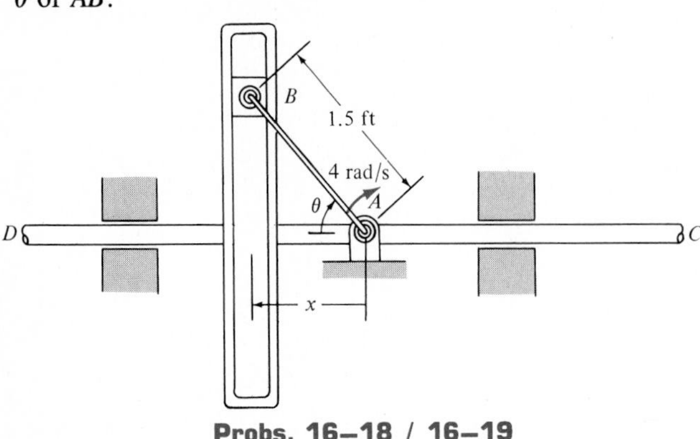

Probs. 16–18 / 16–19

16–19. Determine the velocity and acceleration of rod *CD* in Prob. 16–18 when $\theta = 60°$ if at this instant *AB* has an angular velocity of 4 rad/s and an angular acceleration of 2 rad/s^2.

***16–20.** Rod *CD* presses against *AB*, giving it an angular velocity. If the angular velocity of *AB* is maintained at $\omega = 5$ rad/s, determine the required speed $\mathbf{v}$ of *CD* for any angle θ of rod *AB*.

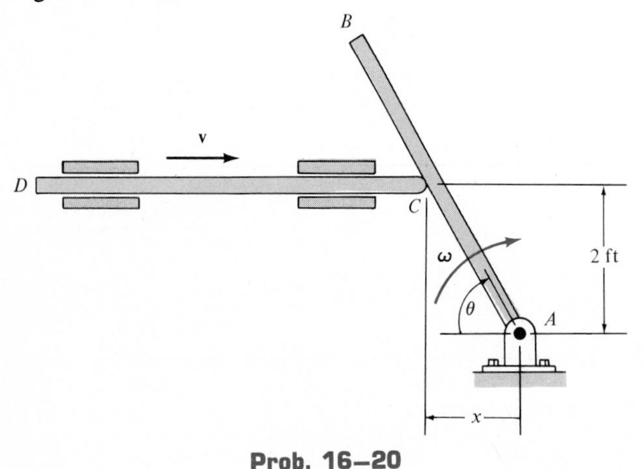

Prob. 16–20

16–21. At the instant $\theta = 50°$ the slotted guide is moving upward with an acceleration of 3 m/s^2 and a velocity of 2 m/s. Determine the angular acceleration and angular velocity of link *AB* at this instant. *Note:* The upward motion of the guide is in the negative *y* direction.

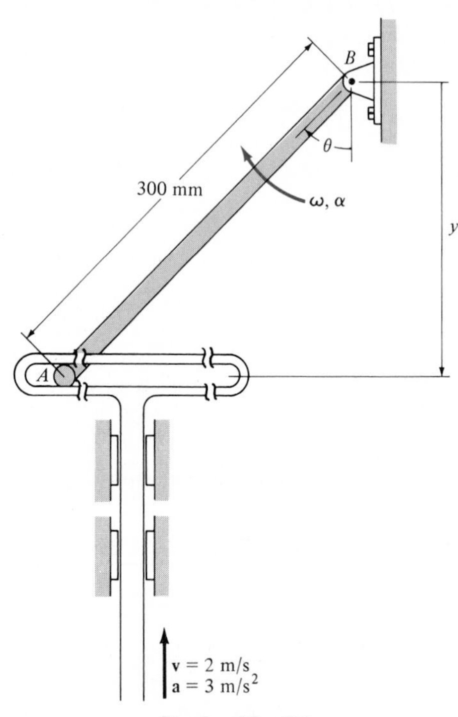

Prob. 16–21

16–22. At the instant shown, $\theta = 60°$, and rod *AB* is subjected to a deceleration of 4 m/s^2 when the velocity is 8 m/s. Determine the angular velocity and angular acceleration of link *CD* at this instant.

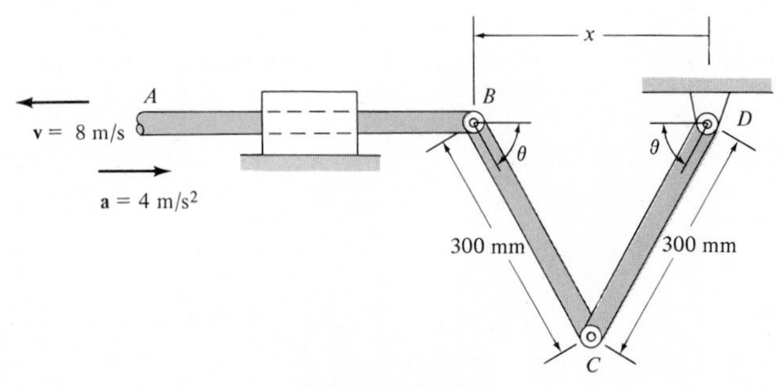

Prob. 16–22

16–23. The 2-m-long bar is confined to move in the horizontal and vertical slots A and B. If the velocity of the slider block at A is 6 m/s, determine the bar's angular velocity and the velocity of block B at the instant $\theta = 60°$.

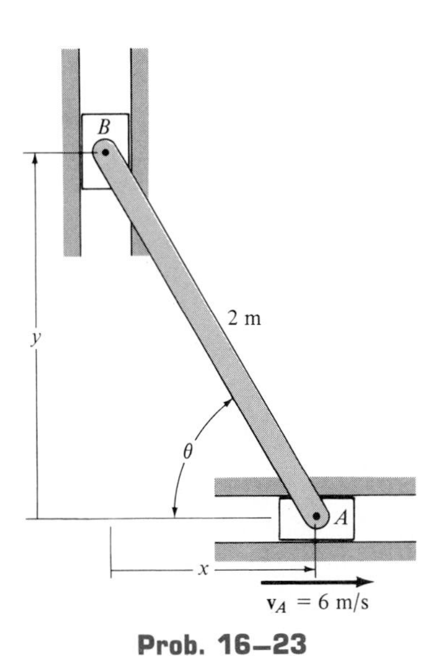

Prob. 16–23

***16–24.** The scaffold S is raised hydraulically by moving the roller at A towards the pin at B. If A is approaching B with a speed of 1.5 ft/s, determine the speed at which the platform is rising as a function of θ. Each link is pin-connected at its midpoint and end points and has a length of 4 ft.

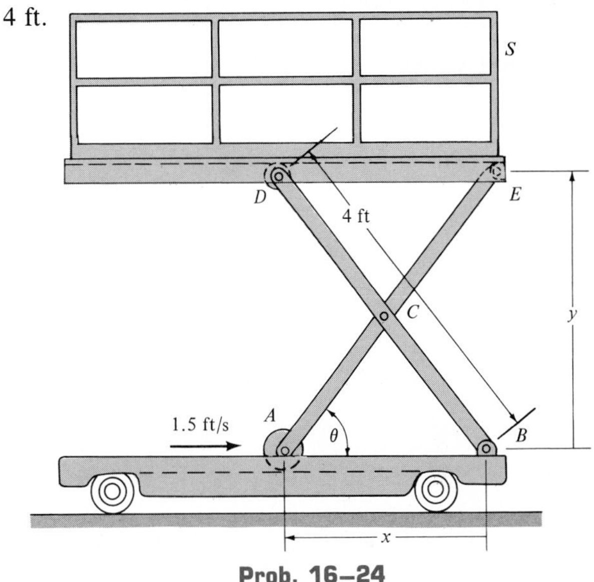

Prob. 16–24

Relative-Motion Analysis: Velocity 16.5

Since general plane motion consists of a combination of translation and rotation of the body, it is often convenient to view each of these "component" motions *separately* using a *relative-motion analysis*. Two sets of coordinate axes will be used to do this. The x, y coordinate system is fixed and therefore it will be used to measure *absolute* positions, velocities, and accelerations of points A and B in the body, Fig. 16–9*a*. The origin of the x', y' coordinate system will be fixed to the selected "base point" A, which generally has a *known* motion. The axes of this coordinate system are *not* fixed to the body; rather they will only be allowed to *translate* with respect to the fixed frame.

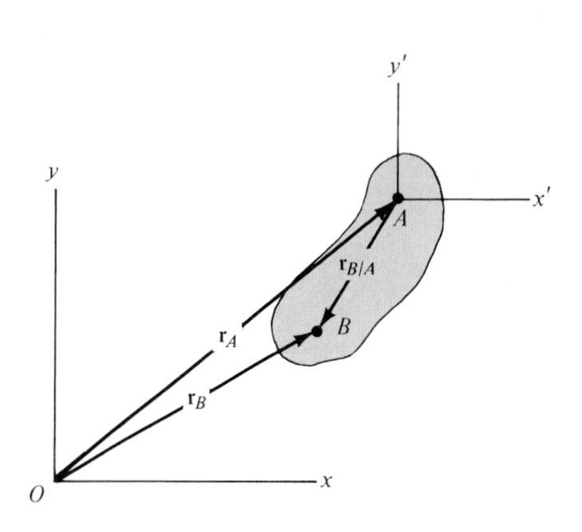

(a)

Fig. 16–9a

Position. The *position vector* $\mathbf{r}_A$ in Fig. 16–9*a* specifies the location of the "base point" A; whereas the *relative-position vector* $\mathbf{r}_{B/A}$ locates point B in the body with respect to A. By vector addition, the *absolute position* of B can be determined from the equation

$$\mathbf{r}_B = \mathbf{r}_A + \mathbf{r}_{B/A} \qquad (16\text{–}10)$$

Velocity. To determine the relationship between the instantaneous velocities of points A and B, it is necessary to take the time derivative of the position equation, i.e., $d\mathbf{r}_B/dt = d\mathbf{r}_A/dt + d\mathbf{r}_{B/A}/dt$. The terms $d\mathbf{r}_B/dt = \mathbf{v}_B$ and

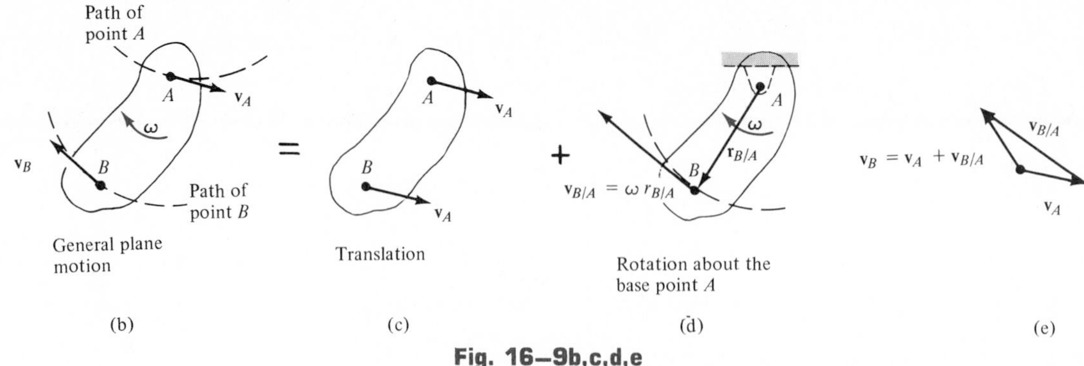

Fig. 16–9b,c,d,e

$d\mathbf{r}_A/dt = \mathbf{v}_A$ are measured from the fixed x, y axes and represent the *absolute velocities* of points B and A, Fig. 16–9*b*. The last term is denoted as the *relative velocity* $\mathbf{v}_{B/A}$, which represents the velocity of B measured by an observer stationed at A and fixed to the translating x', y' axes. Since the body is *rigid*, this observer sees point B move along the dashed *circular arc* that has a radius of curvature of $\mathbf{r}_{B/A}$, Fig. 16–9*d*. In other words, the body moves as if it were *pinned* at A. Consequently, $\mathbf{v}_{B/A}$ has a *magnitude* of $v_{B/A} = \omega r_{B/A}$ and a *direction* which is tangent to the curved path at B. We therefore have

$$\mathbf{v}_B = \mathbf{v}_A + \mathbf{v}_{B/A} \tag{16–11}$$

where

$\mathbf{v}_B$ = absolute velocity of point B
$\mathbf{v}_A$ = absolute velocity of the base point A
$\mathbf{v}_{B/A}$ = relative velocity of "B with respect to A" as measured by a *translating* observer; the *magnitude* is $v_{B/A} = \omega r_{B/A}$ and the *direction* is perpendicular to the line of action of $\mathbf{r}_{B/A}$
$\boldsymbol{\omega}$ = absolute angular velocity of the body
$\mathbf{r}_{B/A}$ = relative-position vector drawn from A to B

Each of the three terms in Eq. 16–11 is represented graphically on the *kinematic diagrams* in Fig. 16–9*b*, *c*, and *d*. Here it is seen that at a given instant the velocity of B, Fig. 16–9*b*, is determined by considering the entire body to translate with a velocity of $\mathbf{v}_A$, Fig. 16–9*c*, and simultaneously rotate about the base point A with an instantaneous angular velocity $\boldsymbol{\omega}$, Fig. 16–9*d*. Vector addition of these two effects, applied to B, yields $\mathbf{v}_B$, as shown in Fig. 16–9*e*.

The velocity equation 16–11 may be used in a practical manner to study the motion of a rigid body which is either pin-connected to or in contact with other moving bodies. To obtain the necessary data when applying this equation, points A and B should generally be selected at joints which are pin-connected, or at points in contact with adjacent bodies which have a *known motion*. For example, both points A and B on link AB, Fig. 16–10*a*, have circular paths of motion. Hence, at the instant shown, the magnitudes of their velocities are determined by the angular motion of the wheel and link BC so

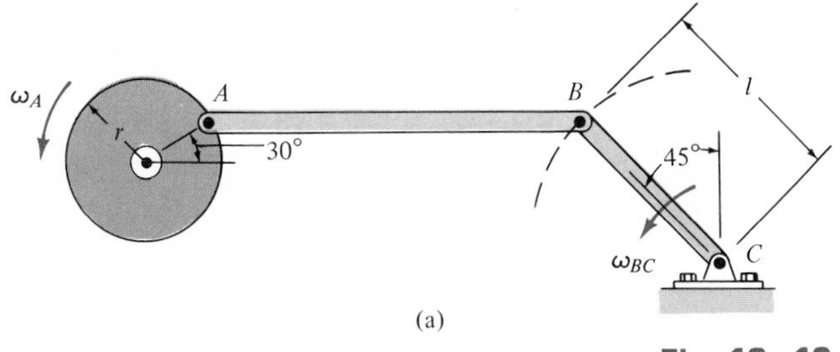

(a)

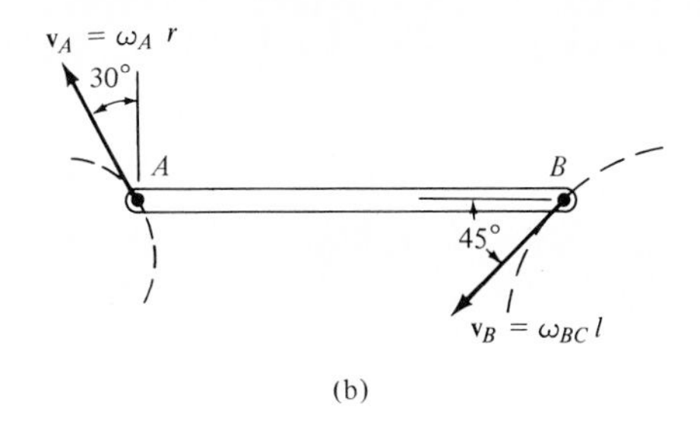

(b)

Fig. 16–10

that $v_A = \omega_A r$ and $v_B = \omega_{BC} l$. The *directions* of $\mathbf{v}_A$ and $\mathbf{v}_B$ are always *tangent* to their paths of motion, Fig. 16–10*b*. In the case of the wheel in Fig. 16–11, which rolls without slipping, point *A* can be selected at the ground. Here *A* has zero velocity since it is in contact with the ground—which, of course, is at rest. Furthermore, the center of the wheel, *B*, moves along a horizontal path so that $\mathbf{v}_B$ is horizontal.

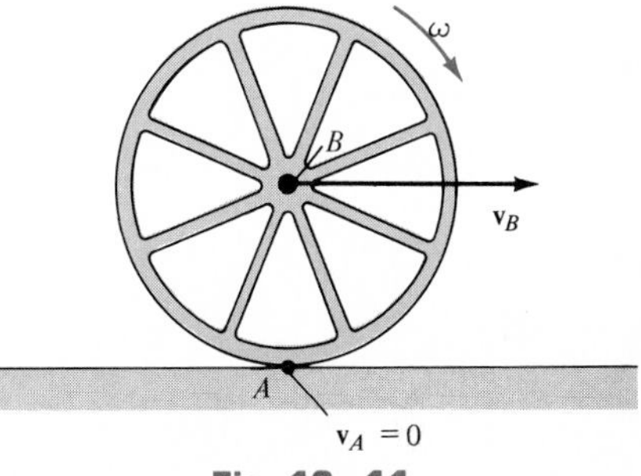

Fig. 16–11

PROCEDURE FOR ANALYSIS

The velocity equation $\mathbf{v}_B = \mathbf{v}_A + \mathbf{v}_{B/A}$ can be applied to any two points *A* and *B* located in the *same* rigid body. For application, it is suggested that the following procedure be used.

Kinematic Diagrams. Draw a kinematic diagram of the body such as shown in Fig. 16–9*b*, and indicate on it the *absolute velocities* $\mathbf{v}_A$ and $\mathbf{v}_B$ of points *A* and *B* and the angular velocity $\boldsymbol{\omega}$. Usually the base point *A* is selected as a point having a known velocity.

Also, draw a kinematic diagram of the body such as shown in Fig. 16–9*d* and indicate on it the *relative velocity* $\mathbf{v}_{B/A}$. Since the body is considered to be pinned at the base point *A*, the *magnitude* of $\mathbf{v}_{B/A}$ is $v_{B/A} = \omega r_{B/A}$. The *direction* is established from the diagram, such that $\mathbf{v}_{B/A}$ acts perpendicular to $\mathbf{r}_{B/A}$ in accordance with the rotational motion $\boldsymbol{\omega}$ of the body.*

Velocity Equation. Write the velocity equation in symbolic form: $\mathbf{v}_B = \mathbf{v}_A + \mathbf{v}_{B/A}$, and underneath each of the terms represent the respective vectors by their magnitudes and directions. To do this, use the data tabulated on the kinematic diagrams. Since motion occurs in the plane, this ''vector'' equation can be expressed in terms of two scalar component equations which provide a solution for at most two unknowns. If the solution yields a *negative* answer for an *unknown,* it indicates its *direction* is *opposite* to that shown on the kinematic diagram.

The following example problems numerically illustrate this scalar method of application.

*Perhaps the notation $\mathbf{v}_B = \mathbf{v}_A + \mathbf{v}_{B/A(\text{pin})}$ helps in recalling that *A* is pinned.

Example 16–5

The link shown in Fig. 16–12*a* is guided by two blocks at *A* and *B*, which move in the fixed slots. If at the instant shown the velocity of *A* is 2 m/s downward, determine the velocity of *B* at this instant.

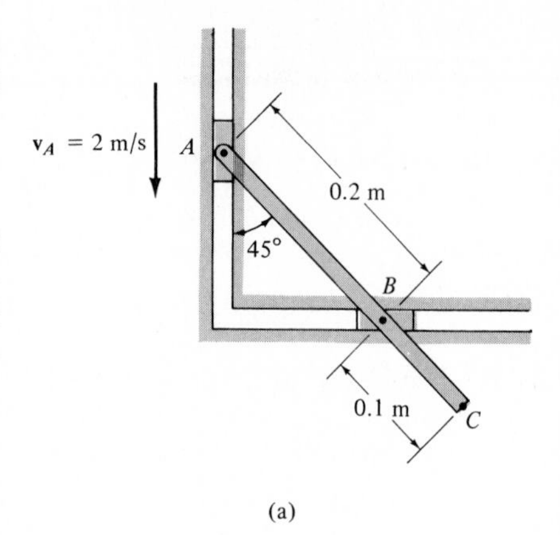

(a)

Solution

Kinematic Diagrams. Since points *A* and *B* are restricted to move along the fixed slots and $\mathbf{v}_A$ is directed downward, the velocity $\mathbf{v}_B$ must be directed horizontally to the right, Fig. 16–12*b*. This motion causes the link to rotate counterclockwise; that is, by the right-hand rule the angular velocity $\boldsymbol{\omega}$ is directed outward, perpendicular to the page (the plane of motion). Knowing the magnitude and direction of $\mathbf{v}_A$ and the lines of action of $\mathbf{v}_B$ and $\boldsymbol{\omega}$, it is possible to apply the velocity equation $\mathbf{v}_B = \mathbf{v}_A + \mathbf{v}_{B/A}$ to points *A* and *B* in order to solve for the two unknown magnitudes v_B and ω.

As shown in Fig. 16–12*c*, the relative velocity $\mathbf{v}_{B/A}$ acts up to the right, *perpendicular* to $\mathbf{r}_{B/A}$, in accordance with the angular velocity $\boldsymbol{\omega}$. The magnitude of $\mathbf{v}_{B/A}$ is

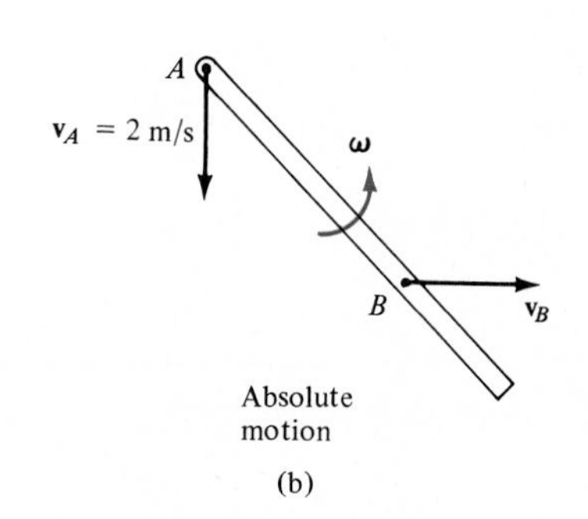

(b)

$$v_{B/A} = \omega r_{B/A} = \omega(0.2 \text{ m})$$

Velocity Equation

$$\mathbf{v}_B = \mathbf{v}_A + \mathbf{v}_{B/A}$$

$$\underset{\rightarrow}{v_B} = \underset{\downarrow}{2 \text{ m/s}} + \underset{\measuredangle 45^\circ}{\omega(0.2 \text{ m})}$$

Equating the horizontal and vertical components, with the assumed positive directions to the right and upward, we have

$$(\overset{+}{\rightarrow}) \qquad v_B = 0 + 0.2\ \omega \cos 45^\circ$$

$$(+\uparrow) \qquad 0 = -2 + 0.2\ \omega \sin 45^\circ$$

Thus,

$$\omega = 14.14 \text{ rad/s}$$

$$v_B = 2 \text{ m/s} \qquad \textit{Ans.}$$

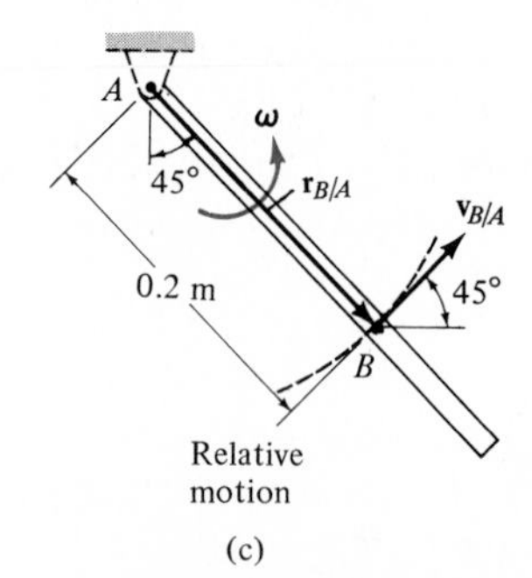

(c)

Fig. 16–12

Since both results are *positive*, the *directions* of $\mathbf{v}_B$ and $\boldsymbol{\omega}$ are indeed *correct* as shown in Fig. 16–12*b*.

Having determined $\boldsymbol{\omega}$, as an exercise apply the velocity equation $\mathbf{v}_C = \mathbf{v}_A + \mathbf{v}_{C/A}$ to points *A* and *C* and show that $v_C = 3.16$ m/s, directed at $\theta = 18.4^\circ$ up from the horizontal.

Example 16–6

The cylinder shown in Fig. 16–13*a* rolls freely on the surface of a conveyor belt which is moving at 2 ft/s. Assuming that no slipping occurs between the cylinder and the belt, determine the velocity of point A. The cylinder has a clockwise angular velocity of $\omega = 15$ rad/s at the instant shown.

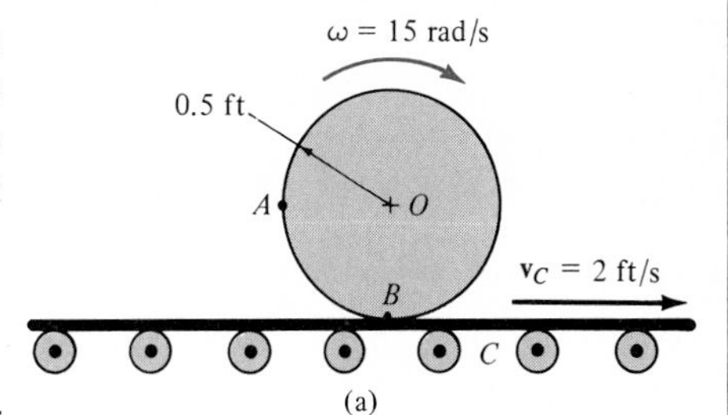

(a)

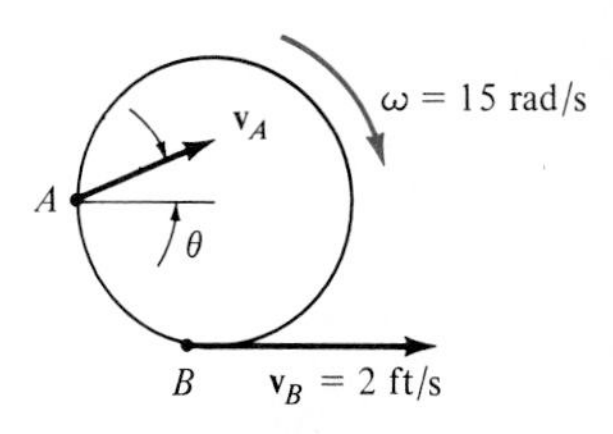

Absolute motion

(b)

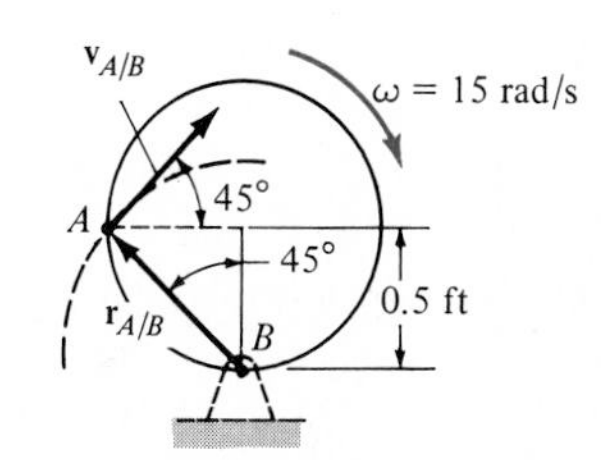

Relative motion

(c)

Fig. 16–13

Solution

Kinematic Diagrams. Since no slipping occurs, point B on the cylinder has the same velocity as the conveyor, Fig. 16–13*b*. Knowing the angular velocity of the cylinder, we will apply the velocity equation to B, the base point, and A to determine the x and y components of $\mathbf{v}_A$, i.e., $\mathbf{v}_A = \mathbf{v}_B + \mathbf{v}_{A/B}$.

The relative velocity $\mathbf{v}_{A/B}$ has a direction as shown in Fig. 16–13*c* and a magnitude of

$$v_{A/B} = \omega r_{A/B} = (15 \text{ rad/s})\left(\frac{0.5 \text{ ft}}{\cos 45^\circ}\right) = 10.6 \text{ ft/s} \quad \measuredangle^{45^\circ}$$

Velocity Equation

$$\mathbf{v}_A = \mathbf{v}_B + \mathbf{v}_{A/B}$$

$$\underset{\rightarrow}{(v_A)_x} + \underset{\uparrow}{(v_A)_y} = \underset{\rightarrow}{2 \text{ ft/s}} + \underset{\measuredangle^{45^\circ}}{10.6 \text{ ft/s}}$$

Equating the horizontal and vertical components gives

$$(\overset{+}{\rightarrow}) \qquad (v_A)_x = 2 + 10.6 \cos 45^\circ = 9.50 \text{ ft/s}$$

$$(+\uparrow) \qquad (v_A)_y = 0 + 10.6 \sin 45^\circ = 7.50 \text{ ft/s}$$

Thus,

$$v_A = \sqrt{(9.50)^2 + (7.50)^2} = 12.10 \text{ ft/s} \qquad \textit{Ans.}$$

$$\theta = \tan^{-1}\left(\frac{7.50}{9.50}\right) = 38.3^\circ \quad \measuredangle_{\theta}^{\mathbf{v}_A}$$

Example 16–7

The bar AB of the linkage shown in Fig. 16–14a has a clockwise angular velocity of 30 rad/s when $\theta = 60°$. Compute the angular velocity of members BC and DC at this instant.

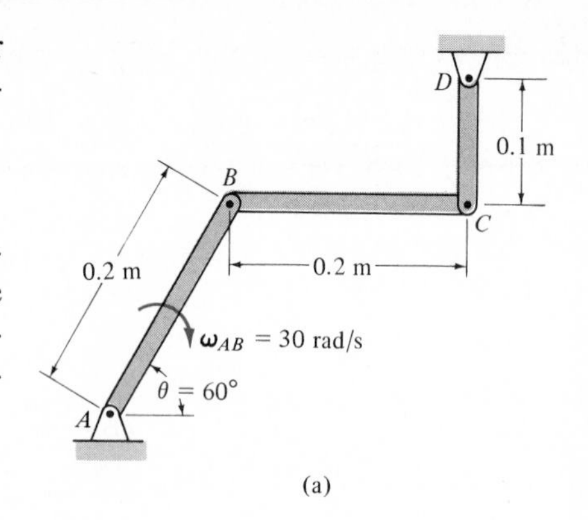

(a)

Solution

Kinematic Diagrams. From the figure, AB and DC are subjected to *rotation about a fixed axis* and BC is subjected to *general plane motion*. The magnitude and direction of $\mathbf{v}_B$ may be determined from the kinematic diagram shown in Fig. 16–14b. Because of the rotation, $\mathbf{v}_B$ *always* acts tangent to the dashed circular path. The magnitude of $\mathbf{v}_B$ is

$$v_B = 30 \text{ rad/s}(0.2 \text{ m}) = 6 \text{ m/s} \quad \searrow^{30°}$$

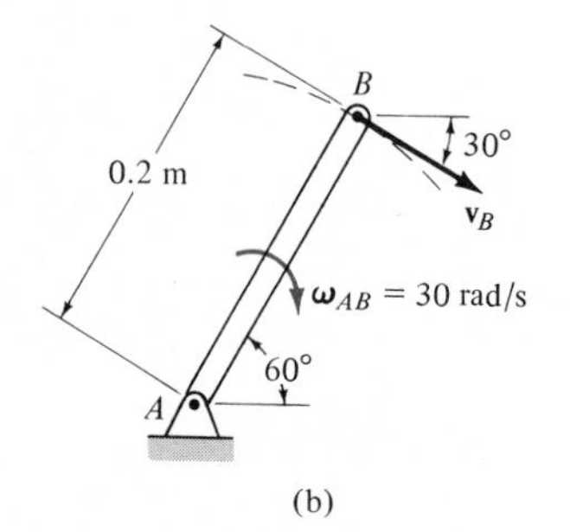

(b)

Likewise, from the kinematic diagram of bar DC, Fig. 16–4c, $\mathbf{v}_C$ acts tangent to its circular path.

Knowing $\mathbf{v}_B$ and the direction of $\mathbf{v}_C$, Fig. 16–14d, we can determine the two unknown magnitudes v_C and ω_{BC} by applying the velocity equation to B, the base point, and C lying on rod BC, i.e., $\mathbf{v}_C = \mathbf{v}_B + \mathbf{v}_{C/B}$.

As shown in Fig. 16–14e, the relative velocity $\mathbf{v}_{C/B}$ acts upward and has a magnitude of

$$v_{C/B} = \omega_{BC}(r_{C/B})$$
$$= \omega_{BC}(0.2 \text{ m}) \uparrow$$

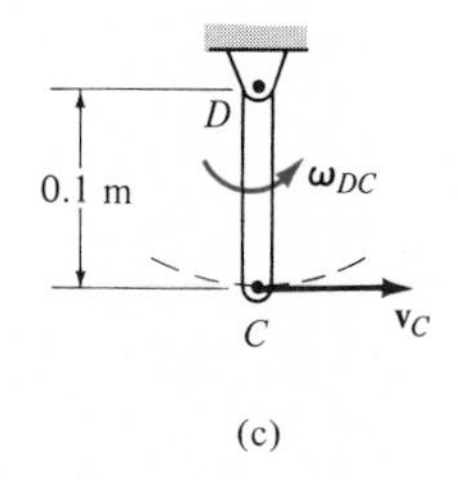

(c)

Velocity Equation

$$\mathbf{v}_C = \mathbf{v}_B + \mathbf{v}_{C/B}$$
$$\underset{\rightarrow}{v_C} = \underset{\searrow^{30°}}{6 \text{ m/s}} + \underset{\uparrow}{\omega_{BC}(0.2 \text{ m})}$$

Equating the horizontal and vertical components yields

$(\overset{+}{\rightarrow})$ $$v_C = 6 \cos 30° + 0 = 5.20 \text{ m/s}$$

$(+\uparrow)$ $$0 = -6 \sin 30° + 0.2\omega_{BC}$$

$$\omega_{BC} = 15 \text{ rad/s} \circlearrowleft \qquad \textit{Ans.}$$

Due to the circular motion of DC, Fig. 16–14c,

$$5.20 \text{ m/s} = \omega_{DC}(0.1 \text{ m})$$
$$\omega_{DC} = 52 \text{ rad/s} \circlearrowleft \qquad \textit{Ans.}$$

Fig. 16–14

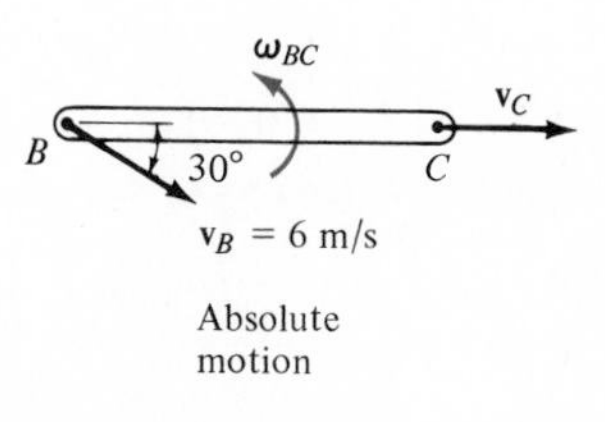

Absolute motion

(d)

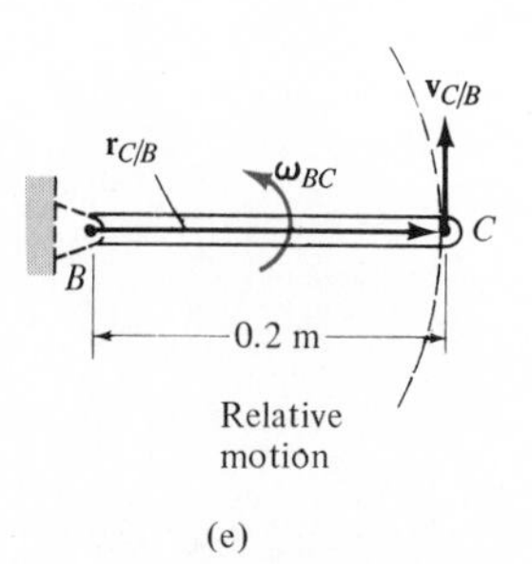

Relative motion

(e)

Problems

16–25. An acrobat is executing a summersault. At the instant shown he has an angular velocity of 1.5 rad/s and his mass center G has a velocity of 0.75 m/s vertically upward. Determine the velocities of his head H and foot F at this instant.

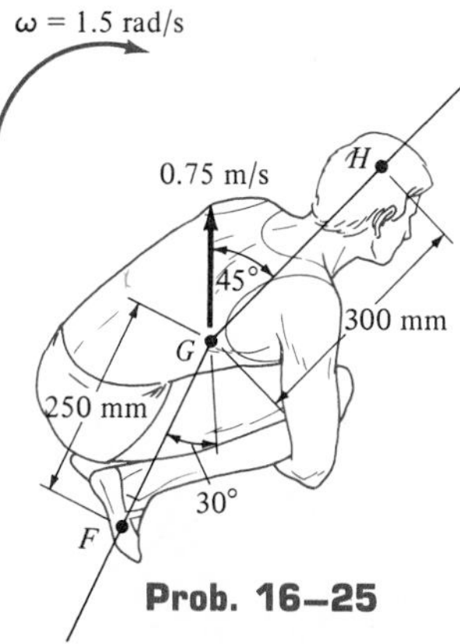

Prob. 16–25

16–26. Due to an engine failure, the missile is rotating at $\omega = 3$ rad/s, while its mass center G is moving upward at 200 ft/s. Determine the velocity of its nose B at this instant.

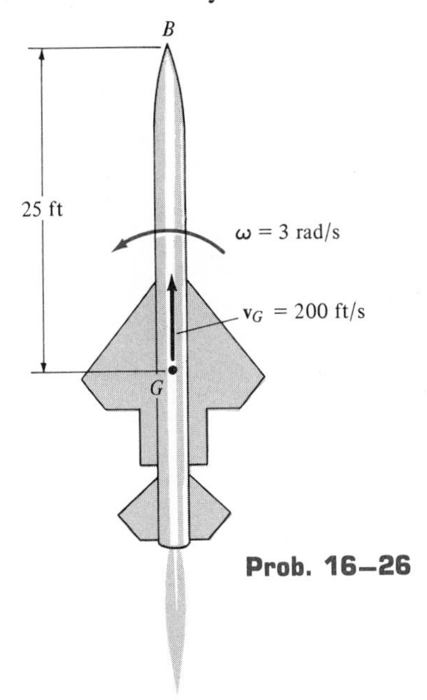

Prob. 16–26

16–27. Knowing that the angular velocity of link CB is $\omega_{CB} = 6$ rad/s, determine the velocity of B and the velocity of the collar A at the instant shown.

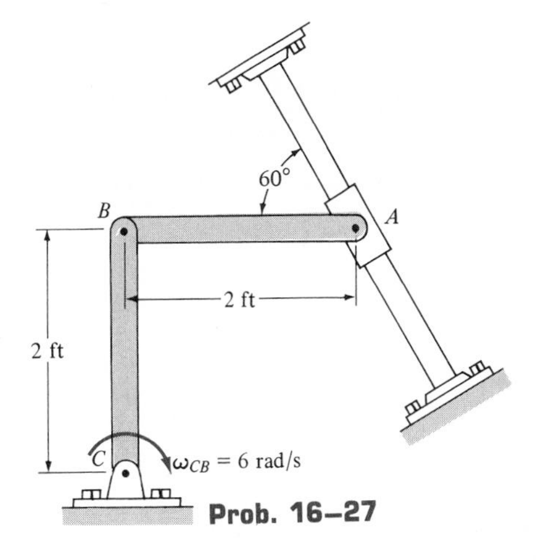

Prob. 16–27

***16–28.** If the block at C is moving downward at 4 ft/s, determine the velocity of B and the angular velocity of bar AB at the instant shown.

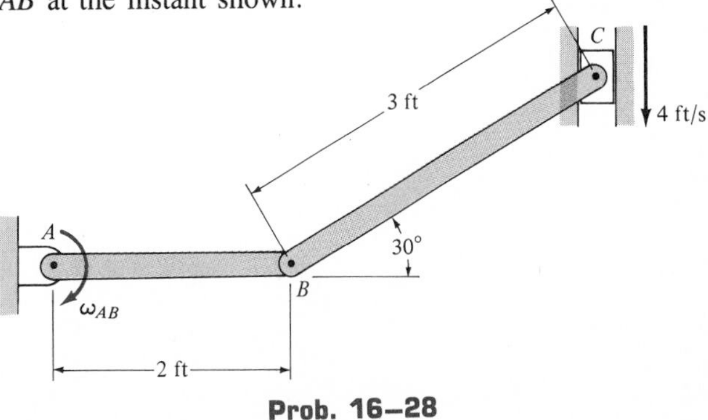

Prob. 16–28

16–29. At the instant shown, the disk is rotating at $\omega = 4$ rad/s. If the end of the cord wrapped around the disk is fixed at D, determine the velocities of points B and C. *Hint:* $v_A = 0$.

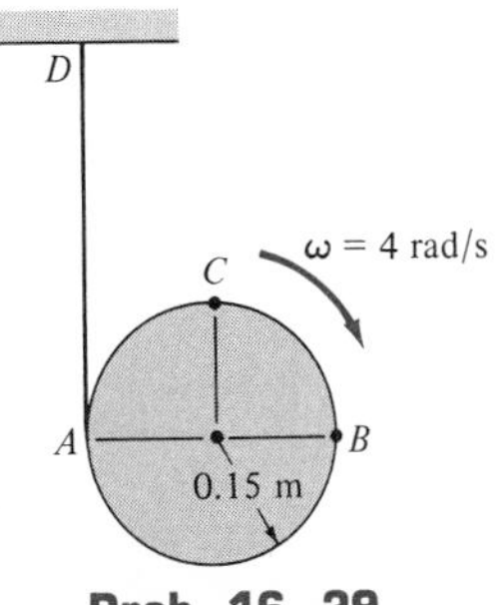

Prob. 16–29

16–30. Knowing the angular velocity of link CD is $\omega_{CD} = 4$ rad/s, determine (a) the velocity of C, (b) the angular velocity of BC, and (c) the velocity of B at the instant shown.

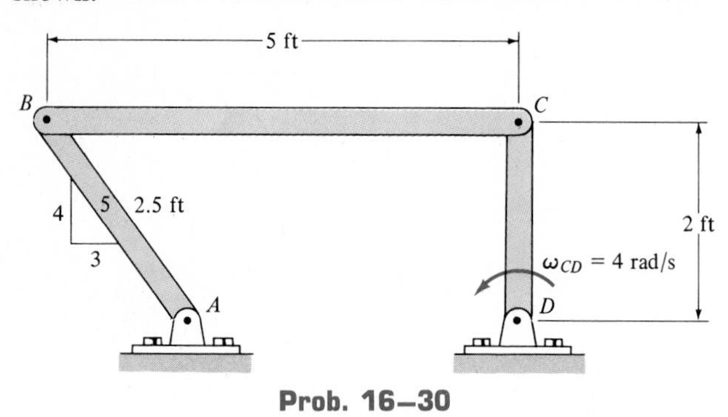

Prob. 16–30

16–31. The bicycle has a velocity of $v = 6$ ft/s, and at the same instant the rear wheel has a clockwise angular velocity of $\omega = 3$ rad/s, which causes it to slip at its contact point A. Determine the velocity of point A. *Hint:* $v_C = 6$ ft/s.

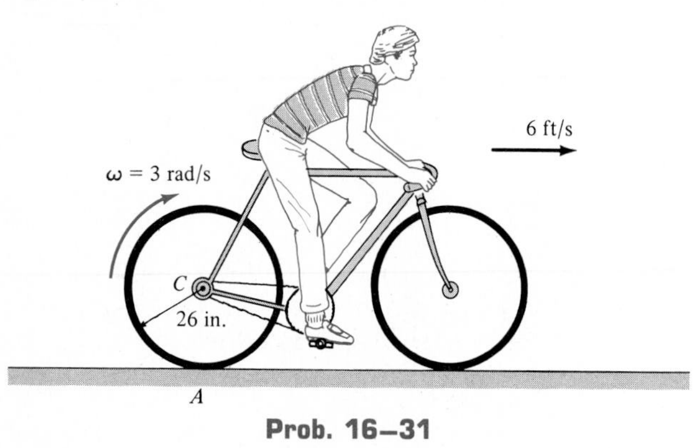

Prob. 16–31

***16–32.** At the instant shown, the truck is moving to the right at 3 m/s, while the spool is rolling at $\omega = 6$ rad/s without slipping at B. Determine the velocity of the spool's center G.

Prob. 16–32

16–33. If rod CD has a downward velocity of 6 ft/s at the instant shown, determine the velocity of the gear rack A at this instant. The rod is pinned at C to gear B.

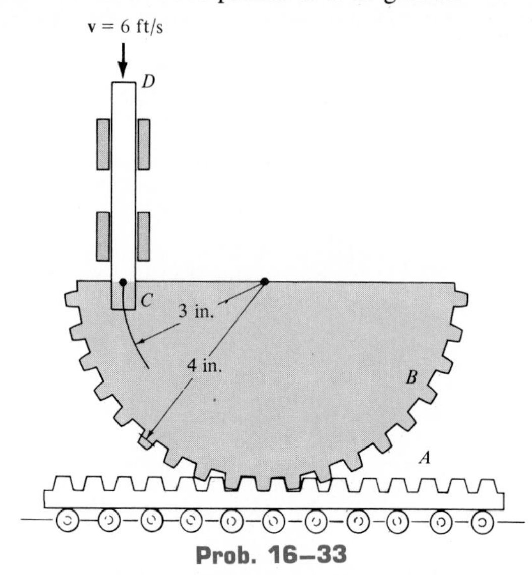

Prob. 16–33

16–34. When the crank on the Chinese windlass is turning, the rope on shaft A unwinds while that on shaft B winds up. Determine the speed at which the block lowers if the crank is turning with an angular velocity of $\omega = 4$ rad/s. What is the angular velocity of the pulley at C? The rope segments on each side of the pulley are both parallel and vertical and the rope does not slip on the pulley.

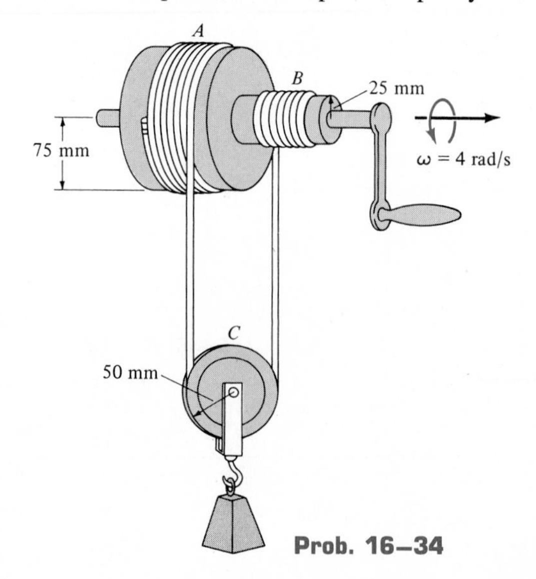

Prob. 16–34

16–35. If the end of the cord is pulled downward with a speed of $v_C = 150$ mm/s, determine the angular velocities of pulleys A and B and the upward speed of block D. Assume that the cord does not slip on the pulleys.

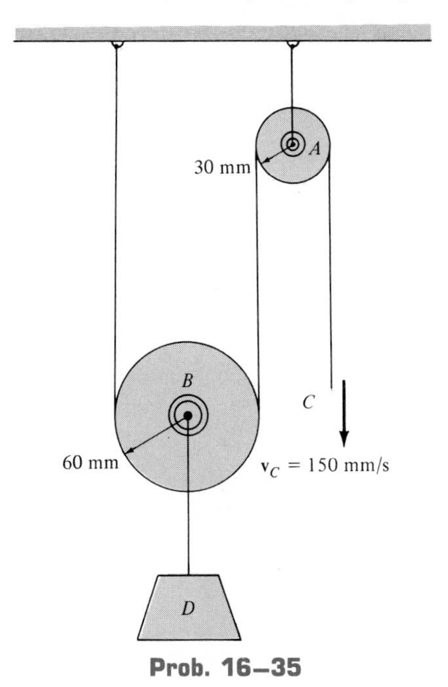

Prob. 16–35

16–37. The device is used to indicate the safe load acting at the end of the boom, B, when it is in any angular position. It consists of a fixed dial plate D and an indicator arm ACE which is pinned to the plate at C and to a short link EF. If the boom is pin-connected to the trunk frame at G and is rotating downward at $\omega_B = 4$ rad/s, determine the velocity of the dial pointer A at the instant shown, i.e., when EF and AC are in the vertical position.

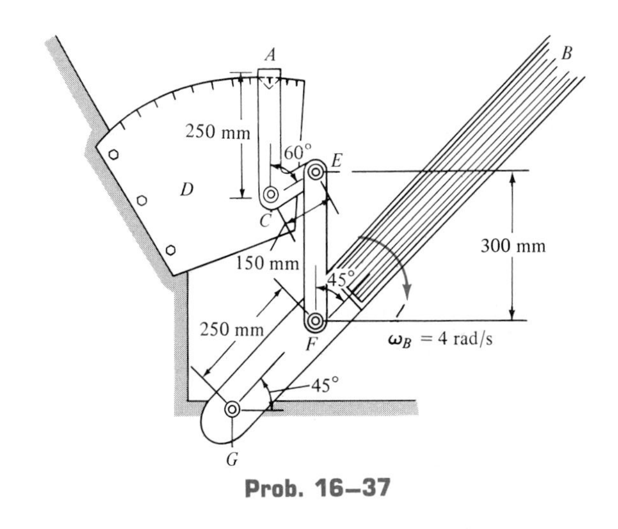

Prob. 16–37

***16–36.** The inner hub of the roller bearing is rotating with an angular velocity of $\omega_i = 8$ rad/s, while the outer hub is rotating in the opposite direction at $\omega_o = 6$ rad/s. Determine the angular velocity of each of the rollers if they roll on the hubs without slipping.

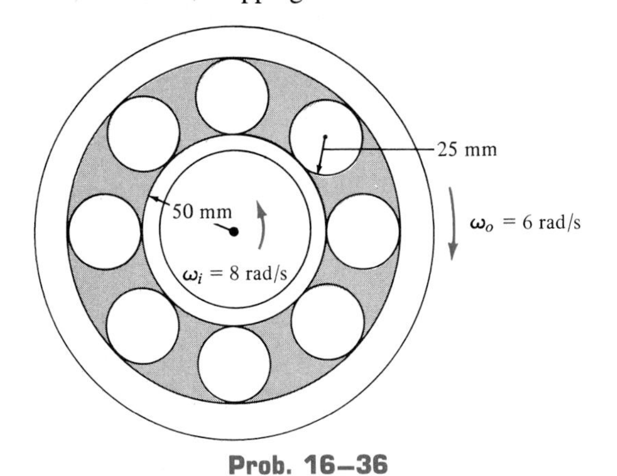

Prob. 16–36

16–38. If roller C and link AB both rotate clockwise with an angular velocity of $\omega = 4$ rad/s, determine the angular velocity of roller D. No slipping occurs between the rollers. Point A is fixed from moving.

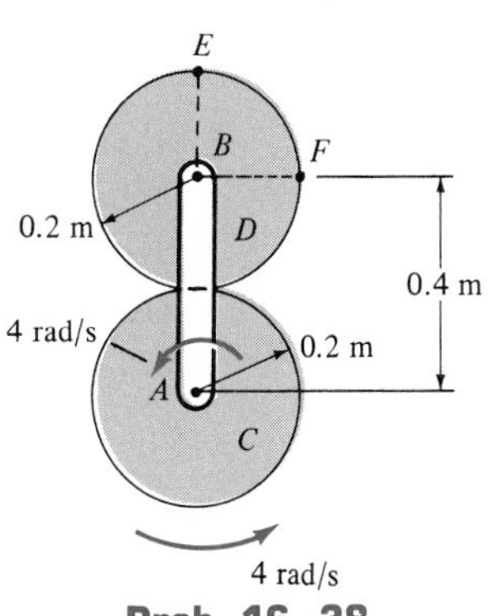

Prob. 16–38

16–39. Determine the velocities of points E and F on the roller D at the instant shown.

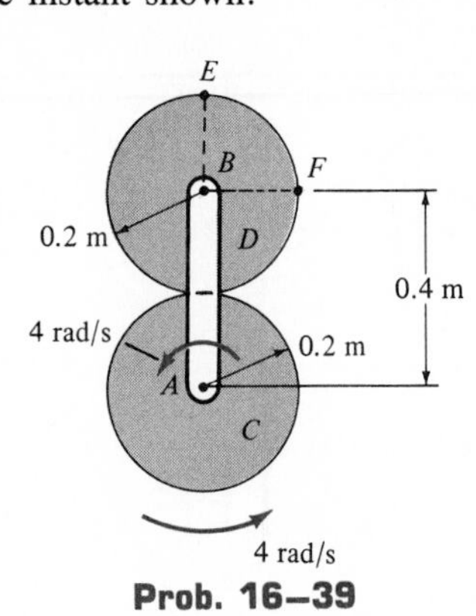

Prob. 16–39

***16–40.** The rotation of link AB creates an oscillating movement of gear F. If AB has an angular velocity of $\omega_{AB} = 8$ rad/s, determine the angular velocity of gear F at the instant shown. Gear E is a part of arm CD and pinned at D to a fixed point.

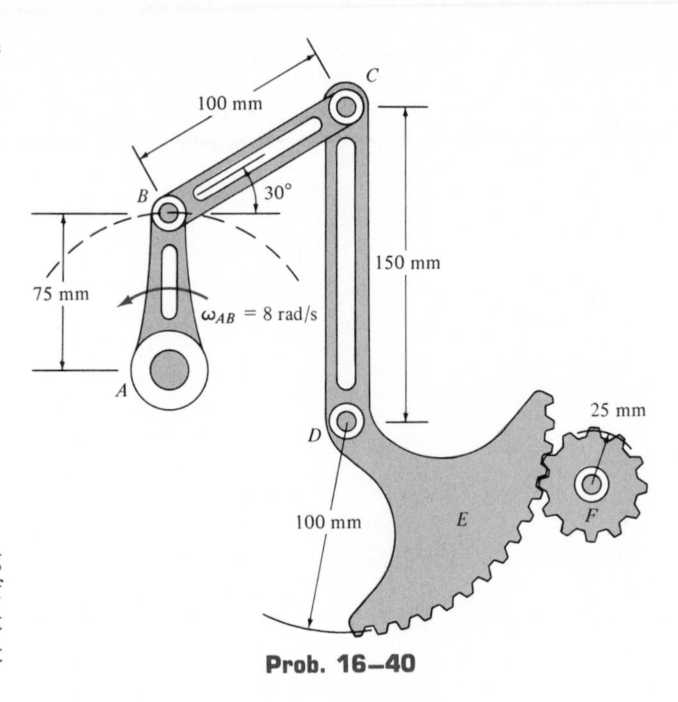

Prob. 16–40

16.6 Instantaneous Center of Zero Velocity

The velocity of any point B located in a rigid body can be obtained in a simple way if one chooses the base point IC to be a point that has *zero velocity* at the instant considered. In this case, $\mathbf{v}_{IC} = \mathbf{0}$, and therefore the velocity equation (Eq. 16–11) becomes $\mathbf{v}_B = \mathbf{v}_{IC} + \mathbf{v}_{B/IC} = \mathbf{v}_{B/IC}$. For a body having general plane motion, point IC so chosen is called the *instantaneous center of zero velocity* and it lies on the *instantaneous axis of zero velocity* (IA). Hence, the IA is always perpendicular to the plane used to represent the motion, and the intersection of the IA with this plane defines the IC. Since $\mathbf{v}_B = \mathbf{v}_{B/IC}$, point B moves momentarily about the IC in a circular path; in other words, the body appears to rotate about the IA, Fig. 16–15a. If the relative-position vector $\mathbf{r}_{B/IC}$ is established, then the *magnitude* of $\mathbf{v}_B$ is simply $v_B = \omega r_{B/IC}$, where ω is the angular velocity of the body. The *direction* of $\mathbf{v}_B$ is *perpendicular* to $\mathbf{r}_{B/IC}$, as shown in Fig. 16–15a.

Although the IC may be conveniently used to determine the velocity of any point in a body, it generally *does not have zero acceleration* and therefore it *should not* be used for finding the accelerations of points in a body.

Location of the *IC*. If the location of the IC is unknown, it may be determined provided one knows the lines of action for the velocities $\mathbf{v}_A$ and $\mathbf{v}_B$ of any two points A and B in the body, Fig. 16–15b. From each of these lines construct at points A and B *perpendicular* line segments which then define the

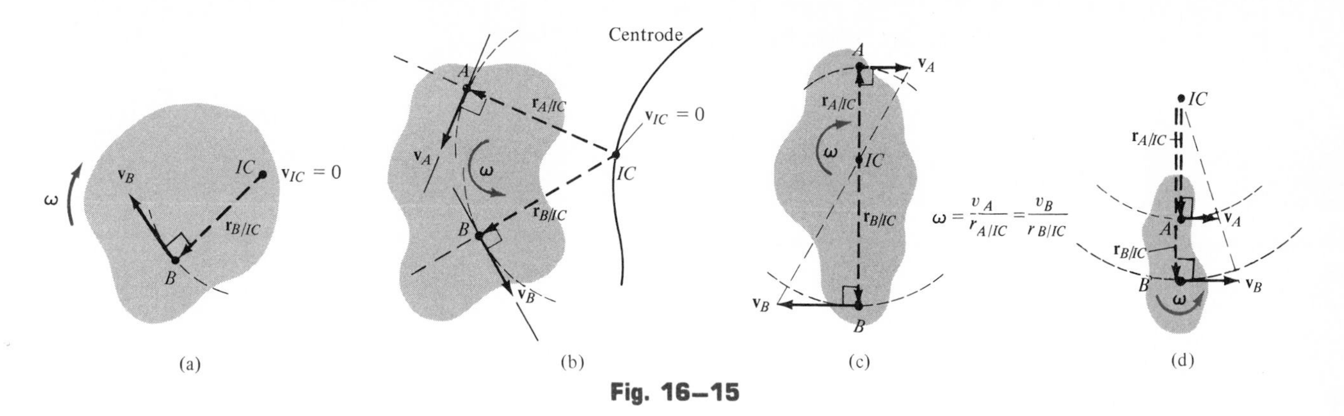

Fig. 16–15

lines of action of $\mathbf{r}_{A/IC}$ and $\mathbf{r}_{B/IC}$, respectively. Extending these perpendiculars to their *point of intersection* as shown locates the *IC* at the instant considered. As a special case, if the lines of action of the velocities $\mathbf{v}_A$ and $\mathbf{v}_B$ are *parallel* to one another, Fig. 16–15*c* or *d*, then, provided one knows the magnitudes of $\mathbf{v}_A$ and $\mathbf{v}_B$, the *IC* can be located using proportional right triangles.

When a body is subjected to general plane motion, the point selected to represent the instantaneous center of zero velocity for the body can only be used for an *instant of time*. Since the body changes its position from one instant to the next, then for each position of the body a unique instantaneous center must be determined. The locus of points which defines the *IC* during various instants of time is called a *centrode,* Fig. 16–15*b*. Hence, each point on the centrode acts as the *IC* for the body only for an instant of time.

PROCEDURE FOR ANALYSIS

The velocity of a point in a body which is subjected to general plane motion can be determined with reference to its instantaneous center of zero velocity, provided the location of the *IC* is first established. This is done by using the method described above. As shown on the kinematic diagram in Fig. 16–16, the body is imagined as "extended and pinned" at the *IC* such that it rotates about this pin with its instantaneous angular velocity $\boldsymbol{\omega}$. The *magnitude* of velocity for the arbitrary points A, B, and C in the body can then be determined by using the equation $v = \omega r$, where r is the radial line drawn from the *IC* to the point. The line of action of each velocity vector is *perpendicular* to its associated radial line, and the velocity has a *direction* which tends to move the point in a manner consistent with the angular rotation of the radial line, Fig. 16–16.

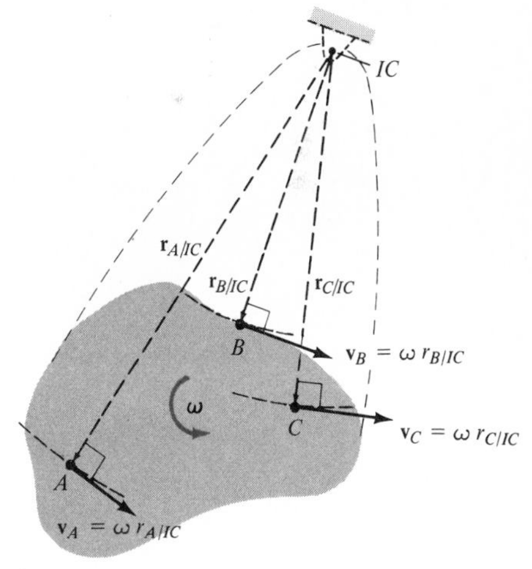

Fig. 16–16

The following examples illustrate the method of determining the *IC* and computing the velocities of points in a rigid body subjected to general plane motion. See also the "velocity analysis" of Examples 16–11 to 16–13, which further illustrates this method of analysis.

Example 16–8

Determine the location of the instantaneous center of zero velocity for (a) the wheel shown in Fig. 16–17*a*, which is rolling without slipping along the ground, (b) the crankshaft *BC* shown in Fig. 16–17*b*, and (c) the link *CB* shown in Fig. 16–17*c*.

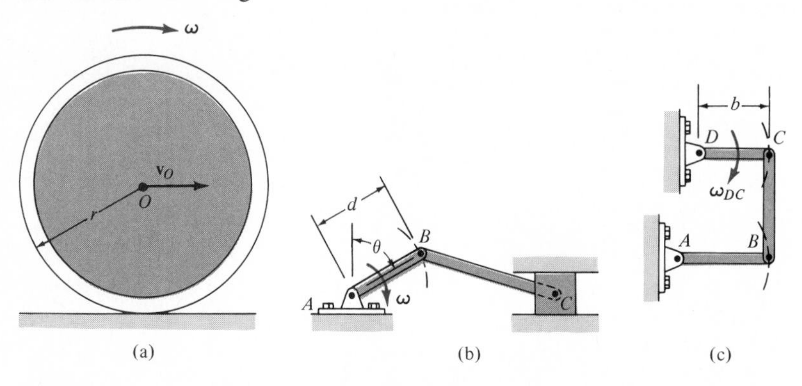

Fig. 16–17

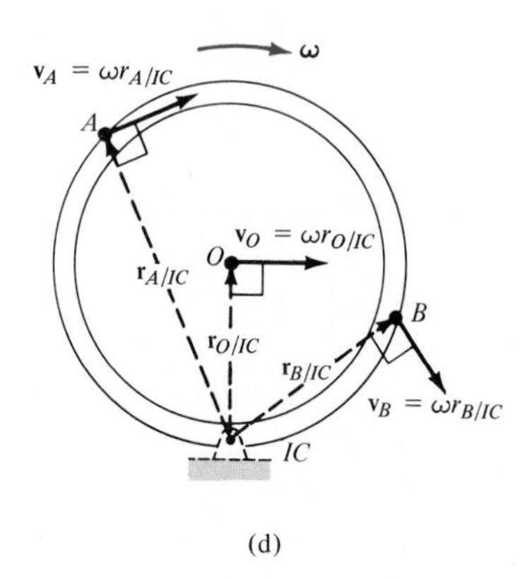

(d)

Solution

Part (a). The wheel rolls without slipping, and therefore the point of *contact* of the wheel with the ground has *zero velocity*. Hence, this point represents the *IC* for the wheel, Fig. 16–17*d*. If it is imagined that the wheel is momentarily pinned at this point, the velocities of points *A*, *B*, *O*, and so on, can be found using $v = \omega r$. As shown, the distances $r_{A/IC}$, $r_{B/IC}$, and $r_{O/IC}$, are determined from the geometry of the wheel.

Part (b). As shown in Fig. 16–17*b*, link *AB* rotates about the fixed pin *A*. Thus, point *B* has a speed of $v_B = \omega d$, caused by the clockwise rotation of *AB*. Also, $\mathbf{v}_B$ is perpendicular to *AB*, so that it acts at an angle θ from the horizontal as shown in Fig. 16–17*e*. The motion of point *B* causes the piston to move forward *horizontally* with a velocity $\mathbf{v}_C$. Consequently, point *C* on the rod moves horizontally with this same velocity. When lines are drawn perpendicular to $\mathbf{v}_B$ and $\mathbf{v}_C$, Fig. 16–17*e*, they intersect at the *IC*. The magnitudes of $\mathbf{r}_{B/IC}$ and $\mathbf{r}_{C/IC}$ are determined strictly from the geometry of construction.

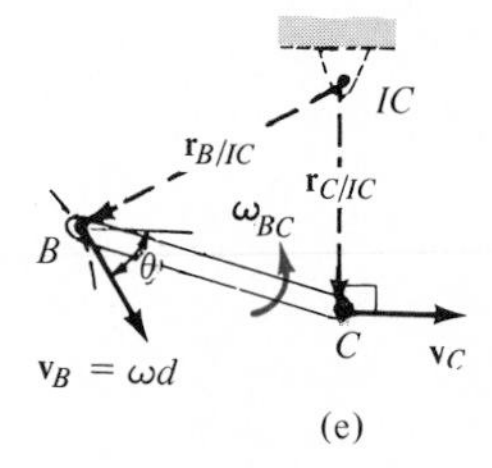

(e)

Part (c). Points *B* and *C* follow circular paths of motion since rods *AB* and *DC* are each subjected to rotation about a fixed axis, Fig. 16–17*c*. In particular, $v_C = \omega_{DC} b$. Since the velocity is always tangent to the path, at the instant considered, $\mathbf{v}_C$ on rod *DC* and $\mathbf{v}_B$ on rod *AB* are both directed vertically downward, along the axis of link *CB*, Fig. 16–17*f*. Furthermore, since *CB* is *rigid,* no relative displacement occurs between points *B* and *C*, so that $\mathbf{v}_B = \mathbf{v}_C$. Radial lines drawn perpendicular to these two velocities form parallel lines which intersect at "infinity;" i.e., $r_{C/IC} \rightarrow \infty$ and $r_{B/IC} \rightarrow \infty$. Since $v_C = r_{C/IC}\omega_{CB}$, then $\omega_{CB} = v_C/r_{C/IC} = (\omega_{DC} b)/\infty = 0$. As a result, rod *CB* momentarily *translates* with a speed of $v_C = \omega_{DC} b$. An instant later, however, *CB* will move to a new position, causing the instantaneous center to *change* to some finite location.

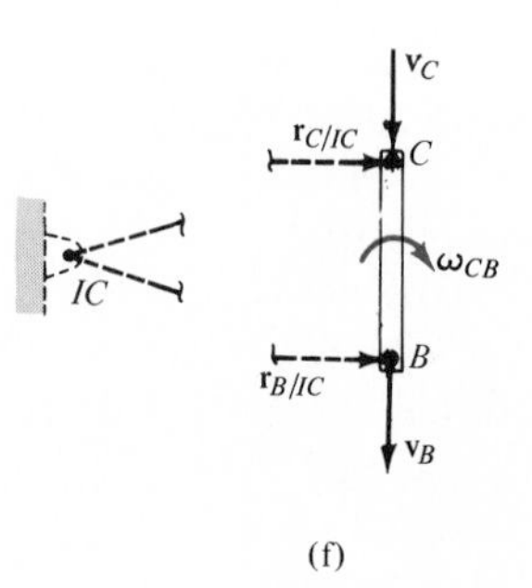

(f)

Example 16–9

Block D shown in Fig. 16–18a moves with a speed of 3 m/s. Determine the angular velocities of links BD and AB, and the velocity of point B at the instant shown.

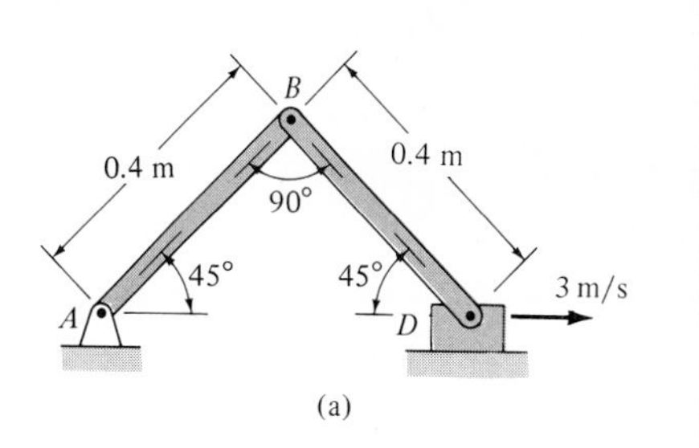

(a)

Solution

Since D moves to the right at 3 m/s, it causes arm AB to rotate about point A in a clockwise direction. Hence, $\mathbf{v}_B$ is directed perpendicular to AB as shown in Fig. 16–18b. The instantaneous center of zero velocity for BD is located at the intersection of the line segments drawn perpendicular to $\mathbf{v}_B$ and $\mathbf{v}_D$, Fig. 16–18b. From the geometry,

$$r_{B/IC} = 0.4 \tan 45^\circ \text{ m} = 0.4 \text{ m}$$

$$r_{D/IC} = \frac{0.4 \text{ m}}{\cos 45^\circ} = 0.566 \text{ m}$$

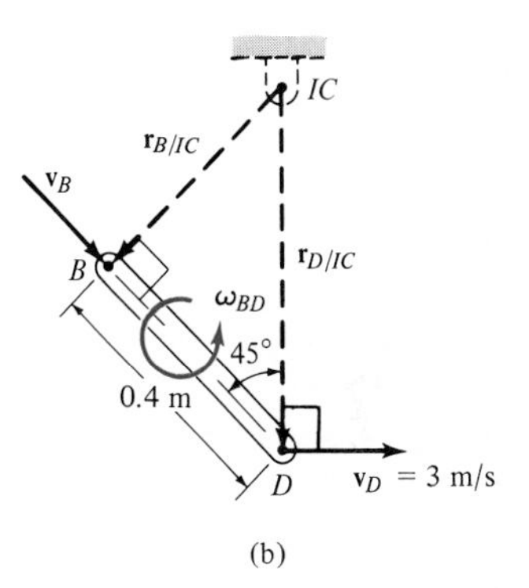

(b)

Since the magnitude of $\mathbf{v}_D$ is known, the angular velocity of the link is

$$\omega_{BD} = \frac{v_D}{r_{D/IC}} = \frac{3 \text{ m/s}}{0.566 \text{ m}} = 5.30 \text{ rad/s} \circlearrowleft \qquad \textit{Ans.}$$

The velocity of B is therefore

$$v_B = \omega_{BD}(r_{B/IC}) = 5.30 \text{ rad/s}(0.4 \text{ m}) = 2.12 \text{ m/s} \qquad \textit{Ans.}$$

Note that link AB is subjected to rotation about a fixed axis passing through A, Fig. 16–18c, and since v_B is known, the angular velocity of AB is

$$\omega_{AB} = \frac{v_B}{r_{B/A}} = \frac{2.12 \text{ m/s}}{0.4 \text{ m}} = 5.30 \text{ rad/s} \circlearrowright \qquad \textit{Ans.}$$

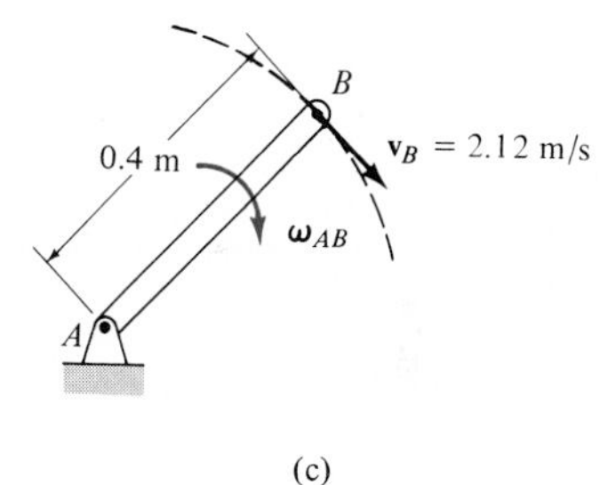

(c)

Fig. 16–18

Example 16–10

The cylinder shown in Fig. 16–19*a* rolls without slipping between the two moving plates *E* and *D*. Determine the angular velocity of the cylinder and the velocity of its center *C* at the instant shown.

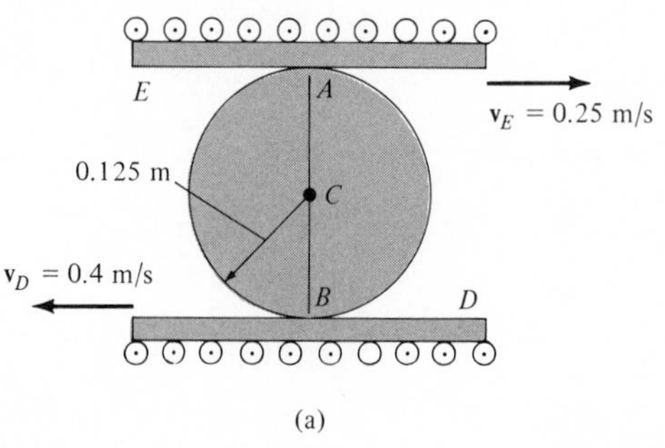

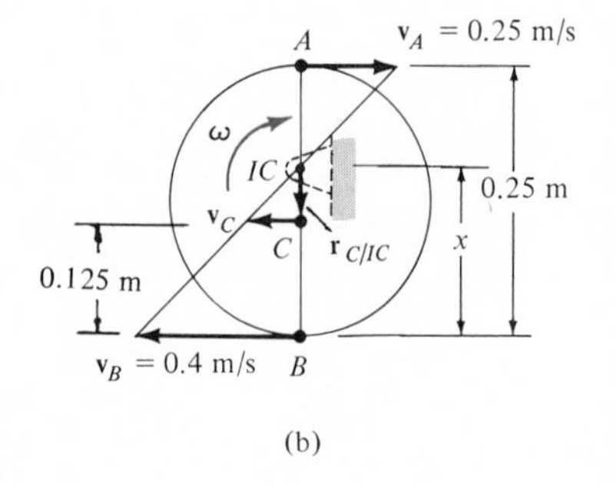

Fig. 16–19

Solution

Since no slipping occurs, the contact points *A* and *B* on the cylinder have the same velocity as the plates *E* and *D*. Furthermore, the velocities $\mathbf{v}_A$ and $\mathbf{v}_B$ are *parallel*, so that by the proportionality of right triangles the *IC* is located at a point on line *AB*, Fig. 16–19*b*. Assuming this point to be a distance x from *B*, we have

$$v_B = \omega x; \qquad 0.4 = \omega x$$

$$v_A = \omega(0.25 - x); \qquad 0.25 = \omega(0.25 - x)$$

Dividing one of these equations into the other eliminates ω and yields

$$0.4(0.25 - x) = 0.25x$$

$$x = \frac{0.1}{0.65} = 0.154 \text{ m}$$

Hence, the angular velocity is

$$\omega = \frac{v_B}{x} = \frac{0.4}{0.154} = 2.60 \text{ rad/s} \qquad \textit{Ans.}$$

The velocity of point *C* is therefore

$$v_C = \omega r_{C/IC} = 2.60(0.154 - 0.125)$$

$$= 0.0754 \text{ m/s} \qquad \textit{Ans.}$$

Problems

16–41. Solve Prob. 16–27 using the method of instantaneous center of zero velocity.

16–42. Solve Prob. 16–28 using the method of instantaneous center of zero velocity.

16–43. Solve Prob. 16–29 using the method of instantaneous center of zero velocity.

***16–44.** Solve Prob. 16–30 using the method of instantaneous center of zero velocity.

16–45. Solve Prob. 16–32 using the method of instantaneous center of zero velocity.

16–46. The automobile with wheels 2.5 ft in diameter is traveling in a straight path at a rate of 60 ft/s. If no slipping occurs, determine the angular velocity of one of the rear wheels and the velocity of the fastest moving point on the wheel.

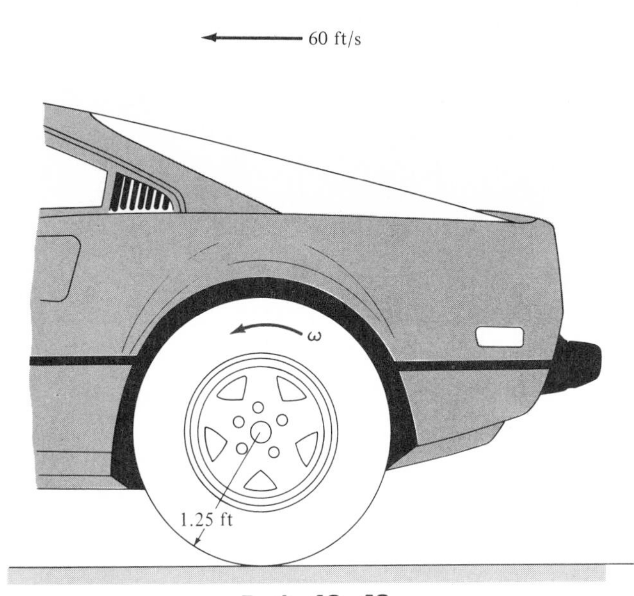

Prob. 16–46

16–47. As the cord unravels from the wheel's inner hub, the wheel is rotating at $\omega = 2$ rad/s at the instant shown. Determine the velocities of points A and B.

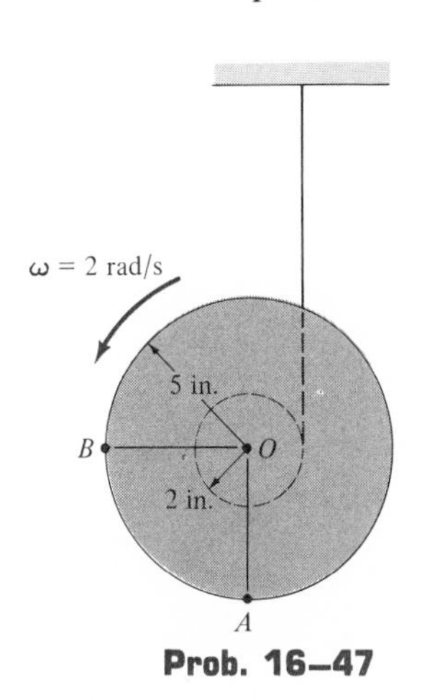

Prob. 16–47

***16–48.** Part of an automatic transmission consists of a *fixed* ring gear R, three equal planet gears P, the sun gear S, and the planet carrier C, which is shaded. If the sun gear is rotating at $\omega_S = 5$ rad/s, determine the angular velocity of the *planet carrier*. Note that C is pin-connected to the center of each of the planet gears.

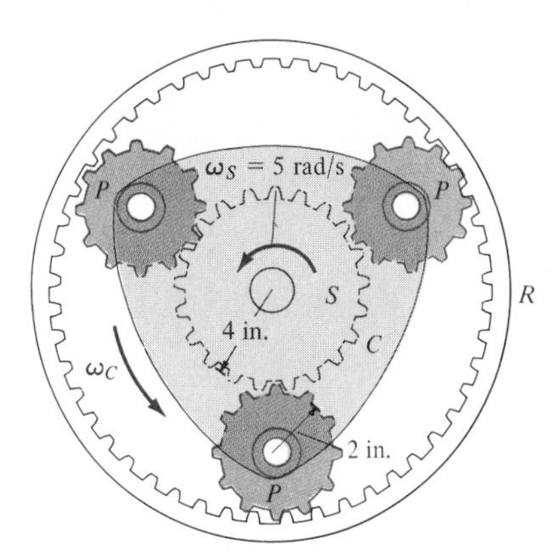

Prob. 16–48

16–49. Show that if the rim of the wheel and its hub maintain contact with the three tracks as the wheel rolls, it is necessary that slipping occurs at the hub A if no slipping occurs at B. Under these conditions, what is the speed at A if the wheel has an angular velocity $\boldsymbol{\omega}$?

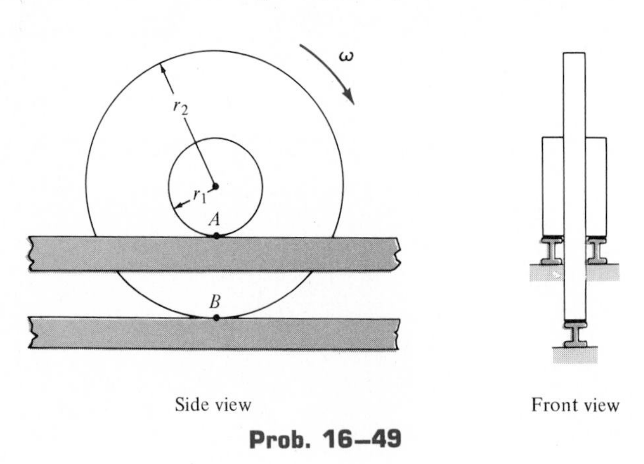

Prob. 16–49

16–50. The wheel rolls on its hub without slipping on the horizontal surface S. If the velocity of the center of the wheel is $v_C = 300$ mm/s to the right, determine the velocities of points A and B at the instant shown.

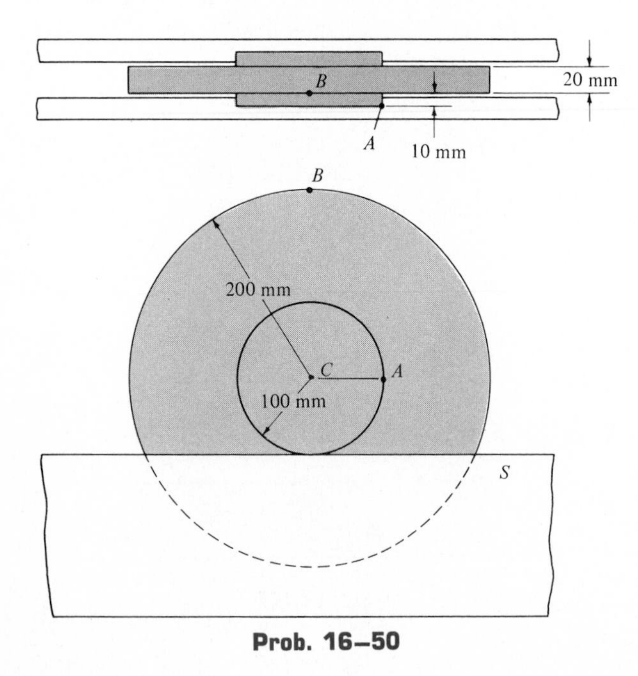

Prob. 16–50

16–51. If link AB is rotating at 3 rad/s, determine the angular velocities of links BC and CD at the instant shown.

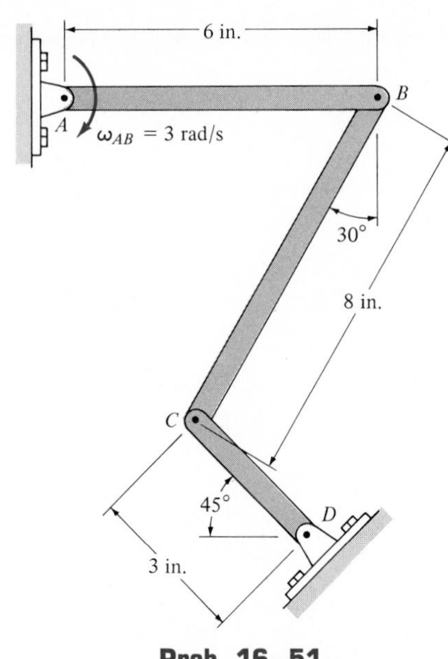

Prob. 16–51

***16–52.** The oil pumping unit consists of a walking beam AB, connecting rod BC, and crank CD. If the crank rotates at a constant rate of 6 rad/s, determine the speed of the rod hanger H at the instant shown. *Hint:* Point B follows a circular path about point E and therefore the velocity of B is *not* vertical.

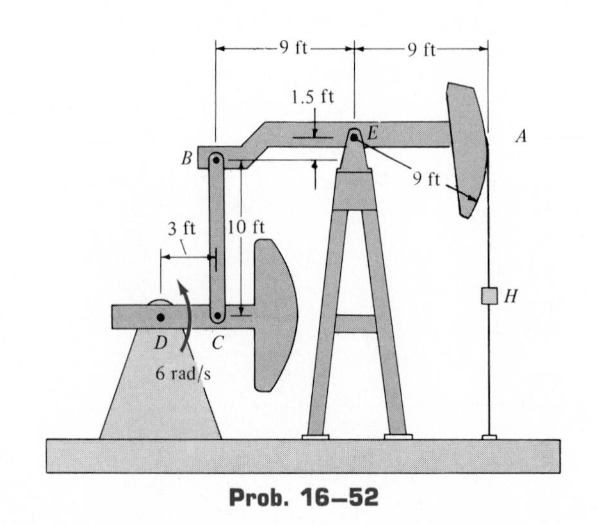

Prob. 16–52

16–53. The mechanism shown is used in a riveting machine. It consists of a driving piston A, three links, and a riveter which is attached to the slider block D. Determine the velocity of C at the instant shown, when the piston at A is traveling at $v_A = 30$ m/s.

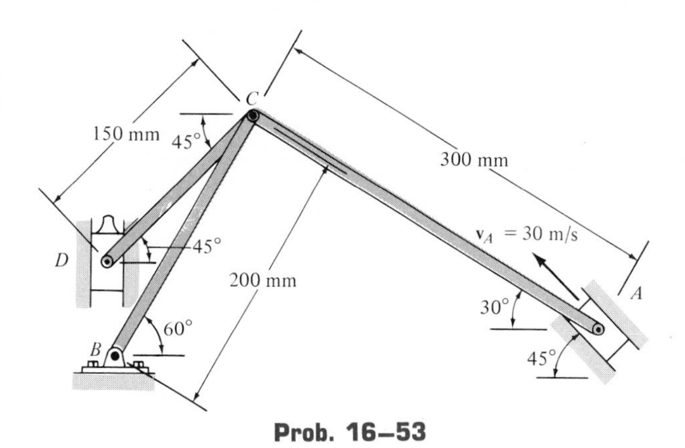

Prob. 16–53

Relative-Motion Analysis: Acceleration 16.7

An equation that relates the accelerations of two points in a rigid body subjected to general plane motion may be determined by differentiating the velocity equation $\mathbf{v}_B = \mathbf{v}_A + \mathbf{v}_{B/A}$ with respect to time. Thus,

$$\frac{d\mathbf{v}_B}{dt} = \frac{d\mathbf{v}_A}{dt} + \frac{d\mathbf{v}_{B/A}}{dt}$$

The terms $d\mathbf{v}_B/dt = \mathbf{a}_B$ and $d\mathbf{v}_A/dt = \mathbf{a}_A$ are measured from a set of *fixed x, y axes* and represent the *absolute accelerations* of points B and A, Fig. 16–20*a*. The last term is denoted as the *relative acceleration* $\mathbf{a}_{B/A}$. This vector represents the acceleration of B measured by an observer stationed at A and fixed to a set of *translating axes*. Since the body is rigid, to this observer point B appears to move along the dashed *circular arc* that has a radius of curvature of $r_{B/A}$, Fig. 16–20*c*. In other words, the body moves as if it were *pinned* at

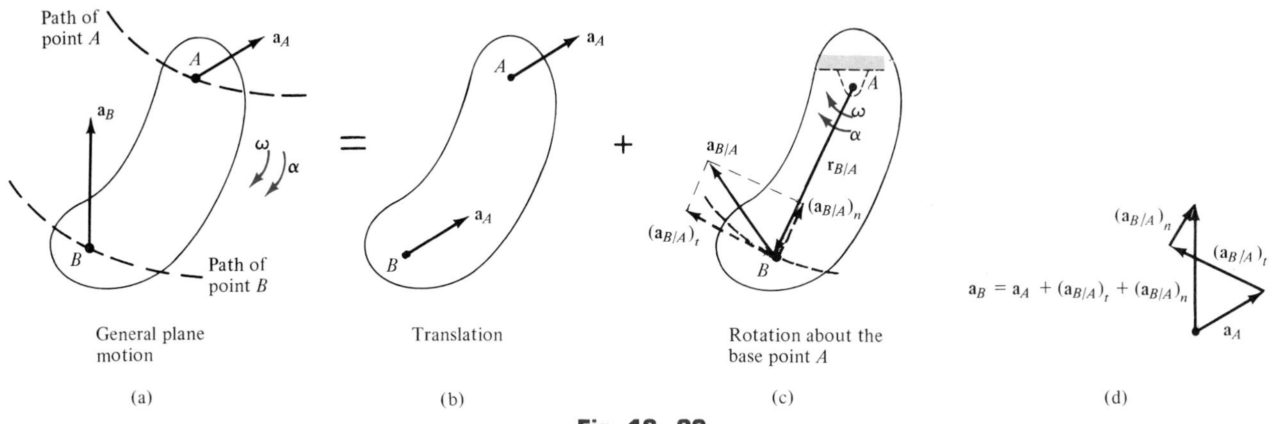

Fig. 16–20

A. Consequently, $\mathbf{a}_{B/A}$ can be expressed in terms of its tangential and normal components of motion; i.e., $\mathbf{a}_{B/A} = (\mathbf{a}_{B/A})_t + (\mathbf{a}_{B/A})_n$, where $(a_{B/A})_t = \alpha r_{B/A}$ and $(a_{B/A})_n = \omega^2 r_{B/A}$. Hence, the above equation can be written in the form

$$\mathbf{a}_B = \mathbf{a}_A + (\mathbf{a}_{B/A})_t + (\mathbf{a}_{B/A})_n \tag{16–12}$$

where

$\mathbf{a}_B$ = absolute acceleration of point B

$\mathbf{a}_A$ = absolute acceleration of point A

$(\mathbf{a}_{B/A})_t$ = relative tangential acceleration component of "B with respect to A;" the *magnitude* is $(a_{B/A})_t = \alpha r_{B/A}$ and the *direction* is perpendicular to the line of action of $\mathbf{r}_{B/A}$

$(\mathbf{a}_{B/A})_n$ = relative normal acceleration component of "B with respect to A;" the *magnitude* is $(a_{B/A})_n = \omega^2 r_{B/A}$ and the *direction* is always from B towards A

$\boldsymbol{\alpha}$ = absolute angular acceleration of the body

$\boldsymbol{\omega}$ = absolute angular velocity of the body

$\mathbf{r}_{B/A}$ = relative-position vector drawn from A to B

Each of the four terms in Eq. 16–12 is represented graphically on the *kinematic diagrams* shown in Fig. 16–20. Here it is seen that at a given instant the acceleration of B, Fig. 16–20a, is determined by considering the body to translate with an acceleration $\mathbf{a}_A$, Fig. 16–20b, and simultaneously rotate about the base point A with an instantaneous angular velocity $\boldsymbol{\omega}$ and angular acceleration $\boldsymbol{\alpha}$, Fig. 16–20c. Vector addition of these two effects, applied to B, yields $\mathbf{a}_B$, as shown in Fig. 16–20d. It should be noted from Fig. 16–20a that points A and B move along *curved paths,* and as a result the accelerations of these points have *both tangential and normal components.* (Recall that the acceleration of a point is *tangent to the path* only when the path is *rectilinear.*)

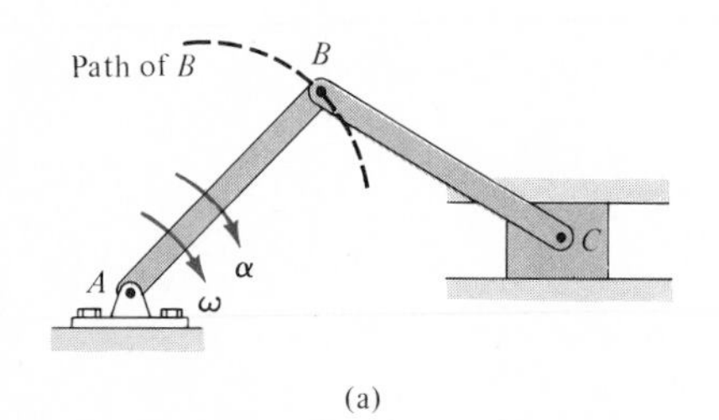

(a)

If Eq. 16–12 is applied in a practical manner to study the accelerated motion of a rigid body which is pin-connected to two other bodies, it should be realized that points which are *coincident at the pin* move with the *same acceleration,* since the path of motion over which they travel is the *same*. For example, point B lying on either rod AB or BC of the crank mechanism shown in Fig. 16–21a has the same acceleration, since the rods are pin-connected at B. Here the motion of B is along a *curved path,* so that $\mathbf{a}_B$ is calculated on the basis of its tangential and normal components $(a_B)_t = \alpha r_{B/A}$ and $(a_B)_n = \omega^2 r_{B/A}$, which are defined by the angular motion of AB. At the other end of rod BC, however, point C moves along a *rectilinear path,* which is defined by the piston. Hence, in this case, the acceleration $\mathbf{a}_C$ is directed along the path, Fig. 16–21b.

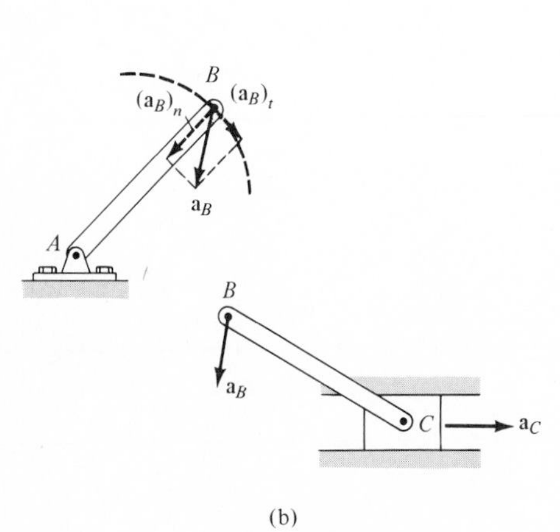

(b)

Fig. 16–21

If two bodies contact one another *without slipping,* and the *points in contact* move along *different paths,* the *tangential components* of acceleration of the points will be the *same;* however, the *normal components* will *not* be the same. For example, consider the two meshed gears in Fig. 16–22a. Point A is located on gear B and A' is located on gear C. Due to the rotational motion, $(\mathbf{a}_A)_t = (\mathbf{a}_{A'})_t$; however, since both points follow different curved paths, $(\mathbf{a}_A)_n \neq (\mathbf{a}_{A'})_n$ and therefore $\mathbf{a}_A \neq \mathbf{a}_{A'}$, Fig. 16–22*b*.

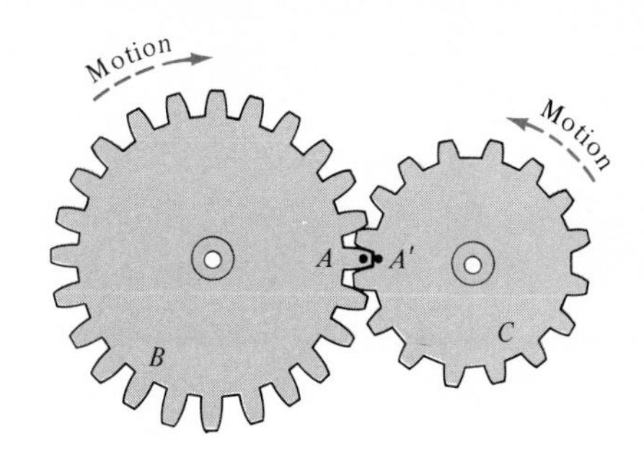

(a)

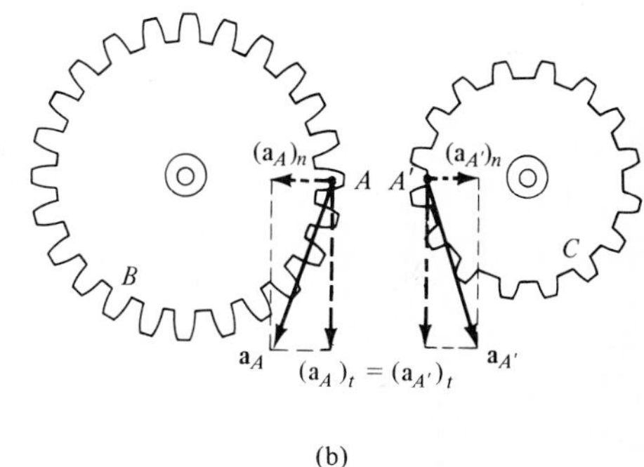

(b)

Fig. 16–22

PROCEDURE FOR ANALYSIS

The acceleration equation $\mathbf{a}_B = \mathbf{a}_A + (\mathbf{a}_{B/A})_t + (\mathbf{a}_{B/A})_n$ applies between any two points located in the *same* rigid body. For application, it is suggested that the following procedure be used.

Velocity Analysis. If the angular velocity $\boldsymbol{\omega}$ of the body is unknown, determine it by using a velocity analysis as discussed in Secs. 16–5 or 16–6. Determine also the velocities $\mathbf{v}_A$ and $\mathbf{v}_B$ of points A and B *if these points move along curved paths.*

Kinematic Diagrams. Draw a kinematic diagram of the body such as shown in Fig. 16–20*a* and indicate on it the absolute accelerations of points A and B, $\mathbf{a}_A$ and $\mathbf{a}_B$, the angular velocity $\boldsymbol{\omega}$, and the angular acceleration $\boldsymbol{\alpha}$. Usually the base point A is selected as a point having a known acceleration. In particular, if points A and B move along *curved paths,* their accelerations should be expressed in terms of their tangential and normal components; i.e., $\mathbf{a}_A = (\mathbf{a}_A)_t + (\mathbf{a}_A)_n$ and $\mathbf{a}_B = (\mathbf{a}_B)_t + (\mathbf{a}_B)_n$. Here $(a_A)_n = (v_A)^2/\rho_A$ and $(a_B)_n = (v_B)^2/\rho_B$, where ρ_A and ρ_B define the radii of curvature of the paths of points A and B, respectively.

Also, draw a kinematic diagram of the body such as shown in Fig. 16–20*c* and indicate on it the two components of the relative acceleration, $(\mathbf{a}_{B/A})_t$ and $(\mathbf{a}_{B/A})_n$. Since the body is considered to be pinned at the base point A, these components have *magnitudes* of $(a_{B/A})_t = \alpha r_{B/A}$ and $(a_{B/A})_n = \omega^2 r_{B/A}$ and their *directions* are established from the diagram such that $(\mathbf{a}_{B/A})_t$ acts perpendicular to $\mathbf{r}_{B/A}$, in accordance with the rotational motion $\boldsymbol{\alpha}$ of the body, and $(\mathbf{a}_{B/A})_n$ is directed from B toward A.*

Acceleration Equation. Write the acceleration equation in symbolic form: $\mathbf{a}_B = \mathbf{a}_A + (\mathbf{a}_{B/A})_t + (\mathbf{a}_{B/A})_n$, and underneath each of the terms represent the respective vectors by their magnitudes and directions. To do this use the data tabulated on the kinematic diagrams. Since motion occurs in the plane, this "vector" equation can be expressed in terms of two scalar component equations which provide a solution for at most two unknowns. If a solution yields a *negative* answer for an *unknown,* it indicates that its *direction* is *opposite* to that shown on the kinematic diagram.

The following example problems numerically illustrate this scalar method of application.

*Perhaps the notation $\mathbf{a}_B = \mathbf{a}_A + (\mathbf{a}_{B/A(\text{pin})})_t + (\mathbf{a}_{B/A(\text{pin})})_n$ helps in recalling that A is pinned.

Example 16–11

The rod AB shown in Fig. 16–23a is confined to move along the inclined planes at A and B. If point A has an acceleration of 3 m/s^2 and a velocity of 2 m/s, both directed down the plane at the instant the rod becomes horizontal, determine the angular acceleration of the rod at this instant.

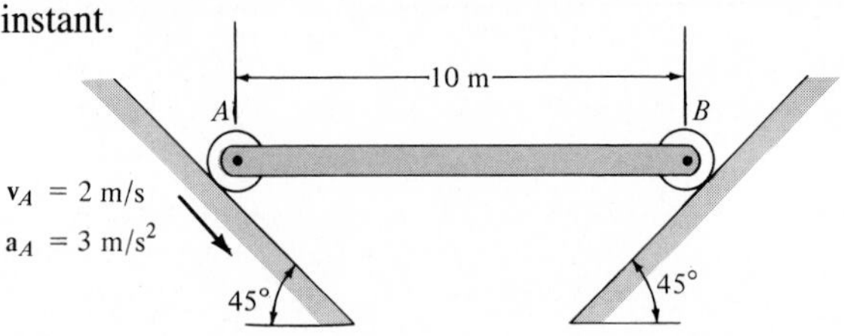

(a)

Solution

We will apply the acceleration equation to points A and B on the rod.

Velocity Analysis. Since the velocity of A is known and the velocity of B is tangent to its path, the instantaneous center of zero velocity for the rod is located as shown in Fig. 16–23b. From the geometry,

$$r_{A/IC} = r_{B/IC} = 10 \cos 45° = 7.07 \text{ m}$$

Thus,

$$\omega = \frac{v_A}{r_{A/IC}} = \frac{2 \text{ m/s}}{7.07 \text{ m}} = 0.283 \text{ rad/s} \circlearrowleft$$

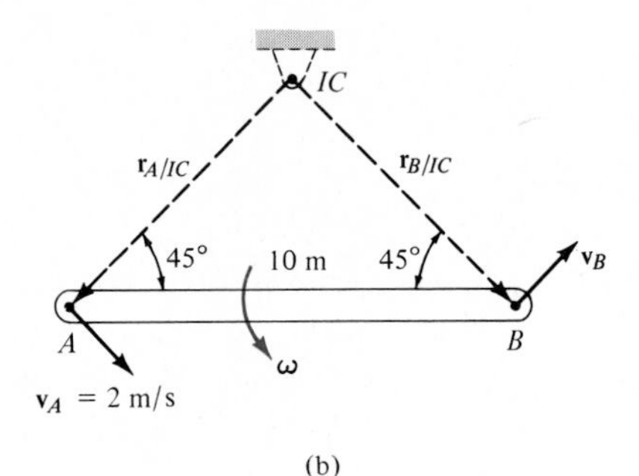

(b)

Kinematic Diagrams. The accelerated motions of points A and B are shown on the kinematic diagram in Fig. 16–23c. Since B moves along a straight-line path, it has an acceleration along the path.* There are two unknowns in Fig. 16–23c, i.e., a_B and α. Point A will be chosen as the base point, so that $\mathbf{a}_B = \mathbf{a}_A + \mathbf{a}_{B/A}$.

The relative accelerated motion of B with respect to A is shown on the kinematic diagram in Fig. 16–23d.

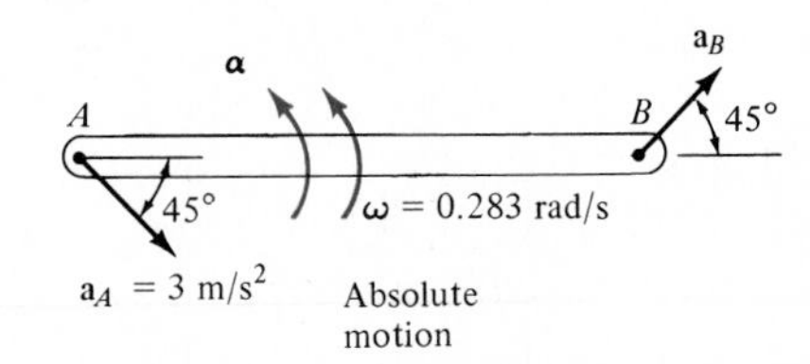

(c)

Acceleration Equation

$$\mathbf{a}_B = \mathbf{a}_A + (\mathbf{a}_{B/A})_t + (\mathbf{a}_{B/A})_n$$

$$\underset{\nearrow 45°}{a_B} = \underset{\searrow 45°}{3 \text{ m/s}^2} + \underset{\uparrow}{\alpha(10 \text{ m})} + \underset{\leftarrow}{(0.283 \text{ rad/s})^2(10 \text{ m})}$$

Equating the horizontal and vertical components yields

$$(\overset{+}{\rightarrow}) \qquad a_B \cos 45° = 3 \cos 45° - (0.283)^2(10)$$

$$(+\uparrow) \qquad a_B \sin 45° = -3 \sin 45° + \alpha(10)$$

Solving, we have

$$a_B = 1.87 \text{ m/s}^2$$

$$\alpha = 0.344 \text{ rad/s}^2 \circlearrowleft \qquad \textit{Ans.}$$

(d)

Fig. 16–23

*If the path were *curved*, B would have both normal and tangential components of acceleration.

Example 16–12

The collar C in Fig. 16–24a is moving downward with an acceleration of 1 m/s^2, and at the instant shown it has a speed of 2 m/s. Determine the angular acceleration of link CB at this instant.

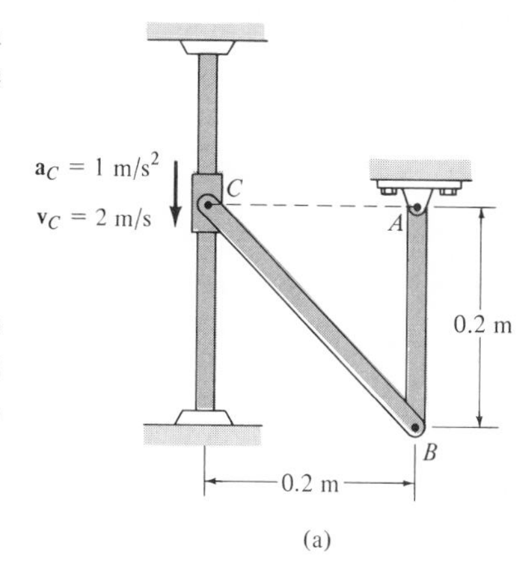

(a)

Solution

We will apply the acceleration equation to points C and B on link CB.

Velocity Analysis. Since $\mathbf{v}_C$ is downward, $\mathbf{v}_B$ is directed horizontally (to the right). The instantaneous center of zero velocity for CB is located as shown in Fig. 16–24b. From the figure, $r_{C/IC} = r_{B/IC} = 0.2$ m and $\boldsymbol{\omega}_{CB}$ is assumed to act counterclockwise. Show that

$$\omega_{CB} = 10 \text{ rad/s} \circlearrowleft \qquad v_B = 2 \text{ m/s} \rightarrow$$

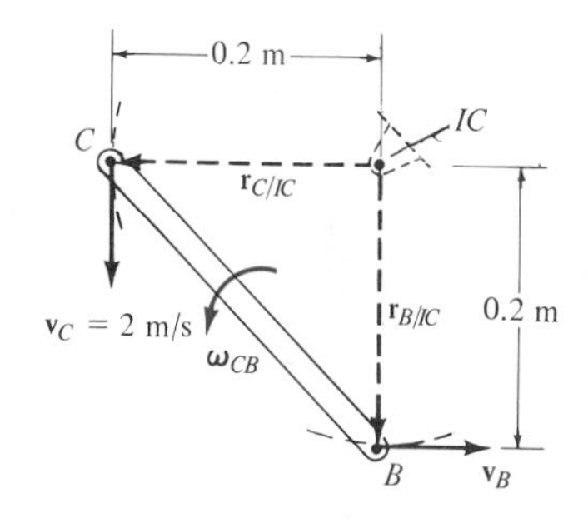

(b)

Kinematic Diagrams. As shown on the kinematic diagram in Fig. 16–24c, the acceleration of point C is directed downward since it moves along a straight-line path. This point will be chosen as the base point since $\mathbf{a}_C$ is known. Point B moves along a *curved path* and therefore $\mathbf{a}_B$ is represented by its normal and tangential components. From bar AB, the normal component is

$$(a_B)_n = \frac{(v_B)^2}{r_{B/A}} = \frac{(2)^2}{0.2} = 20 \text{ m/s}^2$$

There are two unknowns shown on the kinematic diagram in Fig. 16–24c, α_{CB} and $(a_B)_t$.

The relative accelerated motion of B with respect to C is shown on the kinematic diagram in Fig. 16–24d.

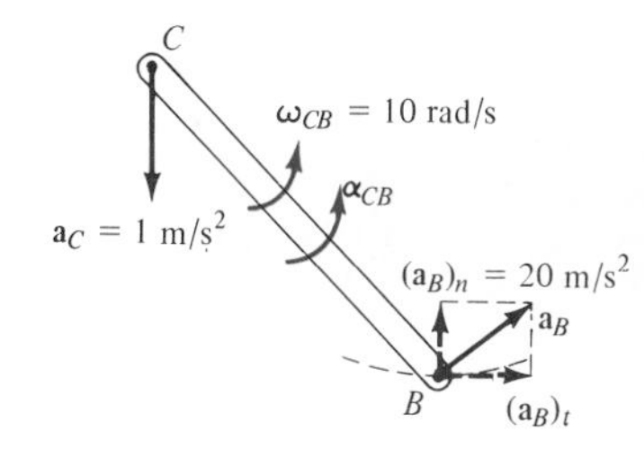

Absolute motion

(c)

Acceleration Equation

$$(\mathbf{a}_B)_t + (\mathbf{a}_B)_n = \mathbf{a}_C + (\mathbf{a}_{B/C})_t + (\mathbf{a}_{B/C})_n$$

$$\underset{\rightarrow}{(a_B)_t} + \underset{\uparrow}{20 \text{ m/s}^2} = \underset{\downarrow}{1 \text{ m/s}^2} + \underset{\angle 45^\circ}{\alpha_{CB}(0.2\sqrt{2} \text{ m})} + \underset{45^\circ \nwarrow}{(10 \text{ rad/s})^2(0.2\sqrt{2} \text{ m})}$$

Equating components in the horizontal and vertical directions gives

$$(\overset{+}{\rightarrow}) \qquad (a_B)_t = \alpha_{CB}(0.2\sqrt{2}) \cos 45^\circ - (10)^2(0.2\sqrt{2}) \cos 45^\circ$$

$$(+\uparrow) \qquad 20 = -1 + \alpha_{CB}(0.2\sqrt{2}) \sin 45^\circ + (10)^2(0.2\sqrt{2}) \sin 45^\circ$$

Solving, we get

$$(a_B)_t = -19 \text{ m/s}^2, \ \alpha_{CB} = 5 \text{ rad/s}^2 \qquad \textit{Ans.}$$

The negative sign indicates that $(\mathbf{a}_B)_t$ acts in the opposite direction to that shown in Fig. 16–24c.

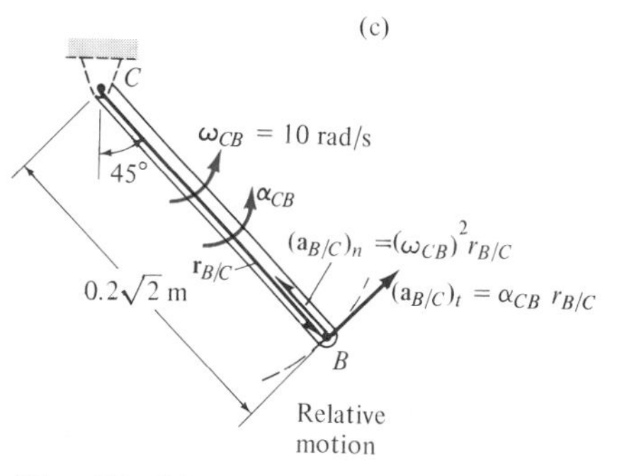

Relative motion

(d)

Fig. 16–24

Example 16–13

The cylinder shown in Fig. 16–25*a* rolls without slipping, such that at the instant shown its center has a velocity of 1.5 ft/s and an acceleration of 4 ft/s². Determine the acceleration of point *B*.

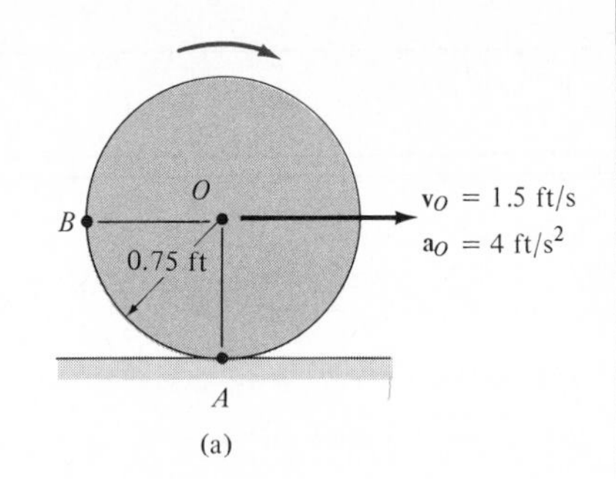

(a)

Solution

We will apply the acceleration equation to points *B* and *O* on the cylinder.

Velocity Analysis. Since no slipping occurs, $\mathbf{v}_A = \mathbf{0}$, so that point *A* represents the instantaneous center of zero velocity, Fig. 16–25*b*. Hence,

$$\omega = \frac{v_O}{r} = \frac{1.5}{0.75} = 2 \text{ rad/s} \curvearrowright$$

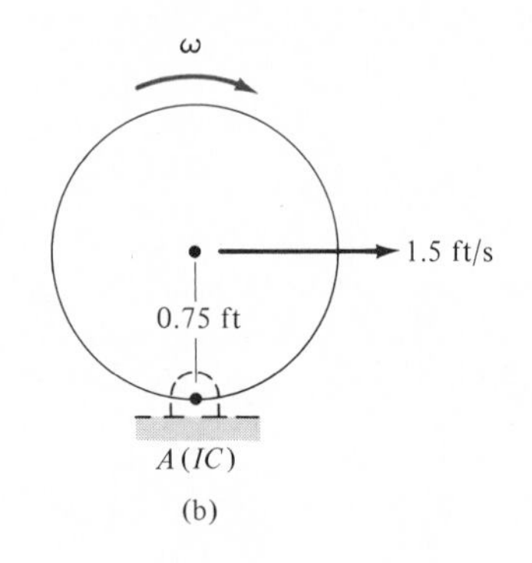

(b)

This result can also be obtained *directly* from the formulation in Example 16–3, i.e., $v_O = r\omega$.

Kinematic Diagrams. Using the result of Example 16–3, $a_O = r\alpha$, the angular acceleration is

$$a_O = r\alpha; \qquad 4 = 0.75\alpha$$
$$\alpha = 5.33 \text{ rad/s}^2 \curvearrowright$$

Since $\mathbf{a}_O$ is known, *O* will be chosen as the base point. Point *B* moves along a *curved path* having an unknown radius of curvature.* Its acceleration will be represented by its unknown *x* and *y* components, Fig. 16–25*c*.

The relative accelerated motion of *B* with respect to *O* is shown on the kinematic diagram in Fig. 16–25*d*.

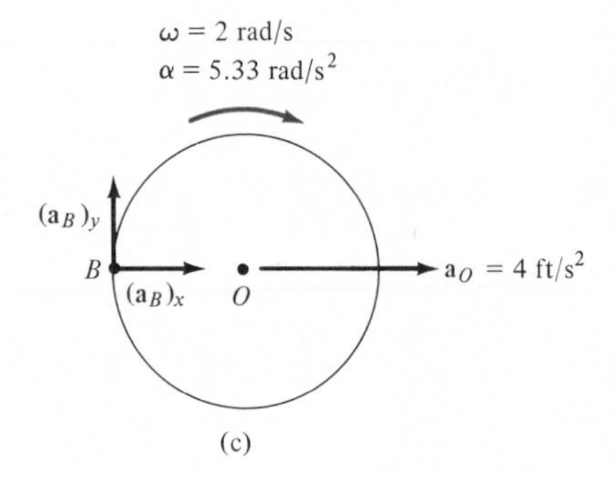

(c)

Acceleration Equation

$$(\mathbf{a}_B)_x + (\mathbf{a}_B)_y = \mathbf{a}_O + (\mathbf{a}_{B/O})_t + (\mathbf{a}_{B/O})_n$$

$$\underset{\rightarrow}{(a_B)_x} + \underset{\uparrow}{(a_B)_y} = \underset{\rightarrow}{4 \text{ ft/s}^2} + \underset{\uparrow}{5.33 \text{ rad/s}^2 (0.75 \text{ ft})} + \underset{\rightarrow}{(2 \text{ rad/s})^2(0.75 \text{ ft})}$$

Equating the horizontal and vertical components yields

$$\overset{+}{\rightarrow} \qquad (a_B)_x = 4 + (2)^2(0.75) = 7.0$$

$$+\uparrow \qquad (a_B)_y = 5.33(0.75) = 4.0$$

Thus,

$$a_B = \sqrt{(7.0)^2 + (4.0)^2} = 8.06 \text{ ft/s}^2 \qquad \textit{Ans.}$$

$$\theta = \tan^{-1}\left(\frac{4.0}{7.0}\right) = 29.7^\circ \qquad \textit{Ans.}$$

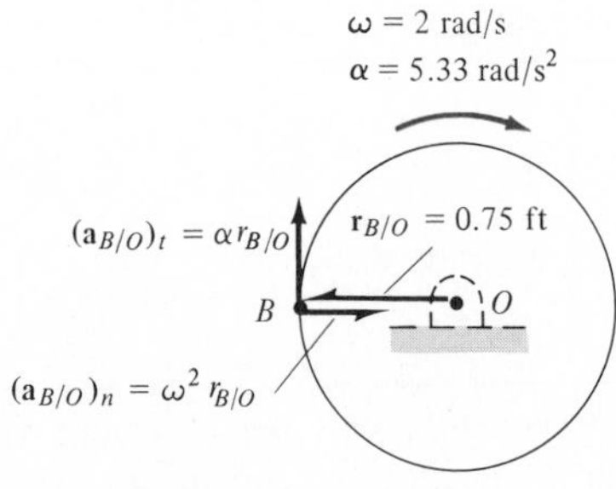

(d)

Fig. 16–25

*The radius of curvature ρ is *not* equal to the radius of the cylinder since the cylinder is *not* rotating about point *O*. Furthermore, ρ is *not* defined as the distance from the *IC* to *B*, since the location of the *IC* depends only on the velocity of a point and *not* the geometry of its path.

Problems

16–54. At a given instant the bottom A of the ladder has the velocity and acceleration shown. Determine the acceleration of the top of the ladder, B, and the ladder's angular acceleration at the same instant. *Hint:* First show that $\omega_{AB} = 0.75$ rad/s.

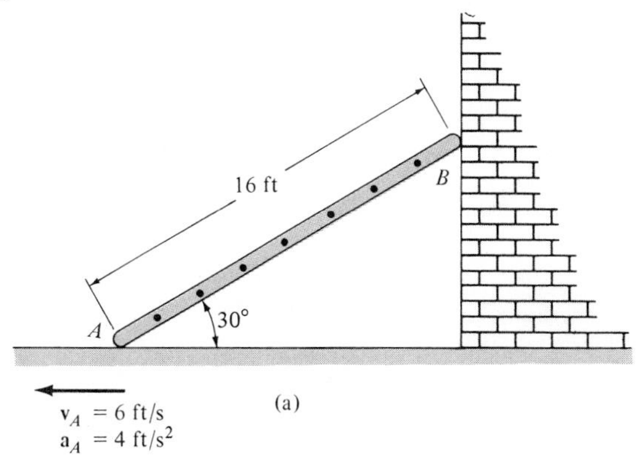

Prob. 16–54

16–55. Determine the angular acceleration of link BC at the instant $\theta = 90°$ if the collar C has an instantaneous velocity of $v_C = 4$ ft/s and deceleration of $a_C = 3$ ft/s² as shown. *Hint:* First show that $\omega_{BC} = \omega_{BA} = 5.66$ rad/s.

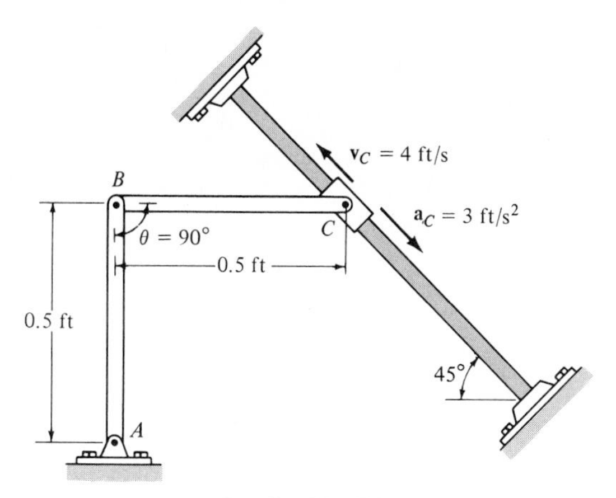

Prob. 16–55

***16–56.** At the instant shown, the disk is rotating with an angular acceleration of $\alpha = 8$ rad/s² and angular velocity of $\omega = 3$ rad/s. Determine the angular acceleration of the link AB and the acceleration of the piston at this instant. *Hint:* First show that $\omega_{AB} = 0$.

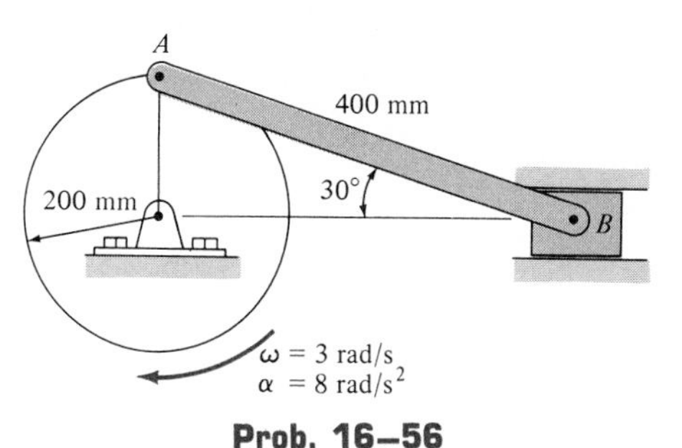

Prob. 16–56

16–57. The pulley is pin-connected to block B at A. As cord CF unwinds from the inner hub with the motion shown, cord DE unwinds from the outer rim. Determine the angular acceleration of the pulley and the acceleration of block B at the instant shown. *Hint:* The IC is at D so $\omega = 40$ rad/s.

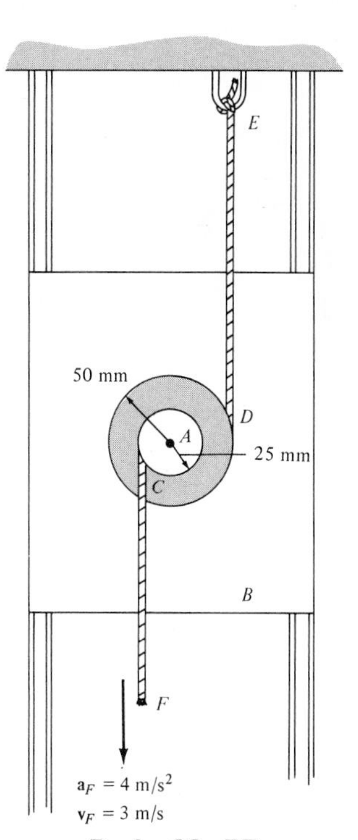

Prob. 16–57

16–58. As the cord unravels from the cylinder, the cylinder has an angular acceleration of $\alpha = 4\ \text{rad/s}^2$ and an angular velocity of $\omega = 2$ rad/s at the instant shown. Determine the accelerations of points A and B at this instant.

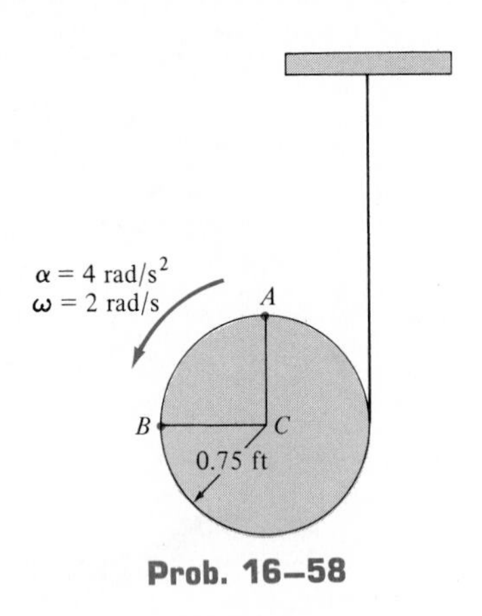

Prob. 16–58

16–59. The wheel is moving to the right such that it has an angular acceleration of $\alpha = 4\ \text{rad/s}^2$ and angular velocity of $\omega = 2$ rad/s at the instant shown. If it does not slip at A, determine the acceleration of point B.

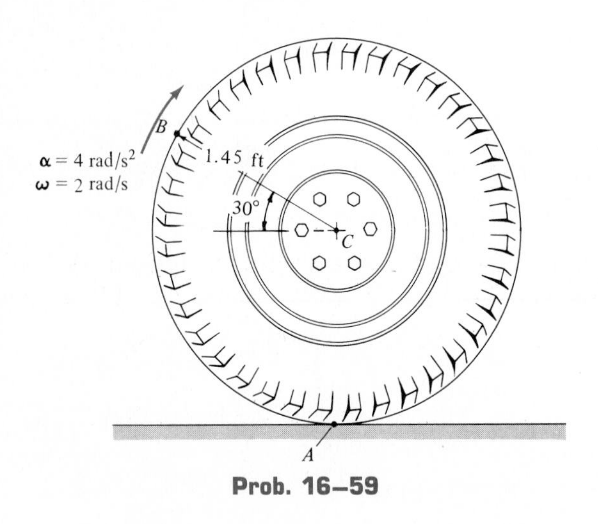

Prob. 16–59

***16–60.** The disk rolls without slipping such that it has an angular acceleration of $\alpha = 4\ \text{rad/s}^2$ and angular velocity of $\omega = 2$ rad/s at the instant shown. Determine the acceleration of points A and B on the link and the link's angular acceleration at this instant. Assume point A lies on the periphery of the disk, 150 mm from C.

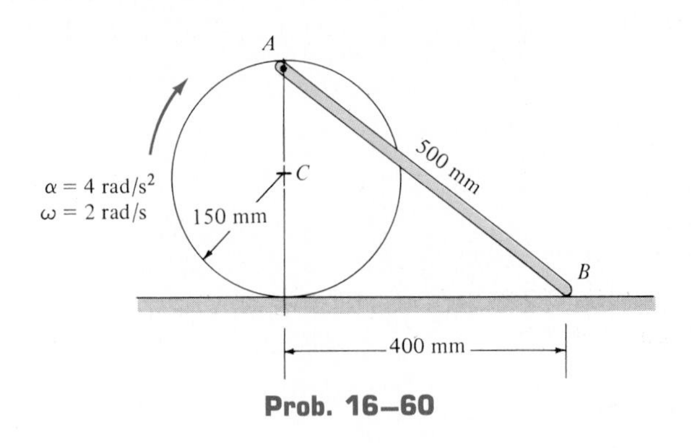

Prob. 16–60

16–61. Gear C is rotating with a constant angular velocity of $\omega_C = 3$ rad/s. Determine the acceleration of the piston A and the angular acceleration of rod AB at the instant $\theta = 90°$. Set $r_C = 0.2$ ft and $r_D = 0.3$ ft. *Hint:* $v_B = 0.6$ ft/s and $\omega_{AB} = 0$.

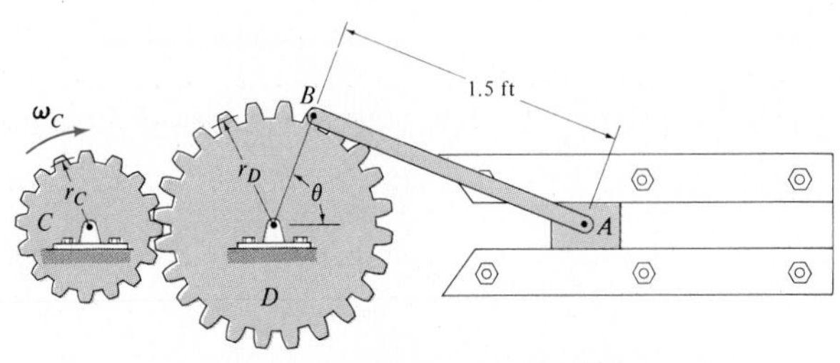

Probs. 16–61 / 16–62

16–62. Determine the acceleration of the piston A and the angular acceleration of rod AB at the instant $\theta = 0°$ in Prob. 16–61. *Hint:* $v_B = 0.6$ ft/s and $\omega_{AB} = 0.4$ rad/s.

16–63. The retractable wing-tip float is used on an airplane operating off water. Determine the angular accelerations $\boldsymbol{\alpha}_{CD}$, $\boldsymbol{\alpha}_{BD}$, and $\boldsymbol{\alpha}_{AB}$ at the instant shown if the trunnion C travels along the horizontal rotating screw with an acceleration of $a_C = 0.5\ \text{ft/s}^2$. In the position shown, $v_C = 0$. Also, points A and E are pin-connected to the wing.

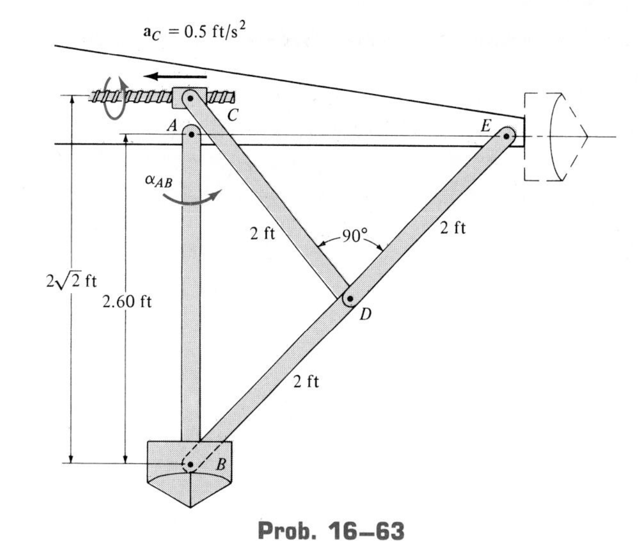

Prob. 16–63

Review Problems

***16–64.** A wheel has an initial clockwise angular velocity of 60 rad/s. Determine the number of revolutions required to bring it to a stop in 10 s. What is the constant angular deceleration?

16–65. Determine the angular velocities of bars BC and CD at the instant shown.

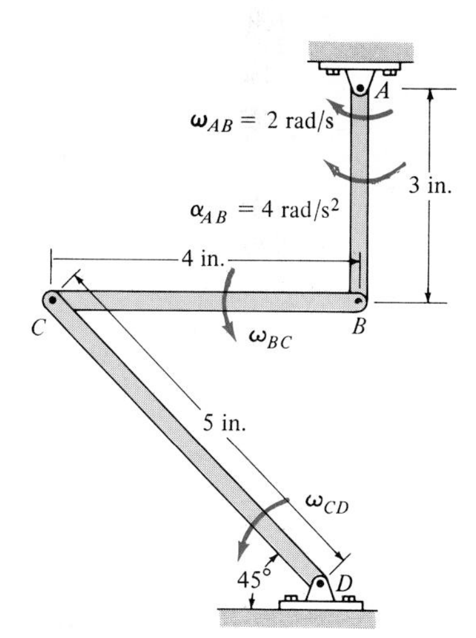

Probs. 16–65 / 16/66

16–66. Determine the angular accelerations of bars BC and CD at the instant shown. $\omega_{BC} = 1.5$ rad/s, $\omega_{CD} = 1.70$ rad/s.

16–67. If a shaft is rotating at 400 rad/s, and it is subjected to an angular deceleration of $\alpha = (-4t^2)\ \text{rad/s}^2$, where t is in seconds, determine the time needed to bring it to a stop.

***16–68.** Compute the velocity of rod R for any angle θ of the cam C if the cam rotates with a constant angular velocity $\boldsymbol{\omega}$. The pin connection at O does not cause an interference with the motion of A on C.

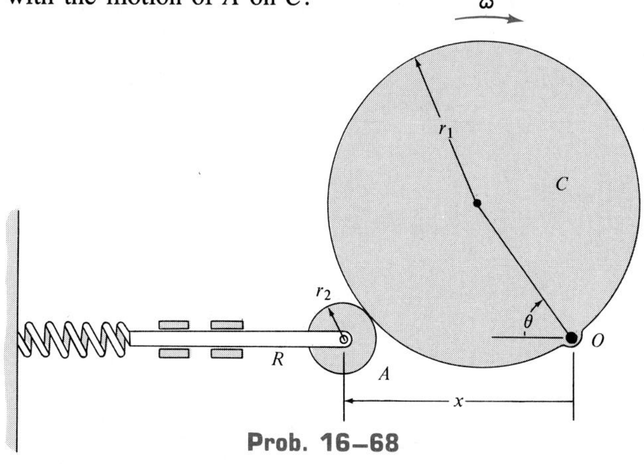

Prob. 16–68

16–69. A turbine is originally at rest and is subjected to an angular acceleration of $\alpha = (5 + \theta)$ rad/s^2, where θ is in radians. Determine the angular velocity of the turbine when it has turned 2 revolutions.

16–70. The tire is slipping on the ground such that point A has a speed of $v_A = 6$ ft/s. If the tire is rotating at $\omega = 10$ rad/s, determine the velocity of point B.

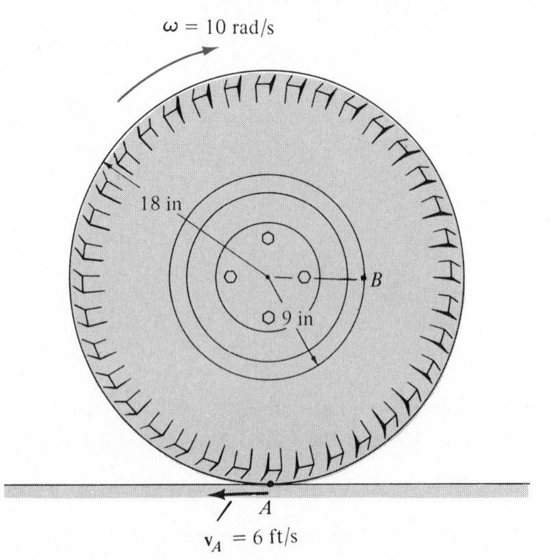

Prob. 16–70

16–71. When the slider block C is in the position shown, link AB has a clockwise angular velocity of $\omega_{AB} = 7$ rad/s. Determine the angular velocity of BC and the velocity of C at this instant.

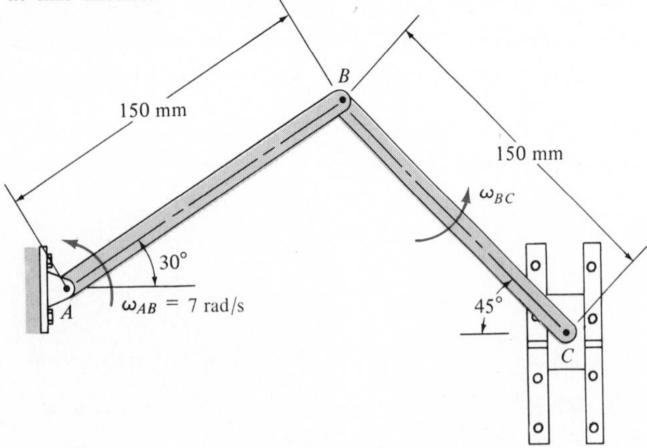

Probs. 16–71 / 16–72

***16–72.** Determine the acceleration of block C at the instant shown. $\omega_{AB} = 7$ rad/s and $\omega_{BC} = 4.95$ rad/s.

16–73. The crankshaft AB has a constant angular velocity of 200 rad/s. Determine the angular velocity of the connecting rod BP and the velocity of the piston when the system is in the position shown.

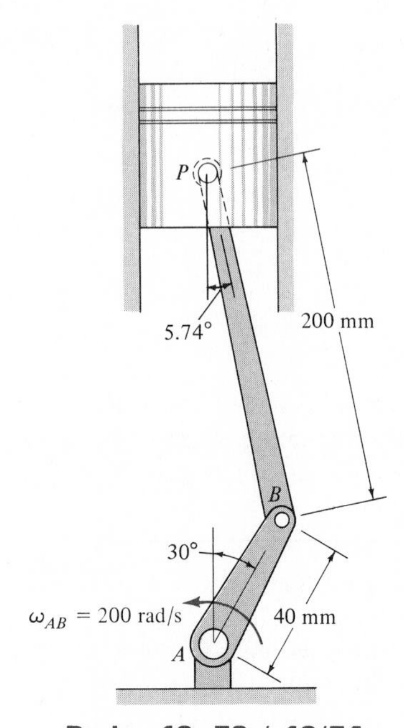

Probs. 16–73 / 16/74

16–74. Determine the acceleration of the piston P in Prob. 16–73 when the system is in the position shown. $\omega_{BP} = 34.8$ rad/s.

Planar Kinetics of a Rigid Body: Force and Acceleration 17

In Chapter 16 the planar kinematics of rigid-body motion was presented in order of increasing complexity, that is, translation, rotation about a fixed axis, and general plane motion. The study of rigid-body kinetics in this chapter will be presented in somewhat the same order. The chapter begins by introducing a property of a body called the mass moment of inertia. Afterward, a derivation of the equations of general plane motion for symmetrical rigid bodies is given, and then these equations are applied to specific problems of rigid-body translation, rotation about a fixed axis, and finally general plane motion. A rigid body subjected to any of these three types of motion may be analyzed in a fixed reference plane, because the path of motion of each particle of the body lies in a plane that is parallel to the reference plane. A kinetic study of these motions is referred to as the kinetics of planar motions or simply *planar kinetics*.

Moment of Inertia 17.1

Since a body has a definite size and shape, an applied nonconcurrent force system may cause the body to both translate and rotate. The translational aspects of the motion were studied in Chapter 13 and are governed by the equation $\mathbf{F} = m\mathbf{a}$. The rotational aspects, caused by the moments $\mathbf{M}$, are governed by an equation of the form $\mathbf{M} = I\boldsymbol{\alpha}$. The symbol I in this equation is termed the moment of inertia. By comparison, the *moment of inertia* is a measure of the resistance of a body to *angular acceleration* ($\mathbf{M} = I\boldsymbol{\alpha}$) in the same way that *mass* is a measure of the body's resistance to *acceleration* ($\mathbf{F} = m\mathbf{a}$).

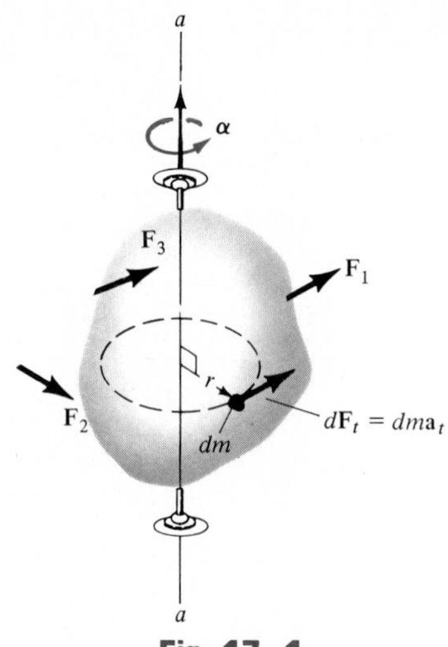

Fig. 17–1

We can formalize a definition of the moment of inertia by considering the rigid body shown in Fig. 17–1. Here the applied external forces cause an unbalanced moment about the *aa* axis which gives the body an angular acceleration $\boldsymbol{\alpha}$. To analyze this motion, consider an element of mass *dm* located at a perpendicular distance *r* from the axis. Due to the external loading, an unbalanced tangential force component $d\mathbf{F}_t$ will be transferred to the element from the surrounding particles (elements). It is this force which creates the angular motion of the element about the *aa* axis. Applying the equation of motion to the element in the tangential direction yields $dF_t = dm\, a_t$. However, $a_t = r\alpha$, so that the moment of dF_t about the *aa* axis is $dM = r\, dF_t = r^2\alpha\, dm$. The moment of the tangential forces acting on *all the elements* of the body is determined by integration, which gives $M = \int_m r^2\alpha\, dm$. Since α is the same for all radial lines *r* extending from the axis to each element *dm*, α may be factored out of the integrand, leaving $M = I\alpha$, where

$$I = \int_m r^2\, dm \tag{17–1}$$

This integral is termed the *moment of inertia*. Since the formulation involves the distance *r*, Fig. 17–1, the value of *I* is *unique* for each axis *aa* about which it is computed. In the study of planar kinetics, however, the axis which is generally chosen for analysis passes through the body's mass center *G* and is *always* perpendicular to the plane of motion. The moment of inertia computed about this axis will be defined as I_G. Since the moment of inertia is an important property, used throughout the study of rigid-body planar kinetics, methods used for its calculation will now be discussed.

PROCEDURE FOR ANALYSIS

In general, when integrating Eq. 17–1, it is best to choose a coordinate system which simplifies the equations that describe the boundary of the body. For example, cylindrical coordinates are generally appropriate when solving problems which involve bodies having circular boundaries.

If the body consists of material having a variable mass density, $\rho = \rho(x, y, z)$, the elemental mass dm of the body may be expressed in terms of its density and volume as $dm = \rho\, dV$. Substituting in Eq. 17–1, the body's moment of inertia is then computed using *volume elements* for integration, i.e.,

$$I = \int_V r^2 \rho \, dV \qquad (17\text{–}2)$$

In the special case of ρ being a *constant,* this term may be factored out of the integral and the integration is then purely a function of geometry,

$$I = \rho \int_V r^2 \, dV \qquad (17\text{–}3)$$

When the elemental volume chosen for integration has differential sizes in all three directions, e.g., $dV = dx\, dy\, dz$, Fig. 17–2*a*, the moment of inertia of the body must be computed using "triple integration." The integration process can, however, be simplified to a *single integration* provided the chosen elemental volume has a differential size or thickness in only *one direction*. Shell or disk elements are often used for this purpose.

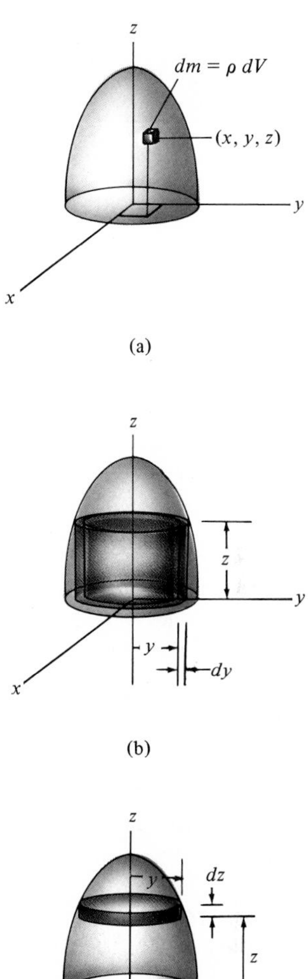

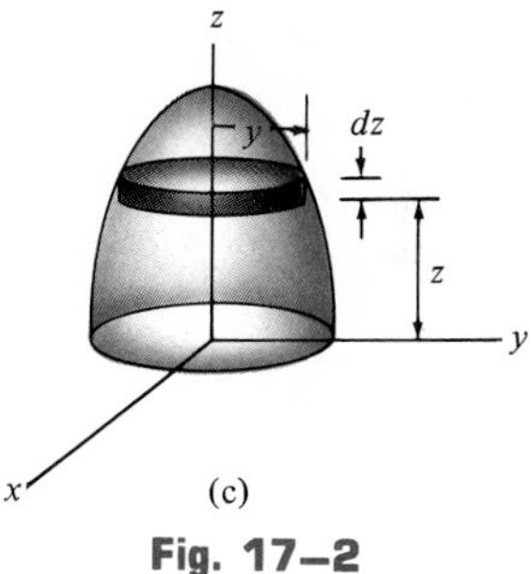

Fig. 17–2

Shell Element. If a *shell element* having a height z, radius $r = y$, and thickness dy is chosen for integration, Fig. 17–2*b*, then the volume is $dV = (2\pi y)(z)\, dy$. This element may be used in Eq. 17–2 or 17–3 for computing the moment of inertia I_z of the body about the z axis, since the *entire element,* due to its "thinness," lies at the *same* perpendicular distance $r = y$ from the z axis (see Example 17–1).

Disk Element. If a disk element having a radius $r = y$ and a thickness dz is chosen for integration, Fig. 17–2*c*, then the volume is $dV = (\pi y^2)\, dz$. In this case, however, the element is *finite* in the radial direction, and consequently parts of it *do not* all lie at the *same radial distance* r from the z axis. As a result, Eq. 17–2 or 17–3 *cannot* be used to determine I_z. Instead, to perform the integration using this element, it is first necessary to determine the moment of inertia *of the element* about the z axis and then integrate this result (see Example 17–2).

Example 17–1

Determine the moment of inertia of the right circular cylinder shown in Fig. 17–3a about the z axis. The mass density ρ of the material is constant.

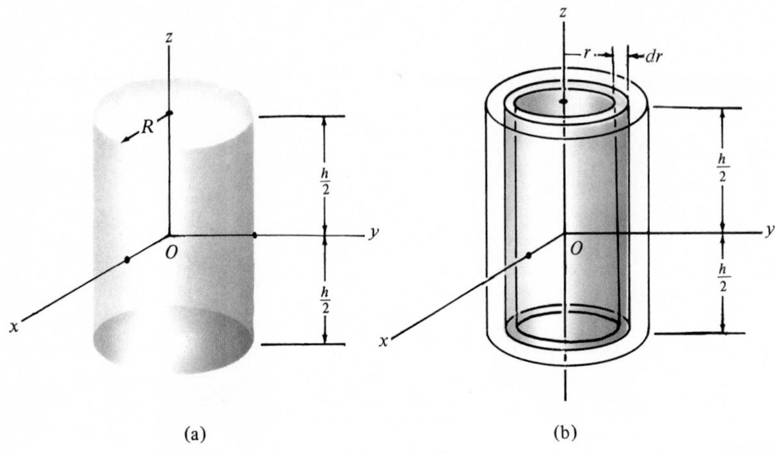

Fig. 17–3

Solution

Shell Element. This problem may be solved using the *shell element* in Fig. 17–3b and single integration. The volume of the element is $dV = (2\pi r)(h)\,dr$, so that the mass is $dm = \rho\,dV = \rho(2\pi hr\,dr)$. Since the *entire element* lies at the same distance r from the z axis, the moment of inertia *of the element* is

$$dI_z = r^2\,dm = \rho 2\pi h r^3\,dr$$

Integrating over the entire region of the cylinder yields

$$I_z = \int_m r^2\,dm = \rho 2\pi h \int_0^R r^3\,dr = \frac{\rho\pi}{2} R^4 h$$

The mass of the cylinder is

$$m = \int_m dm = \rho 2\pi h \int_0^R r\,dr = \rho\pi h R^2$$

so that

$$I_z = \frac{1}{2} m R^2 \qquad \textit{Ans.}$$

Example 17–2

A solid is formed by revolving the shaded area shown in Fig. 17–4*a* about the y axis. If the mass density of the material is 5 slug/ft^3, determine the moment of inertia about the y axis.

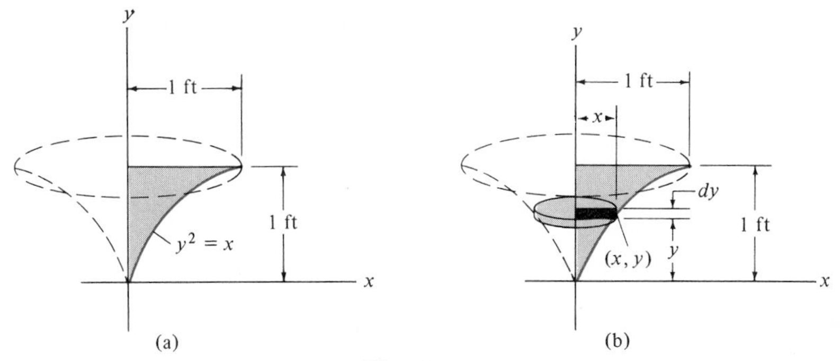

Fig. 17–4

Solution

Disk Element. The moment of inertia will be computed using a *disk element,* as shown in Fig. 17–4*b*. Here the element intersects the curve at the arbitrary point (x, y) and has a mass

$$dm = \rho\, dV = \rho(\pi x^2)\, dy$$

Although all portions of the element are *not* located at the same distance from the y axis, it is still possible to determine the moment of inertia dI_y *of the element* about the y axis. In the preceding example it was shown that the moment of inertia of a cylinder about its longitudinal axis is $I = \frac{1}{2}mR^2$, where m and R are the mass and radius of the cylinder. Since the height of the cylinder is not involved in this formula, the moment of inertia of the disk element in Fig. 17–4*b* is

$$dI_y = \tfrac{1}{2}(dm)x^2 = \tfrac{1}{2}[\rho(\pi x^2)\, dy]x^2$$

Substituting $x = y^2$, $\rho = 5$ slug/ft^3, and integrating with respect to y, from $y = 0$ to $y = 1$ ft, yields the moment of inertia for the entire solid.

$$I_y = \frac{\pi 5}{2}\int_0^1 x^4\, dy = \frac{\pi 5}{2}\int_0^1 y^8\, dy = 0.873 \text{ slug} \cdot \text{ft}^2 \qquad \textit{Ans.}$$

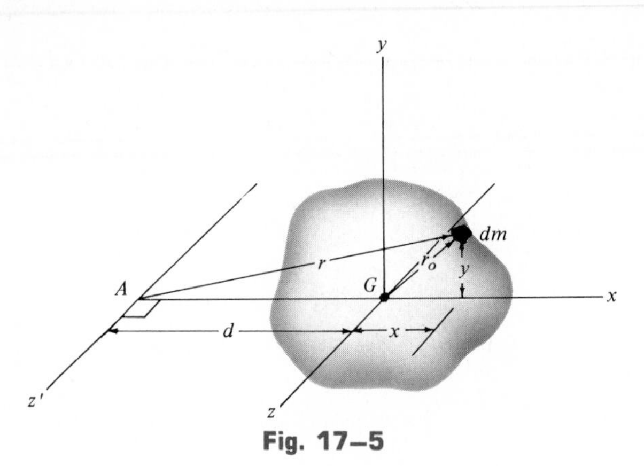

Fig. 17–5

Parallel-Axis Theorem. If the moment of inertia of the body about an axis passing through the body's mass center is known, then the body's moment of inertia may be determined about any other *parallel axis* by using the *parallel-axis theorem*. This theorem can be derived by considering the body shown in Fig. 17–5. The z axis passes through the body's mass center G, whereas the corresponding *parallel axis* z' lies at a constant distance d away. Selecting the differential element of mass dm which is located at point (x, y) and using the Pythagorean theorem, $r^2 = (d + x)^2 + y^2$, we can express the moment of inertia of the body computed about the z' axis as

$$I = \int_m r^2\, dm = \int_m [(d + x)^2 + y^2]\, dm$$

$$= \int_m (x^2 + y^2)\, dm + 2d \int_m x\, dm + d^2 \int_m dm$$

Since $r_o^2 = x^2 + y^2$, the first integral represents I_G. The second integral equals *zero,* since the z axis passes through the body's mass center, so that $\int x\, dm = \bar{x} \int dm$ since $\bar{x} = 0$. Finally, the third integral represents the total mass m of the body. Hence, the moment of inertia about the z' axis can be written as

$$I = I_G + md^2 \tag{17–4}$$

where

I_G = moment of inertia about the z axis passing through the mass center G
m = mass of the body
d = perpendicular distance between the parallel axes

Radius of Gyration. Occasionally, the moment of inertia of a body about a specified axis is reported in handbooks using the *radius of gyration*. When this length, k, and the body's mass m are known, the body's moment of inertia is determined from the equation

$$I = mk^2 \quad \text{or} \quad k = \sqrt{\frac{I}{m}} \tag{17-5}$$

Note the *similarity* between the definition of k in this formula and r in the equation $dI = r^2\, dm$, which defines the moment of inertia of an elemental mass dm of the body about an axis.

Composite Bodies. The parallel-axis theorem is often used to determine the moment of inertia of composite shapes when the moment of inertia I_G of each of the composite parts is either known or can be computed by integration.* For example, if the body is constructed of a number of simple shapes such as disks, spheres, and rods, the moment of inertia of the body about any axis z' can be determined by adding algebraically the moments of inertia of all the composite shapes computed about the z' axis. Algebraic addition is necessary since a composite part must be considered as a negative quantity if it has already been counted as part of another part—for example a "hole" subtracted from a solid plate. The parallel-axis theorem is needed for the calculations if the center of gravity of each composite part does not lie on the z' axis.

*See the table given on the inside back cover of this book.

Example 17–3

If the plate shown in Fig. 17–6a has a density of 8000 kg/m^3 and a thickness of 10 mm, compute its moment of inertia about an axis directed perpendicular to the page and passing through point O.

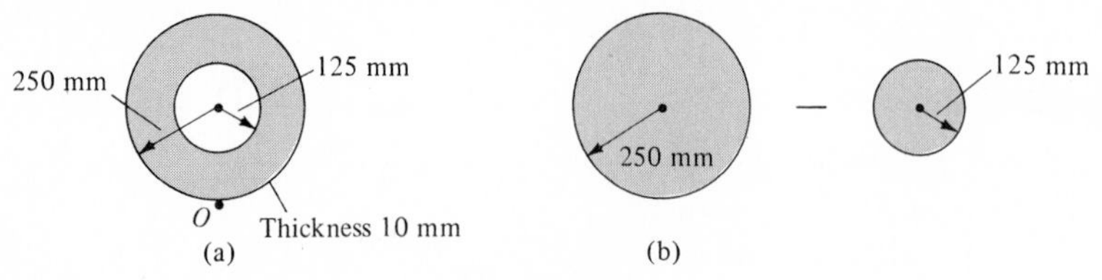

Fig. 17–6

Solution

The plate consists of two composite parts, Fig. 17–6b. The 250-mm-radius disk *minus* a 125-mm-radius disk. The moment of inertia about O can be determined by computing the moment of inertia of each of these parts about O and then *algebraically* adding the results. The computations are performed by using the parallel-axis theorem in conjunction with the data listed in the table on the inside back cover.

Disk. The moment of inertia of a thin disk about an axis perpendicular to the plane of the disk is $I_G = \frac{1}{2}mr^2$. The mass centers of *both* the 250-mm-radius disk and the 125-mm-radius disk (hole) are located at a distance of 0.25 m from point O. For the 250-mm-radius disk, we have

$$m_d = \rho_d V_d = 8000 \text{ kg/m}^3[\pi(0.25 \text{ m})^2(0.01 \text{ m})] = 15.71 \text{ kg}$$

$$\begin{aligned}(I_O)_d &= \tfrac{1}{2}m_d r_d^2 + m_d d^2 \\ &= \tfrac{1}{2}(15.71 \text{ kg})(0.25 \text{ m})^2 + (15.71 \text{ kg})(0.25 \text{ m})^2 \\ &= 1.473 \text{ kg}\cdot\text{m}^2\end{aligned}$$

Hole. For the 125-mm-radius disk (hole), we have

$$m_h = \rho_h V_h = 8000 \text{ kg/m}^3[\pi(0.125 \text{ m})^2(0.01 \text{ m})] = 3.93 \text{ kg}$$

$$\begin{aligned}(I_O)_h &= \tfrac{1}{2}m_h r_h^2 + m_h d^2 \\ &= \tfrac{1}{2}(3.93 \text{ kg})(0.125 \text{ m})^2 + (3.93 \text{ kg})(0.25 \text{ m})^2 \\ &= 0.276 \text{ kg}\cdot\text{m}^2\end{aligned}$$

The moment of inertia of the plate about point O is therefore

$$\begin{aligned}I_O &= (I_O)_d - (I_O)_h \\ &= 1.473 - 0.276 \\ &= 1.197 \text{ kg}\cdot\text{m}^2\end{aligned}$$ *Ans.*

Example 17–4

The pendulum consists of two 10-lb rods suspended from point O as shown in Fig. 17–7. Compute the pendulum's moment of inertia about an axis passing through (a) the pin at O and (b) the mass center G of the pendulum.

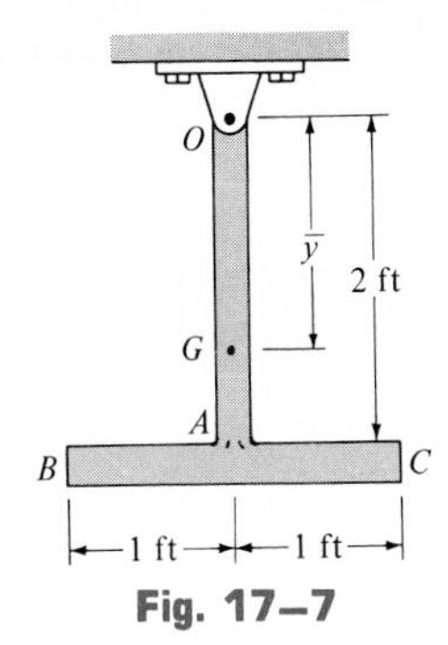

Fig. 17–7

Solution

Part (a). The moment of inertia of rod OA about an axis perpendicular to the page and passing through the end point O of the rod is $I_O = \frac{1}{3}ml^2$. Hence,

$$(I_O)_{OA} = \frac{1}{3}ml^2 = \frac{1}{3}\left(\frac{10}{32.2}\right)(2)^2 = 0.414 \text{ slug} \cdot \text{ft}^2$$

The same value may be computed using $I_G = \frac{1}{12}ml^2$ and the parallel-axis theorem; i.e.,

$$(I_O)_{OA} = \frac{1}{12}ml^2 + md^2 = \frac{1}{12}\left(\frac{10}{32.2}\right)(2)^2 + \frac{10}{32.2}(1)^2$$
$$= 0.414 \text{ slug} \cdot \text{ft}^2$$

For rod BC we have

$$(I_O)_{BC} = \frac{1}{12}ml^2 + md^2 = \frac{1}{12}\left(\frac{10}{32.2}\right)(2)^2 + \left(\frac{10}{32.2}\right)(2)^2$$
$$= 1.346 \text{ slug} \cdot \text{ft}^2$$

The moment of inertia of the pendulum is therefore

$$I_O = 0.414 + 1.346 = 1.76 \text{ slug} \cdot \text{ft}^2 \qquad \textit{Ans.}$$

Part (b). The mass center G will be located relative to the pin at O. Assuming this distance to be $\bar{y}$, Fig. 17–7, and using the formula for determining the mass center, we have

$$\bar{y} = \frac{\Sigma \tilde{y}m}{\Sigma m} = \frac{1\left(\frac{10}{32.2}\right) + 2\left(\frac{10}{32.2}\right)}{\left(\frac{10}{32.2}\right) + \left(\frac{10}{32.2}\right)} = 1.5 \text{ ft}$$

The moment of inertia I_G may be computed in the same manner as I_O, which requires successive applications of the parallel-axis theorem in order to transfer the moments of inertia of rods OA and BC to G. A more direct solution, however, involves applying the parallel-axis theorem using the result for I_O, i.e.,

$$I_O = I_G + md^2; \qquad 1.76 = I_G + \left(\frac{20}{32.2}\right)(1.5)^2$$
$$I_G = 0.362 \text{ slug} \cdot \text{ft}^2 \qquad \textit{Ans.}$$

Problems

17–1. Determine the moment of inertia of the homogeneous sphere having a mass density of $\rho = 6\ \text{slug/ft}^3$ and radius of 2 ft with respect to the y axis.

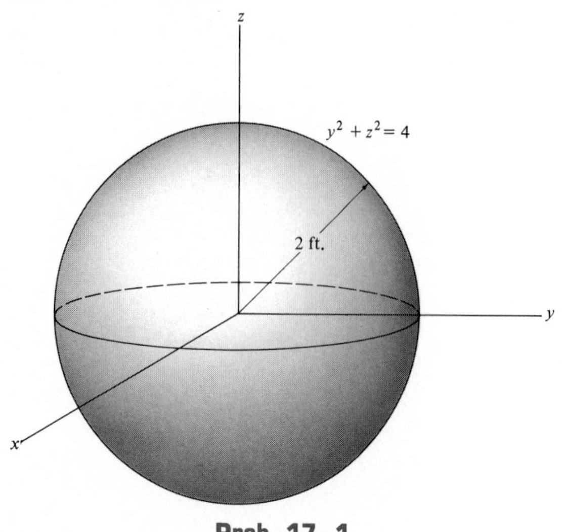

Prob. 17–1

17–2. The right circular cone is formed by revolving the shaded area around the x axis. Determine the moment of inertia I_x. The cone has a density of $4\ \text{Mg/m}^3$.

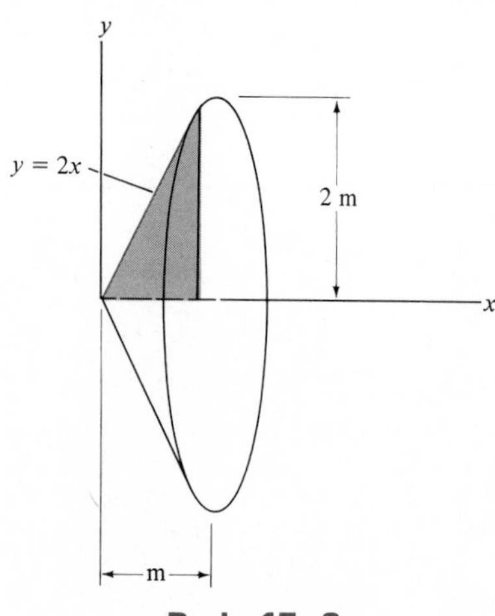

Prob. 17–2

17–3. The solid is formed by revolving the shaded area around the x axis. Determine the moment of inertia about the x axis. The mass density of the material is $\rho = 5\ \text{Mg/m}^3$.

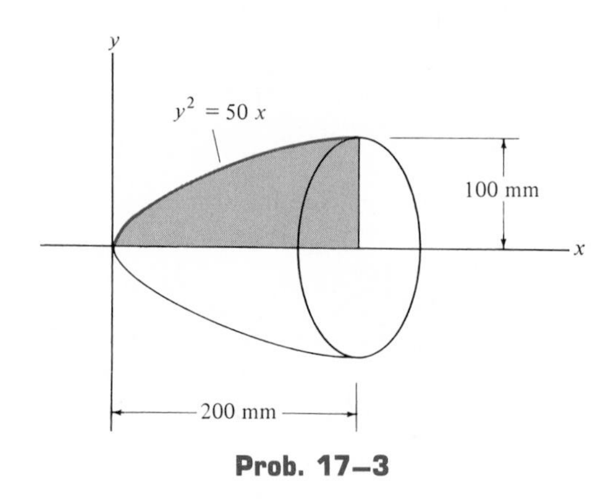

Prob. 17–3

***17–4.** An ellipsoid is formed by rotating the shaded area about the x axis. Determine the moment of inertia of this body with respect to the x axis. The density of the material is $\rho = 4\ \text{Mg/m}^3$.

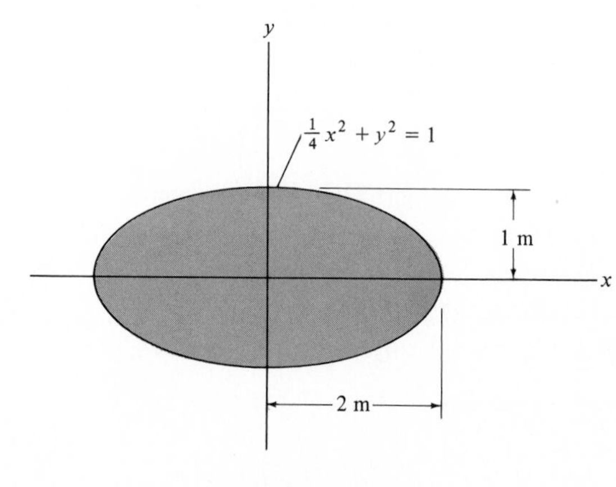

Prob. 17–4

17–5. Determine the moment of inertia of the hemispherical solid about the y axis. The density of the material is $\rho = 4\ \text{slug/ft}^3$.

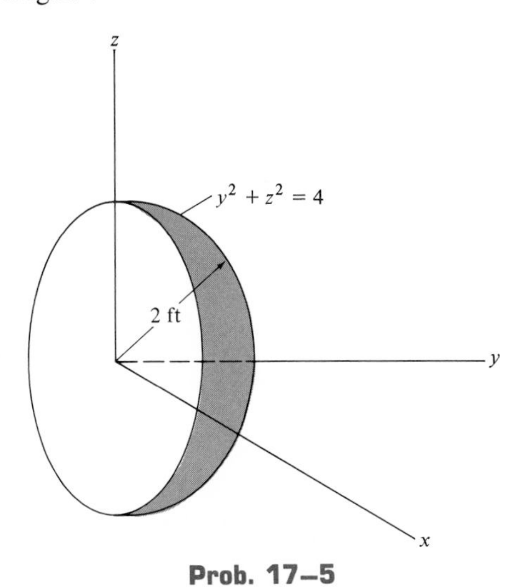

Prob. 17–5

17–6. Determine the moment of inertia of the homogeneous pyramid of mass m with respect to the z axis. The density of the material is ρ. *Suggestion:* Use a rectangular plate element having a volume of $dV = (2x)(2y)\ dz$.

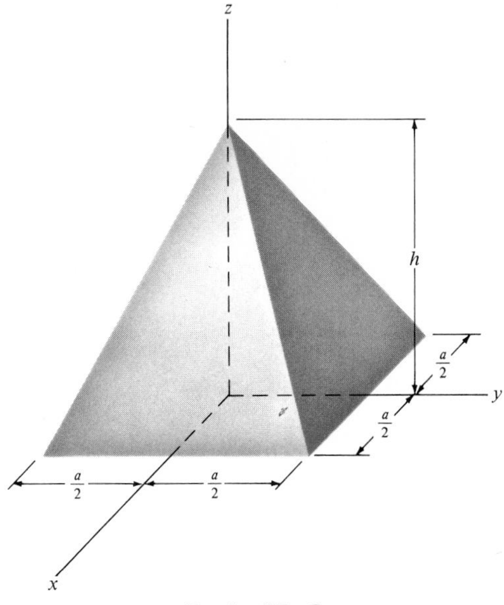

Prob. 17–6

17–7. A concrete solid is formed by rotating the shaded area about the y axis. Determine the moment of inertia I_y. The density of material is $\rho = 150\ \text{lb/ft}^3$.

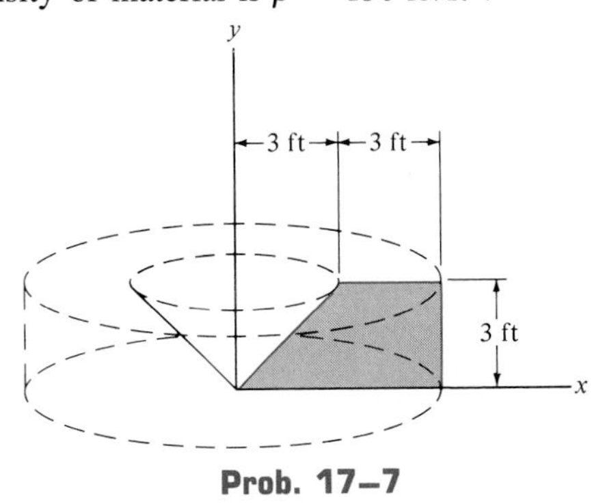

Prob. 17–7

***17–8.** The assembly consists of a uniform rod AB having a mass of 0.2 kg and two solid spheres C and D having a mass of 0.4 kg and 0.6 kg, respectively. Determine $\bar{x}$, which locates the center of mass G, and then calculate the moment of inertia about an axis perpendicular to the page and passing through G.

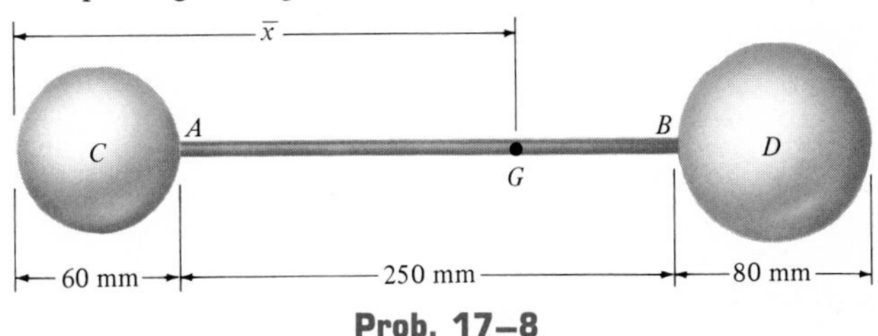

Prob. 17–8

17–9. The pulley has the cross section shown. If the density of the material is $\rho = 400\ \text{lb/ft}^3$, determine the pulley's moment of inertia about the xx axis.

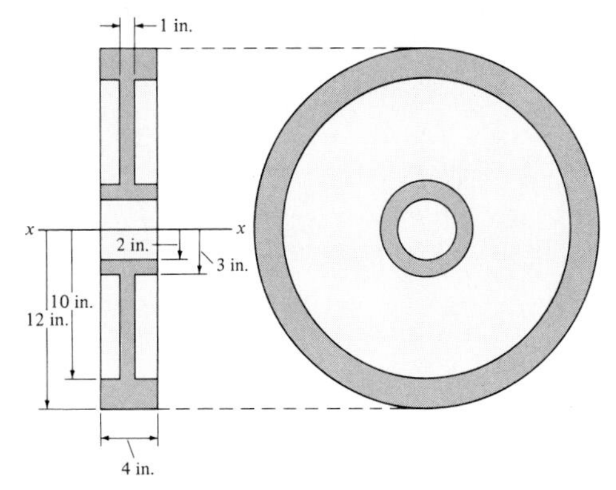

Prob. 17–9

17–10. Determine the moment of inertia of the wheel with respect to the z axis. The rim R and plate P both have a density of $\rho = 600 \text{ lb/ft}^3$.

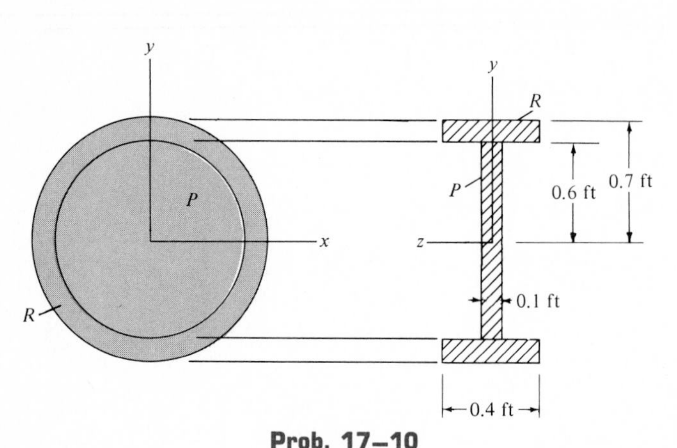

Prob. 17–10

17–11. Determine the moment of inertia of the single-throw crankshaft about the xx axis. The density of material is $\rho = 400 \text{ lb/ft}^3$.

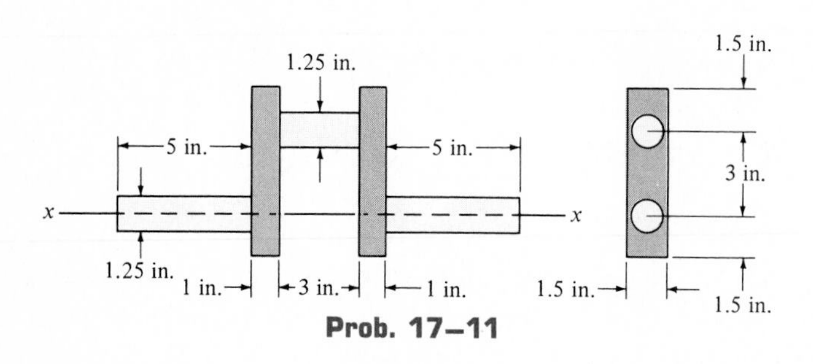

Prob. 17–11

***17–12.** Determine the moment of inertia I_x of the solid having a hemispherical head and conical depression. The density of the material is $\rho = 6 \text{ Mg/m}^3$.

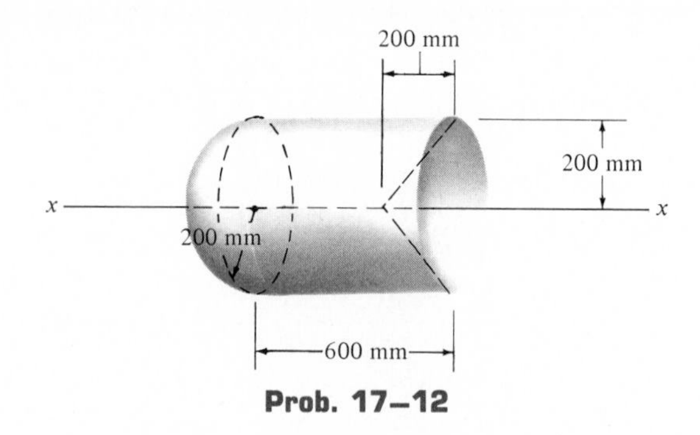

Prob. 17–12

17–13. Locate the center of mass G of the baseball bat by determining $\bar{y}$. Then, compute the moment of inertia of the bat about an axis which passes through G and is perpendicular to the plane of the page. For the calculation, consider the bat to be composed of a truncated cone and cylinder. Neglect the size of the lip at A. The density of wood is $\rho_w = 750 \text{ kg/m}^3$.

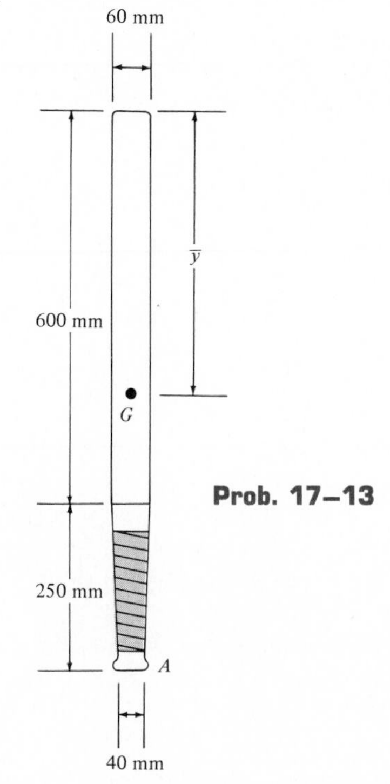

Prob. 17–13

Planar Kinetic Equations of Motion 17.2

Consider the rigid body (slab) shown in Fig. 17–8a, which has a mass m and is subjected to motion viewed in a reference plane. The *inertial frame of reference x,y* has its origin at point P. By definition, these axes do not rotate and are either fixed or translate with constant velocity. At the *instant* considered, the body is rotating with an angular velocity $\boldsymbol{\omega}$, and the applied force system causes the slab or body to have an angular acceleration $\boldsymbol{\alpha}$ while the center of mass has an acceleration $\mathbf{a}_G$.

Equation of Translational Motion. The external forces shown on the body in Fig. 17–8a symbolically represent the effect of gravitational, electrical, magnetic, or contact forces between adjacent bodies. Since this force system has been considered previously in Sec. 13–3, for the analysis of a system of particles, the results may be used here, in which case the particles are contained within the boundary of the body (or slab). Hence, if the equation of motion is applied to each of the "i" particles of the body, and the results added vectorially, it may be concluded that

$$\Sigma \mathbf{F} = m\mathbf{a}_G$$

This equation is referred to as the *equation of translational motion* for a rigid body. It states that *the sum of all the external forces acting on the body is equal to the body's mass times the acceleration of the mass center.*

For motion of the body (or slab) in the x-y plane, the equation of translational motion may be written in the form of two independent scalar equations: namely,

$$\Sigma F_x = m(a_G)_x$$
$$\Sigma F_y = m(a_G)_y$$

Equation of Rotational Motion. We will now determine the effects caused by the moments of the external force system computed about an axis perpendicular to the plane of motion (the z axis) and passing through the arbitrary point P in the body, Fig. 17–8a.* At the instant considered, point P has x and y components of acceleration $(\mathbf{a}_P)_x$ and $(\mathbf{a}_P)_y$, and the body has an angular acceleration $\boldsymbol{\alpha}$ and angular velocity $\boldsymbol{\omega}$. If the acceleration of the ith particle of the body is to be determined, then

$$\mathbf{a}_i = \mathbf{a}_P + \mathbf{a}_{i/P}$$
$$\mathbf{a}_i = \underset{\rightarrow}{(a_P)_x} + \underset{\uparrow}{(a_P)_y} + \underset{\swarrow^{\theta}}{r\omega^2} + \underset{\nwarrow^{\theta}}{r\alpha}$$

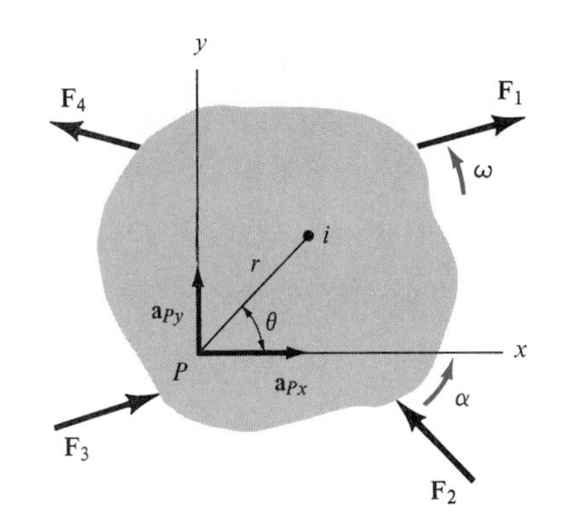

(a)

Fig. 17–8a

*Due to the symmetry of loading on each side of the plane of motion, i.e., along the z axis, moments of the forces about any axis lying in the plane of motion will be zero and therefore not considered here.

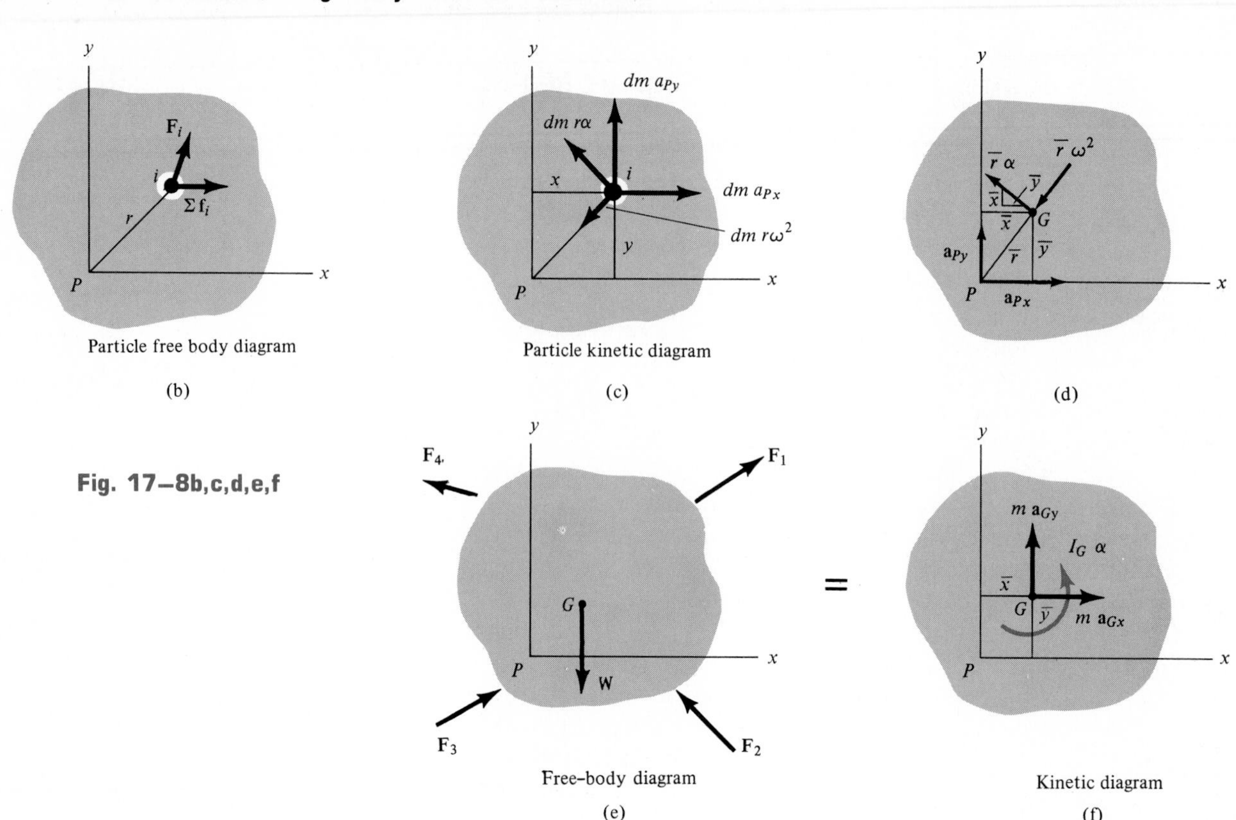

Particle free body diagram

(b)

Particle kinetic diagram

(c)

(d)

Free-body diagram

(e)

Kinetic diagram

(f)

Fig. 17–8b,c,d,e,f

As shown on the free-body diagram, Fig. 17–8*b*, $\mathbf{F}_i$ represents the *resultant* external force acting on the particle, and $\Sigma \mathbf{f}_i$ is the *resultant* of the internal forces caused by interactions with the adjacent particles. Using the components of acceleration listed above and assuming the particle has a mass of dm, the kinetic diagram is constructed as shown in Fig. 17–8*c*. If the particle is at point x, y and moments are summed about point P (the z axis), we have

$$\circlearrowleft + dM_P = -y\, dm(a_P)_x + x\, dm(a_P)_y + r^2\alpha\, dm$$

By integrating with respect to the entire mass m of the body so as to include all the particles contained in the body, we obtain the resultant moment equation

$$\Sigma M_P = -\left(\int_m y\, dm\right)(a_P)_x + \left(\int_m x\, dm\right)(a_P)_y + \left(\int_m r^2\, dm\right)\alpha$$

Here ΣM_P represents only the moment of the *external forces* acting on the body about point P. The resultant moment of the internal forces is zero, since for the entire body they occur in equal and opposite collinear pairs and hence cancel each other out. The integrals in the first and second terms on the right are used to locate the body's center of mass G with respect to P, since $\bar{y}m = \int_m y\, dm$ and $\bar{x}m = \int_m x\, dm$. Also, the last integral represents the body's moment of inertia computed about the z axis, i.e., $I_P = \int r^2\, dm$.

Thus,

$$\circlearrowleft +\Sigma M_P = -\bar{y}m(a_P)_x + \bar{x}m(a_P)_y + I_P\alpha \qquad (17\text{–}6)$$

It is possible to reduce this equation to a simpler form if point P coincides with the mass center G for the body, since then $\bar{x} = \bar{y} = 0$, and therefore

$$\Sigma M_G = I_G\alpha \qquad (17\text{–}7)$$

This *rotational equation of motion states that the sum of the moments of all the external forces computed about the body's mass center G is equal to the product of the moment of inertia of the body about G and the magnitude of the body's angular acceleration.*

Equation 17–6 can also be rewritten in terms of the x and y components of $\mathbf{a}_G$ and the moment of inertia I_G. Using the parallel-axis theorem, $I_P = I_G + m(\bar{x}^2 + \bar{y}^2)$, we get

$$\circlearrowleft + \Sigma M_P = \bar{y}m[-(a_P)_x + \bar{y}\alpha] + \bar{x}m[(a_P)_y + \bar{x}\alpha] + I_G\alpha \qquad (17\text{–}8)$$

Expressing a_P in terms of a_G, Fig. 17–8*d*, we have

$$\mathbf{a}_G = \mathbf{a}_P + \mathbf{a}_{G/P}$$

$$\mathbf{a}_G = \underset{\rightarrow}{(a_P)_x} + \underset{\uparrow}{(a_P)_y} + \bar{r}\omega^2 + \bar{r}\alpha$$

Resolving into x and y components yields

$$\overset{+}{\rightarrow} \qquad (a_G)_x = (a_P)_x - \bar{x}\omega^2 - \bar{y}\alpha$$

$$+\uparrow \qquad (a_G)_y = (a_P)_y - \bar{y}\omega^2 + \bar{x}\alpha$$

From these equations, $[-(a_P)_x + \bar{y}\alpha] = [-(a_G)_x - \bar{x}\omega^2]$ and $[(a_P)_y + \bar{x}\alpha] = [(a_G)_y - \bar{x}\alpha]$. Substituting these results into Eq. 17–8 and simplifying gives

$$\circlearrowleft +\Sigma M_P = -\bar{y}m(a_G)_x + \bar{x}m(a_G)_y + I_G\alpha \qquad (17\text{–}9)$$

This important result indicates that when moments of the external forces acting on the free-body diagram are summed about point P, Fig. 17–8e, they are equivalent to the sum of the "kinetic moments" of the components of m$\mathbf{a}_G$ *about P plus the "kinetic moment" of* $I_G\boldsymbol{\alpha}$*, Fig. 17–8f.* In other words, when these "kinetic moments", ΣM_k, are computed, the vectors $m(\mathbf{a}_G)_x$ and $m(\mathbf{a}_G)_y$ are treated in the same manner as a force, that is, they can act at *any point along their line of action*. In a similar manner, $I_G\boldsymbol{\alpha}$ has the same properties as a couple and can therefore act at *any point* on the kinetic diagram. It is important to keep in mind, however, that $m\mathbf{a}_G$ and $I_G\boldsymbol{\alpha}$ are *not* the same as a force or a couple. Instead, they are *caused* by the effects of forces and moments acting on the body. (See Sec. 13–2.) With this in mind we can therefore write

$$\Sigma M_P = \Sigma(M_k)_P \qquad (17\text{–}10)$$

17.3 Equations of Motion: Translation

When a rigid body undergoes a *translation,* Fig. 17–9*a*, all the particles of the body have the *same acceleration,* so that $\mathbf{a}_G = \mathbf{a}$. Furthermore, $\boldsymbol{\alpha} = \mathbf{0}$, in which case the rotational equation of motion applied at point G reduces to a simplified form, $\Sigma M_G = 0$. Application of this and the translational equations of motion will now be discussed for each of the two types of translation presented in Chapter 16.

Rectilinear Translation. When a body is subjected to *rectilinear translation,* all the particles of the body (slab) travel along parallel straight-line paths. The free-body and kinetic diagrams are shown in Fig. 17–9*b*. At the instant considered, the origin of the inertial reference is located at the center of mass. Since $I_G\boldsymbol{\alpha} = \mathbf{0}$, only $m\mathbf{a}_G$ is shown on the kinetic diagram. Hence, the equations of motion which apply in this case become

$$\begin{aligned} \Sigma F_x &= m(a_G)_x \\ \Sigma F_y &= m(a_G)_y \\ \Sigma M_G &= 0 \end{aligned} \tag{17–11}$$

The last equation requires that the sum of the moments of all the external forces computed about the body's center of mass be equal to zero. It is possible, of course, to sum moments about other points on or off the body, in which case the "kinetic moment" of $m\mathbf{a}_G$ must be taken into account. For example, if point A is chosen, which lies at a perpendicular distance d from the line of action of $m\mathbf{a}_G$, the following moment equation applies:

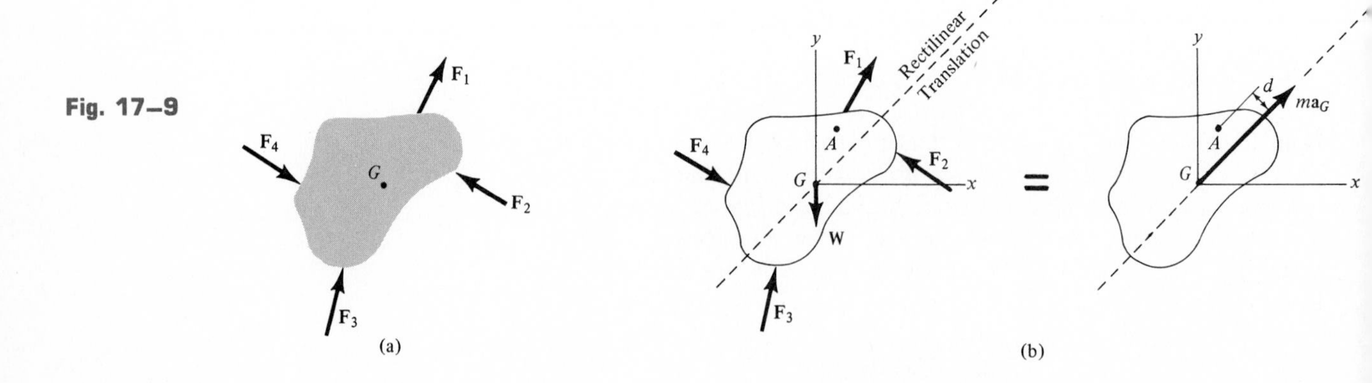

Fig. 17–9

$$\circlearrowleft + \Sigma M_A = \Sigma(M_k)_A; \qquad \Sigma M_A = (ma_G)d$$

Here the moment of the external forces about A (ΣM_A, free-body diagram) equals the "kinetic moment" of $m\mathbf{a}_G$ about A ($\Sigma(M_k)_A$, kinetic diagram).

Curvilinear Translation. When a rigid body is subjected to *curvilinear translation,* all the particles of the body travel along *parallel curved paths.* For analysis, it is often convenient to use an inertial coordinate system having an origin at G and axes which are oriented in the normal and tangential directions of the path of motion, Fig. 17–9c. The three scalar equations of motion are then

$$\begin{aligned} \Sigma F_n &= m(a_G)_n \\ \Sigma F_t &= m(a_G)_t \\ \Sigma M_G &= 0 \end{aligned} \qquad (17\text{–}12)$$

where $(a_G)_t$ and $(a_G)_n$ represent, respectively, the magnitudes of the tangential and normal components of acceleration of point G.

If the moment equation $\Sigma M_G = 0$ is replaced by a moment summation about the arbitrary point B, Fig. 17–9c, it is necessary to account for the "kinetic moments," $\Sigma(M_k)_B$, of the two components $m(\mathbf{a}_G)_n$ and $m(\mathbf{a}_G)_t$ about this point. From the kinetic diagram, h and e represent the perpendicular distances (or "moment arms") from B to the lines of action of the components. If positive moments are assumed to be clockwise, the required moment equation becomes

$$\circlearrowright + \Sigma M_B = \Sigma(M_k)_B; \qquad \Sigma M_B = h[m(a_G)_n] - e[m(a_G)_t]$$

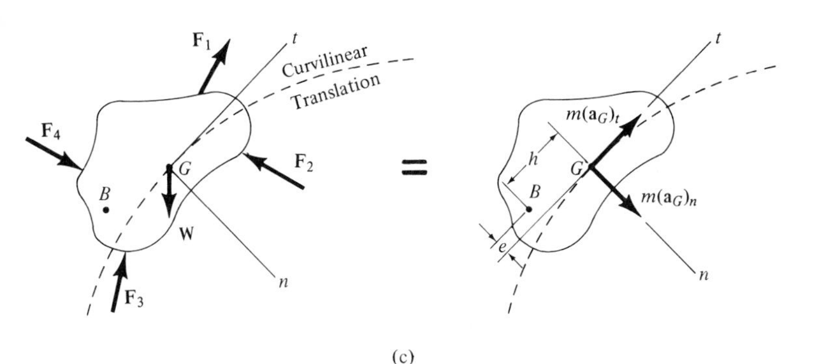

(c)

PROCEDURE FOR ANALYSIS

The following procedure provides a method for solving kinetic problems involving rigid-body translation.

Free-Body and Kinetic Diagrams. Draw the free-body and kinetic diagrams for the body. Recall that the *free-body diagram* is a graphical representation of all the external forces ($\Sigma \mathbf{F}$) which act on the body,* whereas the *kinetic diagram* graphically accounts for the components $m(\mathbf{a}_G)_x$, $m(\mathbf{a}_G)_y$ or $m(\mathbf{a}_G)_t$, $m(\mathbf{a}_G)_n$, Fig. 17–9*b* or *c*.

Equations of Motion. Apply the three equations of motion, Eqs. 17–11 or Eqs. 17–12, with reference to an established x, y or n, t inertial coordinate system. All the terms in these equations are computed directly from the data shown on the free-body and kinetic diagrams. To simplify the analysis, the moment equation $\Sigma M_G = 0$ can be replaced by the more general equation $\Sigma M_P = \Sigma(\mathcal{M}_k)_P$, where point P is usually located at the intersection of the lines of action of as many unknown forces as possible.

If the body is in contact with a *rough surface* and slipping occurs, use the frictional equation $F = \mu_k N$ to relate the normal force $\mathbf{N}$ to its associated frictional force $\mathbf{F}$.†

Kinematics. Use kinematics if a complete solution cannot be obtained strictly from the equations of motion. For *rectilinear translation* with *variable acceleration,* use

$$a_G = \frac{dv_G}{dt} \qquad a_G\, ds_G = v_G\, dv_G \qquad v_G = \frac{ds_G}{dt}$$

For *rectilinear translation* with *constant acceleration,* use

$$v_G = (v_G)_0 + a_G t$$
$$v_G^2 = (v_G)_0^2 + 2a_G[s_G - (s_G)_0]$$
$$s_G = (s_G)_0 + (v_G)_0 t + \tfrac{1}{2}a_G t^2$$

For *curvilinear translation,* use

$$(a_G)_n = \frac{v_G^2}{\rho} = \omega^2 \rho$$

$$(a_G)_t = \frac{dv_G}{dt} \qquad (a_G)_t\, ds_G = v_G\, dv_G \qquad (a_G)_t = \alpha\rho$$

The following examples numerically illustrate application of this procedure.

*See Chapter 5 of *Mechanics for Engineers: Statics* for a discussion of free-body diagrams for rigid bodies.

†Since many of the problems in rigid-body kinetics involve friction, it is suggested that one review the material on friction, covered in Secs. 8–1 and 8–2 of *Mechanics for Engineers: Statics*.

Example 17–5

The car shown in Fig. 17–10*a* has a mass of 2 Mg and a center of mass at G. Determine the car's acceleration if the "driving" wheels in the back are always slipping, whereas the front wheels freely rotate. Neglect the mass of the wheels. The coefficient of kinetic friction between the wheels and the road is $\mu_k = 0.25$.

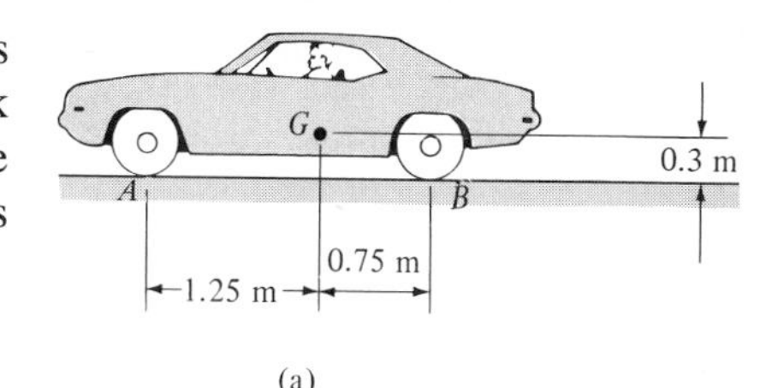

(a)

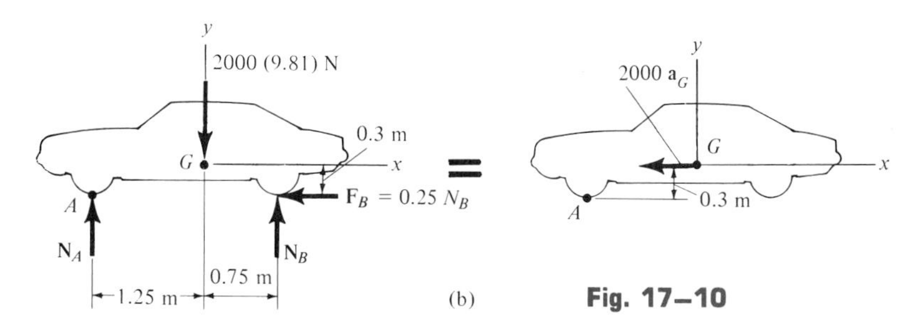

(b)

Fig. 17–10

Solution

Free-Body and Kinetic Diagrams. As shown in Fig. 17–10*b*, the rear-wheel frictional force $\mathbf{F}_B$ pushes the car forward and since *slipping occurs*, this force is related to its associated normal force $\mathbf{N}_B$ by $F_B = 0.25N_B$. The frictional force acting on the *front wheels* is *zero*, since these wheels have negligible mass.*

Equations of Motion

$$\overset{+}{\leftarrow}\Sigma F_x = m(a_G)_x; \qquad 0.25N_B = 2000a_G \qquad (1)$$

$$+\uparrow \Sigma F_y = m(a_G)_y; \quad N_A + N_B - 2000(9.81) = 0 \qquad (2)$$

$$\circlearrowleft +\Sigma M_G = 0; \quad N_A(1.25) + 0.25N_B(0.3) - N_B(0.75) = 0 \qquad (3)$$

The "moment" equation can also be applied at point A, which eliminates N_A from the equation. In this case the moment of the force system about A (free-body diagram) is equivalent to the "kinetic moment" of $m\mathbf{a}_G$ about A (kinetic diagram).

$$\circlearrowleft +\Sigma M_A = \Sigma(M_k)_A; \quad N_B(2) - 2000(9.81)(1.25) = 2000(a_G)(0.3) \qquad (4)$$

Solving Eqs. (1) to (3) or the simpler set of Eqs. (1), (2), and (4) gives

$$a_G = 1.59 \text{ m/s}^2 \qquad \textit{Ans.}$$

$$N_A = 6.88 \text{ kN}$$

$$N_B = 12.74 \text{ kN}$$

*If the mass of the front wheels were to be included in the analysis, the frictional force acting at A would be *directed to the right* to create the counterclockwise rotation of the wheels. The problem solution for this case would be more involved since a general-plane-motion analysis of the wheels would have to be included (see Sec. 17–5).

Example 17–6

A 5-lb force is applied to the center of the roller at A, Fig. 17–11a. If the roller has negligible mass, determine the angle θ which the 6-lb uniform rod makes with the vertical and the speed of the roller after it moves 4 ft starting from rest.

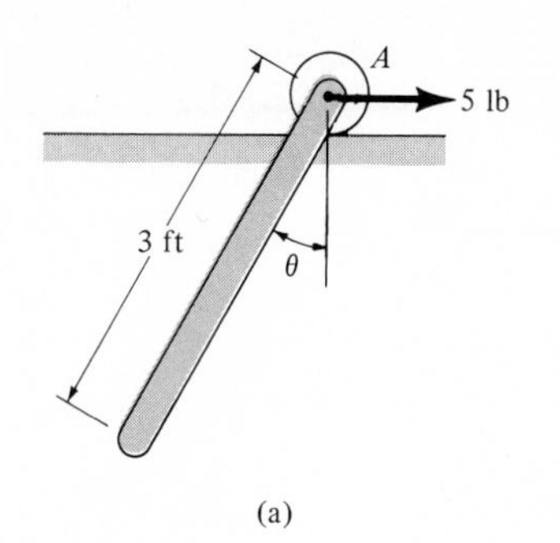

(a)

Solution

Free-Body and Kinetic Diagrams. Since the force acting on the roller is constant, the rod makes a constant angle θ with the vertical and therefore the roller and rod have the *same* acceleration $\mathbf{a}_G$. The free-body and kinetic diagrams of the rod and roller are shown in Fig. 17–11b.

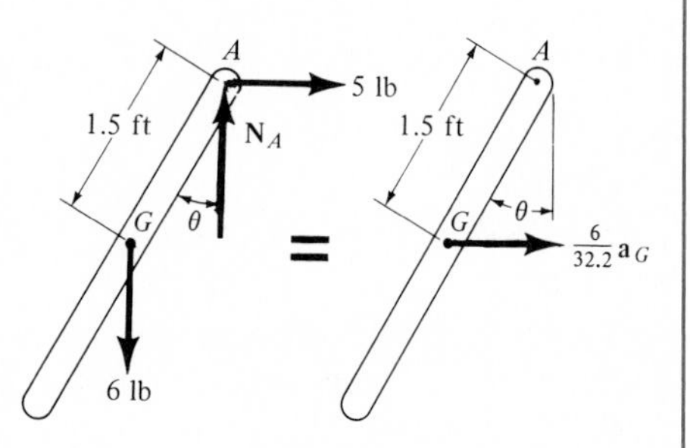

(b)

Fig. 17–11

Equations of Motion

$$\overset{+}{\rightarrow}\Sigma F_x = m(a_G)_x; \qquad 5 = \frac{6}{32.2}a_G \qquad (1)$$

$$+\uparrow \Sigma F_y = m(a_G)_y; \qquad N_A - 6 = 0 \qquad (2)$$

$$\circlearrowright +\Sigma M_G = 0; \qquad 5(1.5\cos\theta) - N_A(1.5\sin\theta) = 0 \qquad (3)$$

Moments can also be summed about A in order to eliminate the unknown normal force which passes through this point.

$$\circlearrowleft +\Sigma M_A = \Sigma(\mathcal{M}_k)_A; \qquad 6(1.5\sin\theta) = \left(\frac{6}{32.2}a_G\right)(1.5\cos\theta) \qquad (4)$$

Solving Eqs. (1) to (3) or Eqs. (1), (2), and (4) yields

$$a_G = 26.8 \text{ ft/s}^2$$
$$N_A = 6 \text{ lb}$$
$$\theta = 39.8^\circ \qquad \textit{Ans.}$$

Kinematics. Since the acceleration is *constant,* the speed after 4 ft of displacement is

$$(\overset{+}{\rightarrow}) \qquad v_G^2 = (v_G)_0^2 + 2a_G[s_G - (s_G)_0]$$
$$v_G^2 = 0 + 2(26.8)[4 - 0]$$
$$v_G = 14.6 \text{ ft/s} \qquad \textit{Ans.}$$

Example 17–7

A uniform 50-kg crate rests on a horizontal surface for which the coefficient of kinetic friction is $\mu_k = 0.2$. Determine the crate's acceleration if a force of $P = 600$ N is applied to the crate as shown in Fig. 17–12*a*.

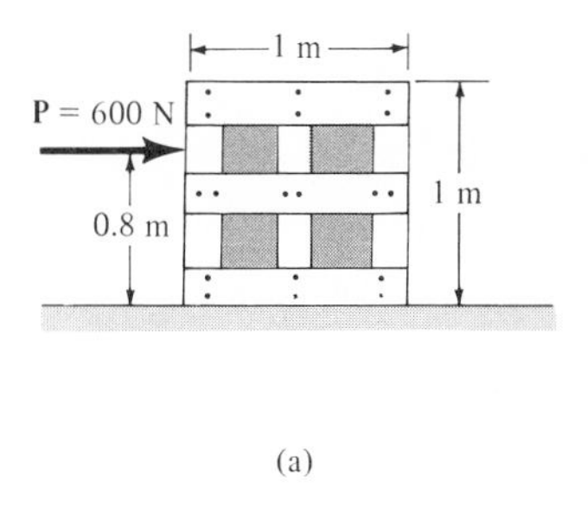

(a)

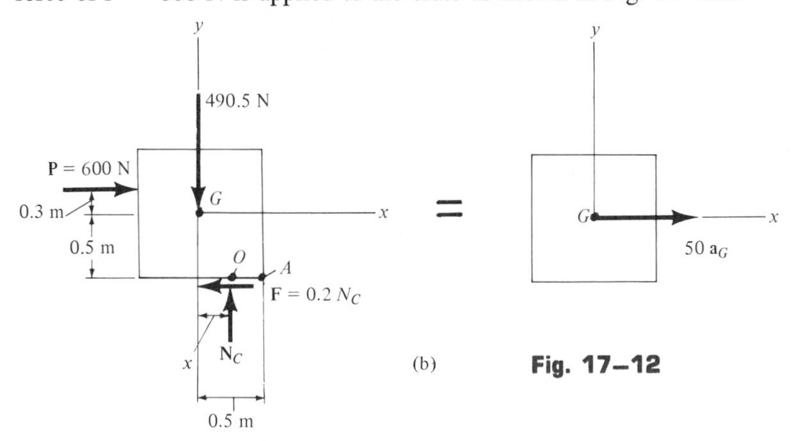

(b)

Fig. 17–12

Solution

Free-Body and Kinetic Diagrams. The force **P** can either cause the crate to slide or to tip over. As shown in Fig. 17–12*b*, it is assumed that the crate slides, so that $F = \mu_k N_C = 0.2N_C$. Also, the resultant normal force $\mathbf{N}_C$ acts at O, a distance x (where $0 < x \leqslant 0.5$ m) from the crate's center line.*

Equations of Motion

$$\overset{+}{\rightarrow}\Sigma F_x = m(a_G)_x; \qquad 600 - 0.2N_C = 50a_G \qquad (1)$$

$$+\uparrow \Sigma F_y = m(a_G)_y; \qquad N_C - 490.5 = 0 \qquad (2)$$

$$\circlearrowleft + \Sigma M_G = 0; \quad -600(0.3) + N_C(x) - 0.2N_C(0.5) = 0 \qquad (3)$$

Solving, we obtain

$$N_C = 490.5 \text{ N}$$

$$x = 0.47 \text{ m}$$

$$a_G = 10.0 \text{ m/s}^2 \qquad \textit{Ans.}$$

Since $x = 0.47$ m < 0.5 m, indeed the crate slides as originally assumed. If the solution had given a value of $x > 0.5$ m, the problem would have to be reworked with the assumption that tipping occurred. If this were the case, $\mathbf{N}_C$ would act at the *corner point A* and $F \leqslant 0.2N_C$, since in general the crate would *not* be on the verge of sliding at the instant it begins to tip.

*The line of action of $\mathbf{N}_C$ does not necessarily pass through the mass center G ($x = 0$), since $\mathbf{N}_C$ must counteract the effect of tipping caused by **P**. See Sec. 8–1 of *Mechanics for Engineers: Statics*.

Example 17–8

The 100-kg beam BD shown in Fig. 17–13a is supported by two rods having negligible mass. Determine the force created in each rod if at the instant $\theta = 0°$ the rods are both rotating with a constant angular velocity of $\omega = 6$ rad/s.

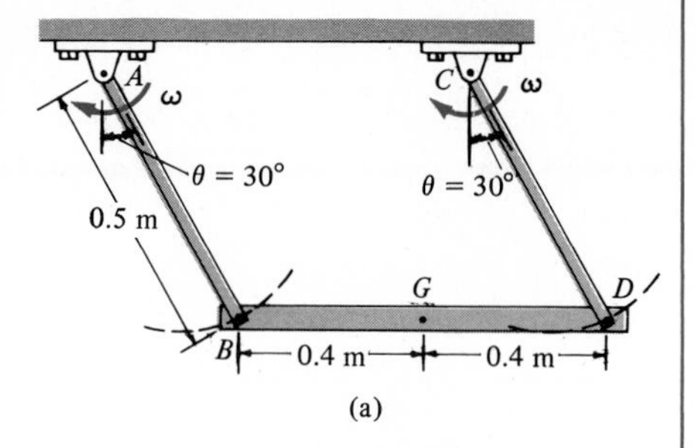

Fig. 17–13

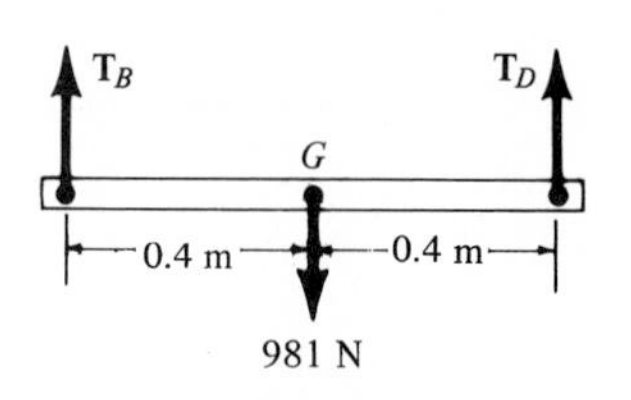

=

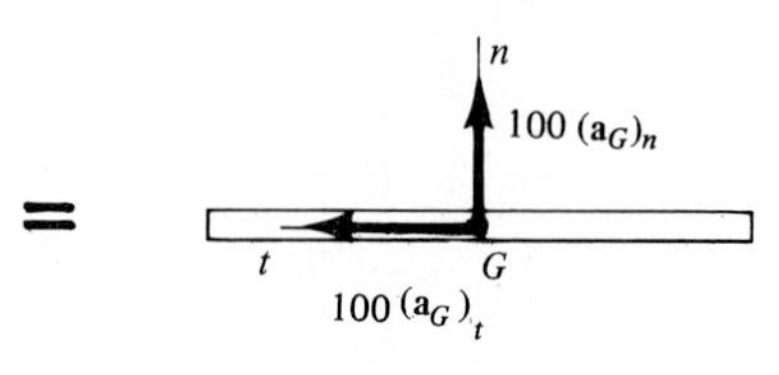

(b)

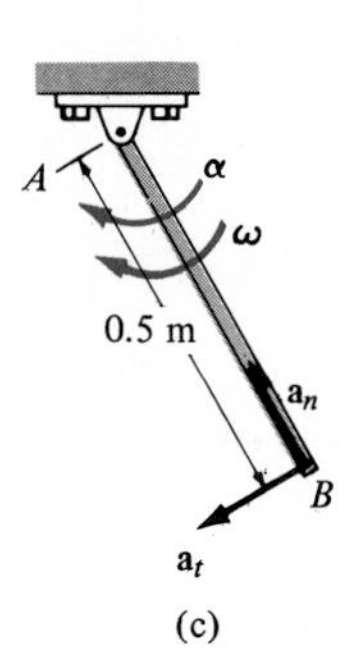

Solution

Free-Body and Kinetic Diagrams. The beam moves with *curvilinear translation* since points B and D and the center of mass G move along circular paths, each path having the same radius of 0.5 m. Using normal and tangential coordinates, the free-body and kinetic diagrams for the beam at $\theta = 0°$ are shown in Fig. 17–13b. Because of the *translation*, G has the *same* motion as the pin at B, which is connected to both the rod and the beam. By studying the angular motion of rod AB, Fig. 17–13c, note that the tangential component of acceleration acts to the left due to the clockwise direction of $\boldsymbol{\alpha}$. Furthermore, the normal component of acceleration is *always* directed toward the center of curvature (toward point A for rod AB). Since the angular velocity of AB is 6 rad/s, then

$$a_n = \omega^2 r = (6)^2(0.5) = 18 \text{ m/s}^2$$

Equations of Motion

$$+\uparrow \Sigma F_n = m(a_G)_n; \quad T_B + T_D - 981 = 100(18) \qquad (1)$$

$$\overset{+}{\leftarrow} \Sigma F_t = m(a_G)_t; \quad 0 = 100(a_G)_t \qquad (2)$$

$$\circlearrowleft + \Sigma M_G = 0; \quad -T_B(0.4) + T_D(0.4) = 0 \qquad (3)$$

Solution of these three equations gives

$$T_B = T_D = 1390.5 \text{ N} \qquad \textit{Ans.}$$

$$(a_G)_t = 0 \text{ m/s}^2$$

These results can also be obtained by applying the rotational equation of motion about point D, in which case

$$\circlearrowright + \Sigma M_D = \Sigma(M_k)_D; \quad T_B(0.8) - 981(0.4) = 100(a_G)_n(0.4) \qquad (4)$$

Solving Eqs. (1), (2), and (4) yields the same results obtained previously.

Problems

Except when stated otherwise, throughout this chapter assume that the coefficients of static and kinetic friction are equal, i.e., $\mu = \mu_s = \mu_k$.

17–14. In order to test its engine, the 2-Mg missile is restrained from being fired by *four* short links, such as AB, which are placed symmetrically around the nozzle, i.e., 90° apart. Determine the force developed in each link, knowing that without these restraints the missile would accelerate upward at 20 m/s^2. Because of symmetry of geometry and loading, the force in each link is the same.

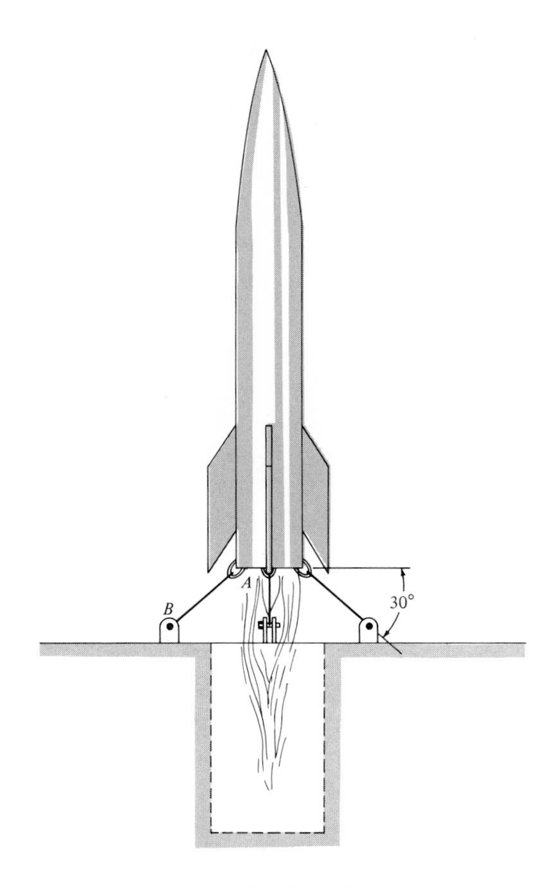

Prob. 17–14

17–15. The door has a weight of 200 lb and a center of gravity at G. Determine how far the door moves in 2 s, starting from rest, if a man pushes on it at C with a horizontal force of 30 lb. Also, find the vertical reactions at the rollers A and B.

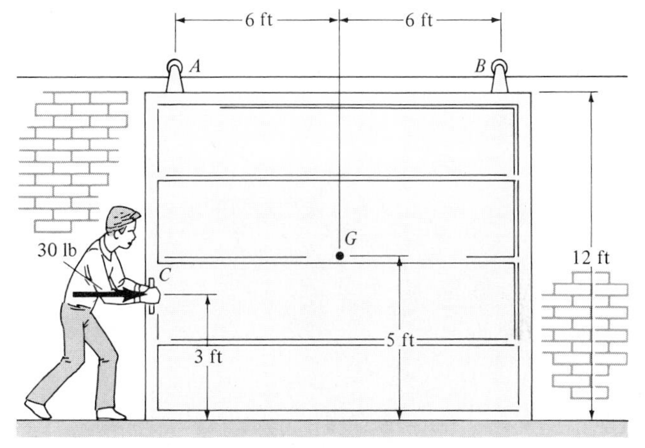

Prob. 17–15

***17–16.** At the *start* of take-off, the propeller on the 2-Mg plane exerts a horizontal thrust of 600 N on the plane. Determine the plane's acceleration and the vertical reactions at the nose wheel A and each of the *two* wing wheels B. Neglect the lifting force of the wings since the plane is originally at rest. The mass center is at G.

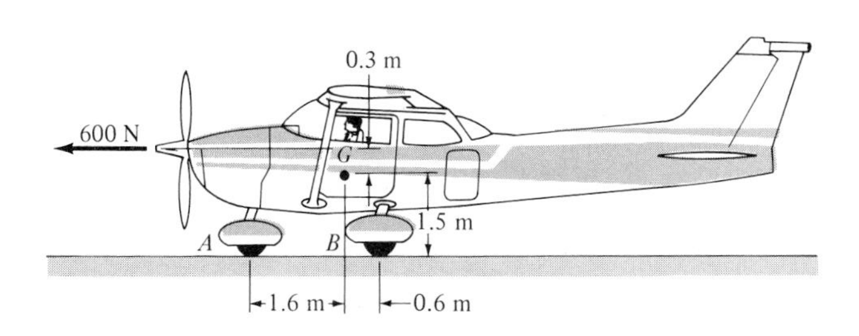

Prob. 17–16

17–17. The sports car has a mass of 1.5 Mg and a center of mass at G. Determine the shortest time it takes for it to reach a speed of 80 km/h, starting from rest, if the engine only drives the rear wheels, whereas the front wheels are free rolling. The coefficient of friction between the wheels and the road is $\mu = 0.2$. Neglect the mass of the wheels for the calculation. If driving power could be supplied to all four wheels, what would be the shortest time for the car to reach a speed of 80 km/h?

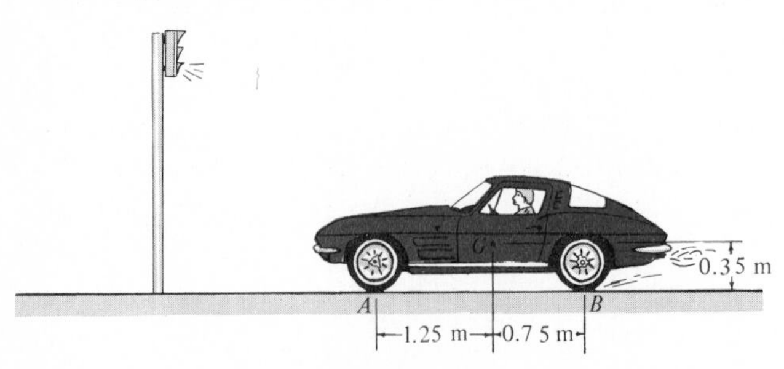

Prob. 17–17

17–18. Bar AB has a weight of 10 lb and is fixed to the carriage at A. Determine the internal axial force $\mathbf{A}_y$, shear force $\mathbf{V}_x$, and moment $\mathbf{M}_A$ at A if the carriage is moving down the plane with an acceleration of 4 ft/s^2.

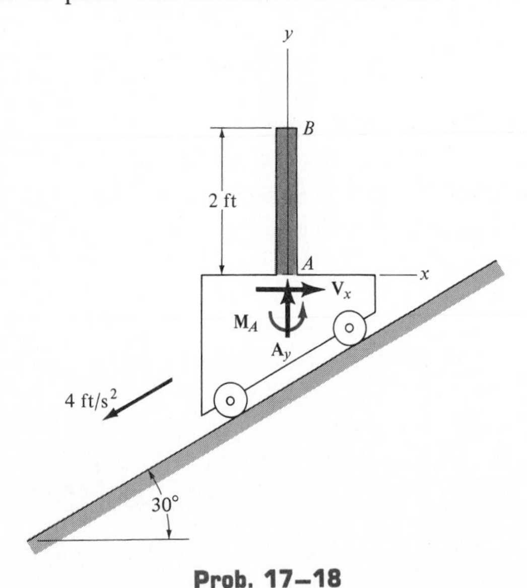

Prob. 17–18

17–19. The jet aircraft has a total mass of 22 Mg and a center of mass at G. If a towing cable is attached to the upper portion of the nose wheel and exerts a force of $T = 400$ N as shown, determine the acceleration of the plane and the normal reactions on the nose wheel and each of the *two* wing wheels located at B. Neglect the size of the wheels in the calculation and any lift caused by the wings.

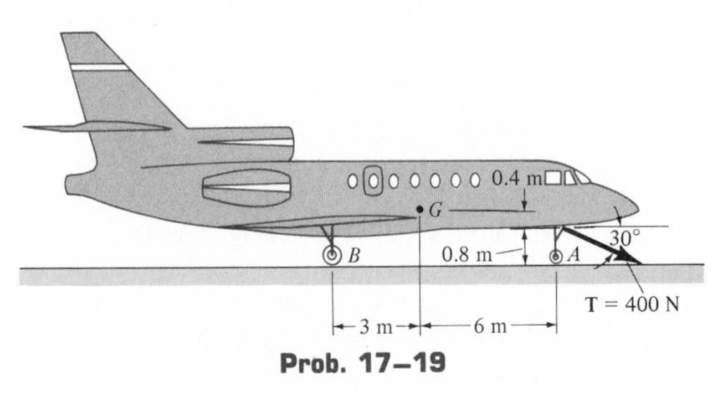

Prob. 17–19

***17–20.** The dragster has a mass of 1.3 Mg and a center of mass at G. If a braking parachute is attached at C and provides a horizontal braking force of $F = 400$ N, determine the dragster's deceleration and the normal reactions of the wheels at A and B. Neglect the mass of the wheels and assume the engine is disengaged so that the wheels are freely rolling.

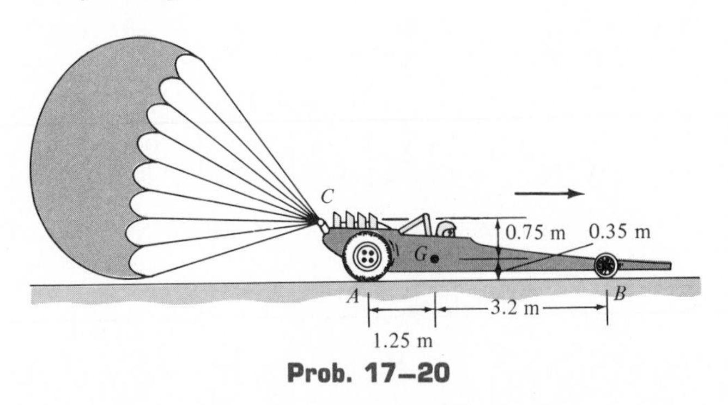

Prob. 17–20

17–21. A motorcyclist is traveling along the horizontally curved road which has a radius of $\rho = 250$ ft. If the coefficient of friction between the tires and the road is $\mu = 0.3$, determine the maximum *constant* speed at which he may round the curve and the corresponding angle θ at which he must lean so as not to tip over or slip. The motorcycle and the rider have a total weight of 400 lb and a center of mass at G; $h = 3.5$ ft.

Prob. 17–21

17–22. The 10-kg block rests on the platform for which $\mu = 0.4$. If at the instant shown link AB has an angular velocity of $\omega = 2$ rad/s, determine the greatest angular acceleration of the link so that the block doesn't slip. *Hint:* the block is subjected to curvilinear translation.

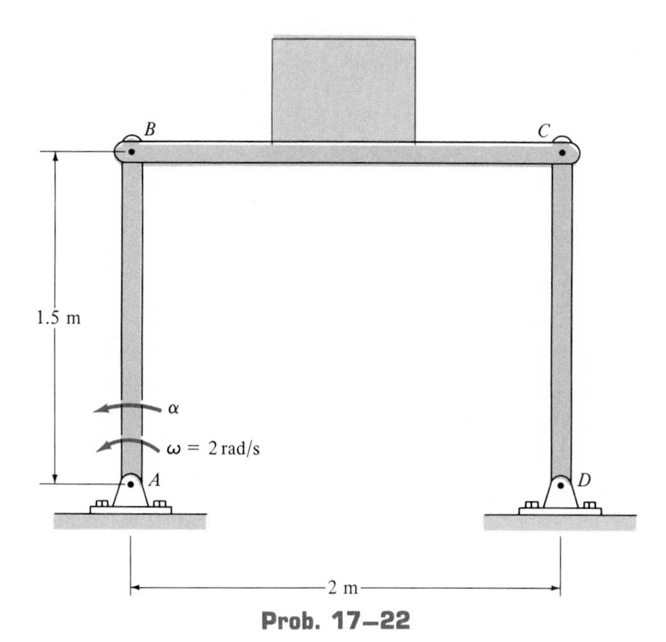

Prob. 17–22

17–23. A 75-kg man and 40-kg boy sit on the seesaw, which has negligible mass. At the instant the man lifts his feet from the ground, determine their accelerations if each sits upright, i.e., they do not rotate. The centers of mass of the man and boy are at G_m and G_b, respectively. *Hint:* Include $m\mathbf{a}$ for the man and boy on the kinetic diagram of the man-boy-seesaw system.

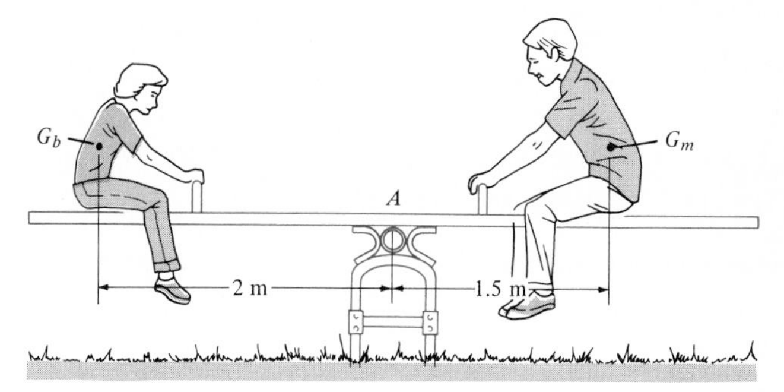

Prob. 17–23

***17–24.** A train, traveling at a constant speed of 15 m/s, is rounding a horizontal curve having a radius of 200 m measured to the center of mass G. Determine the correct banking angle θ of the track so that the wheels of the train exert an equal force on both rails.

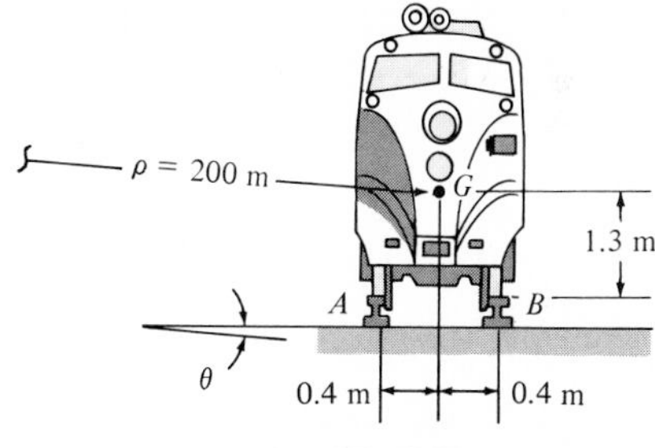

Prob. 17–24

17–25. The car has a weight of 4,000 lb and a center of mass at G. If the coefficient of friction between the wheels and the road is $\mu = 0.30$, determine the minimum constant speed at which it may round the horizontally banked curve without sliding downward. The car travels in the same horizontal plane.

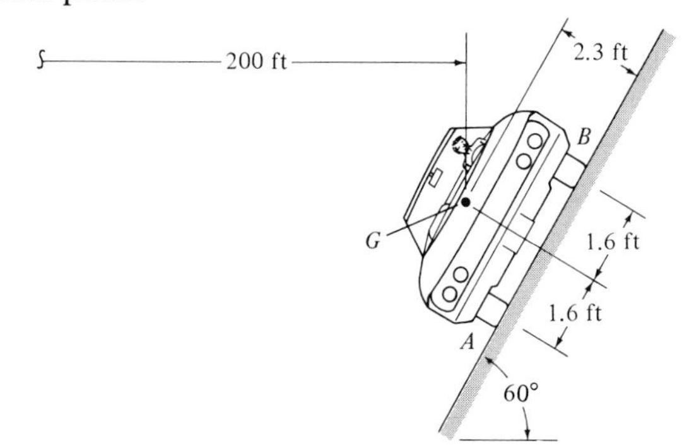

Prob. 17–25

17–26. A car having a weight of 4,000 lb begins to skid and turn with the brakes applied to all four wheels. If the coefficient of friction between the wheels and the road is $\mu = 0.8$, determine the maximum critical height h of the center of gravity G such that the car does not overturn. Tipping will begin to occur after the car rotates 90° from its original direction of motion and, as shown in the figure, undergoes *translation* while skidding. *Hint:* Draw free-body and kinetic diagrams of the car viewed from the front. When tipping occurs, the normal reactions of the wheels on the right side (or passenger side) are zero.

Prob. 17–26

17–27. The motorcycle has a mass of 125 kg and a center of mass at G_1, while the rider has a mass of 75 kg and a center of mass at G_2. If no slipping occurs, determine the minimum driving force $\mathbf{F}_B$ which must be supplied to the rear wheel B in order to create an acceleration of $a = 3\ \text{m/s}^2$. What are the normal reactions of the wheels on the ground? Neglect the mass of the wheels and assume that the front wheel is free to roll. *Hint:* Include $m\mathbf{a}$ for both G_1 and G_2 on the kinetic diagram of the motorcycle and rider.

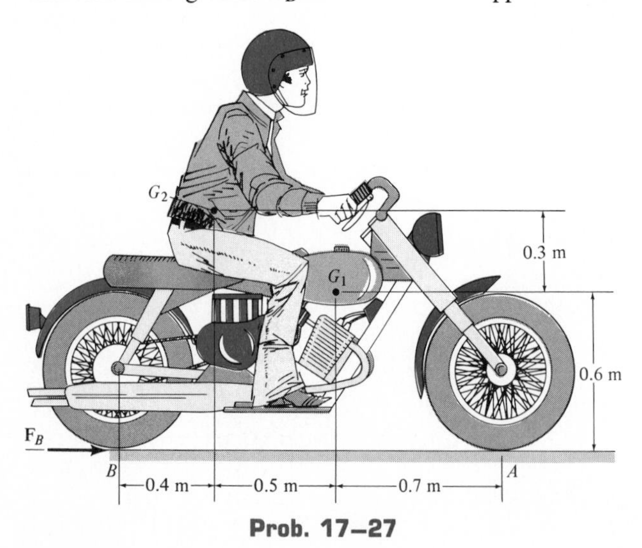

Prob. 17–27

***17–28.** The crate C has a weight of 150 lb and rests on the floor of a truck elevator for which $\mu = 0.4$. Determine the largest initial angular acceleration $\boldsymbol{\alpha}$, starting from rest, which the parallel links AB and DE can have without causing the crate to slip. No tipping occurs.

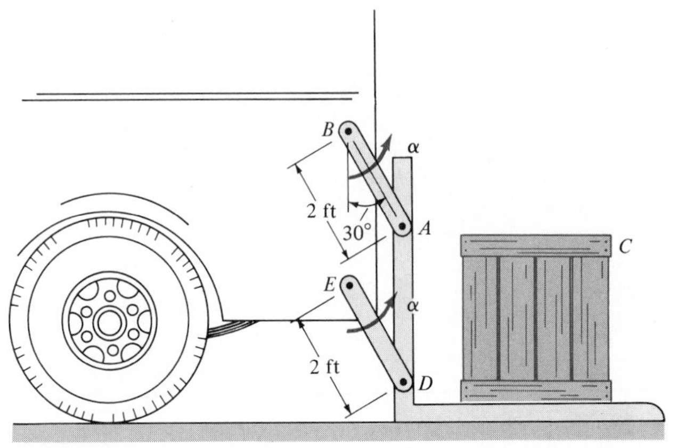

Prob. 17–28

17–29. The trailer portion of a truck has a mass of 4 Mg with a center of mass at G. If a *uniform* crate, having a mass of 800 kg and a center of mass at G_c, rests on the trailer, determine the horizontal and vertical components of reaction at the ball-and-socket joint (pin) A when the truck is decelerating at a constant rate of $a = 3\ \text{m/s}^2$. Assume that the crate does not slip on the trailer and neglect the mass of the wheels. The wheels at B roll freely.

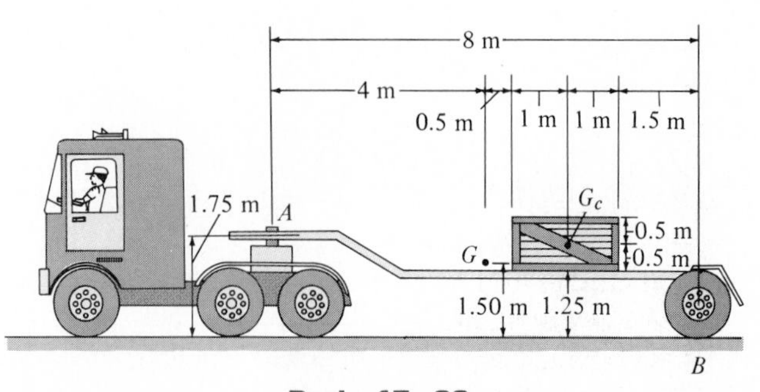

Prob. 17–29

17–30. If the coefficient of friction between the trailer and the crate in Prob. 17–29 is $\mu = 0.3$, determine the maximum allowable deceleration of the truck such that the crate does not slide on the trailer. No tipping occurs.

17–31. Solve Prob. 17–29 assuming that the coefficient of friction between the crate and the trailer is $\mu = 0.15$.

***17–32.** The car, having a mass of 1.4 Mg and mass center at G_c, pulls a loaded trailer having a mass of 0.8 Mg and mass center at G_t. Determine the normal reactions at the car and trailer wheels if the driver applies the car's rear brakes C and causes the car to skid. Take $\mu_C = 0.4$ and assume the hitch at A is a pin or ball-and-socket joint.

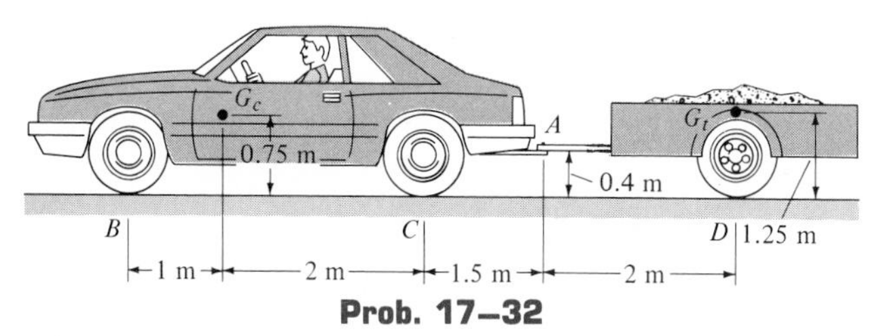

Prob. 17–32

17–33. The traction between two automobiles, A and B, is matched by connecting the rear bumpers with a cable CD. If A has a mass of 2000 kg, with center of mass at G_A, and B has a mass of 1500 kg, with center of mass at G_B, determine the tension developed in the cable and the acceleration of each vehicle. Slipping occurs only at the rear wheels F and H, where in both cases the coefficient of friction is $\mu = 0.3$. Neglect the mass of the wheels and assume that the front wheels at E and I are free to roll.

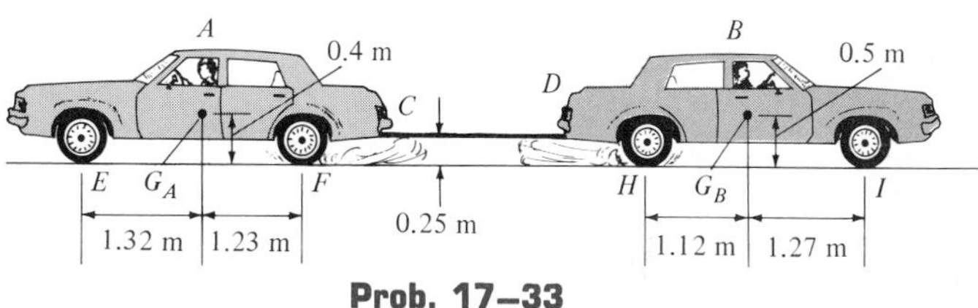

Prob. 17–33

17.4 Equations of Motion: Rotation About a Fixed Axis

Consider the rigid body (or slab) shown in Fig. 17–14*a*, which is constrained to rotate in the vertical plane about a fixed axis perpendicular to the page and passing through the pin at O. The angular velocity and angular acceleration are caused by the external force system acting on the body. Because the body's center of mass G moves in a *circular path,* the acceleration of this point is represented by its tangential and normal components. The *tangential component of acceleration* has a *magnitude* of $(a_G)_t = \alpha r_G$ and must act in a *direction* which is *consistent* with the angular acceleration $\boldsymbol{\alpha}$. The *magnitude* of the *normal component of acceleration* is $(a_G)_n = \omega^2 r_G$. This component is *always directed* from point G to O regardless of the direction of $\boldsymbol{\omega}$.

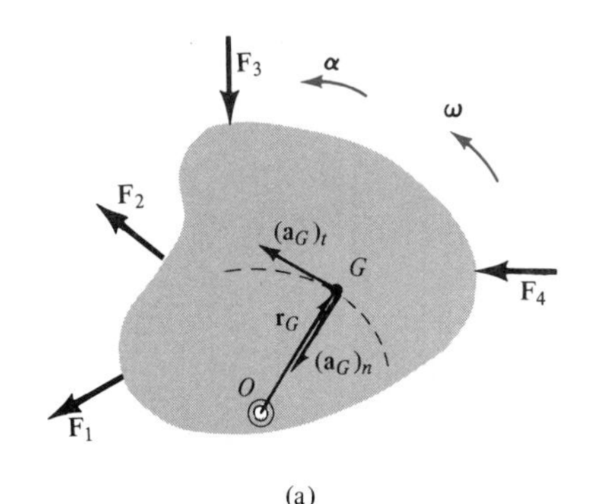

(a)

Fig. 17–14a

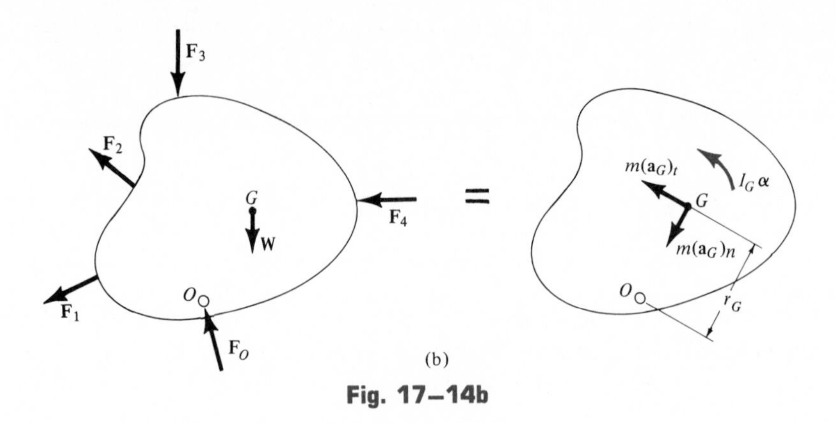

(b)

Fig. 17–14b

The free-body and kinetic diagrams for the body are shown in Fig. 17–14*b*. The weight of the body, $W = mg$, and the pin reaction $\mathbf{F}_O$ are included on the free-body diagram since they represent external forces acting on the body. The two components $m(\mathbf{a}_G)_t$ and $m(\mathbf{a}_G)_n$, shown on the kinetic diagram, are associated with the tangential and normal acceleration components of the body's mass center. These vectors act in the same *direction* as the acceleration components and have *magnitudes* of $m(a_G)_t$ and $m(a_G)_n$. The $I_G\boldsymbol{\alpha}$ vector acts in the same *direction* as $\boldsymbol{\alpha}$ and has a *magnitude* of $I_G\alpha$, where I_G is the body's moment of inertia calculated about an axis which is perpendicular to the page and passing through G. From the derivation given in Sec. 17–2, the equations of motion which apply to the body may be written in the form

$$\begin{aligned} \Sigma F_n &= m(a_G)_n = m\omega^2 r_G \\ \Sigma F_t &= m(a_G)_t = m\alpha r_G \\ \Sigma M_G &= I_G\alpha \end{aligned} \qquad (17\text{–}13)$$

The moment equation may be replaced by a moment summation about any arbitrary point on or off the body provided one accounts for the "kinetic moments" ΣM_k produced by $I_G\boldsymbol{\alpha}$, $m(\mathbf{a}_G)_t$, and $m(\mathbf{a}_G)_n$ about the point. In many problems it is convenient to sum moments about the pin at O in order to eliminate the *unknown* force $\mathbf{F}_O$. From the kinetic diagram, Fig. 17–14*b*, this requires

$$\circlearrowleft + \Sigma M_O = \Sigma(M_k)_O; \qquad \Sigma M_O = r_G m(a_G)_t + I_G\alpha \qquad (17\text{–}14)$$

Note that the "kinetic moment" of $m(\mathbf{a}_G)_n$ is not included in the summation since the line of action of this vector passes through O. Substituting $(a_G)_t = r_G\alpha$, we may rewrite the above equation as $\circlearrowleft + \Sigma M_O = (I_G + mr_G^2)\alpha$. From the parallel-axis theorem, $I = I_G + md^2$, the term in parentheses represents the *moment of inertia of the body about the fixed axis of rotation passing through O*. Denoting this term by I_O, we can write the three equations of motion for the body as*

$$\begin{aligned} \Sigma F_n &= m(a_G)_n = m\omega^2 r_G \\ \Sigma F_t &= m(a_G)_t = m\alpha r_G \\ \Sigma M_O &= I_O\alpha \end{aligned} \tag{17–15}$$

For applications, one should remember that "$I_O\alpha$" accounts for the "kinetic moment" of *both* $(\mathbf{a}_G)_t$ *and* $I_G\boldsymbol{\alpha}$ about point O, Fig. 17–14*b*. In other words, $\Sigma M_O = \Sigma(M_k)_O = I_O\alpha$, as indicated by Eqs. 17–14 and 17–15.

PROCEDURE FOR ANALYSIS

The following procedure provides a method for solving kinetic problems which involve the rotation of a body about a fixed axis.

Free-Body and Kinetic Diagrams. Draw a free-body and kinetic diagram for the body, Fig. 17–14*b*, and compute the moment of inertia I_G or I_O.

Equations of Motion. Apply the three equations of motion, Eqs. 17–13 or 17–15, with reference to an established x, y or n, t inertial coordinate system.

Kinematics. Use kinematics if a complete solution cannot be obtained strictly from the equations of motion. In this regard, if the *angular acceleration is variable,* use

$$\alpha = \frac{d\omega}{dt}, \qquad \alpha\, d\theta = \omega\, d\omega \qquad \omega = \frac{d\theta}{dt}$$

If the *angular acceleration is constant,* use

$$\begin{aligned} \omega &= \omega_0 + \alpha_c t \\ \theta &= \theta_0 + \omega_0 t + \tfrac{1}{2}\alpha_c t^2 \\ \omega^2 &= \omega_0^2 + 2\alpha_c(\theta - \theta_0) \end{aligned}$$

The following examples numerically illustrate application of this procedure.

*The result $\Sigma M_O = I_O\alpha$ can also be obtained *directly* from Eq. 17–6 by selecting point P to coincide with O, realizing that $(\mathbf{a}_P)_x = (\mathbf{a}_P)_y = \mathbf{0}$.

Example 17–9

The 30-kg disk shown in Fig. 17–15*a* is pin-supported at its center. If it starts from rest, determine the number of revolutions it must make to attain an angular velocity of 20 rad/s. Also, what are the reactions at the pin? The disk is acted upon by a constant force $F = 10$ N, which is applied to a cord wrapped around its periphery, and a constant couple $M = 5$ N · m. Neglect the mass of the cord in the calculation.

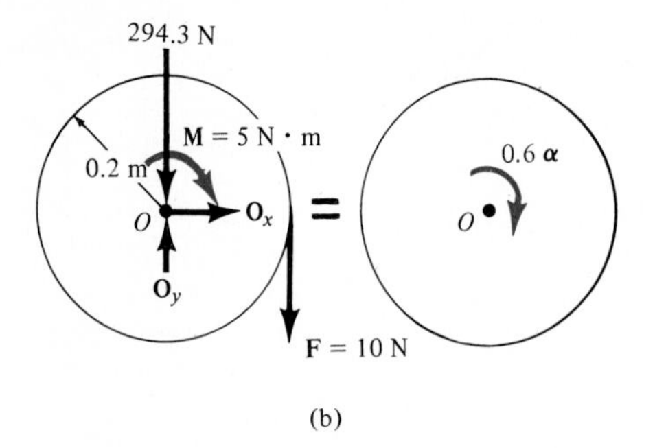

(b)

Fig. 17–15

Solution

Free-Body and Kinetic Diagrams. Since the mass center is not subjected to an acceleration, only $I_O\boldsymbol{\alpha}$ acts on the disk, Fig. 17–15*b*.

The moment of inertia of the disk is

$$I_O = \tfrac{1}{2}mr^2 = \tfrac{1}{2}(30)(0.2)^2 = 0.6 \text{ kg} \cdot \text{m}^2$$

Equations of Motion

$$\overset{+}{\rightarrow}\Sigma F_x = m(a_G)_x; \qquad O_x = 0 \qquad \textit{Ans.}$$

$$+\uparrow \Sigma F_y = m(a_G)_y; \qquad O_y - 294.3 - 10 = 0$$

$$O_y = 304.3 \text{ N} \qquad \textit{Ans.}$$

$$\circlearrowright +\Sigma M_O = I_O\alpha; \qquad 10(0.2) + 5 = 0.6\alpha \qquad \alpha = 11.7 \text{ rad/s}^2$$

Kinematics. Since α is constant, the number of radians the disk must turn to obtain an angular velocity of 20 rad/s is

$$\circlearrowright + \qquad \omega^2 = \omega_0^2 + 2\alpha_c(\theta - \theta_0)$$

$$(20)^2 = 0 + 2(11.7)(\theta - 0)$$

$$\theta = 17.1 \text{ rad}$$

Hence,

$$\theta = 17.1 \text{ rad}\left(\frac{1 \text{ rev}}{2\pi \text{ rad}}\right) = 2.73 \text{ rev} \qquad \textit{Ans.}$$

Example 17–10

The drum shown in Fig. 17–16*a* has a mass of 60 kg and a radius of gyration of $k_O = 0.25$ m. A cord of negligible mass is wrapped around the periphery of the drum and attached to a crate having a mass of 20 kg. If the crate is released, determine the drum's angular acceleration.

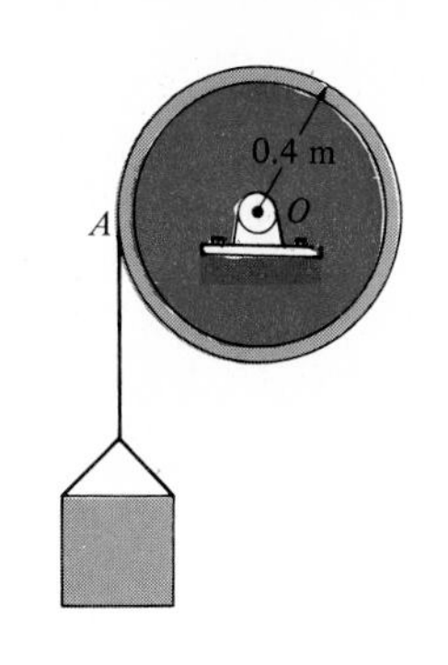

(a)

Solution I

Free-Body and Kinetic Diagrams. Here we will consider the drum and crate separately, Fig. 17–16*b*. Note that the downward acceleration of the crate, **a,** creates a counterclockwise angular acceleration $\boldsymbol{\alpha}$ of the drum.

The moment of inertia of the drum is

$$I_O = mk_O^2 = (60)(0.25)^2 = 3.75 \text{ kg} \cdot \text{m}^2$$

Equations of Motion. By observation, applying the translational equations of motion $\Sigma F_x = m(a_G)_x$ and $\Sigma F_y = m(a_G)_y$ to the drum is of no consequence to the solution, since these equations involve the unknowns O_x and O_y. Thus, for the drum and crate, respectively,

$$\circlearrowleft + \Sigma M_O = I_O\alpha; \qquad T(0.4) = 3.75\alpha \qquad (1)$$

$$+\downarrow \Sigma F_y = m(a_G)_y; \qquad 20(9.81) - T = 20a \qquad (2)$$

Kinematics. Since the point of contact A between the cord and drum has the same tangential component of acceleration, Fig. 17–16*a*, this requires that

$$\circlearrowleft + a = \alpha r; \qquad a = \alpha(0.4) \qquad (3)$$

Solving the above equations

$$T = 105.9 \text{ N}$$
$$a = 4.52 \text{ m/s}^2$$
$$\alpha = 11.3 \text{ rad/s}^2 \qquad \textit{Ans.}$$

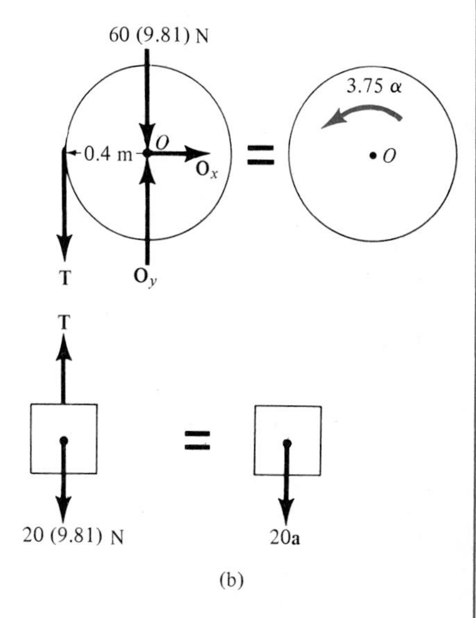

(b)

Solution II

Free-Body and Kinetic Diagrams. The cable tension T can be eliminated from the analysis by considering the drum and crate as a single system, Fig. 17–16*c*.

Equations of Motion. Using Eq. (3) and applying the moment equation about point O to eliminate the unknowns O_x and O_y, we have

$$\circlearrowleft + \Sigma M_O = \Sigma(\mathcal{M}_k)_O; \qquad 20(9.81)(0.4) = 3.75\alpha + [20(0.4\alpha)]0.4$$
$$\alpha = 11.3 \text{ rad/s}^2 \qquad \textit{Ans.}$$

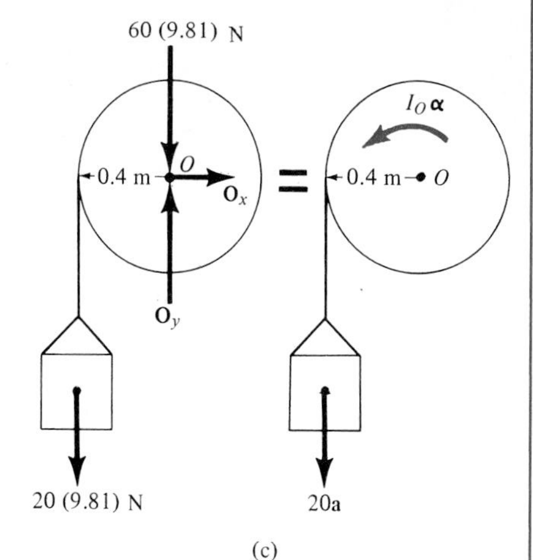

(c)

Fig. 17–16

Example 17–11

The unbalanced 50-lb flywheel shown in Fig. 17–17a has a radius of gyration of $k_G = 0.6$ ft about an axis passing through its mass center G. If it has a clockwise angular velocity of 8 rad/s at the instant shown, determine the horizontal and vertical components of reaction at the pin O.

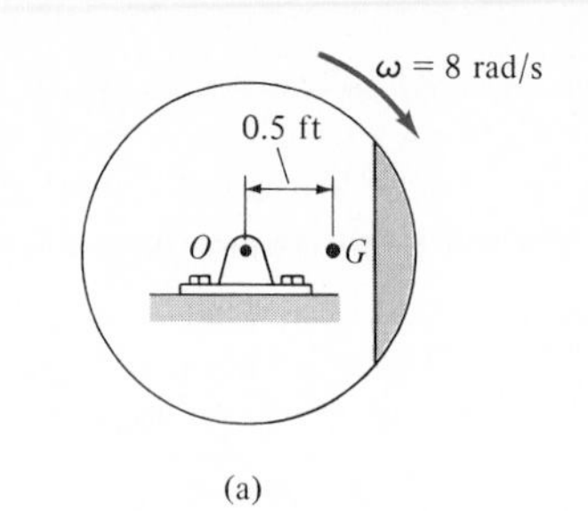

(a)

Solution

Free-Body and Kinetic Diagrams. Since G moves in a *circular path,* $(a_G)_x = \alpha r_G$ and $(a_G)_y = \omega^2 r_G$. By inspection, the weight creates a clockwise angular acceleration $\boldsymbol{\alpha}$, so $I_G\boldsymbol{\alpha}$ acts in a clockwise sense and $m(\mathbf{a}_G)_y$ acts downward, in accordance with the direction of $\boldsymbol{\alpha}$, Fig. 17–17b. Furthermore, $m(\mathbf{a}_G)_x$ acts to the left, toward the center of curvature O.

The moment of inertia of the flywheel about its mass center is determined from the radius of gyration and the mass; i.e., $I_G = mk_G^2 = (50/32.2)(0.6)^2 = 0.559 \text{ slug}\cdot\text{ft}^2$.

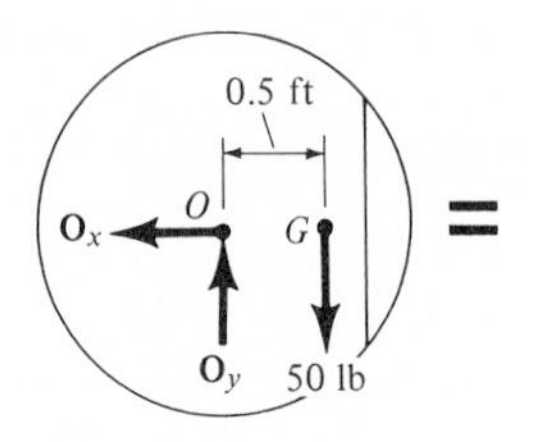

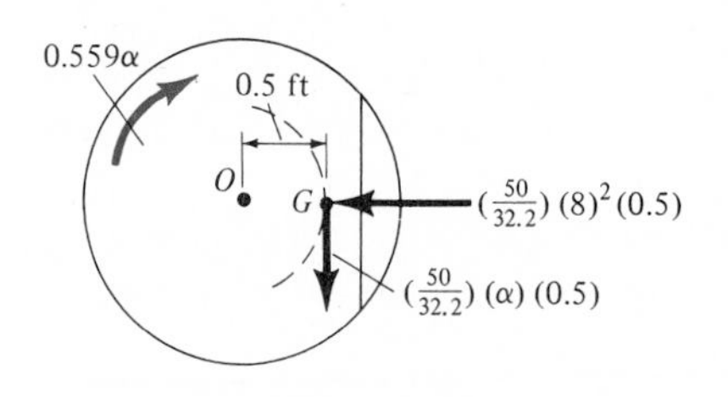

(b)

Fig. 17–17

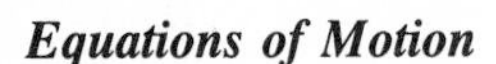

Equations of Motion

$$\overset{+}{\leftarrow}\Sigma F_x = m\omega^2 r_G; \qquad O_x = \left(\frac{50}{32.2}\right)(8)^2(0.5) \qquad (1)$$

$$+\uparrow \Sigma F_y = m\alpha r_G; \qquad O_y - 50 = -\left(\frac{50}{32.2}\right)(\alpha)(0.5) \qquad (2)$$

$$\circlearrowright +\Sigma M_G = I_G\alpha; \qquad O_y(0.5) = 0.559\alpha \qquad (3)$$

Solving,

$$\alpha = 26.4 \text{ rad/s}^2, \qquad O_x = 49.7 \text{ lb}, \qquad O_y = 29.5 \text{ lb} \qquad \textit{Ans.}$$

The moment equation can also be applied at point O in order to eliminate $\mathbf{O}_x$ and $\mathbf{O}_y$ and thereby obtain a *direct solution* for $\boldsymbol{\alpha}$, Fig. 17–17b. This can be done in one of *two* ways, i.e., by using either $\Sigma M_O = \Sigma(\mathcal{M}_k)_O$ or $\Sigma M_O = I_O\alpha$. If the first of these equations is applied, then

$$\circlearrowright +\Sigma M_O = \Sigma(\mathcal{M}_k)_O; \qquad 50(0.5) = 0.559\alpha + \left[\left(\frac{50}{32.2}\right)\alpha(0.5)\right](0.5)$$

$$50(0.5) = 0.947\alpha \qquad (4)$$

If $\Sigma M_O = I_O\alpha$ is applied, then by the parallel-axis theorem the moment of inertia of the flywheel about O is

$$I_O = I_G + mr_G^2 = 0.559 + \left(\frac{50}{32.2}\right)(0.5)^2 = 0.947 \text{ slug}\cdot\text{ft}^2$$

Hence, from the free-body diagram, Fig. 17–17b, we require

$$\circlearrowright +\Sigma M_O = I_O\alpha; \qquad 50(0.5) = 0.947\alpha$$

which is the same as Eq. (4). Solving for α and substituting into Eq. (2) or (3) yields the answer for O_y obtained previously.

Example 17–12

The 20-kg slender rod shown in Fig. 17–18*a* is subjected to a moment of 5 N · m. If the rod is vertical and has an angular velocity of $\omega = 3$ rad/s at the instant shown, determine the horizontal and vertical components of force which the pin at A exerts on the rod at this instant.

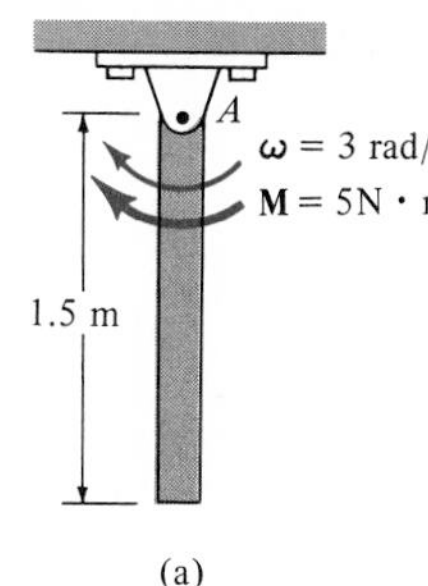

Fig. 17–18

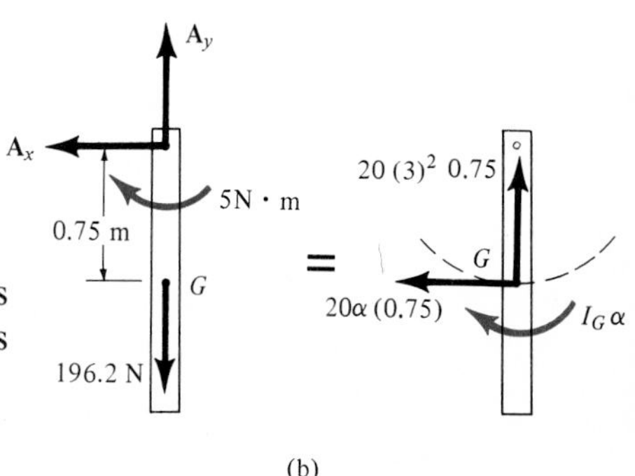

Solution

Free-Body and Kinetic Diagrams. The free-body and kinetic diagrams for the rod are shown, Fig. 17–18*b*. On the kinetic diagram $I_G\boldsymbol{\alpha}$ acts clockwise and $m(\mathbf{a}_G)_t$ acts to the left in accordance with $\boldsymbol{\alpha}$.

The moment of inertia of the rod about point A is

$$I_A = \tfrac{1}{3}ml^2 = \tfrac{1}{3}(20)(1.5)^2 = 15 \text{ kg} \cdot \text{m}^2$$

Equations of Motion. We have

$$\overset{+}{\leftarrow}\Sigma F_x = m(a_G)_x; \qquad A_x = 20\alpha(0.75) \qquad (1)$$

$$+\uparrow \Sigma F_y = m(a_G)_y; \qquad A_y - 196.2 = 20(3)^2(0.75) \qquad (2)$$

Moments will be summed about A in order to eliminate the reactive forces there.

$$\circlearrowright +\Sigma M_A = I_A\alpha; \qquad 5 = 15\alpha \qquad (3)$$

We can also use Eq. 17–14 by accounting for the "kinetic moments" of $I_G\boldsymbol{\alpha}$ and $m(\mathbf{a}_G)_t$ about A, i.e.,

$$\circlearrowright +\Sigma M_A = \Sigma(M_k)_A; \qquad 5 = [\tfrac{1}{12}(20)(1.5)^2]\alpha + [20\alpha(0.75)]0.75 \qquad (4)$$

Furthermore, moments about G could be applied, i.e.,

$$\circlearrowright +\Sigma M_G = I_G\alpha; \qquad 5 - A_x(0.75) = [\tfrac{1}{12}(20)(1.5)^2]\alpha \qquad (5)$$

Either one of the above three moment equations can be used with Eq. (1) to obtain the solution. Obviously, Eq. (3) yields the simplest solution. We have

$$\alpha = 0.333 \text{ rad/s}^2$$

$$A_x = 5.0 \text{ N} \qquad \textit{Ans.}$$

$$A_y = 331.2 \text{ N} \qquad \textit{Ans.}$$

Problems

17–34. A 500-mm-diameter flywheel has a mass of 100 kg and is free to rotate about its center. An inextensible cord is wrapped around the rim of the flywheel, and when a 10-kg block is attached to the cord, the block attains a speed of 200 mm/s after moving downward a distance of 0.5 m starting from rest. Determine the radius of gyration of the flywheel about its central axis.

17–35. The 15-lb rod is pinned at its end and has an angular velocity of $\omega = 5$ rad/s when it is in the horizontal position shown. Determine the rod's angular acceleration and the pin reactions at this instant.

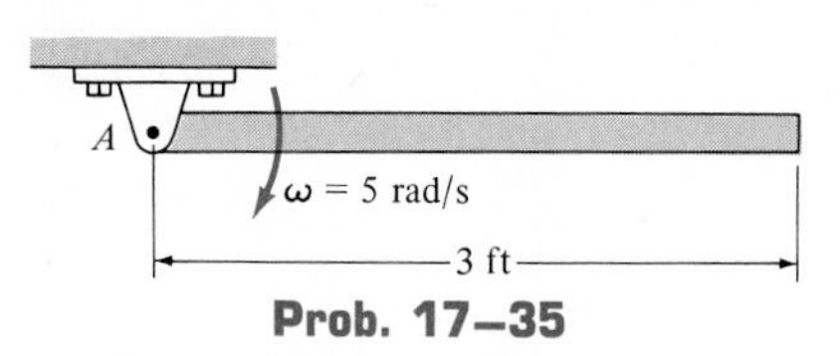

Prob. 17–35

***17–36.** The 10-kg wheel has a radius of gyration of $k_A = 225$ mm and is subjected to a moment of $M = 4$ N · m as shown. Determine its angular velocity in $t = 2$ s, starting from rest, and compute the reactions at the pin A.

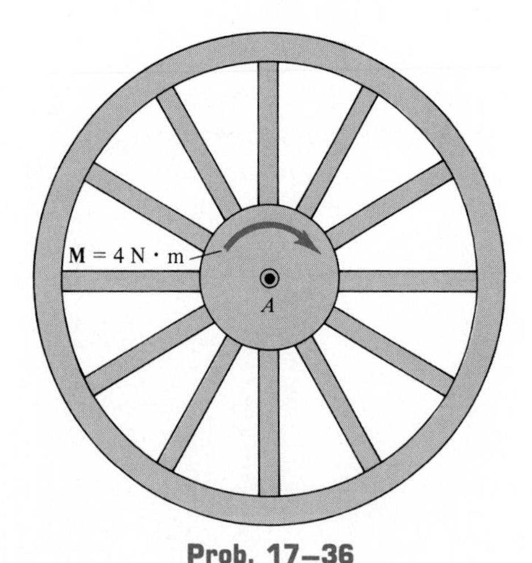

Prob. 17–36

17–37. A cord is wrapped around the inner core of a spool. If the cord is pulled with a constant tension of 30 lb and the spool is originally at rest, determine the spool's angular velocity when $s = 8$ ft of cord have unraveled. Neglect the weight of the cord. The spool and cord have a total weight of 400 lb and the radius of gyration about the axle A is $k_A = 1.30$ ft.

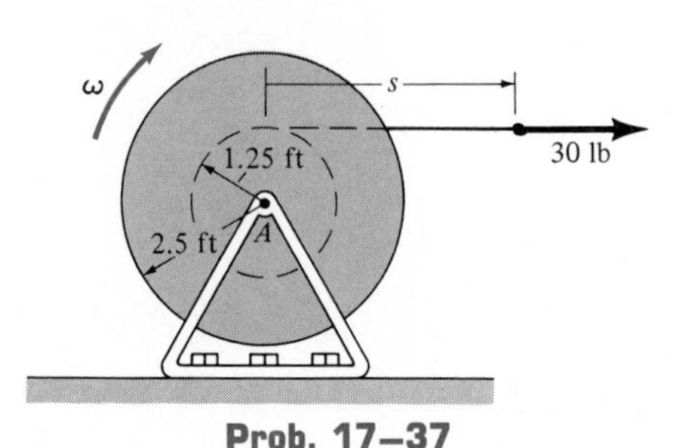

Prob. 17–37

17–38. Determine the force $\mathbf{T}_A$ which must be applied to the cable at A in order to give the 10-kg block B an upward acceleration of 200 mm/s^2. Assume that the cable does not slip over the surface of the 20-kg disk. Compute the tension in the vertical segment of the cord that supports the block and explain why this tension is different from that at A. The disk is pinned at its center C and is free to rotate.

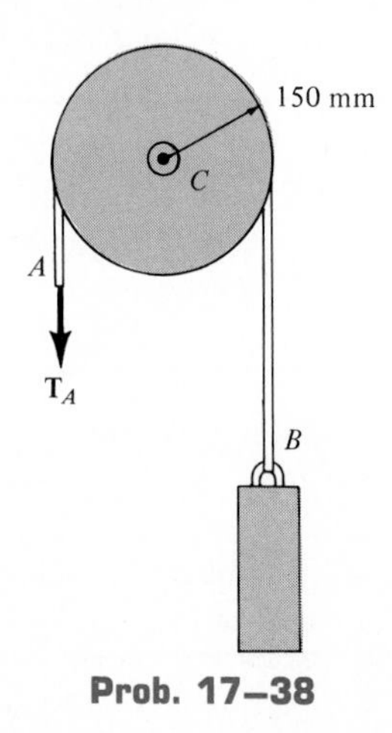

Prob. 17–38

17–39. The pendulum consists of a 20-lb sphere and a 5-lb slender rod. Compute the reaction at the pin O just after it is released from the position shown.

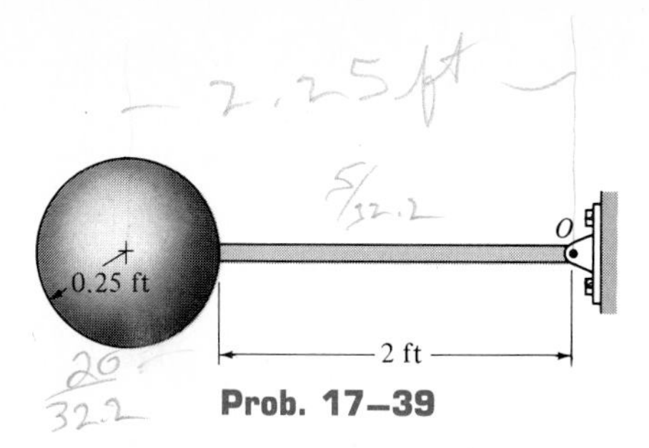

Prob. 17–39

***17–40.** If the support at B is suddenly removed, determine the initial reactions at the pin A. The plate has a weight of 30 lb. *Hint:* At this instant $\omega = 0$.

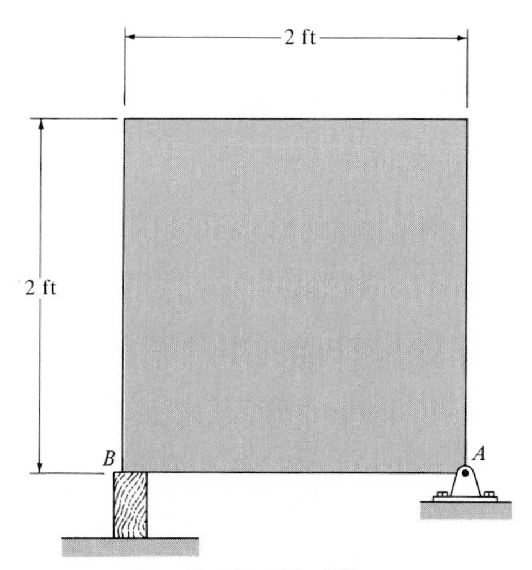

Prob. 17–40

17–41. Determine the angular acceleration of the 25-kg diving board and the horizontal and vertical components of reaction at the pin A the instant the man jumps off. Assume that the board is uniform and rigid and that at the instant he jumps off the spring is compressed a maximum amount of 200 mm and $\boldsymbol{\omega} = \mathbf{0}$.

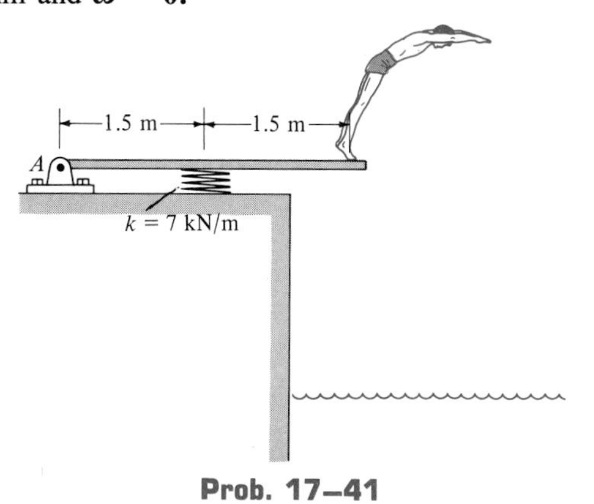

Prob. 17–41

17–42. The disk has a mass of 20 kg and is originally spinning at the end of the strut with an angular velocity of $\omega = 60$ rad/s. If it is then placed against the wall, for which $\mu_A = 0.3$, determine the time required for the motion to stop. What is the force in strut BC during this time?

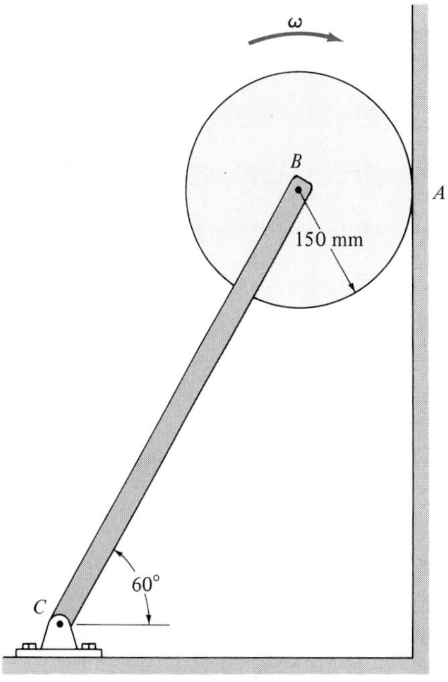

Prob. 17–42

17–43. The kinetic diagram representing the general rotational motion of a rigid body about a fixed axis is shown in the figure. Show that $I_G\boldsymbol{\alpha}$ may be eliminated by moving the vectors $m(\mathbf{a}_G)_t$ and $m(\mathbf{a}_G)_n$ to point P, located a distance $r_{GP} = k_G^2/r_{OG}$ from the center of mass G of the body. Here k_G represents the radius of gyration of the body about G. The point P is called the *center of percussion* of the body.

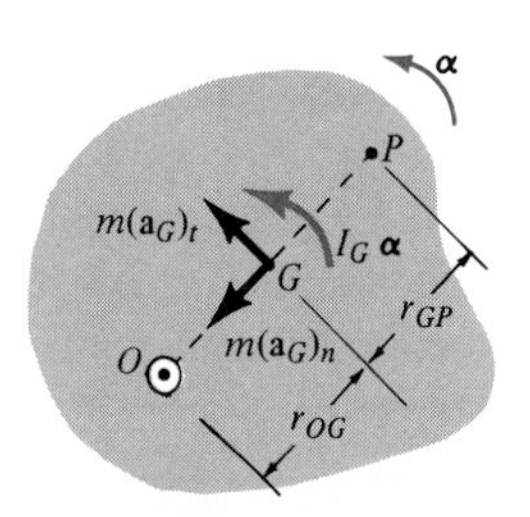

Prob. 17–43

***17–44.** If $\bar{y} = 0.393$ m and $I_G = 0.0773$ kg · m^2, determine the distance h from the center of the grip O to the center of percussion P of the baseball bat. The bat has a mass of 1.65 kg. When the bat strikes a ball at the center of percussion, no stinging effect is felt in the hands of the batter at O. Explain why this is so. *Hint:* See Prob. 17–43 and assume that the point of rotation is at O.

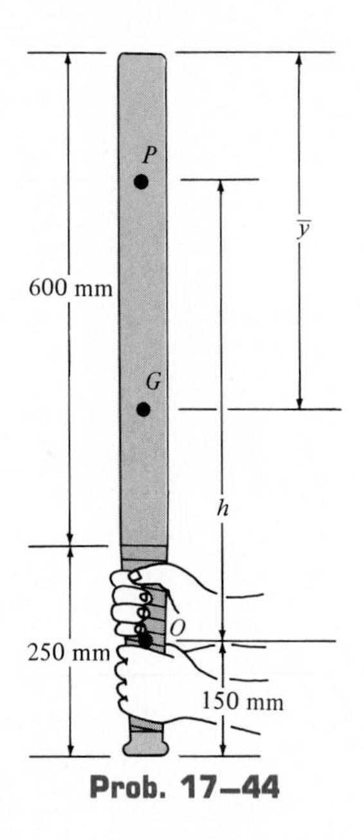

Prob. 17–44

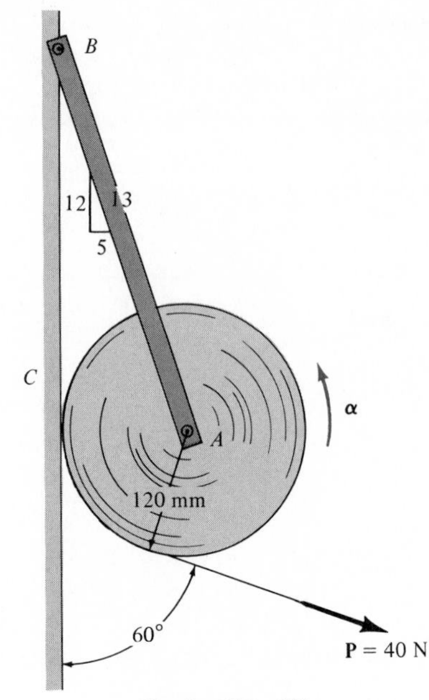

Prob. 17–45

17–45. A 20-kg roll of paper, originally at rest, is pin-supported at its ends to bracket AB. If the roll rests against a wall for which $\mu_C = 0.3$ and a force of 40 N is applied uniformly to the end of the sheet, determine the initial angular acceleration of the roll and the tension in the bracket as the paper unwraps. For the calculation, treat the roll as a cylinder.

17–46. The lid of the vessel has a mass of 15 kg and a radius of gyration about its mass center G of $k_G = 0.2$ m. In order to raise the lid, an operator applies a vertical force of $F = 800$ N at B. Determine the lid's initial angular acceleration and the horizontal and vertical components of reaction which the lid exerts on the hinge at A at the instant it begins to open. The hinge at A is connected to the vessel. Initially $\omega = 0$.

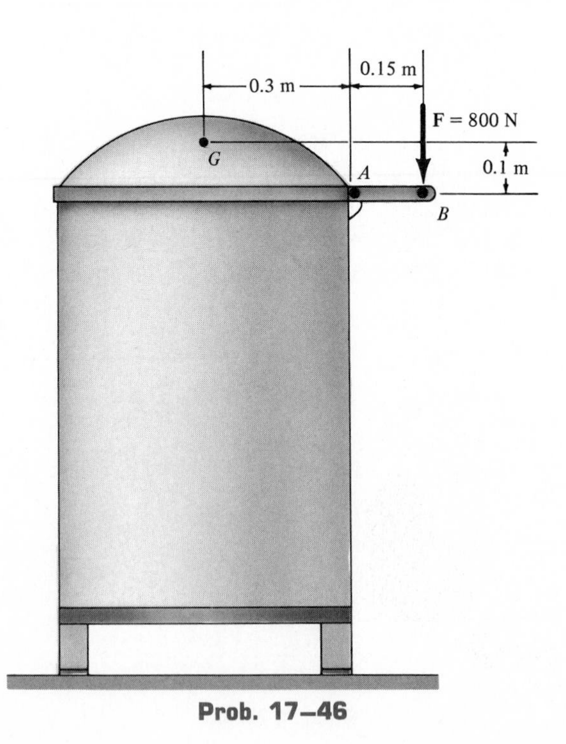

Prob. 17–46

17–47. Disk A has a weight of 5 lb and disk B has a weight of 10 lb. If no slipping occurs between them, determine the couple $\mathbf{M}$ which must be applied to disk A to give it an angular acceleration of 4 rad/s^2.

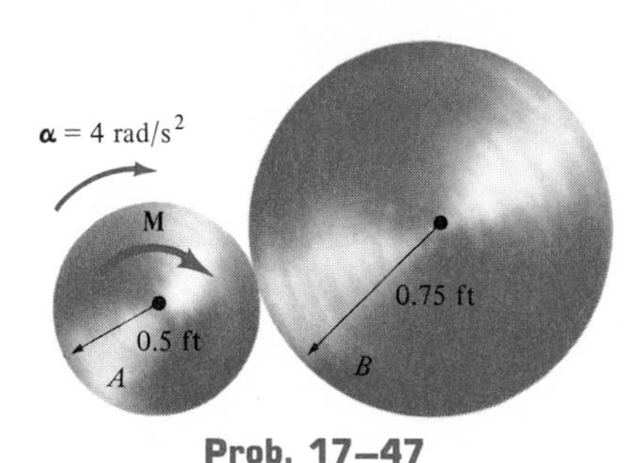

Prob. 17–47

***17–48.** As a man enters a room he pushes on the 30-lb smooth door with a constant force of $F = 3$ lb which always remains perpendicular to the face of the door. If the door is originally at rest when the force is applied at A, determine the door's angular velocity in $t = 1$ s. The moment of inertia of the door about the z axis is $I_z = 3.75$ slug · ft^2. How far has the door rotated during this time?

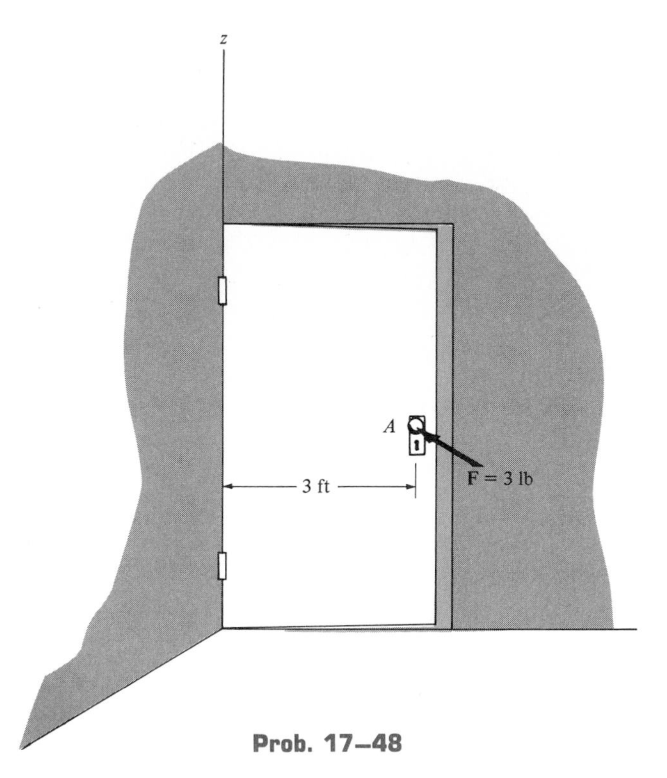

Prob. 17–48

17–49. The relay switch consists of an electromagnet E and a 20-g armature AB (slender bar) which is pinned at A and lies in the vertical plane. When the current is turned off, the armature is held open against the smooth stop at B by the spring CD, which exerts a vertical force on the armature at C of $F_s = 0.85$ N. When the current is turned on, the electromagnet attracts the armature at E with a vertical force of $F = 0.8$ N. Determine the initial angular acceleration of the armature when the contact BF begins to close.

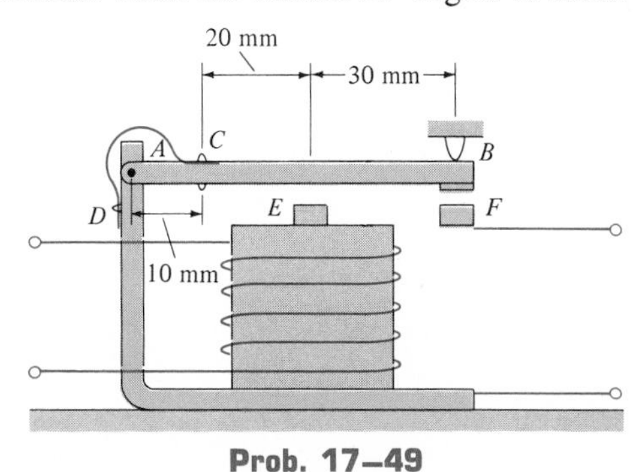

Prob. 17–49

17–50. Cable is unwound from a spool supported on small rollers at A and B by exerting a force of $T = 300$ N on the cable in the direction shown. Compute the time needed to unravel 5 m of cable from the spool if the spool and cable have a total mass of 600 kg and a centroidal radius of gyration of $k_O = 1.2$ m. For the calculation, neglect the mass of the cable being unwound and the mass of the rollers at A and B. The rollers turn with no friction.

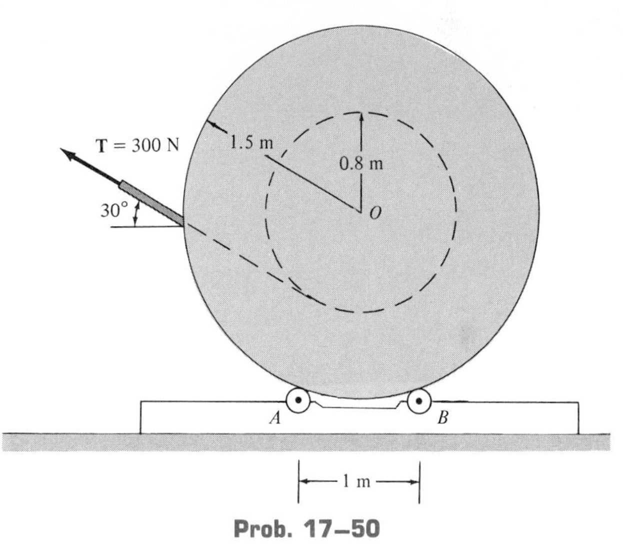

Prob. 17–50

17–51. The armature (slender rod) AB has a mass of 0.2 kg and can pivot about the pin at A. Movement is controlled by the electromagnet E, which exerts an initial horizontal attractive force on the armature at B of $F_B = 4$ N. If the armature lies in the *horizontal plane,* and is originally at rest, determine the initial angular acceleration of the armature at the instant the gap beings to close.

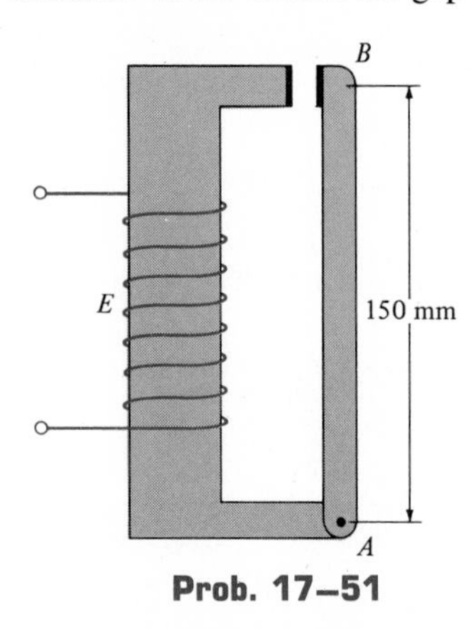

Prob. 17–51

***17–52.** In order to experimentally determine the moment of inertia I_G of a 3-kg connecting rod, the rod is suspended horizontally at A by a cord and at B by a piezoelectric sensor, an instrument used for measuring force. Under these equilibrium conditions, the force at B is measured as 14.6 N. If at the instant the cord is cut the reaction at B is measured as 9.3 N, determine the value of I_G. The support at B does not move when the measurement is taken. For the calculation the horizontal location of G must be determined.

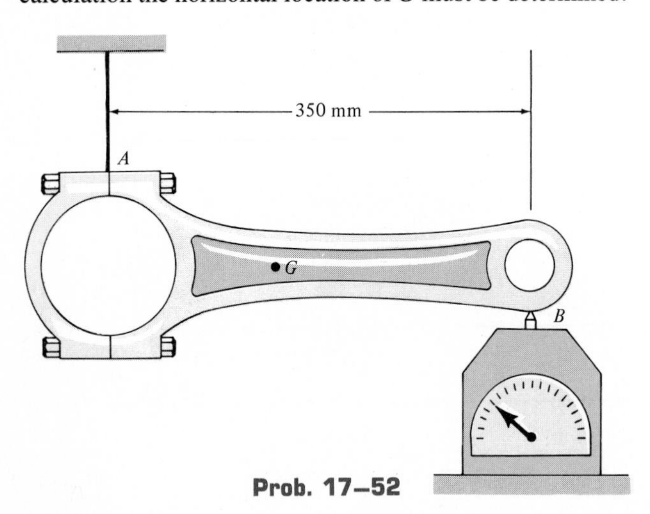

Prob. 17–52

17–53. Two blocks A and B, having a weight of 10 lb and 5 lb, respectively, are attached to the ends of a cord which passes over a 3-lb disk (pulley). If the blocks are released from rest, determine their speed in $t = 0.5$ s. The cord does not slip on the pulley. *Suggestion:* Analyze the "system" consisting of both the blocks and the pulley.

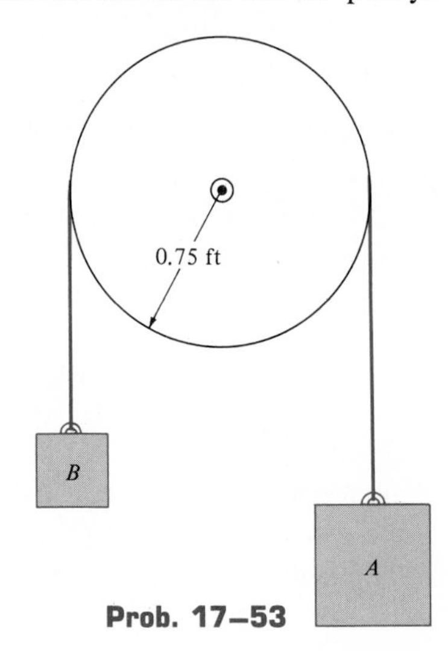

Prob. 17–53

17–54. A clown, mounted on stilts, loses his balance and falls backward. Paralyzed with fear, he remains *rigid* as he falls. His mass including the stilts is 80 kg, the mass center is at G, and the radius of gyration about G is $k_G = 1.2$ m. Determine the frictional force between his shoes and the ground at A when $\theta = 30°$, if at this instant he is rotating clockwise at $\omega = 1.5$ rad/s. What is his angular acceleration?

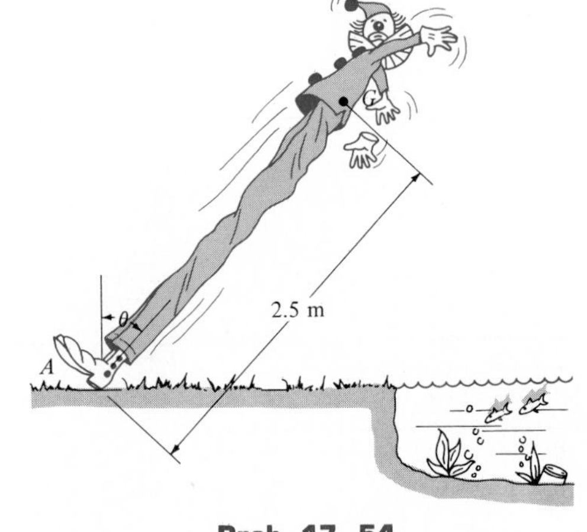

Prob. 17–54

17–55. The "Catherine wheel" is a firework that consists of a coiled tube of powder which is pinned at its center. If the powder burns such that the exhaust gases always exert a force having a constant magnitude of 0.3 N directed tangent to the wheel, determine the angular velocity when $t = 1.5$ s starting from rest. Take $r = 75$ mm and assume that no mass is lost during the "short" time of powder burning. For the calculation, consider the wheel to be a thin disk having a mass of 0.1 kg.

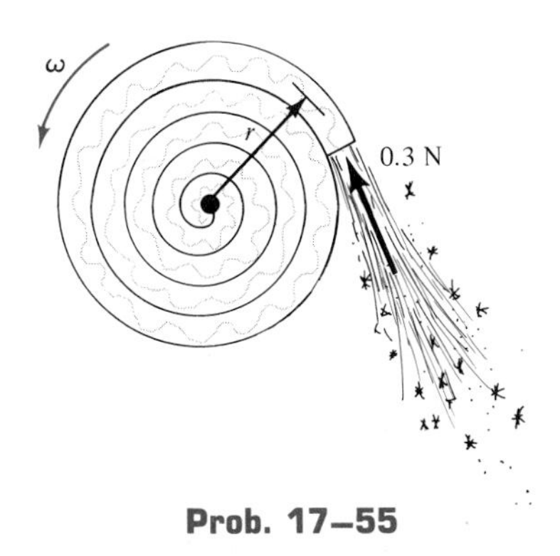

Prob. 17–55

***17–56.** A 2-Mg hippopotamus H is being lifted over its water hole using a gantry which has a mass of 300 kg, a center of mass at G, and a radius of gyration about G of $k_G = 2.4$ m. In order to prevent the gantry from overturning, a man having a mass of 70 kg stands directly over the cross-bar frame at A. By a remote control device, the distance x is slowly increased until the gantry is on the verge of tipping. If the man panics and jumps off the gantry at this instant, determine the initial acceleration of the hippo. Assume that the wheels at B are prevented from slipping, i.e., so that rotation occurs about B. *Note:* The hippo will accelerate downward, whereas point G begins to move in a circular path with center at B.

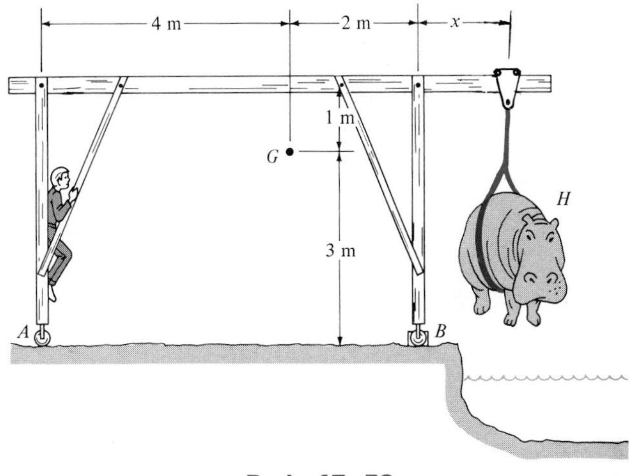

Prob. 17–56

17.5 Equations of Motion: General Plane Motion

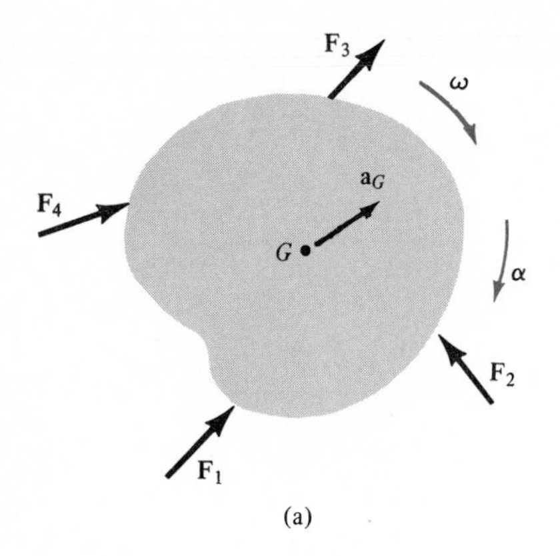

The rigid body (or slab) shown in Fig. 17–19*a* is subjected to general plane motion caused by the externally applied force system. The free-body and kinetic diagrams for the body are shown in Fig. 17–19*b*. The vector $m\mathbf{a}_G$ (shown dashed) has the same *direction* as the acceleration of the body's mass center, and $I_G\boldsymbol{\alpha}$ acts in the same *direction* as the angular acceleration. If an x and y inertial coordinate system is chosen as shown, the three equations of motion may be written as

$$\begin{aligned} \Sigma F_x &= m(a_G)_x \\ \Sigma F_y &= m(a_G)_y \\ \Sigma M_G &= I_G\alpha \end{aligned} \tag{17–16}$$

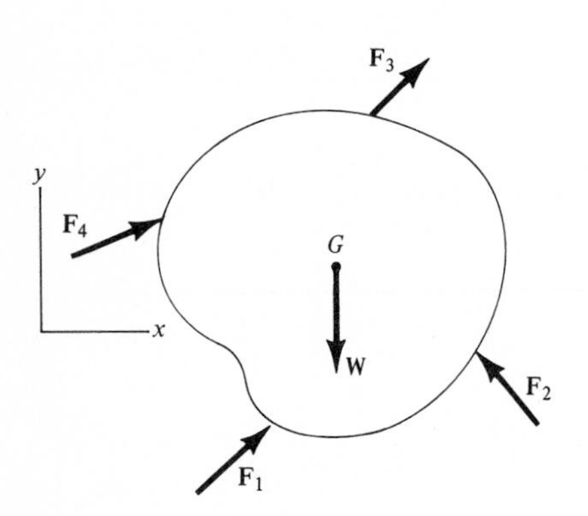

In some problems it may be convenient to sum moments about some point A other than G. This is usually done in order to eliminate unknown forces from the moment summation. When used in this more general sense, the three equations of motion become

$$\begin{aligned} \Sigma F_x &= m(a_G)_x \\ \Sigma F_y &= m(a_G)_y \\ \Sigma M_A &= \Sigma(\mathcal{M}_k)_A \end{aligned} \tag{17–17}$$

where $\Sigma(\mathcal{M}_k)_A$ represents the moment sum of $I_G\boldsymbol{\alpha}$ and $m\mathbf{a}_G$ (or its components) about A as determined by the data on the kinetic diagram.

=

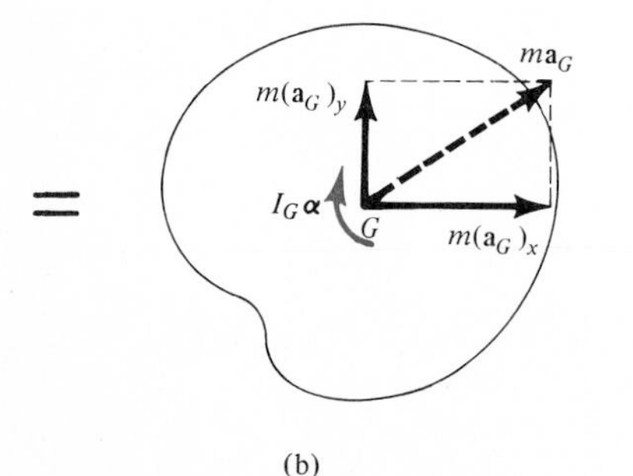

(b)

Fig. 17–19

PROCEDURE FOR ANALYSIS

The following procedure provides a method for solving kinetic problems involving general plane motion of a rigid body.

Free-Body and Kinetic Diagrams. Draw the free-body and kinetic diagrams for the body, Fig. 17–19*b*, and compute the mass moment of inertia I_G.

Equations of Motion. Apply the three equations of motion, Eqs. 17–16 or 17–17, with reference to an established x, y inertial coordinate system.

Kinematics. Use kinematics if a complete solution cannot be obtained strictly from the equations of motion. In particular, if the body's motion is constrained due to its supports, additional equations may be obtained by using $\mathbf{v}_B = \mathbf{v}_A + \mathbf{v}_{B/A}$ and $\mathbf{a}_B = \mathbf{a}_A + \mathbf{a}_{B/A}$, which relate the motions of any two points A and B on the body (see Example 17–16).

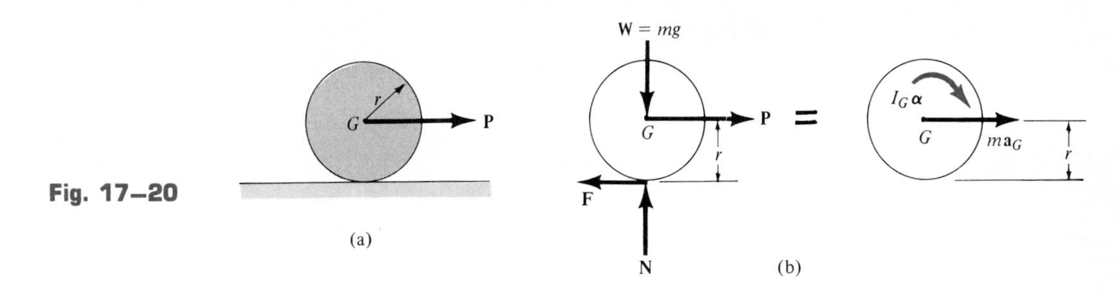

Fig. 17–20

Frictional Rolling Problems. There is a class of planar kinetics problems which deserves special mentioning. These problems involve wheels, cylinders, or bodies of similar shape, which roll on a *rough* plane surface. Because of the applied loadings, it may not be known if the body *rolls without slipping,* or if it *slides as it rolls*. For example, consider the homogeneous disk shown in Fig. 17–20*a*, which has a mass m and is subjected to a known horizontal force **P.** Following the procedure outlined above, the free-body and kinetic diagrams are shown in Fig. 17–20*b*. Applying the three equations of motion yields

$$\overset{+}{\rightarrow}\Sigma F_x = m(a_G)_x; \qquad P - F = ma_G \tag{17–18}$$

$$+\uparrow \Sigma F_y = m(a_G)_y; \qquad N - mg = 0 \tag{17–19}$$

$$\circlearrowright + \Sigma M_G = I_G\alpha; \qquad Fr = I_G\alpha \tag{17–20}$$

A fourth equation is needed since these *three equations* contain *four unknowns: F, N,* α, and a_G.

No Slipping. If the frictional force **F** is great enough to allow the disk to roll *without slipping,* then a_G may be related to α by the *kinematic equation**

$$a_G = \alpha r \tag{17–21}$$

The four unknowns are determined by *solving simultaneously* Eqs. 17–18 to 17–21. When the solution is obtained, the assumption of no slipping must be *checked*. In this regard, recall that no slipping occurs provided $F \leqslant \mu_s N$, where μ_s is the static coefficient of friction. If the inequality is satisfied, the problem is solved. However, if $F > \mu_s N$, the problem must be *reworked,* since then the disk slips as it rolls.

Slipping. In the case of slipping, α and a_G are *independent of one another* so that Eq. 17–21 does not apply. Instead, the magnitude of the frictional force is related to the magnitude of the normal force using the coefficient of kinetic friction μ_k, i.e.,

$$F = \mu_k N \tag{17–22}$$

Here **F** acts to the left on the disk to prevent the slipping motion to the right. In this case Eqs. 17–18 to 17–20 and 17–22 are used for the solution. Examples 17–14 and 17–15 illustrate these concepts numerically.

*Note that the contact point at the ground is the instantaneous center of zero velocity; it does *not* have zero acceleration. Equation 17–21 is derived in Example 16–3.

Example 17–13

The spool in Fig. 17–21a has a mass of 8 kg and a radius of gyration of $k_G = 0.35$ m. If cords of negligible mass are wrapped around its inner hub and outer rim as shown, determine the spool's angular velocity 3 s after it is released from rest.

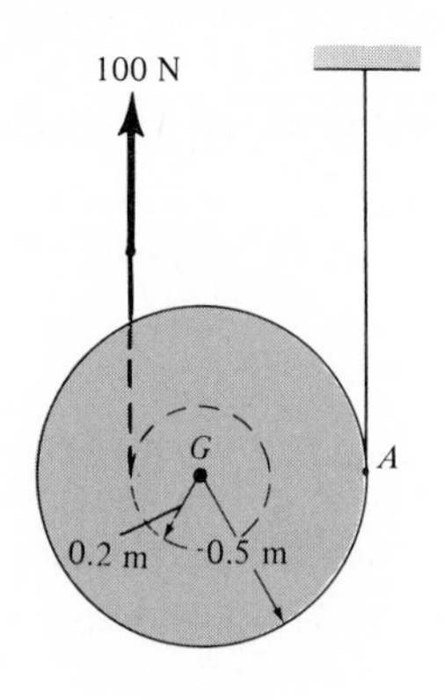

(a)

Solution

Free-Body and Kinetic Diagrams. Since no forces act on the spool in the horizontal direction, $m\mathbf{a}_G$ acts upward, and consequently, $I_G\boldsymbol{\alpha}$ acts clockwise, Fig. 17–21b.

The moment of inertia of the spool about its mass center is

$$I_G = mk_G^2 = 8(0.35)^2 = 0.980 \text{ kg} \cdot \text{m}^2$$

Equations of Motion

$$+\uparrow \Sigma F_y = m(a_G)_y; \qquad T + 100 - 78.48 = 8a_G \qquad (1)$$

$$\circlearrowright + \Sigma M_G = I_G\alpha; \qquad 100(0.2) - T(0.5) = 0.980\alpha \qquad (2)$$

Moments may also be computed about point A in order to eliminate the unknown T. We have

$$\circlearrowright + \Sigma M_A = \Sigma(\mathcal{M}_k)_A; \qquad 100(0.7) - 78.48(0.5) = 0.980\alpha + (8a_G)(0.5) \qquad (3)$$

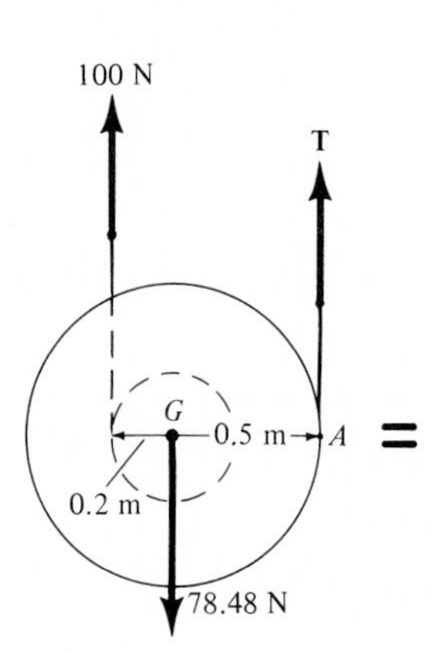

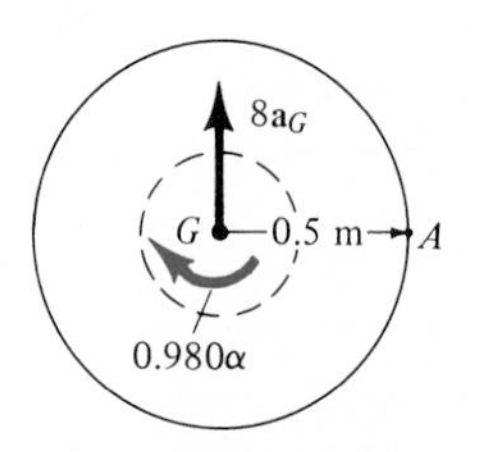

(b)

Fig. 17–21

Kinematics. A complete solution is obtained if kinematics is used to relate a_G to α. In this case the spool "rolls without slipping" on the cord at A. Hence, we can use the results of Example 16–3, so that

$$\circlearrowright + a_G = r\alpha; \qquad a_G = 0.5\alpha \qquad (4)$$

Solving Eqs. (1), (2), and (4) or the simpler set Eqs. (1), (3), and (4), we have

$$\alpha = 10.3 \text{ rad/s}^2$$
$$a_G = 5.16 \text{ m/s}^2$$
$$T = 19.8 \text{ N}$$

Since α is constant, the angular velocity in 3 s is

$$\circlearrowright + \qquad \omega = \omega_0 + \alpha_c t$$
$$= 0 + 10.3(3)$$
$$= 30.9 \text{ rad/s} \qquad \textit{Ans.}$$

Example 17–14

The 50-lb wheel shown in Fig. 17–22*a* has a radius of gyration of $k_G = 0.70$ ft. If a 35-lb · ft couple moment is applied to the wheel, determine the acceleration of its mass center G. The static and kinetic coefficients of friction between the wheel and the plane at A are $\mu_s = 0.3$ and $\mu_k = 0.25$, respectively.

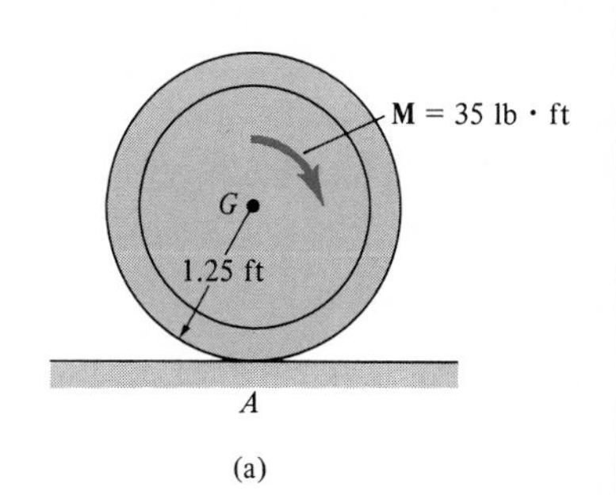

(a)

Fig. 17–22

Solution

Free-Body and Kinetic Diagrams. By inspection, it is seen that the frictional force $\mathbf{F}_A$, acting to the right, gives the wheel its forward acceleration, Fig. 17–22*b*.

The moment of inertia is

$$I_G = mk_G^2 = \frac{50}{32.2}(0.70)^2 = 0.761 \text{ slug} \cdot \text{ft}^2$$

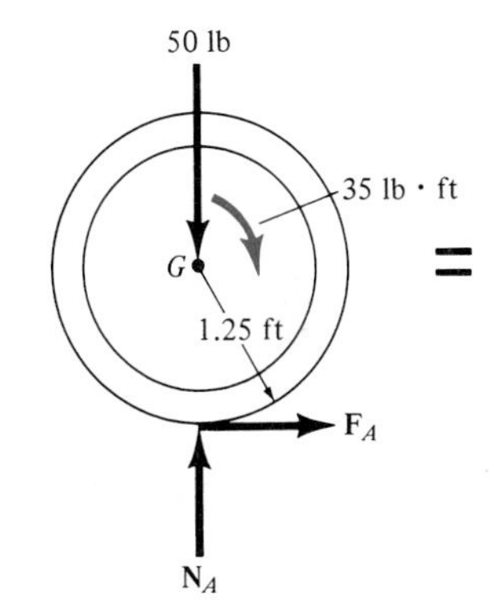

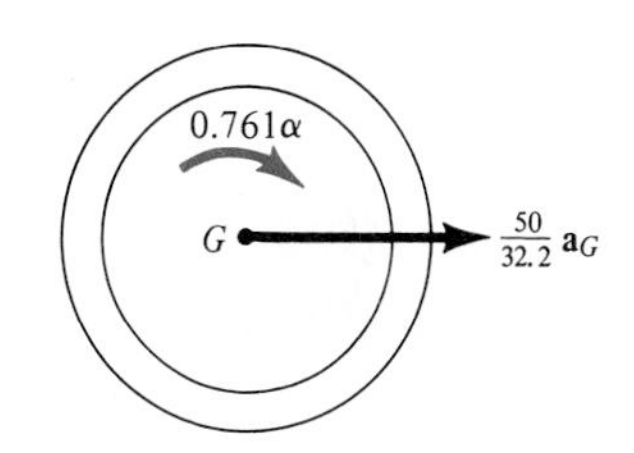

(b)

Equations of Motion

$$\overset{+}{\rightarrow}\Sigma F_x = m(a_G)_x; \qquad F_A = \frac{50}{32.2}a_G \qquad (1)$$

$$+\uparrow \Sigma F_y = m(a_G)_y; \qquad N_A - 50 = 0 \qquad (2)$$

$$\circlearrowright +\Sigma M_G = I_G\alpha; \qquad 35 - 1.25(F_A) = 0.761\alpha \qquad (3)$$

Kinematics (No Slipping). If this assumption is made then

$$a_G = 1.25\alpha \qquad (4)$$

Solving Eqs. (1) to (4),

$$N_A = 50.0 \text{ lb} \qquad F_A = 21.3 \text{ lb}$$
$$\alpha = 11.0 \text{ rad/s}^2 \qquad a_G = 13.7 \text{ ft/s}^2$$

The original assumption of no slipping requires $F_A \leq \mu_s N_A$. However, since 21.3 lb > 0.3(50) = 15 lb, the wheel slips as it rolls.

(Slipping.) This requires $F_A = \mu_k N_A$, or

$$F_A = 0.25N_A \qquad (5)$$

Solving Eqs. (1) to (3) and (5) yields

$$N_A = 50.0 \text{ lb} \qquad F_A = 12.5 \text{ lb}$$
$$\alpha = 25.5 \text{ rad/s}^2$$
$$a_G = 8.0 \text{ ft/s}^2 \qquad \textit{Ans.}$$

Example 17–15

The uniform slender beam shown in Fig. 17–23*a* has a mass of 100 kg and a moment of inertia $I_G = 75\text{ kg}\cdot\text{m}^2$. If the coefficients of static and kinetic friction between the end of the beam and the surface are both equal to $\mu_A = 0.25$, determine the beam's angular acceleration at the instant the 400-N horizontal force is applied. The beam is originally at rest.

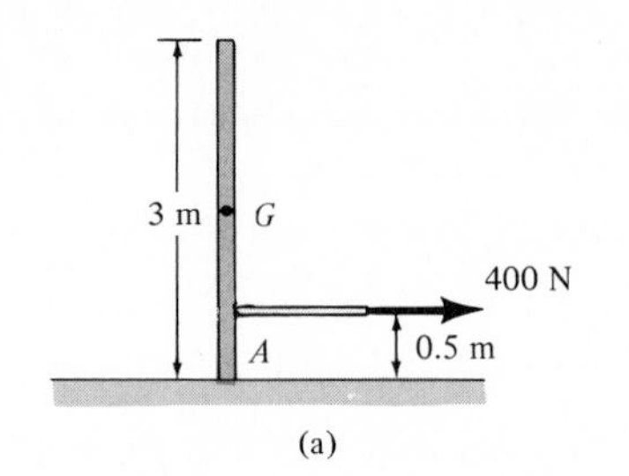

(a)

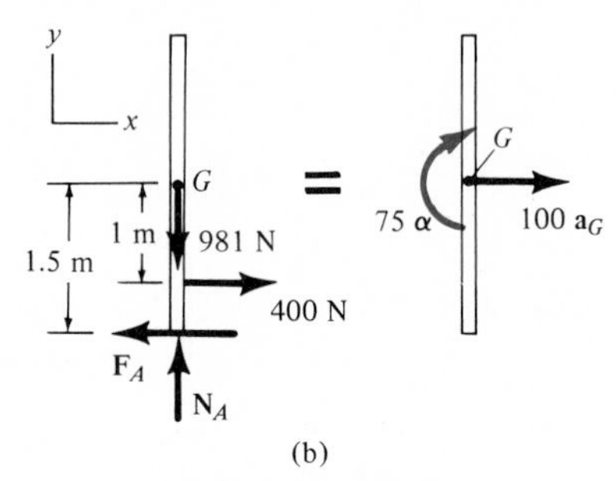

(b)

Fig. 17–23

Solution

Free-Body and Kinetic Diagrams. Figure 17–23*b*. The path of motion of the mass center *G* will be along an unknown curved path having a radius of curvature which is initially parallel to the *y* axis. There is no normal or *y* component of acceleration since the beam is originally at rest, i.e., $\mathbf{v}_G = \mathbf{0}$, so that $a_y = v_G^2/\rho = 0$.

Equations of Motion

$$\overset{+}{\rightarrow}\Sigma F_x = m(a_G)_x; \qquad 400 - F_A = 100a_G \qquad (1)$$

$$+\uparrow \Sigma F_y = m(a_G)_y; \qquad N_A - 981 = 0 \qquad (2)$$

$$\circlearrowright + \Sigma M_G = I_G\alpha; \qquad F_A(1.5) - 400(1) = 75\alpha \qquad (3)$$

A fourth equation is needed for a complete solution.

Kinematics (No Slipping). In this case point *A* acts as a "pivot" so that

$$\circlearrowright + a_G = \alpha r_{AG}; \qquad a_G = 1.5\alpha \qquad (4)$$

Solving Eqs. (1) to (4) yields

$$N_A = 981\text{ N} \qquad F_A = 300\text{ N}$$

$$a_G = 1.0\text{ m/s}^2 \qquad \alpha = 0.667\text{ rad/s}^2$$

Testing the original assumption of no slipping requires $F_A < \mu_A N_A$. However, $300 > 0.25(981) = 245.3$ N.

(Slipping.) For this case Eq. (4) does *not* apply. Instead, the frictional equation $F_A = \mu_A N_A$ is used, i.e.,

$$F_A = 0.25N_A \qquad (5)$$

Solving Eqs. (1) to (3) and (5) simultaneously yields

$$N_A = 981\text{ N} \qquad F_A = 245.3\text{ N} \qquad a_G = 1.55\text{ m/s}^2$$

$$\alpha = -0.427\text{ rad/s}^2 \qquad \textit{Ans.}$$

Because of the negative sign, the angular acceleration of the beam is counterclockwise.

Example 17–16

The 30-kg wheel shown in Fig. 17–24*a* has a mass center at G and a radius of gyration of $k_G = 0.15$ m. If the wheel is originally at rest and released from the position shown, determine its angular acceleration. *No slipping occurs.*

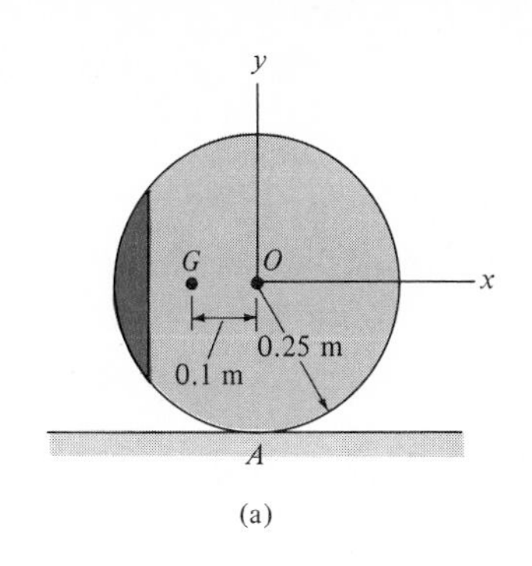

(a)

Solution

Free-Body and Kinetic Diagrams. Since the *path of motion* of G is *unknown,* the two components $m(\mathbf{a}_G)_x$ and $m(\mathbf{a}_G)_y$ must be shown on the kinetic diagram, Fig. 17–24*b*.

The moment of inertia is

$$I_G = mk_G^2 = 30(0.15)^2 = 0.675 \text{ kg} \cdot \text{m}^2$$

Equation of Motion. The unknown normal and frictional forces can be eliminated from the analysis by applying the rotational equation of motion about point A.

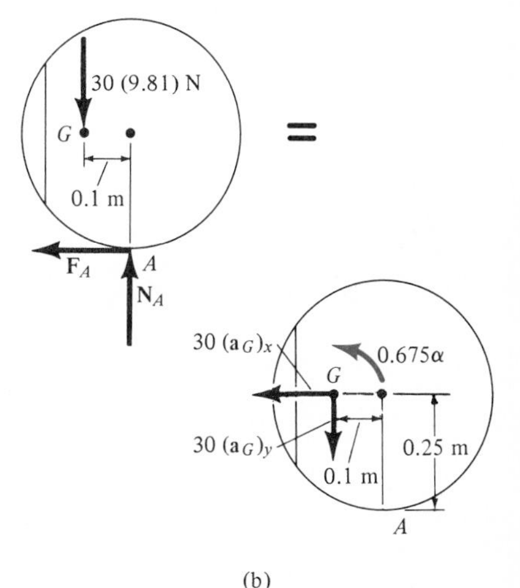

(b)

$\circlearrowleft +\Sigma M_A = \Sigma(M_k)_A;$

$$30(9.81)(0.1) = 0.675\alpha + 30(a_G)_x(0.25) + 30(a_G)_y(0.1) \quad (1)$$

There are three unknowns in this equation: $(a_G)_x$, $(a_G)_y$, and α.

Kinematics. Using kinematics, $(a_G)_x$, $(a_G)_y$, and α can be related. Since no slipping occurs, $a_O = \alpha r = \alpha(0.25)$, directed to the left, Fig. 17–24*c*. Also, $\boldsymbol{\omega} = 0$ since the wheel is originally at rest. Using Eq. 16–12, with point O as the base point, Fig. 17–24*d*, we have

$$\mathbf{a}_G = \mathbf{a}_O + (\mathbf{a}_{G/O})_t + (\mathbf{a}_{G/O})_n$$

$$\underset{\leftarrow}{(a_G)_x} + \underset{\downarrow}{(a_G)_y} = \underset{\leftarrow}{\alpha(0.25)} + \underset{\downarrow}{\alpha(0.1)} + 0$$

Equating the respective horizontal and vertical components, we have

$$(a_G)_x = \alpha(0.25) \quad (2)$$

$$(a_G)_y = \alpha(0.1) \quad (3)$$

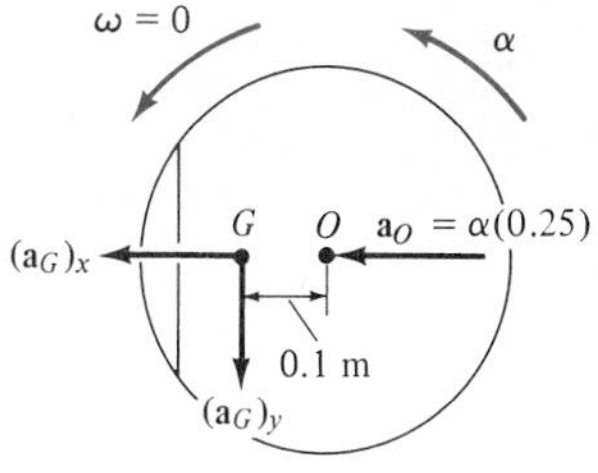

Absolute motion

(c)

Solving Eqs. (1) to (3) yields

$$\alpha = 10.33 \text{ rad/s}^2 \qquad \textit{Ans.}$$

$$(a_G)_x = 2.58 \text{ m/s}^2$$

$$(a_G)_y = 1.03 \text{ m/s}^2$$

As an exercise, show that $F_A = 77.4$ N and $N_A = 263.3$ N.

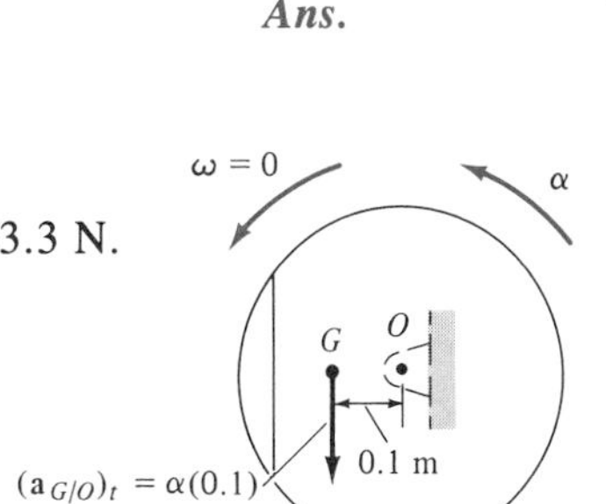

Relative motion

(d)

Fig. 17–24

Problems

17–57. If the disk in Fig. 17–20*a rolls without slipping,* show that when moments are summed about the instantaneous center of zero velocity, *IC*, it is possible to use the moment equation $\Sigma M_{IC} = I_{IC}\alpha$, where I_{IC} represents the moment of inertia of the disk calculated about the instantaneous axis of zero velocity.

17–58. The wheel has a weight of 30 lb, a radius of $r = 0.5$ ft, and a radius of gyration of $k_G = 0.23$ ft. If the wheel does not slip as it rolls, determine its angular acceleration and the normal and frictional forces at the incline. Set $\theta = 12°$.

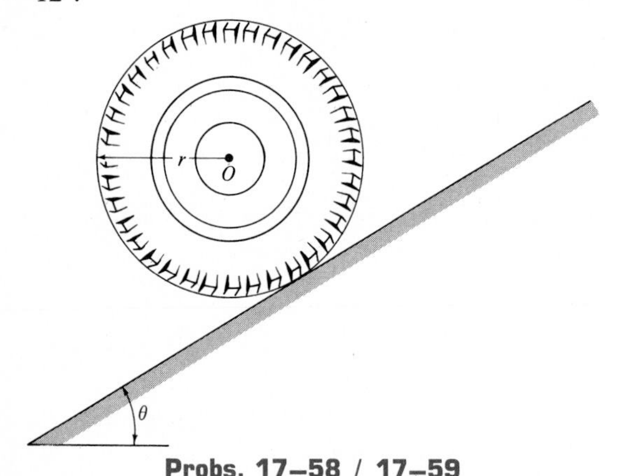

Probs. 17–58 / 17–59

17–59. Solve Prob. 17–58 with $\theta = 30°$.

***17–60.** Two men exert constant vertical forces of 40 lb and 30 lb at the ends *A* and *B* of a uniform plank which has a weight of 50 lb. If the plank is originally at rest, determine the acceleration of its center and its angular acceleration. Assume the plank to be a slender rod.

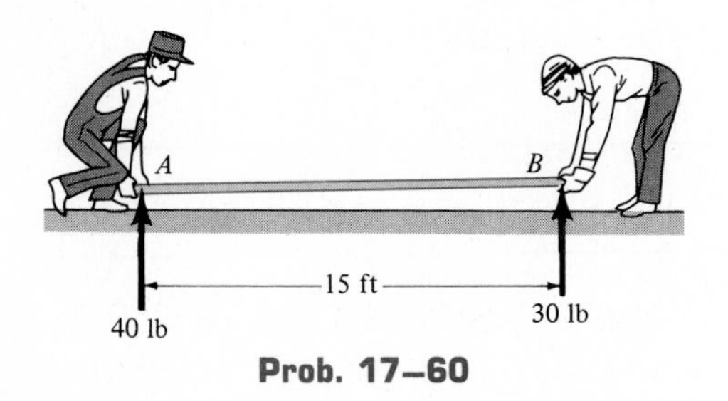

Prob. 17–60

17–61. The slender 200-kg beam is suspended by a cable at its end as shown. If a man pushes on its other end with a horizontal force of 30 N, determine the initial acceleration of its mass center *G*, the beam's angular acceleration, and the tension in the cable *AB*. The beam is originally at rest and can be considered a slender rod.

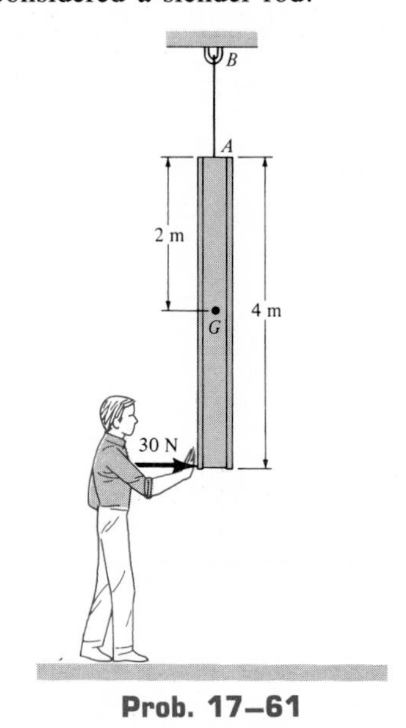

Prob. 17–61

17–62. A woman sits in a rigid position on her rocking chair by keeping her feet on the bottom rungs at *B*. At the instant shown, she has reached an extreme backward position and has zero angular velocity. Determine her forward

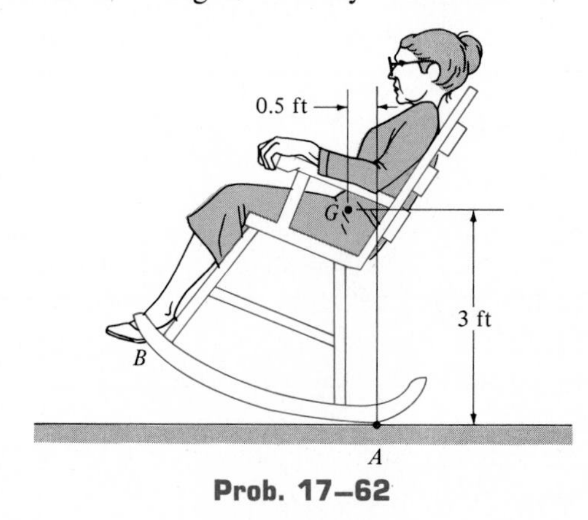

Prob. 17–62

angular acceleration $\boldsymbol{\alpha}$ and the frictional force at A necessary to prevent the rocker from slipping. The woman and the rocker have a combined weight of 180 lb and a radius of gyration about G of $k_G = 2.2$ ft.

17–63. A uniform rod having a weight of 10 lb is pin-supported at A from a roller which rides on a horizontal track. If the rod is originally at rest, and a horizontal force of $F = 15$ lb is applied to the roller, determine the rod's angular acceleration and the acceleration of the rod's mass center. Neglect the mass of the roller and its size d in the computations.

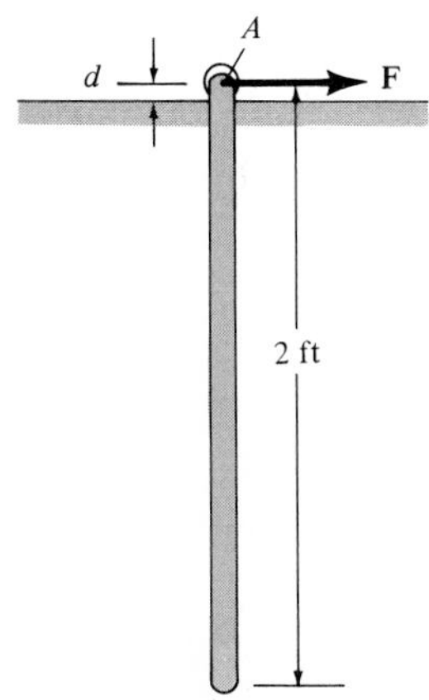

Probs. 17–63 / 17–64

***17–64.** Solve Prob. 17–63 assuming that the roller at A is replaced by a slider block having a negligible mass. The coefficient of friction between the block and the track is $\mu = 0.2$. Neglect the dimension d and the size of the block in the computations.

17–65. A bowling ball has a weight of 16 lb and a radius of 0.375 ft. It is cast horizontally onto an alley such that initially $\omega = 0$ and its mass center has a velocity of $v = 8$ ft/s. If the coefficient of friction between the floor and the ball is $\mu = 0.12$, determine the angular deceleration of the ball before it rolls without slipping. For the calculation, neglect the finger holes in the ball and assume the ball has a uniform density.

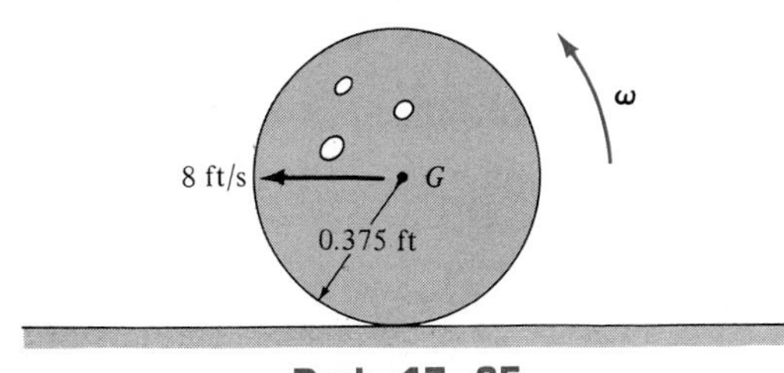

Prob. 17–65

17–66. By pressing down with the finger at B, a thin ring having a weight of 0.04 lb is given an initial velocity of 5 ft/s and a backspin of 40 rad/s when the finger is released. If the coefficient of friction between the table and the ring is $\mu = 0.1$, determine the angular deceleration of the ring before backspinning stops.

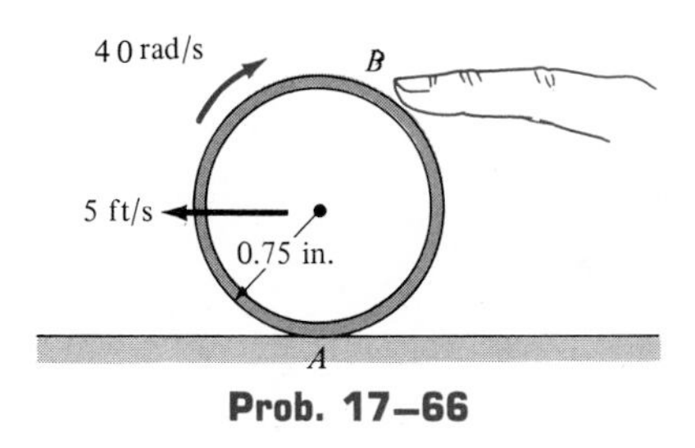

Prob. 17–66

17–67. The lawn roller has a mass of 80 kg and a radius of gyration of $k_G = 0.175$ m. If it is pushed forward with a force of 200 N when the handle is at 45°, determine its angular acceleration. Assume the roller does not slip at the ground.

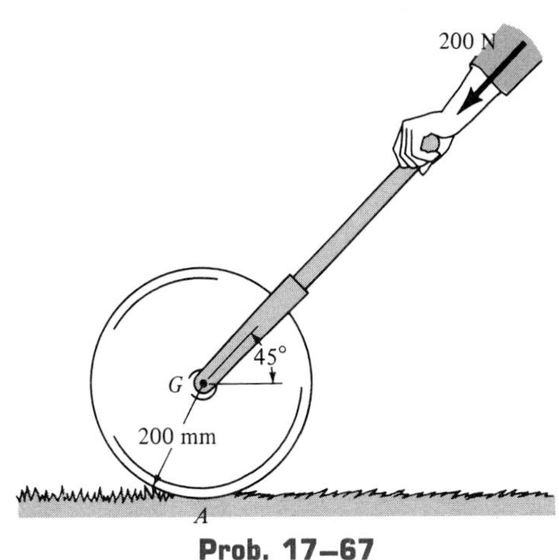

Prob. 17–67

***17–68.** The spool and wire wrapped around its core have a mass of 20 kg and a centroidal radius of gyration of $k_G = 250$ mm. If the coefficient of friction at the ground is $\mu_B = 0.1$, determine the angular acceleration of the spool when the 30-N · m couple is applied.

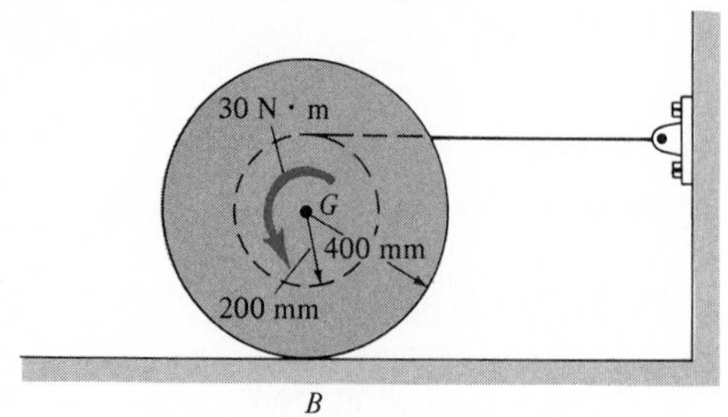

Prob. 17–68

17–69. A spool and the telephone wire wrapped around its core have a total weight of 80 lb and a radius of gyration of $k_G = 0.75$ ft. If the coefficient of friction between the spool and the ground is $\mu_A = 0.4$, determine the angular acceleration of the spool if the end of the cable is subjected to a horizontal force of 30 lb.

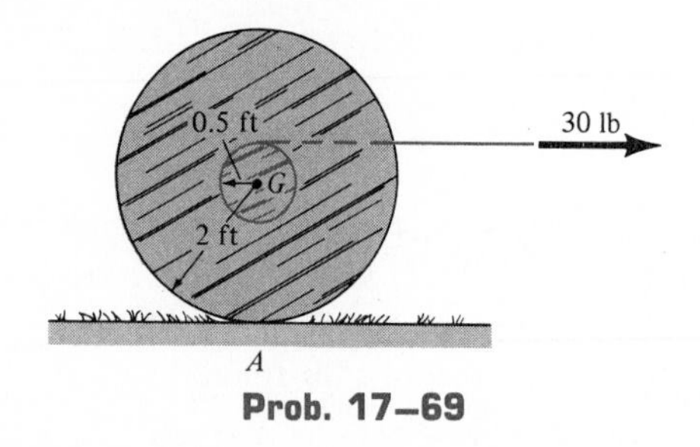

Prob. 17–69

17–70. The spool has a weight of 500 lb, an outer radius of $r = 2$ ft, and a centroidal radius of gyration of $k_G = 1.83$ ft. If it is originally at rest on the bed of a truck, determine its initial angular acceleration if the truck is given an acceleration of $a_T = 1.5$ ft/s^2. The spool rolls without slipping at B. *Hint:* Use kinematics to show that $a_G = 1.5 - 2\alpha$ with $a_B = 1.5$ ft/s^2.

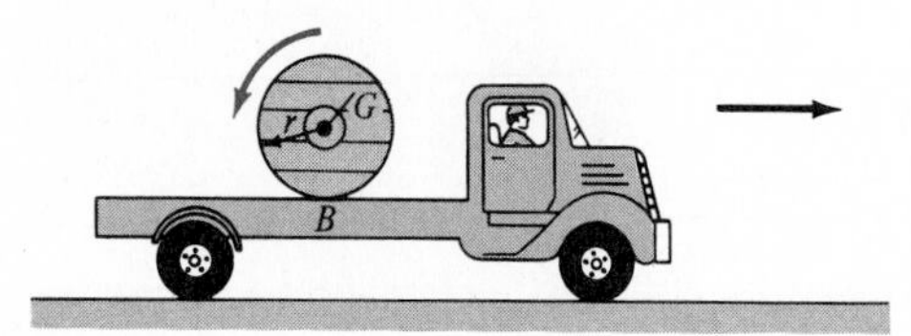

Prob. 17–70

17–71. A long strip of paper is wrapped into two rolls, each having a mass of 10 kg. Roll A is pin-supported at its center, whereas roll B is not centrally supported. If B is brought into contact with A and released from rest, determine the initial tension in the paper between the rolls and the angular acceleration of each roll. For the calculation, assume the rolls to be approximated by cylinders. *Hint:* Show that the necessary kinematic equation is $a_B = 0.09\alpha_A + 0.09\alpha_B$.

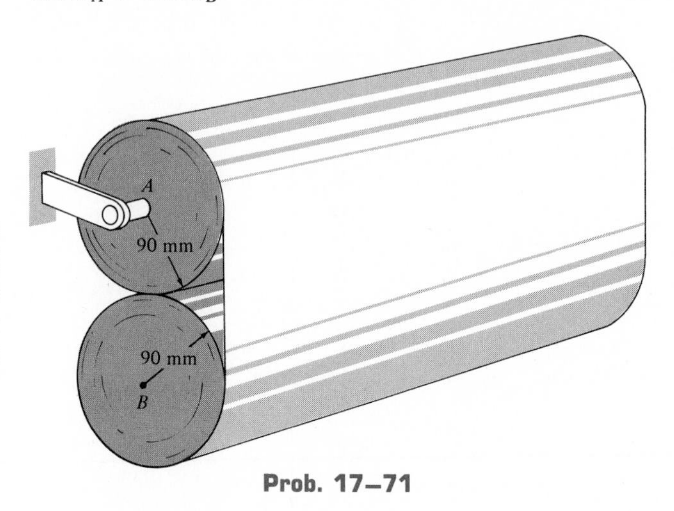

Prob. 17–71

***17–72.** If the cable CB is horizontal and the beam is at rest in the position shown, determine the tension in the cable at the instant the towing force $F = 1500$ N is applied. The coefficient of friction between the beam and the floor at A is $\mu_A = 0.3$. For the calculation, assume that the beam is a uniform slender rod having a mass of 100 kg.

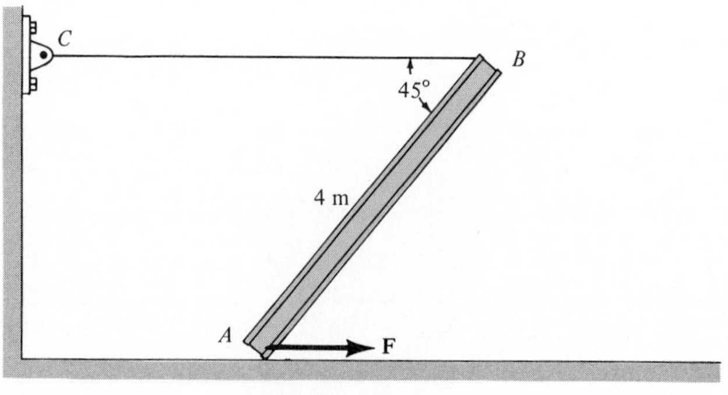

Prob. 17–72

17–73. The cart and its contents have a mass of 40 kg and a mass center at G, excluding the wheels. Each of the two wheels has a mass of 2 kg and a radius of gyration of $k_O = 0.12$ m. If the cart is released from rest from the position shown, determine its speed after it travels 4 m down the incline. The coefficient of friction is $\mu_A = 0.3$ between the incline and A. The wheels roll without slipping at B.

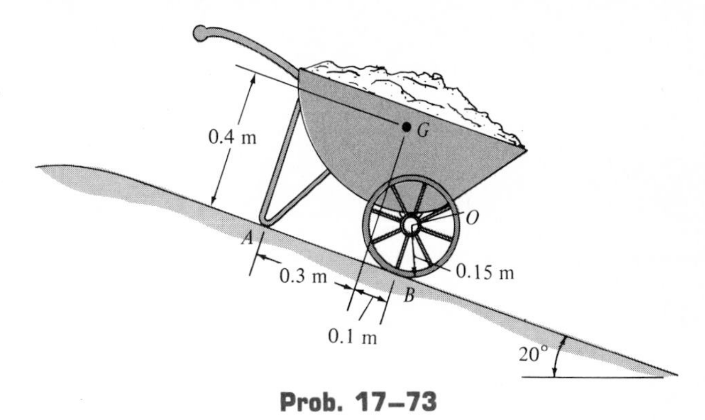

Prob. 17–73

Review Problems

17–74. Two 30-lb disks are *fixed* to the shaft AB, which is free to turn around the y axis. If a cord is wrapped around the periphery of one of the disks and a vertical force of 20 lb is applied, determine the angular velocity of the shaft in $t = 4$ s, starting from rest.

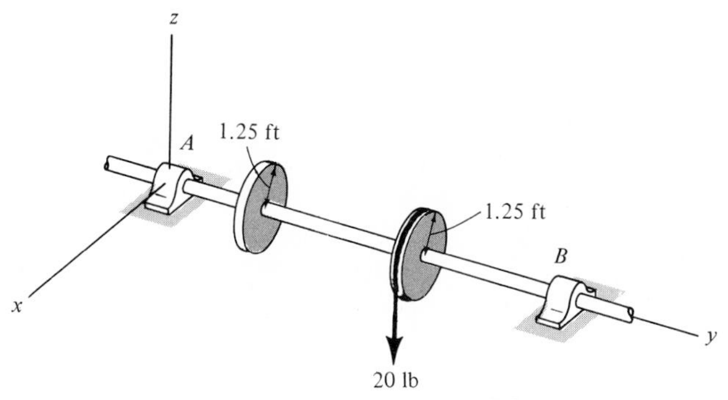

Probs. 17–74 / 17–75

17–75. Solve Prob. 17–74 if a block having a weight of 20 lb is applied to the cord, rather than the 20-lb force.

***17–76.** The uniform slender rod has a mass of 5 kg. Determine the reaction at the pin O just after the cord at A is cut.

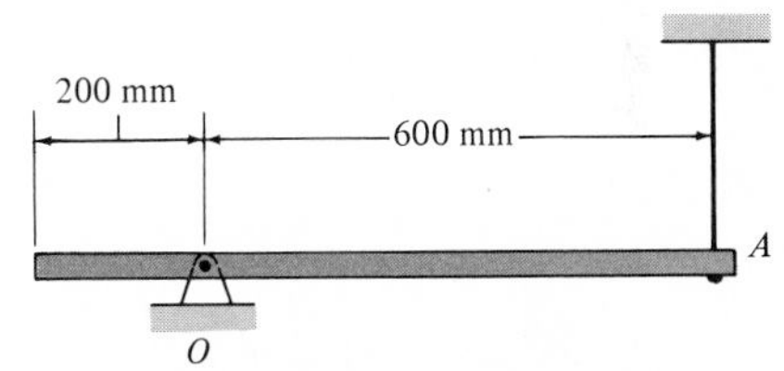

Prob. 17–76

17–77. If the motorcyclist is rounding a banked curve which has a radius of $\rho = 20$ m and angle of inclination of $\phi = 60°$, determine the minimum constant speed at which he must travel, and the angle θ at which he must lean, so that he does not tip over or slip. The coefficient of friction between the tires and the road is $\mu = 0.5$. The motorcycle and rider have a total mass of 200 kg and a center of mass at G.

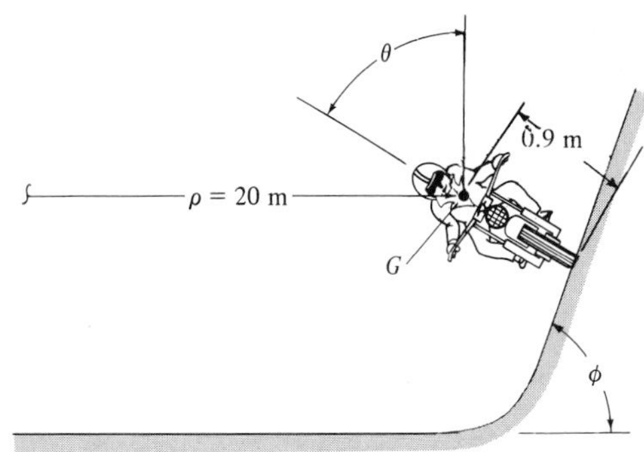

Prob. 17–77

17–78. The automobile has a mass of 2.3 Mg and a mass center at G. Determine the normal reactions of each of the four wheels on the road and the automobile's acceleration. Assume driving power is supplied only to the front wheels and slipping occurs, where $\mu = 0.4$. Neglect the mass of the wheels in the calculation.

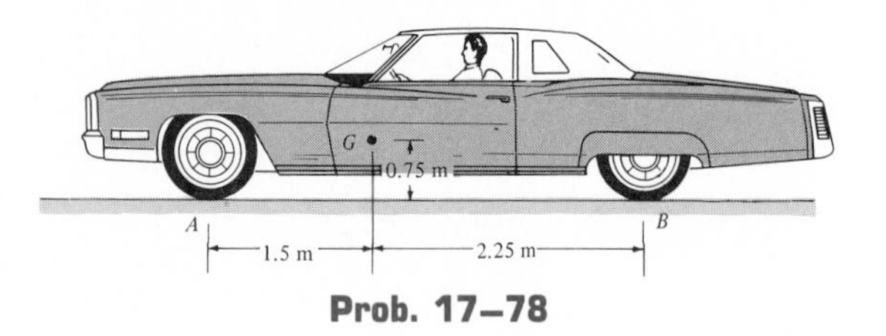

Prob. 17–78

17–79. If the 150-lb crate C is to be given an acceleration of 5 ft/s^2, determine the largest horizontal force **P** which can be applied to it and not cause it to tip or slip on the trolley. The coefficient of friction between the trolley and crate is $\mu = 0.3$. Neglect the mass of the trolley and assume it is free to roll.

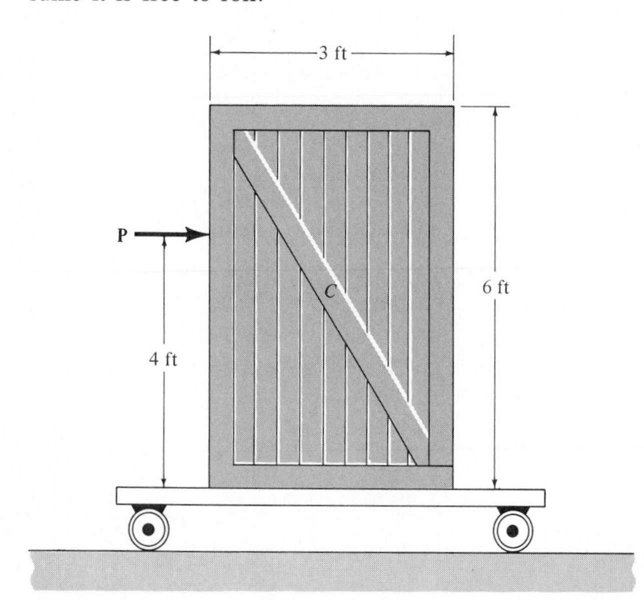

Prob. 17–79

***17–80.** A ring having a mass of 3 kg is given an angular velocity of $\omega = 12$ rad/s and placed on a floor for which the coefficient of friction is $\mu_A = 0.5$. If point O has no initial velocity, determine the ring's angular acceleration while it rolls with slipping.

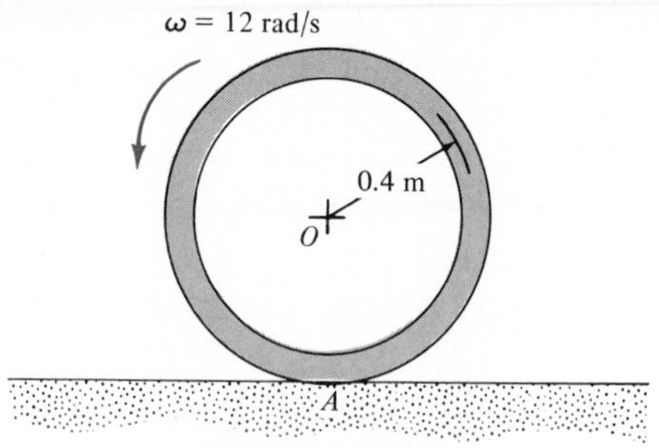

Prob. 17–80

17–81. The 20-lb solid ball is cast on the floor such that it has a backspin of $\omega = 15$ rad/s and its center has an initial horizontal velocity of $v_G = 20$ ft/s. If the coefficient of friction between the floor and the ball is $\mu_A = 0.3$, determine the distance it travels before it stops spinning.

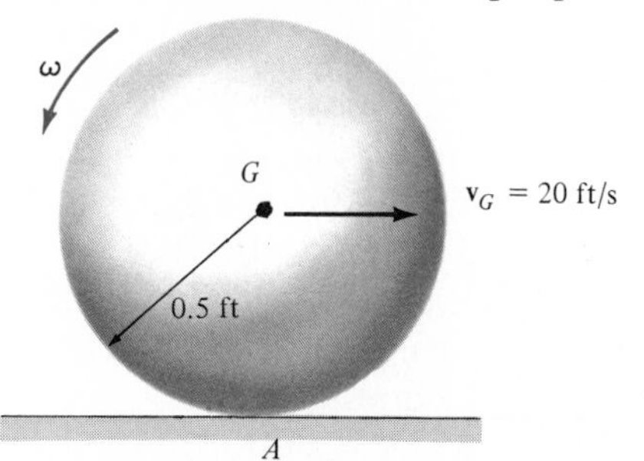

Probs. 17–81 / 17–82

17–82. Determine the backspin $\boldsymbol{\omega}$ which should be given to the 20-lb ball so that when its center is given an initial horizontal velocity of $v_G = 20$ ft/s it stops spinning and translating at the same instant. Take $\mu_A = 0.3$.

17–83. The 5-lb semicylinder is released from rest in the position shown. If it does not slip, determine its angular acceleration and the acceleration of its mass center G. *Hint:* Use the table on the inside back cover of the book for locating G and determining I_G.

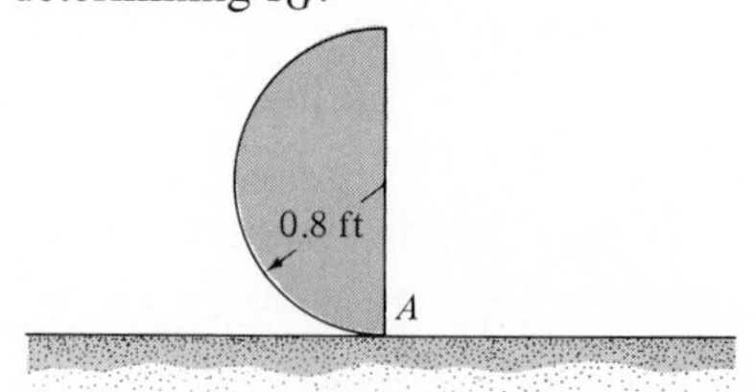

Probs. 17–83 / 17–84

***17–84.** Solve Prob. 17–83 if the coefficient of friction between the semicylinder and the horizontal surface at A is $\mu = 0.2$.

Planar Kinetics of a Rigid Body: Work and Energy

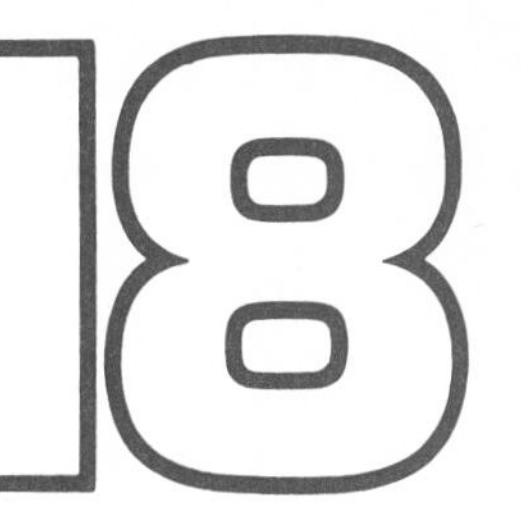

It was shown in Chapter 14 that problems which involve force, velocity, and displacement can conveniently be solved by using the principle of work and energy, or if the force system is "conservative," the conservation of energy theorem. In this chapter we will apply work and energy methods to solve problems involving the planar motion of a rigid body.

Before discussing the principle of work and energy for a body, however, the methods for obtaining the body's kinetic energy when it is subjected to translation, rotation about a fixed axis, or general plane motion will be developed.

Kinetic Energy 18.1

Consider the rigid body shown in Fig. 18–1, which is represented here by a *slab* moving in the reference plane of motion. An arbitrary ith particle of the body, having a mass dm, is located at r from the arbitrary point P. If at the *instant* shown the particle has a velocity $\mathbf{v}_i$, the particle's kinetic energy is $T_i = \frac{1}{2} dm\, v_i^2$. The kinetic energy of the entire body is determined by writing similar expressions for each particle of the body and summing the results, i.e.,

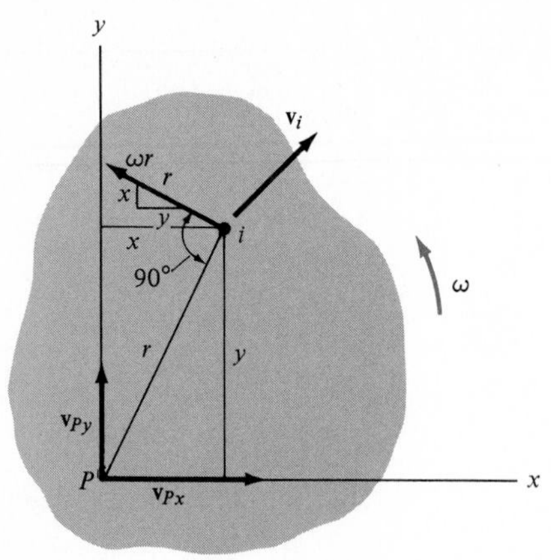

Fig. 18–1

$$T = \tfrac{1}{2} \int_m dm\, v_i^2$$

This equation may be written in another manner by using kinematics to relate $\mathbf{v}_i$ to $\mathbf{v}_P$, the velocity of point P. In this case

$$\mathbf{v}_i = \mathbf{v}_P + \mathbf{v}_{i/P}$$

where $\mathbf{v}_{i/P}$ is the relative velocity of i with respect to P. From Fig. 18–1 we have

$$\mathbf{v}_i = \underset{\rightarrow}{(v_P)_x} + \underset{\uparrow}{(v_P)_y} + \omega r$$

or

$$\mathbf{v}_i = \underset{\rightarrow}{[(v_P)_x - \omega y)]} + \underset{\uparrow}{[(v_P)_y + \omega x]}$$

The square of the magnitude of $\mathbf{v}_i$ is thus

$$\begin{aligned} v_i^2 &= [(v_P)_x - \omega y]^2 + [(v_P)_y + \omega x]^2 \\ &= (v_P^2)_x - 2(v_P)_x \omega y + \omega^2 y^2 + (v_P^2)_y + 2(v_P)_y \omega x + \omega^2 x^2 \\ &= v_P^2 - 2(v_P)_x \omega y + 2(v_P)_y \omega x + \omega^2 r^2 \end{aligned}$$

Substituting into the equation of kinetic energy, realizing that v_P, $(v_P)_x$, $(v_P)_y$ and ω are not functions of position, we have

$$T = \tfrac{1}{2}\left(\int_m dm\right)v_P^2 - (v_P)_x\omega\left(\int_m y\, dm\right) + (v_P)_y\omega\left(\int_m x\, dm\right) + \tfrac{1}{2}\omega^2\left(\int_m r^2\, dm\right)$$

The first integral on the right represents the entire mass m of the body. Since $\bar{x}m = \int_m x\, dm$ and $\bar{y}m = \int_m y\, dm$, the second and third integrals locate the body's center of mass G with respect to P. The last integral represents the body's moment of inertia I_P, computed about the z axis passing through point P. Thus,

$$T = \tfrac{1}{2}mv_P^2 - (v_P)_x\omega\bar{y}\, m + (v_P)_y\omega\bar{x}\, m + \tfrac{1}{2}I_P\omega^2 \qquad (18\text{–}1)$$

This equation reduces to a simpler form if point P coincides with the mass center G for the body, in which case $\bar{x} = \bar{y} = 0$, and therefore

$$T = \tfrac{1}{2}mv_G^2 + \tfrac{1}{2}I_G\omega^2 \qquad (18\text{–}2)$$

Here I_G is the mass moment of inertia of the body about an axis which is perpendicular to the plane of motion and passes through the mass center. Both terms on the right side are always *positive,* since the velocities are squared. Furthermore, it may be verified that these terms have units of length times force, common units being m · N or ft · lb. Recall that in the SI system 1 J (joule) = 1 m · N.

Translation. When a rigid body of mass m is subjected to either rectilinear or curvilinear *translation,* the kinetic energy due to rotation is zero, since $\boldsymbol{\omega} = \mathbf{0}$. From Eq. 18–2, the kinetic energy of the body is therefore

$$T = \tfrac{1}{2} m v_G^2$$

where v_G is the magnitude of the body's translational velocity at the instant considered, Fig. 18–2.

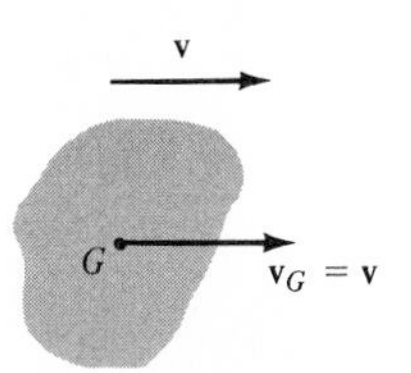

Translation
Fig. 18–2

Rotation About a Fixed Axis. When a rigid body is *rotating about a fixed axis* passing through point O, Fig. 18–3, the body has both *translational* and *rotational* kinetic energy as defined by Eq. 18–2, i.e.,

$$T = \tfrac{1}{2} m v_G^2 + \tfrac{1}{2} I_G \omega^2$$

The body's kinetic energy may be formulated in another manner by noting that $v_G = r_{G/O}\omega$, in which case $T = \frac{1}{2}(I_G + m r_{G/O}^2)\omega^2$. By the parallel-axis theorem, the terms inside the parentheses represent the moment of inertia I_O of the body about an axis perpendicular to the plane of motion and passing through point O. Hence,*

$$T = \tfrac{1}{2} I_O \omega^2 \qquad (18\text{–}3)$$

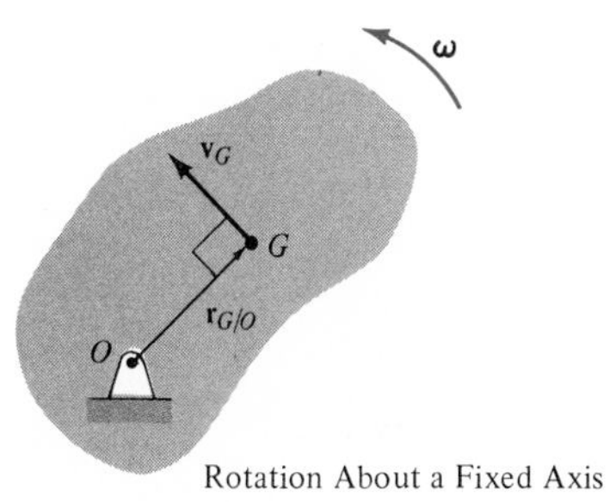

Rotation About a Fixed Axis
Fig. 18–3

From the derivation, this equation may be substituted for Eq. 18–2, since it accounts for *both* the translational kinetic energy of the body's mass center and the rotational kinetic energy of the body computed about the mass center.

General Plane Motion. When a rigid body is subjected to general plane motion, Fig. 18–4, it has an angular velocity $\boldsymbol{\omega}$ and its mass center has a velocity $\mathbf{v}_G$. Hence, the kinetic energy is defined by Eq. 18–2, i.e.,

$$T = \tfrac{1}{2} m v_G^2 + \tfrac{1}{2} I_G \omega^2$$

Here it is seen that the total kinetic energy of the body consists of the *scalar* sum of the body's *translational* kinetic energy, $\frac{1}{2} m v_G^2$, and *rotational* kinetic energy about its mass center, $\frac{1}{2} I_G \omega^2$.

Because energy is a scalar quantity, the total kinetic energy for a system of *connected* rigid bodies is the sum of the kinetic energies of all the moving parts. Depending upon the type of motion, the kinetic energy of *each body* is found by applying Eq. 18–2 or the alternative forms mentioned above.†

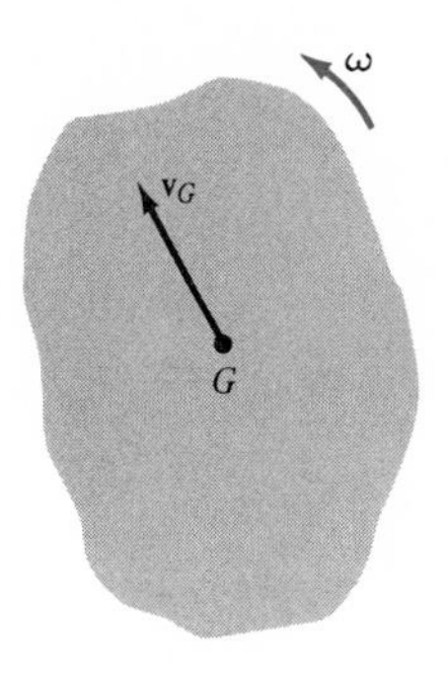

General plane motion
Fig. 18–4

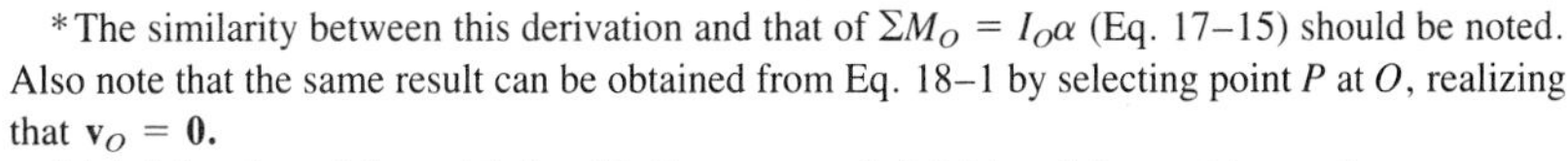
*The similarity between this derivation and that of $\Sigma M_O = I_O\alpha$ (Eq. 17–15) should be noted. Also note that the same result can be obtained from Eq. 18–1 by selecting point P at O, realizing that $\mathbf{v}_O = \mathbf{0}$.

†A brief review of Secs. 16–5 to 16–7 may prove helpful in solving problems, since computations for kinetic energy require a kinematic analysis of velocity.

Example 18–1

The system of three elements shown in Fig. 18–5*a* consists of a 6-kg block *B*, a 10-kg disk *D*, and a 12-kg cylinder *C*. A continuous cord of negligible mass is wrapped around the cylinder, passes over the disk, and is then attached to the block. If the block is moving downward with a speed of 0.8 m/s and the cylinder rolls without slipping, determine the total kinetic energy of the system at this instant.

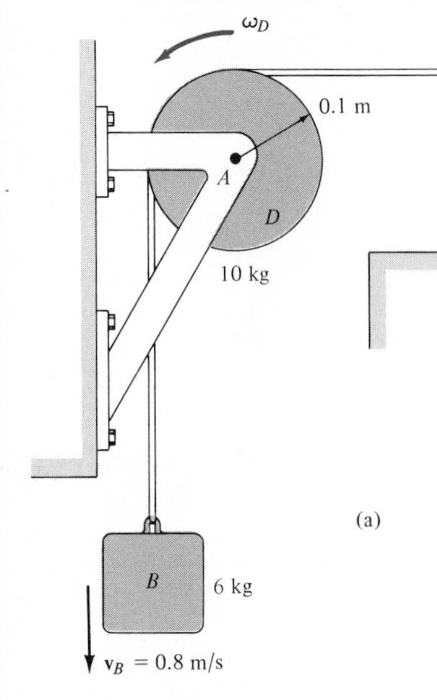

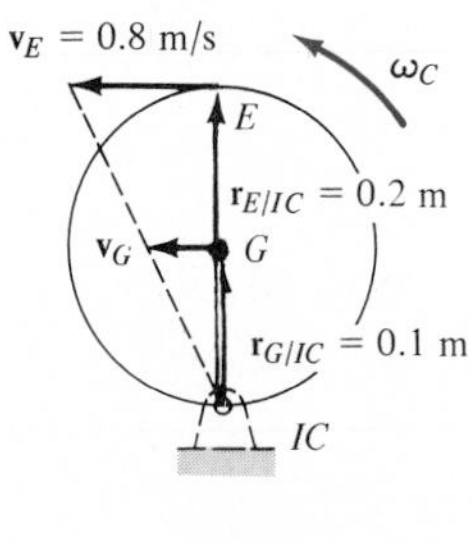

Fig. 18–5

Solution

By inspection, the block is translating, the disk rotates about a fixed axis, and the cylinder has general plane motion. Hence, in order to compute the kinetic energy of the disk and cylinder, it is first necessary to determine ω_D, ω_C, and v_G, Fig. 18–5*a*. From the *kinematics* of the disk,

$$v_B = r_D\omega_D; \quad 0.8 \text{ m/s} = (0.1 \text{ m})\omega_D; \quad \omega_D = 8 \text{ rad/s}$$

Since the cylinder rolls without slipping, the instantaneous center of zero velocity is at the point of contact with the ground, Fig. 18–5*b*, hence,

$$v_E = r_{E/IC}\omega_C; \; 0.8 \text{ m/s} = (0.2 \text{ m})\omega_C; \quad \omega_C = 4 \text{ rad/s}$$

$$v_G = r_{G/IC}\omega_C; \quad v_G = (0.1 \text{ m})(4 \text{ rad/s}) = 0.4 \text{ m/s}$$

The kinetic energy of the block is

$$T_B = \tfrac{1}{2}m_B v_B^2; \quad T_B = \tfrac{1}{2}(6 \text{ kg})(0.8 \text{ m/s})^2 = 1.92 \text{ J}$$

The kinetic energy of the disk is

$$T_D = \tfrac{1}{2}I_D\omega_D^2; \quad T_D = \tfrac{1}{2}(\tfrac{1}{2}m_D r_D^2)\omega_D^2$$
$$= \tfrac{1}{2}[\tfrac{1}{2}(10 \text{ kg})(0.1 \text{ m})^2](8 \text{ rad/s})^2 = 1.60 \text{ J}$$

Finally, the kinetic energy of the cylinder is

$$T_C = \tfrac{1}{2}mv_G^2 + \tfrac{1}{2}I_G\omega_C^2; \quad T_C = \tfrac{1}{2}mv_G^2 + \tfrac{1}{2}(\tfrac{1}{2}m_C r_C^2)\omega_C^2$$
$$= \tfrac{1}{2}(12 \text{ kg})(0.4 \text{ m/s})^2$$
$$+ \tfrac{1}{2}[\tfrac{1}{2}(12 \text{ kg})(0.1 \text{ m})^2](4 \text{ rad/s})^2 = 1.44 \text{ J}$$

The total kinetic energy of the system is therefore

$$T = T_B + T_D + T_C$$
$$= 1.92 \text{ J} + 1.60 \text{ J} + 1.44 \text{ J} = 4.96 \text{ J} \qquad \textit{Ans.}$$

The Work of a Force 18.2

Several types of forces are often encountered in planar kinetics problems involving a rigid body. The work of each of these forces has been derived in Sec. 14–1 and is listed below as a summary.

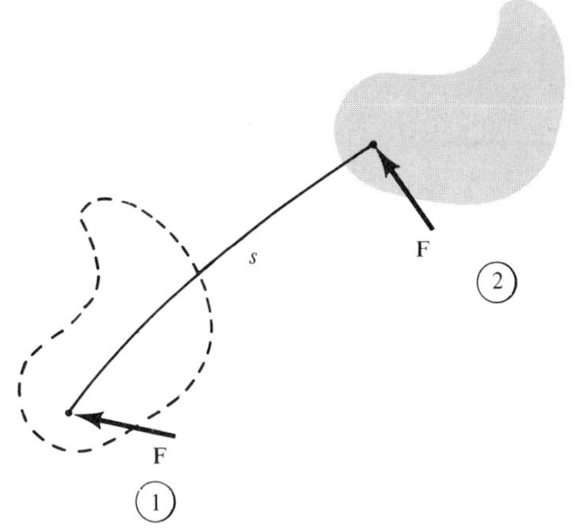

Fig. 18–6

Work of a Variable Force. If an external force **F** acts on a rigid body the work done by the force when it moves along the path s, Fig. 18–6, is defined as

$$U = \int_s F \cos \theta \, ds \tag{18–4}$$

Here θ is the angle between the "tails" of the force vector and the differential path displacement $d\mathbf{s}$. In general, the integration must account for the variation of the force's direction and magnitude. Note that the work of the *internal forces* of the body does not have to be considered. These forces occur in equal but opposite collinear pairs, so as a result, when the body moves, the work of one force cancels that of its counterpart. Furthermore, since the body is rigid, no relative movement between the forces occurs, so that no internal work is done.

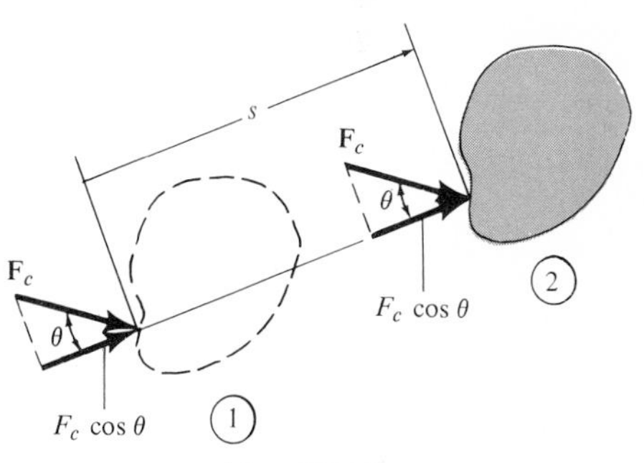

Fig. 18–7

Work of a Constant Force. If an external force $\mathbf{F}_c$ acts on a rigid body, Fig. 18–7, and maintains a constant magnitude F_c and constant direction θ, while the body undergoes a translation s, Eq. 18–4 can be integrated so that the work becomes

$$U_{F_c} = (F_c \cos \theta)s \tag{18–5}$$

Here $F_c \cos \theta$ represents the magnitude of the component of force in the direction of displacement.

Work of a Weight. The weight of a body does work only when the body's center of mass G undergoes a *vertical displacement* y. If this displacement is *downward,* Fig. 18–8, the work is *positive,* since the weight and displacement are in the *same* direction.

$$U_W = Wy \tag{18–6}$$

Here the elevation change is considered to be small so that **W**, which is caused by gravitation, is constant.*

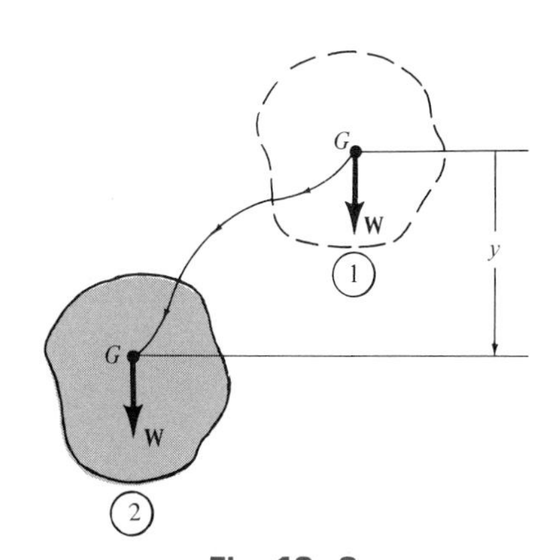

Fig. 18–8

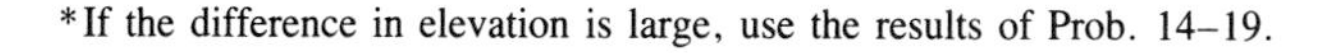

*If the difference in elevation is large, use the results of Prob. 14–19.

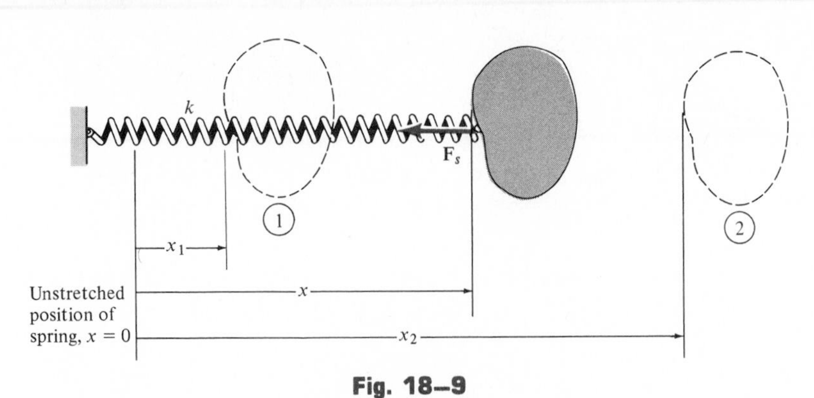

Fig. 18–9

Work of a Spring Force. If a linear elastic spring is attached to a body, the spring force $F_s = kx$ *acting on the body* does work when the spring either stretches or compresses from x_1 to a *further* position x_2. In both cases the work will be *negative* since the *displacement of the body* is always in the opposite direction to the force, Fig. 18–9. The work done is

$$U_s = -\left(\tfrac{1}{2}kx_2^2 - \tfrac{1}{2}kx_1^2\right) \qquad (18\text{–}7)$$

where $|x_2| > |x_1|$.

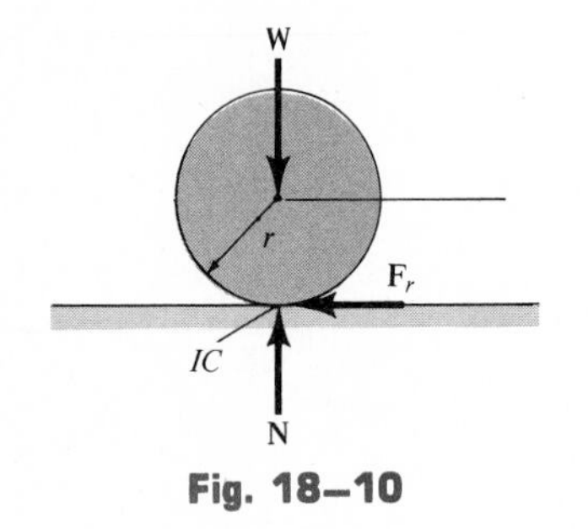

Fig. 18–10

Forces That Do No Work. There are some external forces that do no work when the body is displaced. These forces can act either at *fixed points* on the body or they can have a direction *perpendicular to their displacement*. Examples include the reactions at a pin support about which a body rotates, the normal reaction acting on a body that moves along a surface, and the weight of a body when the center of gravity of the body moves in a *horizontal plane*. A frictional force $\mathbf{F}_r$ acting on a body as it *rolls without slipping* over a rough surface also does no work, Fig. 18–10. This is because, during any *instant of time dt*, $\mathbf{F}_r$ acts at a point having *zero velocity* (instantaneous center, *IC*). In other words, for any differential rolling movement $d\mathbf{s}_{IC}$ of the body's *IC*, the work is $dU = F_r \cdot ds_{IC} = F_r(v_{IC}dt) = 0$, since $v_{IC} = 0$.

18.3 The Work of a Couple

Recall that a *couple* consists of a pair of noncollinear forces which have equal magnitudes and opposite directions. When a body subjected to a couple undergoes general plane motion, the two forces do work *only* when the body undergoes a *rotation*. To show this, consider the body in Fig. 18–11*a*, which is subjected to a couple having a magnitude of $M = Fr$. Any general differential displacement of the body can be considered as a separate translation and rotation. When the body *translates* such that the *component of displacement* along the line of action of the forces is $d\mathbf{s}_t$, Fig. 18–11*b*, clearly the "posi-

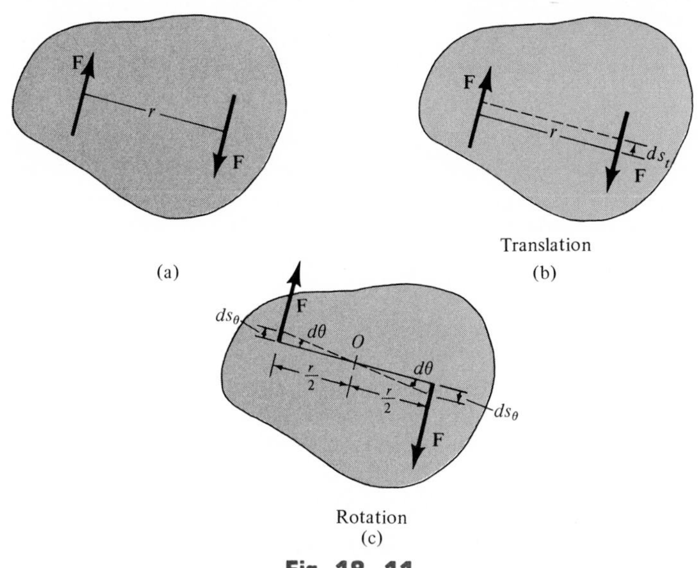

Fig. 18–11

tive'' work of one force *cancels* the ''negative'' work of the other. Consider now a differential rotation $d\boldsymbol{\theta}$ of the body about an axis which is perpendicular to the plane of the couple and intersects the plane at the midpoint O, Fig. 18–11c. (For the derivation any other point in the plane may also be considered.) As shown, each force undergoes a displacement $ds_\theta = (r/2)d\theta$ in the direction of the force; hence, the total work done is

$$dU_M = F\left(\frac{r}{2}d\theta\right) + F\left(\frac{r}{2}d\theta\right) = (Fr)d\theta$$
$$= M\,d\theta$$

Here the line of action of $d\boldsymbol{\theta}$ is parallel to the line of action of **M.** This is *always the case for general plane motion,* since **M** and $d\boldsymbol{\theta}$ are perpendicular to the plane of motion. Furthermore, the resultant work is *positive* when **M** and $d\boldsymbol{\theta}$ are in the *same direction* and *negative* if these vectors are in *opposite directions*.

When the body rotates in the plane through a finite angle θ, from θ_1 to θ_2, the work of a couple is

$$U_M = \int_{\theta_1}^{\theta_2} M\,d\theta \qquad (18\text{–}8)$$

If **M** has a *constant magnitude,* then

$$U_M = M(\theta_2 - \theta_1) \qquad (18\text{–}9)$$

where in all cases the angles θ_1 and θ_2 are measured in radians.

18.4 Principle of Work and Energy

In Sec. 14–2 the principle of work and energy was developed for a particle. By applying this principle to each of the particles of a rigid body and adding the results algebraically, since energy is a scalar, the principle of work and energy for a rigid body may be written as

$$T_1 + \Sigma U_{1-2} = T_2 \tag{18–10}$$

This equation states that the body's initial translational *and* rotational kinetic energy plus the work done by all the external forces and couples acting on the body as the body moves from its initial to its final position is equal to the body's final translational *and* rotational kinetic energy.

When several rigid bodies are pin-connected, connected by inextensible cables, or in mesh with one another, this equation may be applied to the entire system of connected bodies. In all these cases the internal forces, which hold the various members together, do no work and hence are eliminated from the analysis.

PROCEDURE FOR ANALYSIS

The principle of work and energy is used to solve kinetics problems that involve *velocity, force,* and *displacement,* since these terms are involved in the formulation. For application, it is suggested that the following procedure be used.

Kinetic Energy (Kinematic Diagrams). Determine the kinetic-energy terms T_1 and T_2 by applying the equation $T = \frac{1}{2}mv_G^2 + \frac{1}{2}I_G\omega^2$ or an appropriate form of this equation developed in Sec. 18–1. In this regard, *kinematic diagrams* for velocity may be useful for determining v_G and ω, or for establishing a *relationship* between v_G and ω.

Work (Free-Body Diagram). Draw a free-body diagram of the body when it is located at an intermediate point along the path, in order to account for all the forces and couples which do work on the body. The work of each force and couple can be computed using the appropriate relations outlined in Secs. 18–2 and 18–3. Since *algebraic addition* of the work terms is required, it is important that the proper sign of each term be specified. Specifically, work is *positive* when the force (couple) is in the *same direction* as its displacement (rotation); otherwise, it is negative.

Principle of Work and Energy. Apply the principle of work and energy, $T_1 + \Sigma U_{1-2} = T_2$. Since this is a scalar equation, it can be used to solve for only one unknown when it is applied to a single rigid body. This is in contrast to the three scalar equations of motion which may be written for the same body.

The following example problems numerically illustrate application of this procedure.

Example 18–2

The 30-kg disk shown in Fig. 18–12*a* is pin-supported at its center. Determine the number of revolutions it must make to attain an angular velocity of 20 rad/s starting from rest. It is acted upon by a constant force $F = 10$ N, which is applied to a cord wrapped around its periphery, and a constant couple $M = 5\ \text{N}\cdot\text{m}$. Neglect the mass of the cord in the calculation.

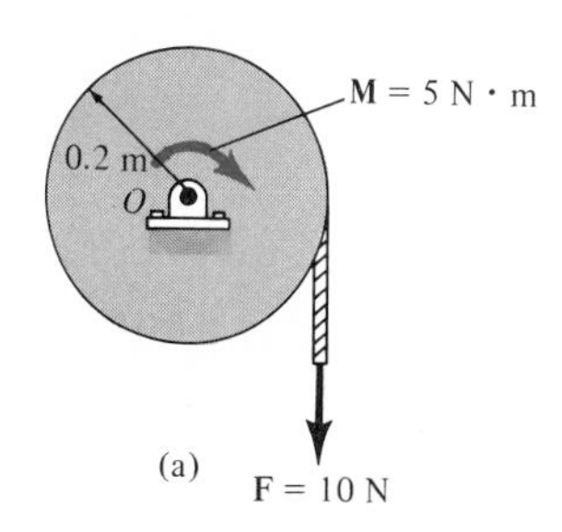

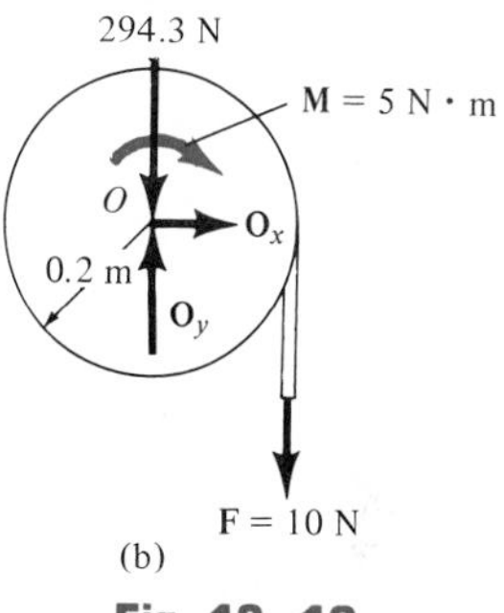

Fig. 18–12

Solution

Kinetic Energy. Since the disk rotates about a fixed axis, the kinetic energy can be computed using $T = \frac{1}{2}I_O\omega^2$, where the moment of inertia is $I_O = \frac{1}{2}mr^2$. Initially, the disk is at rest, so that

$$T_1 = 0$$

$$T_2 = \tfrac{1}{2}I_O\omega_2^2 = \tfrac{1}{2}[\tfrac{1}{2}(30)(0.2)^2](20)^2 = 120\ \text{J}$$

Work (Free-Body Diagram). As shown in Fig. 18–12*b*, the pin reactions $\mathbf{O}_x$ and $\mathbf{O}_y$ and the weight (294.3 N) do no work, since they are not displaced. The *couple,* having a constant magnitude, does positive work $U_M = M\theta$ as the disk *rotates* through a clockwise angle of θ rad, and the *constant force* **F** does positive work $U_{F_c} = Fs$ as the cord *moves* downward $s = \theta r = \theta(0.2)$ m.

Principle of Work and Energy

$$\{T_1\} + \{\Sigma U_{1-2}\} = \{T_2\}$$

$$\{T_1\} + \{M\theta + Fs\} = \{T_2\}$$

$$\{0\} + \{5\theta + (10)\theta(0.2)\} = \{120\}$$

$$\theta = 17.1\ \text{rad} = 17.1\ \text{rad}\left(\frac{1\ \text{rev}}{2\pi\ \text{rad}}\right) = 2.73\ \text{rev} \qquad \textit{Ans.}$$

This problem has also been solved in Example 17–9. Compare the two methods of solution and note that since force, velocity, and displacement θ are involved, a work–energy approach yields a more direct solution.

Example 18–3

The 5-kg bar shown in Fig. 18–13*a* has a center of mass at G. If it is given an initial clockwise angular velocity of $\omega_1 = 10$ rad/s when $\theta = 90°$, compute the spring constant k so that it stops when $\theta = 0°$. The spring deforms 0.1 m when $\theta = 0°$.

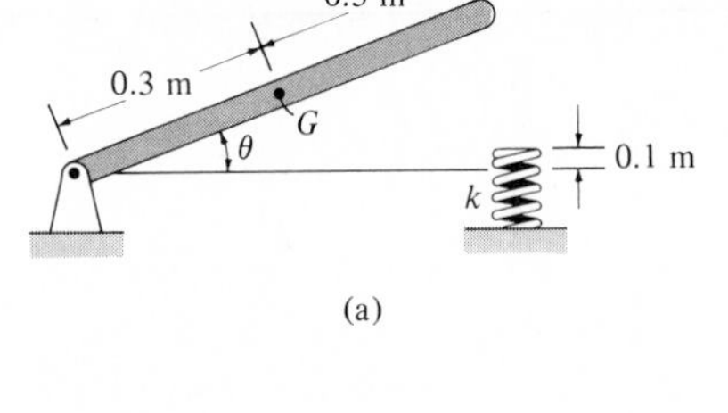

(a)

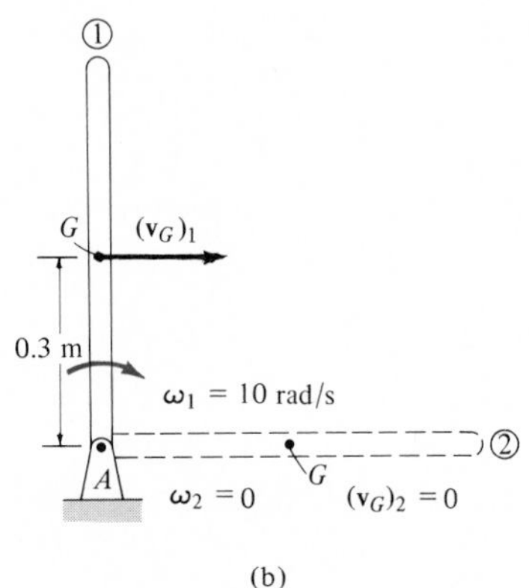

(b)

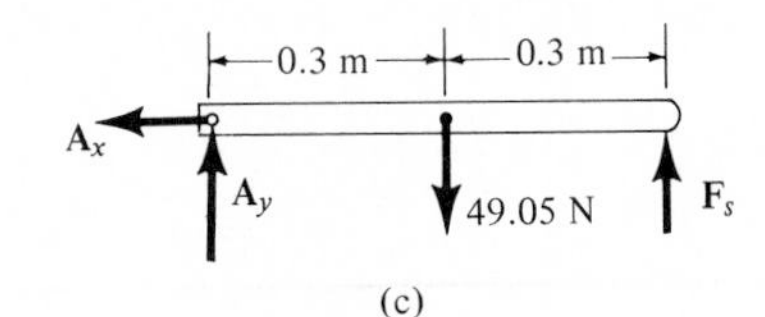

(c)

Fig. 18–13

Solution

Kinetic Energy (Kinematic Diagrams). Two kinematic diagrams of the bar when $\theta = 90°$ (position 1) and $\theta = 0°$ (position 2) are shown in Fig. 18–13*b*. The initial kinetic energy of the bar may be computed with reference to the fixed point of rotation A or the center of mass G.* If A is considered, it is necessary to apply $T_1 = \frac{1}{2}I_A\omega_1^2$. In this case

$$I_A = \tfrac{1}{3}ml^2 = \tfrac{1}{3}(5)(0.6)^2 = 0.60 \text{ kg}\cdot\text{m}^2$$

Thus,

$$T_1 = \tfrac{1}{2}I_A\omega_1^2 = \tfrac{1}{2}(0.60)(10)^2 = 30 \text{ J}$$

Since in the final position $(v_G)_2 = \omega_2 = 0$, the final kinetic energy is

$$T_2 = 0$$

Work (Free-Body Diagram). Figure 18–13*c*. The reactions $\mathbf{A}_x$ and $\mathbf{A}_y$ do no work, since these forces do not move. The 49.05-N weight, centered at G, moves downward through a vertical height of $y = 0.3$ m. Since this displacement is in the *same* direction as the force, the work is *positive*. The spring force $\mathbf{F}_s$ does *negative work on the bar*. Why? This force acts while the spring is being compressed from zero to $x = 0.1$ m. Hence, the work is determined from $U_s = -\frac{1}{2}kx^2$.

Principle of Work and Energy

$$\{T_1\} + \{\Sigma U_{1-2}\} = \{T_2\}$$
$$\{T_1\} + \{Wy - \tfrac{1}{2}kx^2\} = \{T_2\}$$
$$\{30\} + \{49.05(0.3) - \tfrac{1}{2}k(0.1)^2\} = \{0\}$$

Solving for k yields

$$k = 8943 \text{ N/m} = 8.94 \text{ kN/m} \qquad \textit{Ans.}$$

*Apply $T_1 = \frac{1}{2}m(v_G)_1^2 + \frac{1}{2}I_G\omega_1^2$ and show that it gives the same result, $T_1 = 30$ J.

Example 18–4

The wheel shown in Fig. 18–14a weighs 40 lb and has a radius of gyration of $k_G = 0.6$ ft about its mass center G. If it is subjected to a clockwise couple of 15 lb · ft and rolls from rest without slipping, determine its angular velocity after its center G moves 0.5 ft.

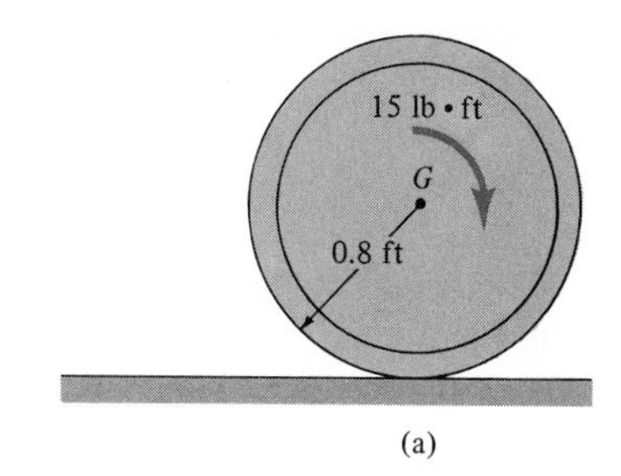

Fig. 18–14

(a)

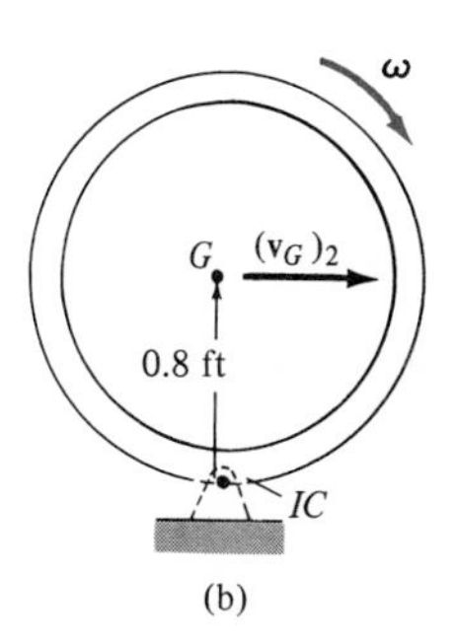

(b)

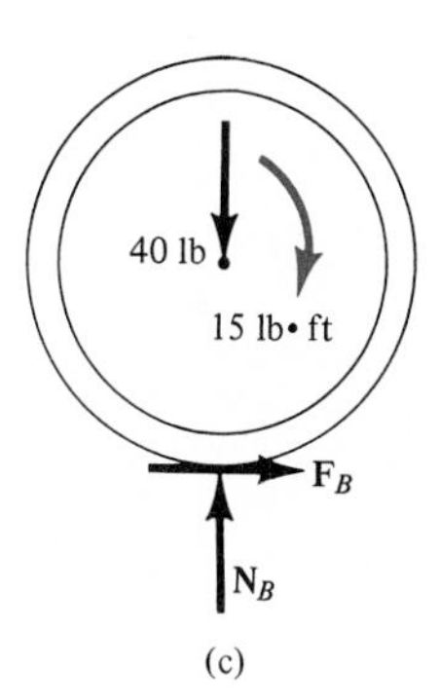

(c)

Solution

Kinetic Energy (Kinematic Diagram). Since the wheel is initially at rest,

$$T_1 = 0$$

The kinematic diagram of the wheel when it is in the final position is shown in Fig. 18–14b. The velocity of the mass center $(\mathbf{v}_G)_2$ can be related to the angular velocity $\boldsymbol{\omega}_2$ by using the instantaneous center of zero velocity (IC), i.e., $(v_G)_2 = 0.8\omega_2$. Hence, the final kinetic energy is

$$\begin{aligned} T_2 &= \frac{1}{2}m(v_G)_2^2 + \frac{1}{2}I_G(\omega_2)^2 \\ &= \frac{1}{2}\left(\frac{40}{32.2}\right)(0.8\omega_2)^2 + \frac{1}{2}\left[\frac{40}{32.2}(0.6)^2\right](\omega_2)^2 \\ &= 0.621(\omega_2)^2 \end{aligned}$$

Work (Free-Body Diagram). As shown in Fig. 18–14c, only the spring force $\mathbf{F}_s$ and the couple do work. The normal force does not move along its line of action and the frictional force does *no work*, since the wheel does not slip as it rolls.

The work of $\mathbf{F}_s$ may be computed using $U_s = -\frac{1}{2}kx^2$. (Why is the work negative?) Since the wheel does not slip when the center G moves 0.5 ft, the wheel rotates $\theta = s/r_{G/IC} = 0.5/0.8 = 0.625$ rad, Fig. 18–14b.

Principle of Work and Energy

$$\begin{aligned} \{T_1\} + \{\Sigma U_{1-2}\} &= \{T_2\} \\ \{T_1\} + \{M\theta\} &= \{T_2\} \\ \{0\} + \{15(0.625)\} &= \{0.621(\omega_2)^2\} \\ \omega_2 &= 3.89 \text{ rad/s} \end{aligned} \qquad \textit{Ans.}$$

Example 18–5

The 10-kg rod shown in Fig. 18–15*a* is constrained so that its ends move along the grooved slots. The rod is initially at rest when $\theta = 0°$. If the slider block at B is acted upon by a horizontal force of $P = 50$ N, determine the angular velocity of the rod at the instant $\theta = 45°$. Neglect the mass of blocks A and B. (Why can the principle of work and energy be used to solve this problem?)

Solution

Kinetic Energy (Kinematic Diagrams). Two kinematic diagrams of the rod, when it is in the initial position 1 and final position 2, are shown in Fig. 18–15*b*. When the rod is in position 1, $T_1 = 0$ since $(\mathbf{v}_G)_1 = \boldsymbol{\omega}_1 = \mathbf{0}$. In position 2 the angular velocity is $\boldsymbol{\omega}_2$ and the velocity of the mass center is $(\mathbf{v}_G)_2$. Hence, the kinetic energy is

$$
\begin{aligned}
T_2 &= \tfrac{1}{2}m(v_G)_2^2 + \tfrac{1}{2}I_G(\omega_2)^2 \\
&= \tfrac{1}{2}(10)(v_G)_2^2 + \tfrac{1}{2}[\tfrac{1}{12}(10)(0.8)^2](\omega_2)^2 \\
&= 5(v_G)_2^2 + 0.267(\omega_2)^2 \qquad (1)
\end{aligned}
$$

The two unknowns $(v_G)_2$ and ω_2 may be related via the instantaneous center of zero velocity for the rod, Fig. 18–15*b*. It is seen that as A moves downward with a velocity $(\mathbf{v}_A)_2$, B moves horizontally to the left with a velocity $(\mathbf{v}_B)_2$. Knowing these directions, the IC may be determined as shown in the figure. Hence,

$$
\begin{aligned}
(v_G)_2 &= r_{G/IC}\omega_2 = (0.4 \tan 45°)\omega_2 \\
&= 0.4\omega_2
\end{aligned}
$$

Substituting into Eq. (1), we have

$$T_2 = 5(0.4\omega_2)^2 + 0.267(\omega_2)^2 = 1.067(\omega_2)^2$$

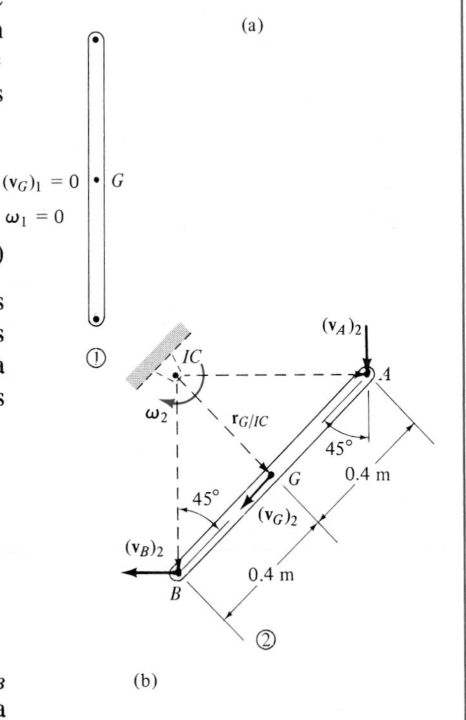

Work (Free-Body Diagram). Fig. 18–15*c*. The normal forces $\mathbf{N}_A$ and $\mathbf{N}_B$ do no work as the rod is displaced. Why? The 98.1-N weight is displaced a vertical distance of $y = (0.4 - 0.4 \cos 45°)$ m; whereas the 50-N force moves a horizontal distance of $s = (0.8 \sin 45°)$ m. The work done by both of these forces is *positive,* since the forces act in the same direction as their corresponding displacement.

Principle of Work and Energy

$$
\begin{aligned}
\{T_1\} + \{\Sigma U_{1-2}\} &= \{T_2\} \\
\{T_1\} + \{Wy + Ps\} &= \{T_2\} \\
\{0\} + \{98.1(0.4 - 0.4 \cos 45°) + 50(0.8 \sin 45°)\} &= \{1.067(\omega_2)^2\}
\end{aligned}
$$

Solving for ω_2 gives

$$\omega_2 = 6.11 \text{ rad/s} \qquad \textit{Ans.}$$

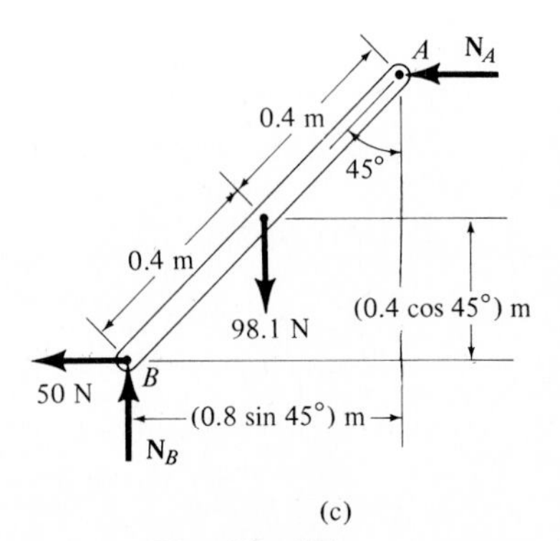

(c)

Fig. 18–15

Problems

Except when stated otherwise, throughout this chapter, assume that the coefficients of static and kinetic friction are equal, i.e., $\mu = \mu_s = \mu_k$.

18–1. At a given instant the body of mass m has an angular velocity $\boldsymbol{\omega}$ and its mass center has a velocity $\mathbf{v}_G$. Show that its kinetic energy can be represented as $T = \frac{1}{2}I_{IC}\omega^2$, where I_{IC} is the moment of inertia of the body computed about the instantaneous axis of zero velocity, located at a distance $r_{G/IC}$ from the mass center as shown.

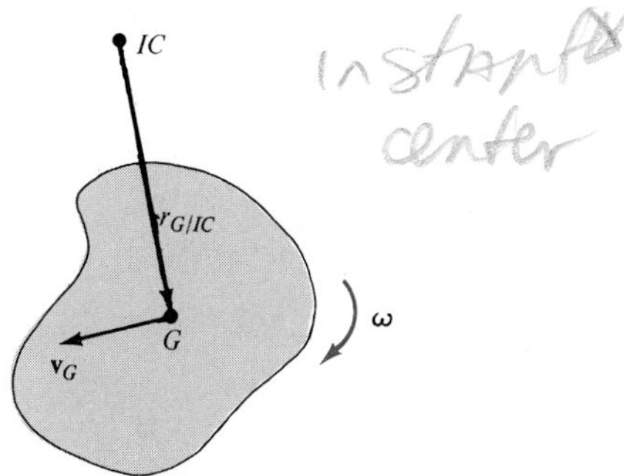

Prob. 18–1

18–2. An 800-lb tree falls from the vertical position such that it pivots about its cut section at A. If the tree can be considered as a uniform rod, pin-supported at A, determine the speed of its top branch B just before it strikes the ground.

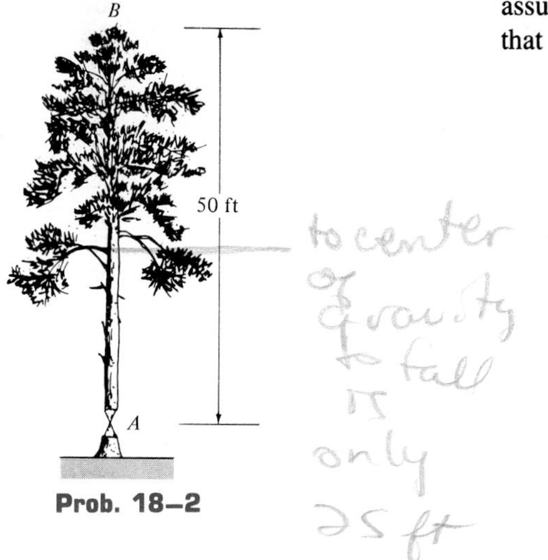

Prob. 18–2

18–3. A motor supplies a constant torque or twist of $M = 120$ lb · ft to the drum. If the drum has a weight of 30 lb and a radius of gyration of $k_O = 0.8$ ft, determine the speed of the 15-lb crate A after it rises $s = 4$ ft starting from rest. Neglect the weight of the cord.

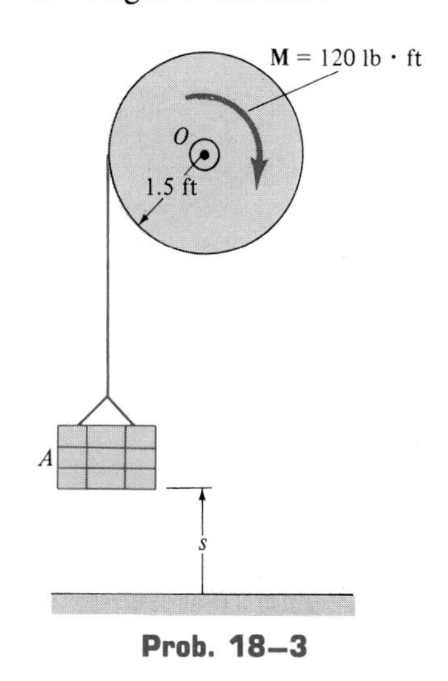

Prob. 18–3

***18–4.** A yo-yo has a weight of 0.3 lb and a radius of gyration of $k_O = 0.06$ ft. If it is released from rest, determine how far it must descend in order to attain an angular velocity of $\omega = 70$ rad/s. Neglect the mass of the string and assume that the string is wound around the central peg such that the mean radius at which it unravels is $r = 0.02$ ft.

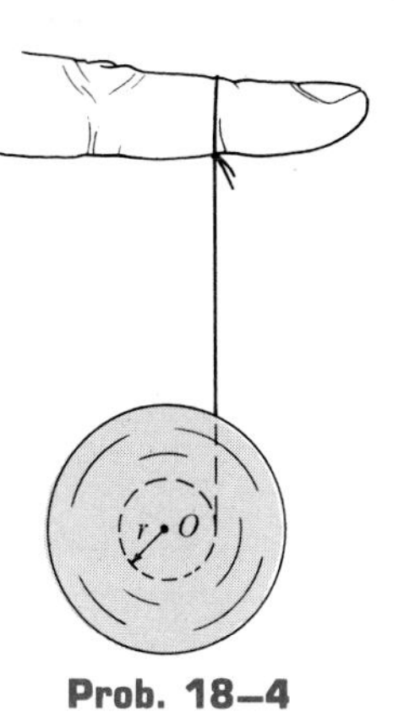

Prob. 18–4

18–5. The 50-kg uniform door is closed by applying a constant handle force of 20 N, which is always directed perpendicular to the plane of the door. If the door is originally at rest and open at $\theta = 90°$, determine the door's angular velocity just before it closes ($\theta = 0°$). Neglect friction at the hinges. For the calculation, assume the door to be approximated by a thin plate. *Hint:* Force **P** undergoes a displacement of $s = (\pi/2)(0.9)$ m.

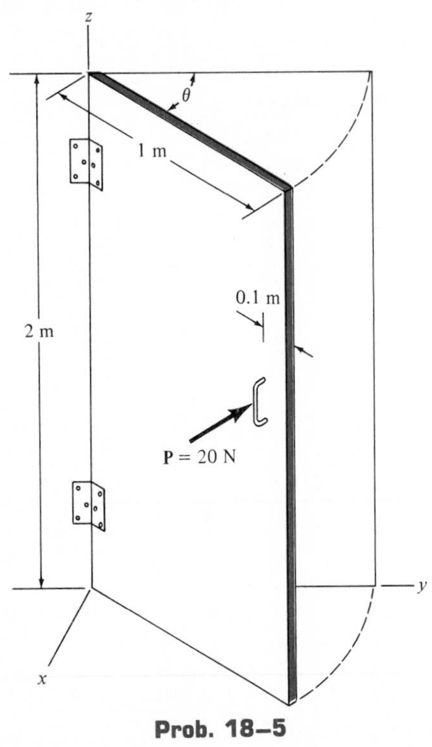

Prob. 18–5

18–6. The 80-kg wheel has a radius of gyration of $k_O = 125$ mm and is rotating at 20 rad/s when the vertical force of $P = 200$ N is applied to the brake handle. If the coefficient of friction between the brake at B and the wheel is $\mu = 0.35$, determine the total number of revolutions the wheel makes before it stops.

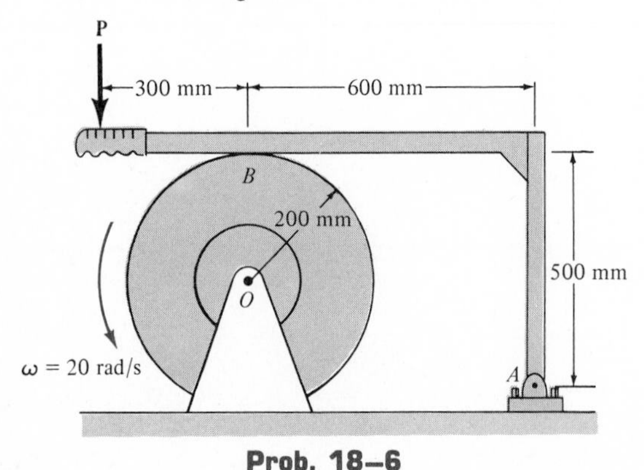

Prob. 18–6

18–7. The spool of cable, originally at rest, has a mass of 200 kg and a radius of gyration of $k_G = 325$ mm. If the spool rests on two small rollers A and B and a constant horizontal force of $P = 400$ N is applied to the end of the cable, compute the angular velocity of the spool when 8 m of cable has been unraveled. Neglect friction and the mass of the rollers and unraveled cable.

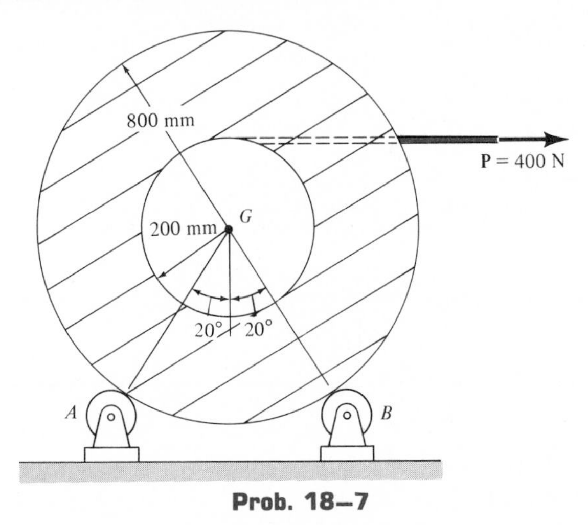

Prob. 18–7

***18–8.** The tub of the mixer has a weight of 70 lb and a radius of gyration about its center of gravity of $k_G = 1.3$ ft.

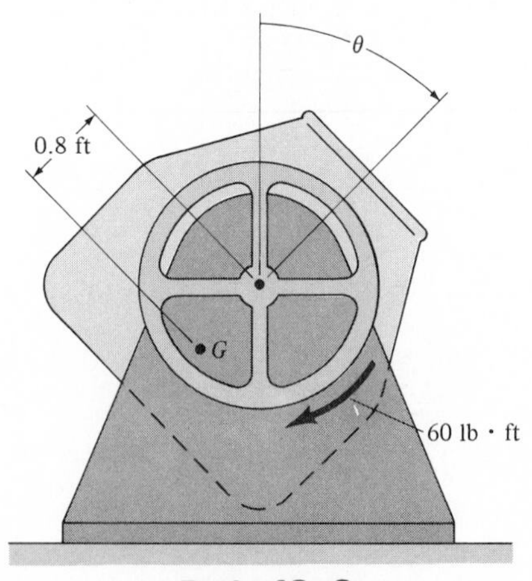

Prob. 18–8

If a constant torque of 60 lb · ft is applied to the dumping wheel, determine the angular velocity of the tub when it has turned $\pi/2$ rad. = 90°. Originally the tub is at rest when $\theta = 0°$.

18–9. A man having a weight of 180 lb sits in a chair of the Ferris wheel, which has a weight of 15,000 lb and a radius of gyration of $k_O = 37$ ft. If a torque of $M = 80(10^3)$ lb · ft is applied about O, determine the angular velocity of the wheel after it has rotated 180°. Neglect the weight of the chairs and note that the man remains in an upright position as the wheel rotates. The wheel starts from rest in the position shown. *Hint:* The kinetic energy of the man must be included in the calculation.

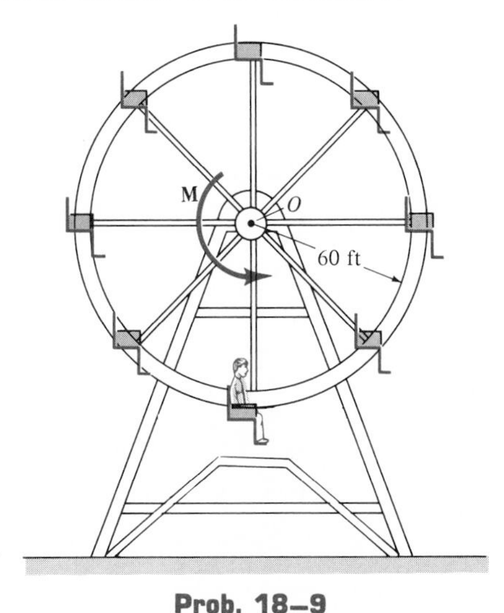

Prob. 18–9

18–10. The 10-kg rod AB is pin-connected at A and subjected to a couple of $M = 15$ N · m. If the rod is released from rest when the spring is unstretched, at $\theta = 30°$, determine the rod's angular velocity at the instant $\theta = 60°$. As the rod rotates, the spring always remains horizontal, because of the roller support at C.

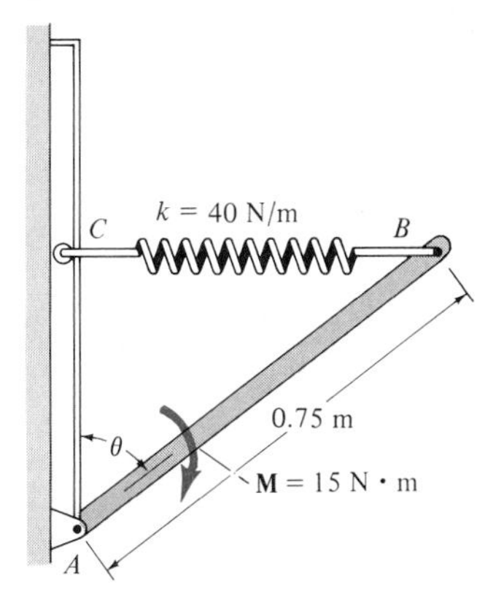

Prob. 18–10

18–11. The wheel has a mass of 20 kg and a radius of gyration of $k_A = 0.13$ m. The attached spring is unstretched and is restricted from stretching more than 300 mm due to the chain. If the wheel is subjected to a torque of $M = 40$ N · m, determine its angular velocity just before the chain becomes taut. Assume no slipping at B. Neglect the mass of the chain.

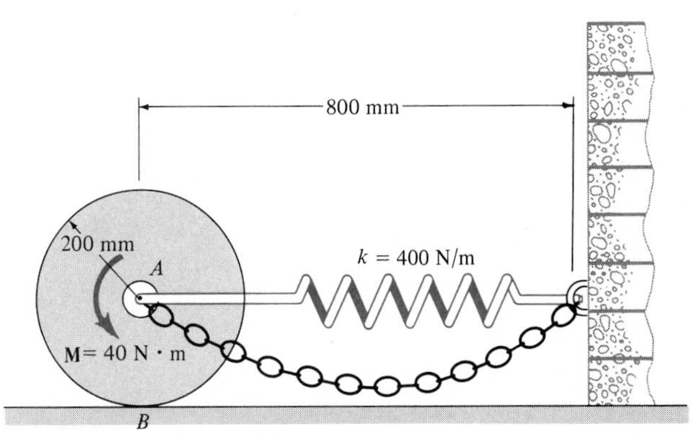

Prob. 18–11

***18–12.** The beam having a weight of 150 lb is supported by two cables. If the cable at end B is cut so that the beam is released from rest when $\theta = 30°$, determine the speed at which end A strikes the wall. Neglect friction at B. Consider the beam to be a thin rod.

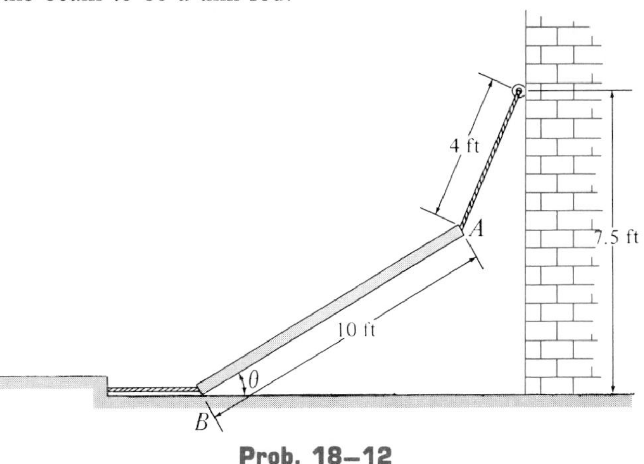

Prob. 18–12

18–13. A man having a weight of 150 lb crouches down at the end of a diving board as shown. In this position the radius of gyration about his center of gravity is $k_G = 1.2$ ft. While holding this position at $\theta = 0°$, he rotates about his toes at A until he loses contact with the board when $\theta = 90°$. If he remains rigid, determine approximately how many times he rolls around (360°) before striking the water after falling 30 ft. *Hint:* First show that his angular velocity at A becomes $\omega = 5.12$ rad/s at $\theta = 90°$.

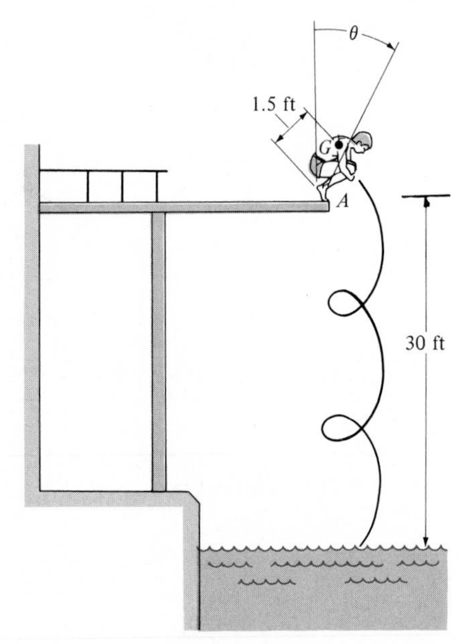

Prob. 18–13

18–14. A uniform ladder having a weight of 30 lb is released from rest when it is in the vertical position, $\theta = 0°$. If it is allowed to fall freely, determine its angular velocity ω when $\theta = 30°$. For the calculation, assume the ladder to be a slender rod which pivots about A.

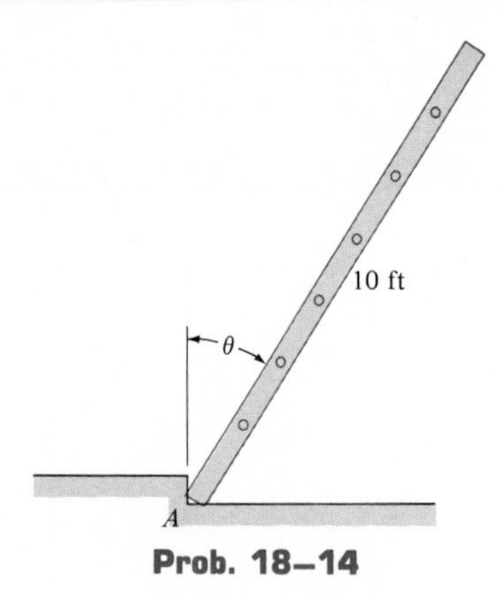

Prob. 18–14

18–15. A vertical force of $P = 1.5$ kN is used to lift the 20-kg wheel and the 50-kg crate B starting from rest. Assuming that the rope does not slip over the wheel, determine the speed of B after the center O of the wheel rises 2 m causing B to rise 4 m. The radius of gyration of the wheel is $k_O = 125$ mm.

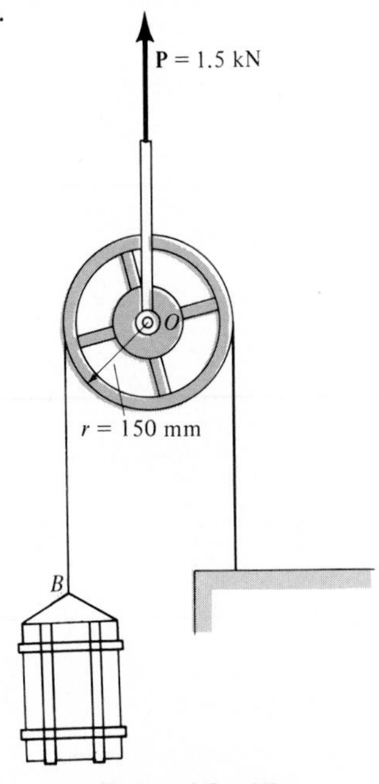

Prob. 18–15

Conservation of Energy 18.5

When a force system acting on a rigid body consists only of *conservative forces,* the conservation of energy theorem may be used to solve a problem which otherwise would be solved using the principle of work and energy. This theorem is often easier to apply since the work of a conservative force is *independent of the path* and only depends upon the initial and final positions of the body. It was shown in Sec. 14–5 that the work of a conservative force may be expressed as the difference in the body's potential energy measured from an arbitrarily selected reference or datum.

Gravitational Potential Energy. Since the weight of a body is concentrated at its center of gravity, the *gravitational potential energy* of the body is determined by knowing the height of the body's center of gravity from a datum plane and the body's weight. We have

$$V_g = +Wy_G \tag{18–11}$$

In this case, the potential energy is *positive,* since the weight has the ability to do *positive work* when the body is moved back to the datum, Fig. 18–16. If the body is located y_G *below* the datum, the gravitational potential energy is *negative,* i.e.,

$$V_g = -Wy_G \tag{18–12}$$

Here the weight does *negative work* when the body is moved back to the datum.

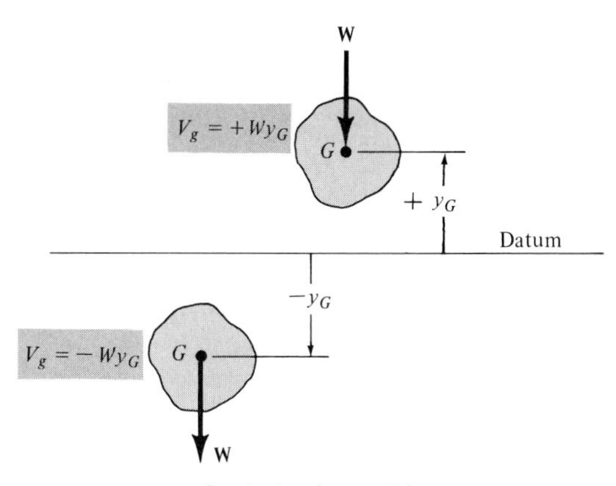

Gravitational potential energy

Fig. 18–16

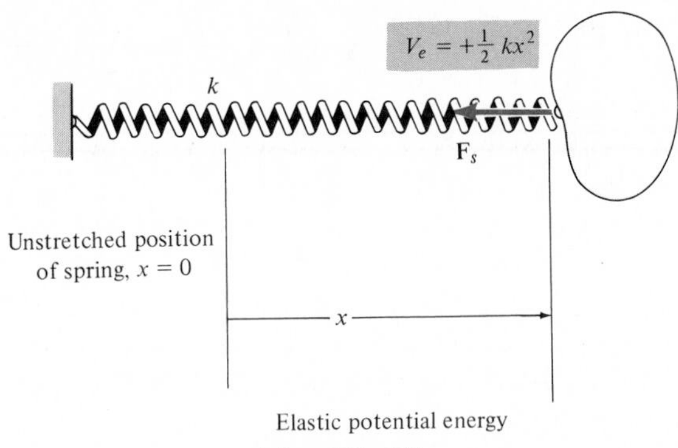

Fig. 18–17

Elastic Potential Energy. The force developed by an elastic spring is also a conservative force. The *elastic potential energy* which a spring imparts to an attached body when the spring is elongated or compressed from an initial unstretched position ($x = 0$) to a final position x, Fig. 18–17, is

$$V_e = +\tfrac{1}{2}kx^2 \tag{18–13}$$

In the deformed position, the spring force acting *on the body* always has the capacity for doing positive work when the spring is returned back to its original undeformed position (see Sec. 14–5).

In general, if a body is subjected to both gravitational and elastic forces, the total *potential energy* is expressed as the algebraic sum

$$V = V_g + V_e \tag{18–14}$$

Here measurement of V depends upon the location of the body with respect to a selected datum in accordance with Eqs. 18–11 to 18–13.

Conservation of Energy Theorem. In Sec. 14–6 the conservation of energy theorem was developed for a particle. By applying this theorem to each of the particles of a rigid body and adding the results algebraically, since energy is a scalar, the conservation of energy theorem for a rigid body may be written as

$$T_1 + V_1 = T_2 + V_2 \tag{18–15}$$

This theorem states that the *sum* of the potential and kinetic energies of the body remains *constant* when the body moves from one position to another. It also applies to a system of smooth, pin-connected rigid bodies, bodies connected by inextensible cords, and bodies in mesh with other bodies. In these cases the forces acting at all points of contact are *eliminated* from the analysis, since they occur in equal and opposite pairs and each pair of forces moves through an equal distance when the system undergoes a small displacement.

PROCEDURE FOR ANALYSIS

The conservation of energy theorem is used to solve problems involving *velocity, displacement,* and *conservative force systems.* For application it is suggested that the following procedure be used.

Potential Energy. Draw two diagrams showing the body located at its initial and final positions along the path. If the center of mass of the body is subjected to a *vertical displacement,* determine where to establish the fixed horizontal datum from which to measure the body's gravitational potential energy V_g. Although the location of the datum is arbitrary, it is advantageous to place it through G when the body is either at the initial or final point of its path, since at the datum $V_g = 0$. Data pertaining to the elevation y of the body's mass center from the datum and the extension or compression of any connecting springs can be determined from the geometry associated with the two diagrams. Recall that the potential energy $V = V_g + V_e$, where $V_g = \pm Wy$ and $V_e = \frac{1}{2}kx^2$.

Kinetic Energy. The kinetic-energy terms T_1 and T_2 are determined from $T = \frac{1}{2}mv_G^2 + \frac{1}{2}I_G\omega^2$ or an appropriate form of this equation as developed in Sec. 18–1. In this regard, kinematic diagrams for velocity may be useful for determining v_G and ω, or for establishing a *relationship* between these quantities.

Conservation of Energy Theorem. Apply the conservation of energy theorem: $T_1 + V_1 = T_2 + V_2$.

It is important to remember that *only problems involving conservative force systems may be solved by using this theorem.* As stated in Sec. 14–5, friction or other drag-resistant forces, which depend upon velocity or acceleration, are nonconservative. The work of such forces is transformed into thermal energy used to heat up the surfaces of contact, and consequently this energy is dissipated into the surroundings and may not be recovered. Therefore, problems involving frictional forces should either be solved by using the principle of work and energy, if it applies, or the equations of motion.

The following example problems numerically illustrate application of the above procedure.

Example 18–6

The uniform 5-kg slender rod AB shown in Fig. 18–18a is attached to a roller at A and smooth collar at B, each having a negligible mass. If the assembly is released from rest when $\theta = 60°$, determine the angular velocity of the rod when $\theta = 0°$. Assume that the roller does not slip.

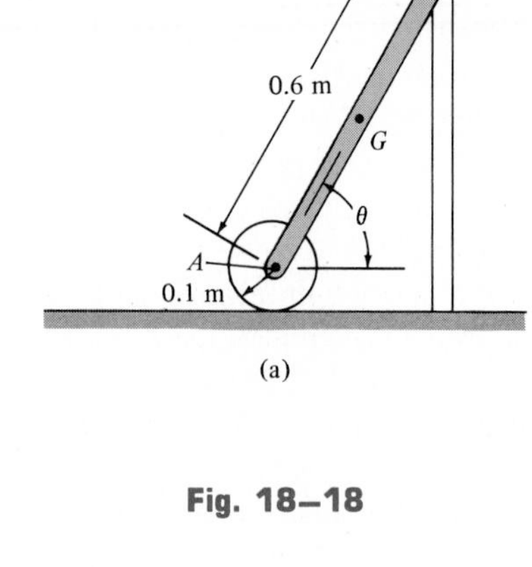

(a)

Fig. 18–18

Solution

Potential Energy. Two diagrams for the rod, when it is located at its initial and final positions, are shown in Fig. 18–18b. For convenience the datum, which is horizontally fixed, passes through point A.

When the system is in position 1, the rod's weight has positive potential energy, since this force is above the datum and therefore has the capacity to do work in moving the rod. Thus,

$$V_1 = W_R y_1 = 49.05(0.3 \sin 60°) = 12.74 \text{ J}$$

When the system is in position 2, the weight of the rod has zero potential energy. Why? Thus,

$$V_2 = 0$$

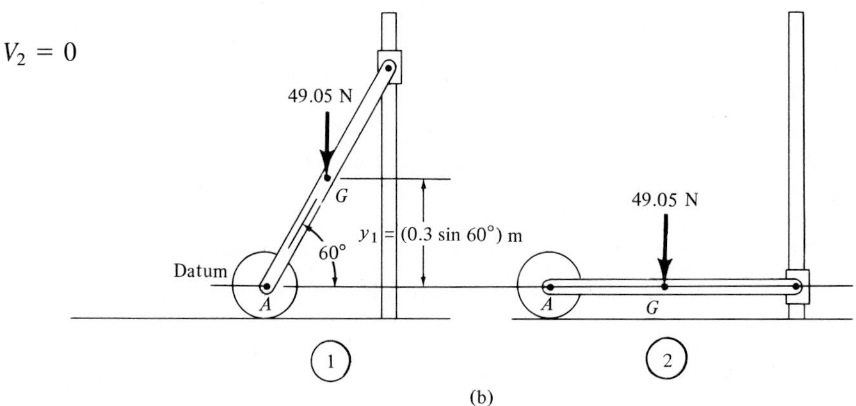

(b)

Kinetic Energy. Since the rod is at rest in the initial position,

$$T_1 = 0$$

In the final position the rod has an angular velocity $(\boldsymbol{\omega}_R)_2$ and its mass center has a velocity $(\mathbf{v}_G)_2$, Fig. 18–18c. For the rod $(\mathbf{v}_G)_2$ can be related to $(\boldsymbol{\omega}_R)_2$ from the instantaneous center of zero velocity, which is located at point A, Fig. 18–18c. Hence, $(v_G)_2 = r_{G/IC}(\omega_R)_2$ or $(v_G)_2 = 0.3(\omega_R)_2$. Thus,

$$\begin{aligned} T_2 &= \tfrac{1}{2}m_R(v_G)_2^2 + \tfrac{1}{2}I_G(\omega_R)_2^2 \\ &= \tfrac{1}{2}(5)(0.3(\omega_R)_2)^2 + \tfrac{1}{2}[\tfrac{1}{12}(5)(0.6)^2](\omega_R)_2^2 \\ &= 0.3(\omega_R)_2^2 \end{aligned}$$

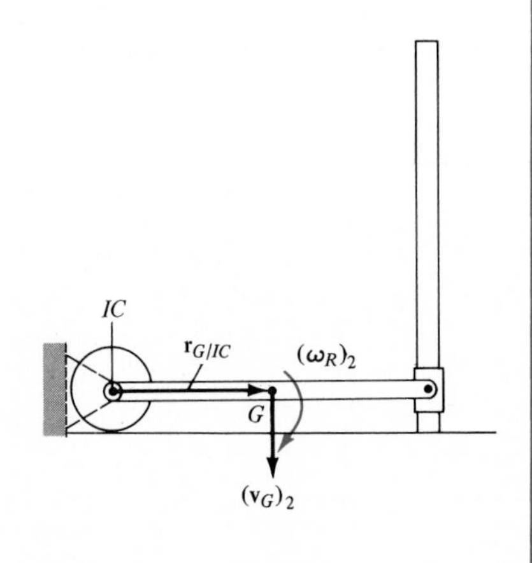

(c)

Conservation of Energy Theorem

$$\{T_1\} + \{V_1\} = \{T_2\} + \{V_2\}$$

$$\{0\} + \{12.74\} = \{0.3(\omega_R)_2^2\} + \{0\}$$

$$(\omega_R)_2 = 6.52 \text{ rad/s} \qquad \textit{Ans.}$$

Example 18–7

The disk shown in Fig. 18–19*a* has a weight of 30 lb and is attached to a spring which has a stiffness of $k = 2$ lb/ft and an unstretched length of 1 ft. If the disk is released from rest in the position shown and rolls without slipping, determine its angular velocity at the instant it is displaced 3 ft.

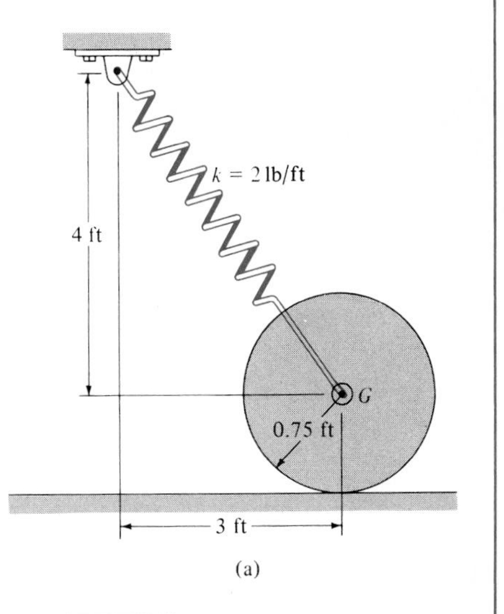

(a)

Solution

Potential Energy. Two diagrams of the disk, when it is located in its initial and final positions, are shown in Fig. 18–19*b*. A gravitational datum is not needed here since the weight is not displaced vertically. From the problem geometry the spring is stretched $x_1 = (\sqrt{3^2 + 4^2} - 1) = 4$ ft and $x_2 = (4 - 1) = 3$ ft in the initial and final positions, respectively. Hence,

$$V_1 = \frac{1}{2}kx_1^2 = \frac{1}{2}(2)(4)^2 = 16 \text{ J}$$

$$V_2 = \frac{1}{2}kx_2^2 = \frac{1}{2}(2)(3)^2 = 9 \text{ J}$$

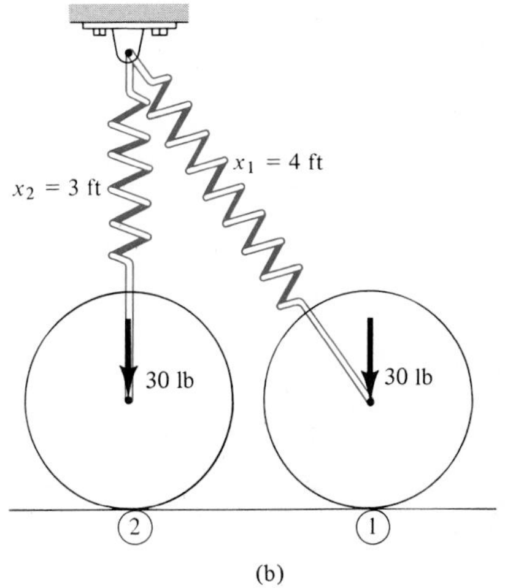

(b)

Kinetic Energy. The disk is released from rest so that $(\mathbf{v}_G)_1 = \mathbf{0}$, $\boldsymbol{\omega}_1 = \mathbf{0}$, and

$$T_1 = 0$$

Since the disk rolls without slipping, $(\mathbf{v}_G)_2$ can be related to $\boldsymbol{\omega}_2$ from the instantaneous center of zero velocity, Fig. 18–19*c*. Hence, $(v_G)_2 = 0.75\omega_2$. Thus,

$$\begin{aligned} T_2 &= \frac{1}{2}m(v_G)_2^2 + \frac{1}{2}I_G(\omega_2)^2 \\ &= \frac{1}{2}\left(\frac{30}{32.2}\right)(0.75\omega_2)^2 + \frac{1}{2}\left[\frac{1}{2}\left(\frac{30}{32.2}\right)(0.75)^2\right](\omega_2)^2 \\ &= 0.393(\omega_2)^2 \end{aligned}$$

Conservation of Energy Theorem

$$\{T_1\} + \{V_1\} = \{T_2\} + \{V_2\}$$

$$\{0\} + \{16\} = \{0.393(\omega_2)^2\} + \{9\}$$

$$\omega_2 = 4.22 \text{ rad/s} \qquad \textit{Ans.}$$

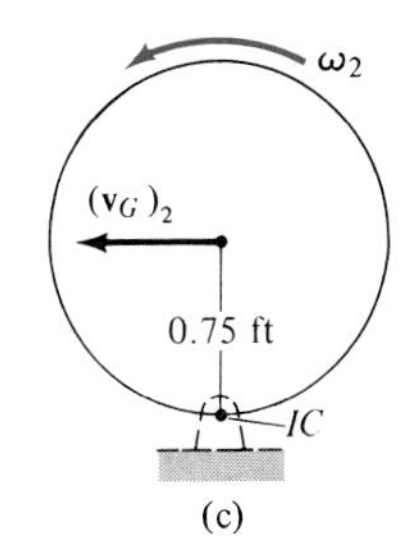

(c)

Fig. 18–19

Example 18–8

The 10-kg rod AB shown in Fig. 18–20a is confined so that its ends move in the horizontal and vertical slots. The spring has a stiffness of $k = 800$ N/m and is unstretched when $\theta = 0°$. Determine the angular velocity of AB when $\theta = 0°$, if AB is released from rest when $\theta = 30°$. Neglect the mass of the slider blocks.

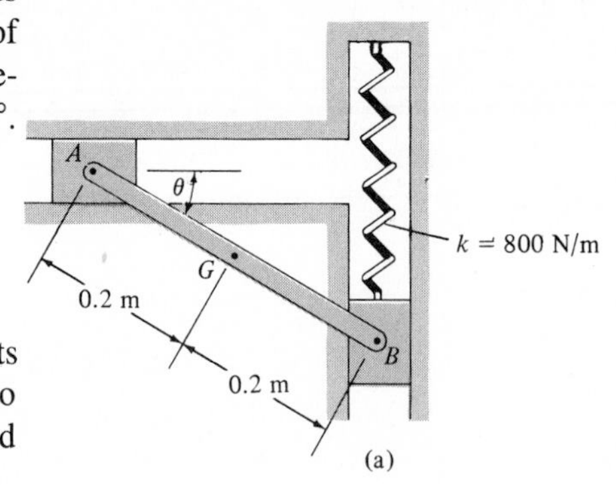

(a)

Solution

Potential Energy. The two diagrams of the rod, when it is located at its initial and final positions, are shown in Fig. 18–20b. The datum, used to measure the gravitational potential energy, is placed in line with the rod when $\theta = 0°$.

When the rod is in position 1, the center of mass G is located *below the datum* so that the gravitational potential energy is *negative*. Furthermore, (positive) elastic potential energy is stored in the spring, since it is stretched a distance of $x_1 = (0.4 \sin 30°)$ m. Thus,

$$V_1 = -Wy_1 + \tfrac{1}{2}kx_1^2$$
$$= -98.1(0.2 \sin 30°) + \tfrac{1}{2}(800)(0.4 \sin 30°)^2 = 6.19 \text{ J}$$

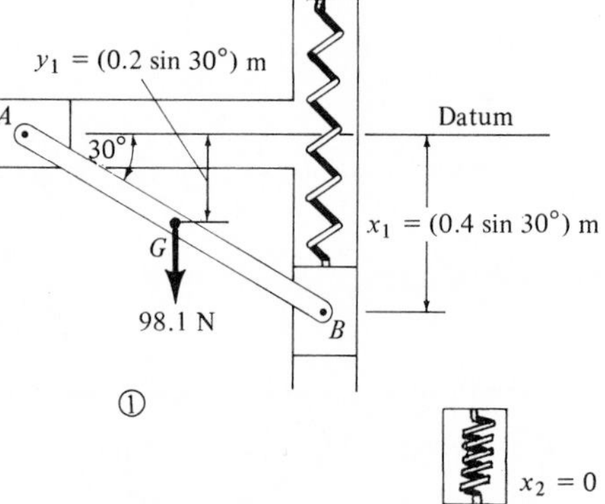

When the rod is in position 2, the potential energy of the rod is zero, since the spring is unstretched, $x_2 = 0$, and the center of mass G is located at the datum. Thus,

$$V_2 = 0$$

Kinetic Energy. The rod is released from rest from position 1, thus $(\mathbf{v}_G)_1 = \mathbf{0}$ and $\boldsymbol{\omega}_1 = \mathbf{0}$, and

$$T_1 = 0$$

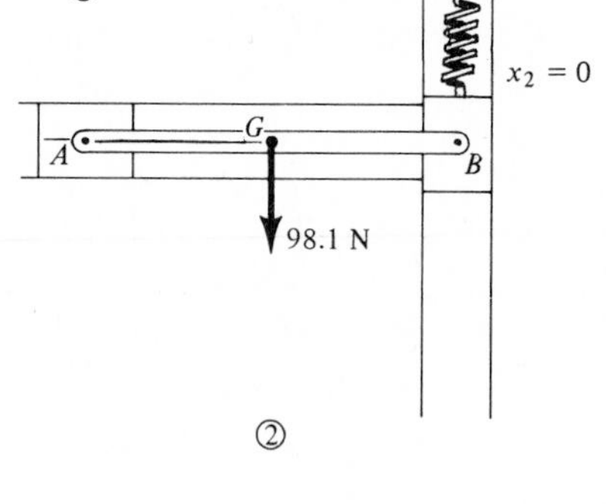

(b)

In position 2 the angular velocity is $\boldsymbol{\omega}_2$ and the rod's mass center has a velocity of $(\mathbf{v}_G)_2$. Using *kinematics*, $(\mathbf{v}_G)_2$ can be related to $\boldsymbol{\omega}_2$ as shown in Fig. 18–20c. At the instant considered, the instantaneous center of zero velocity (IC) for the rod is at point A; hence, $(v_G)_2 = (r_{G/IC})\omega_2 = (0.2)\omega_2$. Thus,

$$T_2 = \tfrac{1}{2}m(v_G)_2^2 + \tfrac{1}{2}I_G(\omega_2)^2$$
$$= \tfrac{1}{2}(10)(0.2\omega_2)^2 + \tfrac{1}{2}[\tfrac{1}{12}(10)(0.4)^2](\omega_2)^2 = 0.267\omega_2^2$$

Conservation of Energy Theorem

$$\{T_1\} + \{V_1\} = \{T_2\} + \{V_2\}$$
$$\{0\} + \{6.19\} = \{0.267\omega_2^2\} + \{0\}$$
$$\omega_2 = 4.82 \text{ rad/s} \qquad \textit{Ans.}$$

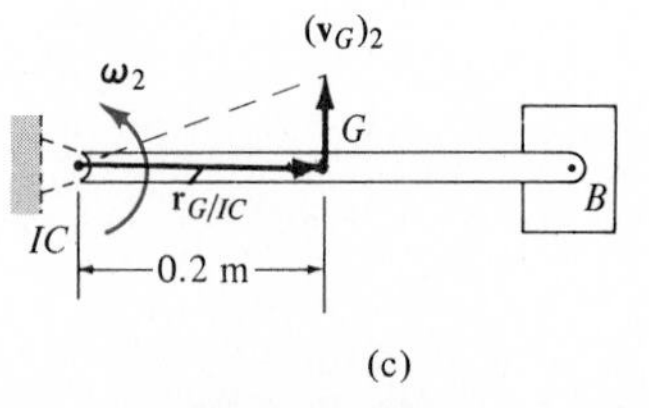

(c)

Fig. 18–20

Problems

***18–16.** Solve Prob. 18–4 using the conservation of energy theorem.

18–17. Solve Prob. 18–13 using the conservation of energy theorem.

18–18. The 500-g rod AB rests along the smooth inner surface of a hemispherical bowl. If the rod is released from rest from the position shown, determine its angular velocity $\boldsymbol{\omega}$ at the instant it swings downward and becomes horizontal. This requires the mass center to move 2.68 mm downward.

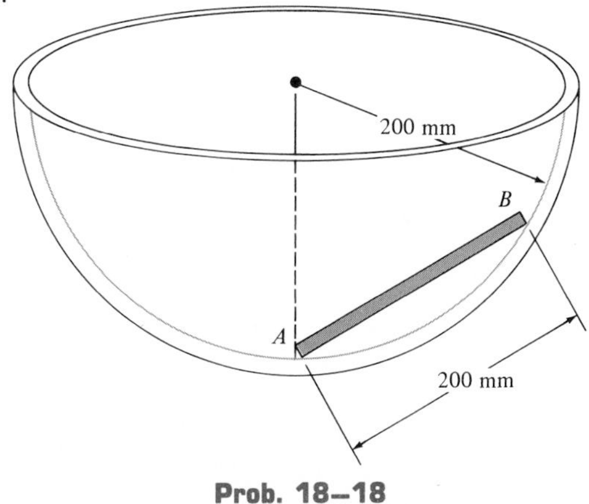

Prob. 18–18

18–19. An automobile tire has a mass of 7 kg and radius of gyration of $k_G = 0.3$ m. If it is released from rest at A on the incline, determine its angular velocity when it reaches the horizontal plane. The tire rolls without slipping.

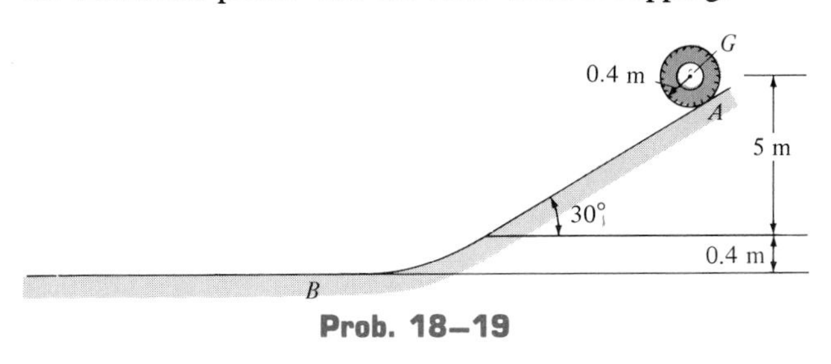

Prob. 18–19

***18–20.** The uniform 150-lb stone (rectangular block) is being turned over on its side by pulling the vertical cable *slowly* upward until the stone begins to tip. If it then falls freely ($\mathbf{T} = \mathbf{0}$) from an essentially balanced at rest position, determine the speed at which the corner A strikes the pad at B. The stone does not slip at its corner C as it falls. Take $I_G = 1.65$ slug · ft^2.

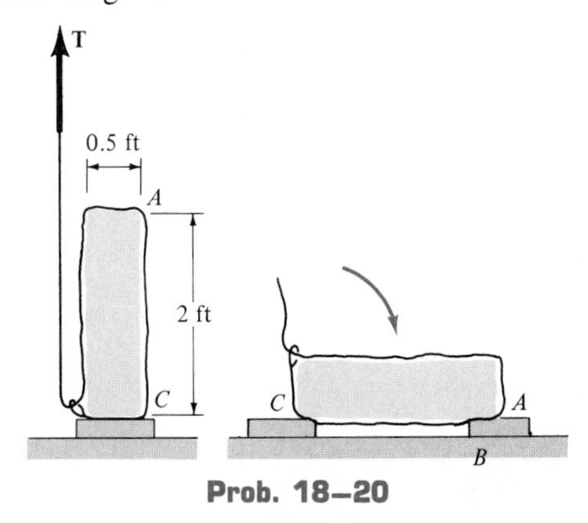

Prob. 18–20

18–21. A girl, having a weight of 110 lb, is sitting on a swing such that when she is at A, $\theta = 30°$, her velocity is equal to zero, and her center of gravity is at G. If she keeps this same fixed (rigid) position as she swings forward, determine the angular velocity of the swing when she reaches the lowest point at B ($\theta = 0°$). Her radius of gyration about an axis passing through G is 1.8 ft.

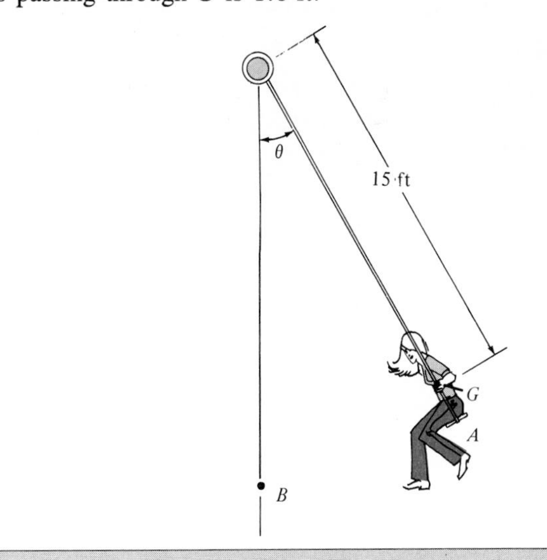

Prob. 18–21

18–22. A chain that has a negligible mass is draped over a sprocket which has a mass of 2 kg and a radius of gyration of $k_O = 50$ mm. If the 4-kg block A is released from rest in the position shown, $s = 1$ m, determine the angular velocity which the chain imparts to the sprocket when $s = 2$ m.

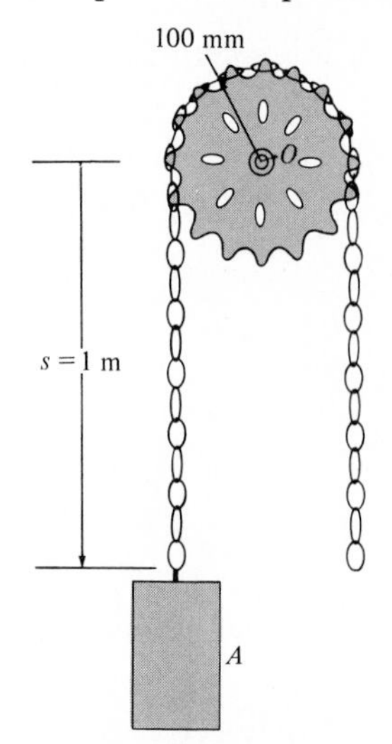

Prob. 18–22

18–23. The jeep and passenger have a total weight of 1,600 lb, not including the four wheels. Each wheel has a weight of 20 lb and a radius of gyration of $k_G = 0.4$ ft about an axis passing through its center. If the jeep starts from rest and coasts down the hill in neutral gear, determine the distance it must travel to attain a speed of 15 ft/s. The tires do not slip on the ground.

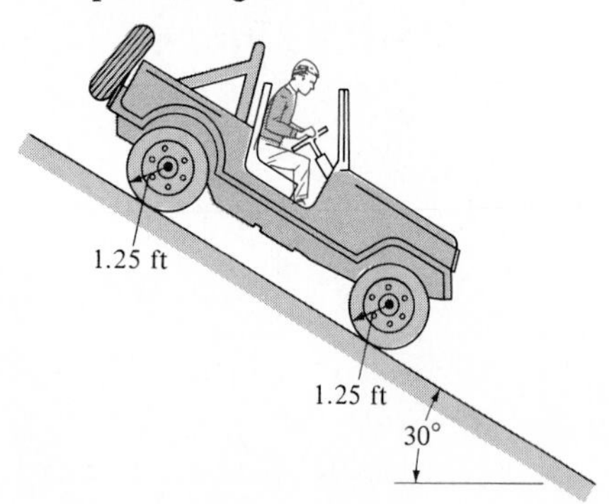

Prob. 18–23

***18–24.** The window AB, shown in side view, has a mass of 10 kg and a radius of gyration of $k_G = 0.2$ m. The attached elastic cable BC has a stiffness of $k = 50$ N/m and an unstretched length of 0.4 m. If the window is released from rest when $\theta \approx 0°$, determine its angular velocity just before it strikes the wall at D, $\theta = 90°$. *Hint:* At $\theta = 0°$, $x_1 = 1.4$ m; and at $\theta = 90°$, $x_2 = 0.880$ m.

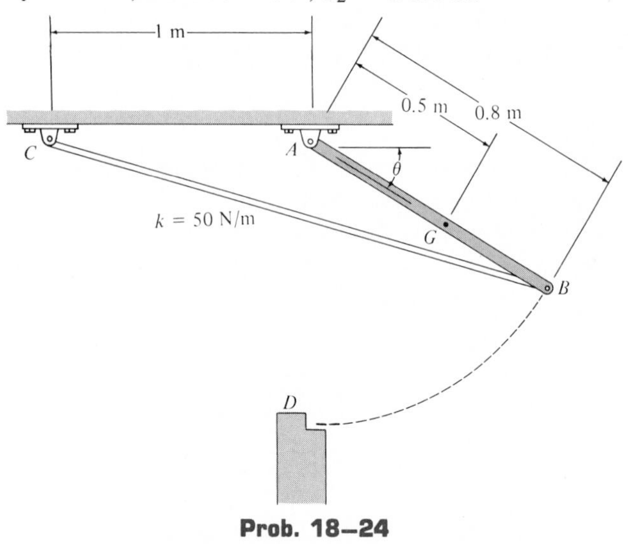

Prob. 18–24

18–25. The small bridge consists of an 1,800-lb uniform deck EF (thin plate), two overhead beams AB (slender rods), each having a weight of 200 lb, and a 2,400-lb counterweight BC, which can be considered as a thin plate having the dimensions shown. The weight of the tie rods AE can be neglected. If the operator lets go of the rope when the bridge is at an at-rest position, $\theta = 45°$, determine the speed at which the end of the deck E hits the roadway step at H, $\theta = 0°$. The bridge is pin-connected at A, D, E, and F.

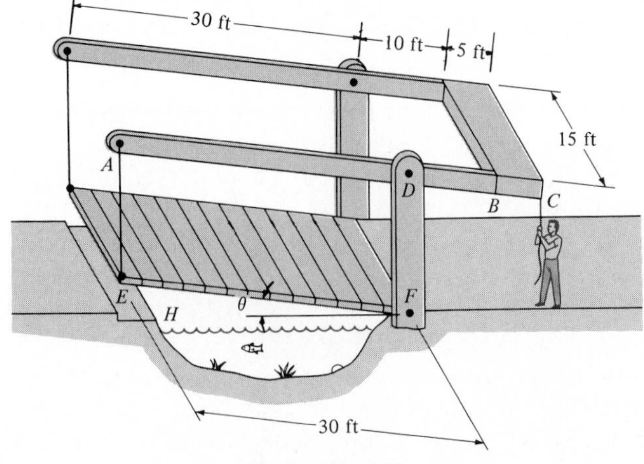

Prob. 18–25

18–26. The window AB, shown in side view, has a weight of 8 lb, a center of gravity at G, and a radius of gyration of $k_A = 0.4$ ft about its hinged axis at A. If the spring is unstretched when $\theta \approx 0°$ and the window is released from rest from this position, determine the velocity with which the end B strikes the wall D, $\theta = 90°$. Neglect friction at the hinge A and at pulley C.

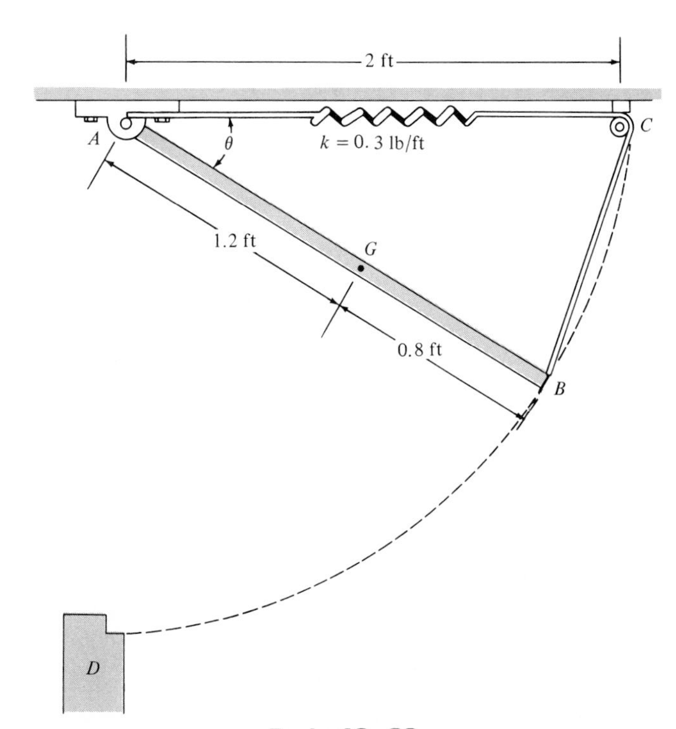

Prob. 18–26

18–27. The uniform window shade AB has a total weight of 0.4 lb. When it is released, it winds around the spring-loaded core O. Motion is caused by a spring within the core, which is coiled so that it exerts a torque on the core of $M = (3(10^{-4})\theta)$ lb · ft. If the shade is released from rest, determine the angular velocity of the core at the instant the shade is completely rolled up, i.e., after 12 revolutions. When this occurs, the spring becomes uncoiled and the radius of gyration of the shade about the axle at O is $k_O = 0.9$ in. *Note:* The elastic potential energy of a torsional spring is $V_e = \frac{1}{2}k\theta^2$, where $M = k\theta$. Here $k = 3(10^{-4})$ lb · ft/rad.

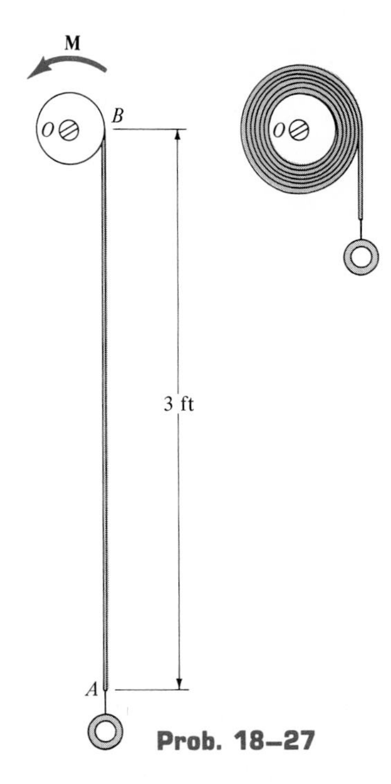

Prob. 18–27

***18–28.** The drum A has a weight of 20 lb and a centroidal radius of gyration of $k_A = 0.6$ ft. Determine the speed of the 20-lb crate C at the instant $s = 10$ ft. Initially, the crate is released from rest when $s = 5$ ft. For the calculation, neglect the mass of this pulley and the cord as it unravels from A.

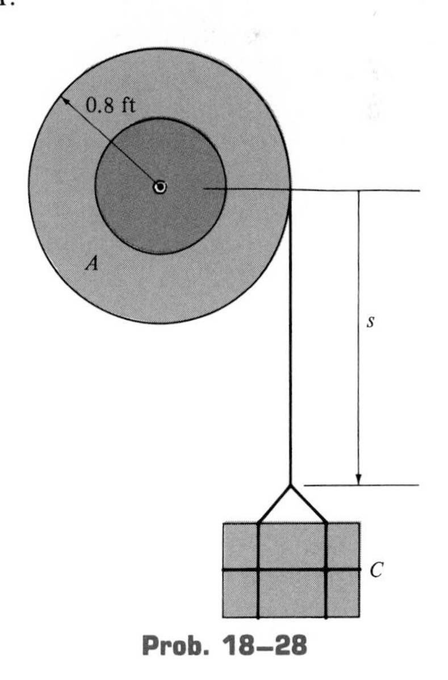

Prob. 18–28

18–29. The uniform bar AB has a mass of 10 kg and is pin-connected at A. If the support at B is removed ($\theta = 90°$), determine the angular velocity of the bar at the instant the bar rotates downward to $\theta = 150°$.

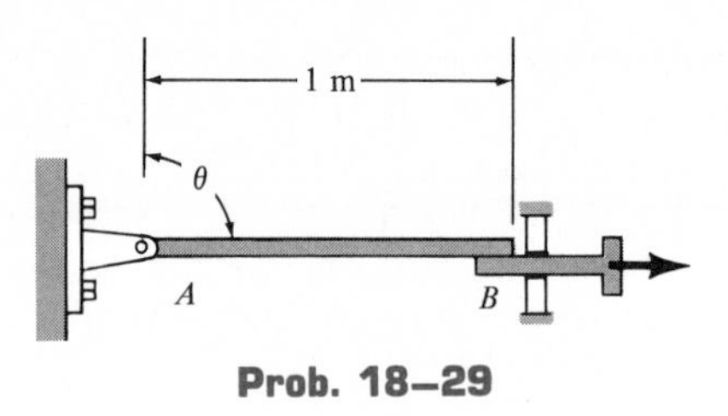

Prob. 18–29

Review Problems

18–30. The spool has a weight of 150 lb and a radius of gyration of $k_O = 2.25$ ft. If a cord is wrapped around its inner core and the end is pulled with a horizontal force of $P = 40$ lb, determine the angular velocity of the spool after the center O has moved 10 ft to the right. The spool starts from rest and does not slip as it rolls. Neglect the mass of the cord.

Probs. 18–30 / 18–31

18–31. Determine the magnitude of the horizontal force **P** that must be applied to the cord which is wrapped around the inner core of the spool, so that the spool attains an angular velocity of $\omega = 5$ rad/s when the center O has moved 10 ft to the right. Neglect the mass of the cord. The spool has a weight of 150 lb and a radius of gyration of $k_O = 2.25$ ft. It starts from rest and does not slip as it rolls.

***18–32.** Determine the reaction at the pin O of the 5-kg rod in Prob. 17–76 at the instant the rod swings down to the vertical position.

18–33. The pendulum of the Charpy impact machine has a mass of 50 kg and a radius of gyration of $k_A = 0.75$ m. If it is released from rest when $\theta = 0°$, determine its angular velocity just before it strikes the specimen S; $\theta = 90°$.

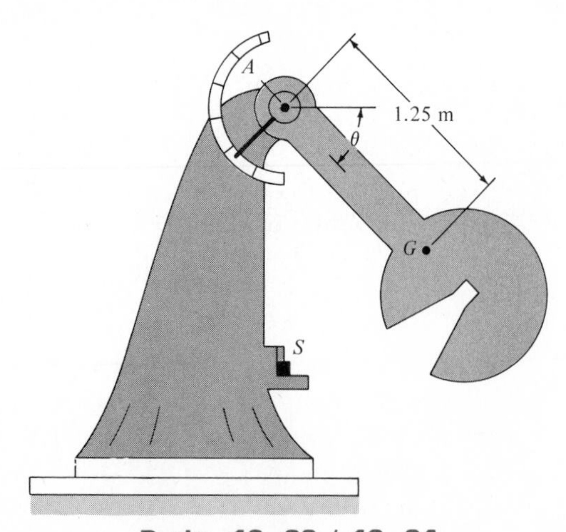

Probs. 18–33 / 18–34

18–34. If the 50-kg pendulum is released from rest when $\theta = 0°$ and after striking the specimen S swings to a maximum value of $\theta = 135°$ before stopping, determine the energy lost when the pendulum strikes the specimen. The radius of gyration of the pendulum is $k_A = 1.5$ m.

18–35. Excluding the four wheels, the car and its contents have a total mass of 600 kg. If each of its *four* wheels is assumed to be a 20-kg disk which rolls without slipping, determine the speed of the car after the horizontal towing force $P = 175$ N moves the car 15 m, starting from rest.

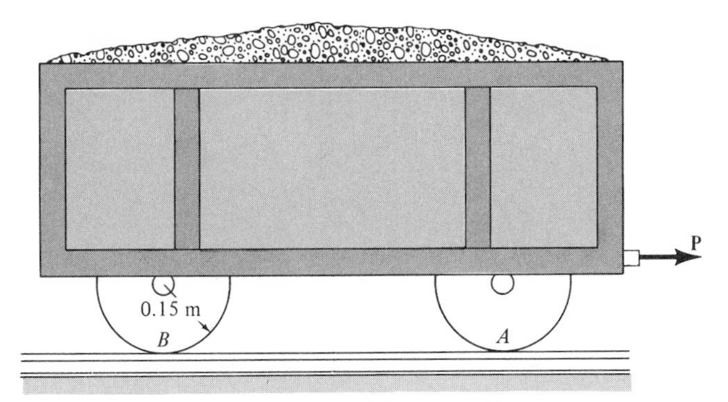

Probs. 18–35 / 18–36

***18–36.** Determine the magnitude of the horizontal towing force **P** which should be applied to the car so that it attains a speed of 6 m/s after traveling 15 m starting from rest. Excluding its wheels the car has a mass of 600 kg. Assume each wheel is a 20-kg disk which rolls without slipping.

18–37. The motor provides a constant torque of 60 N · m to the drum D. If the attached cable is wrapped around the inner core of the spool at A, determine the angular velocity of the spool just after 20 m of cable have been wrapped around the drum. The spool starts from rest and has a mass of 130 kg and a radius of gyration of $k_O = 225$ mm. Neglect the mass of the cable and pulley at B.

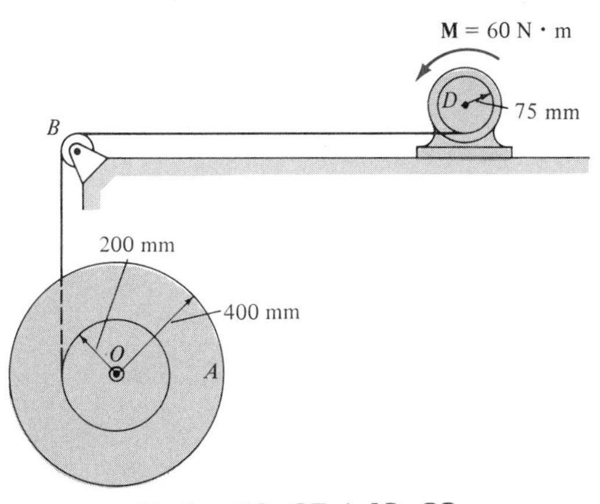

Probs. 18–37 / 18–38

18–38. The motor M provides a constant torque of 60 N · m to the drum D. If the attached cable is wrapped around the inner core of the spool at A, determine the number of revolutions the drum must make before the spool attains an angular velocity of 60 rad/s. The spool starts from rest and has a mass of 130 kg and a radius of gyration of $k_O = 225$ mm.

18–39. If the 3-lb solid ball is released from rest when $\theta = 30°$, determine its angular velocity when $\theta = 0°$, which is the lowest point of the curved path having a radius of 11.5 in. The ball does not slip as it rolls.

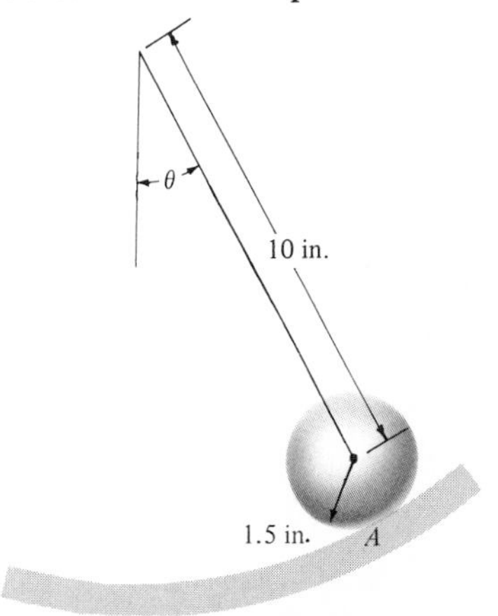

Prob. 18–39

***18–40.** A large roll of paper having a mass of 20 kg and radius $r = 150$ mm is resting over the edge of a table, such that the end of the paper on the roll is attached to the table's surface. If the roll is disturbed slightly from its equilibrium position, determine the angular velocity of the roll at the instant $\theta = 45°$. The centroidal radius of gyration of the roll is $k_G = 75$ mm.

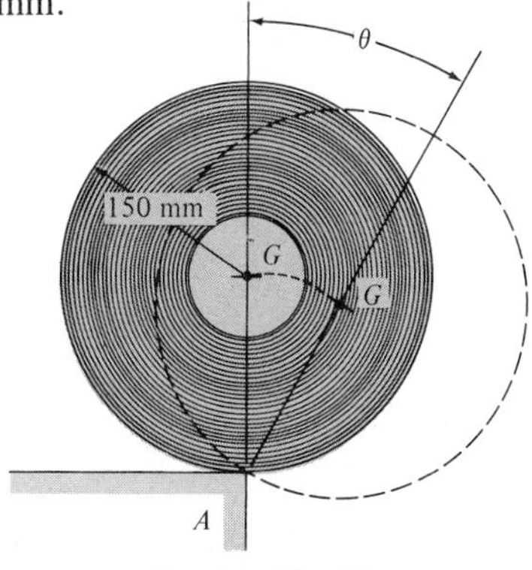

Prob. 18–40

Planar Kinetics of a Rigid Body: Impulse and Momentum

In Chapter 15 it was shown that problems which involve force, velocity, and time can conveniently be solved by using the principle of linear or angular impulse and momentum. In this chapter we will extend these concepts somewhat in order to determine the impulse and momentum relationships that apply to a rigid body. The three planar motions which will be considered are translation, rotation about a fixed axis, and general plane motion. A more general discussion of momentum principles applied to the gyroscopic motion of a rigid body is given at the end of the chapter.

19.1 Linear and Angular Momentum

Linear Momentum. The linear momentum of a rigid body is determined by summing vectorially the linear momenta of all the particles of the body, i.e., $\mathbf{L} = \Sigma m_i \mathbf{v}_i$. We can simplify this expression by noting that $\Sigma m_i \mathbf{v}_i = m\mathbf{v}_G$ (see Sec. 15–2) so that

$$\mathbf{L} = m\mathbf{v}_G \tag{19–1}$$

This equation states that the body's linear momentum is a vector quantity having a *magnitude* mv_G, which is commonly measured in units of kg · m/s or slug · ft/s, and a *direction* defined by $\mathbf{v}_G$, the instantaneous velocity of the mass center.

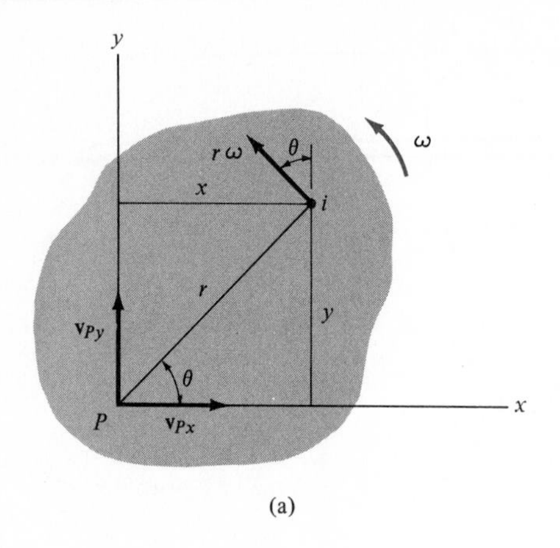

(a)

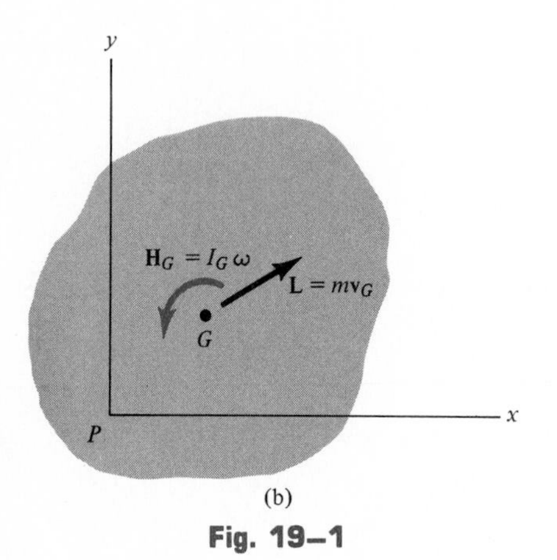

(b)

Fig. 19–1

Angular Momentum. Consider the body in Fig. 19–1*a*, which is subjected to general plane motion. At the instant shown, point *P* has *x* and *y* components of velocity $(\mathbf{v}_P)_x$ and $(\mathbf{v}_P)_y$, and the body has an angular velocity $\boldsymbol{\omega}$. If the motion of the *i*th particle of the body is to be determined, then

$$\mathbf{v}_i = \mathbf{v}_P + \mathbf{v}_{i/P}$$

$$\mathbf{v}_i = \underset{\rightarrow}{(v_P)_x} + \underset{\uparrow}{(v_P)_y} + \underset{\nwarrow \theta}{r\omega}$$

The angular momentum for particle *i* about point *P* is equal to the "moment" of the particle's linear momentum about *P*. From Fig. 19–1*a*, using the above result for $\mathbf{v}_i$, we have

$$\circlearrowleft + \; dH_P = -y\,dm(v_P)_x + x\,dm(v_P)_y + r^2\omega\,dm$$

Integrating over the entire mass *m* of the body, we obtain

$$\circlearrowleft + \; H_P = -\left(\int_m y\,dm\right)(v_P)_x + \left(\int_m x\,dm\right)(v_P)_y + \left(\int_m r^2\,dm\right)\omega$$

Here H_P represents the angular momentum of the body about an axis (the *z* axis) perpendicular to the plane of motion and passing through point *P*. Since $\bar{x}m = \int x\,dm$ and $\bar{y}m = \int y\,dm$, the integrals for the first and second terms on the right are used to locate the body's center of mass *G* with respect to *P*. Also, the last integral represents the body's moment of inertia computed about the *z* axis, i.e., $I_P = \int r^2\,dm$. Thus,

$$\circlearrowleft + \; H_P = -\bar{y}m(v_P)_x + \bar{x}m(v_P)_y + I_P\omega \qquad (19\text{–}2)$$

This equation reduces to a simpler form if point *P* coincides with the mass center *G* for the body, in which case $\bar{x} = \bar{y} = 0$ and therefore

$$H_G = I_G\omega$$

where I_G is the mass moment of inertia of the body about an axis which is perpendicular to the plane of motion and passes through the mass center, and $\boldsymbol{\omega}$ is the instantaneous angular velocity of the body. It is important to realize that $\mathbf{H}_G$ is a vector quantity having a *magnitude* $I_G\omega$, which is commonly measured in units of kg · m²/s or slug · ft²/s, and a *direction* defined by $\boldsymbol{\omega}$, which is always perpendicular to the slab. Since $\boldsymbol{\omega}$ is a free vector, $\mathbf{H}_G$ *can act at any point on the body* (or slab) provided it preserves its same magnitude and direction. Furthermore, since angular momentum is equal to the moment of the linear momentum, the *line of action of* **L** *must pass through the body's mass center* *G* in order to preserve the correct magnitude of $\mathbf{H}_G$ when the "moments" are computed about *G*, Fig. 19–1*b*. In other words, **L** creates *zero* angular momentum about *G*, so that simply $\mathbf{H}_G = I_G\boldsymbol{\omega}$.

Translation. When a rigid body of mass m is subjected to either rectilinear or curvilinear *translation,* Fig. 19–2a, its mass center has a velocity of $\mathbf{v}_G = \mathbf{v}$ and $\boldsymbol{\omega} = \mathbf{0}$. Hence, the linear momentum and the angular momentum, computed about G, become

$$\mathbf{L} = m\mathbf{v}_G \qquad \mathbf{H}_G = \mathbf{0} \tag{19–3}$$

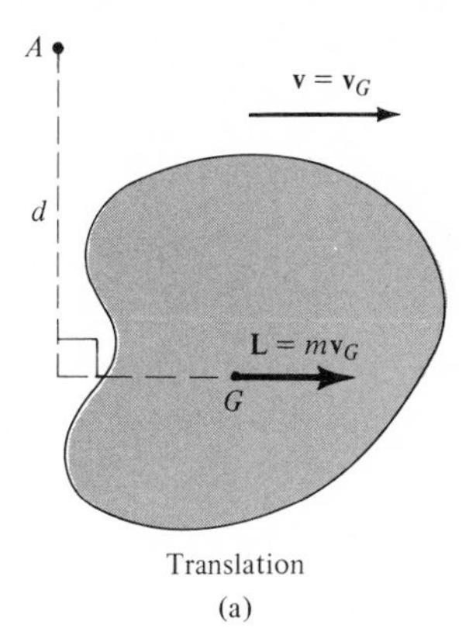

Translation

(a)

If the angular momentum is computed about any other point A on or off the body, Fig. 19–2a, the "moment" of the linear momentum $\mathbf{L}$ must be computed about the point. Since d is the "moment arm" as shown in the figure, then $\circlearrowleft + H_A = (d)(mv_G)$.

Rotation about a Fixed Axis. When a rigid body is *rotating about a fixed axis* passing through point O, Fig. 19–2b, the linear momentum and the angular momentum computed about G are

$$L = mv_G \qquad H_G = I_G\omega \tag{19–4}$$

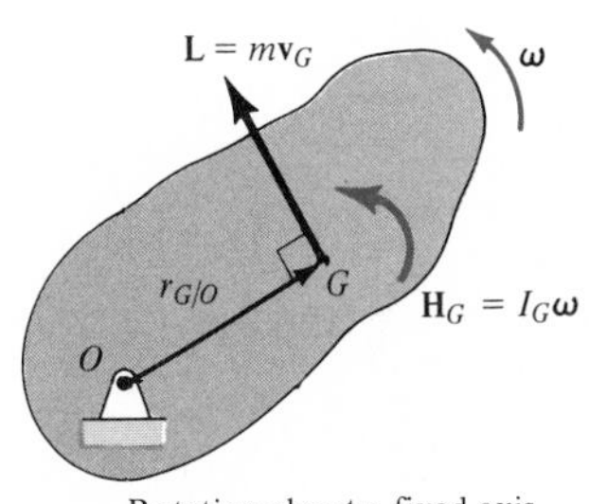

Rotation about a fixed axis

(b)

Fig. 19–2a, b

It is sometimes convenient to compute the angular momentum of the body about point O. In this case it is necessary to account for the "moments" of *both* $\mathbf{L}$ and $\mathbf{H}_G$ about O. Noting that $\mathbf{L}$ (or $\mathbf{v}_G$) is always *perpendicular to* $\mathbf{r}_{G/O}$, we have

$$\circlearrowleft + H_O = I_G\omega + r_{G/O}(mv_G) \tag{19–5}$$

This equation may be *simplified* by first substituting $v_G = r_{G/O}\omega$, in which case $H_O = (I_G + mr_{G/O})\omega$, and, by the parallel-axis theorem, noting that the terms inside the parentheses represent the moment of inertia I_O of the body about an axis perpendicular to the plane of motion and passing through point O. Hence,*

$$H_O = I_O\omega \tag{19–6}$$

For the computation, then, either Eq. 19–5 or 19–6 can be used.

*The similarity between this derivation and that of Eq. 17–15 ($\Sigma M_O = I_O\alpha$) and Eq. 18–3 ($T = \frac{1}{2}I_O\omega^2$) should be noted. Also note that the same result can be obtained from Eq. 19–2 by selecting point P at O, realizing that $(\mathbf{v}_O)_x = (\mathbf{v}_O)_y = 0$.

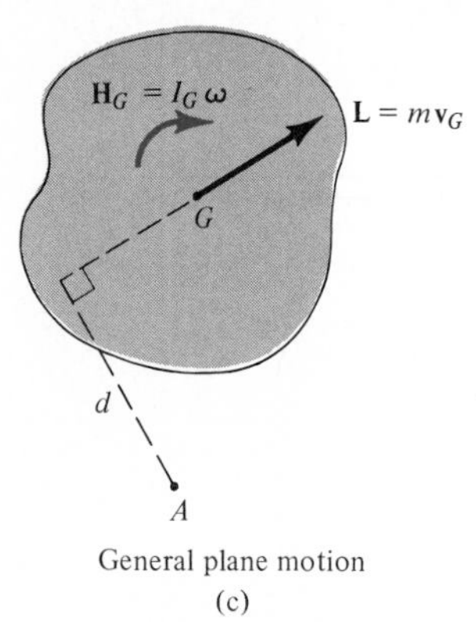

General plane motion
(c)
Fig. 19–2c

General Plane Motion. When a rigid body is subjected to general plane motion, Fig. 19–2*c*, the linear momentum and the angular momentum computed about G become

$$L = mv_G$$
$$H_G = I_G\omega \tag{19–7}$$

If the angular momentum is computed about a point A located either on or off the body, Fig. 19–2*c*, it is necessary to compute the moments of *both* **L** and $\mathbf{H}_G$ about this point. In this case,

$$\circlearrowleft + H_A = I_G\omega + (d)(mv_G)$$

Here d is the moment arm, as shown in the figure.

19.2 Principle of Impulse and Momentum

Principle of Linear Impulse and Momentum. The equation of translational motion for a rigid body can be written as $\Sigma\mathbf{F} = m\mathbf{a}_G = m\,(d\mathbf{v}_G/dt)$. Since the mass of the body is constant,

$$\Sigma\mathbf{F} = \frac{d}{dt}(m\mathbf{v}_G)$$

Multiplying both sides by dt and integrating from $t = t_1$, $\mathbf{v}_G = (\mathbf{v}_G)_1$ to $t = t_2$, $\mathbf{v}_G = (\mathbf{v}_G)_2$ yields

$$\Sigma\int_{t_1}^{t_2} \mathbf{F}\,dt = m(\mathbf{v}_G)_2 - m(\mathbf{v}_G)_1 \tag{19–8}$$

This equation is referred to as the *principle of linear impulse and momentum*. It states that the sum of all the impulses created by the *external force system* which acts on the body during the time interval t_1 to t_2 is equal to the change in the linear momentum of the body during the time interval.

Principle of Angular Impulse and Momentum. If the body is subjected to *general plane motion*, using Eq. 17–7 we can write $\Sigma M_G = I_G\alpha = I_G(d\omega/dt)$. Since the moment of inertia is constant,

$$\Sigma M_G = \frac{d}{dt}(I_G\omega)$$

Multiplying both sides by dt and integrating from $t = t_1$, $\omega = \omega_1$ to $t = t_2$, $\omega = \omega_2$ gives

$$\Sigma\int_{t_1}^{t_2} M_G\,dt = I_G\omega_2 - I_G\omega_1 \tag{19–9}$$

In a similar manner, for *rotation about a fixed axis* passing through point O, Eq. 17–15 ($\Sigma M_O = I_O\alpha$) when integrated becomes

$$\Sigma \int_{t_1}^{t_2} M_O \, dt = I_O\omega_2 - I_O\omega_1 \tag{19–10}$$

Equations 19–9 and 19–10 are referred to as the *principle of angular impulse and momentum*. Both equations state that the sum of the angular impulses acting on the body during the time interval t_1 to t_2 is equal to the change in the body's angular momentum during this time interval. In particular, the angular impulse considered is determined by integrating the moments about point G or O of all the external forces and couples applied to the body.

To summarize the preceding concepts, if motion is occurring in the x-y plane, using impulse and momentum principles the following *three scalar equations* may be written which describe the *planar motion* of the body:

$$\begin{aligned} m(v_{Gx})_1 + \Sigma \int_{t_1}^{t_2} F_x dt &= m(v_{Gx})_2 \\ m(v_{Gy})_1 + \Sigma \int_{t_1}^{t_2} F_y dt &= m(v_{Gy})_2 \\ I_G\omega_1 + \Sigma \int_{t_1}^{t_2} M_G dt &= I_G\omega_2 \end{aligned} \tag{19–11}$$

The first two of these equations represent the principle of linear impulse and momentum in the x-y plane (Eq. 19–8), and the third equation represents the principle of angular impulse and momentum about the z axis (Eq. 19–9).

Equations 19–11 may also be applied to an entire system of connected bodies rather than to each body separately. Doing this eliminates the need to include reactive impulses which occur at the connections since they are *internal* to the system. The resultant equations may be written in symbolic form as

$$\begin{aligned} \left(\Sigma \begin{matrix}\text{syst. linear}\\ \text{momentum}\end{matrix}\right)_{x1} + \left(\Sigma \begin{matrix}\text{syst. linear}\\ \text{impulse}\end{matrix}\right)_{x(1-2)} &= \left(\Sigma \begin{matrix}\text{syst. linear}\\ \text{momentum}\end{matrix}\right)_{x2} \\ \left(\Sigma \begin{matrix}\text{syst. linear}\\ \text{momentum}\end{matrix}\right)_{y1} + \left(\Sigma \begin{matrix}\text{syst. linear}\\ \text{impulse}\end{matrix}\right)_{y(1-2)} &= \left(\Sigma \begin{matrix}\text{syst. linear}\\ \text{momentum}\end{matrix}\right)_{y2} \\ \left(\Sigma \begin{matrix}\text{syst. angular}\\ \text{momentum}\end{matrix}\right)_{O1} + \left(\Sigma \begin{matrix}\text{syst. angular}\\ \text{impulse}\end{matrix}\right)_{O(1-2)} &= \left(\Sigma \begin{matrix}\text{syst. angular}\\ \text{momentum}\end{matrix}\right)_{O2} \end{aligned} \tag{19–12}$$

As indicated, the system's angular momentum and angular impulse must be computed with respect to the *same fixed reference point O* for all the bodies of the system.

PROCEDURE FOR ANALYSIS

Impulse and momentum principles are used to solve kinetics problems that involve *velocity, force,* and *time* since these terms are involved in the formulation. For application it is suggested that the following procedure be used.

Impulse and Momentum Diagrams. Draw the impulse and momentum diagrams for the body or system of bodies. Each of these diagrams represents an outlined shape of the body which graphically accounts for the data required for each of the three terms in Eqs. 19–11 or 19–12.

An appropriate set of impulse and momentum diagrams for a rigid body subjected to general plane motion is shown in Fig. 19–3. Note that the linear-momentum vectors $m\mathbf{v}_G$ are applied at the body's mass center, Fig. 19–3*a* and *c*; whereas the angular-momentum vectors, $I_G\boldsymbol{\omega}$, are free vectors, and therefore, like a couple, they may be applied at any point on the body.

When the impulse diagram is constructed, Fig. 19–3*b*, vectors **F** and **M** which vary with time are indicated by the integrals. However, if **F** and **M** are *constant* from t_1 to t_2, integration of the impulses yields $\mathbf{F}(t_2 - t_1)$ and $\mathbf{M}(t_2 - t_1)$, respectively. Such is the case for the body's weight **W**, Fig. 19–3*b*.

Principle of Impulse and Momentum. Apply the three scalar equations 19–11 (or 19–12) by determining the vector components for each of the terms in these equations directly from the impulse and momentum diagrams. In cases where the body is rotating about a fixed axis, Eq. 19–10 may be substituted for the third of Eqs. 19–11.

Kinematics. If more than three equations are needed for a complete solution, it may be possible to relate the velocity of the body's mass center to the body's angular velocity using *kinematics*. If the motion appears to be complicated, kinematic (velocity) diagrams may be helpful in obtaining the necessary relation.

In general, a method to be used for the solution of a particular type of problem should be decided upon *before* attempting to solve the problem. As stated above, the *principle of impulse and momentum* is most suitable for solving problems which involve *velocity, force,* and *time*. For some problems, however, a combination of the equations of motion and its two integrated forms, the principle of work and energy and the principle of impulse and momentum, will yield the most direct solution to the problem.

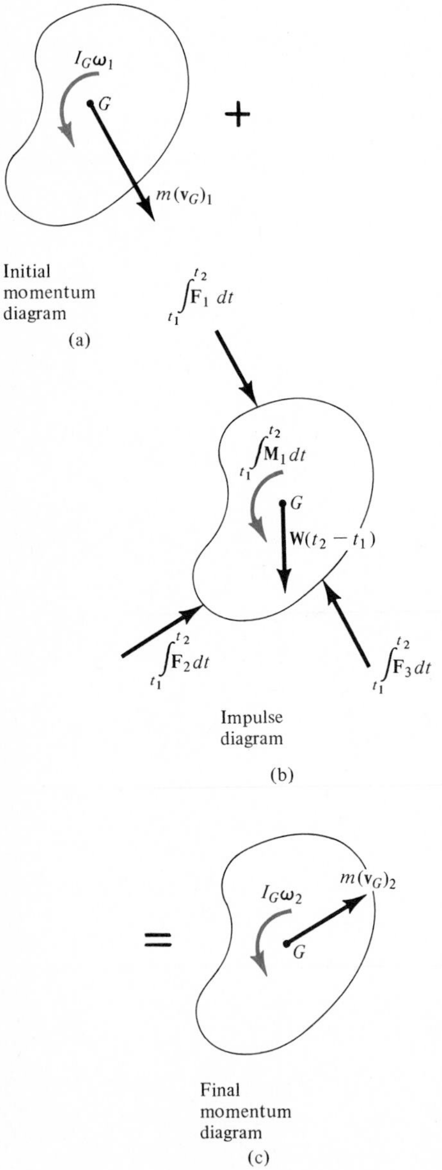

$I_G\omega_1$ G $m(\mathbf{v}_G)_1$ +

Initial momentum diagram (a)

$\int_{t_1}^{t_2}\mathbf{F}_1\,dt$ $\int_{t_1}^{t_2}\mathbf{M}_1\,dt$ G $\mathbf{W}(t_2 - t_1)$ $\int_{t_1}^{t_2}\mathbf{F}_2\,dt$ $\int_{t_1}^{t_2}\mathbf{F}_3\,dt$

Impulse diagram (b)

= $I_G\omega_2$ G $m(\mathbf{v}_G)_2$

Final momentum diagram (c)

Fig. 19–3

Example 19–1

The disk shown in Fig. 19–4*a* weighs 20 lb and is pin-supported at its center. If it is acted upon by a constant couple of 4 lb · ft and a force of 10 lb which is applied to a cord wrapped around its periphery, determine the angular velocity of the disk two seconds after starting from rest. What are the force components of reaction at pin *A*?

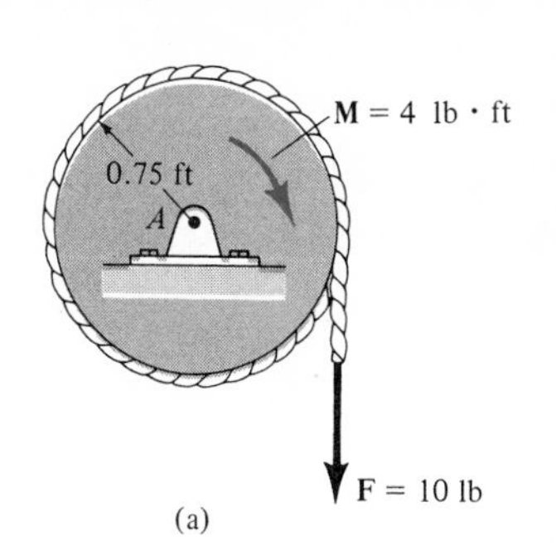

(a)

Solution

Impulse and Momentum Diagrams. Fig. 19–4*b*. Note that all the impulses acting on the impulse diagram are constant since the force and couple causing the motion are constant.

The moment of inertia of the disk about its fixed axis of rotation is

$$I_A = \frac{1}{2}mr^2 = \frac{1}{2}\left(\frac{20}{32.2}\right)(0.75)^2 = 0.175 \text{ slug} \cdot \text{ft}^2$$

Principle of Impulse and Momentum. With reference to the data in Fig. 19–4*b*, we have

$$\overset{+}{\rightarrow} \qquad m(v_{Ax})_1 + \Sigma \int_{t_1}^{t_2} F_x\, dt = m(v_{Ax})_2$$

$$0 + A_x(2) = 0$$

$$+\uparrow \qquad m(v_{Ay})_1 + \Sigma \int_{t_1}^{t_2} F_y\, dt = m(v_{Ay})_2$$

$$0 + A_y(2) - 20(2) - 10(2) = 0$$

$$\circlearrowright + \qquad I_A\omega_1 + \Sigma \int_{t_1}^{t_2} M_A\, dt = I_A\omega_2$$

$$0 + 4(2) + [10(2)](0.75) = 0.175\omega_2$$

Solving these equations yields

$$A_x = 0 \qquad \textit{Ans.}$$

$$A_y = 30 \text{ lb} \qquad \textit{Ans.}$$

$$\omega_2 = 131.6 \text{ rad/s} \qquad \textit{Ans.}$$

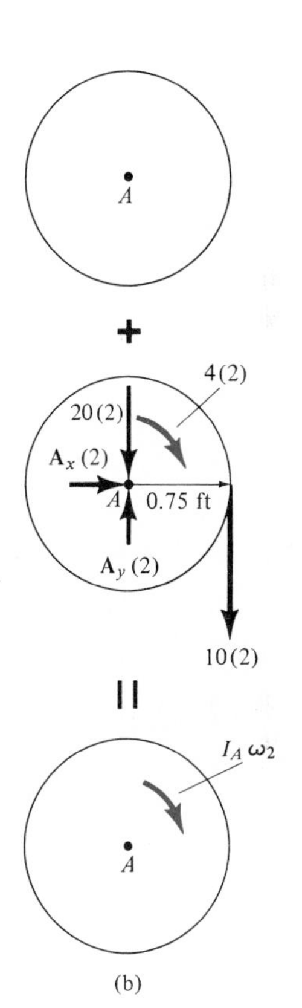

(b)

Fig. 19–4

Example 19–2

The block shown in Fig. 19–5*a* has a mass of 6 kg. It is attached to a cord which is wrapped around the periphery of a 20-kg disk that has a moment of inertia of $I_A = 0.40 \text{ kg} \cdot \text{m}^2$. If the block is initially moving downward with a speed of 2 m/s, determine its speed in 3 s. Neglect the mass of the cord in the calculation.

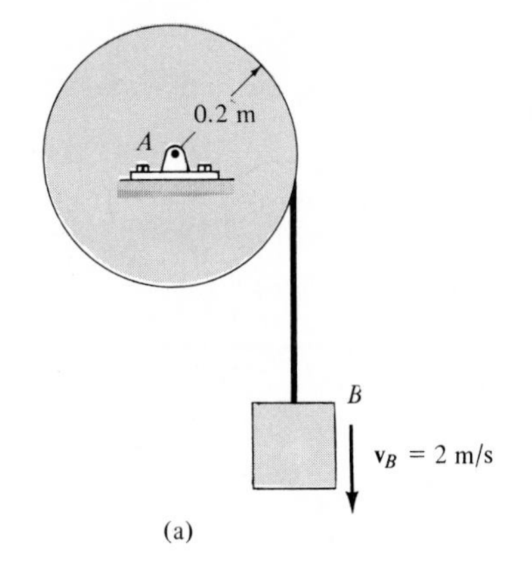

Fig. 19–5

Solution

Impulse and Momentum Diagrams. As shown in Fig. 19–5*b*, we have considered the *system* consisting of the block, the cable, and the disk. The unknown cable tension between the block and disk is thereby eliminated from the analysis, since it acts as an internal force.

Principle of Angular Impulse and Momentum. We can eliminate the unknown impulses acting at A by applying the principle of angular impulse and momentum about this point. It is necessary, however, to account for the ''moments'' of *both* the linear momentum and linear impulses. From Fig. 19–5*b*, we have

$$\left(\Sigma \begin{matrix}\text{syst. angular}\\ \text{momentum}\end{matrix}\right)_{A1} + \left(\Sigma \begin{matrix}\text{syst. angular}\\ \text{impulse}\end{matrix}\right)_{A(1-2)} = \left(\Sigma \begin{matrix}\text{syst. angular}\\ \text{momentum}\end{matrix}\right)_{A2}$$

$$(\circlearrowright +) \quad (6)(2)(0.2) + 0.40\omega_1 + 58.86(3)(0.2) = (6)(v_B)_2(0.2) + 0.40\omega_2$$

Kinematics. Since $\omega = v_B/r$, then $\omega_1 = 2/0.2 = 10$ rad/s and $\omega_2 = (v_B)_2/0.2 = 5(v_B)_2$. Substituting and solving for $(v_B)_2$ yields

$$(v_B)_2 = 13.0 \text{ m/s} \qquad \textit{Ans.}$$

Try solving the problem by applying the principle of impulse and momentum to the block and disk separately.

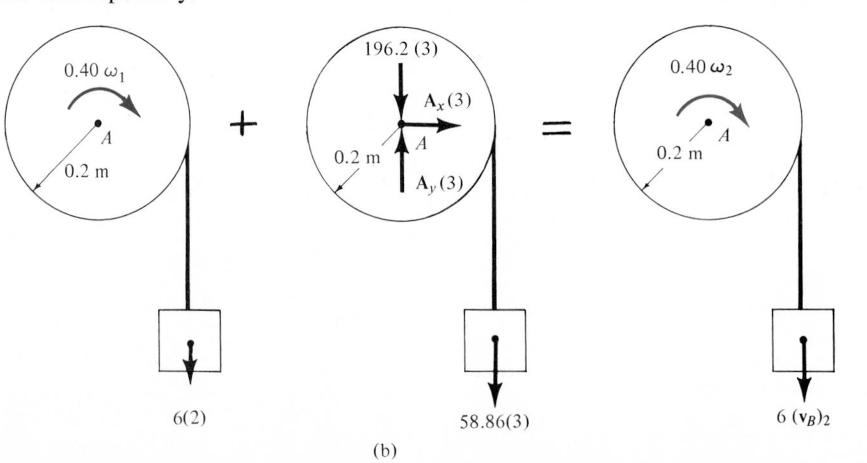

Example 19–3

The 100-kg spool shown in Fig. 19–6*a* has a radius of gyration of $k_G = 0.35$ m. A cable is wrapped around the central hub of the spool and a horizontal force of $P = 10$ N is applied. If the spool is initially at rest, determine its angular velocity in 5 s. Assume that the spool rolls without slipping.

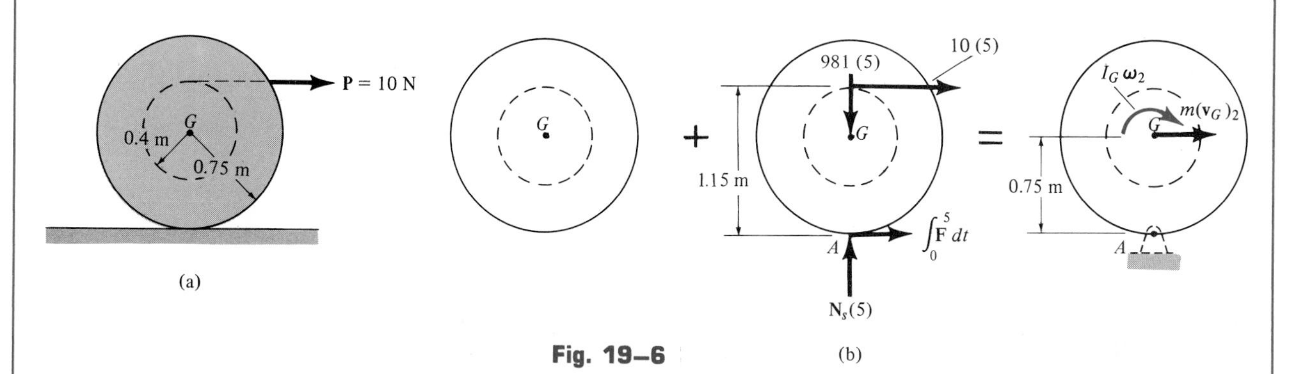

Fig. 19–6

Solution

Impulse and Momentum Diagrams. The impulse and momentum diagrams are shown in Fig. 19–6*b*.

The moment of inertia of the spool about its mass center is

$$I_G = mk_G^2 = (100)(0.35)^2 = 12.25 \text{ kg} \cdot \text{m}^2$$

Principle of Impulse and Momentum. We can eliminate the unknown normal and frictional impulses and thereby obtain a *direct solution* for ω_2 by computing the angular momentum and angular impulses about point A. Using the data on the impulse and momentum diagrams, we have

$$(\circlearrowleft +) \quad I_G\omega_1 + r_{G/A}[m_G(v_G)_1] + \Sigma \int_{t_1}^{t_2} M_A \, dt = I_G\omega_2 + r_{G/A}[m_G(v_G)_2]$$

$$0 + 0 + (10)(5)(1.15) = 12.25\omega_2 + 0.75[100(v_G)_2]$$

Kinematics. Since the spool does not slip, the instantaneous center of zero velocity is at point A. Hence, the velocity of G can be expressed in terms of the spool's angular velocity as $(v_G)_2 = 0.75\omega_2$, Fig. 19–6*b*. Substituting and solving, we obtain

$$\omega_2 = 0.839 \text{ rad/s} \qquad \textit{Ans.}$$

Example 19–4

A monkey having a mass of 2 kg steps off a platform, Fig. 19–7a, and begins to climb up a rope with a constant speed of 1.2 m/s, measured relative to the rope. If the rope is wound around a 100-kg spool, determine the time t required for the spool to rotate at $\omega_2 = 2$ rad/s. Neglect the mass of the rope in the calculation. The spool has a radius of gyration of $k_O = 0.3$ m.

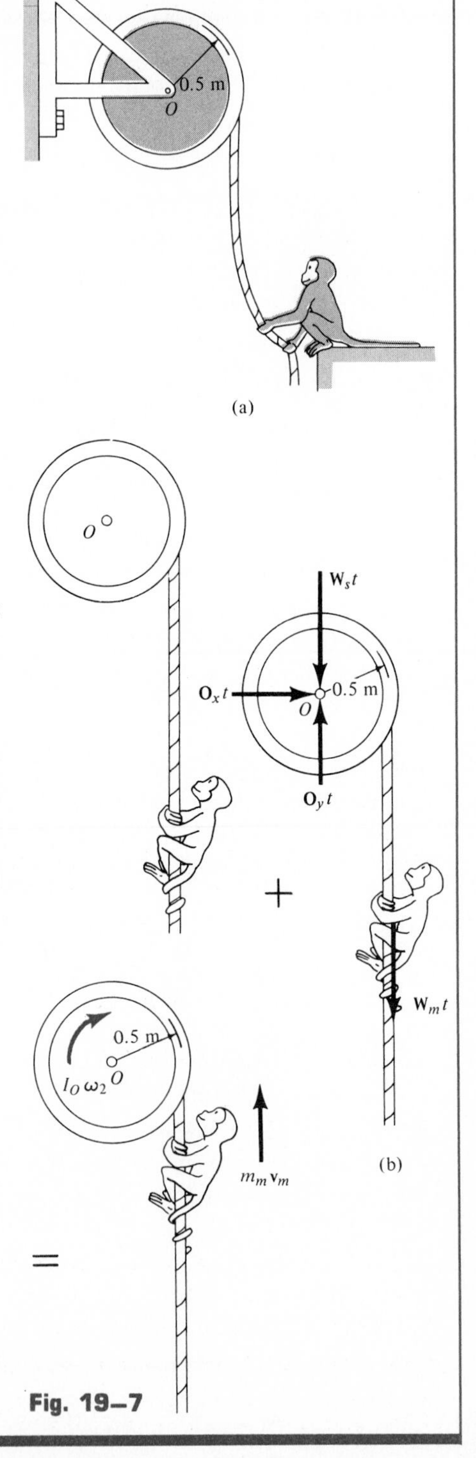

Fig. 19–7

Solution

Impulse and Momentum Diagrams. As shown in Fig. 19–7b, we have considered the monkey and spool as a *single system,* since then the impulses of the monkey on the rope are *internal* and therefore eliminated from the analysis. The reaction at the pin O and the weights of the spool, $\mathbf{W}_s$, and monkey, $\mathbf{W}_m$, create impulses on the system.

The moment of inertia of the spool about O is

$$I_O = mk_O^2 = 100(0.3)^2 = 9 \text{ kg} \cdot \text{m}^2$$

Principle of Angular Impulse and Momentum. Application is considered about point O in order to eliminate the unknown impulsive reaction components at this point. It is required that

$$\left(\sum \begin{matrix}\text{syst. angular}\\ \text{momentum}\end{matrix}\right)_{O1} + \left(\sum \begin{matrix}\text{syst. angular}\\ \text{impulse}\end{matrix}\right)_{O(1-2)} = \left(\sum \begin{matrix}\text{syst. angular}\\ \text{momentum}\end{matrix}\right)_{O2}$$

$$(\circlearrowleft +) \qquad [0] + [W_m t(r)] = [I_O\omega_2 + (m_m v_m)r]$$

$$[0] + [2(9.81)(t)(0.5)] = [9\omega_2 - (2v_m)(0.5)]$$

Kinematics. When the spool has an angular velocity $\omega_2 = 2$ rad/s, the rope is unwinding (downward) at a rate of $v_R = \omega_2 r = 2(0.5) = 1$ m/s. Consequently, if the positive direction is upward, the monkey has a velocity of $v_m = v_R + v_{m/R} = (-1 + 1.2) = 0.2$ m/s (upward), as seen by an observer on the ground (the inertial reference frame). Substituting into the above equation and solving yields

$$t = 1.81 \text{ s} \qquad \textbf{\textit{Ans.}}$$

Problems

Except when stated otherwise, throughout this chapter, assume that the coefficients of static and kinetic friction are equal, i.e., $\mu = \mu_s = \mu_k$.

19–1. Gear A is pinned at B and rotates along the periphery of the circular gear rack R. If A has a weight of 4 lb and a radius of gyration of $k_B = 0.5$ ft, determine the angular momentum of gear A about point C when $\omega_{CB} = 30$ rad/s and (a) $\omega_R = 0$, (b) $\omega_R = 20$ rad/s.

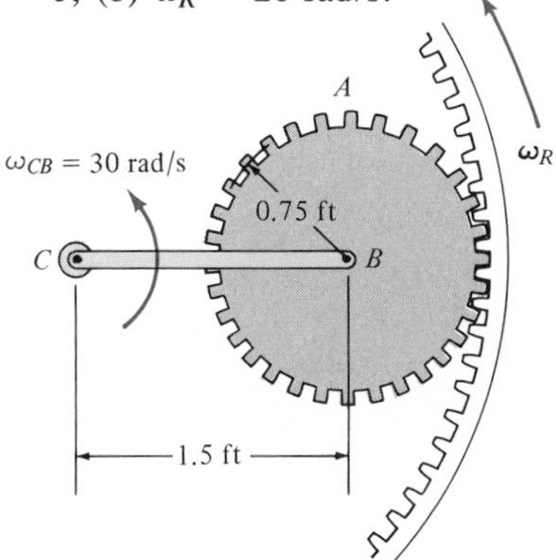

Prob. 19–1

19–2. A cord of negligible mass is wrapped around the outer surface of the 50-lb cylinder and its end is subjected to a constant horizontal force of $P = 2$ lb. If the cylinder rolls without slipping at A, determine its angular velocity in 4 s starting from rest. Neglect the thickness of the cord.

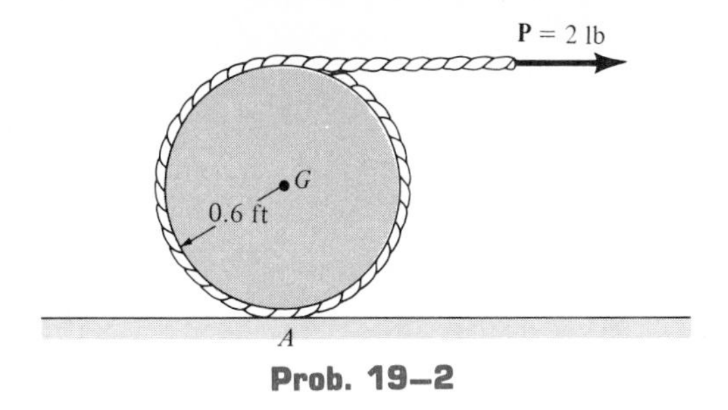

Prob. 19–2

19–3. The car strikes the back of a light pole, which is designed to break away from its base with negligible resistance. From a motion picture taken of the collision it is observed that the pole was given an angular velocity of 60 rad/s when AC was vertical. The pole has a mass of 75 kg, a center of mass at G, and a radius of gyration about an axis perpendicular to the plane of the pole and passing through G of $k_G = 2.25$ m. Determine the horizontal impulse which the car exerts on the pole while AC is essentially vertical, if the impact occurs at B during a time of 0.7 s. *Hint:* The impulses at A are $A_x\,\Delta t \approx 0$ and $A_y\,\Delta t$, where $A_y = 75(9.81)$ N.

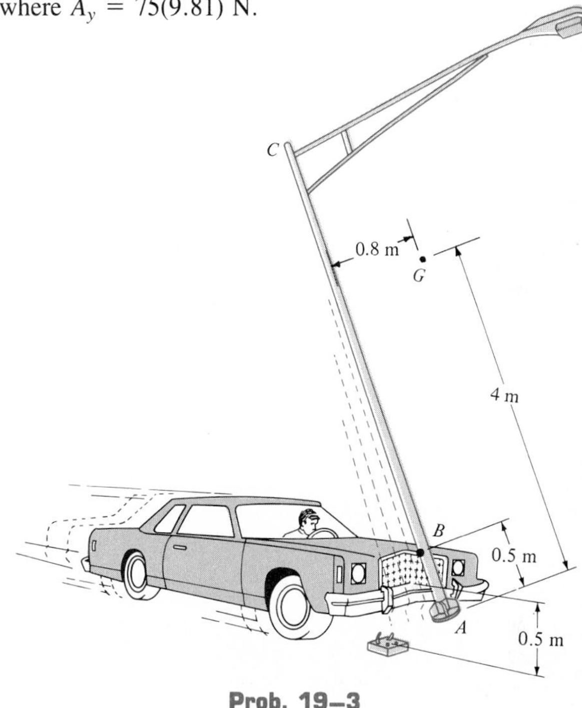

Prob. 19–3

***19–4.** Show that if a slab is rotating about a fixed axis perpendicular to the slab and passing through its mass center G, the angular momentum is the same when computed about any other point P on the slab.

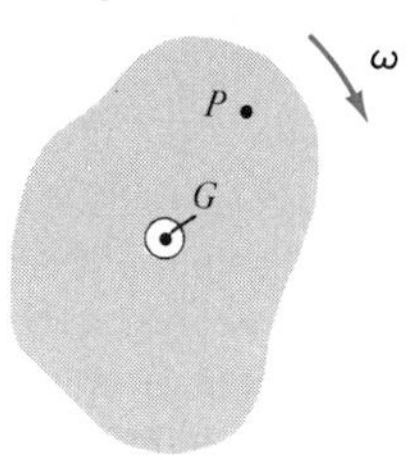

Prob. 19–4

19–5. A cord of negligible mass is wrapped around the outer surface of the 2-kg disk. If the disk is released from rest, determine its angular velocity in 3 s.

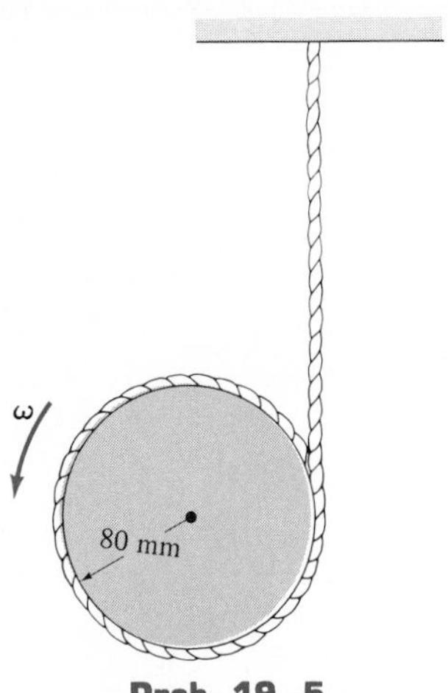

Prob. 19–5

19–6. The spacecraft has a mass of 1500 kg and a moment of inertia about an axis passing through G and directed perpendicular to the page of $I_G = 900 \text{ kg} \cdot \text{m}^2$. If it is traveling forward with a speed of $v_G = 800$ m/s and executes a turn by means of two jets, which provide a constant thrust of 400 N for 0.3 s, determine the spacecraft's angular velocity just after the jets are turned off.

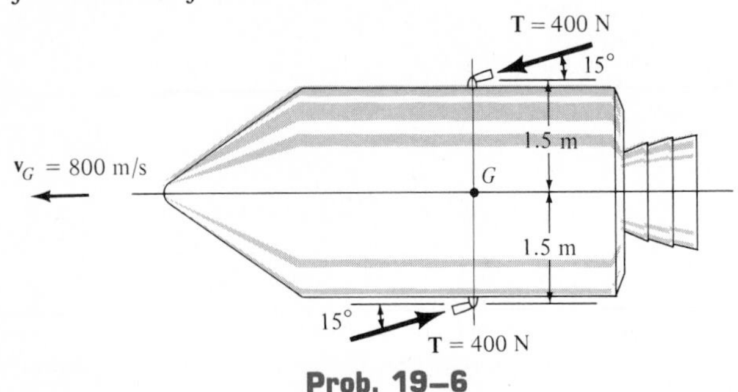

Prob. 19–6

19–7. At a given instant, the body has a linear momentum of $\mathbf{L} = m\mathbf{v}_G$ and an angular momentum computed about its mass center of $\mathbf{H}_G = I_G\boldsymbol{\omega}$. Show that the angular momentum of the body computed about the instantaneous center of zero velocity IC can be expressed as $\mathbf{H}_{IC} = I_{IC}\boldsymbol{\omega}$, where I_{IC} represents the body's moment of inertia computed about the instantaneous axis of zero velocity. As shown, the IC is located at a distance of $r_{G/IC}$ away from the mass center G.

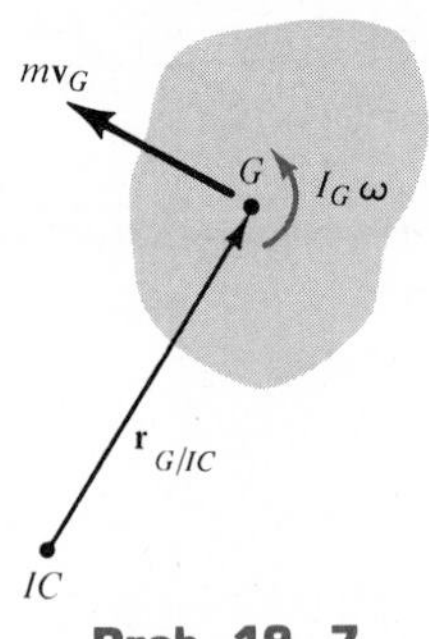

Prob. 19–7

***19–8.** The rigid body (slab) has a mass m and is rotating with an angular velocity $\boldsymbol{\omega}$ about an axis passing through the fixed point O. Show that the momentum of all the particles composing the body can be represented by a single vector having a magnitude of mv_G and acting through point P, called the *center of percussion,* which lies at a distance $r_{P/G} = k_G^2/r_{G/O}$ from the mass center G. Here k_G is the radius of gyration of the body, computed about an axis perpendicular to the plane of motion and passing through G.

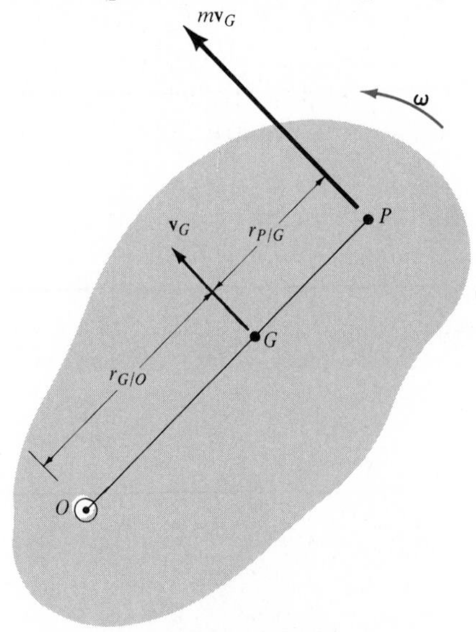

Prob. 19–8

19–9. The space shuttle is located in "deep space," where the effects of gravity can be neglected. It has a mass of 120 Mg, a center of mass at G, and a radius of gyration of $k_G = 14$ m about the x axis. It is originally traveling straight at $v = 3$ km/s when the pilot turns on the engine at

A, creating a thrust of $T = 600$ kN. Determine the shuttle's angular velocity 2 s later.

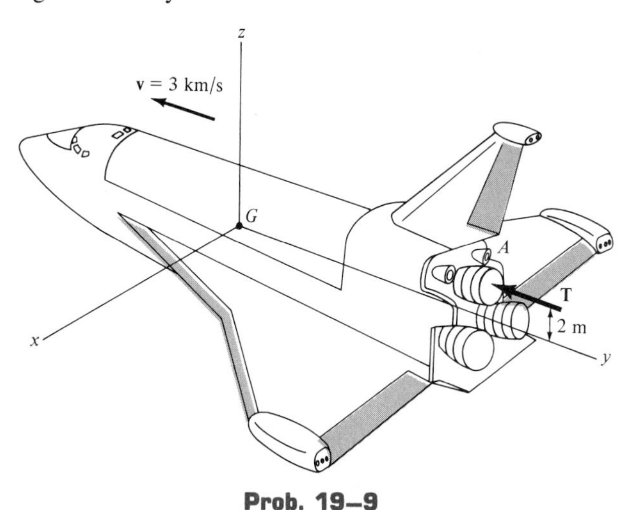

Prob. 19–9

19–10. A 4-kg disk A is mounted on arm BC, which has a negligible mass. If a variable torque or twist of $M = 5$ N · m, where t is measured in seconds, is applied to the arm at C, determine the angular velocity of BC in 2 s starting from rest. Solve the problem assuming that (a) the disk is set in a smooth bearing at B so that it rotates with curvilinear translation, (b) the disk is fixed to the shaft BC, and (c) the disk is given an initial freely spinning angular velocity of $\omega_{D_z} = -80$ rad/s prior to application of the torque.

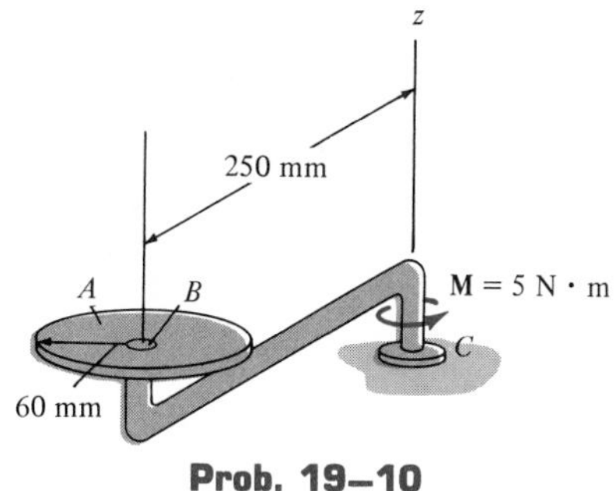

Prob. 19–10

19–11. If the ball has a weight of 15 lb and is thrown onto a *rough surface* with a velocity of 6 ft/s parallel to the surface, determine the amount of backspin, $\boldsymbol{\omega}$, it must be given so that it stops spinning at the same instant that its forward velocity is zero. It is not necessary to know the coefficient of friction at A for the calculation.

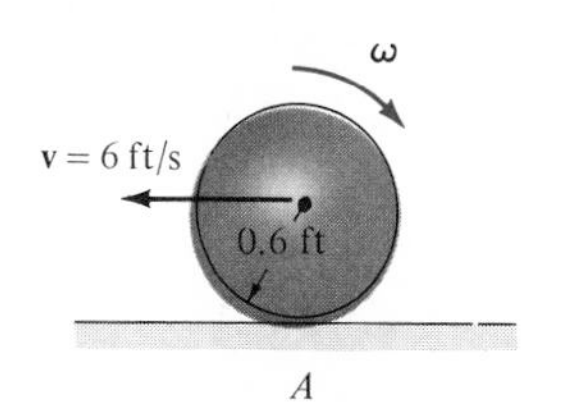

Prob. 19–11

***19–12.** The double pulley consists of two wheels which are attached to one another and turn at the same rate. The pulley has a mass of 15 kg and a radius of gyration of $k_O = 110$ mm. If the block at A has a mass of 40 kg, determine the speed of the block in 3 s after a constant force of 2 kN is applied to the rope wrapped around the inner hub of the pulley. The block is originally at rest.

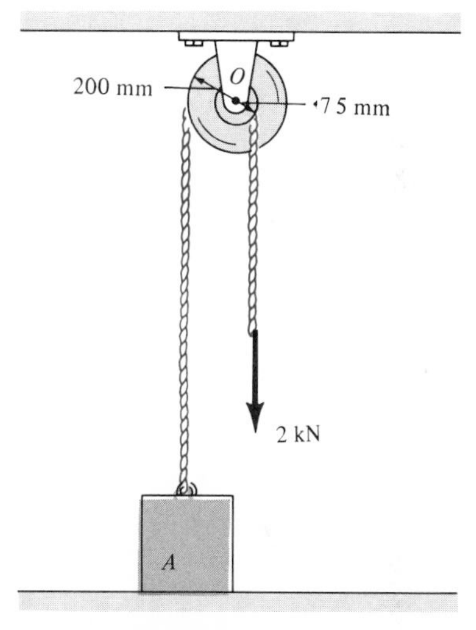

Prob. 19–12

19–13. The 50-kg drum, having a radius of gyration of $k_O = 180$ mm, rolls without slipping at A along an inclined plane for which $\theta = 30°$. If the drum is released from rest, determine its angular velocity in $t = 3$ s.

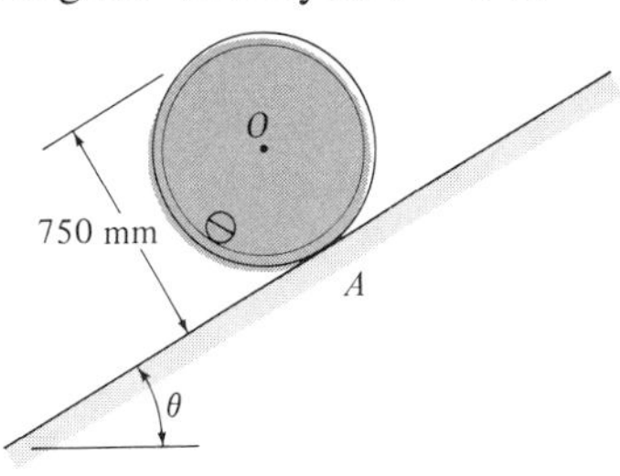

Prob. 19–13

19–14. A constant torque or twist of $M = 0.4\ \text{N}\cdot\text{m}$ is applied to the center gear A. If the system starts from rest, determine the angular velocity of each of the three (equal) smaller gears in 3 s. The smaller gears (B) are pinned at their centers, and the mass and centroidal radii of gyration of the gears are given in the figure.

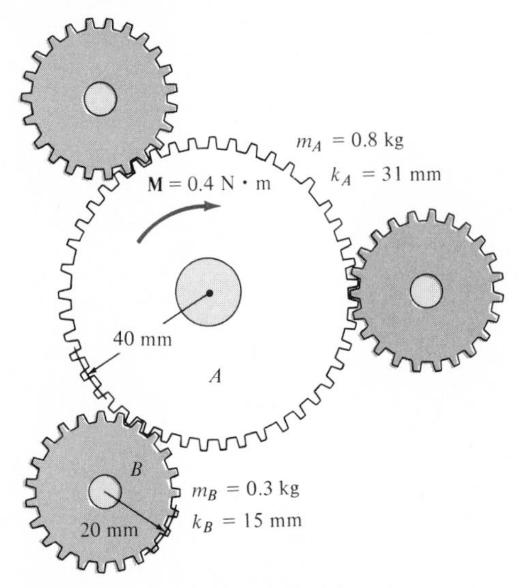

Prob. 19–14

19–15. Gear A has a weight of 1.5 lb, a radius of 0.2 ft, and a radius of gyration of $k_O = 0.13$ ft. The coefficient of friction between the gear rack B and the horizontal surface is $\mu = 0.3$. If the rack has a weight of 0.8 lb and is initially sliding to the left with a velocity of $(v_B)_1 = 4$ ft/s, determine the constant moment **M** which must be applied to the gear to increase the motion of the rack so that in $t = 2.5$ s it will have a velocity of $(v_B)_2 = 8$ ft/s to the left. Neglect friction between the rack and the gear and assume that the gear exerts *only* a horizontal force on the rack.

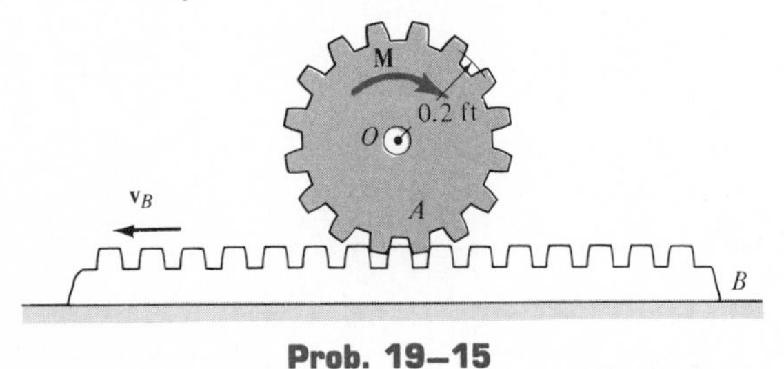

Prob. 19–15

***19–16.** The 12-kg disk has an angular velocity of $\omega = 20$ rad/s. If the brake ABC is applied such that $P = 5\ N$, determine the time needed to stop the disk. The coefficient of friction at B is $\mu = 0.4$.

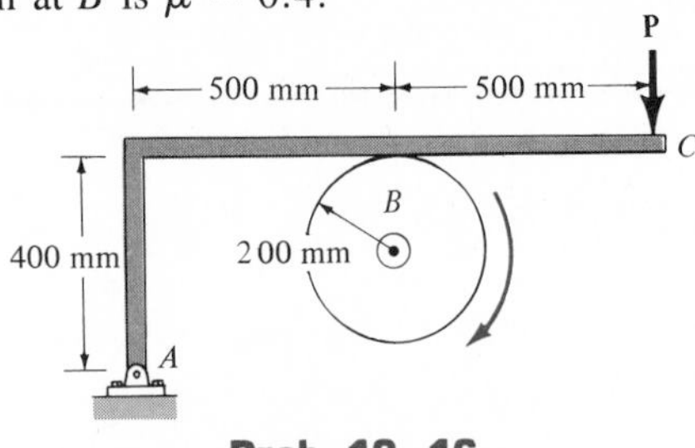

Prob. 19–16

19–17. The flywheel A has a mass of 30 kg and a radius of gyration of $k_C = 95$ mm. Disk B has a mass of 25 kg, is pinned at D, and is coupled to the flywheel using a belt which is subjected to a tension such that it does not slip at its contacting surfaces. If a motor supplies a counterclockwise torque or twist to the flywheel, having a magnitude of $M = 12\ \text{N}\cdot\text{m}$, determine the angular velocity of the disk 3 s after the motor is turned on. Initially, the flywheel is at rest.

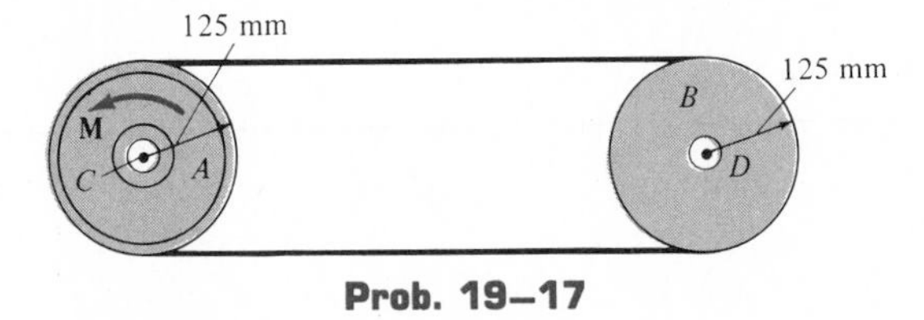

Prob. 19–17

19–18. The drum A has a mass of 150 kg and a centroidal radius of gyration of $k_O = 210$ mm. If a torque or twist of $M = 700\ \text{N}\cdot\text{m}$ is applied to the drum, determine the speed of the 500-kg mine car C in 3 s. The car is originally resting along the stop s and the cable is loose. Neglect friction and the mass of both the pulley B and the car wheels for the calculation.

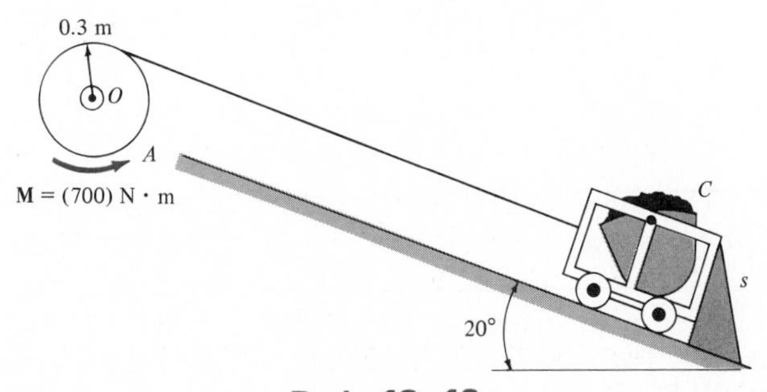

Prob. 19–18

Conservation of Momentum 19.3

Conservation of Linear Momentum. If the sum of all the *linear impulses* acting on a system of connected rigid bodies is *zero,* the linear momentum of the system is constant or conserved. Consequently, the first two of Eqs. 19–12 reduce to the form

$$\left(\Sigma \begin{matrix}\text{syst. linear}\\ \text{momentum}\end{matrix}\right)_1 = \left(\Sigma \begin{matrix}\text{syst. linear}\\ \text{momentum}\end{matrix}\right)_2 \qquad (19\text{–}13)$$

This equation is referred to as the *conservation of linear momentum.*

Without inducing appreciable errors in the computations, it may be possible to apply Eq. 19–13 in a specified direction for which the linear impulses are small or *nonimpulsive*. Specifically, nonimpulsive forces occur when small forces act over very short periods of time. For example, the impulse created by the force of a tennis racket hitting a ball during a very short time interval Δt is large, whereas the impulse of the weight of the ball during this time is small by comparison and may therefore be neglected in the motion analysis of the ball during Δt.

Conservation of Angular Momentum. The angular momentum of a system of connected rigid bodies is conserved about the system's center of mass G, or a fixed point O, when the sum of all the angular impulses created by the external forces acting on the system is zero or appreciably small (nonimpulsive) when computed about these points. The third of Eqs. 19–12 then becomes

$$\left(\Sigma \begin{matrix}\text{syst. angular}\\ \text{momentum}\end{matrix}\right)_{O1} = \left(\Sigma \begin{matrix}\text{syst. angular}\\ \text{momentum}\end{matrix}\right)_{O2} \qquad (19\text{–}14)$$

This equation is referred to as the *conservation of angular momentum.* In the case of a single rigid body, Eq. 19–14 applied to point G becomes $(I_G\omega)_1 = (I_G\omega)_2$. To illustrate an application of this equation, consider a swimmer who executes a somersault after jumping off a diving board. By tucking his arms and legs in close to his chest, he *decreases* his body's moment of inertia and thus *increases* his angular velocity. If he straightens out just before entering the water, his body's moment of inertia is *increased* and his angular velocity *decreases* ($I_G\omega$ must be constant). Since the weight of his body creates a linear impulse during the time of motion, this example also illustrates that the angular momentum of a body is conserved and yet the linear momentum is *not*. Such cases occur whenever the external forces creating the linear impulse pass through either the center of mass of the body or a fixed axis of rotation.

PROCEDURE FOR ANALYSIS

Provided the initial velocity of the body is known, the conservation of linear or angular momentum is used to determine the final velocity of a body *just after* the time period considered. Furthermore, by applying these equations to a *system* of bodies, the internal impulses acting within the system, which may be unknown, are eliminated from the analysis, since they occur in equal but opposite collinear pairs. For application it is suggested that the following procedure be used.

Impulse and Momentum Diagrams. Draw the impulse and momentum diagrams for the body or system of bodies. From the impulse diagram it is possible to classify each of the applied forces as being either "impulsive" or "nonimpulsive." In general, for short times, "nonimpulsive forces" consist of the weight of a body, the force of a slightly deformed spring, or any force that is *known to be small* when compared to the impulsive force.

By inspection of the impulse diagram, it will be possible to tell if the conservation of linear or angular momentum can be applied. Specifically, the *conservation of linear momentum* applies in a given direction when *no* external impulsive forces act on the body or system in that direction; whereas the *conservation of angular momentum* applies about a fixed point O or at the mass center G of a body or system of bodies when all the external impulsive forces acting on the body or system create zero moment (or zero angular impulse) about O or G.

Conservation of Momentum. Apply the conservation of linear or angular momentum in the appropriate directions. Most often, the scalar equations can be formulated by resolving the vector components or summing "moments" *directly* from the impulse and momentum diagrams.

Kinematics. Use *kinematics* if further equations are necessary for the solution of a problem. If the motion appears to be complicated, kinematic (velocity) diagrams may be helpful in obtaining the necessary kinematic relations.

If it is necessary to determine an *internal impulsive force* acting on only one body of a system, the body must be *isolated* and the principle of linear or angular impulse and momentum must be applied *to the body*. After the impulse $\int F\,dt$ is calculated, then, provided the time Δt for which the impulse acts is known, the *average impulsive force* F_{avg} can be determined from $F_{avg} = \int F\,dt/\Delta t$.

The following examples numerically illustrate application of this procedure.

Example 19–5

The 10-kg wheel shown in Fig. 19–8*a* has a moment of inertia of $I_G = 0.156\ \text{kg} \cdot \text{m}^2$. Assuming that the wheel does not slip or rebound, determine the minimum velocity $\mathbf{v}_G$ it must have to just roll over the obstruction at A.

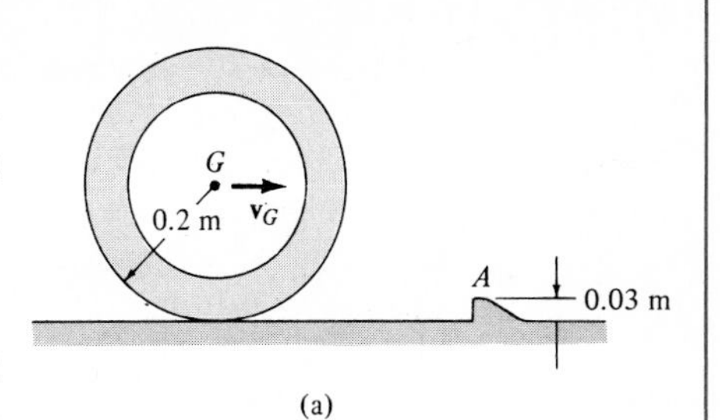

(a)

Solution

Impulse and Momentum Diagrams. Since no slipping or rebounding occurs, the wheel essentially *pivots* about point A during contact. This condition is shown in Fig. 19–8*b*, which indicates, respectively, the momentum of the wheel *just before impact,* the impulses given to the wheel *during impact,* and the momentum of the wheel *just after impact*. Note that only two impulses (forces) act on the wheel. By comparison, the impulse at A is much greater than that caused by the weight ($W = 10(9.81) = 98.1$ N), and since the time of impact is very short, the weight can be considered nonimpulsive. The impulsive force $\mathbf{F}$ at A has both an unknown magnitude and an unknown direction θ. To eliminate this force from the analysis, note that during impact *only* the nonimpulsive weight force creates an *angular impulse about point A*. Furthermore, since $(98.1\Delta t)d \approx 0$, angular momentum about A is essentially *conserved.*

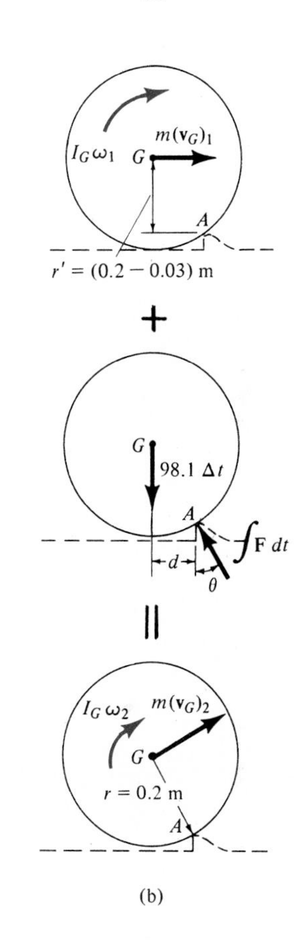

(b)

Conservation of Angular Momentum. With reference to Fig. 19–8*b*,

$$(\circlearrowright +) \qquad (\mathbf{H}_A)_1 = (\mathbf{H}_A)_2$$

$$r'm(v_G)_1 + I_G\omega_1 = rm(v_G)_2 + I_G\omega_2$$

$$(0.2 - 0.03)(10)(v_G)_1 + (0.156)(\omega_1) = (0.2)(10)(v_G)_2 + (0.156)(\omega_2)$$

Kinematics. Since no slipping occurs, $\omega = v_G/r = v_G/0.2 = 5v_G$. Substituting this into the above equation and simplifying yields

$$(v_G)_2 = 0.892(v_G)_1 \qquad (1)$$

Conservation of Energy Theorem. In order to roll over the obstruction, the wheel must pass the dashed position 3 shown in Fig. 19–8*c*. Hence, if $(v_G)_2$ (or $(v_G)_1$) is to be a minimum, it is necessary that the kinetic energy of the wheel at position 2 be equal to the potential energy at position 3. Constructing the datum through the center of mass, as shown in the figure, and applying the conservation of energy theorem, we have

$$\{T_2\} + \{V_2\} = \{T_3\} + \{V_3\}$$

$$\{\tfrac{1}{2}(10)(v_G)_2^2 + \tfrac{1}{2}(0.156)(\omega_2)^2\} + \{0\} = \{0\} + \{(98.1)(0.03)\}$$

Substituting $\omega_2 = 5(v_G)_2$ and Eq. (1) into this equation, and simplifying, yields

$$3.98(v_G)_1^2 + 1.55(v_G)_1^2 = 2.94$$

Thus,

$$(v_G)_1 = 0.729 \text{ m/s} \qquad \textit{Ans.}$$

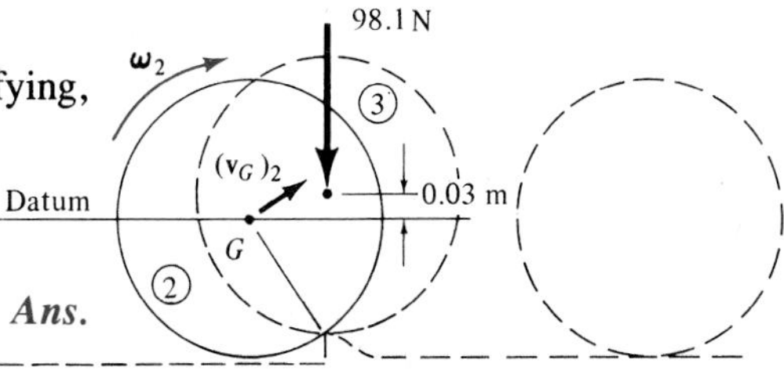

(c)

Example 19–6

The 5-kg slender rod shown in Fig. 19–9*a* is pinned at O and is initially at rest. If a 4-g bullet is fired into the rod with a velocity of 400 m/s, as shown in the figure, determine the angular velocity of the rod just after the bullet becomes embedded in it.

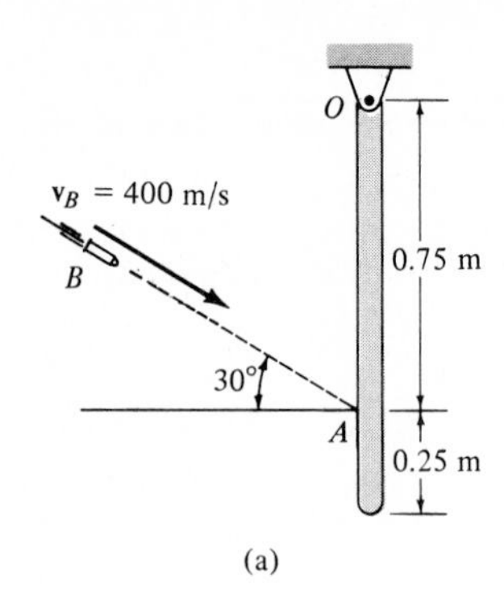

Fig. 19–9

Solution

Impulse and Momentum Diagrams. The impulse which the bullet exerts on the rod can be eliminated from the analysis and the angular velocity of the rod just after impact can be determined by considering the bullet and rod as a single system, Fig. 19–9*b*. The momentum diagrams are drawn *just before and just after impact.* During impact, the bullet and rod exchange equal but opposite *internal impulses* at A. As shown on the impulse diagram, the impulses that are external to the system are due to the reactions at O and the weights of the bullet and rod. Since the time of impact, Δt, is very short, the rod moves only a slight amount and so the "moments" of these impulses about point O are essentially zero, and therefore angular momentum is conserved about this point.

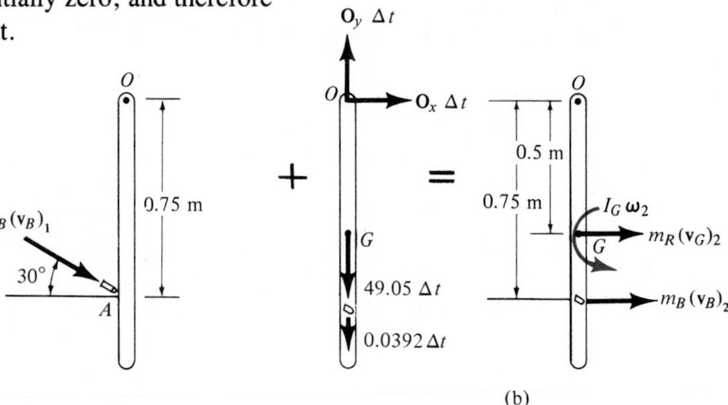

Conservation of Angular Momentum. From Fig. 19–9*b*, we have

$$(\circlearrowleft +) \qquad \Sigma(H_O)_1 = \Sigma(H_O)_2$$

$$m_B(v_B)_1 \cos 30°(0.75 \text{ m}) = m_B(v_B)_2(0.75 \text{ m}) + m_R(v_G)_2(0.5 \text{ m}) + I_G\omega_2$$

$$(0.004)(400 \cos 30°)(0.75) = (0.004)(v_B)_2(0.75) + (5)(v_G)_2(0.5) + [\tfrac{1}{12}(5)(1)^2]\omega_2$$

or

$$1.039 = 0.003(v_B)_2 + 2.50(v_G)_2 + 0.417\omega_2 \qquad (1)$$

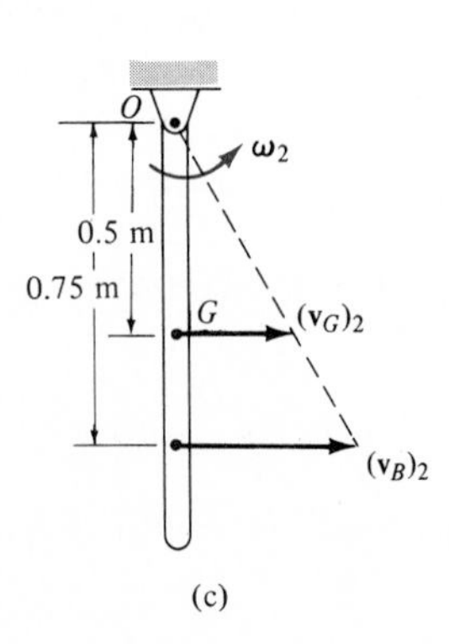

Kinematics. Since the rod is pinned at O, from Fig. 19–9*c* we have

$$(v_G)_2 = 0.5\omega_2, \qquad (v_B)_2 = 0.75\omega_2$$

Substituting into Eq. (1) and solving yields

$$\omega_2 = 0.622 \text{ rad/s} \qquad \textit{Ans.}$$

Problems

19–19. The space satellite has a mass of 125 kg and a moment of inertia of $I_z = 0.940\ \text{kg}\cdot\text{m}^2$, excluding the four solar panels A, B, C, and D. Each solar panel has a mass of 20 kg and can be approximated as a thin plate. If the satellite is originally spinning about the z axis at a constant rate of $\omega_z = 0.5$ rad/s when $\theta = 90°$, determine the rate of spin if all the panels are raised and reach the upward position, $\theta = 0°$, at the same instant.

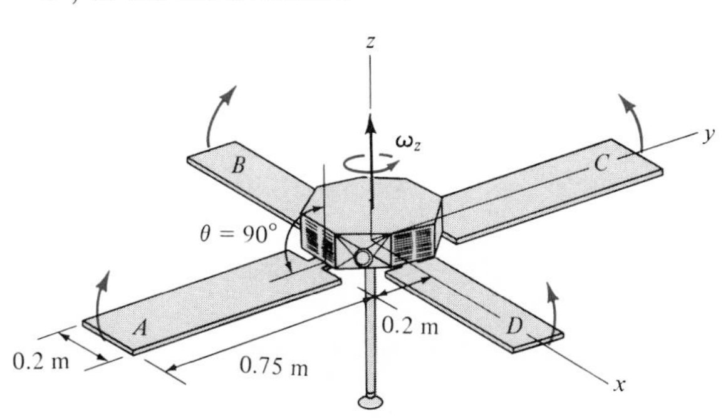

Prob. 19–19

***19–20.** A man has a moment of inertia I_z about the z axis. He is originally at rest and standing on a small platform which can turn freely. If he is handed a wheel which is rotating at $\boldsymbol{\omega}$ and has a moment of inertia I about its spinning axis, determine his angular velocity if (a) he holds the wheel upright as shown, (b) turns the wheel out, $\theta = 90°$, and (c) turns the wheel downward, $\theta = 180°$.

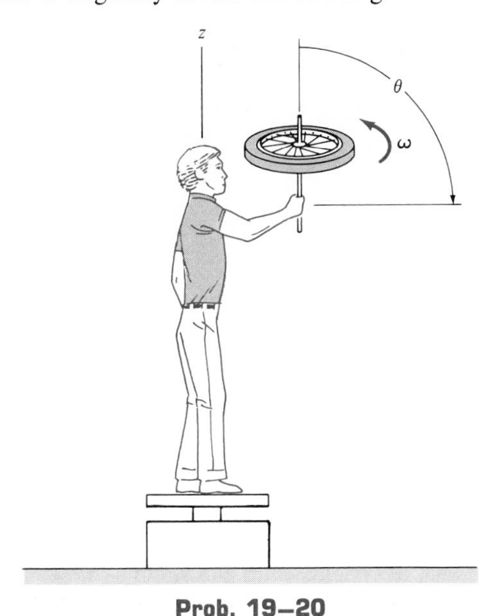

Prob. 19–20

19–21. If the man in Prob. 19–20 is given the wheel when it is at *rest* and he starts it spinning with an angular velocity $\boldsymbol{\omega}$, determine his angular velocity if (a) he holds the wheel upright as shown, (b) turns the wheel out, $\theta = 90°$, and (c) turns the wheel downward, $\theta = 180°$.

19–22. A horizontal circular platform has a weight of 300 lb and a radius of gyration about the z axis passing through its center O of $k_z = 8$ ft. The platform is free to rotate about the z axis and is initially at rest. A man, having a weight of 150 lb, begins to run along the edge in a circular path of radius 10 ft. If he has a speed of 4 ft/s and maintains this speed relative to the platform, compute the angular velocity of the platform.

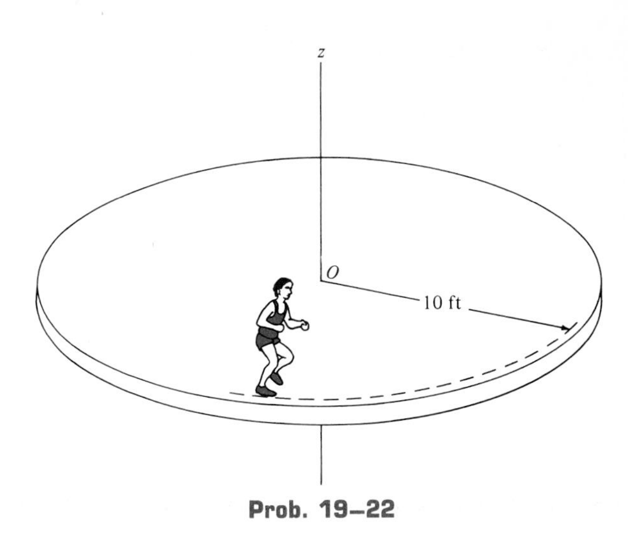

Prob. 19–22

19–23. When a cat falls from an upside-down position she will instinctively rotate her tail so she turns 180° and lands safely on her paws upon reaching the ground. Assuming that a cat's body, excluding its tail, has a mass of 2.5 kg and a radius of gyration of $k_G = 40$ mm about the horizontal *aa* axis passing through the mass center *G*, and furthermore that the 300-mm-long tail has a mass of 0.08 kg and can be considered a slender rod, determine the constant angular speed $\boldsymbol{\omega}$ at which the cat's tail must rotate so that she lands safely when falling accidentally from rest in an upside-down position at a height of $h = 2$ m. This assumes the cat holds her body *rigid* while falling. *Comment:* Time photographs of the motion reveal that a cat *also twists* her body as she falls. The problem specified here is strictly meant to illustrate the mechanics principles involved. Experimental verification is not recommended!

Prob. 19–23

***19–24.** The square plate, where $a = 0.75$ ft, has a weight of 4 lb and is rotating on the smooth surface with a constant angular velocity of $\omega_1 = 10$ rad/s. Determine the new angular velocity of the plate just after its corner strikes the peg *P* and the plate starts to rotate about *P* without rebounding.

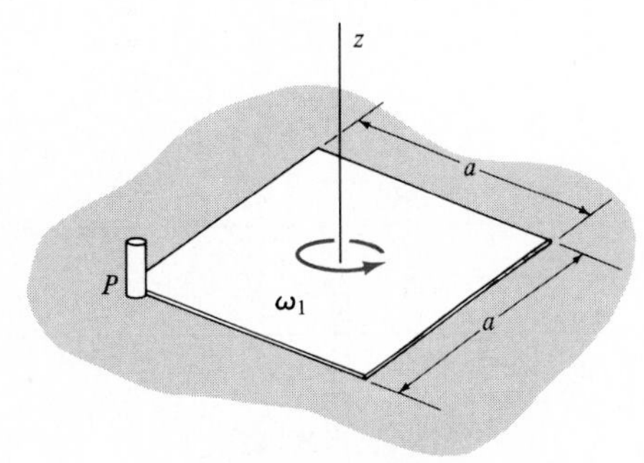

Prob. 19–24

19–25. For safety reasons, the 20-kg supporting leg of a sign is designed to break away with negligible resistance at *B* when the leg is subjected to the impact of a car. Assuming that the leg is pin-supported at *A* and approximates a thin rod, determine the impulse the car bumper exerts on it, if after the impact the leg appears to rotate upward to an angle of $\theta_{max} = 150°$.

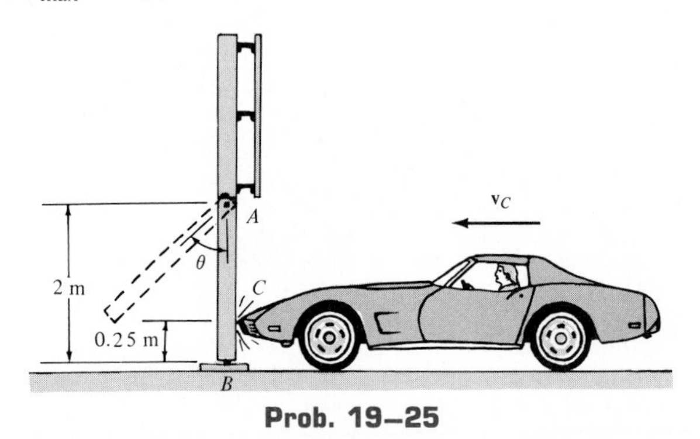

Prob. 19–25

19–26. Determine the height *h* at which a bullet can strike the disk that has a weight of 15 lb and cause it to roll without slipping at *A*. For the calculation this requires that the frictional force at *A* be essentially zero.

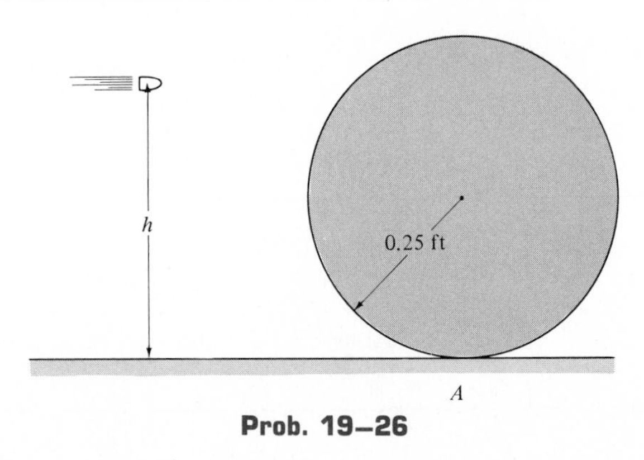

Prob. 19–26

19–27. Determine the height *h* at which a billiard ball of mass *m* must be struck so that no frictional force develops between it and the table at *A*. Assume that the cue *C* only exerts a horizontal force **P** on the ball.

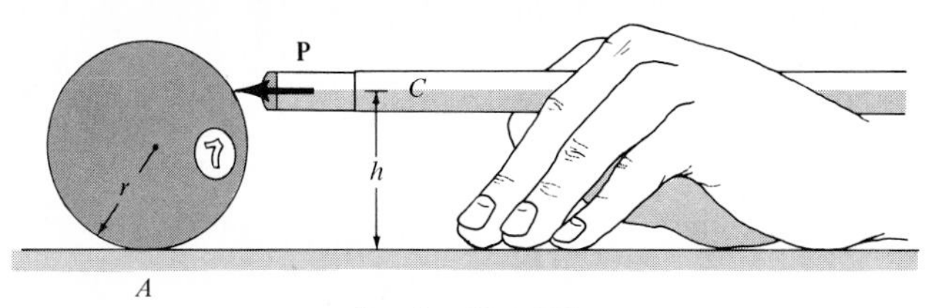

Prob. 19–27

***19–28.** A thin rod having a mass of 4 kg is balanced vertically as shown. Determine the height h at which it can be struck with a horizontal force **F** and not slip at the floor. For the solution this requires that the frictional force at A be essentially zero.

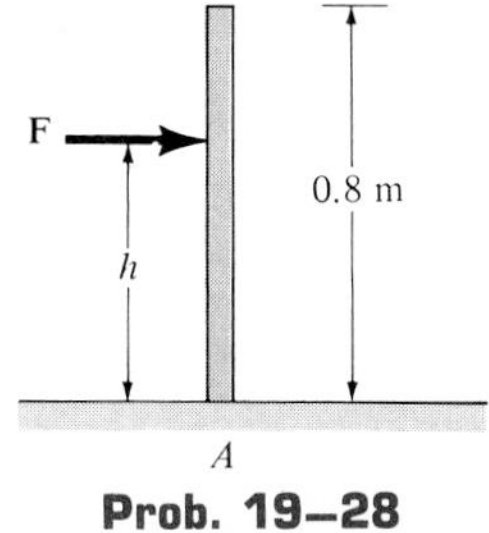

Prob. 19–28

19–29. The uniform rod AB has a weight of 3 lb and is released from rest without rotating from the position shown. As it falls, the end A strikes a hook S, which provides a permanent connection. Determine the speed at which the other end B strikes the wall at C.

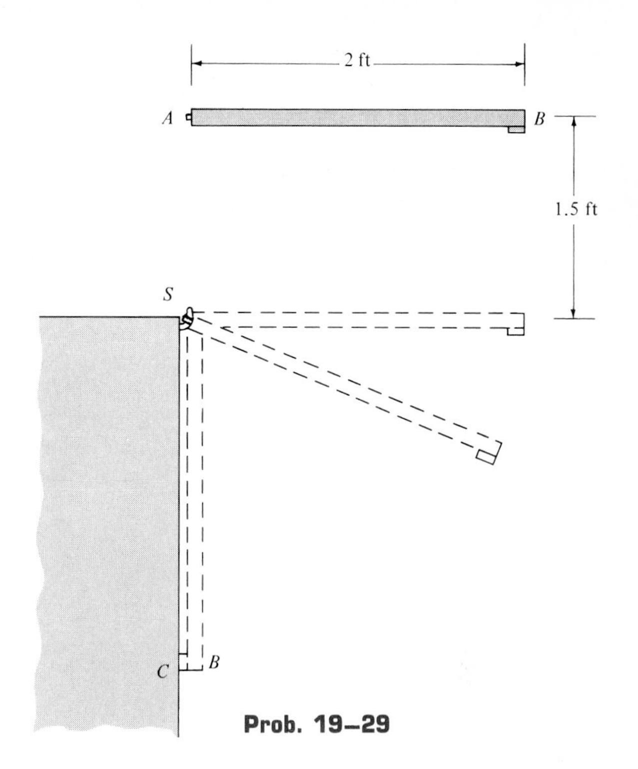

Prob. 19–29

19–30. *Eccentric impact* of two bodies occurs when the line connecting the mass centers of the bodies *does not* coincide with the line of impact. This situation is shown in the figure, where just before impact at C body B is rotating with an angular velocity of $\boldsymbol{\omega}_1$, such that $(v_B)_1 = \omega_1 r$, and the velocity of the contact point on body A is $(\mathbf{u}_A)_1$, such that the component of velocity along the line of impact is $(\mathbf{v}_A)_1$, where $(v_A)_1 > (v_B)_1$. Just after collision, B has an angular velocity of $\boldsymbol{\omega}_2$, such that $(v_B)_2 = \omega_2 r$, and the contact point at A has a velocity of $(\mathbf{u}_A)_2$ with a component of velocity along the line of impact of $(\mathbf{v}_A)_2$. Assuming that body B has a moment of inertia I_O at O and mass m_B, and body A has a mass m_A, apply the conservation of angular momentum for *both* bodies about point O and the principle of angular impulse and momentum to *each* body about O, and show that the above velocities are related by $e = [(v_B)_2 - (v_A)_2]/[(v_A)_1 - (v_B)_1]$, where e is the coefficient of restitution between the bodies. *Suggestion:* Follow the procedure outlined in Sec. 15–4, which was used to derive Eq. 15–11.

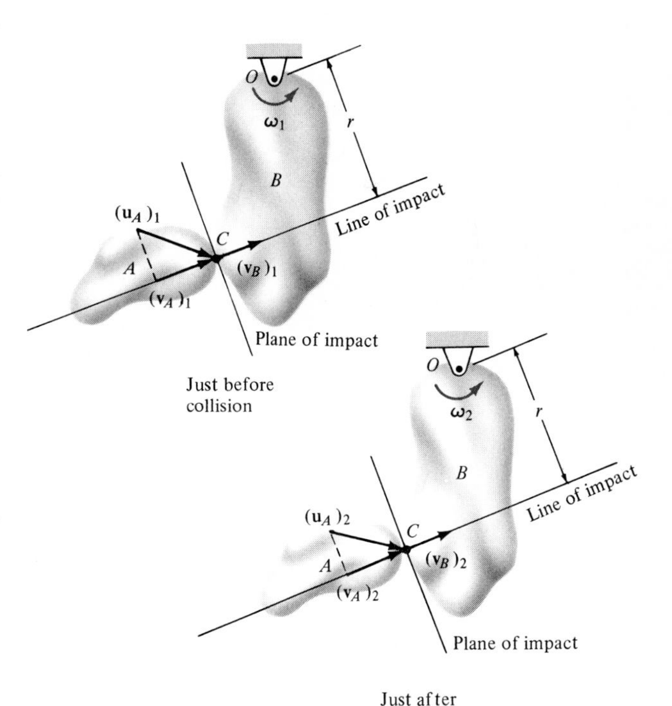

Prob. 19–30

19–31. The disk has a mass of 15 kg. If it is released from rest when $\theta = 30°$, determine the maximum angle θ of rebound after it collides with the wall. The coefficient of restitution between the disk and the wall is $e = 0.6$. When $\theta = 0°$, the disk hangs such that it just touches the wall. Neglect friction at the pin C. *Hint:* Use the equation for e defined in Prob. 19–30.

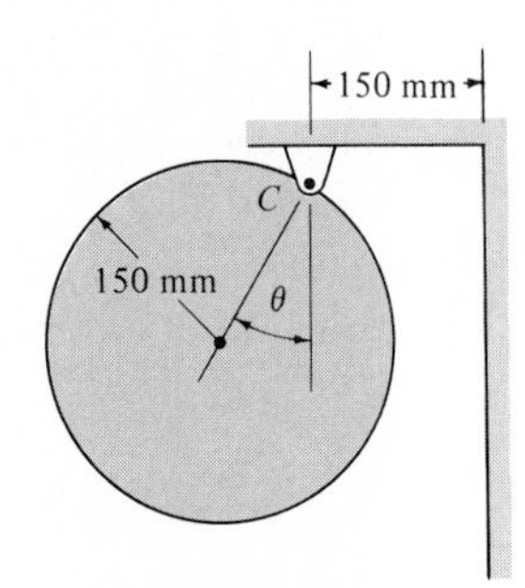

Prob. 19–31

***19–32.** The ball has a weight of 3 lb and rolls without slipping along a horizontal plane with a velocity of $v_O = 0.5$ ft/s. Provided it does not slip or rebound, determine the velocity of its center O as it just starts to roll up the inclined plane.

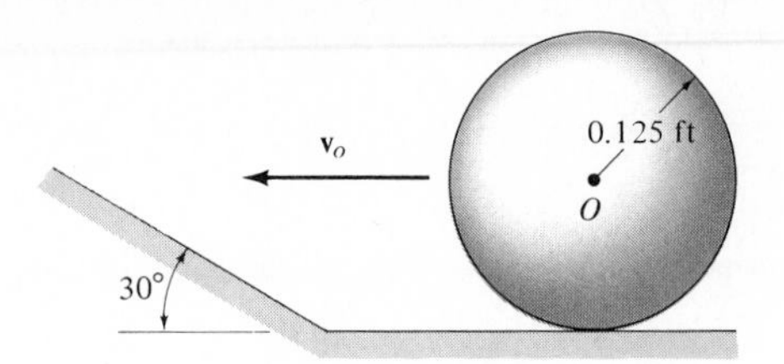

Prob. 19–32

19–33. A ball having a mass of 8 kg and initial speed of $v_1 = 2$ m/s rolls over a 30-mm-long depression. Assuming that the ball does not rebound off the edges of contact, first A, then B, determine its final velocity when it reaches the other side.

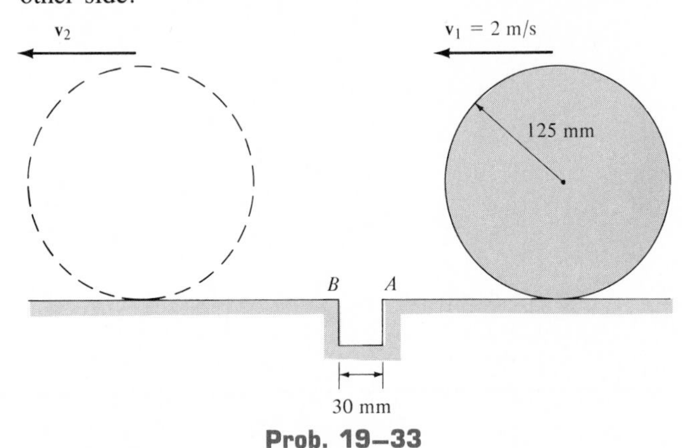

Prob. 19–33

19.4 Gyroscopic Motion

In this section the equation used to analyze the motion of a body (or top) which is symmetrical with respect to an axis and moving about a fixed point lying on the axis will be developed. This equation will then be applied to study the motion of a particularly interesting device, the gyroscope.

Gyroscopic motion can be explained by considering the top shown in Fig. 19–10a. If the top is *spinning* at a very high rate $\boldsymbol{\omega}_z$, the top will *precess* at a *constant angular rate* $\boldsymbol{\Omega}_y$ about the y axis and will *not* fall downward as one would expect instinctively. This unusual phenomenon is often referred to as the *gyroscopic effect*.

Gyroscopic motion can be explained by applying the principle of angular impulse and momentum to the top, written in differential form as $\Sigma \mathbf{M}_O\, dt = d(\mathbf{H}_O)$. Provided the rate of spin $\boldsymbol{\omega}_z$ is considerably greater than its precessional rate of rotation $\boldsymbol{\Omega}_y$, then the top in Fig. 19–10a is referred to as a *gyro*. For all practical purposes, the angular momentum of a gyro can be assumed to be directed along its axis of spin, Fig. 19–10b. Hence $H_O = I_z\omega_z$, where I_z is the moment of inertia of the gyro about the z axis. Consider now the motion during an instant of time dt. We require the angular impulse $\Sigma \mathbf{M}_O\, dt$, created by the weight about the x axis, to be equal to the change in

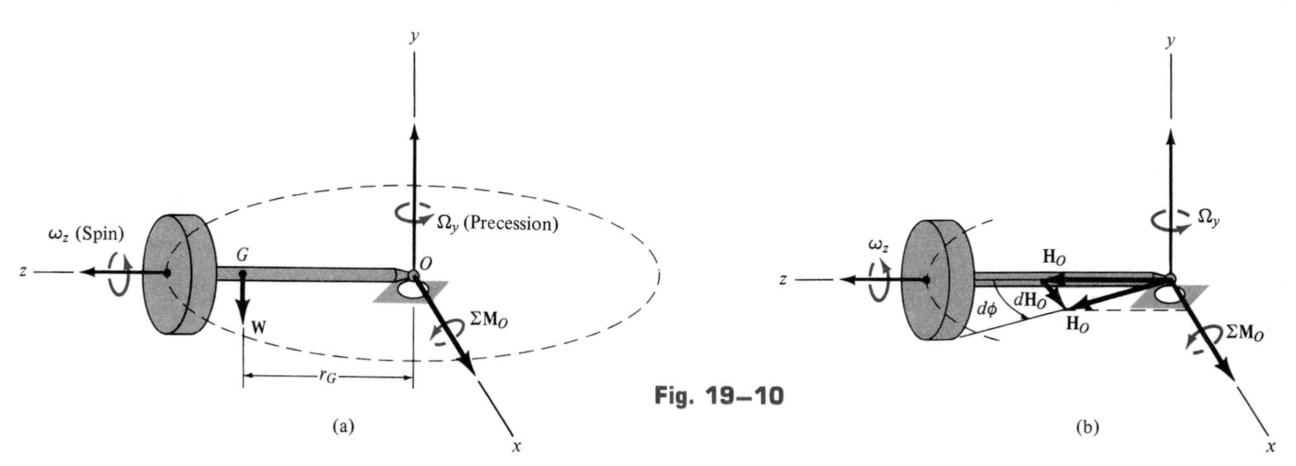

Fig. 19–10

angular momentum $d\mathbf{H}_O$ of the gyro. Since $\mathbf{H}_O$ is constant in magnitude, only the directional change $d\mathbf{H}_O$ occurs, which is due to the precession of $\mathbf{H}_O$ about the y axis, Fig. 19–10b. Hence, $d\mathbf{H}_O$ acts in the direction of the positive x axis and has a magnitude of $dH_O = H_O\,d\phi = I_z\omega_z\,d\phi$, so that

$$\Sigma M_O\,dt = dH_O = I_z\omega_z\,d\phi$$

or

$$\Sigma M_O = I_z\,\omega_z\,\frac{d\phi}{dt}$$

Since $\Omega_y = d\phi/dt$, the final result can be written as

$$\Sigma M_O = I_z\omega_z\Omega_y \qquad (19\text{–}15)$$

It should be noted that the vectors $\Sigma\mathbf{M}_O$, $\boldsymbol{\omega}_z$ (spin) and $\boldsymbol{\Omega}_y$ (precession) are all mutually *perpendicular,* Fig. 19–10a. In this regard, the direction of $\boldsymbol{\Omega}_y$ can be determined if one realizes that $\boldsymbol{\Omega}_y$ always swings $\mathbf{H}_O$ (or $\boldsymbol{\omega}_z$) toward the line of action of $\Sigma\mathbf{M}_O$, Fig. 19–10b.* Hence, it can be seen that the gyro does not fall downward as expected. Instead, if the product $I_z\omega_z\Omega_y$ is correctly chosen to counterbalance the moment $\Sigma M_O = Wr_G$, the gyro remains in the horizontal plane.

When a gyro is mounted in gimbal rings, Fig. 19–11, it becomes *free* of external moments applied to its base. Thus, in theory, its angular momentum $\mathbf{H}_O$ will never precess but, instead, maintain its same fixed orientation along the axis of spin when the base is rotated. This type of gyroscope is called a *free gyro* and is useful as a gyrocompass when the spin axis of the gyro is directed north. In reality, the gimbal mechanism is never completely free of friction, so that such a device is useful only for the local navigation of ships and aircraft. The gyroscopic effect is also useful as a means of stabilizing both the rolling motion of ships at sea and the trajectories of missiles and projectiles. Furthermore, this effect is of significant importance in the design of shafts and bearings for rotors which are subjected to forced precessions.

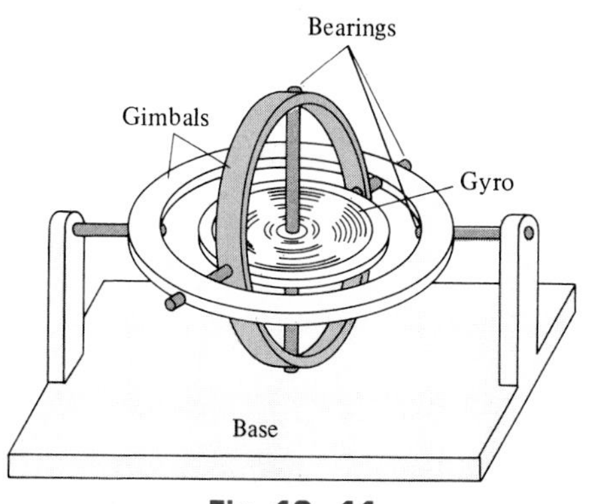

Fig. 19–11

*By the right-hand rule, curl the fingers from $\mathbf{H}_O$ (or $\boldsymbol{\omega}_z$) towards $\Sigma\mathbf{M}_O$. The thumb acts in the direction of $\boldsymbol{\Omega}_y$.

Example 19–7

The 1-kg disk shown in Fig. 19–12*a* is spinning about its axis with a constant angular velocity of $\omega_D = 70$ rad/s. The block at B has a mass of 2 kg, and by adjusting its position s one can change the precession of the disk about its supporting pivot at O. Compute the position s which will enable the disk to have a constant precessional velocity of $\Omega_p = 0.5$ rad/s about the pivot. Neglect the weight of the shaft.

Solution

The free-body diagram of the disk is shown in Fig. 19–12*b*, where **F** represents the force reaction of the shaft on the disk. The origin for the x, y, z coordinate system is located at point O, which represents a *fixed point* for the disk. (Although point O does not lie on the disk, imagine a massless extension of the disk to this point.) In the conventional sense, the y axis is chosen along the axis of precession, and the z axis is along the axis of spin. Hence,

$$\Sigma M_x = I_z \omega_z \Omega_y$$

Substituting the required data gives

$$9.81(0.2) - F(0.2) = [\tfrac{1}{2}(1)(0.05)^2](-70)(0.5)$$

$$F = 10.0 \text{ N}$$

As shown on the free-body diagram of the shaft and block B, Fig. 19–12*c*, summing moments about the x axis requires

$$(19.62)s = (10.0)(0.20)$$

$$= 0.101\ 9 \text{ m} = 101.9 \text{ mm} \qquad \textit{Ans.}$$

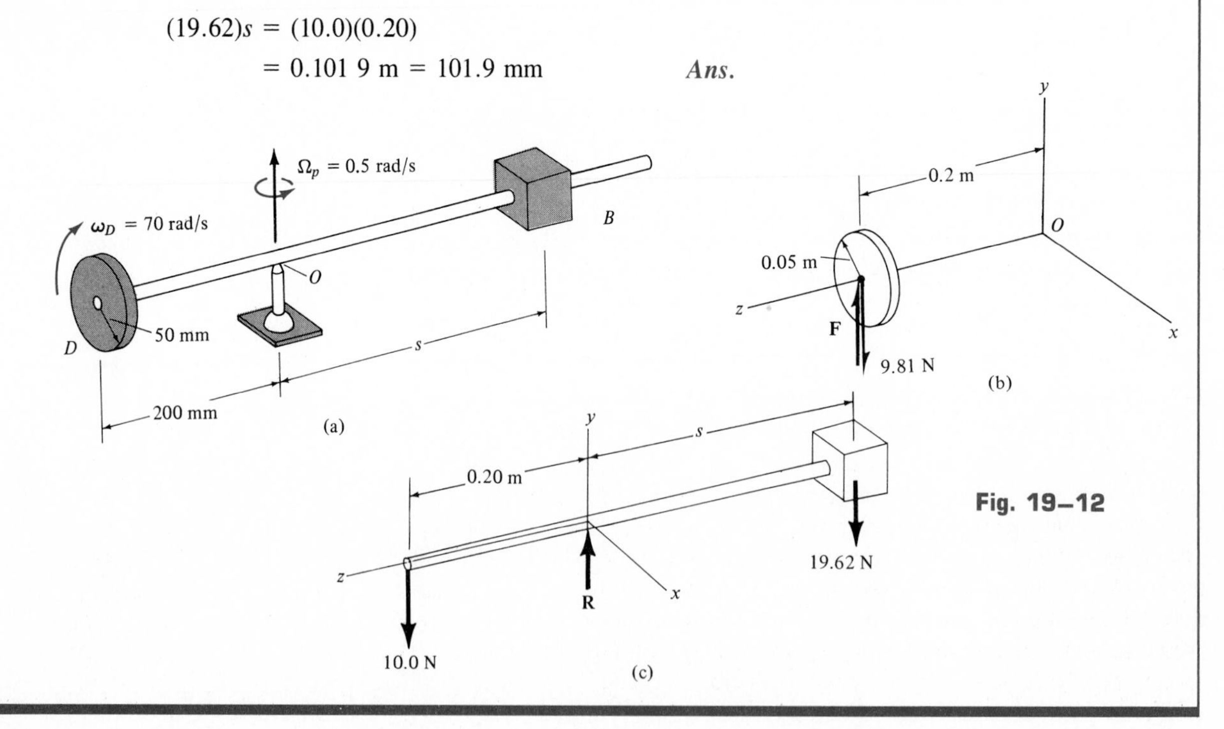

Fig. 19–12

Problems

19–34. The driving armature of a ship's engine may be approximated by the 80-kg cylinder having a radius of 0.150 m. The armature is rotating with an angular velocity of $\omega_s = 250$ rad/s as the ship turns at $\omega_T = 0.3$ rad/s. Determine the vertical reactions at each of the bearings A and B due to this motion.

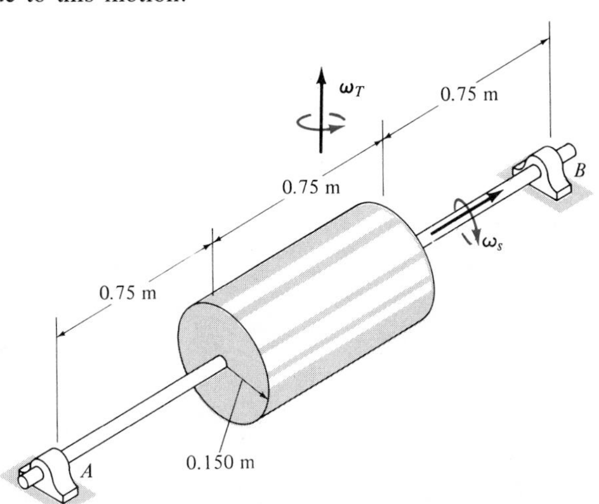

Prob. 19–34

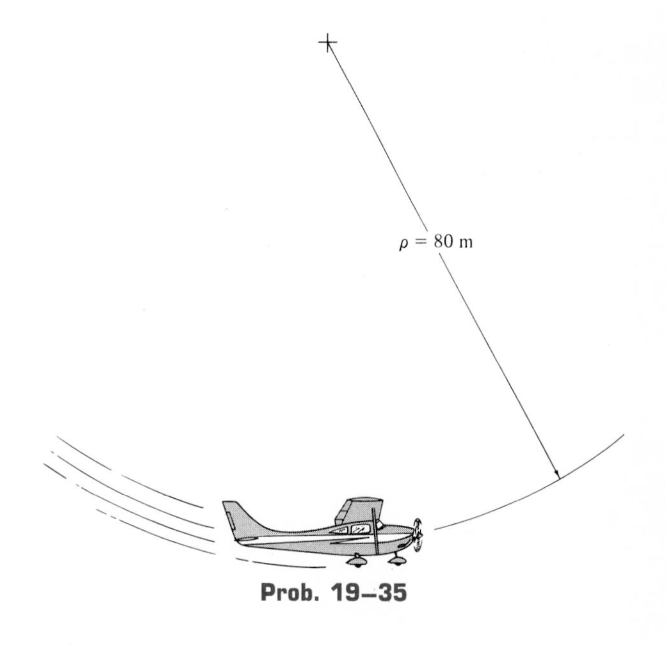

Prob. 19–35

19–35. The propeller on a single-engine airplane has a mass of 15 kg and a centroidal radius of gyration of 0.3 m computed about the axis of spin. When viewed from the front of the airplane, the propeller is turning clockwise at 350 rad/s about the spin axis. If the airplane enters a vertical curve having a radius of $\rho = 80$ m and is traveling at 200 km/h, determine the gyroscopic bending moment which the propeller exerts on the bearings of the engine when the airplane is in its lowest position, as shown.

***19–36.** The car is traveling at $v_C = 100$ km/h around the horizontal curve having a radius of curvature of 80 m. If each wheel has a mass of 16 kg, a radius of gyration of $k_G = 300$ mm about its spinning axis, and a diameter of 400 mm, determine the difference in the normal force between the wheels caused by the gyroscopic effect. The distance between the wheels is 1.30 m.

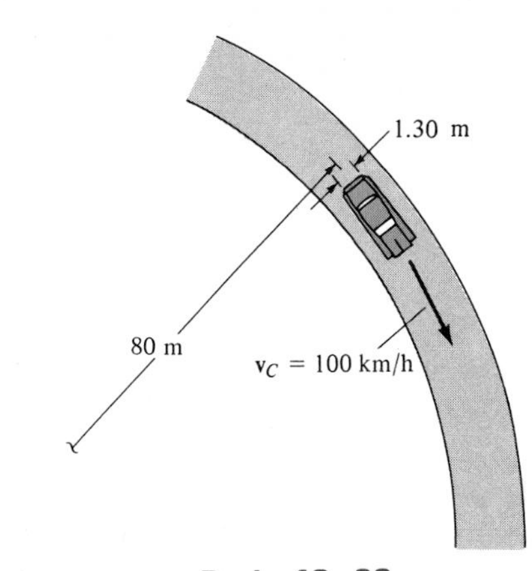

Prob. 19–36

19–37. The mechanics of a Dutch windmill used for grinding grain are shown. If each millstone (disk) has a mass of 400 kg, determine the gyroscopic moment which the millstone at A exerts on the horizontal shaft at B. The mill blade is spinning at 1.5 rad/s and the millstones roll without slipping.

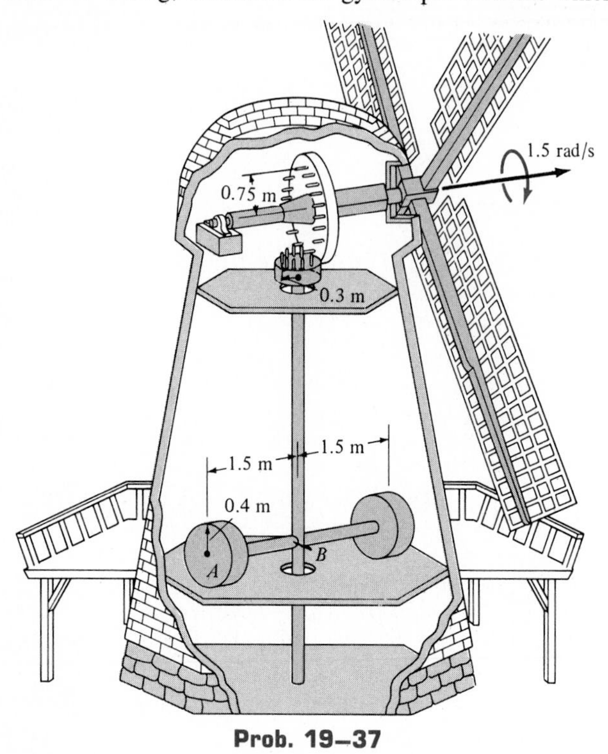

Prob. 19–37

19–38. The gyroscope consists of a uniform 450-g disk D which is attached to an axle AB of negligible mass. The supporting frame has a mass of 180 g and a center of gravity at G. If the disk is rotating about the axle at an angular speed of $\omega_D = 90$ rad/s, determine the constant angular velocity ω_p at which the frame precesses about the pivot point O. The frame moves in the horizontal plane.

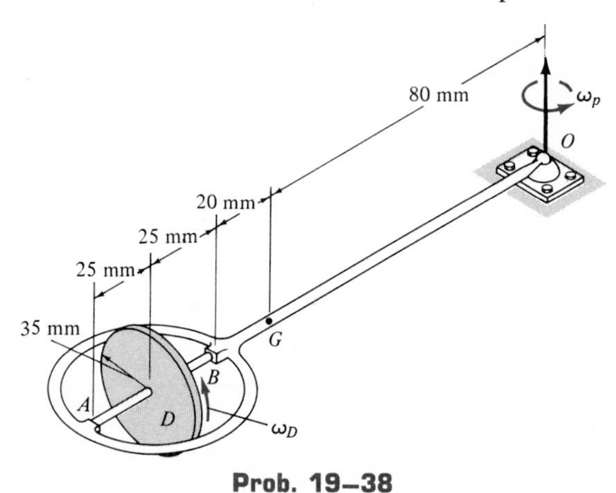

Prob. 19–38

Review Problems

19–39. Solve Prob. 17–74 using impulse and momentum methods.

***19–40.** Solve Prob. 17–75 using impulse and momentum methods.

19–41. The disk has a weight of 10 lb and is pinned at its center O. If a vertical force of $P = 2$ lb is applied to the cord wrapped around its outer rim, determine the angular velocity of the disk in four seconds starting from rest. Neglect the mass of the cable.

19–42. Solve Prob. 19–41 if a 2-lb block is attached to the cord, rather than a 2-lb force **P**.

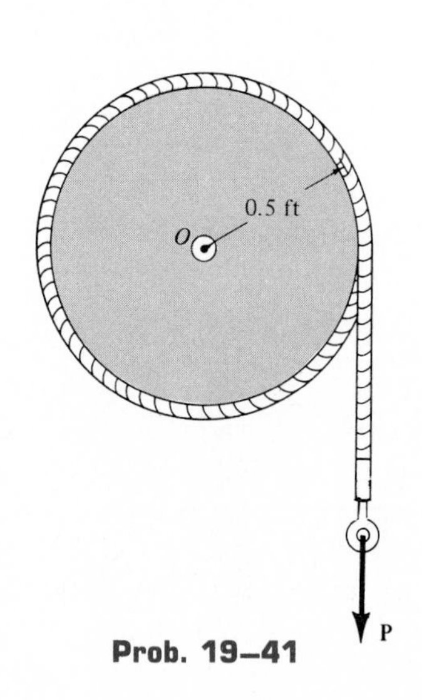

Prob. 19–41

19–43. The 50-kg cylinder has an angular velocity of 30 rad/s when it is brought into contact with the horizontal surface at C. If the coefficient of friction is $\mu_C = 0.2$, determine how long it takes for the cylinder to stop spinning. What force is developed at the pin A during this time? The axis of the cylinder is connected to *two* symmetrical links. (Only AB is shown.) For the computation, neglect the weight of the links.

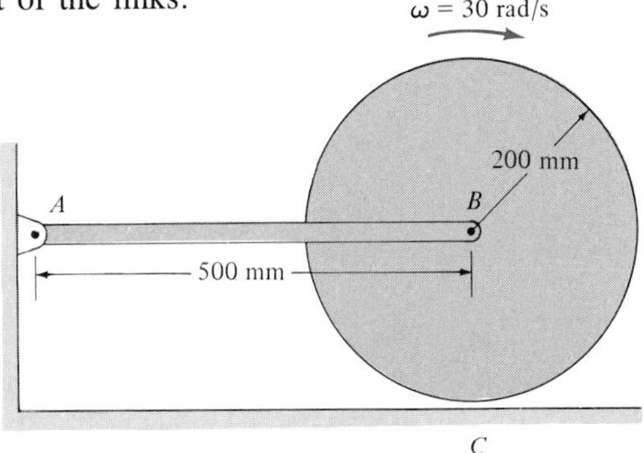

Prob. 19–43

***19–44.** Solve Prob. 19–43 if the cylinder is rotating *counterclockwise* at $\omega = 30$ rad/s.

19–45. The disk has a weight of 15 lb and is pinned at its center O. The spring-loaded plunger exerts a vertical force of 50 lb on the disk, at A. If a moment of $M = (10t)$ lb · ft, where t is in seconds, is applied to the disk, determine the angular velocity of the disk in four seconds starting from rest. The coefficient of friction at A is $\mu_A = 0.3$. *Hint:* First determine the time needed to start the disk rotating.

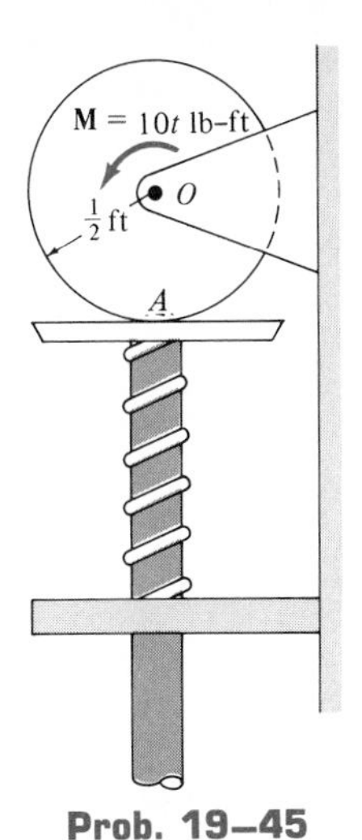

Prob. 19–45

19–46. The spool has a weight of 30 lb and a radius of gyration of $k_O = 1.40$ ft. If a force of 40 lb is applied to the supporting cord at A as shown, determine the angular velocity of the spool in $t = 3$ s starting from rest.

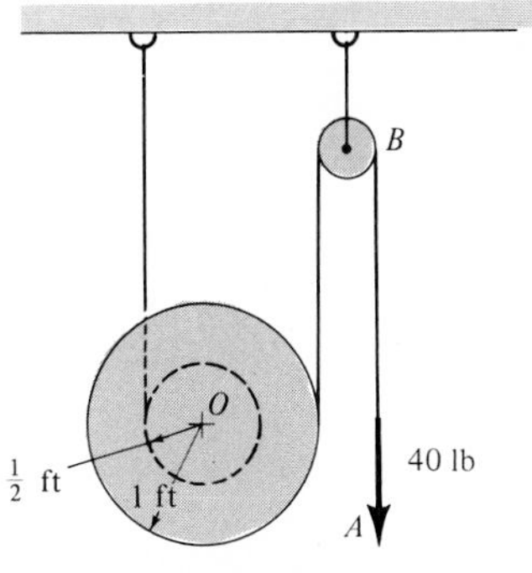

Probs. 19–46 / 19–47

19–47. Solve Prob. 19–46 if a 40-lb block is suspended from the cord at A, rather than applying the 40-lb force.

***19–48.** The two disks A and C, each having a weight of 10 lb and a radius of 0.5 ft, are attached to the crossbar frame. The vertical axis DE of the frame is subjected to an angular velocity of $\omega = 15$ rad/s, causing both disks to roll on the horizontal surface without slipping. If the shaft DE is raised slightly so that the disks leave the surface, determine the gyroscopic bending moment exerted at B on the frame by each of the rotating disks. If the shaft ABC were elastic, would it bend upward or downward because of this moment?

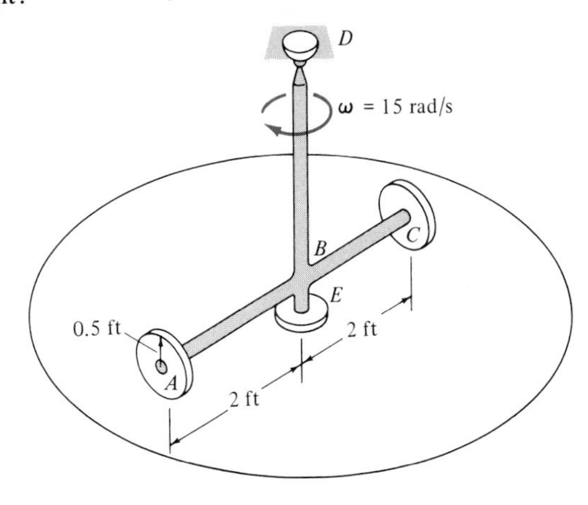

Prob. 19–48

Special Applications

In this chapter we will discuss some applications of dynamics requiring a more advanced mathematical treatment. The chapter begins with application of cylindrical coordinates used to solve kinematic and kinetic problems of particles, followed by a relative motion analysis of particles and rigid bodies using rotating axes. In the last part of the chapter an impulse—momentum analysis is used to analyze systems which are subjected to a steady fluid stream.

20.1* Curvilinear Motion: Cylindrical Components

In some engineering problems it is often convenient to express the path of motion of a particle in terms of cylindrical coordinates r, θ, and z. If motion is restricted to the plane, the polar coordinates r and θ are used.

Polar Coordinates. We can specify the location of particle P shown in Fig. 20–1a using both the *radial coordinate* r, which extends outward from the fixed origin O to the particle, and a *transverse coordinate* θ, which is the counterclockwise angle between a fixed reference line and the r axis. The angle is generally measured in degrees or radians, where 1 rad $= 180°/\pi$. In the following analysis, the velocity and acceleration will be resolved into their radial and transverse components, which act collinear and perpendicular to r. These directions will be *positive* when r and θ are *increasing* as shown in Fig. 20–1a.

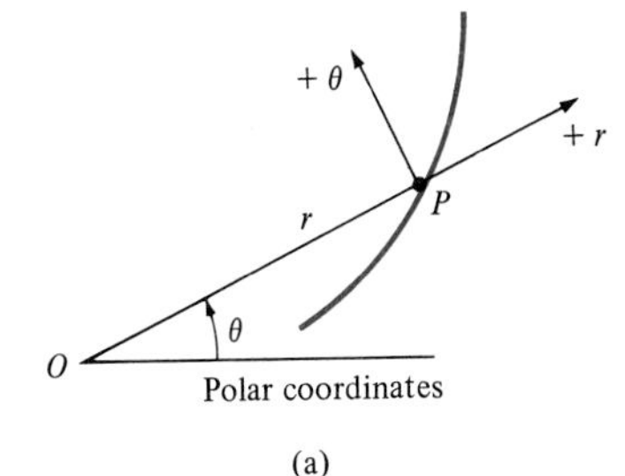

Fig. 20–1a

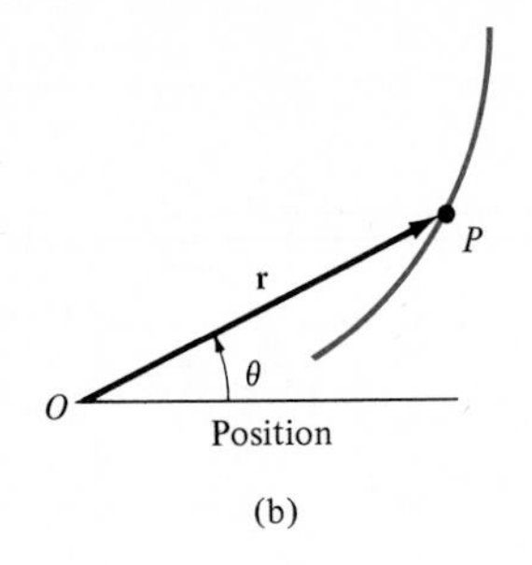

(b)

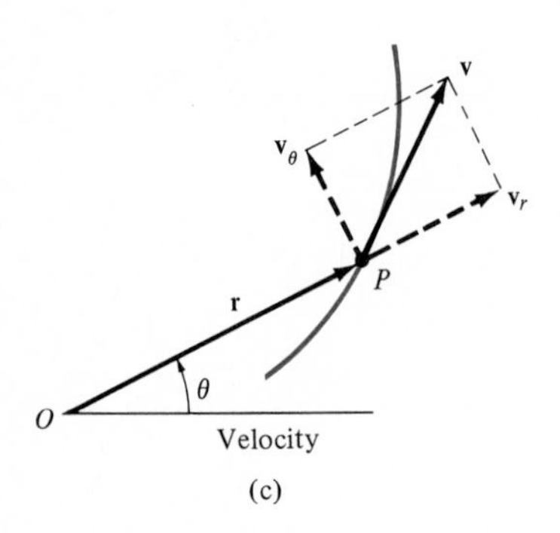

(c)

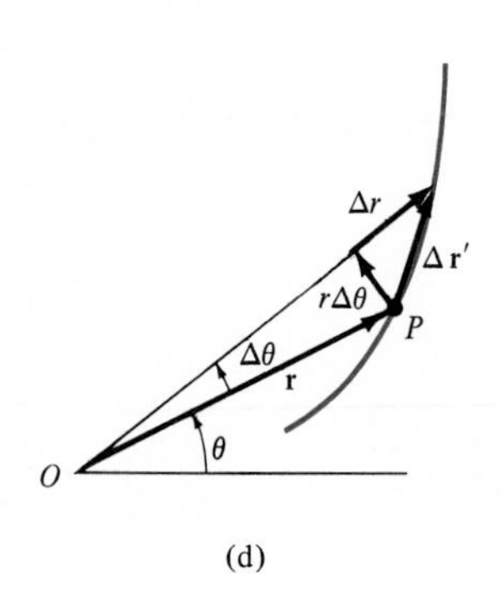

(d)

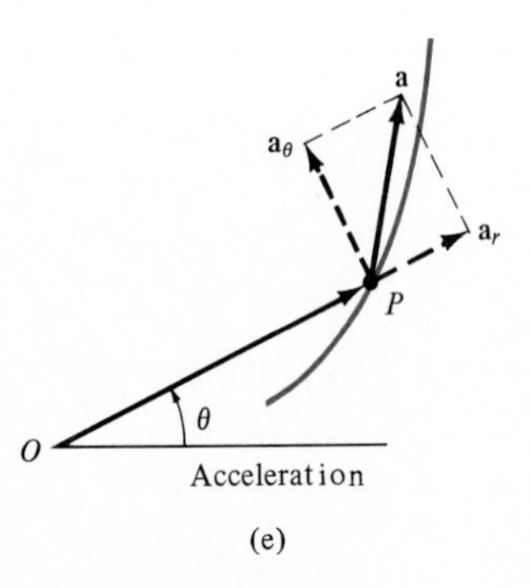

(e)

Fig. 20–1b, c, d, e

Position. At any instant the position of the particle is defined by the position vector **r**, Fig. 20–1*b*.

Velocity. As indicated in Sec. 12–3, the instantaneous velocity **v** is always tangent to the path. Here we will express **v** in terms of its radial and transverse components $\mathbf{v}_r$ and $\mathbf{v}_\theta$, as shown in Fig. 20–1*c*. We require

$$\mathbf{v} = \mathbf{v}_r + \mathbf{v}_\theta \qquad (20\text{–}1)$$

In order to determine these components we must take the time derivative of **r**, while accounting for its changes in the r and θ directions. In Fig. 20–1*d* it is seen that as the particle undergoes the displacement $\Delta\mathbf{r}'$ along the curve in time Δt, the coordinates r and θ change by Δr and $\Delta\theta$. For small angles $\Delta\theta$, $\Delta\mathbf{r}'$ has components of $+\Delta r$ and $+r\,\Delta\theta$ in the r and θ directions, respectively.

We require $\mathbf{v} = \lim_{\Delta t\to 0} \Delta\mathbf{r}'/\Delta t$, or

$$v_r = \lim_{\Delta t\to 0} \frac{\Delta r}{\Delta t} \qquad v_\theta = \lim_{\Delta t\to 0} \frac{r\,\Delta\theta}{\Delta t}$$

which yields

$$\begin{aligned} v_r &= \dot{r} \\ v_\theta &= r\dot{\theta} \end{aligned} \qquad (20\text{–}2)$$

In particular, the term $\dot{\theta} = d\theta/dt$ is called the *angular velocity,* since it provides a measure of the time rate of change of the angle θ. Common units for this measurement are rad/s.

The magnitude of velocity is simply

$$v = \sqrt{(\dot{r})^2 + (r\dot{\theta})^2} \qquad (20\text{–}3)$$

and as stated above, the *direction* of **v** is tangent to the path at point P, Fig. 20–1*c*.

Acceleration. The instantaneous acceleration can be resolved into its radial and tangential components as shown in Fig. 20–1*e*, where

$$\mathbf{a} = \mathbf{a}_r + \mathbf{a}_\theta \qquad (20\text{–}4)$$

In order to determine these components, we must consider the change in the magnitude and direction of the particle's velocity when the particle moves from point P to P' during the time Δt, Fig. 20–1*f*. The changes made in the components of velocity can be determined if we superimpose the r and θ components of the velocities at P and P', which is shown greatly exaggerated in Fig. 20–1*g*. Here it is seen that the change in the radial component is $\Delta\mathbf{v}_r = (\Delta\mathbf{v}_r)_r + (\Delta\mathbf{v}_r)_\theta$, and the change in the transverse component is $\Delta\mathbf{v}_\theta = (\Delta\mathbf{v}_\theta)_r + (\Delta\mathbf{v}_\theta)_\theta$. For small angles $\Delta\theta$, $(\Delta v_r)_\theta = v_r\,\Delta\theta$ and

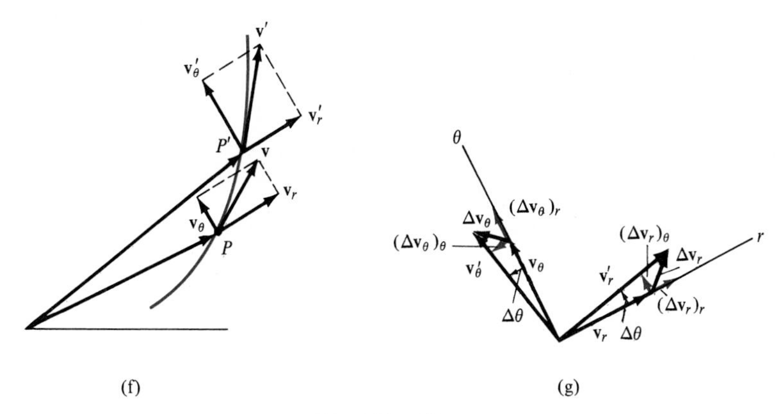

Fig. 20–1f, g

$(\Delta v_\theta)_\theta = v_\theta\,\Delta\theta$, Fig. 20–1$g$, and noting that $(\Delta \mathbf{v}_\theta)_\theta$ points in the negative r direction, we have

$$\lim_{\Delta t\to 0}\frac{\Delta v_r}{\Delta t} = \lim_{\Delta t\to 0}\frac{(\Delta v_r)_r}{\Delta t} - \lim_{\Delta t\to 0} v_r\frac{\Delta\theta}{\Delta t}$$

$$a_r = (\dot{v}_r)_r - v_r\dot{\theta}$$

Also,

$$\lim_{\Delta t\to 0}\frac{\Delta v_\theta}{\Delta t} = \lim_{\Delta t\to 0}\frac{(\Delta v_\theta)_r}{\Delta t} + \lim_{\Delta t\to 0} v_\theta\frac{\Delta\theta}{\Delta t}$$

$$a_\theta = (\dot{v}_\theta)_\theta + v_\theta\dot{\theta}$$

Note that the terms $(\dot{v}_r)_r$ and $(\dot{v}_\theta)_\theta$ represent the time rate of change in the *magnitudes* of $\mathbf{v}_r$ and $\mathbf{v}_\theta$, respectively. From Eqs. 20–2, we have

$$(\dot{v}_r)_r = \ddot{r} \qquad (\dot{v}_\theta)_\theta = \dot{r}\dot{\theta} + r\ddot{\theta}$$

Substituting these and Eqs. 20–2 into the above equations for a_r and a_θ yields the final results

$$\begin{aligned} a_r &= \ddot{r} - r\dot{\theta}^2 \\ a_\theta &= r\ddot{\theta} + 2\dot{r}\dot{\theta} \end{aligned} \tag{20–5}$$

The term $\ddot{\theta} = d^2\theta/dt^2 = d/dt(d\theta/dt)$ is called the *angular acceleration* since it measures the change made in the rate of change of θ during an instant of time. Common units for this measurement are rad/s^2.

The *magnitude* of acceleration is simply

$$a = \sqrt{(\ddot{r} - r\dot{\theta}^2)^2 + (r\ddot{\theta} + 2\dot{r}\dot{\theta})^2} \tag{20–6}$$

The *direction* is determined from the vector addition of its two components. In general, **a** will *not* be tangent to the path, Fig. 20–1e.

PROCEDURE FOR ANALYSIS

Application of the preceding polar equations for computing the particle's velocity and acceleration requires a *straightforward substitution* once r and the time derivatives $\dot{r}$, $\ddot{r}$, $\dot{\theta}$, and $\ddot{\theta}$ have been computed and evaluated at the instant considered. Two types of problems generally occur:

1. If the coordinates are specified as time-parametric equations, $r = r(t)$ and $\theta = \theta(t)$, the derivatives are simply $\dot{r} = dr/dt$, $\ddot{r} = d^2r/dt^2$, $\dot{\theta} = d\theta/dt$, and $\ddot{\theta} = d^2\theta/dt^2$ (see Examples 20–1 and 20–2).
2. If the time-parametric equations of motion are not given, it will be necessary to specify the path $r = f(\theta)$ and compute the relationship between the time derivatives using the chain rule of calculus. In this case,

$$\dot{r} = \frac{df(\theta)}{d\theta}\dot{\theta} \qquad \text{and} \qquad \ddot{r} = \left(\frac{d^2f(\theta)}{d\theta^2}\dot{\theta}\right)\dot{\theta} + \frac{df(\theta)}{d\theta}\ddot{\theta}$$

Thus, if two of the *four* time derivatives $\dot{r}$, $\ddot{r}$, $\dot{\theta}$, and $\ddot{\theta}$ are *known,* the other two can be obtained from these equations (see Example 20–3). In some problems, however, two of these time derivatives may *not* be known; instead, the magnitude of the particle's velocity or acceleration may be specified. If this is the case, $v^2 = \dot{r}^2 + (r\dot{\theta})^2$ and $a^2 = (\ddot{r} - r\dot{\theta}^2)^2 + (r\ddot{\theta} + 2\dot{r}\dot{\theta})^2$ may be used to obtain the necessary relationships involving $\dot{r}$, $\ddot{r}$, $\dot{\theta}$, and $\ddot{\theta}$ (see Example 20–4).

Besides the examples which follow, further examples involving the calculation of a_r and a_θ can be found in the ''kinematics'' of Examples 20–5 to 20–6.

Example 20–1

The rod AB shown in Fig. 20–2 has a constant angular rotation of 5 rad/s. Express the velocity and acceleration of point A on the rod in terms of its polar components.

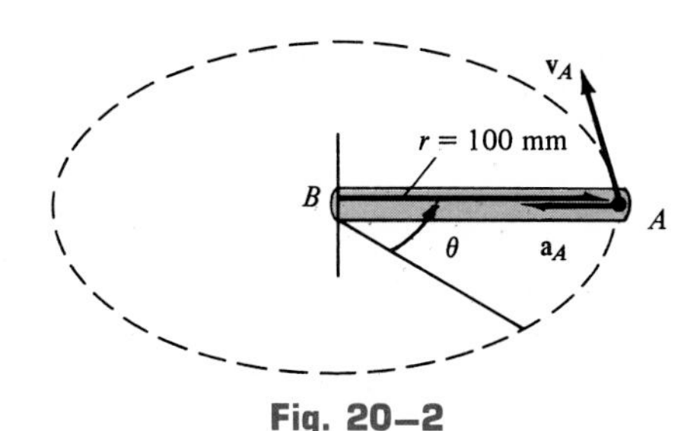

Fig. 20–2

Solution

Time Derivatives. Since Eqs. 20–5 will be used for the solution, it is necessary to obtain the first and second time derivatives of r and θ. Noting that the radius r and the angular rate of rotation $\dot{\theta}$ are *constant*, we have

$$r = 100 \text{ mm} \qquad \dot{r} = 0 \qquad \ddot{r} = 0$$
$$\dot{\theta} = 5 \text{ rad/s} \qquad \ddot{\theta} = 0$$

Velocity

$$v_r = \dot{r} = 0 \qquad \textit{Ans.}$$
$$v_\theta = r\dot{\theta} = 100(5) = 500 \text{ mm/s} \qquad \textit{Ans.}$$

Acceleration

$$a_r = \ddot{r} - r\dot{\theta}^2 = 0 - 100(5)^2 = -2500 \text{ mm/s}^2 \qquad \textit{Ans.}$$
$$a_\theta = r\ddot{\theta} + 2\dot{r}\dot{\theta} = 100(0) + 2(0)(5) = 0 \qquad \textit{Ans.}$$

The path of motion for point A is circular. From the results, the velocity $\mathbf{v}_A$ is tangent to this path and the acceleration $\mathbf{a}_A$ is directed toward B, Fig. 20–2.

Example 20–2

The rod OA, shown in Fig. 20–3a, is rotating in the horizontal x-y plane such that at any instant $\theta = (t^{2/3})$ rad. At the same time, the collar B is sliding outward along OA so that $r = (100t^2)$ mm. If in both cases t is measured in seconds, determine the velocity and acceleration of the collar when $t = 1$ s.

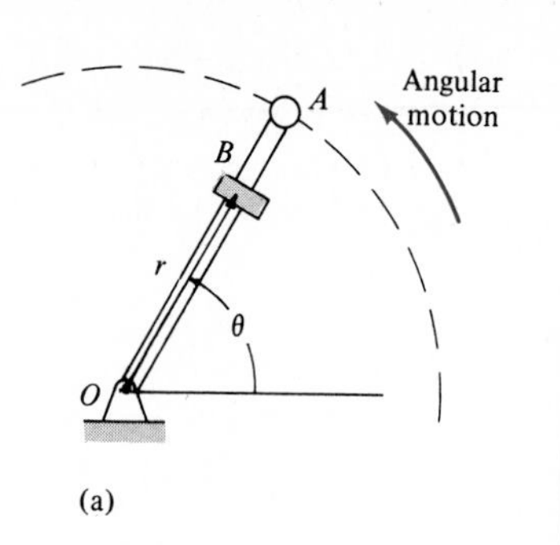

(a)

Solution

Time Derivatives. Computing the time derivatives, and evaluating each at $t = 1$ s, we have

$$r = 100t^2\Big|_{t=1\text{ s}} = 100 \text{ mm} \qquad \theta = t^{2/3}\Big|_{t=1\text{ s}} = 1 \text{ rad} = 57.3^\circ$$

$$\dot{r} = 200t\Big|_{t=1\text{ s}} = 200 \text{ mm/s} \qquad \dot{\theta} = \tfrac{2}{3}t^{-1/3}\Big|_{t=1\text{ s}} = 0.667 \text{ rad/s}$$

$$\ddot{r} = 200\Big|_{t=1\text{ s}} = 200 \text{ mm/s}^2 \qquad \ddot{\theta} = -\tfrac{2}{9}t^{-4/3}\Big|_{t=1\text{ s}} = -0.222 \text{ rad/s}^2$$

Velocity (Fig. 20–3b)

$$v_r = \dot{r} = 200 \text{ mm/s}$$
$$v_\theta = r\dot{\theta} = 100(0.667) = 66.7 \text{ mm/s}$$

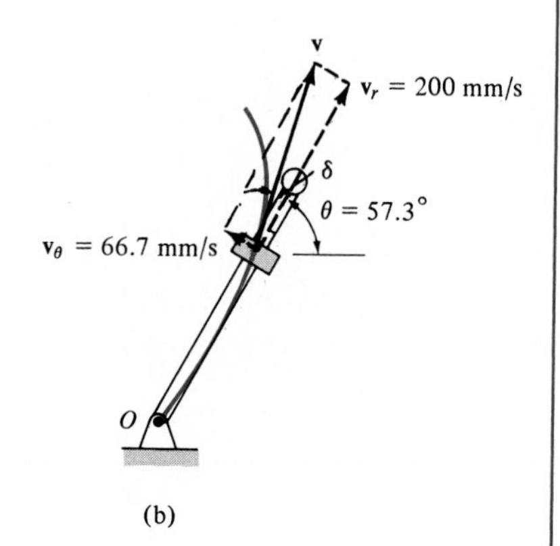

(b)

The magnitude of $\mathbf{v}$ is

$$v = \sqrt{(200)^2 + (66.7)^2} = 210.8 \text{ mm/s} \qquad \textit{Ans.}$$

From Fig. 20–3b,

$$\delta = \tan^{-1}\left(\frac{66.7}{200}\right) = 18.4^\circ \qquad \delta + 57.3^\circ = 75.7^\circ \quad \measuredangle^{\mathbf{v}}_{75.7^\circ} \qquad \textit{Ans.}$$

Acceleration (Fig. 20–3c)

$$a_r = \ddot{r} - r\dot{\theta}^2 = 200 - 100(0.667)^2 = 155.5 \text{ mm/s}^2$$
$$a_\theta = r\ddot{\theta} + 2\dot{r}\dot{\theta} = 100(-0.222) + 2(200)0.667 = 244.4 \text{ mm/s}^2$$

The magnitude of $\mathbf{a}$ is

$$a = \sqrt{(155.5)^2 + (244.4)^2} = 289.7 \text{ mm/s}^2 \qquad \textit{Ans.}$$

From Fig. 20–3c,

$$\phi = \tan^{-1}\left(\frac{244.4}{155.5}\right) = 57.5^\circ \qquad \phi + 57.3^\circ = 114.8^\circ \quad {}^{\mathbf{a}}_{114.8^\circ} \qquad \textit{Ans.}$$

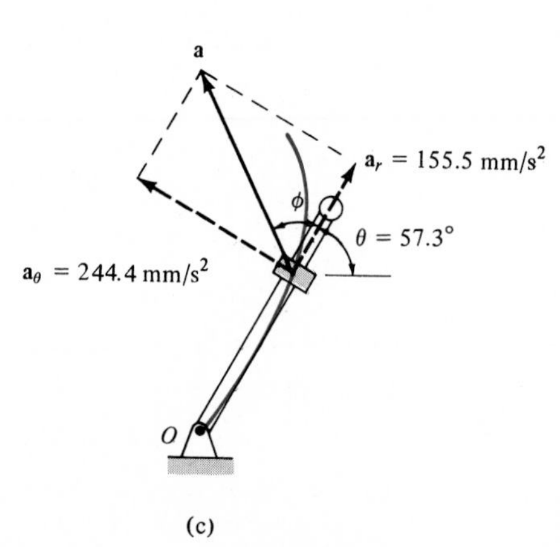

(c)

Fig. 20–3

Example 20–3

The searchlight shown in Fig. 20–4*a* casts a spot of light along the face of a wall that is located 100 m from the searchlight. Determine the magnitudes of the velocity and acceleration at which the spot appears to travel across the wall at the instant $\theta = 45°$. The searchlight is rotating about the z axis at a constant rate of $\dot{\theta} = 4$ rad/s.

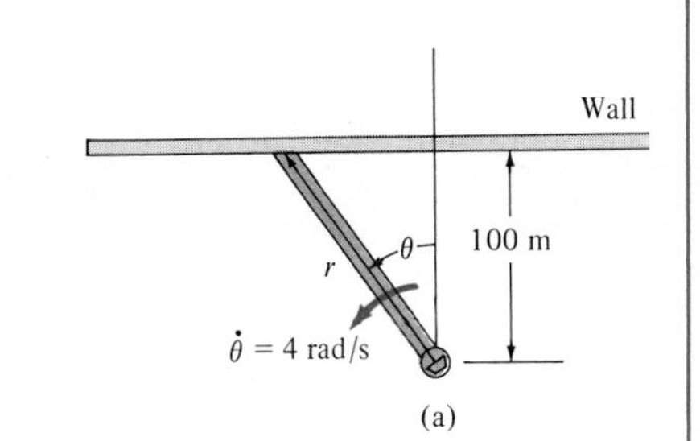

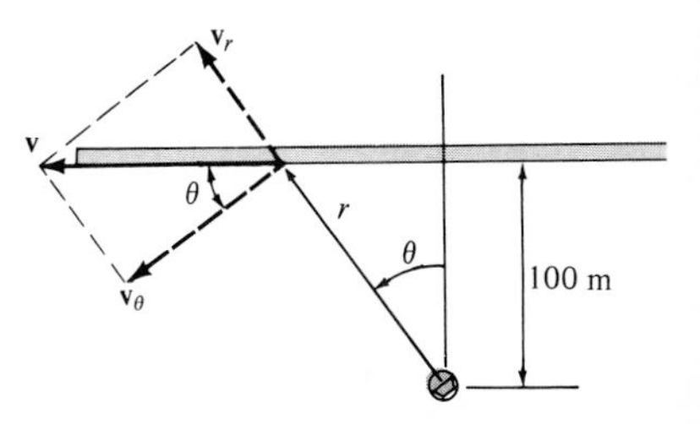

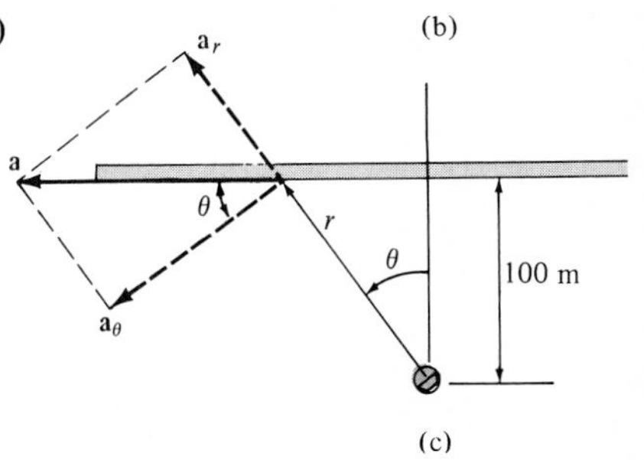

Fig. 20–4

Solution

Time Derivatives. To compute the necessary time derivatives it is first necessary to relate r to θ. From Fig. 20–4*a*, this relation is

$$r = \frac{100}{\cos\theta} = 100 \sec\theta$$

Using the chain rule of calculus, noting that $d(\sec\theta) = \sec\theta\tan\theta\, d\theta$, and $d(\tan\theta) = \sec^2\theta\, d\theta$, we have

$$\dot{r} = 100 \sec\theta \tan\theta\dot{\theta}$$

$$\ddot{r} = 100(\sec\theta\tan\theta\dot{\theta})\tan\theta\dot{\theta} + 100\sec\theta(\sec^2\theta\dot{\theta})\dot{\theta} + 100\sec\theta\tan\theta(\ddot{\theta})$$

$$= 100\sec\theta\tan^2\theta(\dot{\theta})^2 + 100\sec^3\theta(\dot{\theta})^2 + 100\sec\theta\tan\theta\ddot{\theta}$$

Since $\dot{\theta} = 4$ rad/s = constant, then $\ddot{\theta} = 0$, and the above equations, when $\theta = 45°$, become

$$r = 100 \sec 45° = 141.4$$

$$\dot{r} = 400 \sec 45° \tan 45° = 565.7$$

$$\ddot{r} = 1600(\sec 45° \tan^2 45° + \sec^3 45°) = 6788.2$$

Velocity (Fig. 20–4*b*)

$$v_r = \dot{r} = 565.7 \text{ m/s}$$

$$v_\theta = r\dot{\theta} = 141.4(4) = 565.7 \text{ m/s}$$

$$v = \sqrt{v_r^2 + v_\theta^2} = \sqrt{(565.7)^2 + (565.7)^2}$$

$$= 800 \text{ m/s} \qquad \textit{Ans.}$$

Acceleration (Fig. 20–4*c*)

$$a_r = \ddot{r} - r\dot{\theta}^2 = 6788.2 - 141.4(4)^2 = 4525.8 \text{ m/s}^2$$

$$a_\theta = r\ddot{\theta} + 2\dot{r}\dot{\theta} = 141.4(0) + 2(565.7)4 = 4525.8 \text{ m/s}^2$$

$$a = \sqrt{a_r^2 + a_\theta^2} = \sqrt{(4525.8)^2 + (4525.8)^2}$$

$$= 6400 \text{ m/s}^2 \qquad \textit{Ans.}$$

Example 20–4

Due to the rotation of the forked rod, the cylindrical peg A in Fig. 20–5a travels around the slotted path, a portion of which is in the shape of a cardioid, $r = 0.5(1 - \cos\theta)$ ft, where θ is measured in radians. If the peg's velocity is $v = 4$ ft/s and its acceleration is $a = 30$ ft/s^2 at the instant $\theta = 180°$, determine the angular velocity $\dot{\theta}$ and angular acceleration $\ddot{\theta}$ of the fork.

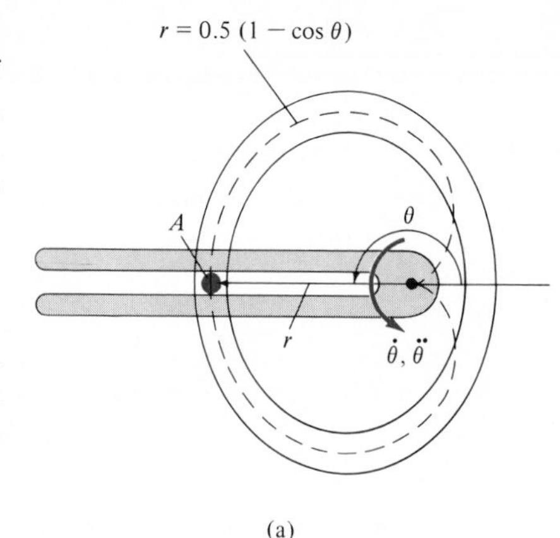

(a)

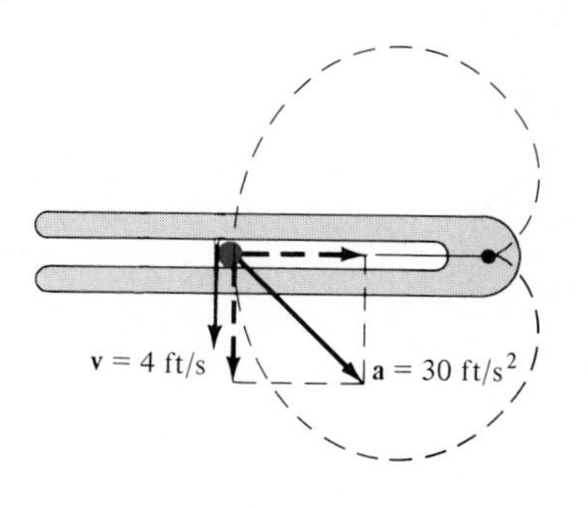

(b)

Fig. 20–5

Solution

Time Derivatives. The magnitudes of the peg's velocity and acceleration are defined in terms of the time derivatives of r and θ by Eqs. 20–3 and 20–6. Computing the time derivatives of r using the chain rule of calculus yields

$$r = 0.5(1 - \cos\theta)$$
$$\dot{r} = 0.5(\sin\theta)\dot{\theta}$$
$$\ddot{r} = 0.5(\cos\theta\dot{\theta})\dot{\theta} + 0.5(\sin\theta)\ddot{\theta}$$

Evaluating these results at $\theta = 180°$, we have

$$r = 1 \text{ ft} \qquad \dot{r} = 0 \qquad \ddot{r} = -0.5\dot{\theta}^2$$

Velocity. Using Eq. 20–3 to determine $\dot{\theta}$ yields

$$v = \sqrt{(\dot{r})^2 + (r\dot{\theta})^2}$$
$$4 = \sqrt{(0)^2 + (1\dot{\theta})^2}$$
$$\dot{\theta} = 4 \text{ rad/s} \qquad \textit{Ans.}$$

Acceleration. In a similar manner, $\ddot{\theta}$ can be found using Eq. 20–6.

$$a = \sqrt{(\ddot{r} - r\dot{\theta}^2)^2 + (r\ddot{\theta} + 2\dot{r}\dot{\theta})^2}$$
$$30 = \sqrt{[-0.5(4)^2 - 1(4)^2]^2 + [1\ddot{\theta} + 2(0)(4)]^2}$$
$$(30)^2 = (-24)^2 + \ddot{\theta}^2$$
$$\ddot{\theta} = 18 \text{ rad/s}^2 \qquad \textit{Ans.}$$

Vectors **a** and **v** are shown in Fig. 20–5b.

Problems

20–1. A particle is moving along a circular path of 2-m radius such that its position as a function of time is given by $\theta = (5t^2)$ rad, where t is in seconds. Determine the magnitude of the particle's acceleration when $\theta = 30°$. The particle starts from rest when $\theta = 0°$.

20–2. A car is traveling along the circular curve of radius $r = 300$ ft. At the instant shown, its angular rate of rotation is $\dot{\theta} = 0.4$ rad/s, which is increasing at the rate of $\ddot{\theta} = 0.2$ rad/s^2. Determine the magnitudes of the car's velocity and acceleration at this instant.

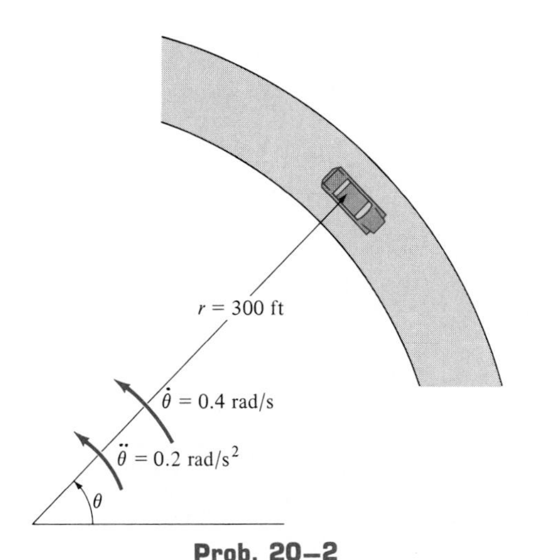

Prob. 20–2

20–3. At the instant shown, the man is twirling a hose over his head with an angular speed of $\dot{\theta} = 2$ rad/s and an angular acceleration of $\ddot{\theta} = 3$ rad/s^2. If it is assumed that the hose lies in a horizontal plane, and water is flowing through it at a constant rate of 3 m/s, determine the magnitudes of velocity and acceleration of a water particle as it exits the open end, $r = 2$ m.

Prob. 20–3

***20–4.** A car is traveling along the circular curve of radius $r = 50$ m with a constant speed of $v = 15$ m/s. Determine the angular rate of rotation $\dot{\theta}$ and the acceleration of the car.

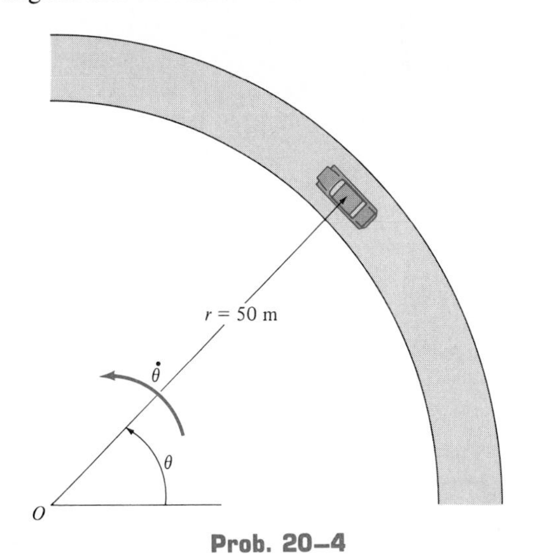

Prob. 20–4

20–5. The double collar C is pin-connected together such that one collar slides over a fixed rod and the other slides over a rotating rod AB. If the angular position of AB is given as $\theta = 4(t^2 + 1)$ rad, where t is measured in seconds, and the path defined by the fixed rod is $r = (0.4 \sin \theta + 0.2)$ m, determine the radial and transverse components of the collar's velocity and acceleration when $t = 2$ s.

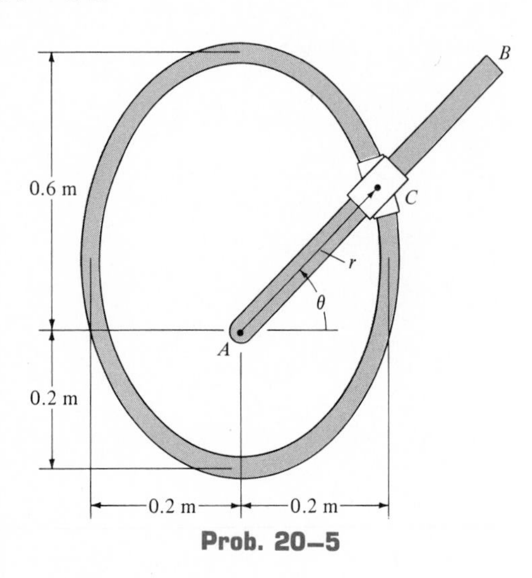

Prob. 20–5

20–6. A particle P moves along the spiral path $r = (10/\theta)$ ft, where θ is in radians. If it maintains a constant speed of $v = 20$ ft/s, determine the magnitudes v_r and v_θ as functions of θ and evaluate each at $\theta = 1$ rad.

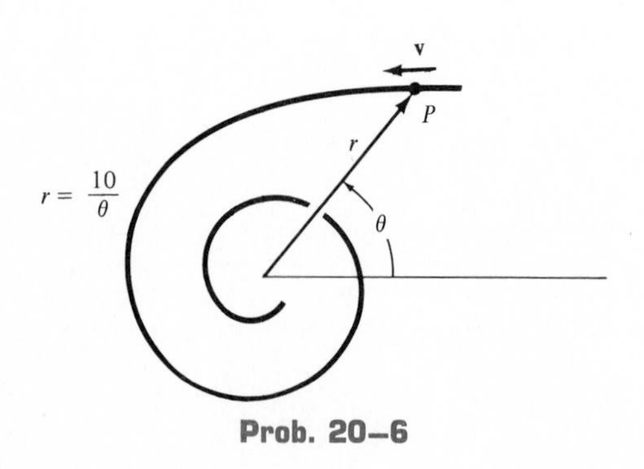

Prob. 20–6

20–7. A cameraman standing at A is following the movement of a race car, B, which is traveling along a straight track at a constant speed of 80 ft/s. Determine the angular rate at which he must turn in order to keep the camera directed on the car at the instant $\theta = 60°$. *Hint:* First show that $r = (100 \csc \theta)$ ft.

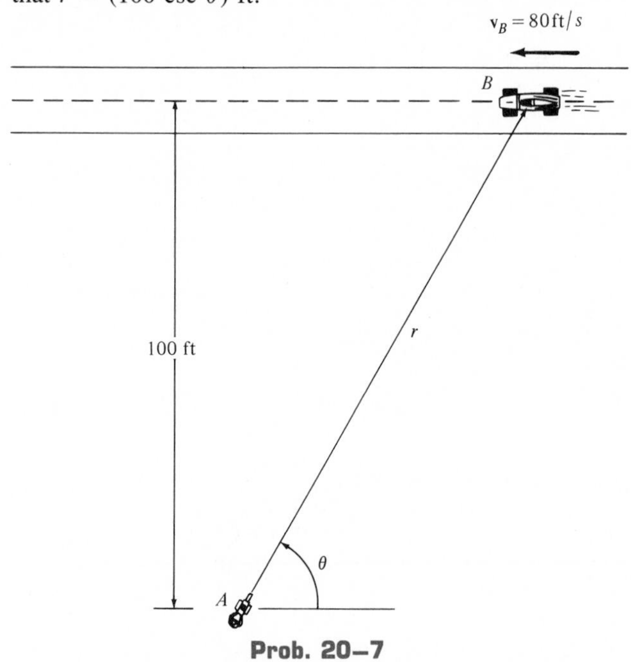

Prob. 20–7

***20–8.** The slotted link is pinned at O, and as a result of rotation it drives the peg P along the horizontal guide. Compute the magnitudes of the velocity and acceleration of P as a function of θ if $\theta = (3t)$ rad, where t is measured in seconds. *Hint:* First show that $r = (500 \sec \theta)$ mm.

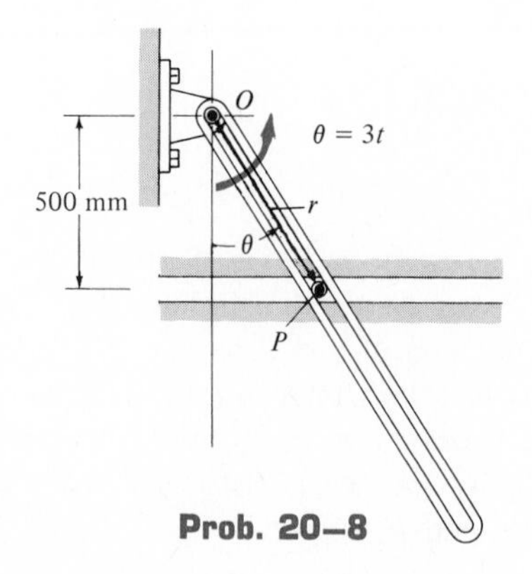

Prob. 20–8

20–9. A cameraman standing at A is following the movement of a race car, B, which is traveling around a curved track at a constant speed of 20 m/s. Determine the angular rate $\dot{\theta}$ at which the man must turn in order to keep the camera directed on the car at the instant $\theta = 30°$. *Hint:* First show that $r = (60 \cos \theta)$ m.

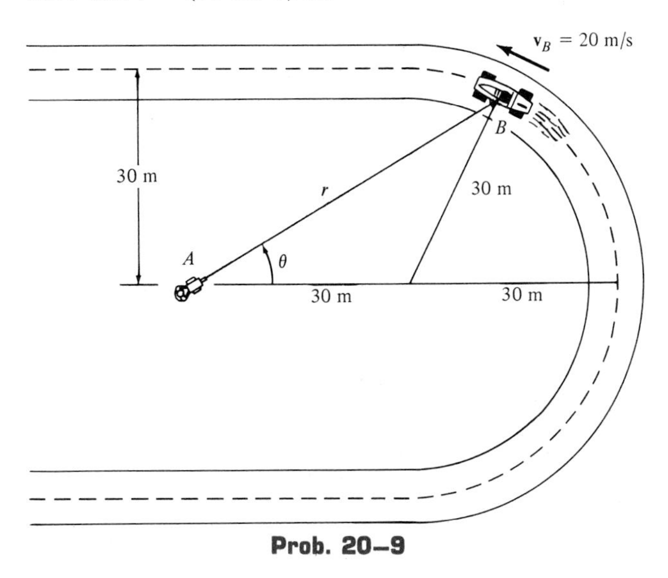

Prob. 20–9

20–10. The slotted arm AB drives the pin C through the spiral groove described by the equation $r = 1.5\theta$, where θ is in radians and r is in feet. If the arm starts from rest when $\theta = 60°$ and is driven at an angular rate of $\dot{\theta} = 4t$ rad/s, determine the radial and transverse components of velocity and acceleration of the pin when $t = 1$ s.

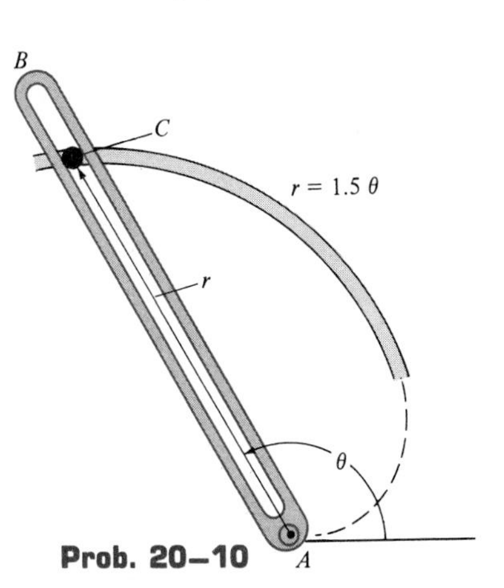

Prob. 20–10

20–11. A double collar C is pin-connected together such that one collar slides over a fixed rod and the other slides over a rotating rod. If the geometry of the fixed rod for a short distance can be defined by a lemniscate, $r^2 = (4 \cos 2\theta)$ ft^2, determine the collar's radial and transverse components of velocity and acceleration at the instant $\theta = 0°$ as shown. Rod OA is rotating at a constant rate of $\dot{\theta} = 6$ rad/s.

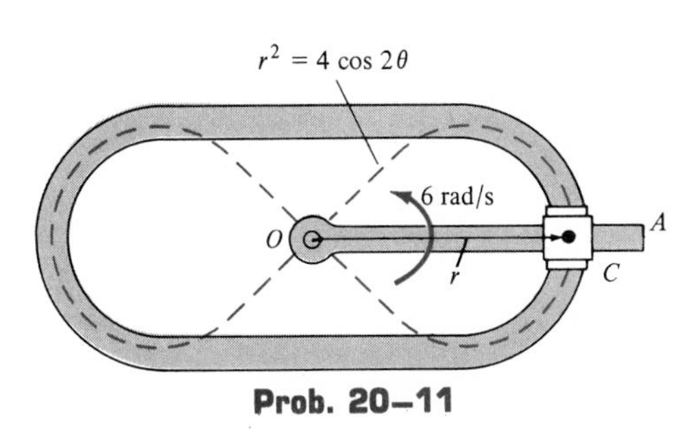

Prob. 20–11

***20–12.** A block rests in the grooved slot of a platform and moves outward along the slot with a speed of $(4t)$ m/s, where t is in seconds. The platform rotates at a constant rate of 6 rad/s. If the block starts from rest at the center, compute the magnitudes of its velocity and acceleration when $t = 1$ s.

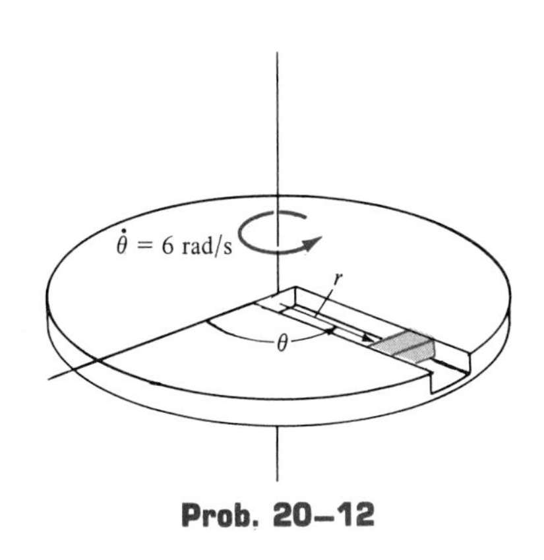

Prob. 20–12

★20.2 Equations of Motion: Cylindrical Coordinates

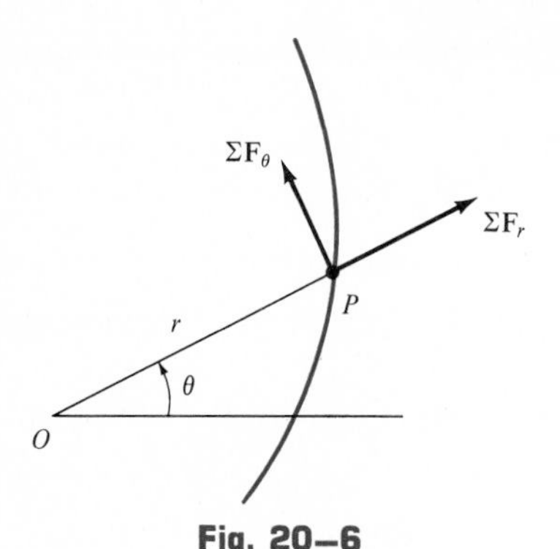

Fig. 20–6

When all the forces acting on a particle are resolved into r, θ components, Fig. 20–6, the equations of motion may be expressed as

$$\Sigma F_r = ma_r \qquad \Sigma F_\theta = ma_\theta \tag{20–7}$$

Tangent and Normal Forces. The most straightforward type of problem involving cylindrical coordinates requires the determination of the resultant force components $\Sigma \mathbf{F}_r$ and $\Sigma \mathbf{F}_\theta$, causing a particle to move with a *known* acceleration. If, however, the particle's accelerated motion is not completely specified at the given instant, then some information regarding the directions or magnitudes of the forces acting on the particle must be known or computed in order to solve Eqs. 20–7. For example, the force **T** causes the particle in Fig. 20–7a to move along a path defined in polar coordinates by $r = f(\theta)$. The *normal force* **N** acting on the particle is always *perpendicular to the tangent of the path;* whereas the frictional force **F** always acts along the tangent in the opposite direction of motion. The *directions* of **N** and **F** can be determined by first computing the angle ψ (psi), Fig. 20–7b, which is defined between the *extended* radial line $r = OP$ and the tangent to the curve. This angle, *always measured counterclockwise,* or in the *positive direction of* θ, is determined from the equation*

$$\tan \psi = r \bigg/ \frac{dr}{d\theta} \tag{20–8}$$

Application is illustrated numerically in Example 20–6.

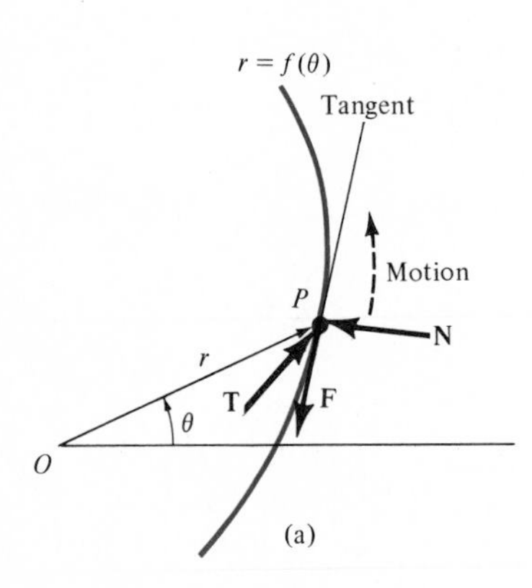

(a)

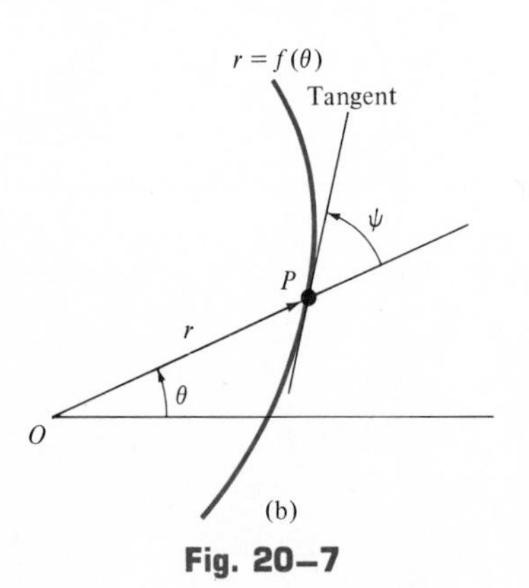

(b)

Fig. 20–7

PROCEDURE FOR ANALYSIS

The procedure given in Sec. 13–4 may be stated as follows when applied to problems involving cylindrical coordinates.

Free-Body and Kinetic Diagrams. Establish the r, θ inertial coordinate system and draw the particle's free-body and kinetic diagrams. When drawing the kinetic diagram assume the vectors $m\mathbf{a}_r$ and $m\mathbf{a}_\theta$ act in the *positive direction* of r and θ if they are unknown.

Equations of Motion. Apply the equations of motion, Eqs. 20–7.

Kinematics. Use the methods of Sec. 20–1 to determine r and the time derivatives $\dot{r}$, $\ddot{r}$, $\dot{\theta}$, and $\ddot{\theta}$, and evaluate the acceleration components $a_r = \ddot{r} - r\dot{\theta}^2$ and $a_\theta = r\ddot{\theta} + 2\dot{r}\dot{\theta}$. If any of these components are computed as negative quantities, it indicates that they act in their negative coordinate directions.

The following examples numerically illustrate application of this procedure.

*The derivation is given in any standard calculus text.

Example 20–5

The smooth 2-kg cylinder C in Fig. 20–8a has a peg P through its center which passes through the slot in arm OA. If the arm rotates in the *vertical plane* at a constant rate of $\dot{\theta} = 0.5$ rad/s, determine the force that the arm exerts on the peg at the instant $\theta = 60°$.

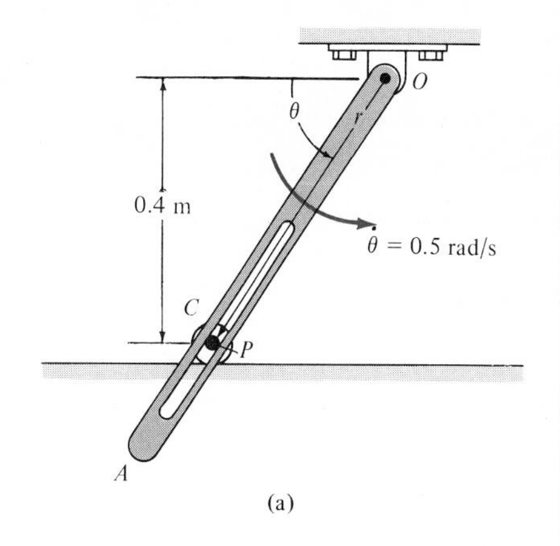

(a)

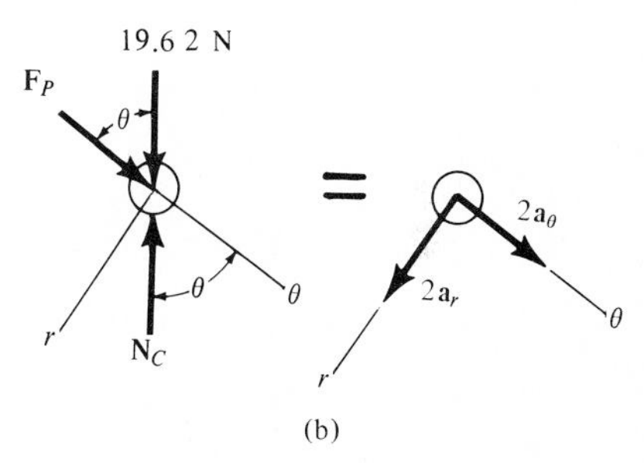

(b)

Fig. 20–8

Solution

Free-Body and Kinetic Diagrams. The free-body and kinetic diagrams for the cylinder are shown in Fig. 20–8b. The force of the peg, $\mathbf{F}_P$, acts perpendicular to the slot in the arm. As required, $2\mathbf{a}_r$ and $2\mathbf{a}_\theta$ are assumed to act in the direction of *positive* r and θ, respectively.

Equations of Motion. Using the data in Fig. 20–8b, we have

$$+\swarrow \Sigma F_r = ma_r; \qquad 19.62 \sin\theta - N_C \sin\theta = 2a_r \qquad (1)$$

$$+\searrow \Sigma F_\theta = ma_\theta; \qquad 19.62 \cos\theta + F_P - N_C \cos\theta = 2a_\theta \qquad (2)$$

Kinematics. From Fig. 20–8a, r can be related to θ by the equation

$$r = \frac{0.4}{\sin\theta} = 0.4 \csc\theta$$

Since $d(\csc\theta) = -(\csc\theta \cot\theta)d\theta$ and $d(\cot\theta) = -(\csc^2\theta)\,d\theta$, then r and the necessary time derivatives become

$$\begin{aligned} \dot{\theta} &= 0.5 & r &= 0.4 \csc\theta \\ \ddot{\theta} &= 0 & \dot{r} &= -0.4 \csc\theta \cot\theta\dot{\theta} \\ & & &= -0.2 \csc\theta \cot\theta \\ & & \ddot{r} &= -0.2(-\csc\theta \cot\theta\dot{\theta}) \cot\theta - 0.2 \csc\theta(-\csc^2\theta\dot{\theta}) \\ & & &= 0.1 \csc\theta(\cot^2\theta + \csc^2\theta) \end{aligned}$$

Evaluating these formulas at $\theta = 60°$, we get

$$\begin{aligned} \dot{\theta} &= 0.5 & r &= 0.462 \\ \ddot{\theta} &= 0 & \dot{r} &= -0.133 \\ & & \ddot{r} &= 0.192 \end{aligned}$$

$$a_r = \ddot{r} - r\dot{\theta}^2 = 0.192 - 0.462(0.5)^2 = 0.0765$$

$$a_\theta = r\ddot{\theta} + 2\dot{r}\dot{\theta} = 0 + 2(-0.133)(0.5) = -0.133$$

Substituting these results into Eqs. (1) and (2) with $\theta = 60°$ and solving yields

$$N_C = 19.44 \text{ N} \qquad F_P = -0.355 \text{ N} \qquad \textit{Ans.}$$

The negative sign indicates that $\mathbf{F}_P$ acts opposite to that shown in Fig. 20–8b.

Example 20–6

A can C, having a mass of 0.5 kg, moves along a grooved horizontal slot shown in Fig. 20–9*a*. The slot is in the form of a spiral, which is defined by the equation $r = (0.1\,\theta)$ m, where θ is measured in radians. If the arm OA is rotating at a constant rate of $\dot{\theta} = 4$ rad/s in the horizontal plane, determine the force it exerts on the can at the instant $\theta = \pi$ rad. Neglect friction and the size of the can.

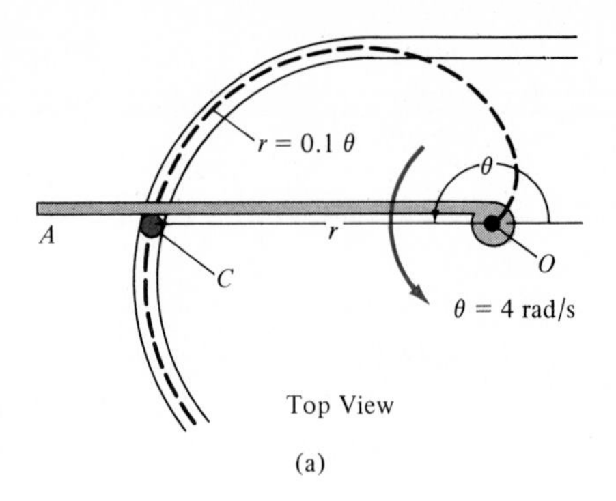

(a)

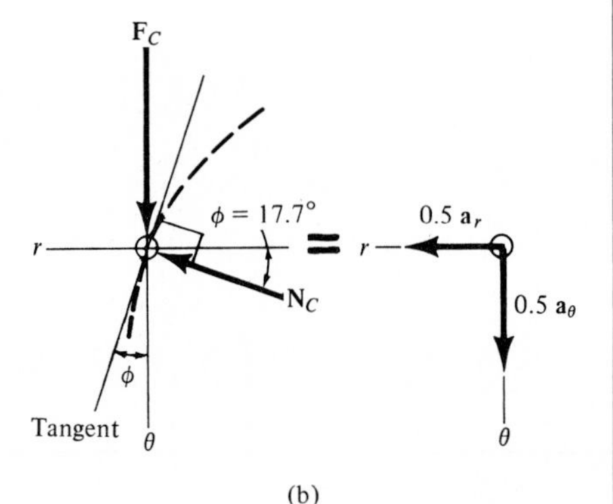

(b)

(c)

Fig. 20–9

Solution

Free-Body and Kinetic Diagrams. As usual, the vectors $0.5\mathbf{a}_r$ and $0.5\mathbf{a}_\theta$ act in the *positive directions* of r and θ, respectively, Fig. 20–9*b*. The driving force $\mathbf{F}_C$ acts perpendicular to the arm OA; whereas the normal force of the wall of the slot on the can, $\mathbf{N}_C$, acts perpendicular to the tangent to the curve at $\theta = \pi$ rad. Since the path is specified, the angle ψ which the extended radial line r makes with the tangent, Fig. 20–9*c*, can be determined from Eq. 20–8. We have $r = 0.1\theta$, so that $dr/d\theta = 0.1$, and therefore

$$\tan \psi = r \Big/ \frac{dr}{d\theta} = \frac{0.1\,\theta}{0.1} = \theta$$

When $\theta = \pi$, $\psi = \tan^{-1} \pi = 72.3°$, so that $\phi = 90° - \psi = 17.7°$, as shown in Fig. 20–9*b* and *c*.

Equations of Motion. Using the data shown in Fig. 20–9*b*, we have

$$\overset{+}{\leftarrow}\Sigma F_r = ma_r; \qquad N_C \cos 17.7° = 0.5a_r \qquad (1)$$

$$+\downarrow \Sigma F_\theta = ma_\theta; \qquad F_C - N_C \sin 17.7° = 0.5a_\theta \qquad (2)$$

Kinematics. Computing the time derivatives of r and θ, we have

$$\dot{\theta} = 4 \text{ rad/s} \qquad r = 0.1\,\theta$$

$$\ddot{\theta} = 0 \qquad \dot{r} = 0.1\,\dot{\theta} = 0.1(4) = 0.4 \text{ m/s}$$

$$\ddot{r} = 0.1\,\ddot{\theta} = 0$$

At the instant $\theta = \pi$ rad,

$$a_r = \ddot{r} - r\dot{\theta}^2 = 0 - 0.1(\pi)(4)^2 = -5.03 \text{ m/s}^2$$

$$a_\theta = r\ddot{\theta} + 2\dot{r}\dot{\theta} = 0 + 2(0.4)(4) = 3.20 \text{ m/s}^2$$

Substituting these results into Eqs. (1) and (2) and solving yields

$$N_C = -2.64 \text{ N}$$

$$F_C = 0.797 \text{ N} \qquad \textit{Ans.}$$

Problems

20–13. A smooth can C, having a mass of 2 kg, is lifted from a feed at A to a ramp at B by a forked rotating rod. If the rod maintains a constant angular motion of $\dot{\theta} = 0.5$ rad/s, determine the force which the rod exerts on the can at the instant $\theta = 30°$. Neglect the effects of friction in the calculation. The ramp from A to B is circular, having a radius of 700 mm. *Hint:* First show that $r = (1400 \cos \theta)$ mm.

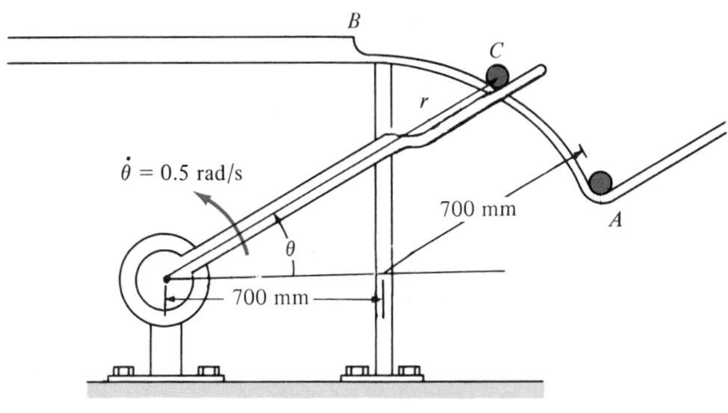

Prob. 20–13

20–14. The spool, which has a weight of 2 lb, slides along the smooth *horizontal* spiral rod, $r = (2\theta)$ ft, where θ is in radians. If its angular rate of rotation is constant and equals $\dot{\theta} = 4$ rad/s, determine the tangential force **P** needed to cause the motion and the normal force that the spool exerts on the rod at the instant $\theta = 90°$.

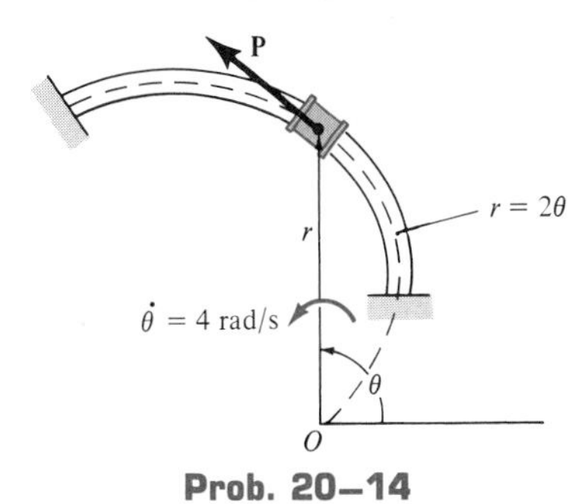

Prob. 20–14

20–15. Solve Prob. 20–14 if the spiral path $r = 2\theta$ is *vertical*.

***20–16.** Rod OA rotates counterclockwise with a constant angular rate of $\dot{\theta} = 5$ rad/s. The double collar B is pin-connected together such that one collar slides over the rotating rod and the other slides over the *horizontal* curved rod, of which the shape is a limaçon described by the equation $r = 1.5(2 - \cos \theta)$ ft. If both collars weigh 0.75 lb, determine the normal force which the curved path exerts on one of the collars, and the force that OA exerts on the other collar at the instant $\theta = 90°$.

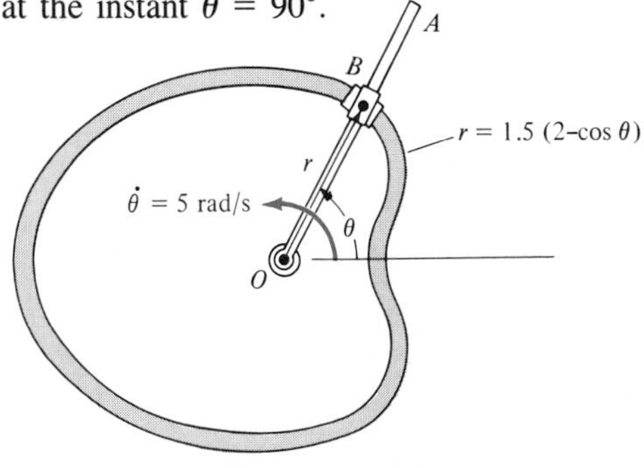

Prob. 20–16

20–17. Solve Prob. 20–16 if the path is *vertical*.

20–18. Using a forked rod, a smooth can C having a mass of 0.4 kg is forced to move along the *vertical slotted* path $r = (0.6\theta)$ m, where θ is measured in radians. If the can has a constant speed of $v_C = 2$ m/s, determine the force of the forked rod on the can and the normal force of the slot on the can at the instant $\theta = \pi$ rad. Assume the can is in contact with only *one edge* of the rod and slot at any instant. *Hint:* To obtain the time derivatives necessary to compute the can's acceleration components a_r and a_θ, take the first and second time derivatives of $r = 0.6\theta$. Then, for further information, use Eq. 20–3 to determine $\dot{r}$ and $\dot{\theta}$. Also, take the time derivative of Eq. 20–3, noting that $\dot{v}_C = 0$, to determine $\ddot{r}$ and $\ddot{\theta}$.

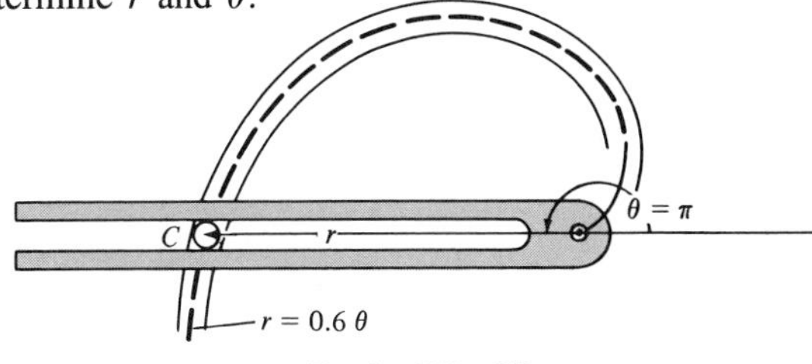

Prob. 20–18

20–19. The pilot of an airplane executes a vertical loop which in part follows the path of a "four-leaved rose," $r = (-600 \cos 2\theta)$ ft, where θ is in radians. If his speed at A is a constant $v_P = 80$ ft/s, determine the vertical reaction he exerts on the seat of the plane when the plane is at A. He weighs 130 lb. *Hint:* To determine the time derivatives necessary to compute the plane's acceleration components a_r and a_θ, take the first and second time derivatives of $r = -600 \cos 2\theta$. Then, for further information, use Eq. 20–3 to determine $\dot{r}$ and $\dot{\theta}$. Also, take the time derivative of Eq. 20–3, noting that $\dot{v}_P = 0$, to determine $\ddot{r}$ and $\ddot{\theta}$.

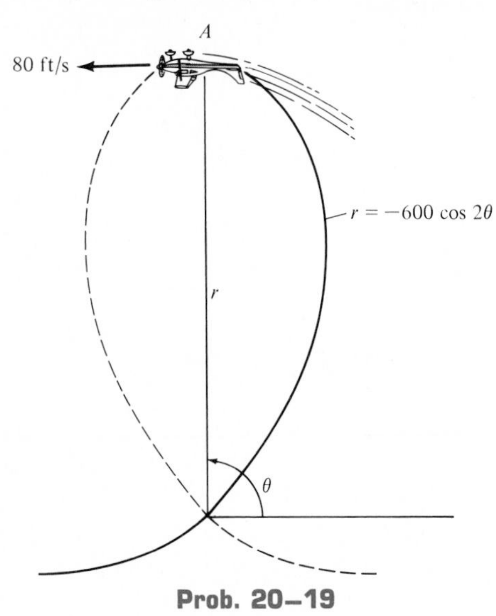

Prob. 20–19

***20–20.** A truck T has a weight of 8,000 lb and is traveling along a portion of a road defined by the lemniscate $r^2 = 0.2(10^6) \cos 2\theta$, where r is measured in meters and θ is in radians. If the truck maintains a constant speed of $v_T = 4$ ft/s, determine the magnitude of the resultant frictional force which must be exerted by all the wheels to maintain the motion when $\theta = 0$. *Hint:* To determine the time derivatives necessary to compute the truck's acceleration components a_r and a_θ, take the first and second time derivatives of $r^2 = 0.2(10^6) \cos 2\theta$. Then, for further information, use Eq. 20–3 to determine $\dot{r}$ and $\dot{\theta}$. Also take the time derivative of Eq. 20–3, noting that $\dot{v}_T = 0$, to determine $\ddot{r}$ and $\ddot{\theta}$.

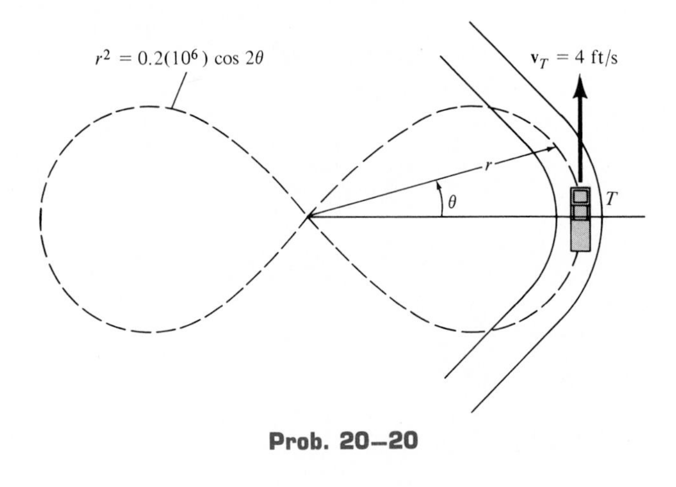

Prob. 20–20

*20.3 Relative-Motion Analysis Using Rotating Axes

In the previous sections, the relative-motion analysis for velocity and acceleration was described using a translating coordinate system. This type of analysis is useful for determining the motion of points on the *same* rigid body, or the motion of points located on several pin-connected rigid bodies. In some problems, however, rigid bodies (mechanisms) are constructed such that *sliding* will occur at their connections. The kinematic analysis for such cases is best performed if the motion is analyzed using a coordinate system which *rotates*. Furthermore, this frame of reference is useful for analyzing the motions of two points on a mechanism which are *not* located on the *same* rigid body and for specifying the kinematics of particle motion when one of the particles is moving along a rotating path.

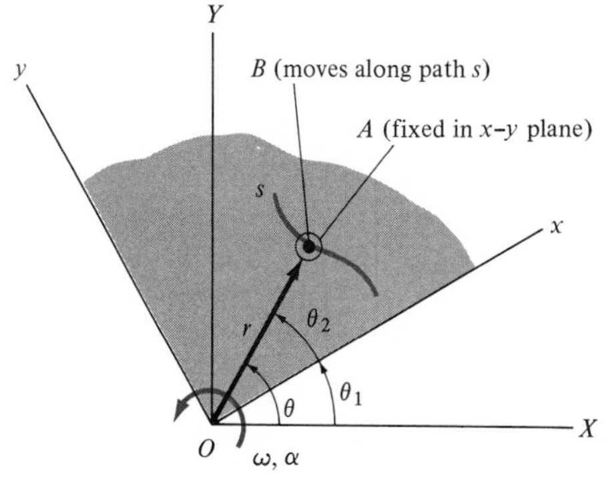

Fig. 20–10

Position. In the following analysis we will consider two sets of axes having the same origin O, Fig. 20–10. The X, Y axes are *fixed*, and at the instant considered the x, y axes are positioned at θ_1 and have an angular velocity $\boldsymbol{\omega}$ and angular acceleration $\boldsymbol{\alpha}$. For most applications these axes are chosen as *fixed in a rotating body,* which is symbolized by the colored shading. Also fixed in this body or plane is the path s. We wish to relate the absolute motion of point B, which *moves along this path,* to the motion of point A, which is *fixed in the x-y plane* and at the instant considered is coincident with B. Both of these points are located at the radial coordinate r, and their transverse coordinates are θ and θ_2, measured from the fixed X or rotating x axis, respectively.

Velocity. Using Eqs. 20–2, the absolute velocity of point B has r, θ components of

$$v_{B_r} = \dot{r} \qquad v_{B_\theta} = r\dot{\theta} = r(\dot{\theta}_1 + \dot{\theta}_2)$$

Since point A is fixed in the x-y plane, $\dot{r} = 0$ and $\dot{\theta}_2 = 0$; thus, its components are

$$v_{A_r} = 0 \qquad v_{A_\theta} = r\dot{\theta}_1$$

To obtain the relative velocity of "B with respect to A," we must consider the x-y plane fixed so that $\dot{\theta}_1 = 0$. Thus,

$$(v_{B/A})_{\text{rel}_r} = \dot{r} \qquad (v_{B/A})_{\text{rel}_\theta} = \dot{\theta}_2 \tag{20–9}$$

When compared, the above three components yield the two component equations

$$v_{B_r} = v_{A_r} + (v_{B/A})_{\text{rel}_r} \qquad v_{B_\theta} = v_{A_\theta} + (v_{B/A})_{\text{rel}_\theta}$$

or the one vector equation

$$\mathbf{v}_B = \mathbf{v}_A + (\mathbf{v}_{B/A})_{\text{rel}} \tag{20–10}$$

where

$\mathbf{v}_B$ = absolute velocity of point B, which moves along the path s

$\mathbf{v}_A$ = absolute velocity of point A, which is fixed on the path and coincides with B at the instant considered

$(\mathbf{v}_{B/A})_{\text{rel}}$ = relative velocity of the concurrent points "B with respect to A," as measured by an observer attached to the *rotating* x-y plane

Acceleration. We will now repeat the above steps with reference to the acceleration. In this regard, the acceleration of point B moving along the path s has r and θ components of acceleration

$$a_{B_r} = \ddot{r} - r\dot{\theta}^2 = \ddot{r} - r(\dot{\theta}_1 + \dot{\theta}_2)^2 = \ddot{r} - r(\dot{\theta}_1^2 + 2\dot{\theta}_1\dot{\theta}_2 + \dot{\theta}_2^2)$$
$$a_{B_\theta} = r\ddot{\theta} + 2\dot{r}\dot{\theta} = r(\ddot{\theta}_1 + \ddot{\theta}_2) + 2\dot{r}(\dot{\theta}_1 + \dot{\theta}_2)$$

For point A, $\dot{r} = \ddot{r} = 0$ and $\dot{\theta}_2 = \ddot{\theta}_2 = 0$, since A is fixed in the x-y plane. Thus,

$$a_{A_r} = -r\dot{\theta}_1^2$$
$$a_{A_\theta} = r\ddot{\theta}_1$$

And lastly, the acceleration of "B with respect to A," which considers the x-y plane fixed so $\dot{\theta}_1 = 0$, gives

$$(a_{B/A})_{\text{rel}_r} = \ddot{r} - r\dot{\theta}_2^2$$
$$(a_{B/A})_{\text{rel}_\theta} = r\ddot{\theta}_2 + 2\dot{r}\dot{\theta}_2$$

When the above components are added, we get the following two scalar equations:

$$a_{B_r} = a_{A_r} + (a_{B/A_r})_{\text{rel}_r} + a_{\text{cor}_r} \qquad (a_B)_\theta = a_{A_\theta} + (a_{B/A_\theta})_{\text{rel}_\theta} + a_{\text{cor}_\theta}$$

Note that the additional components $a_{\text{cor}_r} = -2r\dot{\theta}_1\dot{\theta}_2$ and $a_{\text{cor}_\theta} = 2\dot{r}\dot{\theta}_1$ must be included here. Since $\dot{\theta}_1 = \omega$, and using Eqs. 20–9, we can also write these components as

$$a_{\text{cor}_r} = -2\omega(v_{B/A})_{\text{rel}_\theta} \qquad a_{\text{cor}_\theta} = 2\omega(v_{B/A})_{\text{rel}_r}$$

The resultant acceleration a_{cor} is called the *Coriolis acceleration,* named after the French engineer G. C. Coriolis (1792–1843), who was the first to determine it. This term represents the *difference* in acceleration of B as measured from nonrotating and rotating x, y axes. As shown in Fig. 20–11, its *direction* is found by rotating $\mathbf{a}_{\text{cor}}$ 90° from $(\mathbf{v}_{B/A})_{\text{rel}}$ counterclockwise, in the same direction as $\boldsymbol{\omega}$. Furthermore, this vector has a *magnitude* of $a_{\text{cor}} = 2\omega(v_{B/A})_{\text{rel}}$.

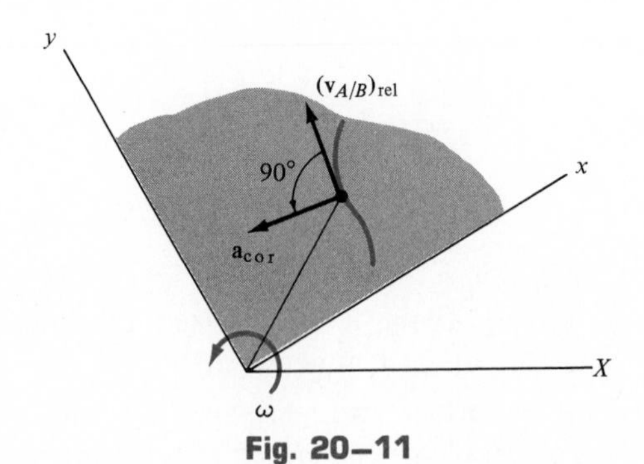

Fig. 20–11

Rewriting the above two scalar equations in vector form, the final result is

$$\mathbf{a}_B = \mathbf{a}_A + (\mathbf{a}_{B/A})_{\text{rel}} + \mathbf{a}_{\text{cor}} \tag{20–11}$$

where

$\mathbf{a}_B$ = absolute acceleration of point B, which moves along the path s

$\mathbf{a}_A$ = absolute acceleration of point A, which is fixed on the path and coincides with B at the instant considered

$(\mathbf{a}_{B/A})_{\text{rel}}$ = relative acceleration of the concurrent points "B with respect to A," as measured by an observer attached to the *rotating* x-y plane

$\mathbf{a}_{\text{cor}}$ = Coriolis acceleration, which has a *magnitude* of $a_{\text{cor}} = 2\omega(v_{B/A})_{\text{rel}}$ and a *direction* found by rotating the vector $(\mathbf{v}_{B/A})_{\text{rel}}$ 90° in the direction of $\boldsymbol{\omega}$

PROCEDURE FOR ANALYSIS

The following procedure provides a method for applying Eqs. 20–10 and 20–11 to the solution of problems involving the planar motion of particles or rigid bodies.

Coordinate Axes. Choose an appropriate location for the origin and proper orientation of the axes for both the fixed X, Y, and rotating x, y reference frames. Most often solutions are easily obtained if at the instant considered the axes are collinear. The rotating frame should be selected fixed to the body or device where the relative motion occurs.

Kinematic Equations. After specifying the moving point B and point A fixed in the x, y references, Eqs. 20–10 and 20–11 should be written in symbolic form

$$\mathbf{v}_B = \mathbf{v}_A + (\mathbf{v}_{B/A})_{\text{rel}}$$
$$\mathbf{a}_B = \mathbf{a}_A + (\mathbf{a}_{B/A})_{\text{rel}} + \mathbf{a}_{\text{cor}}$$

The magnitude and direction of each of the vectors in these equations should be specified from the problem data. The vector components may be selected along either the X, Y axes or the x, y axes. The choice is arbitrary.

Finally, substitute the data into the kinematic equations and solve the resulting scalar component equations.

The following examples numerically illustrate this procedure.

Example 20–7

At the instant $\theta = 60^\circ$, the rod in Fig. 20–12 has an angular velocity of 3 rad/s and an angular acceleration of 2 rad/s^2. At this same instant, the collar C is traveling outward along the rod such that when $x = 0.2$ m the velocity is 2 m/s and the acceleration is 3 m/s^2, both measured relative to the rod. Determine the Coriolis acceleration and the velocity and acceleration of the collar at this instant.

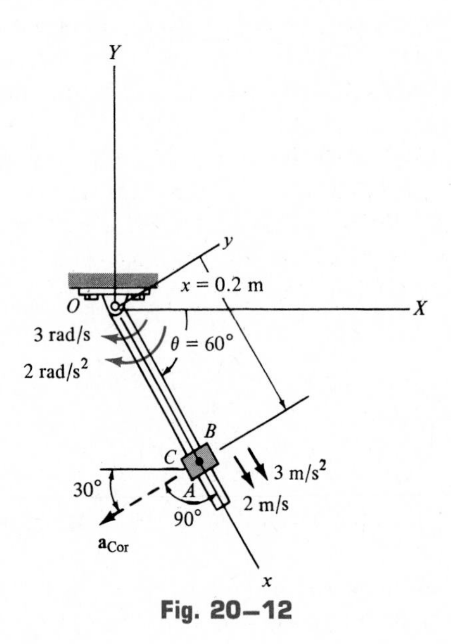

Fig. 20–12

Solution

Coordinate Axes. The origin of both coordinate systems is located at point O, Fig. 20–12. Since the motion of the collar is reported relative to the rod, the rotating x, y frame of reference is *attached* to the rod.

Kinematic Equations. We will specify point B to have the same motion as the collar C, and the concurrent point A to be fixed to the rod, i.e., fixed in the x-y plane, Fig. 20–12. Hence

$$\mathbf{v}_B = \mathbf{v}_A + (\mathbf{v}_{B/A})_{\text{rel}} \tag{1}$$

$$\mathbf{a}_B = \mathbf{a}_A + (\mathbf{a}_{B/A})_{\text{rel}} + \mathbf{a}_{\text{cor}} \tag{2}$$

Point B is traveling along an unknown path, therefore its velocity and acceleration are unknown in magnitude and direction, i.e.,

$$\underset{\rightarrow}{v_{B_X}},\ \underset{\uparrow}{v_{B_Y}},\qquad \underset{\rightarrow}{a_{B_X}},\ \underset{\uparrow}{a_{B_Y}}$$

Point A is fixed to the rod and therefore undergoes rotation about a fixed axis.

$$v_A = \omega r = 3(0.2) = 0.6 \text{ m/s} \quad \overset{30°}{\swarrow}$$
$$(a_A)_n = \omega^2 r = 3^2(0.2) = 1.8 \text{ m/s}^2 \quad \searrow^{30°}$$
$$(a_A)_t = \alpha r = 2(0.2) = 0.4 \text{ m/s}^2 \ {}_{30°}\swarrow$$

An observer fixed on the rod at point A will see point B or the collar C move along the rod. Thus,

$$(v_{B/A})_{\text{rel}} = 2 \text{ m/s} \ \searrow^{30°} \qquad (a_{B/A})_{\text{rel}} = 3 \text{ m/s}^2 \ \searrow^{30°}$$

Lastly, the Coriolis acceleration has a *magnitude* of

$$a_{\text{cor}} = 2\omega(v_{B/A})_{\text{rel}} = 2(3)(2) = 12 \text{ m/s}^2 \ {}_{30°}\swarrow \qquad \textit{Ans.}$$

Its *direction* is 90° apart from $(\mathbf{v}_{B/A})_{\text{rel}}$, where the rotation is in the same direction as $\boldsymbol{\omega}$, Fig. 20–12. Substituting the data into Eqs. (1) and (2), we have

$$\mathbf{v}_B = \mathbf{v}_A + (\mathbf{v}_{B/A})_{\text{rel}}$$
$$\underset{\rightarrow}{v_{B_X}} + \underset{\uparrow}{v_{B_Y}} = \underset{\overset{30°}{\swarrow}}{0.6} + \underset{\searrow_{30°}}{2}$$

$\overset{+}{\rightarrow}$ $$v_{B_X} = -0.6 \cos 30° + 2 \sin 30° = 0.480 \text{ m/s}\rightarrow$$

$+\uparrow$ $$v_{B_Y} = -0.6 \sin 30° - 2 \cos 30° = 2.03 \text{ m/s}\downarrow$$

$$v_B = \sqrt{(0.480)^2 + (2.03)^2} = 2.09 \text{ m/s} \qquad \textit{Ans.}$$

$$\theta = \tan^{-1}\left(\frac{2.03}{0.480}\right) = 76.7° \ \searrow_{30°} \qquad \textit{Ans.}$$

$$\mathbf{a}_B = \mathbf{a}_A + (\mathbf{a}_{B/A})_{\text{rel}} + \mathbf{a}_{\text{cor}}$$
$$\underset{\rightarrow}{a_{B_X}} + \underset{\uparrow}{a_{B_Y}} = \underset{\searrow^{30°}}{1.8} + \underset{\overset{30°}{\swarrow}}{0.4} + \underset{\searrow^{30°}}{3} + \underset{\overset{30°}{\swarrow}}{12}$$

$\overset{+}{\rightarrow}$ $$a_{B_X} = -1.8 \sin 30° - 0.4 \cos 30° + 3 \sin 30° - 12 \cos 30° = 10.14 \text{ m/s}^2 \ \leftarrow$$

$+\uparrow$ $$a_{B_Y} = 1.8 \cos 30° - 0.4 \sin 30° - 3 \cos 30° - 12 \sin 30° = 7.24 \text{ m/s}^2 \ \downarrow$$

$$a_B = \sqrt{(10.14)^2 + (7.24)^2} = 12.5 \text{ m/s}^2 \qquad \textit{Ans.}$$

$$\phi = \tan^{-1}\left(\frac{7.24}{10.14}\right) = 35.5° \ {}_{30°}\swarrow \qquad \textit{Ans.}$$

Example 20–8

The rod AB, shown in Fig. 20–13, rotates clockwise such that it has an angular velocity of $\omega_{AB} = 2$ rad/s when $\theta = 45°$. Determine the angular velocity of rod DE at this instant. The collar at C is pin-connected to AB and slides over rod DE.

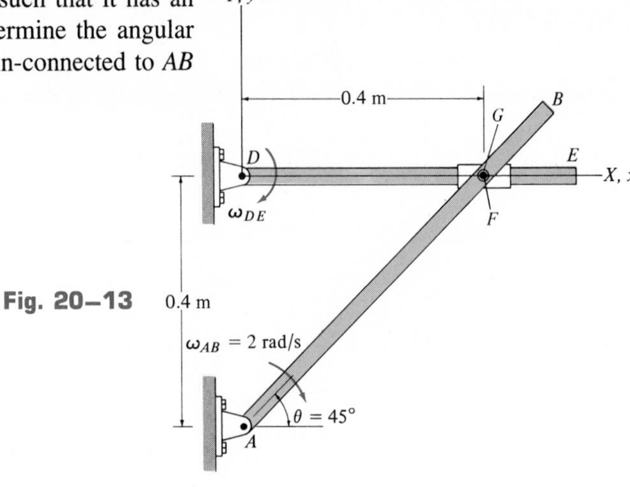

Fig. 20–13

Solution

Coordinate Axes. The origins of both the fixed and rotating frames of reference are located at D, Fig. 20–13. Furthermore, the x, y reference is attached to and rotates with rod DE.

Velocity Equation. Point F is chosen as lying on rod AB or fixed to the collar, and the concurrent point G is fixed in the x-y plane to rod DE. We require

$$\mathbf{v}_F = \mathbf{v}_G + (\mathbf{v}_{F/G})_{\text{rel}} \quad (1)$$

Since point F follows a circular path with center at point A, the velocity of F is

$$v_F = \omega_{AB} r_{AF} = 2(0.4\sqrt{2}) = 1.13 \text{ m/s} \ \searrow^{45°}$$

Point G also follows a circular path with center at point D. Thus,

$$v_G = \omega_{DE} r_{DG} = \omega_{DE}(0.4) \downarrow$$

An observer fixed on rod DE at point G will see point F or the collar C move along rod DE. Thus,

$$(v_{F/G})_{\text{rel}} \rightarrow$$

Substituting the data into Eq. (1) yields

$$\mathbf{v}_F = \mathbf{v}_G + (\mathbf{v}_{F/G})_{\text{rel}}$$

$$\underset{\searrow^{45°}}{1.13} = \underset{\downarrow}{\omega_{DE}(0.4)} + \underset{\rightarrow}{(v_{F/G})_{\text{rel}}}$$

$$\overset{+}{\rightarrow} \qquad 1.13 \cos 45° = (v_{F/G})_{\text{rel}}$$

$$+\downarrow \qquad 1.13 \sin 45° = \omega_{DE}(0.4)$$

Solving,

$$(v_{F/G})_{\text{rel}} = 0.8 \text{ m/s} \rightarrow \qquad \omega_{DE} = 2 \text{ rad/s} \circlearrowright \qquad \textit{Ans.}$$

Problems

20–21. The ball is moving along the slot of a rotating platform such that when it is at C it has a speed of 0.5 m/s, measured relative to the platform. What is the magnitude of the ball's Coriolis acceleration when it is at this point?

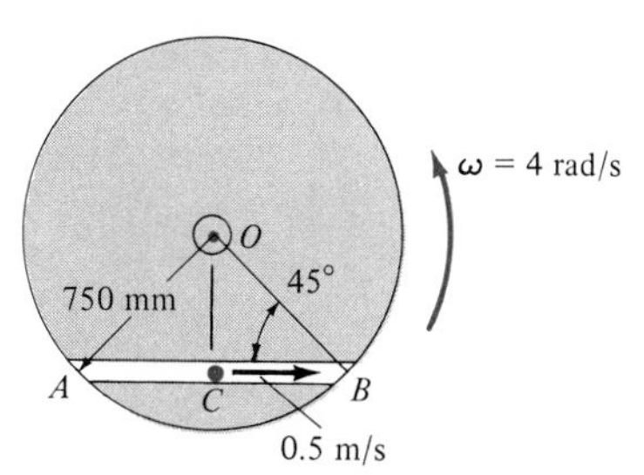

Prob. 20–21

20–22. At the instant shown, the ball B is rolling along the slot in the disk with a velocity of 4 in./s and an acceleration of 5 in./s^2, both measured with respect to the disk and directed away from O. If at the same instant the disk has the angular velocity and angular acceleration shown, determine the velocity and acceleration of the ball at this instant.

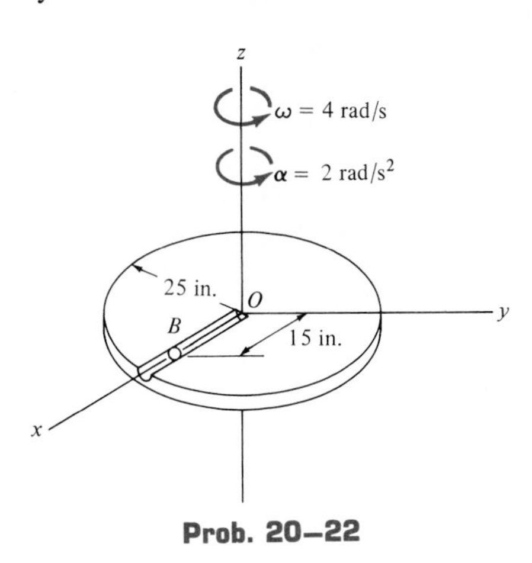

Prob. 20–22

20–23. The slider block B, which is attached to a cord, moves along the slot of a horizontal circular disk. If the cord is pulled down through the central hole A in the disk at a constant rate of $\dot{x} = -2$ m/s, determine the acceleration of the block at the instant $x = 0.1$ m. The disk has a constant angular velocity of $\omega_D = 4$ rad/s.

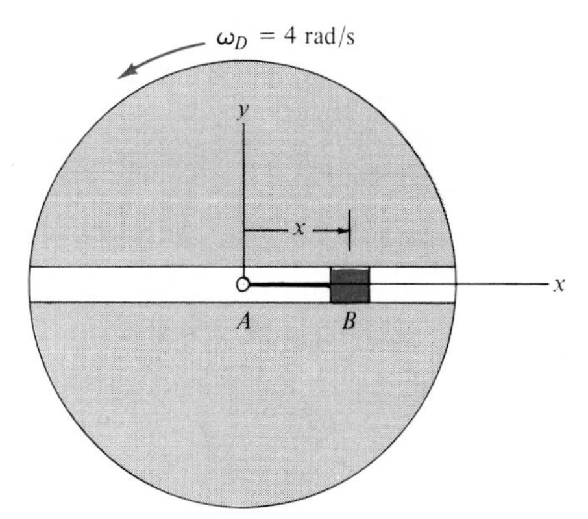

Probs. 20–23 / 20–24

***20–24.** Solve Prob. 20–23 assuming that at the instant $x = 0.1$ m, $\dot{x} = -2$ m/s, $\ddot{x} = 3$ m/s^2, $\omega_D = 4$ rad/s, and the disk has an *angular deceleration* of $\alpha_D = -2$ rad/s^2.

20–25. A girl stands at A on a platform which is rotating with a constant angular velocity of $\omega = 0.5$ rad/s. If she walks at a constant speed of $v = 2$ ft/s on the platform, determine her acceleration (a) when she reaches point D in going along the path ADC and (b) when she reaches point B if she follows the path ABC.

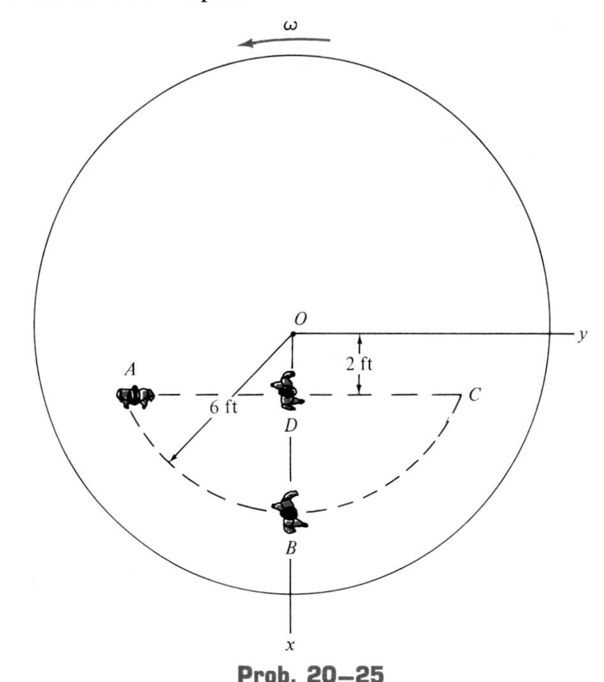

Prob. 20–25

20–26. Block B of the mechanism is confined to move within the slot in member CD. If AB is rotating at a constant rate of $\omega_{AB} = 3$ rad/s, determine the angular velocity and angular acceleration of member CD at the instant shown.

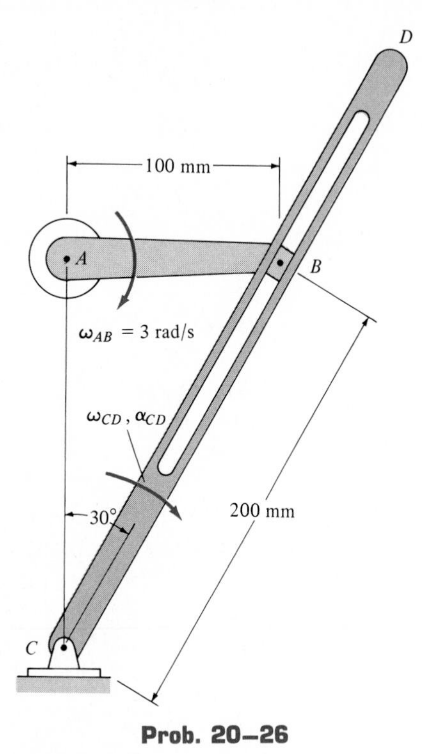

Prob. 20–26

20–27. The two-link mechanism serves to amplify angular motion. Link AB has a pin at B which is confined to move within the slot of link CD. If at the instant shown AB (input) has an angular velocity of $\omega_{AB} = 3$ rad/s and an angular acceleration of $\alpha_{AB} = 2$ rad/s^2, determine the angular velocity and angular acceleration of CD (output) at this instant.

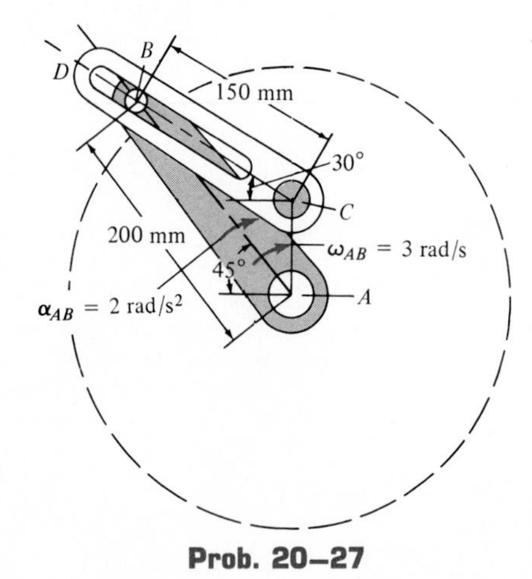

Prob. 20–27

***20–28.** At the instant $\theta = 45°$ link CD has an angular velocity of $\omega_{CD} = 4$ rad/s and an angular acceleration of $\alpha_{CD} = 2$ rad/s^2. Determine the angular velocity and angular acceleration of rod AB at this instant. The collar at C is pin-connected to DC and slides over AB.

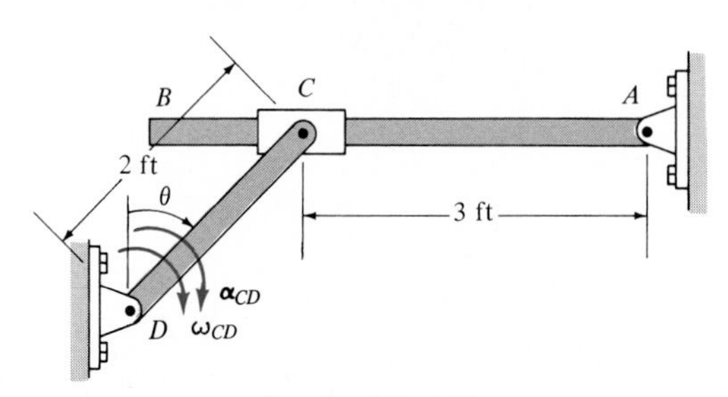

Prob. 20–28

20–29. If the slider block C is fixed to the disk that has a constant counterclockwise angular velocity of 5 rad/s, determine the angular velocity and angular acceleration of the slotted arm AB at the instant shown.

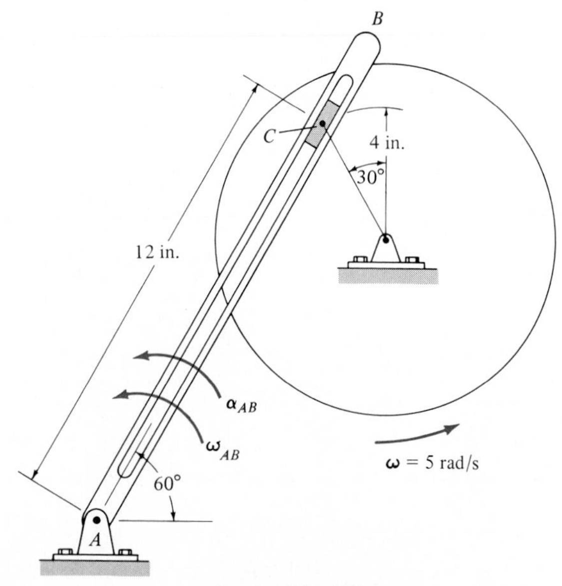

Prob. 20–29

Steady Fluid Streams 20.4*

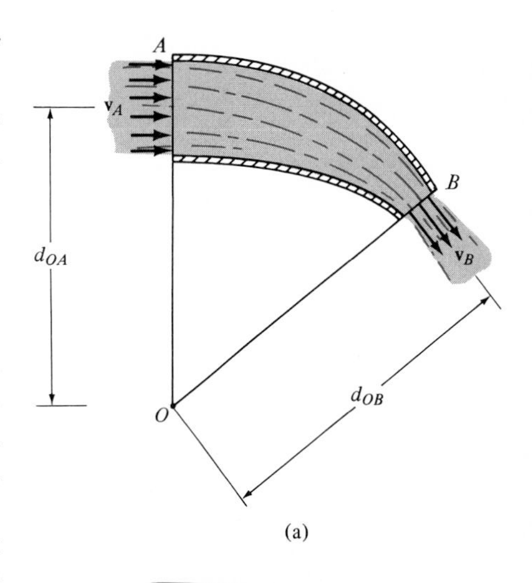

(a)

Knowledge of the forces developed by steadily moving fluid streams is of importance in the design and analysis of turbines, pumps, blades, and fans. To illustrate how the principle of impulse and momentum may be used to determine these forces, consider the diversion of a steady stream of fluid (liquid or gas) by a fixed pipe, Fig. 20–14*a*. The fluid enters the pipe with a velocity $\mathbf{v}_A$ and exits with a velocity $\mathbf{v}_B$. The impulse and momentum diagrams for the fluid stream are shown in Fig. 20–14*b*. The force $\Sigma\mathbf{F}$, shown on the impulse diagram, represents the resultant of all the external forces acting on the fluid stream. It is this loading which gives the fluid stream an impulse whereby the original momentum of the fluid is changed in both its magnitude and direction. Since the flow is steady, $\Sigma\mathbf{F}$ will be *constant* during the time interval dt. During this time the fluid stream is in motion, and as a result a small amount of fluid, having a mass dm, is about to enter the pipe with a velocity $\mathbf{v}_A$ at time t. If this element of mass and the mass of fluid in the pipe is considered as a "closed system," then at time $t + dt$ a corresponding element of mass dm must leave the pipe with a velocity $\mathbf{v}_B$. The fluid stream *within* the pipe section has a mass m and an *average velocity* $\mathbf{v}$ which is constant during the time interval dt. Applying the principle of linear impulse and momentum to the fluid stream, we have

$$dm\,\mathbf{v}_A + m\mathbf{v} + \Sigma\mathbf{F}\,dt = dm\,\mathbf{v}_B + m\mathbf{v}$$

Force Resultant. Solving for the resultant force yields

$$\Sigma\mathbf{F} = \frac{dm}{dt}(\mathbf{v}_B - \mathbf{v}_A) \qquad (20\text{–}12)$$

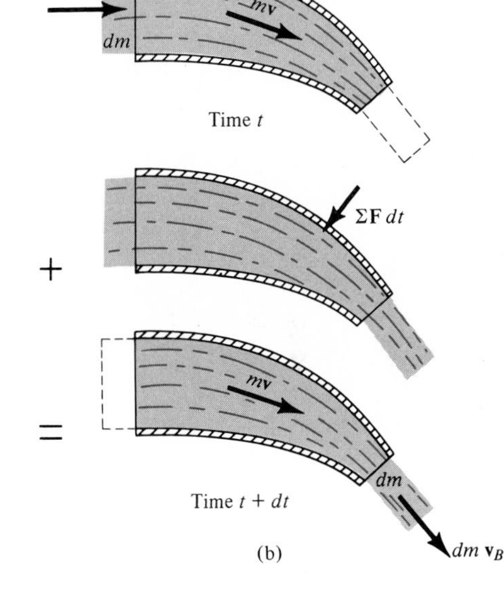

(b)

Provided the motion of the fluid can be represented in the x-y plane, it is usually convenient to express this vector equation in the form of two scalar component equations, i.e.,

$$\Sigma F_x = \frac{dm}{dt}(v_{Bx} - v_{Ax}) \qquad \Sigma F_y = \frac{dm}{dt}(v_{By} - v_{Ay}) \qquad (20\text{–}13)$$

The term dm/dt is called the *mass flow* and indicates the constant amount of fluid which flows either into or out of the pipe per unit of time. If the cross-sectional areas and densities of the fluid at the entrance A and exit B are ρ_A, A_A and ρ_B, A_B, respectively, Fig. 20–14*c*, then continuity of mass requires that $dm = \rho\,dV = \rho_A(ds_A\,A_A) = \rho_B(ds_B\,A_B)$. Hence, during the time dt, since $v_A = ds_A/dt$ and $v_B = ds_B/dt$, we have

$$\frac{dm}{dt} = \rho_A v_A A_A = \rho_B v_B A_B = \rho_A Q_A = \rho_B Q_B \qquad (20\text{–}14)$$

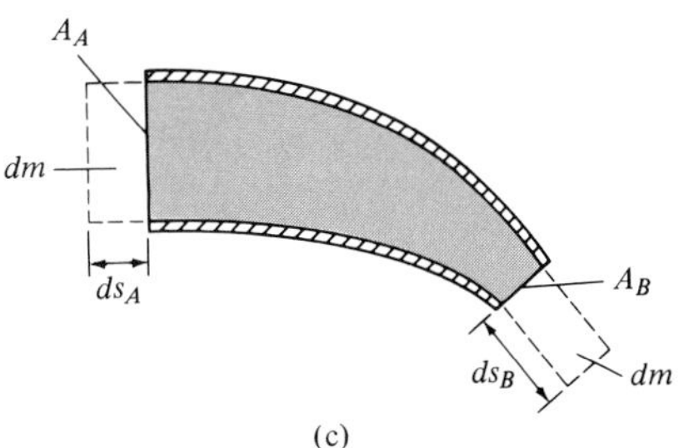

(c)

Fig. 20–14

Here $Q = vA$ is the volumetric *flow rate,* which measures the volume of fluid flowing per unit of time.

Moment Resultant. In some cases it is necessary to obtain the support reactions on the fluid-carrying device. If Eq. 20–12 does not provide enough information to do this, the principle of angular impulse and momentum must be used. The formulation of this principle applied to fluid streams can be obtained from Eq. 15–15, $\Sigma \mathbf{M}_O = d\mathbf{H}_O/dt$, which states that the moment of all the external forces acting on the system about point O is equal to the time rate of change of angular momentum about O. In the case of the pipe shown in Fig. 20–14a, the flow is steady in the x-y plane; hence, we have

$$(\curvearrowleft +) \qquad \Sigma M_O = \frac{dm}{dt}(d_{OB}v_B - d_{OA}v_A) \qquad (20\text{–}15)$$

where the moment arms d_{OB} and d_{OA} are directed from O to the *center* of the openings at A and B.

PROCEDURE FOR ANALYSIS

The following procedure provides a method for solving problems involving steady flow.

Kinematic Diagram. In problems where the device is *moving,* a *kinematic diagram* may be helpful for determining the absolute entrance and exit velocities of the fluid flowing onto the device, since a *relative-motion analysis* of velocity will be involved. The *kinematic diagram* in this case is simply a graphical representation of the velocities showing the vector addition of the relative-motion components. Once the absolute velocity of the fluid flowing onto the device is determined, the mass flow is calculated using Eq. 20–14.

Free-Body Diagram. Draw a free-body diagram of the device which is directing the fluid in order to establish the forces $\Sigma\mathbf{F}$ acting on it. These external forces will include the support reactions, the weight of the device and the fluid contained within it, and the static (gauge) pressure forces of the fluid at the entrance and exit sections of the device.*

Equations of Steady Flow. Applying the equations of steady flow, Eqs. 20–13 and 20–15, using the appropriate components of velocity and force shown on the kinematic and free-body diagrams.

The following examples numerically illustrate application of this procedure.

* In the SI system pressure is measured using the *pascal* (Pa) as the basic unit, where 1 Pa = 1 N/m^2.

Example 20–9

Determine the reaction components which the pipe joint at A exerts on the elbow in Fig. 20–15 if water flowing through the pipe is subjected to a static pressure of 100 kPa at A. The discharge at B is $Q_B = 0.2\ \text{m}^3/\text{s}$. Water is assumed to have a constant density of $\rho_w = 1000\ \text{kg/m}^3$, and the water-filled elbow has a mass of 20 kg and center of gravity at G.

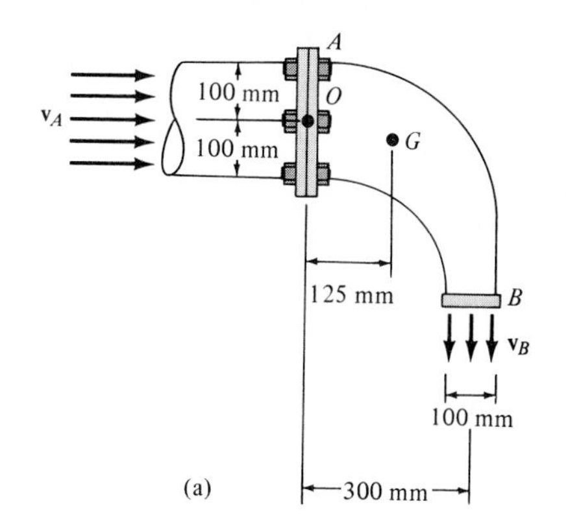

(a)

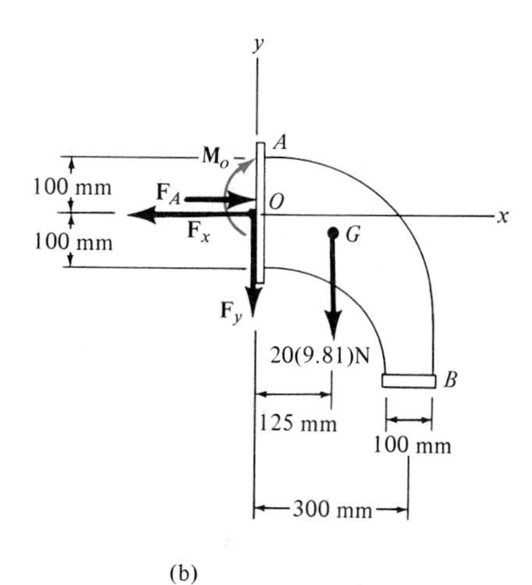

(b)

Fig. 20–15

Solution

The absolute velocity of flow at A and B and the mass flow rate can be obtained from Eq. 20–14. Since the density of water is constant, $Q_B = Q_A = Q$. Hence,

$$\frac{dm}{dt} = \rho_w Q = (1000\ \text{kg/m}^3)(0.2\ \text{m}^3/\text{s}) = 200\ \text{kg/s}$$

$$v_B = \frac{Q}{A_B} = \frac{0.2\ \text{m}^3/\text{s}}{\pi(0.05\ \text{m})^2} = 25.46\ \text{m/s}$$

$$v_A = \frac{Q}{A_A} = \frac{0.2\ \text{m}^3/\text{s}}{\pi(0.1\ \text{m})^2} = 6.37\ \text{m/s}$$

Free-Body Diagram. As shown on the free-body diagram, Fig. 20–15*b*, the *fixed* connection at A exerts a resultant couple $\mathbf{M}_O$ and force components $\mathbf{F}_x$ and $\mathbf{F}_y$ on the elbow. Due to the static pressure of water in the pipe, the pressure force acting on the fluid at A is $F_A = p_A A_A$. Since $1\ \text{kPa} = 1000\ \text{N/m}^2$,

$$F_A = p_A A_A; \quad F_A = (100(10^3)\ \text{N/m}^2)[\pi(0.1\ \text{m})^2] = 3141.6\ \text{N}$$

There is no static pressure acting at B, since the water is discharged at atmospheric pressure; i.e., the pressure measured by a gauge at B is equal to zero, $p_B = 0$.

Equations of Steady Flow

$$(\overset{+}{\rightarrow}) \quad \Sigma F_x = \frac{dm}{dt}(v_{Bx} - v_{Ax}); \quad -F_x + 3141.6 = 200(0 - 6.37)$$

$$F_x = 4.42\ \text{kN} \qquad \textit{Ans.}$$

$$(+\uparrow) \quad \Sigma F_y = \frac{dm}{dt}(v_{By} - v_{Ay}); \quad -F_y - 20(9.81) = 200(-25.46 - 0)$$

$$F_y = 4.90\ \text{kN} \qquad \textit{Ans.}$$

If moments are summed about point O, Fig. 20–15*b*, $\mathbf{F}_x$, $\mathbf{F}_y$, and the static pressure F_A are eliminated, as well as the moment of momentum of the water entering at A, Fig. 20–15*a*. Hence,

$$(\circlearrowleft +) \qquad \Sigma M_O = \frac{dm}{dt}(d_{OB}v_B - d_{OA}v_A)$$

$$M_O + 20(9.81)(0.125) = 200[(0.3)(25.46) - 0]$$

$$M_O = 1.50\ \text{kN}\cdot\text{m} \qquad \textit{Ans.}$$

Example 20–10

A 2-in.-diameter water jet having a velocity of 25 ft/s impinges upon a single moving blade, Fig. 20–16*a*. If the blade is moving at 5 ft/s away from the jet, determine the horizontal and vertical components of force which the blade is exerting on the water, $\rho_w = 62.4\ \text{lb/ft}^3$. What power does the fluid generate on the blade?

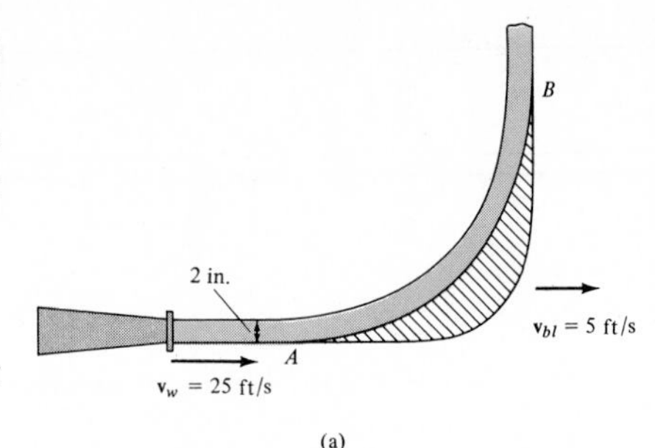

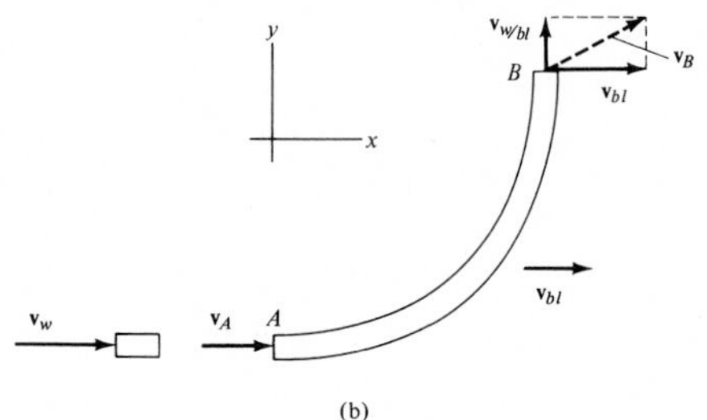

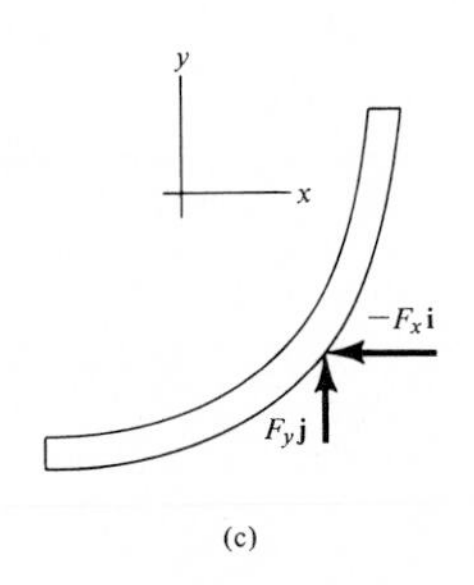

Fig. 20–16

Solution

Kinematic Diagram. As shown in Fig. 20–16*b*, the rate at which water enters the blade is

$$(\overset{+}{\rightarrow}) \qquad v_A = v_w - v_{bl}$$
$$= 25 - 5 = 20\ \text{ft/s}$$

This same *relative-flow velocity* is directed vertically upward on the blade at B. Hence, $v_{w/bl} = 20$ ft/s. Since the blade is moving with a velocity of $v_{bl} = 5$ ft/s to the right, the velocity of flow at B is the vector sum, shown in Fig. 20–16*b*. Thus,

$$v_{B_x} = 5\ \text{ft/s} \qquad v_{B_y} = 20\ \text{ft/s}$$

The mass flow of water onto the blade is

$$\frac{dm}{dt} = \rho_w(v_A)A_A = \frac{62.4}{32.2}(20)\left[\pi\left(\frac{1}{12}\right)^2\right] = 0.846\ \text{slug/s}$$

Free-Body Diagram. The free-body diagram of a section of fluid acting on the blade is shown in Fig. 20–16*c*. The weight of the fluid will be neglected in the calculation, since this force is small compared to the reactive components $\mathbf{F}_x$ and $\mathbf{F}_y$.

Equations of Steady Flow

$$\overset{+}{\rightarrow}\Sigma F_x = \frac{dm}{dt}(v_{B_x} - v_{A_x}); \qquad -F_x = 0.846(5 - 20)$$

$$+\uparrow \Sigma F_y = \frac{dm}{dt}(v_{B_y} - v_{A_y}); \qquad F_y = 0.846(20 - 0)$$

Thus,

$$F_x = 0.846(15) = 12.7\ \text{lb} \qquad \textit{Ans.}$$
$$F_y = 0.846(20) = 16.9\ \text{lb} \qquad \textit{Ans.}$$

The water exerts equal but opposite forces on the blade.

Since the fluid force which causes the blade to move forward horizontally with a velocity of 5 ft/s is $F_x = 12.7$ lb, then from Eq. 14–10 the power is

$$P = (F\cos\theta)v; \qquad P = \frac{12.7(5)}{550} = 0.115\ \text{hp} \qquad \textit{Ans.}$$

Problems

20–30. The nozzle discharges water at the rate of $v_1 = 50$ ft/s against a shield. If the cross-sectional area of the water stream is 2 in.^{2}, determine the force **F** required to hold the shield motionless. $\rho_w = 62.4$ lb/ft.^{3}

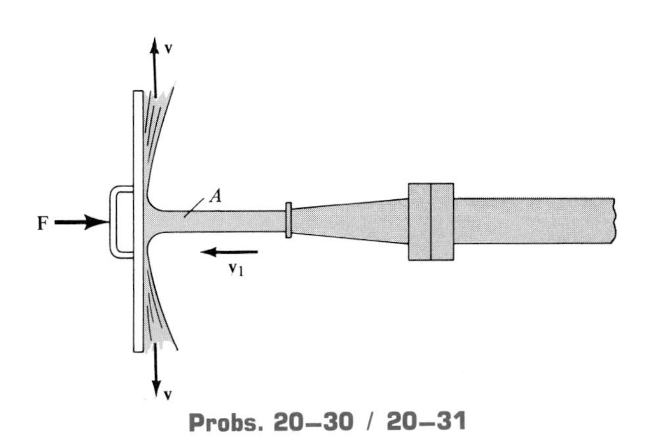

Probs. 20–30 / 20–31

20–31. What force is required to move the shield in Prob. 20–30 forward against the nozzle with a speed of 3 ft/s?

***20–32.** Water is discharged at 16 m/s against the fixed cone diffuser. If the opening diameter of the nozzle is 40 mm, determine the horizontal force exerted by the water on the diffuser. $\rho_w = 1$ Mg/m^3.

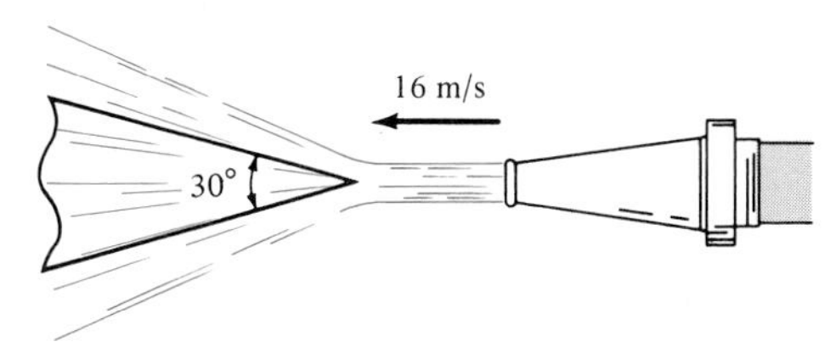

Prob. 20–32

20–33. A jet of water having a cross-sectional area of 4 in.^{2} strikes the fixed blade with a speed of 25 ft/s. Determine the horizontal and vertical components of force which the blade exerts on the water. $\rho_w = 62.4$ lb/ft^3.

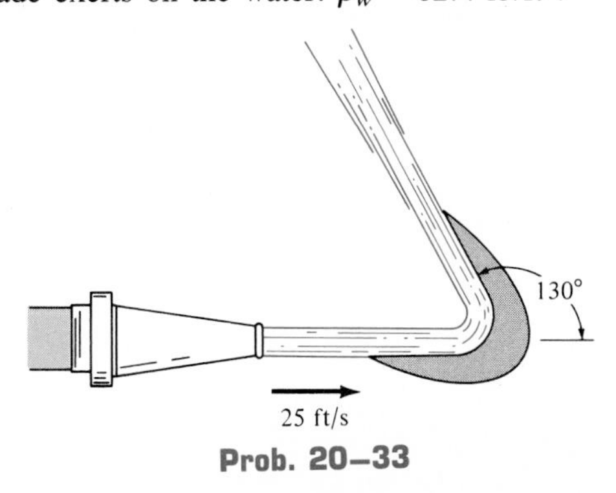

Prob. 20–33

20–34. A power lawn mower hovers very close over the ground. This is done by drawing in air at a speed of 6 m/s through an intake unit A, which has a cross-sectional area of $A_A = 0.25$ m^2, and discharging it at the ground, B, where the cross-sectional area $A_B = 0.35$ m^2. If air at A is subjected only to atmospheric pressure, determine the air pressure which the lawn mower exerts on the ground when the weight of the mower is freely supported and no load is placed on the handle. The mower has a mass of 15 kg with center of mass at G. Assume that air has a constant density of $\rho_a = 1.22$ kg/m^3.

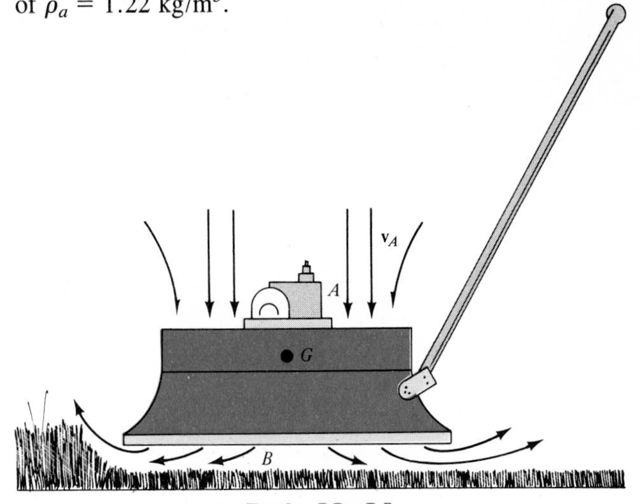

Prob. 20–34

20–35. A snow blower having a scoop S with a cross-sectional area of $A_S = 1.5\ \text{ft}^2$ is pushed into a snow drift with a speed of $v_S = 3$ ft/s. The machine discharges the snow through a tube T that has a cross-sectional area of $A_T = 0.25\ \text{ft}^2$ and is directed 60° from the horizontal. If the density of snow is $\rho_s = 6.5\ \text{lb/ft}^3$, determine the horizontal force **P** required to push the blower forward and the resultant frictional force **F** of the tires on the ground, necessary to prevent the blower from moving sideways. The tires roll freely.

Prob. 20–35

***20–36.** The 200-kg boat is powered by a fan F which develops a slip stream having a diameter of 0.75 m. If the fan ejects air with a speed of 14 m/s, measured relative to the boat, determine the acceleration of the boat if it is initially at rest. Assume that air has a constant density of $\rho_a = 1.22\ \text{kg/m}^3$ and that the entering air is essentially at rest. Neglect the drag resistance of the water.

Prob. 20–36

20–37. Water is flowing from the 150-mm-diameter fire hydrant with a velocity of $v_B = 15$ m/s. Determine the horizontal and vertical components of force and the moment developed at the base joint A if the static (gauge) pressure at A is 50 kPa and the density of water is 1 Mg/m³.

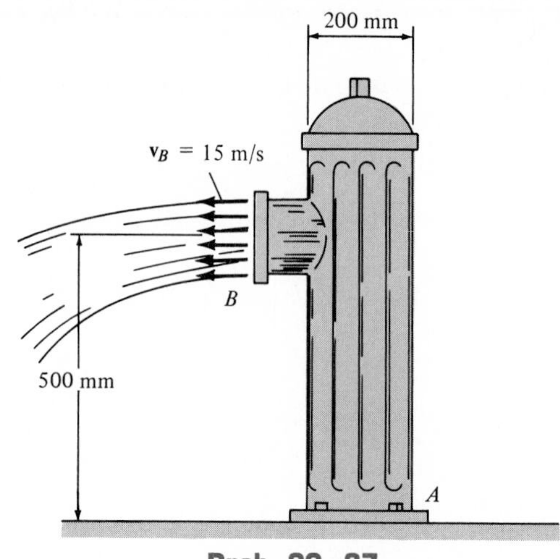

Prob. 20–37

20–38. The air-jet fan discharges air with a speed of $v_B = 15$ m/s into a slip stream having a diameter of 0.5 m. If air has a density of 1.22 kg/m³, determine the horizontal and vertical components of reaction at C and the vertical reaction at each of the two wheels D. The fan and its frame have a mass of 30 kg and a center of mass at G. Due to symmetry, both of the wheels support an equal load. Assume that air entering the fan at A is essentially at rest.

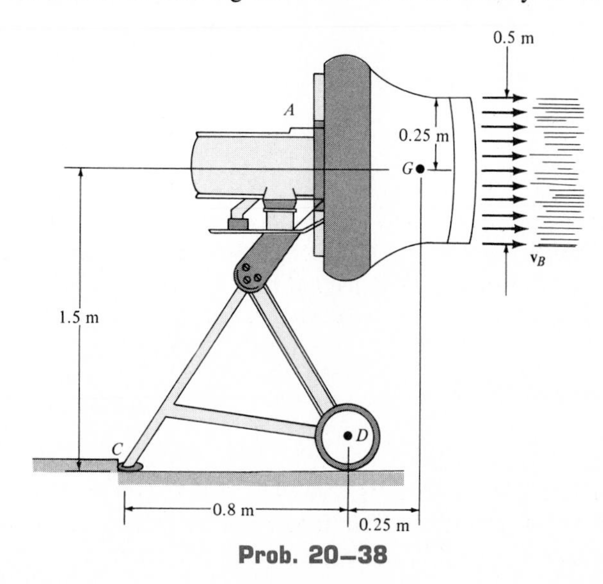

Prob. 20–38

Review Problems

20–39. The motion of the rod B is controlled by the rotation of the grooved link OA. If the link is rotating at a constant angular rate of $\dot{\theta} = 6$ rad/s, determine the magnitudes of the velocity and acceleration of B at the instant $\theta = \pi/2$ rad. The spiral path is defined by the equation $r = (40\theta)$ mm, where θ is measured in radians.

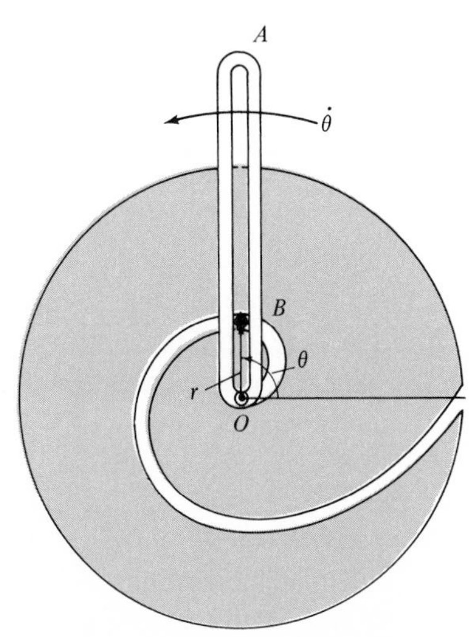

Probs. 20–39 / 20–40

***20–40.** The motion of the rod B is controlled by the rotation of the grooved link OA. If the link is rotating at a constant angular rate of $\dot{\theta} = 6$ rad/s, determine the magnitudes of the velocity and acceleration of the rod when $\theta = \pi$ rad. The spiral path is defined by the equation $r = (40\theta)$ mm, where θ is measured in radians.

20–41. The satellite is traveling with a velocity of 600 mi/h in the Northern Hemisphere at a latitude of 30° from south to north. Determine the satellite's Coriolis acceleration with respect to the center of the earth. The radius of the earth is 3,960 mi, and the satellite is at an altitude of 70 mi.

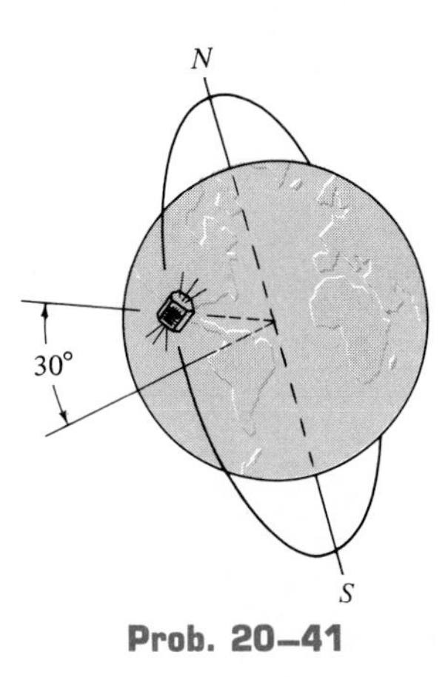

Prob. 20–41

20–42. Sand having a density of 120 lb/ft^3 is discharged onto a chute at A with a velocity of $v_A = 6$ ft/s. If the bin has a 0.2-ft^2 opening, determine the magnitude of the resultant force which the sand exerts on the chute at A.

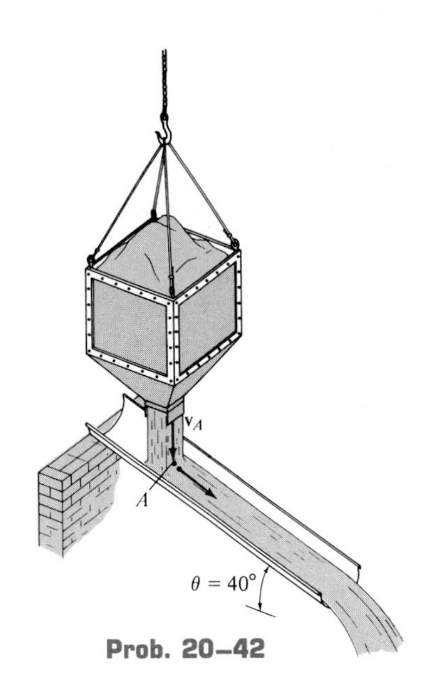

Prob. 20–42

20–43. A ride in an amusement park consists of a cart which is supported by small wheels. Initially the cart is traveling in a circular path of radius $r_0 = 16$ ft such that the angular rate of rotation is $\dot{\theta}_0 = 0.2$ rad/s. If the attached cable OC is drawn inward at a constant speed of $\dot{r} = -0.5$ ft/s, determine the tension it exerts on the cart at the instant $r = 4$ ft. The cart and its passengers have a total weight of 400 lb. Neglect the effects of friction. *Hint:* First show that the equation of motion in the θ direction yields $a_\theta = r\ddot{\theta} + 2\dot{r}\dot{\theta} = d/dt(r^2\dot{\theta}) = 0$. When integrated, $r^2\dot{\theta} = C$, where the constant C is determined from the problem data.

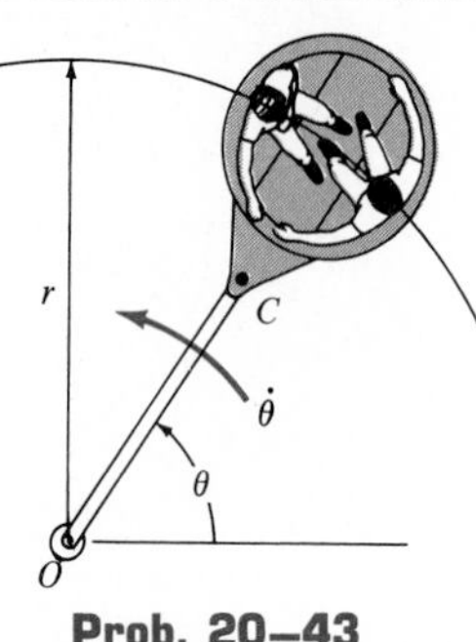

Prob. 20–43

***20–44.** The rod OA rotates counterclockwise with a constant angular rate of $\dot{\theta} = 5$ rad/s. Two pin-connected slider blocks, located at B, move freely on OA and the curved rod whose shape is a limaçon described by the equation $r = (2 - \cos\theta)$ ft. Determine the magnitudes of the velocity and acceleration of the slider blocks at the instant $\theta = 120°$.

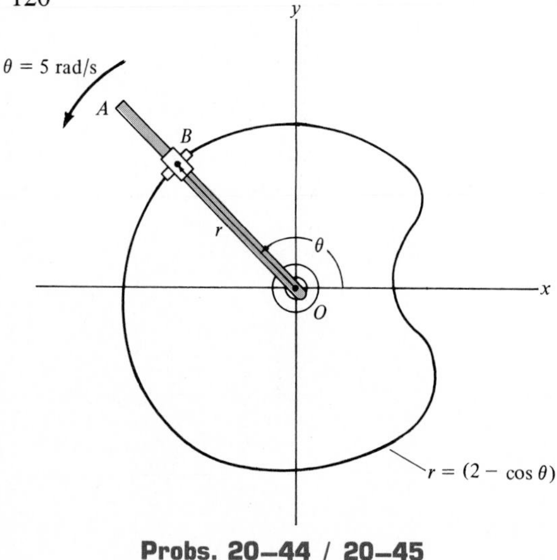

Probs. 20–44 / 20–45

20–45. Two pin-connected slider blocks, located at B, move freely on rod OA and the curved rod whose shape is a limaçon described by the equation $r = (2 - \cos\theta)$ ft. Determine the magnitudes of the velocity and acceleration of the slider blocks when $\theta = 120°$ if at this instant $\dot{\theta} = 5$ rad/s and $\ddot{\theta} = 2$ rad/s^2.

20–46. Sand is deposited from a chute onto a conveyor belt which is moving at 0.5 m/s. If the sand is assumed to fall vertically on the belt at A at the rate of 4 kg/s, determine the required belt tension $\mathbf{F}_B$ to the right of A. The belt is free to move over the conveyor rollers and its tension to the left of A is $F_C = 400$ N.

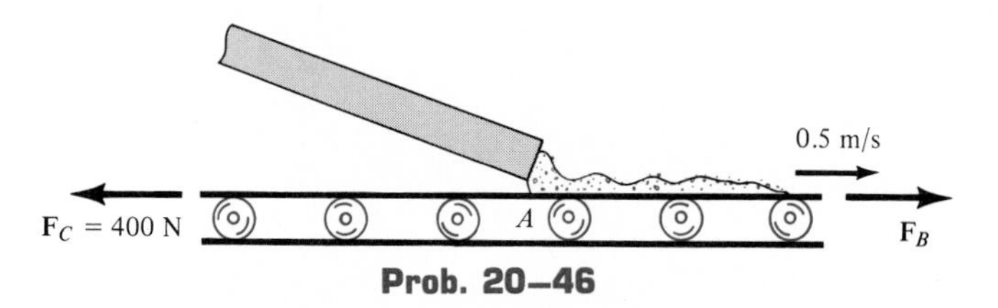

Prob. 20–46

20–47. The rod AB rotates counterclockwise with a constant angular velocity of 2 rad/s. Determine the velocity of point C located on the double collar when $\theta = 45°$. The collar consists of two slider blocks which are constrained to move along the circular shaft and the rod AB.

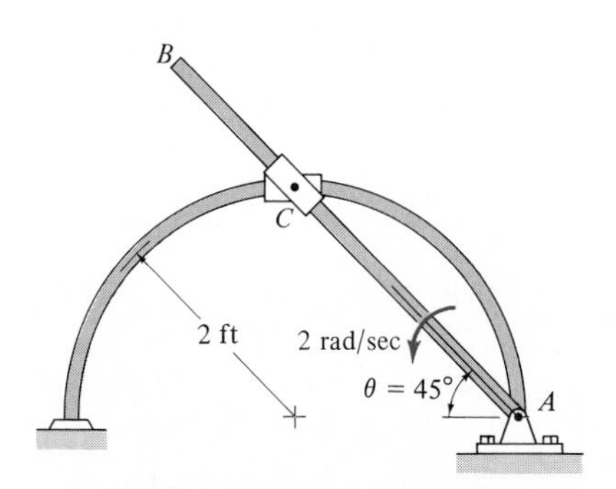

Probs. 20–47 / 20–48

***20–48.** Determine the acceleration of point C in Prob. 20–47.

20–49. A chain is attached to a small 0.5-lb disk that rests on the smooth table. If the disk is given an initial velocity of $v_0 = v_\theta = 2$ ft/s perpendicular to the chain when $\theta = 0°$ and $r_0 = 1.5$ ft, and the chain is pulled downward through a small hole in the center of the table at a constant velocity $\mathbf{v}'$, determine the angular velocity $\dot{\theta}$ of the chain and the force $\mathbf{F}$ at the instant $r = 0.5$ ft. *Hint:* Using the equation of motion in the θ direction, show that $a_\theta = r\ddot{\theta} + 2\dot{r}\dot{\theta} = d/dt(r^2\dot{\theta}) = 0$, which when integrated yields $r^2\dot{\theta} = \text{const}$.

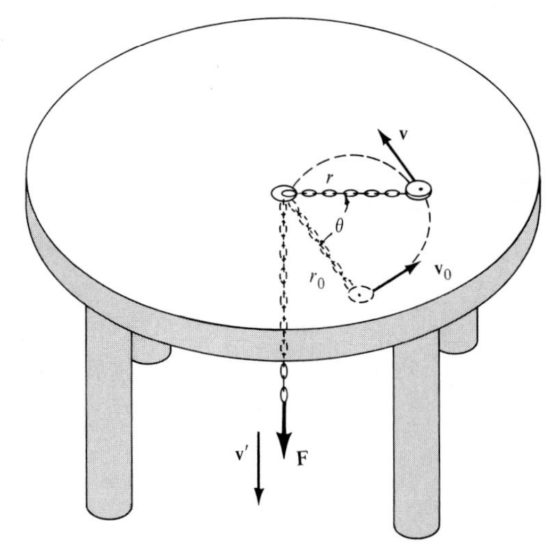

Prob. 20–49

Vibrations

A *vibration* is the periodic motion of a body or system of connected bodies displaced from a position of equilibrium. In general, there are two types of vibration, free and forced. *Free vibration* occurs when the motion is maintained by gravitational or elastic restoring forces, such as the swinging motion of a pendulum or the vibration of an elastic rod. *Forced vibration* is caused by an external periodic or intermittent force applied to the system. Both of these types of vibration may be either damped or undamped. *Undamped* vibrations can continue indefinitely because frictional effects are neglected in the analysis. Since in reality both internal and external frictional forces are present, the motion of all vibrating bodies is actually *damped*.

In this chapter we will study the characteristics of the above types of vibrating motion. The analysis will apply to those bodies which are constrained to move only in one direction. These single-degree-of-freedom systems require only one coordinate to specify completely the position of the system at any time. The analysis of multi-degree-of-freedom systems is based on this simplified case and is thoroughly treated in textbooks devoted to vibrational theory.

Undamped Free Vibration 21.1*

The simplest type of vibrating motion is undamped free vibration, represented by the model shown in Fig. 21–1a. The block has a mass m and is attached to a spring having a stiffness k. Vibrating motion is provided by displacing the block a distance x from its equilibrium position and allowing the spring to restore it to its original position. As the spring pulls on the block, the block

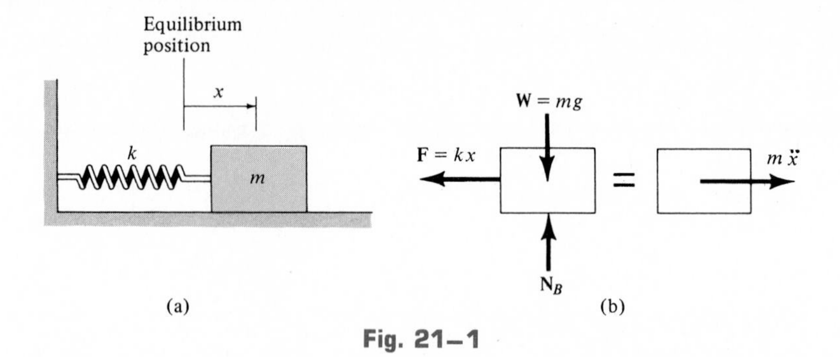

Fig. 21–1

will attain a velocity such that it will proceed to move out of equilibrium when $x = 0$. Provided the supporting surface is smooth, oscillation will continue indefinitely.

The time-dependent path of motion of the block may be determined by applying the equation of motion to the block when it is in the displaced position x. The free-body and kinetic diagrams are shown in Fig. 21–1*b*. The elastic restoring force $F = kx$ is always directed toward the equilibrium position, whereas the acceleration **a** acts in the direction of *positive displacement*. Noting that $a = d^2x/dt^2 = \ddot{x}$, we have

$$\overset{+}{\rightarrow}\Sigma F_x = ma_x; \qquad -kx = m\ddot{x}$$

Here it is seen that the acceleration is proportional to the position. Motion described in this manner is called *simple harmonic motion*. Rearranging the terms into a "standard form" gives

$$\ddot{x} + p^2x = 0 \tag{21-1}$$

The constant p is called the *circular frequency*, expressed in rad/s, and in this case

$$p = \sqrt{\frac{k}{m}} \tag{21-2}$$

Equation 21–1 may also be obtained by considering the block to be suspended, as shown in Fig. 21–2*a*, and measuring the displacement y from the block's *equilibrium position*. The free-body and kinetic diagrams are shown in Fig. 21–2*b*. When the block is in equilibrium, the spring exerts an upward force of $F = W = mg$ on the block. Hence, when the block is displaced a distance y downward from this position, the magnitude of the spring force is $F = W + ky$. Applying the equation of motion gives

$$+\downarrow \Sigma F_y = ma_y; \qquad -W - ky + W = m\ddot{y}$$

or

$$\ddot{y} + p^2y = 0$$

which is the same form as Eq. 21–1, where p is defined by Eq. 21–2.

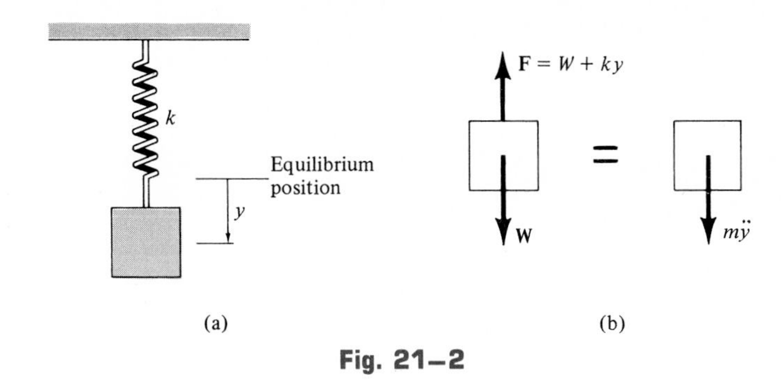

Fig. 21–2

Equation 21–1 is a homogeneous, second-order, linear, differential equation with constant coefficients. It can be shown, using the methods of differential equations, that the general solution of this equation is

$$x = A \sin pt + B \cos pt \tag{21–3}$$

where A and B represent two constants of integration. The block's velocity and acceleration are determined by taking successive time derivatives, which yields

$$v = \dot{x} = Ap \cos pt - Bp \sin pt \tag{21–4}$$

$$a = \ddot{x} = -Ap^2 \sin pt - Bp^2 \cos pt \tag{21–5}$$

When Eqs. 21–3 and 21–5 are substituted into Eq. 21–1, the differential equation is indeed satisfied, and therefore Eq. 21–3 represents the true solution to Eq. 21–1.

The integration constants A and B in Eq. 21–3 are generally determined from the initial conditions of the problem. For example, suppose that the block in Fig. 21–1a has been displaced a distance x_1 to the right from its equilibrium position and given an initial (positive) velocity $\mathbf{v}_1$ directed to the right. Substituting $x = x_1$ at $t = 0$ into Eq. 21–3 yields $B = x_1$. Since $v = v_1$ at $t = 0$, using Eq. 21–4 we obtain $A = v_1/p$. If these values are substituted into Eq. 21–3, the equation describing the motion becomes

$$x = \frac{v_1}{p} \sin pt + x_1 \cos pt \tag{21–6}$$

Equation 21–3 may also be expressed in terms of simple sinusoidal motion. Let

$$A = C \cos \phi \tag{21–7}$$

and

$$B = C \sin \phi \tag{21–8}$$

where C and ϕ are new constants to be determined in place of A and B. Substituting into Eq. 21–3 yields

$$x = C \cos \phi \sin pt + C \sin \phi \cos pt$$

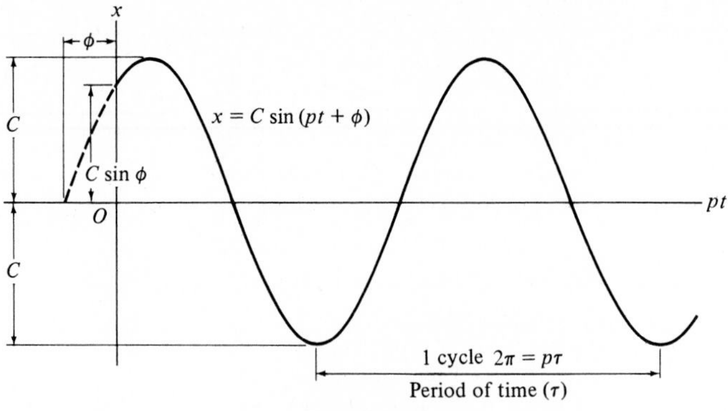

Fig. 21–3

Since $\sin(\theta + \phi) = \sin\theta\cos\phi + \cos\theta\sin\phi$, then

$$x = C\sin(pt + \phi) \tag{21–9}$$

If this equation is plotted on an x versus pt axis, the graph shown in Fig. 21–3 is obtained. The maximum displacement of the block from its equilibrium position is defined as the *amplitude* of vibration. From either the figure or Eq. 21–9 the amplitude is C. The angle ϕ is called the *phase angle* since it represents the amount by which the curve is displaced from the origin when $t = 0$. The constants C and ϕ are related to A and B by Eqs. 21–7 and 21–8. Squaring and adding these two equations, the amplitude becomes

$$C = \sqrt{A^2 + B^2} \tag{21–10}$$

If Eq. 21–8 is divided by Eq. 21–7, the phase angle is

$$\phi = \tan^{-1}\frac{B}{A} \tag{21–11}$$

Note that the sine curve, Eq. 21–9, completes one *cycle*, and hence the cyclic motion of the block is repeated in time $t = \tau$ (tau), so that $p\tau = 2\pi$ or

$$\tau = \frac{2\pi}{p} \tag{21–12}$$

This length of time is called a *period*, Fig. 21–3. Using Eq. 21–2, the period may also be represented as

$$\tau = 2\pi\sqrt{\frac{m}{k}} \tag{21–13}$$

The *frequency* f is defined as the number of cycles completed per unit of time, which is the reciprocal of the period:

$$f = \frac{1}{\tau} = \frac{p}{2\pi} \tag{21–14}$$

or

$$f = \frac{1}{2\pi}\sqrt{\frac{k}{m}} \tag{21–15}$$

The frequency is expressed in cycles/s. This ratio of units is called a *hertz* (Hz), where 1 Hz = 1 cycle/s = 2π rad/s.

When a body or system of connected bodies is given an initial displacement from its equilibrium position and released, it will vibrate with a definite frequency known as the *natural frequency*. This type of vibration is called *free vibration,* since no external forces except gravitational or elastic forces act on the body during the motion. Provided the *amplitude* of vibration remains *constant,* the motion is said to be *undamped*.

The undamped free vibration of a body having a single degree of freedom has the same characteristics as simple harmonic motion of the block and spring discussed above. Consequently, the body's motion is described by a differential equation of the *same form* as Eq. 21–1, i.e.,

$$\ddot{x} + p^2x = 0 \tag{21–16}$$

Hence, if the circular frequency p of the body is known, the period of vibration τ, natural frequency f, and other vibrating characteristics of the body can be established using the above equations.

PROCEDURE FOR ANALYSIS

As in the case of the block and spring, the circular frequency p of a rigid body or system of connected rigid bodies having a single degree of freedom can be determined using the following procedure:

Free-Body and Kinetic Diagrams. Draw the free-body and kinetic diagrams of the body when the body is displaced by a *small amount* from its equilibrium position. Locate the body with respect to its equilibrium position by using an appropriate *position coordinate* q. The vectors $m\mathbf{a}_G$ and $I_G\boldsymbol{\alpha}$, shown on the kinetic diagram, should be directed such that the body is accelerating and thereby causing an *increase* in the position coordinate.

Equation of Motion. Apply the equation of motion to relate the elastic or gravitational restoring forces and couples acting on the body to the body's accelerated motion.

Kinematics. Using kinematics, express the body's accelerated motion in terms of the second time derivative of the position coordinate, $\ddot{q}$. Substitute this result into the equation of motion and determine p by rearranging the terms so that the resulting equation is of the form $\ddot{q} + p^2q = 0$.

The following examples illustrate this procedure.

Example 21–1

Determine the period of vibration for the simple pendulum shown in Fig. 21–4*a*. The bob has a mass m and is attached to a cord of length l. Neglect the size of the bob.

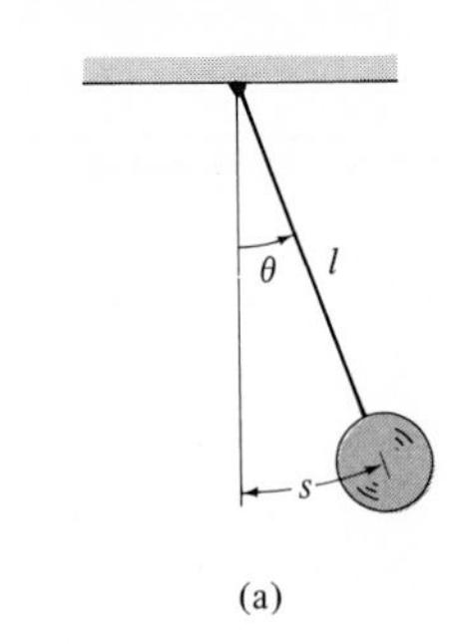

(a)

Solution

Free-Body and Kinetic Diagrams. Motion of the system will be related to the position coordinate $(q =)\theta$, Fig. 21–4*b*. When the bob is displaced by an angle θ, the *restoring force* acting on the bob is created by the *weight component mg* sin θ. Furthermore, note that $m\mathbf{a}_t$ is shown acting in the direction of *increasing* θ.

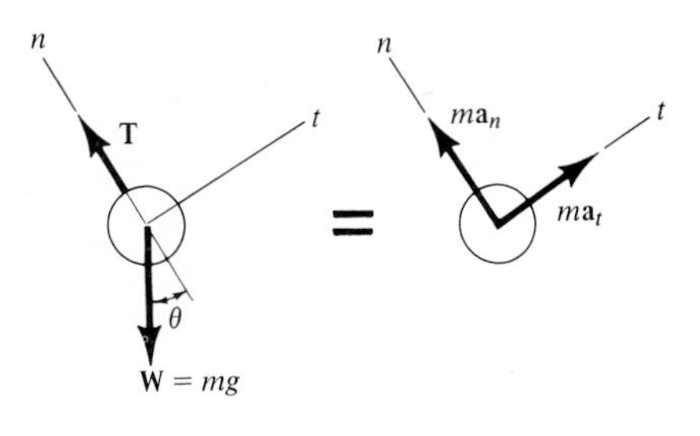

(b)

Fig. 21–4

Equation of Motion. Applying the equation of motion in the *tangential direction,* since it involves the restoring force, yields

$$+\nearrow\Sigma F_t = ma_t; \qquad -mg \sin\theta = ma_t \qquad (1)$$

Kinematics. $a_t = d^2s/dt^2 = \ddot{s}$. Furthermore, s may be related to θ by the equation $s = l\theta$, so that $a_t = l\ddot{\theta}$. Hence, Eq. (1) reduces to the form

$$\ddot{\theta} + \frac{g}{l}\sin\theta = 0 \qquad (2)$$

The solution of this equation involves the use of an elliptic integral. For *small displacements,* however, $\sin\theta \approx \theta$, in which case

$$\ddot{\theta} + \frac{g}{l}\theta = 0 \qquad (3)$$

Comparing this equation with Eq. 21–16 ($\ddot{x} + p^2x = 0$), which is the "standard form" for simple harmonic motion, it is seen that $p = \sqrt{g/l}$. From Eq. 21–12, the period of time required for the bob to make one complete swing is therefore

$$\tau = \frac{2\pi}{p} = 2\pi\sqrt{\frac{l}{g}} \qquad \textit{Ans.}$$

This interesting result, originally discovered by Galileo Galilei through experiment, indicates that the period depends only on the length of the cord and not on the mass of the pendulum bob.

The solution of Eq. (3) is given by Eq. 21–3, where $p = \sqrt{g/l}$ and θ is substituted for x. Like the block and spring, the constants A and B in this problem may be determined if, for example, one knows the displacement and velocity of the bob at a given instant.

Example 21–2

The 10-kg rectangular plate shown in Fig. 21–5*a* is suspended at its center from a rod having a torsional stiffness of $k = 1.5\ \text{N}\cdot\text{m/rad}$. Determine the natural period of vibration of the plate when it is given a small angular displacement θ in the plane of the plate.

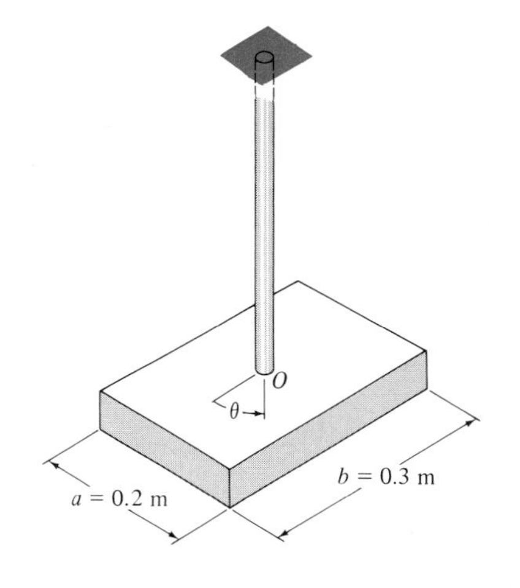

(a)
Fig. 21–5a

Solution

Free-Body and Kinetic Diagrams. Figure 21–5*b*. Since the plate is displaced in its own plane, the torsional *restoring* moment created by the rod is $M = k\theta$. This moment acts in the direction opposite to the angular displacement θ. The vector $I_O\ddot{\theta}$ acts in the direction of *positive* θ.

Equation of Motion

$$\Sigma M_O = I_O\alpha; \qquad -k\theta = I_O\ddot{\theta}$$

or

$$\ddot{\theta} + \frac{k}{I_O}\theta = 0$$

Since this equation is in "standard form," the circular frequency is $p = \sqrt{k/I_O}$.

The moment of inertia of the plate about an axis coincident with the rod is $I_O = \frac{1}{12}m(a^2 + b^2)$. Hence,

$$I_O = \frac{1}{12}[10][(0.2)^2 + (0.3)^2] = 0.108\ \text{kg}\cdot\text{m}^2$$

The natural period of vibration is, therefore,

$$\tau = \frac{2\pi}{p} = 2\pi\sqrt{\frac{I_O}{k}} = 2\pi\sqrt{\frac{0.108}{1.5}} = 1.69\ \text{s} \qquad \textit{Ans.}$$

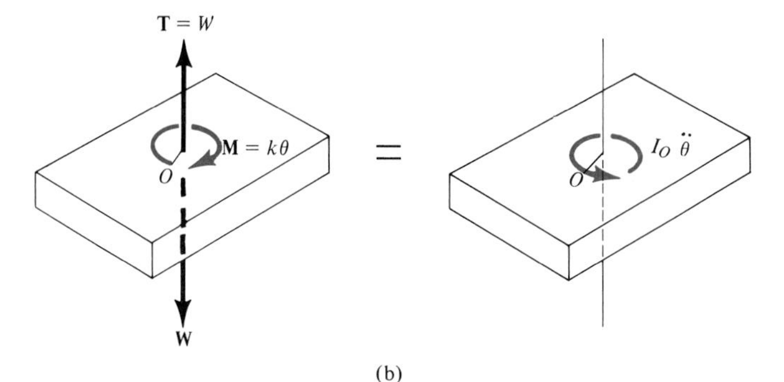

(b)
Fig. 21–5b

Problems

21–1. When a 3-kg block is suspended from a spring, the spring is stretched a distance of 60 mm. Determine the natural frequency and the period of vibration for a 0.2-kg block attached to the same spring.

21–2. A spring has a stiffness of 600 N/m. If a 4-kg block is attached to the spring, pushed 50 mm above its equilibrium position, and released from rest, determine the equation which describes the block's motion. Assume that positive displacement is measured downward.

21–3. A spring is stretched 175 mm by an 8-kg block. If the block is displaced 100 mm downward from its equilibrium position and given a downward velocity of 1.50 m/s, determine the equation which describes the motion. What is the phase angle? Assume that positive displacement is measured downward.

***21–4.** If the block in Prob. 21–3 is given an upward velocity of 4 m/s when it is displaced downward a distance of 60 mm from its equilibrium position, determine the equation which describes the motion. What is the amplitude of the motion? Assume that positive displacement is measured downward.

21–5. A block having a weight of 8 lb is suspended from a spring having a stiffness of $k = 40$ lb/ft. If the block is pushed $y = 0.2$ ft upward from its equilibrium position and then released from rest, determine the equation which describes the motion. What is the amplitude and the natural frequency of the vibration? Assume that positive displacement is measured downward.

21–6. An 8-kg block is suspended from a spring having a stiffness of $k = 80$ N/m. If the block is given an upward velocity of 0.4 m/s when it is 90 mm above its equilibrium position, determine the equation which describes the motion and the maximum upward displacement of the block measured from the equilibrium position. Assume that positive displacement is measured downward.

21–7. A cable is used to suspend the 800-kg safe. If the safe is being lowered at 6 m/s when the motor controlling the cable suddenly jams, determine the maximum tension in the cable and the frequency of vibration of the safe. Neglect the mass of the cable and assume it is elastic such that it stretches 20 mm when subjected to a tension of 4 kN.

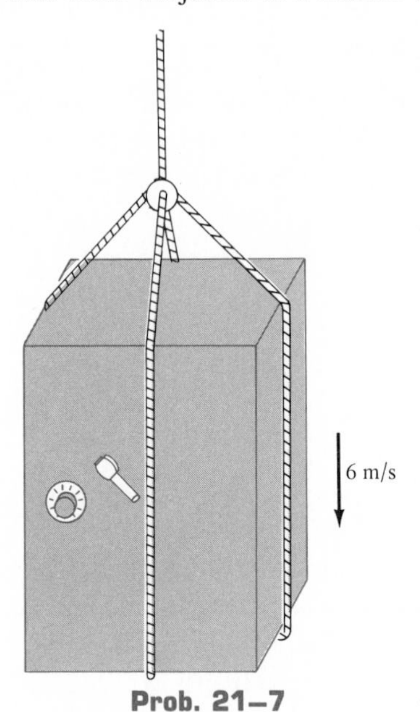

Prob. 21–7

***21–8.** A pendulum has a 0.4-m-long cord and is given a tangential velocity of 0.2 m/s toward the vertical from a position of $\theta = 0.3$ rad. Determine the equation which describes the angular motion.

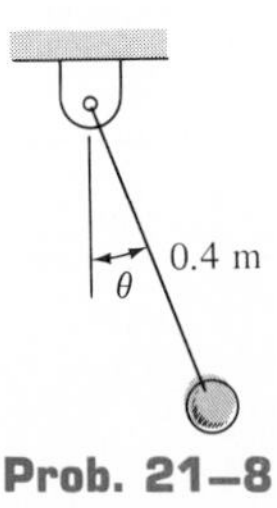

Prob. 21–8

21–9. Determine to the nearest degree the maximum angular displacement of the bob in Prob. 21–8 if it is initially displaced $\theta = 0.2$ rad from the vertical and given a tangential velocity of 0.4 m/s away from the vertical.

21–10. Determine the frequency of vibration for the block-and-spring mechanisms.

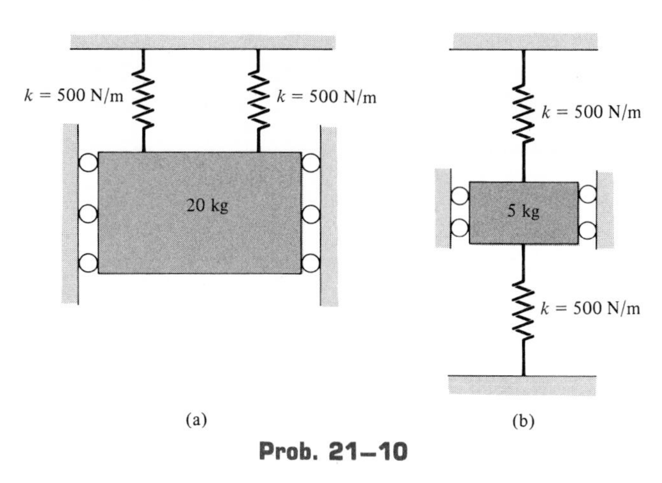

Prob. 21–10

21–11. The semicircular disk has a weight of 15 lb and is pinned at O. Determine the period of oscillation if it is displaced a small amount and released.

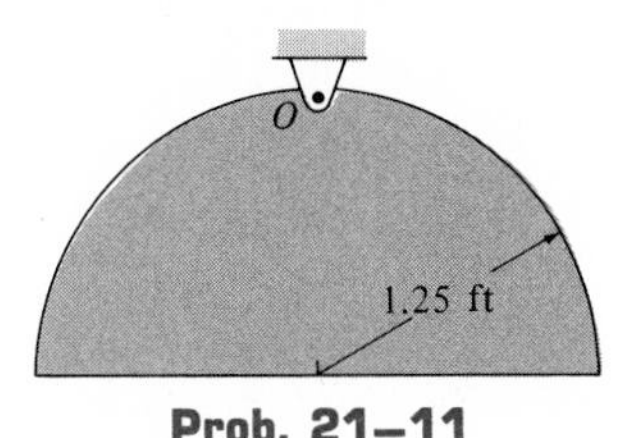

Prob. 21–11

***21–12.** The thin hoop is supported by a knife edge. Determine the period of oscillation for small amplitudes of swing.

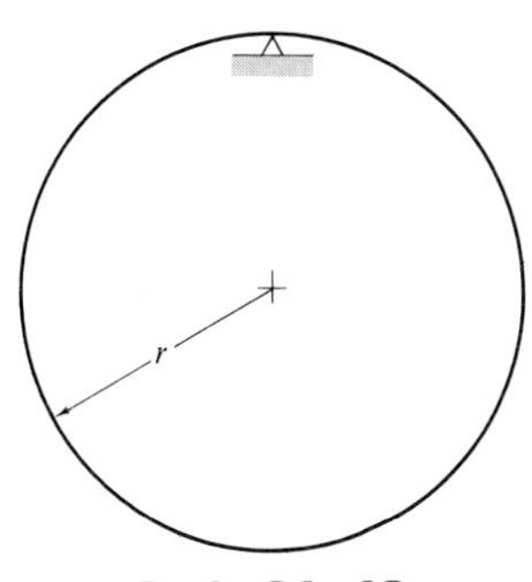

Prob. 21–12

21–13. The uniform beam has a mass of 250 kg/m and a moment of inertia of $I_O = 375 \text{ kg}\cdot\text{m}^2$ calculated about the pin at O. If the lower end is displaced a small amount and released from rest, determine the frequency of vibration. Each spring has a stiffness of $k = 500$ N/m and is unstretched when the beam is hanging from the vertical.

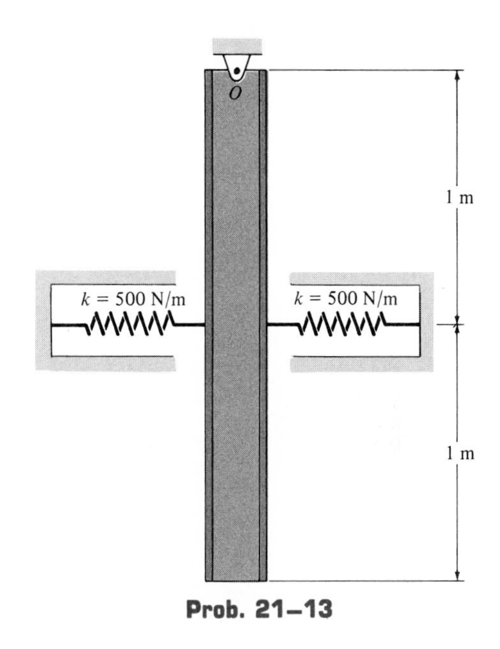

Prob. 21–13

21–14. A platform A, having an unknown mass, is supported by *four* springs, each having the same stiffness k. When nothing is on the platform, the period of vertical vibration is measured as 2.35 s; whereas if a 3-kg block is supported on the platform, the period of vertical vibration is 5.23 s. Compute the mass of a block placed on the (empty) platform which causes the platform to vibrate vertically with a period of 5.62 s. What is the stiffness k of each of the springs?

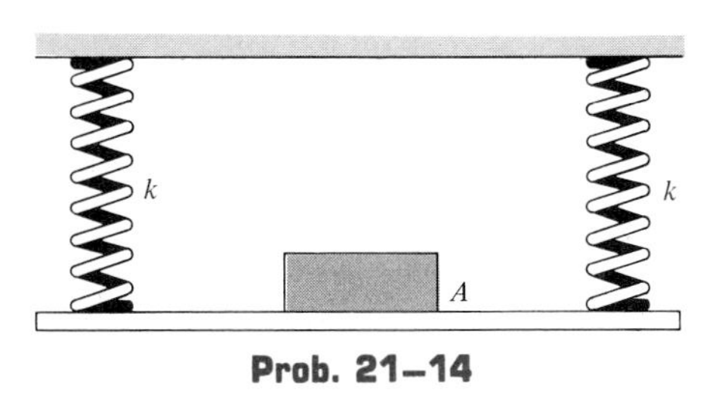

Prob. 21–14

21–15. The body of arbitrary shape has a mass m, mass center at G, and a radius of gyration about G of k_G. If it is displaced by a slight amount θ from its equilibrium position and released, determine the period of vibration.

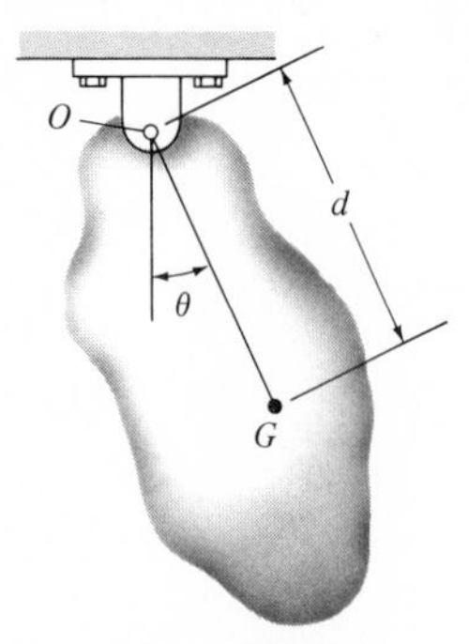

Prob. 21–15

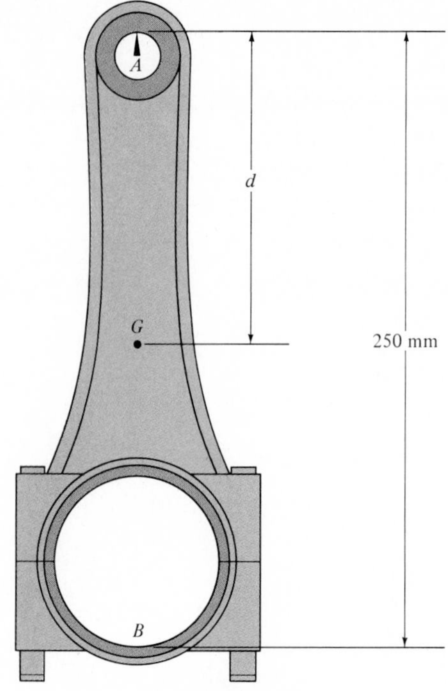

Prob. 21–16

***21–16.** The connecting rod is supported by a knife edge at A and the period of vibration is measured as $\tau_A = 3.38$ s. It is then removed and rotated 180° so that it is supported by the knife edge at B. In this case the period of vibration is measured as $\tau = 3.96$ s. Determine the location d of the center of gravity G, and compute the radius of gyration k_G. *Hint:* See Prob. 21–15.

21–17. The pointer on a metronome supports a 0.4-lb slider A, which is positioned at a fixed distance from the pivot O of the pointer. When the pointer is displaced, a torsional spring at O exerts a restoring moment on the pointer having a magnitude of $M = (1.2\theta)$ lb · ft, where θ represents the angle of displacement from the vertical, measured in radians. Determine the period of vibration when the pointer is displaced a small amount θ and released. Neglect the mass of the pointer.

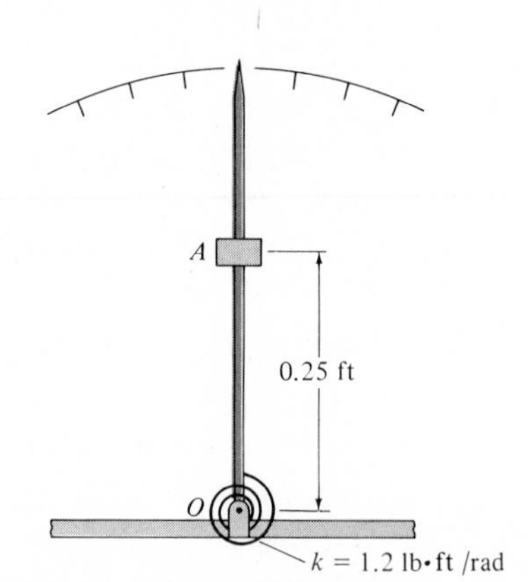

Prob. 21–17

Undamped Forced Vibration 21.2*

Undamped forced vibration is considered to be one of the most important types of vibrating motion in engineering work. The principles which describe the nature of this motion may be applied to the analysis of forces which cause vibration in various types of machines and structures.

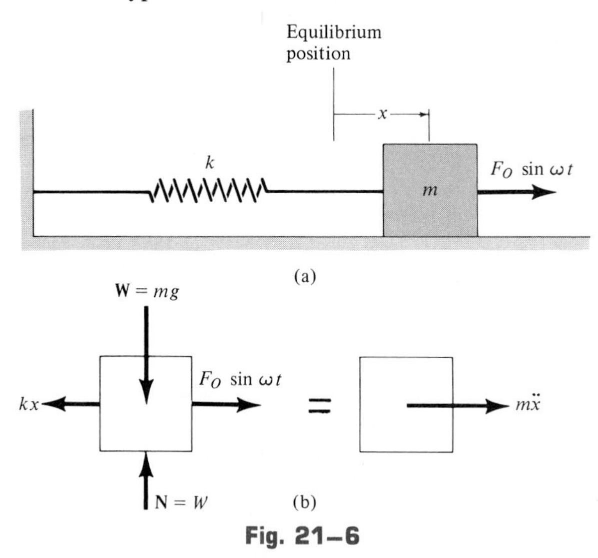

Fig. 21–6

Periodic Force. The block and spring shown in Fig. 21–6*a* provide a convenient "model" which represents the vibrational characteristics of a system subjected to a periodic force $F = F_O \sin \omega t$. This force has a maximum magnitude of F_O and a *forcing frequency* ω. The free-body and kinetic diagrams for the block when it is displaced a distance x are shown in Fig. 21–6*b*. Applying the equation of motion yields

$\overset{+}{\rightarrow}\Sigma F_x = ma_x;$ $\qquad F_O \sin \omega t - kx = m\ddot{x}$

or

$$\ddot{x} + \frac{k}{m}x = \frac{F_O}{m}\sin \omega t \tag{21–17}$$

This equation is referred to as a nonhomogeneous second-order differential equation. The general solution consists of a complementary solution, x_c, *plus* a particular solution, x_p.

The *complementary solution* is determined by setting the term on the right side of Eq. 21–17 equal to zero and solving the resulting homogeneous equation, which is equivalent to Eq. 21–1. The solution is defined by Eq. 21–3, i.e.,

$$x_c = A \sin pt + B \cos pt \tag{21–18}$$

where p is the circular frequency, $p = \sqrt{k/m}$ (Eq. 21–2).

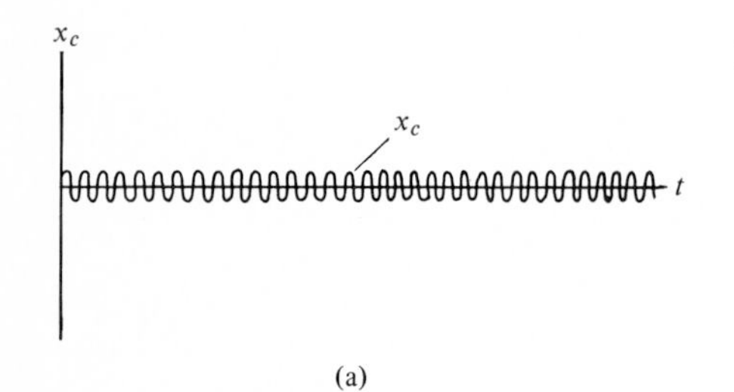

(a)

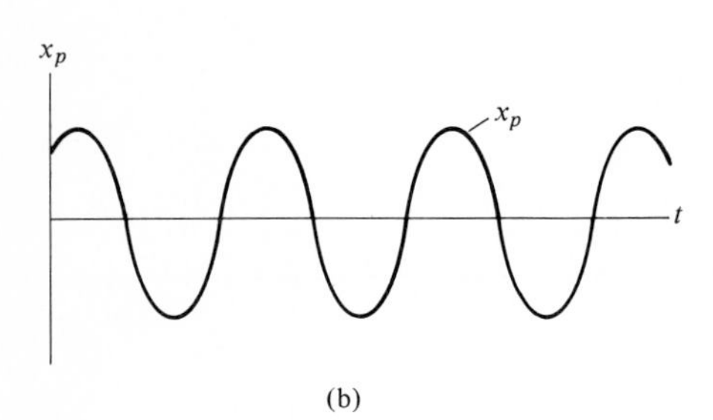

(b)

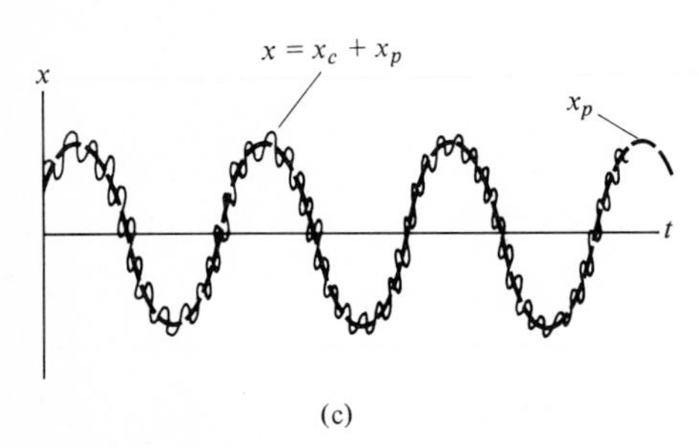

(c)

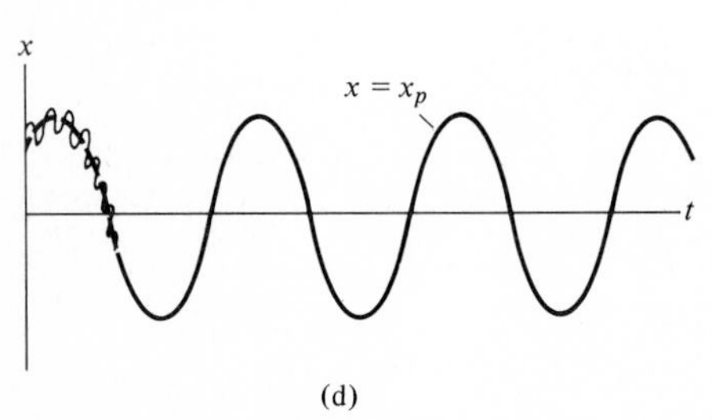

(d)

Fig. 21–7

Since the motion is periodic, the *particular solution* of Eq. 21–17 may be determined by assuming a solution of the form

$$x_p = C \sin \omega t \tag{21–19}$$

where C is a constant. Taking the second time derivative and substituting into Eq. 21–17 yields

$$-C\omega^2 \sin \omega t + \frac{k}{m}(C \sin \omega t) = \frac{F_O}{m} \sin \omega t$$

Factoring out $\sin \omega t$ and solving for C gives

$$C = \frac{F_O/m}{\dfrac{k}{m} - \omega^2} = \frac{F_O/k}{1 - \left(\dfrac{\omega}{p}\right)^2} \tag{21–20}$$

Substituting into Eq. 21–19, we obtain the particular solution

$$x_p = \frac{F_O/k}{1 - \left(\dfrac{\omega}{p}\right)^2} \sin \omega t \tag{21–21}$$

The *general solution* is therefore

$$x = x_c + x_p = A \sin pt + B \cos pt + \frac{F_O/k}{1 - \left(\dfrac{\omega}{p}\right)^2} \sin \omega t \tag{21–22}$$

Here x describes two types of vibrating motion of the block. The *complementary solution* x_c defines the *free vibration,* which depends upon the circular frequency $p = \sqrt{k/m}$ and the constants A and B, Fig. 21–7*a*. Specific values for A and B are obtained by evaluating Eq. 21–22 at a given instant when the displacement and velocity are known. The *particular solution* x_p describes the *forced vibration* of the block caused by the applied force $F = F_O \sin \omega t$, Fig. 21–7*b*. The resultant vibration x is shown in Fig. 21–7*c*. Since all vibrating systems are subject to *friction,* the free vibration, x_c, will in time dampen out. For this reason the free vibration is referred to as *transient,* and the forced vibration is called *steady state,* since it is the only vibration that remains, Fig. 21–7*d*.

From Eq. 21–20 it is seen that the *amplitude* of forced vibration depends upon the *frequency ratio* ω/p. If the *magnification factor* MF is defined as the ratio of the amplitude of steady-state vibration, $(x_p)_{\max}$, to the static deflection F_O/k, which is caused by the amplitude of the periodic force F_O, then, from Eq. 21–20,

$$\text{MF} = \frac{(x_p)_{\max}}{F_O/k} = \frac{1}{1 - \left(\dfrac{\omega}{p}\right)^2} \tag{21–23}$$

This equation is graphed in Fig. 21–8, where it is seen that for $\omega \approx 0$, the MF ≈ 1. In this case, because of the low frequency $\omega < p$, the vibration of the block will be in phase with the applied force **F**. If the force or displacement is applied with a frequency close to the natural frequency of the system, i.e., $\omega/p \approx 1$, the amplitude of vibration of the block becomes extremely large. This condition is called *resonance,* and in practice, resonating vibrations can cause tremendous stress and rapid failure of parts. When the cyclic force $F_O \sin \omega t$ is applied at high frequencies ($\omega > p$), the value of the MF becomes negative, indicating that the motion of the block is out of phase with the force. Under these conditions, as the block is displaced to the right, the force acts to the left, and vice versa. For extremely high frequencies ($\omega \gg p$) the block remains almost stationary, and hence the MF is approximately zero.

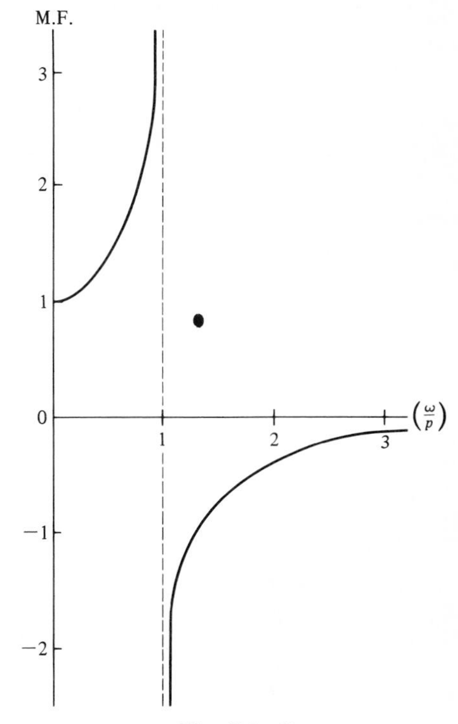

Fig. 21–8

Periodic Support Displacement. Forced vibrations can also arise from the periodic excitation of the support of a system. The model shown in Fig. 21–9*a* represents the periodic vibration of a block which is caused by harmonic movement $\delta = \delta_O \sin \omega t$ of the support. The free-body and kinetic diagrams for the block in this case are shown in Fig. 21–9*b*. The coordinate x is measured from the point of zero displacement of the support, i.e., when the radius vector OA coincides with OB, Fig. 21–9*a*. Therefore, general displacement of the spring is $(x - \delta_O \sin \omega t)$. Applying the equation of motion yields

$$\stackrel{+}{\rightarrow}\Sigma F_x = ma_x; \qquad -k(x - \delta_O \sin \omega t) = m\ddot{x}$$

or

$$\ddot{x} + \frac{k}{m}x = \frac{k\delta_O}{m}\sin \omega t \tag{21–24}$$

By comparison, this equation is identical to the form of Eq. 21–17, *provided* F_O *is replaced* by $k\delta_O$. If this substitution is made into the solutions defined by Eqs. 21–20 to 21–22, the results are appropriate for describing the motion of the block when subjected to the support displacement $\delta = \delta_O \sin \omega t$.

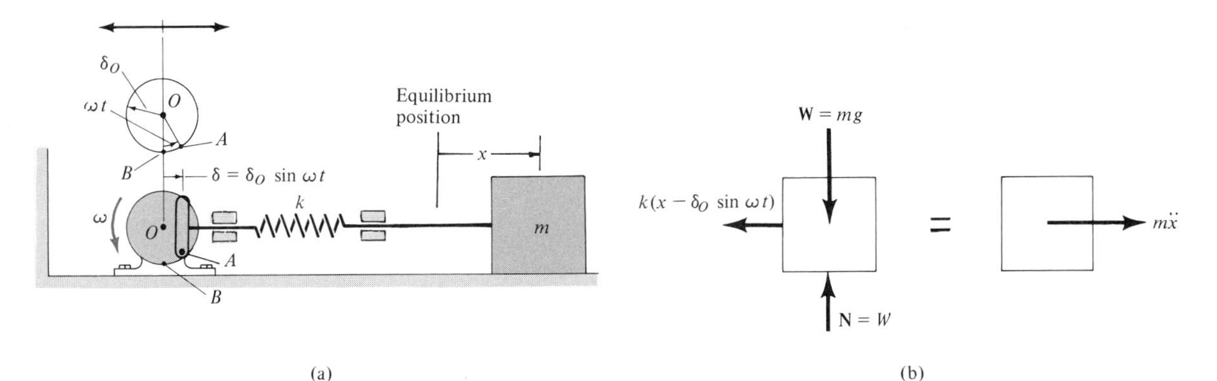

Fig. 21–9

Example 21–3

The instrument shown in Fig. 21–10 is rigidly attached to a platform P, which in turn is supported by *four* springs, each having a stiffness of $k = 800$ N/m. Initially the platform is at rest when the floor is subjected to a displacement of $\delta = 10 \sin (8t)$ mm, where t is measured in seconds. If the instrument is constrained to move vertically, and the total mass of the instrument and platform is 20 kg, determine the vertical displacement y of the platform, measured from the equilibrium position, as a function of time. What would be the floor vibration to cause resonance?

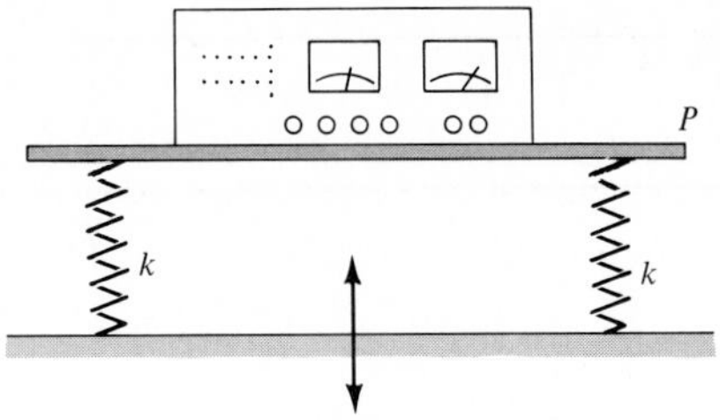

Fig. 21–10

Solution

Since the induced vibration is caused by the displacement of the supports, the motion is described by Eq. 21–22, with F_O replaced by $k\delta_O$, i.e.,

$$y = A \sin pt + B \cos pt + \frac{\delta_O}{1 - \left(\frac{\omega}{p}\right)^2} \sin \omega t \qquad (1)$$

Here $\delta = \delta_O \sin \omega t = 10 \sin (8t)$ mm, so that

$$\delta_O = 10 \text{ mm}, \qquad \omega = 8 \text{ rad/s}$$

$$p = \sqrt{\frac{k}{m}} = \sqrt{\frac{4(800)}{20}} = 12.6 \text{ rad/s}$$

From Eq. 21–21, with $k\delta_O$ replacing F_O, the amplitude of vibration caused by the floor displacement is

$$(y_p)_{\max} = \frac{\delta_O}{1 - \left(\frac{\omega}{p}\right)^2} = \frac{10}{1 - \left(\frac{8}{12.6}\right)^2} = 16.7 \text{ mm} \qquad (2)$$

Hence, Eq. (1) and its time derivative become

$$y = A \sin (12.6t) + B \cos (12.6t) + 16.7 \sin (8t)$$
$$\dot{y} = A(12.6) \cos (12.6t) - B(12.6) \sin (12.6t) + 133.3 \cos (8t)$$

The constants A and B are evaluated from these equations. Since $y = \dot{y} = 0$ at $t = 0$, then

$$0 = 0 + B + 0; \qquad B = 0$$
$$0 = A(12.6) - 0 + 133.3; \qquad A = -10.6$$

The vibrating motion is therefore described by the equation

$$y = -10.6 \sin (12.6t) + 16.7 \sin (8t) \qquad \textit{Ans.}$$

Resonance will occur when the amplitude of vibration caused by the floor displacement approaches infinity. From Eq. (2), this requires that

$$\omega = p = 12.6 \text{ rad/s} \qquad \textit{Ans.}$$

Problems

21–18. If the block is subjected to the impressed force $F = F_O \cos \omega t$, show that the differential equation of motion is $\ddot{y} + (k/m)y = (F_O/m) \cos \omega t$, where y is measured from the equilibrium position of the block. What is the general solution of this equation?

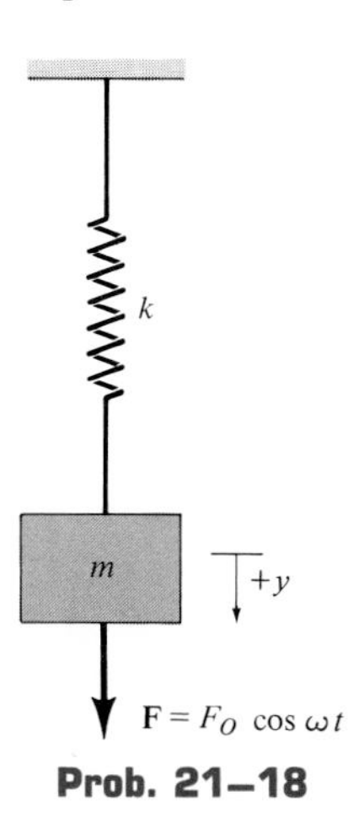

Prob. 21–18

21–19. A 5-kg block is suspended from a spring having a stiffness of 300 N/m. If the block is acted upon by a vertical force of $F = (7 \sin 8t)$ N, where t is measured in seconds, determine the equation which describes the motion of the block when it is pulled down 100 mm from the equilibrium position and released from rest at $t = 0$. Assume that positive displacement is measured downward.

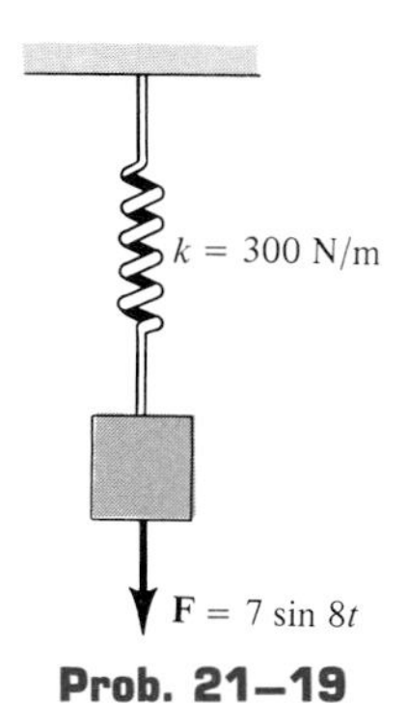

Prob. 21–19

***21–20.** A 7-kg block is suspended from a spring that has a stiffness of $k = 350$ N/m. The block is drawn downward 70 mm from the equilibrium position and released from rest at $t = 0$. If the support moves with an impressed displacement of $\delta = (20 \sin 4t)$ mm, where t is measured in seconds, determine the equation which describes the vertical motion of the block. Assume that positive displacement is measured downward.

21–21. The engine is mounted on a foundation block which is spring-supported. Describe the steady-state vibration of the system if the block and engine have a total weight of $W = 1500$ lb and the engine, when running, creates an impressed force of $F = (50 \sin 2t)$ lb, where t is measured in seconds. Assume that the system vibrates only in the vertical direction, with the positive displacement measured downward, and that the total stiffness of the springs can be represented as $k = 2000$ lb/ft.

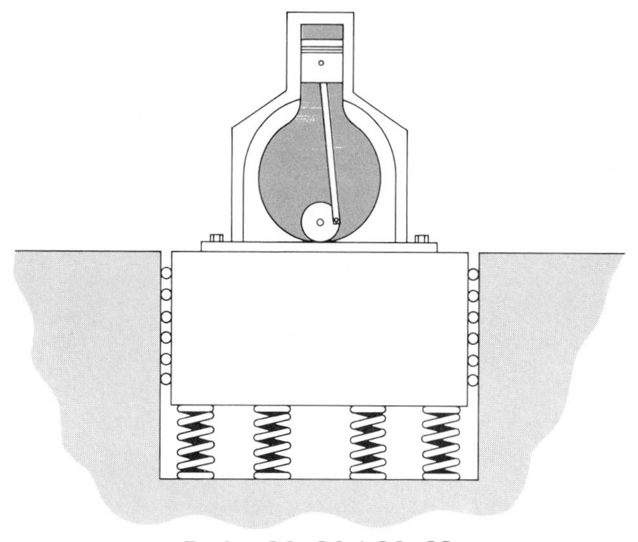

Probs. 21–21 / 21–22

21–22. Determine the rotational speed ω of the engine in Prob. 21–21 which will cause resonance.

21–23. The instrument is centered uniformly on a platform P, which in turn is supported by *four* springs, each spring having a stiffness of $k = 130$ N/m. If the floor is subjected to a vibration of $\omega = 7$ Hz, having a vertical displacement amplitude of $\delta_O = 0.17$ ft, determine the vertical displacement amplitude of the platform and instrument. The instrument and the platform have a total weight of 18 lb.

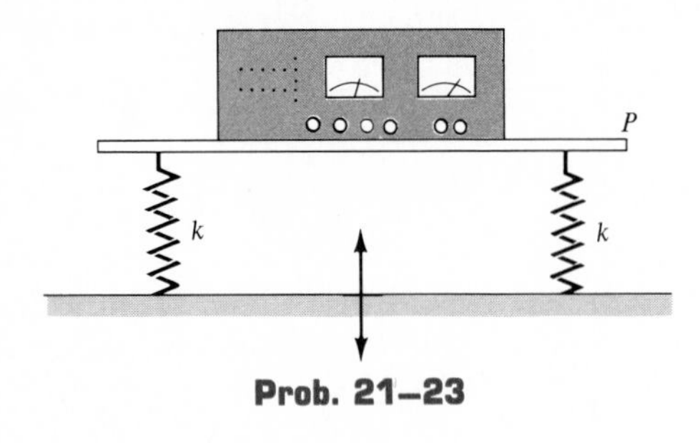

Prob. 21–23

***21–24.** The light elastic rod supports a 4-kg sphere. When an 18-N vertical force is applied to the sphere, the rod deflects 14 mm. If the wall oscillates with a harmonic frequency of 2 Hz and has an amplitude of 15 mm, determine the amplitude of vibration for the sphere.

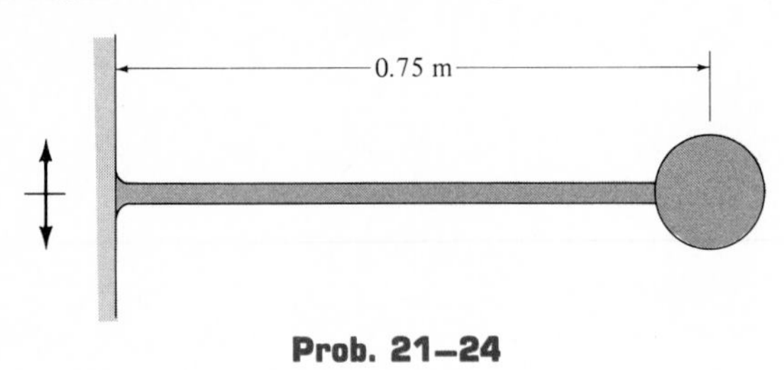

Prob. 21–24

21–25. The 450-kg trailer is pulled with a constant speed over the surface of a bumpy road, which may be approximated by a cosine curve having an amplitude of 50 mm and wave length of 4 m. If the two springs s which support the trailer each have a stiffness of 800 N/m, determine the speed v which will cause the greatest vibration (resonance) of the trailer. Neglect the weight of the wheels.

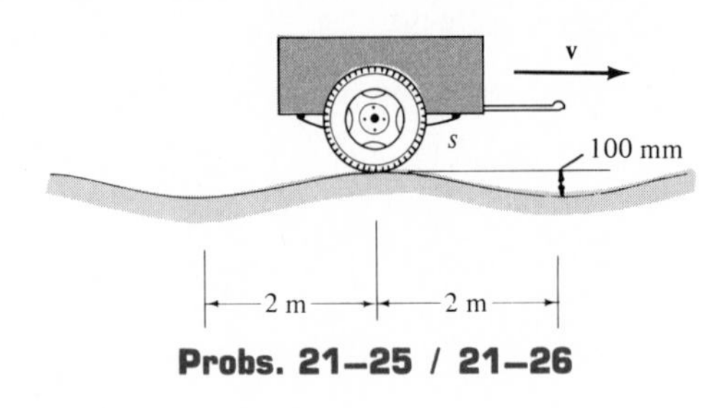

Probs. 21–25 / 21–26

21–26. Determine the amplitude of vibration of the trailer in Prob. 21–25 if the speed $v = 15$ km/h.

*21.3 Viscous Damped Free Vibration

The vibration analysis considered thus far has not included the effects of friction or damping in the system, and as a result, the solutions obtained are only in close agreement with the actual motion. Since all vibrations die out in time, the presence of damping forces should be included in the analysis.

In many cases damping is attributed to the resistance created by the substance, such as water, oil, or air, in which the system vibrates. Provided the body moves slowly through this substance, the resistance to motion is directly

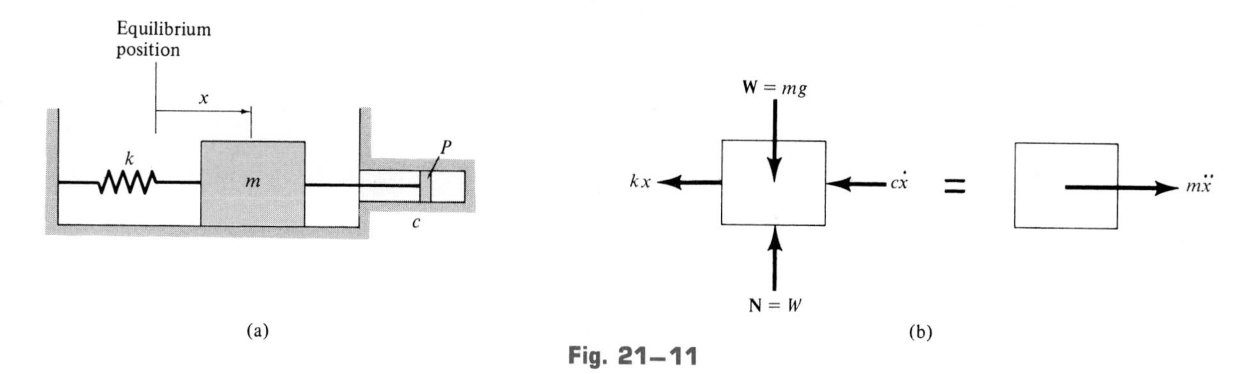

Fig. 21–11

proportional to the body's speed. The type of force developed under these conditions is called a *viscous damping force*. The magnitude of this force may be expressed by an equation of the form

$$F = c\dot{x} \tag{21–25}$$

where the constant c is called the *coefficient of viscous damping* and has units of N · s/m or lb · s/ft.

The vibrating motion of a body or system having viscous damping may be characterized by the block and spring shown in Fig. 21–11*a*. The effect of damping is provided by the *dashpot* connected to the block on the right side. Damping occurs when the piston P moves to the right or left within the enclosed cylinder. The cylinder contains a fluid, and the motion of the piston is retarded since the fluid must flow around or through a small hole in the piston. The dashpot is assumed to have a coefficient of viscous damping c.

If the block is displaced a distance x from its equilibrium position, the resulting free-body and kinetic diagrams are shown in Fig. 21–11*b*. Both the spring force kx and the damping force $c\dot{x}$ oppose the forward motion of the block, so that applying the equation of motion yields

$$\overset{+}{\rightarrow}\Sigma F_x = ma_x; \qquad -kx - c\dot{x} = m\ddot{x}$$

or

$$m\ddot{x} + c\dot{x} + kx = 0 \tag{21–26}$$

This linear, second-order, homogeneous, differential equation has solutions of the form

$$x = e^{\lambda t}$$

where e is the base of the natural logarithm and λ is a constant. The value of λ (lambda) may be obtained by substituting this solution into Eq. 21–26, which yields

$$m\lambda^2 e^{\lambda t} + c\lambda e^{\lambda t} + ke^{\lambda t} = 0$$

or

$$e^{\lambda t}(m\lambda^2 + c\lambda + k) = 0$$

Since $e^{\lambda t}$ is always positive, a solution is possible provided

$$m\lambda^2 + c\lambda + k = 0$$

Hence, by the quadratic formula, the two values of λ are

$$\lambda_1 = -\frac{c}{2m} + \sqrt{\left(\frac{c}{2m}\right)^2 - \frac{k}{m}}, \qquad \lambda_2 = -\frac{c}{2m} - \sqrt{\left(\frac{c}{2m}\right)^2 - \frac{k}{m}} \tag{21–27}$$

The general solution of Eq. 21–26 is therefore a linear combination of exponentials which involves both of these roots. There are three possible combinations of λ_1 and λ_2 which must be considered for the general solution. Before discussing these combinations, however, it is first necessary to consider the definition of the *critical damping coefficient* c_c as the value of c which makes the radical in Eqs. 21–27 equal to zero; i.e.,

$$\left(\frac{c_c}{2m}\right)^2 - \frac{k}{m} = 0$$

or

$$c_c = 2m\sqrt{\frac{k}{m}} = 2mp \tag{21–28}$$

Here the value of p is the circular frequency $p = \sqrt{k/m}$ (Eq. 21–2).

Overdamped System. When $c > c_c$, the roots λ_1 and λ_2 are both real. The general solution of Eq. 21–26 may then be written as

$$x = Ae^{\lambda_1 t} + Be^{\lambda_2 t} \tag{21–29}$$

Motion corresponding to this solution is *nonvibrating*. The effect of damping is so strong that when the block is displaced and released, it simply creeps back to its original position without oscillating. The system is said to be *overdamped*.

Critically Damped System. If $c = c_c$, then $\lambda_1 = \lambda_2 = -c_c/2m = -p$. This situation is known as *critical damping,* since it represents a condition where c has the smallest value necessary to cause the system to be nonvibrating. Using the methods of differential equations, it may be shown that the solution to Eq. 21–26 for critical damping is

$$x = (A + Bt)e^{-pt} \tag{21–30}$$

Underdamped System. Most often $c < c_c$, in which case the system is referred to as *underdamped*. In this case the roots λ_1 and λ_2 are complex numbers and it may be shown that the general solution of Eq. 21–26 can be written as

$$x = D[e^{-(c/2m)t} \sin (p_d t + \phi)] \qquad (21\text{–}31)$$

where D and ϕ are constants generally determined from the initial conditions of the problem. The constant p_d is called the *damped natural frequency* of the system. It has a value of

$$p_d = \sqrt{\frac{k}{m} - \left(\frac{c}{2m}\right)^2} = p\sqrt{1 - \left(\frac{c}{c_c}\right)^2} \qquad (21\text{–}32)$$

where the ratio c/c_c is called the *damping factor*.

The graph of Eq. 21–31 is shown in Fig. 21–12. The initial limit of motion, D, diminishes with each cycle of vibration, since motion is confined within the bounds of the exponential curve. Using the damped natural frequency p_d, the period of damped vibration may be written as

$$\tau_d = \frac{2\pi}{p_d} \qquad (21\text{–}33)$$

Since $p_d < p$ (Eq. 21–32), the period of damped vibration, τ_d, will be greater than that of free vibration, $\tau = 2\pi/p$.

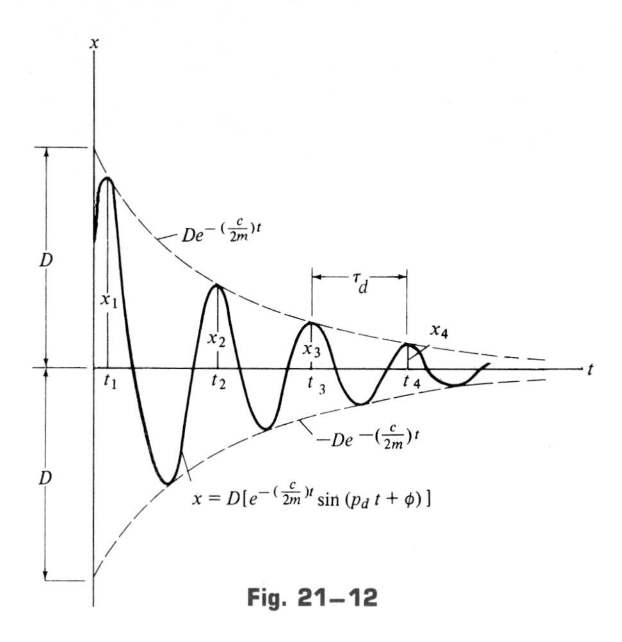

Fig. 21–12

*21.4 Viscous Damped Forced Vibration

The most general case of single-degree-of-freedom vibrating motion occurs when the system includes the effects of forced motion and induced damping. The analysis of this particular type of vibration is of practical value when applied to systems having significant damping characteristics.

If a dashpot is attached to the block and spring shown in Fig. 21–6*a*, the differential equation which describes the motion becomes

$$m\ddot{x} + c\dot{x} + kx = F_O \sin \omega t \qquad (21\text{–}34)$$

A similar equation may be written for a block and spring having a periodic support displacement, Fig. 21–9*a*, which includes the effects of damping. In that case, however, F_O is replaced by $k\delta_O$. Since Eq. 21–34 is nonhomogeneous, the general solution is the sum of a complementary solution, x_c, and a particular solution, x_p. The complementary solution is determined by setting the right side of Eq. 21–34 equal to zero and solving the homogeneous equation, which is equivalent to Eq. 21–26. The solution is therefore given by Eqs. 21–29, 21–30, and 21–31, depending upon the values of λ_1 and λ_2. Because all systems contain friction, however, this solution will dampen out with time. Only the particular solution, which describes the *steady-state vibration* of the system, will remain. Since the applied forcing function is harmonic, the steady-state motion will also be harmonic. Consequently, the particular solution will be of the form

$$x_p = A' \sin \omega t + B' \cos \omega t \qquad (21\text{–}35)$$

The constants A' and B' are determined by taking the necessary time derivatives and substituting them into Eq. 21–34, which after simplification yields

$$[-A'm\omega^2 - cB'\omega + kA'] \sin \omega t + [-B'm\omega^2 + cA'\omega + kB'] \cos \omega t = F_O \sin \omega t$$

Since this equation holds for all time, the constant coefficients of sin ωt and cos ωt may be equated; i.e.,

$$-A'm\omega^2 - cB'\omega + kA' = F_O$$
$$-B'm\omega^2 + cA'\omega + kB' = 0$$

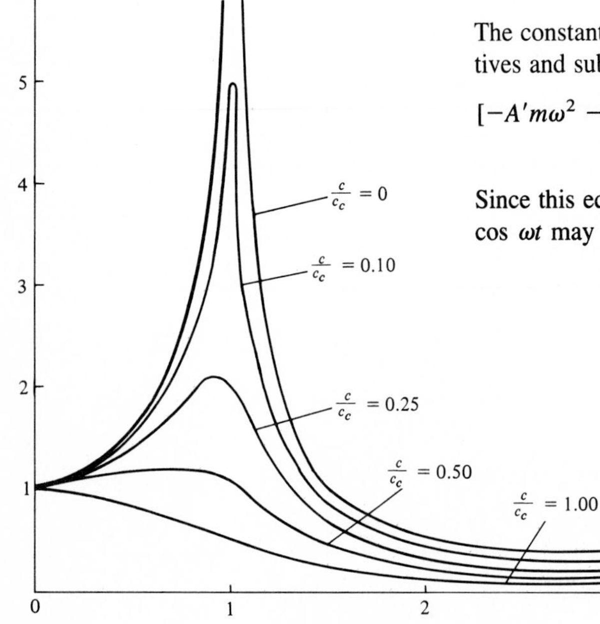

Fig. 21–13

Solving for A' and B', realizing that $p^2 = k/m$, yields

$$A' = \frac{\left(\frac{F_O}{m}\right)(p^2 - \omega^2)}{(p^2 - \omega^2)^2 + \left(\frac{c\omega}{m}\right)^2}$$

$$B' = \frac{-F_O \frac{c\omega}{m^2}}{(p^2 - \omega^2)^2 + \left(\frac{c\omega}{m}\right)^2} \qquad (21\text{–}36)$$

It is also possible to express Eq. 21–35 in a form similar to Eq. 21–9,

$$x_p = C' \sin(\omega t - \phi') \qquad (21\text{–}37)$$

in which case the constants C' and ϕ' are

$$C' = \frac{F_O/k}{\sqrt{\left[1 - \left(\frac{\omega}{p}\right)^2\right]^2 + \left(2\frac{c}{c_c}\frac{\omega}{p}\right)^2}}$$

$$\phi' = \tan^{-1}\left(\frac{c\omega/k}{1 - \left(\frac{\omega}{p}\right)^2}\right) \qquad (21\text{–}38)$$

The angle ϕ' represents the phase difference between the applied force and the resulting steady-state vibration of the damped system.

The *magnification factor* MF has been defined in Sec. 21–2 as the ratio of the amplitude of deflection caused by the forced vibration to the deflection caused by a static force $\mathbf{F}_O$. From Eq. 21–37, the forced vibration has an amplitude of C'; thus,

$$\text{MF} = \frac{C'}{F_O/k} = \frac{1}{\sqrt{\left[1 - \left(\frac{\omega}{p}\right)^2\right]^2 + \left(2\frac{c}{c_c}\frac{\omega}{p}\right)^2}} \qquad (21\text{–}39)$$

The MF is plotted in Fig. 21–13 versus the frequency ratio ω/p for various values of the damping factor c/c_c. It can be seen from this graph that the magnification of the amplitude increases as the damping factor decreases. Resonance obviously occurs only when the damping is zero and the frequency ratio equals 1.

Example 21–4

The 30-kg electric motor shown in Fig. 21–14 is supported by *four* springs, each spring having a stiffness of 200 N/m. If the rotor R is unbalanced such that its effect is equivalent to a 4-kg mass located 60 mm from the axis of rotation, determine the amplitude of vibration when the rotor is turning at $\omega = 10$ rad/s. The damping factor is $c/c_c = 0.15$.

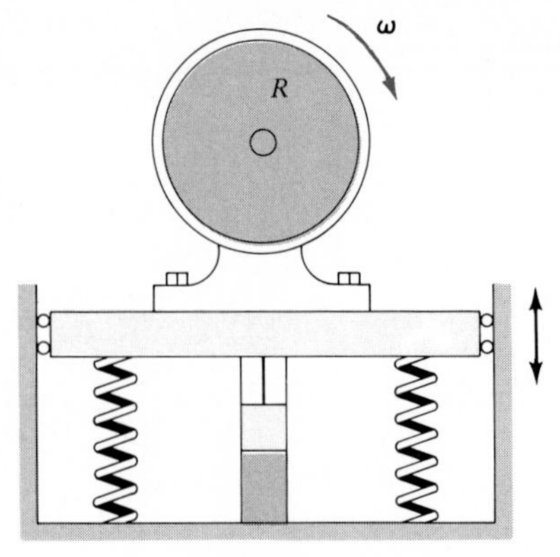

Fig. 21–14

Solution

The periodic force which causes the motor to vibrate is the centrifugal force due to the unbalanced rotor. This force has a constant magnitude of

$$F_O = ma_n = mr\omega^2 = 4\text{ kg}(0.06\text{ m})(10\text{ rad/s})^2 = 24\text{ N}$$

Oscillation in the vertical direction may be expressed in the periodic form $F = F_O \sin \omega t$, where $\omega = 10$ rad/s. Thus,

$$F = 24 \sin 10t$$

The stiffness of the entire system of four springs is $k = 4(200) = 800$ N/m. Therefore, the circular frequency of vibration is

$$p = \sqrt{\frac{k}{m}} = \sqrt{\frac{800}{30}} = 5.16\text{ rad/s}$$

Since the damping factor is known, the steady-state amplitude may be determined from the first of Eqs. 21–38, i.e.,

$$C' = \frac{F_O/k}{\sqrt{\left[1 - \left(\frac{\omega}{p}\right)^2\right]^2 + \left(2\frac{c}{c_c}\frac{\omega}{p}\right)^2}}$$

$$= \frac{24/800}{\sqrt{\left[1 - \left(\frac{10}{5.16}\right)^2\right]^2 + \left[2(0.15)\frac{10}{5.16}\right]^2}}$$

$$= 0.010\,7\text{ m} = 10.7\text{ mm}$$ *Ans.*

Problems

21–27. A block having a mass of 0.8 kg is suspended from a spring having a stiffness of 120 N/m. If a dashpot provides a damping force of 2.5 N when the speed of the block is 0.2 m/s, determine the period of free vibration.

***21–28.** A block having a weight of 7 lb is suspended from a spring having a stiffness of $k = 75$ lb/ft. The support to which the spring is attached is given a simple harmonic motion which may be expressed by $\delta = (0.15 \sin 2t)$ ft, where t is in seconds. If the damping factor is $c/c_c = 0.8$, determine the phase angle ϕ of forced vibration.

21–29. Determine the magnification factor of the block, spring, and dashpot combination in Prob. 21–28.

21–30. The 20-kg block is subjected to the action of the harmonic force $F = (90 \cos 6t)$ N, where t is measured in seconds. Write the equation which describes the steady-state motion.

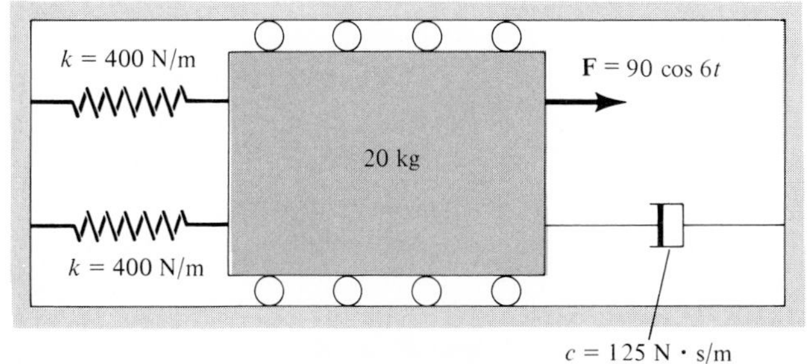

Prob. 21–30

21–31. The barrel of a cannon has a mass of 700 kg, and after firing it recoils a distance of 0.64 m. If it returns to its original position by means of a single recuperator having a damping coefficient of 2 kN · s/m, determine the required stiffness of each of the two springs fixed to the base and attached to the barrel so that the barrel recuperates without vibration.

***21–32.** A block having a mass of 7 kg is suspended from a spring that has a stiffness of $k = 600$ N/m. If it is given an upward velocity of 0.6 m/s from its equilibrium position at $t = 0$, determine the position of the block as a function of time. Assume that positive displacement of the block is downward, and that motion takes place in a medium which furnishes a damping force of $F = (50|v|)$ N, where v is measured in m/s.

21–33. The block, having a weight of 12 lb, is immersed in a liquid such that the damping force acting on the block has a magnitude of $F = (0.7|v|)$ lb, where v is measured in ft/s. If the block is pulled down 0.62 ft and released from rest, describe the motion. The spring has a stiffness of $k = 53$ lb/ft. Assume that positive displacement is measured downward.

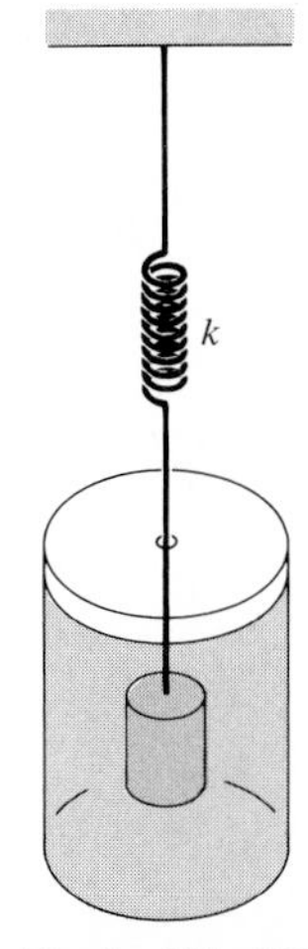

Prob. 21–33

21–34. The 4-kg circular disk is attached to three springs, each spring having a stiffness of $k = 180$ N/m. If the disk is immersed in a fluid and given a downward velocity of 0.3 m/s at the equilibrium position, determine the equation which describes the motion. Assume that positive displacement is measured downward, and that fluid resistance acting on the disk furnishes a damping force having a magnitude of $F = (60|v|)$ N, where v is measured in m/s.

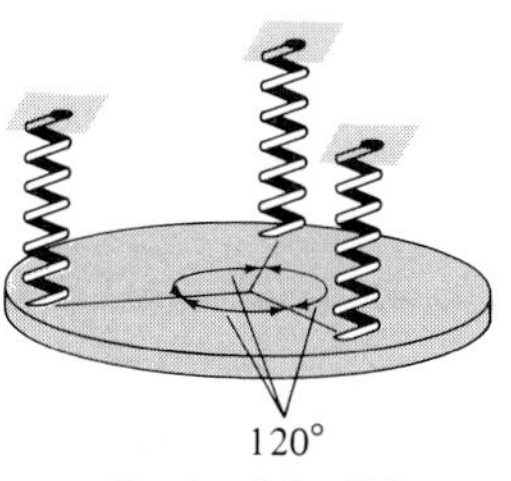

Prob. 21–34

21–35. The damping factor, c/c_c, may be determined experimentally by measuring the successive amplitudes of vibrating motion of a system. If two of these maximum displacements can be approximated by x_1 and x_2, as shown in Fig. 21–12, show that the ratio $\ln x_1/x_2 = 2\pi(c/c_c)/\sqrt{1-(c/c_c)^2}$. The quantity $\ln x_1/x_2$ is called the *logarithmic decrement.*

***21–36.** The block shown in Fig. 21–11 has a mass of 20 kg and the spring has a stiffness of $k = 600$ N/m. When the block is displaced and released, two successive amplitudes are measured as $x_1 = 150$ mm and $x_2 = 87$ mm. Determine the coefficient of viscous damping, c. *Hint:* See Prob. 21–35.

21–37. The 48-kg electric motor is fastened to the midpoint of a pinned supported beam having negligible mass. It is found that the beam deflects 35 mm when the motor is not running. If the motor turns an eccentric flywheel which is equivalent to an unbalanced mass of 4 kg located 100 mm from the axis of rotation, determine the amplitude of steady-state vibration when the motor is turning at 17 rad/s. The damping factor is $c/c_c = 0.20$.

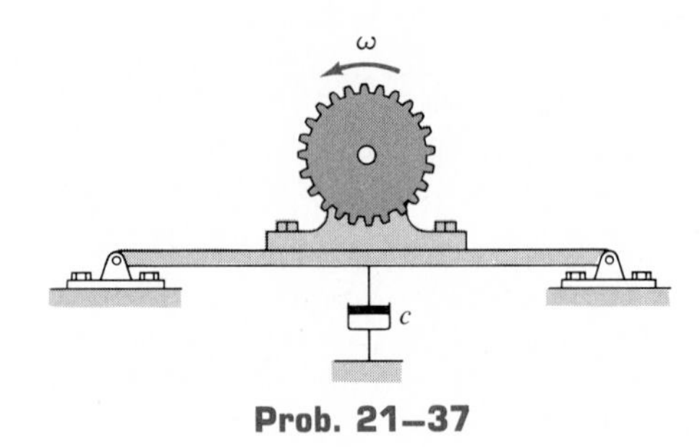

Prob. 21–37

Review Problems

21–38. A spring has a stiffness of 2 lb/in. If a 2-lb weight is attached to the spring, pushed 3 in. above its equilibrium position, and released from rest, determine the equation of motion.

21–39. A 5-lb weight is suspended from a spring having a stiffness of 50 lb/ft. An impressed vertical force of $F = \frac{1}{4}\sin 8t$, where F is given in pounds and t is measured in seconds, is acting on the weight. Determine the equation of motion of the weight when it is pulled down 3 in. from the equilibrium position and released.

***21–40.** The 4-kg sphere is attached to a rod of negligible mass. Use the equation of motion and determine the natural frequency of vibration.

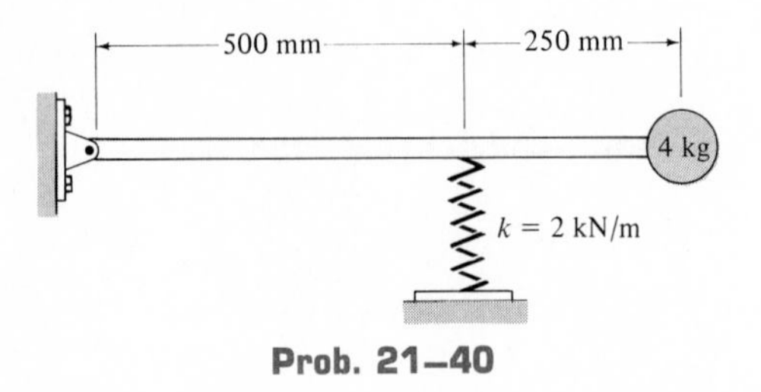

Prob. 21–40

21–41. The spring shown stretches 240 mm when it is attached to a 7-kg block. Determine the equation which describes the motion of the block when it is pulled 50 mm below its equilibrium position and released from rest at $t = 0$. The block is subjected to the impressed force of $F = 200 \sin 2t$, where F is measured in newtons and t in seconds. Assume that positive displacement is measured downward.

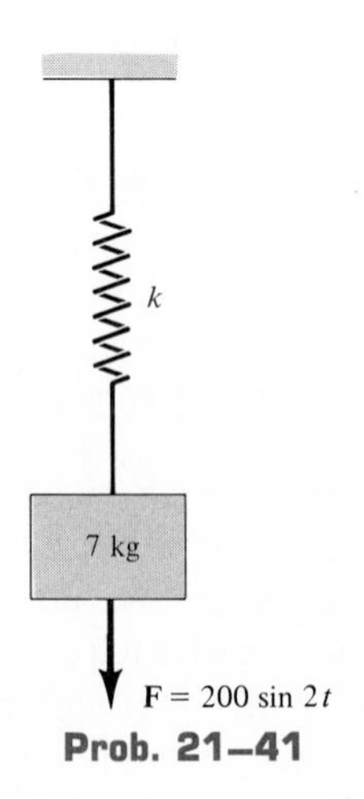

Prob. 21–41

21–42. A 5-lb block is suspended from a spring having a stiffness of 6 lb/in. The support to which the spring is attached is given a displacement which may be expressed by $\delta = 1.5 \sin 2t$, where δ is given in inches and t is measured in seconds. If a dashpot having a damping coefficient of 0.08 lb · s/in. is connected to the block, determine the phase angle ϕ of vibration.

21–43. Determine the magnification factor of the spring and dashpot combination in Prob. 21–42.

***21–44.** The 20-lb block is attached to a spring having a stiffness of 20 lb/ft. If a force of $F = (6 \cos 2t)$ lb, where t is measured in seconds, is applied to the block, determine the maximum velocity of the block after frictional forces cause the free vibrations to dampen out.

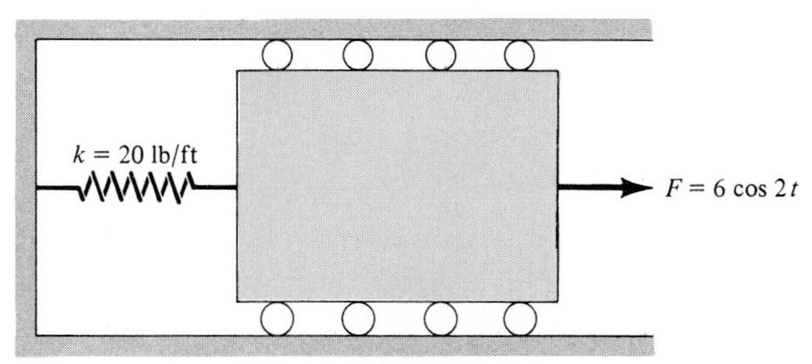

Prob. 21–44

21–45. A 2-lb weight is suspended from a spring having a stiffness of $k = 2$ lb/in. If the weight is pushed 1 in. upward from its equilibrium position and then released, determine the equation which describes the motion. What are the amplitude and the natural frequency of the vibration?

21–46. The 3-kg block is fixed to the end of a rod assembly that has negligible weight. If both springs are unstretched when the assembly is in the position shown, use the equation of motion and determine the natural period of vibration for the block when it is rotated slightly about the pivot at O and released.

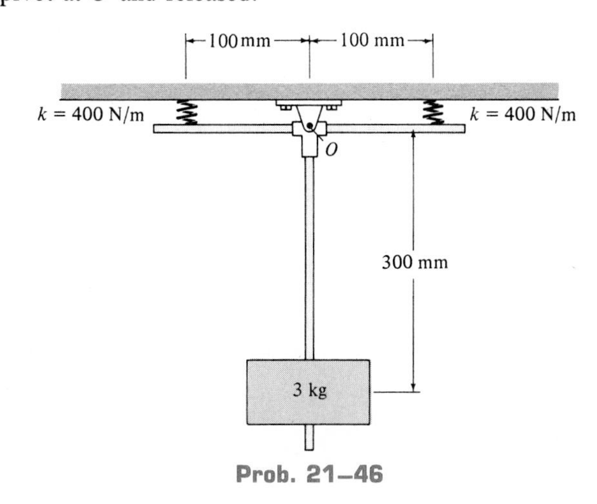

Prob. 21–46

Mathematical Expressions

Quadratic Formula:

If $ax^2 + bx + c = 0$, then $x = \dfrac{-b \pm \sqrt{b^2 - 4ac}}{2a}$

Trigonometric Identities:

$\sin^2 \theta + \cos^2 \theta = 1$

$\sin 2\theta = 2 \sin \theta \cos \theta$

$\cos 2\theta = \cos^2 \theta - \sin^2 \theta$

$\tan \theta = \dfrac{\sin \theta}{\cos \theta}$

$1 + \tan^2 \theta = \sec^2 \theta$

$1 + \cot^2 \theta = \csc^2 \theta$

Integrals:

$\int x^n \, dx = \dfrac{x^{n+1}}{n+1}, \; n \neq -1$

$\int \sin x \, dx = -\cos x$

$\int \cos x \, dx = \sin x$

Derivatives:

$\dfrac{d}{dx}(u^n) = nu^{n-1}\dfrac{du}{dx}$

$\dfrac{d}{dx}(uv) = u\dfrac{dv}{dx} + v\dfrac{du}{dx}$

$\dfrac{d}{dx}\left(\dfrac{u}{v}\right) = \dfrac{v\dfrac{du}{dx} - u\dfrac{dv}{dx}}{v^2}$

$\dfrac{d}{dx}(\sin u) = \cos u\dfrac{du}{dx}$

$\dfrac{d}{dx}(\cos u) = -\sin u\dfrac{du}{dx}$

$\dfrac{d}{dx}(\tan u) = \sec^2 u\dfrac{du}{dx}$

$\dfrac{d}{dx}(\cot u) = -\csc^2 u\dfrac{du}{dx}$

$\dfrac{d}{dx}(\sec u) = \tan u \sec u\dfrac{du}{dx}$

$\dfrac{d}{dx}(\csc u) = -\csc u \cot u\dfrac{du}{dx}$

Answers*

Chapter 12

12–1. 1.86 m/s^2.
12–2. 8 s, 320 ft.
12–3. 445 ft, 214 ft/s, 70 ft/s^2.
12–5. 20 ft.
12–6. **(a)** 135 ft, **(b)** 45 ft/s, **(c)** 18 ft/s^2.
12–7. **(a)** 19.0 ft, **(b)** 1.09 s, **(c)** 3.31 s.
12–9. 29.4 m/s, 88.3 m.
12–10. 0, 80.67 m.
12–11. 60 m, −2 m/s.
12–13. 0.6 ft/s, 2.6 ft/s.
12–14. **(a)** 834 mm, **(b)** 1122 mm/s, 450 mm/s^2.
12–15. 4.4 m/s^2.
12–17. 313.9 m, 72.5 m/s.
12–18. 45.2 ft/s, 13.9 ft/s^2.
12–19. $-5.82\ s^{-2}$, 1.12 ft.
12–21. 40.8 s.
12–22. A is ahead of B 1.45 m.
12–23. 7.90 ft/s.
12–25. 11.17 km/s.
12–26. $v = R\sqrt{\dfrac{2g_0(y_0 - y)}{(R + y)(R + y_0)}}$, 3.52 km/s.
12–27. 450 m.
12–29. 11.25 s.
12–30. 110.48 ft, $v = 4 - 32.2t$.
12–31. $t = 15$ s, $v = 270$ m/s, $s = 2025$ m; $t = 20$ s, $v = 395$ m/s, $s = 3687.5$ m.
12–33. $0 \leq t < 20$ s, $v = 7.5t$, $a = 7.5$; 20 s $< t \leq 40$ s, $v = 150$, $a = 0$.
12–34. 11.1 m/s^2, −25 m/s^2.
12–35. $t = 10$ s, $v = 25$ m/s, $s = 83.3$ m; $t = 22.5$ s, $v = 0$, $s = 239.5$ m.
12–37. 150 ft/s, 34.5 s.
12–38. 125 m, A ahead of B.
12–39. 22.4 ft/s, 38.7 ft/s, 47.4 ft/s.
12–41. 4 ft, 3 ft.
12–42. 24.33 m/s $\measuredangle$ 9.46°, 8 m/s^2 →.
12–43. 12.34 ft/s^2, 13.5°.
12–45. 2 km, 1.92 km, 175.3 m/s, 7.62 m/s^2.
12–46. 52 m/s.
12–47. 16.5 km/s, 64 km/s^2.

*Note: Answers to every fourth problem are omitted.

12–49. 5 km, 7 km.

12–50. 14.14 m/s ↘ 45°.

12–51. $(a_{AB})_{\text{Avg}} = 4.91\ \text{m/s}^2$, $(a_{AC})_{\text{Avg}} = 5\ \text{m/s}^2$.

12–53. 37.7 ft/s, 33.2 ft/s.

12–54. 47.7°.

12–55. 16.8 m/s.

12–57. 2.09 s, 50.9 ft/s.

12–58. 8.21 ft, 3.37 ft.

12–59. 2.40 m.

12–61. 66.2 ft.

12–62. 2.36 m/s^2.

12–63. 1.04 m/s^2.

12–65. 66.6 m.

12–66. 12.0 m/s^2.

12–67. 1.81 ft/s^2.

12–69. 5.66 m/s, 0.961 m/s^2.

12–70. 21.1 m/s^2, 40 m/s^2.

12–71. 35 m/s.

12–73. **(a)** $(a_n)_A = 32.2\ \text{ft/s}^2$, $(a_t)_A = 0$, $\rho_A = 698.8$ ft.
(b) $(a_n)_B = 14.0\ \text{ft/s}^2$, $(a_t)_B = 29.0\ \text{ft/s}^2$, $\rho_B = 8506.7$ ft.

12–74. **(a)** 1 m/s, **(b)** 0.5 m/s.

12–75. 15 m/s.

12–77. **(a)** 60 s, **(b)** 48 s.

12–78. 12 ft/s.

12–79. $v_B = 2$ m/s →, $v_A = 8$ m/s →.

12–81. 56.5°.

12–82. 13.5 ft/s, 74.2 s.

12–83. 13 ft/s ↗ 22.6°, 66.7 ft.

12–85. 5 m/s ↘ 36.9°.

12–86. 48.6°.

12–87. 16.6 km/h, 25.0°.

12–89. 60.8 mi/h, 1426.4 mi/h 82.5° ↘.

12–90. 30.8 m/s ↘ 35.8°, 1.76 m/s^2, ↘ 58.5°.

12–91. 144.2 km/h ↗ 56.3°, 39 773.0 km/h^2 ↘ 84.2°.

12–93. 37.9 ft ←, 1.33 s.

12–94. $x = 0$, $y = -2$ ft, 250 ft/s^2.

12–95. 1.11 s, 668.7 ft.

12–97. 4 mm/s, 0.

12–98. $0 \leq t < 20$, $a = 2.5$, $s = 1.25t^2$, $20 < t \leq 60$, $a = 0.5$, $s = 0.25t^2 + 40t - 400$.

12–99. 9.82 m/s, not constant, $a_t = 2.30\ \text{m/s}^2$.

12–101. 14 ft/s^2.

12–102. 96 ft.

12–103. 4.59 m, ↘ 25.9°.

12–105. $a = \frac{1}{2}t$, $v = \frac{1}{4}t^2$, $s = \frac{1}{12}t^3$.

12–106. 12 ft/s ↑, 18 ft/s ↓.

12–107. 28 s.

Chapter 13

13–1. 41.7 nN.

13–2. $1.92(10^{20})$ N.

13–3. 151.7 N, 40.8 kg, 3.72 m/s^2.

13–5. 0.0278 m/s^2.

13–6. 0.19 m/s^2 ↑.

13–7. 1200 N, 16.67 m.

13–9. 15.23 ft/s.

13–10. 3.06 s.

13–11. 13.1 ft/s.

13–13. 7994.6 lb.

13–14. 17.1 m/s.

13–15. 5.64 m/s.

13–17. 43.1 s.

13–18. 4.95 ft/s^2, 92.3 lb.

13–19. 24 kg.

13–21. 250 mm.

13–22. 4.28 ft/s^2.

13–23. 0.755 m/s^2, 1.51 m/s^2, 90.6 N, 45.3 N.

13–25. **(a)** 6.94 m/s^2, **(b)** 6.94 m/s^2.

13–26. 35.59 lb, 235.59 lb, 677.95 lb · ft.

13–27. 0.391 s.

13–29. 40.1 ft/s.

13–30. 11.3 ft/s, No.

13–31. 9.37 m/s, 339.9 N, 0, 588.6 N.

13–33. 310.3 N, 9.36 m/s^2.

13–34. 408.6 lb, 0.

13–35. 20.9°.

13–37. 1.82 N, 0.844 N.

13–38. 2.83 m/s.

13–39. 1.63 m/s.

13–41. 75.6°.

13–42. 3.13 m/s.

13–43. 120.23 N, 48.1 m/s^2.

13–45. 1,518.6 lb, 424.5 lb.
13–46. 32.2 ft/s, 48.3 ft, 559.0 lb.
13–47. 24.4 m/s.
13–49. 86.9 ft/s^2.
13–50. 78.6 ft/s^2.
13–51. T = 5.67 lb, N_A = 2.27 lb.
13–53. 2.84 rad/s.
13–54. 1271.8 N, 573.2 N.
13–55. 1095.3 N, 493.6 N.

Chapter 14

14–1. 3.2 kN.
14–2. 744.4 N, 14,380.7 ft · lb.
14–3. 347.2 kN.
14–5. 21.5 ft/s.
14–6. 5,525.0 lb · ft, 1,901.9 lb.
14–7. 44.1 kN, 38.2 mm.
14–9. 2.77 m.
14–10. 35.2 ft/s, 41.7 ft.
14–11. 434.7 g.
14–13. 35.9 ft/s.
14–14. 1.14 ft.
14–15. 10.13 ft.
14–17. 38.5 N.
14–18. 17.97 m/s, 12.54 kN.
14–19. $GM_em\left(\frac{1}{r_2} - \frac{1}{r_1}\right)$.
14–21. 762.64 W, 30.5 s.
14–22. 34.0 min.
14–23. 946.4 kW.
14–25. 5.82 hp.
14–26. 0.245.
14–27. 8.86°.
14–29. 70 hp.
14–30. 8.31t MW.
14–31. 22.24 kW.
14–33. 434.7 g.
14–34. 17.97 m/s.
14–35. 25.4 ft/s.
14–37. $h = 24.5$ m, $N_B = 0$, $N_C = 9810$ N.
14–38. 2.34 m/s.
14–39. 49.05 mm.
14–41. 40.12 ft/s.
14–42. 114.5 ft/s, 29.1 lb.
14–43. 17.7 ft/s.
14–45. 56.7 Mm/h.
14–46. 7.66 m/s.
14–47. 0.516 ft, 0.387 ft.
14–49. 214.2 ft/s.
14–50. 155.4 ft/s.
14–51. 29.2 ft/s.
14–53. 1.10 in.
14–54. 6.50 ft.
14–55. 18.9 ft/s.
14–57. 3.33 ft.
14–58. 40.7 s.
14–59. 9.24 hp.

Chapter 15

15–1. 0.373 lb · s.
15–2. 46.6 s.
15–3. 3.11 s.
15–5. 0.810 lb · s.
15–6. 57.5 lb · s, 63 lb · s.
15–7. 1.02 s.
15–9. 1.5 kN.
15–10. 5.68 N · s.
15–11. 525.7 lb.
15–13. 21.5 ft/s.
15–14. 162.2 N · s.
15–15. 8.54 m/s ↑, 25.62 m/s ↓.
15–17. 4.44 kN.
15–18. 22.5 kN · s.
15–19. 0.486 s.
15–21. 4.14 s.
15–22. 0.84 m/s.
15–23. 0.203 m/s.
15–25. 20.7 ft/s, 38.6 lb · s.
15–26. 0.178 m/s, 770.5 N.
15–27. 4.25 ft/s, 120.5 lb.
15–29. 11.01 km/h.
15–30. 0.933 m/s. 0.498 m/s.

15–31. **(a)** 1.25 ft/s, **(b)** 1.125 ft/s.
15–33. 45.4 ft/s.
15–34. 0.22 m/s, 2.22 m/s.
15–35. 0.775.
15–37. 16.04 ft/s, 12.8 ft/s, 2.56 ft.
15–38. 13.9 ft/s, 14.29 ft/s, 5.96 ft/s, 3.52 ft.
15–39. **(a)** 19.7 ft/s, **(b)** 9.43 ft/s, 15.3 ft/s, **(c)** 9.13 ft.
15–42. 80.75 km/h, 9.52 m.
15–43. $v_D = v$, $v_A = 0$, $v_B = 0$, $v_C = v$.
15–45. 4.13 ft/s, 5.89 ft/s.
15–46. 6.90 m/s, 75.6 m/s.
15–47. 2.88 ft/s, 1.77 ft/s.
15–49. 90°.
15–50. **(a)** 1.50 slug · ft²/s, 1.74 slug · ft²/s, 0.396 slug · ft²/s.
15–51. $6.75(10^6)$ kg · m²/s.
15–53. 12.67 m/s.
15–54. 91.8 ft/s.
15–55. 10.2 ft/s, 98.1 ft · lb.
15–57. 13.9 ft/s, 7.89 ft/s.
15–58. 4 m/s, 200 mm.
15–59. 21.9 m/s, 20.9°.
15–61. 2.55 lb · s, 2.40 slug · ft/s, 2.55 slug · ft/s.
15–62. 2.45 slug · ft/s.
15–63. 0.349 m/s.
15–65. 6.99 ft/s, 6.50 in.
15–66. 3.20 ft/s.
15–67. 3.80 m/s ←, 6.51 m/s ⦪ 68.6°.
15–69. 1.5 m/s.

Chapter 16

16–1. **(a)** −2.5 rad/s², **(b)** 79.6 rev.
16–2. 0.375 ft/s², 3600 ft/s², 4500 ft.
16–3. 6.41 rev, 3.5 s.
16–5. 18.8 ft/s.
16–6. 2.07 in.
16–7. 0.225 m/s², 0.202 m/s².
16–9. 8.49 rad/s, 0.6 m/s.
16–10. 2.86 ft/s, 0.273 ft/s².
16–11. 6.93 ft/s, 84.3 ft/s².
16–13. 0.45 ft/s ↑, −0.075 ft/s² ↓.
16–14. 1 m/s.
16–15. 70 ft/s, 4.24 rev.
16–17. 6 ft/s, 36.9 ft/s².
16–18. $v_{CD} = -6 \sin \theta$, $a_{CD} = -24 \cos \theta$.
16–19. −5.20 ft/s, −14.60 ft/s².
16–21. 8.70 rad/s, 50.5 rad/s².
16–22. −15.40 rad/s, −129.23 rad/s².
16–23. 3.46 rad/s ↺, 3.46 m/s ↓.
16–25. 0.536 m/s, ⦣ 53.6°; 0.992 m/s, 70.9° ⦨.
16–26. 213.6 ft/s, ⦨ 69.4°.
16–27. 12 ft/s, 24 ft/s.
16–29. 0, 1.2 m/s ↓, 0.849 m/s ⦩ 45°.
16–30. 8 ft/s, 10 ft/s, 1.2 rad/s.
16–31. 0.5 ft/s ←.
16–33. 8 ft/s.
16–34. 4 rad/s ↺, 100 mm/s ↓.
16–35. $\omega_A = 5$ rad/s, $\omega_B = 1.25$ rad/s, 75 mm/s ↑.
16–37. 1.36 m/s →.
16–38. 4 rad/s.
16–39. 2.4 m/s, 1.79 m/s 26.6° ⦨.
16–41. 12 ft/s, 24 ft/s.
16–42. 4 ft/s, 2 rad/s.
16–43. 0, 1.2 m/s ↓, 0.849 m/s ⦩ 45°.
16–45. 6 m/s →.
16–46. 48 rad/s ↺, 120 ft/s ←.
16–47. 14 in./s ↓, 10.8 in./s ⦪ 21.8°.
16–49. $v_A = \omega(r_2 - r_1)$.
16–50. $v_A = 424.3$ mm/s, ⦪ 45°, $v_B = 900$ mm/s →.
16–51. 1.65 rad/s, 5.38 rad/s.
16–53. 10.64 m/s.
16–54. 1.47 rad/s² ↻, 24.93 ft/s² ↓.
16–55. 27.79 rad/s² ↺.
16–57. 53.3 rad/s² ↻, 2.67 m/s² ↓.
16–58. $a_B = 6.71$ ft/s², ⦪ 63.4°, $a_A = 6.71$ ft/s², 63.4° ⦫.
16–59. 13.88 ft/s² ⦣ 8.78°.
16–61. 0.245 ft/s² →, 0.816 rad/s² ↺.
16–62. $a_A = 1.44$ ft/s², $\alpha_{AB} = 0$.
16–63. $\alpha_{CD} = 0.177$ rad/s², $\alpha_{BD} = 0.177$ rad/s², $\alpha_{AB} = 0$.
16–65. 1.70 rad/s.
16–66. 3.73 rad/s², 8.84 rad/s².
16–67. 6.69 s.
16–69. 16.8 rad/s.
16–70. 11.7 ft/s ⦪ 39.8°.

16–71. 4.95 rad/s, 1.43 m/s.
16–73. 4.70 m/s, 34.8 rad/s.
16–74. 301.6 m/s^2.

Chapter 17

17–1. 321.7 slug · ft^2.
17–2. 20.1 Mg · m^2.
17–3. $52.34(10^6)$ kg · m^2.
17–5. 107.2 slug · ft^2.
17–6. $\frac{1}{10}ma^2$.
17–7. $28.09(10^3)$ slug · ft^2.
17–9. 4.17 slug · ft^2.
17–10. 1.67 slug · ft^2.
17–11. $6.70(10^{-3})$ slug · ft^2.
17–13. 0.393 m, 0.0773 kg · m^2.
17–14. 20 kN.
17–15. 95.0 lb ↑, 105.0 lb ↑, 9.66 ft.
17–17. 2-wheel drive 17.5 s, 4-wheel drive 11.33 s.
17–18. 9.38 N, 1.08 N, 1.08 N · m.
17–19. 0.0157 m/s^2, 72.12 kN, 71.95 kN.
17–21. 49.14 ft/s, 16.7°.
17–22. 1.02 rad/s^2.
17–23. 1.94 m/s^2, 1.45 m/s^2.
17–25. 77.9 ft/s.
17–26. 3.125 ft.
17–27. 600 N, 606.5 N, 1355.5 N.
17–29. 14.4 kN, 21.7 kN.
17–30. 2.94 m/s^2.
17–31. 9.93 kN, 19.35 kN.
17–33. 2.91 kN, 0.337 m/s^2.
17–34. 1.24 m.
17–35. 16.1 ft/s^2, 17.5 lb, 3.75 lb.
17–37. 4.79 rad/s.
17–38. 102.1 N.
17–39. 0, 1.941 lb, 14.85 rad/s^2.
17–41. 0, 288.7 N, 23.1 rad/s^2.
17–42. 193.1 N, 3.11 s.
17–45. 218.20 N, 21.01 rad/s^2 ↰.
17–46. 36.1 rad/s^2, 54.2 N, 1109.7 N.
17–47. 0.233 lb · ft.
17–49. 891.1 rad/s^2.
17–50. 6.71 s.
17–51. 400 rad/s.
17–53. 4.88 ft/s.
17–54. 50.4 kN.
17–55. 80 rad/s^2, 120 rad/s.
17–58. 1.089 lb, 29.34 lb, 11.05 rad/s^2.
17–59. 2.62 lb, 26.0 lb, 26.6 rad/s^2.
17–61. 1962 N, 0.15 m/s^2, 0.225 rad/s^2.
17–62. 1.14 rad/s^2, 19.2 lb, 176.8 lb.
17–63. 48.3 ft/s^2, 144.9 rad/s^2.
17–65. 25.76 rad/s^2.
17–66. 51.52 rad/s^2.
17–67. 5 rad/s^2.
17–69. 6.62 rad/s^2.
17–70. 10.61 lb, 0.4082 rad/s^2.
17–71. 43.6 rad/s^2, 43.6 rad/s^2, 19.62 N.
17–73. 4.77 m/s.
17–74. 68.7 rad/s.
17–75. 41.2 rad/s.
17–77. 11.38 m/s, 33.4°.
17–78. 2.18 m/s^2, 6.27 kN, 5.01 kN.
17–79. 68.3 lb.
17–81. 5.75 ft.
17–82. 100 rad/s.
17–83. 9.90 ft/s^2, 11.39 rad/s^2.

Chapter 18

18–2. 69.5 ft/s.
18–3. 26.7 ft/s.
18–5. 1.84 rad/s.
18–6. 1.34 rev.
18–7. 17.4 rad/s.
18–9. 0.836 rad/s.
18–10. 4.60 rad/s.
18–11. 8.59 rad/s.
18–13. 0.935 rev.
18–14. 1.14 rad/s.
18–15. 4.70 m/s.
18–17. ≈1 rev.
18–18. 3.70 rad/s.
18–19. 19.8 rad/s.
18–21. 0.753 rad/s.

18–22. 41.76 rad/s.
18–23. 7.02 ft.
18–25. 6.36 ft/s.
18–26. 11.7 ft/s ←.
18–27. 85.1 rad/s.
18–29. 5.05 rad/s.
18–30. 4.5 rad/s.
18–31. 49.1 lb.
18–33. 3.30 rad/s.
18–34. 433.5 J.
18–35. 2.70 m/s.
18–37. 69.7 rad/s.
18–38. 31.4 rev.
18–39. 18.1 rad/s.

Chapter 19

19–1. **(a)** 6.52 slug · ft^2/s, **(b)** 8.39 slug · ft^2/s.
19–2. 11.4 rad/s.
19–3. 6.63 kN · s.
19–5. 245.2 rad/s.
19–6. 0.386 rad/s.
19–9. 0.102 rad/s.
19–10. **(a)** 40 rad/s, **(b)** 38.9 rad/s, **(c)** 41.1 rad/s.
19–11. 25 rad/s.
19–13. 31.9 rad/s.
19–14. 1520.1 rad/s.
19–15. 0.0622 lb · ft, 0.280 lb.
19–17. 77.2 rad/s.
19–18. 3.42 m/s.
19–19. 3.68 rad/s.
19–21. **(a)** $(I/I_z)\omega$, **(b)** 0, **(c)** $(I/I_z)\omega$.
19–22. 0.175 rad/s.
19–23. 32.8 rad/s.
19–25. 79.84 N · s.
19–26. 0.375 ft.
19–27. $\frac{7}{5}r$.
19–29. 10.16 ft/s.
19–31. 17.86°.
19–33. 1.96 m/s.
19–34. 422.4 N, 362.4 N.
19–35. 327.9 N · m.
19–37. 1.69 kN · m.
19–38. 27.94 rad/s.
19–39. 68.7 rad/s.
19–41. 103.4 rad/s.
19–42. 73.6 rad/s.
19–43. 1.53 s, $T_{AB}/2 = 49.0$ N.
19–45. 907.0 rad/s.
19–46. 65.6 rad/s.
19–47. 27.8 rad/s.

Chapter 20

20–1. 29.0 m/s^2.
20–2. 120 ft/s, 76.8 ft/s^2.
20–3. 5 m/s, 19.7 m/s^2.
20–5. 2.61 m/s, 9.04 m/s, −236.82 m/s^2, 88.04 m/s^2.
20–6. −14.4 ft/s, 14.14 ft/s.
20–7. 0.6 rad/s.
20–9. 0.33 rad/s.
20–10. 6 ft/s, 18.3 ft/s, −67.12 ft/s^2, 66.28 ft/s^2.
20–11. 0, 12 ft/s, −216 ft/s^2, 0.
20–13. 11.33 N.
20–14. 1.66 lb, 4.78 lb.
20–15. 2.75 lb, 3.10 lb.
20–17. 1.11 lb, 1.25 lb.
20–18. 0.886 N, 3.92 N.
20–19. 85.3 lb.
20–21. 4 m/s^2.
20–22. 60.13 in./s, 243.04 in./s^2.
20–23. 16.08 m/s^2.
20–25. **(a)** 2.5 ft/s^2, **(b)** 4.17 ft/s^2.
20–26. 0.75 rad/s ↻, 1.95 rad/s^2 ↺.
20–27. 3.86 rad/s ↻, 7.46 rad/s^2 ↻.
20–29. 0.833 rad/s ↺, 4.81 rad/s^2 ↻.
20–30. 67.5 lb.
20–31. 75.8 lb.
20–33. 55.3 lb, 25.8 lb.
20–34. 451.8 Pa.
20–35. 8.17 lb, 2.72 lb.
20–37. 3806.6 N, 1988.0 N · m
20–38. 53.9 N, 9.1 N, 142.6 N.
20–39. 446.9 mm/s, 3662.1 mm/s^2.

20–41. 151.1 mi/h^2.

20–42. 22.7 lb.

20–43. 508.8 lb.

20–45. 13.2 ft/s, 87.8 ft/s^2.

20–46. 402 N.

20–47. 8 ft/s.

20–49. 1.12 lb.

Chapter 21

21–1. 7.88 Hz, 0.127 s.

21–2. $x = -0.05 \cos (12.25t)$.

21–3. $x = 0.2 \sin (7.49t) + 0.1 \cos (7.49t)$, 26.6°.

21–5. 2.02 Hz, $y = -0.2 \cos (12.69t)$, 0.2 ft.

21–6. $x = -0.127 \sin (3.162t) - 0.09 \cos (3.162t)$, 0.156 m.

21–7. 2.52 cycles/s, 83 749.3 N.

21–9. 16.3°.

21–10. **(a)** 1.125 Hz, **(b)** 2.25 Hz.

21–11. 1.32 s.

21–13. 0.632 Hz.

21–14. 1.36 N/m, 3.59 kg.

21–15. $2\pi\sqrt{(k_G^2 + d^2)/gd}$.

21–17. 0.167 s.

21–18. $y = A \sin pt + B \cos pt + \dfrac{F_o/m}{\left(\dfrac{k}{m} - \omega^2\right)} \cos \omega t$.

21–19. $y = (361 \sin (7.746t) + 100 \cos (7.746t) - 350 \sin 8t)$ mm.

21–21. $x_P = (0.0276 \sin 2t)$ ft.

21–22. 6.55 rad/s.

21–23. 0.158 ft.

21–25. 1.20 m/s.

21–26. 4.51 mm.

21–27. 0.666 s.

21–29. 0.997.

21–30. $x = 0.119 \cos (6t - 84.0°)$ m.

21–31. 714.3 N/m.

21–33. $y = 0.622e^{-0.939t} \sin (11.89t + 85.5°)$ ft.

21–34. $y = 33.8(e^{-7.5t}\sin (8.88t))$ mm.

21–37. 21.1 mm.

21–38. $x = -0.25 \cos 19.7t$.

21–39. $x = 0.00279 \sin 17.9t + 0.25 \cos 17.9t + 0.00625 \sin 8t$.

21–41. $y = -243 \sin 6.39t + 50 \cos 6.39t + 775 \sin 2t)$ mm.

21–42. 1.54°.

21–43. 1.008.

21–45. 3.13 Hz, $x = -0.0833 \cos (19.7t)$.

21–46. 0.796 s.

Index

DEON
STOCKERT
232-1346

Obituaries

Ex-Moorhead planning director dies at 72

Robert J. Roberts of 302 5th St. S., Moorhead, a planning director and Clay County surveyor known for his highly-visible public battle with the Moorhead City Council and Planning Commission during the city's urban renewal years, died Monday. He was 72.

His funeral will be at 10 Saturday in the Cathedral of St. Paul (Minn.), with burial in Fort Snelling National Cemetery, Minneapolis. A memorial Mass will be at 7 Thursday in St. Joseph Catholic Church, Moorhead.

Mr. Roberts served as the executive director of Moorhead's Housing and Redevelopment Authority for Urban Renewal for eight years, and as the city planning director for 13 years. In addition, he was the director for the Clay County Planning Commission, County Surveyor for 32 years and held engineering jobs in the private sector.

During the late 1960s and early 1970s, Mr. Roberts was best known for his protracted battles with the city council, which tried to force him to resign some of his authority with the city, and his newpaper column that berated city government.

Roberts was born July 24, 1917, in St. Paul. He grew up in St. Paul and graduated from Mechanic Arts High School in 1935. He married Eileen Marie Eichinger in 1945 in White Bear Lake, Minn. He served in the U.S. Army Air Corps from 1942 to 1946. They moved to Moorhead in 1950. They also had lived in Boston and Minneapolis. He had been owner of Robert J. Roberts and Associates Inc. and Lake Agassiz Testing Laboratories in Moorhead.

He is survived by his wife; a son, Robert J., Moorhead; three daughters, Mary Pat Roberts-Kraft, St. Paul; and Eileen M. Roberts and Kate Roberts, both Minneapolis; a brother, George W., White Bear Lake; a sister, Margaret A. Roberts, St. Paul; and a grandchild. (Boulger)

Geometric Properties of Line and Area Elements

Centroid Location

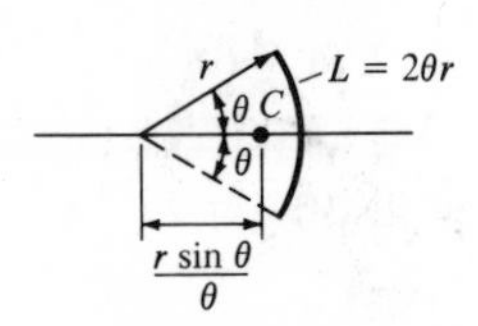

Circular arc segment

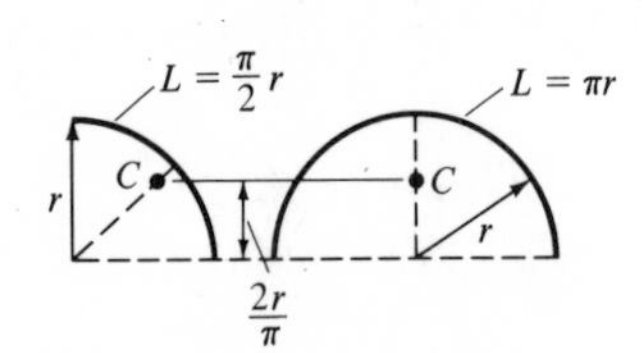

Quarter and semicircular arcs

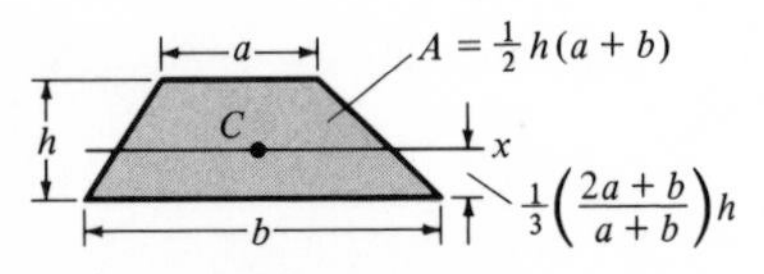

Trapezoidal area

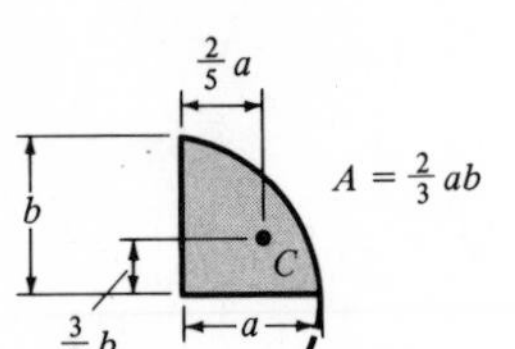

Semiparabolic area

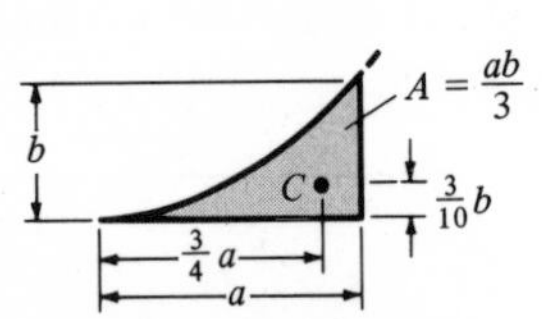

Exparabolic area

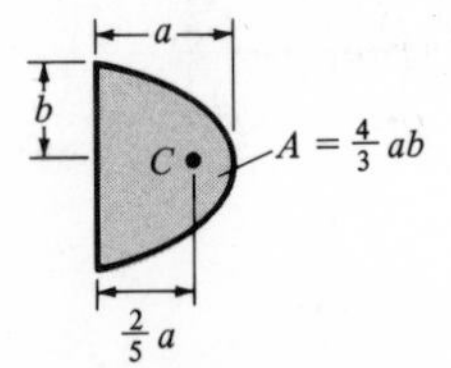

Parabolic area

Centroid Location / Area Moment of Inertia

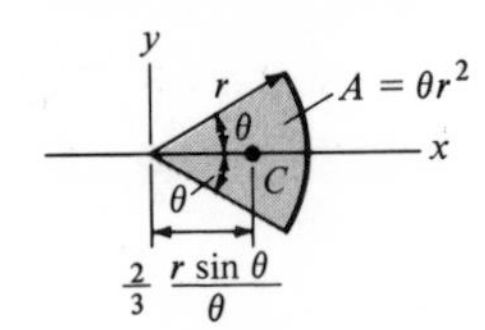

Circular sector area

$$I_x = \tfrac{1}{4} r^4(\theta - \tfrac{1}{2} \sin 2\theta)$$
$$I_y = \tfrac{1}{4} r^4(\theta + \tfrac{1}{2} \sin 2\theta)$$

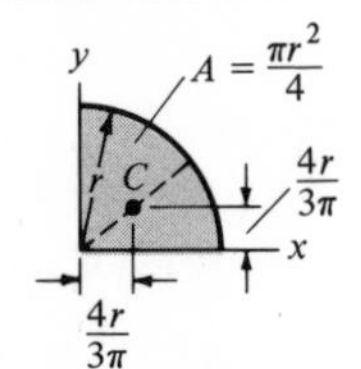

Quarter circular area

$$I_x = \tfrac{1}{16} \pi r^4$$
$$I_y = \tfrac{1}{16} \pi r^4$$

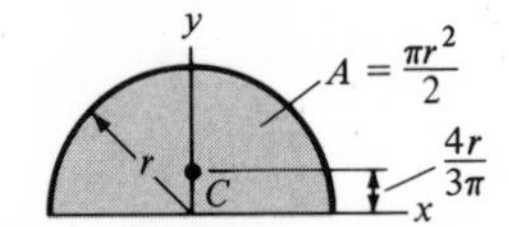

Semicircular area

$$I_x = \tfrac{1}{8} \pi r^4$$
$$I_y = \tfrac{1}{8} \pi r^4$$

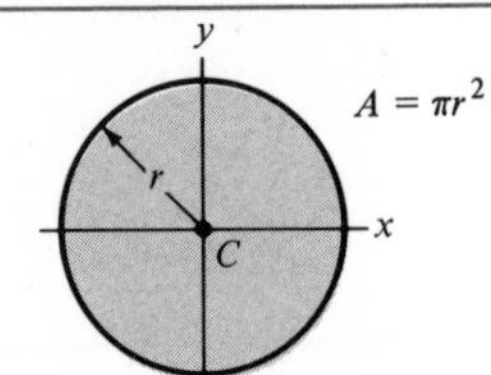

Circular area

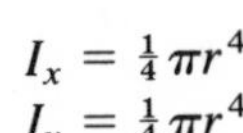

$$I_x = \tfrac{1}{4} \pi r^4$$
$$I_y = \tfrac{1}{4} \pi r^4$$

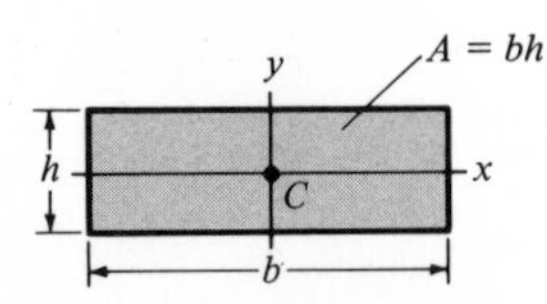

Rectangular area

$$I_x = \tfrac{1}{12} bh^3$$
$$I_y = \tfrac{1}{12} hb^3$$

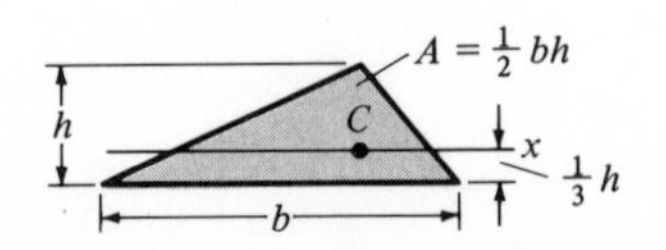

Triangular area

$$I_x = \tfrac{1}{36} bh^3$$